AF249045

Electrical

Level One

Trainee Guide
2014 *NEC*® Revision

PEARSON

Boston Columbus Indianapolis New York San Francisco Upper Saddle River
Amsterdam Cape Town Dubai London Madrid Milan Munich Paris Montreal Toronto
Delhi Mexico City São Paulo Sydney Hong Kong Seoul Singapore Taipei Tokyo

NCCER
President: Don Whyte
Director of Product Development: Daniele Dixon
Electrical Project Managers: Daniele Dixon, Chris Wilson
Production Manager: Tim Davis
Quality Assurance Coordinator: Debie Hicks
Desktop Publishing Coordinator: James McKay
Permissions Specialist: Megan Casey
Production Specialist: Megan Casey
Editors: Chris Wilson, Debie Hicks, James Bassett

Writing and development services provided by Topaz Publications, Liverpool, NY
Lead Writer/Project Manager: Veronica Westfall
Desktop Publisher: Joanne Hart
Art Director: Alison Kenyon
Permissions Editors: Andrea LaBarge, Tonia Burke
Writers: Thomas Burke, Gerald Shannon, Nancy Brown, Charles Rogers, Anna Meade

Pearson Education, Inc.
Director, Global Employability Solutions: Jonell Sanchez
Head of Associations: Andrew Taylor
Editorial Assistant: Douglas Greive
Program Manager: Alexandrina B. Wolf
Project Manager: Janet Portisch
Operations Supervisor: Deidra M. Skahill
Art Director: Diane Ernsberger
Directors of Marketing: David Gesell, Margaret Waples
Field Marketers: Brian Hoehl, Stacey Martinez

Composition: NCCER
Printer/Binder: LSC Communications
Cover Printer: LSC Communications
Text Fonts: Palatino and Univers

Credits and acknowledgments for content borrowed from other sources and reproduced, with permission, in this textbook appear at the end of each module.

---

1 16

ISBN-13: 978-0-13-473028-8
ISBN-10: 0-13-473028-3

Preface

To the Trainee

Electricity powers the applications that make our daily lives more productive and efficient. The demand for electricity has led to vast job opportunities in the electrical field. Electricians constitute one of the largest construction occupations in the United States, and they are among the highest-paid workers in the construction industry. According to the U.S. Bureau of Labor Statistics, job opportunities for electricians are expected to be very good as the demand for skilled craftspeople is projected to outpace the supply of trained electricians.

Electricians install electrical systems in structures. They install wiring and other electrical components, such as circuit breaker panels, switches, and light fixtures. Electricians follow blueprints, the *National Electrical Code®*, and state and local codes. They use specialized tools and testing equipment, such as ammeters, ohmmeters, and voltmeters. Electricians learn their trade through craft and apprenticeship programs. These programs provide classroom instruction and on-the-job learning with experienced electricians.

We wish you success as you embark on your first year of training in the electrical craft, and hope that you will continue your training beyond this textbook. There are more than 700,000 people employed in electrical work in the United States, and, as most of them can tell you, there are many opportunities awaiting those with the skills and desire to move forward in the construction industry.

New with *Electrical Level One*

NCCER and Pearson are pleased to present the eighth edition of *Electrical Level One.* This edition has been updated to meet the 2014 *National Electrical Code®* and includes revisions to the module examinations.

The opening page of each of the twelve modules in this textbook features photos of award-winning construction projects from Associated Builders and Contractors, Inc. and The Associated General Contractors of America. Check out these awesome projects and see where a career as an electrician could take you.

We wish you success as you progress through this training program. If you have any comments on how NCCER might improve upon this textbook, please complete the User Update form located at the back of each module and send it to us. We will always consider and respond to input from our customers.

We invite you to visit the NCCER website at **www.nccer.org** for information on the latest product releases and training, as well as online versions of the *Cornerstone* magazine and Pearson's NCCER product catalog.

Your feedback is welcome. You may email your comments to **curriculum@nccer.org** or send general comments and inquiries to **info@nccer.org**.

NCCER Standardized Curricula

NCCER is a not-for-profit 501(c)(3) education foundation established in 1996 by the world's largest and most progressive construction companies and national construction associations. It was founded to address the severe workforce shortage facing the industry and to develop a standardized training process and curricula. Today, NCCER is supported by hundreds of leading construction and maintenance companies, manufacturers, and national associations. The NCCER Standardized Curricula was developed by NCCER in partnership with Pearson, the world's largest educational publisher.

Some features of NCCER's Standardized Curricula are as follows:

- An industry-proven record of success
- Curricula developed by the industry for the industry
- National standardization providing portability of learned job skills and educational credits
- Compliance with the Office of Apprenticeship requirements for related classroom training *(CFR 29:29)*
- Well-illustrated, up-to-date, and practical information

NCCER also maintains a National Registry that provides transcripts, certificates, and wallet cards to individuals who have successfully completed a level of training within a craft in NCCER's Curricula. *Training programs must be delivered by an NCCER Accredited Training Sponsor in order to receive these credentials.*

Special Features

In an effort to provide a comprehensive, user-friendly training resource, we have incorporated many different features for your use. Whether you are a visual or hands-on learner, this book will provide you with the proper tools to get started in the electrical industry.

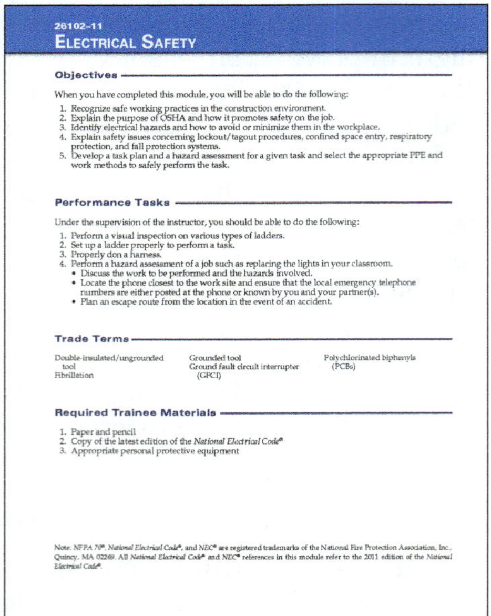

26102-11
ELECTRICAL SAFETY

Objectives

When you have completed this module, you will be able to do the following:

1. Recognize safe working practices in the construction environment.
2. Explain the purpose of OSHA and how it promotes safety on the job.
3. Identify electrical hazards and how to avoid or minimize them in the workplace.
4. Explain safety issues concerning lockout/tagout procedures, confined space entry, respiratory protection, and fall protection systems.
5. Develop a task plan and a hazard assessment for a given task and select the appropriate PPE and work methods to safely perform the task.

Performance Tasks

Under the supervision of the instructor, you should be able to do the following:

1. Perform a visual inspection on various types of ladders.
2. Set up a ladder properly to perform a task.
3. Properly don a harness.
4. Perform a hazard assessment of a job such as replacing the lights in your classroom.
   - Discuss the work to be performed and the hazards involved.
   - Locate the phone closest to the work site and ensure that the local emergency telephone numbers are either posted at the phone or known by you and your partner(s).
   - Plan an escape route from the location in the event of an accident.

Trade Terms

Double-insulated/ungrounded tool
Fibrillation
Grounded tool
Ground fault circuit interrupter (GFCI)
Polychlorinated biphenyls (PCBs)

Required Trainee Materials

1. Paper and pencil
2. Copy of the latest edition of the *National Electrical Code*®
3. Appropriate personal protective equipment

Note: *NFPA 70*®, *National Electrical Code*®, and *NEC*® are registered trademarks of the National Fire Protection Association, Inc., Quincy, MA 02269. All *National Electrical Code*® and *NEC*® references in this module refer to the 2011 edition of the *National Electrical Code*®.

Introduction

This page is found at the beginning of each module and lists the Objectives, Performance Tasks, Trade Terms, and Required Trainee Materials for that module. The Objectives list the skills and knowledge you will need in order to complete the module successfully. The Performance Tasks give you an opportunity to apply your knowledge to the real-world duties that electricians perform. The list of Trade Terms identifies important terms you will need to know by the end of the module. Required Trainee Materials list the materials and supplies needed for the module.

On Site

On Site features provide a head start for those entering the electrical field by presenting technical tips and professional practices from master electricians in a variety of disciplines. The On Site features often include real-life scenarios similar to those you might encounter on the job site.

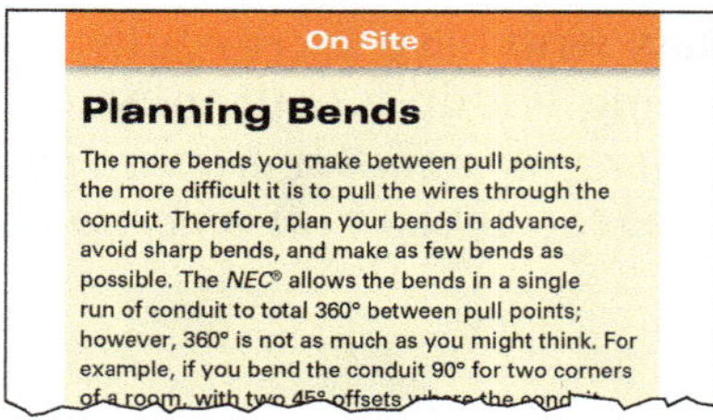

Color Illustrations and Photographs

Full-color illustrations and photographs are used throughout each module to provide vivid detail. These figures highlight important concepts from the text and provide clarity for complex instructions. Each figure reference is denoted in the text in *italic type* for easy reference.

Figure 2 Pushing down on the bender to complete the bend.

Notes, Cautions, and Warnings

Safety features are set off from the main text in highlighted boxes and are organized into three categories based on the potential danger of the issue being addressed. Notes simply provide additional information on the topic area. Cautions alert you of a danger that does not present potential injury but may cause damage to equipment. Warnings stress a potentially dangerous situation that may cause injury to you or a co-worker.

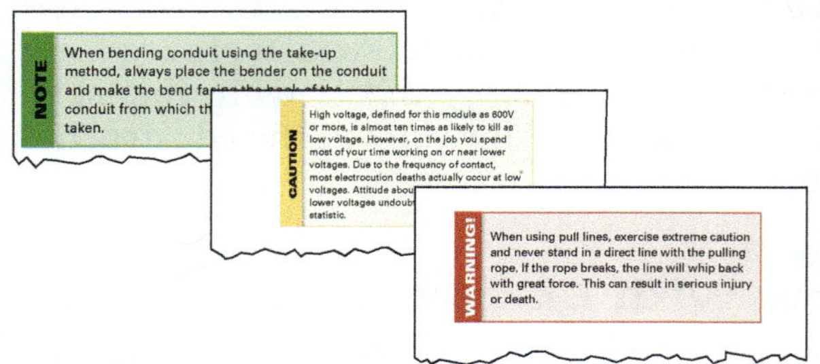

Going Green

Going Green looks at ways to preserve the environment, save energy, and make good choices regarding the health of the planet. Through the introduction of new construction practices and products, you will see how the "greening of America" has already taken root.

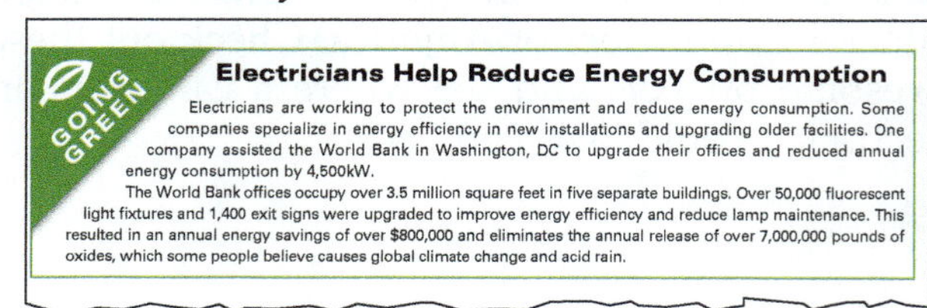

Case History

Case History features emphasize the importance of safety by citing examples of the costly (and often devastating) consequences of ignoring *National Electrical Code®* or OSHA regulations.

Case History

Powder-Actuated Tools

A 22-year-old apprentice was killed when he was struck in the head by a nail fired from a powder-actuated tool in an adjacent room. The tool operator was attempting to anchor plywood to a hollow wall and fired the gun, causing the nail to pass through the wall, where it traveled nearly thirty feet before striking the victim. The tool operator had never received training in the proper use of the tool, and none of the employees in the area were wearing personal protective equipment.

The Bottom Line: Never use a powder-actuated tool to secure fasteners into easily penetrated materials; these tools are designed primarily for installing fasteners into masonry. The use of powder-actuated tools requires special training and certification. In addition, all personnel in the area must be aware that the tool is in use and should be wearing appropriate personal protective equipment.

What's wrong with this picture?

What's wrong with this picture? features include photos of actual code violations for identification and encourage you to approach each installation with a critical eye.

What's wrong with this picture?

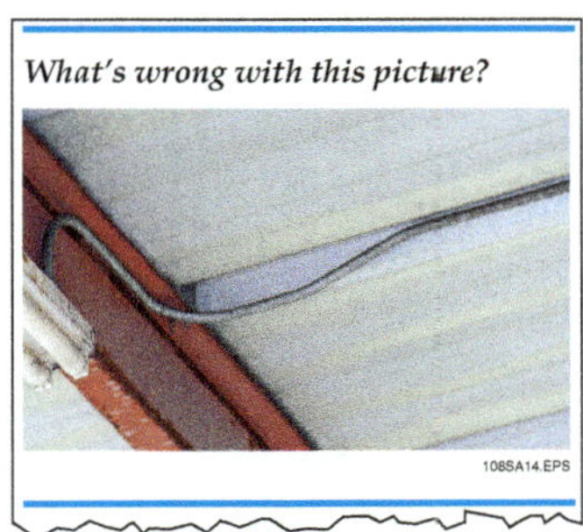

108SA14.EPS

Think About It

Think About It features use "What if?" questions to help you apply theory to real-world experiences and put your ideas into action.

Think About It

Residential Wiring

Make a mental wiring tour of your home. Picture several rooms, including the kitchen and utility/laundry room. How is each device connected to the power source, and what is the probable amperage and overcurrent protection? What other devices might or might not be included in the circuit? How many branch circuits serve each room? Later, examine the panelboard and exposed wiring to see how accurately you identified the branch circuits.

Step-by-Step Instructions

Step-by-step instructions are used throughout to guide you through technical procedures and tasks from start to finish. These steps show you not only how to perform a task but also how to do it safely and efficiently.

are energized until you have verified that the circuit is de-energized. This is called a live-dead-live test. Follow these steps to verify that a circuit is de-energized:

Step 1 Ensure that the circuit is properly tagged and locked out *(CFR 1910.333/1926.417)*.

Step 2 Verify the test instrument operation on a known source using the appropriately rated tester.

Step 3 Using the test instrument, check the circuit to be de-energized. The voltage should be zero.

Step 4 Verify the test instrument operation, once again on a known power source.

Trade Terms

Each module presents a list of Trade Terms that are discussed within the text and defined in the Glossary at the end of the module. These terms are denoted in the text with **blue bold type** upon their first occurrence. To make searches for key information easier, a comprehensive Glossary of Trade Terms from all modules is located at the back of this book.

Electricity is all about cause and effect. The presence of voltage (volts) in a closed circuit will cause current (amps) to flow. The more voltage you apply, the more current will flow. However, the amount of current flow is also determined by how much **resistance** (**ohms**) the load offers to the flow of current. In order to convert electrical energy into work, the load consumes energy. The amount of energy a device consumes is called **power**, and is expressed in **watts (W)**. **Volts (V)**, amps, ohms, and watts are related in such a way that if any one of them changes, the others are proportionally affected. This relationship can be seen using basic math principles that you will learn in this module. You will also learn how electricity is produced and how test instruments are used to measure electricity.

Review Questions

Review Questions are provided to reinforce the knowledge you have gained. This makes them a useful tool for measuring what you have learned.

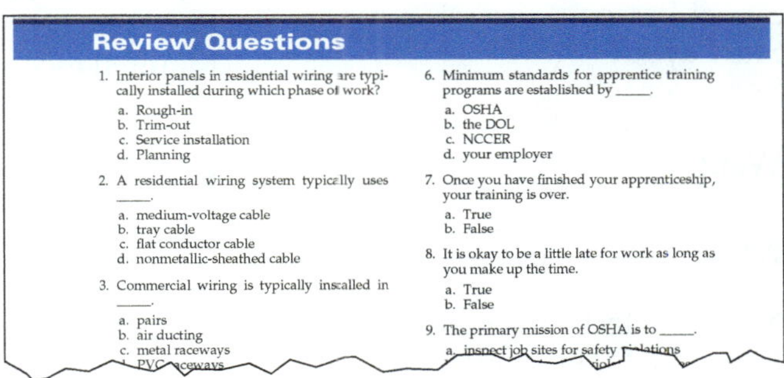

Review Questions

1. Interior panels in residential wiring are typically installed during which phase of work?
   a. Rough-in
   b. Trim-out
   c. Service installation
   d. Planning
2. A residential wiring system typically uses _____.
   a. medium-voltage cable
   b. tray cable
   c. flat conductor cable
   d. nonmetallic-sheathed cable
3. Commercial wiring is typically installed in _____.
   a. pairs
   b. air ducting
   c. metal raceways
   d. PVC raceways
6. Minimum standards for apprentice training programs are established by _____.
   a. OSHA
   b. the DOL
   c. NCCER
   d. your employer
7. Once you have finished your apprenticeship, your training is over.
   a. True
   b. False
8. It is okay to be a little late for work as long as you make up the time.
   a. True
   b. False
9. The primary mission of OSHA is to _____.
   a. inspect job sites for safety violations

NCCER Standardized Curricula

NCCER's training programs comprise more than 80 construction, maintenance, pipeline, and utility areas and include skills assessments, safety training, and management education.

Boilermaking
Cabinetmaking
Carpentry
Concrete Finishing
Construction Craft Laborer
Construction Technology
Core Curriculum: Introductory Craft Skills
Drywall
Electrical
Electronic Systems Technician
Heating, Ventilating, and Air Conditioning
Heavy Equipment Operations
Highway/Heavy Construction
Hydroblasting
Industrial Coating and Lining Application Specialist
Industrial Maintenance Electrical and Instrumentation Technician
Industrial Maintenance Mechanic
Instrumentation
Insulating
Ironworking
Masonry
Millwright
Mobile Crane Operations
Painting
Painting, Industrial
Pipefitting
Pipelayer
Plumbing
Reinforcing Ironwork
Rigging
Scaffolding
Sheet Metal
Signal Person
Site Layout
Sprinkler Fitting
Tower Crane Operator
Welding

Maritime

Maritime Industry Fundamentals
Maritime Pipefitting
Structural Fitter

Green/Sustainable Construction

Building Auditor
Fundamentals of Weatherization
Introduction to Weatherization
Sustainable Construction Supervisor
Weatherization Crew Chief
Weatherization Technician
Your Role in the Green Environment

Energy

Alternative Energy
Introduction to the Power Industry
Introduction to Solar Photovoltaics
Introduction to Wind Energy
Power Industry Fundamentals
Power Generation Maintenance Electrician
Power Generation I&C Maintenance Technician
Power Generation Maintenance Mechanic
Power Line Worker
Power Line Worker: Distribution
Power Line Worker: Substation
Power Line Worker: Transmission
Solar Photovoltaic Systems Installer
Wind Turbine Maintenance Technician

Pipeline

Control Center Operations, Liquid
Corrosion Control
Electrical and Instrumentation
Field Operations, Liquid
Field Operations, Gas
Maintenance
Mechanical

Safety

Field Safety
Safety Orientation
Safety Technology

Management

Fundamentals of Crew Leadership
Project Management
Project Supervision

Supplemental Titles

Applied Construction Math
Careers in Construction
Tools for Success

Spanish Translations

Basic Rigging (Principios Básicos de Maniobras)
Carpentry Fundamentals (Introducción a la Carpintería, Nivel Uno)
Carpentry Forms (Formas para Carpintería, Nivel Trés)
Concrete Finishing, Level One (Acabado de Concreto, Nivel Uno)
Core Curriculum: Introductory Craft Skills (Currículo Básico: Habilidades Introductorias del Oficio)
Drywall, Level One (Paneles de Yeso, Nivel Uno)
Electrical, Level One (Electricidad, Nivel Uno)
Field Safety (Seguridad de Campo)
Insulating, Level One (Aislamiento, Nivel Uno)
Ironworking, Level One (Herrería, Nivel Uno)
Masonry, Level One (Albañilería, Nivel Uno)
Pipefitting, Level One (Instalación de Tubería Industrial, Nivel Uno)
Reinforcing Ironwork, Level One (Herreria de Refuerzo, Nivel Uno)
Safety Orientation (Orientación de Seguridad)
Scaffolding (Andamios)
Sprinkler Fitting, Level One (Instalación de Rociadores, Nivel Uno)

Acknowledgments

This curriculum was revised as a result of the farsightedness and leadership of the following sponsors:

ABC Heart of America Chapter
ABC New Orleans/ Bayou Chapter
ABC of Iowa
Baltimore City Community College
Beacon Electrical Contractors
Center for Employment Training
Cianbro Corporation
Duck Creek Engineering, Inc.
Exelon Generation
Faith Technologies, Inc.
Lamphear Electric
Lee College
Lee Company
Madison Comprehensive High School
Pumba Electric LLC
Robins and Morton
Satellite Broadcasting and Communication
The College of Southern Maryland
TIC – The Industrial Company
Tri-City Electrical Contractors
Trident Technical College

This curriculum would not exist were it not for the dedication and unselfish energy of those volunteers who served on the Authoring Team. A sincere thanks is extended to the following:

Chuck Ackland
Ed Cockrell
Tim Dean
Tim Ely
John Evans
Scott Haldiman
Steven Hill
Dan Lamphear
L.J. LeBlanc
Neil Matthes
Brenton Miller
Scott Mitchell
Jim Mitchem
John Mueller
Nick Musmeci
Mike Owens
Mike Powers
Gregory Pratt
Raymond Saldivar
Alton Smith
Wayne Stratton
Joel Timpanaro

NCCER Partners

American Fire Sprinkler Association
Associated Builders and Contractors, Inc.
Associated General Contractors of America
Association for Career and Technical Education
Association for Skilled and Technical Sciences
Carolinas AGC, Inc.
Carolinas Electrical Contractors Association
Center for the Improvement of Construction Management and Processes
Construction Industry Institute
Construction Users Roundtable
Construction Workforce Development Center
Design Build Institute of America
GSSC – Gulf States Shipbuilders Consortium
Manufacturing Institute
Mason Contractors Association of America
Merit Contractors Association of Canada
NACE International
National Association of Minority Contractors
National Association of Women in Construction
National Insulation Association
National Ready Mixed Concrete Association
National Technical Honor Society
National Utility Contractors Association
NAWIC Education Foundation
North American Technician Excellence
Painting & Decorating Contractors of America
Portland Cement Association
Skills USA
Steel Erectors Association of America
U.S. Army Corps of Engineers
University of Florida, M. E. Rinker School of Building Construction
Women Construction Owners & Executives, USA

Table of Contents

Note: NFPA 70®, National Electrical Code® and NEC® are registered trademarks of the National Fire Protection Association, Inc., Quincy, MA 02269. All National Electrical Code® and NEC® references in this textbook refer to the 2014 edition of the National Electrical Code®.

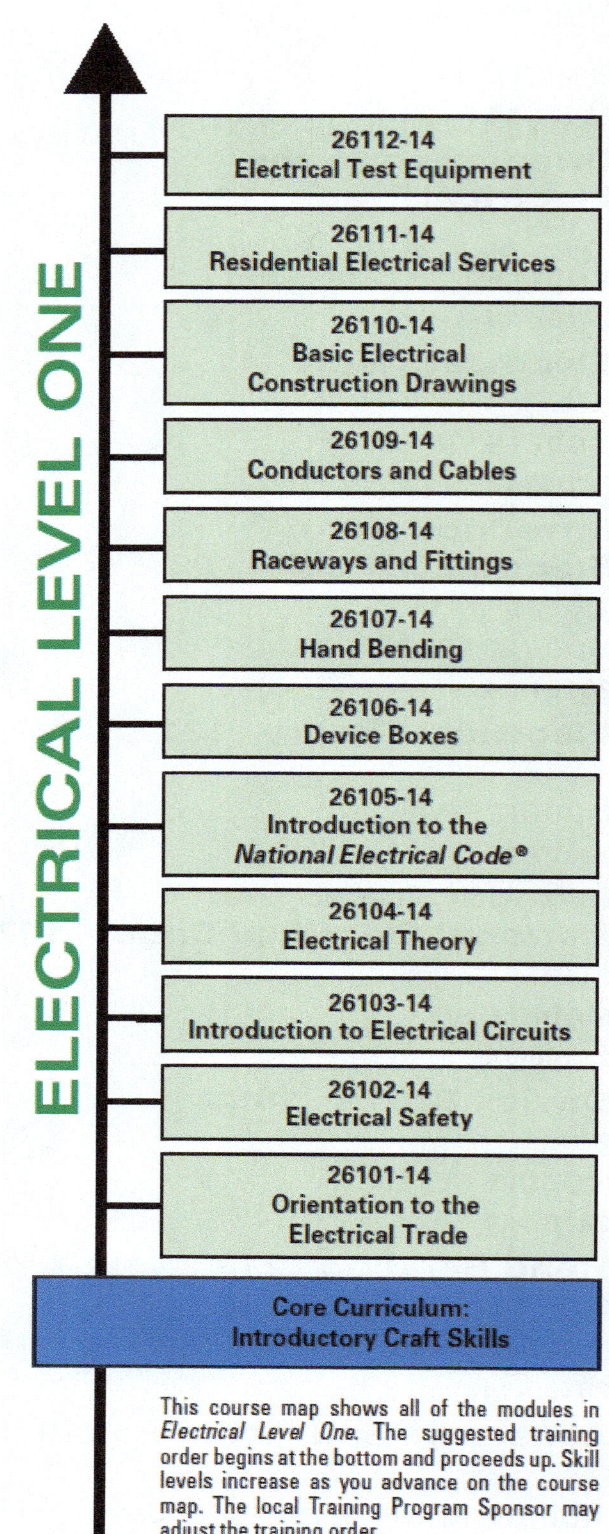

This course map shows all of the modules in *Electrical Level One*. The suggested training order begins at the bottom and proceeds up. Skill levels increase as you advance on the course map. The local Training Program Sponsor may adjust the training order.

Orientation to the Electrical Trade

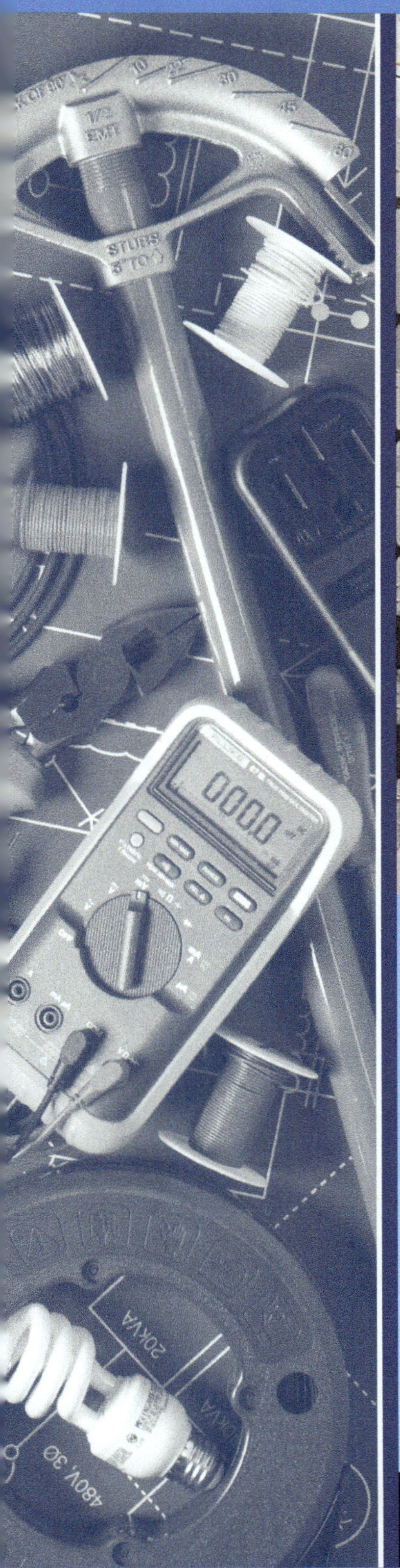

Pet Food Facility

With an extensive safety program and involvement in the project's pre-planning stages, Interstates Companies performed a complete electrical design-build, including all electrical engineering, construction, and automation programming, clocking 69,000 hours without a recordable incident.

26101-14

Trainees with successful module completions may be eligible for credentialing through NCCER's National Registry. To learn more, go to **www.nccer.org** or contact us at **1.888.622.3720.** Our website has information on the latest product releases and training, as well as online versions of our *Cornerstone* magazine and Pearson's product catalog.

Your feedback is welcome. You may email your comments to **curriculum@nccer.org,** send general comments and inquiries to **info@nccer.org,** or fill in the User Update form at the back of this module.

This information is general in nature and intended for training purposes only. Actual performance of activities described in this manual requires compliance with all applicable operating, service, maintenance, and safety procedures under the direction of qualified personnel. References in this manual to patented or proprietary devices do not constitute a recommendation of their use.

26101-14

Orientation to the Electrical Trade

Objectives

When you have completed this module, you will be able to do the following:

1. Describe the apprenticeship/training process for electricians.
2. Describe various career paths/opportunities one might follow in the electrical trade.
3. Define the various sectors of the electrical industry.
4. State the tasks typically performed by an electrician.
5. Explain the responsibilities and aptitudes of an electrician.

Performance Tasks

This is a knowledge-based module. There are no performance tasks.

Trade Terms

Electrical service
Occupational Safety and Health Administration (OSHA)
On-the-job learning (OJL)
Raceway systems
Rough-in
Trim-out

Required Trainee Materials

1. Paper and pencil
2. Copy of the latest edition of the *National Electrical Code*®

Note:
NFPA 70®, *National Electrical Code*®, and *NEC*® are registered trademarks of the National Fire Protection Association, Inc., Quincy, MA 02269. All *National Electrical Code*® and *NEC*® references in this module refer to the 2014 edition of the *National Electrical Code*®.

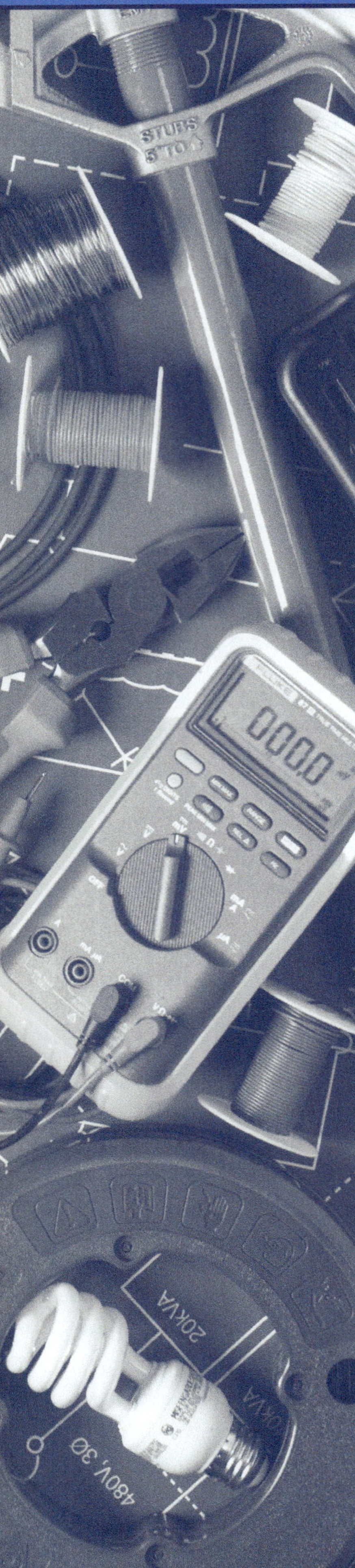

Contents

Topics to be presented in this module include:

Figures

1.0.0 Introduction

The electrical field can be divided into three broad categories: residential, commercial, and industrial systems. As you study to become an electrician, you will reach a point at which you must decide which area of electrical work you want to pursue. Many skilled electricians become comfortable in residential or commercial wiring, while others feel at home in large industrial facilities such as petrochemical plants, installing or maintaining huge electrical systems including motors and control devices. Later on in your career, you may decide to start your own company or teach others the trade.

1.1.0 Residential Wiring

Components of a residential electrical system include an electrical supply, **electrical service**, nonmetallic-sheathed cable, nail-on device boxes, panelboards, and fixtures. Phases of residential electrical wiring include **rough-in**, **trim-out**, testing, and troubleshooting. *Figure 1* shows examples of some primary components of residential wiring, including:

- Pad-mounted transformer
- Electrical service
- Nail-on device box
- Nonmetallic-sheathed cable
- Interior panel (subpanel)
- Luminaire (lighting fixture)

Interior panel enclosures, such as the one shown in *Figure 1(E)*, are typically installed and partially terminated during the rough-in stage.

1.2.0 Commercial Wiring

Electrical installations in commercial structures contain many of the same elements as residential installations. One major exception, however, is that in commercial and industrial electrical installations, conductors are typically installed in metal raceways, requiring the installing electricians to be skilled in conduit bending. A well-trained electrician has the ability to install a metal **raceway system** with little or no waste in conduit, while an inexperienced beginner will typically go through several pieces of conduit before acquiring the necessary bend. Practice makes perfect in conduit bending.

Figure 2 shows some elements that make up a commercial electrical system. These include a pad-mounted commercial transformer, the electrical service, conduit, a fire alarm system, and lighting.

1.3.0 Industrial Wiring

Because of the hazardous materials that exist in many industrial facilities, the installation and maintenance of electrical systems in these volatile environments must follow rigid requirements as governed by the *National Electrical Code® (NEC®)*. For similar reasons, commercial and residential installations also have strict code requirements that must be followed.

Conduit systems in volatile environments must be sealed to outside vapors and gases, and any potential sparking or arcing device must be enclosed within a special enclosure or casing to prevent the ignition of hazardous vapors that might be present.

Industrial electricians are generally split into two groups: installers and maintenance personnel. In many large industrial facilities, electrical systems are typically installed by contract electricians who do not work directly for the facility in which they are working, but for a contractor hired by the facility. It is the responsibility of these electricians to install conduit systems, conductors, motors, and equipment. They then turn the completed system over to maintenance electricians who work directly for the facility. These electricians maintain the system once it is energized and operating. In smaller facilities, industrial plant electricians may both install and maintain the electrical equipment.

Figure 3 illustrates some of the electrical equipment that may be found in industrial facilities, including distribution switchgear, rigid metallic conduit (RMC), and a motor control center.

(A) PAD-MOUNTED TRANSFORMER

(B) RESIDENTIAL ELECTRICAL SERVICE

(C) DEVICE BOX

(D) NONMETALLIC-SHEATHED CABLE

(E) INTERIOR PANEL (SUBPANEL)

(F) LUMINAIRE

26101-14_F01.EPS

Figure 1 Primary components of residential wiring.

(A) PAD-MOUNTED COMMERCIAL TRANSFORMER

(B) COMMERCIAL ELECTRICAL SERVICE

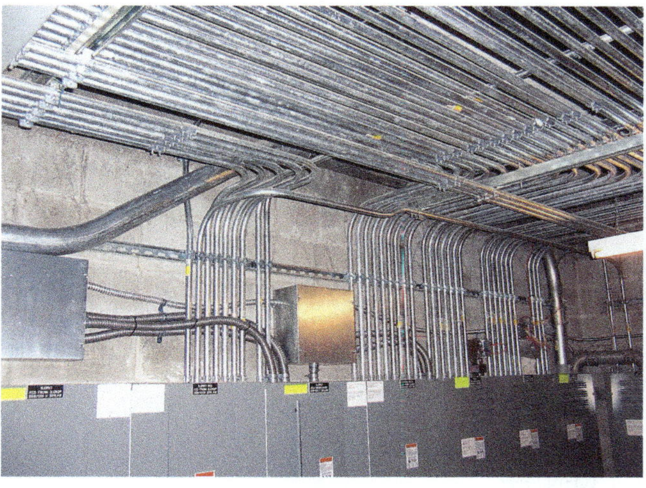
(C) CONDUIT SYSTEM

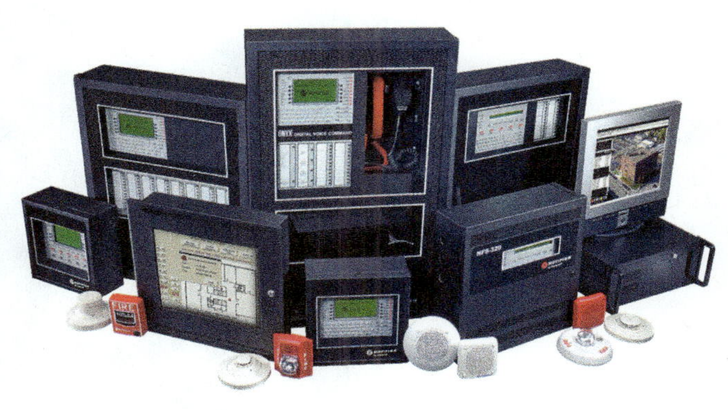
(D) FIRE ALARM SYSTEM

(E) OFFICE LIGHTING

(F) OUTDOOR LIGHTING

26101-14_F02.EPS

Figure 2 Commercial electrical systems.

(A) DISTRIBUTION SWITCHGEAR

(B) RIGID CONDUIT SYSTEM

(C) MOTOR CONTROL CENTER

26101-14_F03.EPS

Figure 3 Typical industrial electrical equipment.

2.0.0 Career Opportunities in the Electrical Field

Try to visualize anything operating without electricity. We live in a world of electrical dependency, with most of us taking its availability for granted. We have all experienced the unexpected power failures and power outages due to weather conditions, brownouts, or even blackouts.

It takes a small army of electrically skilled individuals to generate, transmit, distribute, and maintain electrical systems and equipment in order for us to have the convenience of continuous and quality electrical energy at our fingertips. Examples of electrical occupations include residential electrician, commercial electrician, industrial electrician, and electrical maintenance technician.

2.1.0 What You Can Do with Electrical Training and Experience

Wherever you might go, there is a very good chance that you will come in contact with the effects of electricity. Driving through the city and suburbs you see buildings, stores, and signs lighted by electricity. The buildings you see have various kinds of electrical systems installed for cooling, heating, lighting, and for powering all the equipment in the building. This is true of a small home as well as the largest buildings in the world. You will find it difficult to do anything in today's world without having electricity involved. Getting ready for work, cooking, shopping, seeing a movie, surfing the Web, using power tools, and most other things you might do during a typical day require electricity.

On Site

Lighting the New York Skyline

The first searchlight on top of the Empire State Building heralded the election of Franklin D. Roosevelt in 1932. A series of floodlights were installed in 1964 to illuminate the top 30 floors of the building. Today, the color of the lights is changed to mark various events. For example, yellow marks the U.S. Open and red, white, and blue marks Independence Day.

The building is lit from the 72nd floor to the base of the TV antenna by 204 metal halide lamps and 310 fluorescent lamps. In 1984, a color-changing apparatus was added in the uppermost mooring mast. There are 880 vertical and 220 horizontal fluorescent lights. The colors can be changed with the flick of a switch.

26101-14_SA01.EPS

Once you begin to look, you will see the effects of electricity everywhere. All of the electrical systems in the United States and all over the world must be installed and maintained by someone qualified to do the work. That means that you have a very big opportunity for work that is rewarding both personally and financially. Growth in the industry, new technology, upgrading and retrofitting existing equipment, and retirement of current workers create opportunities for trained and skilled electricians.

The demand for skilled electricians is high. There is a huge amount of construction work underway. New homes, schools, office buildings, malls, airports, industrial plants, and many other types of structures are being constructed every

On Site

Wide World of Sports

Every fall, 600 men and women from 47 countries participate in the U.S. Open tennis tournament held in New York. The U.S.T.A. National Tennis Center includes 33 outdoor courts, 9 indoor courts, and 3 stadium-style courts. It is the largest public tennis facility in the world. When the tennis center was expanded, electricians laid over 415 miles of high-performance cable with over 80,000 terminations and 5 miles of fiber optic cable. The scope of work included multimedia systems covering network broadcast, internal cable TV distribution, scoring systems, and audio capabilities. The broadcast system includes 6 TV studios, 3 interview rooms, and 30 broadcast booths. During the Open, the scores from all the courts are posted live through the complex and to the U.S. Open website.

day. According to the Bureau of Labor Statistics, this new construction creates a very large demand for new workers every year.

2.1.1 Earn While You Learn

While you are learning the trade, typically through an **on-the-job learning (OJL)** program, you are making money. You are getting paid to learn the trade. Compare that with a typical college student who may be paying a large tuition bill while attending class. Another advantage of learning on the job is the hands-on experience you get. There is no substitute in any job for hands-on experience.

Pay in the construction industry is very good, and the pay for electricians is close to the top of the scale for all occupations.

2.1.2 A Diversified and Interesting Career

There are many ways to increase your skills and grow professionally in construction. There are many opportunities to try different types of jobs and earn more money within the electrical trade. You will find that electrical work is demanding but fulfilling. There is a large variety of work to be done. You may be indoors installing boxes or hooking up motor controllers, or you may be outside climbing a ladder or running conduit on a pipe rack high in the air.

Electricity and electrical equipment is needed everywhere. This means that jobs for electricians are also needed everywhere.

On Site

Electricians Connect the New England Patriots to the 21st Century

It took a team of 200 electricians to build Gillette Stadium™ for the New England Patriots in 2000. The main scope of work was a $30 million contract for the stadium's power, lighting, voice and data systems, and aboveground electrical infrastructure. In 16 months, electricians installed 4,000,000 feet of wire and cable, 900,000 feet of raceway, 310 panel boards, 11,000 interior light fixtures, and 728 sports lighting fixtures. The project also included a 2,000kV generator and 5 substations for emergency power systems.

2.2.0 Residential Electrician

The primary goal of a residential electrician is to provide a complete electrical system in a residential structure. Elements of a residential wiring installation include installing the electrical service entrance equipment, branch circuit conductors, device boxes, panel enclosures, overcurrent protective devices (circuit breakers), and the fixtures such as lighting and smoke detectors.

Residential electrical contractors are the primary employers of residential electricians. Various methods of employment are available, depending on the policies of the contractor. Residential electricians may work directly for the contractor, or may be hired as individual contractors who are responsible for filing their own taxes and providing their own insurance, as well as supplying their own equipment. These types of electricians are generally paid by piece work, meaning that they are paid a fixed price for each house they complete. Tract homes, similar to those shown in *Figure 4*, are frequently wired by residential electricians.

26101-14_F04.EPS

Figure 4 Tract homes (subdivisions).

2.3.0 Commercial Electrician

Commercial electricians install power, light, and control wiring in a variety of locations including apartment buildings, stores, offices, service stations, and hospitals.

Employers of commercial electricians are generally electrical contractors who work as subcontractors for a general contractor. Many electrical contractors install both residential and commercial wiring, such as in metal frame commercial buildings, as shown in *Figure 5*. However, they often rely on electricians that are specialists in one or the other type of wiring.

26101-14_F05.EPS

Figure 5 Metal frame commercial building.

2.4.0 Industrial Electrician

Electricians who specialize in installing electrical systems in industrial facilities require additional training due to the amount of specialty equipment that must be installed and tested. Electricians working in hazardous locations must understand the special code requirements associated with these locations. They must differentiate between the hazardous classes and divisions, and know the requirements for each class and division. They must be familiar with three-phase power, motors, and motor control systems. Industrial electricians may also be responsible for installing the conduit and wiring for process control instrumentation, such as the installation shown in *Figure 6*. They must be able to troubleshoot any of these systems should they fail during initial testing.

Contractors who specialize in designing and building industrial facilities are typically the employers of industrial electricians.

26101-14_F06.EPS

Figure 6 Instrumentation installation.

2.5.0 Electrical Maintenance Technician

Electrical maintenance technicians can be found in commercial and industrial facilities. They normally work directly for the owner or management of the facility, and in large facilities are members of a maintenance group supervised by a maintenance manager or supervisor.

In industrial facilities, maintenance electricians are frequently responsible for both the electrical and instrumentation systems and equipment, and are referred to as E&I technicians. Instrumentation expertise is a trade of its own, and requires additional training over and above the electrical skills training found in this course.

Electrical maintenance electricians are usually employees of the facility; however, there are contract maintenance groups that provide maintenance personnel who work side-by-side with full-time plant personnel in the maintenance of the electrical and instrumentation systems.

One common component that all electrical maintenance personnel must be familiar with in industrial facilities is the magnetic motor starter, as illustrated in *Figure 7*. Industrial maintenance electricians should be able to disassemble and reassemble, troubleshoot, and repair magnetic motor starters, as these components frequently fail in these environments.

26101-14_F07.EPS

Figure 7 Magnetic motor starter.

3.0.0 Your Training Program

The Department of Labor (DOL) Office of Apprenticeship sets the minimum standards for training programs across the country. These programs rely on mandatory classroom instruction and on-the-job learning (OJL). They require at least 144 hours of classroom instruction per year and 2,000 hours of OJL per year. In a typical electrical apprenticeship program, trainees spend a total of at least 576 hours in classroom instruction and 8,000 hours in OJL before receiving journeyman certificates issued by registered apprenticeship programs.

To address the training needs of the professional communities, NCCER developed a four-year electrical training program. NCCER uses the minimum Department of Labor standards as a foundation for comprehensive curricula that provide trainees with in-depth classroom and OJL experience.

The standardized NCCER curriculum provides trainees with industry-driven training and education. It adopts a purely competency-based teaching approach. This means that trainees must show the instructor that they possess the knowledge and skills needed to safely perform the hands-on tasks that are covered in each module.

When the instructor is satisfied that a trainee has the required knowledge and skills for a given module, that information is sent to NCCER and kept in the National Registry. The National Registry can then confirm training and skills for workers as they move from state to state, company to company, or even within a company. See the *Appendix* for examples of the credentials issued by NCCER.

Whether you enroll in an NCCER program or another apprenticeship program, make sure you work for an employer or sponsor who supports a nationally standardized training program that includes credentials to confirm your skill development.

3.1.0 Apprenticeship

Apprentice training goes back thousands of years. Its basic principles have not changed over time. First, it is a means for a person entering the craft to learn from those who have mastered the craft. Second, it focuses on learning by doing; real skills versus theory. Some theory is presented in the classroom. However, it is always presented in a way that helps the trainee understand the purpose behind the skill that is to be learned.

3.1.1 Apprenticeship Standards

All apprenticeship standards prescribe certain work-related or on-the-job learning. This OJL is broken down into specific tasks in which the apprentice receives hands-on training. In addition, a specified number of hours is required in each task. The total number of OJL hours for an apprenticeship program is traditionally 8,000, which amounts to four years of training. In a competency-based program, it may be possible to shorten this time by testing out of specific tasks through a series of performance exams.

In a traditional program, the required OJL may be acquired in increments of 2,000 hours per year.

The apprentice must log all work time and turn it in to the apprenticeship committee so that accurate time control can be maintained. After each 1,000 hours of related work, the apprentice will receive a pay increase as prescribed by the apprenticeship standards.

Informal OJL provided by employers is usually less thorough than that provided through a formal apprenticeship program. The degree of training and supervision in this type of program often depends on the size of the employing firm. A small contractor may provide training in only one area, while a large company may be able to provide training in several areas.

For those entering an apprenticeship program, a high school or technical school education is desirable. Courses in shop, mechanical drawing, and general mathematics are helpful. Manual dexterity, good physical conditioning, and quick reflexes are important. The ability to solve problems quickly and accurately and to work closely with others is essential. You must also have a high concern for safety.

The prospective apprentice must submit certain information to the apprenticeship committee. This may include the following:

- Aptitude test (General Aptitude Test Battery or GATB Form Test) results (usually administered by the local Employment Security Commission)
- Proof of educational background (candidate should have school transcripts sent to the committee)
- Letters of reference from past employers and friends
- Proof of age
- If the candidate is a veteran, a copy of Form DD214
- A record of technical training received that relates to the construction industry and/or a record of any pre-apprenticeship training

NOTE

Some companies have physical activity requirements that must be met by apprentices. These requirements vary from company to company.

The apprentice must do the following:

- Wear proper safety equipment on the job
- Purchase and maintain tools of the trade as needed and required by the contractor
- Submit a monthly on-the-job training report to the committee
- Report to the committee if a change in employment status occurs
- Attend classroom-related instruction and adhere to all classroom regulations such as attendance requirements

3.1.2 Youth Apprenticeship Program

A Youth Apprenticeship Program is also available that allows students to begin their apprentice training while still in high school. A student entering the program in eleventh grade may complete as much as two years of the NCCER standardized four-year curriculum by high school graduation. In addition, the program, in cooperation with local craft employers, allows students to work in the trade and earn money while still in school. Upon graduation, the student can enter the industry at a higher level and with more pay than someone just starting the apprenticeship program.

This training program is similar to the one used by NCCER learning centers, contractors, and colleges across the country. Students are recognized through official transcripts and can enter the next year of the program wherever it is offered. They may also have the option of applying the credits at a two-year or four-year college that offers degree or certification programs in the construction trades.

3.1.3 Licensing

After you complete your training, you will probably want to take your state or local licensing exam. The purpose of licensing is to provide assurance that you are qualified to install and/or maintain electrical systems. You will be able to work independently and earn a higher income. As a licensed electrician, you are not only responsible for your work, but are liable for that work as well. If someone is working for you, then you are also responsible and liable for that person's work.

Licensing requirements vary from state to state, and may also vary by municipality. Contact your local building department for the requirements in your area. After you receive your license, your state or locality may require continuing education in order to renew your license.

Think About It

Licensing

What are the licensing requirements in your area?

4.0.0 Responsibilities of the Employee

In order to be successful, you must be able to use current trade materials, tools, and equipment to finish the task quickly and efficiently. You must keep abreast of technical advancements and continually gain the skills to use them. A professional never takes chances with regard to personal safety or the safety of others.

4.1.0 Professionalism

The word *professionalism* is a broad term that describes the desired overall behavior and attitude expected in the workplace. Professionalism is too often absent from the construction site and the various trades. Most people would argue that professionalism must start at the top in order to be successful. It is true that management support of professionalism is important to its success in the workplace, but it is just as important that individuals recognize their own responsibility for professionalism.

Professionalism includes honesty, productivity, safety, civility, cooperation, teamwork, clear and concise communication, being on time, and coming prepared to work. It can be demonstrated in a variety of ways every minute you are in the workplace.

Professionalism is a benefit to both the employer and the employee. It is a personal responsibility. The construction industry is what each individual chooses to make of it; choose professionalism and the industry image will follow.

4.2.0 Honesty

Honesty and personal integrity are important traits of the successful professional. Professionals pride themselves on performing a job well, and on being punctual and dependable. Each job is completed in a professional way, never by cutting corners or reducing materials. A valued professional maintains work attitudes and ethics that protect property such as tools and materials belonging to employers, customers, and other trades from damage or theft at the shop or job site.

Honesty and success go hand-in-hand for both the employer and the professional electrician. It is not simply a choice between good and bad, but a choice between success and failure. Dishonesty will always catch up with you. Whether you are stealing materials, tools, or equipment from the job site or simply lying about your work, it will not take long for your employer to find out. Of course, you can always go and find another employer, but this option will ultimately run out on you.

If you plan to be successful and enjoy continuous employment, consistency of earnings, and being sought after as opposed to seeking employment, then start out with the basic understanding of honesty in the workplace and you will reap the benefits.

Honesty means more, however, than just not taking things that do not belong to you. It also means giving a fair day's work for a fair day's pay. Employers place a high value on employees who display honesty.

4.3.0 Loyalty

Employees expect employers to look out for their interests, to provide them with steady employment, and to promote them to better jobs as openings occur. Employers feel that they, too, have a right to expect their employees to be loyal to them—to keep their interests in mind, to speak well of them to others, to keep any minor troubles strictly within the plant or office, and to keep absolutely confidential all matters that pertain to the business. Both employers and employees should keep in mind that loyalty is not something to be demanded; rather, it is something to be earned.

4.4.0 Willingness to Learn

Every company and job site has its own way of doing things. Employers expect their workers to be willing to learn these ways. You must be

On Site

Electricians Are Key Players for the NBA and NHL

Electricians are critical players in building modern sports complexes for professional sports franchises. The Pepsi Center was built in Denver, Colorado for the NBA's Denver Nuggets, the NHL's Colorado Avalanche, and several other teams. During the two-year construction cycle, electricians laid over 120 miles of conduit and 569 miles of wire, and installed 13,000 fixtures and 280 panels. The lighting systems are computer controlled and can be preset for basketball games and concerts. In addition to quality sound for concerts, the state-of-the-art sound and security system includes CCTV monitoring and card access security controls.

26101-14_SA02.EPS

willing to adapt to change and learn new methods and procedures as quickly as possible. Sometimes, a change in safety regulations or the purchase of new equipment makes it necessary for even experienced employees to learn new methods and operations. Successful people take every opportunity to learn more about their trade.

4.5.0 Willingness to Take Responsibility

Most employers expect their employees to see what needs to be done, then go ahead and do it. Once an assignment is received and the procedure and safety guidelines are fully understood, you should assume the responsibility for that task without further reminders.

4.6.0 Willingness to Cooperate

To cooperate means to work together. In our modern business world, cooperation is the key to getting things done. Learn to work as a member of a team with your employer, supervisor, and fellow workers in a common effort to get the work done efficiently, safely, and on time.

4.7.0 Rules and Regulations

People can work together well only if there is some understanding about what work is to be done, when and how it will be done, and who will do it. Rules and regulations are a necessity in any work situation and must be followed by all employees.

4.8.0 Tardiness and Absenteeism

Tardiness means being late for work, and absenteeism means being off the job for one reason or another. Consistent tardiness and frequent absences are an indication of poor work habits, unprofessional conduct, and a lack of commitment.

On Site

Ethical Principles for Members of the Construction Trades

Honesty: Be honest and truthful in all dealings. Conduct business according to the highest professional standards. Faithfully fulfill all contracts and commitments. Do not deliberately mislead or deceive others.

Integrity: Demonstrate personal integrity and the courage of your convictions by doing what is right even where there is pressure to do otherwise. Do not sacrifice your principles because it seems easier.

Loyalty: Be worthy of trust. Demonstrate fidelity and loyalty to companies, employers and sponsors, co-workers, trade institutions, and other organizations.

Fairness: Be fair and just in all dealings. Do not take undue advantage of another's mistakes or difficulties. Fair people are open-minded and committed to justice, equal treatment of individuals, and tolerance for and acceptance of diversity.

Respect for others: Be courteous and treat all people with equal respect and dignity.

Obedience: Abide by laws, rules, and regulations relating to all personal and business activities.

Commitment to excellence: Pursue excellence in performing your duties, be well-informed and prepared, and constantly try to increase your proficiency by gaining new skills and knowledge.

Leadership: By your own conduct, seek to be a positive role model for others.

On Site

Tips for a Positive Attitude

Here is a short checklist of things that you should keep in mind to help you develop and maintain a positive attitude:

- Remember that your attitude follows you wherever you go.
- Helpful suggestions and compliments are much more effective than negative ones.
- Look for the positive characteristics of your co-workers and supervisors.

Although workers don't get paid when they are absent or tardy, there is still a cost to the employer. For example, the worker's health care insurance must still be paid, even though the worker is not on site. Also, jobs are bid and scheduled based on a certain work force size. If you are not there, work is not getting done and schedules are not being met. It is important for you to be at work, on time, every day.

If it is necessary to stay home, phone the office early in the morning so that your supervisor can find another worker for the day.

5.0.0 Responsibilities of the Employer

Just as the employee has responsibilities on the job, the employer also has responsibilities. These are set out in the *Occupational Safety and Health Act* of 1970. The job of the **Occupational Safety and Health Administration (OSHA)** is to set occupational safety and health standards for all places of employment, enforce these standards, ensure that employers provide and maintain a safe workplace for all employees, and provide research and educational programs to support safe working practices.

OSHA was adopted with the stated purpose to assure as far as possible every working man and woman in the nation safe and healthful working conditions and to preserve our human resources.

OSHA requires each employer to provide a safe and hazard-free working environment. OSHA also requires that employees comply with OSHA rules and regulations that relate to their conduct on the job. To gain compliance, OSHA can perform spot inspections of job sites, impose fines for violations, and even stop any more work from proceeding until the job site is safe.

According to OSHA standards, you are entitled to on-the-job safety training. Your employer must do the following:

- Show you how to do each job safely
- Provide you with the required personal protective equipment
- Warn you about specific hazards
- Supervise you for safety while performing the work

The enforcement for this act of Congress is provided by the federal and state safety inspectors, who have the legal authority to impose fines for safety violations. The law allows states to have their own safety regulations and agencies to enforce them, but they must first be approved by the U.S. Secretary of Labor. For states that do not develop such regulations and agencies, federal OSHA standards must be obeyed.

These standards are listed in *OSHA Safety and Health Standards for the Construction Industry (29 CFR, Part 1926)*, sometimes called *OSHA Standards 1926*. Other safety standards that apply to construction are published in *OSHA Safety and Health Standards for General Industry (29 CFR, Parts 1900 to 1910)*, in *NFPA 70E®*, and the *NEC®*.

The most important general requirements that OSHA places on employers in the construction industry are as follows:

- The employer must post, in an easily seen area, signs informing employees of their rights and responsibilities.
- The employer must ensure that there are no serious hazards on the job site, and make sure that the workplace is in compliance with OSHA rules and regulations.
- Warning signs, posters, and labels must be posted in all required areas.
- The employer must perform frequent and regular job site inspections of equipment.
- The employer must instruct all employees to recognize and avoid unsafe conditions, and to know the regulations that pertain to the job so they may control or eliminate any hazards.
- No one may use any tools, equipment, machines, or materials that do not comply with *OSHA Standards 1926*.
- The employer must ensure that only qualified individuals operate tools, equipment, and machines.
- The employer must provide medical training and examinations when required by OSHA.
- Employers with greater than 10 employees must keep records of work-related injuries and illnesses. These records must be available to employees.
- Employers must not discriminate against employees who are exercising their rights under OSHA regulations.

Additional employer responsibilities are described in the *Americans with Disabilities Act of 1990 (ADA)*. If a worker has a disability but is qualified to do the job, that worker has equal rights to employment. The U.S. Equal Employment Opportunity Commission, along with state and local civil rights agencies, enforce ADA regulations.

6.0.0 Safety

In exchange for the benefits of your employment and your own well-being, you are obligated to work safely. You are also obligated to make sure anyone you supervise or work with is working safely. Your employer is obligated to maintain a safe workplace for all employees. Safety is everyone's responsibility.

You have a responsibility to maintain a safe working environment. This means two things:

- Follow your company's rules for proper working procedures and practices.
- Report any unsafe equipment and conditions directly to your supervisor.

On the job, if you see something that is not safe, report it! Do not ignore it. It will not correct itself. In the long run, even if you do not think an unsafe condition affects you, it does. Always report unsafe conditions. Do not think your employer will be angry because your productivity suffers while the condition is being reported. On the contrary, your employer will be more likely to criticize you for not reporting a problem.

Your employer knows that the short time lost in making conditions safe again is nothing compared with shutting down the whole job because of a major disaster. If that happens, you are out of work anyway. In fact, OSHA regulations require you to report hazardous conditions.

This applies to every part of the construction industry. Whether you work for a large contractor or a small contractor, you are obligated to report unsafe conditions.

In addition to the OSHA standards, there are specific standards related to electrical systems and devices. The *National Electrical Code® (NEC®)* sets the minimum standards for the safe installation of electrical systems. You will become very familiar with the *NEC®* as you progress through your training. *NEC®* temporary power requirements are more stringent than OSHA standards and can be enforced by the inspector and OSHA.

Another standard you should become familiar with is *NFPA 70E®, Standard for Electrical Safety in the Workplace.* This standard covers safe work practices that must be applied when working on or near exposed energized parts, and describes in detail the steps required for putting a circuit or electrical system into an electrically safe working condition. It also covers safe approach distances to exposed energized parts, and introduces the little understood but very dangerous reality of arc flash hazard. Proper personal protective equipment (PPE) required to protect oneself from both the hazard of electrical shock and the hazard of arc flash is extensively covered. *Figure 8* shows a higher level of PPE (arc flash hood) that will be required when exposed to a potentially high level of arc flash.

26101-14_F08.EPS

Figure 8 Arc flash hood.

GOING GREEN

Electricians Help Reduce Energy Consumption

Electricians are working to protect the environment and reduce energy consumption. Some companies specialize in energy efficiency in new installations and upgrading older facilities. One company assisted the World Bank in Washington, DC to upgrade their offices and reduced annual energy consumption by 4,500kW.

The World Bank offices occupy over 3.5 million square feet in five separate buildings. Over 50,000 fluorescent light fixtures and 1,400 exit signs were upgraded to improve energy efficiency and reduce lamp maintenance. This resulted in an annual energy savings of over $800,000 and eliminates the annual release of over 7,000,000 pounds of oxides, which some people believe causes global climate change and acid rain.

Summary

The electrical trade can be divided into three main areas of expertise: residential, commercial, and industrial. Each of these areas requires specific skills and training.

Regardless of the area of expertise, all electricians must understand the fundamentals of electricity, the hazards associated with electrical shock and arc flash, and the practices that must be applied in order to work safely.

Job opportunities in the electrical field can provide an above-average income and a rewarding yet challenging career with many levels of advancement possible.

Review Questions

1. Interior panels in residential wiring are typically installed during which phase of work?
   a. Rough-in
   b. Trim-out
   c. Service installation
   d. Planning

2. A residential wiring system typically uses _____.
   a. medium-voltage cable
   b. tray cable
   c. flat conductor cable
   d. nonmetallic-sheathed cable

3. Commercial wiring is typically installed in _____.
   a. pairs
   b. air ducting
   c. metal raceways
   d. PVC raceways

4. Which of the following would most likely require special knowledge of hazardous locations?
   a. Residential wiring
   b. Industrial wiring
   c. Commercial wiring
   d. Service work

5. RMC is a type of _____.
   a. electrical service
   b. conduit
   c. transformer
   d. motor control

6. Minimum standards for apprentice training programs are established by _____.
   a. OSHA
   b. the DOL
   c. NCCER
   d. your employer

7. Once you have finished your apprenticeship, your training is over.
   a. True
   b. False

8. It is okay to be a little late for work as long as you make up the time.
   a. True
   b. False

9. The primary mission of OSHA is to _____.
   a. inspect job sites for safety violations
   b. fine companies that violate safety regulations
   c. distribute safety equipment to workers
   d. ensure that employers maintain a safe workplace

10. If you see a safety violation at your job site, you should ignore it unless it affects you directly.
    a. True
    b. False

Module 26101-14
Supplemental Exercises

1. Phases of residential electrical wiring include ______________________________

____

2. True or False? A major difference between residential and commercial wiring is that commercial wiring is usually installed in metal conduit.

3. Industrial electricians are usually split into two groups: ____________________ and ____________________.

4. Your apprenticeship program requires ________ hours of OJL per year.
   a. 500
   b. 1,000
   c. 2,000
   d. 4,000

5. True or False? Competency-based training means that you must demonstrate the skills necessary to perform hands-on tasks before advancing to the next stage of the curriculum.

6. A ______________________ allows you to complete as much as two years of your apprenticeship before you have finished high school.

7. True or False? It's okay to take scrap pieces home as long as you don't think they'll be needed on the job.

8. Many employees feel that sick days can be treated as floating holidays to be taken whenever they would like a day off. How is this unfair to an employer or co-workers?

____

____

____

9. True or False? OSHA requires that employers provide a safe and hazard-free job site.

10. In addition to the OSHA safety standards, name two other standards that relate to electrical systems and devices.

____

____

Trade Terms Quiz

Fill in the blank with the correct term that you learned from your study of this module.

1. ___________ is the federal government agency established to ensure a safe and healthy environment in the workplace.
2. Job-related learning acquired while working is known as ___________.
3. The main panelboard enclosure would probably be installed in the ___________ stage.
4. The ___________ connects the commercial power to the premises wiring system.
5. Devices and fixtures would be installed during ___________.
6. Conduit is part of the ___________.

Trade Terms

Electrical service
Occupational Safety and Health Administration (OSHA)
On-the-job learning (OJL)
Raceway systems
Rough-in
Trim-out

Cornerstone of Craftsmanship

Tim Dean

Electrician/Electrical Trades Instructor
Central Ohio ABC/Madison
Comprehensive High School

Provide a summary of how you got started in the construction industry.
Upon exiting The University of Toledo, I took a job with my brother-in-law, who worked as an electrician in Akron, Ohio. I knew little or nothing about electricity but needed a job to support myself and my wife.

Who inspired you to enter the industry? Why?
I suppose my brother-in-law, Tom Argenio, was my inspiration. He possessed a work ethic and craftsmanship that is very rare in our society today. He taught me not only how to be a good tradesman but to understand the pride of quality workmanship.

What do you enjoy most about your job?
In the early days of my career, I simply appreciated having a job but as time passed and my experience grew, I acquired a thirst for knowledge and understanding. Knowing the why's, when's, where's, and how's brought new meaning to the skills I was attaining.

Do you think training and education are important in construction? If so, why?
The electrical trade is one of the most diverse and challenging of all the construction trades. Training and education are not only important but mandatory to stay safe, efficient, qualified, and prepared for the challenges of new technology.

How important are NCCER credentials to your career?
Being a part of NCCER has been one of the most rewarding experiences of my life. Acquiring credentials for performing as a Subject Matter Expert has benefited not only me personally by recognition and professional development but the organization I work for as well. I enjoy the challenges required to stay abreast of the continual evolution of our industry and NCCER provides the vehicle and resources for me to keep up with cutting-edge technological changes.

How has training/construction impacted your life and you career?
When I entered the trade, I was clueless. I had no idea what I was getting myself into but as I continued, I found a challenge in seeking an understanding of how and why things worked the way they did. I was engaged in the process and discovered that the more I knew and learned, the more I was worth to my employer.

Would you suggest construction as a career to others? If so, why?
I would recommend a career in the electrical trade to anyone who has a desire to learn and a thirst for understanding the new technologies that continue to evolve in our society. The electrical construction trade is both rewarding and challenging. Being a part of this great industry opens doors to new and exciting opportunities in many different areas. One only has to look around at the world we live in to know that being an electrician is much more than a job. Career opportunities abound and there is truly no limit to what you can attain.

How do you define craftsmanship?
To me craftsmanship is the result of acquiring knowledge, developing skills, infusing moral and ethical values, and blending personal pride to construct a product or process of recognizable and enduring quality.

Samples of NCCER Training Credentials

26101-14_A01.EPS

THE STANDARD FOR DEVELOPING CRAFT PROFESSIONALS

13614 Progress Blvd • Alachua, Florida 32615 • p. 888.622.7320 f. 386.518.6255 • www.nccer.org

Official Training Transcript

Below are your credentials from NCCER's National Registry. These industry-recognized credentials give you flexibility in planning your career and ensure your achievements follow you wherever you go.

To access your training online via the Automated National Registry (ANR) web site, go to **https://anr.nccer.org**, click the **Individuals** button, and enter your NCCER card number and PIN *(Note: Your card number is below, and also on your NCCER wallet card. The default PIN is the last four digits of your SSN, you can change it after you log in.)* The first time you log in you'll need to answer a few security questions.

NCCER Card #: 01234567
Trainee Name: Sample Student
Sponsor: YouthBuild USA/YouthBuild
Address: 58 Day St
Somerville, MA 02144

Current Employer/School:

Module	Description	Instructor	Training Location	Completed
00101-04	Basic Safety	James Harris	YouthBuild USA/YouthBuild International Quad Area YouthBuild	2/26/2013
00102-04	Introduction to Construction Math	James Harris	YouthBuild USA/YouthBuild International Quad Area YouthBuild	2/26/2013
00103-04	Introduction to Hand Tools	James Harris	YouthBuild USA/YouthBuild International Quad Area YouthBuild	2/26/2013
00104-04	Introduction to Power Tools	James Harris	YouthBuild USA/YouthBuild International Quad Area YouthBuild	2/26/2013
00105-04	Introduction to Blueprints	James Harris	YouthBuild USA/YouthBuild International Quad Area YouthBuild	2/26/2013
00106-04	Basic Rigging	James Harris	YouthBuild USA/YouthBuild International Quad Area YouthBuild	2/26/2013
00107-04	Basic Communication Skills	James Harris	YouthBuild USA/YouthBuild International Quad Area YouthBuild	12/4/2012
00108-04	Basic Employability Skills	James Harris	YouthBuild USA/YouthBuild International Quad Area YouthBuild	2/26/2013

NO ENTRIES BELOW THIS LINE

Page 1

Printed 6/17/201

26101-14_A02.EPS

26101-14_A03.EPS

Trade Terms Introduced in This Module

Electrical service: The electrical components that are used to connect the commercial power to the premises wiring system.

Occupational Safety and Health Administration (OSHA): The federal government agency established to ensure a safe and healthy environment in the workplace.

On-the-job learning (OJL): Job-related learning acquired while working.

Raceway systems: Conduit, fittings, boxes, and enclosures that house the conductors in an electrical system.

Rough-in: The installation of the raceway system (including conduit, boxes, and enclosures), wiring, or cable.

Trim-out: After rough-in, the installation and termination of devices and fixtures.

Additional Resources

This module presents thorough resources for task training. The following resource material is suggested for further study.

National Electrical Code® Handbook, Latest Edition. Quincy, MA: National Fire Protection Association.

Figure Credits

Associated Builders and Contractors, Inc., Module opener

Tim Dean, Figure 1A

Topaz Publications, Inc., Figures 1B–E, 2A, 3–8

Hubbell Incorporated, Figures 1F, 2E, 2F

Al Hamilton, Figure 2B

Mike Powers, Figure 2C

Honeywell Fire Systems, Americas, Figure 2D

Dumond Chemicals, Inc., SA01

Pepsi Center, Denver, Colorado, SA02

NCCER CURRICULA — USER UPDATE

NCCER makes every effort to keep its textbooks up-to-date and free of technical errors. We appreciate your help in this process. If you find an error, a typographical mistake, or an inaccuracy in NCCER's curricula, please fill out this form (or a photocopy), or complete the online form at **www.nccer.org/olf**. Be sure to include the exact module ID number, page number, a detailed description, and your recommended correction. Your input will be brought to the attention of the Authoring Team. Thank you for your assistance.

Instructors – If you have an idea for improving this textbook, or have found that additional materials were necessary to teach this module effectively, please let us know so that we may present your suggestions to the Authoring Team.

NCCER Product Development and Revision
13614 Progress Blvd., Alachua, FL 32615
Email: curriculum@nccer.org
Online: www.nccer.org/olf

❑ Trainee Guide ❑ AIG ❑ Exam ❑ PowerPoints Other ______________________

Craft / Level: Copyright Date:

Module ID Number / Title:

Section Number(s):

Description:

Recommended Correction:

Your Name:

Address:

Email: Phone:

Electrical Safety

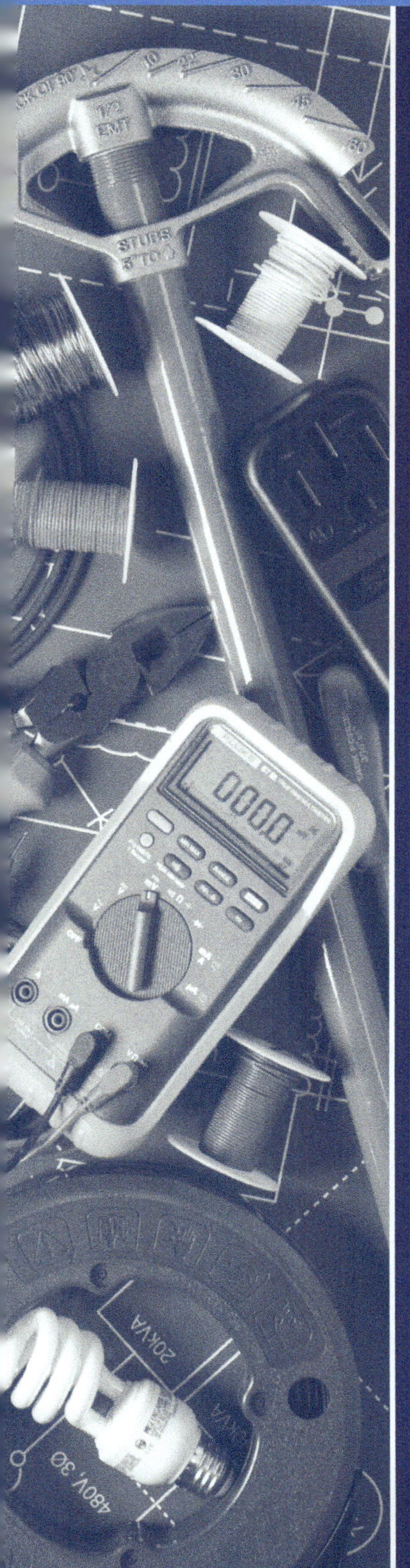

Ponnequin Wind Farm

The Ponnequin Wind Farm generates electrical power from the wind and is located on the plains of eastern Colorado just south of the Wyoming border. It consists of 44 wind turbines and can generate up to 30 megawatts of electricity. Each turbine weighs nearly 100 tons and stands 181 feet tall.

26102-14

Trainees with successful module completions may be eligible for credentialing through NCCER's National Registry. To learn more, go to **www.nccer.org** or contact us at **1.888.622.3720.** Our website has information on the latest product releases and training, as well as online versions of our *Cornerstone* magazine and Pearson's product catalog.

Your feedback is welcome. You may email your comments to **curriculum@nccer.org,** send general comments and inquiries to **info@nccer.org,** or fill in the User Update form at the back of this module.

This information is general in nature and intended for training purposes only. Actual performance of activities described in this manual requires compliance with all applicable operating, service, maintenance, and safety procedures under the direction of qualified personnel. References in this manual to patented or proprietary devices do not constitute a recommendation of their use.

26102-14
ELECTRICAL SAFETY

Objectives

When you have completed this module, you will be able to do the following:

1. Recognize safe working practices in the construction environment.
2. Explain the purpose of OSHA and how it promotes safety on the job.
3. Identify electrical hazards and how to avoid or minimize them in the workplace.
4. Explain safety issues concerning lockout/tagout procedures, confined space entry, respiratory protection, and fall protection systems.
5. Develop a task plan and a hazard assessment for a given task and select the appropriate PPE and work methods to safely perform the task.

Performance Tasks

Under the supervision of the instructor, you should be able to do the following:

1. Perform a visual inspection on various types of ladders.
2. Set up a ladder properly to perform a task.
3. Properly don a harness.
4. Perform a hazard assessment of a job such as replacing the lights in your classroom.
   - Discuss the work to be performed and the hazards involved.
   - Locate the phone closest to the work site and ensure that the local emergency telephone numbers are either posted at the phone or known by you and your partner(s).
   - Plan an escape route from the location in the event of an accident.

Trade Terms

Double-insulated/ungrounded tool
Fibrillation
Grounded tool
Ground fault circuit interrupter (GFCI)
Polychlorinated biphenyls (PCBs)

Required Trainee Materials

1. Paper and pencil
2. Copy of the latest edition of the *National Electrical Code®*
3. Appropriate personal protective equipment

Note:
NFPA 70®, *National Electrical Code®*, and *NEC®* are registered trademarks of the National Fire Protection Association, Inc., Quincy, MA 02269. All *National Electrical Code®* and *NEC®* references in this module refer to the 2014 edition of the *National Electrical Code®*.

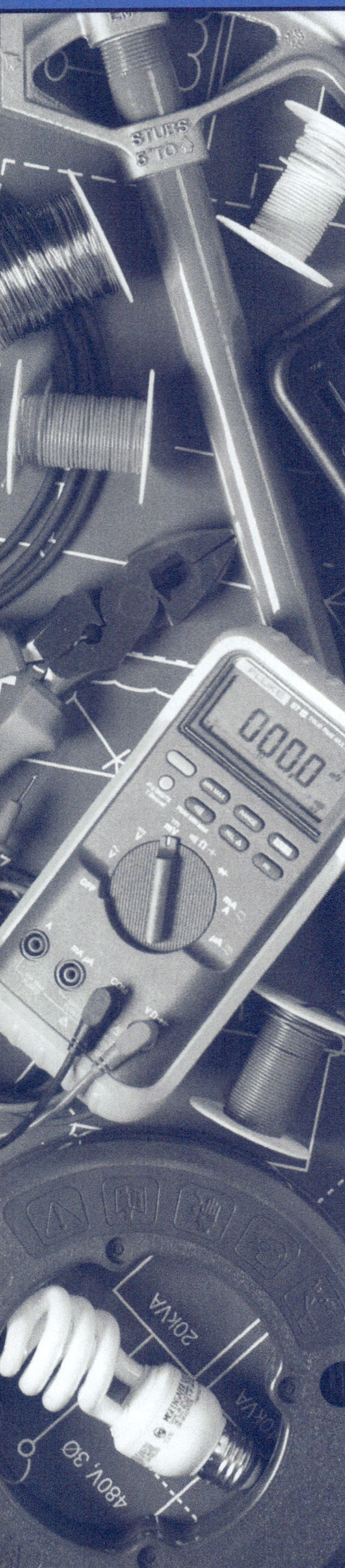

Contents

Topics to be presented in this module include:

Figures and Tables

1.0.0 Introduction

In order to be safe, you need to be aware of potential hazards and stay constantly alert to these hazards. You must take the proper precautions and practice the basic rules of safety. You must be safety-conscious at all times and report any unsafe conditions to your supervisor and co-workers. Safety should become a habit. Keeping a safe attitude on the job will go a long way in reducing the number and severity of accidents. Remember that your safety is up to you.

As an apprentice electrician, you need to be especially careful. You should only work under the direction of experienced personnel who are familiar with the various job-site hazards and the means of avoiding them.

The most life-threatening hazards on a construction site are:

- Falls when you are working in high places
- The possibility of being crushed by falling materials or equipment
- Electric shock and arc-related burns caused by coming into contact with live electrical circuits
- The possibility of being struck by flying objects or moving equipment/vehicles such as trucks, forklifts, and construction equipment

These exposures account for approximately 90% of fatalities on construction sites. Other hazards include cuts, burns, back sprains, and getting chemicals or objects in your eyes. Most injuries, both those that are life-threatening and those that are less severe, are preventable if the proper precautions are taken.

What's wrong with this picture?

26102-14_SA01.EPS

2.0.0 Electrical Shock

Electricity can be described as a potential that results in the movement of electrons in a conductor. This movement of electrons is called electrical current. Some substances, such as silver, copper, steel, and aluminum, are excellent conductors. The human body is also a conductor. The conductivity of the human body greatly increases when the skin is wet or moistened with perspiration.

Electrical current flows along any path in which the voltage can overcome the resistance. If the human body contacts an electrically energized point and is also in contact with the ground or another point in the circuit, the human body becomes a path for the current. *Table 1* shows the effects of current passing through the human body. One mA is one milliamp, or one one-thousandth of an ampere.

Table 1 Current Level Effects on the Human Body

Current Value	Typical Effects
1mA	Perception level. Slight tingling sensation.
5mA	Slight shock. Involuntary reactions can result in serious injuries such as falls from elevations.
6 to 30mA	Painful shock, loss of muscular control.
50 to 150mA	Extreme pain, respiratory arrest, severe muscular contractions. Death possible.
1000mA to 4300mA	Ventricular fibrillation, severe muscular contractions, nerve damage. Typically results in death.

Source: Occupational Safety and Health Administration

26102-14_T01.EPS

On Site

Electrical Safety in the Workplace

Each year in the U.S., there are approximately 20,000 electricity-related accidents at home and in the workplace. In a recent year, these accidents resulted in 700 deaths. Electrical accidents are the third leading cause of death in the workplace.

Think About It

Severity of Shock

In *Table 1,* how many milliamps separate a mild shock from a potentially fatal one? What is the fractional equivalent of this in amps? How many amps are drawn by a 60W light bulb?

A primary cause of death from electrical shock is when the heart's rhythm is overcome by an electrical current. Normally, the heart's operation uses a very low-level electrical signal to cause the heart to contract and pump blood. When an abnormal electrical signal, such as current from an electrical shock, reaches the heart, the low-level heartbeat signals are overcome. The heart begins twitching in an irregular manner and goes out of rhythm with the pulse. This twitching is called **fibrillation**. The use of cardiopulmonary resuscitation (CPR) can keep oxygen flowing to the body, but unless the normal heartbeat is restored using special defibrillation equipment (paddles), the individual will die. Other effects of electrical shock may include immediate heart stoppage and burns. In addition, the body's reaction to the shock can cause a fall or other accident. Delayed internal problems can also result. This is why it is critical for you to have a medical exam if you receive even a minor shock.

2.1.0 The Effect of Current

The amount of current measured in amperes that passes through a body determines the outcome of an electrical shock. The higher the voltage, the greater the chance for a fatal shock. In a one-year study in California, the following results were observed by the State Division of Industry Safety:

- Thirty percent of all electrical accidents were caused by contact with conductors. Of these accidents, 66% involved low-voltage conductors (those carrying 600 volts [V] or less).
- Portable, electrically operated hand tools made up the second largest number of injuries (15%). Almost 70% of these injuries happened when the frame or case of the tool became energized. These injuries could have been prevented by following proper safety practices, using properly maintained grounded or **double-insulated/ungrounded tools**, and using **ground fault circuit interrupter (GFCI)** protection.

NOTE

Electric shocks or burns are a major cause of accidents in the construction industry. According to the National Institute for Occupational Safety and Health, workers in the construction industry are four times more likely to be electrocuted at work than all other industries combined.

In one ten-year study, investigators found 9,765 electrical injuries in the U.S. A little more than 13% of the high-voltage injuries (over 600V) resulted in death. These high-voltage totals included limited-amperage contacts, which are often found on

CAUTION

High voltage, defined for this module as 600V or more, is almost ten times as likely to kill as low voltage. However, on the job you spend most of your time working on or near lower voltages. Due to the frequency of contact, most electrocution deaths actually occur at low voltages. Attitude about the harmlessness of lower voltages undoubtedly contributes to this statistic.

On Site

Dangers of Electricity

Never underestimate the power of electricity. For example, the current through a 25W light bulb is more than enough to kill you.

Think About It

Electrocution

Why can a bird perch safely on an electric wire? Squirrels are a common cause of shorts at substations; why does a squirrel get electrocuted when a bird does not?

electronic equipment. When tools or equipment touch high-voltage overhead lines, the chance that a resulting injury will be fatal climbs to 28%. Of the low-voltage injuries, 1.4% were fatal.

These statistics have been included to help you gain respect for the environment where you work and to stress how important safe working habits really are.

2.1.1 Body Resistance

Electricity travels in closed circuits, and its normal route is through a conductor. Shock occurs when the body becomes part of the electric circuit (*Figure 1*). The current must enter the body at one point and leave at another. Shock normally occurs in one of three ways: the person must come in contact with both wires of the electric circuit; one wire of the electric circuit and the ground; or a metallic part that has become live by being in contact with an energized wire while the person is also in contact with the ground.

To fully understand the harm done by electrical shock, you need to understand something about the physiology of certain body parts: the skin, the heart, and muscles.

Skin covers the body and is made up of three layers. The most important layer, as far as electric shock is concerned, is the outer layer of dead cells referred to as the horny layer. This layer is composed mostly of a protein called keratin, and it is the keratin that provides the largest percentage of the body's electrical resistance. When it is dry, the outer layer of skin may have a resistance of several thousand ohms. When it is moist, there is a radical drop in resistance, as is also the case if there is a cut or abrasion that pierces the horny layer. The amount of resistance provided by the skin will vary widely from person to person. A worker with a thick horny layer will have a much higher resistance than a child. The resistance will also vary widely at different parts of the body. For instance, the worker with high-resistance hands may have low-resistance skin on the back of his calf.

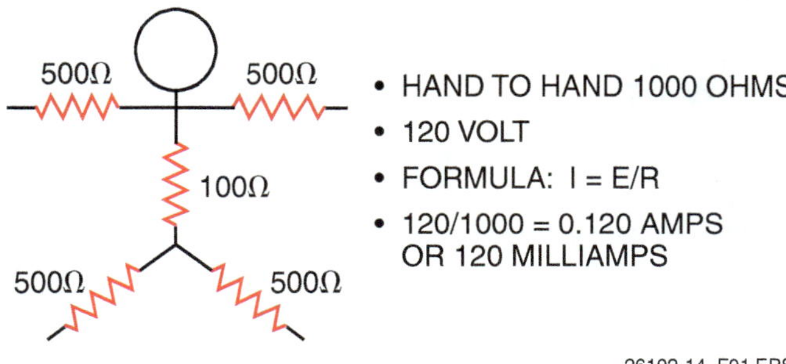

Figure 1 Body resistance.

The heart is the pump that sends life-sustaining blood to all parts of the body. The blood flow is caused by the contractions of the heart muscle, which is controlled by electrical impulses. The electrical impulses are delivered by an intricate system of nerve tissue with built-in timing mechanisms, which make the chambers of the heart contract at exactly the right time. An outside electric current of as little as 75 milliamperes can upset the rhythmic, coordinated beating of the heart by disturbing the nerve impulses. When this happens, the heart is said to be in fibrillation, and the pumping action stops. Death will occur quickly if the normal beat is not restored. Remarkable as it may seem, what is needed to defibrillate the heart is a shock of an even higher intensity.

The other muscles of the body are also controlled by electrical impulses delivered by nerves. Electric shock can cause loss of muscular control, resulting in the inability to let go of an electrical conductor. Electric shock can also cause injuries of an indirect nature in which involuntary muscle reaction from the electric shock can cause bruises, fractures, and even death resulting from collisions or falls.

The severity of shock received when a person becomes a part of an electric circuit is affected by three primary factors: the amount of current flowing through the body (measured in amperes), the path of the current through the body, and the length of time the body is in the circuit. Other factors that may affect the severity of the shock are the frequency of the current, the phase of the heart cycle when shock occurs, and the general health of the person prior to the shock. Effects can range from a barely perceptible tingle to immediate cardiac arrest. Although there are no absolute limits, or even known values that show the exact injury at any given amperage range, *Table 1* lists the general effects of electric current on the body for different current levels. As this table illustrates, a difference of only 100 milliamperes exists between a current that is barely perceptible and one that is likely to kill you.

A severe shock can cause considerably more damage to the body than is visible. For example, a person may suffer internal hemorrhages and destruction of tissues, nerves, and muscle. In addition, shock is often only the beginning in a chain of events. The final injury may well be from a fall, cuts, burns, or broken bones.

2.1.2 Burns

The most common shock-related injury is a burn. Burns suffered in electrical accidents may be of three types: electrical burns, arc burns, and thermal contact burns.

> **Think About It**
>
> **Bodily Harm**
>
> What factors affect the amount of damage to the body during an electric shock?

Electrical burns are the result of electric current flowing through the tissues or bones. Tissue damage is caused by the heat generated by the current flow through the body. An electrical burn is one of the most serious injuries you can receive, and should be given immediate attention. Since the most severe burning is likely to be internal, what may appear at first to be a small surface wound could, in fact, be an indication of severe internal burns.

Arc burns make up a substantial portion of the injuries from electrical malfunctions. The electric arc between metals can be up to 35,000°F, which is about four times hotter than the surface of the sun. Workers several feet from the source of the arc can receive severe or fatal burns. Since most electrical safety guidelines recommend safe working distances based on shock considerations, workers can be following these guidelines and still be at risk from arc. Electric arcs can occur due to poor electrical contact or failed insulation. Electrical arcing is caused by the passage of substantial amounts of current through the vaporized terminal material (usually metal or carbon).

> **CAUTION**
>
> Since the heat of the arc is dependent on the short circuit current available at the arcing point and the time it takes to clear (trip), arcs generated by low-voltage systems can be just as dangerous as those generated at 13,000V.

The third type of burn is a thermal contact burn. It is caused by contact with objects thrown during the blast associated with an electric arc. This blast comes from the pressure developed by the near-instantaneous heating of the air surrounding the arc, and from the expansion of the metal as it is vaporized. (Copper expands by a factor in excess of 65,000 times in boiling.) The pressure wave can be great enough to hurl people, switchgear, and cabinets considerable distances. It can stop your heart, impale you with shrapnel, blow off limbs, cause deafness, and cause you to inhale vaporized metal. Another hazard associated with the blast is the hurling of molten metal droplets, which can also cause thermal contact burns and associated damage.

3.0.0 Reducing Your Risk

There are many things that can be done to greatly reduce the chance of receiving an electrical shock. Always comply with your company's safety policy and all applicable rules and regulations, including job-site rules. In addition, the Occupational Safety and Health Administration (OSHA) publishes the *Code of Federal Regulations (CFR). CFR Part 1910* covers the OSHA standards for general industry and *CFR Part 1926* covers the OSHA standards for the construction industry.

Do not approach any electrical conductors closer than indicated in *Table 2* unless they are de-energized and your company has designated you as a qualified individual for that task. The values given in the table are minimum safe clearance distances; your company may have more restrictive requirements. *Table 2* is based on information provided in *NEC 70E®, Standard for Electrical Safety in the Workplace, Table 130.4(C)(a).*

3.1.0 Protective Equipment

You must also become familiar with common personal protective equipment. In particular, know the voltage rating of each piece of equipment. Rubber gloves are used to prevent the skin from coming into contact with energized circuits. A separate leather cover protects the rubber glove from punctures and other damage (see *Figure 2*). The leather protectors also provide the wearer with a certain level of protection against arc flash. OSHA addresses the use of protective equipment, apparel, and tools in *CFR 1910.335(a)*. This article is divided into two sections: *Personal Protective Equipment* and *General Protective Equipment and Tools.*

Table 2 Limited Approach Boundaries to Live Parts

Nominal System Voltage Range (Phase-to-Phase)	Limited Approach Boundary from Fixed Circuit Component
50 to 300	3 ft 6 in
301 to 750	3 ft 6 in
751 to 15kV	5 ft 0 in
15.1kV to 36kV	6 ft 0 in
36.1kV to 46kV	8 ft 0 in
46.1kV to 72.5kV	8 ft 0 in
72.6kV to 121kV	8 ft 0 in
138kV to 145kV	10 ft 0 in
161kV to 169kV	11 ft 8 in
230kV to 242kV	13 ft 0 in
345kV to 362kV	15 ft 4 in
500kV to 550kV	19 ft 0 in
765kV to 800kV	23 ft 9 in

26102-14_T02.EPS

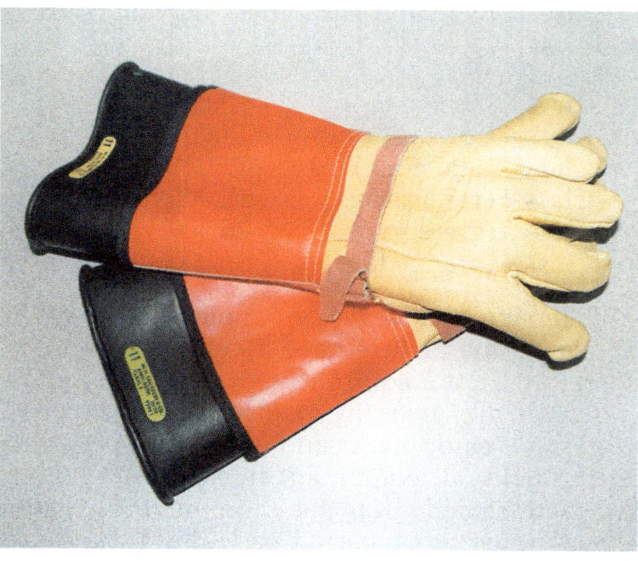

26102-14_F02.EPS

Figure 2 Rubber gloves and leather protectors.

The first section, *Personal Protective Equipment,* includes the following requirements:

- Employees working in areas where there are potential electrical hazards shall be provided with, and shall use, electrical protective equipment that is appropriate for the specific parts of the body to be protected and for the work to be performed.
- Protective equipment shall be maintained in a safe, reliable condition and shall be periodically inspected or tested, as required by *CFR 1910.137(b)/1926.95(a).*
- If the insulating capability of protective equipment may be subject to damage during use, the insulating material shall be protected.
- Employees shall wear nonconductive head protection wherever there is a danger of head injury from electric shock or burns due to contact with exposed energized parts.
- Employees shall wear protective equipment for the eyes and face wherever there is danger of injury to the eyes or face from electric arcs or flashes or from flying objects resulting from an electrical explosion.

Case History

Arc Blast

An electrician in Louisville, KY was cleaning a high-voltage switch cabinet. He removed the padlock securing the switch enclosure and opened the door. He used a voltage meter to verify absence of voltage on the three load phases at the rear of the cabinet. However, he did not test all potentially energized parts within the cabinet, and some components were still energized. He was wearing standard work boots and safety glasses, but was not wearing protective rubber gloves or an arc suit. As he used a paintbrush to clean the switch, an arc blast occurred, which lasted approximately one-sixth of a second. The electrician was knocked down by the blast. He suffered third-degree burns and required skin grafts on his arms and hands. The investigation determined that the blast was caused by debris, such as a cobweb, falling across the open switch.

The Bottom Line: Always wear the appropriate personal protective equipment and test all components for voltage before working in or around electrical devices.

The second section, *General Protective Equipment and Tools,* includes the following requirements:

- When working near exposed energized conductors or circuit parts, each employee shall use insulated tools or handling equipment if the tools or handling equipment might make contact with such conductors or parts. If the insulating capability of insulated tools or handling equipment is subject to damage, the insulating material shall be protected.
- Fuse pullers, insulated for the circuit voltage, shall be used to remove or install fuses.

WARNING!

Fuses should never be installed or removed when energized. This could cause an electrical arc, which can result in death or a serious injury.

- Ropes and handlines used near exposed energized parts shall be nonconductive.
- Protective shields, protective barriers, or insulating materials shall be used to protect each employee from shock, burns, or other electrically related injuries while that employee is working near exposed energized parts that might be accidentally contacted or where dangerous electric heating or arcing might occur. When normally enclosed live parts are exposed for maintenance or repair, they shall be guarded to protect unqualified persons from contact with the live parts.

The types of electrical safety equipment, protective apparel, and protective tools available for use are quite varied. This module will discuss the most common types of safety equipment. These include the following:

- Currently tested rubber protective equipment, including gloves and blankets
- Protective apparel
- Natural fiber clothing
- Hot sticks
- Fuse pullers
- Shorting probes
- Safety glasses
- Face shields

3.1.1 Rubber Protective Equipment

All electrical workers may be exposed to energized circuits or equipment. Two of the most important articles of protection for electrical workers are insulated rubber gloves and rubber blankets, which must be matched to the voltage rating for the circuit or equipment. Rubber protective equipment is designed for the protection of the user. If it fails during use, a serious injury could occur.

Rubber protective equipment is available in two types. Type 1 designates rubber protective equipment that is manufactured of natural or synthetic rubber that is properly vulcanized, and Type 2 designates equipment that is ozone resistant, made from any elastomer or combination of elastomeric compounds. Ozone is a form of oxygen that is produced from electricity and is present in the air surrounding a conductor under high voltages. Normally, ozone is found at voltages of 10kV and higher, such as those found in electric utility transmission and distribution systems. Type 1 protective equipment can be damaged by corona cutting, which is the cutting action of ozone on natural rubber when it is under mechanical stress. Type 1 rubber protective equipment can also be damaged by ultraviolet rays. However, it is very important that the rubber protective equipment in use today be natural rubber (Type 1). Type 2 rubber protective equipment is very stiff and is not as easily worn as Type 1 equipment.

Various classes – The American National Standards Institute (ANSI) and the American Society for Testing and Materials International (ASTM) have designated a specific classification system for rubber protective equipment. The maximum AC use voltage and equipment tag colors are as follows:

- Class 00 (beige tag) 500V
- Class 0 (red tag) 1,000V
- Class 1 (white tag) 7,500V
- Class 2 (yellow tag) 17,000V
- Class 3 (green tag) 26,500V
- Class 4 (orange tag) 36,000V

Referring back to *Figure 2*, note that the gloves shown in the picture have a yellow tag and are therefore Class 2 equipment.

Inspection of protective equipment – Before rubber protective equipment can be worn by personnel in the field, all equipment must have a current test date stenciled on the equipment. Insulating gloves must be inspected each day by the user before they can be used. They must also be electrically tested every six months and any time the insulating value is in question. Because rubber protective equipment is used for personal protection and serious injury could result from its misuse or failure, it is important that an adequate safety factor be provided between the voltage on which it is to be used and the voltage at which it was tested.

All rubber protective equipment must be marked with the appropriate voltage rating and last inspection date. The markings that are required to be on rubber protective equipment must be applied in a manner that will not interfere with the protection that is provided by the equipment.

WARNING!

Never work on anything energized without direct instruction from your employer and appropriate safety equipment.

Gloves – Both high- and low-voltage rubber gloves are of the gauntlet type and are available in various sizes. To get the best possible protection and service life, here are a few general rules that apply whenever they are used in electrical work:

- Always wear leather protectors over your gloves as they provide burn protection not provided by the gloves themselves. Any direct contact with sharp or pointed objects may cut, snag, or puncture the gloves and take away the protection you are depending on.

- Always wear rubber gloves right side out (serial number and size to the outside).
- Always keep the gauntlets up. Rolling them down sacrifices a valuable area of protection. Tuck the sleeves of your shirt or protective suit under the glove cuffs to prevent an arc blast from coming inside your clothing.
- Always inspect and field check gloves before using them. Always check the inside for any debris.
- Use light amounts of manufacturer-approved glove dust or cotton liners with the rubber gloves. This gives the user more comfort, and it also helps to absorb some of the perspiration that can damage the gloves over years of use.
- Wash the rubber gloves in lukewarm, clean, fresh water after each use. Dry the gloves inside and out prior to returning to storage. Never use any type of cleaning solution on the gloves.
- Once the gloves have been properly cleaned, inspected, and tested, they must be properly stored. Store them in a cool, dry, dark place that is free from ozone, chemicals, oils, solvents, or other materials that could damage the gloves. Do not store gloves near hot pipes or in direct sunlight. Store both gloves and sleeves in their natural shape in a bag or box inside their leather protectors. They should be undistorted, right side out, and unfolded.
- Gloves can be damaged by many different chemicals, especially petroleum-based products such as oils, gasoline, hydraulic fluid inhibitors, hand creams, pastes, and salves. If contact is made with these or other petroleum-based products, the contaminant should be wiped off immediately. If any signs of physical damage or chemical deterioration are found (e.g., swelling, softness, hardening, stickiness, ozone deterioration, or sun checking), the protective equipment must not be used.
- Never wear watches or rings while wearing rubber gloves; this can cause damage from the inside out and defeats the purpose of using rubber gloves. Never wear anything conductive.
- Rubber gloves must be electrically tested every six months by a certified testing laboratory. Always check the inspection date before using gloves.
- Use rubber gloves only for their intended purpose, not for handling chemicals or other work. This also applies to the leather protectors.

Before rubber gloves are used, a visual inspection and an air test should be made. This should be done prior to use and as many times during the day as you feel necessary. To perform a visual inspection, stretch a small area of the glove, checking to see that no defects exist, such as:

- Embedded foreign material
- Deep scratches
- Pinholes or punctures
- Snags or cuts

Gloves and sleeves can be inspected by rolling the outside and inside of the protective equipment between the hands. This can be done by squeezing together the inside of the gloves or sleeves to bend the outside area and create enough stress to the inside surface to expose any cracks, cuts, or other defects. When the entire surface has been checked in this manner, the equipment is then turned inside out, and the procedure is repeated. It is very important not to leave the rubber protective equipment inside out as that places stress on the preformed rubber.

Remember, any damage at all reduces the insulating ability of the rubber glove. Look for signs of deterioration from age, such as hardening and slight cracking. Also, if the glove has been exposed to petroleum products, it should be considered suspect because deterioration can be caused by such exposure. If the gloves are suspect, turn them in for evaluation. If the gloves are defective, turn them in for disposal. Never leave a damaged glove lying around; someone may think it is a good glove and not perform an inspection prior to using it.

After visually inspecting the glove, other defects may be observed by applying the air test (*Figure 3*).

Step 1 Stretch the glove and look for any defects.

Step 2 Twirl the glove around quickly or roll it down from the glove gauntlet to trap air inside.

Step 3 Trap the air by squeezing the gauntlet with one hand. Use the other hand to squeeze the palm, fingers, and thumb to check for weaknesses and defects.

Step 4 Hold the glove up to your ear to try to detect any escaping air.

Step 5 If the glove does not pass this inspection, it must be turned in for disposal.

CAUTION

Never blow the gloves up like a balloon or use compressed gas for the air test as this can damage the glove.

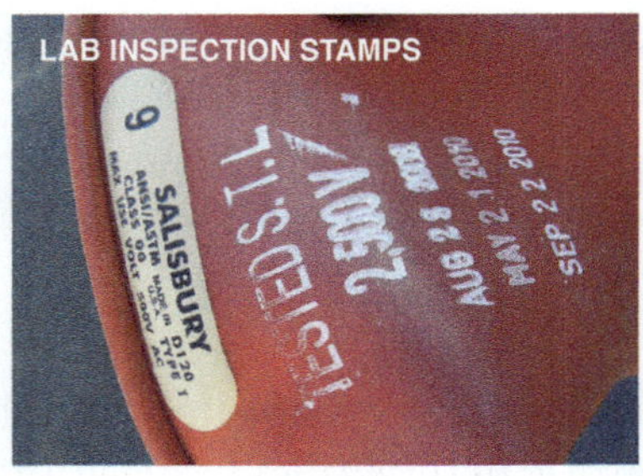

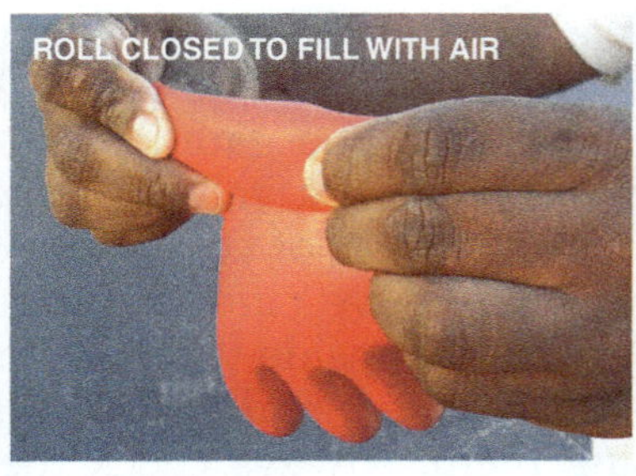

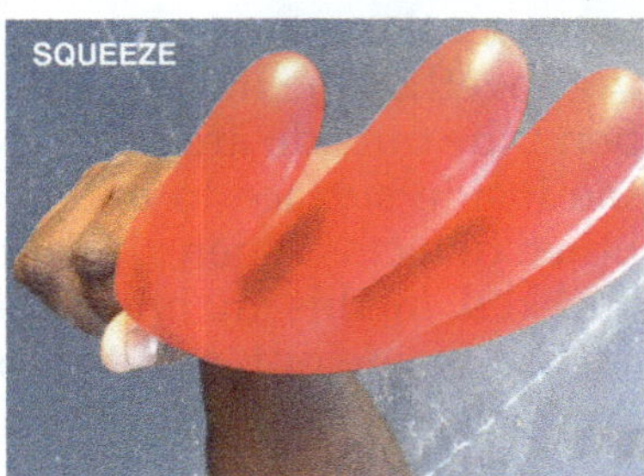

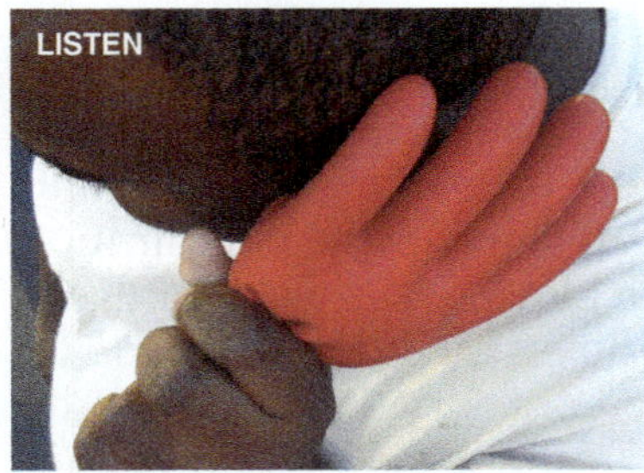

26102-14_F03.EPS

Figure 3 Glove inspection.

Insulating blankets – An insulating blanket is a versatile cover-up device best suited for the protection of maintenance technicians against accidental contact with energized electrical equipment.

These blankets are designed and manufactured to provide insulating quality and flexibility for use in covering. Insulating blankets are designed only for covering equipment and should not be used on the floor. Special rubber floor mats, called switchboard matting, are available for floor use. Use caution when installing these over sharp edges or when covering pointed objects.

Blankets must be tested yearly and inspected before each use. To check rubber blankets, place the blanket on a flat surface and roll the blanket from one corner to the opposite corner. If there are any irregularities in the rubber, this method will expose them. After the blanket has been rolled from each corner, it should then be turned over and the procedure repeated.

Insulating blankets are cleaned in the same manner as rubber gloves. Once the protective equipment has been properly cleaned, inspected, and tested, it must be properly stored. It should be stored in a cool, dry, dark place that is free from ozone, chemicals, oils, solvents, or other materials that could damage the equipment. Such storage should not be in the vicinity of hot pipes or direct sunlight. Blankets may be stored rolled in containers that are designed for this use; the inside diameter of the roll should be at least two inches.

On Site

Field-Marked Flash Protection Signs

In other than dwelling units, *NEC Section 110.16* requires that all switchboards, switchgear, panelboards, industrial control panels, meter socket enclosures, and motor control centers be clearly marked to warn qualified persons of potential electric arc flash hazards. *NEC Section 110.24(A)* requires service equipment at these locations to be marked with the maximum fault current available and the date of the calculation.

3.1.2 Protective Apparel

Besides rubber gloves, there are other types of special application protective apparel, such as fire suits, face shields, and rubber sleeves.

Manufacturing plants should have other types of special application protective equipment available for use, such as high-voltage sleeves, high-voltage boots, nonconductive protective helmets, nonconductive eyewear and face protection, and switchboard blankets.

All equipment must be inspected before use and during use, as necessary. The equipment used and the extent of the precautions taken depend on each individual situation; however, it is better to be overprotected than underprotected when you are trying to prevent electric shock, arc blast, and burns.

When working with energized equipment, flash suits may be required in some applications.

Face shields must also be worn during all switching operations where arcs are a possibility. Always use a rated face shield.

3.1.3 Personal Clothing

Any individual who will perform work in an electrical environment or in plant substations should dress accordingly. Never wear synthetic-fiber

Think About It

Dressing for Safety

How could minor flaws in clothing cause harm? What about metal components, such as the rivets in jeans, or synthetic materials, such as polyester? How are these dangerous? What about rings, watches, earrings, or other body piercings? How does protective apparel prevent accidents?

clothing on a job site; these types of materials will melt when exposed to high temperatures and because they burn hotter, will actually increase the severity of a burn. Wear cotton clothing, fiberglass-toe boots or shoes, safety glasses, and hard hats. Use hearing protection where needed.

3.1.4 Hot Sticks

Hot sticks are insulated tools designed for the manual operation of disconnecting switches, fuse removal and insertion, and the application and removal of temporary grounds.

A hot stick is made up of two parts, the head or hood and the insulating rod. The head can be made of metal or hardened plastic, while the insulating section may be wood, plastic, laminated wood, or other effective insulating materials. There are also telescoping sticks available.

Most plants have hot sticks available for different purposes. Select a stick of the correct type and size for the application and inspect it before use. Look for signs of obvious damage, deep scratches, dust, or surface contaminants. Never use a damaged hot stick.

Storage of hot sticks is important. They should be hung up vertically on a wall to prevent any damage. They should also be stored away from direct sunlight and prevented from being exposed to petroleum products.

3.1.5 Fuse Pullers

Use the plastic or fiberglass style of fuse puller for removing and installing low-voltage cartridge fuses. All fuse pulling and replacement operations must be done using fuse pullers.

WARNING!

Fuses should only be pulled when they are safely de-energized. Failure to do so can result in a serious injury or death.

The best type of fuse puller is one that has a spread guard installed. This prevents the puller from opening if resistance is met when installing fuses.

3.1.6 Shorting Probes

Before working on de-energized circuits that have capacitors installed, you must discharge the capacitors using a safety shorting probe. This is a procedure that requires special training and may only be performed by qualified individuals.

3.1.7 Eye and Face Protection

NFPA 70E® and OSHA require that you wear safety glasses at all times whenever you are working on or near energized circuits. Face protection is worn over your safety glasses, and is required whenever there is danger of electrical arcs or flashes, or from flying or falling objects resulting from an electrical explosion. *NFPA 70E®* is discussed in detail later in this module.

3.2.0 Verify That Circuits Are De-Energized

You should always assume that all the circuits are energized until you have verified that the circuit is de-energized. This is called a live-dead-live test. Follow these steps to verify that a circuit is de-energized:

Step 1 Ensure that the circuit is properly tagged and locked out *(CFR 1910.333/1926.417)*.

Step 2 Verify the test instrument operation on a known source using the appropriately rated tester.

Step 3 Using the test instrument, check the circuit to be de-energized. The voltage should be zero.

Step 4 Verify the test instrument operation, once again on a known power source.

3.3.0 Other Precautions

There are several other precautions you can take to help make your job safer. For example:

- Always remove all jewelry (e.g., rings, watches, body piercings, bracelets, and necklaces) before working on electrical equipment. Most jewelry is made of conductive material and wearing it can result in a shock, as well as other injuries if the jewelry gets caught in moving components.

On Site

Using a Phasing Tester

Using a phasing tester or other voltage detection equipment can expose you to a considerable arc flash hazard. Always take the time to use the appropriate personal protective equipment. Remember, you only have one chance to protect yourself.

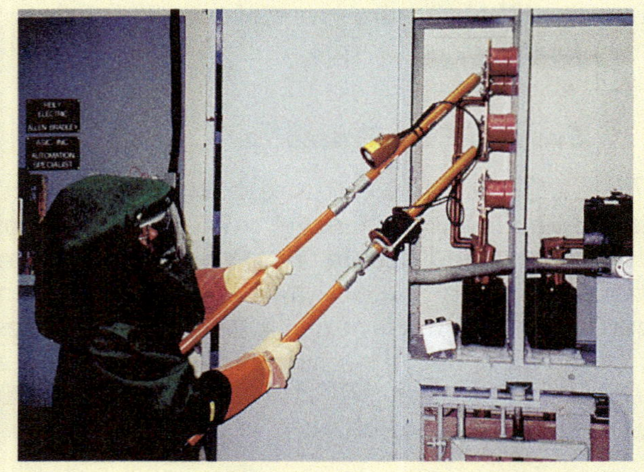

26102-14_SA02.EPS

- When working on energized equipment, it is safer to work in pairs. In doing so, if one of the workers experiences a harmful electrical shock, the other worker can call for help.
- Plan each job before you begin it. Make sure you understand exactly what it is you are going to do. If you are not sure, ask your supervisor.

WARNING!

The first time you perform a specific task, it cannot be done energized.

- You will need to look over the appropriate prints and drawings to locate isolation devices and potential hazards. Never defeat safety interlocks. Remember to plan your escape route before starting work. Know where the nearest phone is and the emergency number to dial for assistance.
- If you realize that the work will go beyond the scope of what was planned, stop and get instructions from your supervisor before continuing. Do not attempt to plan as you go.
- It is critical that you stay alert. Workplaces are dynamic, and situations relative to safety are always changing. If you leave the work area to pick up material, take a break, or have lunch, reevaluate your surroundings when you return. Remember, plan ahead.

4.0.0 OSHA

The purpose of OSHA is "to ensure safe and healthful working conditions for working men and women." OSHA is authorized to enforce standards and assist and encourage the states in their efforts to ensure safe and healthful working conditions. OSHA assists states by providing for research, information, education, and training in the field of occupational safety and health.

The law that established OSHA specifies the duties of both the employer and employee with respect to safety. Some of the key requirements are outlined below. This list does not include everything, nor does it override the procedures called for by your employer.

- Employers shall provide a place of employment free from recognized hazards likely to cause death or serious injury.
- Employers shall comply with OSHA standards.
- Employers shall be subject to fines and other penalties for violation of those standards.

CAUTION

OSHA states that employees have a duty to follow the safety rules laid down by the employer. Additionally, some states can reduce the amount of benefits paid to an injured employee if that employee was not following known, established safety rules. Your company may also terminate you if you violate an established safety rule.

4.1.0 Safety Standards

The OSHA standards are split into several sections. As discussed earlier, the two that affect you the most are *CFR 1926,* construction specific, and *CFR 1910,* which is the standard for general industry. Either or both may apply depending on where you are working and what you are doing. Your company may also have its own policies and procedures. In addition, you will be required to follow the safety procedures of any plant or facility where you are working.

On Site

Safety on the Job Site

Uncovered openings present several hazards:

- Workers may trip over them.
- If they are large enough, workers may fall through them.
- If there is a work area below, tools or other objects may fall through them, causing serious injury to workers below.

Any hole deeper than 6 feet and more than 2 inches in any direction must be protected with a cover or with guardrails, as shown here. A cover must be able to handle twice the anticipated load, be secured in place, and have the word "HOLE" or "COVER" on it.

26102-14_SA03.EPS

On Site

Working Around Openings

Did you know that OSHA could cite your company for working around the open pipe in the photo shown here, even if you weren't responsible for creating it? If you are working near uncovered openings, you have only two options:

- Cover the opening properly (either you may do it or you may have the responsible contractor cover it).
- Stop work and leave the job site until the opening has been covered.

26102-14_SA04.EPS

4.2.0 Safety Philosophy and General Safety Precautions

The most important piece of safety equipment required when performing work in an electrical environment is common sense. All areas of electrical safety precautions and practices draw upon common sense and attention to detail. One of the most dangerous conditions in an electrical work area is a poor attitude toward safety.

WARNING!

Only qualified individuals may work on or near electrical equipment. Your employer will determine who is qualified. Remember, your employer's safety rules must always be followed.

As stated in *CFR 1910.333(a)/1926.416*, safety-related work practices shall be employed to prevent electric shock or other injuries resulting from either direct or indirect electrical contact when work is performed near or on equipment or circuits that are or may be energized. The specific safety-related work practices shall be consistent with the nature and extent of the associated electrical hazards. The following are considered some of the basic and necessary attitudes and electrical safety precautions that lay the groundwork for a proper safety program. Before going on any electrical work assignment, these safety precautions should be reviewed and adhered to.

Think About It

Working on Energized Systems

Some electricians commonly work on energized systems because they think it's too much trouble to turn off the power. What practices have you seen around your home or workplace that could be deadly?

- *All work on electrical equipment should be done with circuits de-energized and cleared or grounded* – It is obvious that working on energized equipment is much more dangerous than working on equipment that is de-energized. Work on energized electrical equipment should be avoided if at all possible. *CFR 1910.333(a)(1)* states that live parts to which an employee may be exposed shall be de-energized before the employee works on or near them, unless the employer can demonstrate that de-energizing introduces additional or increased hazards or is not possible because of equipment design or operational limitations. Live parts that operate at less than 50 volts to ground need not be de-energized if there will be no increased exposure to electrical burns or to explosion due to electric arcs.
- *All conductors, buses, and connections should be considered energized until proven otherwise* – As stated in *1910.333(b)(1)*, conductors and parts of electrical equipment that have not been locked out or tagged out in accordance with this section should be considered energized. Routine operation of the circuit breakers and disconnect switches contained in a power distribution system can be hazardous if not approached in the right manner. Several basic precautions that can be observed in switchgear operations are:
  - Wear proper clothing made of natural fiber or fire-resistant fabric in accordance with *NFPA 70E®*.
  - Wear eye, face, and head protection.
  - Whenever operating circuit breakers in low-voltage or medium-voltage systems, always stand off to the side of the unit.
  - Always try to operate disconnect switches and circuit breakers under a no-load condition.
  - Never intentionally force an interlock on a system or circuit breaker.

Often, a circuit breaker or disconnect switch is used for providing lockout on an electrical system. Perform the following procedures when using the device as a lockout point:

- Breakers must always be locked out and tagged whenever you are working on a circuit that is tied to an energized breaker. Always follow the standard rack-out and removal procedures that were supplied with the switchgear. Once removed, a sign must be hung on the breaker identifying its use as a lockout point, and approved safety locks must be installed when the breaker is used for isolation. Breakers equipped with closing springs should be discharged to release all stored energy in the breaker mechanism.

WARNING!

The manufacturers recommend that racking be done de-energized and in compliance with *NFPA 70E®*.

- Some of the circuit breakers used are equipped with keyed interlocks (kirklocks) for protection during operation. These locks are used to ensure proper sequence of operation only. They are not used to lock out a circuit or system. When opening or closing a disconnect manually, it should be done quickly with a positive force. Lockout/tagout should be used when the disconnects are open.
- Whenever performing switching or fuse replacements, always use the protective equipment necessary to ensure personnel safety. Always prepare for the worst case accident when performing switching.
- Whenever re-energizing circuits following maintenance or removal of a faulted component, use extreme care. Always verify that the equipment is in a condition to be re-energized safely. All connections should be insulated and all covers and screws installed. Have all personnel stand clear of the area for the initial re-energization. Never assume everything is in perfect condition. Verify the conditions.

The following procedure is provided as a guideline for ensuring that equipment and systems will not be damaged by reclosing low-voltage circuit breakers into faults. If a low-voltage circuit breaker has opened for no apparent reason, perform the following:

Step 1 Verify that the equipment being supplied is not physically damaged and shows no obvious signs of overheating or fire.

Step 2 Make all appropriate tests to locate any faults.

Step 3 Reclose the feeder breaker. Stand off to the side when closing the breaker.

Step 4 If the circuit breaker trips again, do not attempt to reclose the breaker. In a plant environment, Electrical Engineering should be notified, and the cause of the trip must be isolated and repaired.

The same general procedure should be followed for fuse replacement, with the exception of transformer fuses. If a transformer fuse blows, the transformer and feeder cabling should be inspected and tested before re-energizing. A blown fuse to a transformer is very significant because it normally indicates an internal fault. Transformer failures are catastrophic in nature and can be extremely dangerous. If applicable, contact the in-plant Electrical Engineering Department prior to commencing any effort to re-energize a transformer.

Power must always be removed from a circuit when removing and installing fuses. The air break disconnects (or quick disconnects) provided on the upstream side of a large transformer must be opened prior to removing the transformer's fuses. Otherwise, severe arcing will occur as the fuse is removed. This arcing can result in personnel injury and equipment damage.

To replace fuses servicing circuits below 600 volts:

- Turn off the power to the disconnect.
- Verify that the fuses are de-energized.
- Remove the blown fuse.
- Install the new fuse. Push it in firmly and verify that it is seated properly.
- Turn the power back on.

When replacing fuses servicing systems above 600 volts:

- Open and lock out the disconnect switches.
- Unlock the fuse compartment.
- Verify that the fuses are de-energized.
- Attach the fuse removal hot stick to the fuse and remove it.

4.3.0 Electrical Regulations

OSHA has certain regulations that apply to job-site electrical safety. These regulations include:

- All electrical work shall be in compliance with the latest *NEC*® and OSHA standards.
- The noncurrent-carrying metal parts of fixed, portable, and plug-connected equipment shall be grounded. Choose either **grounded tools** or double-insulated tools. *Figure 4* shows an example of a double-insulated tool.
- Extension cords shall be the three-wire type, shall be protected from damage, and if hung overhead, shall not be fastened with staples or bare wire or hung in a manner that could cause damage to the outer jacket or insulation. Never run an extension cord through a doorway or window that can pinch the cord. Also, never allow vehicles or equipment to drive over cords.
- Exposed lamps in temporary lights shall be guarded to prevent accidental contact, except where lamps are deeply recessed in the reflector. Temporary lights shall not be suspended, except in accordance with their listed labeling.
- Receptacles for attachment plugs shall be of an approved type and properly installed. Installation of the receptacle will be in accordance with the listing and labeling for each receptacle.
- *NEC Section 590.6* states that all 125V, single-phase, 15A, 20A, and 30A receptacle outlets for temporary or permanent power shall be ground fault protected.
- Each disconnecting means for motors and appliances and each service feeder or branch circuit at the point where it originates shall be legibly marked with its purpose and voltage.
- Flexible cords shall be used in continuous lengths (no splices) and shall be of a type listed in *NEC Table 400.4*.
- Receptacles other than 125V, single-phase, 15A, 20A, and 30A should be ground fault protected if possible. If that is not possible, a written assured equipment grounding conductor program must be continuously enforced at the site for all cord sets per *NEC Section 590.6(B)*. *Figure 5* shows a typical ground fault circuit interrupter (GFCI).

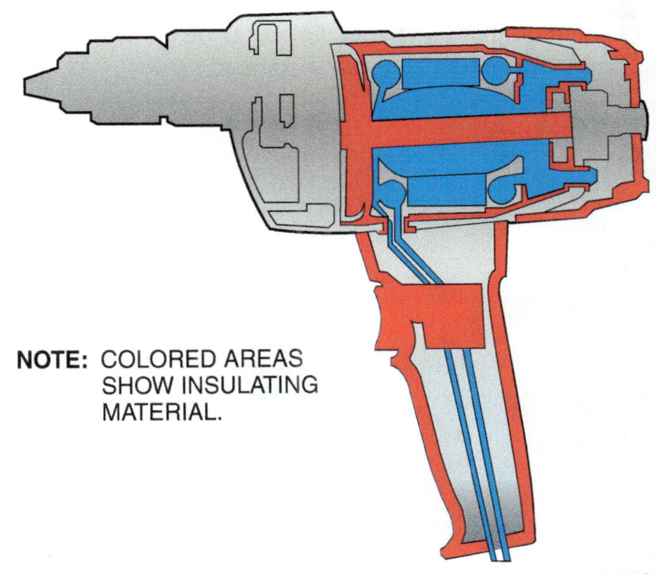

Figure 4 Double-insulated electric drill.

Case History

Potential Hazards

A self-employed builder was using a metal cutting tool on a metal carport roof and was not using GFCI protection. The male and female plugs of his extension cord partially separated, and the active pin touched the metal roofing. When the builder grounded himself on the gutter of an adjacent roof, he received a fatal shock.

The Bottom Line: Always use GFCI protection and be on the lookout for potential hazards.

4.3.1 OSHA Lockout/Tagout Rule

The OSHA lockout/tagout rule covers the specific procedure to be followed for the "servicing and maintenance of machines and equipment in which the unexpected energization or startup of the machines or equipment, or releases of stored energy, could cause injury to employees." This standard establishes minimum performance requirements for the control of such hazardous energy.

The first step to be completed before working on a circuit is to ensure that equipment is isolated from all potentially hazardous energy (for example, electrical, mechanical, hydraulic, chemical, or thermal), and tagged and locked out before employees perform any servicing or maintenance activities in which the unexpected energization, startup, or release of stored energy could cause injury. All employees shall be instructed in the lockout/tagout procedure.

CAUTION

Although 99% of your work may be electrical, be aware that you may also need to lock out mechanical and other types of energy equipment.

26102-14_F05.EPS

Figure 5 Typical GFCI receptacle.

Think About It

GFCIs

Explain how GFCIs protect people. Where should a GFCI be installed in the circuit to be most effective?

The following is an example of a lockout/tagout procedure. Make sure to use the procedure that is specific to your employer or job site.

WARNING!

This procedure is provided for your information only. The OSHA procedure provides only the minimum requirements for lockouts/tagouts. Consult the lockout/tagout procedure for your company and the plant or job site at which you are working. Remember that your life could depend on the lockout/tagout procedure. It is critical that you use the correct procedure for your site. The *NEC*® requires that remote-mounted motor disconnects be permanently equipped with a lockout feature.

I. *Introduction*

A. This lockout/tagout procedure has been established for the protection of personnel from potential exposure to hazardous energy sources during construction, installation, service, and maintenance of electrical energy systems.

B. This procedure applies to and must be followed by all personnel who may be potentially exposed to the unexpected startup or release of hazardous energy (e.g., electrical, mechanical, pneumatic, hydraulic, chemical, or thermal).

Exception: This procedure does not apply to process and/or utility equipment or systems with cord and plug power supply systems when the cord and plug are the only source of hazardous energy, are removed from the source, and remain under the exclusive control of the authorized employee.

Exception: This procedure does not apply to troubleshooting (diagnostic) procedures and installation of electrical equipment and systems when the energy source cannot be de-energized because continuity of service is essential or shutdown of the system is impractical. Additional personal protective equipment for such work is required and the safe work practices identified for this work must be followed.

II. *Definitions*
- *Affected employee* – Any person working on or near equipment or machinery when maintenance or installation tasks are being performed by others during lockout/ tagout conditions.
- *Appointed authorized employee* – Any person appointed by the job-site supervisor to coordinate and maintain the security of a group lockout/tagout condition.
- *Authorized employee* – Any person authorized by the job-site supervisor to use lockout/tagout procedures while working on electrical equipment.
- *Authorized supervisor* – The assigned job-site supervisor who is in charge of coordination of procedures and maintenance of security of all lockout/tagout operations at the job site.
- *Energy isolation device* – An approved electrical disconnect switch capable of accepting approved lockout/tagout hardware for the purpose of isolating and securing a hazardous electrical source in an open or safe position.
- *Lockout/tagout hardware* – A combination of padlocks, danger tags, and other devices designed to attach to and secure electrical isolation devices.

III. *Training*
- A. Each authorized supervisor, authorized employee, and appointed authorized employee shall receive initial and as-needed user-level training in lockout/tagout procedures.
- B. Training is to include recognition of hazardous energy sources, the type and magnitude of energy sources in the workplace, and the procedures for energy isolation and control.
- C. Retraining will be conducted on an as-needed basis whenever lockout/tagout procedures are changed or there is evidence that procedures are not being followed properly.

IV. *Protective Equipment and Hardware*
- A. Lockout/tagout devices shall be used exclusively for controlling hazardous energy sources.
- B. All padlocks must be numbered and assigned to one employee only.
- C. No duplicate or master keys will be made available to anyone except the site supervisor.
- D. A current list with the lock number and authorized employee's name must be maintained by the site supervisor.
- E. Danger tags must be of the standard white, red, and black *DANGER—DO NOT OPERATE* design and shall include the authorized employee's name, the date, and the appropriate network company (use permanent markers).
- F. Danger tags must be used in conjunction with padlocks, as shown in *Figure 6*.

V. *Procedures*
- A. Preparation for lockout/tagout:
  1. Check the procedures to ensure that no changes have been made since you last used a lockout/tagout.
  2. Identify all authorized and affected employees involved with the pending lockout/tagout.
- B. Sequence for lockout/tagout:
  1. Notify all authorized and affected personnel that a lockout/tagout is to be used and explain the reason why.
  2. Shut down the equipment or system using the normal Off or Stop procedures.
  3. Lock out energy sources and test disconnects to be sure they cannot be moved to the On position and open the control cutout switch. If there is no cutout switch, block the magnet in the Switch Open position before working on electrically operated equipment/apparatus such as motors, relays, etc. Remove the control wire.
  4. Lock and tag the required switches in the Open position. Each authorized employee must affix a separate lock and tag. An example is shown in *Figure 7*.
  5. Dissipate any stored energy by attaching the equipment or system to ground.
  6. Verify that the test equipment is functional via a known power source.
  7. Confirm that all switches are in the Open position and use test equipment to verify that all parts are de-energized.
  8. If it is necessary to temporarily leave the area, upon returning, retest to ensure that the equipment or system is still de-energized.
- C. Restoration of energy:
  1. Confirm that all personnel and tools, including shorting probes, are accounted for and removed from the equipment or system.

(A) ELECTRICAL LOCKOUT

(B) PNEUMATIC LOCKOUT

26102-14_F06.EPS

Figure 6 Lockout/tagout devices.

2. Completely reassemble and secure the equipment or system.
3. Replace and/or reactivate all safety controls.
4. Remove locks and tags from isolation switches. Authorized employees must remove their own locks and tags.

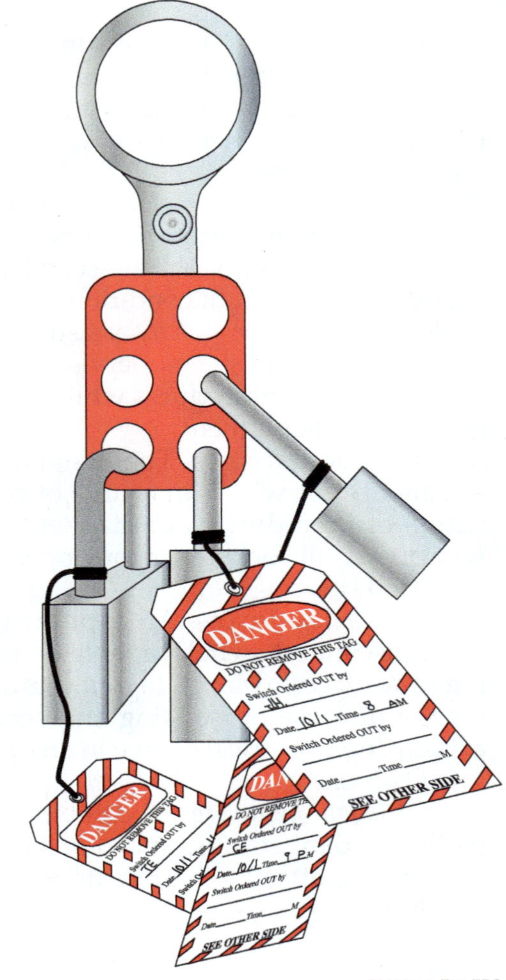

26102-14_F07.EPS

Figure 7 Multiple lockout/tagout device.

5. Notify all affected personnel that the lockout/tagout has ended and the equipment or system is energized.
6. Operate or close isolation switches to restore energy.

VI. *Emergency Removal Authorization*

A. In the event a lockout/tagout device is left secured, and the authorized employee is absent, or the key is lost, the authorized supervisor can remove the lockout/tagout device.

B. The authorized employee must be informed that the lockout/tagout device has been removed.

C. Written verification of the action taken, including informing the authorized employee of the removal, must be recorded in the job journal.

What's wrong with this picture?

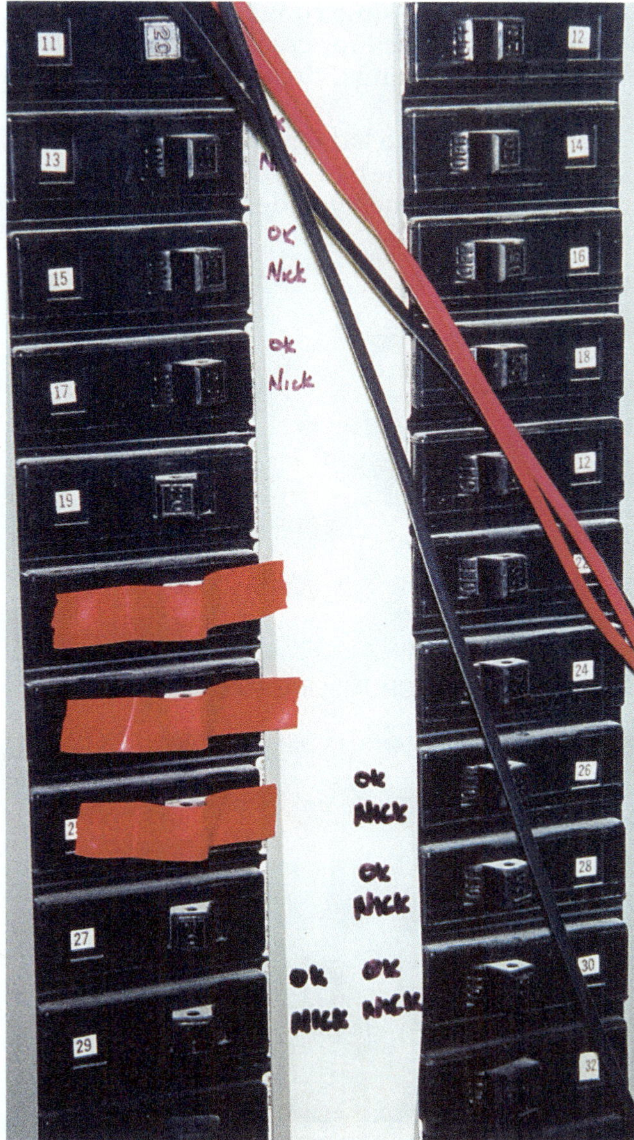

26102-14_SA05.EPS

Case History

Lockout/Tagout Dilemma

In Georgia, electricians found energized switches after the lockout of a circuit panel in an older system that had been upgraded several times. The existing wiring did not match the current site drawings. A subsequent investigation found many such situations in older facilities.

The Bottom Line: Never rely solely on drawings. It is mandatory that the circuit be tested after lockout to verify that it is de-energized.

Think About It

Lockout/Tagout – Who Does It and When?

What situations are likely to require lockout/tagout? Who is responsible for performing the lockout/tagout? When would more than one person be responsible?

4.4.0 Other OSHA Regulations

There are other OSHA regulations that you need to be aware of on the job site. For example:

- OSHA requires the posting of hard hat areas. Be alert to those areas and always wear your hard hat properly, with the bill in front. Hard hats should be worn whenever overhead hazards exist, or there is the risk of exposure to electric shock or burns.
- You should wear safety shoes on all job sites. Keep them in good condition.
- Do not wear clothing with exposed metal zippers, buttons, or other metal fasteners. Avoid wearing loose-fitting or torn clothing.
- Protect your eyes. Your eyesight is threatened by many activities on the job site. Always wear safety glasses with full side shields. In addition, the job may also require protective equipment such as face shields or goggles.

5.0.0 *NFPA 70E®*

In addition to the *NEC®*, the National Fire Protection Association also publishes the *Standard for Electrical Safety in the Workplace (NFPA 70E®)*. The *NEC®* specifies the minimum installation provisions necessary to safeguard persons and property from electrical hazards, while *NFPA 70E®* addresses the hazards that arise during the installation, operation, and maintenance of electrical equipment. In other words, the *NEC®* applies to installations, while *NFPA 70E®* applies to workplaces. *NFPA 70E®* is a national consensus standard, which means that it is a standard developed by the same people it affects and is then adopted by a nationally recognized institution. Other national consensus standards include those developed by ASTM International and the American National Standards Institute (ANSI).

The electrical safety requirements in *OSHA 1910.331* to *1910.335* are to use appropriate safe work practices, including de-energization of equipment, and the use of PPE appropriate to the hazard. *NFPA 70E®* is really the first consensus standard to provide practical guidance in hazard assessment, training of qualified persons, and management of electrical hazards. For many, *NFPA 70E*® provided the first functional understanding of the electrical arc flash hazard, direction for establishing flash protection boundaries, and levels of protection when qualified persons must work within the flash protection boundary. *NFPA 70E*® is arranged as follows:

- *Chapter 1, Safety-Related Work Practices* – This chapter covers the basic safety requirements for working on or near electrical equipment. It covers safety training programs, explains the difference between qualified and unqualified individuals, and includes requirements for analysis of both the electrical shock hazard and the electrical arc flash/blast hazard for a given task. The analysis includes assessment of the presence and extent of hazardous voltage and the available fault current for arc flash. The end result of each hazard analysis is also selection of safe work practices and appropriate PPE to be used in performing the task.
- *Chapter 2, Safety-Related Maintenance Requirements* – This chapter covers the maintenance requirements for various types of electrical equipment. This includes premises wiring, various types of electric equipment and devices, and personal safety and protective equipment.
- *Chapter 3, Safety Requirements for Special Equipment* – This chapter covers electrolytic cells, batteries, lasers, and various types of power electronic equipment, such as radio and television transmitters, UPS systems, arc welding equipment, and lighting controllers.
- *Annexes A through P* – The annexes provide various technical resources. These include safe approach distances, determination of flash protection boundaries and incident energy exposure from arc flash, a sample electrical lockout/tagout procedure, and other useful information.
- *Part Two* – This part contains supplements that cover the *NEC*® requirements asssociated with safe work practices, electrical preventive maintenance programs, and a case history of an arc-flash incident.

6.0.0 Ladders and Scaffolds

Ladders and scaffolds account for about half of the injuries from workplace electrocutions. The involuntary recoil that can occur when a person is shocked can cause the person to be thrown from a ladder or high place.

6.1.0 Ladders

Many job-site accidents involve the misuse of ladders. Make sure to follow these general rules every time you use any ladder. Following these rules can prevent serious injury or even death.

- Before using any ladder, inspect it. Look for loose or missing rungs, cleats, bolts, or screws. Also check for cracked, bent, broken, or badly worn rungs, cleats, or side rails. See *Figure 8*.
- Before you climb a ladder, make sure you clear any debris from the base of the ladder so you do not trip over it when you descend.

(A) CRUMBLING RAIL

(B) CRACKED STEPLADDER

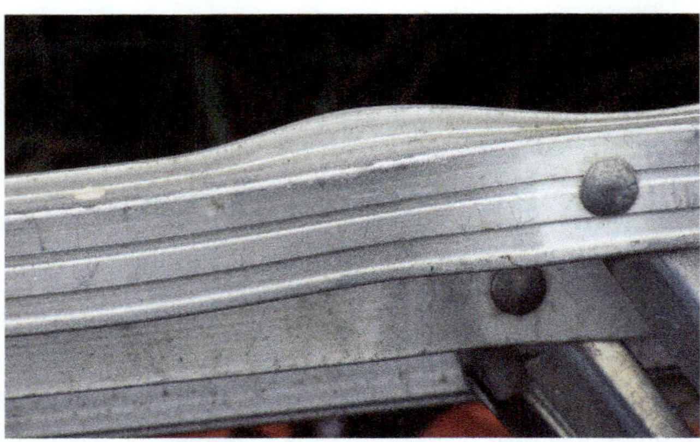

(C) BENT BACK BRACE

26102-14_F08.EPS

Figure 8 Types of ladder damage.

> **WARNING!**
> When performing electrical work, always use ladders made of nonconductive material.

- If you find a ladder in poor condition, do not use it. Report it and tag it for repair or disposal.
- Never modify a ladder by cutting it or weakening its parts.
- Do not set up ladders where they may be run into by others, such as in doorways or walkways. If it is absolutely necessary to set up a ladder in such a location, protect the ladder with barriers.
- Do not increase a ladder's reach by putting it in a mechanical lift or standing it on boxes, barrels, or anything other than a flat, solid surface.
- Check your shoes for grease, oil, or mud before climbing a ladder. These materials could make you slip.
- Always face the ladder and maintain three-point contact with the ladder (either have both feet and one hand on the ladder or both hands and one foot as you climb).
- Never lean out from the ladder. Keep your belt buckle centered between the rails. If something is out of reach, get down and move the ladder.

6.1.1 Straight and Extension Ladders

There are some specific rules to follow when working with straight and extension ladders:

- Always place a straight ladder at the proper angle. The horizontal distance from the ladder feet to the base of the wall or support should be about one-fourth the working height of the ladder. See *Figure 9*.
- Secure straight ladders to prevent slipping. Use ladder shoes or hooks at the top and bottom. Another method is to secure a board to the floor against the ladder feet. For brief jobs, someone can hold the straight ladder.
- Side rails should extend above the top support point by at least 36 inches.
- It takes two people to safely extend and raise an extension ladder. Extend the ladder only after it has been raised to an upright position.
- Never carry an extended ladder.
- Never use two ladders spliced together.
- Ladders should not be painted because paint can hide defects.

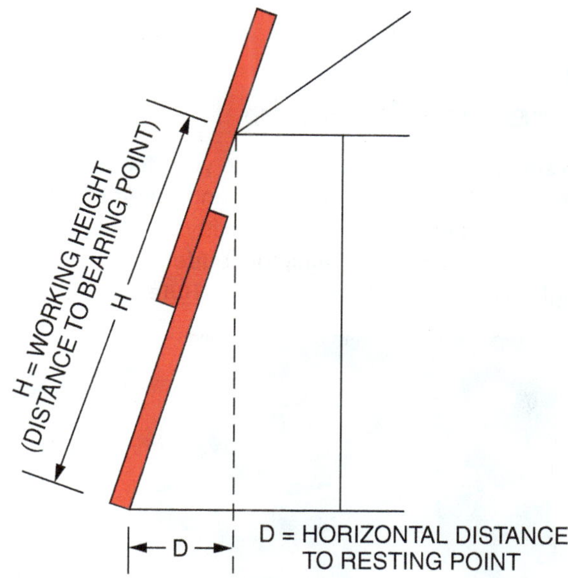

Figure 9 Straight ladder positioning.

What's wrong with this picture?

26102-14_SA07.EPS

On Site

Fireman's Rule

To set up a straight ladder, place your feet against the side rails, stand straight up, and put your hands at right angles to your body directly in front of you. If you can grab the side rails, the ladder is at the correct angle. This is called the Fireman's Rule.

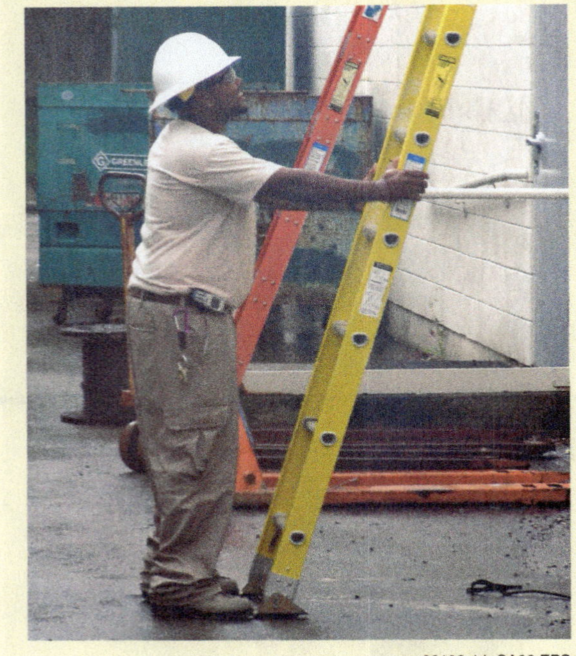

26102-14_SA06.EPS

6.1.2 Stepladders

There are also a few specific rules to use with a stepladder:

- Always open the stepladder all the way and lock the spreaders to avoid collapsing the ladder accidentally.
- Use a stepladder that is high enough for the job so that you do not have to reach. Get someone to hold the ladder if it is more than 10 feet high.
- Never use a stepladder as a straight ladder.
- Never stand on or straddle the top two rungs of a stepladder.
- Ladders are not shelves.

WARNING!

Do not leave tools or materials on a stepladder.

26102-14_F10.EPS

Figure 10 Scaffolding.

Sometimes you will need to move or remove protective equipment, guards, or guardrails to complete a task using a ladder. Remember, always replace what you moved or removed before leaving the area.

6.2.0 Scaffolds

Working on scaffolds (*Figure 10*) also involves being safe and alert to hazards. In general, keep scaffold platforms clear of unnecessary material or scrap. These can become deadly tripping hazards or falling objects. Carefully inspect each part of the scaffold as it is erected. Your life may depend on it! Makeshift scaffolds have caused many injuries and deaths on job sites. Use only scaffolding and planking materials designed and marked for their specific use. When working on a scaffold, follow the established specific requirements set by OSHA for the use of fall protection. When appropriate, wear an approved harness with a lanyard properly anchored to the structure.

The following requirements represent a compilation of the more stringent requirements of both *CFR 1910* and *CFR 1926.*

The following are some of the basic OSHA rules for working safely on scaffolds:

- Scaffolds must be erected on sound, rigid footing (referred to as a mud sill) that can carry the maximum intended load. If the scaffolding is not erected on concrete or another firm surface, you can use 2 × 10 or 2 × 12 scaffold grade lumber to support the base plates of the scaffolding.
- Scaffolds must be erected straight and plumb with no bent or deformed pieces. Correctly

Case History

Scaffolds and Electrical Hazards

Remember that scaffolds are excellent conductors of electricity. Recently, a maintenance crew needed to move a scaffold and although time was allocated in the work order to dismantle and rebuild the scaffold, the crew decided to push it instead. They did not follow OSHA recommendations for scaffold clearance and did not perform a job site survey. During the move, the five-tier scaffold contacted a 12,000V overhead power line. All four members of the crew were killed and the crew chief received serious injuries.

The Bottom Line: Never take shortcuts when it comes to your safety and the safety of others. Trained safety personnel should survey each job site prior to the start of work to assess potential hazards. Safe working distances should be maintained between scaffolding and power lines.

erected scaffolding will be symmetrical with the same parts on both sides (unless it has an outrigger attachment).

- Guardrails and toe boards must be installed on the open sides and ends of platforms higher than ten feet above the ground or floor.
- There must be a screen of ½-inch maximum openings between the toe board and the midrail where persons are required to work or pass under the scaffold.
- Scaffold planks must extend over their end supports not less than 6 inches nor more than 12 inches and must be properly blocked.
- If the scaffold does not have built-in ladders that meet the standard, then it must have an attached ladder access.
- All employees must be trained to erect, dismantle, and use scaffold(s).
- Unless it is impossible, fall protection must be worn while building or dismantling all scaffolding.
- Work platforms must be completely decked for use by employees.
- Use trash containers or other similar means to keep debris from falling and never throw or sweep material from above.

7.0.0 Lifts, Hoists, and Cranes

On the job, you may be working in the operating area of lifts, hoists, or cranes. The following safety rules are for those who are working in the area with overhead equipment but are not directly involved in its operation.

- Stay alert and pay attention to the warning signals from operators.
- Never stand or walk under a load, regardless of whether it is moving or stationary.
- Always warn others of moving or approaching overhead loads.
- Never attempt to distract signal persons or operators of overhead equipment.
- Obey warning signs.
- Do not use equipment that you are not qualified to operate.
- Cranes that are operated in areas with places in which a person can become trapped or pinched must have barricades placed around them to warn away workers.
- Hoists rigged in shafts or the outside of buildings must be secured to prevent them from being pulled down.
- Never overload a hoist, lift, or crane. Always follow lift ratings—there is no such thing as a safety factor that can be used to cheat a load.
- Cranes, lifts, and hoists must be inspected daily.
- Never lift a person using a crane, hoist, or material lift.
- Only people who have been trained in rigging should ever do rigging for a lift. It is extremely easy for a load to fall out of inappropriately rigged lifts.
- Personnel hoists require gates on every landing which can only be opened from the hoist side. This ensures that they are not opened onto a fall hazard.

What's wrong with this picture?

26102-14_SA08.EPS

On Site

Vertical Towers

This electrician is installing a light fixture using a vertical tower or Genie® lift. This lift is designed to fold up small enough so that it can be maneuvered through house doorways. Some models can be driven.

26102-14_SA09.EPS

On Site

Lifting

If you bend from the waist to pick up a 50-pound object, you are applying 10 times the amount of pressure (500 pounds) to your lower back. Lower back injuries are one of the most common workplace injuries because it's so easy to be careless about lifting, especially when you are in a hurry. Remember, it is much easier to ask for help than it is to nurse an injured back.

8.0.0 Lifting

Back injuries cause many lost working hours every year. That is in addition to the misery felt by the person with the hurt back! Learn how to lift properly and size up the load. To lift, first stand close to the load. Then, squat down and keep your back straight. Get a firm grip on the load and keep the load close to your body. Lift by straightening your legs. Make sure that you lift with your legs and not your back. Do not be afraid to ask for help if you feel the load is too heavy. See *Figure 11* for an example of proper lifting.

Keep the following precautions in mind when lifting:

- Make the lift smoothly and under control.
- Move your feet to pivot; do not twist or you may injure yourself.
- Constantly scan the path ahead for obstructions. If you cannot see your path over or around the object being carried, then you must have help to transport the object.
- Avoid lifting objects over your head.
- Never lift over the side or tailgate of a pickup truck.
- Don't twist your body when lifting up or setting an object down.
- Never reach over an obstacle to lift a load.
- Don't step over objects in your way.

What's wrong with this picture?

26102-14_SA10.EPS

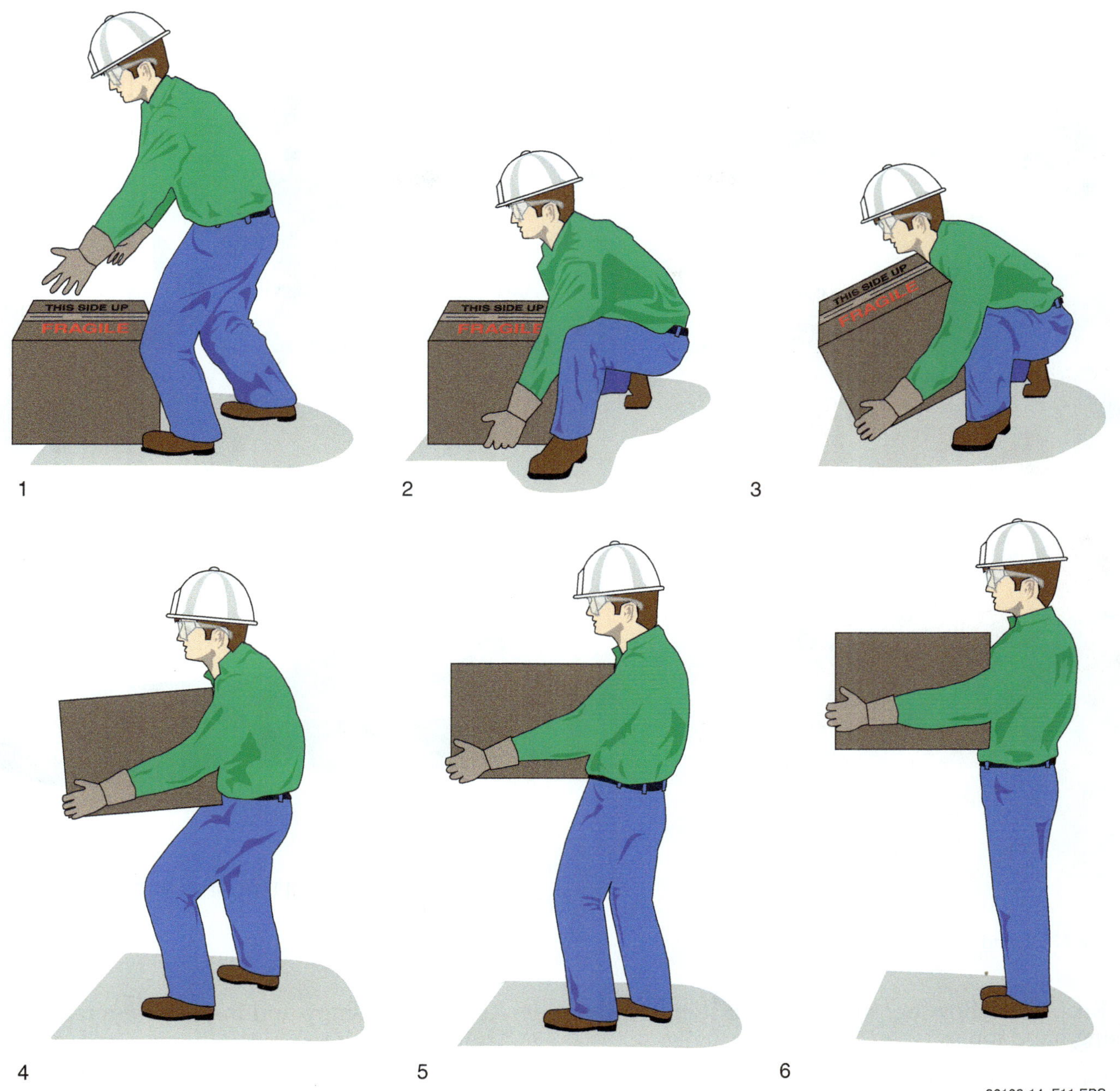

Figure 11 Proper lifting.

On Site

Lift Safely to Preserve Your B.A.C.K.

B – Balance: keep your stance wide, get a good grip on the object.

A – Alignment: keep your back relaxed and upright.

C – Contract and close: contract your stomach muscles and hold loads close.

K – Knees: make sure you bend them, not your waist.

9.0.0 Basic Tool Safety

When using any tools for the first time, read the operator's manual to learn the recommended safety precautions. If you are not certain about the operation of any tool, ask the advice of a more experienced worker. Before using a tool, you should know its function and how it works.

9.1.0 Hand Tool Safety

Hand tools are non-powered tools and may include anything from screwdrivers to cable strippers (*Figure 12*). Hand tools are dangerous if they are misused or improperly maintained.

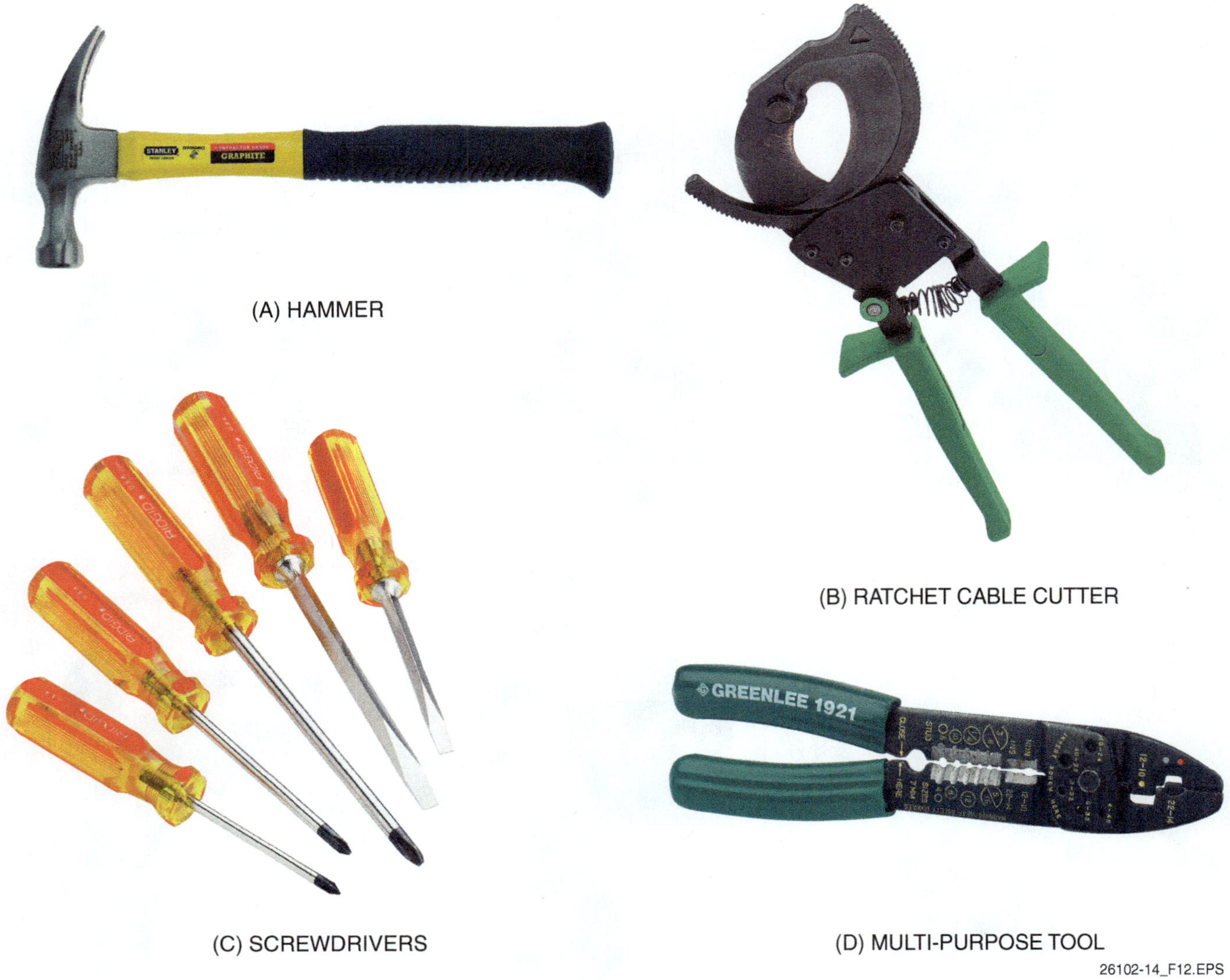

(A) HAMMER

(B) RATCHET CABLE CUTTER

(C) SCREWDRIVERS

(D) MULTI-PURPOSE TOOL

26102-14_F12.EPS

Figure 12 Hand tools.

Keep the following precautions in mind when using hand tools:

- Only use tools for their designated purpose.
- Always maintain your tools properly. If the wooden handle on a tool such as an ax or hammer is loose, splintered, or cracked, the head of the tool may fly off and strike the user or another person.
- Repair or replace damaged or worn tools. A wrench with its jaws sprung might easily slip, causing hand injuries. If the wrench flies, it may strike the user or another person.
- Impact tools such as chisels, wedges, and drift pins are unsafe if they have mushroomed heads. The heads might shatter when struck, sending sharp fragments flying.
- Use bladed tools with the blades and points aimed away from yourself and other people.
- Store bladed tools properly; use the sheath or protective covering if there is one.
- Keep blades sharp and inspect them regularly. Dull blades are difficult to use and control and can be far more dangerous than well-maintained blades.
- Never leave tools on top of ladders or scaffolding.

In general, your risks are greatly reduced by inspecting and maintaining tools regularly and always wearing appropriate personal protective equipment such as safety goggles, hard hats, and filtering masks. Also, keep floors dry and clean to prevent accidental slips that can result in injuries caused by the tools you may be using.

9.2.0 Power Tool Safety

Power tools can be hazardous when they are improperly used or not well maintained. Most of the risks associated with hand tools are also risks

when using power tools. When you add a power source to a tool, however, the risk factors increase.

Power tools are powered by different sources. Some examples of power sources for power tools include:

- Electricity
- Pneumatics (air pressure)
- Liquid fuel (gasoline or propane)
- Hydraulics (fluid pressure)

You must know the safety rules and proper operating procedures for each tool you use. Specific operating procedures and safety rules for using a tool are provided in the operator's/user's manual supplied by the manufacturer. Before operating any power tool for the first time, always read the manual to familiarize yourself with the tool. If the manual is missing, contact the manufacturer for a replacement.

WARNING!

Never use a tool if you are unsure of how to use it correctly.

Follow these general guidelines in order to prevent accidents and injury:

- Inspect all tools for damage before use. Remove damaged tools from use and tag DO NOT USE.
- Never carry or lower a tool by the cord or hose.
- Keep cords and hoses away from heat, oil, and sharp edges.
- Do not attempt to operate any power tool before being checked out by your instructor on that particular tool.
- Always wear eye protection, a hard hat, and any other required personal protective equipment when operating power tools.
- Wear face and hearing protection when required.
- Wear proper respiratory equipment when necessary.
- Wear the appropriate clothing for the job being done. Wear close-fitting clothing that cannot become caught in moving tools. Roll up or button long sleeves, tuck in shirttails, and tie back long hair. Do not wear any jewelry, including watches or rings.
- Do not distract others or let anyone distract you while operating a power tool.
- Do not engage in horseplay.
- Do not run or throw objects.
- Consider the safety of others, as well as yourself. Observers should be kept at a safe distance away from the work area.
- Never leave a power tool running while it is unattended.
- Assume a safe and comfortable position before using a power tool. Be sure to maintain good footing and balance in order to respond to kickbacks, jumps, or sudden shifts.
- Secure work with clamps or a vise, freeing both hands to safely operate the tool.
- To avoid accidental starting, never carry a tool with your finger on the switch.
- Be sure that a power tool is properly grounded and connected to a ground fault circuit interrupter (GFCI) before using it.
- Ensure that power tools are disconnected before performing maintenance or changing accessories.
- Use a power tool only for its intended use.
- Keep your feet, fingers, and hair away from the blade and/or other moving parts of a power tool.
- Never use a power tool with guards or safety devices removed or disabled.
- Never operate a power tool if your hands or feet are wet.
- Keep the work area clean at all times.
- Become familiar with the correct operation and adjustments of a power tool before attempting to use it.
- Keep tools sharp and clean for the best performance.
- Follow the instructions in the user's manual for lubricating and changing accessories.
- Keep a firm grip on the power tool at all times.
- Use electric extension cords of sufficient size to service the particular power tool you are using.
- Do not run extension cords across walkways where they will pose a tripping hazard.
- Report unsafe conditions to your instructor or supervisor.
- Tools that shoot nails (*Figure 13*), rivets, or staples, and operate at pressures greater than 100 pounds per square inch (psi), must be equipped with a safety device that won't allow fasteners to be shot unless the muzzle is pressed against a work surface.
- Compressed-air guns should never be pointed toward anyone and the muzzle should never be pressed against a person.
- The use of powder-actuated tools requires special training and certification.
- Never use explosive or flammable materials around powder-actuated tools.
- Never point powder-actuated tools at anybody.

Case History

Don't Remove the Grounding Prong

An employee was climbing a metal ladder to hand an electric drill to the journeyman installer on a scaffold about 5 feet above him. When the victim reached the third rung from the bottom of the ladder, he received an electric shock that killed him. The investigation revealed that the extension cord had a missing grounding prong and that a conductor on the green grounding wire was making intermittent contact with the energized black wire, thereby energizing the entire length of the grounding wire and the drill's frame. The drill was not double insulated.

The Bottom Line: Do not disable any safety device on a power tool. A ground fault can be deadly.

Source: The Occupational Safety and Health Administration (OSHA)

26102-14_F13.EPS

Figure 13 Pneumatic nail gun.

- Never pick up an unattended powder-actuated tool. Instead, tell your supervisor that a powder-actuated tool has been left unattended.
- Never play with powder-actuated tools. These tools are as dangerous as a loaded gun.

10.0.0 Confined Space Entry Procedures

Occasionally, you may be required to do your work in a manhole or vault. If this is the case, there are some special safety considerations that you need to be aware of. For details on the subject of working in manholes and vaults, refer to *CFR 1910.146/1926.21(a)(6)(i) and (ii)*. The general precautions are listed in the following paragraphs.

10.1.0 General Guidelines

A confined space includes (but is not limited to) any of the following: a manhole (*Figure 14*), boiler, tank, trench (four feet or deeper), tunnel, hopper, bin, sewer, vat, pipeline, vault, pit, air duct, or vessel. A confined space is identified as follows:

- It has limited entry and exit.
- It is not intended for continued human occupancy.
- It has poor ventilation.
- It has the potential for entrapment/engulfment.
- It has the potential for accumulating a dangerous atmosphere.

Entry into a confined space occurs when any part of the body crosses the plane of entry. No employee shall enter a confined space unless the employee has been trained in confined space entry procedures. Other requirements for confined spaces include the following:

- All hazards must be eliminated or controlled before a confined space entry is made.
- The air quality in the confined space must be continually monitored.
- All appropriate personal protective equipment shall be worn at all times during confined space entry and work. The minimum required equipment includes a hard hat, safety glasses, full body harness, and life line.
- Ladders used for entry must be secured.
- A rescue retrieval system must be in use when entering confined spaces and while working

26102-14_F14.EPS

Figure 14 Manhole.

in permit-required confined spaces (discussed later). Each employee must be capable of being rescued by the retrieval system.
- Only no-entry rescues will be performed by company personnel. Entry rescues will be performed by trained rescue personnel identified on the entry permit.
- The area outside the confined space must be properly barricaded, and appropriate warning signs must be posted.
- Entry permits can only be issued and signed by a qualified person such as the job-site supervisor. Permits must be kept at the confined space while work is being conducted. At the end of the shift, the entry permits must be made part of the job journal and retained for one year.

10.2.0 Confined Space Hazard Review

Before determining the proper procedure for confined space entry, a hazard review shall be performed. The hazard review shall include, but not be limited to, the following conditions:

- The past and current uses of the confined space
- The physical characteristics of the space including size, shape, air circulation, etc.
- Proximity of the space to other hazards
- Existing or potential hazards in the confined space, such as:
  - Atmospheric conditions (oxygen levels, flammable/explosive levels, and/or toxic levels)
  - Presence/potential for liquids
  - Presence/potential for particulates
- Potential for mechanical/electrical hazards in the confined space (including work to be done)

Once the hazard review is completed, the supervisor, in consultation with the project managers and/or safety manager, shall classify the confined space as one of the following:

- A nonpermit confined space
- A permit-required confined space controlled by ventilation
- A permit-required confined space

Once the confined space has been properly classified, the appropriate entry and work procedures must be followed.

> **WARNING!**
>
> Only qualified and trained individuals may enter a confined space.

11.0.0 First Aid

You should be prepared in case an accident does occur on the job site or anywhere else. First aid training that includes certification classes in CPR and artificial respiration could be the best insurance you and your fellow workers ever receive. Make sure that you know where first aid is available at your job site. Also, make sure you know the accident reporting procedure. Each job site should also have a first aid manual or booklet giving easy-to-find emergency treatment procedures for various types of injuries. Emergency first aid telephone numbers should be readily available to everyone on the job site. Refer to *CFR 1910.151/ 1926.23* and *1926.50* for specific requirements.

12.0.0 Solvents and Toxic Vapors

The solvents that are used by electricians may give off vapors that are toxic enough to make people temporarily ill or even cause permanent injury. Many solvents are skin and eye irritants. Solvents can also be systemic poisons when they are swallowed or absorbed through the skin.

Solvents in spray or aerosol form are dangerous in another way. Small aerosol particles or solvent vapors mix with air to form a combustible mixture with oxygen. The slightest spark could cause an explosion in a confined area because the mix is perfect for fast ignition. There are procedures and methods for using, storing, and disposing of most solvents and chemicals. These procedures are normally found in the safety data sheets (SDSs)/ material safety data sheets (MSDSs) available at your facility.

An SDS is required for all materials that could be hazardous to personnel or equipment. These sheets contain information on the material, such as the manufacturer and chemical makeup. As much information as possible is kept on the hazardous material to prevent a dangerous situation; or, in the event of a dangerous situation, the information is used to rectify the problem in as safe a manner as possible. See *Figure 15* for an example of SDS information you may find on the job.

12.1.0 Precautions When Using Solvents

It is always best to use a nonflammable, nontoxic solvent whenever possible. However, any time solvents are used, it is essential that your work area be adequately ventilated and that you wear the appropriate personal protective equipment:

- Wear a chemical face shield with chemical goggles to protect the eyes and skin from sprays and splashes.

H336: May cause drowsiness or dizziness	P403+P233: Store in a well ventilated place. Keep container tightly closed
EUH019: May form explosive peroxides	P501: Dispose of contents/container in accordance with local regulation

SECTION 3 - COMPOSITION/INFORMATION ON INGREDIENTS

	CAS#	EINECS #	REACH Pre-registration Number	CONCENTRATION % by Weight
Tetrahydrofuran (THF)	109-99-9	203-726-8	05-2116297729-22-0000	25 - 50
Methyl Ethyl Ketone (MEK)	78-93-3	201-159-0	05-2116297728-24-0000	5 - 36
Cyclohexanone	108-94-1	203-631-1	05-2116297718-25-0000	15 - 30

All of the constituents of this adhesive product are listed on the TSCA inventory of chemical substances maintained by the US EPA, or are exempt from that listing.
* Indicates this chemical is subject to the reporting requirements of Section 313 of the Emergency Planning and Community Right-to-Know Act of 1986 (40CFR372).
# indicates that this chemical is found on Proposition 65's List of chemicals known to the State of California to cause cancer or reproductive toxicity.

SECTION 4 - FIRST AID MEASURES

Contact with eyes: Flush eyes immediately with plenty of water for 15 minutes and seek medical advice immediately.
Skin contact: Remove contaminated clothing and shoes. Wash skin thoroughly with soap and water. If irritation develops, seek medical advice.
Inhalation: Remove to fresh air. If breathing is stopped, give artificial respiration. If breathing is difficult, give oxygen. Seek medical advice.
Ingestion: Rinse mouth with water. Give 1 or 2 glasses of water or milk to dilute. Do not induce vomiting. Seek medical advice immediately.

SECTION 5 - FIREFIGHTING MEASURES

Suitable Extinguishing Media: Dry chemical powder, carbon dioxide gas, foam, Halon, water fog.
Unsuitable Extinguishing Media: Water spray or stream.
Exposure Hazards: Inhalation and dermal contact
Combustion Products: Oxides of carbon, hydrogen chloride and smoke

	HMIS	NFPA	
			0-Minimal
Health	2	2	1-Slight
Flammability	3	3	2-Moderate
Reactivity	0	0	3-Serious
PPE	B		4-Severe

Protection for Firefighters: Self-contained breathing apparatus or full-face positive pressure airline masks.

SECTION 6 - ACCIDENTAL RELEASE MEASURES

Personal precautions: Keep away from heat, sparks and open flame.
Provide sufficient ventilation, use explosion-proof exhaust ventilation equipment or wear suitable respiratory protective equipment.
Prevent contact with skin or eyes (see section 8).
Environmental Precautions: Prevent product or liquids contaminated with product from entering sewers, drains, soil or open water course.
Methods for Cleaning up: Clean up with sand or other inert absorbent material. Transfer to a closable steel vessel.
Materials not to be used for clean up: Aluminum or plastic containers

SECTION 7 - HANDLING AND STORAGE

Handling: Avoid breathing of vapor, avoid contact with eyes, skin and clothing.
Keep away from ignition sources, use only electrically grounded handling equipment and ensure adequate ventilation/fume exhaust hoods.
Do not eat, drink or smoke while handling.
Storage: Store in ventilated room or shade below 44°C (110°F) and away from direct sunlight.
Keep away from ignition sources and incompatible materials: caustics, ammonia, inorganic acids, chlorinated compounds, strong oxidizers and isocyanates.
Follow all precautionary information on container label, product bulletins and solvent cementing literature.

SECTION 8 - PRECAUTIONS TO CONTROL EXPOSURE / PERSONAL PROTECTION

EXPOSURE LIMITS:

Component	ACGIH TLV	ACGIH STEL	OSHA PEL	OSHA STEL:
Tetrahydrofuran (THF)	50 ppm	100 ppm	200 ppm	
Methyl Ethyl Ketone (MEK)	200 ppm	300 ppm	200 ppm	
Cycloh			50 ppm	

26102-14_F15.EPS

Figure 15 Portion of an SDS.

On Site

MSDS Centers

MSDSs are required to be in an accessible location. Many job sites provide MSDS centers, such as that shown here.

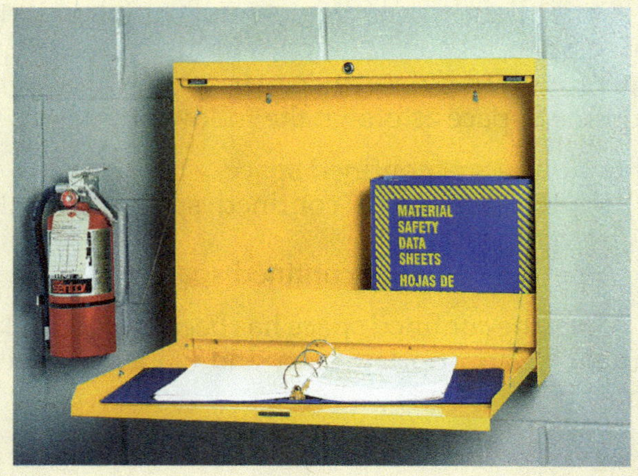

26102-14_SA11.EPS

- Wear a chemical apron to protect your body from sprays and splashes. Remember that some solvents are acid-based. If they come into contact with your clothes, solvents can eat through your clothes to your skin.
- A paper filter mask does not stop vapors; it is used only for nuisance dust. In situations where a paper mask does not supply adequate protection, chemical cartridge respirators might be needed. These respirators can stop many vapors if the correct cartridge is selected. In areas where ventilation is a serious problem, a self-contained breathing apparatus (SCBA) must be used.
- Make sure that you have been given a full medical evaluation and that you are properly trained in using respirators at your site.

12.2.0 Respiratory Protection

The best respiratory protection is to avoid the hazard entirely. Off-shift work, ventilation, and/or rescheduled work schedules should always be used to eliminate the need for working in areas with poor air quality. For example, in an area where hazardous solvents are used, the electrical work can be done off schedule when the solvents are not being used. When this cannot be done, protection against high concentrations of dust, mist, fumes, vapors, and gases is provided by appropriate respirators.

Appropriate respiratory protective devices should be used for the hazardous material involved and the extent and nature of the work performed.

An air-purifying respirator is, as its name implies, a respirator that removes contaminants from air inhaled by the wearer. The respirators may be divided into the following types: particulate-removing (mechanical filter), gas- and vapor-removing (chemical filter), and a combination of particulate-removing and gas- and vapor-removing.

Particulate-removing respirators are designed to protect the wearer against the inhalation of particulate matter in the ambient atmosphere. They may be designed to protect against a single type of particulate, such as pneumoconiosis-producing and nuisance dust, toxic dust, metal fumes or mist, or against various combinations of these types.

Gas- and vapor-removing respirators are de--signed to protect the wearer against the inhalation of gases or vapors in the ambient atmosphere. They are designated as gas masks, chemical cartridge respirators (nonemergency gas respirators), and self-rescue respirators. They may be designed to protect against a single gas such as chlorine; a single type of gas, such as acid gases; or a combination of types of gases, such as acid gases and organic vapors.

If you are required to use a respiratory protective device, you must be evaluated by a physician to ensure that you are physically fit to use a respirator. You must then be fitted and thoroughly instructed in the respirator's use.

Any employee whose job entails having to wear a respirator must keep his face free of facial hair in the seal area.

WARNING!

Do not use any respirator unless you have been fitted for it and thoroughly understand its use. As with all safety rules, follow your employer's respiratory program and policies.

Respiratory protective equipment must be inspected regularly and maintained in good condition. Respiratory equipment must be properly cleaned on a regular basis and stored in a sanitary, dustproof container.

On Site

Chemical Safety

The first line of defense with chemicals is to read and follow the directions found on the container. If you follow these instructions, you should be safe from chemical exposure. Be aware that everyone reacts differently to chemicals and you may be hypersensitive to a particular chemical that does not bother your co-workers. Leave the area at the first sign of an allergic reaction and seek medical attention.

Case History

Altered Respiratory Equipment

A self-employed man applied a solvent-based coating to the inside of a tank. Instead of wearing the proper respirator, he used nonstandard air supply hoses and altered the face mask. All joints and the exhalation slots were sealed with tape. He collapsed and was not discovered for several hours.

The Bottom Line: Never alter or improvise safety equipment.

13.0.0 Asbestos

Asbestos is a mineral-based material that is resistant to heat and corrosive chemicals. Depending on the chemical composition, asbestos fibers may range in texture from coarse to silky. The properties that make asbestos fibers so valuable to industry are its high tensile strength, flexibility, heat and chemical resistance, and good frictional properties.

Asbestos fibers enter the body by inhalation of airborne particles or by ingestion and can become embedded in the tissues of the respiratory or digestive systems. Exposure to asbestos can cause numerous disabling or fatal diseases. Among these diseases are asbestosis, an emphysema-like condition; lung cancer; mesothelioma, a cancerous tumor that spreads rapidly in the cells of membranes covering the lungs and body organs; and gastrointestinal cancer. The use of asbestos was banned in 1978.

Because asbestos was still in the manufacturing pipeline for a while after it was banned, you need to assume that any facility constructed before 1980 has asbestos in it. The owner must have a survey with any work rules needed to work safely around the asbestos. Common products that contain asbestos include thermal pipe insulation, mastic for ducts and insulation, spray-on fireproofing, floor tiles, ceiling tiles, roof insulation, exterior building sheathing, old wire insulation, and even pipe. As an electrician, you must not drill through or otherwise work with asbestos—you can only be trained to work around it when it can be done safely. Asbestos work is a trade that requires special training and protective equipment.

The following signs must be placed in areas containing asbestos.

DANGER

ASBESTOS
CANCER AND LUNG DISEASE HAZARD
AUTHORIZED PERSONNEL ONLY
RESPIRATORS AND PROTECTIVE CLOTHING ARE REQUIRED IN THIS AREA

DANGER

CONTAINS ASBESTOS FIBERS
AVOID CREATING DUST
CANCER AND LUNG DISEASE HAZARD

14.0.0 Batteries

Working around wet cell batteries can be dangerous if the proper precautions are not taken. Batteries often give off hydrogen gas as a byproduct. When hydrogen mixes with air, the mixture can be explosive in the proper concentration. For this reason, smoking is strictly prohibited in battery rooms, and only insulated tools should be used. Proper ventilation also reduces the chance of explosion in battery areas. Follow your company's procedures for working near batteries. Also, ensure that your company's procedures are followed for lifting heavy batteries.

WARNING!

Battery-powered scissor lifts may have unsealed batteries that require water levels to be checked and topped off. Never use a flame to look inside a battery—it can cause an explosion. Charging batteries with low water levels can damage them.

14.1.0 Acids

Batteries also contain acid, which will eat away human skin and many other materials. Personal protective equipment for battery work typically includes chemical aprons, sleeves, gloves, face shields, and goggles to prevent acid from contacting skin and eyes. Follow your site procedures for dealing with spills of these materials. Also, know the location of first aid when working with these chemicals.

14.2.0 Wash Stations

Because of the chance that battery acid may contact someone's eyes or skin, wash stations are located near battery rooms. Do not connect or disconnect batteries without proper supervision. Everyone who works in the area should know where the nearest wash station is and how to use it. Battery acid should be flushed from the skin and eyes with large amounts of water or with a neutralizing solution.

CAUTION

If you come in contact with battery acid, flush the affected area with water and report it immediately to your supervisor.

15.0.0 PCBs and Vapor Lamps

Polychlorinated biphenyls (PCBs) are chemicals that were marketed under various trade names as a liquid insulator/cooler in older transformers. In addition to being used in older transformers, PCBs are also found in some large capacitors and in the small ballast transformers used in street lighting and ordinary fluorescent light fixtures. Disposal of these materials is regulated by the Environmental Protection Agency (EPA) and must be done through a regulated disposal company; use extreme caution and follow your facility procedures.

WARNING!

Do not come into contact with PCBs. They present a variety of serious health risks, including lung damage and cancer.

In addition, any vapor lamps, such as fluorescent, halide, or mercury vapor lamps, must be recycled. The tubes must be packaged and handled carefully to avoid breakage.

16.0.0 Lead Safety

In 2010, the Environmental Protection Agency (EPA) enacted the Renovation, Repair, and Painting (RRP) rule. This standard is designed primarily to protect young children living in or spending time in buildings containing lead-based paint. It requires lead-safe certification and work practices to be followed during renovation work in all homes or other child-occupied facilities built prior to 1978. The rule is triggered by disturbing more than 6 square feet per room of an interior painted surface or more than 20 square feet of an exterior painted surface.

Housing and Urban Development (HUD) projects have a lower trigger of only 2 square feet of interior lead-based paint, more than 20 square feet of exterior lead-based paint, or 10 percent of the total surface area on an interior or exterior painted component that contains lead-based paint. This can include relatively small areas such as window sills, baseboard, and trim.

The RRP rule contains the following basic requirements:

- At least one RRP certified renovator is required at each job site. Certification involves lead-safe work training by an EPA-accredited training provider.
- In addition to individual certification, the contracting firm or agency must also be certified.
- Contractors must give the client a copy of the "Renovate Right" pamphlet (available for download at www.epa.gov).

17.0.0 Fall Protection

All employees must receive documented training before working in any area where there is the possibility of exposure to a fall of 6 feet or more. This training must be renewed annually and must include ladder safety. The 6-foot rule does not apply to ladders, scaffolds, and mechanical lifts, which have their own standards.

17.1.0 Fall Protection Procedures

Fall protection must be used when employees are on a walking or working surface that is 6 feet or more above a lower level and has an unprotected edge or side. The areas covered include, but are not limited to the following:

- Finished and unfinished floors or mezzanines
- Temporary or permanent walkways/ramps
- Finished or unfinished roof areas
- Elevator shafts and hoistways
- Floor, roof, or walkway holes
- Working 6 feet or more above dangerous equipment (*Exception:* If the dangerous equipment is unguarded, fall protection must be used at all heights regardless of the fall distance.)

Figure 16 shows a simple temporary guardrail installed on the stairway of a residence under

26102-14_F16.EPS

Figure 16 Temporary guardrail on stairs.

construction. *Figure 17* shows a complex guardrail system used during the construction of a department store. The large central opening will eventually house the building escalators.

> **NOTE**
> Walking/working surfaces do not include ladders, scaffolds, vehicles, or trailers. Also, an unprotected edge or side is an edge/side where there is no guardrail system at least 39 inches high.

According to OSHA, an employee can never be exposed to a fall of more than 6 feet. This is called 100% fall protection. The employee must be protected by one of the following in this order of preference:

1. Guardrail systems
2. Personal fall arrest systems (PFAS)
3. Controlled access zone or other administrative system

26102-14_F17.EPS

Figure 17 Complex guardrail system.

17.1.1 Guardrail Systems

Guardrail systems must be constructed as follows:

- The top rails must be 42 inches (±3 inches) and must be capable of withstanding 200 lbs.

What's wrong with this picture?

26102-14_SA12.EPS

- The mid-rails must be at 21 inches (±3 inches) and must be capable of withstanding 150 lbs.
- The toe board must be 4 inches tall and no more than ¼ inch above the floor for drainage.
- Guardrails can be made of 2 × 4s, pipes, chains, or cables.
- Chains and cables require flags every 6 feet. If cables are used, they must be secured to avoid deflection greater than 3 inches in, out, or down from the 42-inch requirement.
- Banding material is not allowed for guardrail construction. Cable guardrails require the use of cable clamps on the cable. Clamps must be forged and not malleable. They must be torqued and installed properly. Ensure that the clamp does not damage the loadbearing line.

> **NOTE**
> These ratings apply to all portions of the rail system such as anchors, anchorage material, and clamps. Guardrails cannot be used to secure a PFAS unless they are designed to support 5,000 lbs per person.

Warning lines, signs, or barricades must be installed back from the edge at least 6 feet. That way, even if someone falls over the barrier, they will not fall to the level below.

17.1.2 Personal Fall Arrest Systems (PFAS)

PFAS provide fall arrest after an employee falls. This equipment must be selected, inspected, donned, anchored, and maintained to be effective. The complete system usually consists of a full-body harness, lanyard, and anchorage device.

Full-body harnesses – Full-body harnesses are the only acceptable equipment to wear for PFAS. Select the appropriate harness based on size and gender. Inspect the equipment before use. Harnesses must be worn snug (but not tight) with all required straps attached. When properly applied, you should be able to slide two fingers under the straps with little difficulty. The D-ring in the back of the harness must be centered between the shoulder blades. After donning the harness, have a co-worker pull sharply up on the D-ring. You should feel the grab around the thighs, chest, and buttocks. Jobs that require positioning must be accomplished using a full-body harness with side D-rings. Safety belts are not allowed.

Lanyards – Lanyards are used to connect the harness to the attachment point. As no employee can be exposed to a fall of more than 6 feet, standard lanyards must be no longer than 6 feet. You can be exposed to 1,800 lbs of force in a properly worn harness. The use of shock absorbing lanyards or retractable lanyards can reduce that force to as low as 400 to 600 lbs. Shock absorbers work by slowing the employee to a stop by ripping stitches while elongating up to 42 inches. All lanyards must have locking snap hooks. Never attach two locking snap hooks to the same D-ring as they can foul each other, causing the relatively weak gates to break. *Figure 18* shows an electrician making an adjustment to an outdoor wallwasher lighting fixture while hanging out of a 15th-story window. For extra protection, he is attached to two lanyards on two separate D-rings.

What's wrong with this picture?

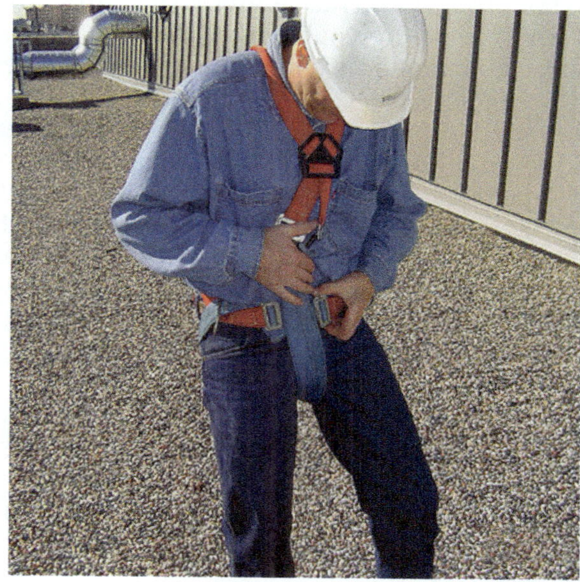

26102-14_SA13.EPS

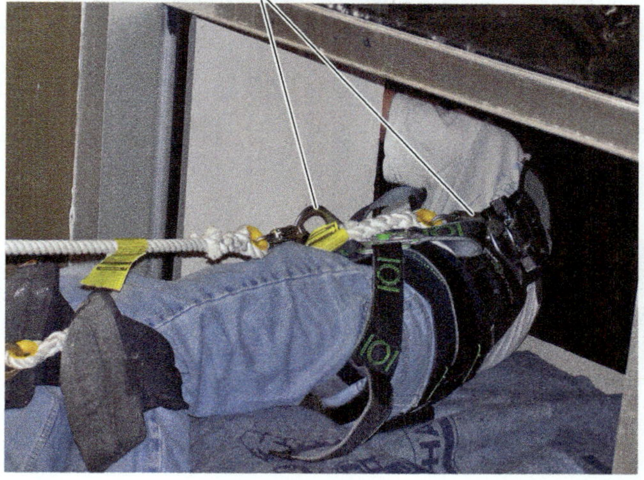

26102-14_F18.EPS

Figure 18 Electrician tied off to two lanyards.

A twin-tailed lanyard is required when climbing. While climbing, you cannot unhook your lanyard to move it to another anchorage and still have 100% fall protection. Thus, with two lanyards, you can "walk" to where you are working.

Retractable lanyards come in a variety of sizes from 6 feet to over 150 feet. *Figure 19* shows a worker tied off to a retractable lanyard when working on a boom lift. Retractable lanyards are used where movement or close proximity with the ground will render standard lanyards ineffective or inefficient. Retractable lanyards can be longer than 6 feet because when a fall occurs they grab hold within 2 feet. This quick reaction also eliminates the need for a shock absorber.

WARNING!

Do not put a shock absorber in line with a retractable lanyard because it may interfere with the quick response of the retractable lanyard.

26102-14_F19.EPS

Figure 19 Proper fall protection on a boom lift.

What's wrong with this picture?

26102-14_SA14.EPS

OSHA standards require PFAS to control falls for any working surface 6 feet or more above the ground. If you are forced to attach below the D-ring on your harness, you will fall that distance plus the 6-foot length of a standard lanyard. This could result in a fall of up to 12 feet. A PFAS is not designed to meet those forces and will transfer much higher impacts when it arrests the fall. To combat this, the American National Standards Institute (ANSI) has changed the equipment manufacturing standards for fall protection. The new standard is commonly referred to as the *12-foot lanyard*. The lanyard is still only 6 feet long, but the allowable deployment (extendable length) of the shock absorber has been increased. This lessens the force transferred to the body through the harness, reducing the likelihood of injury.

Before using a PFAS, you must examine the space below any potential fall point. Make sure it is clear of any hazards or obstructions. When assessing the required free length of a PFAS, you must account for the deployment of the shock absorber. In addition, PFAS manufacturers require up to a 3-foot safety factor. Taking this safety factor into account, *Table 3* indicates the minimum distance required between any hazard/obstruction and the worker or attachment point.

As you can see, a typical single-story building may not provide enough height to use standard lanyards. Retractable lanyards that engage and stop falls in less than 2 feet are usually the best choice for fall protection when used below 20 feet or on any type of lift equipment.

Table 3 Required Free Lengths for Various Lanyards

Lanyard Type	Lanyard Length	Shock Absorber Length	Average Worker Height	Safety Factor	Required Free Length
Standard 6' lanyard with 3.5' shock absorber	6'	3.5'	6'	3'	18.5'
New ANSI 6' lanyard with 4' shock absorber	6'	4'	6'	3'	19'
New ANSI 12' lanyard with 5' shock absorber	6' (12' freefall)	5'	N/A (accounted for by lanyard)	3'	20'

ANSI has also addressed the locking mechanism (gate) on snap hooks. OSHA requires the gate to be rated for 350 pounds when the rest of the system is rated at 5,000 pounds. This creates a weak link in the PFAS. ANSI issued a new standard increasing the strength of these gates to 3,600 pounds.

NOTE

Ladders, scaffolds, stairs, and similar types of equipment have other fall protection standards. Make sure you know and follow your company safety policies when using this equipment.

Anchorage devices – Anchorage devices and points are the interface between the PFAS and the structure to which they are attached. This point must hold 5,000 lbs (this is the equivalent of a full-size extended cab pickup truck).

A lanyard cannot be wrapped around an anchorage and then attached to itself unless it is specially designed with cross arm straps. Cross arm straps are made of webbing, two inches wide, of any necessary length, with two different size D-rings. The cross arm straps are passed over whatever object you are going to attach them to and wrapped around to reduce the length of the lanyard. Beam clamps, wire hangers, trolleys, and other manufactured devices are also used for specific applications.

Equipment inspection process – All PFAS must be inspected when received and before each use. Check the manufacturer's tag for the manufacturer's inspection date. If the fall equipment has no date, it should be disposed of immediately. Carefully look over the webbing. If you observe any burns, ripped stitches, color marker threads, distorted grommets, bent or cracked buckle tongues, distorted D-rings, or bent, cracked, or pulled fabrics, remove the PFAS from service. Retractable lanyards must undergo the same inspection process as other PFAS. In addition, you must also pull out the entire lanyard for inspection, let it back in, and then pull out 2 to 4 feet of lanyard and give it a swift tug to see if it properly engages. If any of these inspections fail, the equipment must be destroyed immediately or tagged DO NOT USE and returned to the shop.

WARNING!

All fall protection equipment that is involved in a fall must be taken out of service and destroyed. Any employees involved in a fall must receive medical attention, even if they do not feel they have been injured. Falls can cause internal injuries that are not readily apparent to the victim.

Rescue – Never pull anyone up by their fall protection; always rescue them with ladders or equipment from below. If the standard equipment is not available to provide rescue, a plan must be created before work can proceed. Rescues must be accomplished from below using ladders, lifts, and/or scaffolds.

WARNING!

Unless a person is in immediate danger, never attempt to lift them up by their lanyard. This could cause an additional drop for the fallen worker and/or injure the rescuers.

Immediately summon the fire department to assist in the rescue effort unless you can rescue the person without assistance. Rescue must take place as quickly as possible, as hanging from a harness presents additional hazards. If you fall, continue to move your limbs while awaiting rescue. This will help maintain circulation in your lower extremities.

> **WARNING!**
>
> Per OSHA, any employee whose weight exceeds 310 pounds, including their tools, cannot wear fall protection.

PFAS selection – The type of system selected depends on the fall hazards associated with the work to be performed. First, a hazard analysis must be conducted by the job site supervisor prior to the start of work. Based on the hazard analysis, the job site supervisor and project manager, in consultation with the safety manager, will select the appropriate fall protection system. All employees must be instructed in the use of the fall protection system before starting work.

17.1.3 Controlled Access Zones

There are times when a guardrail cannot be attached to the building. In these cases, a controlled access zone may be the solution. A controlled access zone may consist of guards, barricades, badge systems, or other administrative measures.

Figure 20 shows a controlled access zone on a roof. Note that the barricade is located 6 feet from the edge. This way, even if someone trips over the barricade, they won't fall off the roof.

Figure 20 Controlled access zone.

> **Think About It**
>
> **Putting It All Together**
>
> This module has described a professional approach to electrical safety. How does this professional outlook differ from an everyday attitude? What do you think are the key features of a professional philosophy of safety?

SUMMARY

Safety must be your concern at all times so that you do not become either the victim of an accident or the cause of one. Safety requirements and safe work practices are provided by OSHA and your employer. It is essential that you adhere to all safety requirements and follow your employer's safe work practices and procedures. Also, you must be able to identify the potential safety hazards of your job site. The consequences of unsafe job site conduct can often be expensive, painful, or even deadly. Report any unsafe act or condition immediately to your supervisor. You should also report all work-related accidents, injuries, and illnesses to your supervisor immediately. Remember, proper construction techniques, common sense, and a good safety attitude will help to prevent accidents, injuries, and fatalities.

Review Questions

1. The most life-threatening hazards on a construction site typically include all of the following *except* _____.
   a. falls
   b. electric shock
   c. being crushed or struck by falling or flying objects
   d. chemical burns

2. If a person's heart begins to fibrillate due to an electrical shock, the solution is to _____.
   a. leave the person alone until the fibrillation stops
   b. immerse the person in ice water
   c. use the Heimlich maneuver
   d. have a qualified person use emergency defibrillation equipment

3. Low-voltage conductors rarely cause injuries.
   a. True
   b. False

4. Class 00 rubber gloves are used when working with voltages less than _____.
   a. 500 volts
   b. 1,000 volts
   c. 5,000 volts
   d. 7,500 volts

5. An important use of a hot stick is to _____.
   a. replace busbars
   b. test for voltage
   c. replace fuses
   d. test for continuity

6. Which of these statements correctly describes a double-insulated power tool?
   a. There is twice as much insulation on the power cord.
   b. It can safely be used in place of a grounded tool.
   c. It is made entirely of plastic or other nonconducting material.
   d. The entire tool is covered in rubber.

7. Which of the following applies in a lockout/tagout procedure?
   a. Only the supervisor can install lockout/tagout devices.
   b. If several employees are involved, the lockout/tagout equipment is applied only by the first employee to arrive at the disconnect.
   c. Lockout/tagout devices applied by one employee can be removed by another employee as long as it can be verified that the first employee has left for the day.
   d. Lockout/tagout devices are installed by every authorized employee involved in the work.

8. The *NEC*® provides requirements for safe electrical installations, while *NFPA 70E*® provides guidance for establishing safe electrical workplaces.
   a. True
   b. False

9. What is the proper distance from the feet of a straight ladder to the wall?
   a. one-fourth the working height of the ladder
   b. one-half the height of the ladder
   c. three feet
   d. one-fourth of the square root of the height of the ladder

10. What are the minimum and maximum distances (in inches) that a scaffold plank can extend beyond its end support?
    a. 4; 8
    b. 6; 10
    c. 6; 12
    d. 8; 12

11. All of the following are usually considered confined spaces *except* a _____.
    a. 3 ft trench
    b. sewer
    c. pipeline
    d. manhole

12. The best way to protect yourself from solvent hazards is to _____.
    a. always wear vinyl gloves and a paper filter mask
    b. ask a co-worker for instructions
    c. ask the supplier
    d. read and follow all instructions on the product's SDS/MSDS

13. Asbestos was banned in _____; therefore, you must assume that any facility constructed before _____ has asbestos in it.
    a. 1943; 1945
    b. 1964; 1966
    c. 1978; 1980
    d. 1988; 1990

14. You may throw vapor lamps out with regular trash as long as you wrap them carefully to avoid breakage.
    a. True
    b. False

15. A PFAS anchorage point for a 6-foot lanyard must be able to hold _____.
    a. 250 lbs
    b. 500 lbs
    c. 1,000 lbs
    d. 5,000 lbs

Trade Terms Quiz

Fill in the blank with the correct term that you learned from your study of this module.

1. A life-threatening condition of the heart in which the muscle fibers contract irregularly is called ____________.
2. Any tool that has a case made of nonconductive material and that has been constructed so that the case is insulated from electrical energy is a ____________.
3. ____________ are chemicals often found in liquids that use a tube used to cool certain types of large transformers and capacitors.
4. A ____________ will de-energize a circuit or a portion of it if the current to ground exceeds some predetermined value.
5. A ____________ has a three-prong plug at the end of its power cord or some other means to ensure that stray current travels to ground without passing through the body of the operator.

Trade Terms

Double-insulated/ ungrounded tool
Fibrillation
Grounded tool
Ground fault circuit interrupter (GFCI)
Polychlorinated biphenyls (PCBs)

Module 26102-14
Supplemental Exercises

1. *CFR Part 1910* covers the ____.

2. *CFR Part 1926* covers the ____.

3. True or False? You should use a compressor for an air test on high-voltage gloves because it has more pressure and is much faster.

4. There are six classes of rubber protective equipment. List the six classes and their voltage ratings.

   ____

   ____

   ____

   ____

   ____

   ____

5. All conductors, buses, and connections should be considered ____________________ until proven otherwise.

6. The distance from the ladder feet to the base of the wall or support should be about ______________ the vertical distance from the bottom to the top of the ladder.

7. As it is erected, each part of a scaffolding shall be carefully ____________________.

8. List the general guidelines used in identifying a confined space.

   ____

   ____

   ____

   ____

   ____

9. Before determining the proper procedure for confined space entry, a(n) ____________________ ____________________ must be performed.

10. Fall protection must be used when employees are on a walking or working surface that is ____________________ or more above a lower level and has an unprotected edge or side.

Cornerstone of Craftsmanship

Michael J. Powers

Tri-City Electrical Contractors, Inc.

How did you choose a career in the electrical field?
My father was an electrician and after I "burned out" with a career in fast-food management, I decided to choose a completely different field.

Tell us about your apprenticeship experience.
It was excellent! I worked under several very knowledgeable electricians and had a pretty good selection of teachers. Over my four-year apprenticeship, I was able to work on a variety of jobs, from photomats to kennels to colleges.

What positions have you held and how did they help you to get where you are now?
I have been an electrical apprentice, a licensed electrician, a job-site superintendent, a master electrician, and am currently a corporate safety and training director. The knowledge I acquired in electrical theory in apprenticeship school and preparing for my licensing exams, as well as the practical on-the-job experience over thirty years in the trade, were wonderful training for my current position.

I also serve on the authoring team for NCCER's Electrical curricula, which has provided me with not only the opportunity to share what I have learned, but is also a great way to keep current in other areas by meeting with electricians from a variety of disciplines (we have commercial, residential, and industrial electricians on the team, as well as instructors).

What would you say was the single greatest factor that contributed to your success?
Choosing a company that recognized and rewarded competent, hard workers and provided them with the support and guidance to allow them to develop and succeed in the industry.

What does your current job entail?
I am responsible for safe work practices and procedures through the job-site management team at Tri-City Electrical Contractors, Inc. I also assist in developing, delivering, and administering the training program, from apprenticeship to in-house to outsourced training.

What advice do you have for trainees?
Training in all its aspects is the key to your success and advancement in the industry. Any time you are given a training opportunity, take it, even if it might not appear relevant at the time. Eventually, all knowledge can be applied to some situation.

Most importantly, have fun! The construction industry is composed of good people. I firmly believe that construction workers, as a group, are much more honest and direct than any other comparable group. Wait—did I say comparable group? That's a misstatement—there is no comparable group. Construction workers build America!

Trade Terms Introduced in This Module

Double-insulated/ungrounded tool: An electrical tool that is constructed so that the case is insulated from electrical energy. The case is made of a nonconductive material.

Fibrillation: Very rapid irregular contractions of the muscle fibers of the heart that result in the heartbeat and pulse going out of rhythm with each other.

Grounded tool: An electrical tool with a three-prong plug at the end of its power cord or some other means to ensure that stray current travels to ground without passing through the body of the user. The ground plug is bonded to the conductive frame of the tool.

Ground fault circuit interrupter (GFCI): A protective device that functions to de-energize a circuit or portion thereof within an established period of time when a current to ground exceeds some predetermined value. This value is less than that required to operate the overcurrent protective device of the supply circuit.

Polychlorinated biphenyls (PCBs): Toxic chemicals that may be contained in liquids used to cool certain types of large transformers and capacitors.

Additional Resources

This module presents thorough resources for task training. The following resource material is suggested for further study.

29 CFR Parts 1900–1910, Standards for General Industry. Occupational Safety and Health Administration, U.S. Department of Labor.

29 CFR Part 1926, Standards for the Construction Industry. Occupational Safety and Health Administration, U.S. Department of Labor.

Managing Electrical Hazards, Latest Edition. Upper Saddle River, NJ: Pearson Education, Inc.

National Electrical Code® Handbook, Latest Edition. Quincy, MA: National Fire Protection Association.

Standard for Electrical Safety in the Workplace, Latest Edition. Quincy, MA: National Fire Protection Association.

Figure Credits

TIC, The Industrial Company, Module opener

Topaz Publications, Inc., Figures 2, 5, 6A

Panduit Corp., Figure 6B

Mike Powers, Figures 3, 8–10, 14, 16–20, SA01, SA03–SA10, SA12–14

The Stanley Works, Figures 12A, 13

Greenlee/A Textron Company, Figures 12B, 12D

RIDGID®, Figure 12C

Tim Ely, SA02

Weld-On Adhesives, Inc., a division of IPS Corporation, Figure 15

Courtesy of Datum Filing Systems at www.datumfiling.com., SA11

U.S. Department of Labor Tables 1, 2

NCCER CURRICULA — USER UPDATE

NCCER makes every effort to keep its textbooks up-to-date and free of technical errors. We appreciate your help in this process. If you find an error, a typographical mistake, or an inaccuracy in NCCER's curricula, please fill out this form (or a photocopy), or complete the online form at **www.nccer.org/olf**. Be sure to include the exact module ID number, page number, a detailed description, and your recommended correction. Your input will be brought to the attention of the Authoring Team. Thank you for your assistance.

Instructors – If you have an idea for improving this textbook, or have found that additional materials were necessary to teach this module effectively, please let us know so that we may present your suggestions to the Authoring Team.

NCCER Product Development and Revision

13614 Progress Blvd., Alachua, FL 32615

Email: curriculum@nccer.org

Online: www.nccer.org/olf

❑ Trainee Guide ❑ AIG ❑ Exam ❑ PowerPoints Other ______________

Craft / Level: Copyright Date:

Module ID Number / Title:

Section Number(s):

Description:

Recommended Correction:

Your Name:

Address:

Email: Phone:

Introduction to Electrical Circuits

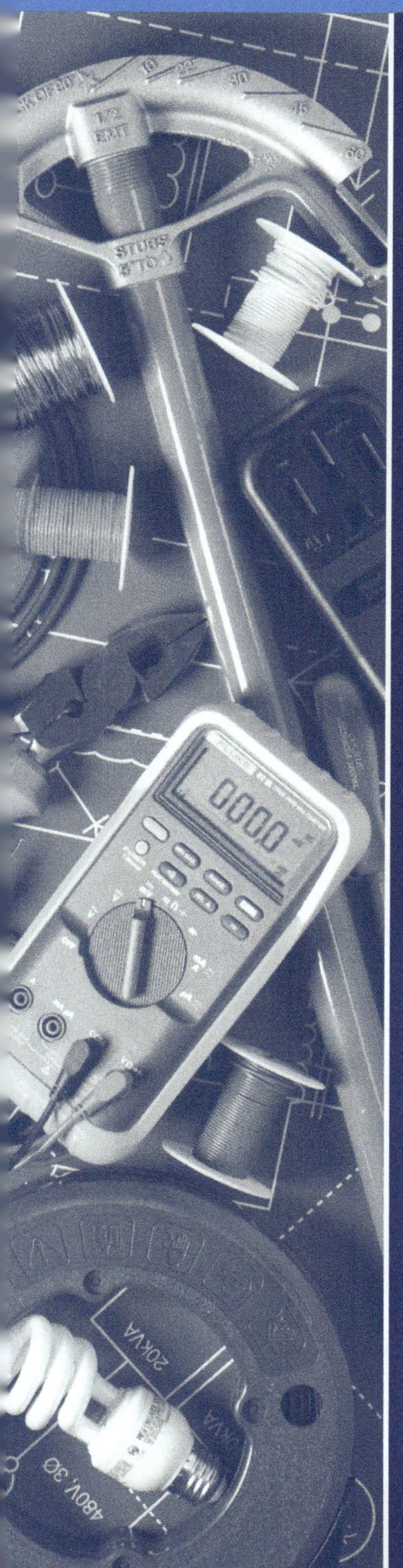

Reno Transportation Rail Access Corridor

The Reno Transportation Rail Access Corridor (ReTRAC) is the largest public works project ever undertaken in Nevada and includes the design and construction of a 2.1-mile long, 54-foot wide, 33-foot deep train trench running through downtown Reno. ReTRAC was built using the design-build method, resulting in shorter construction times, lessened traffic impacts, and lower costs.

26103-14

Trainees with successful module completions may be eligible for credentialing through NCCER's National Registry. To learn more, go to **www.nccer.org** or contact us at **1.888.622.3720.** Our website has information on the latest product releases and training, as well as online versions of our *Cornerstone* magazine and Pearson's product catalog.

Your feedback is welcome. You may email your comments to **curriculum@nccer.org,** send general comments and inquiries to **info@nccer.org,** or fill in the User Update form at the back of this module.

This information is general in nature and intended for training purposes only. Actual performance of activities described in this manual requires compliance with all applicable operating, service, maintenance, and safety procedures under the direction of qualified personnel. References in this manual to patented or proprietary devices do not constitute a recommendation of their use.

26103-14

Introduction to Electrical Circuits

Objectives

When you have completed this module, you will be able to do the following:

1. Define voltage and identify the ways in which it can be produced.
2. Explain the difference between conductors and insulators.
3. Define the units of measurement that are used to measure the properties of electricity.
4. Identify the meters used to measure voltage, current, and resistance.
5. Explain the basic characteristics of series and parallel circuits.

Performance Tasks

This is a knowledge-based module. There are no performance tasks.

Trade Terms

Ammeter
Ampere (A)
Atom
Battery
Circuit
Conductor
Coulomb
Current
Electron
Insulator
Joule (J)
Kilo
Matter
Mega
Neutrons
Nucleus
Ohm (Ω)
Ohmmeter
Ohm's law
Power
Protons
Relay
Resistance
Resistor
Schematic
Series circuit
Solenoid
Transformer
Valence shell
Volt (V)
Voltage
Voltage drop
Voltmeter
Watt (W)

Required Trainee Materials

1. Paper and pencil
2. Appropriate personal protective equipment
3. Calculator

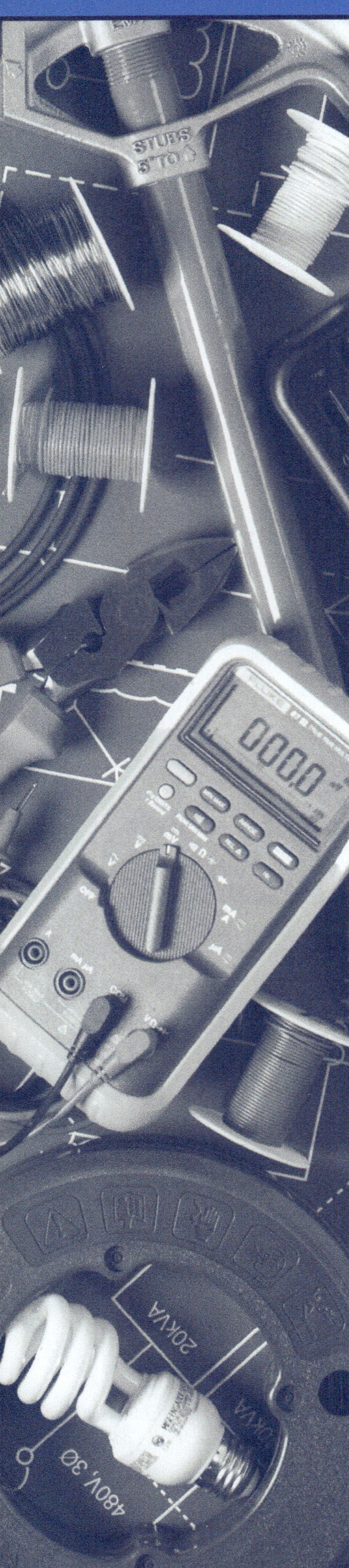

Contents

Topics to be presented in this module include:

Figures and Tables

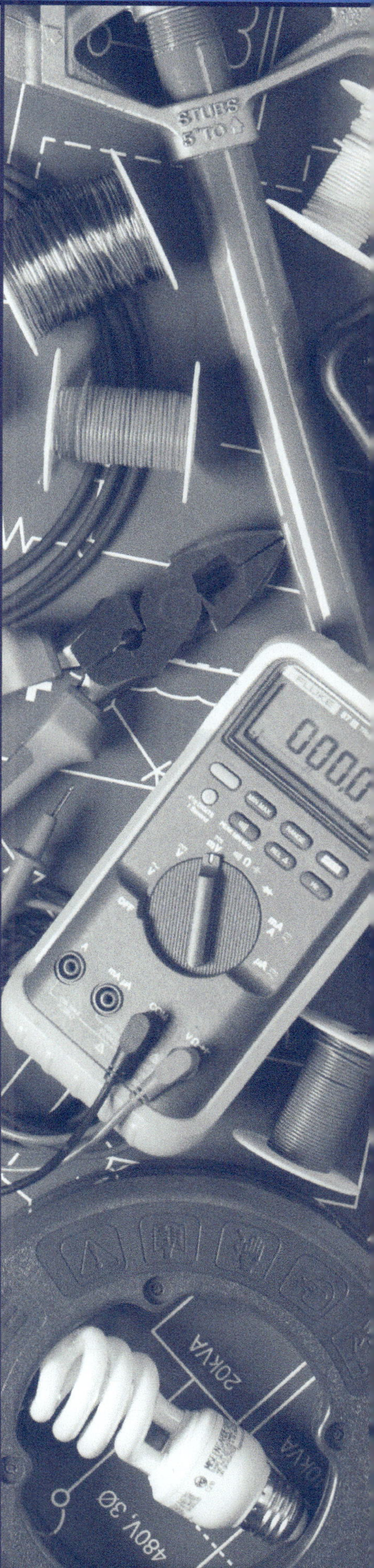

1.0.0 Introduction

Electricity is a form of energy that can be used by electrical devices such as motors, lights, TVs, heaters, and numerous other devices to perform work. Electricity is also used to control non-electrical devices that perform work. For example, your car is driven by a gasoline engine, but you wouldn't be able to start it or turn it off without the electrical system. In order to work with electricity, you need to know how it is produced and how it acts in electrical **circuits.**

You will hear the term *circuit* throughout your training. An electrical circuit contains, at minimum, a **voltage** source, a load, and conductors (wires) to carry the electrical **current** (*Figure 1*). The circuit should also have a means to stop and start the current, such as a switch.

Electricity is all about cause and effect. The presence of voltage (volts) in a closed circuit will cause current (amps) to flow. The more voltage you apply, the more current will flow. However, the amount of current flow is also determined by how much **resistance** (**ohms**) the load offers to the flow of current. In order to convert electrical energy into work, the load consumes energy. The amount of energy a device consumes is called **power**, and is expressed in **watts (W). Volts (V)**, amps, ohms, and watts are related in such a way that if any one of them changes, the others are proportionally affected. This relationship can be seen using basic math principles that you will learn in this module. You will also learn how electricity is produced and how test instruments are used to measure electricity.

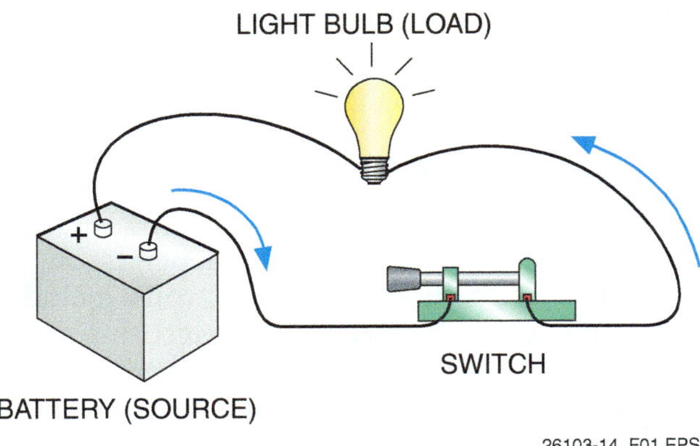

Figure 1 Basic electrical circuit.

2.0.0 Atomic Theory

In order to understand electrical theory, you must first understand the basic concepts of atomic theory. Atomic theory explains the construction and behavior of atoms, including the transfer of electrons that results in current flow.

2.1.0 The Atom

The **atom** is the smallest part of an element that enters into a chemical change, but it does so in the form of a charged particle. These charged particles are called ions, and are of two types—positive and negative. A positive ion may be defined as an atom that has become positively charged. A negative ion may be defined as an atom that has become negatively charged. One of the properties of charged ions is that ions of the same charge tend to repel one another, whereas ions of unlike charge will attract one another. The term *charge* can be taken to mean a quantity of electricity that is either positive or negative.

The structure of an atom is best explained by a detailed analysis of the simplest of all atoms, that of the element hydrogen. The hydrogen atom in *Figure 2* is composed of a **nucleus** containing one **proton** and a single orbiting **electron**. As the electron revolves around the nucleus, it is held in this orbit by two counteracting forces. One of these forces is called centrifugal force, which is the force that tends to cause the electron to fly outward as it travels around its circular orbit. The second force acting on the electron is electrostatic force. This force tends to pull the electron in toward the

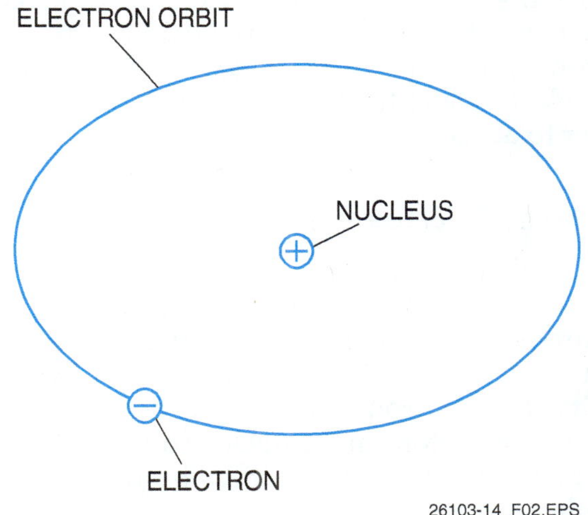

Figure 2 Hydrogen atom.

On Site

Why Bother Learning Theory?

Many trainees wonder why they need to bother learning the theory behind how things operate. They figure, why should I learn how it works as long as I know how to install it? The answer is, if you only know how to install something (e.g., run wire, connect switches, etc.), that's all you are ever going to be able to do. For example, if you don't know how your car operates, how can you troubleshoot it? The answer is, you can't. You can only keep changing out the parts until you finally hit on what is causing the problem. (How many times have you seen people do this?) Remember, unless you understand not only how things work but why they work, you'll only be a parts changer. With theory behind you, there is no limit to what you can do.

nucleus and is caused by the mutual attraction between the positive nucleus and the negative electron. At some given radius, the two forces will balance each other, providing a stable path for the electron.

- A proton (+) repels another proton (+).
- An electron (–) repels another electron (–).
- A proton (+) attracts an electron (–).

Basically, an atom contains three types of subatomic particles that are of interest in electricity: electrons, protons, and **neutrons**.

The protons and neutrons are located in the center, or nucleus, of the atom, and the electrons travel around the nucleus in orbits.

Because protons are relatively heavy, the repulsive force they exert on one another in the nucleus of an atom has little effect.

The attracting and repelling forces on charged materials occur because of the electrostatic lines of force that exist around the charged materials. In a negatively charged object, the lines of force of the excess electrons combine to produce an electrostatic field that has lines of force coming into the object from all directions. In a positively charged object, the lines of force of the excess protons combine to produce an electrostatic field that has lines of force going out of the object in all directions. The electrostatic fields either aid or oppose each other to attract or repel.

2.1.1 The Nucleus

The nucleus is the central part of the atom. It is made up of heavy particles called protons and neutrons. The proton is a charged particle containing the smallest known unit of positive electricity. The neutron has no electrical charge. The number of protons in the nucleus determines how the atom of one element differs from the atom of another element.

Although a neutron is actually a particle by itself, it is generally thought of as an electron and proton combined, and is electrically neutral. Since neutrons are electrically neutral, they are not considered relevant to the electrical nature of atoms.

Think About It

Electrical Charges

Think about the things you come in contact with every day. Where do you see or find examples of electrostatic attraction?

2.1.2 Electrical Charges

The negative charge of an electron is equal but opposite to the positive charge of a proton. The charges of an electron and a proton are called electrostatic charges. The lines of force associated with each particle produce electrostatic fields. Because of the way these fields act together, charged particles can attract or repel one another. The Law of Electrical Charges states that particles with like charges repel each other and those with unlike charges attract each other. This is shown in *Figure 3*.

2.2.0 Conductors and Insulators

The difference between atoms, with respect to chemical activity and stability, depends on the number and position of the electrons included within the atom. In general, the electrons reside in groups of orbits called shells. The shells are arranged in steps that correspond to fixed energy levels.

The outer shell of an atom is called the **valence shell**, and the electrons contained in this shell are called valence electrons (*Figure 4*). The number of valence electrons determines an atom's ability to gain or lose an electron, which in turn determines the chemical and electrical properties of the atom. An atom that is lacking only one or two electrons from its outer shell will easily gain electrons to complete its shell, but a large amount of energy is required to

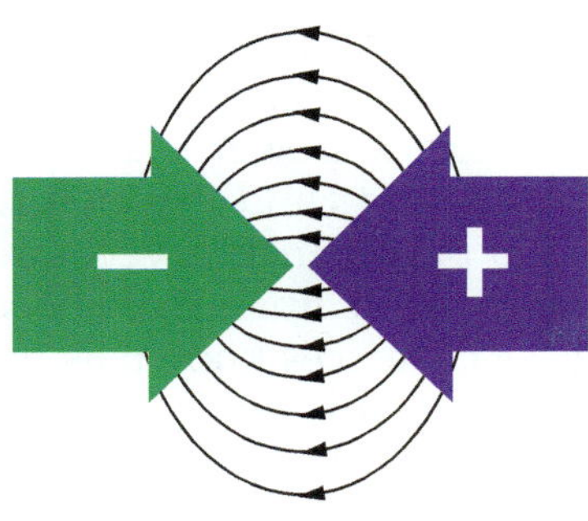

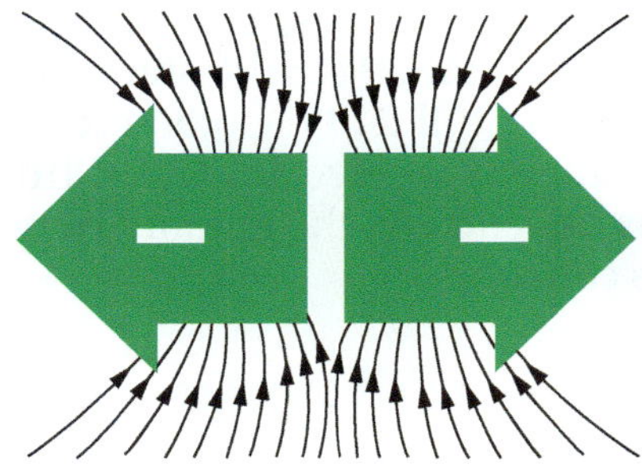

Figure 3 Law of electrical charges.

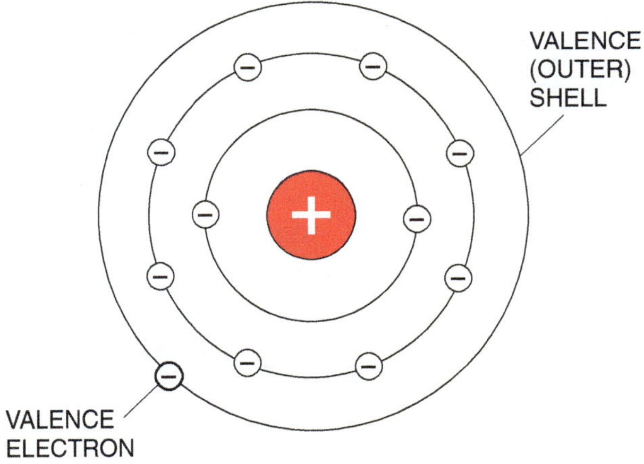

Figure 4 Valence shell and electrons.

free any of its electrons. An atom having a relatively small number of electrons in its outer shell in comparison to the number of electrons required to fill the shell will easily lose these valence electrons.

It is the valence electrons that we are most concerned with in electricity. These are the electrons that are easiest to break loose from their parent atom. Normally, a **conductor** has three or less valence electrons, an **insulator** has five or more valence electrons, and semiconductors usually have four valence electrons.

All the elements of which **matter** is made may be placed into one of three categories: conductors, insulators, and semiconductors.

Conductors, for example, are elements such as copper and silver that will conduct a flow of electricity very readily. Because of their good conducting abilities, they are formed into wire and used whenever it is desired to transfer electrical energy from one point to another.

Insulators, on the other hand, do not conduct electricity to any great degree and are used when it is desirable to prevent the flow of electricity. Compounds such as porcelain and plastic are good insulators.

Materials such as germanium and silicon are not good conductors but cannot be used as insulators either, since their electrical characteristics fall between those of conductors and those of insulators. These in-between materials are classified as semiconductors. As you will learn later in your training, semiconductors play a crucial role in electronic circuits.

2.3.0 Magnetism

The operation of many electrical components relies on the power of magnetism. Motors, **relays, transformers**, and **solenoids** are examples. Magnetized iron generates a magnetic field consisting of magnetic lines of force, also known as magnetic flux lines (*Figure 5*). Magnetic objects within the field will be attracted or repelled by the magnetic field. The more powerful the magnet, the more powerful the magnetic field around it. Each magnet has a north pole and a south pole. Opposing poles attract each other; like poles repel each other.

Electricity also produces magnetism. Current flowing through a conductor produces a small magnetic field around the conductor. If the conductor is coiled around an iron bar, the result is an electromagnet (*Figure 6*) that attracts and repels other magnetic objects just like an iron magnet. This is the basis on which electric motors and other components operate.

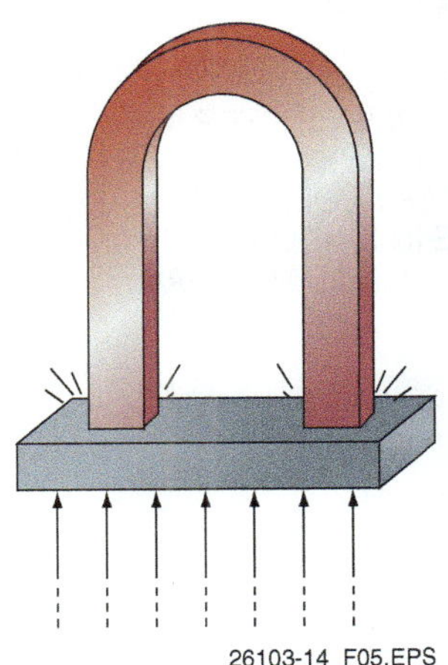

Figure 5 Magnetism.

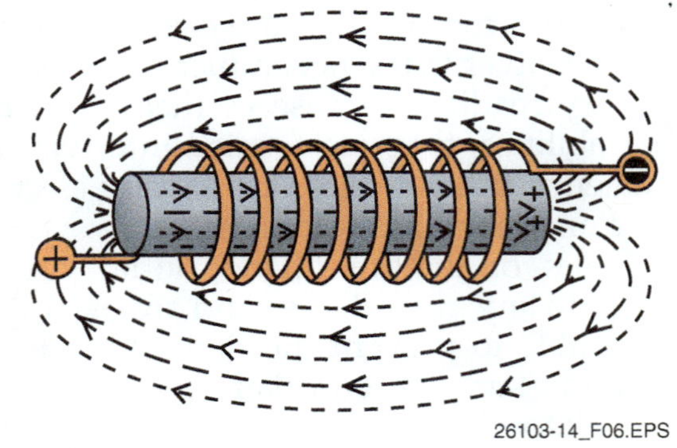

Figure 6 Electromagnet.

3.0.0 Electrical Power Generation and Distribution

Electricity comes from electrical generating plants (*Figure 7*) operated by utilities like your local power company. Steam from coal-burning or nuclear power plants is used to power huge generators called turbines, which generate electricity. There are also hydroelectric power plants, solar power generating plants, and wind-driven turbines.

The electrical power that travels through long-distance transmission lines may be as high as 750,000 volts (V). Devices known as transformers are used to step the voltage down to lower levels as it reaches electrical substations and eventually homes, offices, and factories. The voltage you receive at home is usually about 240V. At the wall outlet where you plug in small appliances such as televisions and toasters, the voltage is about 120V (*Figure 8*). Electric stoves, clothes dryers, water heaters, and central air conditioning systems usually require the full 240V. Commercial buildings and factories may receive anywhere from 208V to 575V. This depends on the amount of power their machines consume.

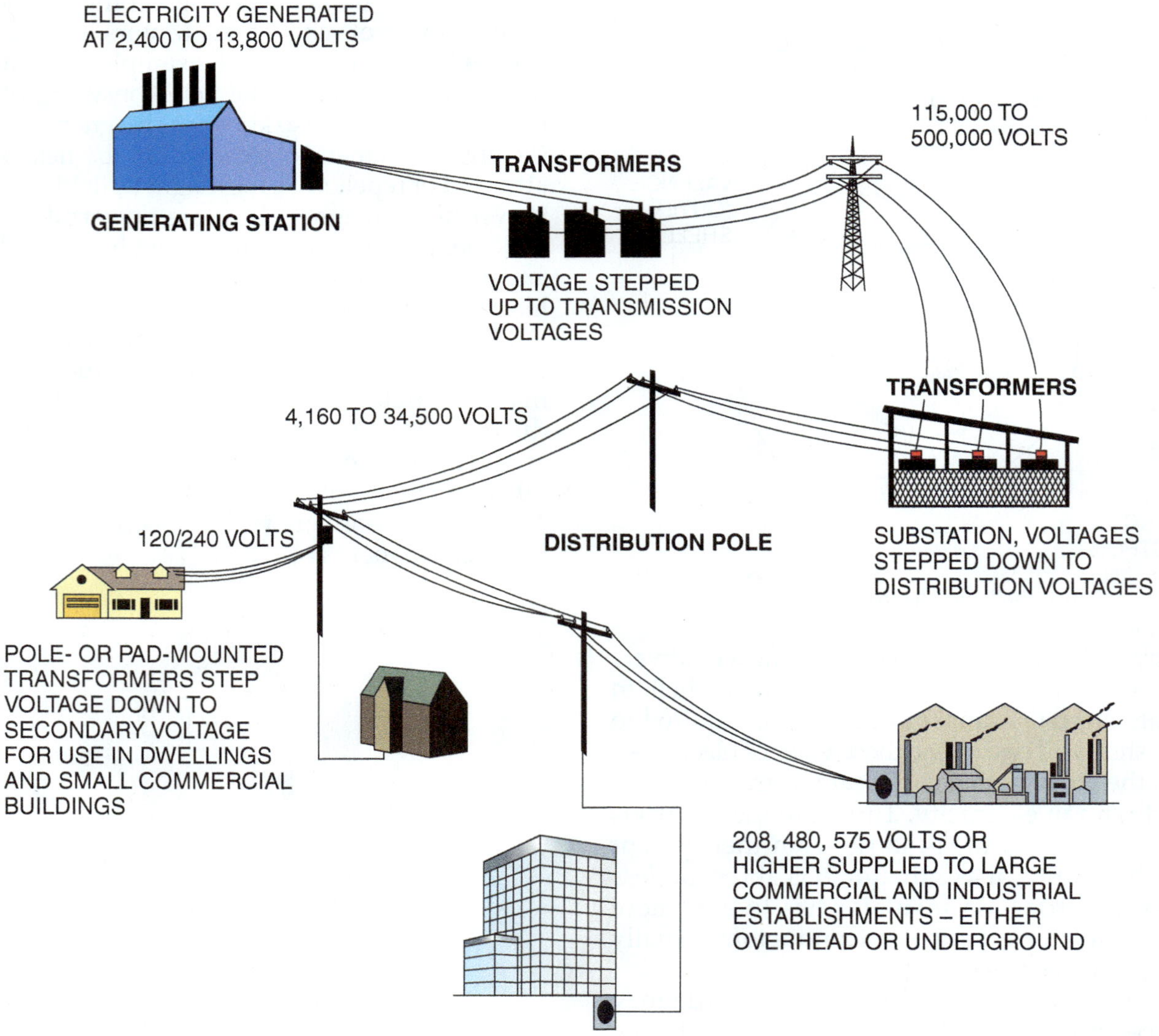

ELECTRICITY GENERATED AT 2,400 TO 13,800 VOLTS
GENERATING STATION
TRANSFORMERS
VOLTAGE STEPPED UP TO TRANSMISSION VOLTAGES
115,000 TO 500,000 VOLTS
TRANSFORMERS
SUBSTATION, VOLTAGES STEPPED DOWN TO DISTRIBUTION VOLTAGES
4,160 TO 34,500 VOLTS
DISTRIBUTION POLE
120/240 VOLTS
POLE- OR PAD-MOUNTED TRANSFORMERS STEP VOLTAGE DOWN TO SECONDARY VOLTAGE FOR USE IN DWELLINGS AND SMALL COMMERCIAL BUILDINGS
208, 480, 575 VOLTS OR HIGHER SUPPLIED TO LARGE COMMERCIAL AND INDUSTRIAL ESTABLISHMENTS – EITHER OVERHEAD OR UNDERGROUND
26103-14_F07.EPS

Figure 7 Electrical power distribution.

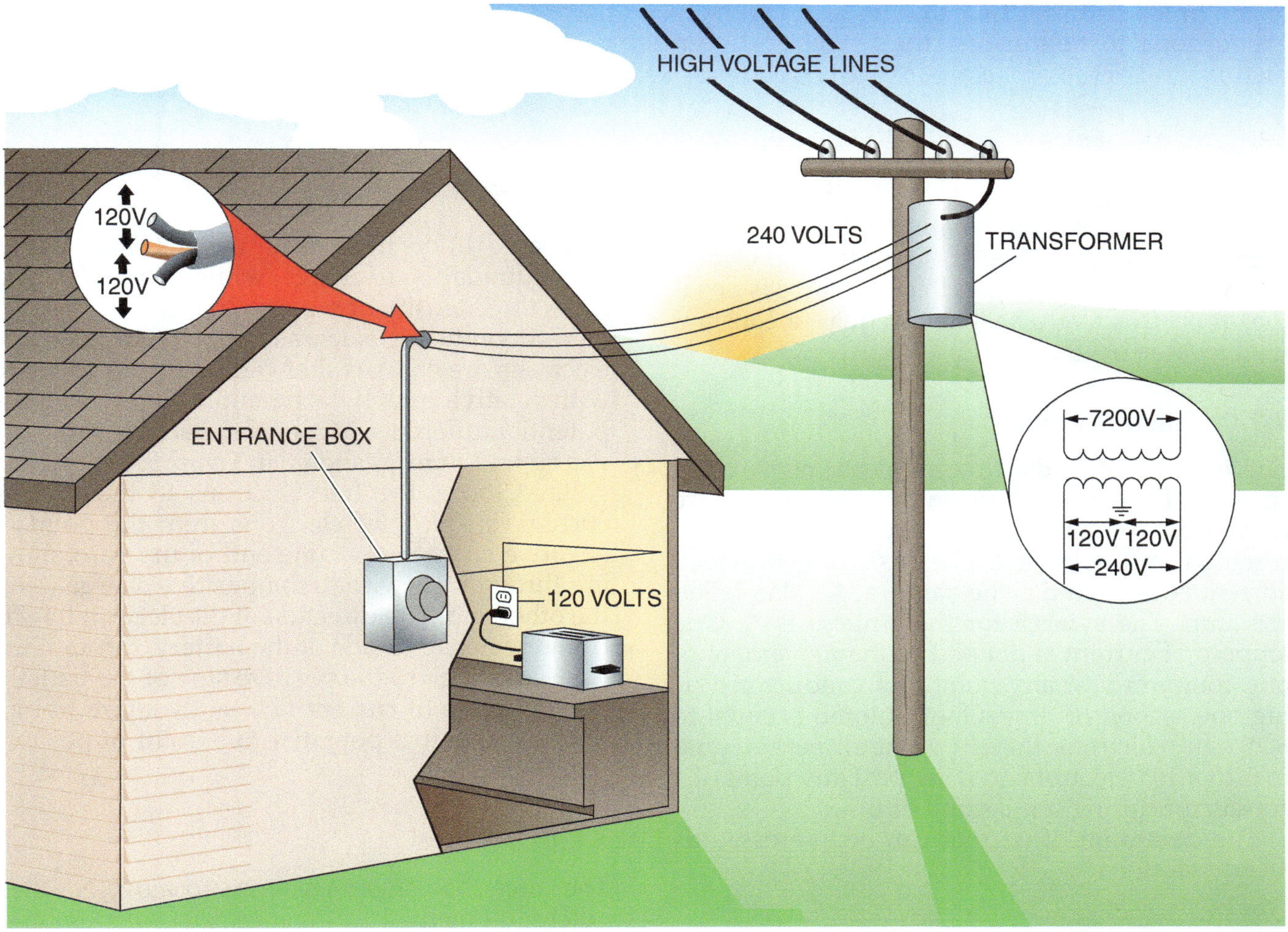

26103-14_F08.EPS

Figure 8 Internal power distribution.

GOING GREEN

Hydroelectric Plants

Hydroelectric plants use the power generated by water to drive turbines that produce electricity.

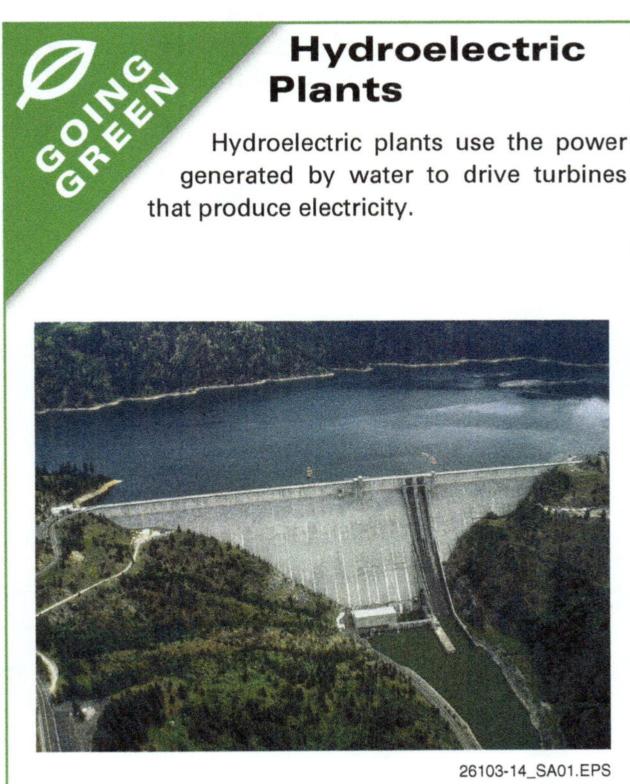

26103-14_SA01.EPS

4.0.0 Electric Charge and Current

An electric charge has the ability to do the work of moving another charge by attraction or repulsion. The ability of a charge to do work is called its potential. When one charge is different from another, there is a difference in potential between them. The sum of the difference of potential of all the charges in the electrostatic field is referred to as electromotive force (emf) or voltage. Voltage is frequently represented by the letter *E*.

Electric charge is measured in **coulombs**. An electron has 1.6×10^{-19} coulombs of charge. Therefore, it takes 6.25×10^{18} electrons to make up one coulomb of charge, as shown below.

$$\frac{1}{1.6 \times 10^{-19}} = 6.25 \times 10^{18} \text{ electrons}$$

If two particles, one having charge Q_1 and the other charge Q_2, are a distance (d) apart, then the force between them is given by Coulomb's law, which states that the force is directly proportional to the product of the two charges and inversely proportional to the square of the distance between them:

$$\text{Force} = \frac{k \times Q_1 \times Q_2}{d^2}$$

If Q_1 and Q_2 are both positive or both negative, then the force is positive; it is repulsive. If Q_1 and Q_2 are of opposite charges, then the force is negative; it is attractive. The letter k equals a constant with a value of 10^9.

4.1.0 Current Flow

The movement of the flow of electrons is called current. To produce current, the electrons are moved by a potential difference. Current is represented by the letter *I*. The basic unit in which current is measured is the **ampere (A)**, also called the amp. The symbol for the ampere is *A*. One ampere of current is defined as the movement of one coulomb past any point of a conductor during one second of time. One coulomb is equal to 6.25×10^{18} electrons; therefore, one ampere is equal to 6.25×10^{18} electrons moving past any point of a conductor during one second of time.

The definition of current can be expressed as an equation:

$$I = \frac{Q}{T}$$

Where:

I = current (amperes)
Q = charge (coulombs)
T = time (seconds)

Charge differs from current in that charge (Q) is an accumulation of charge, while current (I) measures the intensity of moving charges.

In a conductor, such as copper wire, the free electrons are charges that can be forced to move with relative ease by a potential difference. If a potential difference is connected across two ends of a copper wire, as shown in *Figure 9*, the applied voltage forces the free electrons to move. This current is a flow of electrons from the point of negative charge (–) at one end of the wire, moving through the wire to the positive charge (+) at the other end. The direction of the electron flow is from the negative side of the **battery**, through the wire, and back to the positive side of the battery. The direction of current flow is therefore from a point of negative potential to a point of positive potential.

4.2.0 Voltage

The force that causes electrons to move is called voltage, potential difference, or electromotive force (emf). One volt is the potential difference between two points for which one coulomb of electricity will do one **joule (J)** of work. A battery is one of several means of creating voltage. It chemically creates a large reserve of free electrons at the

On Site

Transformers

Large distribution transformers at power substations step down the power to the level required for local distribution. Pole transformers like the one shown here step it down further to the voltages needed for homes and businesses.

26103-14_SA02.EPS

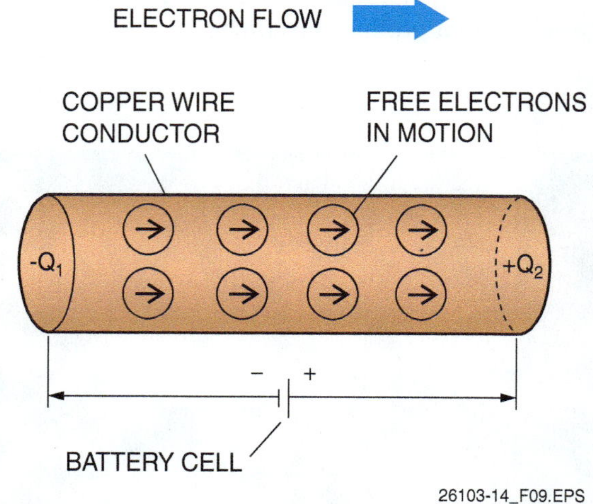

Figure 9 Potential difference causing electric current.

On Site

Units of Electricity and Volta

A disagreement with a fellow scientist over the twitching of a frog's leg eventually led 18th-century physicist Alessandro Volta to theorize that when certain objects and chemicals come into contact with each other, they produce an electric current. Believing that electricity came from contact between metals only, Volta coined the term *metallic electricity*. To demonstrate his theory, Volta placed two discs, one of silver and the other of zinc, into a weak acidic solution. When he linked the discs together with wire, electricity flowed through the wire. Thus, Volta introduced the world to the battery, also known as the Voltaic pile. Now Volta needed a term to measure the strength of the electric push or the flowing charge; the volt is that measure.

negative (–) terminal. The positive (+) terminal has electrons chemically removed and will therefore accept them if an external path is provided from the negative (–) terminal. When a battery is no longer able to chemically deposit electrons at the negative (–) terminal, it is said to be dead, or in need of recharging. Batteries are normally rated in volts. Large batteries are also rated in ampere-hours, where one ampere-hour is a current of one amp supplied for one hour.

4.3.0 Resistance

Resistance is directly related to the ability of a material to conduct electricity. All conductors have very low resistance; insulators have very high resistance.

On Site

Law of Electrical Force and de Coulomb

In the 18th century, a French physicist named Charles de Coulomb was concerned with how electric charges behaved. He watched the repelling forces electric charges exerted by measuring the twist in a wire. An object's weight acted as a turning force to twist the wire, and the amount of twist was proportional to the object's weight. After many experiments with opposing forces, de Coulomb proposed the Inverse Square Law, later known as the Law of Electrical Force.

Think About It

The Magic of Electricity

The effect of the flow of electrons occurs at close to the speed of light, about 186,000 miles per second. How long does it take the light from the end of a flashlight to reach the floor? If you ran a light circuit from Maine to California and flipped the switch, how long would it take for the light to come on?

4.3.1 Characteristics of Resistance

Resistance can be defined as the opposition to current flow. To add resistance to a circuit, electrical components called **resistors** are used. A resistor is a device whose resistance to current flow is a known, specified value. Resistance is measured in ohms and is represented by the symbol *R* in equations. One ohm is defined as the amount of resistance that will limit the current in a conductor to one ampere when the voltage applied to the conductor is one volt. The symbol for an ohm is Ω.

The resistance of a wire is proportional to the length of the wire, inversely proportional to the cross-sectional area of the wire, and dependent upon the kind of material of which the wire is made. The relationship for finding the resistance of a wire is:

$$R = \rho \frac{L}{A}$$

Where:

R = resistance (ohms)
L = length of wire (feet)
A = area of wire (circular mils, CM, or cm^2)
ρ = specific resistance (ohm-CM/ft or microhm-CM)

A mil equals 0.001 inch; a circular mil is the cross-sectional area of a wire one mil in diameter.

The specific resistance is a constant that depends on the material of which the wire is made. *Table 1* shows the properties of various wire conductors.

Think About It

Current Flow

Why do you need two wires to use electrical devices? Why can't current simply move to a lamp and be released as light energy?

On Site

Joule's Law

While other scientists of the 19th century were experimenting with batteries, cells, and circuits, James Joule was theorizing about the relationship between heat and energy. He discovered, contrary to popular belief, that work did not just move heat from one place to another; work, in fact, generated heat. Furthermore, he demonstrated that over time, a relationship existed between the temperature of water and electric current. These ideas formed the basis for the concept of energy. In his honor, the modern unit of energy was named the joule.

Table 1 shows that at 75°F, a one-mil diameter, pure annealed copper wire that is one foot long has a resistance of 10.351 ohms; while a one-mil diameter, one-foot-long aluminum wire has a resistance of 16.758 ohms. Temperature is important in determining the resistance of a wire. The hotter a wire, the greater its resistance.

Table 1 Conductor Properties

Metal	Specific Resistance (Resistance of 1 CM/ft in ohms)	
	32°F or 0°C	75°F or 23.8°C
Silver, pure annealed	8.831	9.674
Copper, pure annealed	9.390	10.351
Copper, annealed	9.590	10.505
Copper, hard-drawn	9.810	10.745
Gold	13.216	14.404
Aluminum	15.219	16.758
Zinc	34.595	37.957
Iron	54.529	62.643

26103-14_T01.EPS

5.0.0 Ohm's Law

Ohm's law defines the relationship between current, voltage, and resistance. There are three ways to express Ohm's law mathematically.

- The current in a circuit is equal to the voltage applied to the circuit divided by the resistance of the circuit:

$$I = \frac{E}{R}$$

- The resistance of a circuit is equal to the voltage applied to the circuit divided by the current in the circuit:

$$R = \frac{E}{I}$$

On Site

The Visual Language of Electricity

Learning to read circuit diagrams is like learning to read a book—first you learn to read the letters, then you learn to read the words, and before you know it, you are reading without paying attention to the individual letters anymore. Circuits are the same way—you will struggle at first with the individual pieces, but before you know it, you will be reading a circuit without even thinking about it. Studying the table below will help you to understand the fundamental language of electricity.

What's Measured	Unit of Measurement	Symbol	Ohm's Law Symbol
Amount of current	Amp	A	I
Electrical power	Watt	W	P
Force of current	Volt	V	E
Resistance to current	Ohm	Ω	R

- The applied voltage to a circuit is equal to the product of the current and the resistance of the circuit:

$$E = I \times R = IR$$

Where:

I = current (amperes)
R = resistance (ohms)
E = voltage or emf (volts)

If any two of the quantities E, I, or R are known, the third can be calculated.

The Ohm's law equations can be memorized and practiced effectively by using an Ohm's law circle, as shown in *Figure 10*. To find the equation for E, I, or R when two quantities are known, cover the unknown third quantity. The other two quantities in the circle will indicate how the covered quantity may be found.

Example 1:

Find I when E = 120V and R = 30Ω.

$$I = \frac{E}{R}$$

$$I = \frac{120V}{30\Omega}$$

$$I = 4A$$

This formula shows that in a DC circuit, current (I) is directly proportional to voltage (E) and inversely proportional to resistance (R).

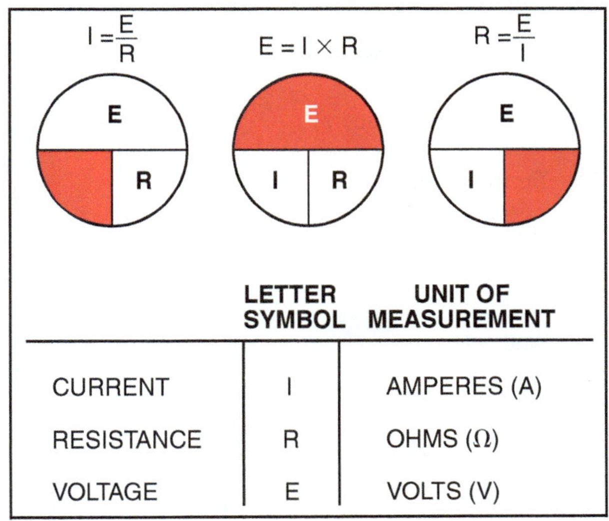

	LETTER SYMBOL	UNIT OF MEASUREMENT
CURRENT	I	AMPERES (A)
RESISTANCE	R	OHMS (Ω)
VOLTAGE	E	VOLTS (V)

26103-14_F10.EPS

Figure 10 Ohm's law circle.

Example 2:

Find R when E = 240V and I = 20A.

$$R = \frac{E}{I}$$

$$R = \frac{240V}{20A}$$

$$R = 12\Omega$$

Example 3:

Find E when I = 15A and R = 8Ω.

$$E = I \times R$$

$$E = 15A \times 8\Omega$$

$$E = 120V$$

6.0.0 Schematic Representation of Circuit Elements

The simple electric circuit shown earlier is shown in both pictorial and schematic forms in *Figure 11*. The schematic diagram is a shorthand way to draw an electric circuit, and circuits are usually represented in this way. In addition to the connecting wire, three components are shown symbolically: the battery, the switch, and the lamp. Note the positive (+) and negative (–) markings in both

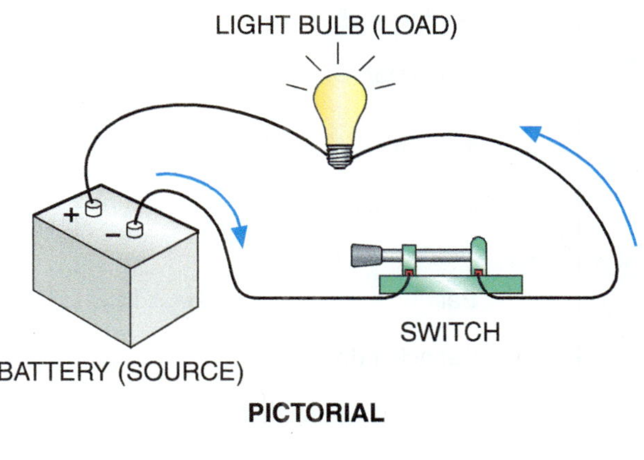

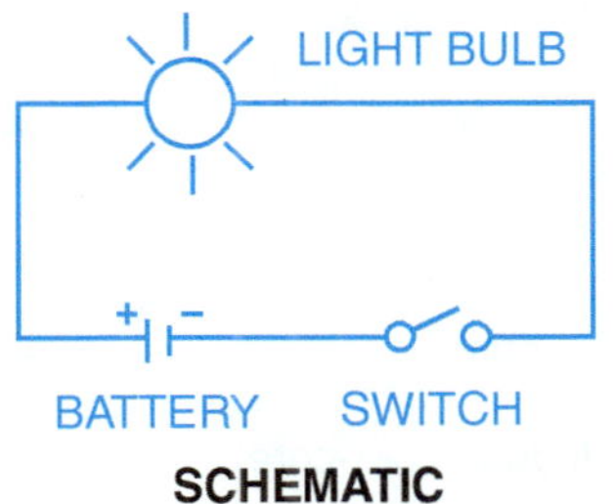

26103-14_F11.EPS

Figure 11 Electrical circuit.

Think About It

Voltage Matters

Standard household voltage is different the world over, from 100V in Japan to 600V in Bombay, India. Many countries have no standard voltage; for example, France varies from 110V to 360V. If you were to plug a 120V hair dryer into England's 240V, you would burn out the dryer. Use basic electric theory to explain exactly what would happen to destroy the hair dryer.

the pictorial and schematic representations of the battery. The schematic components represent the pictorial components in a simplified manner. A schematic diagram is one that shows, by means of graphic symbols, the electrical connections and functions of the different parts of a circuit.

The standard graphic symbols for commonly used electrical and electronic components are shown in *Figure 12*.

Think About It

Drawing a Schematic

Draw a schematic diagram showing a voltage source, switch, motor, and fuse.

7.0.0 Resistors

The function of a resistor is to offer a particular resistance to current flow. For a given current and known resistance, the change in voltage across the component, or **voltage drop**, can be predicted using Ohm's law. Voltage drop refers to a specific amount of voltage used, or developed, by that component. An example is a very basic circuit of a 10V battery and a single resistor in a **series circuit**. The voltage drop across that resistor is 10V because it is the only component in the circuit and all voltage must be dropped across that resistor. Similarly, for a given applied voltage, the current that flows may be predetermined by selection of the resistor value.

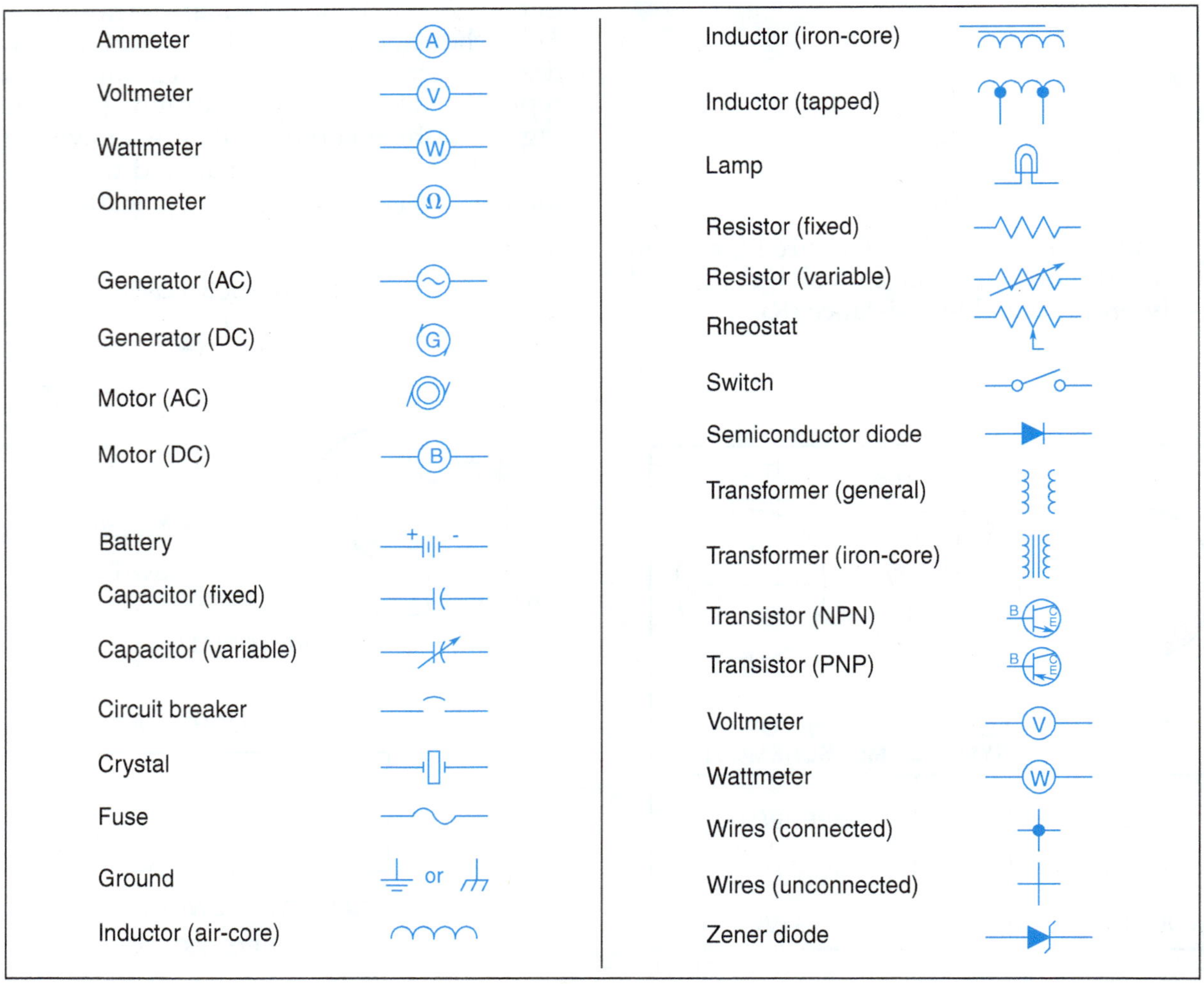

Figure 12 Standard schematic symbols.

On Site

Using Your Intuition

Learning the meanings of various electrical symbols may seem overwhelming, but if you take a moment to study *Figure 12*, you will see that most of them are intuitive—that is, they are shaped (in a symbolic way) to represent the actual object. For example, the battery shows + and –, just like an actual battery. The motor has two arms that suggest a spinning rotor. The transformer shows two coils. The resistor has a jagged edge to suggest pulling or resistance. Connected wires have a black dot that reminds you of solder. Unconnected wires simply cross. The fuse stretches out in both directions as though to provide extra slack in the line. The circuit breaker shows a line with a break in it. The capacitor shows a gap. The variable resistor has an arrow like a swinging compass needle. As you learn to read schematics, take the time to make mental connections between the symbol and the object it represents.

The required power dissipation largely dictates the construction and physical size of a resistor.

The two most common types of electronic resistors are wire-wound and carbon composition construction. A typical wire-wound resistor consists of a length of nickel wire wound on a ceramic tube and covered with porcelain. Low-resistance connecting wires are provided, and the resistance value is usually printed on the side of the component. *Figure 13* illustrates the

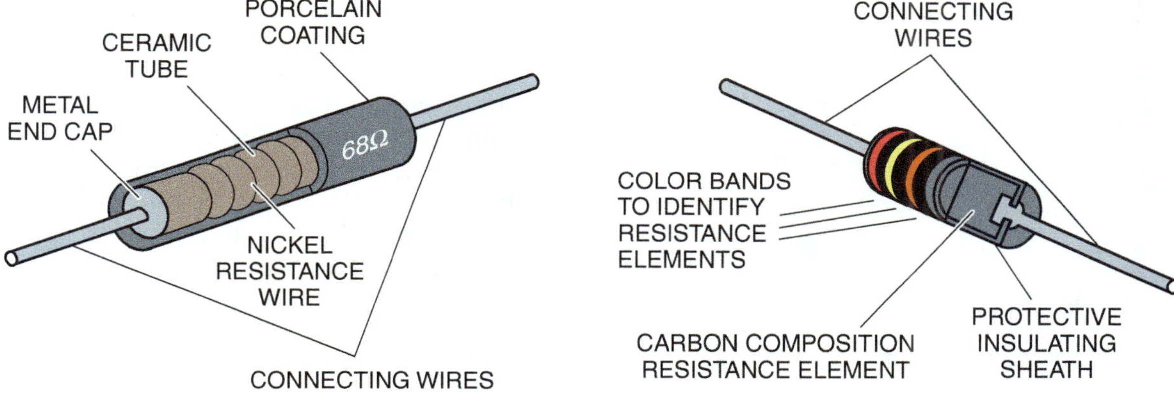

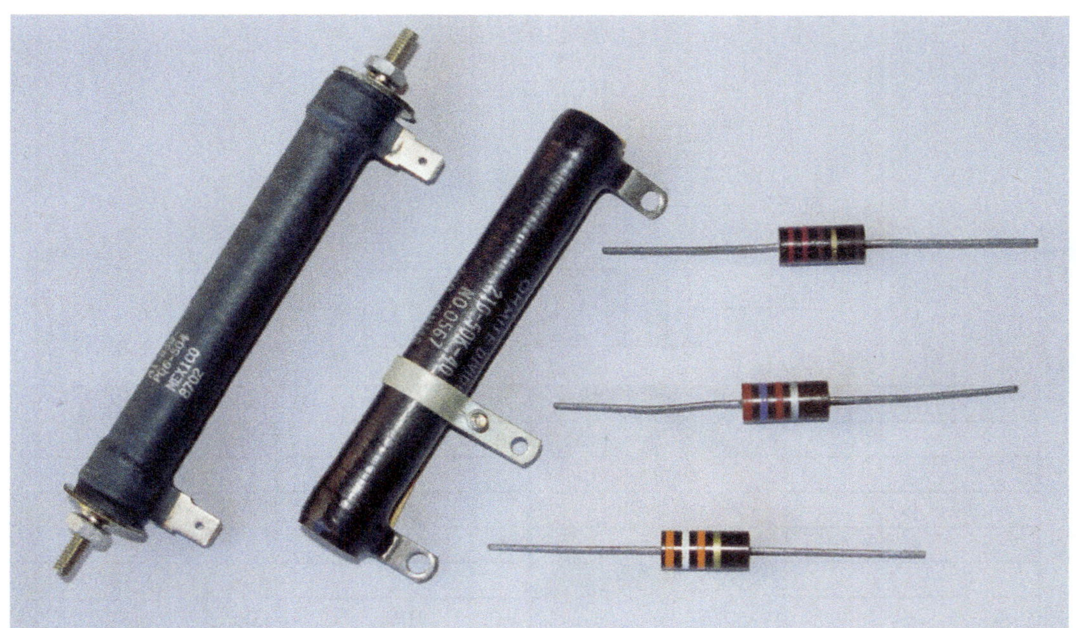

26103-14_F13.EPS

Figure 13 Common resistors.

construction of typical resistors. Carbon composition resistors are constructed by molding mixtures of powdered carbon and insulating materials into a cylindrical shape. An outer sheath of insulating material affords mechanical and electrical protection, and copper connecting wires are provided at each end. Carbon composition resistors are smaller and less expensive than the wire-wound type. However, the wire-wound type is the more rugged of the two and is able to survive much larger power dissipations than the carbon composition type.

Most resistors have standard fixed values, so they can be termed fixed resistors. Variable resistors, also known as adjustable resistors, are used a great deal in electronics. Two common symbols for a variable resistor are shown in *Figure 14*.

A variable resistor consists of a coil of closely wound insulated resistance wire formed into a partial circle. The coil has a low-resistance terminal at each end, and a third terminal is connected to a movable contact with a shaft adjustment facility. The movable contact can be set to any point on a connecting track that extends over one (uninsulated) edge of the coil.

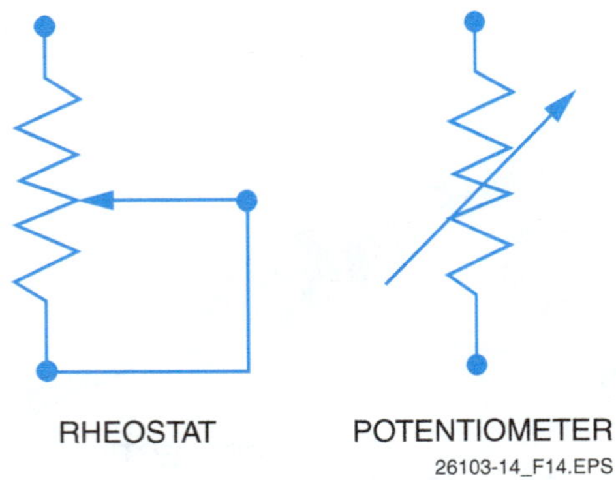

Figure 14 Symbols used for variable resistors.

Using the adjustable contact, the resistance from either end terminal to the center terminal may be adjusted from zero to the maximum coil resistance.

Another type of variable resistor is known as a decade resistance box. This is a laboratory component that contains precise values of switched series-connected resistors.

7.1.0 Resistor Color Codes

Because carbon composition resistors are physically small (some are less than 1 cm in length), it is not convenient to print the resistance value on the side. Instead, a color code in the form of colored bands is employed to identify the resistance value and tolerance. The color code is illustrated in *Figure 15*. Starting from one end of the resistor, the first two bands identify the first and second digits of the resistance value, and the third band indicates the number of zeros. An exception to this is when the third band is either silver or gold, which indicates a 0.01 or 0.1 multiplier, respectively. The fourth band is always either silver or gold, and in this position, silver indicates a ± 10% tolerance and gold indicates a ± 5% tolerance. Where no fourth band is present, the resistor tolerance is ± 20%.

You can put this information to practical use by determining the range of values for the carbon resistor in *Figure 16*.

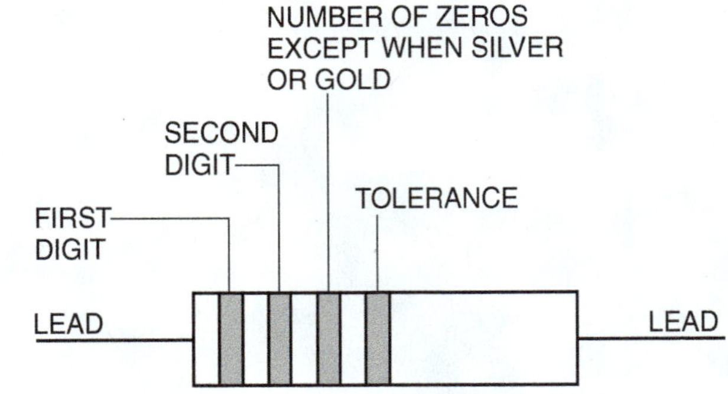

0	BLACK	7	VIOLET
1	BROWN	8	GRAY
2	RED	9	WHITE
3	ORANGE	0.1	GOLD
4	YELLOW	0.01	SILVER
5	GREEN	5%	GOLD – TOLERANCE
6	BLUE	10%	SILVER – TOLERANCE

26103-14_F15.EPS

Figure 15 Resistor color codes.

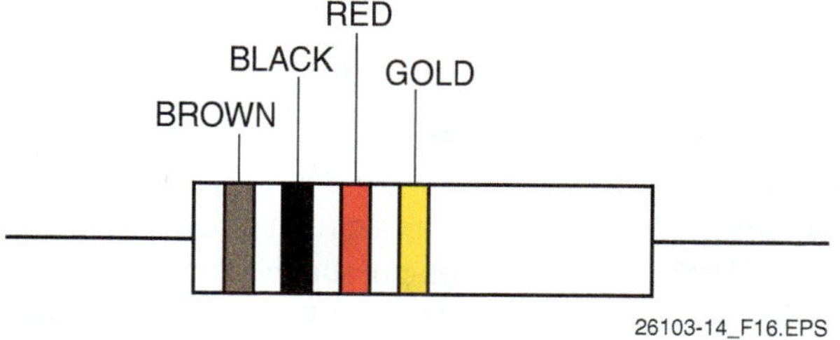

Figure 16 Sample color codes on a fixed resistor.

The color code for this resistor is as follows:

- Brown = 1, black = 0, red = 2, gold = a tolerance of ± 5%
- First digit = 1, second digit = 0, number of zeros (2) = 1,000Ω

Since this resistor has a value of 1,000Ω ± 5%, the resistor can range in value from 950Ω to 1,050Ω.

8.0.0 Electrical Circuits

You will often hear the terms *series circuit* and *parallel circuit* during your training. When you hear these terms, keep in mind that they refer to the way loads are connected in the circuit.

8.1.0 Series Circuits

A series circuit provides only one path for current flow and is a voltage divider. The total resistance of the circuit is equal to the sum of the individual resistances. The 12V series circuit in *Figure 17* has two 30Ω loads. The total resistance is therefore 60Ω. The amount of current flowing in the circuit is 0.2A.

$$I = \frac{E}{R} = \frac{12V}{60\Omega} = 0.2A$$

If there were five 30Ω loads, the total resistance would be 150Ω. The current flow is the same through all the loads. The voltage measured across any load (voltage drop) depends on the resistance of that load. The sum of the voltage drops equals the total voltage applied to the circuit. Circuits containing loads in series are uncommon. An important trait of a series circuit is that if the circuit is open at any point, no current will flow. For example, if you have five light bulbs connected in series and one of them blows, all five lights will go off.

8.2.0 Parallel Circuits

In a parallel circuit, each load is connected directly to the voltage source; therefore, the voltage drop is the same through all loads

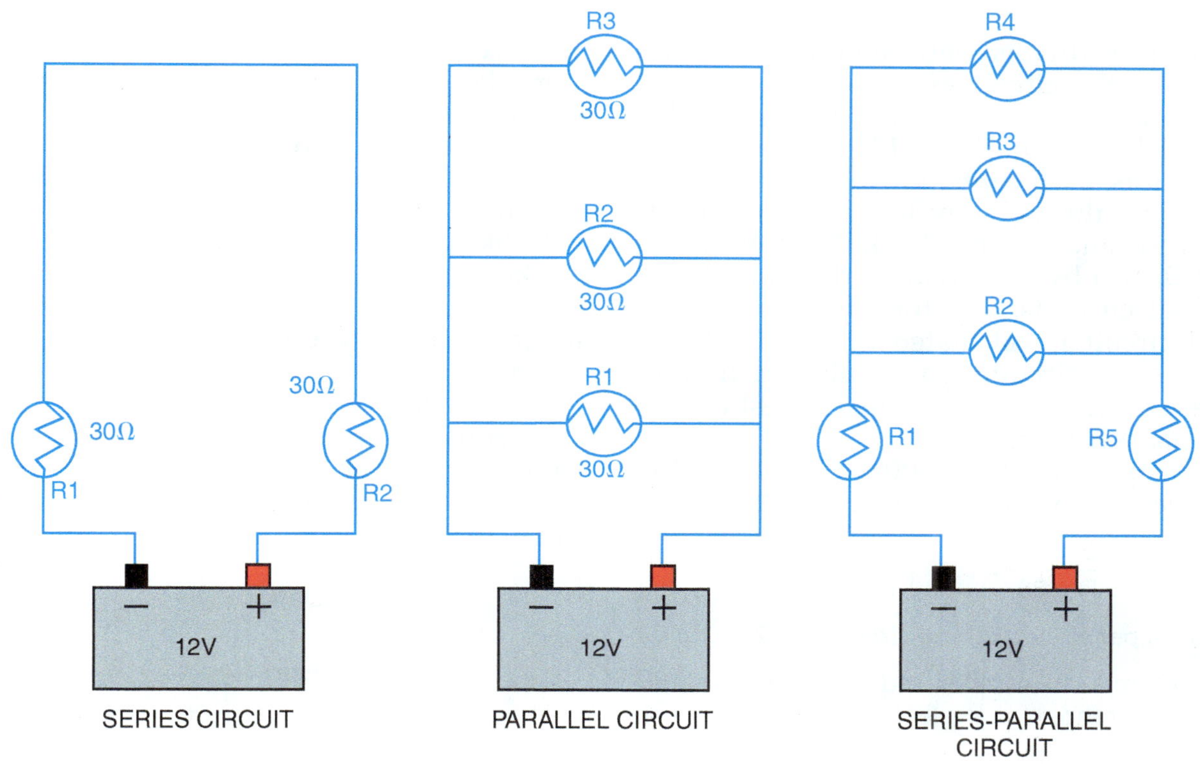

Figure 17 Types of circuits.

On Site

Is It a Series Circuit?

When the term *series circuit* is used, it refers to the way the loads are connected. The same is true for parallel and series-parallel circuits. You will rarely, if ever, find loads connected in series, or in a series-parallel arrangement. The simple circuit shown here illustrates this point. At first glance, you might think it is a series-parallel circuit. On closer examination, you can see that there are only two loads—the relay and the contactor—and they are connected in parallel. Therefore, it is a parallel circuit. The control devices are wired in series with the loads, but only the loads are considered in determining the type of circuit.

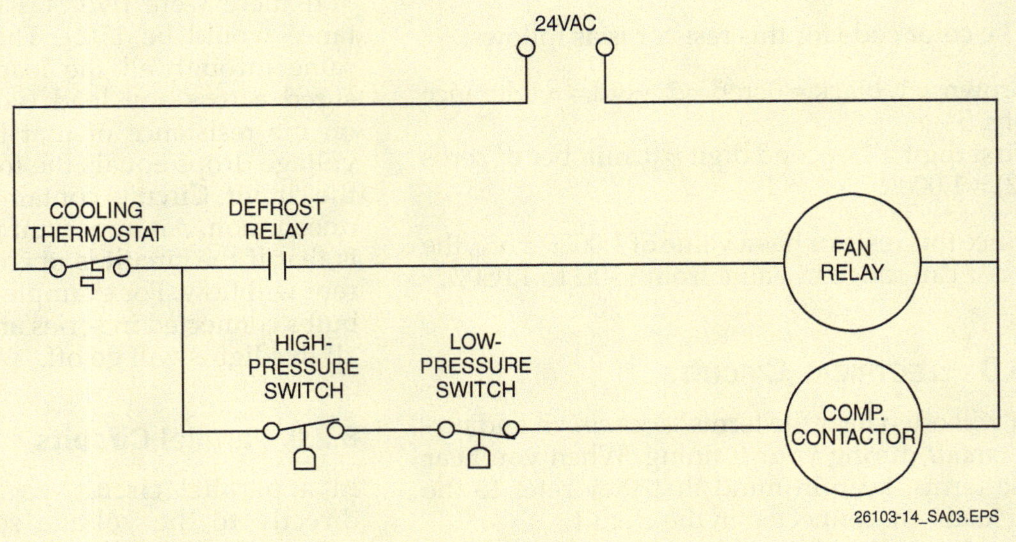

and current is divided between the loads. The source sees the circuit as two or more individual circuits containing one load each. In the parallel circuit in *Figure 17*, the source sees three circuits, each containing a 30Ω load. The current flow through any load is determined by the resistance of that load. Thus the total current drawn by the circuit is the sum of the individual currents. The total resistance of a parallel circuit is calculated differently from that of a series circuit. In a parallel circuit, the total resistance is less than the smallest of the individual resistances.

For example, each of the 30Ω loads draws 0.4A at 12V; therefore, the total current is 1.2A:

$$I = \frac{E}{R} = \frac{12V}{30} = 0.4A \text{ per circuit}$$

$$0.4A \text{ per circuit} \times \text{three circuits} = 1.2A$$

Now, Ohm's law can be used again to calculate the total resistance:

$$R = \frac{E}{I} = \frac{12V}{1.2A} = 10\Omega$$

This one was simple because all the resistances were the same value. The process is the same when the resistances are different, but the current calculation has to be done for each load. The individual currents are added to get the total current.

Unlike series circuits, parallel circuits continue working even if one circuit opens. Household circuits are wired in parallel. In fact, almost all the load circuits you encounter will be parallel circuits.

Either of the following formulas can be used to convert parallel resistances to a single resistance value. The first one is used when there are two resistances in parallel. The second is used when there are three or more.

$$\text{Total resistance} = \frac{R1 \times R2}{R1 + R2}$$

$$\text{Total resistance} = \frac{1}{\frac{1}{R1} + \frac{1}{R2} + \frac{1}{R3}}$$

Example:

1. The total resistance of the parallel circuit below is 6Ω.

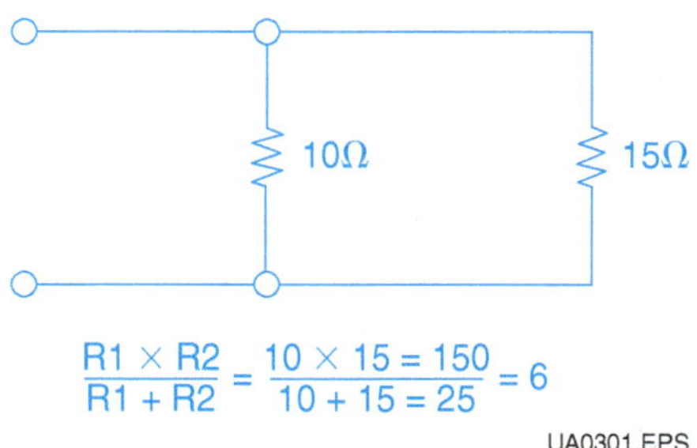

2. The total resistance of the parallel circuit below is 4.76Ω.

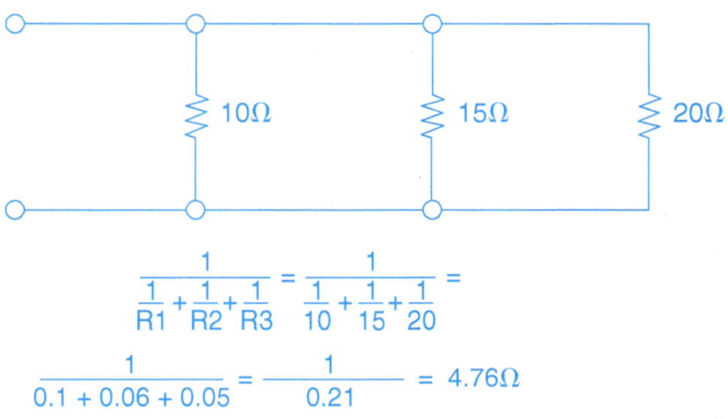

10Ω
15Ω
20Ω
1 / (1/R1 + 1/R2 + 1/R3) = 1 / (1/10 + 1/15 + 1/20) =
1 / (0.1 + 0.06 + 0.05) = 1 / 0.21 = 4.76Ω
UA0302.EPS

8.3.0 Series-Parallel Circuits

Electronic circuits often contain a hybrid arrangement known as a series-parallel circuit (*Figure 17*). It is unlikely, however, that you will ever have to determine the electrical characteristics of one of these circuits. If it becomes necessary, the parallel loads must be converted to their equivalent series resistance. Then the load resistances are added to determine total circuit resistance.

9.0.0 Electrical Measuring Instruments

Electricians frequently use test meters to measure voltage, current, and resistance. The most common test meter is the volt-ohm-milliammeter (VOM), also called a multimeter. *Figure 18* shows both digital and analog multimeters. The analog meter is so-called because the pointer moves in proportion to the value being measured. The person using the meter must then interpret the scale to determine the measured value. Digital meters display the result numerically on the screen.

Multimeters are commonly used to measure AC and DC voltage, DC current, and resistance. They can also be used to measure AC current in the milliamp range. For larger current values, it is usually necessary to use a clamp-on **ammeter** (*Figure 19*).

WARNING!

Only qualified individuals may use these meters. Consult your company's safety policy for applicable rules.

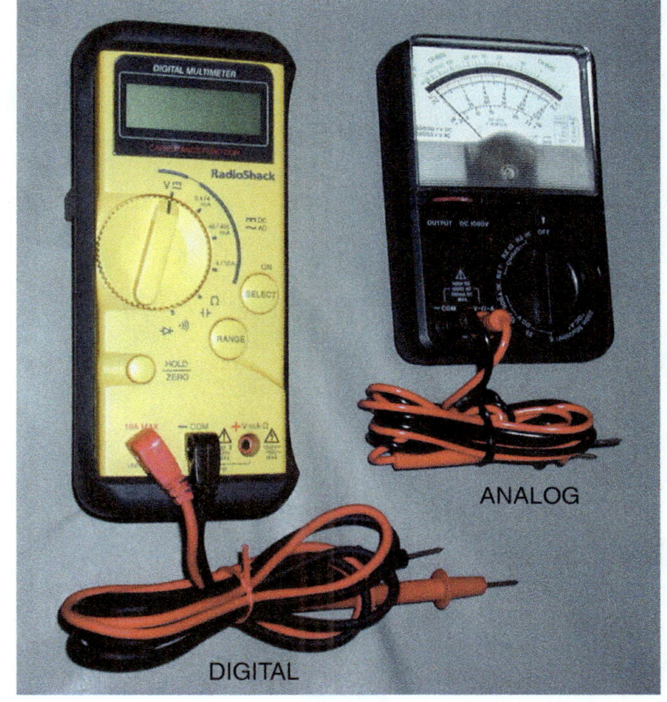

26103-14_F18.EPS

Figure 18 Digital and analog meters.

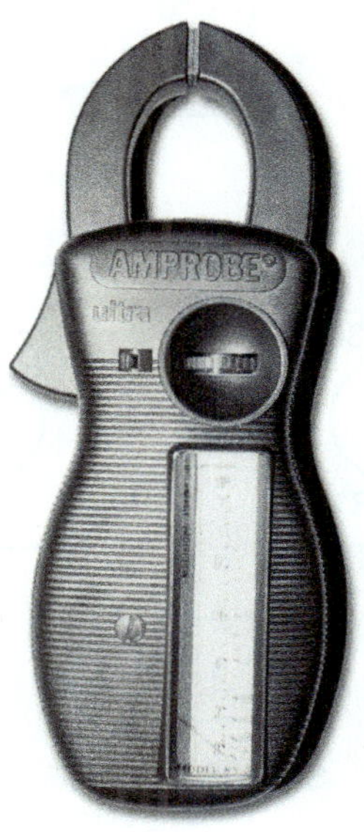

26103-14_F19.EPS

Figure 19 Clamp-on ammeter.

9.1.0 Measuring Current

A clamp-on ammeter is used to measure current. The jaws of the ammeter are placed around a single conductor (*Figure 20*). Current flowing through the wire creates a magnetic field, which induces a proportional current in the ammeter jaws. This current is read by the meter movement and appears as a direct readout or, on an analog meter, as a deflection of the meter needle.

In-line ammeters (*Figure 21*) are less common. This type of meter must be connected in series with the circuit, which means that the circuit must be opened.

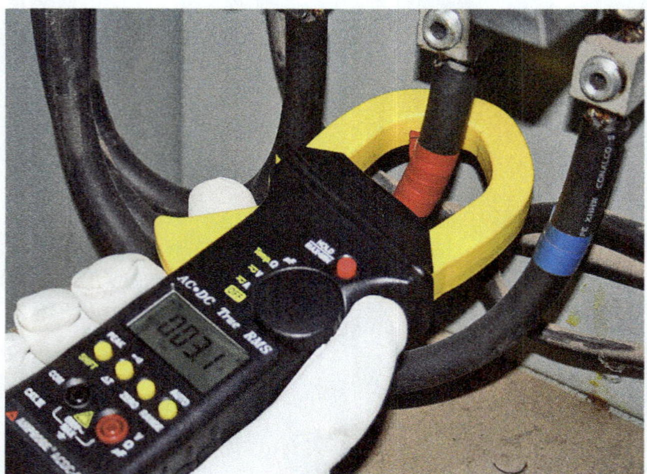

26103-14_F20.EPS

Figure 20 Clamp-on ammeter in use.

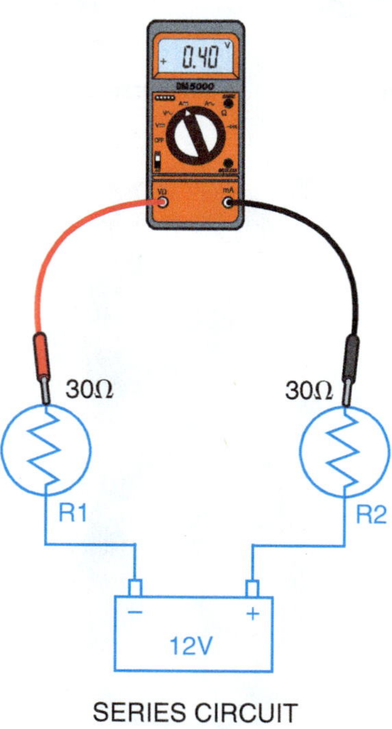

SERIES CIRCUIT

26103-14_F21.EPS

Figure 21 In-line ammeter test setup.

Aside from following good safety practices, there are a few things to remember when measuring current:

- If the ammeter jaws are dirty or misaligned, a meter will not read correctly.
- When using an analog meter, always start at a high range and work down to avoid damaging the meter.
- Do not clamp the meter jaws around two different conductors at the same time, or an inaccurate reading will result.

9.2.0 Measuring Voltage

A **voltmeter** must be connected in parallel with (across) the component or circuit to be tested (*Figure 22*). If a circuit function is not operating, the voltmeter can be used to determine if the correct voltage is available to the circuit. Voltage must be checked with power applied.

9.3.0 Measuring Resistance

An **ohmmeter** contains an internal battery that acts as a voltage source. Therefore, resistance measurements are always made with the system power shut off. Sometimes, an ohmmeter is used to measure

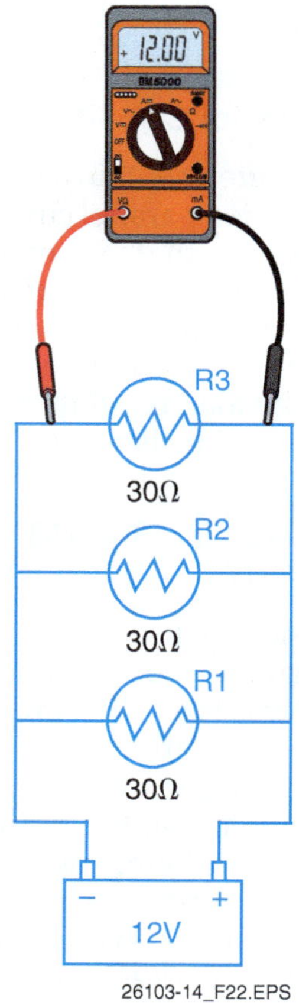

26103-14_F22.EPS

Figure 22 Voltmeter connection.

resistance in a load; motor windings are a good example. More often, an ohmmeter is used to check continuity in a circuit. A wire or closed switch offers negligible resistance. With the ohmmeter connected as in *Figure 23*, and the three switches closed, the current produced by the ohmmeter battery will flow unopposed and the meter will show zero resistance. The circuit has continuity; that is, it is continuous. If a switch is open, however, there is no path for current and the meter will see infinite resistance; that is, a lack of continuity.

A continuity tester (*Figure 24*) is a simple device consisting mainly of a battery and either an audible or visual indicator. It can be used in place of an ohmmeter to test the continuity of a wire and to identify individual wires contained in a conduit or other raceway. To test the continuity of a wire, strip the insulation off the end of the wire to be tested at one end of the conduit run, then connect (short) the wire to the metal conduit. At the other end of the conduit run, clip the alligator clip lead of the tester to the conduit and touch the probe to the end of the wire under test. If the tester audible alarm sounds or the indicator light comes on, there is continuity. Note that this only indicates that there is continuity between the two points being tested; it does not indicate the actual value of the resistance. If there is no indication, the wire is open.

To identify individual wires in a conduit run, touch the tester probe to the wires in the conduit one at a time until the tester audible alarm sounds or the indicator lights. Then, put matching identification tags on both ends of the wire.

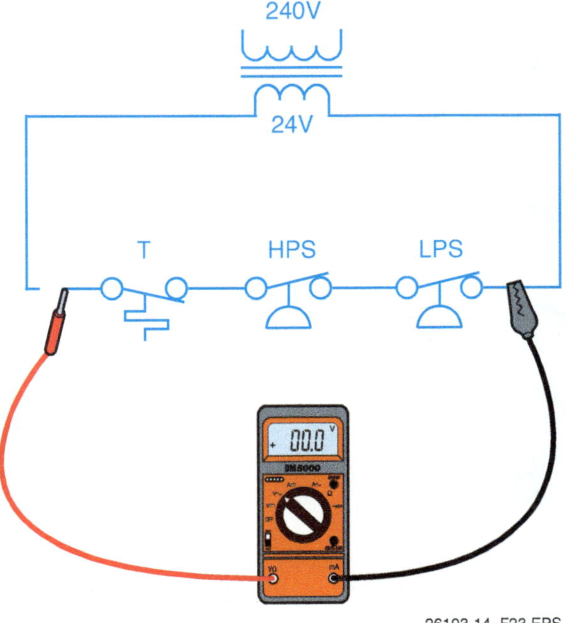

Figure 23 Ohmmeter connection for continuity testing.

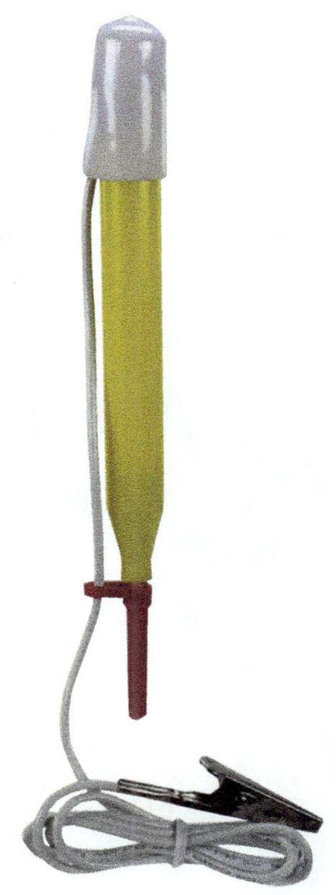

Figure 24 Continuity tester.

Continue this procedure until all the wires have been identified.

9.4.0 Voltage Testers

Figure 25 shows one of the wide variety of devices available for checking for the presence of voltage. It can be used as a troubleshooting tool and as a safety device to make sure the voltage is turned off before touching any terminals or conductors. When the probes are touched to the circuit, the light on the instrument will turn

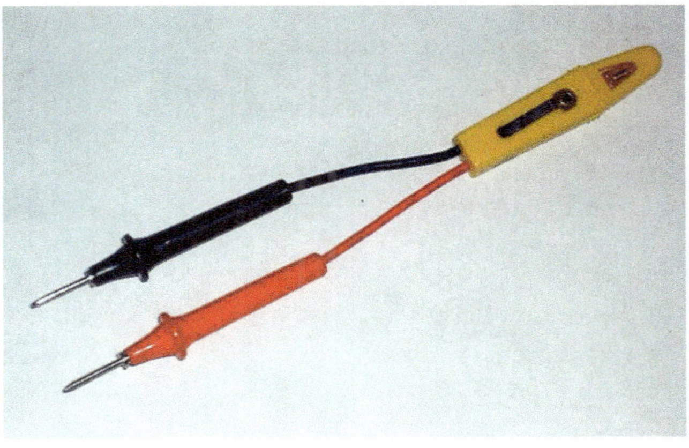

Figure 25 Voltage tester.

On Site

Test Instruments—Old and New

Early electricians used individual meters to test circuit parameters. Today, those instruments seem primitive given the availability of all-purpose instruments like the combination multimeter and clamp-on ammeter with its direct digital readout shown here.

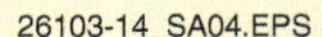

26103-14_SA04.EPS

26103-14_SA05.EPS

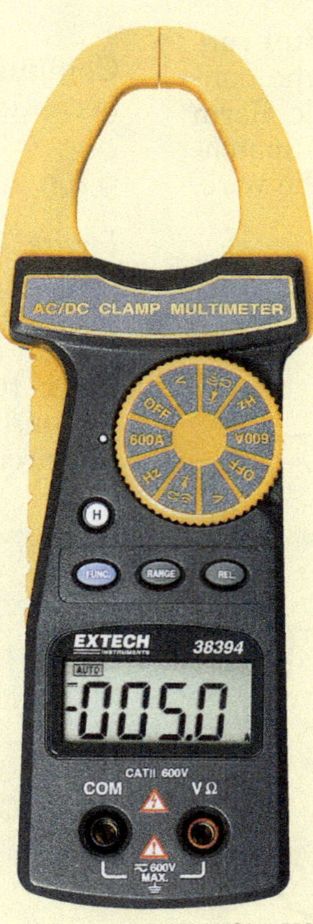

26103-14_SA06.EPS

On Site

Meter Applications

Different electrical systems may have differing levels of available current as well as different voltages. Keep in mind that the meter must be matched to the application. For example, a meter that has enough protective features to be used safely on interior branch circuits may not be safe to use on the service feeders.

on if a voltage is present. Instruments like these are available in several voltage ranges, so it is important to know something about the circuit you are checking.

10.0.0 Electrical Power

Power is defined as the rate of doing work. This is equivalent to the rate at which energy is used or dissipated. Electrons passing through a resistance dissipate energy in the form of heat. In electrical circuits, power is measured in units called watts (W). The power in watts equals the rate of energy conversion. One watt of power equals the work done in one second by one volt of potential difference in moving one coulomb of charge. One coulomb per second is an ampere; therefore, power in watts equals the product of amperes times volts.

The work done in an electrical circuit can be useful work or it can be wasted work. In both cases, the rate at which the work is done is still measured in power. The turning of an electric motor is useful work. On the other hand, the heating of wires or resistors in a circuit is wasted work, since no useful function is performed by the heat.

The unit of electrical work is the joule. This is the amount of work done by one coulomb flowing through a potential difference of one volt. Thus, if five coulombs flow through a potential difference of one volt, five joules of work are done. The time it takes these coulombs to flow through the potential difference has no bearing on the amount of work done.

It is more convenient when working with circuits to think of amperes of current rather than coulombs. As previously discussed, one ampere equals one coulomb passing a point in one second. Using amperes, one joule of work is done in one second when one ampere moves through a potential difference of one volt. This rate of one joule of work in one second is the basic unit of power, and is called a watt. Therefore, a watt is the power used when one ampere of current flows through a potential difference of one volt, as shown in *Figure 26*.

Mechanical power is usually measured in units of horsepower (hp). To convert from horsepower to watts, multiply the number of horsepower by 746. To convert from watts to horsepower, divide the number of watts by 746. Conversions for common units of power are given in *Table 2*.

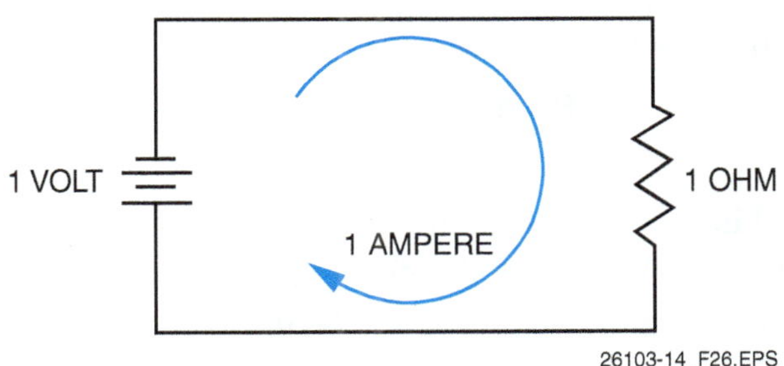

Figure 26 One watt.

On Site

Using an Ohmmeter

An ohmmeter has its own battery to test the resistance or continuity of a circuit. Therefore, the circuit must be de-energized because the ohmmeter is calibrated for its own power source.

On Site

Ohmmeter Batteries

When you use an ohmmeter to measure resistance or test continuity, the power to the circuit under test must be turned off. The internal multimeter battery provides a small DC current that is used to measure resistance. In order to ensure accurate readings, it is important to replace or recharge the battery in accordance with the meter manufacturer's instructions.

Table 2 Conversion Table

1,000 watts (W)	=	1 kilowatt (kW)
1,000,000 watts (W)	=	1 megawatt (MW)
1,000 kilowatts (kW)	=	1 megawatt (MW)
1 watt (W)	=	0.00134 horsepower (hp)
1 horsepower (hp)	=	746 watts (W)

26103-14_T02.EPS

The kilowatt-hour (kWh) is commonly used for large amounts of electrical work or energy. (The prefix **kilo** means one thousand.) The amount is calculated simply as the product of the power in kilowatts multiplied by the time in hours during which the power is used. If a light bulb uses 300W or 0.3kW for 4 hours, the amount of energy is 0.3×4, which equals 1.2kWh.

Very large amounts of electrical work or energy are measured in megawatts (MW). (The prefix **mega** means one million.)

Electricity usage is figured in kilowatt-hours of energy. The power line voltage is fairly constant at 120V. Suppose the total load current in the main line equals 20A. Then the power in watts from the 120V line is:

$$P = 120V \times 20A$$

$$P = 2{,}400W \text{ or } 2.4kW$$

If this power is used for five hours, then the energy of work supplied equals:

$$2.4 \times 5 = 12kWh$$

10.1.0 Power Equation

When one ampere flows through a difference of two volts, two watts must be used. In other words, the number of watts used is equal to the number of amperes of current times the potential difference. This is expressed in equation form as:

$$P = I \times E \text{ or } P = IE$$

Think About It

Power

We take electrical power for granted, never stopping to think how surprising it is that a flow of submicroscopic electrons can pump thousands of gallons of water or illuminate a skyscraper. Our lives now constantly rely on the ability of the electron to do work. Think about your day up to this moment. How has electrical power shaped your experience?

Where:

P = power used in watts

I = current in amperes

E = potential difference in volts

The equation is sometimes called Ohm's law for power, because it is similar to Ohm's law. This equation is used to find the power consumed in a circuit or load when the values of current and voltage are known. The second form of the equation is used to find the voltage when the power and current are known:

$$E = \frac{P}{I}$$

The third form of the equation is used to find the current when the power and voltage are known:

$$I = \frac{P}{E}$$

Using these three equations, the power, voltage, or current in a circuit can be calculated whenever any two of the values are already known.

Example 1:

Calculate the power in a circuit where the source of 100V produces 2A in a 50Ω resistance.

$$P = IE$$

$$P = 2 \times 100$$

$$P = 200W$$

This means the source generates 200W of power while the resistance dissipates 200W in the form of heat.

Example 2:

Calculate the source voltage in a circuit that consumes 1,200W at a current of 5A.

$$E = \frac{P}{I}$$

$$E = \frac{1{,}200}{5}$$

$$E = 240V$$

Example 3:

Calculate the current in a circuit that consumes 600W with a source voltage of 120V.

$$I = \frac{P}{E}$$

$$I = \frac{600}{120}$$

$$I = 5A$$

Components that use the power dissipated in their resistance are generally rated in terms of power. The power is rated at normal operating voltage, which is usually 120V. For instance, an appliance that draws 5A at 120V would dissipate 600W. The rating for the appliance would then be 600W/120V.

To calculate I or R for components rated in terms of power at a specified voltage, it may be convenient to use the power formula in different forms. There are actually three basic power formulas, but each can be rearranged into two other forms for a total of nine combinations:

$$P = IE \qquad P = I^2R \qquad P = \frac{E^2}{R}$$

$$I = \frac{P}{E} \qquad R = \frac{P}{I^2} \qquad R = \frac{E^2}{P}$$

$$E = \frac{P}{I} \qquad I = \sqrt{\frac{P}{R}} \qquad E = \sqrt{PR}$$

Note that all of these formulas are based on Ohm's law (E = IR) and the power formula (P = I × E). *Figure 27* shows all of the applicable power, voltage, resistance, and current equations.

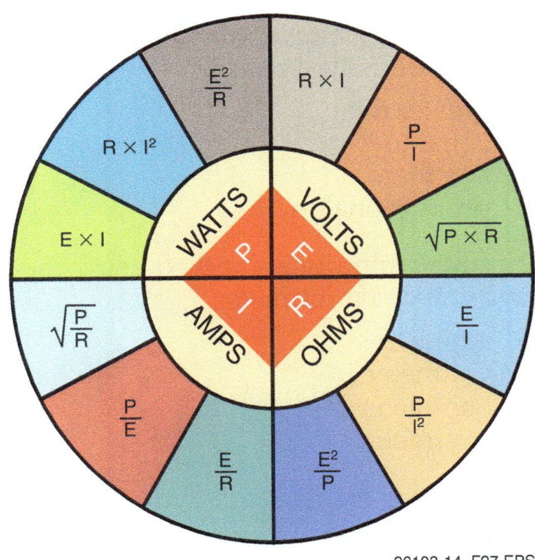

Figure 27 Expanded Ohm's law circle.

10.2.0 Power Rating of Resistors

If too much current flows through a resistor, the heat caused by the current will damage or destroy the resistor. This heat is caused by I^2R heating, which is power loss expressed in watts. Therefore, every resistor is given a wattage, or power rating, to show how much I^2R heating it can take before it burns out. This means that a resistor with a power rating of one watt will burn out if it is used in a circuit where the current causes it to dissipate heat at a rate greater than one watt.

If the power rating of a resistor is known, the maximum current it can carry is found by using an equation derived from $P = I^2R$:

$$P = I^2R$$

$$I^2 = P/R$$

$$I = \sqrt{P/R}$$

Using this equation, find the maximum current that can be carried by a 1Ω resistor with a power rating of 4W:

$$I = \sqrt{P/R} = \sqrt{4/1} = 2 \text{ amperes}$$

If such a resistor conducts more than 2 amperes, it will dissipate more than its rated power and burn out.

Power ratings assigned by resistor manufacturers are usually based on the resistors being mounted in an open location where there is free air circulation, and where the temperature is not higher than 104°F (40°C). Therefore, if a resistor is mounted in a small, crowded, enclosed space, or where the temperature is higher than 104°F, there is a good chance it will burn out even before its power rating is exceeded. Also, some resistors are designed to be attached to a chassis or frame that will carry away the heat.

Think About It

Putting It All Together

Notice the common electrical devices in the building you're in. What is their wattage rating? How much current do they draw? How would you test their voltage or amperage?

Summary

Electricians often test and troubleshoot electrical circuits. This work can be done safely and more effectively if you know the theory of electricity and the interrelationships of voltage, current, resistance, and power. The basic tool for understanding these relationships is Ohm's law.

Testing and troubleshooting of electrical circuits involves the use of test instruments such as the multimeter, or VOM. The multimeter combines the voltmeter, ammeter, and ohmmeter into a single instrument. In analog meters, a pointer moves across a scale in proportion to the current flowing though the meter. In a digital meter, the measured value is displayed directly on the screen of the meter.

Review Questions

1. An electrical circuit contains, at minimum, a(n) _____.
   a. voltage source, load, and switch
   b. ammeter, load, and voltage source
   c. voltage source, load, and conductors
   d. conductor, switch, and load

2. A type of subatomic particle with a positive charge is a(n) _____.
   a. proton
   b. neutron
   c. electron
   d. nucleus

3. Which of the following substances is considered an insulator?
   a. Gold
   b. Copper
   c. Silver
   d. Porcelain

4. The voltage commonly supplied to a residence by the local utility is _____.
   a. 120V
   b. 240V
   c. 480V
   d. 208V

5. Another term used for voltage is _____.
   a. emf
   b. coulomb
   c. current
   d. joule

6. In order to calculate the current flowing in a circuit, you would multiply voltage by resistance.
   a. True
   b. False

7. The color band that represents tolerance on a resistor is the _____.
   a. 4th band
   b. 3rd band
   c. 2nd band
   d. 1st band

8. In a parallel circuit, the total resistance is _____ the smallest resistance.
   a. greater than
   b. equal to
   c. less than
   d. proportional to

9. Circuit continuity is checked using the _____ function of a multimeter.
   a. ammeter
   b. voltmeter
   c. ohmmeter
   d. wattmeter

10. The power in a circuit with 120 volts and 5 amps is _____.
    a. 24 watts
    b. 600 watts
    c. 6,000 watts
    d. $1/24$ watt

Trade Terms Quiz

Fill in the blank with the correct term that you learned from your study of this module.

1. A(n) ____________ is an instrument for measuring electrical current.
2. Measured in amperes, ____________ is the flow of electrons in a circuit.
3. A(n) ____________ is the force required to produce a current of one ampere through a resistance of one ohm.
4. Voltage is measured with a(n) ____________.
5. One volt applied across one ohm of resistance causes a current flow of one ____________.
6. One volt is the potential difference between two points for which one coulomb of electricity will do one ____________ of work.
7. A(n) ____________ is the common unit used for specifying the size of a given charge.
8. ____________ is the driving force that makes current flow in a circuit.
9. The basic unit of measurement for electrical power is the ____________.
10. The ____________ is the smallest particle of an element that will still retain the properties of that element.
11. The ____________ is the center of an atom.
12. Found in the nuclei of atoms, ____________ are electrically positive particles and ____________ are electrically neutral particles.
13. The outermost ring of electrons orbiting the nucleus of an atom is known as the ____________.
14. A(n) ____________ is a negatively charged particle that orbits the nucleus of an atom.
15. ____________ is any substance that has mass and occupies space.
16. The prefix used to indicate one thousand is ____________.
17. The prefix used to indicate one million is ____________.
18. Consisting of two or more cells, ____________ convert chemical energy into electrical energy.
19. A(n) ____________ is a complete path for current flow.
20. A material through which it is relatively easy to maintain an electric current is a(n) ____________.
21. A(n) ____________ is a material through which it is difficult to conduct an electric current.
22. The basic unit of measurement for resistance is the ____________.
23. The instrument that is used to measure resistance is called a(n) ____________.
24. ____________ is a statement of the relationship between current, voltage, and resistance in an electrical circuit.
25. ____________ is the rate of doing work or the rate at which energy is used or dissipated.
26. Measured in ohms, ____________ is the electrical property that opposes the flow of current through a circuit.
27. A(n) ____________ is a component that normally opposes current flow in a DC circuit.
28. A(n) ____________ is a drawing in which symbols are used to represent the components in a system.
29. A(n) ____________ circuit has only one route for current flow.
30. The change in voltage across a component is called ____________.
31. A(n) ____________ is an electromechanical component used as a switching device.
32. A device containing one or more coils of wire wrapped around a common core is called a(n) ____________.
33. An electromagnetic device used to control a mechanical device such as a valve is called a(n) ____________.

Trade Terms

- Ammeter
- Ampere (A)
- Atom
- Battery
- Circuit
- Conductor
- Coulomb
- Current
- Electron
- Insulator
- Joule (J)
- Kilo
- Matter
- Mega
- Neutrons
- Nucleus
- Ohm (Ω)
- Ohmmeter
- Ohm's law
- Power
- Protons
- Relay
- Resistance
- Resistor
- Schematic
- Series circuit
- Solenoid
- Transformer
- Valence shell
- Volt (V)
- Voltage
- Voltage drop
- Voltmeter
- Watt (W)

Module 26103-14
Supplemental Exercises

1. An atom that is missing two or fewer electrons from its outer shell will ______________________.

2. Current is measured in units called _____.

   a. joules
   b. coulombs
   c. amperes
   d. volt-amperes

3. Joules are _____.

   a. units of work
   b. the potential difference between two points
   c. the difference between EMF and current
   d. the rate of current flow

4. All conductors have _____.

   a. current flow
   b. some resistance
   c. EMF
   d. voltage potential

5. To find voltage when both current and resistance are known, use the formula _____.

   a. $E = I \times R$
   b. $E = I \div R$
   c. $E = \frac{R}{I}$
   d. $E = I + R$

6. A resistor with a color code of yellow, orange, red, and silver has a tolerance of ____________.

7. Ammeters measure _____.

   a. voltage
   b. resistance
   c. current
   d. power

8. True or False? When using an in-line ammeter, it is important to connect the meter in series.

9. If a toaster draws 6.2A of current at 120V, how many kilowatt-hours of energy will be used in 3.5 hours? ______________________

10. If a battery sends a current of 10A through a circuit for one hour, how many coulombs will flow through the circuit? _______________

11. The charged particles of an atom are called ______________________.

12. Conductors have ______________________ or fewer valence electrons.

13. The sum of the difference in potential of all the charges in an electrostatic field is called __________ ____.

14. Electric charge is measured in ______________________.

15. Resistance is measured in ______________________.

16. Find the resistance when the voltage is 120V and the current is 6A. ____________________

17. Give the resistance value and tolerance of a resistor where the color bands are red, yellow, brown, and silver. ______________________

18. What is the power in a 120V circuit with a current of 12.5A?________________________

19. What is the voltage in a 30Ω circuit with a power rating of 480W? ____________________

20. An ohmmeter is used to measure resistance and check for ______________________.

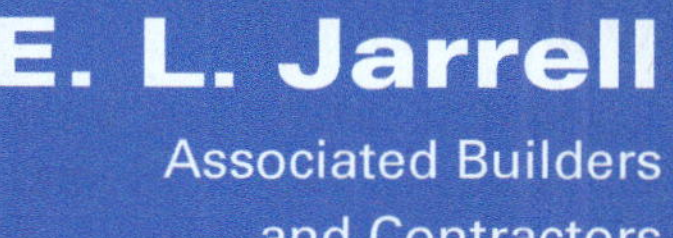

E. L. Jarrell

Associated Builders and Contractors

Eurlin Layne (E. L.) Jarrell is another prime example of a master electrician giving back to the electrical community by teaching and mentoring.

After serving in the United States Army, E. L. went to work for Cities Services, now known as CITGO. He stayed at CITGO for 38 years, finally retiring in 1995. It was during his employment at CITGO that he first received apprenticeship training in the electrical field.

While at CITGO, E. L. worked as a process unit operator before moving to the electrical department. While there, he worked as a trainee electrician for three years until he became a first-class electrician. A few years later, he was promoted to temporary supervisor, planning and scheduling shut-down maintenance. In 1983, he took and passed the Block Master Electrician test for the City of Lake Charles, Louisiana. In 1997, E. L. became involved with Associated Builders and Contractors (ABC).

E. L. is currently the Electrical Department Head for the ABC Training Center, where he works in the lab, overseeing students doing hands-on electrical work. During his first semester teaching at the ABC Training Center, it became clear to E. L. that many students simply had no time to study because they worked 10-hour days, drove over 100 miles to work, and had family obligations. In response, E. L. began an in-class study guide. He encouraged students to form study groups, and he gave students time to study in class.

E. L. was an instrumental member of NCCER's Technical Review Committee, which completely rewrote all four levels of NCCER's Electrical curriculum. In addition, E. L. is currently a member of both NCCER's National Skills Assessment Written Test Committee and the Performance Verification Packet for Industrial Electricians Committee.

E. L. has decided to give back to the electrical community with his expertise and mentoring. Many of E. L.'s students have become his personal friends. He says, "At this point in my life, I just want to continue being the best electrical instructor that I can be and share some of my knowledge and experience with my students and hope that I can make a difference in their lives and careers."

Trade Terms Introduced in This Module

Ammeter: An instrument for measuring electrical current.

Ampere (A): A unit of electrical current. For example, one volt across one ohm of resistance causes a current flow of one ampere.

Atom: The smallest particle to which an element may be divided and still retain the properties of the element.

Battery: A DC voltage source consisting of two or more cells that convert chemical energy into electrical energy.

Circuit: A complete path for current flow.

Conductor: A material through which it is relatively easy to maintain an electric current.

Coulomb: An electrical charge equal to 6.25×10^{18} electrons or 6,250,000,000,000,000,000 electrons. A coulomb is the common unit of quantity used for specifying the size of a given charge.

Current: The movement, or flow, of electrons in a circuit. Current (I) is measured in amperes.

Electron: A negatively charged particle that orbits the nucleus of an atom.

Insulator: A material through which it is difficult to conduct an electric current.

Joule (J): A unit of measurement that represents one newton-meter (Nm), which is a unit of measure for doing work.

Kilo: A prefix used to indicate one thousand; for example, one kilowatt is equal to one thousand watts.

Matter: Any substance that has mass and occupies space.

Mega: A prefix used to indicate one million; for example, one megawatt is equal to one million watts.

Neutrons: Electrically neutral particles (neither positive nor negative) that have the same mass as a proton and are found in the nucleus of an atom.

Nucleus: The center of an atom. It contains the protons and neutrons of the atom.

Ohm (Ω): The basic unit of measurement for resistance.

Ohmmeter: An instrument used for measuring resistance.

Ohm's law: A statement of the relationships among current, voltage, and resistance in an electrical circuit: current (I) equals voltage (E) divided by resistance (R). Generally expressed as a mathematical formula: $I = E/R$.

Power: The rate of doing work or the rate at which energy is used or dissipated. Electrical power is the rate of doing electrical work. Electrical power is measured in watts.

Protons: The smallest positively charged particles of an atom. Protons are contained in the nucleus of an atom.

Relay: An electromechanical device consisting of a coil and one or more sets of contacts. Used as a switching device.

Resistance: An electrical property that opposes the flow of current through a circuit. Resistance (R) is measured in ohms.

Resistor: Any device in a circuit that resists the flow of electrons.

Schematic: A type of drawing in which symbols are used to represent the components in a system.

Series circuit: A circuit with only one path for current flow.

Solenoid: An electromagnetic coil used to control a mechanical device such as a valve.

Transformer: A device consisting of one or more coils of wire wrapped around a common core. It is commonly used to step voltage up or down.

Valence shell: The outermost ring of electrons that orbit about the nucleus of an atom.

Volt (V): The unit of measurement for voltage (electromotive force or difference of potential). One volt is equivalent to the force required to produce a current of one ampere through a resistance of one ohm.

Voltage: The driving force that makes current flow in a circuit. Voltage (E) is also referred to as electromotive force or difference of potential.

Voltage drop: The change in voltage across a component that is caused by the current flowing through it and the amount of resistance opposing it.

Voltmeter: An instrument for measuring voltage. The resistance of the voltmeter is fixed. When the voltmeter is connected to a circuit, the current passing through the meter will be directly proportional to the voltage at the connection points.

Watt (W): The basic unit of measurement for electrical power.

Additional Resources

This module presents thorough resources for task training. The following resource material is suggested for further study.

Electronics Fundamentals: Circuits, Devices, and Applications, Thomas L. Floyd. New York: Prentice Hall.

Principles of Electric Circuits, Thomas L. Floyd. New York: Prentice Hall.

Figure Credits

AGC of America, Module opener

Topaz Publications, Inc., Figures 13 (photo), 18, 19, 25, SA02, SA04, SA05

Tim Dean, Figure 20

Amprobe, Figure 24

U.S. Army Corps of Engineers, SA01

Courtesy of Extech Instruments, a FLIR Company, SA06

NCCER CURRICULA — USER UPDATE

NCCER makes every effort to keep its textbooks up-to-date and free of technical errors. We appreciate your help in this process. If you find an error, a typographical mistake, or an inaccuracy in NCCER's curricula, please fill out this form (or a photocopy), or complete the online form at **www.nccer.org/olf**. Be sure to include the exact module ID number, page number, a detailed description, and your recommended correction. Your input will be brought to the attention of the Authoring Team. Thank you for your assistance.

Instructors – If you have an idea for improving this textbook, or have found that additional materials were necessary to teach this module effectively, please let us know so that we may present your suggestions to the Authoring Team.

NCCER Product Development and Revision

13614 Progress Blvd., Alachua, FL 32615

Email: curriculum@nccer.org

Online: www.nccer.org/olf

❑ Trainee Guide ❑ AIG ❑ Exam ❑ PowerPoints Other ____________________________

Craft / Level: Copyright Date:

Module ID Number / Title:

Section Number(s):

Description:

Recommended Correction:

Your Name:

Address:

Email: Phone:

Electrical Theory

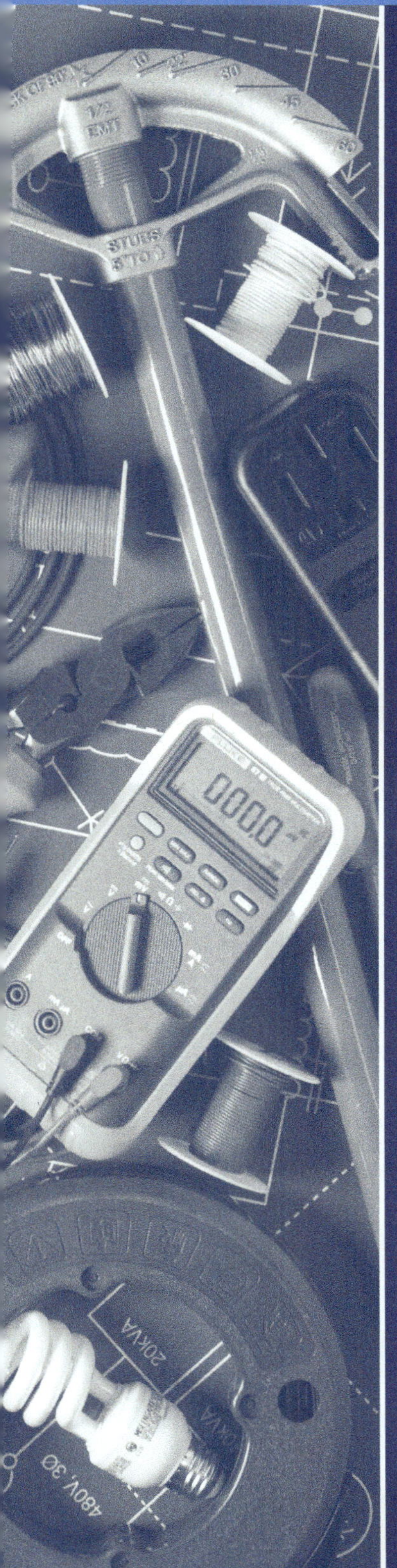

GM Lansing Delta Township Assembly Complex

This project consisted of converting 375 acres of 1,100 acres of farm land into a campus-style automotive assembly facility totaling 2.4 million square feet. In 20 months, a greenfield site was transformed into a major complex where raw steel enters at one end and finished vehicles are shipped to dealers at the opposite end. This project was the first design-build/guaranteed maximum price new manufacturing facility that General Motors has created.

26104-14

Trainees with successful module completions may be eligible for credentialing through NCCER's National Registry. To learn more, go to **www.nccer.org** or contact us at **1.888.622.3720.** Our website has information on the latest product releases and training, as well as online versions of our *Cornerstone* magazine and Pearson's product catalog.

Your feedback is welcome. You may email your comments to **curriculum@nccer.org,** send general comments and inquiries to **info@nccer.org,** or fill in the User Update form at the back of this module.

This information is general in nature and intended for training purposes only. Actual performance of activities described in this manual requires compliance with all applicable operating, service, maintenance, and safety procedures under the direction of qualified personnel. References in this manual to patented or proprietary devices do not constitute a recommendation of their use.

26104-14

Electrical Theory

Objectives

When you have completed this module, you will be able to do the following:

1. Explain the basic characteristics of combination circuits.
2. Calculate, using Kirchhoff's voltage law, the voltage drop in series, parallel, and series-parallel circuits.
3. Calculate, using Kirchhoff's current law, the total current in parallel and series-parallel circuits.
4. Using Ohm's law, find the unknown parameters in series, parallel, and series-parallel circuits.

Performance Tasks

This is a knowledge-based module. There are no Performance Tasks.

Trade Terms

Kirchhoff's current law
Kirchhoff's voltage law
Parallel circuits
Series circuits
Series-parallel circuits

Required Trainee Materials

1. Paper and pencil
2. Appropriate personal protective equipment

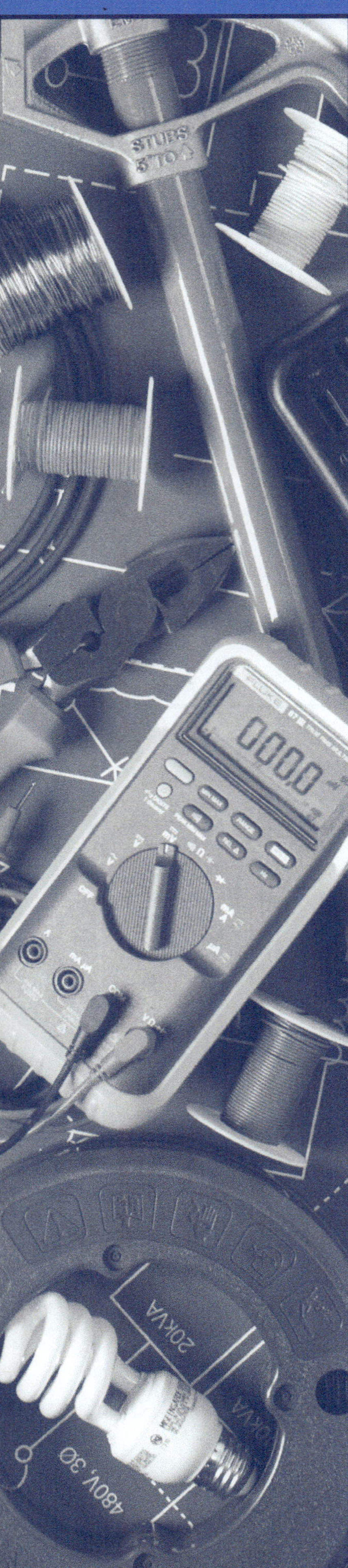

Contents

Topics to be presented in this module include:

Figures

1.0.0 Introduction

Ohm's law was explained in the module *Introduction to Electrical Circuits.* This fundamental concept is now going to be used to analyze more complex series circuits, parallel circuits, and combination series-parallel circuits. This module will explain how to calculate resistance, current, and voltage in these complex circuits. Ohm's law will be used to develop a new law for voltage and current determination. This law, called Kirchhoff's law, will become the new foundation for analyzing circuits.

2.0.0 Resistive Circuits

Resistance is calculated in different ways, depending on whether it is a series or parallel circuit. Resistance is calculated in ohms.

2.1.0 Resistances in Series

A series circuit is a circuit in which there is only one path for current flow. Resistance is measured in ohms (Ω). In the series circuit shown in *Figure 1* the current (I) is the same in all parts of the circuit. This means that the current flowing through R_1 is the same as the current flowing through R_2 and R_3, and it is also the same as the current supplied by the battery.

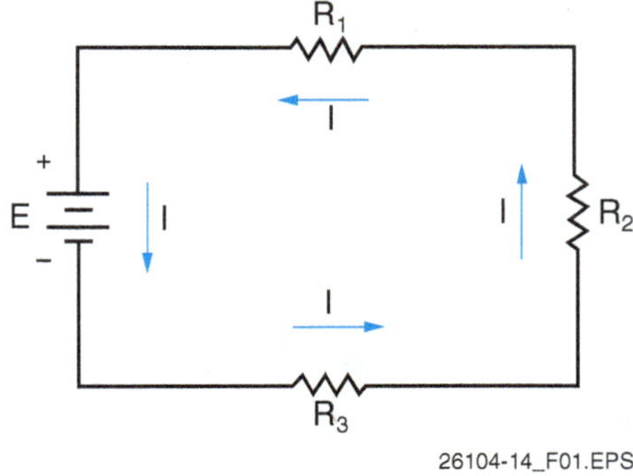

Figure 1 Series circuit.

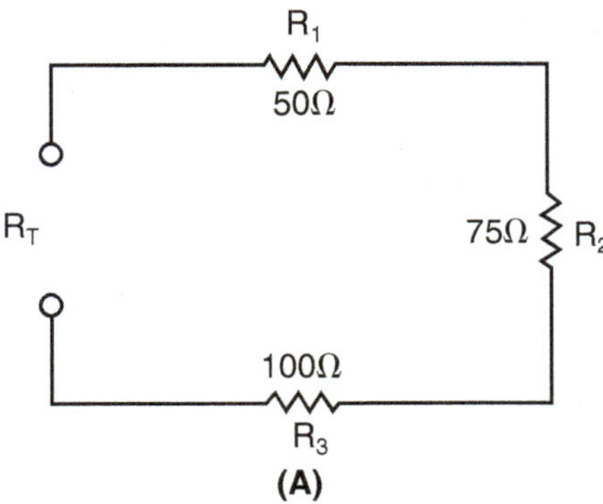

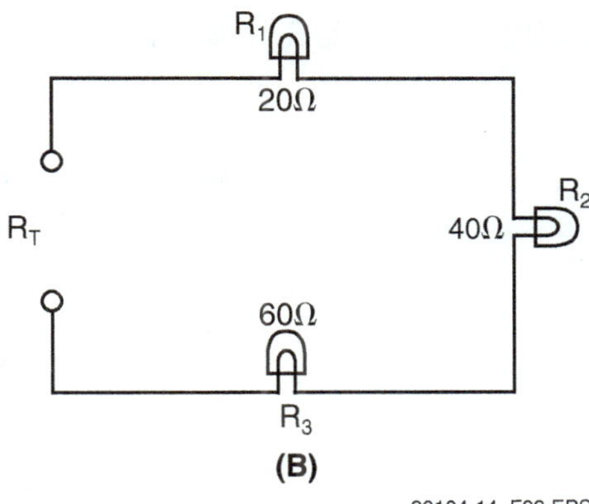

Figure 2 Total resistance.

When resistances are connected in series as in this example, the total resistance in the circuit is equal to the sum of the resistances of all the parts of the circuit:

$$R_T = R_1 + R_2 + R_3$$

Where:

R_T = total resistance

$R_1 + R_2 + R_3$ = resistances in series

Example 1:

The circuit shown in *Figure 2(A)* has 50Ω, 75Ω, and 100Ω resistors in series. Find the total resistance of the circuit.

Add the values of the three resistors in series:

$$R_T = R_1 + R_2 + R_3 = 50 + 75 + 100 = 225\Omega$$

Example 2:

The circuit shown in *Figure 2(B)* has three lamps connected in series with the resistances shown. Find the total resistance of the circuit.

Add the values of the three lamp resistances in series:

$$R_T = R_1 + R_2 + R_3 = 20 + 40 + 60 = 120\Omega$$

2.2.0 Resistances in Parallel

The total resistance in a parallel resistive circuit is given by the formula:

$$R_T = \frac{1}{\frac{1}{R_1} + \frac{1}{R_2} + \frac{1}{R_3} + \frac{1}{R_n}}$$

Where:

R_T = total resistance in parallel

R_1, R_2, R_3, and R_n = branch resistances

Think About It

Series Circuits

Simple series circuits are seldom encountered in practical wiring. The only simple series circuit you may recognize is older strands of Christmas lights, in which the entire string went dead when one lamp burned out. Think about what the actual wiring of a series circuit would look like in household receptacles. How would the circuit physically be wired? What kind of illumination would you get if you wired your household receptacles in series and plugged half a dozen lamps into those receptacles?

Example 1:

Find the total resistance of the 2Ω, 4Ω, and 8Ω resistors in parallel shown in *Figure 3*.

Write the formula for the three resistances in parallel:

$$R_T = \frac{1}{\frac{1}{R_1} + \frac{1}{R_2} + \frac{1}{R_3}}$$

Substitute the resistance values:

$$R_T = \frac{1}{\frac{1}{2} + \frac{1}{4} + \frac{1}{8}}$$

$$R_T = \frac{1}{0.5 + 0.25 + 0.125}$$

$$R_T = \frac{1}{0.875}$$

$$R_T = 1.14\Omega$$

Note that when resistances are connected in parallel, the total resistance is always less than the resistance of any single branch.

In this case:

$R_T = 1.14\Omega < R_1 = 2\Omega$, $R_2 = 4\Omega$, and $R_3 = 8\Omega$

Example 2:

Add a fourth parallel resistor of 2Ω to the circuit in *Figure 3*. What is the new total resistance, and what is the net effect of adding another resistance in parallel?

Write the formula for four resistances in parallel:

$$R_T = \frac{1}{\frac{1}{R_1} + \frac{1}{R_2} + \frac{1}{R_3} + \frac{1}{R_4}}$$

Substitute values:

$$R_T = \frac{1}{\frac{1}{2} + \frac{1}{4} + \frac{1}{8} + \frac{1}{2}}$$

$$R_T = \frac{1}{0.5 + 0.25 + 0.125 + 0.5}$$

$$R_T = \frac{1}{1.375}$$

$$R_T = 0.73\Omega$$

The net effect of adding another resistance in parallel is a reduction of the total resistance from 1.14Ω to 0.73Ω.

2.2.1 Simplified Formulas

The total resistance of equal resistors in parallel is equal to the resistance of one resistor divided by the number of resistors:

$$R_T = \frac{R}{N}$$

Where:

R_T = total resistance of equal resistors in parallel
R = resistance of one of the equal resistors
N = number of equal resistors

If two resistors with the same resistance are connected in parallel, the equivalent resistance is half of that value, as shown in *Figure 4*.

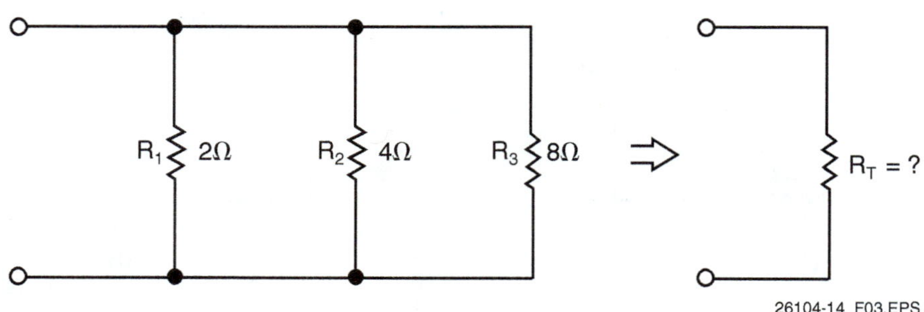

Figure 3 Parallel branch.

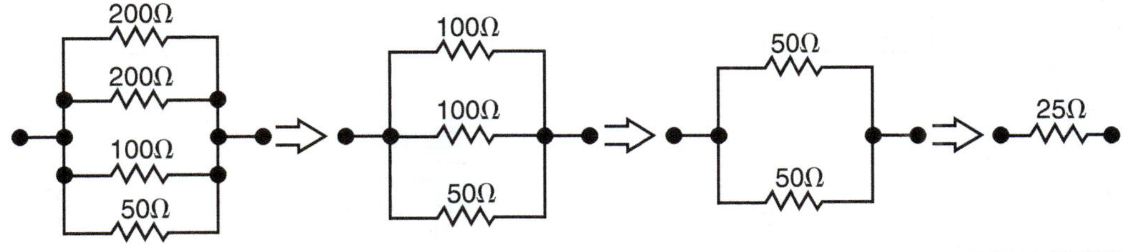

Figure 4 Equal resistances in a parallel circuit.

The two 200Ω resistors in parallel are the equivalent of one 100Ω resistor; the two 100Ω resistors are the equivalent of one 50Ω resistor; and the two 50Ω resistors are the equivalent of one 25Ω resistor.

When any two unequal resistors are in parallel, it is often easier to calculate the total resistance by multiplying the two resistances and then dividing the product by the sum of the resistances:

$$R_T = \frac{R_1 \times R_2}{R_1 + R_2}$$

Where:

R_T = total resistance of unequal resistors in parallel

R_1, R_2 = two unequal resistors in parallel

Example 1:

Find the total resistance of a 6Ω (R_1) resistor and an 18Ω (R_2) resistor in parallel:

$$R_T = \frac{R_1 \times R_2}{R_1 + R_2} = \frac{6 \times 18}{6 + 18} = \frac{108}{24} = 4.5\Omega$$

Example 2:

Find the total resistance of a 100Ω (R_1) resistor and a 150Ω (R_2) resistor in parallel:

$$R_T = \frac{R_1 \times R_2}{R_1 + R_2} = \frac{100 \times 150}{100 + 150} = \frac{15,000}{250} = 60\Omega$$

2.3.0 Series-Parallel Circuits

To find current, voltage, and resistance in series circuits and parallel circuits is fairly easy. When working with either type, use only the rules that apply to that type. In a series-parallel circuit, some parts of the circuit are series connected and other parts are parallel connected. Thus, in some parts the rules for series circuits apply, and in other parts, the rules for parallel circuits apply. To analyze or solve a problem involving a series-parallel circuit, it is necessary to recognize which parts of the circuit are series connected and which parts are parallel connected. This is obvious if the circuit is simple. Many times, however, the circuit must be redrawn, putting it into a form that is easier to recognize.

In a series circuit, the current is the same at all points. In a parallel circuit, there are one or more points where the current divides and flows in separate branches. In a series-parallel circuit, there are both separate branches and series loads. The easiest way to find out whether a circuit is a series, parallel, or series-parallel circuit is to start at the negative terminal of the power source and trace the path of current through the circuit back to the positive terminal of the power source. If the current does not divide anywhere, it is a series circuit. If the current divides into separate branches, but there are no series loads, it is a parallel circuit. If the current divides into separate branches and there are also series loads, it is a series-parallel circuit. *Figure 5* shows electric lamps connected in series, parallel, and series-parallel circuits.

After determining that a circuit is series-parallel, redraw the circuit so that the branches and the series loads are more easily recognized. This is especially helpful when computing the total resistance of the circuit. *Figure 6* shows resistors connected in a series-parallel circuit and the equivalent circuit redrawn to simplify it.

Think About It

Parallel Circuits

An interesting fact about circuits is the drop in resistance in a parallel circuit as more resistors are added. But this fact does not mean that you can add an endless number of devices, such as lamps, in a parallel circuit. Why not?

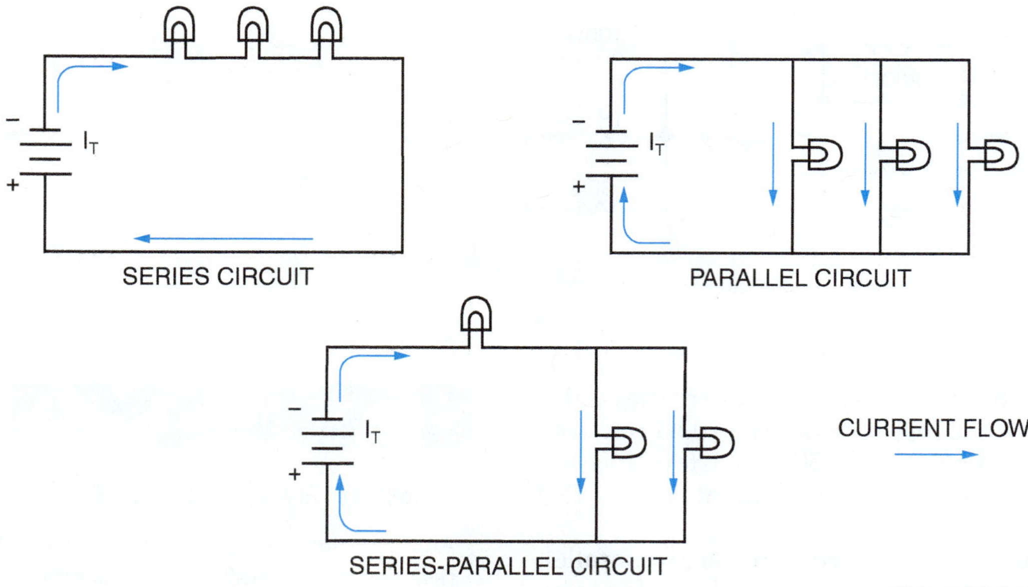

Figure 5 Series, parallel, and series-parallel circuits.

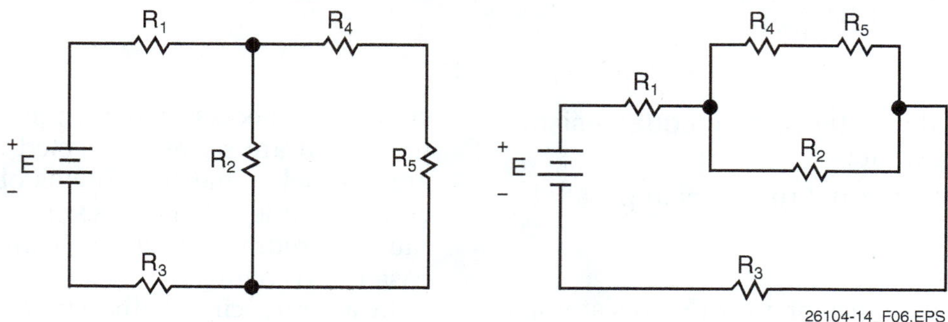

Figure 6 Redrawing a series-parallel circuit.

2.3.1 Reducing Series-Parallel Circuits

Very often, all that is known about a series-parallel circuit is the applied voltage and the values of the individual resistances. To find the voltage drop across any of the loads or the current in any of the branches, the total circuit current must usually be known. But to find the total current, the total resistance of the circuit must be known. To find the total resistance, reduce the circuit to its simplest form, which is usually one resistance that forms a series circuit with the voltage source. This simple series circuit has the equivalent resistance of the series-parallel circuit it was derived from, and also has the same total current. There are four basic steps in reducing a series-parallel circuit:

- If necessary, redraw the circuit so that all parallel combinations of resistances and series resistances are easily recognized.
- For each parallel combination of resistances, calculate its effective resistance.
- Replace each of the parallel combinations with one resistance whose value is equal to the effective resistance of that combination. This provides a circuit with all series loads.
- Find the total resistance of this circuit by adding the resistances of all the series loads.

Examine the series-parallel circuit shown in *Figure 7* and reduce it to an equivalent series circuit.

Think About It

Series-Parallel Circuits

Explain *Figure 6*. Which resistors are in series and which are in parallel?

On Site

Parallel Circuits

Most practical circuits are wired in parallel, like the pole lights shown here.

26104-14_SA01.EPS

In this circuit, resistors R_2 and R_3 are connected in parallel, but resistor R_1 is in series with both the battery and the parallel combination of R_2 and R_3. The current I_T leaving the negative terminal of the voltage source travels through resistor R_1 before it is divided at the junction of resistors R_1, R_2, and R_3 (Point A) to go through the two branches formed by resistors R_2 and R_3.

Given the information in *Figure 7*, calculate the resistance of R_2 and R_3 in parallel and the total resistance of the circuit, R_T.

The total resistance of the circuit is the sum of R_1 and the equivalent resistance of R_2 and R_3 in parallel. To find R_T, first find the resistance of R_2 and R_3 in parallel. Because the two resistances have the same value of 20Ω, the resulting equivalent resistance is 10Ω. Therefore, the total resistance (R_T) is 15Ω (5Ω + 10Ω).

2.4.0 Applying Ohm's Law

In resistive circuits, unknown circuit parameters can be found by using Ohm's law and the techniques for determining equivalent resistance.

2.4.1 Voltage and Current in Series Circuits

Ohm's law may be applied to an entire series circuit or to the individual parts of the circuit. When it is used on a particular part of a circuit, the voltage across that part is equal to the current in that part multiplied by the resistance of that part.

For example, given the information in *Figure 8*, calculate the total resistance (R_T) and the total current (I_T).

To find R_T:

$$R_T = R_1 + R_2 + R_3$$
$$R_T = 20 + 50 + 120$$
$$R_T = 190\Omega$$

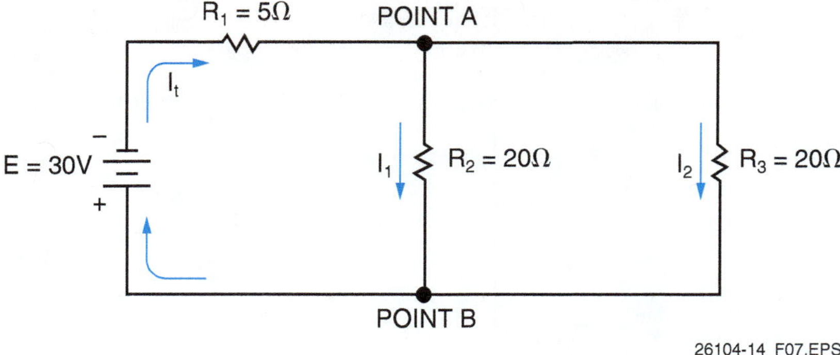

Figure 7 Reducing a series-parallel circuit.

To find I_T using Ohm's law:

$$I_T = \frac{E_T}{R_T}$$

$$I_T = \frac{95}{190}$$

$$I_T = 0.5A$$

Find the voltage across each resistor. In a series circuit, the current is the same; that is, I = 0.5A through each resistor:

$$E_1 = IR_1 = 0.5(20) = 10V$$
$$E_2 = IR_2 = 0.5(50) = 25V$$
$$E_3 = IR_3 = 0.5(120) = 60V$$

The voltages E_1, E_2, and E_3 found for *Figure 8* are known as voltage drops or IR drops. Their effect is to reduce the voltage that is available to be applied across the rest of the components in the circuit. The sum of the voltage drops in any series circuit is always equal to the voltage that is applied to the circuit. The total voltage (E_T) is the same as the applied voltage and can be verified in this example ($E_T = 10 + 25 + 60$ or 95V).

Figure 8 Calculating voltage drops.

2.4.2 Voltage and Current in Parallel Circuits

A parallel circuit is a circuit in which two or more components are connected across the same voltage source, as illustrated in *Figure 9*. The resistors R_1, R_2, and R_3 are in parallel with each other and with the battery. Each parallel path is then a branch with its own individual current. When the total current I_T leaves the voltage source E, part I_1 of the current I_T will flow through R_1, part I_2 will flow through R_2, and the remainder I_3 will flow through R_3. The branch currents I_1, I_2, and I_3 can be different. However, if a voltmeter is connected across R_1, R_2, and R_3, the respective voltages E_1, E_2, and E_3 will be equal to the source voltage E.

The total current I_T is equal to the sum of all branch currents.

This formula applies for any number of parallel branches whether the resistances are equal or unequal.

Using Ohm's law, each branch current equals the applied voltage divided by the resistance between the two points where the voltage is

Think About It

Voltage Drops

Calculating voltage drops is not just a schoolroom exercise. It is important to know the voltage drop when sizing circuit components. What would happen if you sized a component without accounting for a substantial voltage drop in the circuit?

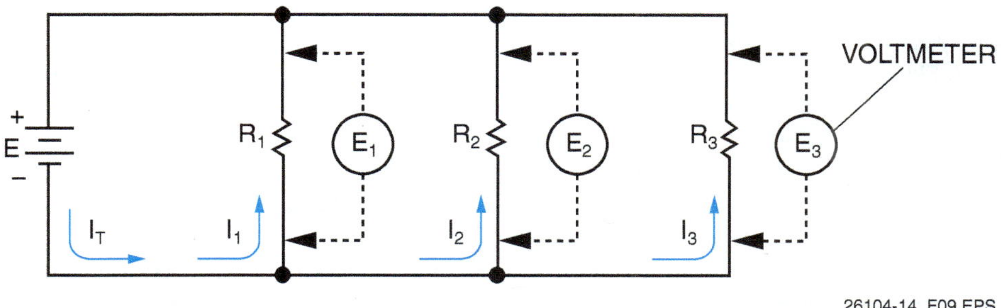

Figure 9 Parallel circuit.

applied. Hence, for each branch in *Figure 9* we have the following equations:

$$\text{Branch 1: } I_1 = \frac{E_1}{R_1} = \frac{E}{R_1}$$

$$\text{Branch 2: } I_2 = \frac{E_2}{R_2} = \frac{E}{R_2}$$

$$\text{Branch 3: } I_3 = \frac{E_3}{R_3} = \frac{E}{R_3}$$

With the same applied voltage, any branch that has less resistance allows more current through it than a branch with higher resistance.

Example 1:

The two branches R_1 and R_2, shown in *Figure 10(A)* across a 110V power line draw a total line current of 20A. Branch R_1 takes 12A. What is the current I_2 in branch R_2?

Transpose to find I_2 and then substitute given values:

$$I_T = I_1 + I_2$$
$$I_2 = I_T - I_1$$
$$I_2 = 20 - 12 = 8A$$

Example 2:

As shown in *Figure 10(B)* the two branches R_1 and R_2 across a 240V power line draw a total line current of 35A. Branch R_2 takes 20A. What is the current I_1 in branch R_1?

Transpose to find I_1 and then substitute given values:

$$I_T = I_1 + I_2$$
$$I_1 = I_T - I_2$$
$$I_1 = 35 - 20 = 15A$$

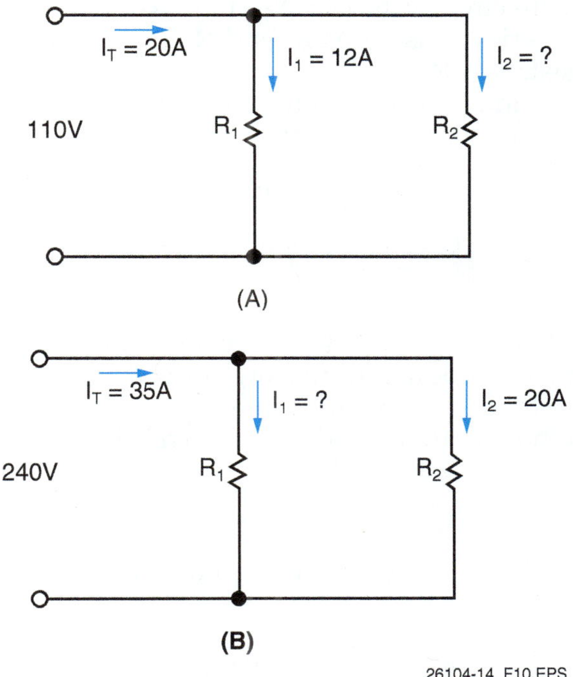

Figure 10 Solving for an unknown current.

2.4.3 Voltage and Current in Series-Parallel Circuits

Series-parallel circuits combine the elements and characteristics of both the series and parallel configurations. By properly applying the equations and methods previously discussed, the values of individual components of the circuit can be determined. *Figure 11* shows a simple series-parallel circuit with a 1.5V battery.

The current and voltage associated with each component can be determined by first simplifying the circuit to find the total current, and then working across the individual components.

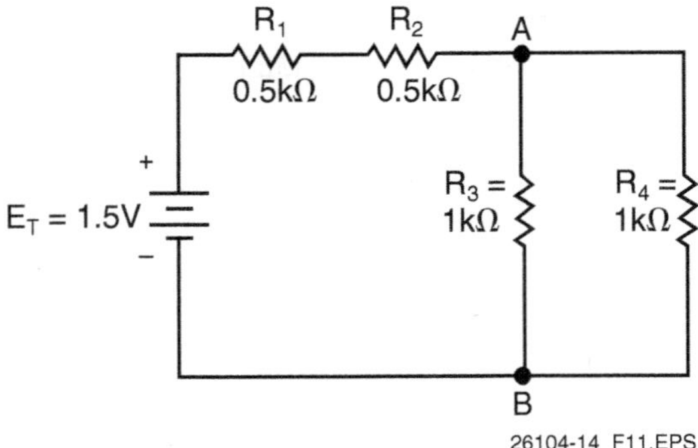

Figure 11 Series-parallel circuit.

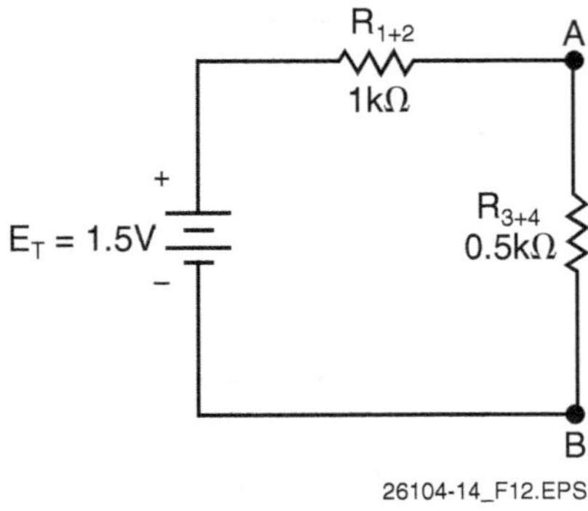

Figure 12 Simplified series-parallel circuit.

This circuit can be broken into two components: the series resistances R_1 and R_2, and the parallel resistances R_3 and R_4.

R_1 and R_2 can be added together to form the equivalent series resistance R_{1+2}:

$$R_{1+2} = R_1 + R_2$$
$$R_{1+2} = 0.5k\Omega + 0.5k\Omega$$
$$R_{1+2} = 1k\Omega$$

R_3 and R_4 can be totaled using either the general reciprocal formula or, since there are two resistances in parallel, the product over sum method. Both methods are shown below.

$$R_{3+4} = \frac{1}{\frac{1}{R_3} + \frac{1}{R_4}} = \frac{1}{\frac{1}{1k\Omega} + \frac{1}{1k\Omega}}$$

$$= \frac{1}{\frac{2}{1{,}000\Omega}} = \frac{1}{0.002} = 0.5k\Omega$$

$$R_{3+4} = \frac{R_3 \times R_4}{R_3 + R_4} = \frac{1k\Omega \times 1k\Omega}{1k\Omega + 1k\Omega}$$

$$= \frac{1{,}000{,}000\Omega}{2{,}000\Omega} = 0.5k\Omega$$

The equivalent circuit containing the R_{1+2} resistance of 1kΩ and the R_{3+4} resistance of 0.5kΩ is shown in *Figure 12*.

Using the Ohm's law relationship that total current equals voltage divided by circuit resistance, the circuit current can be determined. First, however, total circuit resistance must be found. Since the simplified circuit consists of two resistances in series, they are simply added together to obtain total resistance.

$$R_T = R_{1+2} + R_{3+4}$$
$$R_T = 1k\Omega + 0.5k\Omega$$
$$R_T = 1.5k\Omega$$

Applying this to the current/voltage equation:

$$I_T = \frac{E_T}{R_T}$$

$$I_T = \frac{1.5V}{1.5k\Omega}$$

$$I_T = 1mA \text{ or } 0.001A$$

Now that the total current is known, voltage drops across individual components can be determined:

$$E_{R1} = I_T R_1 = 1mA \times 0.5K\Omega = 0.5V$$
$$E_{R2} = I_T R_2 = 1mA \times 0.5K\Omega = 0.5V$$

Since the total voltage equals the sum of all voltage drops, the voltage drop from A to B can be determined by subtraction:

$$E_T = E_{R1} + E_{R2} + E_{A+B}$$
$$E_T - E_{R1} - E_{R2} = E_{A+B}$$
$$1.5V - 0.5V - 0.5V = E_{A+B} = 0.5V$$

Since R_3 and R_4 are in parallel, some of the total current must pass through each resistor. R_3 and R_4 are equal, so the same current should flow through each branch. Using the relationship:

$$I = \frac{E}{R}$$

$$I_{R3} = \frac{E_{R3}}{R_3} \qquad I_{R4} = \frac{E_{R4}}{R_4}$$

$$I_{R3} = \frac{0.5V}{1k\Omega} \qquad I_{R4} = \frac{0.5V}{1k\Omega}$$

$$I_{R3} = 0.5mA \qquad I_{R4} = 0.5mA$$

$$0.5mA + 0.5mA = 1mA$$

Therefore, the total current for the circuit passes through R_1 and R_2 and is evenly divided between R_3 and R_4.

3.0.0 Kirchhoff's Laws

Kirchhoff's laws provide a simple, practical method of solving for unknown parameters in a circuit.

3.1.0 Kirchhoff's Current Law

In its most general form, **Kirchhoff's current law** states that at any point in a circuit, the total current entering that point must equal the total current leaving that point. For parallel circuits, this implies that the current in a parallel circuit is equal to the sum of the currents in each branch.

When using Kirchhoff's laws to solve circuits, it is necessary to adopt conventions that determine the algebraic signs for current and voltage terms. A convenient system for current is to consider all current flowing into a branch point as positive, and all current directed away from that point as negative.

As an example, in *Figure 13* the currents can be written as:

$$I_A + I_B - I_C = 0$$

or

$$5A + 3A - 8A = 0$$

Currents I_A and I_B are positive terms because these currents flow into P, but I_C, directed out of P, is negative.

For a circuit application, refer to Point C at the top of the diagram in *Figure 14*. The 6A I_T into Point C divides into the 2A I_3 and 4A $I_{4/5}$, both directed out. Note that $I_{4/5}$ is the current through R_4 and R_5. The algebraic equation is:

$$I_T - I_3 - I_{4/5} = 0$$

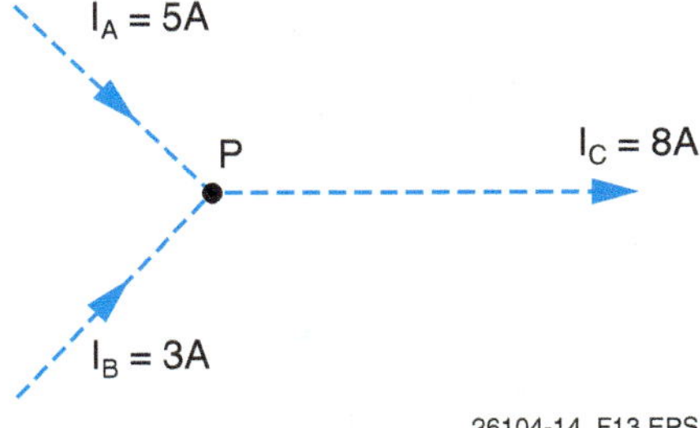

Figure 13 Kirchhoff's current law.

Substituting the values for each current:

$$6A - 2A - 4A = 0$$

For the opposite direction, refer to Point D at the bottom of *Figure 14*. Here, the branch currents into Point D combine to equal the mainline current I_T returning to the voltage source. Now, I_T is directed out from Point D, with I_3 and $I_{4/5}$ directed in. The algebraic equation is:

$$I_3 + I_{4/5} - I_T = 0$$
$$2A + 4A - 6A = 0$$

Note that at either Point C or Point D, the sum of the 2A and 4A branch currents must equal the 6A total line current. Therefore, Kirchhoff's current law can also be stated as:

$$I_{IN} = I_{OUT}$$

For *Figure 14*, the equations for current can be written as shown below.

At Point C:

$$6A = 2A + 4A$$

At Point D:

$$2A + 4A = 6A$$

Kirchhoff's current law is really the basis for the practical rule in parallel circuits that the total line current must equal the sum of the branch currents.

3.2.0 Kirchhoff's Voltage Law

Kirchhoff's voltage law states that the algebraic sum of the voltages around any closed path is zero.

Referring to *Figure 15*, the sum of the voltage drops around the circuit must equal the voltage applied to the circuit:

$$E_A = E_1 + E_2 + E_3$$

Where:

E_A = voltage applied to the circuit

E_1, E_2, and E_3 = voltage drops in the circuit

Another way of stating this law is that the algebraic sum of the voltage rises and voltage drops must be equal to zero. A voltage source is considered a voltage rise; a voltage across a resistor is a voltage drop. (For convenience in labeling, letter subscripts are shown for voltage sources and numerical subscripts are used for voltage drops.)

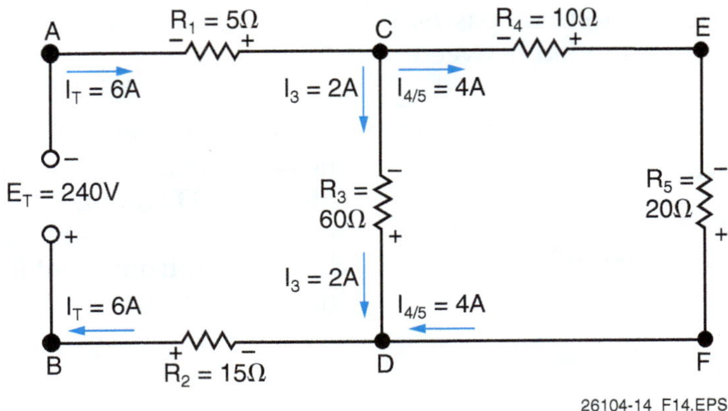

Figure 14 Application of Kirchhoff's current law.

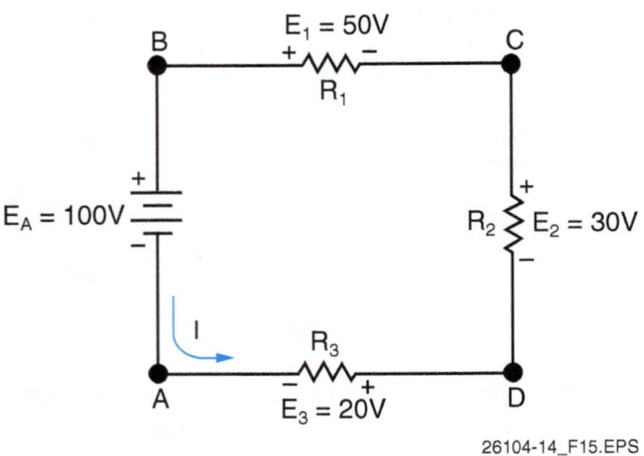

Figure 15 Kirchhoff's voltage law.

Did You Know?

Electrical Laws

As you might have guessed, Ohm's law and Kirchhoff's laws were named after real people. George Ohm first published his theory on the mathematical relationships of current, voltage, and resistance in 1826. Gustav Kirchhoff followed 19 years later with his work on the calculation of currents, voltages, and resistances in electrical networks. As a result of their breakthrough research, both men went on to head mathematics departments at major German universities.

This form of the law can be written by transposing the right members to the left side:

Voltage applied – sum of voltage drops = 0

Substitute letters:

$$E_A - E_1 - E_2 - E_3 = 0$$
$$E_A - (E_1 + E_2 + E_3) = 0$$

3.3.0 Loop Equations

Any closed path for current flow is called a loop. A loop equation specifies the voltages around the loop. Refer to *Figure 16*.

Consider the inside loop through A, C, D, and B. This includes the voltage drops E_1, E_3, and E_2, and the source E_T. In a clockwise direction, starting at Point A, the algebraic sum of the voltages is:

$$-E_1 - E_3 - E_2 + E_T = 0$$

or

$$-30V - 120V - 90V + 240V = 0$$

Voltages E_1, E_3, and E_2 have a negative value, because there is a decrease in voltage seen across each of the resistors in a clockwise direction. However, the source E_T is a positive term because an increase in voltage is seen in that same direction.

For the opposite direction, going counterclockwise in the same loop from Point B, E_T is negative while E_1, E_2, and E_3 have positive values. Therefore:

$$-E_T + E_2 + E_3 + E_1 = 0$$

or

$$-240V + 90V + 120V + 30V = 0$$

When the negative term is transposed, the equation becomes:

$$240V = 90V + 120V + 30V$$

In this form, the loop equation shows that Kirchhoff's voltage law is really the basis for the practical rule in series circuits that the sum of the voltage drops must equal the applied voltage.

For example, determine the voltage E_B for the circuit shown in *Figure 17*. The direction of the

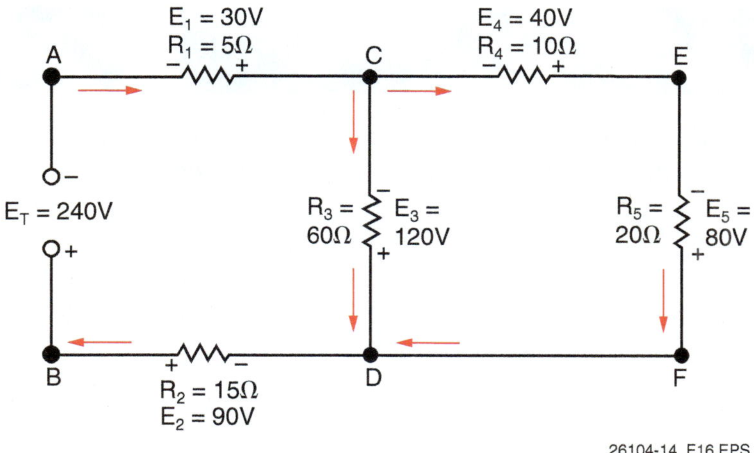

Figure 16 Loop equation.

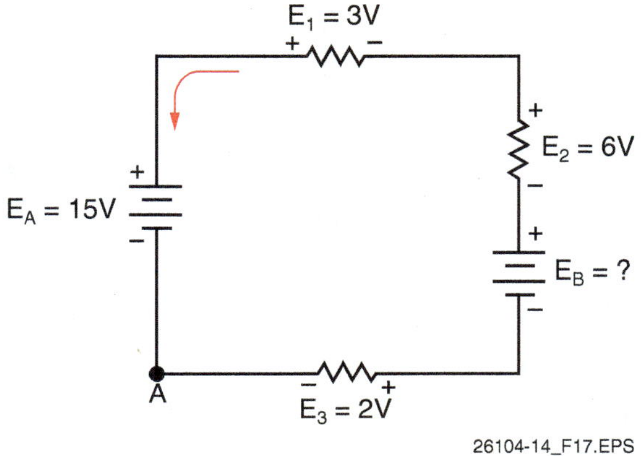

Figure 17 Applying Kirchhoff's voltage law.

current flow is shown by the arrow. First mark the polarity of the voltage drops across the resistors and trace the circuit in the direction of the current flow starting at Point A. Then write the voltage equation around the circuit:

$$-E_3 - E_B - E_2 - E_1 + E_A = 0$$

Solve for E_B:

$$E_B = E_A - E_3 - E_2 - E_1$$
$$E_B = 15V - 2V - 6V - 3V$$
$$E_B = 4V$$

Since E_B was found to be positive, the assumed direction of current is in fact the actual direction of current.

In its most general form, Kirchhoff's voltage law states that the algebraic sum of all the potential differences in a closed loop is equal to zero. A closed loop means any completely closed path consisting of wire, resistors, batteries, or other components. For series circuits, this implies that the sum of the voltage drops around the circuit is equal to the applied voltage. For parallel circuits, this implies that the voltage drops across all branches are equal.

Think About It

Putting It All Together

Draw four 60W lamps in parallel with a 120V power source. What is the amperage in the circuit? What would happen to the amperage if you doubled the voltage?

Summary

The relationships among current, voltage, resistance, and power in Ohm's law are the same for both DC series and DC parallel circuits. Understanding and being able to apply these concepts is necessary for effective circuit analysis and troubleshooting. DC series-parallel circuits also have these fundamental relationships. Since DC series-parallel circuits are a combination of simple series and parallel circuits, Kirchhoff's voltage and current laws will apply. Calculating I, E, R, and P for series-parallel circuits is no more difficult than calculating these values for simple series or parallel circuits. However, for series-parallel circuits, these calculations require more careful circuit analysis in order to use Ohm's law correctly.

Review Questions

1. The formula for calculating the total resistance in a series circuit with three resistors is _____.
   a. $R_T = R_1 + R_2 + R_3$
   b. $R_T = R_1 - R_2 - R_3$
   c. $R_T = R_x R_x R_3$
   d. 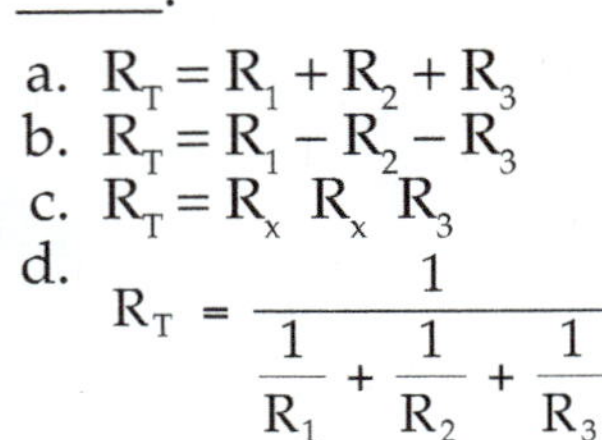
   $$R_T = \frac{1}{\frac{1}{R_1} + \frac{1}{R_2} + \frac{1}{R_3}}$$

2. Find the total resistance in a series circuit with three resistances of 10Ω, 20Ω, and 30Ω.
   a. 1Ω
   b. 15Ω
   c. 20Ω
   d. 60Ω

3. The formula for calculating the total resistance in a parallel circuit with three resistors is _____.
   a. $R_T = R_1 + R_2 + R_3$
   b. $R_T = R_1 - R_2 - R_3$
   c. $R_T = R_x R_x R_3$
   d. 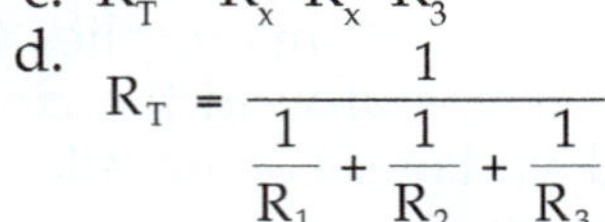
   $$R_T = \frac{1}{\frac{1}{R_1} + \frac{1}{R_2} + \frac{1}{R_3}}$$

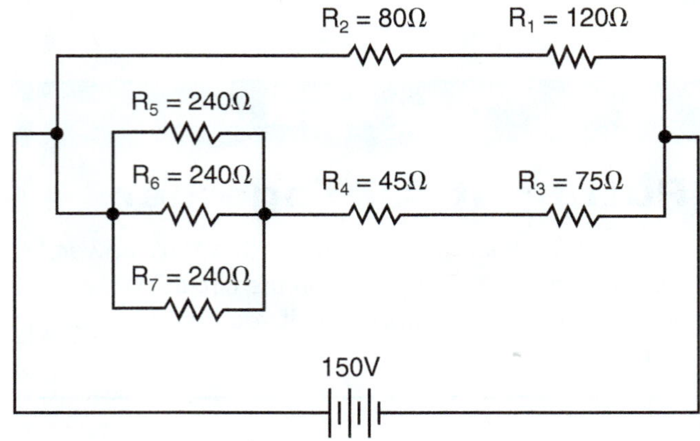

Figure 1

4. The total resistance in *Figure 1* is _____.
   a. 100Ω
   b. 129Ω
   c. 157Ω
   d. 1,040Ω

5. In a parallel circuit, the voltage across each path is equal to the _____.
   a. total circuit resistance times path current
   b. source voltage minus path voltage
   c. path resistance times total current
   d. applied voltage

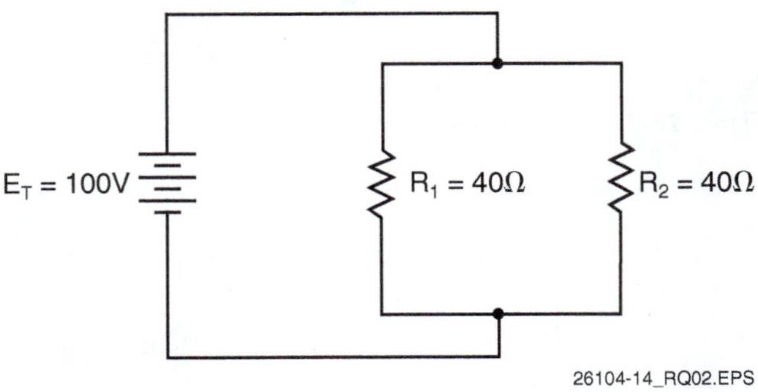

Figure 2

6. The value for total current in *Figure 2* is _____.
   a. 1.25A
   b. 2.50A
   c. 5A
   d. 10A

7. A resistor of 32Ω is in parallel with a resistor of 36Ω, and a 54Ω resistor is in series with the pair. When 350V is applied to the combination, the current through the 54Ω resistor is _____.
   a. 2.87A
   b. 3.26A
   c. 4.93A
   d. 5.86A

8. A 242Ω resistor is in parallel with a 180Ω resistor, and a 420Ω resistor is in series with the combination in a 27V circuit. A current of 22mA flows through the 242Ω resistor. The current through the 180Ω resistor is _____.
   a. 29.6mA
   b. 36.4mA
   c. 59.4mA
   d. 60.3mA

9. Kirchhoff's voltage law states that the algebraic sum of the voltages around any closed current path is _____.
   a. infinity
   b. zero
   c. twice the current
   d. always less than the individual voltages due to voltage drop

10. Two 24Ω resistors are in parallel, and a 42Ω resistor is in series with the combination. When 78V is applied to the three resistors, the voltage drop across the 42Ω resistor is about _____.
    a. 49.8V
    b. 55.8V
    c. 60.7V
    d. 65.3V

Trade Terms Quiz

Fill in the blank with the correct term that you learned from your study of this module.

1. _________ states that the total amount of current flowing through a parallel circuit is equal to the sum of the amounts of current flowing through each current path.
2. _________ states that the sum of all the voltage drops in a circuit is equal to the source voltage of the circuit.
3. _________ contain both series and parallel current paths.
4. _________ contain two or more parallel paths through which current can flow.
5. _________ contain only one path for current flow.

Trade Terms

Kirchhoff's current law
Kirchhoff's voltage law
Parallel circuits
Series circuits
Series-parallel circuits

Module 26104-14

Supplemental Exercises

1. Which formula does *not* apply to a parallel circuit?

   a. $R_T = \frac{R}{N}$

   b. $R_T = \frac{1}{\frac{1}{R_1} + \frac{1}{R_2} + \frac{1}{R_3} + \frac{1}{R_n}}$

   c. $R_T = R_1 + R_2 + R_3$

   d. $R_T = \frac{R_1 \times R_2}{R_1 + R_2}$

2. If two or more resistors are connected in parallel, the _____.

   a. total resistance is higher than any single resistor
   b. total resistance is lower than any single resistor
   c. current flow varies based on voltage fluctuation
   d. resistance depends on the power rating of the circuit

3. True or False? You can use Ohm's law to find the value of each branch current in a parallel circuit.

4. True or False? Parallel circuits are current dividers.

5. Which statement is *not* correct relative to Kirchhoff's current law?

   a. $I_A + I_B - I_C = 0$
   b. The total current entering a point must equal total current leaving a point.
   c. The total line current must equal the sum of the branch currents.
   d. Kirchhoff's laws apply to current only.

6. True or False? In a series circuit, E_T must equal the sum of the individual voltage drops.

7. When calculating series-parallel combination circuits, it is important to _____.

   a. reduce to series loads, then add the loads
   b. replace parallel combinations with one value
   c. reduce to simpler circuits where possible
   d. All of the above.

8. True or False? Kirchhoff's voltage law can be used to determine voltage drops.

9. Two resistors are connected in parallel; R_1 is 90 ohms, R_2 is 45 ohms. What is the total resistance of R_1 and R_2 in parallel? ____________________

10. Two resistors are connected in parallel; R_2 is 90 ohms, R_3 is 45 ohms. A third resistor (R_1, which is 20 ohms) is connected in series with R_2 and R_3.

   a. What is the total resistance of this circuit? ______________________

   b. If the voltage source of this circuit is 150V, what is the total current flow? ______________

   (Hint: sketch the circuit before solving.)

CIRCUIT

Cornerstone of Craftsmanship

James Mitchem

TIC—The Industrial Company

Jim Mitchem serves as a troubleshooter for a large electrical contractor. During his career in the electrical industry, he worked his way up from apprentice to technical services manager.

How did you become an electrician?
Quite by accident. A couple of years after college, I was working as a relief operator in a plant when the lead electrician retired, creating a vacancy. I liked the idea that electricians were expected to use their knowledge and initiative to keep the place running. I applied and was accepted as a trainee.

How did you get your training?
I took an electrical apprenticeship course by correspondence, and I was fortunate enough to work with good people who helped me along. I worked in an environment that exposed me to a variety of equipment and applications, and just about everyone I've ever worked with has taught me something. Now I'm passing my knowledge on to others.

What kinds of work have you done in your career?
I've worked as an apprentice, journeyman, instrument and controls technician, instrument fitter, foreman, general foreman, superintendent, and startup engineer. Each of these positions required that I learn new skills, both technical and managerial. My experience in many disciplines and types of projects has given me a high level of credibility with my employer and our clients.

Now I act as a technical resource and troubleshooter for job sites and in-house functions such as safety, quality assurance, and training. I visit job sites to help solve problems and help out with commissioning and startup.

What factor or factors have contributed the most to your success?
There are several factors. Two very important ones have been a desire to learn and a willingness to do whatever is asked of me. I also keep an eye on the big picture. When I'm on a job, I'm not just pulling wire, I'm building a power plant or whatever the project is. I also think it has helped me to remain with the same employer for 18 years.

Any advice for apprentices just beginning their careers?
Keep learning! And don't depend on others to train you. Take the initiative to buy or borrow books and trade journals. Take licensing tests and do whatever is necessary to keep your licenses current. Finally, make sure you know your own personal and professional values and work with a company that shares those values.

Trade Terms Introduced in This Module

Kirchhoff's current law: The statement that the total amount of current flowing through a parallel circuit is equal to the sum of the amounts of current flowing through each current path.

Kirchhoff's voltage law: The statement that the sum of all the voltage drops in a circuit is equal to the source voltage of the circuit.

Parallel circuits: Circuits containing two or more parallel paths through which current can flow.

Series circuits: Circuits with only one path for current flow.

Series-parallel circuits: Circuits that contain both series and parallel current paths.

Additional Resources

This module presents thorough resources for task training. The following resource material is suggested for further study.

Electronics Fundamentals: Circuits, Devices, and Applications, Thomas L. Floyd. New York: Prentice Hall.

Principles of Electric Circuits, Thomas L. Floyd. New York: Prentice Hall.

Figure Credits

AGC of America, Module opener

Topaz Publications, Inc., SA01

NCCER CURRICULA — USER UPDATE

NCCER makes every effort to keep its textbooks up-to-date and free of technical errors. We appreciate your help in this process. If you find an error, a typographical mistake, or an inaccuracy in NCCER's curricula, please fill out this form (or a photocopy), or complete the online form at **www.nccer.org/olf**. Be sure to include the exact module ID number, page number, a detailed description, and your recommended correction. Your input will be brought to the attention of the Authoring Team. Thank you for your assistance.

Instructors – If you have an idea for improving this textbook, or have found that additional materials were necessary to teach this module effectively, please let us know so that we may present your suggestions to the Authoring Team.

NCCER Product Development and Revision

13614 Progress Blvd., Alachua, FL 32615

Email: curriculum@nccer.org
Online: www.nccer.org/olf

❑ Trainee Guide ❑ AIG ❑ Exam ❑ PowerPoints Other ______________

Craft / Level: Copyright Date:

Module ID Number / Title:

Section Number(s):

Description:

Recommended Correction:

Your Name:

Address:

Email: Phone:

Introduction to the *National Electrical Code*®

Brick Nanotechnology Center at Purdue University

The Gaylor Group's team of nearly 60 electricians followed strict cleanroom protocols to install electrical systems in this leading-edge nanotechnology facility while facing shared-space challenges and work restrictions.

26105-14

Trainees with successful module completions may be eligible for credentialing through NCCER's National Registry. To learn more, go to **www.nccer.org** or contact us at **1.888.622.3720.** Our website has information on the latest product releases and training, as well as online versions of our *Cornerstone* magazine and Pearson's product catalog.

Your feedback is welcome. You may email your comments to **curriculum@nccer.org,** send general comments and inquiries to **info@nccer.org,** or fill in the User Update form at the back of this module.

This information is general in nature and intended for training purposes only. Actual performance of activities described in this manual requires compliance with all applicable operating, service, maintenance, and safety procedures under the direction of qualified personnel. References in this manual to patented or proprietary devices do not constitute a recommendation of their use.

26105-14
Introduction to the *National Electrical Code*®

Objectives

When you have completed this module, you will be able to do the following:

1. Explain the purpose and history of the *NEC*®.
2. Describe the layout of the *NEC*®.
3. Demonstrate how to navigate the *NEC*®.
4. Describe the purpose of the National Electrical Manufacturers Association and the NFPA.
5. Explain the role of nationally recognized testing laboratories.

Performance Tasks

Under the supervision of the instructor, you should be able to do the following:

1. Use ***NEC Article 90*** to determine the scope of the *NEC*®. State what is covered by the *NEC*® and what is not.
2. Find the definition of the term *feeder* in the *NEC*®.
3. Look up the *NEC*® specifications that you would need to follow if you were installing an outlet near a swimming pool.
4. Find the minimum wire bending space required for two No. 1/0 AWG conductors installed in a junction box or cabinet and entering opposite the terminal.

Trade Terms

Articles
Chapters
Exceptions
Informational Note
National Electrical Manufacturers Association (NEMA)
National Fire Protection Association (NFPA)
Nationally Recognized Testing Laboratories (NRTLs)
Parts
Sections

Required Trainee Materials

1. Paper and pencil
2. Copy of the latest edition of the *National Electrical Code*®

Note:
NFPA 70®, *National Electrical Code*®, and *NEC*® are registered trademarks of the National Fire Protection Association, Inc., Quincy, MA 02269. All *National Electrical Code*® and *NEC*® references in this module refer to the 2014 edition of the *National Electrical Code*®.

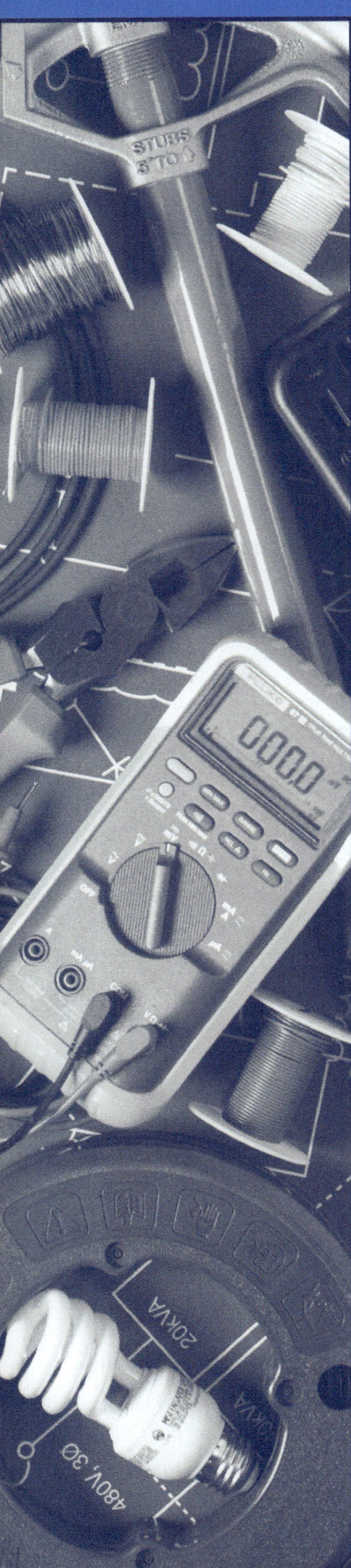

Contents

Topics to be presented in this module include:

Figures

1.0.0 Introduction

The *National Electrical Code*® (*NEC*®) is published by the National Fire Protection Association (NFPA). The *NEC*® is one of the most important tools for the electrician. When used together with the electrical code for your local area, the *NEC*® provides the minimum requirements for the installation of electrical systems. Unless otherwise specified, always use the latest edition of the *NEC*® as your on-the-job reference. It specifies the minimum provisions necessary for protecting people and property from electrical hazards. In some areas, however, local laws may specify different editions of the *NEC*®, so be sure to use the edition specified by your employer. Also, bear in mind that the *NEC*® only specifies minimum requirements, so local or job requirements may be more stringent.

2.0.0 Purpose and History of the *NEC*®

The primary purpose of the *NEC*® is the practical safeguarding of persons and property from hazards arising from the use of electricity [*NEC Section 90.1(A)*]. A thorough knowledge of the *NEC*® is one of the first requirements for becoming a trained electrician. The *NEC*® is probably the most widely used and generally accepted code in the world. It has been translated into several languages. It is used as an electrical installation, safety, and reference guide in the United States. Compliance with *NEC*® standards increases the safety of electrical installations—the reason the *NEC*® is so widely used.

Although *NEC Section 90.1(A)* states, "This Code is not intended as a design specification or an instruction manual for untrained persons," it does provide a sound basis for the study of electrical installation procedures—under the proper guidance. The *NEC*® has become the standard reference of the electrical construction industry. Anyone involved in electrical work should obtain an up-to-date copy and refer to it frequently.

Think About It

The *NEC*®

Why do you think it's necessary to have a standard set of procedures for electrical installations? Find out who does the electrical inspection in your area. Who determines what will be inspected, when it will be inspected, and who will do the inspection?

All electrical work must comply with the currently adopted *NEC*® and all local ordinances. Like most laws, the *NEC*® is easier to work with once you understand the language and know where to look for the information you need.

NOTE

This module is not a substitute for the *NEC*®. You need to acquire a copy of the most recent edition and keep it handy at all times. The more you know about the *NEC*®, the better an electrician you will become.

2.1.0 History

In 1881, the National Association of Fire Engineers met in Richmond, Virginia. From this meeting came the idea to draft the first *National Electrical Code*. The first nationally recommended electrical code was published by the National Board of Fire Underwriters (now the American Insurance Association) in 1895.

In 1896, the National Electric Light Association (NELA) was working to make the requirements of the fire insurance organizations and electrical utilities fit together. NELA succeeded in promoting a conference that would result in producing a standard national code. The NELA code would serve the interests of the insurance industry, operating concerns, manufacturing, and industry.

The conference produced a set of requirements that was unanimously accepted. In 1897, the first edition of the *NEC*® was published, and the *NEC*® became the first cooperatively produced national code. The organization that produced the *NEC*® was known as the *National Conference on Standard Electrical Rules*. This group became a permanent organization, and its job was to develop the *NEC*®.

In 1911, the NFPA took over administration and control of the *NEC*®. However, the National Board of Fire Underwriters continued to publish the *NEC*® until 1962. From 1911 until now, the *NEC*® has experienced several major changes, as well as regular three-year updates. In 1923, the *NEC*® was rearranged and rewritten, and in 1937, it was editorially revised. In 1959, the *NEC*® adopted a numbering system in which each section of an article was identified with an article/section number.

2.1.1 Who Is Involved?

The creation of a universally accepted set of rules is an involved and complicated process. Rules made by a committee have the advantage that they usually do not leave out the interests of any of the groups represented on the committee. However,

What's wrong with these pictures?

26105-14_SA01.EPS

26105-14_SA02.EPS

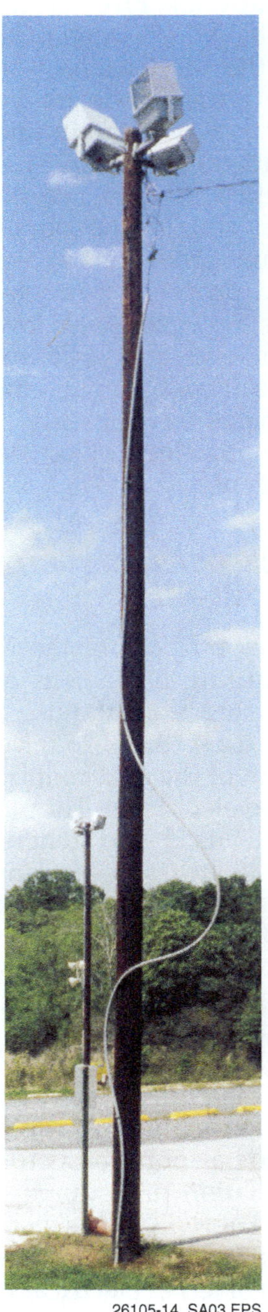

26105-14_SA03.EPS

since the rules must represent the interests and requirements of an assortment of groups, they are often quite complicated and wordy.

In 1949, the NFPA reorganized the *NEC®* into its present structure. The present structure consists of a Correlating Committee and Code Making Panels (CMPs). The Correlating Committee consists of principal voting members and alternates. The principal function of the Correlating Committee is to ensure that:

- No conflict of requirements exists
- Correlation has been achieved
- NFPA regulations governing committee projects have been followed
- A practical schedule of revision and publication is established and maintained

Each of the CMPs have members who are experts on particular subjects and have been assigned certain articles to supervise and revise as

Think About It

Code Changes

This photograph shows the first code book and the current code book. Code changes occur every three years. Who can suggest changes to the *National Electrical Code®*? What might be reasons for submitting changes?

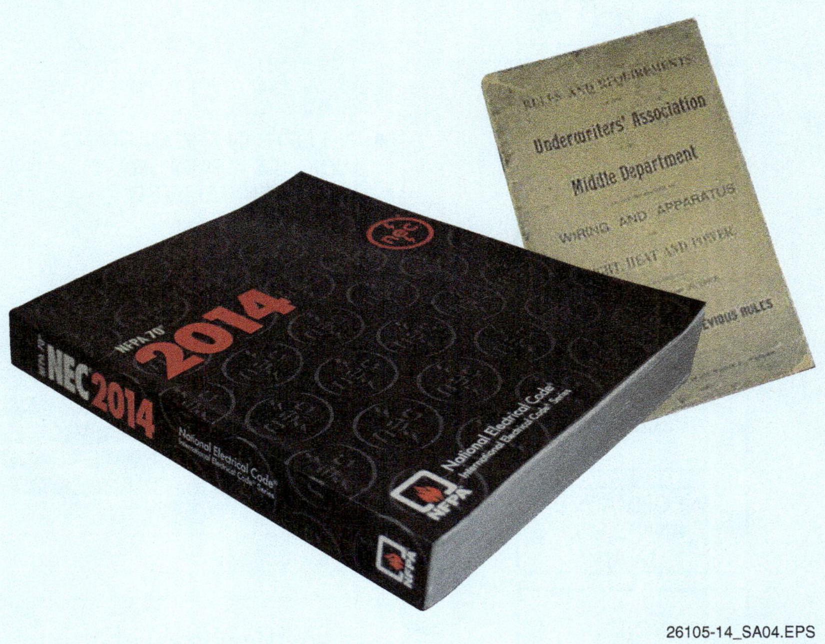

26105-14_SA04.EPS

required. Members of the CMPs represent special interest groups such as trade associations, electrical contractors, electrical designers and engineers, electrical inspectors, electrical manufacturers and suppliers, electrical testing laboratories, and insurance organizations.

Each panel is structured so that not more than one-third of its members are from a single interest group. The members of the *NEC®* Committee create or revise requirements for the *NEC®* through probing, debating, analyzing, weighing, and reviewing new input. Anyone, including you, can submit proposals to amend the *NEC®*. Sample forms for this purpose may be obtained from the Secretary of the Standards Council at NFPA Headquarters. The *NEC®* describes the code-making process in detail.

The NFPA membership is drawn from the fields listed above. In addition to publishing the *NEC®*, the duties of the NFPA include the following:

- Developing, publishing, and distributing standards that are intended to minimize the possibility and effects of fire and explosion
- Conducting fire safety education programs for the general public
- Providing information on fire protection, prevention, and suppression
- Compiling annual statistics on causes and occupancies of fires, large-loss fires (over one million dollars), fire deaths, and firefighter casualties
- Providing field service by specialists on electricity, flammable liquids and gases, and marine fire problems
- Conducting research projects that apply statistical methods and operations research to develop computer models and data management systems

3.0.0 The Layout of the *NEC®*

The *NEC®* begins with a brief history. *Figure 1* shows how the *NEC®* is organized.

3.1.0 Types of Rules

There are two basic types of rules in the *NEC®*: mandatory rules and permissive rules. It is important to understand these rules as they are defined in *NEC Section 90.5*. Mandatory rules contain the words "shall" or "shall not" and must be adhered to. Permissive rules identify actions that are allowed but not required and typically cover options or alternative methods. Permissive rules are indicated by the phrases "shall be permitted"

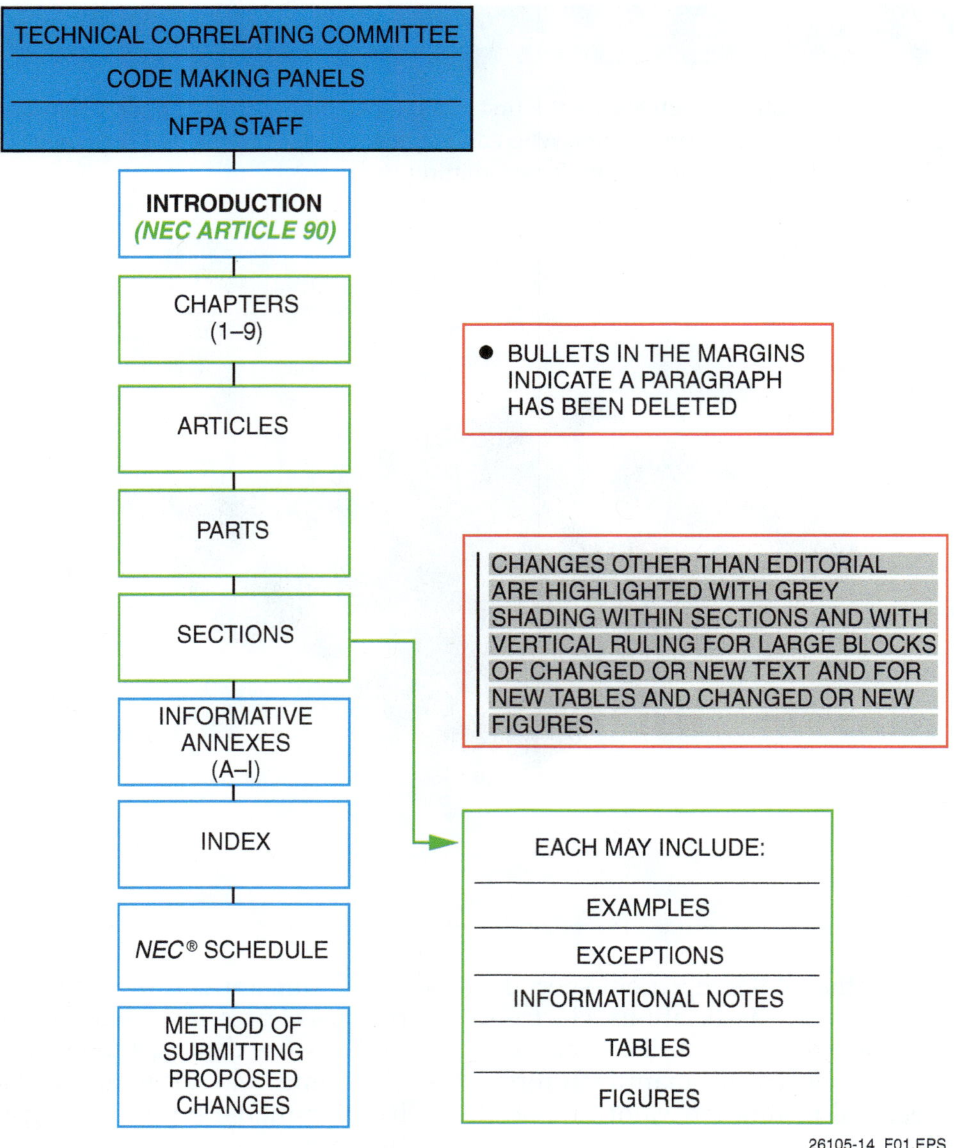

Figure 1 The layout of the *NEC®*.

or "shall not be required." Be aware that local ordinances may amend requirements of the *NEC®*. This means that a city or county may have additional requirements or prohibitions that must be followed in that jurisdiction.

3.2.0 *NEC®* Introduction

The main body of the text begins with an Introduction entitled *NEC Article 90*. This introduction gives you an overview of the *NEC®* (see *Figure 1*). Items included in this section are:

- Purpose of the *NEC®*
- Scope of the code book
- Code arrangement
- Enforcement
- Mandatory rules and explanatory material
- Formal interpretation
- Examination of equipment for safety
- Wiring planning
- Metric units of measurement

3.3.0 The Body of the *NEC®*

The remainder of the code is organized into nine chapters, followed by informative annexes, the index, a schedule for the next code cycle, and information on submitting proposed changes.

NEC Chapters 1 through 8 are subdivided into articles. Each chapter focuses on a general category of electrical application, such as *NEC Chapter 2, Wiring and Protection*. Each article emphasizes a more specific part of that category, such as *NEC Article 210, Branch Circuits, Part I General Provisions*. Each section gives examples of a specific application of the *NEC®*, such as *NEC Section 210.4, Multiwire*

On Site

NEC® Layout

Remember, chapters contain a group of articles relating to a broad category. An article is a specific subject within that category, such as ***NEC Article 250****, Grounding and Bonding,* which is in Chapter 2 relating to wiring and protection. When an article applies to different installations in the same category, it will be divided into parts using Roman numerals. Any specific requirements in any of the articles may also have exception(s) to the main rules.

Branch Circuits. *NEC Chapter 9* contains tables that are referenced by any of the articles in *NEC Chapters 1 through 8*. *Informative Annexes A through J* provide informational material and examples that are helpful when applying *NEC®* requirements.

The chapters of the *NEC®* are organized into four major categories:

- *NEC Chapters 1, 2, 3, 4* – The first four chapters present the rules for the design and installation of electrical systems. They generally apply to all electrical installations.
- *NEC Chapters 5, 6, and 7* – These chapters are concerned with special occupancies, equipment, and conditions. Rules in these chapters may modify or amend those in the first four chapters.
- *NEC Chapter 8* – This chapter covers communications systems, such as the telephone and television receiving equipment. It may also reference other articles, such as the installation of grounding electrode conductor connections as covered in *NEC Section 250.52*.
- *NEC Chapter 9* – This chapter contains tables that are applicable when referenced by other chapters in the *NEC®*.
- *Informative Annexes A through J* – These annexes contain helpful information that is not mandatory.
  - *Informative Annex A* contains a list of product safety standards. These standards provide further references for requirements that are in addition to the *NEC®* requirements for the electrical components mentioned.
  - *Informative Annex B* contains information for determining ampacities of conductors under engineering supervision.
  - *Informative Annex C* contains the conduit fill tables for multiple conductors of the same size and type within the accepted raceways.
  - *Informative Annex D* contains examples of calculations for branch circuits, feeders, and services as well as load calculations.
  - *Informative Annex E* contains information on types of building construction.
  - *Informative Annex F* provides information on critical operations power systems.
  - *Informative Annex G* is for informational purposes and covers supervisory control and data acquisition (SCADA) systems.
  - *Informative Annex H* contains the requirements for administration and enforcement of the *NEC®*.
  - *Informative Annex I* contains the recommended torque tables from *UL Standard 486-A–B*.
  - *Informative Annex J* contains the ADA Standards for accessible design.

3.4.0 Text in the *NEC®*

When you open the *NEC®*, you will notice several different types of text or printing used. Here is an explanation of each type of text:

- *Bold black letters* – Headings for each *NEC®* application are written in **bold** black letters.
- *Exceptions* – These explain the circumstances under which a specific part of the *NEC®* does not apply. Exceptions are written in *italics* under that part of the *NEC®* to which they pertain.
- *Informational Notes* – These explain something in an application, suggest other sections to read about the application, or provide tips

On Site

The *NEC®* as a Reference Tool

The *NEC®*, although carefully organized, is a highly technical document; as such, it uses many technical words and phrases. It's crucial for you to understand their meaning—if you don't, you'll have trouble understanding the rules themselves. When using the *NEC®* as a reference tool, you may have to refer to a number of articles to find your answer(s). *NEC®* definitions are covered in ***NEC Article 100***. Many issues concerning the *NEC®* may be resolved by simply reviewing the definitions.

about the application. These are defined in the text by the term *Informational Note* before an indented paragraph.

- *Figures* – These may be included with explanations to give you a picture of what your application may look like.
- *Tables* – These are often included when there is more than one possible application of the *NEC*®. You would use a table to look up the specifications of your application.

4.0.0 Navigating the *NEC*®

To locate information for a particular procedure being performed, use the following steps:

Step 1 Familiarize yourself with *NEC Articles 90, 100, and 110* to gain an understanding of the material covered in the *NEC*® and the definitions used in it.

Step 2 Turn to the *Table of Contents* at the beginning of the *NEC*®.

Step 3 Locate the chapter that focuses on the desired category.

Step 4 Find the article pertaining to your specific application.

Step 5 Turn to the page indicated. Each application will begin with a bold heading.

NOTE

An index is provided at the end of the *NEC*®. The index lists specific topics and provides a reference to the location of the material within the *NEC*®. The index is helpful when you are looking for a specific topic rather than a general category.

Once you are familiar with *NEC Articles 90, 100, and 110,* you can move on to the rest of the *NEC*®. There are several key sections used often in servicing electrical systems.

4.1.0 *NEC Chapter 1, General*

NEC Article 100 contains a list of common definitions used in the *NEC*®. Definitions specific to a single article are included in the second section of each article. For example, *NEC Section 240.2* contains definitions specific to overcurrent protection.

Two definitions from *NEC Article 100* you should become familiar with are:

- *Labeled* – "Equipment or materials to which has been attached a label, symbol, or other identifying mark of an organization that is acceptable to the authority having jurisdiction and concerned with product evaluation, that maintains periodic inspection of production of labeled equipment or materials, and by whose labeling the manufacturer indicates compliance with appropriate standards or performance in a specified manner."
- *Listed* – "Equipment, materials, or services included in a list published by an organization that is acceptable to the authority having jurisdiction and concerned with evaluation of products or services, that maintains periodic inspection of production of listed equipment or materials, or periodic evaluation of services, and whose listing states that either the equipment, material, or service meets appropriate designated standards or has been tested and found suitable for a specified purpose."

Think About It

Junction Boxes

Find the rule in the *NEC*® that explains whether a junction box without devices can be supported solely by two or more lengths of rigid metal conduit (RMC). Explain the technical terminology in everyday language.

Besides installation rules, you will also have to be concerned with the type and quality of materials that are used in electrical wiring systems. Nationally recognized testing laboratories are product safety certification laboratories. Underwriters Laboratories, Inc., also called UL, is one such laboratory. These laboratories establish and operate product safety certification programs to make sure that items produced under the service are safeguarded against reasonable foreseeable risks. Some of these organizations maintain a worldwide network of field representatives who make unannounced visits to manufacturing facilities to counter-check products bearing their seal of approval. The UL label is shown in *Figure 2*.

NEC Article 110 gives the general requirements for electrical installations. It is important for you to be familiar with this general information and the definitions.

Think About It

***NEC* Article 90**

After you've familiarized yourself with *NEC Article 90,* explain its intent. What part of electrical installation does it not cover?

26105-14_F02.EPS

Figure 2 Underwriters Laboratories label.

4.2.0 *NEC Chapter 2, Wiring and Protection*

NEC Chapter 2 discusses wiring design and protection, the information electrical technicians need most often. It covers the use and identification of grounded conductors, branch circuits, feeders, calculations, services, overcurrent protection, and grounding. This is essential information for all types of electrical systems. If you run into a problem related to the design or installation of a conventional electrical system, you can probably find a solution for it in this chapter.

NEC Article 200 contains important information on the use and identification of grounded conductors.

NEC Articles 210 and 215 cover the provisions and requirements for branch circuits, their ratings and required outlets, and installation and overcurrent protection requirements for feeders.

NEC Article 220 contains the requirements for calculating branch circuit, feeder, and service loads, including the calculation methods for farm loads.

NEC Articles 225 and 230 cover outside branch circuit, feeder, and service installation requirements.

NEC Article 240 covers the requirements for overcurrent protection and overcurrent protective devices, including the standard ratings for fuses and fixed-trip circuit breakers. *NEC Article 240, Part IX* covers overcurrent protection for installations over 1,000 volts.

NEC Article 250 covers the requirements for grounding and bonding electrical systems. It lists the systems required, permitted, and not permitted to be grounded and provides the requirements for grounding connection locations. It also covers accepted methods of grounding and bonding, including types and sizes of grounding and bonding conductors and electrodes.

On Site

Underwriters Laboratories®

The Chicago World's Fair was opened in 1893 and thanks to Edison's introduction of the electric light bulb, included a light display called the Palace of Electricity. But all was not perfect—wires soon sputtered and crackled, and, ironically, the Palace of Electricity caught fire. The fair's insurance company called in a troubleshooting engineer, who, after careful inspection, found faulty and brittle insulation, worn-out and deteriorated wiring, bare wires, and overloaded circuits. This engineer, William Henry Merrill, called for standards in the electrical industry and then set up a testing laboratory above a Chicago firehouse to do just that. An independent testing organization, Underwriters Electrical Bureau, was born. UL is now an internationally recognized authority on product safety testing and safety certification and standards development.

Think About It

Other Testing Laboratories

In addition to UL, there are several other recognized testing laboratories. How many can you name?

4.3.0 *NEC Chapter 3, General Requirements for Wiring Methods and Materials*

NEC Chapter 3 lists the rules on wiring methods and materials. The materials and procedures to use on a particular system depend on the type of building construction, the type of occupancy, the location of the wiring in the building, the type of atmosphere in the building or in the area surrounding the building, mechanical factors, and the relative costs of different wiring methods. The general requirements for conductors and wiring methods that form an integral part of manufactured equipment are not included in the requirements of the code per *NEC Article 300.1(B)*.

NEC Article 300 provides the general requirements for all wiring methods and materials, including information such as minimum burial depths and permitted wiring methods for areas above suspended ceilings.

NEC Article 310 contains a description of acceptable conductors for the wiring methods contained in *NEC Chapter 3*.

NEC Articles 312 and 314 give rules for boxes, cabinets, conduit bodies, and raceway fittings. Outlet boxes vary in size and shape, depending on their use, the size of the raceway, the number of conductors entering the box, the type of building construction, and the atmospheric conditions of the area. These articles should answer most questions on the selection and use of these items.

The *NEC®* does not describe in detail all types and sizes of outlet boxes. However, the manufacturers of outlet boxes provide excellent catalogs showing their products. Collect these catalogs, since these items are essential to your work.

NEC Articles 320 through 340 cover sheathed cables of two or more conductors, such as nonmetallic-sheathed and metal-clad cable.

NEC Articles 342 through 356 cover conduit wiring systems, such as rigid and flexible metal and nonmetallic conduit.

NEC Articles 358 through 362 cover tubing wiring methods, such as electrical metallic and nonmetallic tubing.

NEC Articles 366 through 390 cover other wiring methods, such as busways and wireways. *NEC Article 392* covers cable trays. *NEC Article 393* covers low-voltage suspended ceiling power distribution systems.

4.4.0 *NEC Chapter 4, Equipment for General Use*

NEC Article 400 covers the use and installation of flexible cords and cables, including the trade name, type letter, wire size, number of conductors, conductor insulation, outer covering, and use of each. *NEC Article 402* covers fixture wires, again giving the trade name, type letter, and other important details.

NEC Article 404 covers the requirements for the uses and installation of switches, switching devices, and circuit breakers where used as switches.

NEC Article 406 gives the rules for the ratings, types, and installation of receptacles, cord connectors, and attachment plugs (cord caps).

NEC Article 408 covers the requirements for switchboards, switchgear, and panelboards used to control light and power circuits.

Think About It

Disconnects

How would you proceed to find the *NEC®* rule for the maximum number of disconnects permitted for a service?

NEC Article 410 on luminaires is especially important. It gives installation procedures for luminaires in specific locations. For example, it covers luminaires near combustible material and luminaires in closets. However, the *NEC®* does not describe how many luminaires will be needed in a given area to provide a certain amount of illumination.

NEC Article 422 covers the use of electric appliances in any occupancy. It includes kitchen appliances, various heating appliances, cord-and-plug connected equipment, and other appliances.

NEC Article 424 covers fixed electric space-heating equipment. It includes heating cable, unit heaters, boilers, central systems, and other approved equipment.

NEC Article 430 covers electric motors, including electrical connections, motor controls, and overload protection.

NEC Articles 440 through 460 cover air conditioning and refrigerating equipment, generators, transformers, phase converters, and capacitors.

NEC Article 480 provides requirements related to battery-operated electrical systems. Storage batteries are seldom thought of as part of a conventional electrical system, but they often provide standby emergency lighting service. They may also supply power to security systems that are separate from the main AC electrical system.

4.5.0 *NEC Chapter 5, Special Occupancies*

NEC Chapter 5 covers special occupancy areas. These are areas where the sparks generated by electrical equipment may cause an explosion or fire. The hazard may be due to the atmosphere of the area or the presence of a volatile material in the area. Commercial garages, aircraft hangars, and service stations are typical special occupancy locations.

NEC Article 500 covers the different types of special occupancy atmospheres where an explosion is possible. The atmospheric groups were established to make it easy to test and approve equipment for various types of uses.

NEC Articles 501.10, 502.10, and 503.10 cover the installation of explosion-proof wiring. An explosion-proof system is designed to prevent the ignition of a surrounding explosive atmosphere when arcing occurs within the electrical system.

There are three main classes of special occupancy location:

- *Class I* – Areas containing flammable gases or vapors in the air. Class I areas include paint spray booths, dyeing plants where hazardous

Think About It

Sealing Requirements

Study *NEC Section 501.15*. Summarize key points. Discuss your interpretation with the rest of the class.

liquids are used, and gas generator rooms *(NEC Article 501)*.

- *Class II* – Areas where combustible dust is present, such as grain-handling and storage plants, dust and stock collector areas, and sugar-pulverizing plants *(NEC Article 502)*. These are areas where, under normal operating conditions, there may be enough combustible dust in the air to produce explosive or ignitable mixtures.
- *Class III* – Areas that are hazardous because of the presence of easily ignitable fibers or other particles in the air, although not in large enough quantities to produce ignitable mixtures *(NEC Article 503)*. Class III locations include cotton mills, rayon mills, and clothing manufacturing plants.

NEC Articles 511 and 514 regulate garages and fuel dispensing locations where volatile or flammable liquids are used. While these areas are not always considered critically hazardous locations, there may be enough danger to require special precautions in the electrical installation. In these areas, the *NEC®* requires that volatile gases be confined to an area not more than eighteen inches above the floor or tank. See *NEC Table 514.3(B)(2)*. So in most cases, conventional raceway systems are permitted above this level. If the area is judged to be critically hazardous, explosion-proof wiring (including seal-offs) may be required.

NEC Article 520 regulates theaters and similar occupancies where fire and panic can cause hazards to life and property. Projection rooms and adjacent areas must be properly ventilated and wired for the protection of operating personnel and others using the area.

NEC Chapter 5 also covers floating buildings, marinas, agricultural buildings, service stations, bulk storage plants, health care facilities, mobile homes and parks, and temporary installations.

4.6.0 *NEC Chapter 6, Special Equipment*

NEC Article 600 covers electric signs and outline lighting. *NEC Article 610* applies to cranes and hoists. *NEC Article 620* covers the majority of the electrical work involved in the installation and operation of elevators, dumbwaiters, escalators, and moving walks. The manufacturer is responsible for most of this work. The electrician usually just furnishes a feeder terminating in a disconnect means in the bottom of the elevator shaft. The electrician may also be responsible for a lighting circuit to a junction box midway in the elevator shaft for connecting the elevator cage lighting cable and exhaust fans. The articles in this chapter list most of the requirements for these installations.

NEC Article 630 regulates electric welding equipment. It is normally treated as a piece of industrial power equipment requiring a special power outlet, but there are special conditions that apply to the circuits supplying welding equipment. These are outlined in detail in this chapter.

NEC Article 640 covers wiring for sound recording and similar equipment. This type of equipment normally requires low-voltage wiring. Special outlet boxes or cabinets are usually provided with the equipment, but some items may be mounted in or on standard outlet boxes. Some sound recording systems require direct current. It is supplied from rectifying equipment, batteries, or motor generators. Low-voltage alternating current comes from relatively small transformers connected on the primary side to a 120V circuit within the building.

Other items covered in *NEC Chapter 6* include X-ray equipment *(NEC Article 660)*, induction and dielectric heat-generating equipment *(NEC Article 665)*, industrial machinery *(NEC Article 670)*, swimming pools and fountains *(NEC Article 680)*, and solar photovoltaic (PV) systems *(NEC Article 690)*.

4.7.0 *NEC Chapter 7, Special Conditions*

In most commercial buildings, the *NEC®* and local ordinances require a means of lighting public rooms, halls, stairways, and entrances. There must be enough light to allow the occupants to exit from the building if the general building lighting is interrupted. Exit doors must be clearly indicated by illuminated exit signs.

NEC Chapter 7 covers the installation of emergency and legally required standby systems. These circuits should be arranged so that they can automatically transfer to an alternate source

On Site

Different Interpretations

For as many trainees in class with you today, there will be as many different interpretations of the *NEC®*. However, a difference of opinion is not always a problem—discussing the *NEC®* will allow you to expand your own understanding of it. In your previous discussion about *NEC Section 501.15*, what differences of interpretation did you have to resolve?

of current, usually storage batteries or gasoline-driven generators. As an alternative in some types of occupancies, you can connect them to the supply side of the main service, so disconnecting the main service switch would not disconnect the emergency circuits. This chapter also covers fire alarms and a variety of other equipment, systems, and conditions that are not easily categorized elsewhere in the *NEC®*.

4.8.0 *NEC Chapter 8, Communications Systems*

NEC Chapter 8 is a special category for wiring associated with electronic communications systems including telephone, radio and TV, satellite dish, network-powered broadband systems, and community antenna systems.

4.9.0 Examples of Navigating the *NEC®*

Using your copy of the 2014 *NEC®*, follow along with these sample scenarios to familiarize yourself with the layout of the *NEC®*.

4.9.1 Installing Type SE Cable

Suppose you are installing Type SE (service-entrance) cable on the side of a home. You know that this cable must be secured, but you are not sure of the spacing between cable clamps. To find out this information, use the following procedure:

Step 1 Look in the *NEC® Table of Contents* and follow down the list until you find an appropriate category. (Or you can use the index at the end of the book.)

Step 2 *NEC Article 230* will probably catch your eye first, so turn to the page where it begins.

Step 3 Scan down through the section numbers until you come to *NEC Section 230.51, Mounting Supports.* Upon reading this section, you will find in paragraph *(A) Service-Entrance Cables* that "service-entrance cables shall be supported by straps or other approved means within 300 mm (12 in) of every service head, gooseneck, or connection to a raceway or enclosure and at intervals not exceeding 750 mm (30 in)."

After reading this section, you will know that a cable strap is required within 12 inches of the service head and within 12 inches of the meter base. Furthermore, the cable must be secured in between these two termination points at intervals not exceeding 30 inches.

4.9.2 Installing Track Lighting

Assume that you are installing track lighting in a residential occupancy. The owners want the track located behind the curtain of their sliding glass patio doors. To determine if this is an *NEC®* violation, follow these steps:

Step 1 Look in the *NEC® Table of Contents* and find the chapter that contains information about the general application you are working on. *NEC Chapter 4, Equipment for General Use,* covers track lighting.

Step 2 Now look for the article that fits the specific category you are working on. In this case, *NEC Article 410* covers luminaires, lampholders, and lamps.

Step 3 Next locate the section within *NEC Article 410* that deals with the specific application. For this example, refer to *Part XIV, Lighting Track.*

Step 4 Turn to the page listed.

Step 5 Read *NEC Section 410.2, Definitions,* to become familiar with track lighting. Then go to *NEC Section 410.151* and read the information contained therein. Note that paragraph *(C) Locations Not Permitted* under *NEC Section 410.151* states the following: "Lighting track shall not be installed in the following locations: (1) where likely to be subjected to physical damage; (2) in wet or damp locations; (3) where subject to corrosive vapors; (4) in storage battery rooms; (5) in hazardous (classified) locations; (6) where concealed; (7) where extended through walls or partitions; (8) less than 1.5 m (5 ft) above the finished floor except where protected from physical damage or track operating at less than 30 volts rms open-circuit voltage; (9) where prohibited by *NEC Section 410.10(D)*."

Step 6 Read *NEC Section 410.151(C)* carefully. Do you see any conditions that would violate any *NEC®* requirements if the track lighting is installed in the area specified? In checking these items, you will probably note condition (6), "where concealed." Since the track lighting is to be installed behind a curtain, this sounds like an *NEC®* violation. You need to check further.

Step 7 You need the *NEC®* definition of *concealed*. Therefore, turn to *NEC Article 100, Definitions* and find the main term *concealed*. It reads: "**Concealed**. Rendered inaccessible by the structure or finish of the building."

On Site

NFPA Codes and Standards

In addition to the *NEC®*, NFPA also publishes many other codes and standards. Two that are of interest to the electrician are *NFPA 70E®, Standard for Electrical Safety in the Workplace* and *NFPA 70B, Recommended Practice for Electrical Equipment Maintenance. NFPA 70E®* provides direction on the safe installation, operation, and maintenance of electrical equipment. *NFPA 70B* provides direction for performing electrical tests, inspection, and maintenance procedures. Both documents are excellent resources for the electrical professional.

Step 8 Although the track lighting may be out of sight if the curtain is drawn, it will still be readily accessible for maintenance. Consequently, the track lighting is really not concealed according to the *NEC®* definition.

When using the *NEC®* to determine electrical installation requirements, keep in mind that you will nearly always have to refer to more than one section. Sometimes the *NEC®* itself refers the reader to other articles and sections. In some cases, the user will have to be familiar enough with the *NEC®* to know what other sections pertain to the installation at hand. It can be a confusing situation, but time and experience in using the *NEC®* will make it much easier. A pictorial road map of some *NEC®* topics is shown in *Figure 3*.

5.0.0 Other Organizations

In support of the requirements of the *NEC®* there are other organizations that identify conformance standards for the manufacture and/or use of electrical products.

5.1.0 Nationally Recognized Testing Laboratories

Nationally Recognized Testing Laboratories (NRTLs) are product safety certification laboratories. These laboratories perform extensive testing of new products to make sure they are built to established standards for electrical and fire safety. They establish and operate product safety certification programs to make sure that items produced under the service are safeguarded against reasonably foreseeable risks. NRTLs maintain a worldwide network of field representatives who make unannounced visits to factories to check products bearing their safety marks.

Think About It

Conformance and Electrical Equipment

What other resources are available for finding information about the use of electrical equipment and materials?

5.2.0 National Electrical Manufacturers Association

The **National Electrical Manufacturers Association (NEMA)** was founded in 1926. It is made up of companies that manufacture equipment used for generation, transmission, distribution, control, and utilization of electric power. The objectives of NEMA are to maintain and improve the quality and reliability of products; to ensure safety standards in the manufacture and use of products; and to develop product standards covering such matters as naming, ratings, performance, testing, and dimensions. NEMA participates in developing the *NEC®* and advocates its acceptance by state and local authorities.

Think About It

Putting It All Together

Look around you at the electrical components and products used and the quality of the work. Do you see any components or products that have not been listed or labeled? If so, how might these devices put you in harm's way? Do you see any code violations?

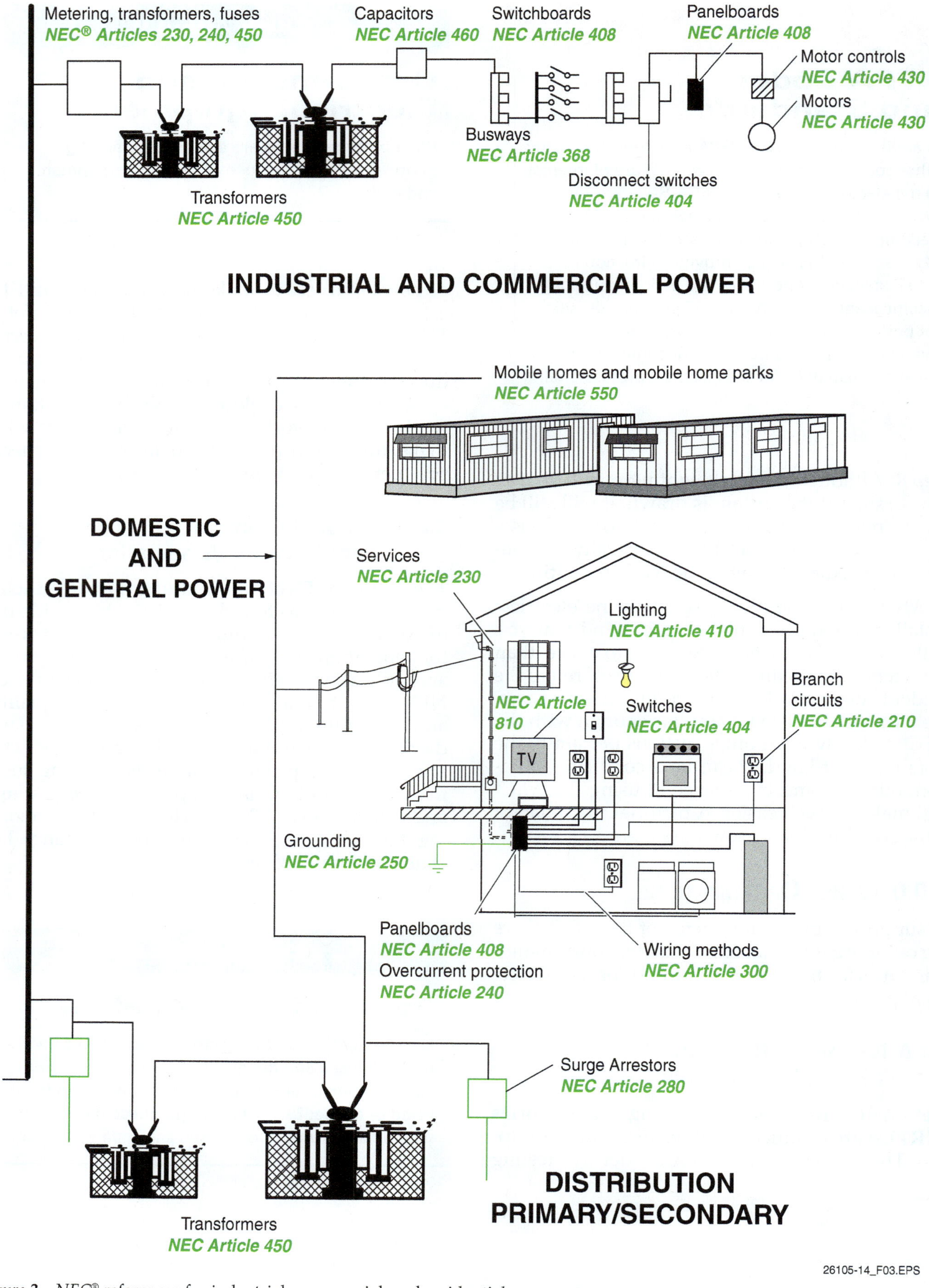

Metering, transformers, fuses
NEC® Articles 230, 240, 450

Capacitors
NEC Article 460

Switchboards
NEC Article 408

Panelboards
NEC Article 408

Motor controls
NEC Article 430

Motors
NEC Article 430

Busways
NEC Article 368

Disconnect switches
NEC Article 404

Transformers
NEC Article 450

INDUSTRIAL AND COMMERCIAL POWER

Mobile homes and mobile home parks
NEC Article 550

DOMESTIC AND GENERAL POWER

Services
NEC Article 230

Lighting
NEC Article 410

Branch circuits
NEC Article 210

NEC Article 810

TV

Switches
NEC Article 404

Grounding
NEC Article 250

Panelboards
NEC Article 408
Overcurrent protection
NEC Article 240

Wiring methods
NEC Article 300

Surge Arrestors
NEC Article 280

DISTRIBUTION PRIMARY/SECONDARY

Transformers
NEC Article 450

26105-14_F03.EPS

Figure 3 NEC® references for industrial, commercial, and residential power.

SUMMARY

The *NEC*® specifies the minimum provisions necessary for protecting people and property from hazards arising from the use of electricity and electrical equipment. As an electrician, you must be aware of how to use and apply the *NEC*® on the job site. Using the *NEC*® will help you to safely install and maintain the electrical equipment and systems you come into contact with.

Review Questions

1. What word or phrase best describes the *NEC*® requirements for the installation of electrical systems?
   a. Minimum
   b. Most stringent
   c. Design specification
   d. Complete

2. All of the following groups are usually represented on the Code Making Panels *except* _____.
   a. trade associations
   b. electrical inspectors
   c. insurance organizations
   d. government lobbyists

3. Mandatory and permissive rules are defined in _____.
   a. *NEC Article 90*
   b. *NEC Article 100*
   c. *NEC Article 110*
   d. *NEC Article 200*

4. The general design and installation of electrical systems is covered in _____.
   a. *NEC Chapters 1, 2, and 7*
   b. *NEC Chapters 1, 2, 3, and 4*
   c. *NEC Chapters 6, 7, and 8*
   d. *NEC Chapters 5, 6, 7, and 9*

5. Devices such as radios, televisions, and telephones are covered in _____.
   a. *NEC Chapter 8*
   b. *NEC Chapter 7*
   c. *NEC Chapter 6*
   d. *NEC Chapter 5*

6. Examples of branch circuit calculations can be found in _____.
   a. *NEC Informative Annex A*
   b. *NEC Informative Annex C*
   c. *NEC Informative Annex D*
   d. *NEC Informative Annex G*

7. *NEC Article 110* covers _____.
   a. branch circuits
   b. definitions
   c. general requirements for electrical installations
   d. wiring design and protection

8. Cable trays are covered in _____.
   a. *NEC Article 330*
   b. *NEC Article 342*
   c. *NEC Article 368*
   d. *NEC Article 392*

Questions 9 through 12 are two-part questions that ask you to first locate an article in the *NEC*®, and then answer a question using the article.

9. Where in the *NEC*® would you find information on conductors for general wiring?
   a. *NEC Article 280*
   b. *NEC Article 310*
   c. *NEC Article 404*
   d. *NEC Article 340*

10. Within the article found in Question 9, locate information on stranded conductors. What size conductors must be stranded when installed in raceways?
    a. 14 AWG or larger
    b. 12 AWG or larger
    c. 10 AWG or larger
    d. 8 AWG or larger

11. Where in the *NEC*® would you find information on switchboards and panelboards?
    a. *NEC Article 285*
    b. *NEC Article 300*
    c. *NEC Article 408*
    d. *NEC Article 540*

12. Within the article found in Question 11, find the section that covers switchboard clearances. Assuming the switchboard does not have a noncombustible shield, what is the minimum clearance between the top of the switchboard and a combustible ceiling?
    a. 12 inches
    b. 18 inches
    c. 24 inches
    d. 36 inches

13. Installation procedures for luminaires are provided in _____.
    a. *NEC Article 410*
    b. *NEC Article 408*
    c. *NEC Article 366*
    d. *NEC Article 460*

14. Theaters are covered in _____.
    a. *NEC Article 338*
    b. *NEC Article 110*
    c. *NEC Article 430*
    d. *NEC Article 520*

15. *NEC Article 600* covers _____.
    a. track lighting
    b. electric signs and outline lighting
    c. X-ray equipment
    d. emergency lighting systems

Trade Terms Quiz

Fill in the blank with the correct term that you learned from your study of this module.

1. ___________ are the main topics of the *NEC®*.
2. Nine ___________ form the broad structure of the *NEC®*.
3. Certain articles in the *NEC®* are subdivided into lettered ___________.
4. Parts and articles are subdivided into numbered ___________.
5. Although they follow the applicable sections of the *NEC®*, ___________ allow alternative methods to be used under specific conditions.
6. A(n) ___________ is explanatory material that follows specific *NEC®* sections.
7. The ___________ is an organization that maintains and improves the quality and reliability of electrical products.
8. The ___________ publishes the *NEC®*; it also develops standards to minimize the possibility and effects of fire.
9. ___________ are organizations that are responsible for testing and certifying electrical equipment.

Trade Terms

Articles
Chapters
Exceptions
Informational Note
National Electrical Manufacturers Association (NEMA)
National Fire Protection Association (NFPA)
Nationally Recognized Testing Laboratories (NRTLs)
Parts
Sections

Module 26105-14
Supplemental Exercises

1. The *NEC*® is made up of ___________ chapters.

2. The *NEC*® provides the ___________ requirements for the installation of electrical systems.

3. *NEC Chapter 2* covers ___________.
   a. wiring and protection
   b. lighting
   c. occupancy
   d. emergency

4. Which of the following covers electric motors?
   a. *NEC Article 230*
   b. *NEC Article 330*
   c. *NEC Article 430*
   d. *NEC Article 130*

5. *NEC Article 520* covers ___________ and similar occupancies where fire and panic can cause hazards to life and property.

6. To size grounding conductors, refer to ___________.
   a. *NEC Chapter 1*
   b. *NEC Chapter 2*
   c. *NEC Chapter 3*
   d. *NEC Chapter 4*

7. Wiring methods are covered in ___________.
   a. *NEC Chapter 1*
   b. *NEC Chapter 2*
   c. *NEC Chapter 3*
   d. *NEC Chapter 4*

8. Gas stations are covered in ___________.
   a. *NEC Chapter 5*
   b. *NEC Chapter 6*
   c. *NEC Chapter 7*
   d. *NEC Chapter 8*

9. Emergency systems are covered in ___________.
   a. *NEC Chapter 6*
   b. *NEC Chapter 7*
   c. *NEC Chapter 8*
   d. *NEC Chapter 9*

10. Fire alarm systems are covered in ___________.
   a. *NEC Chapter 6*
   b. *NEC Chapter 7*
   c. *NEC Chapter 8*
   d. *NEC Chapter 9*

11. Definitions are covered in ___________.

12. Metric units of measurement are covered in ___________.

13. X-ray equipment is covered in ___________.

    a. *NEC Article 665*
    b. *NEC Article 660*
    c. *NEC Article 670*
    d. *NEC Article 650*

14. ___________ contains tables that specify the properties of conductors.

    a. *NEC Chapter 8*
    b. *NEC Chapter 9*
    c. *NEC Chapter 7*
    d. *NEC Chapter 6*

15. Electric signs are covered in ___________.

    a. *NEC Chapter 8*
    b. *NEC Chapter 7*
    c. *NEC Chapter 6*
    d. *NEC Chapter 9*

Christine Thorstensen Porter

Intertek Testing Services

How did you choose a career in the electrical field?
It was "in the blood," literally. My father had his own electrical contracting company and was an instructor for our apprenticeship/training program. One day, my husband, who also worked for my dad, gave me a tool belt and sent me through the Construction Industry Training Council's (CITC) apprenticeship program in Seattle, Washington. Since my father taught many of the classes, the pressure was on. I felt that I had to perform well. My father was a major influence on me both as an electrician, and as a student leader. Eventually, I went into the estimating side of things in my dad's business.

In the fourth year of my apprenticeship program I was taking a lot of estimating and project management courses and applying the skills I gained in the field. My dad's business was high-end residential and some small commercial work. We'd have a crew on site from six months to a year until the projects were complete. A lot of the installations we did were data systems and computer-controlled security systems. When I took the Administrator's exam, that gave me my qualifications to supervise.

What positions have you held in the industry?
In 1981 I was a teaching assistant. That teacher started a new review class for electricians needing to pass journeyman and administrative exams; I took over his second-year apprenticeship/trainee class. To better prepare myself for my new teaching role, I took various classes offered by the International Association of Electrical Inspectors (IAEI), but I was dismayed and disappointed with the instructor because he merely read the *NEC*® to us, instead of giving us any insights as to how the *NEC*® could impact our installations, or why any particular *NEC*® rule was in place. So I also ended up getting involved with teaching classes for the IAEI.

My experience with IAEI gave me a solid grounding in the Code. It also led to more local involvement. In fact, I'm now the chairman of the city's Electrical Code Advisory Committee. I have also served on the Technical Advisory Committee for our state amendments to the *NEC*®.

What does your current job involve?
I teach the second-year electrical training program at CITC. I teach courses for IAEI. I perform inspections for Intertek Testing Services, and I serve on a variety of committees. In addition to chairing the Electrical Code Advisory Committee, I'm a Senior Associate member of the Puget Sound Chapter of IAEI; I sit on the city of Seattle's Construction Codes Advisory Board and am a subject matter expert for NCCER's *Contren*® Electrical curriculum.

Do you have any advice for someone just entering the trade?
I'd tell them that the electrical trade is extremely rich and diverse. There is so much that you can do. From installing all different types of systems, to project management, and to estimating, the field is wide open. The job never stays dull. The field changes with all the changes in technology. Because of that, there are so many new and emerging jobs within the trade. As far as becoming a success, I'd say to keep abreast of changes—changes in the Code and in technology—and stay connected. A good way to do that is to join associations such as IAEI. Network and continue your training. It never ends in this field. Also, spend time mentoring others. You learn more that way, by giving yourself.

Trade Terms Introduced in This Module

Articles: The articles are the main topics of the *NEC®*, beginning with *NEC Article 90, Introduction,* and ending with *NEC Article 840, Premises-Powered Broadband Communications Systems.*

Chapters: Nine chapters form the broad structure of the *NEC®*.

Exceptions: Exceptions follow the applicable sections of the *NEC®* and allow alternative methods to be used under specific conditions.

Informational Note: Explanatory material that follows specific *NEC®* sections.

National Electrical Manufacturers Association (NEMA): The association that maintains and improves the quality and reliability of electrical products.

National Fire Protection Association (NFPA): The publishers of the *NEC®*. The NFPA develops standards to minimize the possibility and effects of fire.

Nationally Recognized Testing Laboratories (NRTLs): Product safety certification laboratories that are responsible for testing and certifying electrical equipment.

Parts: Certain articles in the *NEC®* are subdivided into parts. Parts have letter designations (e.g., Part A).

Sections: Parts and articles are subdivided into sections. Sections have numeric designations that follow the article number and are preceded by a period (e.g., 501.4).

Additional Resources

This module presents thorough resources for task training. The following resource material is suggested for further study.

National Electrical Code® Handbook, Latest Edition. Quincy, MA: National Fire Protection Association.

Figure Credits

Associated Builders and Contractors, Inc., Module opener

John Traister, Figure 3

Topaz Publications, Inc., SA01, SA02, SA04

Tim Ely, SA03

NCCER CURRICULA — USER UPDATE

NCCER makes every effort to keep its textbooks up-to-date and free of technical errors. We appreciate your help in this process. If you find an error, a typographical mistake, or an inaccuracy in NCCER's curricula, please fill out this form (or a photocopy), or complete the online form at **www.nccer.org/olf**. Be sure to include the exact module ID number, page number, a detailed description, and your recommended correction. Your input will be brought to the attention of the Authoring Team. Thank you for your assistance.

Instructors – If you have an idea for improving this textbook, or have found that additional materials were necessary to teach this module effectively, please let us know so that we may present your suggestions to the Authoring Team.

NCCER Product Development and Revision
13614 Progress Blvd., Alachua, FL 32615

Email: curriculum@nccer.org
Online: www.nccer.org/olf

❑ Trainee Guide ❑ AIG ❑ Exam ❑ PowerPoints Other ____________________

Craft / Level: Copyright Date:

Module ID Number / Title:

Section Number(s):

Description:

Recommended Correction:

Your Name:

Address:

Email: Phone:

Device Boxes

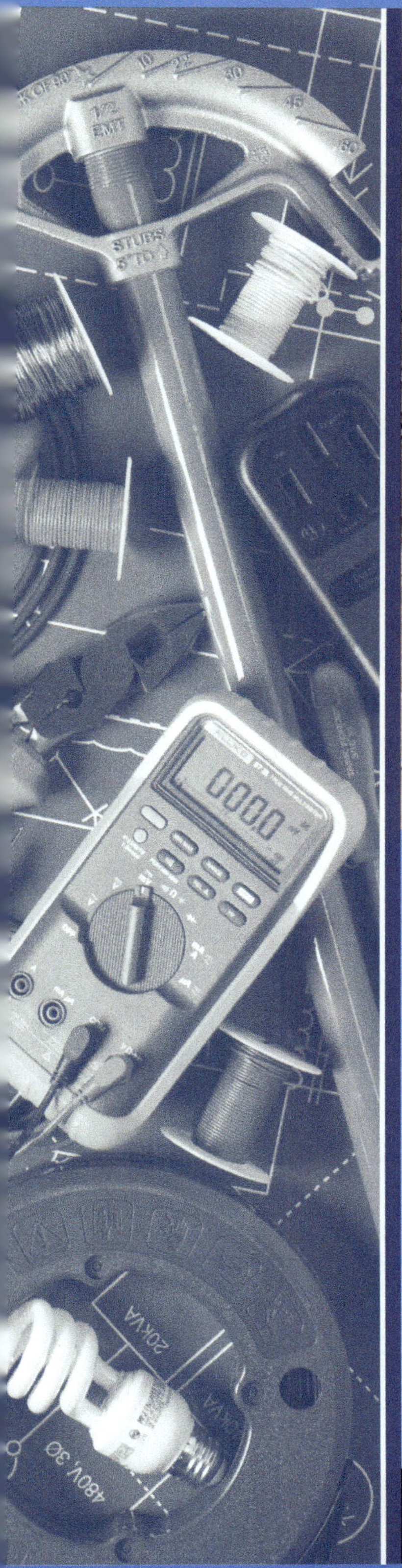

Shell Chemical Norco Subsystems Project

The Automation Group (TAG) and Yokogawa Corporation of America Industrial Specialty Contractors performed a complete reinstrumentation, replacing old-fashioned pneumatic instruments and controls with modern electronics and fiber optics at Shell Chemical's Norco facility.

26106-14

Trainees with successful module completions may be eligible for credentialing through NCCER's National Registry. To learn more, go to **www.nccer.org** or contact us at **1.888.622.3720.** Our website has information on the latest product releases and training, as well as online versions of our *Cornerstone* magazine and Pearson's product catalog.

Your feedback is welcome. You may email your comments to **curriculum@nccer.org,** send general comments and inquiries to **info@nccer.org,** or fill in the User Update form at the back of this module.

This information is general in nature and intended for training purposes only. Actual performance of activities described in this manual requires compliance with all applicable operating, service, maintenance, and safety procedures under the direction of qualified personnel. References in this manual to patented or proprietary devices do not constitute a recommendation of their use.

26106-14

Device Boxes

Objectives

When you have completed this module, you will be able to do the following:

1. Describe the different types of nonmetallic and metallic boxes.
2. Calculate the *NEC*® fill requirements for boxes under 100 cubic inches.
3. Identify the appropriate box type and size for a given application.
4. Select and demonstrate the appropriate method for mounting a given box.

Performance Tasks

Under the supervision of the instructor, you should be able to do the following:

1. Identify the appropriate box type and size for a given application.
2. Select the minimum size pull or junction box for the following applications:
   - Conduit entering and exiting for a straight pull.
   - Conduit entering and exiting at an angle.

Trade Terms

Connector
Explosion-proof
Handy box
Junction box
Outlet box
Pull box
Raintight
Waterproof
Watertight
Weatherproof

Required Trainee Materials

1. Pencil and paper
2. Appropriate personal protective equipment
3. Copy of the latest edition of the *National Electrical Code*®

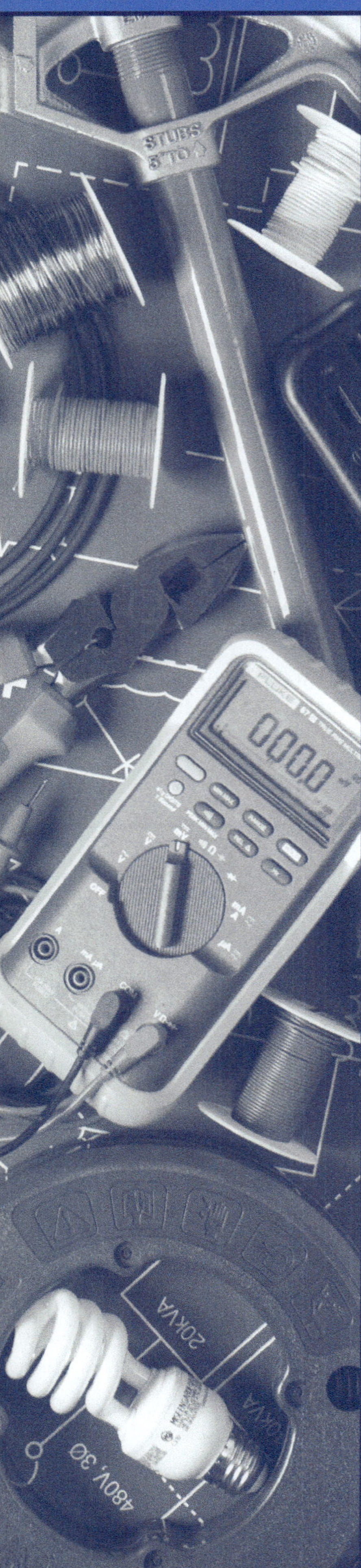

Contents

Topics to be presented in this module include:

Figures

1.0.0 Introduction

On every job, boxes are required for outlets, switches, pull boxes, and junction boxes. All of these must be sized, installed, and supported to meet current *NEC®* requirements. Since the *NEC®* limits the number of conductors, fittings, and devices allowed in each outlet or switch box according to its size, you must install boxes that are large enough to accommodate the number of conductors that must be spliced in the box or fed through it. Therefore, a knowledge of the various types of boxes and the volume of each is essential.

Besides being able to calculate the required box sizes, you must also know how to select the proper type of box for any given application. For example, metallic boxes used in concrete deck pours are different from those used as device boxes in residential or commercial buildings. Boxes used for the support of lighting fixtures or for securing devices in outdoor installations will be different from the two types just mentioned. Boxes for use in certain hazardous locations will further differ in construction; many must be rated as being explosion-proof.

You must also know what fittings are available for terminating the various wiring methods in these boxes.

Electrical drawings rarely indicate the exact types of outlet boxes to be used in a given area, with the possible exception of boxes used in hazardous locations. Electricians who lack practical on-the-job experience may not always choose the best box for a given application. The use of improper boxes and other ill-adapted materials will cause excessive time to be taken on the job. For example, outlet boxes for use with only Type AC cable should contain built-in clamps; many times boxes will be ordered with knockouts only. This latter case requires the use of additional connectors, which may only take a few additional seconds per connection, but when these are added up over the period of a large project, much additional time is wasted. Because the cost of labor is an expensive item, any excess labor required will more than offset any savings gained from the use of inadequate materials.

Labor is often wasted due to obstructions that enter raceways during the general construction work. Most of these obstructions can be avoided by plugging all raceway openings with capped bushings or similar means of protection (*Figure 1*). Care should also be taken to thoroughly tighten all fittings, couplings, and so on. All openings in boxes must be closed per *NEC Section 314.17(A)*. Knockout closures are used for this purpose (*Figure 1*).

On Site

Outlet Boxes in Poured Concrete

When installing outlet boxes in poured concrete structures, stuff the inside of the boxes with newspaper and/or cover the openings with duct tape and use capped bushings to keep concrete out of the raceway system.

Figure 1 Conduit caps and knockout closures.

2.0.0 Types of Boxes

Outlet boxes normally fall into three categories:

- Pressed steel boxes with knockouts of various sizes for raceway or cable entrances
- Cast iron, aluminum, or brass boxes with threaded hubs of various sizes and locations for raceway entrances
- Nonmetallic boxes

On Site

Special Considerations

Before you can effectively plan and route conductors to their termination points and then select and install the correct types and sizes of boxes, you will need to have the following information:

- Length and number of conductors
- Conductor ampacity
- Allowances for voltage drops
- Environment

Pressed steel boxes also fall into two categories:

- Boxes with conduit, electric metallic tubing, and cable
- Boxes designed for use with specific types of surface metal raceways

Outlet boxes vary in size and shape depending upon their use, the size of the raceway, the number of conductors entering the box, the type of building construction, the atmospheric conditions of the area, and special requirements.

Outlet box covers are usually required to adapt the box to the particular use it is to serve. For example, a 4" square box is adapted to one-gang or two-gang switches or receptacles by the use of either one-gang or two-gang flush device covers. A one-gang cast hub box can be adapted to provide a vapor-proof switch or a vapor-proof receptacle cover. Special outlet box hangers are available to facilitate their installation, particularly in frame building construction.

The types of enclosures used as outlet and device boxes for the support of fixtures, or for securing devices such as switches, receptacles, or other equipment on the same yoke or strap, are available in various sizes and shapes. These enclosures may be used in the one-gang, two-gang, three-gang, or four-gang types. Ceiling outlet boxes are available in various shapes. Device boxes are those that are usually installed to support receptacles and switches.

Boxes installed for the support of a lighting fixture are required to be listed for the purpose. Most device boxes are typically not designed or listed for use to support lighting fixtures. The use of a device box to support fixtures is addressed in *NEC Section 314.27(A)*.

A floor box that is listed specifically for installation in a floor is required where receptacles or junction boxes are installed in a floor. Listed floor boxes are provided with covers and gaskets to exclude surface water and cleaning compounds.

A box used at fan outlets is not permitted to be used as the sole support for ceiling (paddle) fans, unless it is listed for the application as the sole means of support. Where a ceiling fan does not exceed 70 pounds in weight, it is permitted to be supported by outlet boxes listed and identified for such use. Boxes designed to support more than 35 pounds must be marked with the maximum weight to be supported. See *NEC Section 314.27(C)*. These boxes must be rigidly supported from a structural member of the building. A paddle fan box and its related accessories are shown in *Figure 2*.

Per *NEC Section 314.27(D)*, boxes used for support of utilization equipment other than ceiling-suspended (paddle) fans shall meet the requirements of *NEC Section 314.27(A)* for support of a luminaire that is the same size and weight.

On Site

Identifying Boxes

After installing all the boxes in a complex installation, you can save yourself much time and trouble later on if you label each box location according to type. Using a can of spray paint, paint small dots on the floor below each box. Use one dot for a plug receptacle, two dots for a switch box, and three dots for a wall light. After the drywall installers are finished, you'll be able to easily identify any wall box.

2.1.0 Octagon and Round Boxes

Octagon boxes are available with knockouts for use with either conduit (using locknuts and bushings) or cable box connectors. They are also available with both Type AC and NM cable clamps. The standard width of octagon boxes is four inches, with depths available in 1¼", 1½", or 2⅛". Extension rings are also available for increasing the depth.

Round boxes are available in the same dimensions, but *NEC Section 314.2* prevents using such boxes where conduit or connectors—requiring locknuts and bushings—are connected to the side of the box. The conduit or box connector must terminate in the top of such boxes. Round boxes with cable clamps, however, are permitted for use with Type AC and NM cables that may terminate in either the side or top of the box. Nonmetallic round boxes are permitted only with open wiring on insulators, concealed knob-and-tube wiring, Type NM cable, and nonmetallic raceways.

Figure 3 shows typical metallic octagon boxes and an octagon extension ring. *Figure 3(A)* shows a box with concentric knockouts for conduit or box connectors, while *Figure 3(B)* shows a box that utilizes cable clamps. *Figure 3(C)* shows the extension ring. *Figure 4* shows a nonmetallic round box

On Site

Locating Boxes on Walls

The Uniform Building Code (UBC) requires boxes on opposite sides of walls between occupancies to be separated by at least 24" in order to maintain the fire rating. (See ***NEC Section 300.21, Informational Note.***)

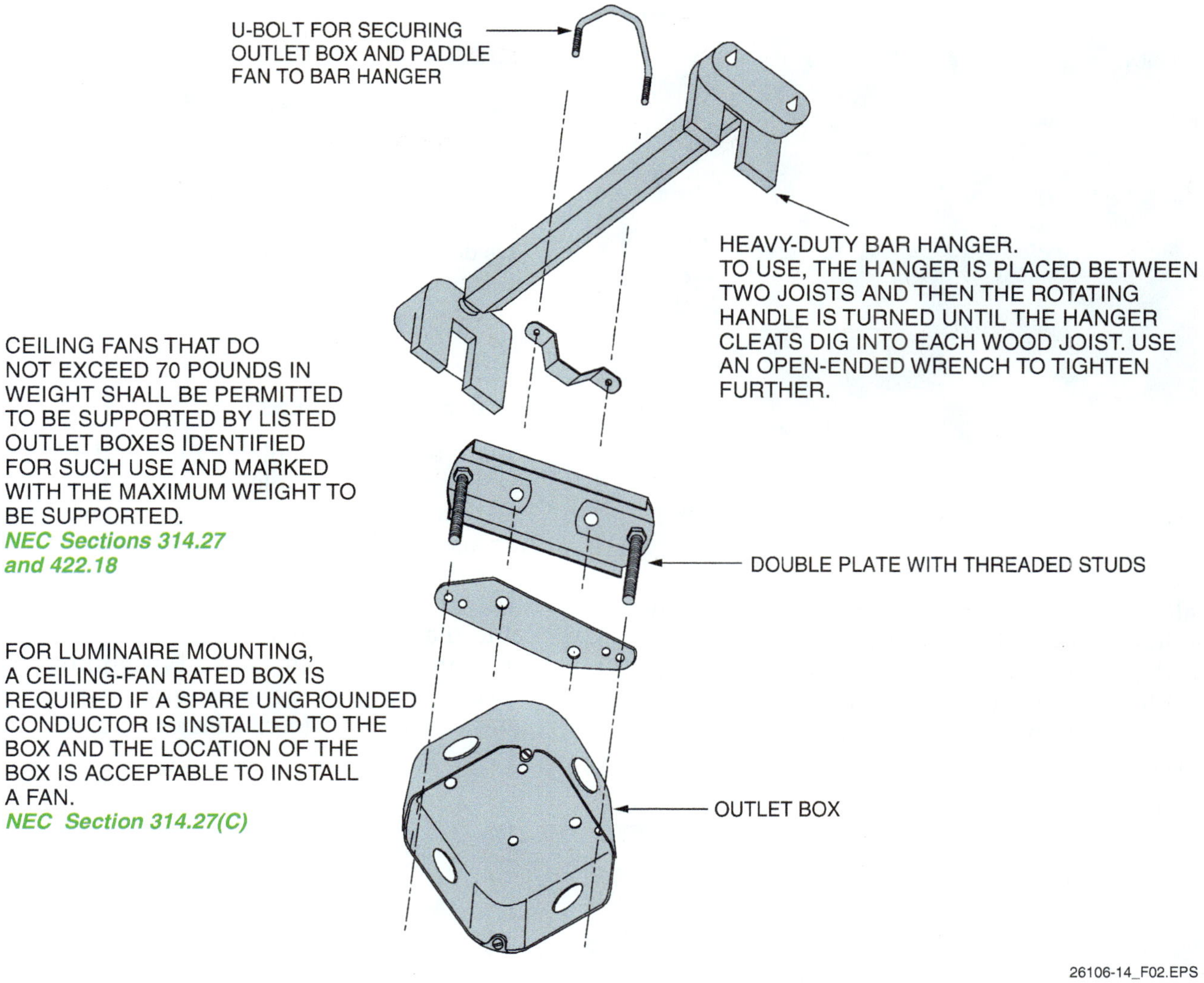

U-BOLT FOR SECURING OUTLET BOX AND PADDLE FAN TO BAR HANGER

HEAVY-DUTY BAR HANGER. TO USE, THE HANGER IS PLACED BETWEEN TWO JOISTS AND THEN THE ROTATING HANDLE IS TURNED UNTIL THE HANGER CLEATS DIG INTO EACH WOOD JOIST. USE AN OPEN-ENDED WRENCH TO TIGHTEN FURTHER.

CEILING FANS THAT DO NOT EXCEED 70 POUNDS IN WEIGHT SHALL BE PERMITTED TO BE SUPPORTED BY LISTED OUTLET BOXES IDENTIFIED FOR SUCH USE AND MARKED WITH THE MAXIMUM WEIGHT TO BE SUPPORTED. *NEC Sections 314.27 and 422.18*

DOUBLE PLATE WITH THREADED STUDS

FOR LUMINAIRE MOUNTING, A CEILING-FAN RATED BOX IS REQUIRED IF A SPARE UNGROUNDED CONDUCTOR IS INSTALLED TO THE BOX AND THE LOCATION OF THE BOX IS ACCEPTABLE TO INSTALL A FAN. *NEC Section 314.27(C)*

OUTLET BOX

26106-14_F02.EPS

Figure 2 Typical fan box for installation in an existing ceiling.

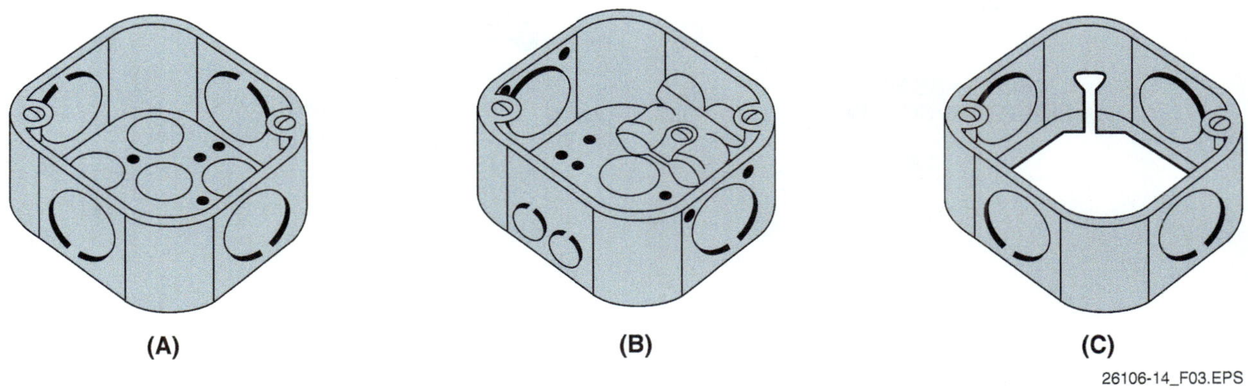

Figure 3 Typical octagon boxes and octagon extension ring.

and fixture ring with a bar hanger for mounting between studs or joists.

Octagon and round boxes are used for wall-mounted lighting fixtures (luminaires). However, covers are available for octagon boxes that will support receptacles and switches. Blank covers are also available when the box is used as a junction box.

Figure 5 shows a cross section of a 3½" round shallow box used for supporting lighting fixtures that have integral wire termination space. *NEC Section 314.24(A)* requires that this and

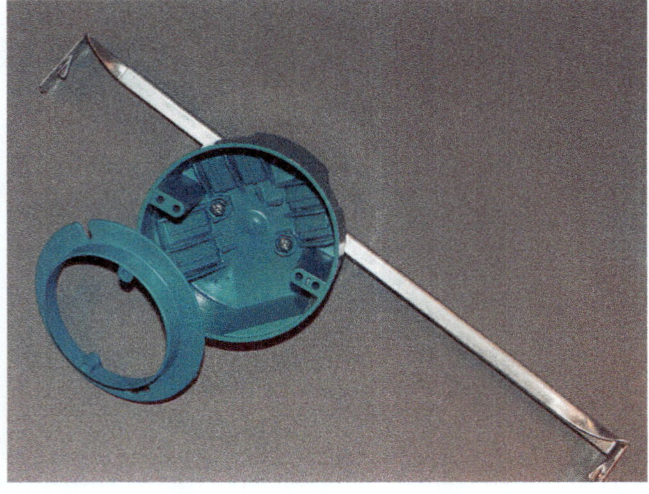

26106-14_F04.EPS

Figure 4 Nonmetallic round box and fixture ring with bar hanger.

all boxes have a minimum depth of ½". Boxes intended to enclose flush devices must not have a depth of less than 15⁄16" for conductors No. 14 AWG and smaller, 1³⁄16" for conductors No. 12 or 10 AWG, and 2¹⁄16" for conductors No. 8, 6, or 4 AWG per *NEC Section 314.24(B)*.

On Site

Use of Round Boxes Is Limited

Round boxes are not to be used in cases where conduit must enter from the side of the box because it is difficult to make a good connection with a locknut or bushing on a rounded surface.

2.2.0 Square Boxes

Square boxes are available in 4" and 4¹¹⁄16" square sizes. Both are available in depths of 1¼", 1½", and 2⅛". Extension rings are also available to further increase the depth. These boxes are available with or without mounting brackets (for fastening to structural members). Boxes designed for use with cable may have either Type AC or NM clamps for securing the cable at the box entrance points. Square boxes may be used with a single or two-gang device ring (e.g., plaster ring or tile ring)

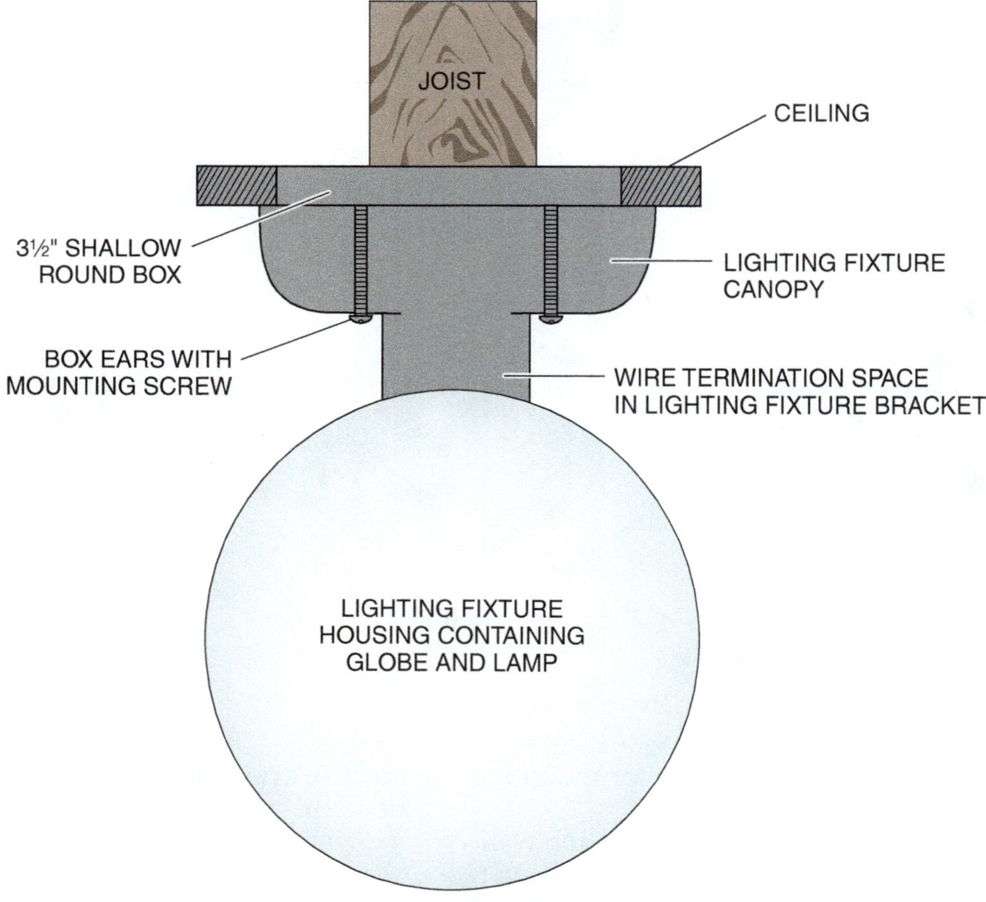

26106-14_F05.EPS

Figure 5 Shallow round box used for mounting a lighting fixture.

for mounting receptacles or switches. A ring with a round opening is also available for mounting lighting fixtures. Blank covers are available when the boxes are used as junction boxes.

Figure 6 shows a square box with a rectangular extension ring that is used to bring the box to the finished surface. Square boxes can be used as junction boxes, and when the number of conductors warrants more capacity than is available in other types of boxes.

2.3.0 Device Boxes

Device boxes are designed for flush mounting mainly in residential and some commercial applications. They are available with or without cable clamps and brackets for mounting to wooden structural members. This type of box is also available with plaster ears for installation in finished wall partitions. Three types of device boxes are shown in *Figure 7*. As you can see, some boxes include integral nails for direct installation into wall studs. Boxes used in metal stud environments are installed using self-tapping sheet metal screws, while those installed in concrete may require the use of special powder-actuated fasteners. These materials are discussed in more detail in a later module.

A special single-gang box with the mounting ears on the inside of the box is called a **handy box** (also known as a utility box). Such boxes are available in depths of 1½", 1⅞", and 2⅛". Care must be exercised when using these boxes as their limited volume restricts the number of conductors permitted in the box.

26106-14_F06.EPS

Figure 6 Square box with extension ring.

2.4.0 Masonry Boxes

Special boxes known as masonry or concrete boxes are used in flat-slab construction jobs. These boxes consist of a sleeve with external ears and a plate that is attached after the sleeve is nailed to the deck.

Masonry boxes are manufactured in different heights and care should be taken to use boxes of sufficient height to allow the knockouts to come well above the reinforcing rods. This eliminates the need for offsets in the conduit where it enters a box. *Figure 8* shows a practical application of a masonry box.

On Site

Knockouts

When completing building renovations, it may be difficult to remove a knockout from a previously installed box without dislodging the box. One way to do it is to drill into the knockout and partially insert a self-tapping screw. Then use diagonal or side-cut pliers to pull the knockout from the box.

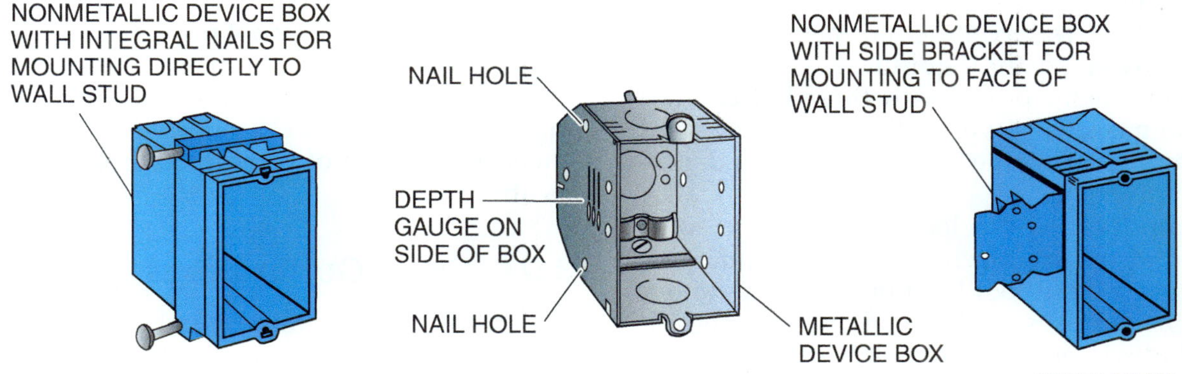

Figure 7 Typical device boxes.

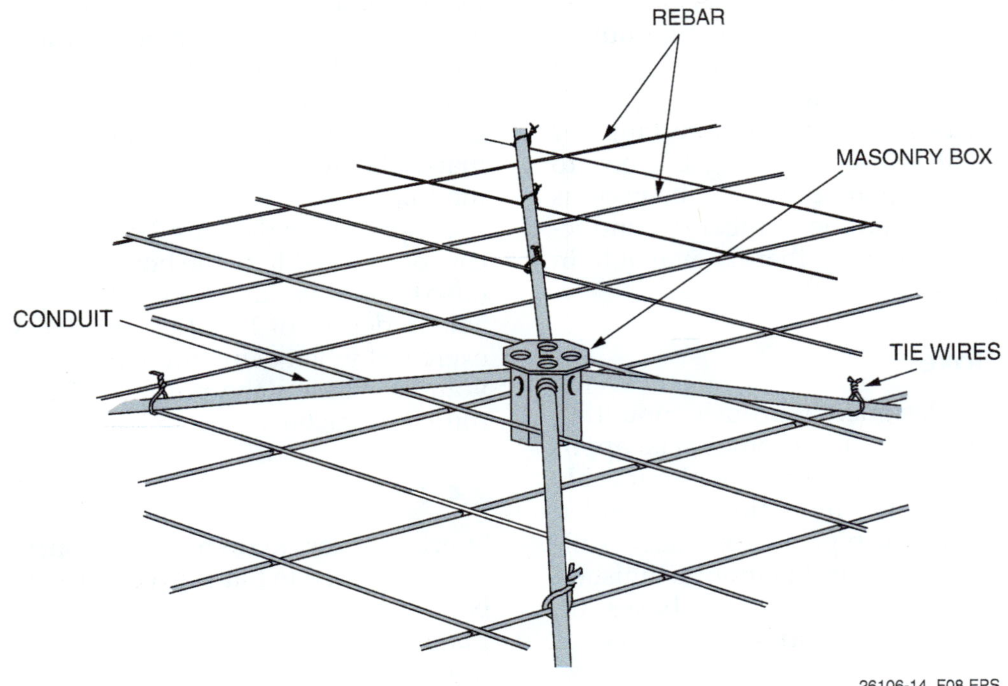

Figure 8 Masonry box.

2.5.0 Boxes for Damp and Wet Locations

In damp or wet locations, boxes and fittings must be placed or equipped to prevent moisture or water from entering and accumulating within the box or fitting. It is recommended that approved boxes of nonconductive material be used with nonmetallic sheathed cable or approved nonmetallic conduit when the cable or conduit is used in locations where there is likely to be occasional moisture present. Boxes installed in wet locations must be listed for such use per *NEC Section 314.15*.

A wet location is any location subject to saturation with water or other liquids, such as locations exposed to weather or water, washrooms, garages, and interiors that might be hosed down. Underground installations or those in concrete slabs or masonry in direct contact with the earth must be considered to be wet locations. **Raintight**, **waterproof**, or **watertight** equipment (including fittings) may satisfy the requirements for **weatherproof** equipment. Boxes with threaded conduit hubs and gasketed covers will normally prevent water from entering the box except for condensation within the box.

A damp location is a location subject to some degree of moisture. Such locations include partially protected outdoor locations—such as under canopies, marquees, and roofed open porches. It also includes interior locations subject to moderate degrees of moisture—such as some basements, some barns, and cold storage warehouses.

On Site

Outdoor Boxes

Outdoor wiring must be able to resist the entry of water. Outdoor boxes are either drip tight, which means sealed against falling water from above, or watertight, which means sealed against water from any direction. Drip-tight boxes simply have lids that deflect rain; they are not waterproof. Watertight boxes are sealed with gaskets to prevent the entry of water from any angle.

Weatherproof covers for outdoor receptacles must be chosen with care. All 15A and 20A 125–250V receptacles in wet locations require a cover that is weatherproof regardless of whether or not the receptacle is in use. Other receptacles may or may not require these covers, depending on the rating. *NEC Sections 406.9(A) and (B)* cover installation of receptacles in damp or wet locations.

3.0.0 Sizing Outlet Boxes

In general, the maximum number of conductors permitted in standard outlet boxes is listed in *NEC Table 314.16(A)*. These figures apply where no fittings or devices such as fixture studs, cable

clamps, switches, or receptacles are contained in the box and where no grounding conductors are part of the wiring within the box. Obviously, in all modern residential wiring systems there will be one or more of these items contained in the outlet box. Therefore, where one or more of the above-mentioned items are present, the number of conductors is reduced by one less than that shown in the table for each type of fitting and by two for each device strap. For example, a deduction of two conductors must be made for each strap containing a device such as a switch or duplex receptacle; a further deduction of one conductor shall be made for one or more grounding conductors entering the box. For example, a 3" × 2" × 3½" box is listed in the table as containing a maximum number of eight No. 12 wires. If the box contains a cable clamp and a duplex receptacle, three wires will have to be deducted from the total of eight—providing for only five No. 12 wires. If a ground wire is used, only four No. 12 wires may be used, which might be the case when a three-wire cable with ground is used to feed a three-way wall switch. Also, each looped conductor over 12" used in the box counts as two conductors.

A pictorial definition of stipulated conditions as they apply to *NEC Section 314.16(A)* is shown in *Figures 9* through *11*. *Figure 9* illustrates an assortment of raised covers and outlet box extensions. These components, when combined with the appropriate outlet boxes, serve to increase the usable space. Each type is marked with its cubic inch capacity, which may be added to the figures in *NEC Table 314.16(A)* to calculate the increased number of conductors allowed.

Figure 10 shows typical wiring configurations, which must be counted as conductors when calculating the total capacity of outlet boxes. A wire passing through the box without a splice or tap is counted as one conductor. Therefore, a cable containing two wires that passes in and out of an outlet box with a splice or tap is counted as two conductors. However, a wire that enters a box and is either spliced or connected to a terminal, and then exits again, is counted as two conductors. In the case of two cables that each have two wires, the total conductors counted will be four. Wires that enter and terminate in the same box are counted as individual conductors and in this case, the total count would be two conductors. Remember, when one or more grounding wires enter the box and are joined, a deduction of only one is required, regardless of their number.

Further components that require deduction adjustments from those specified in *NEC Table 314.16(A)* include fixture studs, hickeys, and fixture stud extensions *[NEC Section 314.16(B)(3)]*. One conductor must be deducted from the total for each type of fitting used. Two conductors must be deducted for each strap-mounted device, such

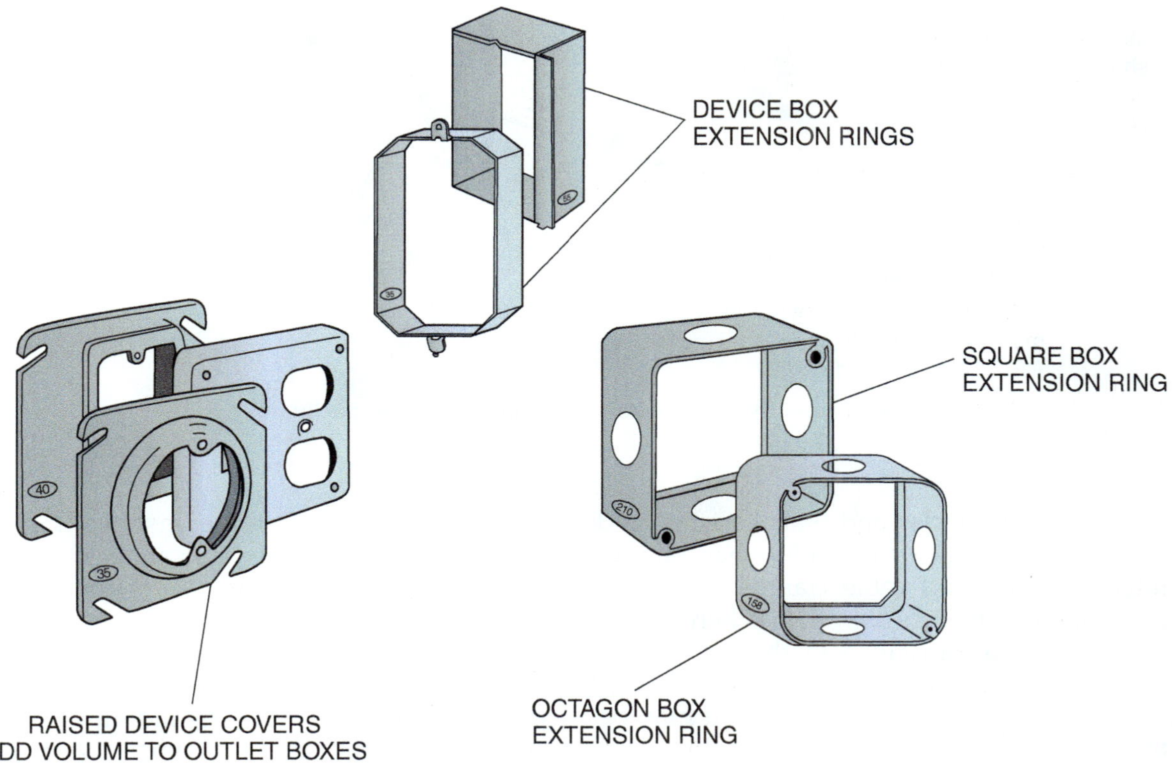

Figure 9 Devices or components that add to outlet box capacity.

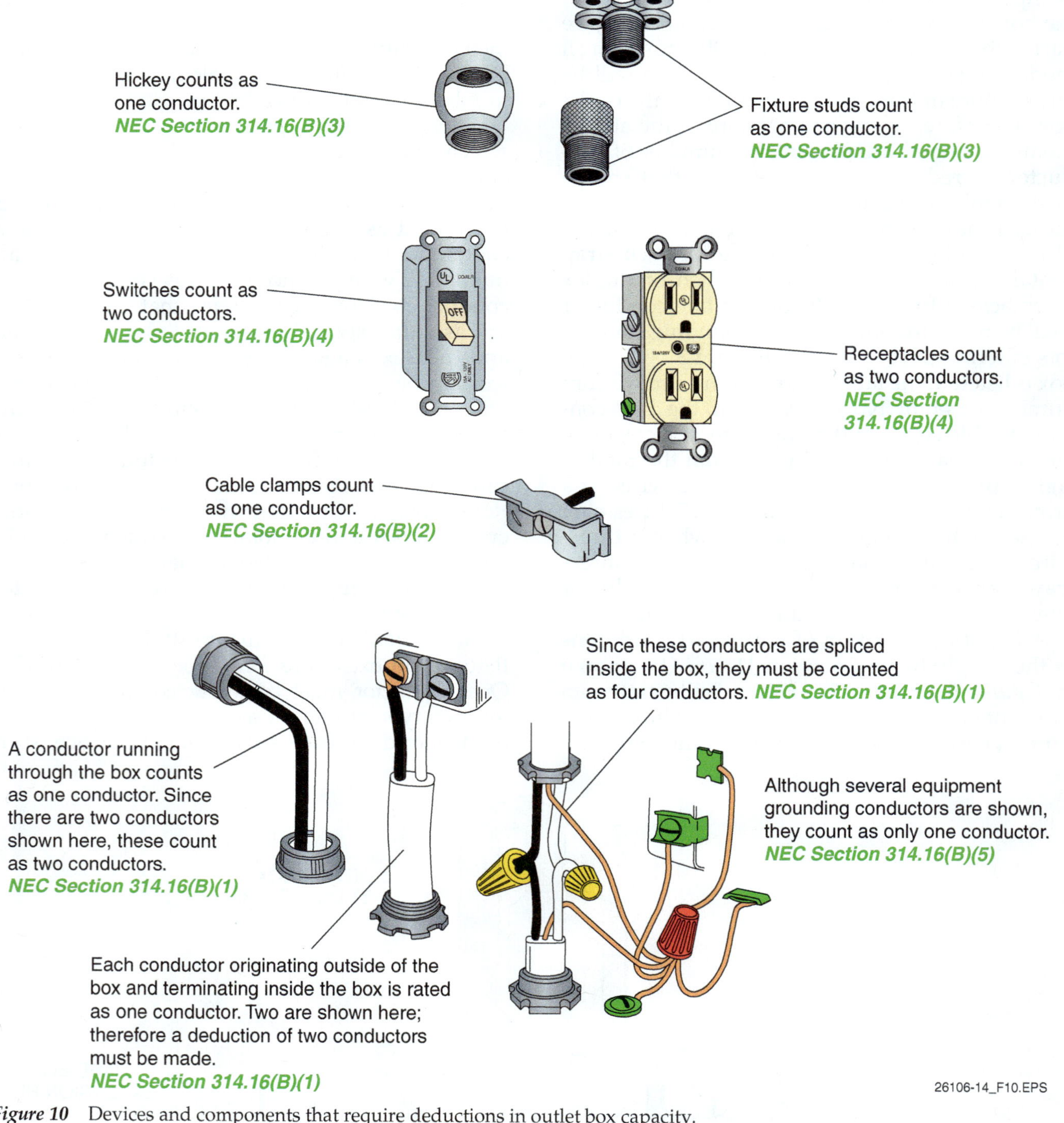

Figure 10 Devices and components that require deductions in outlet box capacity.

as duplex receptacles and wall switches; a deduction of one conductor is made when one or more internally mounted cable clamps are used *[NEC Section 314.16(B)(2) and (4)]*. When mixed conductor sizes are installed in the box, these items must be counted as the largest wire size in the box.

Figure 11 shows components that may be used in outlet boxes without affecting the total number of conductors. Such items include grounding clips and screws, wire nuts, and cable connectors when the latter are inserted through knockout holes in the outlet box and secured with locknuts. Prewired fixture wires are not counted against the total number of allowable conductors in an outlet box; neither are conductors originating and ending in the box, such as pigtails.

To better understand how outlet boxes are sized, use the following as an example: Two No. 12 AWG conductors are installed in ½" EMT and terminate into a metallic outlet box

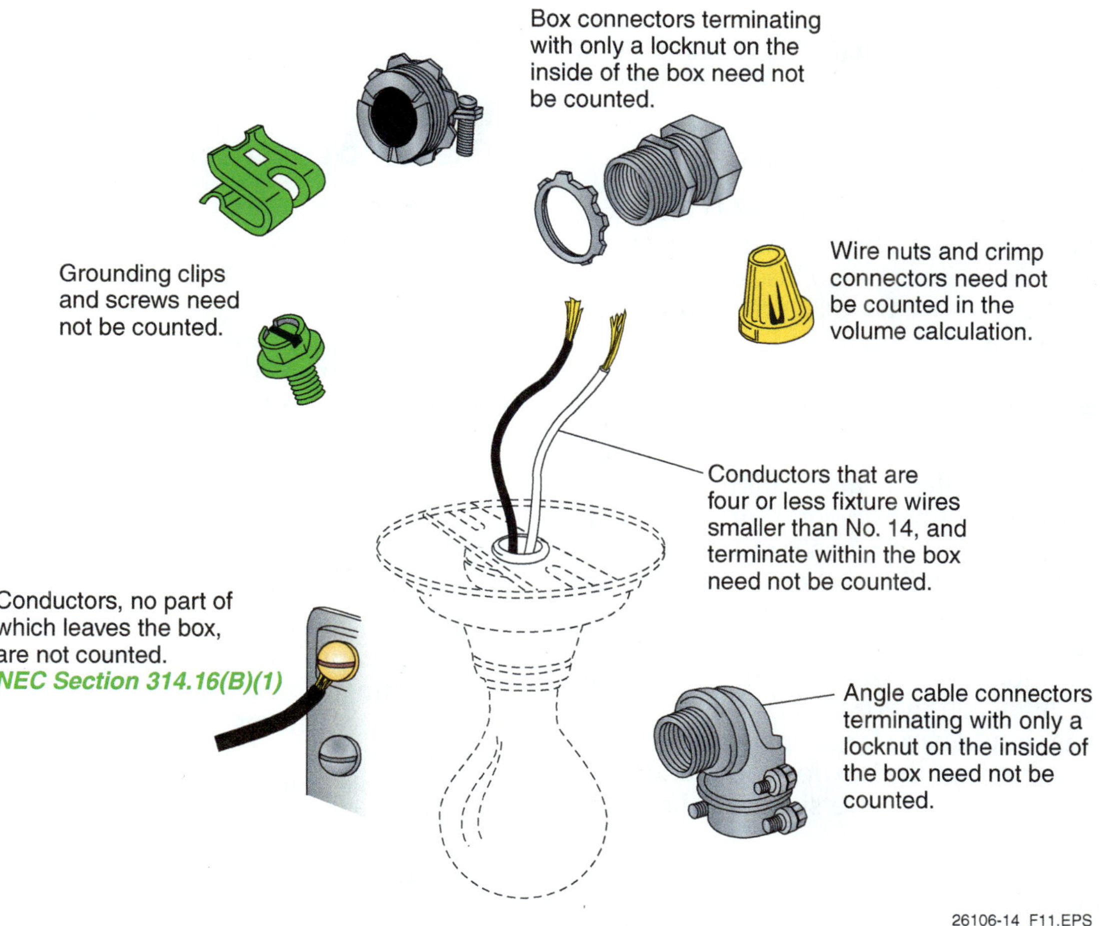

Figure 11 Items that may be disregarded when calculating outlet box capacity.

containing one duplex receptacle. What size outlet box will meet *NEC®* requirements?

The first step is to count the total number of conductors and equivalents that will be used in the box *(NEC Section 314.16)*.

On Site

Allocating Space in a Box

Here is a summary of the *NEC®* rules for determining the capacity of a box. Add the number of conductor equivalents indicated for each device:

Each conductor	1
Each looped conductor over 12"	2
Each strap containing a device	2
Each fixture stud or hickey	1
One or more grounding conductors	1
One or more cable clamps	1

Also add one extra allowance for an additional grounding conductor when it is an isolating ground conductor.

Step 1 Calculate the total number of conductors and their equivalents:

One receptacle = 2
Two #12 conductors = 2
Total #12 conductors = 4

Step 2 Determine the amount of space required for each conductor. *NEC Table 314.16(B)* gives the box volume required for each conductor:

No. 12 AWG = 2.25 cubic inches

Step 3 Calculate the outlet box space required by multiplying the number of cubic inches required for each conductor by the number of conductors found in Step 1 above.

$4 \times 2.25 = 9.00$ cubic inches

Step 4 Once you have determined the required box capacity, again refer to *NEC Table 314.16(A)* and note that a 3" × 2" × 2" box comes closest to our requirements. This box size is rated for 10.0 cubic inches.

For another example, if four No. 12 conductors enter the box, two additional No. 12 conductors must be added to our previous count for a total of six conductors.

$$6 \times 2.25 = 13.5 \text{ cubic inches}$$

Again, refer to *NEC Table 314.16(A)* and note that a 3" × 2" × 2¾"device box with a rated capacity of 14.0 cubic inches is the closest device box that meets *NEC*® requirements. Of course, any box with a larger capacity is permitted.

4.0.0 Pull and Junction Boxes

Pull and junction boxes are provided in an electrical installation to facilitate the installation of conductors, or to provide a junction point for the connection of conductors, or both. In some instances, the location and size of pull boxes are designated on the drawings. In most cases, however, the electricians on the job will have to determine the proper number, location, and sizes of pull or junction boxes to facilitate conductor installation.

Pull boxes should be as large as possible. Workers need space within the box for both hands and in the case of the larger wire sizes, workers will need room for their arms to feed the wire. *NEC Sections 314.28(A) through (E)* specify that pull and junction boxes must provide adequate space and dimensions for the installation of conductors. For raceways containing conductors of No. 4 or larger, and for cables containing conductors of No. 4 or larger, the minimum dimensions of pull or junction boxes installed in a raceway or cable run shall comply with the following:

- In straight pulls, the length of the box shall not be less than eight times the trade diameter of the largest raceway.
- Where angle or U pulls are made, the distance between each raceway entry inside the box and the opposite wall of the box shall not be less than six times the trade diameter of the largest raceway in a row. This distance shall be increased for additional entries by the amount of the sum of the trade diameters of all other raceway entries in the same row on the same wall of the box. Each row shall be calculated individually, and the single row that provides the maximum distance shall be used.
- Also where angle or U pulls are made, the distance between raceway entries enclosing the same conductor shall not be less than six times the trade diameter of the larger raceway.
- When transposing cable size into raceway size, the minimum trade size raceway required for the number and size of conductors in the cable shall be used.

On Site

Metal Boxes Must Be Grounded

Metal boxes are good conductors. Therefore, when metal boxes are used, they must be grounded to the circuit grounding system.

Long runs of conductors should not be made in one pull. Pull boxes, installed at convenient intervals, will relieve much of the strain on the conductors. The length of the pull, in many cases, is left to the judgment of the workers or their supervisor, and the condition under which the work is installed.

The installation of pull boxes may seem to cause a great deal of extra work and trouble, but they save a considerable amount of time and hard work when pulling conductors. Properly placed, they eliminate bends and elbows and do away with the necessity of fishing from both ends of a conduit run.

If possible, pull boxes should be installed in a location that allows electricians to work easily and conveniently. For example, in an installation where the conduit comes up a corner of a wall and changes direction at the ceiling, a pull box that is installed too high will force the electrician to stand on a ladder when feeding conductors, and will allow no room for supporting the weight of the wire loop or for the cable-pulling tools.

Unless the contract drawings or project engineer state otherwise, it is just as easy for the pull boxes to be placed at a convenient height that allows workers to stand on the floor with sufficient room for both wire loop and tools.

In some electrical installations, a number of junction boxes must be installed to route the conduit in the shortest, most economical way. The *NEC*® requires all junction boxes to be readily accessible. This means that a person must be able to get to the conductors inside the box without removing plaster, wall covering, or any other part of the building.

Junction boxes or pull boxes must be securely fastened in place on walls or ceilings or adequately suspended.

While certain sizes of factory-constructed boxes are available with concentric knockouts, in many instances it will be necessary to have them custom built to meet the job requirements. When it is not possible to accurately anticipate the raceway entrance requirements, it will be necessary to cut the required knockouts on the job.

In the case of large pull boxes and troughs, shop drawings should be prepared prior to the

construction of these items with all required knockouts accurately indicated in relation to the conduit run requirements.

4.1.0 Sizing Pull and Junction Boxes

Figure 12 shows a junction box with four runs of conduit. Since this is a straight pull, and 4" conduit is the largest size in the group, the minimum length required for the box can be determined by the following calculation:

Trade size of conduit × 8 [per ***NEC Section 314.28(A)(1)***] = minimum length of box

or:

$$4" \times 8 = 32"$$

Therefore, this particular pull box must be at least 32" in length. The width of the box, however, need only be of sufficient size to enable locknuts and bushings to be installed on all the conduits or connectors entering the enclosure.

Junction or pull boxes in which the conductors are pulled at an angle (*Figure 13*) must have a distance of not less than six times the trade diameter of the largest conduit *[NEC Section 314.28(A)(2)]*. The distance must be increased for additional conduit entries by the amount of the sum of the diameter of all other conduits entering the box on the same side. The distance between raceway entries enclosing the same conductors must not be less than six times the trade diameter of the largest conduit.

Since the 4" conduit is the largest size in this case:

$$L_1 = 6 \times 4" + (3 + 2) = 29"$$

Since the same conduit runs are located on the adjacent wall of the box, L_2 is calculated in the same way; therefore, $L_2 = 29"$.

The distance (D) = 6 × 4" or 24". This is the minimum distance permitted between conduit entries enclosing the same conductor.

The depth of the box need only be of sufficient size to permit locknuts and bushings to be properly installed. In this case, a 6"-deep box would suffice.

On Site

Using Pull Boxes

Pull boxes make it easier to install conductors. They may also be installed to avoid having more than 360° worth of bends in a single run. (Remember, if a pull box is used, it is considered the end of the run for the purposes of the *NEC*® 360° rule.)

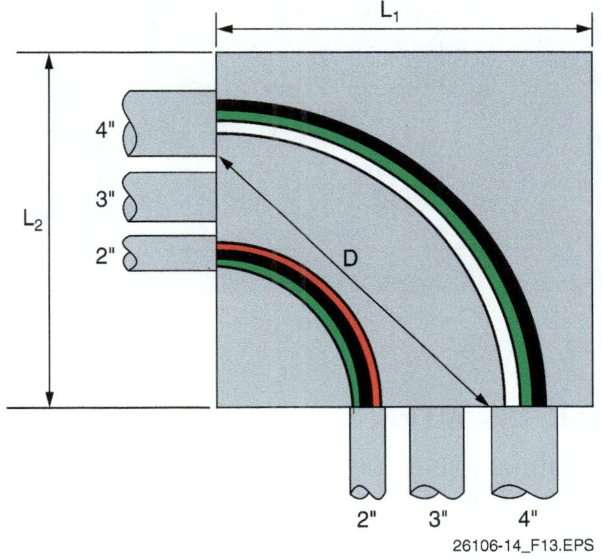

Figure 13 Pull box with conduit runs entering at right angles.

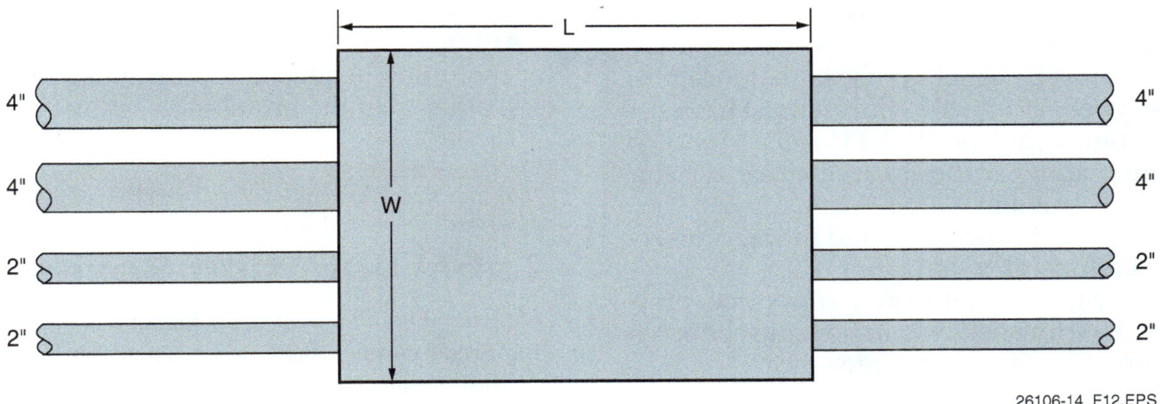

Figure 12 Pull box with two 4" and two 2" conduit runs.

On Site

Knockout Punch Kit

Often, conduit must enter boxes, cabinets, or panels that do not have precut knockouts. In these cases, a knockout punch can be used to make a hole for the conduit connection.

26106-14_SA01.EPS

5.0.0 Installing Boxes

In addition to box fill, there are a number of other considerations when installing boxes. These include additional *NEC®* requirements and making the actual wiring connections.

Some of the general *NEC®* requirements are as follows:

- The box selected must be listed for the given application (for example, a box used in a wet location must be listed for use in that location).
- As discussed previously, the box must have sufficient volume, as listed in *NEC Table 314.16(A)*, and must allow sufficient free space for conductors, as listed in *NEC Table 314.16(B)*.
- Conductors entering boxes and fittings must be protected from abrasion.
- Boxes must be installed and supported properly, and the finished installation must be accessible for later repair or maintenance.

You must also consider the type of box cover or canopy to be used, as well as the type of box (for example, pull/junction or outlet), and in some cases, the direction of the pull. Refer to *NEC Sections 314.15 through 314.30* for additional details. Some of the *NEC®* requirements for receptacle locations are discussed in the following section.

5.1.0 *NEC®* Requirements for Receptacle Locations

NEC Section 210.52 states the minimum requirements for the location of receptacles in dwelling units. It specifies that in each kitchen, family room, and dining room, receptacle outlets shall be installed so that no point along the floor line in any wall space is more than six feet, measured horizontally, from an outlet in that space, including any wall space two feet or more in width (three feet for foyers) and the wall space occupied by fixed panels in exterior walls, but excluding sliding panels (*Figure 14*). When spaced in this manner, a six-foot extension cord will reach a receptacle from any point along the wall line. Receptacle outlets shall, insofar as practicable, be spaced equal distances apart. Receptacle outlets in floors shall not be counted as part of the required number of receptacle outlets unless located within 18" of the wall.

The *NEC®* defines wall space as a wall that is unbroken along the floor line by doorways, fireplaces, or similar openings. Each wall space that is two feet or more in width must be treated individually and separately from other wall spaces within the room. This minimizes the use of cords across doorways, fireplaces, and similar openings.

At least one receptacle is required in each laundry area, within three feet of bathroom sinks, on the outside of the building at the front and back (GFCI protected), on any balcony, deck, or porch accessible from the inside of the house, in each basement, in each attached and detached garage, in each hallway ten feet or more in length, and at an accessible location for servicing any HVAC equipment.

Although no actual *NEC®* requirements exist for mounting heights and positioning of receptacles, other than the prohibition against mounting

On Site

Label Junction Boxes

It's a good idea to label every junction cover with the circuit number, the panel it came from, and its destination. The next person to service the installation will be grateful for this extra help.

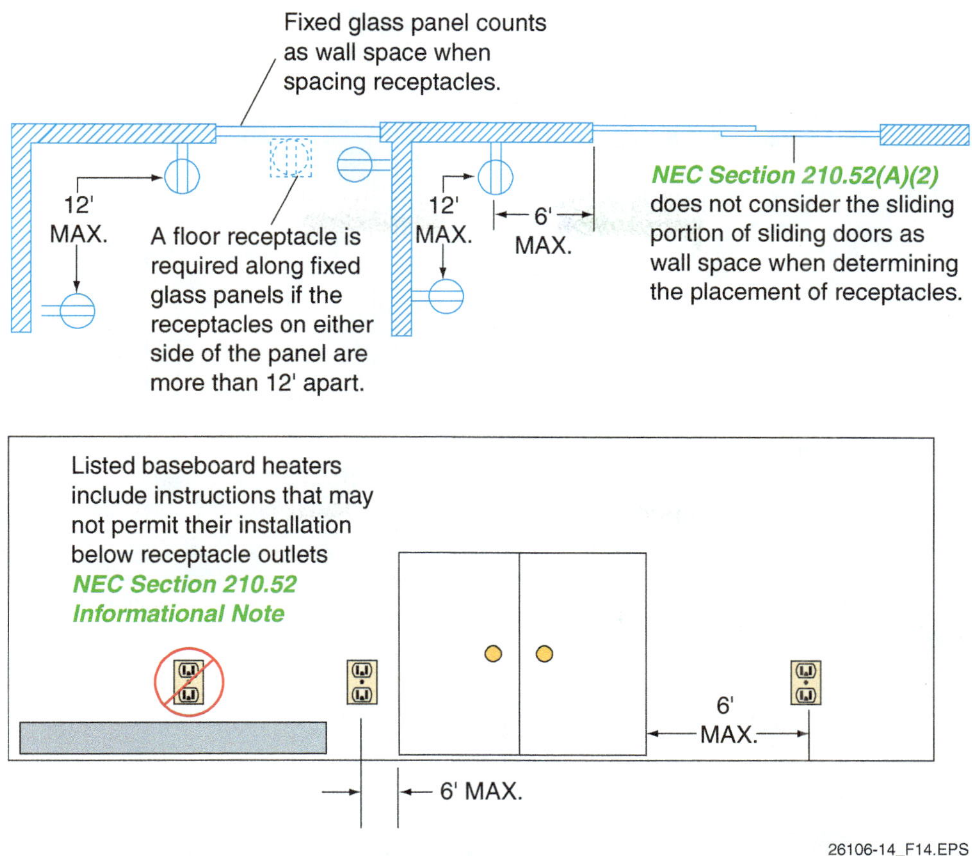

Figure 14 *NEC®* requirements for receptacle locations in dwelling units.

receptacles face up on countertops and similar work surfaces, there are certain *NEC®* requirements regarding receptacle placement. For example, *NEC Section 210.52(C)(5)* states that receptacles shall be not more than 20" above a kitchen countertop. Also, where allowed, receptacles may not be located more than 12" below the countertop surface. See *NEC Section 210.52(C)(5) Exception*. In addition to these *NEC®* guidelines, certain installation methods have become standard in the electrical industry. *Figure 15* shows common mounting heights of duplex receptacles used on conventional residential and small commercial installations. However, these dimensions are frequently varied to suit the building structure. For example, ceramic tile might be placed above a kitchen or bathroom countertop. If the dimensions in *Figure 15* put the receptacle part of the way out of the tile, the mounting height should be adjusted to place the receptacle either completely in the tile or completely out of the tile, as shown in *Figure 16*.

Refer again to *Figure 15* and note that the mounting heights are given to the bottom of the outlet box. Many dimensions on electrical drawings are given to the center of the outlet box or receptacle. However, during the actual installation, workers installing the outlet boxes can mount them more accurately (and in less time) by lining up the bottom of the box with a chalk mark rather than trying to eyeball this mark to the center of the box.

A decade or so ago, most electricians mounted receptacle outlets 12" from the finished floor to the center of the outlet box. However, a recent survey taken of over 500 homeowners shows that they prefer a mounting height of 15" from the finished floor to the bottom of the outlet box. It is easier to plug and unplug the cord assemblies at this height. However, always check the working drawings, written specifications, details of construction, and local codes for measurements that may affect the mounting height of a particular receptacle outlet. For example, those confined to wheelchairs may require more specific receptacle height locations to fit their individual needs.

NEC Section 314.20 requires all outlet boxes installed in walls or ceilings of concrete, tile, or other noncombustible material, such as plaster or drywall, to be installed in such a manner that the front edge of the box or fitting is not set back from the finished surface by more than ¼". Where walls and ceilings are constructed of wood or other combustible materials, outlet boxes and fittings must be flush with the finished surface of the wall.

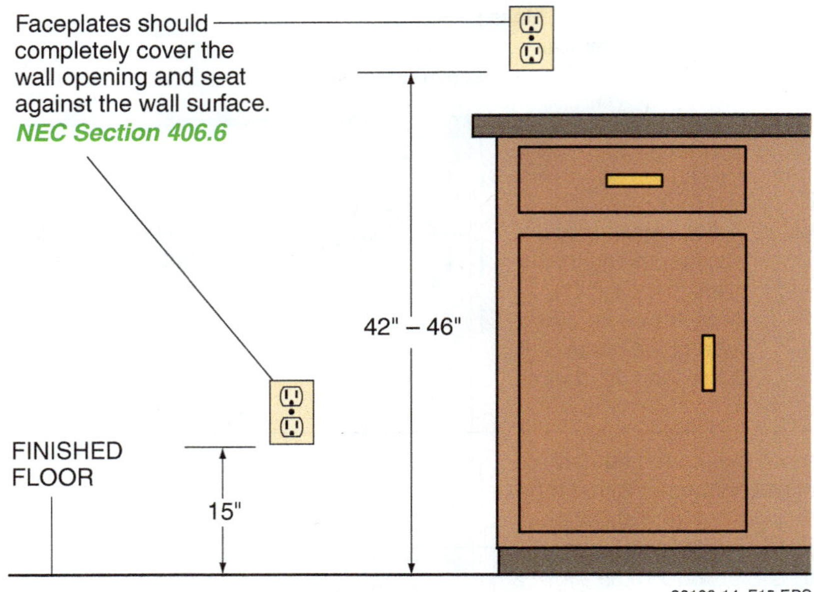

Figure 15 Mounting heights of duplex receptacles.

Figure 16 Adjusting mounting heights.

Wall surfaces such as drywall or plaster that contain wide gaps or are broken, jagged, or otherwise damaged, must be repaired so there will be no gaps or open spaces greater than ⅛" between the outlet box and the wall material (*NEC Section 314.21*). These repairs should be made prior to installing the faceplate. Such repairs are best made using a noncombustible caulking or spackling compound. See *Figure 17*.

Many industrial/commercial outlet boxes are surface-mounted and connect directly to the electrical raceway. These boxes are often mounted to the building steel or to a support channel or strut that is welded or clamped to the steel. The installation of outlet boxes or raceways should be coordinated so as not to interfere with piping, ductwork, fireproofing, and other items that may or may not already be in place.

5.2.0 Making Connections

Before any actual installation begins, study the electrical floor plan and consult with the builder or architect to ensure that no last-minute changes have been made in the electrical system. Install all boxes in accordance with the electrical drawings. The spacing should be as even as possible. The box center is the midpoint on the vertical dimension of the box. Make sure that you check the door swing direction so that the switches are not installed behind a door. Measure the height of the switch boxes from the floor so that they will be at

Gaps or openings around outlet box must not be greater than ⅛"; repair if necessary.
NEC Section 314.21

26106-14_F17.EPS

Figure 17 Gaps or openings around outlet boxes must be repaired.

the proper height when installed. After the boxes are installed, the wires must be spliced.

Insulated spring connectors, commonly called wire nuts or Wirenuts®, are solderless connectors made in various color-coded sizes that allow for splicing the hundreds of different solid or stranded wire combinations typically encountered in branch-circuit and fixture splicing applications. See *Figure 18*. Several varieties of wire nuts are available, but the following are the ones used most often:

- Those for use on wiring systems 300V and under
- Those for use on wiring systems 600V and under (1,000V in lighting fixtures and signs)

Some types of wire nuts have thin wings on each side of the connector to facilitate their installation. Wire nuts are normally made in sizes to accommodate conductors as small as No. 22 AWG up to as large as No. 10 AWG, with practically any combination of those sizes in between.

Always check the listing requirements on the package prior to use. Some wire nuts are only listed for specific applications, such as copper to copper only. The maximum temperature rating is 105°C (221°F).

The general procedure for splicing wires with wire nuts is as follows:

Step 1 Select the proper size wire nut to accommodate the wires being spliced. Wire nut packages contain charts that list the allowable combinations of wires by size. Refer

26106-14_F18.EPS

Figure 18 Wire nuts.

On Site

Making Connections

Conductor connections can be made using either push-in connectors or twist-on connectors, as shown here. Get in the habit of making your connections in the following order: first the grounding conductors, then the grounded conductors, and finally, the ungrounded conductors. If you approach each job in a systematic manner, it will soon become second nature to you and you'll be less likely to make mistakes.

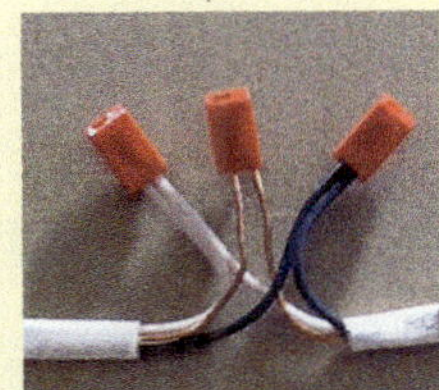

(A) NO. 14/2 NM-B CABLE WITH PUSH-IN WIRE CONNECTORS

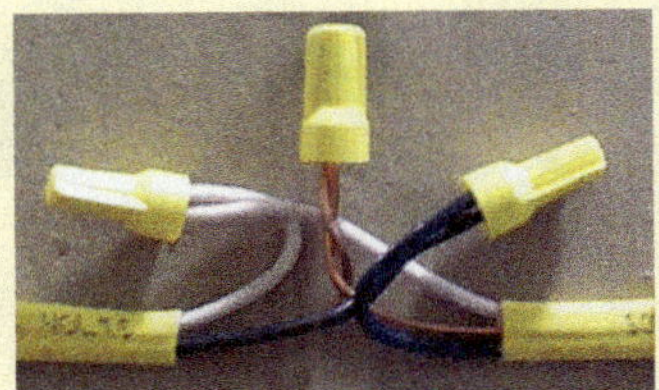

(B) NO. 12/2 NM-B CABLE WITH TWIST-ON WIRE NUTS

26106-14_SA02.EPS

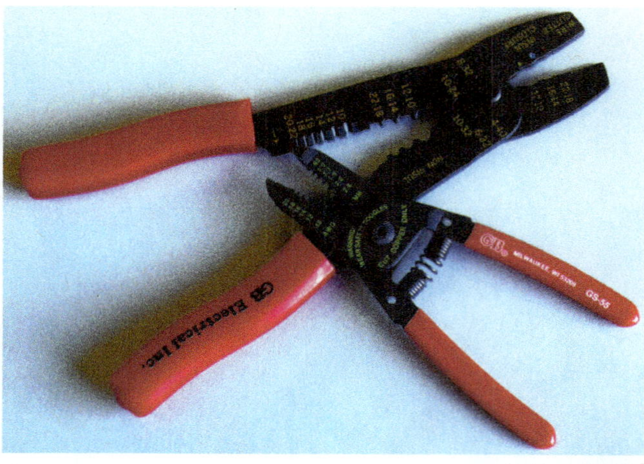

26106-14_F19.EPS

Figure 19 Stripping tools.

to the label on the wire nut box or container for this information.

Step 2 Select the appropriate tool (*Figure 19*), and then strip the insulation from the ends of the wires to be spliced. The length of insulation stripped off is typically about ½" (*Figure 20*), but may vary depending on the wire size and the wire nut being used. Follow the manufacturer's directions given on the wire nut package.

Step 3 Stick the ends of the wires into the wire nut and turn clockwise until tight. The wire nut draws the conductors and insulation into the body of the connector. See *Figure 21*.

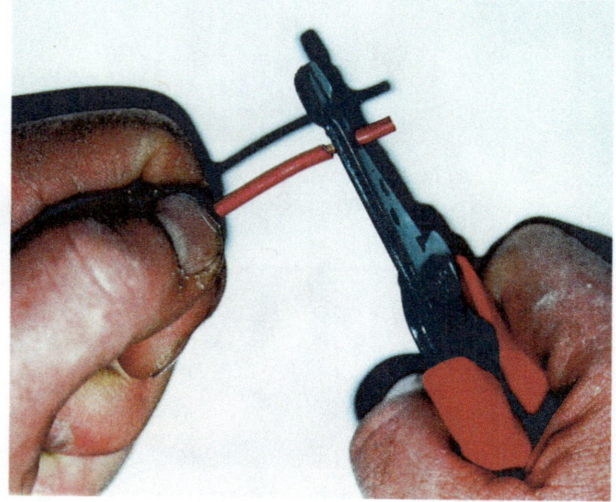

26106-14_F20.EPS

Figure 20 Stripping the insulation.

On Site

Wire Nuts for Wet Locations

Specially designed wire nuts are made for use in wet locations and/or direct burial applications. These wire nuts have a water repellent, non-hardening sealant inside the body that completely seals out moisture to protect the conductors against moisture, fungus, and corrosion. The sealant remains in a gel state and will not melt or run out of the wire nut body throughout the life of the connection. Unlike other types of wire nuts, this type can be used one time only. The wire nut can be backed off, eliminating the need to cut the wires for future or retrofit applications, but once removed, it must be discarded.

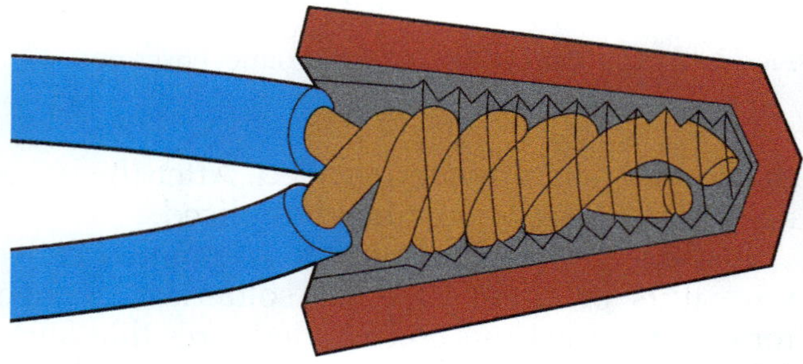

26106-14_F21.EPS

Figure 21 Wires installed in wire nut.

NOTE

Some manufacturers of wire nuts require that the wires be pre-twisted before screwing on the nut. Also, some manufacturers recommend using a nut driver to tighten the wire nut. Always follow the manufacturer's instructions.

Step 4 After making the connections within a box, tuck the wires neatly into the back of the box. That way, when the painters and plasterers come along to finish the walls, the wires will not be covered in drywall compound or paint.

Think About It

Putting It All Together

Turn off the power in one area of your home and then remove some of the switch and receptacle plates. Examine the wiring inside each box. Is the box adequately sized for the number of wires and devices?

SUMMARY

Electricians work with device boxes almost every day on every project. Consequently, you must have a thorough knowledge of the types of boxes available and their applications. *NEC Article 314* covers the installation of boxes.

One of the best ways to learn about boxes and fittings is to study manufacturers' catalogs. You will find a wealth of information in each of these, many of which include detailed instructions on installation techniques. Some of these catalogs also provide simplified *NEC®* explanations of the use of the manufacturer's products.

In this module, you learned that outlet boxes must be sized, installed, and supported to meet current *NEC®* requirements. Since the *NEC®* limits the number of conductors in a given box size, boxes with adequate volume must be selected. The boxes must also be the proper type for the application. Each type of box was discussed, along with references to applicable portions of *NEC Article 314*.

The maximum number of conductors permitted in standard outlet boxes is listed in *NEC Table 314.16(A)*. Certain devices and components require deductions in the number of conductors allowed in the box.

In addition to box sizing, there are various other factors to consider when installing boxes. These requirements are discussed in *NEC Sections 314.15 through 314.30*.

Review Questions

1. The maximum weight allowed by the *NEC®* when ceiling fans are mounted directly to an approved, unmarked outlet box is _____.
   a. 25 pounds
   b. 35 pounds
   c. 45 pounds
   d. 55 pounds

2. Square outlet boxes are available in sizes of 4" and _____.
   a. 4¹¹⁄₁₆"
   b. 5"
   c. 5¼"
   d. 6"

3. A box installed under a roofed open porch is considered a _____.
   a. dry location
   b. wet location
   c. damp location
   d. weatherproof location

4. How many conductor(s) must be deducted for each strap-mounted device in a device box?
   a. One
   b. Two
   c. Three
   d. Four

5. The capacity of an outlet box can be increased by using _____.
   a. fixture studs
   c. strap-mounted devices
   b. wire nuts
   d. raised device covers

6. A conductor under 12" long running through a box counts as two conductors.
   a. True
   b. False

7. If there are four equipment grounding conductors in a box, you must deduct _____ conductor(s) from the total allowed.
   a. 1
   b. 2
   c. 3
   d. 4

8. If the largest trade diameter of a raceway entering a pull box is 3", and it is a straight pull, the minimum size box allowed is _____.
   a. 20"
   b. 24"
   c. 30"
   d. 36"

9. The depth of a pull box with conduit runs entering at right angles need be only of sufficient depth to permit locknuts and bushings to be properly installed.
   a. True
   b. False

10. Factors to consider when installing boxes are covered in _____.
    a. *NEC Sections 314.1 through 314.15*
    b. *NEC Sections 314.15 through 314.30*
    c. *NEC Sections 314.40 through 314.44*
    d. *NEC Sections 314.70 through 314.72*

Trade Terms Quiz

Fill in the blank with the correct term that you learned from your study of this module.

1. ___________ means constructed or protected so that exposure to beating rain will not result in the entrance of water under specified test conditions.
2. A box designed and constructed to withstand an internal explosion without creating an external explosion or fire is called ___________.
3. A box constructed so that water will not enter the enclosure under specified test conditions is ___________.
4. A(n) ___________ is an enclosure used to facilitate the installation of cables from point to point in long runs.
5. A box constructed or protected so that exposure to the weather will not interfere with successful operation is referred to as ___________.
6. A(n) ___________ is a metallic or nonmetallic box installed in an electrical wiring system from which current is taken to supply some apparatus or device.
7. A single-gang outlet box used for surface mounting to enclose receptacles or wall switches on concrete or concrete block construction is called a(n) ___________.
8. A box constructed so that moisture will not interfere with successful operation is said to be ___________.
9. A(n) ___________ is an enclosure where one or more raceways or cables enter, and in which electrical conductors may be spliced.
10. A(n) ___________ is a device used to physically connect conduit or cable to an outlet box, cabinet, or other enclosure.

Trade Terms

Connector
Explosion-proof
Handy box
Junction box
Outlet box
Pull box
Raintight
Waterproof
Watertight
Weatherproof

Module 26106-14
Supplemental Exercises

1. All boxes must be sized, installed, and supported to meet the current ________________ requirements.
2. Blank covers are used when boxes serve as ________________.
3. Each conductor originating outside of the box and terminating inside of the box counts as ________________ conductor(s).
4. When sizing outlet boxes, deduct ________________ conductor(s) for each switch.
5. For a straight pull, the minimum length of the box must be at least ________________ times the trade diameter of the largest raceway.
6. The minimum length for a junction box in which the conductors are pulled at an angle and that contains two 3" conduits and one 4" conduit entering each side is ________________.
7. Octagon and round boxes are used mostly for ________________
________________.
8. When sizing outlet boxes, ________________ conductor(s) must be deducted for each strap-mounted device.
9. An extension ring is used to increase the ________________ of the box.
10. What is the maximum distance between receptacles along the floor line in any wall space?
________________.

Cornerstone of Craftsmanship

Gary Edgington

Baker Electric

What made you decide to become an electrician?
I had an early fascination with electricity and used to work on motorized cars, lights, or about anything else that would plug into the wall or run from a battery. My interest continued through high school. After watching an electrician wire a house, I made up my mind to be in the electrical field somewhere.

How did you learn the trade?
After graduating high school in 1972, I called an electrical contractor, Kinsey Electric, on a regular basis until he hired me. Harry was a seasoned electrical contractor, who did residential, commercial, and agricultural wiring. Because it was a one-man operation, I had an advantage in working with the owner and getting some good hands-on training along with acquiring his good work ethics. Harry was a well-respected person in the community, with a reputation of being the person to call if you wanted it done right. He was my first role model. Along with the electrical training, Harry taught me that if you're going to do a job, be the best at it and take pride in what you do.

What factor or factors have contributed most to your success?
After working with Harry for a few years, I went to work at a manufacturing facility as an industrial electrician. I was fortunate to meet two more role models who greatly influenced me in my career. One was a retired Air Force electrician, who taught me the importance of knowledge and the need to be able to find answers to what you don't know. The second was an electrician from Manchester, England. He helped me recognize my abilities and inspired me to acquire more. Above everything, he gave me confidence in myself which has helped me throughout my life.

What kind of jobs did you hold on the way to your current position?
In addition to my job as an electrician helper/ electrician for Kinsey and as an industrial electrician, I became a licensed electrical contractor in 1978. After being in the electrical contracting industry for 22 years, and employing up to fifty people at times, I sold my business and became an electrical consultant. As the owner of the business I have found myself doing everything from electrical work in the field to estimating to project and business management. During this time and since then I have taught electrical apprentice classes and *National Electrical Code*® classes for thirteen years.

What does an electrical consultant do?
As a consultant I have taught electrical classes and code updates, been involved in the electrical design process, done code review on projects, and have been an expert in court cases.

What advice would you give to someone entering the electrical field?
My advice to someone entering the field is to embrace every opportunity to learn something new, to develop good work practices and ethics, and to think safety all of the time.

What advice do you have for trainees?
As a trainee in this field you need to practice at your profession. Be accurate in your work and pursue knowledge. You will find the ability and the confidence you need to succeed in this field.

What would you say was the single greatest factor that contributed to your success?
The single greatest factor for me has been getting close to a professional in the field and trying to follow in their footsteps.

Trade Terms Introduced in This Module

Connector: Device used to physically connect conduit or cable to an outlet box, cabinet, or other enclosure.

Explosion-proof: Designed and constructed to withstand an internal explosion without creating an external explosion or fire.

Handy box: Single-gang outlet box used for surface mounting to enclose receptacles or wall switches on concrete or concrete block construction of industrial and commercial buildings; nongangable; also made for recessed mounting; also known as a utility box.

Junction box: An enclosure where one or more raceways or cables enter, and in which electrical conductors can be, or are, spliced.

Outlet box: A metallic or nonmetallic box installed in an electrical wiring system from which current is taken to supply some apparatus or device.

Pull box: A sheet metal box-like enclosure used in conduit runs to facilitate the pulling of cables from point to point in long runs, or to provide for the installation of conduit support bushings needed to support the weight of long riser cables, or to provide for turns in multiple-conduit runs.

Raintight: Constructed or protected so that exposure to a beating rain will not result in the entrance of water under specified test conditions.

Waterproof: Constructed so that moisture will not interfere with successful operation.

Watertight: Constructed so that water will not enter the enclosure under specified test conditions.

Weatherproof: Constructed or protected so that exposure to the weather will not interfere with successful operation.

Additional Resources

This module presents thorough resources for task training. The following resource material is suggested for further study.

National Electrical Code® Handbook, Latest Edition. Quincy, MA: National Fire Protection Association.

Figure Credits

Associated Builders and Contractors, Inc., Module opener

Topaz Publications, Inc., Figures 1, 4, 6, 17–20, SA01, SA02

John Traister, Figures 9–12, 14

NCCER CURRICULA — USER UPDATE

NCCER makes every effort to keep its textbooks up-to-date and free of technical errors. We appreciate your help in this process. If you find an error, a typographical mistake, or an inaccuracy in NCCER's curricula, please fill out this form (or a photocopy), or complete the online form at **www.nccer.org/olf**. Be sure to include the exact module ID number, page number, a detailed description, and your recommended correction. Your input will be brought to the attention of the Authoring Team. Thank you for your assistance.

Instructors – If you have an idea for improving this textbook, or have found that additional materials were necessary to teach this module effectively, please let us know so that we may present your suggestions to the Authoring Team.

NCCER Product Development and Revision

13614 Progress Blvd., Alachua, FL 32615

Email: curriculum@nccer.org

Online: www.nccer.org/olf

❑ Trainee Guide ❑ AIG ❑ Exam ❑ PowerPoints Other ____________________

Craft / Level: Copyright Date:

Module ID Number / Title:

Section Number(s):

Description:

Recommended Correction:

Your Name:

Address:

Email: Phone:

Hand Bending

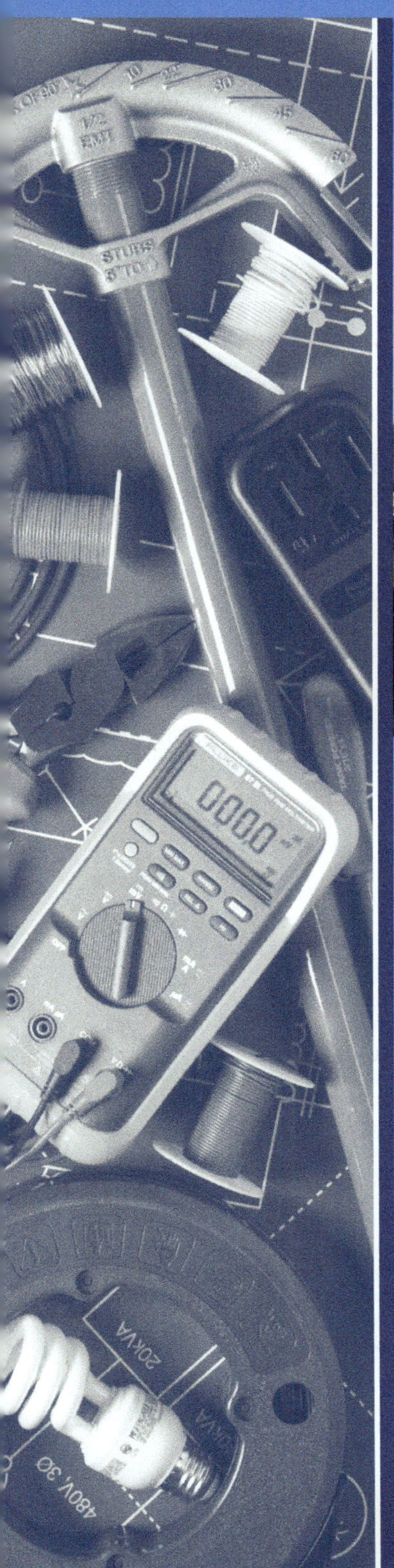

National Air and Space Museum

The Steven F. Udvar-Hazy Center near Washington Dulles International Airport is the companion facility to the Museum on the National Mall. The building opened in December, 2003, and provides enough space for the Smithsonian to display the thousands of aviation and space artifacts that cannot be exhibited on the National Mall.

26107-14

Trainees with successful module completions may be eligible for credentialing through NCCER's National Registry. To learn more, go to **www.nccer.org** or contact us at **1.888.622.3720.** Our website has information on the latest product releases and training, as well as online versions of our *Cornerstone* magazine and Pearson's product catalog.

Your feedback is welcome. You may email your comments to **curriculum@nccer.org,** send general comments and inquiries to **info@nccer.org,** or fill in the User Update form at the back of this module.

This information is general in nature and intended for training purposes only. Actual performance of activities described in this manual requires compliance with all applicable operating, service, maintenance, and safety procedures under the direction of qualified personnel. References in this manual to patented or proprietary devices do not constitute a recommendation of their use.

26107-14

Hand Bending

Objectives

When you have completed this module, you will be able to do the following:

1. Identify the methods for hand bending and installing conduit.
2. Determine conduit bends.
3. Make 90-degree bends, back-to-back bends, offsets, kicks, and saddle bends using a hand bender.
4. Cut, ream, and thread conduit.

Performance Tasks

Under the supervision of the instructor, you should be able to do the following:

1. Make 90° bends, back-to-back bends, offsets, kicks, and saddle bends using a hand bender.
2. Cut, ream, and thread conduit.

Trade Terms

90° bend
Back-to-back bend
Concentric bends
Developed length
Gain
Offsets
Rise
Segment bend
Stub-up

Required Trainee Materials

1. Paper and pencil
2. Copy of the latest edition of the *National Electrical Code*®
3. Appropriate personal protective equipment

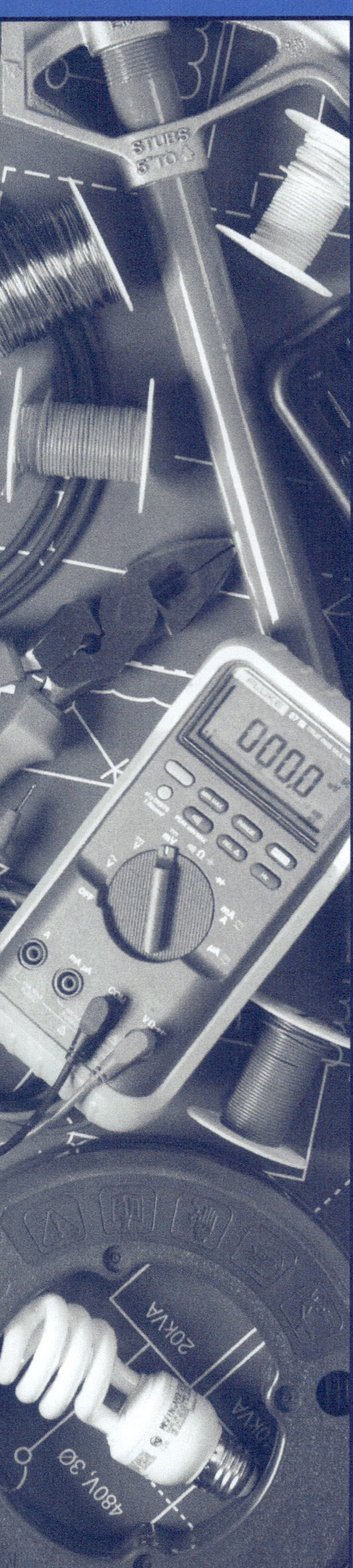

Contents

Topics to be presented in this module include:

Figures and Tables

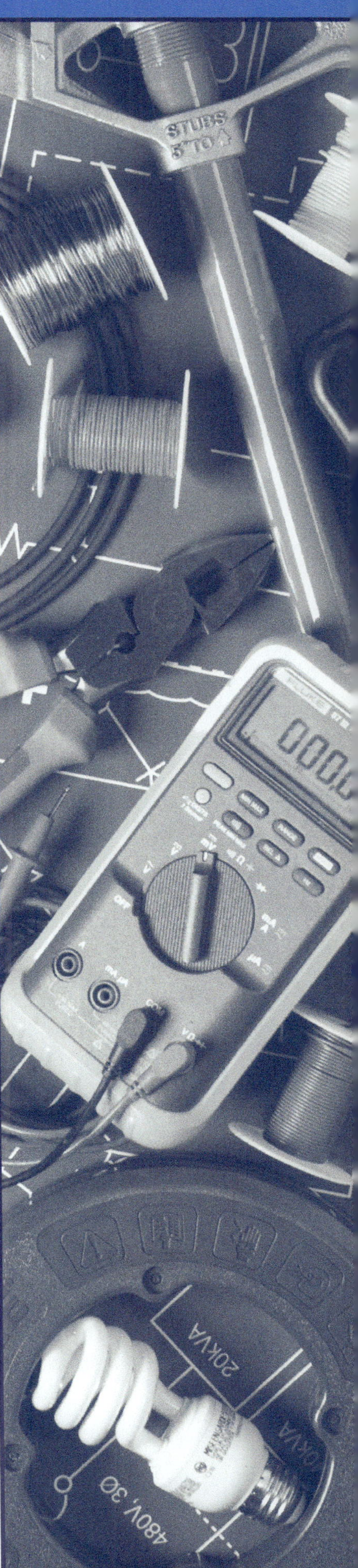

1.0.0 Introduction

The art of conduit bending is dependent upon the skills of the electrician and requires a working knowledge of basic terms and proven procedures. Practice, knowledge, and training will help you gain the skills necessary for proper conduit bending and installation. You will be able to practice conduit bending in the lab and in the field under the supervision of experienced co-workers. In this module, the techniques for using hand-operated and step conduit benders such as the hand bender and the hickey will be covered. The processes of hand bending, cutting, reaming, and threading conduit will also be explained.

2.0.0 Hand Bending Equipment

Figure 1 shows hand benders. Hand benders are convenient to use on the job because they are portable and no electrical power is required. Hand benders have a shape that supports the walls of the conduit being bent.

These benders are used to make various bends in smaller-size conduit (½" to 1¼"). Most hand benders are sized to bend rigid conduit and electrical metallic tubing (EMT) of corresponding sizes. For example, a single hand bender can bend either ¾" EMT or ½" rigid conduit. The next larger size of hand bender will bend either 1" EMT or ¾" rigid conduit. This is because the corresponding sizes of conduit have nearly equal outside diameters.

The first step in making a good bend is familiarizing yourself with the bender. The manufacturer of the bender will typically provide documentation indicating starting points, distance between **offsets**, **gains**, and other important values associated with that particular bender. There is no substitute for taking the time to review this information. It will make the job go faster and result in better bends.

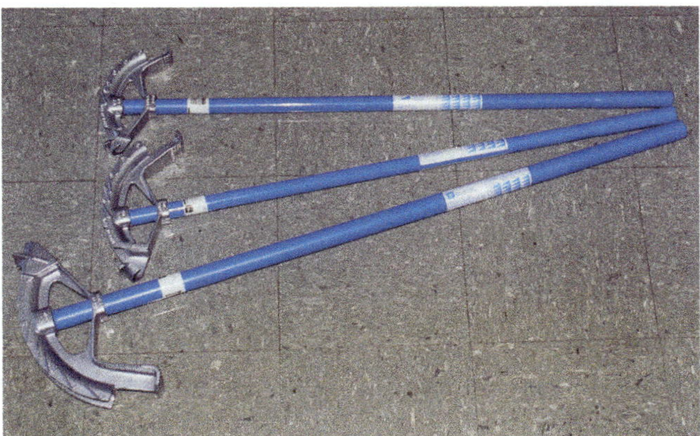

26107-14_F01.EPS

Figure 1 Hand benders.

> **On Site**
>
> **Working with Conduit**
>
> Unprotected electrical cable is susceptible to physical damage; therefore, protect the wiring with conduit.

> **CAUTION**
>
> When making bends, be sure you have a firm grip on the handle to avoid slippage and possible injury.

When performing a bend, it is important to keep the conduit on a stable, firm, flat surface for the entire duration of the bend. Hand benders are designed to have force applied using one foot and the hands. See *Figure 2*. It is important to use constant foot pressure as well as force on the handle to achieve uniform bends. Allowing the conduit to rise up or performing the bend on soft ground can result in distorting the conduit outside the bender.

> **NOTE**
>
> Bends should be made in accordance with the guidelines of *NEC Article 342* (intermediate metal conduit or IMC), *Article 344* (rigid metal conduit or RMC), *Article 352* (rigid polyvinyl chloride conduit or PVC), or *Article 358* (electrical metallic tubing or EMT).

A hickey (*Figure 3*) should not be confused with a hand bender. The hickey, which is used for RMC and IMC only, functions quite differently.

When you use a hickey to bend conduit, you are forming the bend as well as the radius. When

26107-14_F02.EPS

Figure 2 Pushing down on the bender to complete the bend.

On Site

Bending Conduit

A good way to practice bending conduit is to use a piece of No. 10 or No. 12 solid wire and bend it to resemble the bends you need. This gives you some perspective on how to bend the conduit and it will also help you to anticipate any problems with the bends.

using a hickey, be careful not to flatten or kink the conduit. Hickeys should only be used with RMC and IMC because very little support is given to the walls of the conduit being bent.

A hickey is a segment bending device. First, a small bend of about 10° is made. Then, the hickey is moved to a new position and another small bend is made. This process is continued until the bend is completed. A hickey can be used for conduit **stub-ups** in slabs and decks.

On Site

Proper Bends

Kinks are created by bending too small a radius using a hickey.

26107-14_SA01.EPS

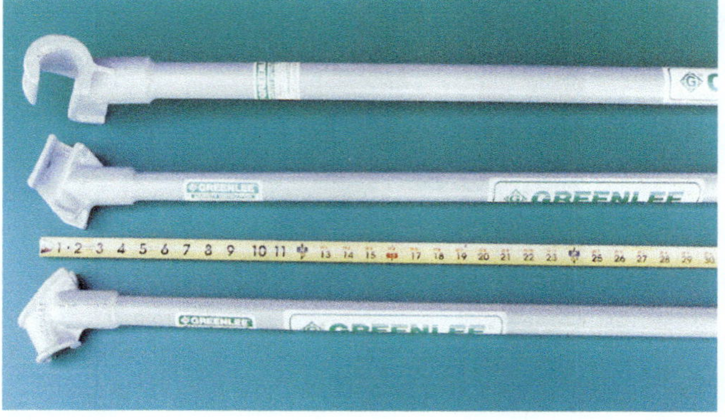

26107-14_F03.EPS

Figure 3 Hickeys.

26107-14_F04.EPS

Figure 4 Typical PVC heating units.

Polyvinyl chloride (PVC) conduit is bent using a heating unit (*Figure 4*). The PVC must be rotated regularly while it is in the heater so that it heats evenly. Once heated, the PVC is removed, and the bending is performed by hand. Some units use an electric heating element, while others use liquid propane (LP). After bending, a damp sponge or cloth is often used so that the PVC sets up faster.

CAUTION

Avoid contact with the case of the heating unit; it can become very hot and cause burns. Also, to avoid a fire hazard, ensure that the unit is cool before storage. If using an LP unit, keep a fire extinguisher nearby.

When bending PVC that is 2" or larger in diameter, there is a risk of wrinkling or flattening the bend. A plug set eliminates this problem (*Figure 5*). A plug is inserted into each end of the piece of PVC being bent. Then, a hand pump is used to pressurize the conduit before bending it. The pressure is about 3 to 5 psi.

NOTE

The plugs must remain in place until the pipe is cooled and set.

2.1.0 Geometry Required to Make a Bend

Bending conduit requires that you use some basic geometry. You may already be familiar with most of the concepts needed; however, here is a review of the concepts directly related to this task.

A right triangle is defined as any triangle with a 90° angle. The side directly opposite the 90° angle is called the hypotenuse, and the side on which the triangle sits is the base. The vertical side is called the height. On the job, you will apply the

What's wrong with this picture?

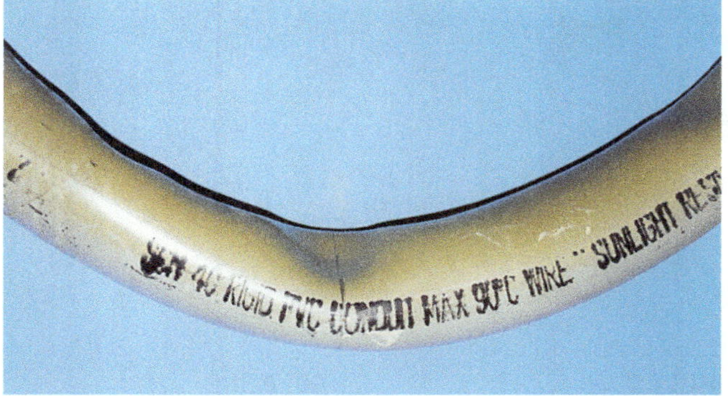

26107-14_SA02.EPS

26107-14_F05.EPS

Figure 5 Typical plug set.

relationships in a right triangle when making an offset bend. The offset forms the hypotenuse of a right triangle (*Figure 6*).

> **On Site**
>
> **Practical Bending**
>
> When making offset bends of 45° to step conduit up to another level, square floor tiles make a convenient grid to gauge the distance and angles.

> **NOTE**
>
> There are reference tables for sizing offset bends based on these relationships (see *Appendix A*).

A circle is defined as a closed curved line whose points are all the same distance from its center. The distance from the center point to the edge of the circle is called the radius. The length from one edge of the circle to the other edge through the center point is the diameter. The distance around the circle is called the circumference. A circle can be divided into four equal quadrants. Each quadrant accounts for 90°, making a total of 360°. When you make a **90° bend**, you will use ¼ of a circle, or one quadrant (see *Figure 7*).

Concentric circles are circles that have a common center but different radii. The concept of concentric circles can be applied to **concentric bends** in conduit. The angle of each bend is 90°. Such bends have the same center point, but the radius of each is different.

To calculate the circumference of a circle, use the following formula:

$$C = \pi \times D \text{ or } C = \pi D$$

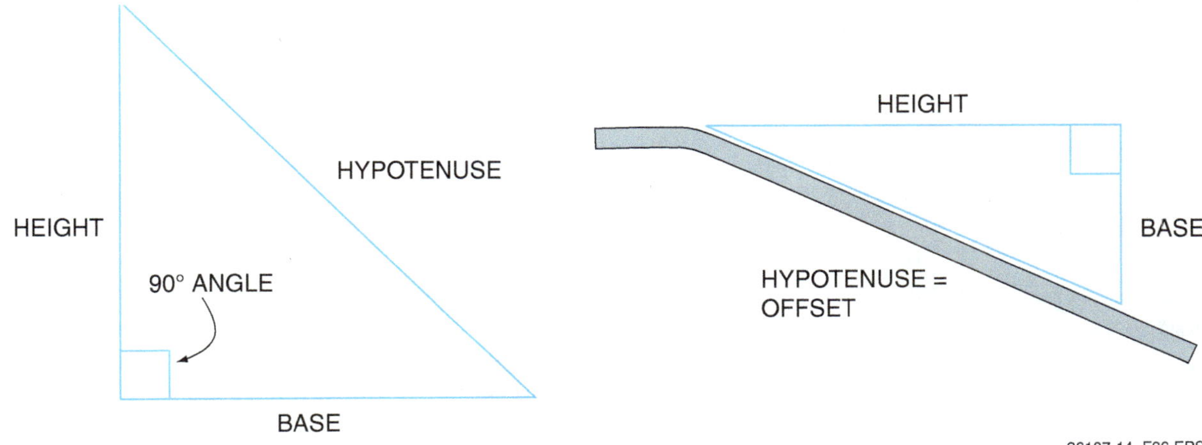

Figure 6 Right triangle and offset bend.

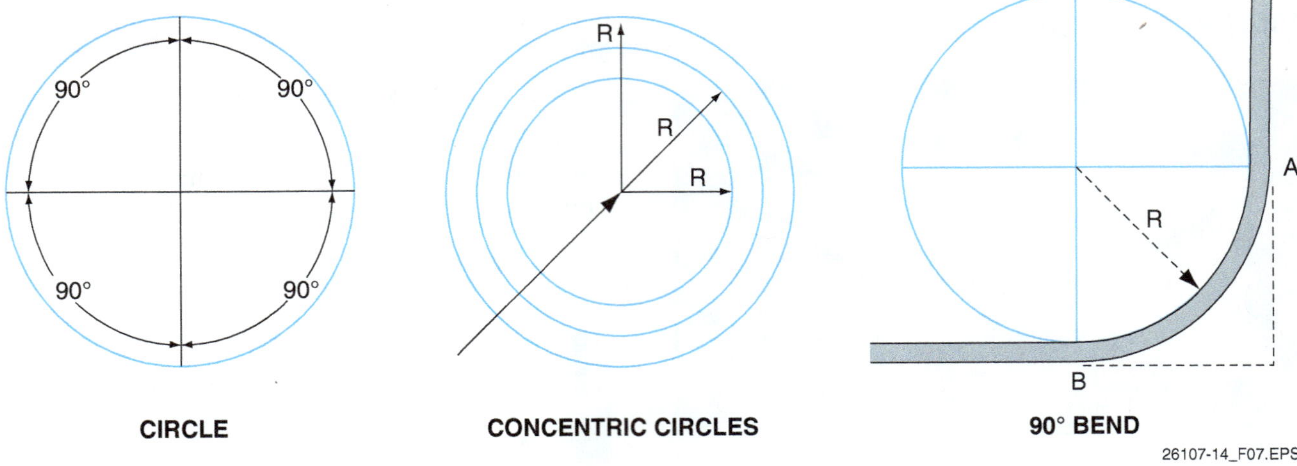

Figure 7 Circles and 90° bends.

In this formula, C = circumference, D = diameter, and π = 3.14. Another way of stating the formula for circumference is C = 2πR, where R equals the radius or ½ the diameter. To figure the arc of a quadrant use:

$$\text{Length of arc} = (0.25)\,2\pi R = 1.57R$$

For this formula, the arc of a quadrant equals ¼ the circumference of the circle or 1.57 times the radius.

A bending radius table is included in *Appendix B*.

2.2.0 Making a 90° Bend

The 90° stub bend is probably the most basic bend of all. The stub bend is used much of the time, regardless of the type of conduit being installed. Before beginning to make the bend, you need to know two measurements:

- Desired rise or stub-up
- Take-up distance of the bender

The desired rise is the height of the stub-up. The take-up is the amount of conduit the bender will use to form the bend. Take-up distances are usually listed in the manufacturer's instruction manual. Typical bender take-up distances are shown in *Table 1*.

Table 1 Typical Bender Take-Up Distances

EMT	Rigid/IMC	Take-Up
½"	—	5"
¾"	½"	6"
1"	¾"	8"
1¼"	1"	11"

Once you have determined the take-up, subtract it from the stub-up height. Mark that distance on the conduit (all the way around) at that distance from the end. The mark will indicate the point at which you will begin to bend the conduit. Line up the starting point on the conduit with the starting point on the bender. Most benders have a mark, such as an arrow, to indicate this point. *Figure 8* shows the take-up required to achieve an 18" stub-up on a piece of ½" EMT.

Once you have lined up the bender, use one foot to hold the conduit steady. Keep your heel on the floor for balance. Apply constant pressure on the bender foot pedal with your other foot. Make sure you hold the bender handle perpendicular to the floor, and as far up as possible, to get maximum leverage. Then, bend the conduit in one smooth motion, pulling as steadily as possible. Avoid overstretching.

On Site

Matching Bends in Parallel Runs

Suppose you are running 1" rigid conduit along with a 2" rigid conduit in a rack and you come to a 90° bend. If you used a 1" shoe, the radius would not match that of the 2" conduit bend. To match the 2" 90° bend, take your 1" conduit and put it in the 2" shoe of your bender, then bend as usual. This 1" 90° bend will now have the same radius as the 2" 90° bend. This trick will only work on rigid conduit. If done with EMT, it will flatten the pipe.

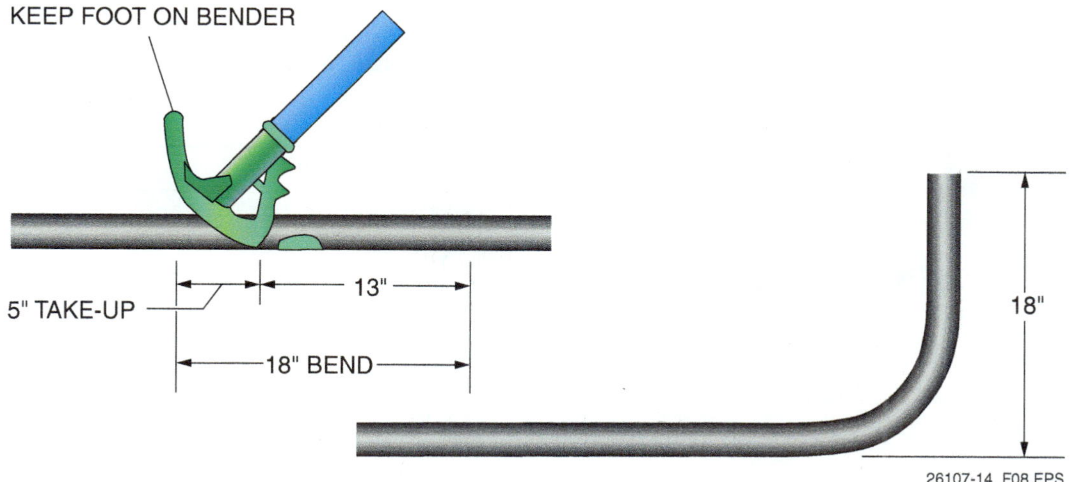

Figure 8 Bending an 18-inch stub-up.

> **NOTE**
> When bending conduit using the take-up method, always place the bender on the conduit and make the bend facing the hook of the conduit from which the measurements were taken.

After finishing the bend, check to make sure you have the correct angle and measurement. Use the following steps to check a 90° bend:

Step 1 With the back of the bend on the floor, measure to the end of the conduit stub-up to make sure it is the right length.

Step 2 Check the 90° angle of the bend with a square or at the angle formed by the floor and a wall. A torpedo level may also be used.

> **NOTE**
> If you overbend a conduit slightly past the desired angle, you can use the bender to bend the conduit back to the correct angle.

The above procedure will produce a 90° one-shot bend. That means that it took a single bend to form the conduit bend. A **segment bend** is any bend that is formed by a series of bends of a few degrees each, rather than a single one-shot bend. A shot is actually one bend in a segment bend. Segment or sweep bends must conform to the provisions of the *NEC*®.

2.3.0 Gain

The gain is the distance saved by the arc of a 90° bend. Knowing the gain can help you to precut, ream, and prethread both ends of the conduit before you bend it. This will make your work go more quickly because it is easier to work with conduit while it is straight. *Figure 9* shows that the overall **developed length** of a piece of conduit with a 90° bend is less than the sum of the horizontal and vertical distances when measured square to the corner. This is shown by the following equation:

$$\text{Developed length} = (A + B) - \text{gain}$$

An example of a manufacturer's gain table is also shown in *Figure 9*. These tables are used to determine the gain for a certain size conduit.

On Site

Take-Up Method

When bending conduit using the take-up method, always place the bender on the conduit and make the bend facing the end of the conduit from which the measurements were taken. It helps to make a narrow mark with a soft lead pencil or marker completely around the conduit. This is called girdling.

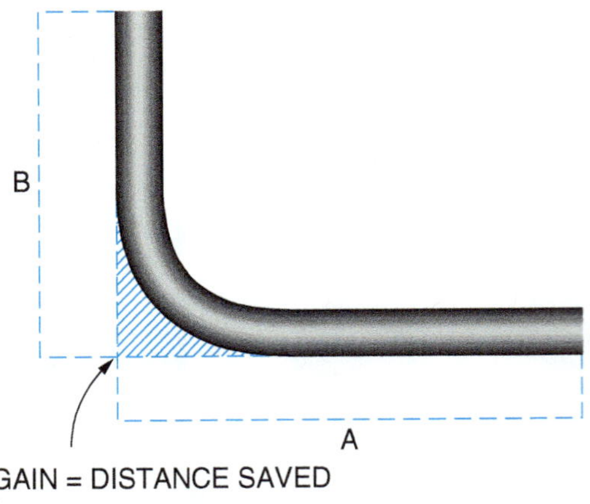

Conduit Size	NEC® Radius	90° Gain
½"	4"	2⅝"
¾"	5"	3¼"
1"	6"	4"
1¼"	8"	5⅝"

TYPICAL GAIN TABLE

26107-14_F09.EPS

Figure 9 Gain.

2.4.0 Back-to-Back 90° Bends

A **back-to-back bend** consists of two 90° bends made on the same piece of conduit and placed back-to-back (*Figure 10*).

To make a back-to-back bend, make the first bend (labeled *X* in *Figure 10*) in the usual manner. To make the second bend, measure the required distance between the bends from the back of the first bend. This distance is labeled *L* in the figure. Reverse the bender on the conduit, as shown in *Figure 10*. Place the bender's back-to-back indicating mark at point *Y* on the conduit. Note that outside measurements from point *X* to point *Y* are used. Holding the bender in the reverse position and properly aligned, apply foot pressure and complete the second bend.

> **On Site**
>
> **Checking Vertical Rise**
>
> Use a torpedo level to check for plumb on a vertical rise.
>
> 26107-14_SA03.EPS

> **Think About It**
>
> **Gain**
>
> What is the difference between the gain and the take-up of a bend?

2.5.0 Making an Offset

Many situations require that the conduit be bent so that it can pass over objects such as beams and other conduits, or enter meter cabinets and junction boxes. Bends used for this purpose are called offsets (kicks). To produce an offset, two

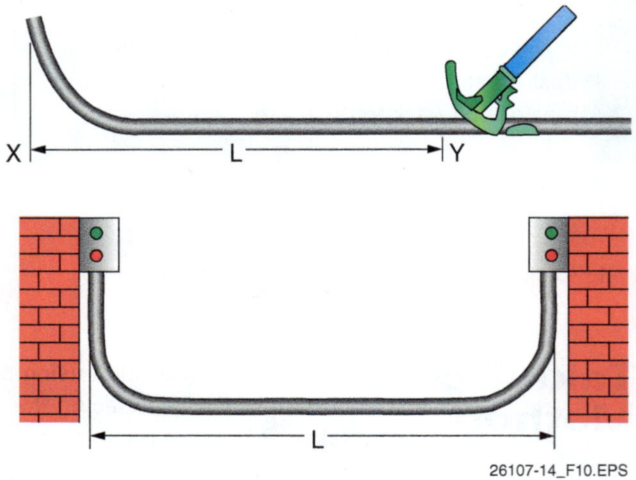

26107-14_F10.EPS

Figure 10 Back-to-back bends.

Think About It

Smooth Bends

Why are smooth bends so important?

equal bends of less than 90° are required, a specified distance apart, as shown in *Figure 11*.

Offsets are a trade-off between space and the effort it will take to pull the wire. The larger the degree of bend, the harder it will be to pull the wire. The smaller the degree of bend, the easier it will be to pull the wire. Use the shallowest degree of bend that will still allow the conduit to bypass the obstruction and fit in the given space.

When conduit is offset, some of the conduit length is used. If the offset is made into the area, an allowance must be made for this shrinkage. If the offset angle is away from the obstruction, the shrinkage can be ignored.

Table 2 shows the amount of shrinkage per inch of rise for common offset angles.

The formula for figuring the distance between bends is as follows:

Distance between bends = depth of offset × multiplier

The distance between the offset bends can generally be found in the manufacturer's documentation for the bender. *Table 3* shows the distance between bends for the most common offset angles.

Calculations related to offsets are derived from the branch of mathematics known as trigonometry, which deals with triangles. The multipliers shown in *Table 2* represent the cosecant (CSC) of the related offset angle. The multiplier is determined by dividing the hypotenuse of the triangle created by the offset by the depth of the offset (*Figure 11*).

On Site

Offset Benders

This offset bender can be used for wall-mounted boxes with exposed conduit. It automatically matches the offset to the box knockout position and is a great timesaver when making multiple offsets.

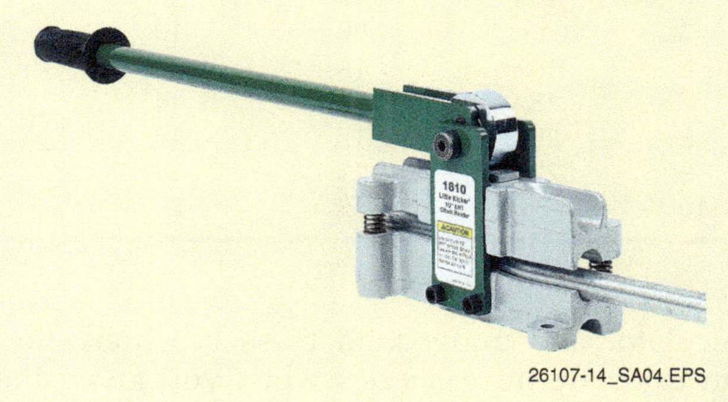

26107-14_SA04.EPS

Table 2 Shrinkage Calculation

Offset Angle	Multiplier	Shrinkage (per inch of rise)
10° × 10°	6.0	1/16"
22½° × 22½°	2.6	3/16"
30° × 30°	2.0	¼"
45° × 45°	1.4	⅜"
60° × 60°	1.2	½"

Basic trigonometry (trig) functions are briefly covered in *Appendix A*. As you will see in the next section, the tangent (TAN) of the offset angle is also used in calculating parallel offsets. Understanding trig functions will help you understand how offsets are determined. If you have a scientific

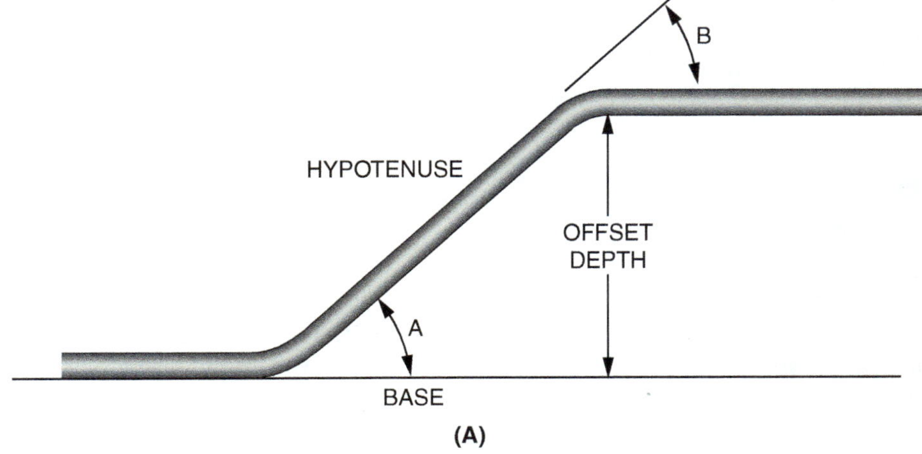

(A)

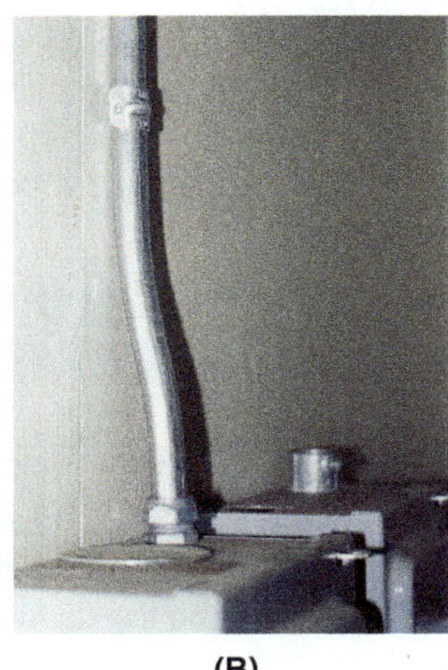

(B)

26107-14_F11.EPS

Figure 11 Offsets.

Table 3 Common Offset Factors (in Inches)

	22½°		30°		45°		60°	
Offset Depth	**Between Bends**	**Shrinkage**	**Between Bends**	**Shrinkage**	**Between Bends**	**Shrinkage**	**Between Bends**	**Shrinkage**
2	5¼	⅜	—	—	—	—	—	—
3	7¾	9⁄16	6	¾	—	—	—	—
4	10½	¾	8	1	—	—	—	—
5	13	15⁄16	10	1¼	7	1⅞	—	—
6	15½	1⅛	12	1½	8½	2¼	7¼	3
7	18¼	1 5⁄16	14	1¾	9¾	2⅝	8⅜	3½
8	20¾	1½	16	2	11¼	3	9⅝	4
9	23½	1¾	18	2¼	12½	3⅜	10⅞	4½
10	26	1⅞	20	2½	14	3¾	12	5

calculator and understand these functions, you can calculate offset angles when you know the dimensions of the triangle created by the offset and the obstacle.

2.6.0 Parallel Offsets

Often, multiple pieces of conduit must be bent around a common obstruction. In this case, parallel offsets are made. Since the bends are laid out along a common radius, an adjustment must be made to ensure that the ends do not come out uneven, as shown in *Figure 12*.

The center of the first bend of the innermost conduit is found first, as shown in *Figure 13*. Each successive conduit must have its centerline moved farther away from the end of the pipe, as shown in *Figure 14*. The amount to add is calculated as follows:

Amount added = center-to-center spacing × tangent (TAN) of ½ offset angle

Tangents can be found using the trig tables provided in *Appendix A*.

For example, *Figure 15* shows three pipes laid out as parallel and offset. The angle of the offset is 30°. The center-to-center spacing is 3". The start of the innermost pipe's first bend is 12".

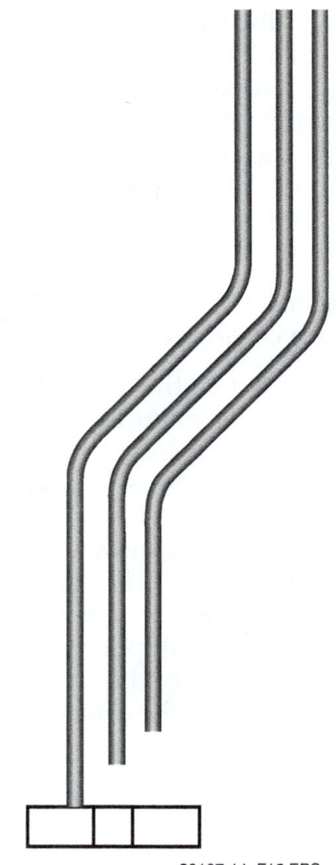

Figure 12 Incorrect parallel offsets.

Think About It

Calculating Shrinkage

You're making a 30° by 30° offset to clear a 6" obstruction. What will be the distance between bends? What will be the developed length shrink? Make the same calculations for a 10" offset with 45° bends.

The starting point of the second pipe will be:

12" + [center-to-center spacing × TAN (½ offset angle)]

12" + (3" × TAN 15°) = 12" + (3" × 0.2679) = 12" + 0.8037"

This is approximately 12 13⁄16".

The starting point for the outermost pipe is:

12 13⁄16" + 13⁄16" = 13⅝"

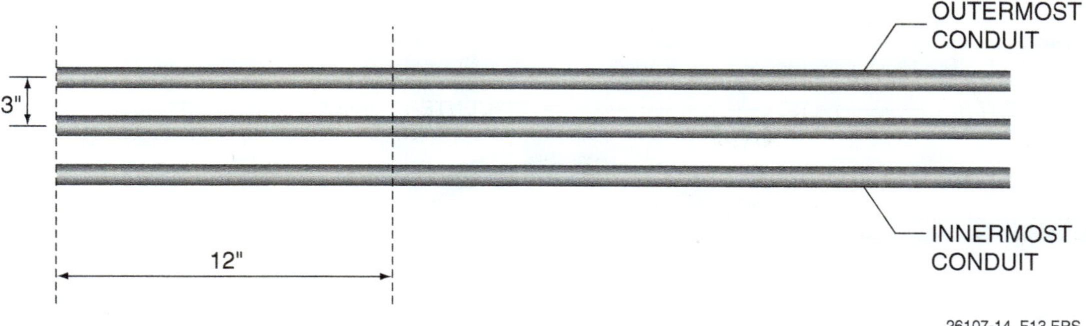

Figure 13 Center of first bend.

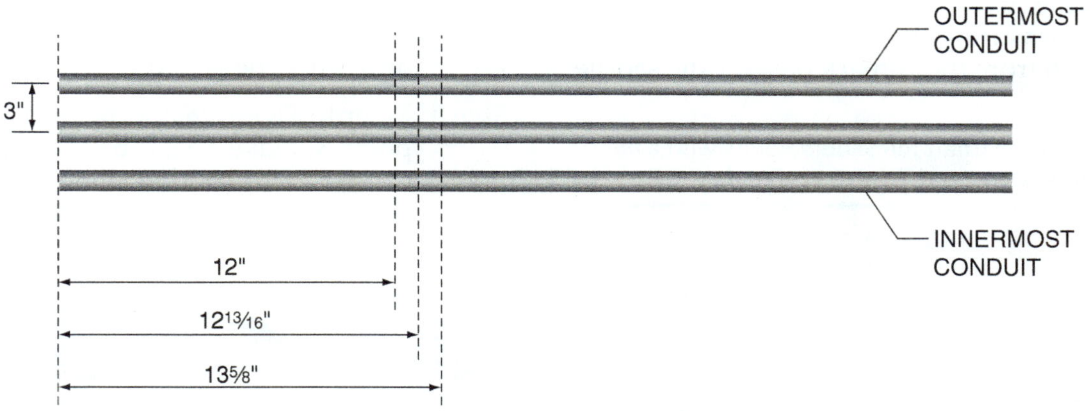

Figure 14 Successive centerlines.

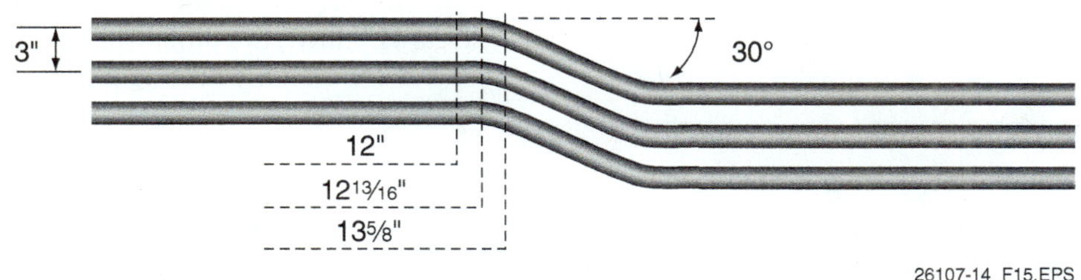

Figure 15 Parallel offset pipes.

Think About It

Equal Angles

Why is it important that the angles be identical when making an offset bend?

2.7.0 Saddle Bends

A saddle bend is used to go around obstructions. *Figure 16* illustrates an example of a saddle bend that is required to clear a pipe obstruction. Making a saddle bend will cause the center of the saddle to shorten ³⁄₁₆" for every inch of saddle depth (see *Table 4*). For example, if the pipe diameter is 2 inches, this would cause a ⅜" shortening of the conduit on each side of the bend. When making saddle bends, the following steps should apply:

Step 1 Locate the center mark A on the conduit by using the size of the obstruction (i.e., pipe diameter) and calculate the shrink rate of the obstruction (for example, if the pipe diameter is 2 inches, ⅜" of conduit will be lost on each side of the bend for a total shrinkage of ¾"). This figure will be added to the measurement from the end of the conduit to the centerline of the obstruction (for example, if the distance measured from the conduit end to the obstruction centerline was 15", the distance to A would be 15⅜").

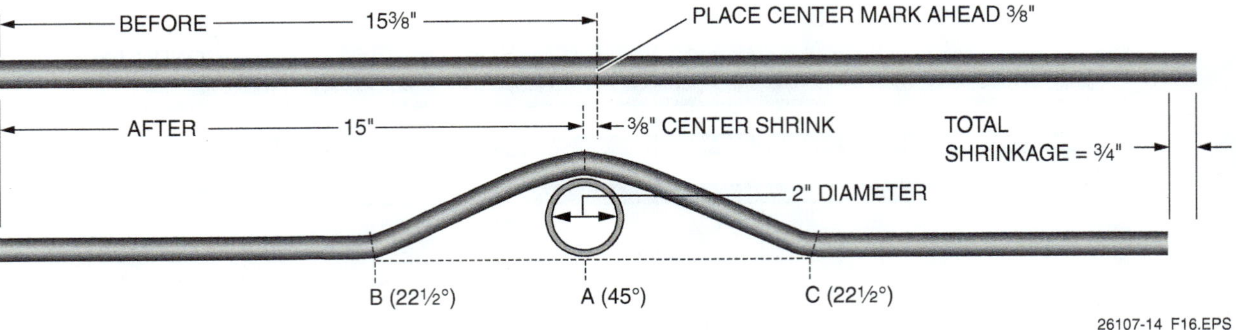

Figure 16 Saddle measurement.

Step 2 Locate marks B and C on the conduit by measuring 2½" for every 1" of saddle depth from the A mark (i.e., for the saddle depth of 2 inches, the B mark would be 5" before the A mark and the C mark would be 5" after the A mark). See *Figure 17*.

Step 3 Refer to *Figure 18* and make a 45° bend at point A, make a 22½° bend at point B, and make a 22½° bend at point C. (Be sure to check the manufacturer's specifications.)

Table 4 Shrinkage Chart for Saddle Bends with a 45° Center Bend and Two 22 ½° Bends

Obstruction Depth	Shrinkage Amount (Move Center Mark Forward)	Make Outside Marks from New Center Mark
1	3⁄16"	2½"
2	⅜"	5"
3	9⁄16"	7½"
4	¾"	10"
5	15⁄16"	12½"
6	1⅛"	15"
For each additional inch, add	3⁄16"	2½"

Think About It

Calculating Parallel Offsets

You're making parallel offsets of 45° and the lengths of conduit are spaced 4" center to center. If the offset starts 12" down the pipe, what is the starting point for the bend on the second pipe?

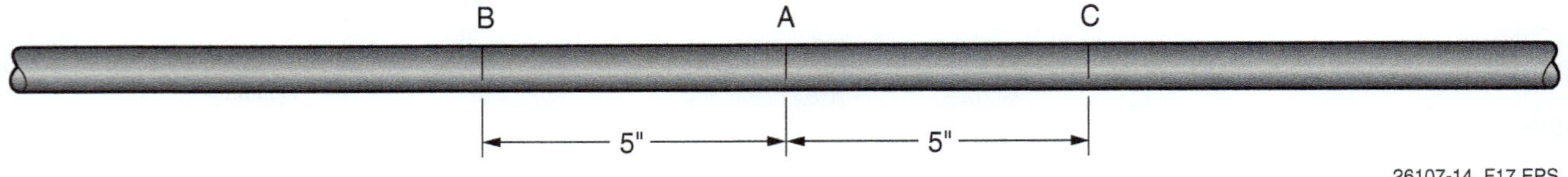

Figure 17 Measurement locations.

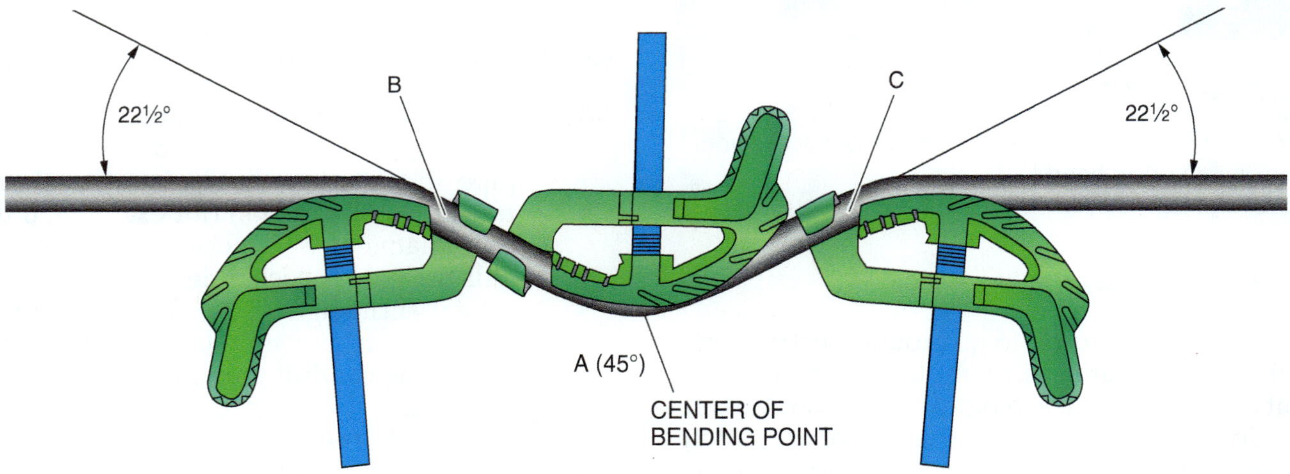

Figure 18 Location of bends.

2.8.0 Four-Bend Saddles

Four-bend saddles can be difficult. The reason is that four bends must be aligned exactly on the same plane. Extra time spent laying it out and performing the bends will pay off in not having to scrap the whole piece and start over.

Figure 19 illustrates that the four-bend saddle is really two offsets formed back-to-back. Working left to right, the procedure for forming this saddle is as follows:

Step 1 Determine the height of the offset.

Step 2 Determine the correct spacing for the first offset and mark the conduit.

Step 3 Bend the first offset.

Step 4 Mark the start point for the second offset at the trailing edge of the obstruction.

Step 5 Mark the spacing for the second offset.

Step 6 Bend the second offset.

Using *Figure 20* as an example, a four-bend saddle using ½" EMT is laid out as follows:

- Height of the box = 6"
- Width of the box = 8"
- Distance to the obstruction = 36"

Two 30° offsets will be used to form the saddle. It is created as follows:

Step 1 See *Figure 21*. Working from left to right, calculate the start point for the first bend. The distance to the obstruction is 36", the offset is 6", and the 30° multiplier from *Table 2* is 2.0:

Distance to the obstruction – (offset × constant for the angle) + shrinkage = distance to the first bend

36" – (6" × 2.0) + 1½" = 25½"

Step 2 Determine where the second bend will end to ensure the conduit clears the obstruction. See *Figure 22*.

Distance to the first bend + distance to second bend + shrinkage = total length of the first offset

25½" + 12" + 1½" = 39"

Step 3 Determine the start point of the second offset. The width of the box is 8"; therefore, the start point of the second offset should be 8" beyond the end of the first offset:

8" + 39" = 47"

Step 4 Determine the spacing for the second offset. Since the first and second offsets have the same rise and angle, the distance between bends will be the same, or 12".

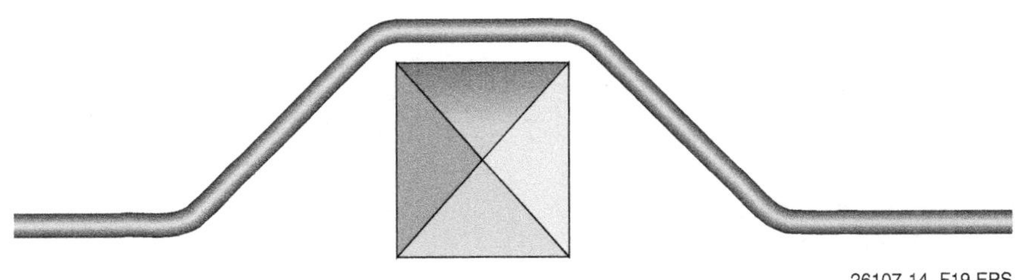

Figure 19 Typical four-bend saddle.

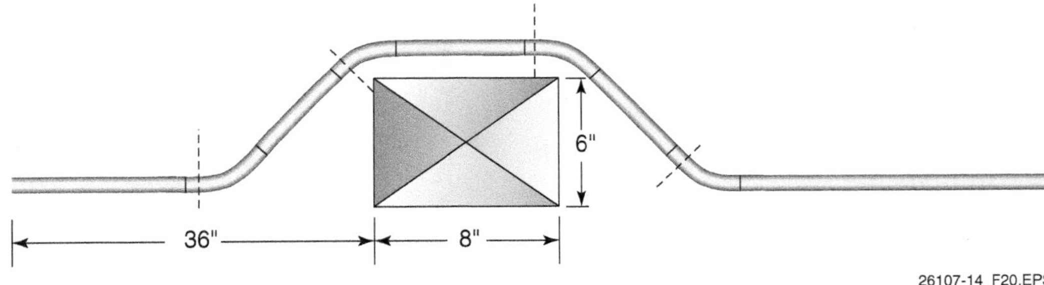

Figure 20 Four-bend saddle.

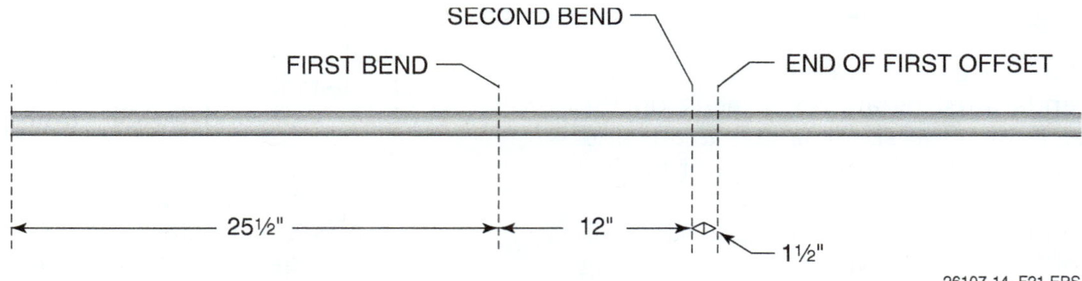

Figure 21 Four-bend saddle measurements.

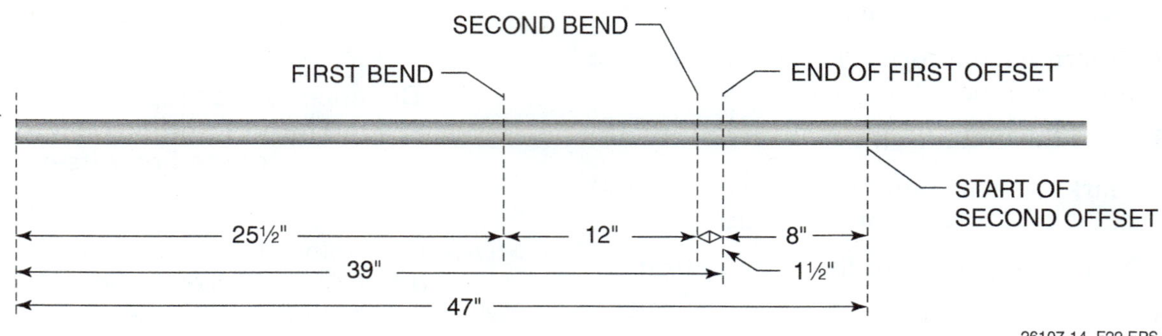

Figure 22 Bend and offset measurements.

On Site

Planning Bends

The more bends you make between pull points, the more difficult it is to pull the wires through the conduit. Therefore, plan your bends in advance, avoid sharp bends, and make as few bends as possible. The *NEC®* allows the bends in a single run of conduit to total 360° between pull points; however, 360° is not as much as you might think. For example, if you bend the conduit 90° for two corners of a room, with two 45° offsets where the conduit connects to a panelboard and junction box, you've used up your 360°.

3.0.0 Cutting, Reaming, and Threading Conduit

RMC, IMC, and EMT are available in standard 10-foot lengths. When installing conduit, it is cut to fit the job requirements.

3.1.0 Hacksaw Method of Cutting Conduit

Conduit is normally cut using a hacksaw. To cut conduit with a hacksaw:

Step 1 Inspect the blade of the hacksaw and replace it, if needed. A blade with 18, 24, or 32 cutting teeth per inch is recommended for conduit. Use a higher tooth count for EMT and a lower tooth count for rigid conduit and IMC. If the blade needs to be replaced, point the teeth toward the front of the saw when installing the new blade.

Step 2 Secure the conduit in a pipe vise.

Step 3 Rest the middle of the hacksaw blade on the conduit where the cut is to be made. Position the saw so the end of the blade is pointing slightly down and the handle is pointing slightly up. Push forward gently until the cut is started. Make even strokes until the cut is finished.

CAUTION

To avoid bruising your knuckles on the newly cut pipe, use gentle strokes for the final cut.

Step 4 Check the cut. The end of the conduit should be straight and smooth. *Figure 23* shows correct and incorrect cuts. Ream the conduit.

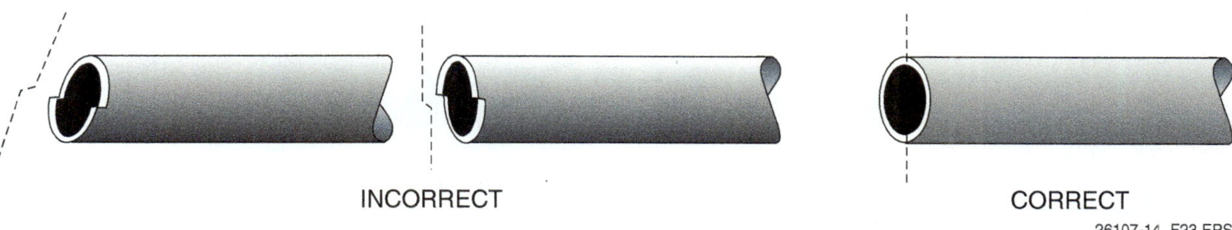

Figure 23 Conduit ends after cutting.

On Site

Using Your Bender Head to Secure Conduit

To secure conduit while cutting, insert the conduit into the bender head, brace your foot against the bender to secure it, then proceed to cut the conduit.

26107-14_SA05.EPS

3.2.0 Pipe Cutter Method

A pipe cutter can also be used to cut RMC and IMC. To use a pipe cutter:

Step 1 Secure the conduit in a pipe vise and mark a place for the cut.

Step 2 Open the cutter and place it over the conduit with the cutter wheel on the mark.

Step 3 Tighten the cutter by rotating the screw handle.

CAUTION

Do not overtighten the cutter. Overtightening can break the cutter wheel and distort the wall of the conduit.

Step 4 Rotate the cutter counterclockwise to start the cut. *Figure 24* shows the proper way to rotate the cutter.

Step 5 Tighten the cutter handle ¼ turn for each full turn around the conduit. Again, make sure that you do not overtighten it.

On Site

EMT Reaming Tools

There are specialty tools available for reaming EMT. One example is shown here. This tool slips over the end of a square-shank screwdriver and is secured in place with setscrews. The tool is inserted into the end of the EMT and rotated back and forth to deburr the conduit.

26107-14_SA06.EPS

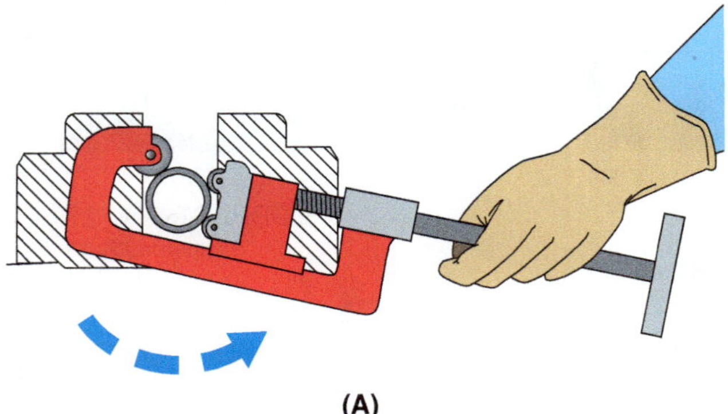

26107-14_F24.EPS

Figure 24 Cutter rotation.

Step 6 Add a few drops of cutting oil to the groove and continue cutting. Avoid skin contact with the oil.

Step 7 When the cut is almost finished, stop cutting and snap the conduit to finish the cut. This reduces the ridge that can be formed on the inside of the conduit.

Step 8 Clean the conduit and cutter with a shop towel rag.

Step 9 Ream the conduit.

3.3.0 Reaming Conduit

When the conduit is cut, the inside edge is sharp. This edge will damage the insulation of the wire when it is pulled through. To avoid this damage, the inside edge must be smoothed or reamed using a reamer (*Figure 25*).

26107-14_F25.EPS

Figure 25 Rigid conduit reamer.

On Site

Using Tape

Use a piece of tape as a guide for marking your cutting lines around the conduit. This ensures a straight cut.

To ream the inside edge of a piece of conduit using a hand reamer, proceed as follows:

Step 1 Place the conduit in a pipe vise.

Step 2 Insert the reamer tip in the conduit.

Step 3 Apply light forward pressure and start rotating the reamer. *Figure 26* shows the proper way to rotate the reamer. It should be rotated using a downward motion. The reamer can be damaged if you rotate it in the wrong direction. The reamer should bite as soon as you apply the proper pressure.

Step 4 Remove the reamer by pulling back on it while continuing to rotate it. Check the progress and then reinsert the reamer. Rotate the reamer until the inside edge is smooth. You should stop when all burrs have been removed.

NOTE

If a conduit reamer is not available, use a half-round file (the tang of the file must have a handle attached). EMT may be reamed using the nose of diagonal cutters or small hand reamers.

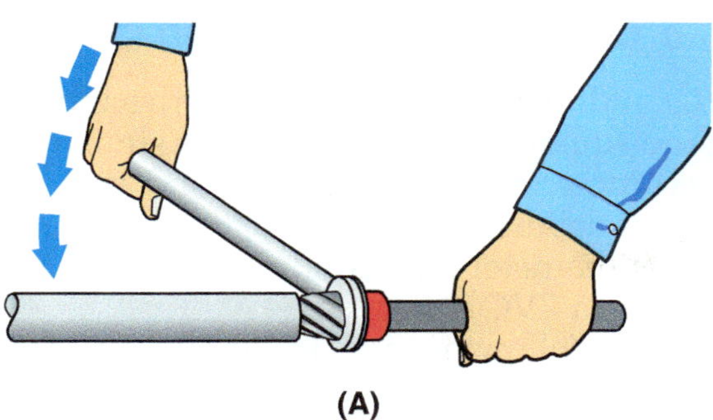
(A)

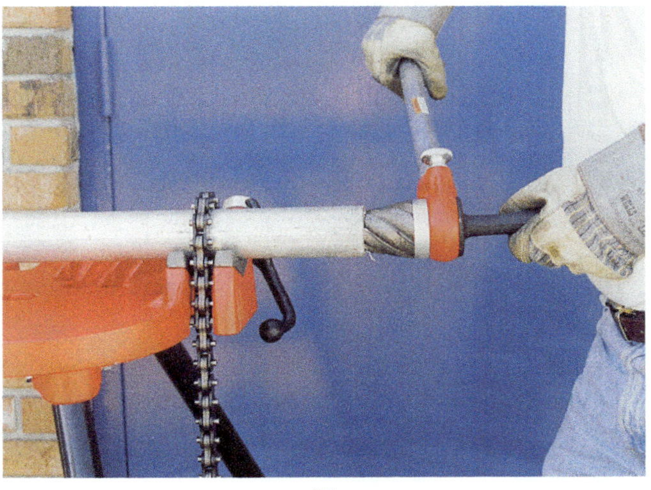
(B)

26102-14_F26.EPS

Figure 26 Reamer rotation.

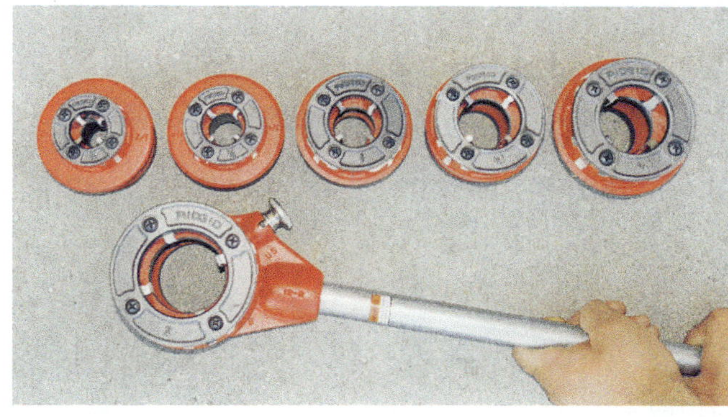
26107-14_F27.EPS

Figure 27 Hand-operated ratchet threader.

3.4.0 Threading Conduit

After conduit is cut and reamed, it is usually threaded so it can be properly joined. Only RMC and IMC have walls thick enough for threading.

The tool used to cut threads in conduit is called a die. Conduit dies are made to cut a taper of ¾ inch per foot. The number of threads per inch varies from 8 to 18, depending upon the diameter of the conduit. A thread gauge is used to measure how many threads per inch are cut.

The threading dies are contained in a die head. The die head can be used with a hand-operated ratchet threader (*Figure 27*) or with a portable power drive.

To thread conduit using a hand-operated threader, proceed as follows:

Step 1 Insert the conduit in a pipe vise. Make sure the vise is fastened to a strong surface. Place supports, if necessary, to help secure the conduit.

Step 2 Determine the correct die and head. Inspect the die for damage such as broken teeth. Never use a damaged die.

Step 3 Insert the die securely in the head. Make sure the proper die is in the appropriately numbered slot on the head.

Step 4 Determine the correct thread length to cut for the conduit size used (match the manufacturer's thread length).

Step 5 Lubricate the die with cutting oil at the beginning and throughout the threading operation. Avoid skin contact with the oil.

Step 6 Cut threads to the proper length. Make sure that the conduit enters the tapered side of the die. Apply pressure and start turning the head. You should back off the head each quarter-turn to clear away chips.

Step 7 Remove the die when the proper cut is made. Threads should be cut only to the length of the die. Overcutting will leave the threads exposed to corrosion.

Step 8 Inspect the threads to make sure they are clean, sharp, and properly made. Use a thread gauge to measure the threads. The finished end should allow for a wrench-tight fit with one or two threads exposed.

NOTE

The conduit should be reamed again after threading to remove any burrs and edges. Cutting oil must be swabbed from the inside and outside of the conduit. Use a sandbox or drip pan under the threader to collect drips and shavings.

On Site

Threading Conduit

The key to threading conduit is to start with a square cut. If you don't get it right, the conduit won't thread properly.

On Site

Oiling the Threader

For smoother operation, oil the threader often while threading the conduit.

26107-14_SA07.EPS

Die heads can also be used with portable power drives. You will follow the same steps when using a portable power drive. Threading machines are often used on larger conduit and where frequent threading is required. Threading machines hold and rotate the conduit while the die is fed onto the conduit for cutting. When using a threading machine, make sure you secure the legs properly and follow the manufacturer's instructions.

3.5.0 Cutting and Joining PVC Conduit

PVC conduit may be easily cut with a handsaw. To ensure square cuts, a miter box or similar device is recommended for cutting 2" and larger PVC. You can deburr the cut ends using a pocket knife. Smaller diameter PVC conduit, up to 1½", may be cut using a PVC cutter.

Use the following steps to join PVC conduit sections or attachments to plastic boxes:

CAUTION

Solvents and cements used with PVC are hazardous. Wear gloves and eye protection, and always follow the product instructions. Ensure that the area is well ventilated.

Step 1 Wipe all the contacting surfaces clean and dry.

Step 2 Apply a coat of cement (a brush or aerosol can is recommended) on both ends of the conduit.

NOTE

Cementing the PVC must be done quickly. The aerosol spray cans of cement or the cement/brush combination are usually provided by the PVC manufacturer. Make sure you use the recommended cement.

On Site

PVC Cutters

A nylon string can be used to cut PVC in place in awkward locations. However, it is best to use a PVC cutter to cut smaller trade sizes of PVC.

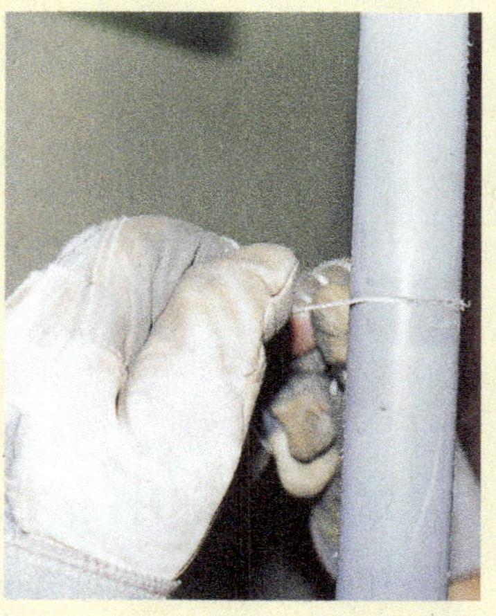

26107-14_SA08.EPS

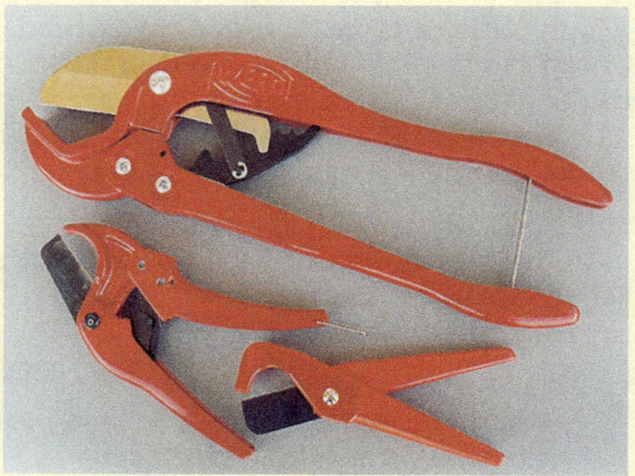

26107-14_SA09.EPS

Step 3 Press the conduit and fitting together and rotate about a half-turn to evenly distribute the cement.

Forming PVC in the field requires a special tool called a hot box or other specialized methods. PVC may not be threaded when it is used for electrical applications.

Think About It

Putting It All Together

This module has stressed the precision necessary for creating accurate and uniform bends. Why is this important? What practical problems can result from sloppy or inaccurate bends?

Summary

You must choose a conduit bender to suit the kind of conduit being installed and the type of bend to be made. Some knowledge of the geometry of right triangles and circles needs to be mastered to make the necessary calculations. You must be able to calculate, lay out, and perform bending operations on a single run of conduit and also on two or more parallel runs of conduit. At times, data tables for the figures may be consulted for the calculations. All work must conform to the requirements of the *NEC*®.

Review Questions

1. The field bending of PVC requires a _____.
   a. hickey
   b. heating unit
   c. segmented bender
   d. one-shot bender

2. A hickey can be used to bend _____.
   a. RMC
   b. EMT
   c. PVC
   d. HDPE

3. What is the key to accurate bending with a hand bender?
   a. Correct size and length of handle
   b. Constant foot pressure on the back piece
   c. Using only the correct brand of bender
   d. Correct inverting of the conduit bender

4. In a right triangle, the side directly opposite the 90° angle is called the _____.
   a. right side
   b. hypotenuse
   c. altitude
   d. base

5. Prior to making a 90° bend, what two measurements must be known?
   a. Length of conduit and size of conduit
   b. Desired rise and length of conduit
   c. Size of bender and size of conduit
   d. Stub-up distance and take-up distance

6. A back-to-back bend is _____.
   a. a two-shot 90° bend
   b. two 90° bends made back-to-back
   c. an offset with four bends back-to-back
   d. a segmented bend

7. To prevent the ends of the conduit from being staggered, what additional information must be used when making parallel offset bends?
   a. Center-to-center spacing and tangent of ½ the offset angle
   b. Length of conduit and size of conduit
   c. Stub-up distance and take-up distance
   d. Offset angle and length of conduit

8. When making a saddle bend, the center of the saddle will cause the conduit to shrink _____ for every inch of saddle depth.
   a. ⅜"
   b. 3⁄16"
   c. ¾"
   d. 3⁄32"

9. When using a pipe cutter, start the cut by rotating the cutter _____.
   a. in a clockwise direction
   b. with the grain
   c. in a counterclockwise direction
   d. against the grain

10. EMT is threaded using a die.
    a. True
    b. False

Trade Terms Quiz

Fill in the blank with the correct term that you learned from your study of this module

1. A right-angle bend is also called a(n) ____________.
2. The rise in a section of conduit is called a(n) ____________.
3. The ____________ length is the actual length of the conduit that will be bent.
4. Also called a kick, a(n) ____________ is two bends placed in a piece of conduit in order to navigate around obstructions.
5. ____________ bends are large bends that are formed by multiple short bends or shots.
6. Two 90° bends with a straight section of conduit between them constitute a(n) ____________ bend.
7. ____________ bends are 90° bends made in two or more parallel sections of conduit, where the radius of each bend in conduit after the inside bend is respectively increased.
8. ____________ is the distance that is saved by the arc of a 90° bend.
9. ____________ is the length of the bent section of conduit measured from the bottom, centerline, or top of the straight section to the end of the bent section.

Trade Terms

90° bend	Developed length	Offsets	Segment bend
Back-to-back bend	Gain	Rise	Stub-up
Concentric bends			

Module 26107-14
Supplemental Exercises

1. The sizes of conduit that can be bent using a hand bender are ____________________.
   a. ½ inch through ¾ inch
   b. ½ inch through 1¼ inch
   c. ½ inch through 1½ inch
   d. ½ inch through 2 inch

2. If a bender can be used to bend ¾-inch RMC, then it can also be used to bend ________ EMT.
   a. 1-inch
   b. 2-inch
   c. 2½-inch
   d. 3-inch

3. The take-up on 1-inch EMT is ____________________.

4. The take-up on ½-inch RMC is ____________________.

5. The typical gain on ½-inch RMC bent at 90° is ____________________.

6. On an offset using 30° bends and a depth of 6 inches, the conduit shrink is ____________________.
   a. 1⁄16 inch
   b. 1½ inches
   c. 1 inch
   d. 2¼ inches

7. On an offset using 30° bends and a depth of 6 inches, the distance between bends is ______________.
   a. 6 inches
   b. 7 inches
   c. 10 inches
   d. 12 inches

8. The conduit shrink is ________ per inch of offset when using 30° bends.
   a. 1⁄16 inch
   b. ⅛ inch
   c. ¼ inch
   d. ½ inch

9. The multiplier for determining the distance between bends is ________ when bending offsets using 30° bends.
   a. 1.4
   b. 2.0
   c. 2.6
   d. 6.0

10. The multiplier for determining the distance between bends is ________ when bending offsets using 45° bends.
    a. 1.2
    b. 1.4
    c. 2.6
    d. 2.8

Cornerstone of Craftsmanship

Timothy Ely

Beacon Electric Company

Tim Ely is a man who believes in giving something back to the industry that nurtured his successful career. Despite working in a demanding executive position, he serves on many industry committees and was instrumental in the development of the NCCER Electrical Program.

What made you decide to become an electrician?
During my last two years of high school, I worked for a do-it-all construction company. We laid concrete, installed roofs, hung drywall, installed plumbing, and did electrical work. I liked the electrical work the best.

How did you learn the trade?
I learned through on-the-job training, hard work, and studying on my own. I had good teachers who were patient with me and took the time to help me succeed.

What kinds of jobs did you hold on the way to your current position?
I started out wiring houses and did that for the first two years. Then I switched over to commercial and industrial work, and worked as an apprentice in that area for two more years before becoming a journeyman. From there, I served as a lead electrician, then foreman, then city superintendent, then finally general superintendent before being promoted to my current job as vice president of construction.

What factor or factors have contributed most to your success?
Hard work helps a lot. I also try to bring a positive attitude to work with me every day. My family and friends have supported me throughout my career.

What does a vice president of construction do in your company?
In my job, I have responsibility for all the job sites, as well as the warehouse and service trucks. I also have responsibility for employee hiring, safety training, job planning and scheduling, quality control, and licensing. I personally hold 28 different state and city licenses, and I firmly believe that getting the training to obtain your licenses and then doing the in-service training to keep your licenses current are important factors in an electrician's success. For example, an electrical contractor can bid on jobs in a wide geographical area. Electricians working for that contractor can work on projects in different cities, even different states. Every place you go will require you to have a valid license.

What advice would you give to someone entering the electrical trade?
Work hard, treat people with respect, and keep an open mind. Be careful how you deal with people. Someone you offend today may wind up being your boss or a potential customer tomorrow.

Trade Terms Introduced in This Module

90° bend: A bend that changes the direction of the conduit by 90°.

Back-to-back bend: Any bend formed by two 90° bends with a straight section of conduit between the bends.

Concentric bends: 90° bends made in two or more parallel runs of conduit with the radius of each bend increasing from the inside of the run toward the outside.

Developed length: The actual length of the conduit that will be bent.

Gain: Because a conduit bends in a radius and not at right angles, the length of conduit needed for a bend will not equal the total determined length. Gain is the distance saved by the arc of a 90° bend.

Offsets: An offset (kick) is two bends placed in a piece of conduit to change elevation to go over or under obstructions or for proper entry into boxes, cabinets, etc.

Rise: The length of the bent section of conduit measured from the bottom, centerline, or top of the straight section to the end of the bent section.

Segment bend: A large bend formed by multiple short bends or shots.

Stub-up: Another name for the rise in a section of conduit. Also, a term used for conduit penetrating a slab or the ground.

Appendix A

Using Trigonometry to Determine Offset Angles and Multipliers

You do not have to be a mathematician to use trigonometry. Understanding the basic trig functions and how to use them can help you calculate unknown distances or angles. Assume that the right triangle below represents a conduit offset. If you know the length of one side and the angle, you can calculate the length of the other sides, or if you know the length of any two of the sides of the triangle, you can then find the offset angle using one or more of these trig functions. You can use a trig table such as that shown on the following pages or a scientific calculator to determine the offset angle. For example, if the cosecant of angle A is 2.6, the trig table tells you that the offset angle is 22½°.

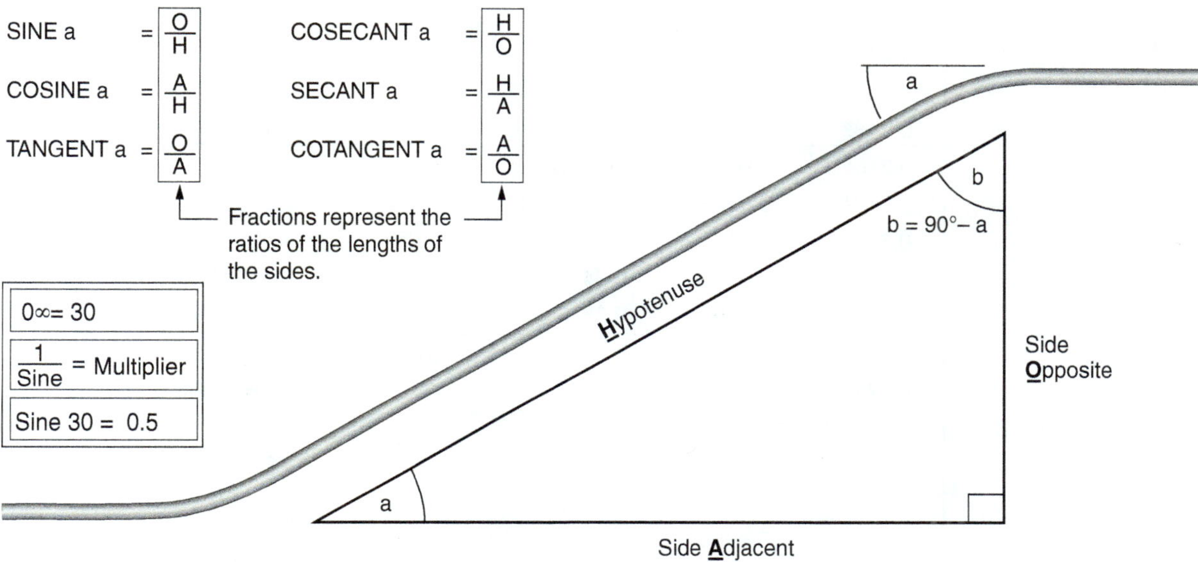

To determine the multiplier for the distance between bends in an offset:

1. Determine the angle of the offset: 30°
2. Find the sine of the angle: 0.5
3. Find the inverse (reciprocal) of the sine: $\frac{1}{0.5}$ = 2. This is also listed in trig tables as the cosecant of the angle.
4. This number multiplied by the height of the offset gives the hypotenuse of the triangle, which is equal to the distance between bends.

26107-14_A01.EPS

ANGLE	SINE	COSINE	TANGENT	COTANGENT	COSECANT
1	0.0175	0.9998	0.0175	57.3000	57.3065
2	0.0349	0.9994	0.0349	28.6000	28.6532
3	0.0523	0.9986	0.0524	19.1000	19.1058
4	0.0698	0.9976	0.0699	14.3000	14.3348
5	0.0872	0.9962	0.0875	11.4000	11.4731
6	0.1045	0.9945	0.1051	9.5100	9.5666
7	0.1219	0.9925	0.1228	8.1400	8.2054
8	0.1392	0.9903	0.1405	7.1200	7.1854
9	0.1564	0.9877	0.1584	6.3100	6.3926
10	0.1736	0.9848	0.1763	5.6700	5.7587
11	0.1908	0.9816	0.1944	5.1400	5.2408
12	0.2079	0.9781	0.2126	4.7000	4.8097
13	0.2250	0.9744	0.2309	4.3300	4.4454
14	0.2419	0.9703	0.2493	4.0100	4.1335
15	0.2588	0.9659	0.2679	3.7300	3.8636
16	0.2756	0.9613	0.2867	3.4900	3.5915
17	0.2924	0.9563	0.3057	3.2700	3.4203
18	0.3090	0.9511	0.3249	3.0800	3.2360
19	0.3256	0.9455	0.3443	2.9000	3.0715
20	0.3420	0.9397	0.3640	2.7500	2.9238
21	0.3584	0.9336	0.3839	2.6100	2.7904
22	0.3746	0.9272	0.4040	2.4800	2.6694
23	0.3907	0.9205	0.4245	2.3600	2.5593
24	0.4067	0.9135	0.4452	2.2500	2.4585
25	0.4226	0.9063	0.4663	2.1400	2.3661
26	0.4384	0.8988	0.4877	2.0500	2.2811
27	0.4540	0.8910	0.5095	1.9600	2.2026
28	0.4695	0.8829	0.5317	1.8800	2.1300
29	0.4848	0.8746	0.5543	1.8000	2.0626
30	0.5000	0.8660	0.5774	1.7300	2.0000
31	0.5150	0.8572	0.6009	1.6600	1.9415
32	0.5299	0.8480	0.6249	1.6000	1.8870
33	0.5446	0.8387	0.6494	1.5400	1.8360
34	0.5592	0.8290	0.6745	1.4800	1.7883
35	0.5736	0.8192	0.7002	1.4300	1.7434
36	0.5878	0.8090	0.7265	1.3800	1.7012
37	0.6018	0.7986	0.7536	1.3300	1.6616
38	0.6157	0.7880	0.7813	1.2800	1.6242
39	0.6293	0.7771	0.8098	1.2300	1.5890
40	0.6428	0.7660	0.8391	1.1900	1.5557
41	0.6561	0.7547	0.8693	1.1500	1.5242
42	0.6691	0.7431	0.9004	1.1100	1.4944
43	0.6820	0.7314	0.9325	1.0700	1.4662
44	0.6947	0.7193	0.9657	1.0400	1.4395
45	0.7071	0.7071	1.0000	1.0000	1.4142

26107-14_A02.EPS

ANGLE	SINE	COSINE	TANGENT	COTANGENT	COSECANT
46°	0.7193	0.6947	1.0355	0.9660	1.4395
47°	0.7314	0.6820	1.0724	0.9330	1.3673
48°	0.7431	0.6691	1.1106	0.9000	1.3456
49°	0.7547	0.6561	1.1504	0.8690	1.3250
50°	0.7660	0.6428	1.1918	0.8390	1.3054
51°	0.7771	0.6293	1.2349	0.8100	1.2867
52°	0.7880	0.6157	1.2799	0.7810	1.2690
53°	0.7986	0.6018	1.3270	0.7540	1.2521
54°	0.8090	0.5878	1.3764	0.7270	1.2360
55°	0.8192	0.5736	1.4281	0.7000	1.2207
56°	0.8290	0.5592	1.4826	0.6750	1.2062
57°	0.8387	0.5446	1.5399	0.6490	1.1923
58°	0.8480	0.5299	1.6003	0.6250	1.1791
59°	0.8572	0.5150	1.6643	0.6010	1.1666
60°	0.8660	0.5000	1.7321	0.5770	1.1547
61°	0.8746	0.4848	1.8040	0.5540	1.1433
62°	0.8829	0.4695	1.8807	0.5320	1.1325
63°	0.8910	0.4540	1.9626	0.5100	1.1223
64°	0.8988	0.4384	2.0503	0.4880	1.1126
65°	0.9063	0.4226	2.1445	0.4660	1.1033
66°	0.9135	0.4067	2.2460	0.4450	1.0946
67°	0.9205	0.3907	2.3559	0.4240	1.0863
68°	0.9272	0.3746	2.4751	0.4040	1.0785
69°	0.9336	0.3584	2.6051	0.3840	1.0711
70°	0.9397	0.3420	2.7475	0.3640	1.0641
71°	0.9455	0.3256	2.9042	0.3440	1.0576
72°	0.9511	0.3090	3.0777	0.3250	1.0514
73°	0.9563	0.2924	3.2709	0.3060	1.0456
74°	0.9613	0.2756	3.4874	0.2870	1.0402
75°	0.9659	0.2588	3.7321	0.2680	1.0352
76°	0.9703	0.2419	4.0108	0.2490	1.0306
77°	0.9744	0.2250	4.3315	0.2310	1.0263
78°	0.9781	0.2079	4.7046	0.2130	1.0223
79°	0.9816	0.1908	5.1446	0.1940	1.0187
80°	0.9848	0.1736	5.6713	0.1760	1.0154
81°	0.9877	0.1564	6.3138	0.1580	1.0124
82°	0.9903	0.1392	7.1154	0.1410	1.0098
83°	0.9925	0.1219	8.1443	0.1230	1.0075
84°	0.9945	0.1045	9.5144	0.1050	1.0055
85°	0.9962	0.0872	11.4301	0.0880	1.0038
86°	0.9976	0.0698	14.3007	0.0700	1.0024
87°	0.9986	0.0523	19.0811	0.0520	1.0013
88°	0.9994	0.0349	28.6363	0.0350	1.0006
89°	0.9998	0.0175	57.2900	0.0180	1.0001
90°	1.0000	0.0000	—	0.0000	1.0000

26107-14_A03.EPS

Appendix B

Bending Radius Table

Radius (Inches)	Radius Increments (Inches)									
	0	**1**	**2**	**3**	**4**	**5**	**6**	**7**	**8**	**9**
0	0.00	1.57	3.14	4.71	6.28	7.85	9.42	10.99	12.56	14.13
10	15.70	17.27	18.84	20.41	21.98	23.85	25.12	26.69	28.26	29.83
20	31.40	32.97	34.54	36.11	37.68	39.25	40.82	42.39	43.96	45.83
30	47.10	48.67	50.24	51.81	53.38	54.95	56.52	58.09	59.66	61.23
40	62.80	64.37	65.94	67.50	69.03	70.65	72.22	73.79	75.36	76.93
50	87.50	80.07	81.64	83.21	84.78	86.35	87.92	89.49	91.06	92.63
60	94.20	95.77	97.34	98.91	100.48	102.05	103.62	105.19	106.76	108.33
70	109.90	111.47	113.04	114.61	116.18	117.75	119.32	120.89	122.46	124.03
80	125.60	127.17	128.74	130.31	131.88	133.45	135.02	136.59	138.16	139.73
90	141.30	142.87	144.44	146.01	147.58	149.15	150.72	–	–	–

To find the developed length for the following angles, use a fraction of the 90° chart.

For	15°	22½°	30°	45°	60°	67½°	75°	90°
Take	1/6	1/4	1/3	1/2	2/3	3/4	5/6	See Chart

For any other degrees: Developed length = 0.01744 × radius × degrees.

26107-14_A04.EPS

Additional Resources

This module presents thorough resources for task training. The following resource material is suggested for further study.

Benfield Conduit Bending Manual, 2nd Edition. Overland Park, KS: EC&M Books.

National Electrical Code® Handbook, Latest Edition. Quincy, MA: National Fire Protection Association.

Tom Henry's Conduit Bending Package (includes video, book, and bending chart). Winter Park, FL: Code Electrical Classes, Inc.

Figure Credits

Associated Builders and Contractors, Inc., Module opener

Topaz Publications, Inc., Figures 1, 3–5, 11B (photo), 24B (photo), 25, 26B (photo), 27, SA01–SA03, SA07–SA09

John Traister, Table 1, Figures 8–10

Greenlee/A Textron Company, SA04

Tim Ely, SA05

Klein Tools, Inc., SA06

NCCER CURRICULA — USER UPDATE

NCCER makes every effort to keep its textbooks up-to-date and free of technical errors. We appreciate your help in this process. If you find an error, a typographical mistake, or an inaccuracy in NCCER's curricula, please fill out this form (or a photocopy), or complete the online form at **www.nccer.org/olf**. Be sure to include the exact module ID number, page number, a detailed description, and your recommended correction. Your input will be brought to the attention of the Authoring Team. Thank you for your assistance.

Instructors – If you have an idea for improving this textbook, or have found that additional materials were necessary to teach this module effectively, please let us know so that we may present your suggestions to the Authoring Team.

NCCER Product Development and Revision

13614 Progress Blvd., Alachua, FL 32615

Email: curriculum@nccer.org

Online: www.nccer.org/olf

❑ Trainee Guide ❑ AIG ❑ Exam ❑ PowerPoints Other ______________________________

Craft / Level: Copyright Date:

Module ID Number / Title:

Section Number(s):

Description:

Recommended Correction:

Your Name:

Address:

Email: Phone:

Raceways and Fittings

The Beauvallon

The Beauvallon, a high-end, mixed-use project in Denver's Golden Triangle features two 14-story towers with a total of 210 luxury, for-sale condos. The project's European flavor comes from what designers call its "modified French Baroque style," which features wide balconies, two-story mansard roofs, ornate cornices, urban landscaping, and other Old World architectural elements.

26108-14

Trainees with successful module completions may be eligible for credentialing through NCCER's National Registry. To learn more, go to **www.nccer.org** or contact us at **1.888.622.3720.** Our website has information on the latest product releases and training, as well as online versions of our *Cornerstone* magazine and Pearson's product catalog.

Your feedback is welcome. You may email your comments to **curriculum@nccer.org,** send general comments and inquiries to **info@nccer.org,** or fill in the User Update form at the back of this module.

This information is general in nature and intended for training purposes only. Actual performance of activities described in this manual requires compliance with all applicable operating, service, maintenance, and safety procedures under the direction of qualified personnel. References in this manual to patented or proprietary devices do not constitute a recommendation of their use.

26108-14

Raceways and Fittings

Objectives

When you have completed this module, you will be able to do the following:

1. Identify and select various types and sizes of raceways and fittings for a given application.
2. Identify various methods used to fabricate (join) and install raceway systems.
3. Identify uses permitted for selected raceways.
4. Demonstrate how to install a flexible raceway system.
5. Terminate a selected raceway system.
6. Identify the appropriate conduit body for a given application.

Performance Tasks

Under the supervision of the instructor, you should be able to do the following:

1. Identify and select various types and sizes of raceways, fittings, and fasteners for a given application.
2. Demonstrate how to install a flexible raceway system.
3. Terminate a selected raceway system.
4. Identify the appropriate conduit body for a given application.

Trade Terms

Accessible
Approved
Bonding wire
Cable trays
Conduit
Exposed location
Kick
Raceways
Splice
Tap
Trough
Underwriters Laboratories, Inc. (UL)
Wireways

Required Trainee Materials

1. Paper and pencil
2. Copy of the latest edition of the *National Electrical Code®*
3. Appropriate personal protective equipment

Note:
NFPA 70®, *National Electrical Code®*, and *NEC®* are registered trademarks of the National Fire Protection Association, Inc., Quincy, MA 02269. All *National Electrical Code®* and *NEC®* references in this module refer to the 2014 edition of the *National Electrical Code®*.

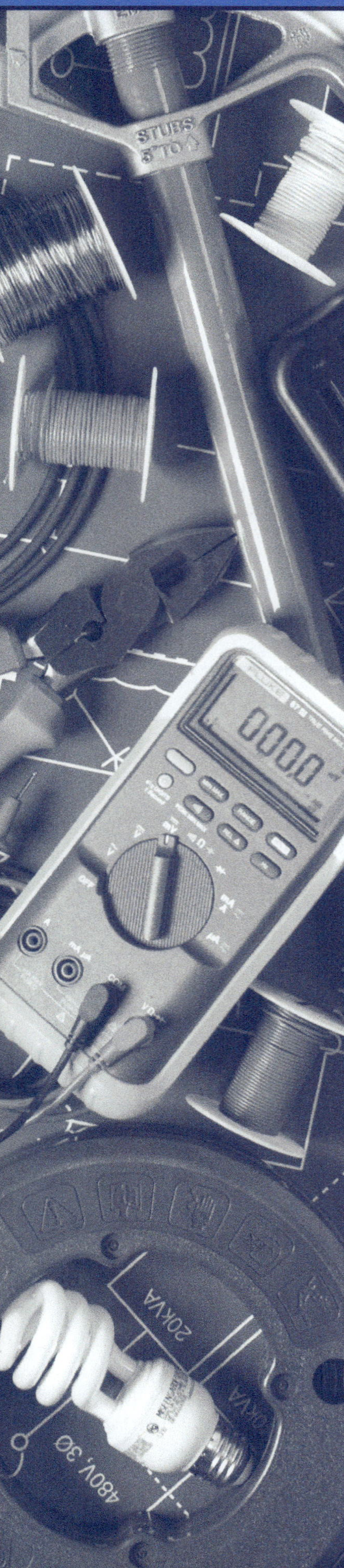

Contents

Topics to be presented in this module include:

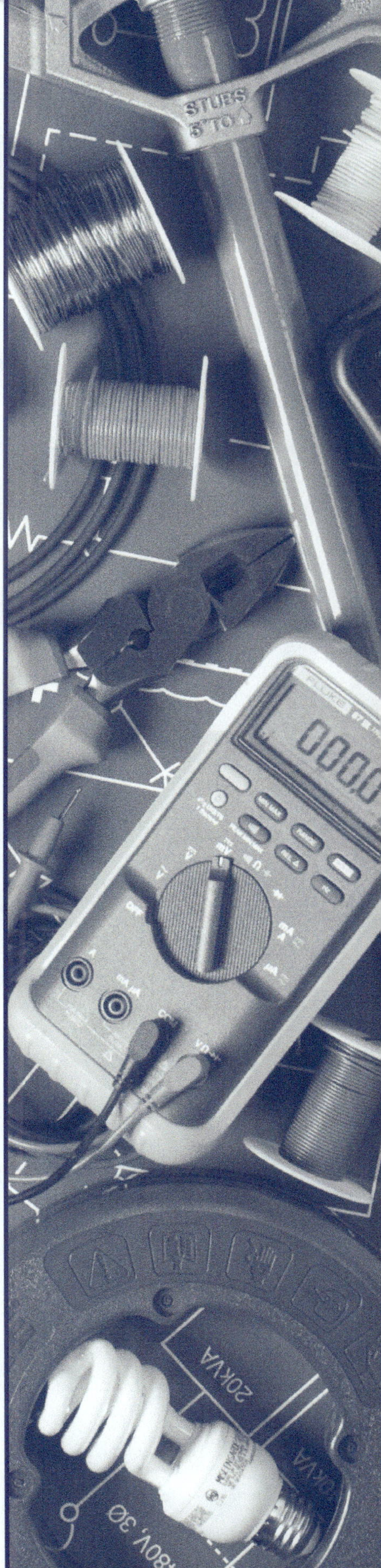

20kVA
480V, 3Ø

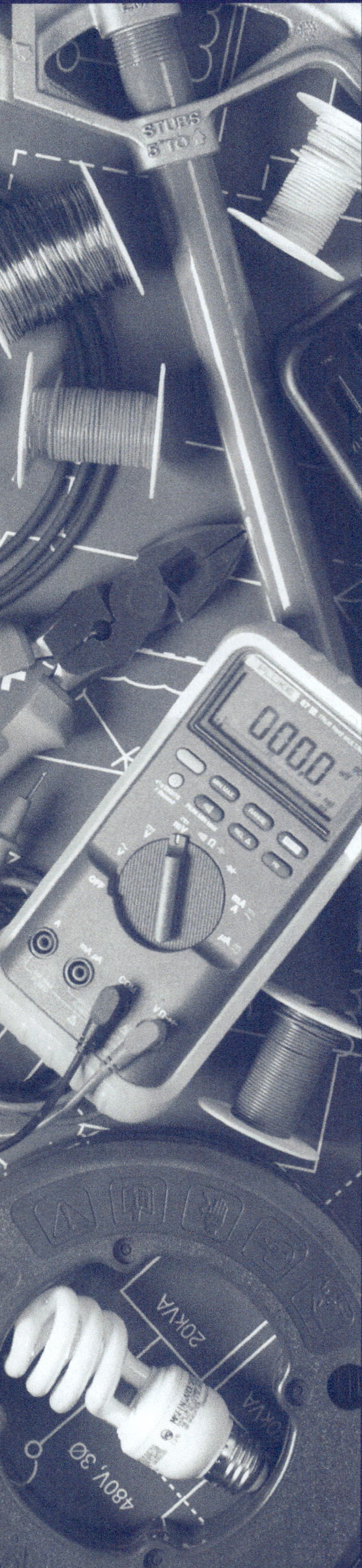

Figures and Tables

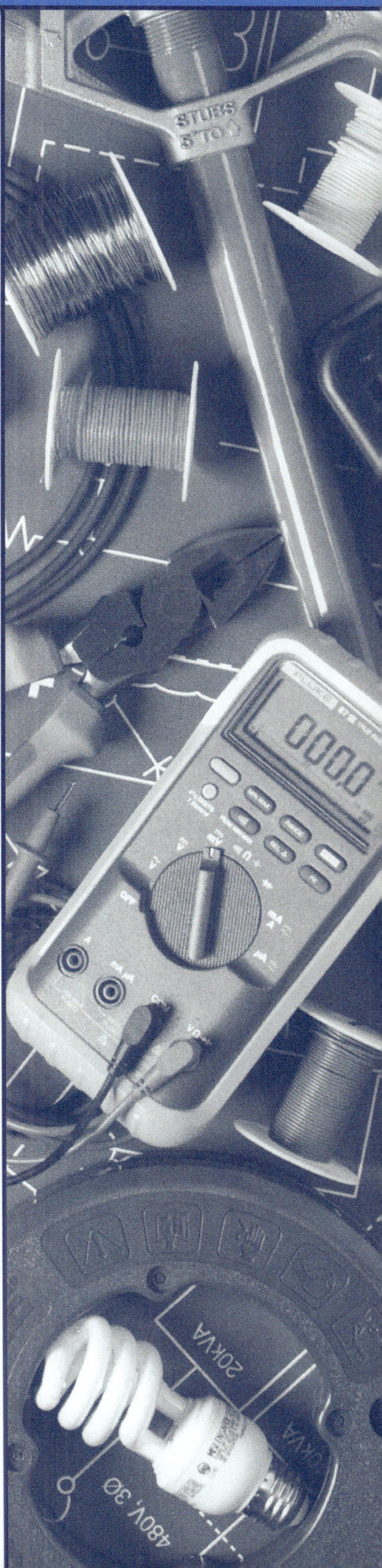

STUBS
5"TO
0000
20kVA
480V, 3Ø

1.0.0 Introduction

Electrical raceways present challenges and requirements involving proper installation techniques, general understanding of raceway systems, and applications of the *NEC®* to raceway systems. Acquiring quality installation skills for raceway systems requires practice, knowledge, and training.

A presentation of the various types of raceway systems and fittings, basic raceway installation skills, and *NEC®* requirements applicable to raceway systems is included in this module. This module also covers raceway supports and environmental considerations for raceway systems, as well as general raceway information.

Along with the study of this module, the following *NEC®* Articles should be referenced:

- *NEC Article 250 – Grounding and Bonding*
- *NEC Article 342 – Intermediate Metal Conduit: Type IMC*
- *NEC Article 344 – Rigid Metal Conduit: Type RMC*
- *NEC Article 348 – Flexible Metal Conduit: Type FMC*
- *NEC Article 350 – Liquidtight Flexible Metal Conduit: Type LFMC*
- *NEC Article 352 – Rigid Polyvinyl Chloride Conduit: Type PVC*
- *NEC Article 353 – High-Density Polyethylene Conduit: Type HDPE*
- *NEC Article 356 – Liquidtight Flexible Nonmetallic Conduit: Type LFNC*
- *NEC Article 358 – Electrical Metallic Tubing: Type EMT*
- *NEC Article 360 – Flexible Metallic Tubing: Type FMT*
- *NEC Article 362 – Electrical Nonmetallic Tubing: Type ENT*
- *NEC Article 376 – Metal Wireways*
- *NEC Article 378 – Nonmetallic Wireways*
- *NEC Article 392 – Cable Trays*

2.0.0 Raceways

Raceway is a general term referring to a wide range of circular and rectangular enclosed channels used to house electrical wiring. Raceways can be metallic or nonmetallic and come in different shapes. Depending on the particular purpose for which they are intended, raceways include enclosures such as underfloor raceways, flexible metal conduit, tubing, wireways, surface metal raceways, surface nonmetallic raceways, and support systems such as cable trays.

3.0.0 Conduit

Conduit is a raceway with a circular cross section, similar to pipe, that contains wires or cables. Conduit is used to provide protection for conductors and route them from one place to another. Metal conduit also provides a permanent electrical path to ground. This equipment must be listed per the *NEC®*.

3.1.0 Conduit as a Ground Path

For safety reasons, most equipment that receives electrical power and has a metallic frame is bonded. In order to bond the equipment, an electrical connection must be made to connect the metal frame of the electrically powered equipment to the grounding point at the service-entrance equipment. This is usually done in one or both of the following ways:

- The frame of the equipment is connected to a wire (equipment grounding conductor), which is directly connected to the ground point at the grounding terminal.
- The frame of the equipment is connected (bonded) to a metal conduit or other type of raceway system, which provides an uninterrupted and low-impedance circuit to the ground point at the service-entrance equipment. The metal raceway or conduit acts as the equipment grounding conductor.

According to *NEC Section 250.96,* metal raceways, cable trays, cable armor, cable sheath, enclosures, frames, fittings, and other metal noncurrent-carrying parts that are to serve as equipment grounding conductors, with or without the use of supplementary equipment grounding conductors, shall be bonded where necessary to ensure electrical continuity and the capacity to safely conduct any fault current likely to be imposed on them. Any nonconductive paint, enamel, or similar coating shall be removed at threads, contact points, and contact surfaces or be connected by means of fittings designed so as to make such removal unnecessary.

The purpose of the equipment grounding conductor is to provide a low-resistance path to ground for all equipment that receives power. This is done so that if an ungrounded conductor comes in contact with the frame of a piece of equipment, the circuit overcurrent device immediately acts to open the circuit. It also reduces the voltage to ground that would be present on the faulted equipment if a person came in contact with the equipment frame.

3.2.0 Types of Conduit and Tubing

There are many types of conduit used in the construction industry. The size of conduit to be used is determined by engineering specifications, local codes, and the *NEC*®. Refer to *NEC Chapter 9, Tables 1 through 4 and Informative Annex C* for conduit fill with various conductors. There are several common types of conduit to examine.

3.2.1 Electrical Metallic Tubing

Electrical metallic tubing (EMT) is the lightest duty tubing available for enclosing and protecting electrical wiring. EMT is widely used for residential, commercial, and industrial wiring systems. It is lightweight, easily bent and/or cut to shape, and is the least costly type of metallic conduit. Because the wall thickness of EMT is less than that of rigid conduit, it is often referred to as thinwall conduit. A comparison of inside and outside diameters of EMT to rigid metal conduit (RMC) and intermediate metal conduit (IMC) is shown in *Figure 1*.

NEC Section 358.10(A) permits the installation of EMT for either exposed or concealed work. Per *NEC Section 358.10(B)*, ferrous or nonferrous EMT, elbows, couplings, and fittings shall be permitted to be installed in concrete, in direct contact with the earth, or in areas subject to severe corrosive influences where protected by corrosion protection and approved as suitable for the condition. Per *NEC Section 358.10(C)*, supports, bolts, straps, screws, and so forth shall be of corrosion-resistant materials or protected against corrosion by corrosion-resistant materials.

EMT shall not be used (1) where, during installation or afterward, it will be subject to severe physical damage; (2) where protected from corrosion solely by enamel; (3) in cinder concrete or cinder fill where subject to permanent moisture unless protected on all sides by a layer of non-cinder concrete at least 2 inches thick or unless the tubing is at least 18 inches under the fill; (4) in any hazardous (classified) locations except as permitted by *NEC Sections 502.10(B)(2), 503.10, and 504.20;* or (5) for the support of fixtures or other equipment.

In a wet area, EMT and other conduit must be installed to prevent water from entering the conduit system. In locations where walls are subject to regular wash-down [see *NEC Section 300.6(D)*], the entire conduit system must be installed to provide a ¼-inch air space between it and the wall or supporting surface. The entire conduit system is considered to include conduit, boxes, and fittings. To ensure resistance to corrosion caused by wet environments, EMT is galvanized. The term *galvanized* is used to describe the procedure in which the interior and exterior of the conduit are coated with a corrosion-resistant zinc compound.

EMT, being a good conductor of electricity, may be used as an equipment grounding conductor. In order to qualify as an equipment grounding conductor [see *NEC Section 250.118(4)*], the conduit system must be tightly connected at each joint and provide a continuous grounding path from each electrical load to the service equipment. The connectors used in an EMT system ensure electrical and mechanical continuity throughout the system (see *NEC Sections 250.96, 300.10, and 358.42*).

NOTE

Support requirements for EMT are also covered in *NEC Section 358.30*. The types of supports will be discussed later in this module.

Because EMT is too thin for threads, fittings listed for EMT must be used. For wet or damp locations, compression fittings such as those shown in *Figure 2* are used. These fittings contain

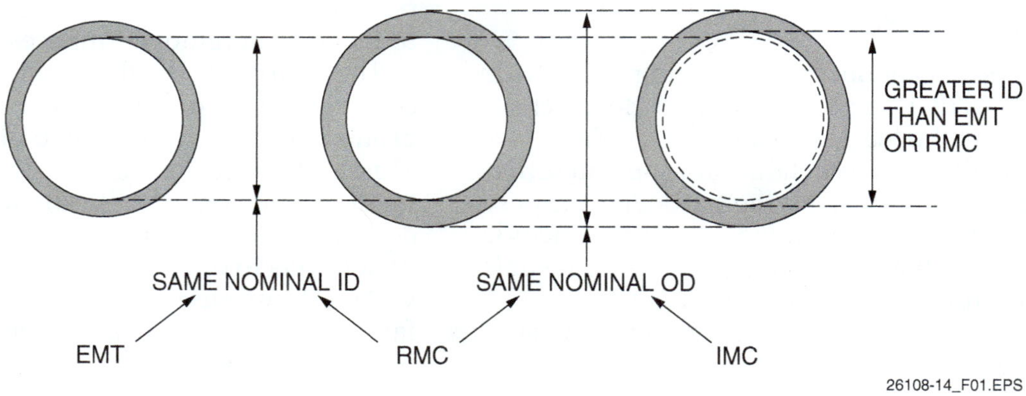

Figure 1 Conduit comparison.

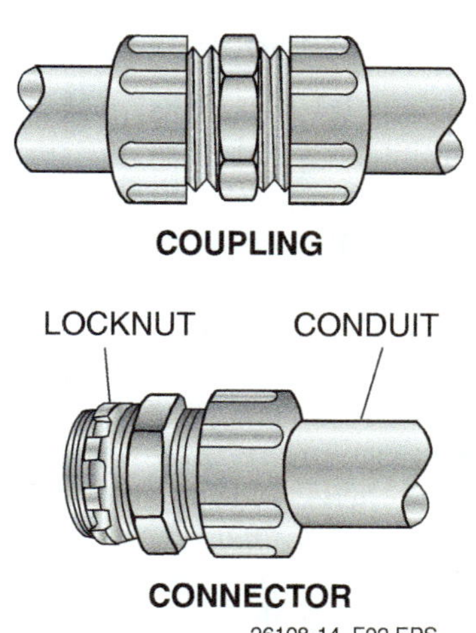

Figure 2 Compression fittings.

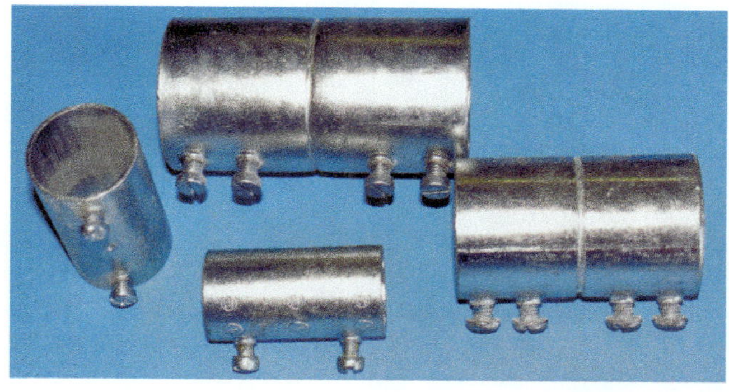

Figure 3 Setscrew fittings.

Think About It

EMT Use

Where would you use EMT? Are there any circumstances where EMT cannot be run through a suspended ceiling? What are some differences between EMT and rigid conduit?

a compression ring made of metal that forms a raintight seal.

When EMT compression couplings are used, they must be securely tightened, and when installed in masonry concrete, they must be of the concrete-tight type. If installed in a wet location, they must be the raintight type. Refer to *NEC Section 358.42.*

EMT fittings for dry locations can be either the setscrew type or the indenting type. To use the setscrew type, the ends of the EMT are inserted into the sleeve and the setscrews are tightened to make the connection. Various types of setscrew fittings are shown in *Figure 3*.

EMT sizes of 2½ inches and larger have the same outside diameter as corresponding sizes of galvanized RMC. RMC threadless connectors may be used to connect EMT.

NOTE

EMT connectors smaller than 2½ inches, although they are the same size as RMC threadless connectors, may not be used to connect RMC.

Both setscrew and compression couplings are available in die-cast or steel construction. Steel couplings are stronger, but may not seal as well.

Support requirements for EMT are presented in *NEC Section 358.30.* As with most other metal conduit, EMT must be supported at least every 10 feet and within 3 feet of each outlet box, junction box, cabinet, fitting, or terminating end of the conduit. An exception to *NEC Section 358.30(A), Exception 1* allows the fastening of unbroken lengths of EMT to be increased to a distance of 5 feet where structural members do not readily permit fastening within 3 feet.

Electrical nonmetallic tubing (ENT) is also available. It provides an economical alternative to EMT, but it can only be used in certain applications. See *NEC Article 362.*

3.2.2 Rigid Metal Conduit

Rigid metal conduit (RMC) is conduit that is constructed of metal of sufficient thickness to permit the cutting of pipe threads at each end. Specific information on RMC may be found in *NEC Article 344*. RMC provides the best physical protection for conductors of any of the various types of conduit. RMC is supplied in 10-foot lengths including a threaded coupling on one end.

RMC may be made from steel or aluminum. Rigid metal steel conduit may be stainless, galvanized, or enamel-coated inside and out. Because of its threaded fittings, RMC provides an excellent equipment grounding conductor as defined in *NEC Section 250.118(2).* A piece of RMC is shown in *Figure 4(A)*. The support requirements for RMC are presented in *NEC Section 344.30(A) and (B)* and *NEC Table 344.30(B)(2).*

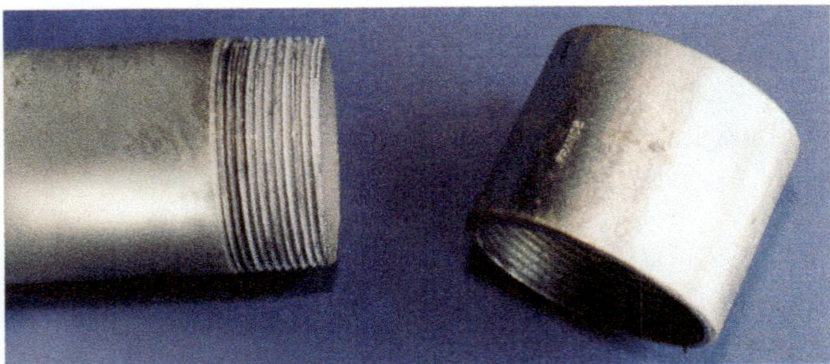

(A) RIGID METAL CONDUIT (RMC)

(B) PLASTIC-COATED RMC

26108-14_F04.EPS

Figure 4 Types of rigid metal conduit (RMC).

RMC is mostly used in industrial applications. RMC is heavier than EMT and IMC. It is more difficult to cut and bend, usually requires threading of cut ends, and has a higher purchase price than EMT and IMC. As a result, the cost of installing RMC is generally higher than the cost of installing EMT and IMC.

3.2.3 Plastic-Coated RMC

Plastic-coated RMC has a thin coating of polyvinyl chloride (PVC) over the RMC. See *Figure 4(B)*. This combination is useful when an environment calls for the ruggedness of RMC along with the corrosion resistance of rigid nonmetallic conduit (PVC). Typical installations where plastic-coated RMC may be required are:

- Chemical plants
- Food plants
- Refineries
- Fertilizer plants
- Paper mills
- Wastewater treatment plants

Plastic-coated RMC requires special threading and bending techniques.

3.2.4 Aluminum Conduit

Aluminum conduit has several characteristics that distinguish it from steel conduit. Because it has better resistance to wet environments and some chemical environments, aluminum conduit generally requires less maintenance in installations such as sewage treatment plants.

NEC Section 300.6(B) states that aluminum conduit used in concrete or in direct contact with soil requires supplementary corrosion protection. According to Underwriters Laboratories *Electrical Construction Equipment Directory* (UL Green Book), examples of supplementary protection are paints approved for the purpose (such as bitumastic paint), tape wraps approved for the purpose, or PVC-coated conduit.

NOTE

Caution must be exercised to avoid burial of aluminum conduit in soil or concrete that contains calcium chloride. Calcium chloride may interfere with the corrosion resistance of aluminum conduit. Calcium chloride and similar materials are often added to concrete to speed concrete setting. It is important to determine if chlorides are to be used in the concrete prior to installing aluminum conduit. If chlorides are to be used, aluminum conduit must be avoided. Check with local authorities regarding this type of usage.

3.2.5 Black Enamel Steel Conduit

Rigid black enamel steel conduit (often called black conduit) is steel conduit that is coated with a black enamel. In the past, this type of conduit was used exclusively for indoor wiring. Black enamel steel conduit is no longer manufactured for sale in the United States. It is mentioned only because it may still be found in existing installations.

3.2.6 Intermediate Metal Conduit

Intermediate metal conduit (IMC) has a wall thickness that is less than RMC but greater than that

On Site

RMC Installations

Use RMC in hazardous environments, such as feed mills, or in areas where there is a chance of physical abuse or extreme moisture, such as outdoor environments. The *NEC®* does allow EMT to be buried in the ground or in concrete, but galvanized RMC is more commonly used.

On Site

EMT Installation

When installing EMT, hook your index finger up through the box to check that the conduit is seated in the connector. If you feel a lip between the conduit and the connector, the conduit is not properly seated.

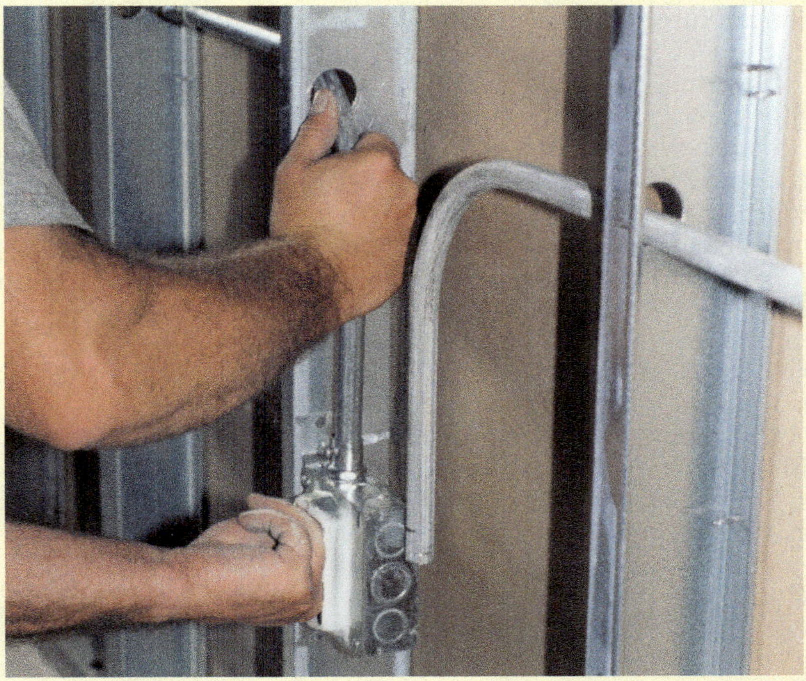

26108-14_SA01.EPS

of EMT. The weight of IMC is approximately ⅔ that of RMC. Because of its lower purchase price, lighter weight, and thinner walls, IMC installations are generally less expensive than comparable RMC installations. However, IMC installations still have high strength ratings.

NOTE

Additional information on IMC may be found in *NEC Article 342.*

The outside diameter of a given size of IMC is the same as that of the comparable size of RMC. Therefore, RMC fittings may be used with IMC. Since the threads on IMC and RMC are the same size, no special threading tools are needed to thread IMC. Some electricians feel that threading IMC is more difficult than threading RMC because IMC is somewhat harder.

The internal diameter of a given size of IMC is somewhat larger than the internal diameter of the same size of RMC because of the difference in wall thickness. Bending IMC is considered easier than bending RMC because of the reduced wall thickness. However, bending is sometimes complicated by kinking, which may be caused by the increased hardness of IMC.

The *NEC®* requires that IMC be identified along its length at 5-foot intervals with the letters IMC. *NEC Sections 110.21 and 342.120* describe this marking requirement.

Like RMC, IMC is permitted to act as an equipment grounding conductor, as defined in *NEC Section 250.118(3).* The use of IMC may be restricted in some jurisdictions. It is important to

On Site

Use of Aluminum Conduit

Aluminum conduit is used for special purposes such as high-cycle lines (400 cycles or above); around cooling towers, food service areas, and other applications in which corrosion is a factor; or where magnetic induction is a concern, such as near magnetic resonance imaging (MRI) equipment in hospitals.

investigate the requirements of each jurisdiction before selecting any materials.

3.2.7 Rigid Polyvinyl Chloride Conduit

The most common type of rigid nonmetallic conduit is manufactured from polyvinyl chloride (PVC). Because PVC is noncorrosive, chemically inert, and non-aging, it is often used for installation in wet or corrosive environments. Corrosion problems found with steel and aluminum RMC do not occur with PVC. However, PVC may deteriorate under some conditions, such as extreme sunlight, unless marked sunlight resistant.

All PVC is marked according to standards established by the National Electrical Manufacturers Association (NEMA) or Underwriters Laboratories, Inc. (UL). A section of PVC is shown in *Figure 5*.

Since PVC is lighter than steel or aluminum rigid conduit, IMC, or EMT, it is considered easier to handle. PVC can usually be installed much faster than other types of conduit because the joints are made up with cement and require no threading.

PVC contains no metal. This characteristic reduces the voltage drop of conductors carrying alternating current in PVC compared to identical conductors in steel conduit.

Because PVC is nonconducting, it cannot be used as an equipment grounding conductor. An equipment grounding conductor sized in accordance with *NEC Table 250.122* must be pulled in each PVC conductor run (except for underground service-entrance conductors).

PVC is available in lengths up to 20 feet. However, some jurisdictions require it to be cut to 10-foot lengths prior to installation. PVC is subject to expansion and contraction directly related to the difference in temperature, plus any radiating effects on the conduit. In moderate climates, even a 10-foot installation of PVC would require an expansion joint per the *NEC*®. Each straight section of conduit run must be treated independently from other sections when connected by elbows. To avoid damage to PVC caused by temperature changes, expansion couplings are used. See *Figure 6*. The inside of the coupling is sealed with one or more O-rings. This type of coupling may allow up to six inches of movement. Check the requirements of the local jurisdiction prior to installing PVC.

PVC is manufactured in two types:

- *Type EB* – Thin wall for underground use only when encased in concrete. Also referred to as Type I.
- *Type DB* – Thick wall for underground use without encasement in concrete. Also referred to as Type II.

Type DB is available in two wall thicknesses, Schedule 40 and Schedule 80.

- Schedule 40 is heavy wall for direct burial in the earth and aboveground installations.
- Schedule 80 is extra heavy wall for direct burial in the earth, aboveground installations for general applications, and installations where the conduit is subject to physical damage.

PVC is affected by higher-than-usual ambient temperatures. Support requirements for PVC are found in *NEC Section 352.30(B) and Table 352.30.* As with other conduit, it must be supported within three feet of each termination, but the maximum spacing between supports depends upon the size of the conduit. Some of the regulations for the maximum spacing of supports are:

- ½- to 1-inch conduit: every 3 feet
- 1¼- to 2-inch conduit: every 5 feet
- 2½- to 3-inch conduit: every 6 feet
- 3½- to 5-inch conduit: every 7 feet
- 6-inch conduit: every 8 feet

Think About It

Applications of PVC

What installations would be suitable for the use of PVC? For what situations would PVC be a poor choice?

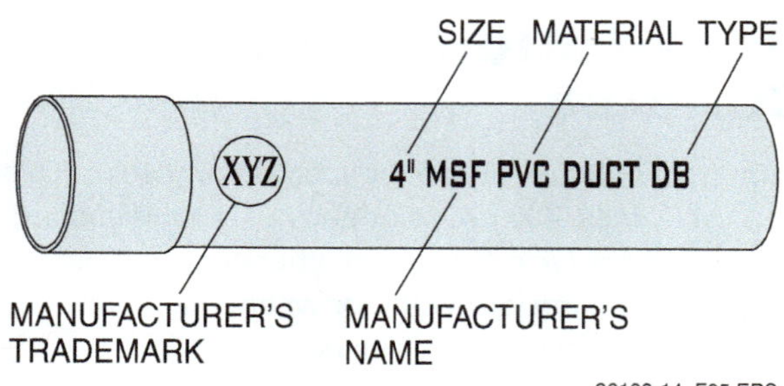

Figure 5 Rigid nonmetallic conduit.

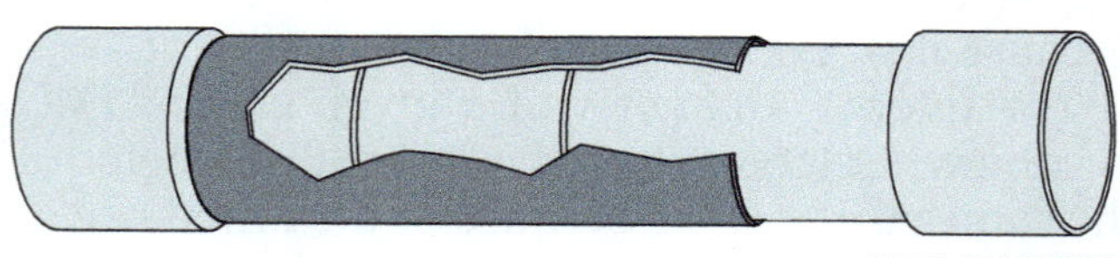

Figure 6 PVC expansion coupling.

3.2.8 *High-Density Polyethylene Conduit*

High-density polyethylene conduit (HDPE) is a rigid nonmetallic conduit listed for underground installations. (It is not listed for aboveground use.) See *NEC Article 353.* It is suitable for direct burial or where encased in concrete. In many signaling and communications applications, it is provided on reels with conductors pre-installed and may be laid in a trench or plowed into the earth.

3.2.9 *Liquidtight Flexible Nonmetallic Conduit*

Liquidtight flexible nonmetallic conduit (LFNC) was developed as a raceway for industrial equipment where flexibility was required and protection of conductors from liquids was also necessary. This is covered by *NEC Article 356.* Usage of LFNC has been expanded from industrial applications to outside and direct burial usage where listed.

Several varieties of LFNC have been introduced. The first product (LFNC-A) is commonly referred to as hose. It consists of an inner and outer layer of neoprene with a nylon reinforcing web between the layers. A second-generation product (LFNC-B), and most widely used, consists of a smooth wall, flexible PVC with a rigid PVC integral reinforcement rod. The third product (LFNC-C) is a nylon corrugated shape without any integral reinforcements. These three permitted LFNC raceway designs must be flame resistant with fittings approved for installation of electrical conductors. Nonmetallic connectors are listed for use and some liquidtight metallic flexible conduit connectors are dual-listed for both metallic and nonmetallic liquidtight flexible conduit.

LFNC is sunlight-resistant and suitable for use at conduit temperatures of 80°C dry and 60°C wet. It is available in ⅜-inch through 4-inch sizes. *NEC Section 356.12* states that LFNC cannot be used where subject to physical damage or in lengths longer than 6 feet, except where properly secured, where flexibility is required, or as permitted by *NEC Section 356.10.* Also, it cannot be used in any hazardous (classified) locations except as specified in other articles of the *NEC®*.

Liquidtight flexible metal conduit is a raceway of circular cross section having an outer liquidtight, nonmetallic, sunlight-resistant jacket over an inner flexible metal core with associated couplings and connectors covered by *NEC Article 350.*

Compression connectors are used to connect liquidtight flexible conduit to boxes or equipment.

On Site

Liquidtight Conduit

Liquidtight conduit protects conductors from vapors, liquids, and solids. Liquidtight conduit that includes an inner metal core is widely used in commercial and industrial construction.

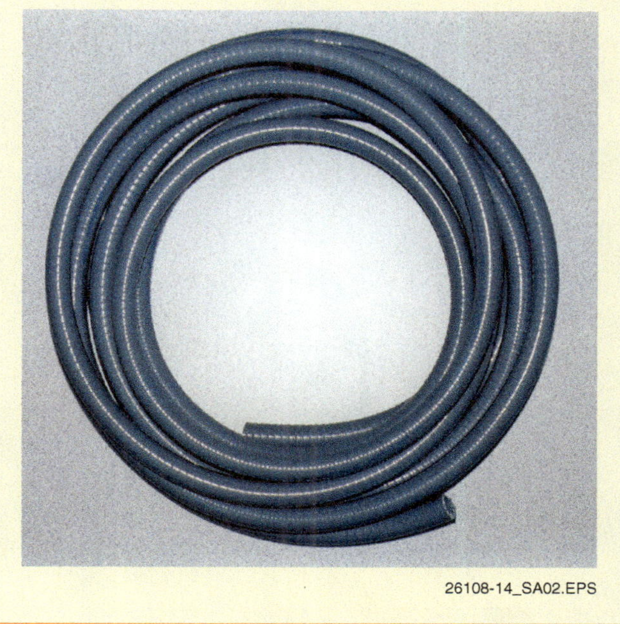

26108-14_SA02.EPS

They are available in straight, 45°, and 90° configurations (*Figure 7*).

3.2.10 Flexible Metal Conduit

Flexible metal conduit, also called flex, may be used for many kinds of wiring systems. Flexible metal conduit is made from a single strip of steel or aluminum, wound and interlocked. It is typically available in sizes from ⅜ inch to 4 inches in diameter. An illustration of flexible metal conduit is shown in *Figure 8*.

Flexible metal conduit is often used to connect equipment or machines that vibrate or move slightly during operation. Also, final connection to equipment having an electrical connection point that is marginally accessible is often accomplished with flexible metal conduit.

Flexible metal conduit is easily bent, but the minimum bending radius is the same as for other types of conduit. It should not be bent more than the equivalent of four quarter bends (360° total) between pull points (e.g., conduit bodies and boxes). It can be connected to boxes with a flexible conduit connector and to rigid conduit or EMT by using a combination coupling.

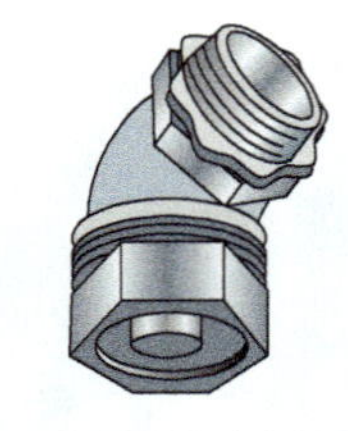

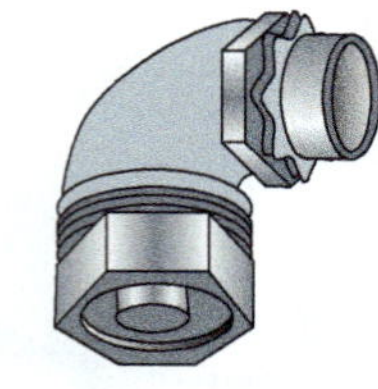

Figure 7 Liquidtight flex connectors.

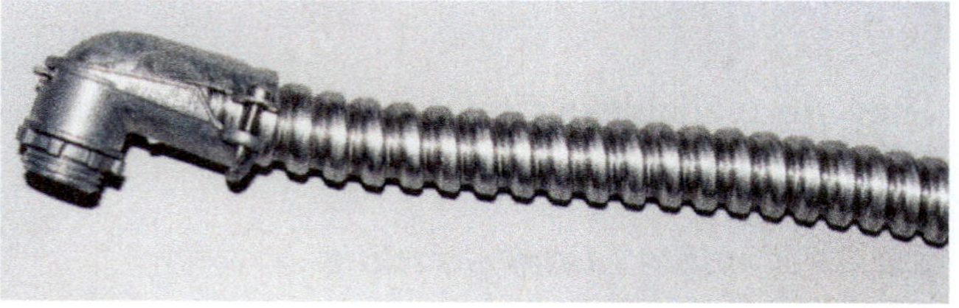

Figure 8 Flexible metal conduit.

Two types of combination couplings are shown in *Figure 9*.

Flexible metal conduit is generally available in two types: nonliquidtight and liquidtight. *NEC Articles 348 and 350* cover the uses of flexible metal conduit.

Liquidtight flexible metal conduit has an outer covering of liquidtight, sunlight-resistant flexible material that acts as a moisture seal. It is intended for use in wet locations. It is used primarily for equipment and motor connections when movement of the equipment is likely to occur. The number of bends, size, and support requirements for liquidtight conduit are the same as for all flexible conduit. Fittings used with liquidtight conduit must also be of the liquidtight type.

Support requirements for flexible metal conduit are found in *NEC Sections 348.30 and 350.30.*

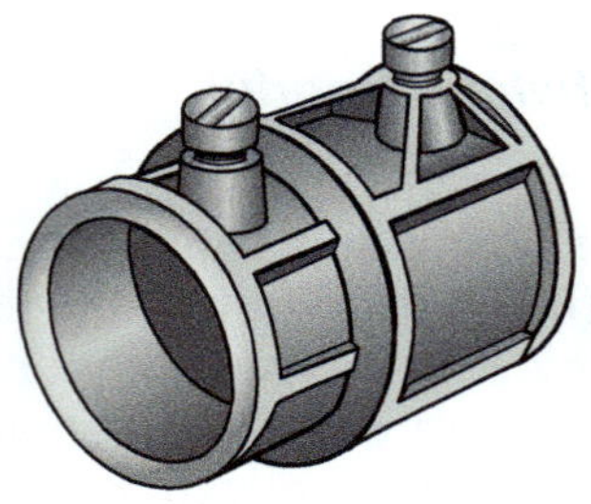

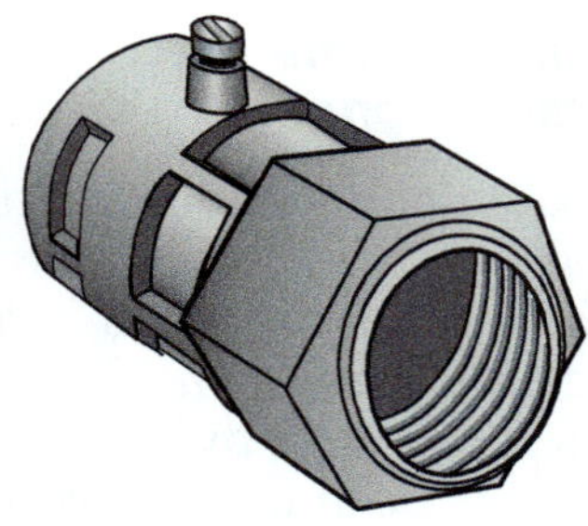

Figure 9 Combination couplings.

Straps or other means of securing the flexible metal conduit must be spaced every 4½ feet and within 12 inches of each end. (This spacing is closer together than for rigid conduit.) However, at terminals where flexibility is necessary, lengths of up to 36 inches without support are permitted. Failure to provide proper support for flexible conduit can make pulling conductors difficult.

4.0.0 Metal Conduit Fittings

A large variety of conduit fittings are available to do electrical work. Manufacturers design and construct fittings to permit a multitude of applications. The type of conduit fitting used in a particular application depends upon the size and type of conduit, the type of fitting needed for the application, the location of the fitting, and the installation method. The requirements and proper applications of boxes and fittings (conduit bodies) are found in *NEC Section 300.15.* Some of the more common types of fittings are examined in the following sections.

NOTE

When using a combination coupling, be sure the flexible conduit is pushed as far as possible into the coupling. This covers the end and protects the conductors from damage.

4.1.0 Couplings

Couplings are sleeve-like fittings that are typically threaded inside to join two male threaded pieces of rigid conduit or IMC. A piece of conduit with a coupling is shown in *Figure 10*.

Other types of couplings may be used depending upon the location and type of conduit. Several types are shown in *Figure 11*.

4.2.0 Conduit Bodies

Conduit bodies, also called condulets, are a separate portion of a conduit or tubing system that provide access through a removable cover(s) to the interior of the system at a junction of two or more sections of the system, a pull point, or at a terminal point of the system. They are usually cast and are significantly higher in cost than the stamped steel boxes permitted with EMT. However, there are situations in which conduit bodies are preferable, such as in outdoor locations, for appearance's sake in an **exposed location**, or to change types or sizes of raceways. Also, conduit bodies do not have to be supported, as do stamped steel boxes. They are also used when elbows or bends would not be appropriate.

NEC Section 314.16(C)(2) states that conduit bodies cannot contain **splices**, **taps**, or devices unless they are durably and legibly marked by the manufacturer with their cubic inch capacity and wire size. The maximum number of conductors permitted in a conduit body is found using *NEC Table 314.16(B)*. (See *Table 1*.)

4.2.1 Type C Conduit Bodies

Type C conduit bodies may be used to provide a pull point in a long conduit run or a conduit run that has bends totaling more than 360°. A Type C conduit body is shown in *Figure 12*.

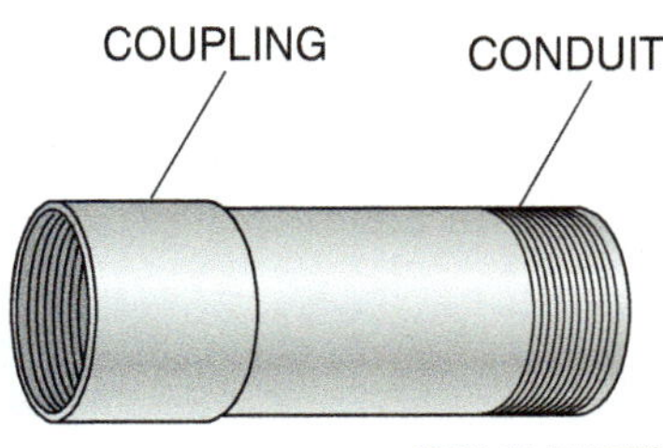

Figure 10 Conduit and coupling.

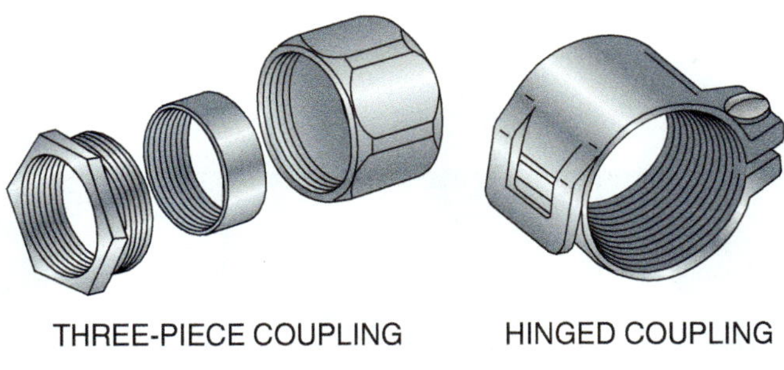

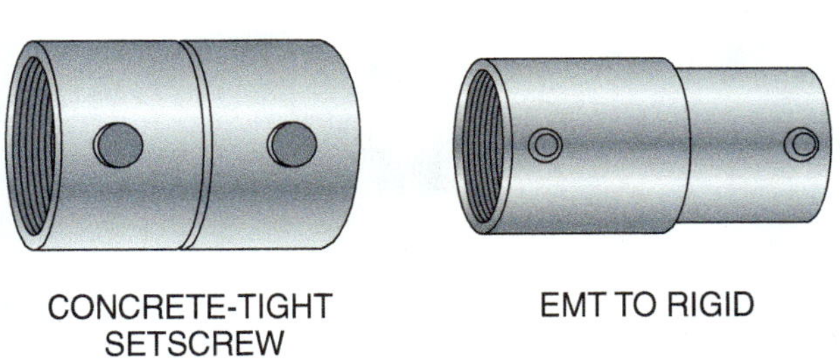

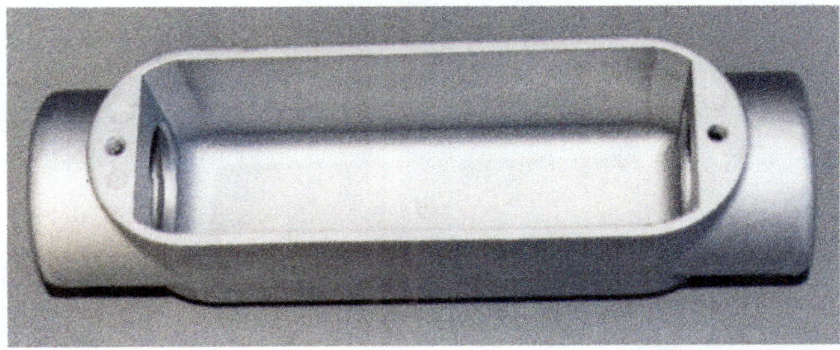

Figure 11 Metal conduit couplings.

Table 1 Volume Required per Conductor [Data from *NEC Table 314.16(B)*]

Size of Conductor	Free Space Within Box for Each Conductor
No. 18	1.5 cubic inches
No. 16	1.75 cubic inches
No. 14	2.0 cubic inches
No. 12	2.25 cubic inches
No. 10	2.5 cubic inches
No. 8	3.0 cubic inches
No. 6	5.0 cubic inches

26108-14_F12.EPS

Figure 12 Type C conduit body.

4.2.2 Type L Conduit Bodies

When referring to conduit bodies, the letter L represents an elbow. A Type L conduit body is used as a pulling point for conduit that requires a 90° change in direction. The cover is removed, then the wire is pulled out, coiled on the ground or floor, reinserted into the other conduit body's opening, and pulled. The cover and its associated gasket are then replaced. Type L conduit bodies are available with the cover on the back (Type LB), on the sides (Type LL or LR), or on both sides (Type LRL). Type L conduit bodies are shown in *Figure 13*.

NOTE

The cover and gasket must be ordered separately. Do not assume that these parts come with conduit bodies when they are ordered.

To identify Type L conduit bodies, use the following method:

Step 1 Hold the body like a pistol.

Step 2 Locate the opening on the body:

- If the opening is to the left, it is a Type LL.
- If the opening is to the right, it is a Type LR.
- If the opening is on top (back), it is a Type LB.
- If there are openings on both the left and the right, it is a Type LRL.

4.2.3 Type T Conduit Bodies

Type T conduit bodies are used to provide a junction point for three intersecting conduits and are used extensively in conduit systems. A Type T conduit body is shown in *Figure 14*.

4.2.4 Type X Conduit Bodies

Type X conduit bodies are used to provide a junction point for four intersecting conduits. The removable cover provides access to the interior of the X so that wire pulling and splicing may be performed. A Type X conduit body is shown in *Figure 15*.

4.2.5 Threaded Weatherproof Hub

Threaded weatherproof hubs are used for conduit entering a box in a wet location. *Figure 16* shows typical threaded weatherproof hubs.

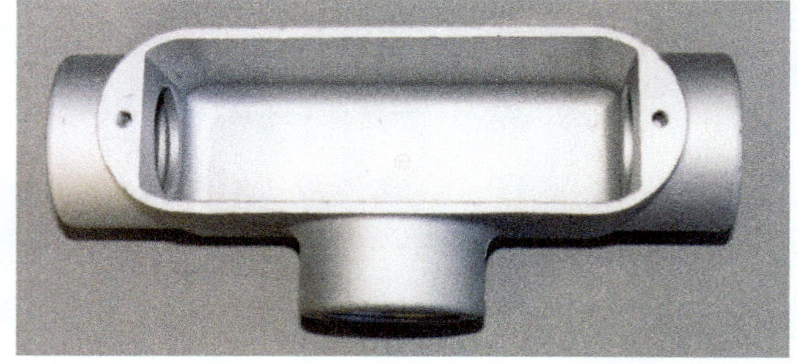

26108-14_F14.EPS

Figure 14 Type T conduit body.

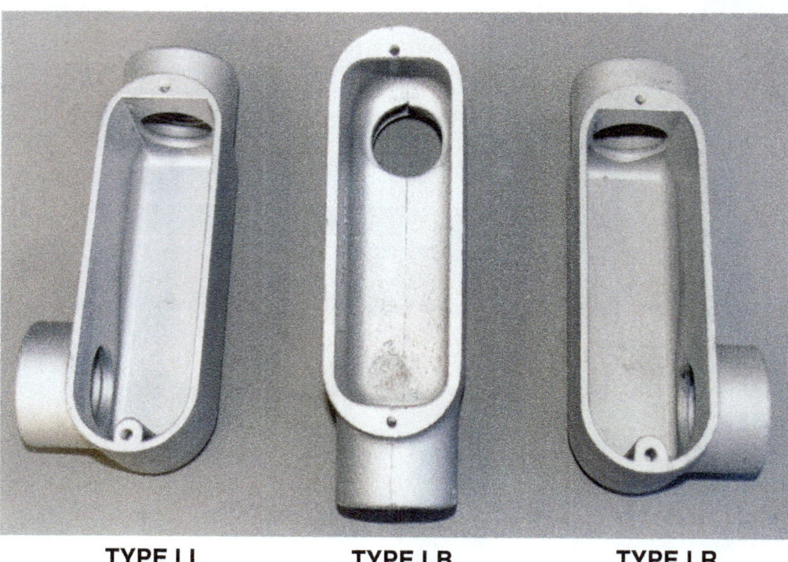

26108-14_F13A.EPS

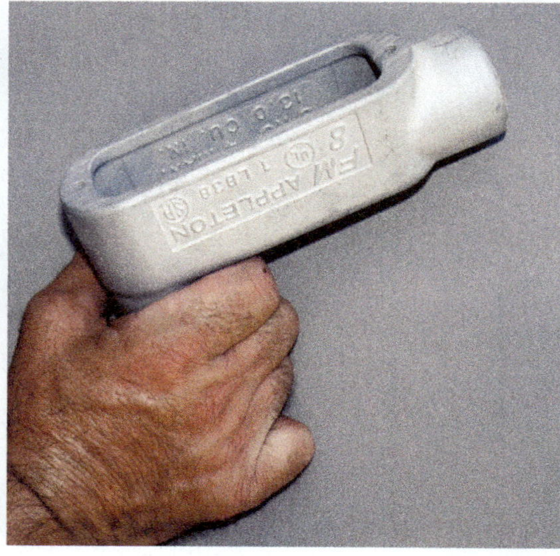

26108-14_F13B.EPS

Figure 13 Type L conduit bodies and how to identify them.

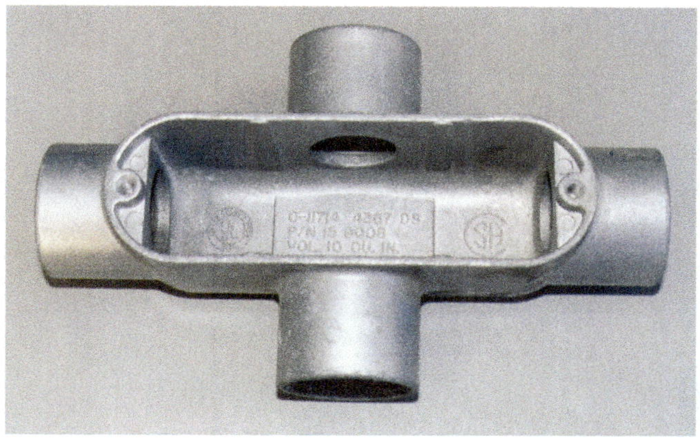

26108-14_F15.EPS

Figure 15 Type X conduit body.

26108-14_F16.EPS

Figure 16 Threaded weatherproof hubs.

4.3.0 Insulating Bushings

An insulating bushing is either nonmetallic or has an insulated throat. Insulating bushings are installed on the threaded end of conduit that enters a sheet metal enclosure.

4.3.1 Nongrounding Insulating Bushings

The purpose of a nongrounding insulating bushing is to protect the conductors from being damaged by the sharp edges of the threaded conduit end. *NEC Section 300.15(C)* states that where a conduit enters a box, fitting, or other enclosure, a fitting must be provided to protect the wire from abrasion unless the design of the box, fitting, or enclosure is such as to afford equivalent protection. *NEC Section 312.6(C)* references *Section 300.4(G),* which states that where ungrounded conductors of No. 4 or larger enter a raceway in a cabinet or box enclosure, the conductors shall be protected by a substantial fitting providing a smoothly rounded insulating surface, unless the conductors are separated from the raceway fitting by substantial insulating material securely fastened in place. An exception is where threaded hubs or bosses that are an integral part of a cabinet, box enclosure, or raceway provide a smoothly rounded or flared entry for conductors. Insulating bushings are shown in *Figure 17*.

4.3.2 Grounding Insulating Bushings

Grounded insulating bushings, usually called grounding bushings, are used to protect conductors and also have provisions for connection of an equipment grounding conductor. The ground wire, once connected to the grounding bushing, may be connected to the enclosure to which the conduit is connected. Grounding insulating bushings are shown in *Figure 18*.

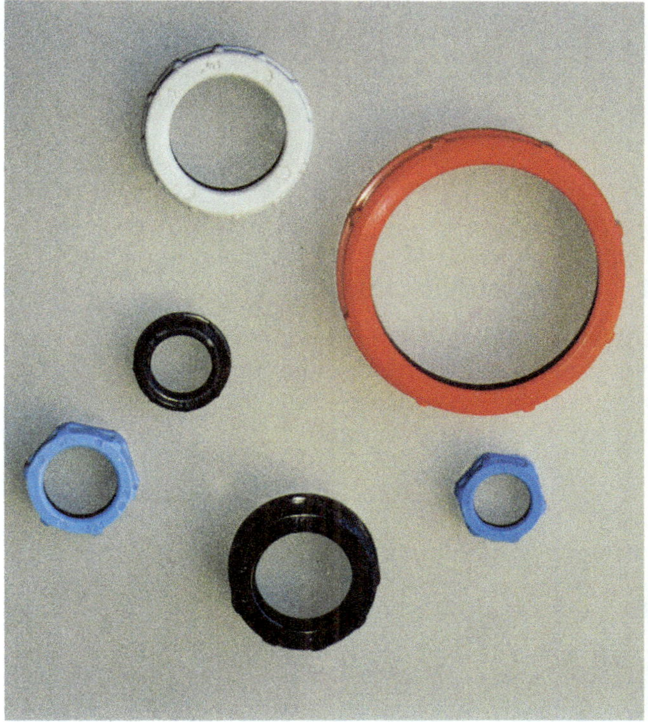

26108-14_F17.EPS

Figure 17 Insulating bushings.

26108-14_F18.EPS

Figure 18 Grounding insulating bushings.

On Site

Installation of Conduit Bodies

It will be much easier to identify conduit bodies once you begin to see them in use. Here you see liquidtight nonmetallic conduit entering a Type T conduit body (A) and a Type LB conduit body in an outdoor commercial application (B).

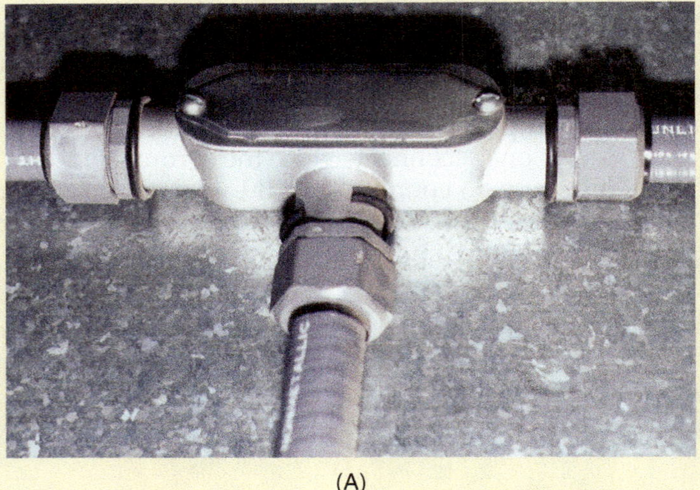

(A)

26108-14_SA03.EPS

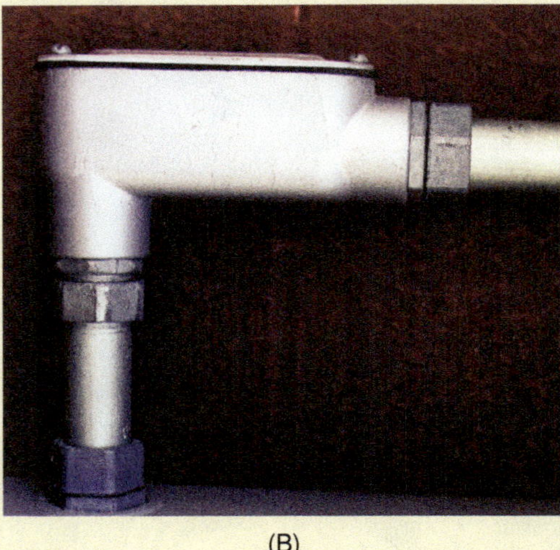

(B)

26108-14_SA04.EPS

4.4.0 Offset Nipples

Offset nipples are used to connect two pieces of electrical equipment in close proximity where a slight offset is required. They come in sizes ranging from ½" to 2" in diameter. See *Figure 19*.

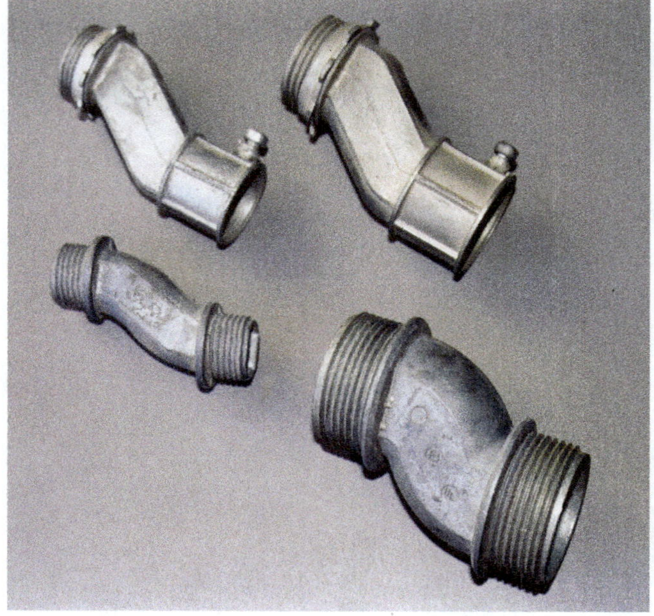

26108-14_F19.EPS

Figure 19 Offset nipples.

5.0.0 Making a Conduit-To-Box Connection

Conduit is joined to boxes by connectors, adapters, threaded hubs, or locknuts.

Bushings protect the wires from the sharp edges of the conduit. As previously discussed, bushings are usually made of plastic or metal. Some metal bushings have a grounding screw to permit a **bonding wire** to be installed.

Locknuts (*Figure 20*) are used on the inside and outside walls of the box to which the conduit is connected. A grounding locknut may be needed

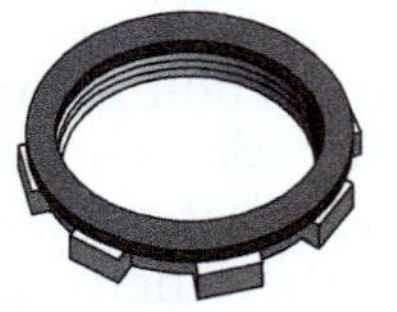

SEALING LOCKNUT

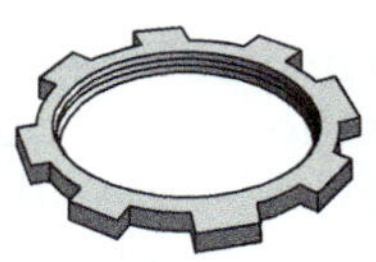

STANDARD LOCKNUT

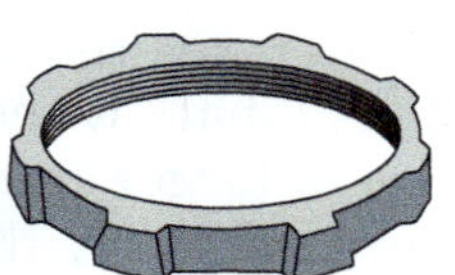

STANDARD LOCKNUT

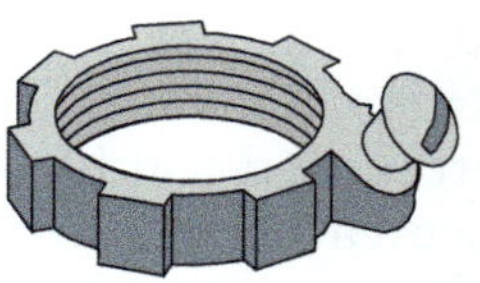

GROUNDING LOCKNUT

26108-14_F20.EPS

Figure 20 Locknuts.

if a bonding wire is to be installed. Special sealing locknuts are also used in wet locations.

When joining metal conduit to metal boxes, a means must be provided in each metal box for the connection of an equipment grounding conductor. The means shall be permitted to be a tapped hole or equivalent per *NEC Section 314.40(D)*.

A proper conduit-to-box connection is shown in *Figure 21*.

In order to make a good connection, use the following procedure:

Step 1 Thread the external locknut onto the conduit. Run the locknut to the bottom of the threads.

Step 2 Insert the conduit into the box opening.

Step 3 If an inside locknut or grounding locknut is required, screw it onto the conduit inside the box opening.

Step 4 Screw the bushing onto the threads projecting into the box opening. Make sure the bushing is tightened as much as possible.

Step 5 Tighten the external locknut to secure the conduit to the box.

It is important that the bushings and locknuts fit tightly. For this reason, the conduit must enter straight into the box. This may require that a box offset or **kick** be made in the conduit.

6.0.0 Sealing Fittings

Hazardous locations in manufacturing plants and other industrial facilities involve a wide variety of flammable gases and vapors and ignitable dusts. These hazardous substances have widely different flash points, ignition temperatures, and flammable limits requiring fittings that can be sealed. Sealing fittings are installed in conduit runs to minimize the passage of gases, vapors, or flames through the conduit and reduce the accumulation of moisture. They are required by *NEC Article 500* in hazardous locations where explosions may occur. They are also required where conduit passes from a hazardous location of one classification to another or to an unclassified location. Several types of sealing fittings are shown in *Figure 22*.

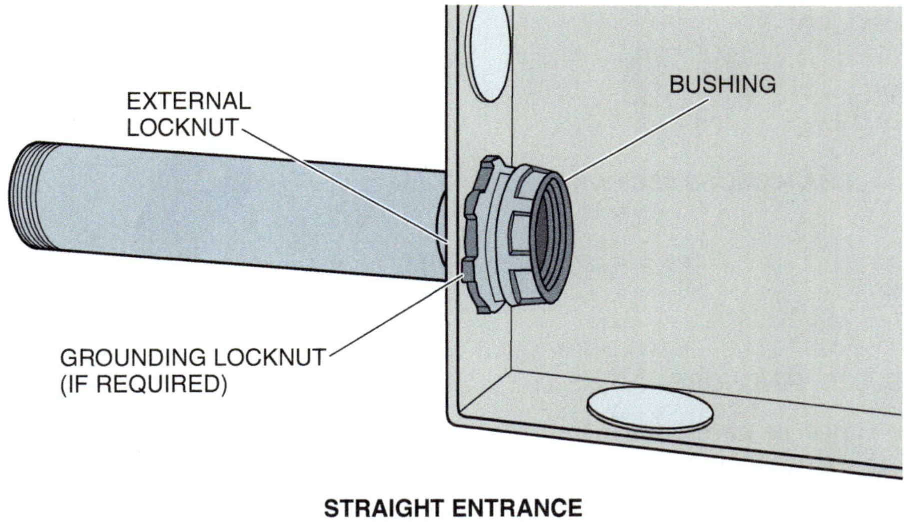

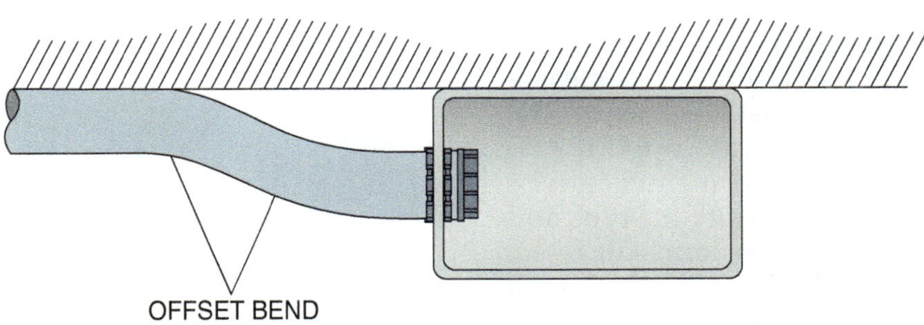

Figure 21 Conduit-to-box connections.

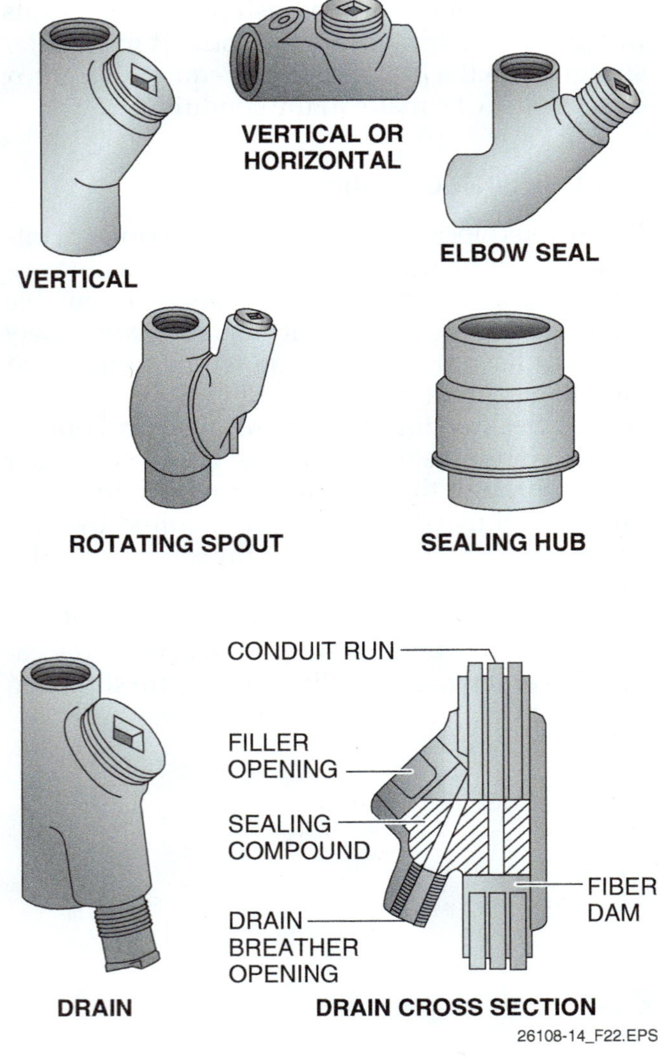

Figure 22 Sealing fittings.

7.0.0 Fasteners and Anchors

Conduit and other types of raceways used to carry wiring and cables must be properly supported. This generally means attaching the raceway to the building structure. Depending on the type of construction, the raceways may have to be attached to wood, concrete, or metal. Each of these materials requires the use of fasteners designed for the specific use. Using the wrong fastener, or installing the right fastener incorrectly, can lead to a failure of the raceway support.

The project specifications and manufacturer's installation instructions may specify the type and size of fasteners to use and how to install them. In other instances, the electrician will be expected to select the right type of fastener for a given application. It is therefore important that every electrician be familiar with the different types of fasteners, their uses, and their limitations.

7.1.0 Tie Wraps

A tie wrap is a one-piece, self-locking cable tie, usually made of nylon, that is used to fasten a bundle of wires and cables together. Tie wraps can be quickly installed either manually or using a special installation tool. Black tie wraps resist ultraviolet light and are recommended for outdoor use.

Tie wraps are made in standard, cable strap and clamp, and identification configurations (*Figure 23*). All types function to clamp bundled

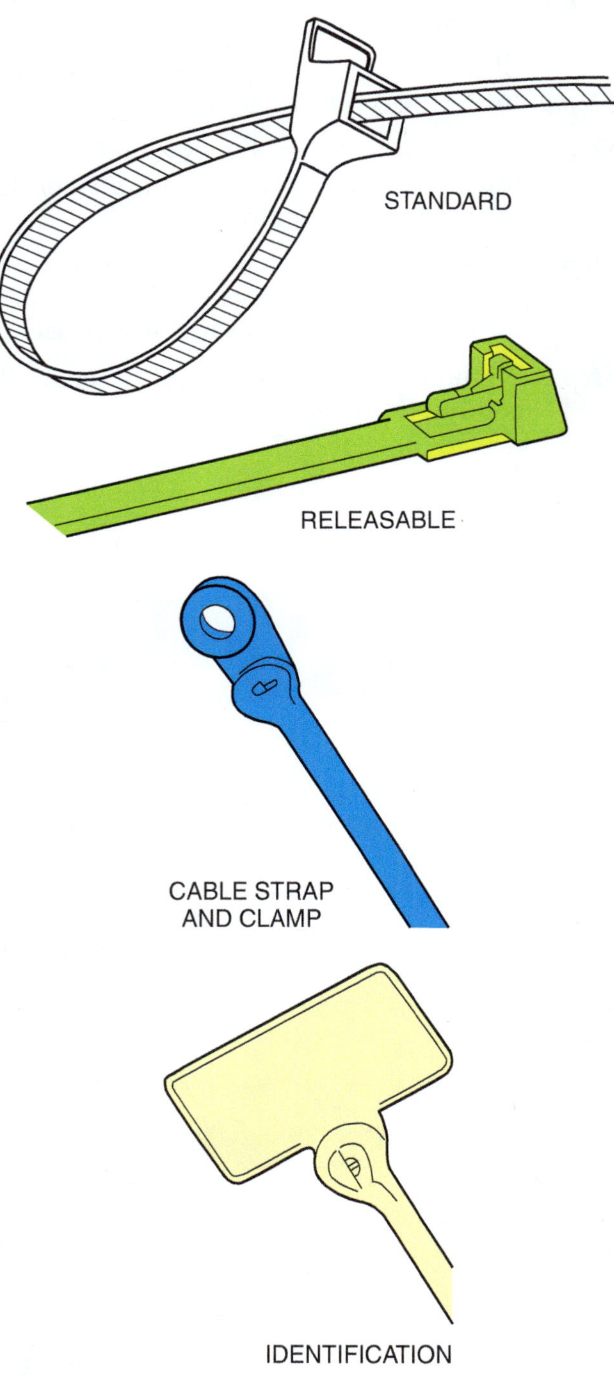

Figure 23 Tie wraps.

On Site

Installing Sealing Fittings

These fittings must be sealed after the wires are pulled. A fiber dam is first packed into the base of the fitting between and around the conductors, then the liquid sealing compound is poured into the fitting.

26108-14_SA05.EPS

wires or cables together. In addition, the cable strap and clamp has a molded mounting hole in the head used to secure the tie with a rivet, screw, or bolt after the tie wrap has been installed around the wires or cable. Identification tie wraps have a large flat area provided for imprinting or writing cable identification information. There is also a releasable version available. It is a non-permanent tie used for bundling wires or cables that may require frequent additions or deletions. Cable ties are made in various lengths ranging from about 3" to 30", allowing them to be used for fastening wires and cables into bundles with diameters ranging from about ½" to 9", respectively. Tie wraps can also be attached to a variety of adhesive mounting bases made for that purpose.

7.2.0 Screws

Screws are made in a variety of shapes and sizes for different fastening jobs. The finish or coating used on a screw determines whether it is for interior or exterior use, corrosion resistant, etc. Screws of all types have heads with different shapes and slots. Some have machine threads and are self-drilling. The size or diameter of a screw body or shank is given in gauge numbers ranging from No. 0 to No. 24, and in fractions of an inch for screws with diameters larger than ¼". The higher the gauge number, the larger the diameter of the shank. Screw lengths range from ¼" to 6", measured from the tip to the part of the head that is flush to the surface when driven in. When choosing a screw for an application, you must consider the type and thickness of the materials to be fastened, the size of the screw, the material it is made of, the shape of its head, and the type of driver. Because of the wide diversity in the types of screws and their application, always follow the manufacturer's recommendation to select the right screw for the job. To prevent damage to the screw head or the material being fastened, always use a screwdriver or power driver bit with the proper size and shape tip to fit the screw.

Some of the more common types of screws are:

- Wood screws
- Lag screws
- Masonry/concrete screws
- Thread-forming and thread-cutting screws
- Deck screws
- Drywall screws
- Drive screws

On Site

Tie Wraps

Tie wraps are available in a wide variety of colors that can be used to color code different cable bundles.

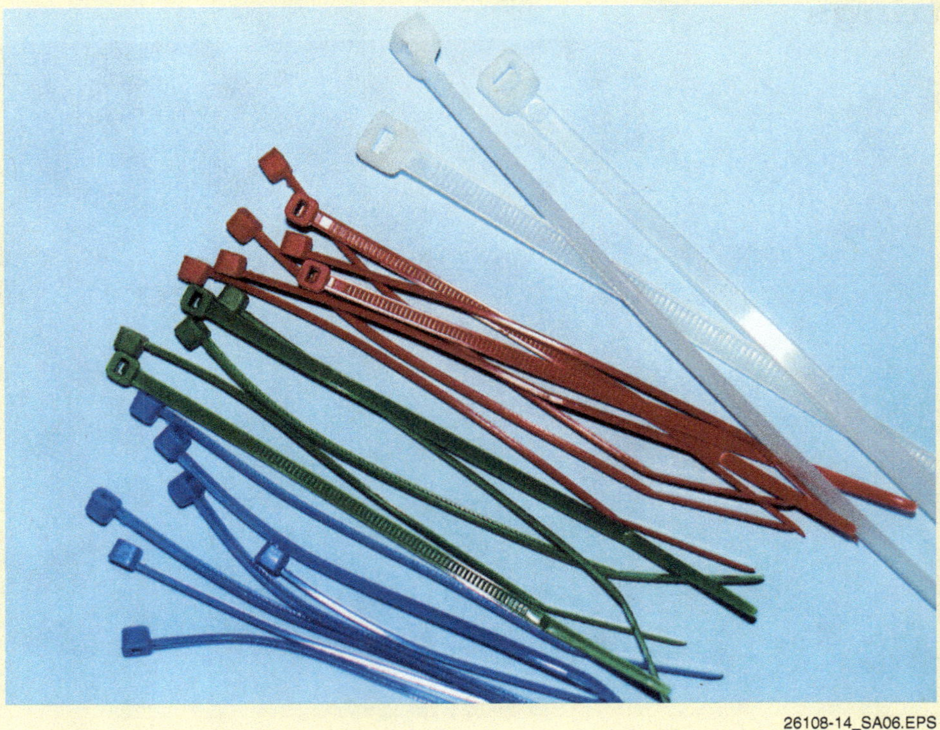

26108-14_SA06.EPS

7.2.1 Wood Screws

Wood screws (*Figure 24*) are typically used to fasten boxes, panel enclosures, etc. to wood framing or structures where greater holding power is needed than can be provided by nails. They are also used to fasten equipment to wood in applications where it may occasionally need to be unfastened and removed. Wood screws are commonly made in lengths from ¼" to 4", with shank gauge sizes ranging from 0 to 24. The shank size used is normally determined by the size hole provided in the box, panel, etc. to be fastened. When determining the length of a wood screw to use, a good rule of thumb is to select screws long enough to allow about ⅔ of the screw length to enter the piece of wood that is being gripped.

On Site

Screws

In most applications, either threaded or non-threaded fasteners such as nails could be used. However, threaded fasteners are sometimes preferred because they can usually be tightened and removed without damaging the surrounding material.

7.2.2 Lag Screws and Shields

Lag screws (*Figure 25*) or lag bolts are heavy-duty wood screws with square- or hex-shaped heads that provide greater holding power. Lag screws with diameters ranging between ¼" and ½" and lengths ranging from 1" to 6" are common. They are typically used to fasten heavy equipment to wood, but can also be used to fasten equipment to concrete when a lag shield is used.

A lag shield is a tube that is split lengthwise but remains joined at one end. It is placed in a predrilled hole in the concrete. When a lag screw is screwed into the lag shield, the shield expands in the hole, firmly securing the lag screw. In hard masonry, short lag shields (typically 1" to 2" long) may be used to minimize drilling time. In soft or weak masonry, long lag shields (typically 1½" to 3" long) should be used to achieve maximum holding strength.

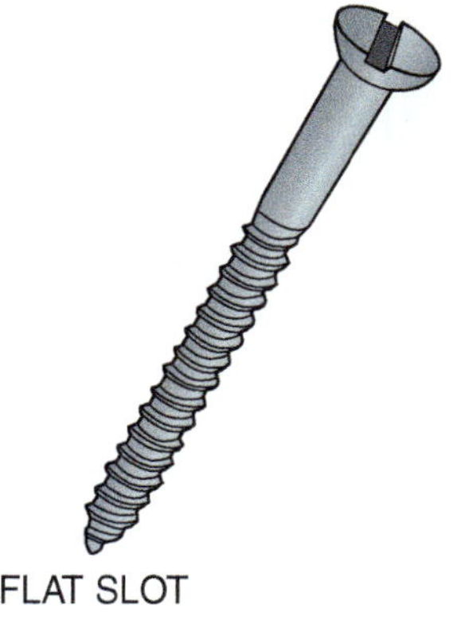

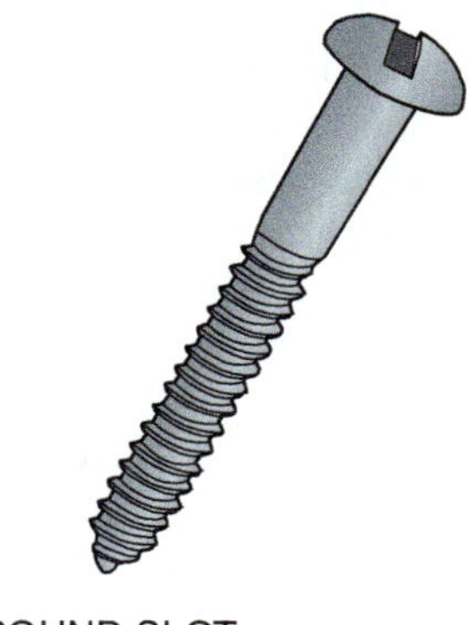

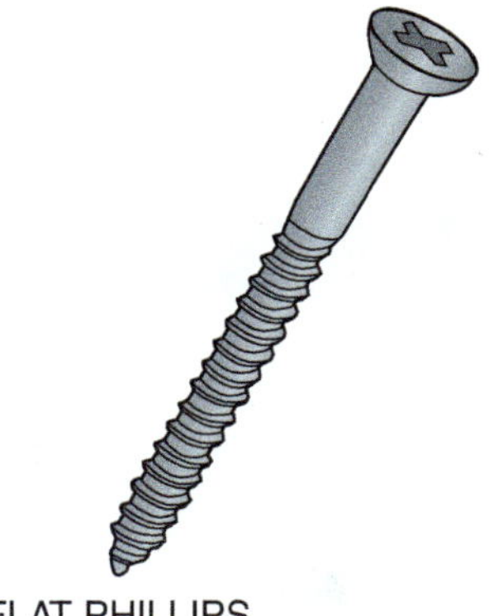

Figure 24 Wood screws.

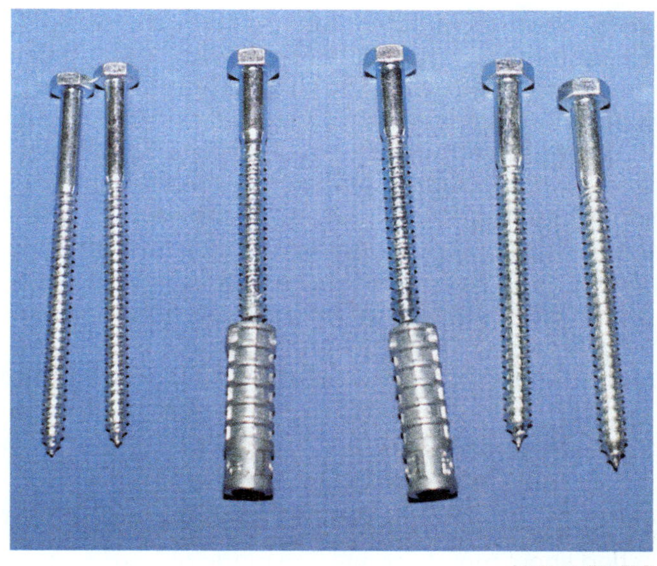

Figure 25 Lag screws and shields.

Make sure to use the proper length lag screw to achieve proper expansion. The length of the lag screw used should be equal to the thickness of the component being fastened plus the length of the lag shield. Also, drill the hole in the masonry to a depth approximately ½" longer than the shield being used. If the head of a lag screw rests directly on wood when installed, a flat washer should be placed under the head to prevent the head from digging into the wood as the lag screw is tightened down. Be sure to take the thickness of any washers used into account when selecting the length of the screw.

7.2.3 Concrete/Masonry Screws

Concrete/masonry screws (*Figure 26*), commonly called self-threading anchors, are used to fasten a device or fixture to concrete, block, or brick. No anchor is needed. To provide a matched tolerance anchoring system, the screws are installed using specially designed carbide drill bits and

On Site

Installing Wood Screws

To maintain holding power, be careful not to drill your pilot hole too large. It's wise to drill a pilot hole deep enough to equal about two-thirds of the length of the threaded portion of the screw. Additionally, to lubricate screw threads, use soap, which makes the screw easier to drive.

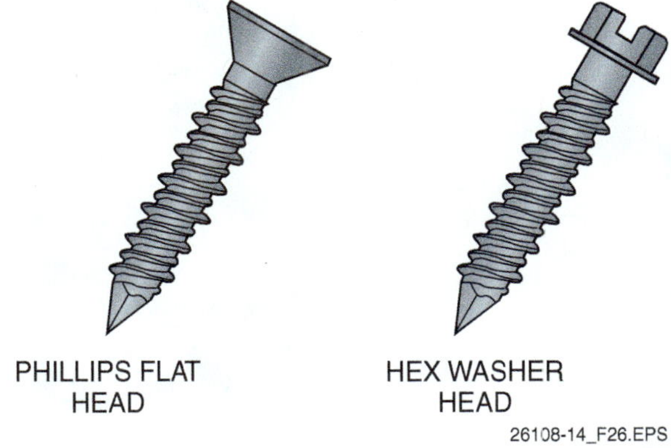

Figure 26 Concrete screws.

installation tools made for use with the screws. These tools are typically used with a standard rotary drill hammer. The installation tool, along with an appropriate drive socket or bit, is used to drive the screws directly into predrilled holes that have a diameter and depth specified by the screw manufacturer. When being driven into the concrete, the widely spaced threads on the screws cut into the walls of the hole to provide a tight friction fit. Most types of concrete/masonry screws can be removed and reinstalled to allow for shimming and leveling of the fastened device.

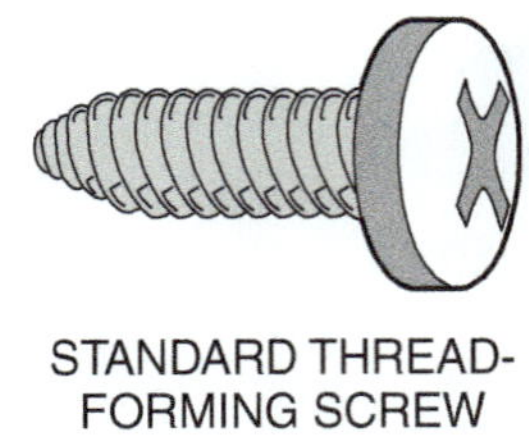

Figure 27 Thread-forming screws.

7.2.4 Thread-Forming and Thread-Cutting Screws

Thread-forming screws (*Figure 27*), commonly called sheet metal screws, are made of hard metal. They form a thread as they are driven into the work. This thread-forming action eliminates the need to tap a hole before installing the screw. To achieve proper holding, it is important to make sure to use the proper size bit when drilling pilot holes for thread-forming screws. The correct drill bit size used for a specific size screw is usually marked on the box containing the screws. Some types of thread-forming screws also drill their own holes, eliminating drilling, punching, and aligning parts. Thread-forming screws are primarily used to fasten light-gauge metal parts together. They are made in the same diameters and lengths as wood screws.

Hardened steel thread-cutting metal screws with blunt points and fine threads (*Figure 28*) are used to join heavy-gauge metals, metals of different gauges, and nonferrous metals. They are also used to fasten sheet metal to building structural members. These screws are made of hardened steel that is harder than the metal being tapped. They cut threads by removing and cutting a portion of the metal as they are driven into a pilot hole and through the material.

Think About It

Self-Drilling Screws

Can you name an electrical application for self-drilling screws?

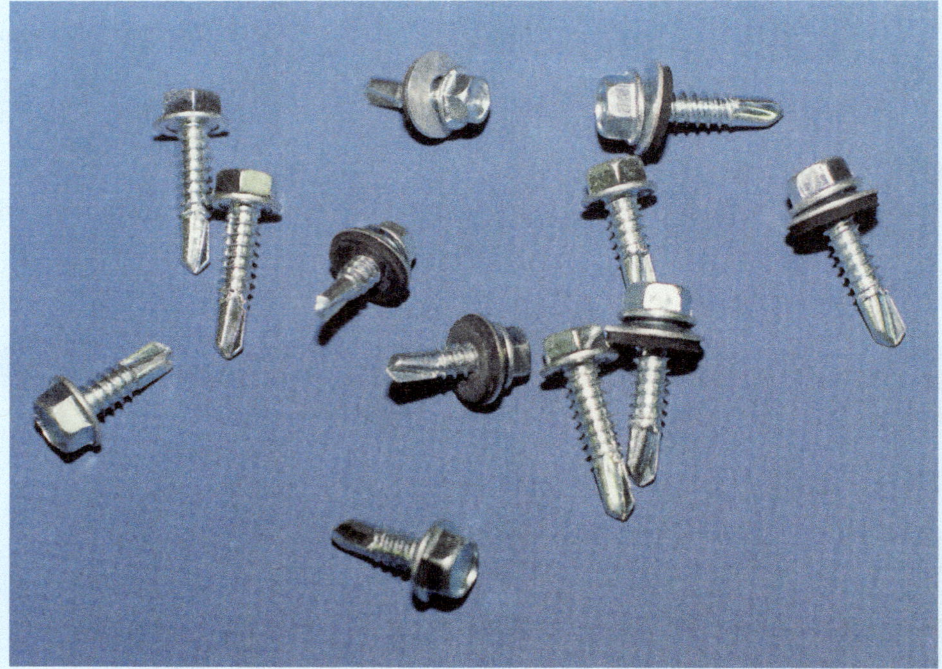

26108-14_SA07.EPS

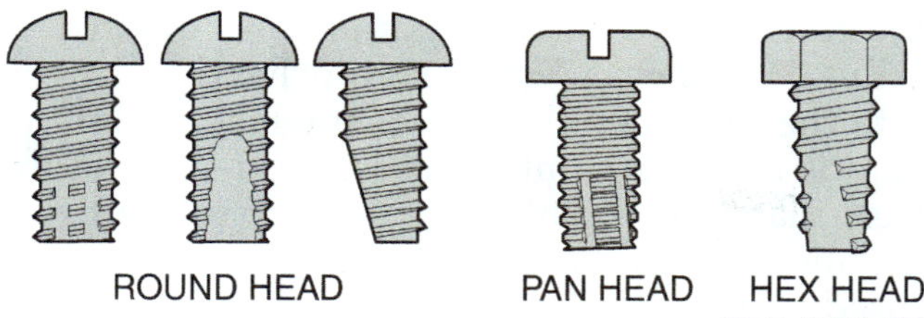

Figure 28 Thread-cutting screws.

7.2.5 Drywall Screws

Drywall screws (*Figure 29*) are thin, self-drilling screws with bugle-shaped heads. Depending on the type of screw, it cuts through the wallboard and anchors itself into wood and/or metal studs, holding the wallboard tight to the stud. Coarse thread screws are normally used to fasten wallboard to wood studs. Fine thread and high-and-low thread types are generally used for fastening to metal studs. Some screws are made for use in either wood or metal. A Phillips or Robertson drive head allows the drywall screw to be countersunk without tearing the surface of the wallboard.

Figure 29 Drywall screws.

7.2.6 Drive Screws

Drive screws do not require that the hole be tapped. They are installed by hammering the screw into a drilled or punched hole of the proper size. Drive screws are mostly used to fasten parts that will not be exposed to much pressure. A typical use of drive screws is to attach permanent name plates on electric motors and other types of equipment. *Figure 30* shows typical drive screws.

7.3.0 Hammer-Driven Pins and Studs

Hammer-driven pins or threaded studs (*Figure 31*) can be used to fasten wood or steel to concrete or block without the need to predrill holes. The pin or threaded stud is inserted into a hammer-driven tool designed for its use. The pin or stud is inserted in the tool point end out with the washer seated in the recess. The pin or stud is then positioned against the base material where it is to be fastened and the drive rod of the tool tapped lightly until the striker pin contacts the pin or stud. Following this, the tool's drive rod is struck using heavy blows with about a two-pound engineer's hammer. The force of the hammer blows is transmitted through the tool directly to the head of the fastener, causing it to be driven into the concrete or block. For best results, the drive pin or stud should be embedded a minimum of ½" in hard concrete to 1¼" in softer concrete block.

7.4.0 Powder-Actuated Tools and Fasteners

Powder-actuated tools (*Figure 32*) can be used to drive a wide variety of specially designed pin and threaded stud-type fasteners into masonry and steel. These tools look and fire like a gun and use

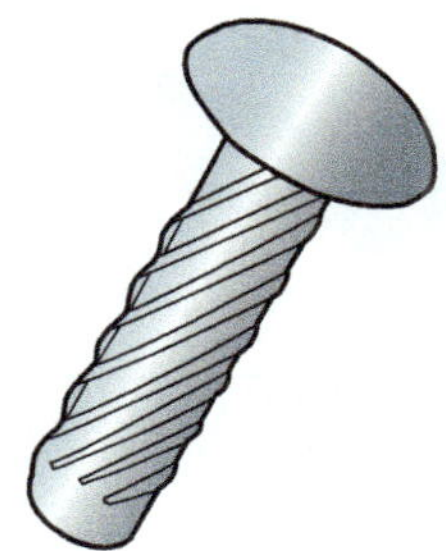

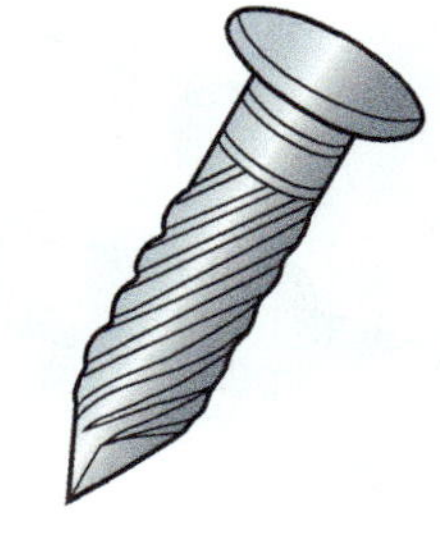

Figure 30 Drive screws.

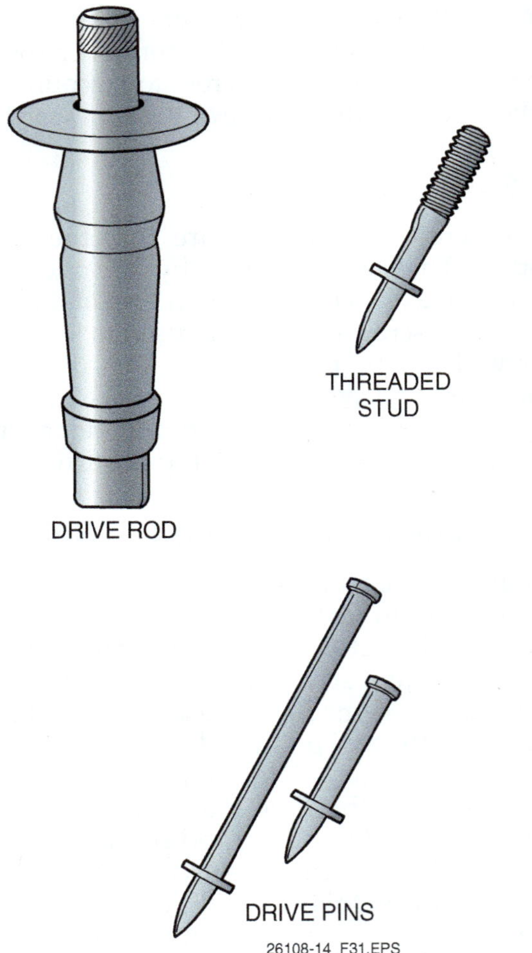

Figure 31 Hammer-driven pins and installation tool.

the force of a detonated gunpowder load (typically .22, .25, or .27 caliber) to drive the fastener into the material. The depth to which the pin or stud is driven is controlled by the density of the base material in which the pin or stud is being installed and by the power level or strength of the cased powder load.

Powder loads and their cases are designed for use with specific types and/or models of powder-actuated tools and are not interchangeable. Typically, powder loads are made in 12 increasing power or load levels used to achieve the proper penetration. The different power levels are identified by a color-code system and load case types. Note that different manufacturers may use different color codes to identify load strength. Power level 1 is the lowest power level while 12 is the highest. Higher number power levels are used when driving into hard materials or when a deeper penetration is needed. Powder loads are available as single-shot units for use with single-shot tools. They are also made in multi-shot strips or disks for semiautomatic tools.

> **WARNING!**
>
> Powder-actuated fastening tools are to be used only by trained and licensed operators and in accordance with the tool operator's manual. You must carry your license with you whenever you are using a powder-actuated tool.

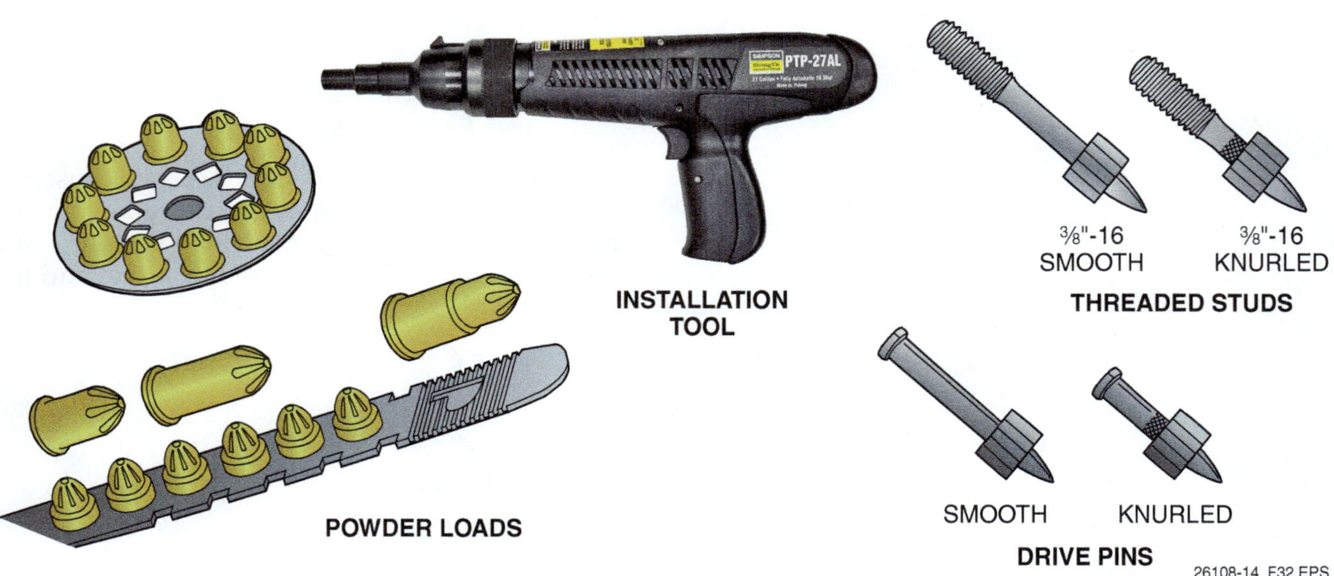

Figure 32 Powder-actuated installation tools and fasteners.

OSHA Standard 29 CFR 1926.302(e) governs the use of powder-actuated tools and states that only those individuals who have been trained in the operation of the particular powder-actuated tool in use be allowed to operate it. Authorized instructors available from the various powder-actuated tool manufacturers generally provide such training and licensing. Trained operators must take precautions to protect both themselves and others in the area when using a powder-actuated driver tool:

- Always use the tool in accordance with the published tool operation instructions. The instructions should be kept with the tool. Never attempt to override the safety features of the tool.
- Never place your hand or other body parts over the front muzzle end of the tool.
- Use only fasteners, powder loads, and tool parts specifically made for use with the tool. Use of other materials can cause improper and unsafe functioning of the tool.
- Operators and bystanders must wear eye and hearing protection along with hard hats. Other personal safety gear, as required, must also be used.
- Always post warning signs that state Powder-Actuated Tool in Use within 50 feet of the area where tools are used.

Case History

Powder-Actuated Tools

A 22-year-old apprentice was killed when he was struck in the head by a nail fired from a powder-actuated tool in an adjacent room. The tool operator was attempting to anchor plywood to a hollow wall and fired the gun, causing the nail to pass through the wall, where it traveled nearly thirty feet before striking the victim. The tool operator had never received training in the proper use of the tool, and none of the employees in the area were wearing personal protective equipment.

The Bottom Line: Never use a powder-actuated tool to secure fasteners into easily penetrated materials; these tools are designed primarily for installing fasteners into masonry. The use of powder-actuated tools requires special training and certification. In addition, all personnel in the area must be aware that the tool is in use and should be wearing appropriate personal protective equipment.

- Before using a tool, make sure it is unloaded and perform a proper function test. Check the functioning of the unloaded tool as described in the published tool operation instructions.
- Do not guess before fastening into any base material; always perform a center punch test.
- Always make a test firing into a suitable base material with the lowest power level recommended for the tool being used. If this does not set the fastener, try the next higher power level. Continue this procedure until the proper fastener penetration is obtained.
- Always point the tool away from operators or bystanders.
- Never use the tool in an explosive or flammable area.
- Never leave a loaded tool unattended. Do not load the tool until you are prepared to complete the fastening. Should you decide not to make a fastening after the tool has been loaded, always remove the powder load first, then the fastener. Always unload the tool before cleaning or servicing, when changing parts, prior to work breaks, and when storing the tool.
- Always hold the tool perpendicular to the work surface and use the spall (chip or fragment) guard or stop spall whenever possible.
- Always follow the required spacing, edge distance, and base material thickness requirements.
- Never fire through an existing hole or into a weld area.
- In the event of a misfire, always hold the tool depressed against the work surface for at least 30 seconds. If the tool still does not fire, follow the published tool instructions. Never carelessly discard or throw unfired powder loads into a trash receptacle.
- Always store the powder loads and unloaded tool under lock and key.

7.5.0 Mechanical Anchors

Mechanical anchors are devices used to give fasteners a firm grip in a variety of materials, where the fasteners by themselves would otherwise have a tendency to pull out. Anchors can be classified in many ways by different manufacturers. In this module, anchors have been divided into five broad categories:

- One-step anchors
- Bolt anchors
- Screw anchors
- Self-drilling anchors
- Hollow-wall anchors

7.5.1 One-Step Anchors

One-step anchors are designed so that they can be installed through the mounting holes in the component to be fastened. This is because the anchor and the drilled hole into which it is installed have the same size diameter. They come in various diameters ranging from ¼" to 1¼" with lengths ranging from 1¾ to 12". Wedge, stud, sleeve, one-piece, screw, and nail anchors (*Figure 33*) are common types of one-step anchors.

- *Wedge anchors* – Wedge anchors are heavy-duty anchors supplied with nuts and washers. The drill bit size used to drill the hole is the same diameter as the anchor. The depth of the hole is not critical as long as the minimum length recommended by the manufacturer is drilled. After the hole is blown clean of dust and other material, the anchor is inserted into the hole and driven with a hammer far enough so that at least six threads are below the top surface of the component. Then, the component is fastened by tightening the anchor nut to expand the anchor and tighten it in the hole.
- *Stud bolt anchors* – Stud bolt anchors are heavy-duty threaded anchors. Because this type of anchor is made to bottom in its mounting hole, it is a good choice to use when jacking or leveling of the fastened component is needed. The depth of the hole drilled in the masonry must be as specified by the manufacturer in order to achieve proper expansion. After the hole is blown clean of dust and other material, the anchor is inserted in the hole with the expander plug end down. Following this, the anchor is driven into the hole with a hammer (or setting tool) to expand the anchor and tighten it in the hole. The anchor is fully set when it can no longer be driven into the hole. The component is fastened using the correct size and thread bolt for use with the anchor stud.
- *Sleeve anchors* – Sleeve anchors are multi-purpose anchors. The depth of the anchor hole is not critical as long as the minimum length recommended by the manufacturer is drilled. After the hole is blown clean of dust and other material, the anchor is inserted into the hole and tapped until flush with the component. Then, the anchor nut or screw is tightened to expand the anchor and tighten it in the hole.
- *One-piece anchors* – One-piece anchors are multi-purpose anchors. They work on the principle that as the anchor is driven into the hole, the spring force of the expansion mechanism is compressed and flexes to fit the size of the hole. Once set, it tries to regain its original shape. The depth of the hole drilled in the masonry must be at least ½" deeper than the required embedment. The proper depth is crucial. Overdrilling is as bad as underdrilling. After the hole is blown clean of dust and other material, the anchor is inserted through the component and driven with a hammer into the hole until the head is firmly seated against the component. It is important to make sure that the anchor is driven to the proper embedment depth. Note that manufacturers also make specially designed drivers and manual tools that are used instead of a hammer to drive one-piece anchors. These tools allow the anchors to be installed in confined spaces and help prevent damage to the component from stray hammer blows.
- *Hammer-set anchors* – Hammer-set anchors are made for use in concrete and masonry. There are two types: nail and screw. An advantage of the screw-type anchors is that they are removable. Both types have a diameter the same size as the anchoring hole. For both types, the anchor hole must be drilled to the diameter of

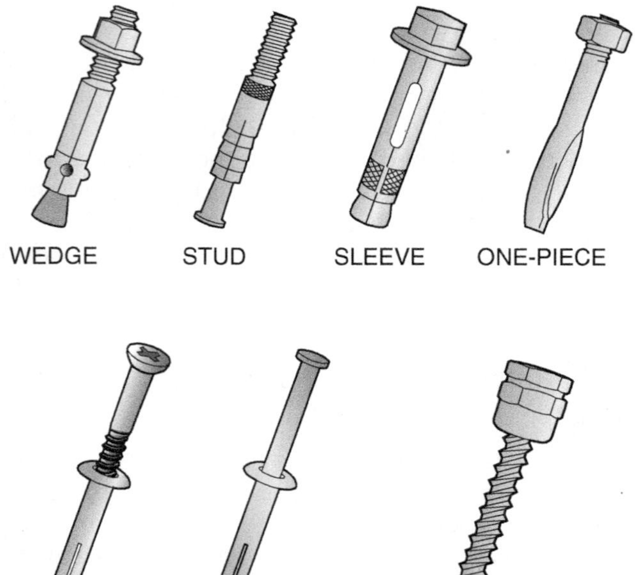

Figure 33 One-step anchors.

the anchor and to a depth of at least ¼" deeper than that required for embedment. After the hole is blown clean of dust and other material, the anchor is inserted into the hole through the mounting holes in the component to be fastened; then the screw or nail is driven into the anchor body to expand it. It is important to make sure that the head is seated firmly against the component and is at the proper embedment.

- *Threaded-rod anchors* – Threaded-rod anchors, such as the Sammy® anchor, are available for installation in concrete, steel, or wood. The anchor is designed to support a threaded rod, which is screwed into the head of the anchor after the anchor is installed. A special nut driver is available for installing the screws.

7.5.2 Bolt Anchors

Bolt anchors are designed to be installed flush with the surface of the base material. They are used in conjunction with threaded machine bolts or screws. In some types, they can be used with threaded rod. Drop-in, single and double expansion, and caulk-in anchors (*Figure 34*) are commonly used types of bolt anchors.

- *Drop-in anchors* – Drop-in anchors are typically used as heavy-duty anchors. There are two types of drop-in anchors. The first type, made for use in solid concrete and masonry, has an internally threaded expansion anchor with a preassembled internal expander plug. The anchor hole must be drilled to the specific diameter and depth specified by the manufacturer. After the hole is blown clean of dust and other material, the anchor is inserted into the hole and tapped until it is flush with the surface. Following this, a setting tool supplied with the anchor is driven into the anchor to expand it. The component to be fastened is positioned in place and fastened by threading and tightening the correct size machine bolt or screw into the anchor.

  The second type, called a hollow set drop-in anchor, is made for use in hollow concrete and masonry base materials. Hollow set drop-in anchors have a slotted, tapered expansion sleeve and a serrated expansion cone. They come in various lengths compatible with the outer wall thickness of most hollow base materials. They can also be used in solid concrete and masonry. The anchor hole must be drilled to the specific diameter specified by the manufacturer. When installed in hollow base materials, the hole is drilled into the cell or void. After the hole is blown clean of dust and other material, the anchor is inserted into the hole and tapped until it is flush with the surface. Following this, the component to be fastened is positioned in place; then the proper size machine bolt or screw is threaded into the anchor and tightened to expand the anchor in the hole.
- *Single- and double-expansion anchors* – Single- and double-expansion anchors are both made for use in concrete and other masonry. The double-expansion anchor is used mainly when fastening into concrete or masonry of questionable strength. For both types, the anchor hole must be drilled to the specific diameter and depth specified by the manufacturer. After the hole is blown clean of dust and other material, the anchor is inserted into the hole, threaded cone end first. It is then tapped until it is flush with the surface. Following this, the component to be fastened is positioned in place; then the proper size machine bolt or screw is threaded into the anchor and tightened to expand the anchor in the hole.

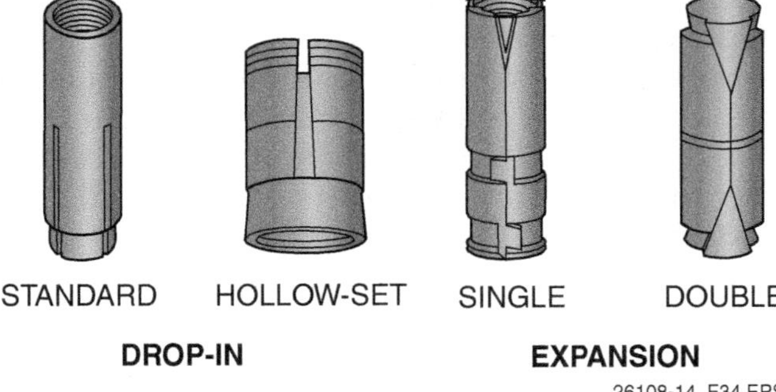

Figure 34 Bolt anchors.

7.5.3 Screw Anchors

Screw anchors are lighter-duty anchors made to be installed flush with the surface of the base material. They are used in conjunction with sheet metal, wood, or lag screws depending on the anchor type. Fiber and plastic anchors are common types of screw anchors (*Figure 35*). The lag shield anchor used with lag screws was described earlier in this module.

Fiber and plastic anchors are typically used in concrete and masonry. Plastic anchors are also

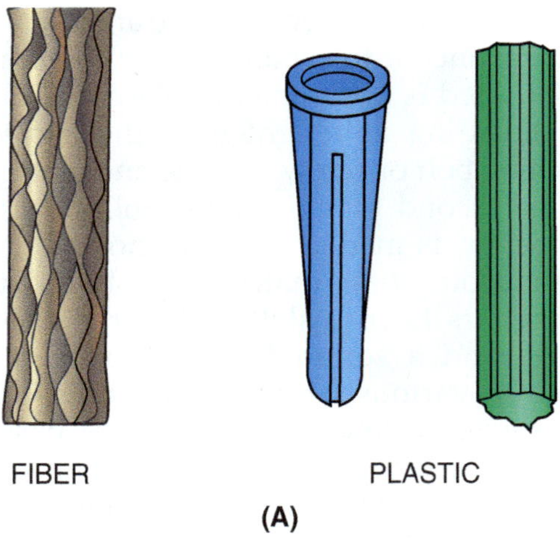

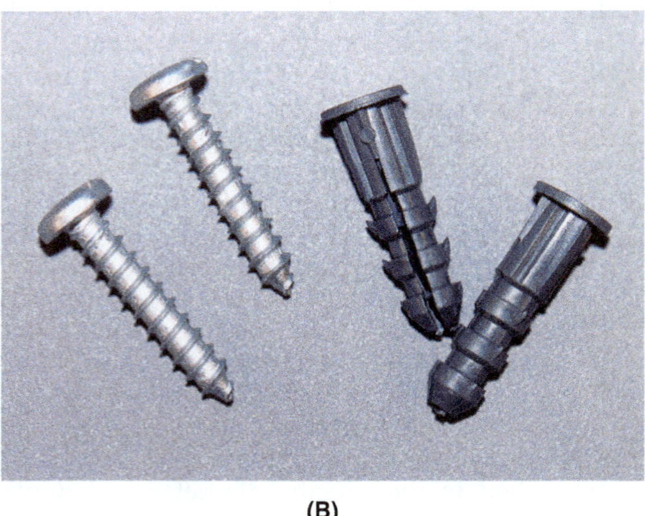

Figure 35 Screw anchors and screws.

commonly used in wallboard and similar base materials. The installation of all types is simple. The anchor hole must be drilled to the diameter specified by the manufacturer. The minimum depth of the hole must equal the anchor length. After the hole is blown clean of dust and other material, the anchor is inserted into the hole and tapped until it is flush with the surface. Following this, the component to be fastened is positioned in place; then the proper type and size screw is driven through the component mounting hole and into the anchor to expand the anchor in the hole.

7.5.4 Self-Drilling Anchors

Some anchors made for use in masonry are self-drilling anchors. *Figure 36* is typical of those in common use. This fastener has a cutting sleeve that is first used as a drill bit and later becomes the expandable fastener itself. A rotary hammer is used to drill the hole in the concrete using the anchor sleeve as the drill bit. After the hole is drilled, the anchor is pulled out and the hole cleaned. This is followed by inserting the anchor's expander plug into the cutting end of the sleeve. The anchor sleeve and expander plug are driven back into the hole with the rotary hammer until they are flush with the surface of the concrete. As the fastener is hammered down, it hits the bottom, where the tapered expander causes the fastener to expand and lock into the hole. The anchor is then snapped off at the shear point with a quick lateral movement of the hammer. The component to be fastened can then be attached to the anchor using the proper size bolt.

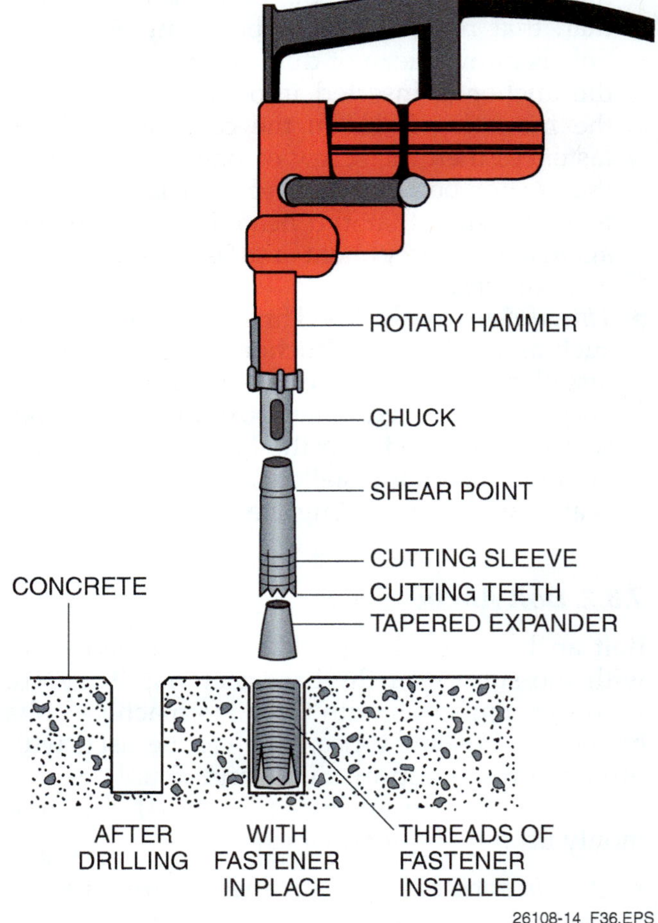

Figure 36 Self-drilling anchor.

7.6.0 Guidelines for Drilling Anchor Holes in Hardened Concrete or Masonry

When selecting masonry anchors, regardless of the type, always take into consideration and follow the manufacturer's recommendations pertaining

to hole diameter and depth, minimum embedment in concrete, maximum thickness of material to be fastened, and the pullout and shear load capacities.

When installing anchors and/or anchor bolts in hardened concrete, make sure the area where the equipment or component is to be fastened is smooth so that it will have solid footing. Uneven footing might cause the equipment to twist, warp, not tighten properly, or vibrate when in operation. Before starting, carefully inspect the rotary hammer or hammer drill and the drill bit(s) to ensure they are in good operating condition. Be sure to use the type of carbide-tipped masonry or percussion drill bits recommended by the drill/hammer or anchor manufacturer because these bits are made to take the higher impact of the masonry materials. Also, it is recommended that the drill or hammer tool depth gauge be set to the depth of the hole needed. The trick to using masonry drill bits is not to force them into the material by pushing down hard on the drill. Use a little pressure and let the drill do the work. For large holes, start with a smaller bit, then change to a larger bit.

The methods for installing the different types of anchors in hardened concrete or masonry have been briefly described. Always install the selected anchors according to the manufacturer's directions. Here is an example of a typical procedure used to install many types of expansion anchors in hardened concrete or masonry.

On Site

Safety

Be sure to wear safety goggles whenever you tackle any fastening project, regardless of how small the job may seem. Remember, you can never replace lost eyesight.

Refer to *Figure 37* as you study the procedure.

WARNING!

Drilling in concrete generates noise, dust, and flying particles. Always wear safety goggles, ear protectors, and gloves. Make sure other workers in the area also wear protective equipment.

Step 1 Drill the anchor bolt hole the same size as the anchor bolt. The hole must be deep enough for six threads of the bolt to be below the surface of the concrete (see *Figure 37, Step 1*). Clean out the hole using a squeeze bulb.

Step 2 Drive the anchor bolt into the hole using a hammer (*Figure 37, Step 2*). Protect the threads of the bolt with a nut that does not allow any threads to be exposed.

Step 3 Put a washer and nut on the bolt, and tighten the nut with a wrench until the anchor is secure in the concrete (*Figure 37, Step 3*).

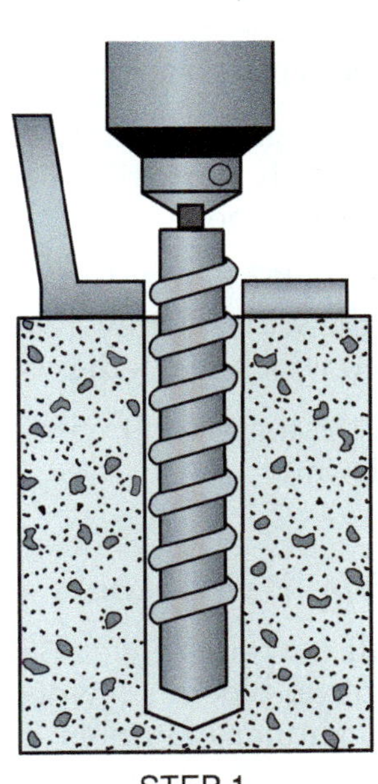

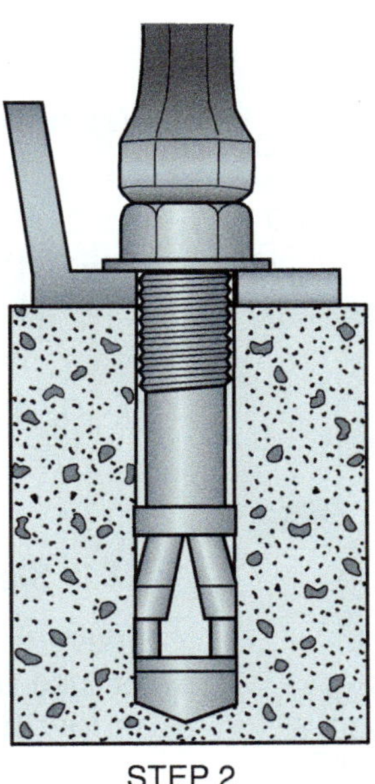

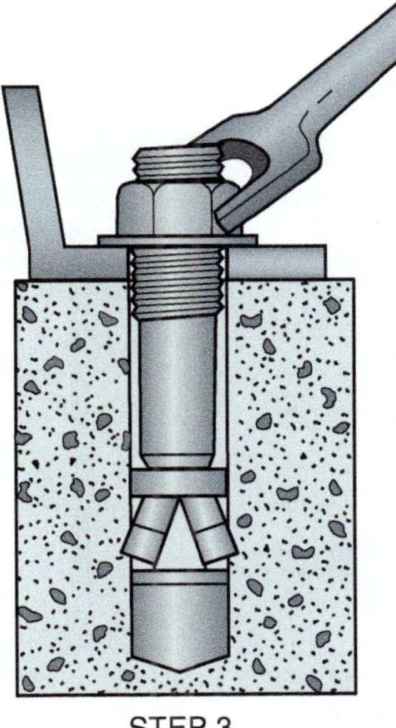

Figure 37 Installing an anchor bolt in hardened concrete.

7.7.0 Hollow-Wall Anchors

Hollow-wall anchors are used in hollow materials such as concrete plank, block, structural steel, wallboard, and plaster. Some types can also be used in solid materials. Toggle bolts, sleeve-type wall anchors, wallboard anchors, and metal drive-in anchors are common anchors used when fastening to hollow materials.

When installing anchors in hollow walls or ceilings, regardless of the type, always follow the manufacturer's recommendations pertaining to use, hole diameter, wall thickness, grip range (thickness of the anchoring material), and the pullout and shear load capacities.

7.7.1 Toggle Bolts

Toggle bolts (*Figure 38*) are used to fasten equipment, hangers, supports, and similar items into hollow surfaces such as walls and ceilings. They consist of a slotted bolt or screw and spring-loaded wings. When the bolt is inserted through the item to be fastened, then through a predrilled hole in the wall or ceiling, the wings spring apart and provide a firm hold on the inside of the hollow wall or ceiling as the bolt is tightened. Note that the hole drilled in the wall or ceiling should be just large enough for the compressed winghead to pass through. Once the toggle bolt is installed, be careful not to completely unscrew the bolt because the wings will fall off, making the fastener useless. Screw-actuated plastic toggle bolts are also made. These are similar to metal toggle bolts, but they come with a pointed screw and do not require as large a hole. Unlike the metal version, the plastic wings remain in place if the screw is removed.

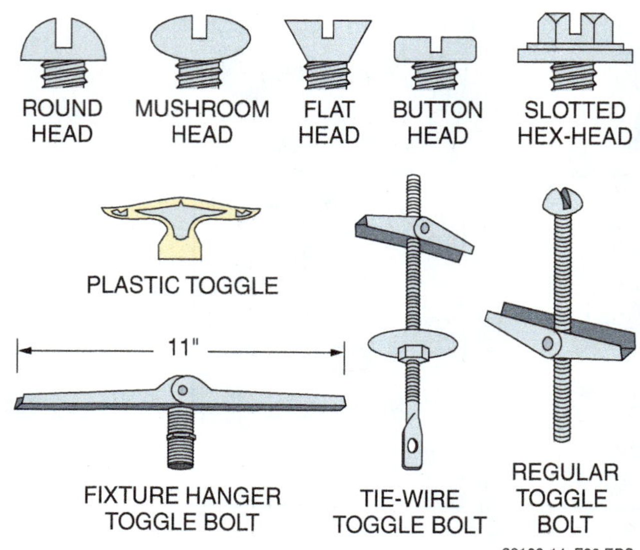

Figure 38 Toggle bolts.

Think About It

Toggle Bolts

What will happen if you fasten a toggle bolt too tightly?

Toggle bolts are used to fasten a part to hollow block, wallboard, plaster, panel, or tile. The following general procedure can be used to install toggle bolts.

WARNING!

Follow all safety precautions when using an electric drill.

Step 1 Select the proper size drill bit or punch and toggle bolt for the job.

Step 2 Check the toggle bolt for damaged or dirty threads or a malfunctioning wing mechanism.

Step 3 Drill a hole completely through the surface to which the part is to be fastened.

Step 4 Insert the toggle bolt through the opening in the item to be fastened.

Step 5 Screw the toggle wing onto the end of the toggle bolt, ensuring that the flat side of the toggle wing is facing the bolt head.

Step 6 Fold the wings completely back and push them through the drilled hole until the wings spring open.

Step 7 Pull back on the item to be fastened in order to hold the wings firmly against the inside surface to which the item is being attached.

Step 8 Tighten the toggle bolt with a screwdriver until it is snug.

7.7.2 Sleeve-Type Wall Anchors

Sleeve-type wall anchors (*Figure 39*) are suitable for use in concrete, block, plywood, wallboard, hollow tile, and similar materials. The two types made are standard and drive. The standard type is commonly used in walls and ceilings and is installed by drilling a mounting hole to the

Case History

Installation Requirements

In a college dormitory, battery-powered emergency lights were anchored to sheetrock hallway ceilings with sheetrock screws, with no additional support. These fixtures weigh 8-10 pounds each and might easily have fallen out of the ceiling, causing severe injury. When the situation was discovered, the contractor had to remove and replace dozens of fixtures.

The Bottom Line: Incorrect anchoring methods can be both costly and dangerous.

required diameter. The anchor is inserted into the hole and tapped until the gripper prongs embed in the base material. Following this, the anchor's screw is tightened to draw the anchor tight against the inside of the wall or ceiling. Note that the drive-type anchor is hammered into the material without the need for drilling a mounting hole. After the anchor is installed, the anchor screw is removed, the component being fastened is positioned in place, then the screw is reinstalled through the mounting hole in the component and into the anchor. The screw is tightened into the anchor to secure the component.

7.7.3 Wallboard Anchors

Wallboard anchors (*Figure 39*) are self-drilling medium- and light-duty anchors used for fastening in wallboard. The anchor is driven into the wall with a Phillips head manual or cordless screwdriver until the head of the anchor is flush with the wall or ceiling surface. Following this, the component being fastened is positioned over the anchor, then secured with the proper size sheet metal screw driven into the anchor.

Think About It

Sleeve-Type Drive Anchors

What happens when you remove the screw when a sleeve-type drive anchor is in place?

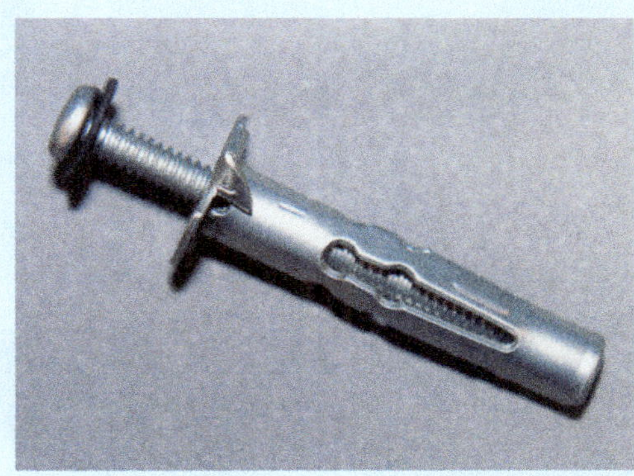

26108-14_SA08.EPS

7.7.4 Metal Drive-In Anchors

Metal drive-in anchors (*Figure 39*) are used to fasten light to medium loads to wallboard. They have two pointed legs that stay together when the anchor is hammered into a wall and spread out against the inside of the wall when a No. 6 or 8 sheet metal screw is driven in.

7.8.0 Epoxy Anchoring Systems

Epoxy resin compounds can be used to anchor threaded rods, dowels, and similar fasteners in solid concrete, hollow wall, and brick. For one manufacturer's product, a two-part epoxy is packaged in a two-chamber cartridge that keeps the resin and hardener ingredients separated until use. This cartridge is placed into a special tool

Figure 39 Sleeve-type, wallboard, and metal drive-in anchors.

Think About It

Ceiling Installations

In the dormitory problem discussed earlier, which of the following fasteners could have been used to safely secure the emergency lights?

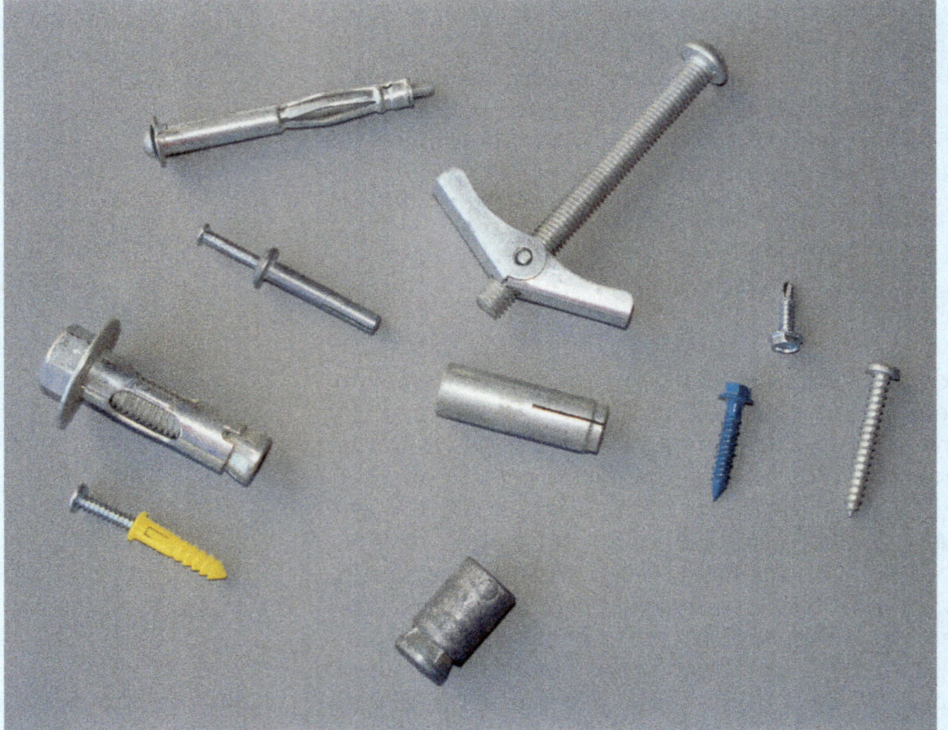

26108-14_SA09.EPS

similar to a caulking gun. When the gun handle is pumped, the epoxy resin and hardener components are mixed within the gun; then the epoxy is ejected from the gun nozzle.

To use the epoxy to install an anchor in solid concrete (*Figure 40*), a hole of the proper size is drilled in the concrete and cleaned using a nylon (not metal) brush. Following this, a small amount of epoxy is dispensed from the gun to make sure that the resin and hardener have mixed properly. This is indicated by the epoxy being of a uniform color. The gun nozzle is then placed into the hole, and the epoxy is injected into the hole until half the depth of the hole is filled. Following this, the selected fastener is pushed into the hole with a slow twisting motion to make sure that the epoxy fills all voids and crevices, then is set to the required plumb (or level) position. After the recommended cure time for the epoxy has elapsed, the fastener nut can be tightened to secure the component or fixture in place.

The procedure for installing a fastener in a hollow wall or brick using epoxy is basically the same as described above. The difference is that the epoxy is first injected into an anchor screen to fill the screen, then the anchor screen is installed into the drilled hole. Use of the anchor screen is necessary to hold the epoxy intact in the hole until the anchor is inserted into the epoxy.

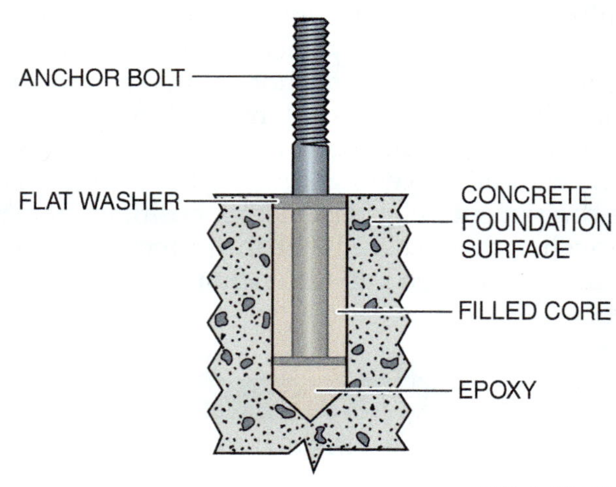

Figure 40 Fastener anchored in epoxy.

On Site

Using Epoxy

Once mixed, epoxy has a limited working time. Therefore, mix exactly what you need and work quickly. After the working time is up, epoxy requires a specific curing time. Always give epoxy its recommended curing time; since epoxy is so strong and sets so quickly, you'll be tempted to stress the bond before it's fully cured.

8.0.0 Raceway Supports

Raceway supports are available in many types and configurations. This section discusses the most common conduit supports found in electrical installations. *NEC Section 300.11* discusses the requirements for branch circuit wiring that is supported from above suspended ceilings. Electrical equipment and raceways must have their own supporting methods and may not be supported by the supporting hardware of a fire-rated roof/ceiling assembly.

8.1.0 Straps

Straps are used to support conduit to a surface (see *Figure 41*). The spacing of these supports must conform to the minimum support spacing requirements for each type of conduit. One- and two-hole straps are used for all types of conduit: EMT, RMC, IMC, PVC, and flex. The straps can be flexible or rigid. Two-part straps are used to secure conduit to electrical framing channels (struts). Parallel and right angle beam clamps are also used to support conduit from structural members.

Clamp back straps can also be used with a backplate to maintain the ¼-inch spacing from the surface required for installations in wet locations.

On Site

Use the Proper Tool for the Application

To avoid damaging fasteners, use the correct tool for the job. For example, don't use pliers to install bolts, and don't use a screwdriver that is too large or too small.

On Site

Wall-Mounted Supports

This wall-mounted support has been fabricated to hold the conduit away from the metal building. While it is shown with only one raceway, additional raceways can be added to the framing channel.

26108-14_SA10.EPS

8.2.0 Standoff Supports

The standoff support, often referred to as a Minerallac® (the name of a manufacturer of this type of support), is used to support conduit away from the supporting structure. In the case of the one-hole and two-hole straps, the conduit must be offset wherever a fitting occurs. If standoff supports are used, the conduit is held away from the supporting surface, and no offsets are required in the conduit at the fittings. Standoff supports may be used to support all types of conduit including RMC, IMC, EMT, PVC, and flex, as well as tubing installations. A standoff support is shown in *Figure 42*.

8.3.0 Electrical Framing Channels

Electrical framing channels or other similar framing materials are used together with Unistrut®-type conduit clamps to support conduit (see *Figure 43*). They may be attached to a ceiling, wall, or other surface or be supported from a trapeze hanger.

8.4.0 Beam Clamps

Beam clamps are used with suspended hangers. The raceway is attached to or laid in the hanger.

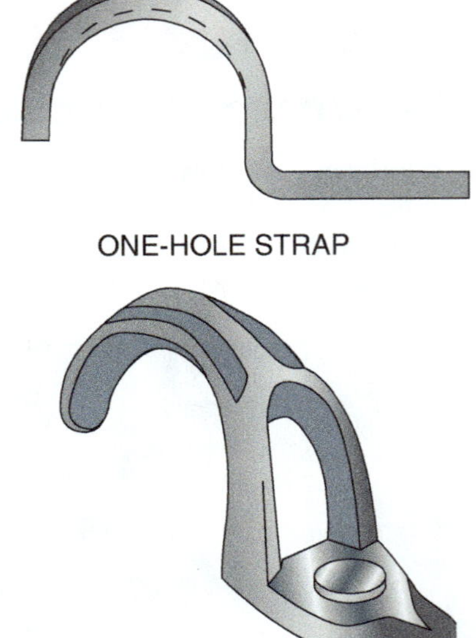

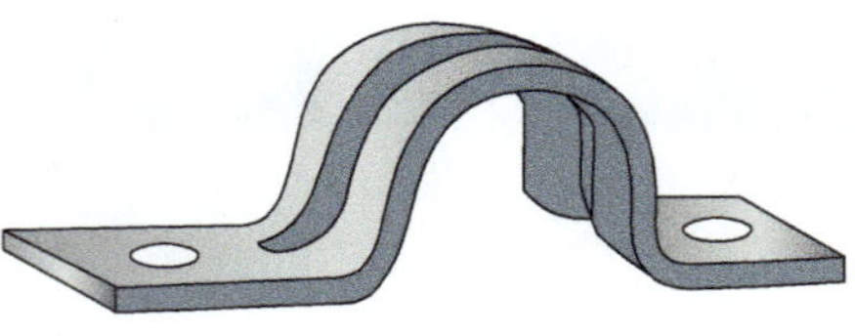

Figure 41 Straps.

The hanger is suspended by a threaded rod. One end of the threaded rod is attached to the hanger and the other end is attached to a beam clamp. The beam clamp is then attached to a beam. A beam clamp with wireway support assembly is shown in *Figure 44*.

On Site

Bundling Conductors

When conductors are bundled together in a wireway their magnetic fields tend to cancel, thus minimizing inductive heating in the conductors.

Figure 42 Standoff support.

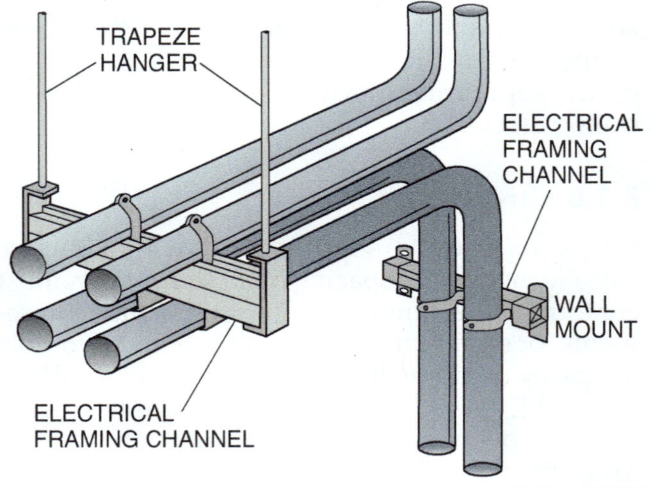

Figure 43 Electrical framing channels.

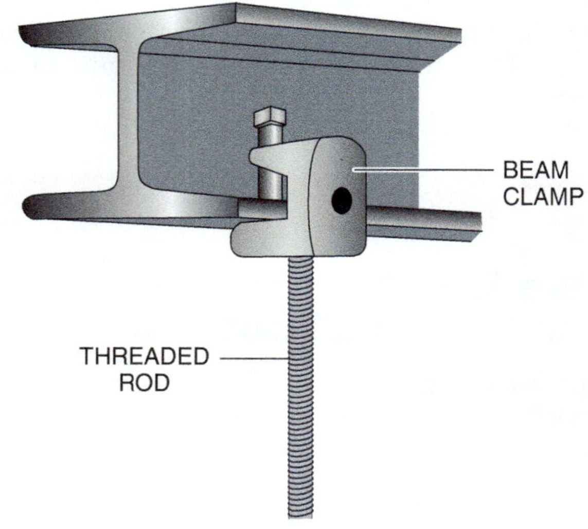

Figure 44 Beam clamp.

9.0.0 WIREWAYS

Wireways are sheet metal troughs provided with hinged or screw-on removable covers. Like other types of raceways, wireways are used for housing electric wires and cables. Wireways are available in various lengths, including 1, 2, 3, 4, 5, and 10 feet. The availability of various lengths allows runs of any exact number of feet to be made without cutting the wireway ducts. Wireways are dealt with specifically in *NEC Article 376.*

As listed in *NEC Section 376.22(A),* the sum of the cross-sectional areas of all contained conductors at any cross section of a wireway shall not exceed 20% of the interior cross-sectional area of the wireway. The derating factors in *NEC Table 310.15(B)(3)(a)* shall be applied only where the number of current-carrying conductors exceeds 3, including neutral conductors classified as current-carrying under the provisions of *NEC Section 310.15(B)(5).* Conductors for signaling or controller conductors between a motor and its starter used only for starting duty shall not be considered current-carrying conductors.

It is also noted in *NEC Section 376.56(A)* that conductors, together with splices and taps, must not fill the wireway to more than 75% of its cross-sectional area. No conductor larger than that for which the wireway is designed shall be installed in any wireway. Be sure to check *NEC Article 378* for the requirements of nonmetallic wireways.

NEC Section 376.23(A) requires that the dimensions of *NEC Table 312.6(A)* be applied where insulated conductors are deflected in a wireway. *NEC Section 376.23(B)* requires that the provisions of *NEC Section 314.28* apply where wireways are used as pull boxes.

9.1.0 Auxiliary Gutters

Strictly speaking, an auxiliary gutter is a wireway that is intended to add to wiring space at switchboards, meters, and other distribution locations. Auxiliary gutters are dealt with specifically in *NEC Article 366.* Even though the component parts of wireways and auxiliary gutters are identical, you should be familiar with the differences in their use. Auxiliary gutters are used as parts of complete assemblies of apparatus such as switchboards, distribution centers, and control equipment. However, an auxiliary gutter may only contain conductors or busbars, even though it looks like a surface metal raceway that may contain devices and equipment. Unlike auxiliary gutters, wireways represent a type of wiring because they are used to carry conductors between points located considerable distances apart.

The allowable ampacities for insulated conductors in wireways and gutters are given in *NEC Tables 310.15(B)(16) and 310.15(B)(18).* It should be noted that these tables are used for raceways in general. These *NEC®* tables and the notes are often used to determine if the correct materials are on hand for an installation. They are also used to determine if it is possible to add conductors in an existing wireway or gutter.

In many situations, it is necessary to make extensions from the wireways to wall receptacles and control devices. In these cases, *NEC Section 376.70* specifies that these extensions be made using any wiring method presented in *NEC Chapter 3* that includes a means for equipment grounding. Finally, as required in *NEC Section 376.120,* wireways must be marked in such a way that their manufacturer's name or trademark will be visible.

As you can see in *Figure 45,* a wide range of fittings is required for connecting wireways to one another and to fixtures such as switchboards, power panels, and conduit.

9.2.0 Types of Wireways

Rectangular duct-type wireways come as either hinged-cover or screw-cover troughs. Typical lengths are 1, 2, 3, 4, 5, and 10 feet. Shorter lengths are also available. Raintight troughs are permitted to be used in environments where moisture is not permitted within the raceway. However, the raintight trough should not be confused with the raintight lay-in wireway, which has a hinged cover. *Figure 46* shows a raintight trough with a removable side cover.

Wireway troughs are exposed when first installed. Whenever possible, they are mounted on the ceilings or walls, although they may sometimes be suspended from the ceiling. Note that in *Figure 47,* the trough has knockouts similar to those found on junction boxes. After the wireway system has been installed, branch circuits are brought from the distribution panels using conduit. The conduit is joined to the wireway at the most convenient knockout possible.

Wireway components such as trough crosses, 90° internal elbows, and tee connectors serve the same function as fittings on other types of

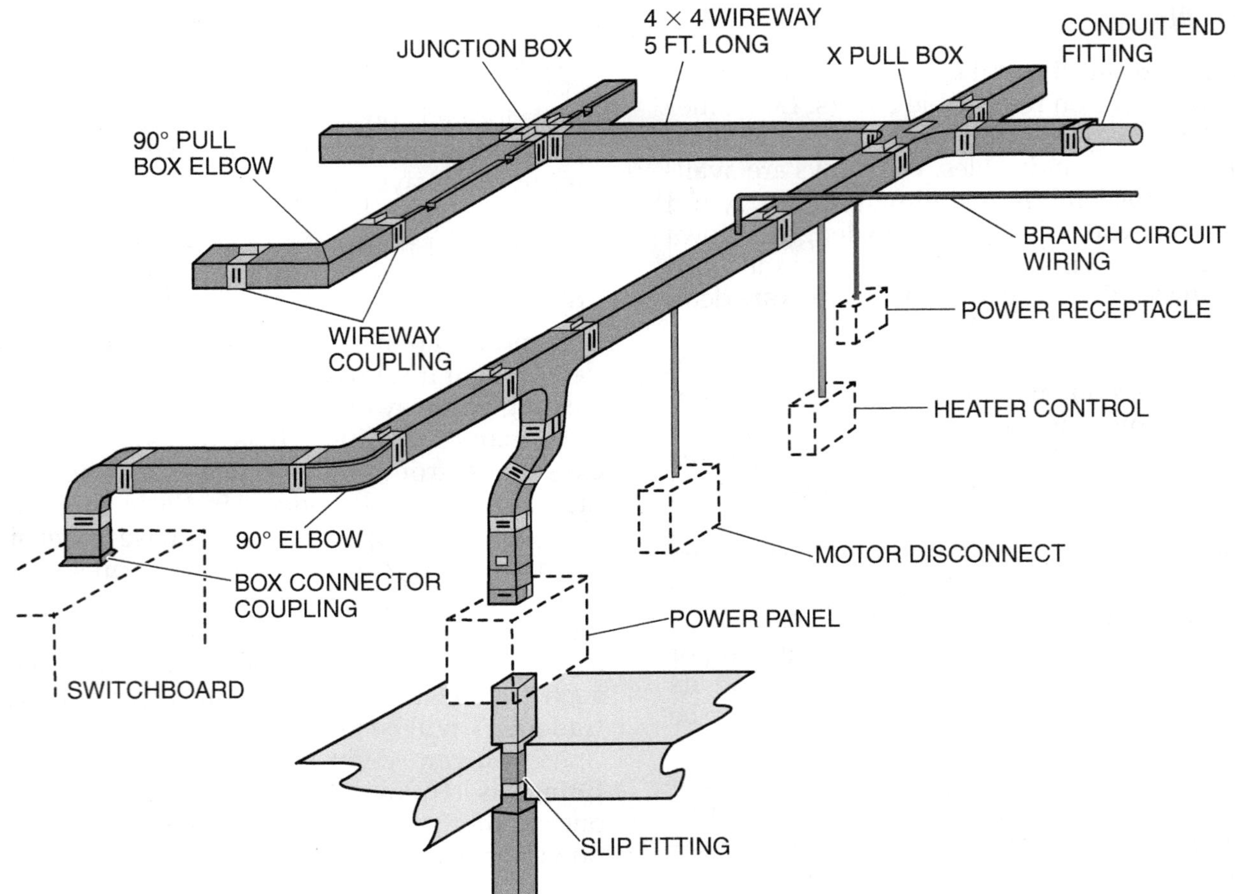

Figure 45 Wireway system layout.

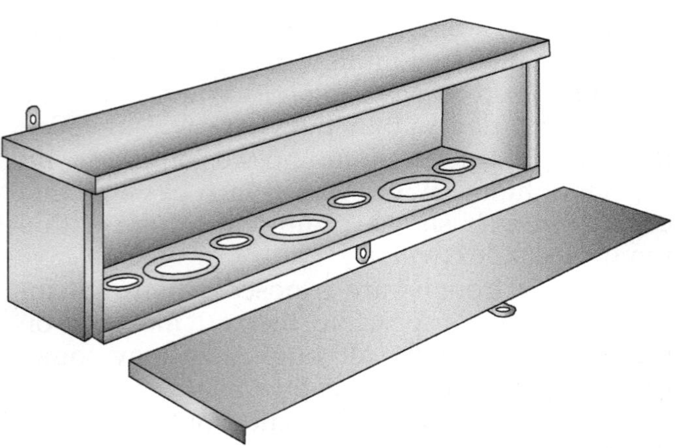

Figure 46 Raintight trough.

raceways. The fittings are attached to the duct using slip-on connectors. All attachments are made with nuts and bolts or screws. When assembling wireways, always place the head of the bolt on the inside and the nut on the outside so that the conductors will not be resting against a sharp edge. It is usually best to assemble sections of the wireway system on the floor, and then raise the sections into position. An exploded view of a section of wireway is shown in *Figure 48*. Both the wireway fittings and the duct come with screw-on, hinged, or snap-on covers to permit conductors to be laid in or pulled through.

The *NEC*® specifies that wireways may be used only for exposed work. Therefore, they cannot be used in underfloor installations. If they are used for outdoor work, they must be of an approved raintight construction. It is important to note that wireways must not be installed where they are subject to severe physical damage, corrosive vapors, or hazardous locations.

Wireway troughs must be installed so that they are supported at distances not exceeding 5 feet. When specially approved supports are used, the distance between supports must not exceed 10 feet.

9.2.1 Wireway Fittings

Many different types of fittings are available for wireways, especially for use in exposed, dry locations. The following sections explain fittings commonly used in the electrical craft.

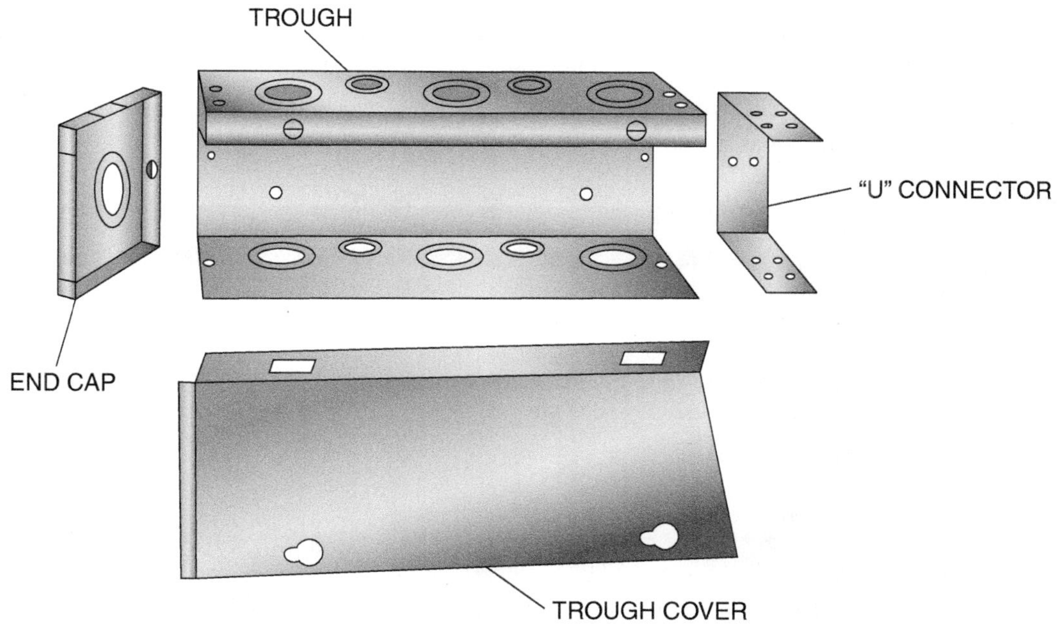

Figure 47 Trough.

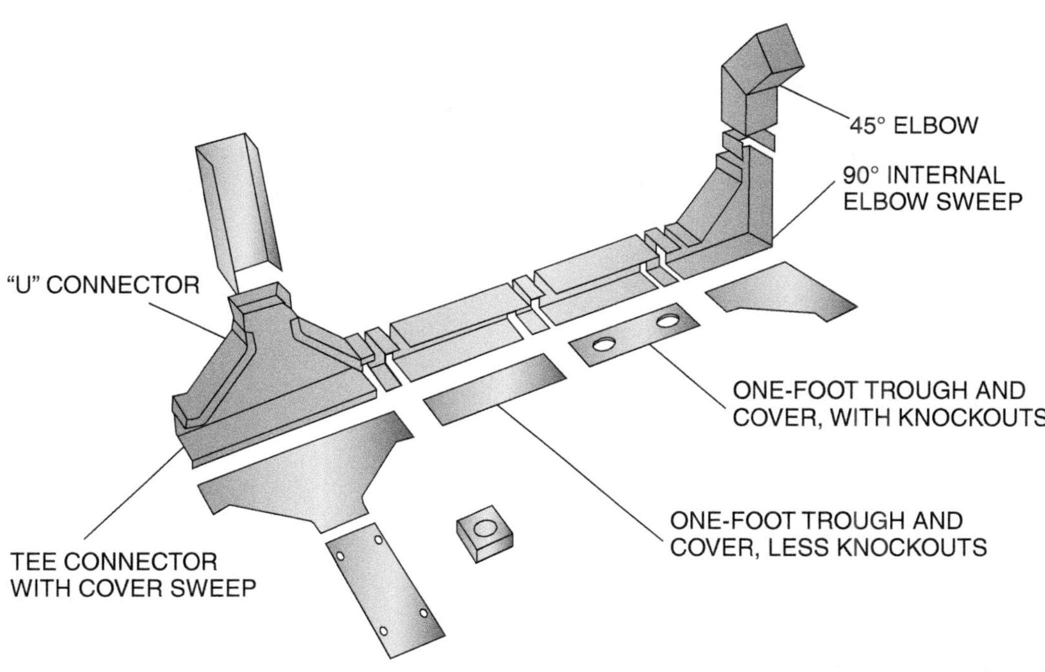

Figure 48 Wireway sections.

9.2.2 Connectors

Connectors (*Figure 49*) are used to join wireway sections and fittings. Connectors are slipped inside the end of a wireway section and are held in place by small bolts and nuts. Alignment slots allow the connector to be moved until it is flush with the inside surface of the wireway. After the connector is in position, it can be bolted to the wireway. This helps to ensure a strong rigid connection. Connectors have a friction hinge that helps hold the wireway cover open when needed.

On Site

Wireway or Trough?

A raintight lay-in wireway has a hinged cover, as shown here. A raintight trough simply has a removable cover.

26108-14_SA11.EPS

9.2.3 End Plates

End plates, or closing plates (*Figure 50*), are used to seal the ends of wireways. They are inserted into the end of the wireway and fastened by screws and bolts. End plates contain knockouts so that conduit or cable may be extended from the wireway.

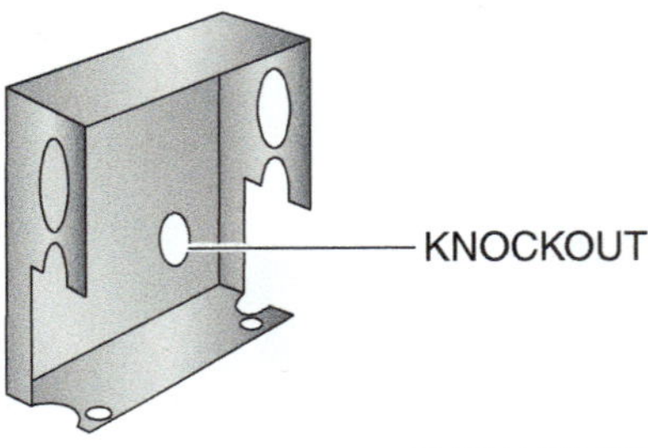

26108-14_F50.EPS

Figure 50 End plate.

9.2.4 Tees

Tee fittings (*Figure 51*) are used when a tee connection is needed in a wireway system. A tee connection is used where circuit conductors may branch in different directions. The tee fitting's covers and

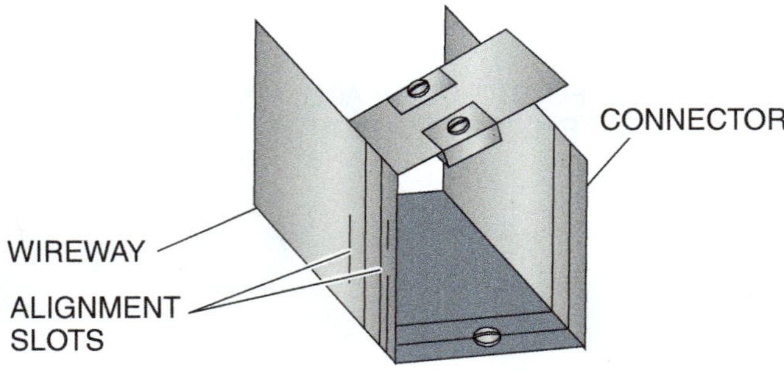

26108-14_F49.EPS

Figure 49 Connector.

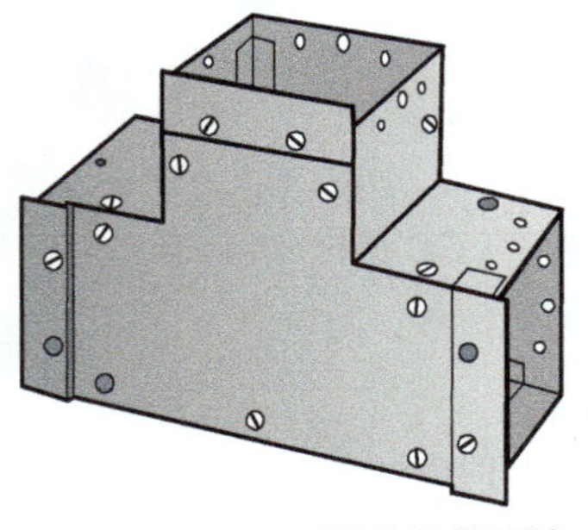
26108-14_F51.EPS

Figure 51 Tee.

sides can be removed for access to splices and taps. Tee fittings are attached to other wireway sections using standard connectors.

9.2.5 Crosses

Crosses (*Figure 52*) have four openings and are attached to other wireway sections with standard connectors. The cover is held in place by screws and can be easily removed for laying in wires or for making connections.

9.2.6 Elbows

Elbows are used to make a bend in the wireway. They are available in angles of 22½°, 45°, or 90°, and are either internal or external. They are attached to wireway sections with standard connectors. Covers and sides can be removed for wire installation. The inside corners of elbows are rounded to prevent damage to conductor insulation. An inside elbow is shown in *Figure 53*.

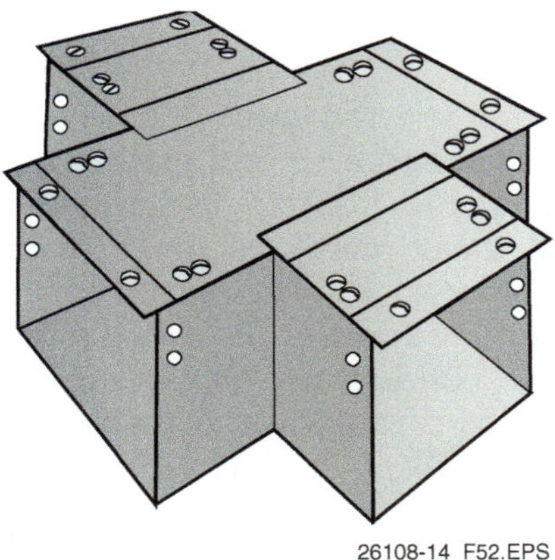

Figure 52 Cross.

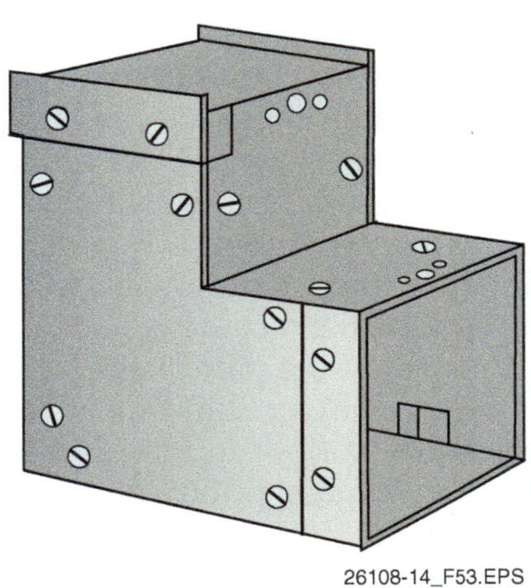

Figure 53 90° inside elbow.

9.2.7 Telescopic Fittings

Telescopic or slip fittings may be used between lengths of wireway. Slip fittings are attached to standard lengths by setscrews and usually adjust from ½ inch to 11½ inches. Slip fittings have a removable cover for installing wires and are similar in appearance to a nipple.

9.3.0 Wireway Supports

Horizontal wireway runs must be securely supported at each end and at intervals of no more than 5 feet or for individual lengths greater than 5 feet at each end or joint, unless listed for other support intervals. In no case shall the support distance be greater than 10 feet, in accordance with *NEC Section 376.30(A)*. If possible, wireways can be mounted directly to a surface. Otherwise, wireways are supported by hangers or brackets.

9.3.1 Suspended Hangers

In many cases, the wireway is supported from a ceiling, beam, or other structural member. In such installations, a suspended hanger (*Figure 54*) may be used to support the wireway.

The wireway is attached to or laid in the hanger. The hanger is suspended by a threaded rod. One end of the rod is attached to the hanger with hex nuts. The other end of the rod is attached to a beam clamp or anchor.

9.3.2 Gusset Brackets

Another type of support used to mount wireways is a gusset bracket (*Figure 55*). This is an L-type bracket that is mounted to a wall. The wireway rests on the bracket and is attached by screws or bolts.

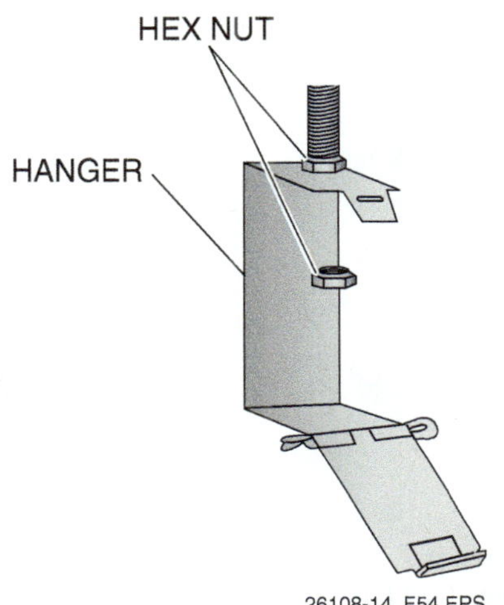

Figure 54 Suspended hanger.

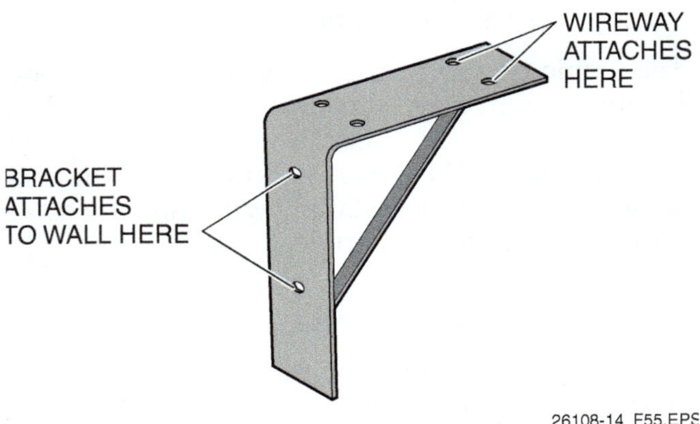

Figure 55 Gusset bracket.

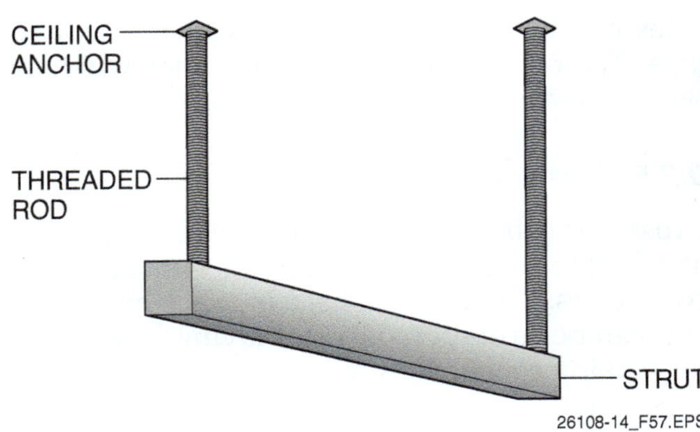

Figure 57 Wireway hanger.

9.3.3 Standard Hangers

Standard hangers (*Figure 56*) are made in two pieces. The two pieces are combined in different ways for different installation requirements. The wireway is attached to the hanger by bolts and nuts.

9.3.4 Wireway Hangers

When a larger wireway must be suspended, a wireway hanger may be used. A wireway hanger is made by suspending a piece of strut from a ceiling, beam, or other structural member. The strut is suspended by threaded rods attached to beam clamps or other ceiling anchors, as shown in *Figure 57*.

9.4.0 Other Types of Raceways

In this section, other types of raceways will be discussed. Depending on the particular purpose for which they are intended, raceways include enclosures such as surface metal and nonmetallic raceways, and underfloor raceways.

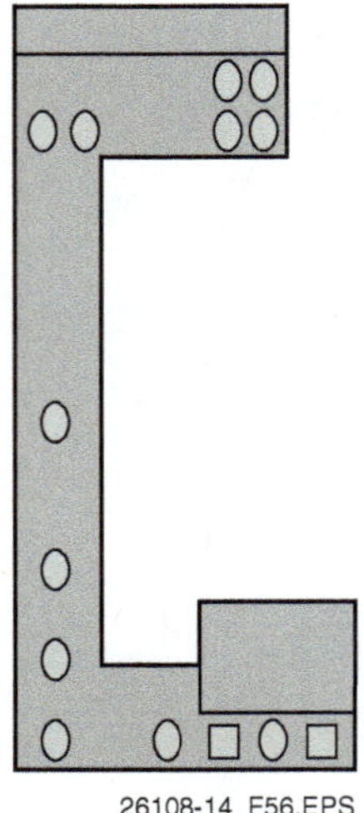

Figure 56 Standard hanger.

9.4.1 Surface Metal and Nonmetallic Raceways

Surface metal raceways consist of a wide variety of special raceways designed primarily to carry power and communications wiring to locations on the surface of ceilings or walls of building interiors.

Installation specifications of both surface metal raceways and surface nonmetallic raceways are listed in detail in *NEC Articles 386 and 388,* respectively. All these raceways must be installed in dry, interior locations. The number of conductors, their amperage, and the allowable cross-sectional area of the conductors, as well as regulations for combination raceways, are specified in *NEC Tables 310.15(B)(16) and 310.15(B)(18)* and *NEC Articles 386 and 388.*

One use of surface metal raceways is to protect conductors that run to non-accessible outlets.

Surface metal and nonmetallic raceways are divided into subgroups based on the specific purpose for which they are intended. There are three small surface raceways that are primarily used for extending power circuits from one point to another. In addition, there are six larger surface raceways that have a much wider range of applications. Typical cross sections of the first three smaller raceways are shown in *Figure 58*.

Additional surface metal raceway designs are referred to as pancake raceways, because their flat cross sections resemble pancakes. Their primary use is to extend power, lighting, telephone, or signal wire across a floor to locations away from the walls of a room without embedding them under the floor. A pancake raceway is shown in *Figure 59*.

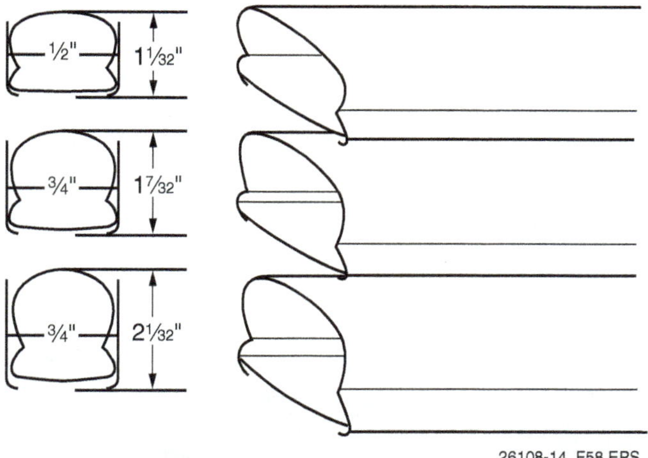

Figure 58 Smaller surface raceways.

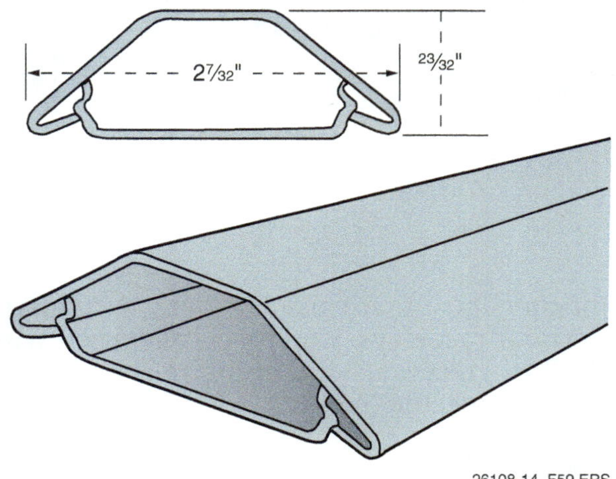

Figure 59 Pancake raceway.

There are also surface metal raceways available that house two or three different conductor raceways. These are referred to as twinduct or tripleduct. These raceways permit different circuits, such as power and signal, to be placed within the same raceway.

Nonmetallic raceways come in a variety of styles. The perimeter raceways shown in *Figure 60* are available in sizes ranging from ¾ to more than 7" wide. Many of these raceways contain barriers that allow them to carry both low voltage and power wiring.

The number and types of conductors permitted to be installed and the capacity of a particular surface raceway must be calculated and matched with *NEC®* requirements, as discussed previously. *NEC Tables 310.15(B)(16) through 310.15(B)(18)* are used for surface raceways in the same manner in which they are used for wireways. For surface raceway installations with more than three conductors in each raceway, particular reference must be made to *NEC Table 310.15(B)(3)(a).*

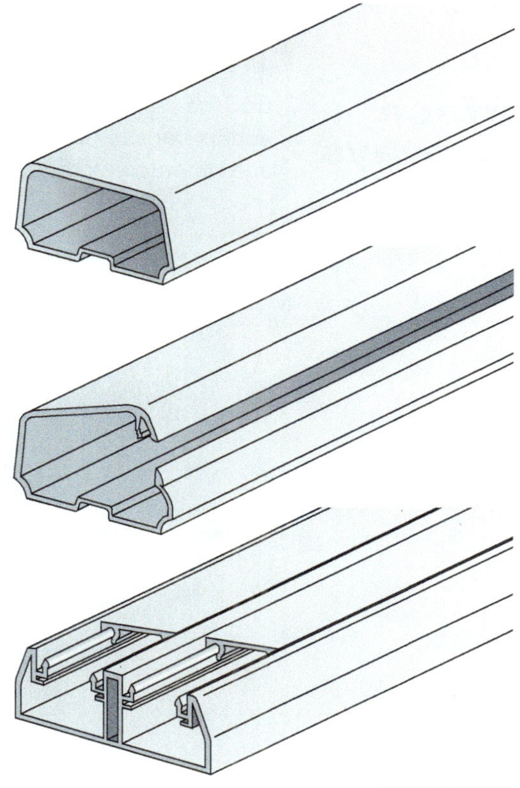

Figure 60 Examples of surface raceway.

9.4.2 Multi-Outlet Assemblies

Manufacturers offer a wide variety of multi-outlet surface raceways. Their function is to hold receptacles and other devices within the raceway. When surface raceways are used in this manner, the assembly is referred to as a multi-outlet assembly. Multi-outlet assemblies (*Figure 61*) are covered in *NEC Article 380.* Multi-outlet systems are either wired in the field or come pre-wired from the factory.

Figure 61 Multi-outlet assembly.

On Site

Surface Raceways

Surface raceways with multiple channels are commonly used in computer networking applications to provide conductors for AC power to the computers, as well as telephone and other low-voltage wiring.

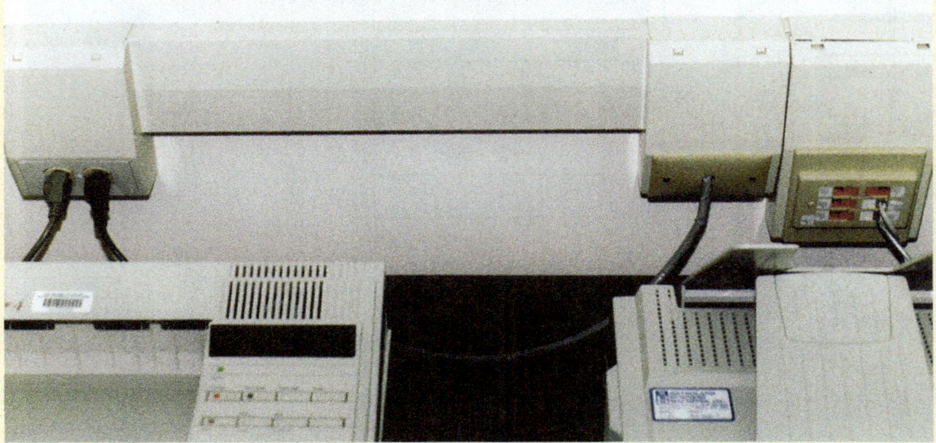

26108-14_SA12.EPS

9.4.3 Pole Systems

There are many situations in which power and other electric circuits have to be carried from overhead wiring systems to devices that are not located near existing wall outlets or control circuits. This type of wiring is typically used in open office spaces where cubicles are provided by temporary dividers. Poles are used to accomplish this. The poles usually come in lengths suitable for 10-, 12-, or 15-foot ceilings. *Figure 62* shows a typical pole base.

9.4.4 Underfloor Systems

Underfloor raceway systems were developed to provide a practical means of bringing conductors for lighting, power, and signaling to cabinets and consoles. Underfloor raceways are available in 10-foot lengths and widths of 4 and 8 inches. The sections are made with inserts spaced every 24 inches. The inserts can be removed for outlet installation. These are explained in *NEC Article 390.*

NOTE

Inserts must be installed so that they are flush with the finished grade of the floor.

Junction boxes are used to join sections of underfloor raceways. Conduit is also used with underfloor raceways by using a raceway-to-conduit connector (conduit adapter). A typical

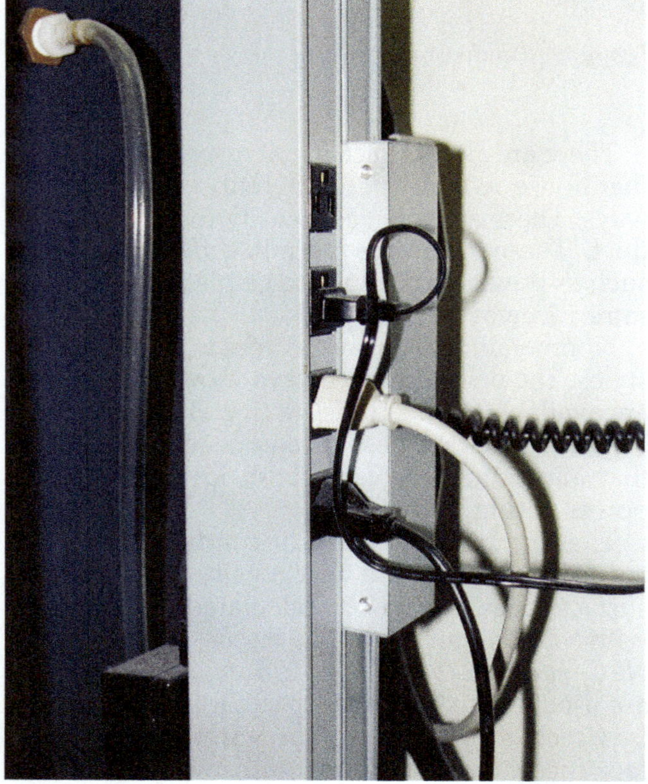

26108-14_F62.EPS

Figure 62 Power pole.

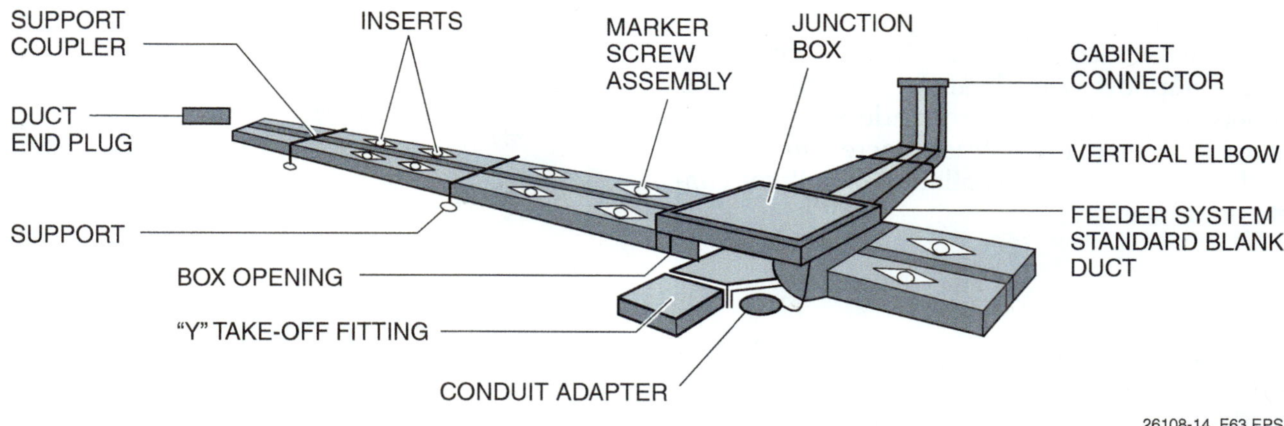

Figure 63 Underfloor raceway duct.

underfloor raceway duct with fittings is shown in *Figure 63*.

This wiring method makes it possible to place a desk or table in any location where it will always be over, or very near to, a duct line. The wiring method for lighting and power between cabinets and the raceway junction boxes may be conduit, underfloor raceway, wall elbows, and cabinet connectors. *NEC Article 390* covers the installation of underfloor raceways.

9.4.5 Cellular Metal Floor Raceways

A cellular metal floor raceway is a type of floor construction designed for use in steel-frame buildings. In these buildings, the members supporting the floor between the beams consist of sheet steel rolled into shapes. These shapes are combined to form cells, or closed passageways, which extend across the building. The cells are of various shapes and sizes, depending upon the structural strength required. The cells of this type of floor construction form the raceways, as shown in *Figure 64*.

Connections to the cells are made using headers that extend across the cells. A header connects only to those cells to be used as raceways for conductors. A junction box or access fitting is necessary at each joint where a header connects to a cell. Two or three separate headers, connecting to different sets of cells, may be used for different systems. For example, light and power, signaling systems, and public telephones would each have a separate header. A special elbow fitting is used to extend the headers up to the distribution equipment on a wall or column. *NEC Article 374* covers the installation of cellular metal floor raceways.

9.4.6 Cellular Concrete Floor Raceways

The term *precast cellular concrete floor* refers to a type of floor used in steel-frame, concrete-frame, and wall-bearing construction. In this type of system, the floor members are precast with hollow voids that form smooth, round cells. The cells form raceways, which can be adapted, using fittings, for use as underfloor raceways. A precast cellular concrete floor is fire-resistant and requires

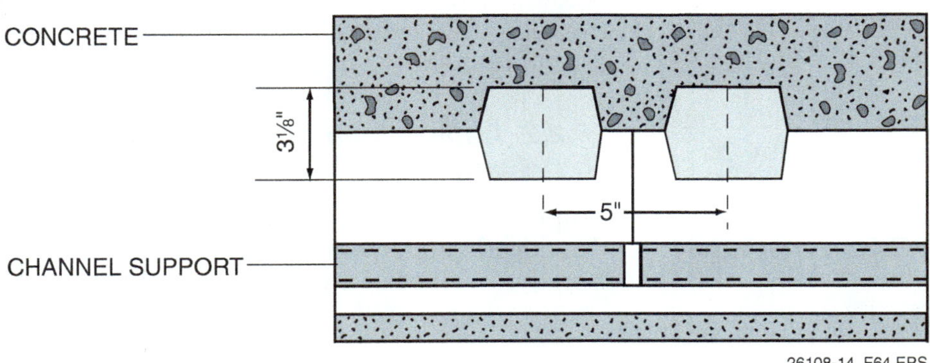

Figure 64 Cross section of a cellular floor.

no further fireproofing. The precast reinforced concrete floor members form the structural floor and are supported by beams or bearing walls. Connections to the cells are made with headers that are secured to the precast concrete floor. *NEC Article 372* covers the installation of cellular concrete floor raceways.

10.0.0 Cable Trays

Cable trays function as a support for conductors and tubing (see *NEC Article 392*). A cable tray has the advantage of easy access to conductors, and thus lends itself to installations where the addition or removal of conductors is a common practice. Cable trays are fabricated from aluminum, steel, and fiberglass. Cable trays are available in two basic forms: ladder and trough. Ladder tray, as the name implies, consists of two parallel channels connected by rungs. Trough consists of two parallel channels (side rails) having a corrugated, ventilated bottom, or a corrugated, solid bottom. There is also a special center rail cable tray available for use in light-duty applications such as telephone and sound wiring.

Cable trays are commonly available in 12- and 24-foot lengths. They are usually available in widths of 6, 9, 12, 18, 24, 30, and 36 inches, and load depths of 4, 6, and 8 inches.

Cable trays may be used in most electrical installations. Cable trays may be used in air handling ceiling space, but only to support the wiring methods permitted in such spaces by *NEC Section 300.22(C)(1)*. Also, cable trays may be used in Class 1, Division 2 locations according to *NEC Section 501.10(B)*. Cable trays may also be used above a suspended ceiling that is not used as an air handling space. Some manufacturers offer an aluminum cable tray that is coated with PVC for installation in caustic environments. A typical cable tray system with fittings is shown in *Figure 65*.

Wire and cable installation in cable trays is defined by the *NEC*®. Read *NEC Article 392* to become familiar with the requirements and restrictions made by the *NEC*® for safe installation of wire and cable in a cable tray.

Metallic cable trays that support electrical conductors must be grounded as required by *NEC Article 250*. Where steel and aluminum cable tray systems are used as an equipment grounding conductor, all of the provisions of *NEC Section 392.60* must be complied with.

WARNING!

Do not stand on, climb in, or walk on a cable tray.

10.1.0 Cable Tray Fittings

Cable tray fittings are part of the cable tray system and provide a means of changing the direction or dimension of the different trays. Some of the uses of horizontal and vertical tees, horizontal and vertical bends, horizontal crosses, reducers, barrier strips, covers, and box connectors are shown in *Figure 65*.

Think About It

Cable Trays and Wireways

What is the difference between a wireway and a cable tray? What kinds of conductors would you expect to find in a cable tray as compared to a wireway?

10.2.0 Cable Tray Supports

Cable trays are usually supported in one of five ways: direct rod suspension, trapeze mounting, center hung, wall mounting, and pipe rack mounting.

10.2.1 Direct Rod Suspension

The direct rod suspension method of supporting cable tray uses threaded rods and hanger clamps. One end of the threaded rod is connected to the ceiling or other overhead structure. The other end is connected to hanger clamps that are attached to the cable tray side rails. A direct rod suspension assembly is shown in *Figure 66*.

10.2.2 Trapeze Mounting and Center Hung Support

Trapeze mounting of cable tray is similar to direct rod suspension mounting. The difference is in the method of attaching the cable tray to the

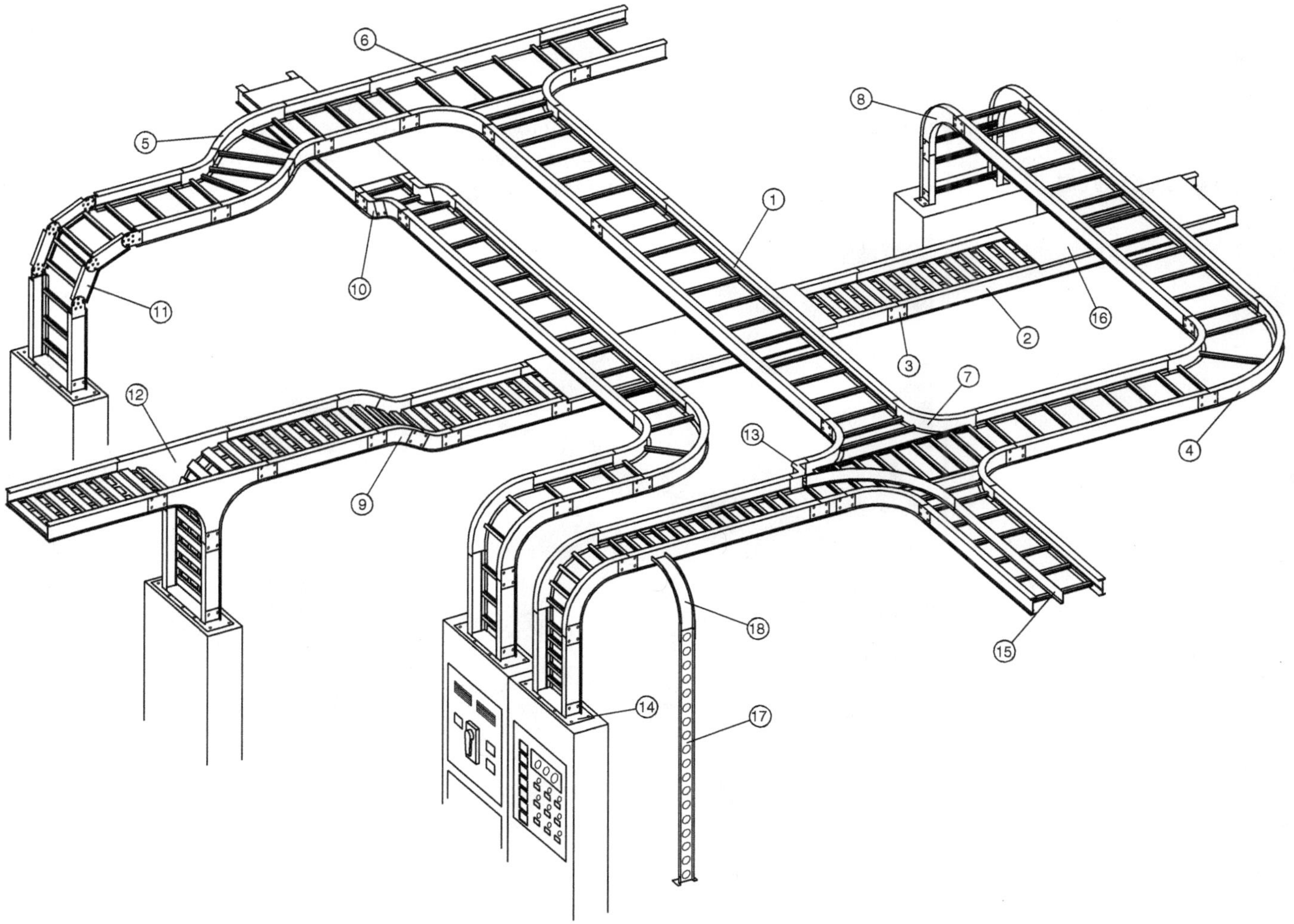

Legend

1. LADDER TYPE CABLE TRAY
2. VENTILATED TROUGH TYPE CABLE TRAY
3. STRAIGHT SPLICE PLATE
4. 90° HORIZONTAL BEND, LADDER TYPE CABLE TRAY
5. 45° HORIZONTAL BEND, LADDER TYPE CABLE TRAY
6. HORIZONTAL TEE, LADDER TYPE CABLE TRAY
7. HORIZONTAL CROSS, LADDER TYPE CABLE TRAY
8. 90° VERTICAL OUTSIDE BEND, LADDER TYPE CABLE TRAY
9. 45° VERTICAL OUTSIDE BEND, VENTILATED TYPE CABLE TRAY
10. 30° VERTICAL INSIDE BEND, LADDER TYPE CABLE TRAY
11. VERTICAL BEND SEGMENT (VBS)
12. VERTICAL TEE DOWN, VENTILATED TROUGH TYPE CABLE TRAY
13. LEFT HAND REDUCER, LADDER TYPE CABLE TRAY
14. FRAME TYPE BOX CONNECTOR
15. BARRIER STRIP STRAIGHT SECTION
16. SOLID FLANGED TRAY COVER
17. VENTILATED CHANNEL STRAIGHT SECTION
18. CHANNEL CABLE TRAY, 90° VERTICAL OUTSIDE BEND

26108-14_F65.EPS

Figure 65 Cable tray system.

threaded rods. A structural member, usually a steel channel or strut, is connected to the vertical supports to provide an appearance similar to a swing or trapeze. The cable tray is mounted to the structural member. Often, the underside of the channel or strut is used to support conduit. A trapeze mounting assembly is shown in *Figure 67*.

A method that is similar to trapeze mounting is a center hung tray support. In this case, only one rod is used and it is centered between the cable tray side rails.

On Site

Cable Tray Systems

Cable tray systems must be continuous and grounded. One of the advantages of using a cable tray system is that it makes it easy to expand or modify the wiring system following installation. Unlike conduit systems, wires can be added or changed by simply laying them into (or lifting them out of) the tray.

26108-14_SA13.EPS

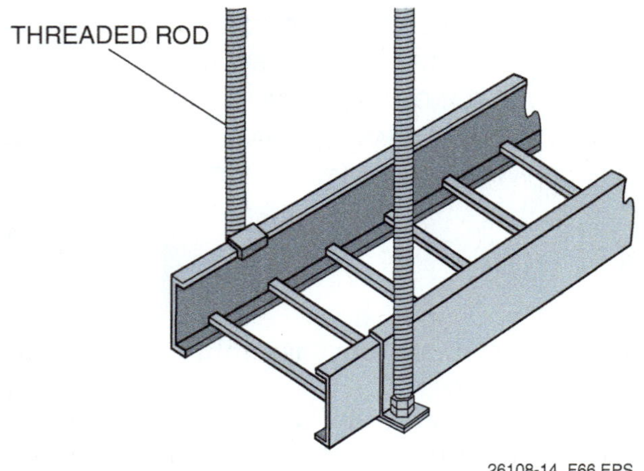

26108-14_F66.EPS

Figure 66 Direct rod suspension.

10.2.3 Wall Mounting

Wall mounting is accomplished by supporting the cable tray with structural members attached to the wall (*Figure 68*). This method of support is often used in tunnels and other underground or sheltered installations where large numbers of conductors interconnect equipment that is separated by long distances.

10.2.4 Pipe Rack Mounting

Pipe racks are structural frames used to support piping that interconnects equipment in outdoor industrial facilities. Usually, some space on the rack is reserved for conduit and cable tray. Pipe

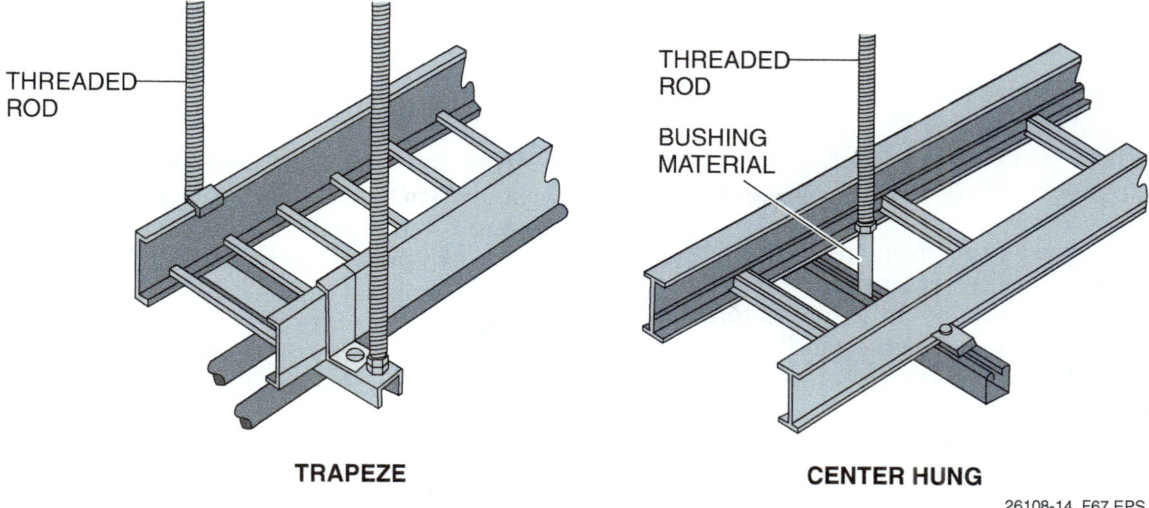

Figure 67 Trapeze mounting and center hung support.

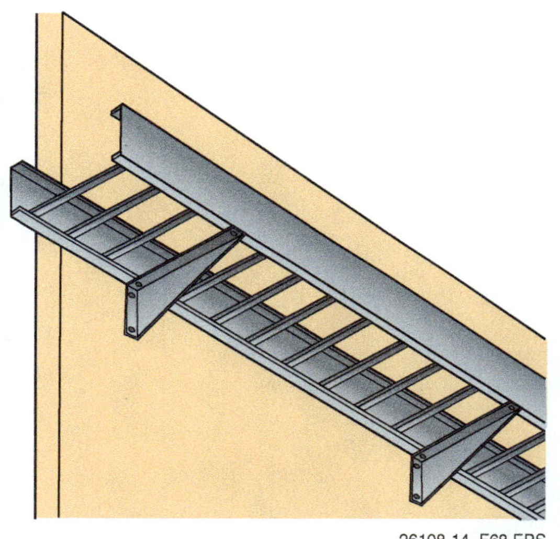

Figure 68 Wall mounting.

rack mounting of cable tray is often used when power distribution and electrical wiring is routed over a large area.

11.0.0 Storing Raceways

Proper and safe methods of storing conduit, wireways, raceways, and cable trays may sound like a simple task, but improper storage techniques can result in wasted time and damage to the raceways, as well as personal injury. There are correct ways to store raceways that will help avoid costly damage, save time in identifying stored raceways, and reduce the chance of personal injury.

Pipe racks are commonly used for storing conduit. The racks provide support to prevent bending, sagging, distorting, scratching, or marring of conduit surfaces. Most racks have compartments where different types and sizes of conduit can be separated for ease of identification and selection. The storage compartments in racks are usually elevated to help avoid damage that might occur at floor level. Conduit that is stored at floor level is easily damaged by people and other materials or equipment in the area.

The ends of stored conduit should be sealed to help prevent contamination and damage. Conduit ends can be capped, taped, or plugged.

Always inspect raceway before storing it to make sure that it is clean and not damaged. It is discouraging to get raceway for a job and find that it is dirty or damaged. Also, make sure that the raceway is stored securely so that when someone comes to get it for a job, it will not fall in any way that could cause injury.

To prevent contamination and corrosion of stored raceway, it should be covered with a tarpaulin or other suitable covering. It should also be separated from noncompatible materials such as hazardous chemicals.

Wireways, surface metal raceways, and cable trays should always be stored off the ground on boards in an area where people will not step on it and equipment will not run over it. Stepping on or running over raceway bends the metal and makes it unusable.

12.0.0 Handling Raceways

Raceway is made to strict specifications. It can be easily damaged by careless handling. From the time raceway is delivered to a job site until the installation is complete, use proper and safe handling techniques. These are a few basic guidelines for handling raceway that will help avoid damaging or contaminating it:

- Never drag raceway off a delivery truck or off other lengths of raceway.
- Never drag raceway on the ground or floor. Dragging raceway can cause damage to the ends.
- Keep the thread protection caps on when handling or transporting conduit raceway.
- Keep raceway away from any material that might contaminate it during handling.
- Flag the ends of long lengths of raceway when transporting it to the job site.
- Never drop or throw raceway when handling it.
- Never hit raceway against other objects when transporting it.
- Always use two people when carrying long pieces of raceway. Make sure that you both stay on the same side and that the load is balanced. Each person should be about one-quarter of the length of the raceway from the end. Lift and put down the raceway at the same time.

13.0.0 Ducting

In the common vocabulary of the electrical trade, a duct is a single enclosed raceway, or runway, through which conductors or cables can be led. Basically, ducting is a system of ducts. However, underground duct systems include manholes, transformer vaults, and risers.

There are several reasons for running power lines underground rather than overhead. In some situations, an overhead high-voltage line would be dangerous, or the space may not be adequate. For aesthetic reasons, architectural plans may require buried lines throughout a subdivision or a planned community. Tunnels may already exist, or be planned, for carrying steam or water lines. In any of these situations, underground installations are appropriate. Underground cables may be buried directly in the ground or run through tunnels or raceways, including conduit and recognized ducts.

In underground construction, a duct system provides a safe passageway for power lines, communication cables, or both. In buildings, underfloor raceways and cellular floor raceways are built to provide ducting so that electricity will be available throughout a large area. As an electrician, you need to know the approved methods of constructing underground ducting. You also need to know how to avoid potential electrical hazards in both original construction and maintenance. It is essential to understand the requirements and limitations imposed on running wires through underfloor and cellular floor raceways and ducts.

13.1.0 Underground Ducts

A duct consists of conduit or an approved duct system (such as HDPE) placed in a trench and covered with earth or concrete. The minimum depth at which the duct will be placed is determined using *NEC Table 300.5.* Encasing the duct in concrete or other materials provides mechanical strength and helps dissipate heat. *Figure 69* shows a duct bank in place and ready for backfill. In this case, it will be covered in concrete.

Manholes are set at intervals in an underground duct run. *Figure 70* shows a manhole with pull strings installed and tied off in preparation for the conductor installation. Manholes provide access through throats (sometimes called chimneys). At ground level, or street surface level, a manhole

26108-14_F69.EPS

Figure 69 Duct bank.

26108-14_F70.EPS

Figure 70 Manhole.

cover closes off the manhole area tightly. A duct line may consist of a single conduit or several, each carrying a cable length from one manhole to the next.

Manholes provide room for conductor installation and maintenance. Workers enter a manhole from above. In a two-way manhole, cables enter and leave in only two directions. There are also three-way and four-way manholes. Often manholes are located at the intersection of two streets so that they can be used for cables leaving in four directions. Manholes are usually constructed of brick or concrete. Their design must provide room for drainage and for workers to move around inside them. A similar opening known as a handhole is sometimes provided for splicing on lateral two-way duct lines.

Transformer vaults house power transformers, voltage regulators, network protectors, meters, and circuit breakers. A cable may end at a transformer vault. Other cables end at a customer's substation or terminate as risers that connect with overhead lines.

13.2.0 Duct Materials

Underground duct lines can be made of fiber, vitrified tile, rigid metal or nonmetallic conduit, or poured concrete. The inside diameter of the ducting for a specific job is determined by the size of the cable that will be drawn into the duct. Sizes from two to six inches (inside diameter) are available for most types of ducting.

WARNING!

Be careful when working with unfamiliar duct materials. In older installations, asbestos/cement duct may have been used. You must be certified to remove or disturb asbestos.

Rigid nonmetallic conduit may be made of PVC (polyvinyl chloride), PE (polyethylene), or styrene. Since this type of conduit is available in lengths up to 20 feet, fewer couplings are needed than with other types of ducting. PVC is popular because it is easy to install, requires less labor than other types of conduit, and is low in cost.

13.3.0 Monolithic Concrete Duct

Monolithic concrete duct is poured at the job site. Multiple duct lines can be formed using rubber tubing cores on spacers. The cores may be removed after the concrete has set. A die containing steel tubes, known as a boat, can also be used to form ducts. It is pulled slowly through the trench on a track as concrete is poured from the top. Poured concrete ducting made by either method is relatively expensive, but offers the advantage of creating a very clean duct interior with no residue that can decay. The rubber core method is especially useful for curving or turning part of a duct system.

13.4.0 Cable-in-Duct

One of the most popular duct types is the cable-in-duct. This type of duct comes from the manufacturer with cables already installed. The duct comes in a reel and can be laid in the trench with ease. The installed cables can be withdrawn in the future, if necessary. This type of duct, because of the form in which it comes, reduces the need for fittings and couplings. It is most frequently used for street lighting systems.

14.0.0 Construction Methods

Conduit and box installation varies with the type of construction. This section discusses some special requirements for masonry and concrete, metal framing, wood, and structural steel construction.

14.1.0 Masonry and Concrete Flush-Mount Construction

In a reinforced concrete construction environment, the conduit and boxes must be embedded in the concrete to achieve a flush surface. Ordinary boxes may be used, but special concrete boxes are preferred and are available in depths up to six inches. These boxes have special ears by which they are nailed to the wooden forms for the concrete. When installing them, stuff the boxes tightly with paper to prevent concrete from seeping in. *Figure 71* shows an installed box.

Flush construction can also be done on existing concrete walls, but this requires chiseling a channel and box opening, anchoring the box and conduit, and then resealing the wall.

To achieve flush construction with masonry walls, the most acceptable method is for the electrician to work closely with the mason laying the blocks. When the construction blocks reach the convenience outlet elevation, boxes are made up as shown in *Figure 72*. The figure shows a raised tile ring or box device cover.

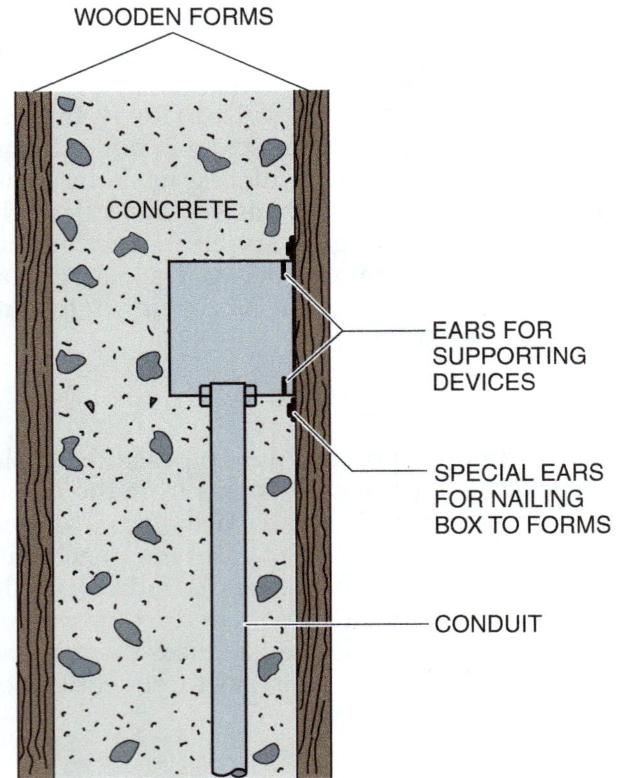

Figure 71 Concrete flush-mount installation.

Figure 72 Box with raised ring.

Figure 73 shows a masonry box that needs no extension or deep plaster ring to bring it to the surface.

> **NOTE**
> The electrician must work with the mason to ensure the box is properly grouted and sealed.

Figure 73 Three-gang concrete box.

Sections of conduit are then coupled in short (4- or 5-foot) lengths. This is done because it is impractical for the mason to maneuver blocks over 10-foot sections of conduit.

14.2.0 Metal Stud Environment

Metal stud walls are a popular method of construction for the interior walls of commercial buildings. Metal stud framing consists of relatively thin metal channel studs, usually constructed of galvanized steel and with an overall dimension the same as standard 2 × 4 wooden studs. Wiring in this type of construction is relatively easy when compared to masonry.

EMT conduit and MC cable are the most common type of wiring methods for metal stud environments. Metal studs usually have some number of pre-punched holes that can be used to route the conduit. If a pre-punched hole is not located where it needs to be, holes can be easily punched in the metal stud with a hole cutter or knockout punch (*Figure 74*).

> **WARNING!**
> Cutting or punching metal studs can create sharp edges. Avoid contact that can result in cuts.

Boxes can be secured to the metal stud using self-tapping screws or one of the many types of box supports available. EMT conduit is supported by the metal studs using conduit straps or other approved

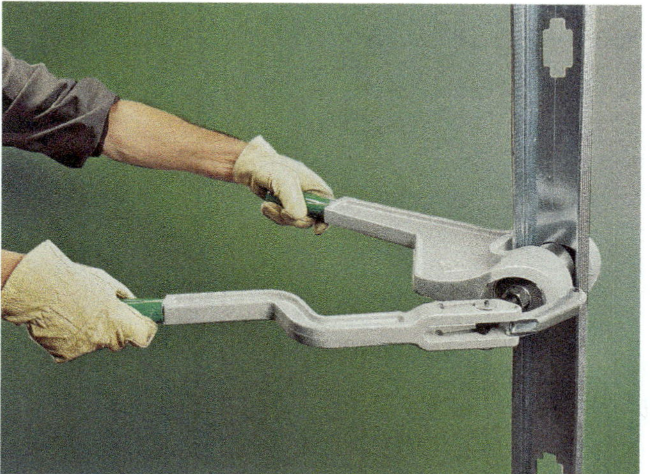

26108-14_F74.EPS

Figure 74 Metal stud punch.

26108-14_F75.EPS

Figure 75 NM cable protected by grommets.

methods. It is important that the conduit be properly supported to facilitate pulling the conductors through the tubing. Boxes are mounted on the metal studs so that the box will be flush with the finished walls. You must know what the finished wall thickness is going to be to properly secure the boxes to the metal studs. For example, if the finished wall will be ⅝-inch drywall, then the box must be fastened so that it protrudes ⅝ of an inch from the metal stud.

WARNING!

When using a screw gun or cordless drill to mount boxes to studs, keep the hand holding the box away from the gun/drill to avoid injury.

Per *NEC Section 300.4(B)(1),* NM cable run through metal studs must be protected by listed bushings or listed grommets (*Figure 75*). This protects the cables from the friction of pulling during installation and from the weight of the cable and vibrations following the installation.

14.3.0 Wood Frame Environment

At one time, the use of rigid conduit in partitions and ceilings was a time-consuming operation. Thinwall conduit makes an easier and quicker job, largely because of the types of fittings that are specially adapted to it.

Figure 76 shows two methods of running thinwall conduit in these locations: boring timbers and notching them. When boring, holes must be drilled large enough for the tubing to be inserted between the studs. The tubing is cut rather short, calling for multiple couplings. EMT can be bowed quite a bit while threading through holes in studs. Boring is the preferred method.

WARNING!

Always wear safety goggles when boring wood.

NEC Section 300.4 addresses the requirements to prevent physical damage to conductors and cabling in wood members. By keeping the edge of the drilled hole 1¼" from the closest edge of the stud, nails are not likely to penetrate the stud far enough to damage the cables. The building codes provide maximum requirements for bored or notched holes in studs.

NEC Section 300.4(A)(1) requires the use of a steel plate or bushing at least 1⁄16" thick or a listed steel nail plate where wiring is installed through bored wooden members less than 1¼" from the nearest edge. See *Figure 77*. Nail plates are also required to protect the conductors in all notched wooden members per *NEC Section 300.4(A)(2)*.

The exception in the *NEC*® permits IMC, RMC, PVC, and EMT to be installed through bored holes

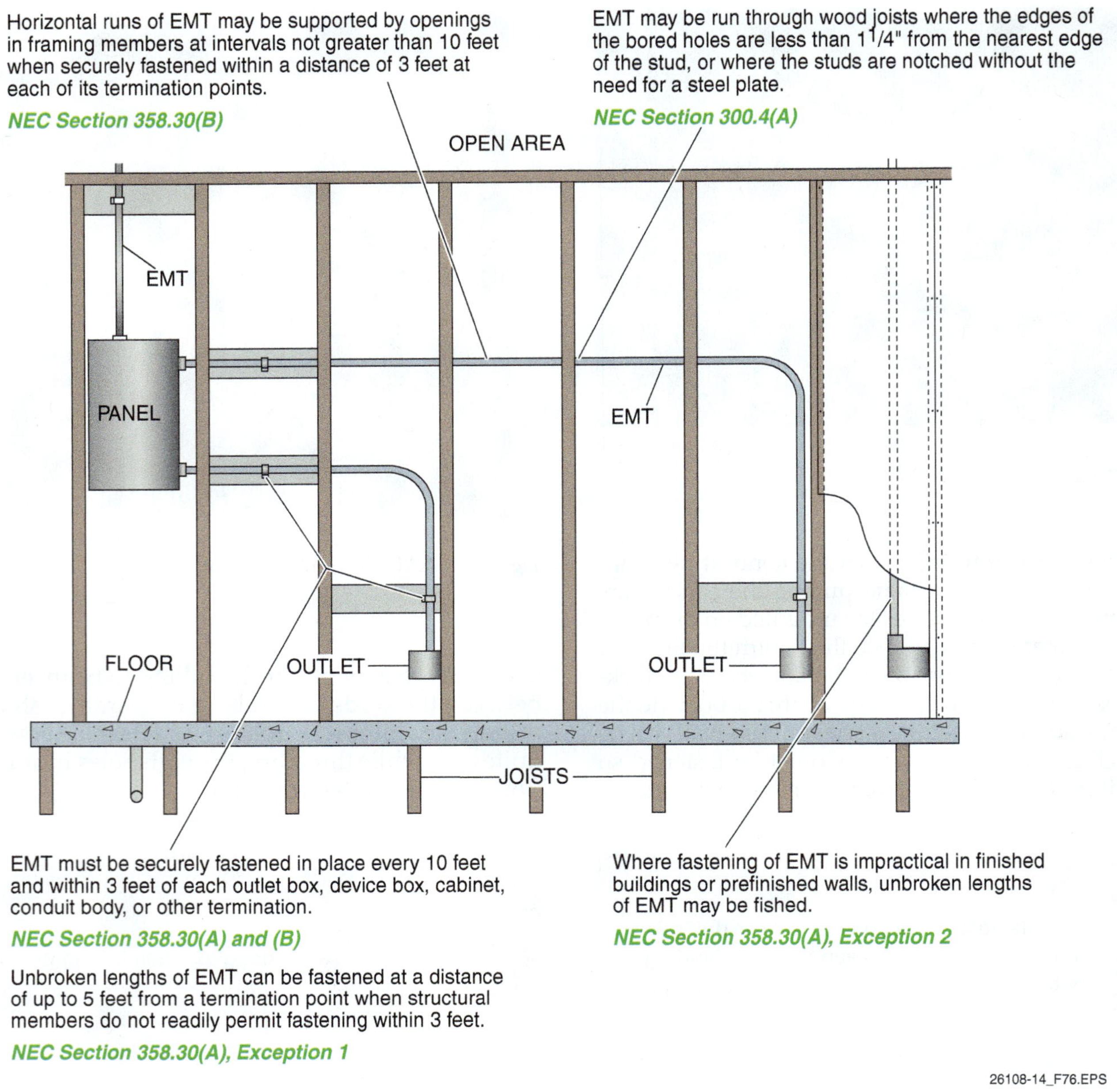

Figure 76 Installing wire or conduit in a wood-frame building.

or laid in notches less than 1¼" from the nearest edge without a steel plate or bushing.

Because of its weakening effect upon the structure, notching should be resorted to only where absolutely necessary. Notches should be as narrow as possible and in no case deeper than 1⁄16 the stock of a bearing timber. A bearing timber supports floor joists or other weight.

> **NOTE**
> Always check with the architect before notching or drilling.

Some wood I-beams are manufactured with perforated knockouts in their web, approximately 12" apart. Never notch or drill through the beam flange or cut other openings in the web without checking the manufacturer's specification sheet.

Also, do not drill or notch other types of engineered lumber without first checking the specification sheets.

14.4.0 Metal Buildings

Many commercial and industrial buildings are prefabricated structures with steel structural supports, and roofing and siding made of light-gauge metal sheets (*Figure 78*). Conduit can be

Figure 77 Steel nail plate.

Figure 78 Metal building.

routed across the structural members that support the roof. *NEC Section 300.4(E)* states that a cable, raceway, or box in exposed or concealed locations under metal-corrugated sheet roof decking must be installed and supported so the nearest outside surface of the cable or raceway is not less than 1½" from the nearest surface of the roof decking. The roof structure can consist of beams and purlins (*Figure 79*) or open-web steel joists (*Figure 80*).

Beams and purlins should not be drilled through; consequently, the conduit is supported

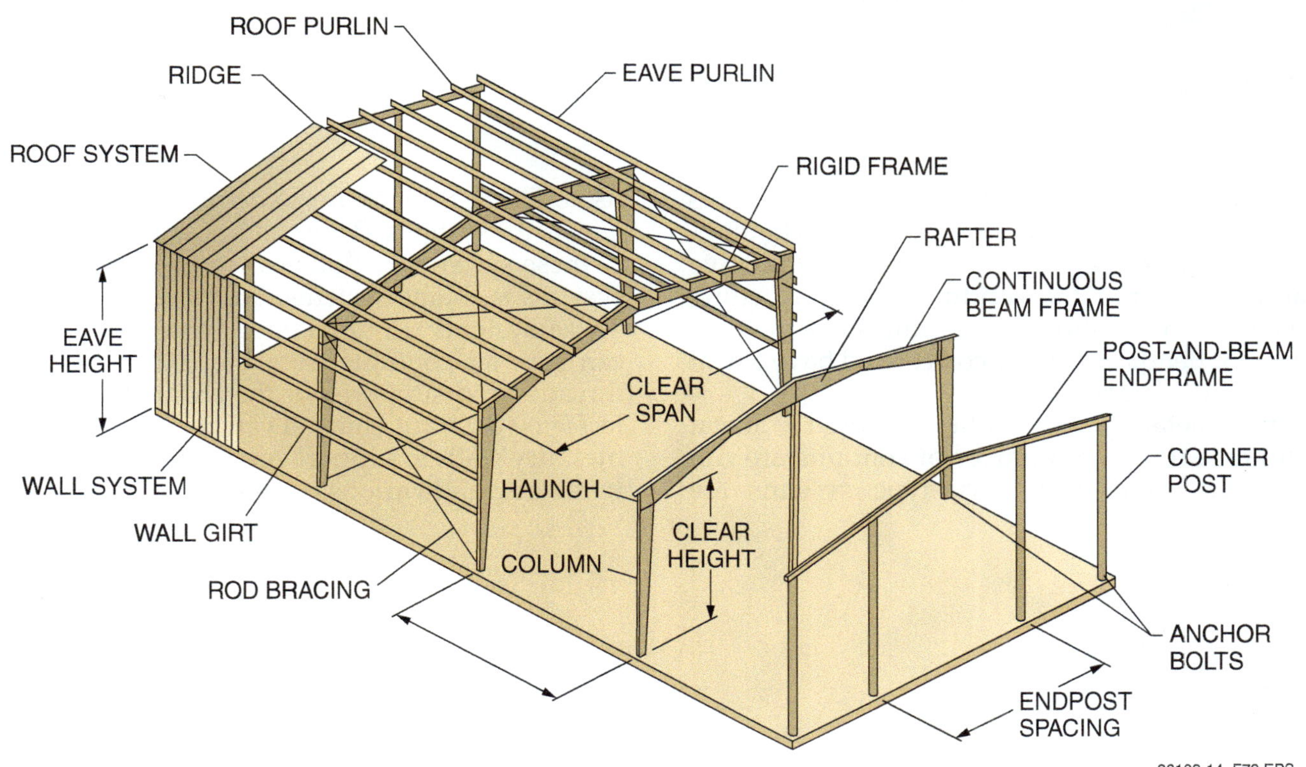

Figure 79 Beam and purlin roof system.

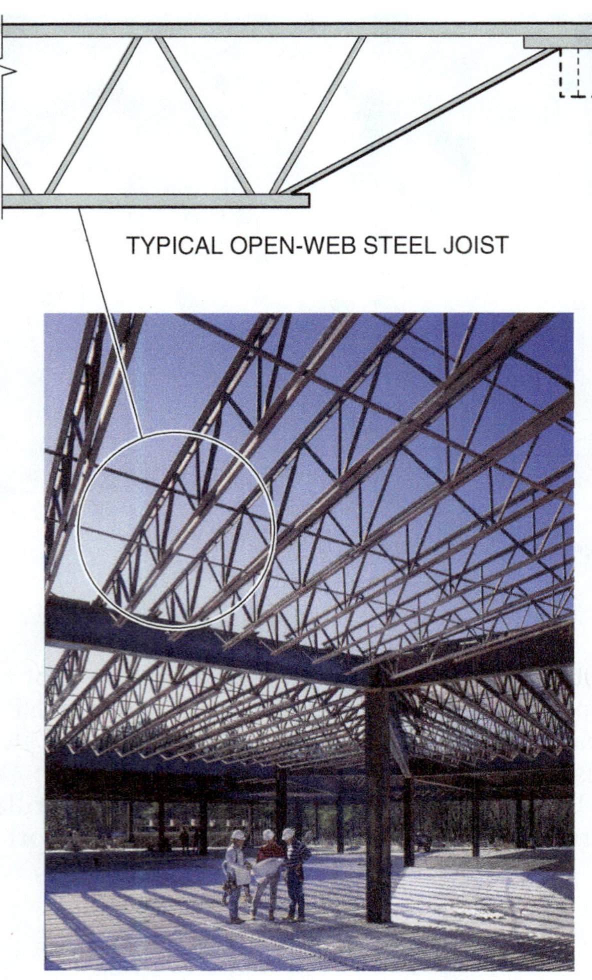

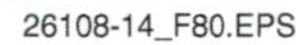

26108-14_F80.EPS

Figure 80 Open-web steel joist roof supports.

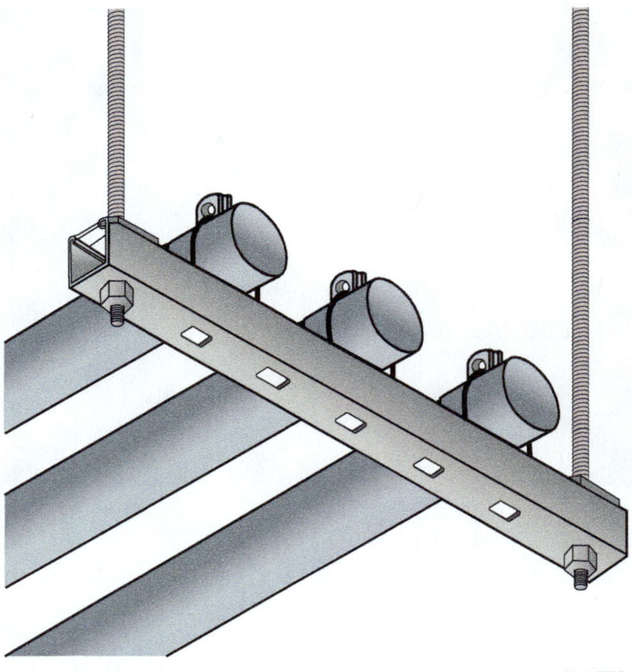

26108-14_F81.EPS

Figure 81 Steel strut system.

from the metal beams by anchoring devices designed especially for that purpose. The supports attach to the beams or supports and have clamps to secure the conduit to the structure. All conduit runs should be plumb since they are exposed. Bends should be correct and have a neat and orderly appearance.

Rigid metal conduit is often required in metal buildings. If a large number of conduits are run along the same path, strut-type systems are used. These systems are sometimes referred to as Unistrut® systems (Unistrut® is a manufacturer of these systems). Another manufacturer of strut systems is B-Line systems. Both are very similar. These systems use a channel-type member that can support conduits from the ceiling by using threaded rod supports for the channel, as shown in *Figure 81*. Strut channel can also be secured to masonry walls to support vertical runs of conduit, wireways, and various types of boxes.

What's wrong with this picture?

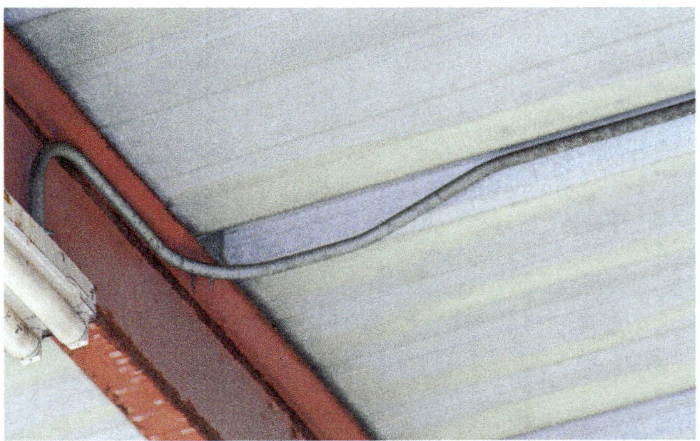

26108-14_SA14.EPS

Think About It

Putting It All Together

Think about the effort that goes into the design of a large industrial installation. If you were to design a large complex, such as the one shown here, where would you start and why?

26108-14_SA15.EPS

SUMMARY

This module discussed the various types of raceways, boxes, and fittings, including their uses and procedures for installation. The primary purpose of raceways is to house electric wire used for power distribution, communication, or electronic signal transmission. Raceways provide protection to the wiring and even a means of identifying one type of wire from another when run adjacent to each other. This process requires proper planning to allow for current needs, future expansion, and a neat and orderly appearance.

Review Questions

1. The lightest duty and most widely used non-flexible metal conduit is _____.
   a. electrical metallic tubing
   b. rigid metal conduit
   c. aluminum conduit
   d. plastic-coated RMC

2. EMT and RMC connectors are interchangeable.
   a. True
   b. False

3. Which of the following is often referred to as thinwall conduit?
   a. IMC
   b. EMT
   c. RMC
   d. Galvanized rigid steel conduit

4. RMC is made of _____.
   a. cast iron
   b. steel or aluminum
   c. copper or aluminum
   d. PVC

5. A Type LB conduit body has a cover on _____.
   a. the left
   b. the right
   c. the back
   d. both sides

6. The fitting used to protect conductors from the sharp edges of conduit where it enters a box is called a _____.
   a. bushing
   b. locknut
   c. coupling
   d. nipple

7. Hammer-set anchors are designed for use in _____.
   a. wood
   b. metal studs
   c. concrete
   d. structural steel

8. *NEC Section 376.56(A)* limits wireway fill to no more than _____ of the cross-sectional area of the wireway.
   a. 40%
   b. 60%
   c. 75%
   d. 90%

9. Raceways designed to extend conductors across a floor without embedding it in the floor are called _____.
   a. cellular raceways
   b. raceway ducts
   c. cellular ductways
   d. pancake raceways

10. Which of the following regulations applies to drilling of wood joists and girders?
    a. A joist cannot be drilled unless approved by the manufacturer.
    b. They can only be drilled in the center third.
    c. The hole must be at least 1 inch from an edge.
    d. The hole diameter must not exceed one-half of the depth of the girder or joist.

Trade Terms Quiz

Fill in the blank with the correct term that you learned from your study of this module.

1. A(n) ____________ area is one that can be reached for service or repair.
2. When something is in a(n) ____________, it is not permanently closed in by the structure or finish of a building.
3. When materials meet a regulatory agency's requirements, the material is then said to be ____________.
4. ____________ is a regulatory agency that evaluates and approves electrical components and equipment.
5. A(n) ____________ is used to make a continuous grounding path between equipment and ground.
6. ____________ are rigid structures, either suspended or mounted, that are used to support electrical conductors.
7. Similar to pipe, ____________ is a round raceway that houses conductors.
8. A(n)____________ is a bend made in a piece of conduit to alter its course.
9. ____________ are enclosed channels that are used to house wires and cables.
10. ____________ are steel troughs designed to carry electrical wire and cable.
11. A(n) ____________ is the connection of two or more conductors.
12. An intermediate point on a main circuit where another wire is connected to supply electrical current to another circuit is called a(n) ____________.
13. Electrical connectors that could be exposed to the environment are housed in long, narrow boxes, or ____________.

Trade Terms

Accessible
Approved
Bonding wire
Cable trays
Conduit
Exposed location
Kick
Raceways
Splice
Tap
Trough
Underwriters Laboratories, Inc. (UL)
Wireways

Module 26108-14
Supplemental Exercises

1. Most electrical equipment that has a metal frame must be ______________________.

2. When installing EMT in a wet location, what type of fittings must be used?
   a. Setscrew
   b. Compression
   c. Raintight
   d. Steel

3. Conductors, along with splices and taps, must not fill a wireway to more than _________________ of its cross-sectional area.
   a. 20%
   b. 30%
   c. 75%
   d. 90%

4. Which of the following covers the installation requirements for cable tray?
   a. *NEC Article 333*
   b. *NEC Article 338*
   c. *NEC Article 392*
   d. *NEC Article 394*

5. True or False? Two-inch IMC conduit has the same internal diameter as two-inch RMC.

6. Which of the following cannot be used as an equipment grounding conductor?
   a. Rigid metal conduit
   b. Rigid nonmetallic conduit
   c. Electrical metallic tubing
   d. Intermediate metal conduit

7. Flexible metal conduit can be connected to rigid metal conduit using a(n) _________________ coupling.

8. Which of the following covers grounding provisions for metal boxes?
   a. *NEC Section 500.8(A)*
   b. *NEC Section 250.30(A)*
   c. *NEC Section 250.53(C)*
   d. *NEC Section 314.40(D)*

9. What type of rigid nonmetallic conduit may be installed where exposed to physical damage?
   a. Type DB Schedule 80
   b. Type DB Schedule 40
   c. Type EB
   d. Type 1

10. RMC is commonly used in ______________________ locations.

11. What type of conduit body is used to provide a junction point for three intersecting conduits?
    a. Type LL
    b. Type LR
    c. Type C
    d. Type T

12. When installing conduit in a wet area, a(n) ____________________ air space must be provided between it and the supporting surface.

13. Conductors installed in PVC are subject to a(n) ____________________.
    a. increase in operating temperature
    b. decrease in operating temperature
    c. increase in ampacity
    d. reduced voltage drop

14. Flexible metal conduit must be supported within ____________________ inches of each end.
    a. 8
    b. 10
    c. 12
    d. 18

15. When installing a cable system near a metal corrugated sheet roofing deck, a spacing of ________________ must be maintained from any point of the roof system.
    a. 1¼"
    b. 1⅜"
    c. 1½"
    d. 1¾"

Cornerstone of Craftsmanship

Leonard "Skip" Layne

Rust Constructors Inc.

How did you choose a career in the electrical field?
I think the electrical field chose me. My father was a contractor for several years before closing shop and accepting a job as an electrical superintendent with the Rust Engineering Company. That happened when I was nine years old. After being moved around the country for the next several years and working as an apprentice on Dad's projects during my college summers, I couldn't think of anything that I would rather do.

Tell us about your apprenticeship experience.
I've never attended a formal apprenticeship school. There are probably several in our group who might say that they suspected this. My electrical education came from field work exposure and several electrical and engineering courses and seminars I've attended over the years.

I'm happy to say that I'm still learning and I've learned a great deal while working on the NCCER Electrical Committee and from my association with the other subject matter experts.

What positions have you held in the industry?
I started as a field apprentice on a tire plant in Madison, Tennessee, in 1959. I've held field positions as an apprentice, journeyman, field engineer, start-up manager, and superintendent. I spent a number of years estimating work, and I established the material control department for another major open-shop contractor several years ago. I managed the project controls group on a nuclear project for another open-shop contractor. I even spent a few years as vice president with an underground utility/treatment plant contractor.

What would you say is the primary factor in achieving success?
Keep learning. Work hard. I've had to work sixteen-hour days on the job site and in the office in order to meet the schedule and incorporate changes. Do what is asked of you and do it well.

What does your current job involve?
My job title says that I'm the Construction Engineering Manager for Rust, but the lack of a definitive job title means that I do whatever the company needs me to do at the time. I qualify the company's electrical licenses in seventeen states where we work.

Recently, Rust volunteered my services to the Gulf Coast Workforce Initiative, a business roundtable initiative to train 20,000 new construction workers for the Gulf Coast area devastated by hurricanes Katrina and Rita.

Do you have any advice for someone just entering the trade?
Get all of the classroom learning you can. Go through all four levels of the Electrical program while working in the field. Ask questions and try to get assigned to as many new and different tasks as you can. All of our larger ABC contractors have excellent supervisory training programs and you need to get into those after your craft training. Be adaptable and keep learning.

Cornerstone of Craftsmanship

Al Hamilton

Willmar Electric Service

How did you become an electrician?

I was in college and met Ed, an electrician who told me about wiring buildings. I went to work for him part time at first and liked it which led to my becoming his apprentice. He was tough taskmaster who cared about his apprentices and he was an excellent communicator and teacher.

How did you get your training?

My training was 100% on the job. Everything was learned "hands on."

I was very fortunate to work for Ed and other experienced electricians who were true craftsmen.

What factor or factors have contributed to your success?

Work ethic, relationships, and reading. A good work ethic has allowed me to overcome mistakes and keep working to learn our craft. I was able to learn about electrical work, business, and life through relationships. I have always looked for successful people to listen and learn from. I discovered that many successful people are happy to share their ideas and that proved to be the key to personal growth. I was told long ago that if I could force myself to read just 10 pages a day that I could read a book a month because the average book is 300 pages. I have done this for many years and I read books on many subjects including the Bible, business, history, biographies, and money and, of course, the *National Electrical Code.* Through reading you can educate yourself and become a more knowledgeable and interesting person who others will look to for advice.

What does your current job entail?

I work for a leading electrical company, Willmar Electric Service Corp. and my responsibilities are business development and estimating. This includes maintaining relationships with our existing customers and finding new customers to work for. Our estimating team uses estimating software to bid jobs. We do competitive bidding to win jobs and we also estimate for design/build projects and projects that are negotiated with customers.

Any advice for apprentices just beginning their careers?

Work for an organization that shares your values and will recognize and reward your efforts. Don't ever give up! Who you become will depend largely on the people you meet and the books that you read. Find honest, ethical people who have demonstrated success and get to know them. If you have not been reading, start today—make yourself do it. Get involved in helping others by becoming a leader in your company, helping out at church, and passing on what you have learned.

Trade Terms Introduced in This Module

Accessible: Able to be reached, as for service or repair.

Approved: Meeting the requirements of an appropriate regulatory agency.

Bonding wire: A wire used to make a continuous grounding path between equipment and ground.

Cable trays: Rigid structures used to support electrical conductors.

Conduit: A round raceway, similar to pipe, that houses conductors.

Exposed location: Not permanently closed in by the structure or finish of a building; able to be installed or removed without damage to the structure.

Kick: A bend in a piece of conduit, usually less than 45°, made to change the direction of the conduit.

Raceways: Enclosed channels designed expressly for holding wires, cables, or busbars, with additional functions as permitted in the *NEC®*.

Splice: Connection of two or more conductors.

Tap: Intermediate point on a main circuit where another wire is connected to supply electrical current to another circuit.

Trough: A long, narrow box used to house electrical connections that could be exposed to the environment.

Underwriters Laboratories, Inc. (UL): An agency that evaluates and approves electrical components and equipment.

Wireways: Steel troughs designed to carry electrical wire and cable.

Additional Resources

This module presents thorough resources for task training. The following resource material is suggested for further study.

Benfield Conduit Bending Manual, 2nd Edition. Overland Park, KS: EC&M Books.

National Electrical Code® Handbook, Latest Edition. Quincy, MA: National Fire Protection Association.

Figure Credits

Associated Builders and Contractors, Inc., Module opener

Topaz Publications, Inc., Figures 3, 4A, 8, 12–19, 25, 35B (photo), 42, 62, 78, SA01–SA09, SA11, SA12, SA15

Tim Dean, Figures 4B, 61, 77, SA10, SA14

Photo courtesy of Simpson Strong-Tie Company, Inc., Figure 32 (photo)

Wiremold/Legrand, Figures 58, 59

Panduit Corp., Figure 60

Cooper B-Line, Figure 65

Jim Mitchem, Figures 69, 70, SA13

Greenlee/A Textron Company, Figures 74, 75

VP Buildings, Figure 79

Nucor Corporation - Vulcraft Group, Figure 80 (photo)

Reprinted with permission from NFPA 70-2014, *National Electrical Code®*, Copyright © 2013, National Fire Protection Association, Quincy, MA. This reprinted material is not the complete and official position of the NFPA on the referenced subject, which is represented only by the standard in its entirety., Table 1

NCCER CURRICULA — USER UPDATE

NCCER makes every effort to keep its textbooks up-to-date and free of technical errors. We appreciate your help in this process. If you find an error, a typographical mistake, or an inaccuracy in NCCER's curricula, please fill out this form (or a photocopy), or complete the online form at **www.nccer.org/olf**. Be sure to include the exact module ID number, page number, a detailed description, and your recommended correction. Your input will be brought to the attention of the Authoring Team. Thank you for your assistance.

Instructors – If you have an idea for improving this textbook, or have found that additional materials were necessary to teach this module effectively, please let us know so that we may present your suggestions to the Authoring Team.

NCCER Product Development and Revision
13614 Progress Blvd., Alachua, FL 32615

Email: curriculum@nccer.org
Online: www.nccer.org/olf

❑ Trainee Guide ❑ AIG ❑ Exam ❑ PowerPoints Other ____________

Craft / Level: Copyright Date:

Module ID Number / Title:

Section Number(s):

Description:

Recommended Correction:

Your Name:

Address:

Email: Phone:

Conductors and Cables

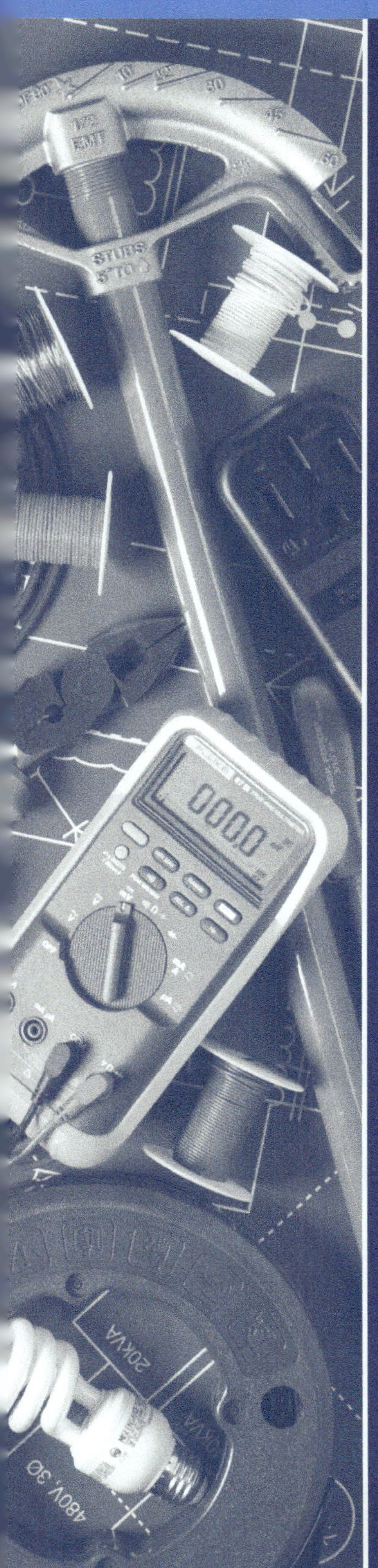

Forest Park-DeBaliviere Station

This MetroLink expansion project in downtown St. Louis used concrete in prolific ways, using more than 500 feet of concrete sewer bridges, installing three cut-and-cover tunnels totaling 1,400 linear feet, and constructing a pedestrian tunnel under a major thoroughfare. The expansion, which opened in August 2006, provides MetroLink customers with a new, eight-mile light rail line.

26109-14

Trainees with successful module completions may be eligible for credentialing through NCCER's National Registry. To learn more, go to **www.nccer.org** or contact us at **1.888.622.3720.** Our website has information on the latest product releases and training, as well as online versions of our *Cornerstone* magazine and Pearson's product catalog.

Your feedback is welcome. You may email your comments to **curriculum@nccer.org,** send general comments and inquiries to **info@nccer.org,** or fill in the User Update form at the back of this module.

This information is general in nature and intended for training purposes only. Actual performance of activities described in this manual requires compliance with all applicable operating, service, maintenance, and safety procedures under the direction of qualified personnel. References in this manual to patented or proprietary devices do not constitute a recommendation of their use.

26109-14

Conductors and Cables

Objectives

When you have completed this module, you will be able to do the following:

1. From the cable markings, describe the insulation and jacket material, conductor size and type, number of conductors, temperature rating, voltage rating, and permitted uses.
2. Determine the allowable ampacity of a conductor for a given application.
3. Identify the *NEC*® requirements for color coding of conductors.
4. Install conductors in a raceway system.

Performance Task

Under the supervision of the instructor, you should be able to do the following:

1. Install conductors in a raceway system.

Trade Terms

Ampacity
Capstan
Fish tape
Mouse
Wire grip

Required Trainee Materials

1. Paper and pencil
2. Copy of the latest edition of the *National Electrical Code*®
3. Appropriate personal protective equipment

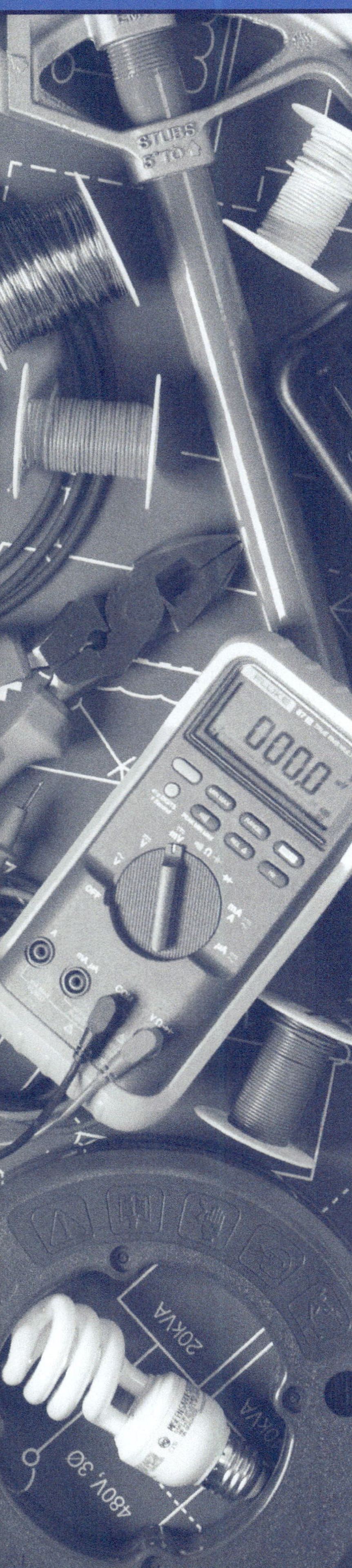

Contents

Topics to be presented in this module include:

Figures and Tables

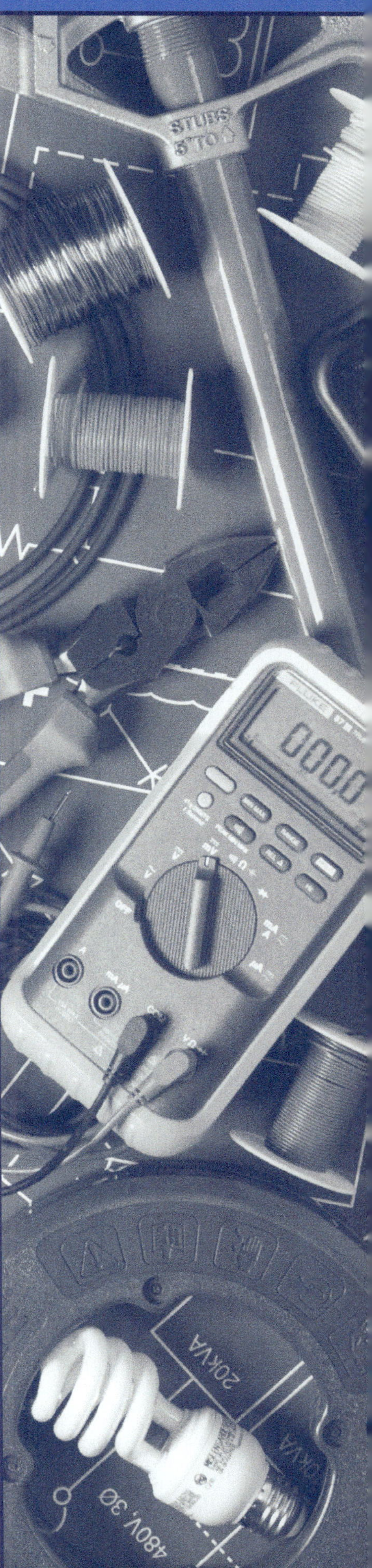

1.0.0 Introduction

As an electrician, you will be required to select the proper wire and/or cable for a job. You will also be required to pull this wire or cable through conduit runs in order to terminate it. This module describes the different types of conductors and conductor insulation. It also explains how these conductors are rated and classified by the *NEC®*. Different methods used for pulling these conductors through conduit runs are also discussed.

2.0.0 Conductors and Insulation

The term *conductor* is used in two ways. It is used to describe the current-carrying portion of a wire or cable, and it is used to describe a wire or cable composed of the current-carrying portion and an outer covering (insulation). In this module, the term *conductor*, if not specified otherwise, is used to describe the wire assembly, which includes the insulation and the current-carrying portion of the wire.

Conductors are uniquely identified by size and insulation material. Size refers to the cross-sectional area of the current-carrying portion of the wire. The ampacity is affected by the conductor material and size, insulation, and installation location.

2.1.0 Wire Size

Wire sizes are expressed in gauge numbers. The standard system of wire sizes in the United States is the American Wire Gauge (AWG) system.

2.1.1 AWG System

The AWG system uses numbers to identify the different sizes of wire and cable (*Figure 1*). The larger the number, the smaller the cross-sectional area of the wire. The larger the cross-sectional area of the current-carrying portion of a conductor, the higher the amount of current the wire can conduct. The AWG numbers range from 50 to 1; then 0, 00, 000, and 0000 (one aught [1/0], two aught [2/0], three aught [3/0], and four aught [4/0]). Any wire larger than 0000 is identified by its area in circular mils. Wire sizes smaller than No. 18 AWG are usually solid, but may be stranded in some cases. Wire sizes of No. 6 AWG or larger are stranded.

For wire sizes larger than No. 16 AWG, the wire size is marked on the insulation (*Figure 2*).

NEC Chapter 9, Table 8 has descriptive information on wire sizes. Again, note that all wires smaller than No. 6 are available as solid or stranded. Wire sizes of No. 6 or larger are shown only as stranded. Solid wire larger than No. 6 is manufactured; however, the *NEC®* only permits the use of solid wire in a raceway for sizes smaller than No. 8 per *NEC Section 310.106(C)*.

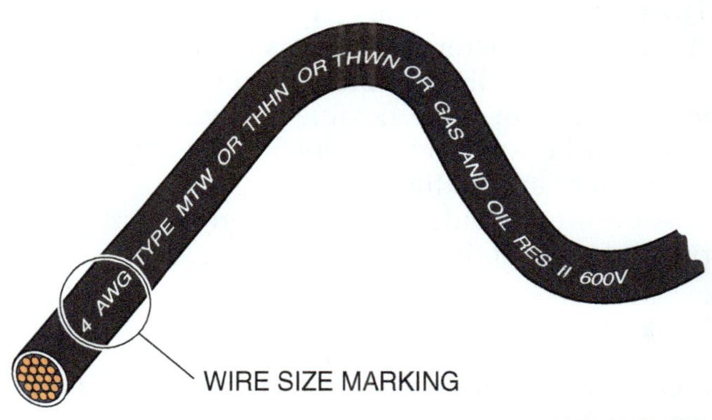

Figure 2 Wire size marking.

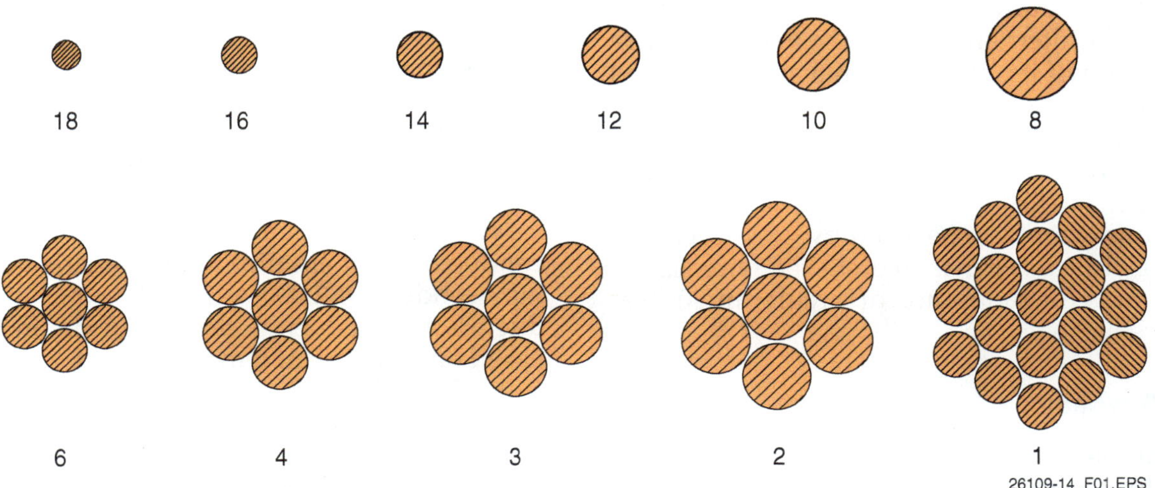

Figure 1 Comparison of wire sizes (enlarged) from No. 18 to No. 1 AWG.

Think About It

Wire Size

Why is wire size a critical factor in a wiring system? Other than load, what other factors may dictate wire size? What can happen when a wire isn't properly sized for the load?

2.1.2 Stranding

According to *NEC Chapter 9, Table 8,* wire sizes No. 18 to No. 2 have seven strands; wire sizes No. 1 to No. 4/0 have 19 strands; and wire sizes between 250 kcmil and 500 kcmil have 37 strands. The purpose of stranding is to increase the flexibility of the wire. Terminating solid wire sizes larger than No. 8 in pull boxes, disconnect switches, and panels would not only be very difficult, but might also result in damage to equipment and wire insulation.

Pulling solid wire in conduit around bends could pose a major problem and cause damage to equipment for wire sizes larger than No. 8. The reason for choosing 7, 19, and 37 strands for stranded conductors is that it is necessary to provide a flexible, almost round conductor. In order for a conductor to be flexible, individual strands must not be too large. *Figure 3* shows how these conductors are configured.

Aluminum conductors are designed with compact (compressed) stranding. See *Figure 4*.

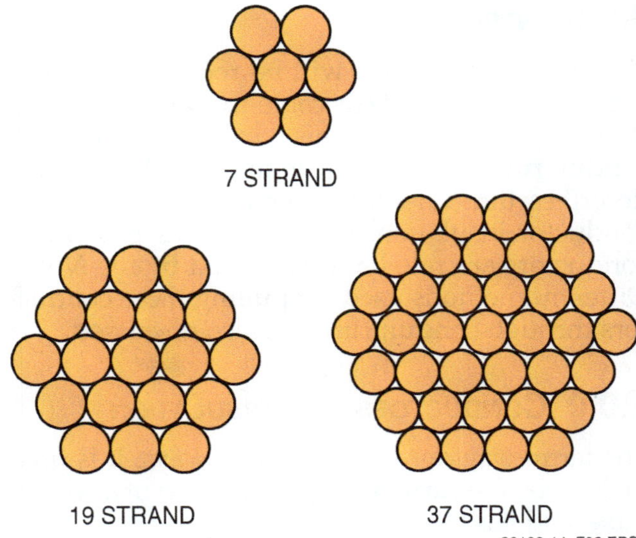

Figure 3 Strand configurations.

Figure 4 Aluminum conductors.

2.1.3 Circular Mils

A circular mil is a circle that has a diameter of 1 mil. A mil is 0.001 inch. When a wire size is 250 kcmil, the cross-sectional area of the current-carrying portion of the wire is the same as 250,000 circles having a diameter of 0.001 inch. This may seem to be a rather clumsy way of sizing wire at first; however, the alternative would be to size the wire as a function of its cross-sectional area expressed in square inches.

According to *NEC Chapter 9, Table 8,* the cross-sectional area of a 250 kcmil conductor is 0.260 square inch.

If a conductor is to be sized by cross-sectional area, it is much easier to express the wire size in circular mils (or thousands of circular mils) than in square inches.

2.2.0 Ampacity

Ampacity is the maximum current in amperes a conductor can carry continuously under the conditions of use without exceeding its temperature rating. The ampacity of conductors for given conditions are listed in *NEC Tables 310.15(B)(16) through 310.15(B)(21). NEC Table 310.15(B)(16)* covers conductors rated 2,000 volts and below where not more than three conductors are installed in a raceway or cable or are directly buried in the earth, based on an ambient temperature of 30°C (86°F).

NEC Table 310.15(B)(17) covers both copper conductors and aluminum or copper-clad aluminum conductors 2,000 volts and below where conductors are used as single conductors in free air, based on an ambient temperature of 30°C (86°F).

NEC Tables 310.15(B)(18) and 310.15(B)(19) apply to conductors rated at 150°C to 250°C (302°F to 482°F), used either in raceway or cable or as single conductors in free air, based on an ambient temperature of 40°C (104°F).

Example:

Determine the ampacity of a No. 12 Cu (copper) THW conductor.

Solution:

25 amps [from *NEC Table 310.15(B)(16)*].

2.3.0 Conductor Material

The most common conductor material is copper. Copper is used because of its excellent conductivity (low resistance), ease of use, and value. The value of a material as an ingredient for wire is determined by several factors, including conductivity, cost, availability, and workability.

2.3.1 Conductivity

Conductivity is a word that describes the ease (or difficulty) of travel presented to an electric current by a conductor. If a conductor has a low resistance, it has a high conductivity. Silver is one of the best conductors since it has very low resistance and high conductivity. Copper has high conductivity and a lower price than silver. Aluminum, another material with good conductivity, is also a good choice for conductor material. The conductivity of aluminum is approximately two-thirds that of copper.

2.3.2 Cost

Cost is always an issue that contributes to the selection of a material to be used for a given application. Often, a material that has low cost may be selected as a conductor material even though it has physical properties that are inferior to the more expensive material. Such is the case in the selection of copper over platinum. Here, the cost of platinum is very high, and very little thinking is required to determine that copper is a better choice. The choice between copper and aluminum is often more difficult to make.

2.3.3 Availability

The availability of some material is often a concern when selecting components for a job. As applied to wire, the mining industry often controls the availability of raw materials, which could produce shortages of some material. The availability of a substance such as copper or aluminum affects the price of the finished product (copper or aluminum wire).

2.3.4 Workability

It is a good idea to select a material that requires less expense for tools and is easier to work with. Aluminum conductors are lighter than copper conductors of the same size. They are also much more flexible than copper conductors and, in general, are easier to work with. However, terminating aluminum conductors often requires special tools and treatment of termination surfaces with an anti-oxidation material. Splicing and terminating aluminum conductors often requires a higher degree of training on the part of the electrician than do similar efforts with copper wire. This is partly due to the fact that aluminum expands and contracts with heat more than copper.

2.4.0 Conductor Insulation

The first attempt to insulate wire was made in the early 1800s during the development of the telegraph. This insulation was designed to provide physical protection rather than electrical protection. Electrical insulation was not an important issue because the telegraph operated at low-voltage DC. This early form of insulation was a substance composed of tarred hemp or cotton fiber and shellac and was used primarily for weatherproofing long-distance distribution lines to mines, industrial sites, and railroads.

Some early electrical distribution systems utilized the knob-and-tube technique of installing wire. The wire was often bare and was pulled between and wrapped around ceramic knobs that were affixed to the building structure. When it was necessary to pull wire through structural members, it was pulled through ceramic tubes. The structural member (usually wood) was drilled, the tube was pressed into the hole, and the wire was pulled through the hole in the tube. As dangerous as this may appear, older homes still exist that have knob-and-tube wiring that was installed in the early 1900s and is still operational. Knob-and-tube wiring was revised to use insulated conductors and was in use up to 1957 in some areas.

The grounded or neutral conductor in overhead services may be bare. Furthermore, the concentric grounded conductor in Type SE cable may be bare when used as a service-entrance cable. However, all current-carrying conductors (including the grounded conductor) must be insulated when

Think About It

Overheating

What happens when insulation is overheated? What affects the ampacity of an insulated conductor?

used on the inside of buildings, or after the first overcurrent protection device.

NEC Table 310.104(A) presents application and construction data on the wide range of 600-volt insulated, individual conductors recognized by the *NEC*®, with the appropriate letter designation used to identify each type of insulated conductor.

2.4.1 Thermoplastic

Thermoplastic is a popular and effective insulation material. The following thermoplastics are widely used:

- *Polyvinyl chloride (PVC)* – The base material used for the manufacture of TW and THW insulation.
- *Polyethylene (PE)* – An excellent weatherproofing material used primarily for insulation of control and communications wiring. It is not used for high-voltage conductors (those exceeding 5,000 volts).
- *Cross-linked polyethylene (XLP)* – An improved PE with superior heat- and moisture-resistant qualities. Used for THHN, THWN, and THHW wiring as well as many high-voltage cables.
- *Nylon* – Primarily used as jacketing material. THHN building wire has an outer coating of nylon.
- *Teflon*® – A high-temperature insulation. Widely used for telephone wiring in a plenum (where other insulated conductors require conduit routing).

2.4.2 Thermoset

Many thermoplastic materials deform when heated. Thermoset materials maintain their form when heated. Thermoset insulations include RHH, RHW, XHH, XHHW, and SIS.

On Site

Terminating Aluminum Wire

Care must be taken to use listed connectors when terminating aluminum wire. All aluminum connections also require the use of anti-oxidizing compound. Some connectors are precoated with compound; others require the addition of it. Be sure to check the connectors before beginning the installation.

2.4.3 Letter Coding

Conductor insulation as applied to building wire is coded by letters. The letters generally, but not always, indicate the type of insulation or its environmental rating. The types of conductor insulation described in this module will be those indicated at the top of *NEC Table 310.15(B)(16)*. The various insulation designations are shown in *Table 1*.

NOTE

Any conductor used in a wet location (see definition under Location, Wet, in *NEC Article 100*) must be listed for use in wet locations. Any conduit run underground is assumed to be subject to water infiltration and is, therefore, in a wet location.

Table 1 Insulation Coding

Letter	Description
B	Braid
E	Ethylene or Entrance
F	Fluorinated or Feeder
H	Heat-Rated or Flame-Retardant
N	Nylon
P	Propylene
R	Rubber
S	Silicon or Synthetic
T	Thermoplastic
U	Underground
W	Weather-Rated
X	Cross-Linked Polyethylene
Z	Modified Ethylene Tetrafluoroethylene
TW	Weather-Rated Thermoplastic (60°C/140°F)
FEP	Fluorinated Ethylene Propylene
FEPB	Fluorinated Ethylene Propylene with Glass Braid
MI	Mineral Insulation
MTW	Moisture, Heat, and Oil-Resistant Thermoplastic
PFA	Perfluoroalkoxy
RHH	Flame-Retardant Heat-Rated Rubber
RHW	Weather-Rated, Heat-Rated Rubber (75°C/167°F)
SA	Silicon
SIS	Synthetic Heat-Resistant
TBS	Thermoplastic Braided Silicon
TFE	Extended Polytetrafluoroethylene
THHN	Heat-Resistant Thermoplastic
THHW	Moisture and Heat-Resistant Thermoplastic
THW	Moisture and Heat-Resistant Thermoplastic
THWN	Weather-Rated, Heat-Rated Thermoplastic with Nylon Cover
UF	Underground Feeder
USE	Underground Service Entrance
XHH	Thermoset
XHHW	Heat-Rated, Flame-Retardant, Weather-Rated Thermoset
ZW	Weather-Rated Modified Ethylene Tetrafluoroethylene

26109-14_T01.EPS

> **Think About It**
>
> ## Conductor Insulation
>
> What are the functions of conductor insulation? Under what conditions does the NEC® allow uninsulated conductors?

2.4.4 Color Coding

A color code is used to help identify wires by the color of the insulation. This makes it easier to install and properly connect the wires. A typical color code is as follows:

- *Two-conductor cable* – One white or gray wire, one black wire, and a grounding wire (usually bare)
- *Three-conductor cable* – One white or gray, one black, one red, and a grounding wire
- *Four-conductor cable* – Same as three-conductor cable plus fourth wire (blue)
- *Five-conductor cable* – Same as four-conductor cable plus fifth wire (yellow)

The grounding conductor may be bare, green, or green with a yellow stripe. Power cable color codes are shown in *Figure 5*.

> **Think About It**
>
> ## Insulation Types
>
> Use ***NEC Table 310.104(A)*** to identify two types of insulation that are suitable for use in wet locations.

The *NEC*® does not require color coding of ungrounded conductors except where more than one nominal voltage system is present [*NEC Section 210.5(C)*]. The ungrounded conductors may be any color with the exception of white, gray, or green; however, it is a good practice to color code conductors as described here. In fact, many construction specifications require color coding. Furthermore, on a four-wire, delta-connected secondary where the midpoint of one phase is grounded to supply lighting and similar loads, the phase conductor having the higher voltage to ground must be identified by an outer finish that is orange in color, by tagging, or by other effective means. Such identification must be placed at each point where a connection is made if the grounded conductor is also present. In most cases, orange tape is used at all termination points when such a condition exists.

2.4.5 Wire Ratings

A critical factor in selecting conductors is the conductor's maximum operating temperature. Consider how and where a conductor will be used so that the conductor's limiting (maximum) temperature rating will not be exceeded. A conductor's operating temperature is determined by the ambient temperature, current flow in the conductor (including harmonic current), current flow in bundled conductors (which raises the ambient temperature), and how fast or slow heat is dissipated into the surrounding medium (which is affected by the conductor insulation).

Another significant factor to consider when selecting and installing conductors is where the conductor will be terminated. The temperature

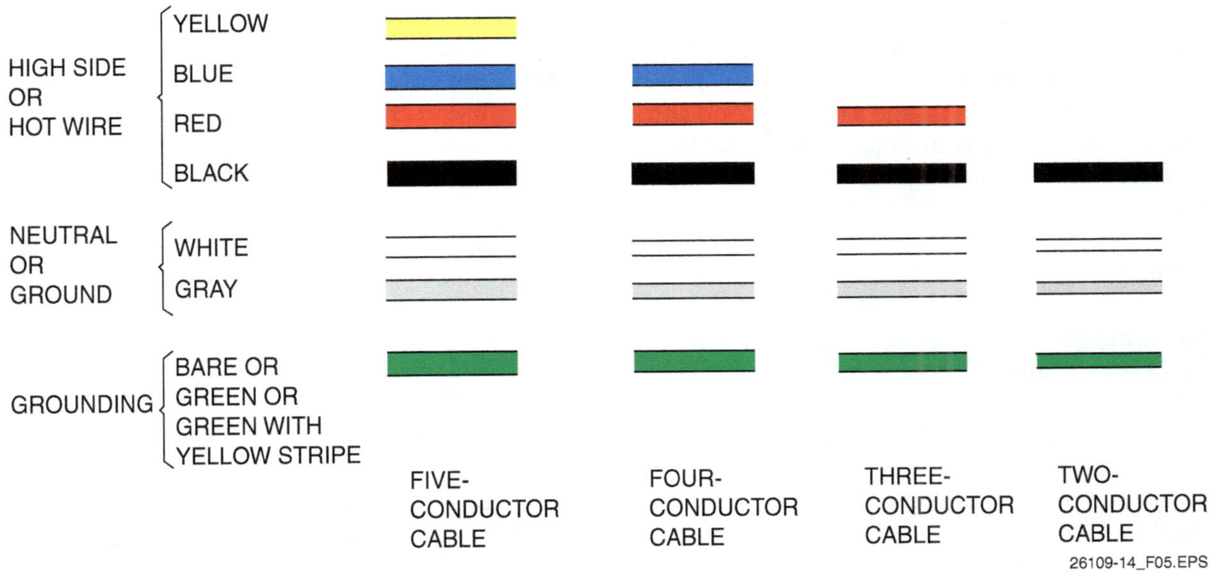

Figure 5 Typical power cable insulation color codes.

Think About It

Color Coding Ungrounded Conductors

Although the *NEC*® does not require the use of color-coded ungrounded current-carrying conductors, why might it be a good idea to use them anyway?

On Site

Color Coding

Color designations are good indicators, but never trust your life to them. Always protect yourself by testing circuits with a voltmeter.

rating of the termination may limit the allowable ampacity of the conductor.

The amount of current a conductor can safely carry, and thus the maximum safe temperature the conductor can reach, is determined in general by conductor size (diameter in circular mils), the ambient (surrounding) temperature, the number of conductors in a bundle, and where the conductors are installed (raceways, conduit, ducts, underground, etc.).

Conductor selection is based largely on the temperature rating of the wire. This requirement is extremely important and is the basis of safe operation for insulated conductors. As shown in *NEC Table 310.104(A),* conductors have various ratings (60°C, 75°C, 90°C, etc.). Since *NEC Tables 310.15(B)(16) and 310.15(B)(17)* are based on an assumed ambient temperature of 30°C (86°F), conductor ampacities are based on the ambient temperature plus the heat (I^2R) produced by the conductor while carrying current. Therefore, the type of insulation used on the conductor is the first consideration in determining the maximum permitted conductor ampacity.

For example, a No. 3/0 THW copper conductor for use in a raceway has an ampacity of 200 according to *NEC Table 310.15(B)(16).* In a 30°C ambient temperature, the conductor is subjected to this temperature when it carries no current. Since a THW-insulated conductor is rated at 75°C, this leaves 45°C (75 – 30) for increased temperature due to current flow. If the ambient temperature exceeds 30°C, the conductor maximum load-current rating must be reduced proportionally per *NEC Table 310.15(B)(2)(a)* so that the total temperature (ambient plus conductor temperature rise due to current flow) will not exceed the temperature rating of the conductor insulation (60°C, 75°C, etc.). For the same reason, the allowable ampacity must be reduced when more than three conductors are contained in a raceway or cable. See *NEC Section 310.15(B)(3)(a).*

Using the ampacity tables – An important step in circuit design is the selection of the type of conductor to be used (TW, THW, THWN, RHH, THHN, XHHW, etc.). The various types of conductors are covered in *NEC Article 310,* and the ampacities of conductors are given in *NEC Tables 310.15(B)(16) through 310.15(B)(21)* for the varying conditions of use (e.g., in a raceway, in open air, at normal or higher-than-normal ambient temperatures). Conductors must be used in accordance with the data in these tables and notes.

2.5.0 Fixture Wires

Fixture wire is used for the interior wiring of fixtures and for wiring fixtures to a power source. Guidelines concerning fixture wire are given in *NEC Article 402.* The list of approved types of fixture wire is given in *NEC Table 402.3. Figure 6* shows one example of fixture wire. The wires are composed of insulated conductors with or without an outer jacket. The conductors range in size from No. 18 to No. 10 AWG.

The decision of which fixture wire to use depends primarily upon the operating temperature that is expected within the fixture. Therefore, it is the character of the insulation that will determine the wire selected. For instance, fixture wires insulated with perfluoroalkoxy (PFA) or extruded polytetrafluoroethylene (PTF) would be selected if the operating temperature of the fixture is expected to reach a maximum of 482°F. This is the highest operating temperature allowed for any fixture wire.

As indicated by *NEC Section 402.3,* fixture wires are suitable for service at 600 volts unless otherwise specified in *NEC Table 402.3.* The allowed ampacities of fixture wire are given in *NEC Table 402.5.*

Although the primary use for fixture wire is the internal wiring of fixtures, several of the wires listed in *NEC Table 402.3* may be used for wiring remote-control, signaling, or power-limited

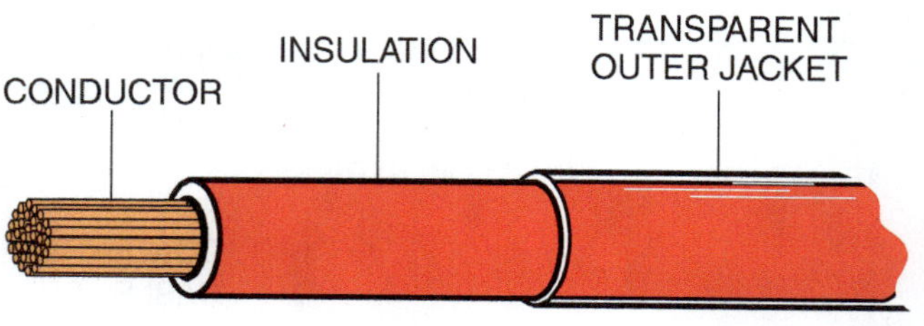

Figure 6 Fixture wire.

On Site

Cable Selection

There are two factors to be considered when determining the type of cable to be used for a specific application: the type of conductor insulation and the cable jacket. Both must be appropriate for the application.

circuits in accordance with *NEC Section 725.49.* Fixture wires may never be used as substitutes for branch circuit conductors.

2.6.0 Cables

Cables are two or more insulated wires and may contain a grounding wire covered by an outer jacket or sheath. Cable is usually classified by the type of covering it has, either nonmetallic (plastic) or metallic, also called armored cable.

Cable may also be classified according to where it can be used (see *NEC Table 400.4*). Because water is such a good conductor of electricity, moisture on conductors can cause power loss or short circuits. For this reason, cables are classified for either dry, damp, or wet locations. Cables can also be classified regarding exposure to sunlight and rough use.

2.6.1 Cable Markings

All cables are marked to show important properties and uses. Cable markings show the wire size, number of conductors, cable type, and voltage rating. In addition, a marking may be included to signify approved service or applications. This information is printed on nonmetallic cable (*Figure 7*).

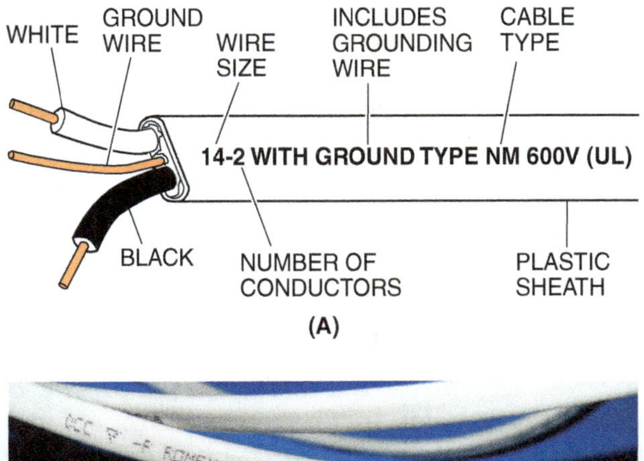

Figure 7 Nonmetallic cable markings.

On metallic cable, marking information is usually included on a tag.

2.6.2 Nonmetallic-Sheathed Cable

Nonmetallic-sheathed cable (Type NM and Type NMC) is widely used for branch circuits and feeders in residential and commercial systems. See *Figure 8*. Both types are commonly called Romex®, even though the cable manufacturer only calls Type NM cable Romex®. Guidelines for the use of nonmetallic-sheathed cable are given in *NEC Article 334.* This cable consists of two or three insulated conductors and one bare conductor enclosed in a nonmetallic sheath. The conductors may be wrapped individually with paper, and the spaces between the conductors may be filled with jute, paper, or other material to protect the conductors and help the cable keep its shape. The sheath covering both Type NM cable and Type NMC cable is flame-retardant and moisture-resistant. The sheath covering Type NMC cable has the additional characteristics of being fungus- and corrosion-resistant.

NEC Article 334 lists the allowed and prohibited uses for Type NM cable and Type NMC cable. Both are allowed to be installed in either exposed or concealed work. The primary difference in their applied uses is that Type NM cable is suitable for dry locations only, whereas Type NMC is permitted for dry, moist, damp, or corrosive locations.

Types NM and NMC cables may be used in one- and two-family dwellings, and in certain multifamily dwellings, depending on the type of construction. See *NEC Section 334.10*.

In general, NM and NMC cable cannot be used in ducts or plenums because toxic gases from burning cable insulation would be spread throughout the structure. NM and NMC cables cannot be installed exposed in the space above suspended ceilings. They cannot be used as service-entrance cable, embedded in concrete, or in hazardous locations. There are many other requirements for NM and NMC specified in *NEC Article 334*. Before installing this type of cable, be sure you read and understand the applicable sections of the *NEC*®.

2.6.3 Type UF Cable

Guidelines for the use of Type UF (underground feeder and branch circuit) cable are given in *NEC Article 340.* Type UF cable is very similar in appearance, construction, and use to Type NMC cable. The main difference between these two cables is that Type UF cable is suitable for direct burial, whereas Type NMC cable is not.

Some of the permitted uses of Type UF cable are: underground and direct burial; as a single-conductor cable; in wet, dry, or corrosive

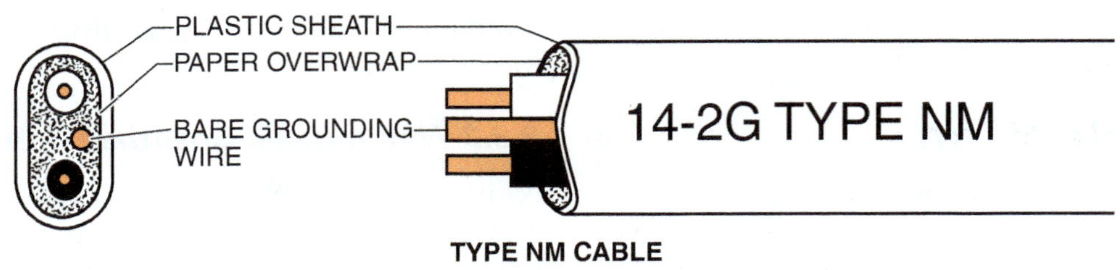

TYPE NM CABLE

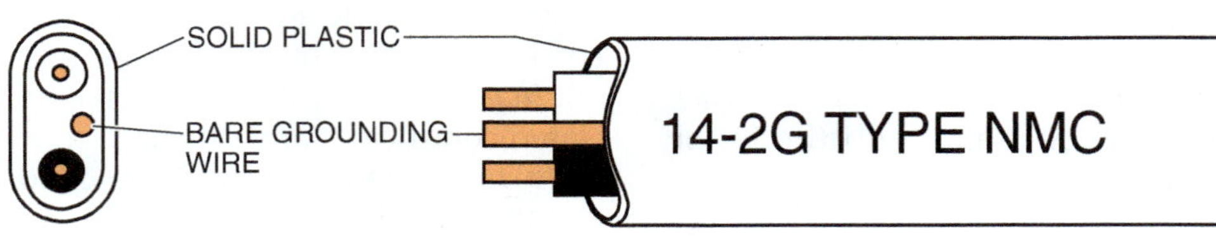

TYPE NMC CABLE

(A)

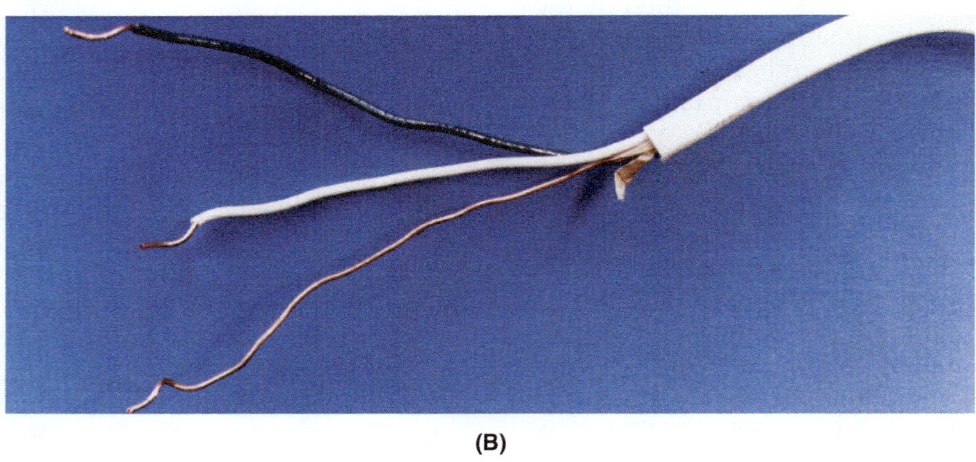

(B)

26109-14_F08.EPS

Figure 8 Nonmetallic-sheathed cable.

conditions; as a nonmetallic-sheathed cable; in solar photovoltaic systems; and in cable trays.

Typically, Type UF cable may not be used as service-entrance cable, in commercial garages, in theaters, in hoistways or elevators, or in hazardous locations. Type UF cable cannot generally be embedded in poured cement, concrete, or aggregate, exposed to sunlight (unless designed for that use), or used as an overhead cable. Refer to *NEC Article 340* for specifics on where and when to use Type UF cable.

2.6.4 Type NMS Cable

Refer to *NEC Sections 334.10* and *334.12* for the applications of Type NMS cable. Type NMS cable is a form of nonmetallic-sheathed cable that contains a factory assembly of power, communications, and signaling conductors enclosed within a moisture-resistant, flame-retardant sheath.

2.6.5 Type MV Cable

Type MV (medium-voltage) cable is covered in *NEC Article 328.* It consists of one or more insulated conductors encased in an outer jacket. This cable is suitable for use with voltages ranging from 2,001 to 35,000 volts. It may be installed in wet and dry locations and may be buried directly in the earth. See *Figure 9.*

2.6.6 Type MC Cable

Type MC (metal clad) cable consists of one or more insulated conductors encased in a metal tape or a metallic sheath. *NEC Article 330* covers Type MC cable. Further information can be found in *UL 1569, Standard for Metal Clad Cables.*

MC cable is used in a wide variety of applications, from small instrumentation cable up to medium voltage feeders. The conductors are

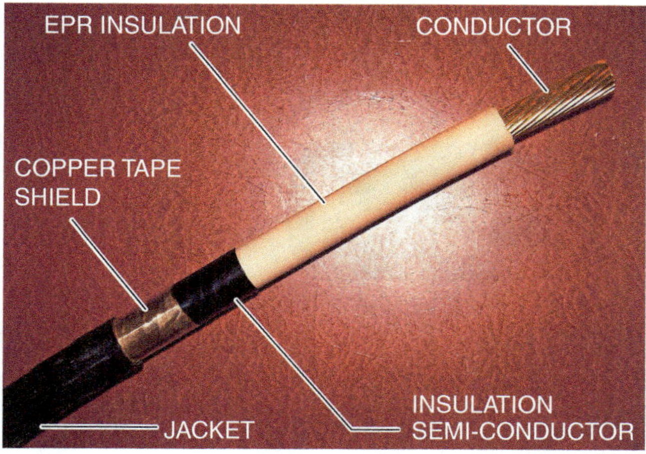

26109-14_F09.EPS

Figure 9 Type MV cable.

coated with a thermoset or thermoplastic insulation. Type MC cable can also be a composite of electrical conductors and optical fiber conductors.

Typical markings on the cable include the maximum rated voltage, AWG size (or circular mil area), and insulation type. If the outer covering will not accept markings, the markings will be on a tape inside the cable along the entire length of the cable. If on the outside, the markings typically have a 24-inch spacing.

The three types of MC cable are: interlocked metal tape, corrugated metal tube, and smooth metal tube. Cables with special uses will be marked accordingly. The outer covering may be a nonmetallic jacket over the metal sheath. One type of MC cable is shown in *Figure 10*.

Some of the typical uses for the three types of MC cable are for services, feeders, and branch circuits; for power, lighting, control, and signal circuits; indoors or outdoors; exposed or concealed; direct burial (if identified for that use); in any raceway; and other uses specified in *NEC Article 330*.

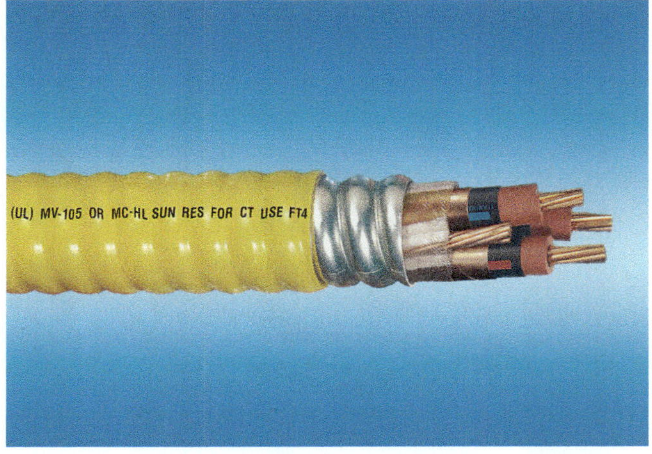

26109-14_F10.EPS

Figure 10 Type of MC cable.

Type MC cable may not be used in corrosive or damaging conditions unless the metal cladding protects the conductors, or some other protective material is used. Uses typically not permitted are in areas where the cable is subject to physical damage, direct burial, in concrete, or where subject to caustic materials.

Typically, UL does not recognize the interlocking metal cladding of Type MC cable as the only means of grounding equipment. For this reason, Type MC cable is not allowed in certain applications, such as patient care areas in hospitals.

Both armored (AC) and MC cables provide advantages during installation. The flexible metal sheath protects the conductors and allows them to bend around corners without kinking or damage to the conductor. Also, since the conductors are already protected by the sheathing, there is no need to pull conductors into a raceway, nor is there concern about conductor contact with pipes or other hard surfaces. Other advantages of metal clad cables are their relatively easy installation without the need for wire pullers, fish tapes, or lubricants.

There are some fundamental differences between Types AC and MC cables. The significant differences are:

- AC cable has a maximum of four conductors, plus a grounding conductor, and comes in sizes from 14 AWG to 1 AWG. Conversely, MC cable has no limitations on the number of conductors, and is sized from 18 AWG to 2,000 kcmil.
- AC cable has a bonding strip (16 AWG). This strip is in constant contact with the armor and, with the armor, forms an equipment ground. MC cable has no bonding strip. The MC cladding is not a ground, although it can supplement the ground.
- AC cable uses moisture-resistant and fire-retardant paper wraps on individual conductors. MC cables have no such paper wrap, but do incorporate a polyester tape used on the assembly.

2.6.7 High-Voltage Shielded Cable

Shielding of high-voltage cables protects the conductor assembly against surface discharge or burning due to corona discharge in ionized air, which can be destructive to the insulation and jacketing.

Electrostatic shielding of cables makes use of both nonmetallic and metallic materials (*Figures 11* and *12*).

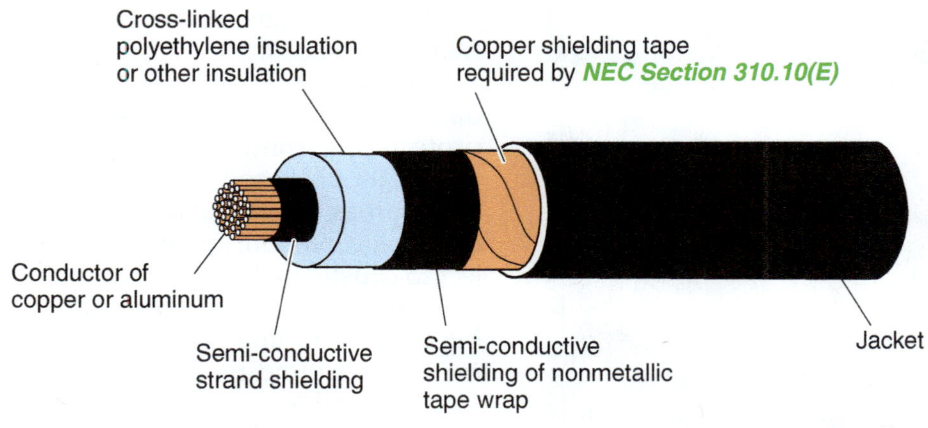

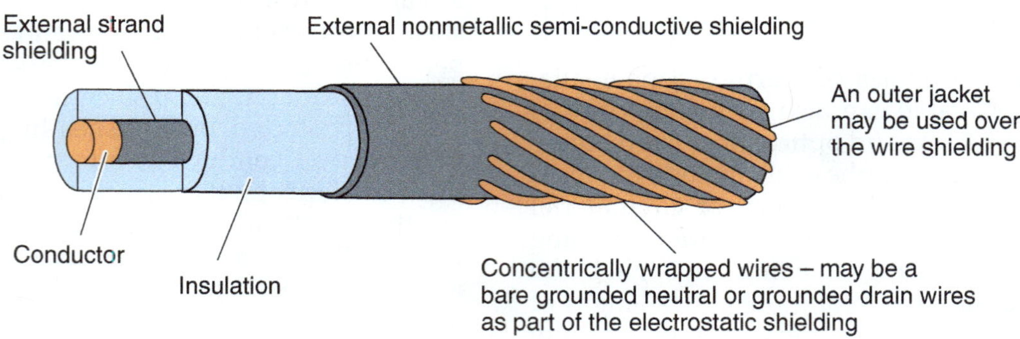

26109-14_F11.EPS

Figure 11 Various types of shielding.

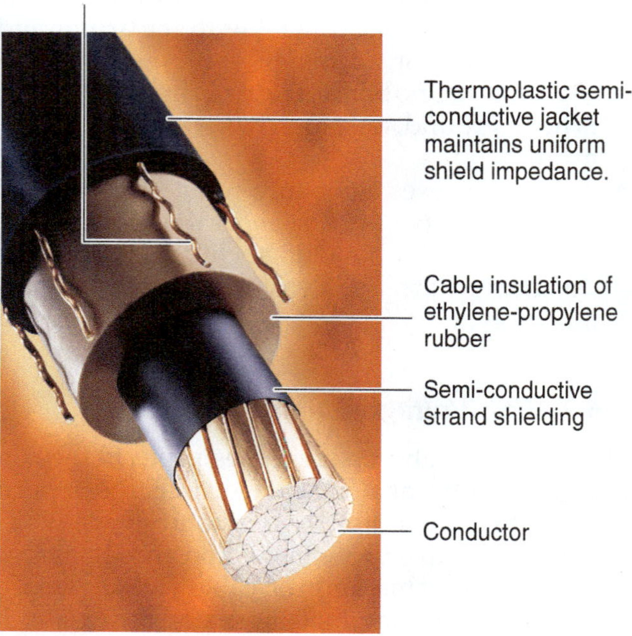

26109-14_F12.EPS

Figure 12 Corrugated drain wire shielding.

On Site

Type MC Cable

Metal clad cable (Type MC) is a type of cable that is widely used in both commercial and industrial environments. It is available in many configurations, with or without an outer jacket. Some of the special applications of MC cable include homerun cables, super neutrals, direct burial, and fire alarm cable.

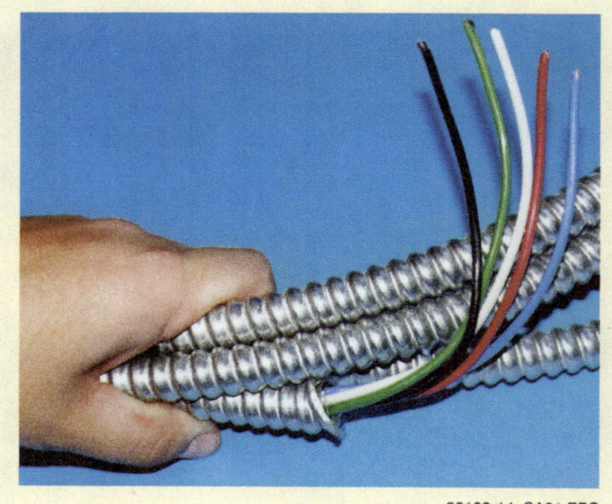

26109-14_SA01.EPS

2.6.8 Channel Wire Assemblies

Channel wire assemblies (Type FC) comprise an entire wiring system, which includes the cable, cable supports, splicers, circuit taps, fixture hangers, and fittings (*Figure 13*). Guidelines for the use of this system are given in *NEC Article 322.* Type FC cable is a flat cable assembly with three or four parallel No. 10 special stranded copper conductors. The assembly is installed in an approved U-channel surface metal raceway with one side open. Tap devices can be inserted anywhere along the run. Connections from the tap devices to the flat cable assembly are made by pin-type contacts when the tap devices are fastened in place. The pin-type contacts penetrate the insulation of the cable assembly and contact the multi-stranded conductors in a matched phase sequence. These taps can then be wired to lighting fixtures or power outlets (*Figure 14*).

As indicated in *NEC Section 322.10,* this wiring system is suitable for branch circuits that only supply small appliances and lights. This system is suitable for exposed wiring only and may not be concealed within the building structure. It is ideal for quick branch circuit wiring at field installations.

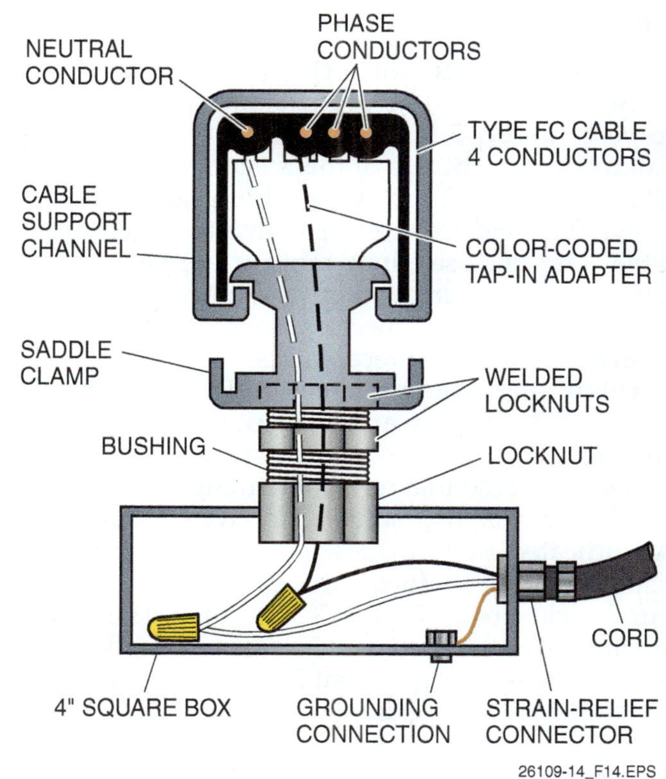

Figure 14 Type FC connection.

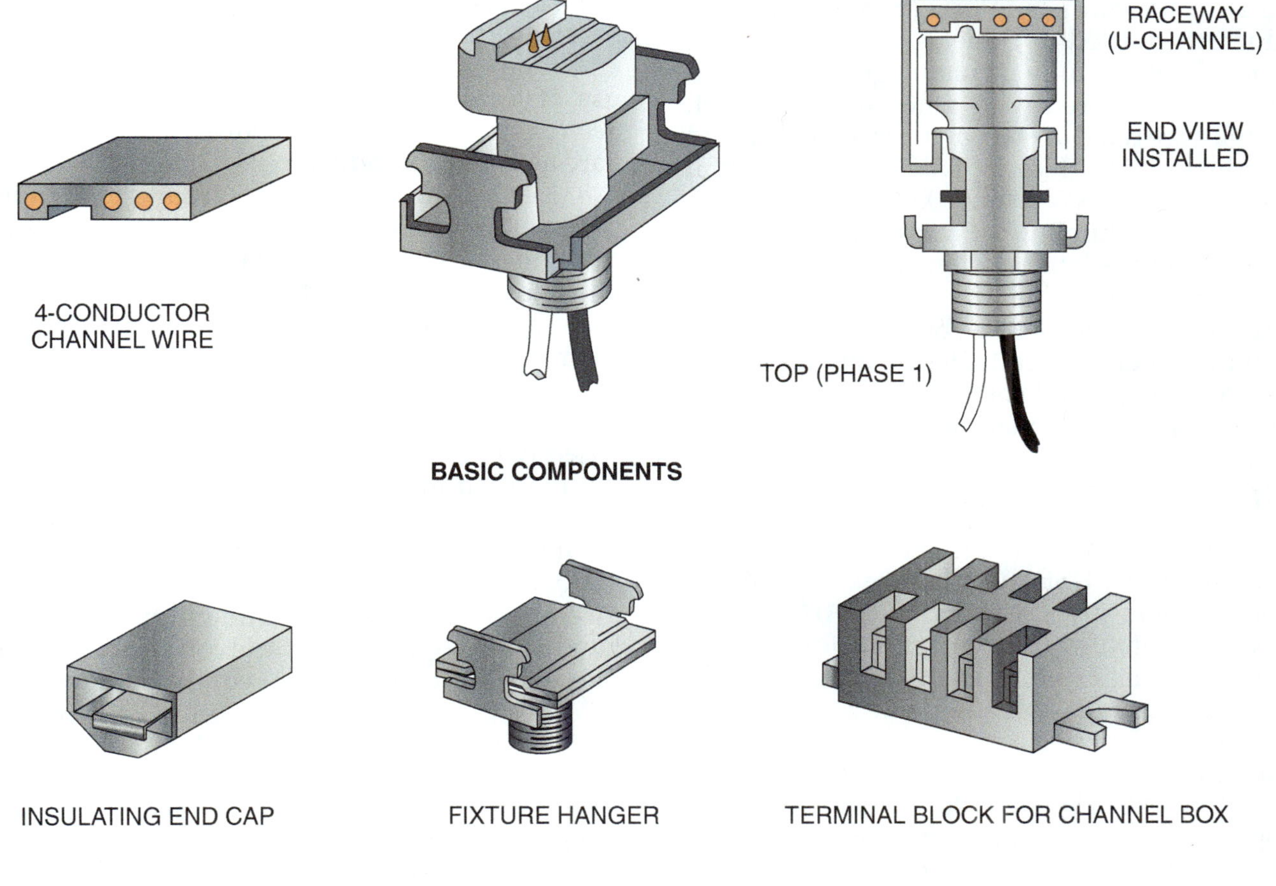

Figure 13 Channel wire components and accessories.

2.6.9 Flat Conductor Cable

Type FCC (flat conductor) cable comprises an entire branch wiring system similar in many respects to Type FC flat conductor assemblies. Guidelines for the use of this system are given in *NEC Article 324.* Type FCC cable consists of three to five flat conductors placed edge-to-edge, separated, and enclosed in a moisture-resistant and flame-retardant insulating assembly. Accessories include cable connectors, terminators, power source adapters, and receptacles.

This wiring system has been designed to supply floor outlets in office areas and other commercial and institutional interiors. It is meant to be run under carpets so that no floor drilling is required. This system is also suitable for wall mounting. As indicated in *NEC Article 324,* telephone and other communications circuits may share the same enclosure as Type FCC flat cable. The main advantage of the system is its ease of installation. It is the ideal wiring system for use when remodeling or expanding existing office facilities.

2.6.10 Type TC Cable

Guidelines for the use of Type TC (power and control tray) cable are given in *NEC Article 336.* Type TC cable consists of two or more insulated conductors twisted together, with or without associated bare or fully insulated grounding conductors, and covered with a nonmetallic jacket. The cables are rated at 600 volts. The cable is listed in conductor sizes No. 18 AWG to 2,000 kcmil copper or No. 12 AWG to 2,000 kcmil aluminum or copper-clad aluminum (*Figure 15*).

As the T in the letter designator indicates, this cable is tray cable. It can be used in cable trays and raceways. It may also be buried directly if the sheathing material is suitable for this use. Type TC cable is also good for use in sunlight when indicated by the cable markings.

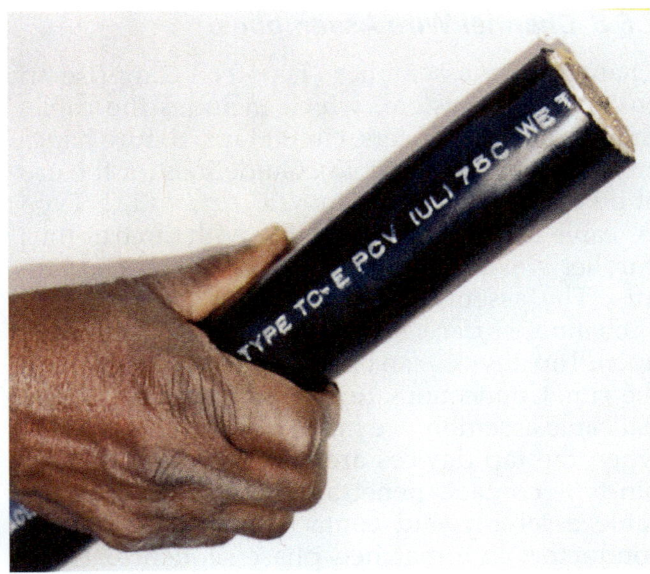

26109-14_F15.EPS

Figure 15 Type TC cable.

2.6.11 SE and USE Cable

Guidelines for the use of Types SE (service-entrance) and USE (underground service-entrance) cable are given in *NEC Article 338.* The *NEC*® contains no specifications for the construction of this cable; it is left to UL to determine what types of cable should be approved for this purpose. Currently, service-entrance cable is labeled in sizes No. 12 AWG and larger for copper, and No. 10 AWG and larger for aluminum or copper-clad aluminum, with Types RH, RHW, RHH, or XHHW conductors. If the type designation for the conductor is marked on the outside surface of the cable, the temperature rating of the cable corresponds to the rating of the individual conductor. When this marking does not appear, the temperature rating of the cable is 75°C (167°F). Type SE cable is for aboveground installation only.

When used as a service-entrance cable, Type SE must be installed as specified in *NEC Article 230.* Service-entrance cable may also be used as feeder and branch circuit cable. Guidelines for the use of service-entrance cable are given in *NEC Section 338.10. Figure 16* shows SE cable with a bare aluminum conductor.

Type USE cable is for underground installation including burial directly in the earth. Type USE cable in sizes No. 4/0 AWG and smaller with all conductors insulated is suitable for all of the underground uses for which Type UF cable is permitted by the *NEC*®.

Type USE cable may consist of either single conductors or a multi-conductor assembly provided with a moisture-resistant covering, but it is not required to have a flame-retardant covering. This type of cable may have a bare copper conductor cabled with the assembly. Furthermore, Type USE single, parallel, or cabled conductor assemblies recognized for underground use may have a bare copper concentric conductor applied. These constructions do not require an outer overall covering. Guidelines for the use of Type USE cable are specified in *NEC Article 338.*

When used as a service-entrance cable, Type USE cable must be installed as specified in *NEC Article 230.* Take the time to read *NEC Article 230* to ensure proper installation. Type USE service-entrance cable may also be used as feeder and branch circuit cable.

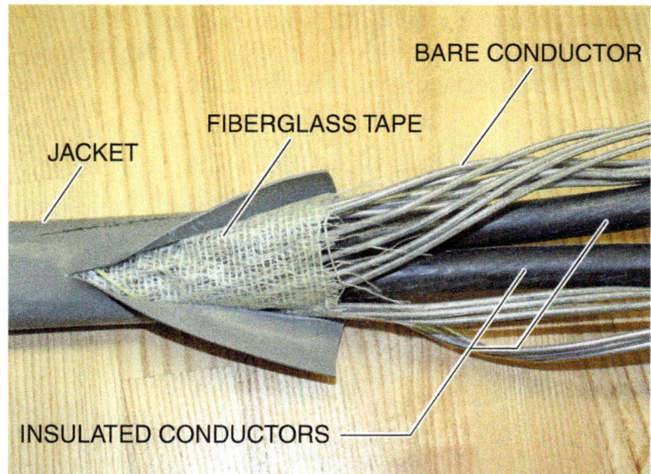

26109-14_F16.EPS

Figure 16 SE cable.

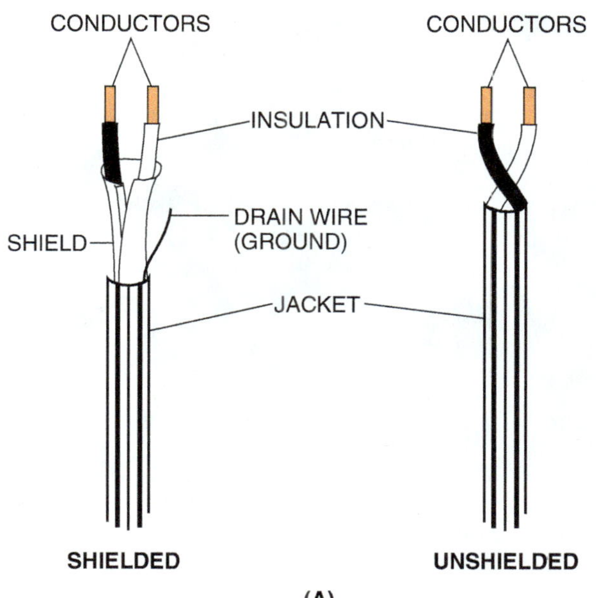

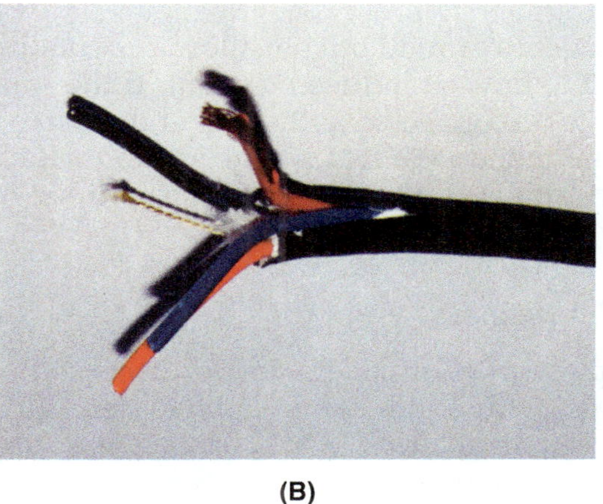

26109-14_F17.EPS

Figure 17 Instrumentation control cable.

2.7.0 Instrumentation Control Wiring

Instrumentation control wiring links the field-sensing, controlling, printout, and operating devices that form an electronic instrumentation control system. The style and size of instrumentation control wiring must be matched to a specific job.

Instrumentation control wiring usually has two or more insulated conductor wires. These wires may also have a shield and a ground wire. An outer layer called the jacket protects the wiring (*Figure 17*). Instrumentation conductor wires come in pairs. The number of pairs in a multi-conductor cable depends on the size of the wire used. A multi-pair cable typically has 12, 24, or 36 pairs of conductors.

2.7.1 Shields

Shields are provided on instrumentation control wiring to protect the electrical signals traveling through the conductors from electrical interference or noise. Shields are usually constructed of aluminum foil bonded to a plastic film (*Figure 18*). If the wiring is not properly shielded, electrical noise may cause erratic or erroneous control signals, false indications, and improper operation of control devices.

2.7.2 Shield Drain

A shield drain is a bare copper wire used in continuous contact with a specified grounding terminal. A shield drain allows connection of all the instruments within a loop to a common grounding point. Always refer to the loop diagram to determine whether or not the shield is to be terminated.

Typically, the shielding in instrumentation circuits is grounded at one end of the conductor only. The purpose of this is to drain induced

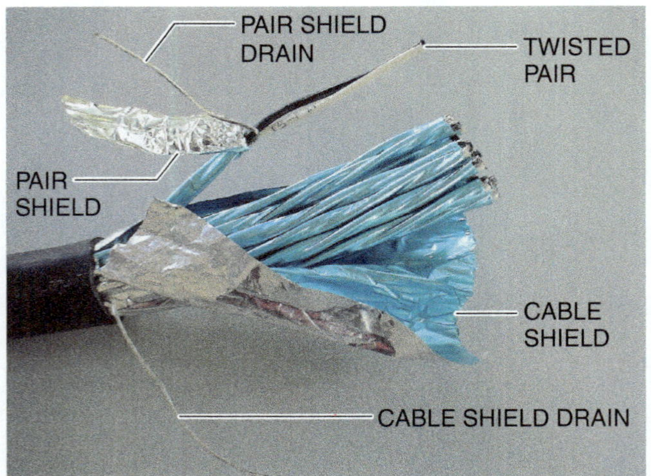

Figure 18 Multi-conductor instrumentation control cable with overall cable shield and individually shielded pairs.

charges to ground but not allow a circulating path for the flow of induced current. If the ground is not to be connected at the end of the wire you are installing, do not remove the ground wire. Fold it back and tape it to the cable. This is called floating the ground.

2.7.3 Jackets

A plastic jacket covers and protects the components within the wire. Polyethylene (PE) and polyvinyl chloride (PVC) jackets are the most commonly used (*Figure 19*). Some jackets have a nylon rip cord that allows the jacket to be peeled back without the use of a knife or cable cutter. This eliminates nicking of the conductor insulation when preparing for termination.

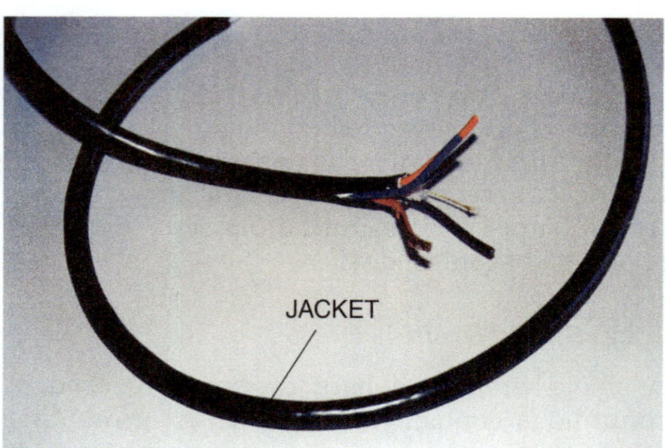

Figure 19 Wire jacket.

3.0.0 Installing Conductors in Conduit Systems

Conductors are installed in all types of conduit by pulling them through the conduit. This is done by using fish tape, pull lines, and pulling equipment.

3.1.0 Fish Tape

Fish tape can be made of flexible steel or nylon and is available in coils of 25 to 200 feet. It should be kept on a reel to avoid twisting. Fish tape has a hook or loop on one end to attach to the conductors to be pulled (*Figure 20*). Broken or damaged fish tape should not be used. To prevent electrical shock, fish tape should not be used near or in live circuits.

Fish tape is fed through the conduit from its reel. The tape usually enters at one outlet or junction box and is fed through to another outlet or junction box (*Figure 21*).

Sometimes fish tape can get hung up in very long conduit runs. These situations call for a rigid fishing tool known as a rodder. Rodders are available in various sizes and in lengths up to 1,000 feet. A typical rodder is shown in *Figure 22*.

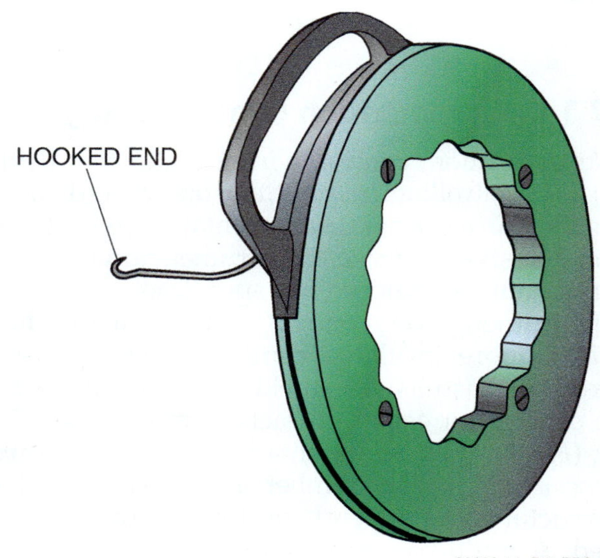

Figure 20 Fish tape.

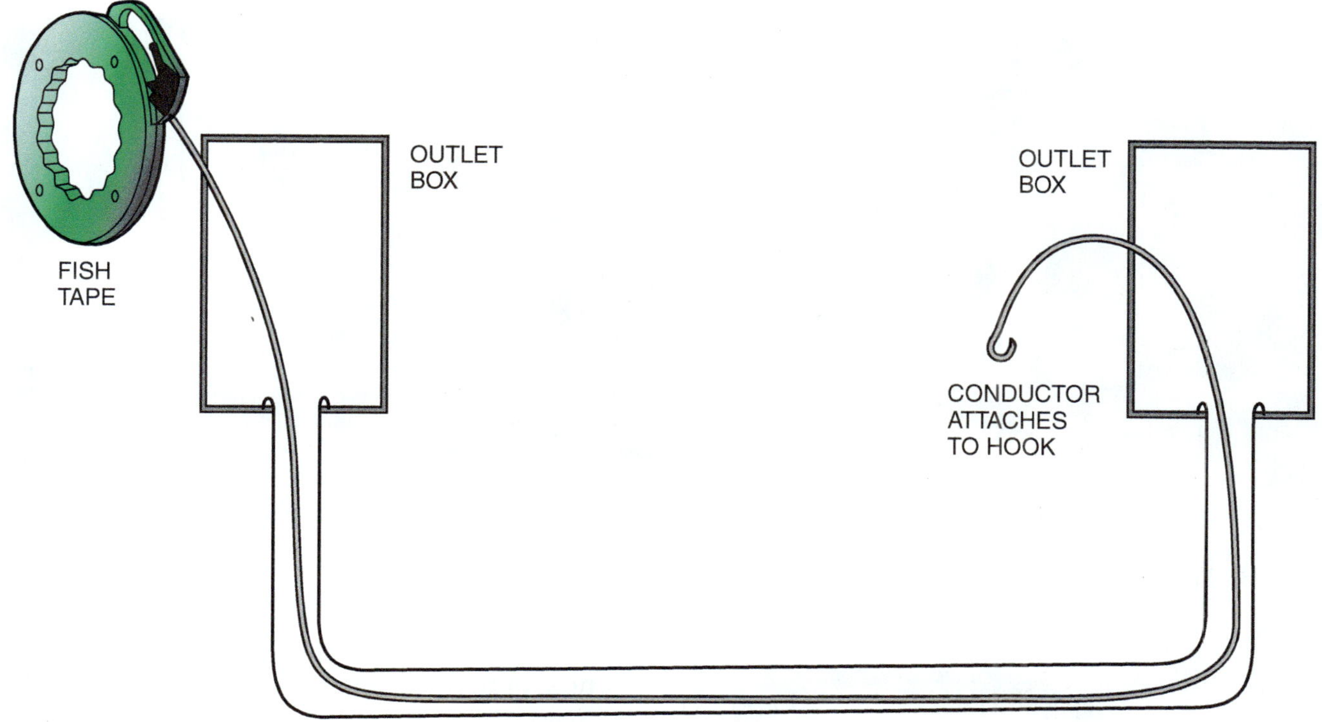

Figure 21 Fish tape installation.

Figure 22 Rodder.

3.1.1 Power Conduit Fishing Systems

String lines can be installed by using different types of power systems. The power system is similar to an industrial vacuum cleaner and pulls a string or rope attached to a piston-like plug (sometimes called a **mouse**) through the conduit. Once the string emerges at the opposite end, either the conductor or a pull rope is then attached and pulled through the conduit, either manually or with power tools. See *Figure 23*.

The hose connection on these vacuum systems can also be reversed to push the mouse through the conduit. In other words, the system can either suck or blow the mouse through the conduit, depending on which method is best in a given situation. In either case, a fish tape is then attached to the string for retrieving through the conduit.

3.1.2 Connecting Wire to a String Line

Once the string is installed in the conduit run, a fish tape is connected to it and pulled back through the conduit. Conductors are then attached to the hooked end of the fish tape or else connected to a basket grip. In most cases, all required conductors are pulled at one time.

3.2.0 Wire Grips

Wire grips are used to attach the cable to the pull tape. One type of wire grip used is a basket grip (sometimes called Chinese Fingers). A basket grip is a steel mesh basket that slips over the end of a large wire or cable (*Figure 24*). The fish tape hooks onto the end and the pull on the fish tape tightens the basket over the conductor.

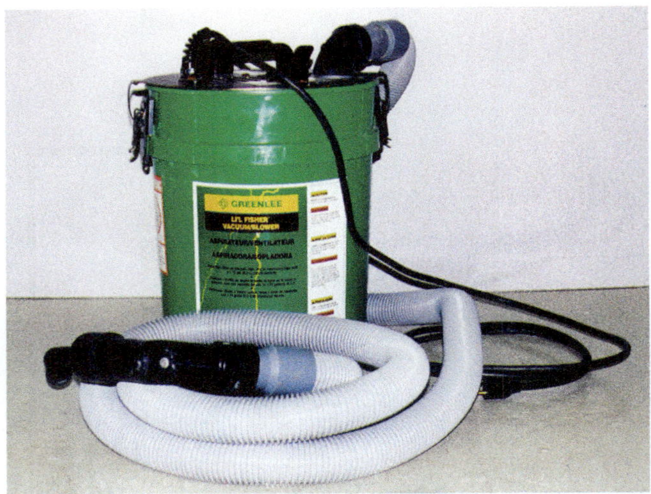

VACUUM BLOWER UNIT

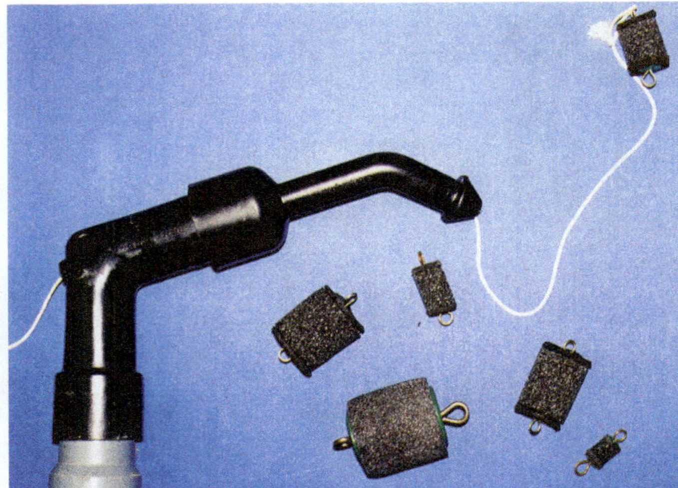

FOAM PLUGS

26109-14_F23.EPS

Figure 23 Power fishing system.

26109-14_F24.EPS

Figure 24 Basket grip.

3.3.0 Pull Lines

If a pull is going to be difficult because of bends in the conduit, the size of the conductors, or the length of the pull, a pull line should be used.

> **WARNING!**
>
> When using pull lines, exercise extreme caution and never stand in a direct line with the pulling rope. If the rope breaks, the line will whip back with great force. This can result in serious injury or death.

A pull line is usually made of nylon or some other synthetic fiber. It is made with a factory-spliced eye for easy connection to fish tape or conductors.

3.4.0 Safety Precautions

The following are several important safety precautions that will help to reduce the chance of being injured while pulling cable.

- To avoid electrical shock, never use fish tape near or in live circuits.
- Read and understand both the operating and safety instructions for the pull system before pulling cable.
- When moving reels of cable, use mechanical lifts for longer spools. For smaller spools, avoid back strain by using your legs to lift (rather than your back) and asking for help with heavy loads. Also, when manually pulling wire, spread your legs to maintain your balance and do not stretch.
- Be careful to avoid any pinchpoints in the capstans and sheaves.
- Select a rope that has a pulling load rating greater than the estimated forces required for the pull.
- Use only low-stretch rope such as multiplex and double-braided polyester for cable pulling. High-stretch ropes store energy much like a stretched rubber band. If there is a failure of the rope, pulling grip, conductors, or any other component in the pulling system, this potential energy will suddenly be unleashed. The whipping action of a rope can cause considerable damage, serious injury, or death.
- Inspect the rope thoroughly before use. Make sure there are no cuts or frays in the rope. Remember, the rope is only as strong as its weakest point.
- When designing the pull, keep the rope confined in conduit wherever possible. Should the rope break or any other part of the pulling system fail, releasing the stored energy in the rope, the confinement in the conduit will work against the whipping action of the rope by playing out much of this energy within the conduit.

On Site

Fish Tape Selection

Metal fish tape (A) generally comes in longer lengths and is the type used most often. Nylon fish tape (B) generally comes in shorter lengths and is more flexible than metal fish tape.

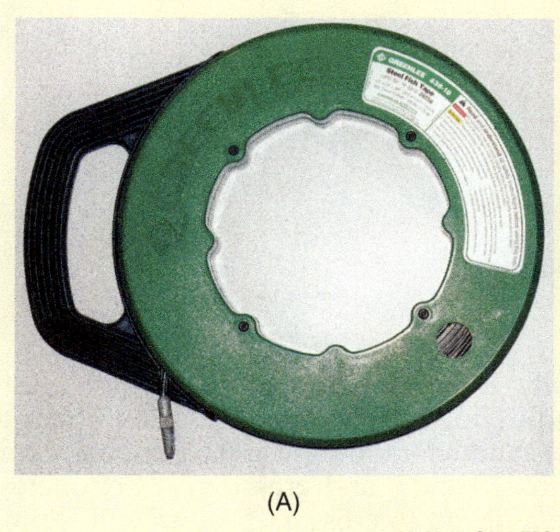

(A)

26109-14_SA02.EPS

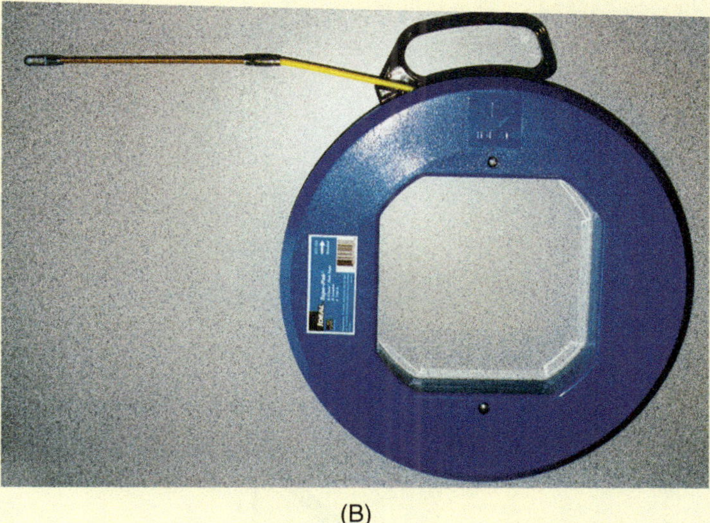

(B)

26109-14_SA03.EPS

- Do not stand in a direct line with the pulling rope. Some equipment is designed so that you may stand to one side for safety.
- Wrap up the pulling rope after use to prevent others from tripping over it.

3.5.0 Pulling Equipment

Many types of pulling equipment are available to help pull conductors through conduit. Pulling equipment can be operated both manually and electrically (*Figure 25*). A manually operated puller is used mainly for smaller pulling jobs where hand pulling is not possible or practical. It is also used in many locations where hand pulling would put an unnecessary strain on the conductors because of the angle of the pull involved.

Electrically driven power pullers are used where long runs, several bends, or large conductors are involved.

The main parts of a power puller are the electric motor, the chain or sprocket drive, the capstan, the sheave, and the pull line.

The pull line is routed over the sheave to ensure a straight pull. The pull line is wrapped around the capstan two or three times to provide a good grip on the capstan. The capstan is driven by the electric motor and does the actual pulling. The pull line is pulled by hand at the same speed at which the capstan is pulling. This eliminates the need for a large spool on the puller to wind the pull line.

Attachments to power pullers, such as special application sheaves and extensions, are available for most pulling jobs. Follow the manufacturer's instructions for setup and operation of the puller.

CAUTION

Before using power pullers, a qualified person must verify the amount of pull or tension that will be exerted on the conductors being pulled.

3.6.0 Feeding Conductors into Conduit

After the fish tape or pull line is attached to the conductors, they must be pulled back through the conduit. As the fish tape is pulled, the attached conductors must be properly fed into the conduit.

On Site

Straightening a Bent Fish Tape

To straighten a bent fish tape, drive five 16-penny (16d) nails into a 2 × 4 about 1 inch apart in a straight line. Then wind the fish tape through the nails in a slalom fashion, and pull it through. This will straighten the tape.

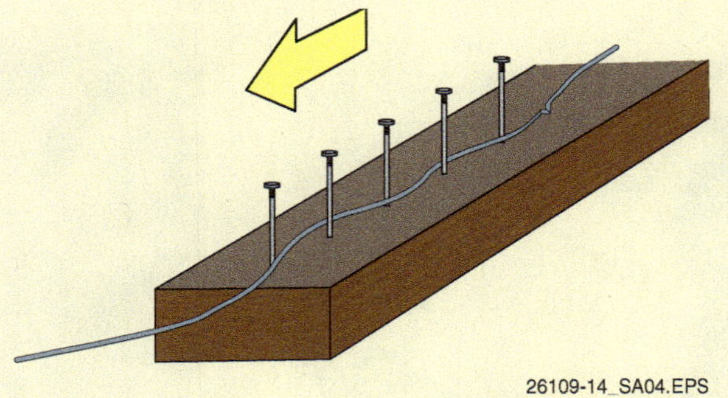

26109-14_SA04.EPS

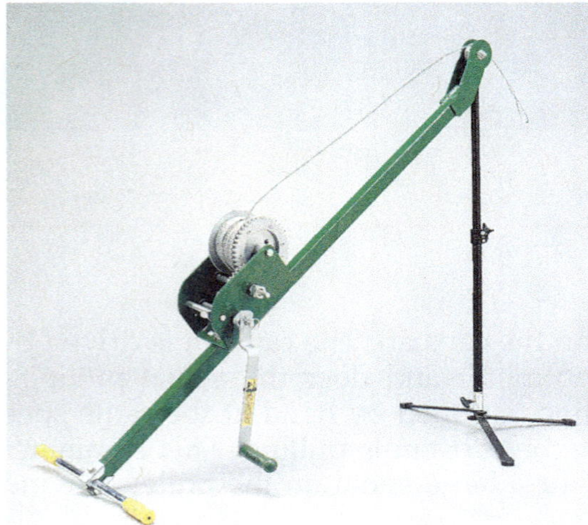

(A) MANUAL WIRE PULLER

(B) POWER PULLER

26109-14_F25.EPS

Figure 25 Pulling equipment.

Usually, more than one conductor is fed into the conduit during a wire pull. It is important to keep the conductors straight and parallel, and free from kinks, bends, and crossovers. Conductors that are allowed to cross each other will form a bulge and make pulling difficult. This could also damage the conductors.

Spools and rolls of conductors must be set up so that they unwind easily, without kinks and bends.

When several conductors must be fed into the conduit at the same time, a reel cart is used (*Figure 26*). The reel cart will allow the spools to turn freely and help prevent the wires from tangling.

3.7.0 Conductor Lubrication

When conductors are fed into long runs of conduit or conduit with several bends, both the conduit and the wires are lubricated with a compound designed for wire lubrication.

Several types of formulated compounds designed for wire lubrication are available in either dry powder, paste, or gel form. These compounds

On Site

Replacing the Hook on a Metal Fish Tape

The hook on a fish tape can be replaced using a propane torch. IMPORTANT: Always work in an appropriate environment and wear the necessary protective equipment. Hold the fish tape securely in a pair of pliers, heat the end with a propane torch until it softens, then use a second set of pliers to form a hook in the fish tape. Allow it to air cool.

26109-14_F26.EPS

Figure 26 Reel cart.

must be noncorrosive to the insulation material of the conductor and to the conduit itself. The compounds are applied by hand to the conductors as they are fed into the conduit. Battery-operated pumps are also available to lubricate the conduit prior to installing the conductors.

3.8.0 Conductor Termination

The amount of free conductor at each junction or outlet box must meet certain *NEC®* specifications. For example, there must be sufficient free conductor so that bends or terminations inside the box, cabinet, or enclosure may be made to a radius as specified in the *NEC®*. The *NEC®* specifies a minimum of six inches for connections made to wiring devices or for splices. Where conductors pass through junction or pull boxes, enough slack should be provided for splices at a later date.

When a box is used as a pull box, the conductors are not necessarily spliced. They may merely enter the pull box via one conduit run and exit via another conduit run. The purpose of a pull box, as the name suggests, is to facilitate pulling conductors on long runs. A junction box, however, is not only used to facilitate pulling conductors through the raceway system, but it also provides an enclosure for splices in the conductors.

On Site

Cleanup

A professional always takes the time to clean up the work area, removing excess lubricant from the boxes, bushings, and conductors. Completing a job in a professional manner is not only good business, it is also an *NEC®* requirement ***(NEC Section 110.12).***

Think About It

Putting It All Together

Think about the design of conductor installations. How does the location of pull points affect the ease of the pull?

Summary

This module introduced key concepts about conductors and cables.

Before selecting a conductor, you must understand wire ratings and why operating temperature is important. The conductor selection process requires that you also consider how and where the conductor will be terminated. The *NEC®* ampacity tables are based on operating temperature. The operating temperature of a conductor is affected by the conductor size (in circular mils), ambient temperature, number of conductors in the bundle, and where the conductors are installed.

Pulling wires and cables through conduit systems is an important part of your job as an electrician. The more you learn about the concepts involved in pulling cable, the safer and more efficient you will be.

Review Questions

1. A conductor's size refers to the cross-sectional area of both the current-carrying wire and the insulation.
   a. True
   b. False

2. The maximum size solid wire that should be terminated in pull boxes and disconnects is _____.
   a. No. 6
   b. No. 8
   c. No. 10
   d. No. 12

3. Compact stranding is used _____.
   a. when installing larger conduit sizes
   b. when decreasing the ampacity of an existing service
   c. with aluminum conductors
   d. in corrosive environments

4. The ampacity ratings of conductors can be found in _____.
   a. *NEC Chapter 1*
   b. *NEC Articles 348 through 352*
   c. *NEC 310.15(B)(16) through 310.15(B)(21)*
   d. *NEC Chapter 9*

5. To determine the ampacity of copper conductors where conductors are used as a single conductor in free air at ambient temperatures of 30°C, use _____.
   a. *NEC Table 310.15(B)(16)*
   b. *NEC Table 310.15(B)(17)*
   c. *NEC Table 310.15(B)(18)*
   d. *NEC Table 310.15(B)(19)*

6. All of the following help to determine how good a material will be for the construction of wire *except* _____.
   a. availability
   b. cost
   c. conductivity
   d. molecular weight

7. A type of thermoset insulation is _____.
   a. RHW
   b. THW
   c. THHW
   d. PVC

8. Polyethylene (PE) is used primarily for _____.
   a. the manufacture of TW and THW insulation
   b. insulation of control and communication wiring
   c. high-voltage cables
   d. high-temperature insulation

9. What letter must be included in the marking of a conductor if it is to be used in a wet, outdoor application?
   a. D
   b. O
   c. I
   d. W

10. What service conductor may be bare, green, or green with a yellow stripe?
    a. The grounding conductor of a multi-conductor cable
    b. The neutral conductor of a multi-conductor cable
    c. The ungrounded conductor of a multi-conductor cable
    d. The high leg of a four-wire, delta-connected secondary

11. What are the colors of insulation on the conductors for a three-conductor NM cable?
    a. One white or gray, one red, and one black
    b. Two white or gray and one red
    c. One white or gray, one red, one black, and a grounding conductor
    d. One green, one white or gray, one blue, and a grounding conductor

12. A conductor's operating temperature is *not* determined by _____.
    a. ambient temperature
    b. current flow
    c. current flow in bundled conductors
    d. the number of conductor strands

13. In some cases, fixture wires may be used as a substitute for branch circuit conductors.
    a. True
    b. False

14. Type NMC cable is suitable for all of the following *except* _____.

    a. dry locations
    b. damp locations
    c. corrosive locations
    d. embedding in concrete

15. The difference between AC cable and MC cable is _____.

    a. AC cable has a maximum of four conductors; MC has a maximum of six
    b. AC cable has a bonding strip; the MC ground is its metal cladding
    c. AC cable has fire-retardant wraps on individual conductors; MC cable does not
    d. AC cable is sized from 10 AWG to 4/0; MC is sized from 18 AWG to 2,000 kcmil

16. Type USE cable can be used for _____.

    a. aboveground installation only
    b. underground installation within a special PVC pipeline
    c. underground installation including direct burial
    d. indoor applications only

17. To prevent unwanted ground loops, instrumentation cable shielding is _____.

    a. not grounded at both ends of the wire
    b. grounded at both ends of the wire
    c. floated on both ends
    d. always ungrounded

18. A pulling line is usually made of _____.

    a. stranded wire
    b. wire rope
    c. steel tape
    d. nylon or other synthetic material

19. When feeding conductors into long runs of conduit, it is important to apply lubricant to the conductors only, and *not* to the conduit.

    a. True
    b. False

20. The *NEC*® specifies a minimum of _____ for connections made to wiring devices or for splices.

    a. 4"
    b. 6"
    c. 8"
    d. 10"

Trade Terms Quiz

Fill in the blank with the correct term that you learned from your study of this module.

1. A conductor's ____________ is the current the conductor can carry continuously without exceeding its temperature rating.
2. On a cable puller, the ____________ is the part on which the pulling rope is wrapped and pulled.
3. During a pull, cable may be attached to the pulling rope using a(n) ____________.
4. A(n) ____________ is a manually operated device that is used to pull a wire through conduit.
5. Composed of foam rubber, a(n) ____________ fits inside a piece of conduit and is propelled by compressed air or vacuumed through the conduit run, pulling a line or tape.

Trade Terms

Ampacity
Capstan
Fish tape
Mouse
Wire grip

Module 26109-14
Supplemental Exercises

1. The designation for one thousand circular mils is ______________________.

2. ______________________ is the maximum current in amperes a conductor can carry continuously under the conditions of use without exceeding its temperature rating.

3. True or False? *NEC Table 310.15(B)(16)* covers conductors rated up to 2,000V where not more than three conductors are installed in a raceway or cable.

4. True or False? Type NMC is suitable for direct burial.

5. Which of the following represents the largest wire size?
   a. 50
   b. 10
   c. 5
   d. 4/0

6. A(n) ______________________ is a circle which has a diameter of 1 mil.

7. True or False? All current-carrying conductors (including the grounded conductor) must be insulated when used on the inside of buildings.

8. True or False? Polyethylene is the base material used for the manufacture of TW and THW insulation.

9. Polyethylene is used primarily for insulation of ______________________.
   a. high-voltage conductors
   b. control and communications wiring
   c. THHN wiring
   d. XHHW wiring

10. What does the letter R stand for with regard to insulation coding?______________________

11. When conductors are installed where the ambient temperature is above ______________________, the conductor ampacity must be reduced proportionally with the increase in temperature.
    a. 25°C
    b. 30°C
    c. 45°C
    d. 60°C

12. Type UF cable ______________________.
    a. is commonly called *Romex*®
    b. is suitable for direct burial
    c. is commonly referred to as *BX*®
    d. consists of five conductors

13. Type MV cable is suitable for ______________________.

    a. use in wet locations
    b. use in dry locations
    c. direct burial
    d. all of the above.

14. Some jackets have a(n) ______________________ that allows the jacket to be peeled back without the use of a knife or cable cutter.

15. A(n) ______________________ is a piston-like plug that is attached to a string or rope for pulling through conduit using a power fishing system.

Cornerstone of Craftsmanship

L. J. LeBlanc

Senior Electrician/Instructor
Pumba Electrical/Baton Rouge Community College

Provide a summary of how you got started in the construction industry.
I took electrical jobs to pay for college and attended night classes at trade school for four years.

Who inspired you to enter the industry? Why?
My dad was an electrician. I was working with him and wiring houses at an early age.

What do you enjoy most about your job?
Electrical work is constantly changing. I love the challenge of troubleshooting. By passing on the knowledge I've obtained, I have truly made a difference in my student's lives.

Do you think training and education are important in construction? If so, why?
Education is extremely important. One must stay abreast of the changes involved in the code.

How important are NCCER credentials to your career?
If you are going to train, you need to have credentials to show you are qualified to teach.

How has training/construction impacted your life and your career?
Pay scales are governed by education and accomplishments.

Would you suggest construction as a career to others? If so, why?
Construction jobs are high paying and the sky is the limit in this field if you are skilled and work hard.

How do you define craftsmanship?
A person signs his name with his work. The job you do not only represents you but your company as well. Installations should be done in a workmanlike manner.

Trade Terms Introduced in This Module

Ampacity: The maximum current in amperes a conductor can carry continuously under the conditions of use without exceeding its temperature rating.

Capstan: The turning drum of the cable puller on which the rope is wrapped and pulled.

Fish tape: A hand device used to pull a wire through a conduit run.

Mouse: A cylinder of foam rubber that fits inside the conduit and is then propelled by compressed air or vacuumed through the conduit run, pulling a line or tape.

Wire grip: A device used to link pulling rope to cable during a pull.

Additional Resources

This module presents thorough resources for task training. The following resource material is suggested for further study.

National Electrical Code® Handbook, Latest Edition. Quincy, MA: National Fire Protection Association.

Figure Credits

AGC of America, Module opener

Tim Dean, Figures 4, 16

Topaz Publications, Inc., Figures 7B (photo), 8B (photo), 15, 17B (photo), 18, 19, 23, 24, 26, SA01–SA03

Jim Mitchem, Figure 9

The Okonite Company, Figure 10

General Cable, Figure 12

Greenlee/A Textron Company, Figures 22, 25

NCCER CURRICULA — USER UPDATE

NCCER makes every effort to keep its textbooks up-to-date and free of technical errors. We appreciate your help in this process. If you find an error, a typographical mistake, or an inaccuracy in NCCER's curricula, please fill out this form (or a photocopy), or complete the online form at **www.nccer.org/olf**. Be sure to include the exact module ID number, page number, a detailed description, and your recommended correction. Your input will be brought to the attention of the Authoring Team. Thank you for your assistance.

Instructors – If you have an idea for improving this textbook, or have found that additional materials were necessary to teach this module effectively, please let us know so that we may present your suggestions to the Authoring Team.

NCCER Product Development and Revision

13614 Progress Blvd., Alachua, FL 32615

Email: curriculum@nccer.org

Online: www.nccer.org/olf

❑ Trainee Guide ❑ AIG ❑ Exam ❑ PowerPoints Other ______________________________

Craft / Level: Copyright Date:

Module ID Number / Title:

Section Number(s):

Description:

Recommended Correction:

Your Name:

Address:

Email: Phone:

Basic Electrical Construction Drawings

Fenway Park Pavilion Seat Expansion and EMC/State Street Club Project

This expansion illustrates the accomplishment of an ambitious concept in a sensitive historical environment. The $45 million project was completed within the six-month winter off-season and used precise phasing and logistics to ensure complete protection of Fenway Field. The contractor structurally lifted and shored the historic facility to accommodate the installation of a new ring of columns that would ultimately support the significant addition.

26110-14

Trainees with successful module completions may be eligible for credentialing through NCCER's National Registry. To learn more, go to **www.nccer.org** or contact us at **1.888.622.3720.** Our website has information on the latest product releases and training, as well as online versions of our *Cornerstone* magazine and Pearson's product catalog.

Your feedback is welcome. You may email your comments to **curriculum@nccer.org,** send general comments and inquiries to **info@nccer.org,** or fill in the User Update form at the back of this module.

This information is general in nature and intended for training purposes only. Actual performance of activities described in this manual requires compliance with all applicable operating, service, maintenance, and safety procedures under the direction of qualified personnel. References in this manual to patented or proprietary devices do not constitute a recommendation of their use.

26110-14

Basic Electrical Construction Drawings

Objectives

When you have completed this module, you will be able to do the following:

1. Explain the basic layout of a set of construction drawings.
2. Describe the information included in the title block of a construction drawing.
3. Identify the types of lines used on construction drawings.
4. Using an architect's scale, state the actual dimensions of a given drawing component.
5. Interpret electrical drawings, including site plans, floor plans, and detail drawings.
6. Interpret equipment schedules found on electrical drawings.
7. Describe the type of information included in electrical specifications.

Performance Tasks

Under the supervision of the instructor, you should be able to do the following:

1. Using an architect's scale, state the actual dimensions of a given drawing component.
2. Make a material takeoff of the lighting fixtures specified in Performance Profile Sheet 2 using the drawing provided on Performance Profile Sheet 3. The takeoff requires that all lighting fixtures be counted, and where applicable, the total number of lamps for each fixture type must be calculated.

Trade Terms

Architectural drawings
Block diagram
Blueprint
Detail drawing
Dimensions
Electrical drawing
Elevation drawing
Floor plan
One-line diagram
Plan view
Power-riser diagram
Scale
Schedule
Schematic diagram
Sectional view
Shop drawing
Site plan
Written specifications

Required Trainee Materials

1. Paper and pencil
2. Copy of the latest edition of the *National Electrical Code*®
3. Appropriate personal protective equipment

Note:
NFPA 70®, *National Electrical Code*®, and *NEC*® are registered trademarks of the National Fire Protection Association, Inc., Quincy, MA 02269. All *National Electrical Code*® and *NEC*® references in this module refer to the 2014 edition of the *National Electrical Code*®.

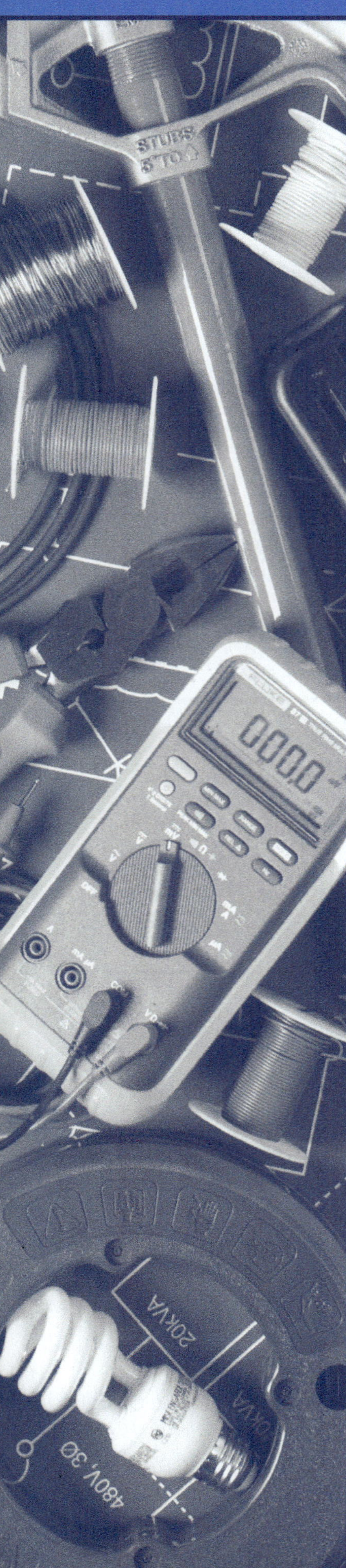

Contents

Topics to be presented in this module include:

Figures

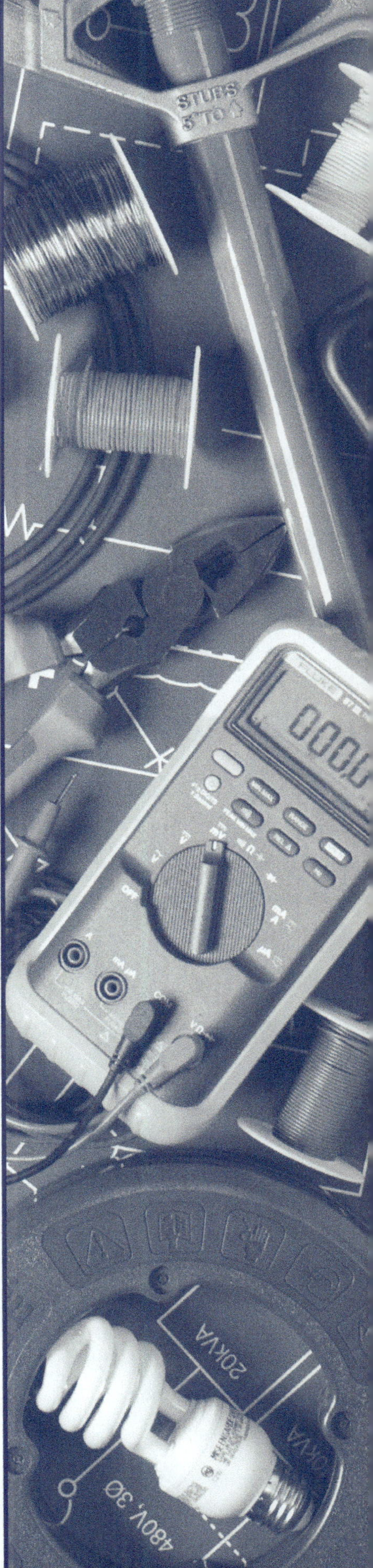

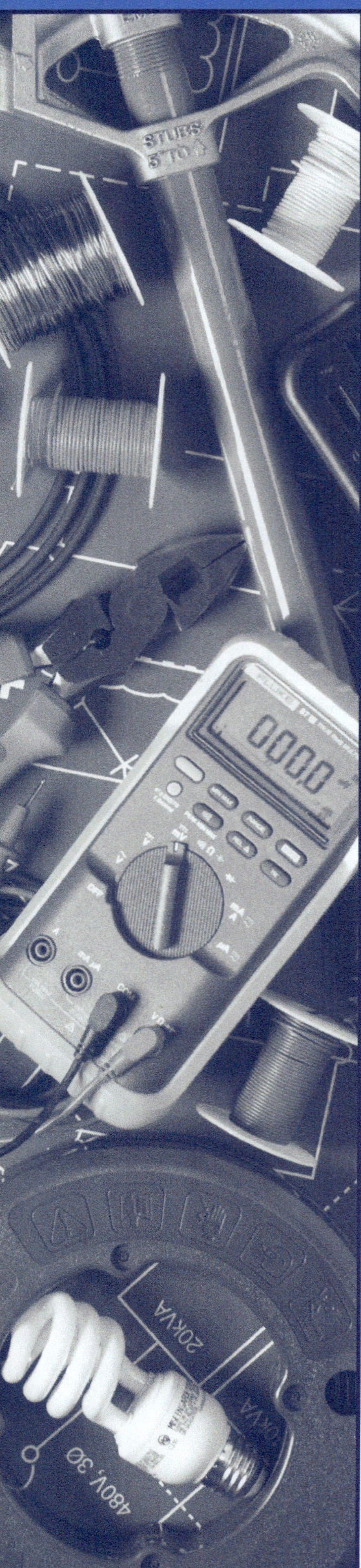

STUBS
5"TO
0.0.0.0
20KVA
480V, 3Ø

1.0.0 Introduction to Construction Drawings

In all large construction projects and in many of the smaller ones, an architect is commissioned to prepare complete working drawings and specifications for the project. These drawings usually include:

- A **site plan** indicating the location of the building on the property.
- **Floor plans** showing the walls and partitions for each floor or level.
- Elevations of all exterior faces of the building.
- Several vertical cross sections to indicate clearly the various floor levels and details of the footings, foundation, walls, floors, ceilings, and roof construction.
- Large-scale **detail drawings** showing such construction details as may be required.

For projects of any consequence, the architect usually hires consulting engineers to prepare structural, electrical, and mechanical drawings, with the latter encompassing pipefitting, instrumentation, plumbing, and heating, ventilating, and air conditioning drawings.

1.1.0 Site Plan

This type of plan of the building site looks as if the site is viewed from an airplane and shows the property boundaries, the existing contour lines, the new contour lines (after grading), the location of the building on the property, new and existing roadways, all utility lines, and other pertinent details. The drawing **scale** is also shown. Descriptive notes may also be found on the site (plot) plan listing names of adjacent property owners, the land surveyor, and the date of the survey. A legend or symbol list is also included so that anyone who must work with the site plan can readily read the information. See *Figure 1*.

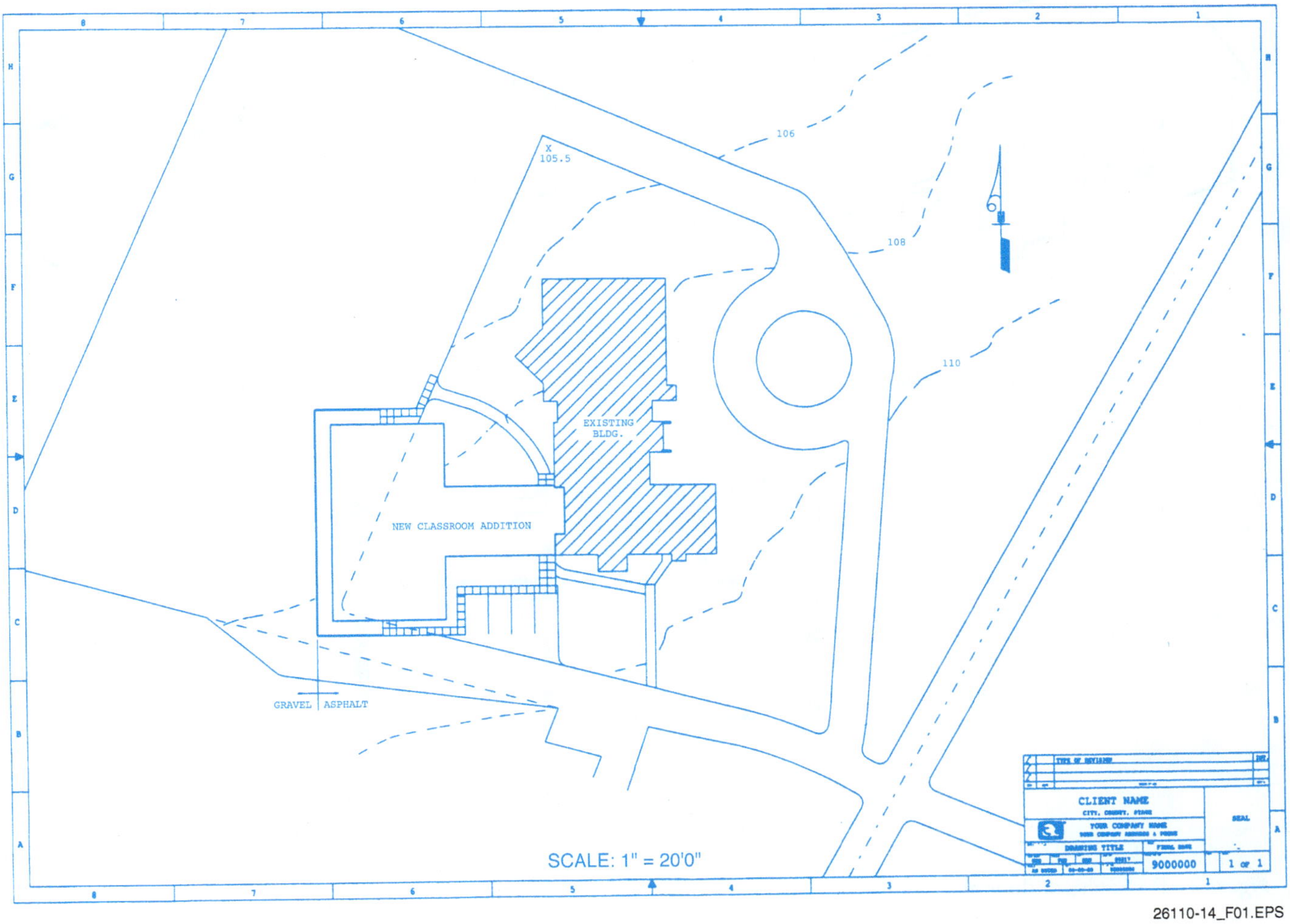

SCALE: 1" = 20'0"

26110-14_F01.EPS

Figure 1 Typical site plan.

1.2.0 Floor Plans

The **plan view** of any object is a drawing showing the outline and all details as seen when looking directly down on the object. It shows only two **dimensions,** length and width. The floor plan of a building is drawn as if a horizontal cut were made through the building—at about window height—and then the top portion removed to reveal the bottom part. See *Figure 2*.

If a plan view of a home's basement is needed, the part of the house above the middle of the basement windows is imagined to be cut away. By looking down on the uncovered portion, every detail and partition can be seen. Likewise, imagine the part above the middle of the first floor windows being cut away. A drawing that looks straight down at the remaining part would be called the first floor plan or lower level. A cut through the second floor windows would be called the second floor plan or upper level. See *Figure 3*.

> **On Site**
>
> **Then and Now**
>
> Years ago, blueprints were created by placing a hand drawing against light-sensitive paper and then exposing it to ultraviolet light. The light would turn the paper blue except where lines were drawn on the original. The light-sensitive paper was then developed, and the resulting print had white lines against a blue background. Modern blueprints usually have blue or black lines against a white background and are generated using computer-aided design programs. Newer programs offer three-dimensional modeling and other enhanced features.

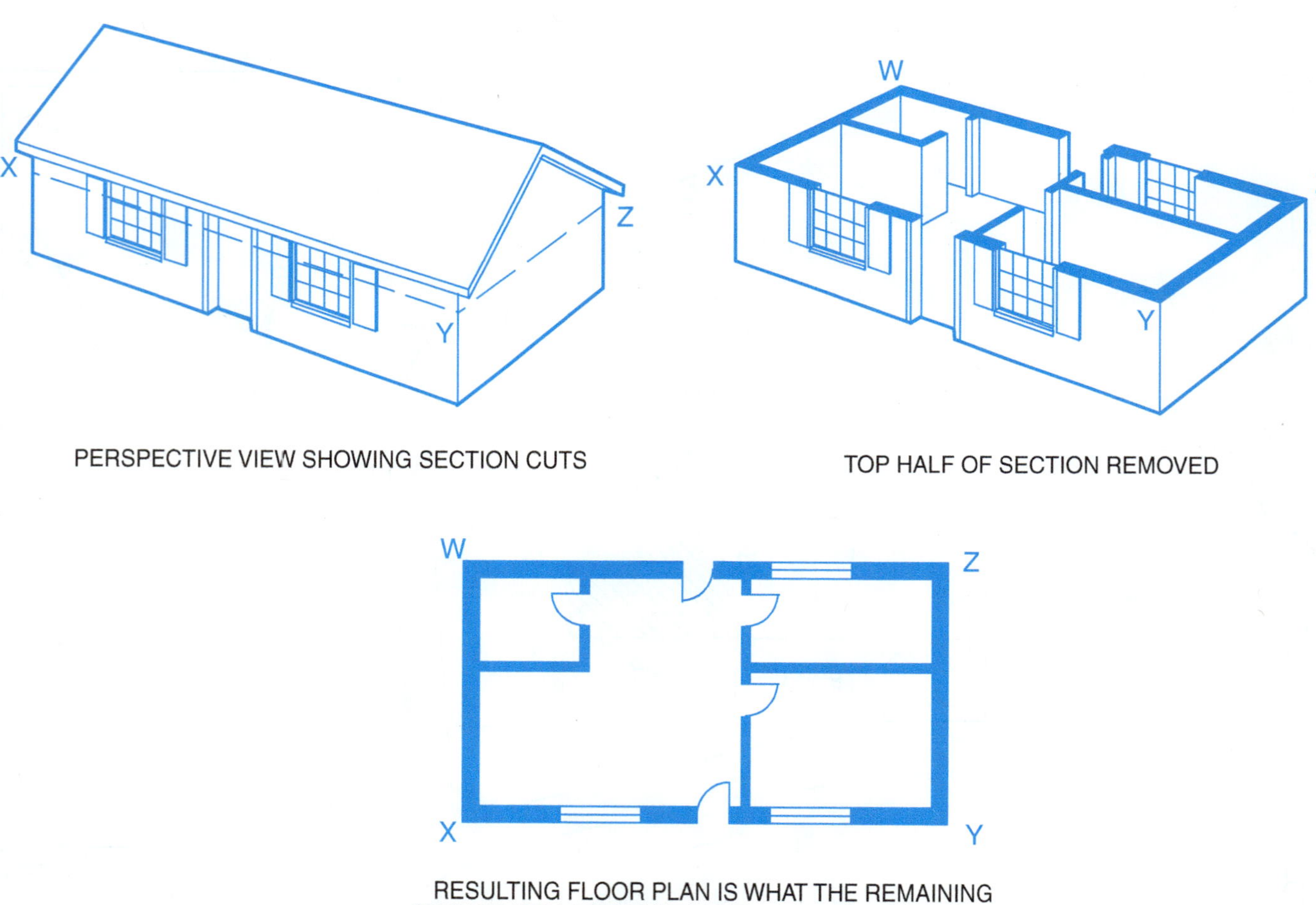

Figure 2 Principles of floor plan layout.

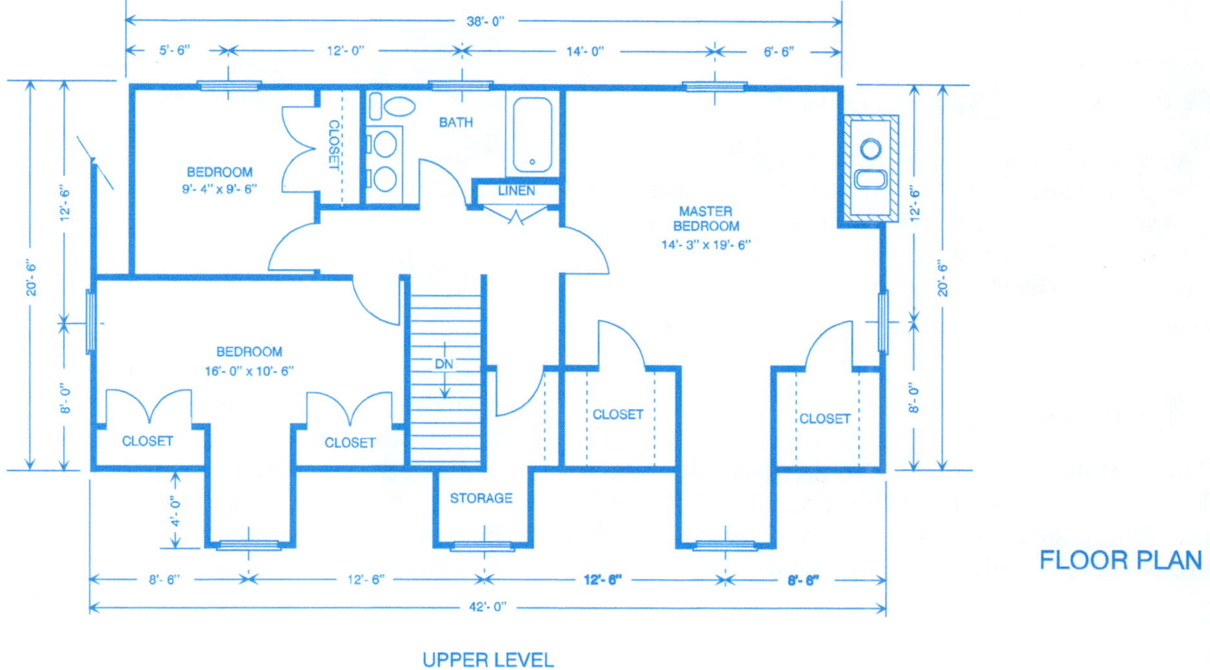

Figure 3 Floor plans of a building.

On Site

Using a Drawing Set

Always treat a drawing set with care. It is best to keep two sets, one for the office and one for field use. Be sure to use the most current revision. After you use a sheet from a set of drawings, refold the sheet with the title block facing up.

1.3.0 Elevations

The elevation is an outline of an object that shows heights and may show the length or width of a particular side, but not depth. *Figures 4* and *5* show **elevation drawings** for a building.

NOTE

Elevation drawings may also show the heights of windows, doors, and porches, the pitch of roofs, etc., because all of these measurements cannot be shown conveniently on floor plans.

1.4.0 Sections

A section or **sectional view** (*Figure 6*) is a cutaway view that allows the viewer to see the inside of a structure. The point on the plan or elevation showing where the imaginary cut has been made is indicated by the section line, which is usually a dashed line. The section line shows the location of the section on the plan or elevation. It is necessary to know which of the cutaway parts is represented in the sectional drawing. To show

FRONT ELEVATION

REAR ELEVATION

26110-14_F04.EPS

Figure 4 Front and rear elevations.

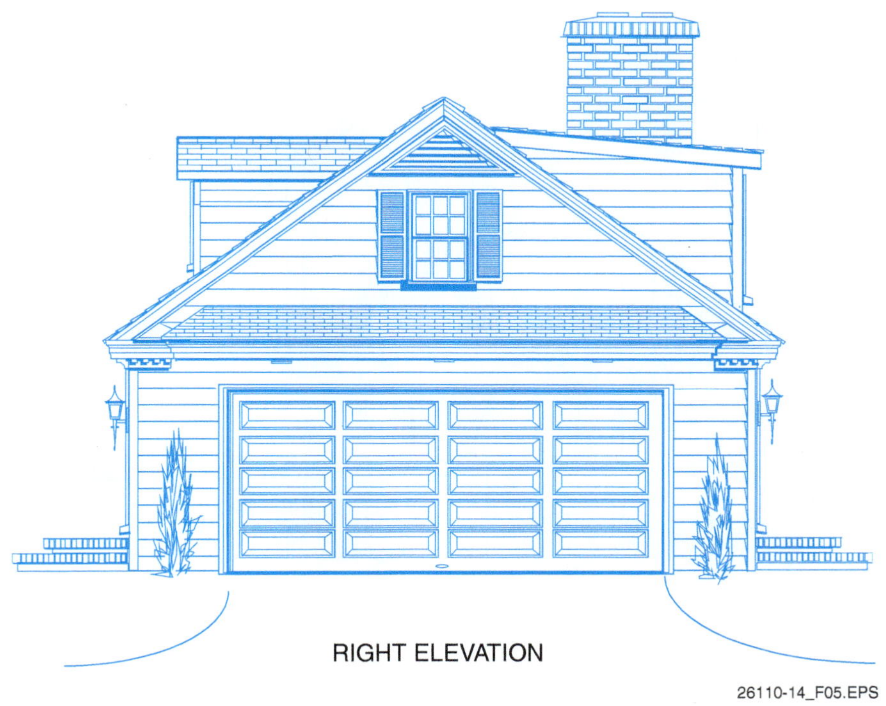

Figure 5 Left and right elevations.

this, arrow points are placed at the ends of the section lines.

In **architectural drawings,** it is often necessary to show more than one section on the same drawing. The different section lines must be distinguished by letters, numbers, or other designations placed at the ends of the lines. These section letters are generally large so as to stand out on the drawings. To further avoid confusion, the same letter is usually placed at each end of the section line. The section is named according to these letters (e.g., Section A-A, Section B-B, and so forth).

A longitudinal section is taken lengthwise while a cross section is usually taken straight across the width of an object. Sometimes, however, a section is not taken along one straight line. It is often taken along a zigzag line to show important parts of the object.

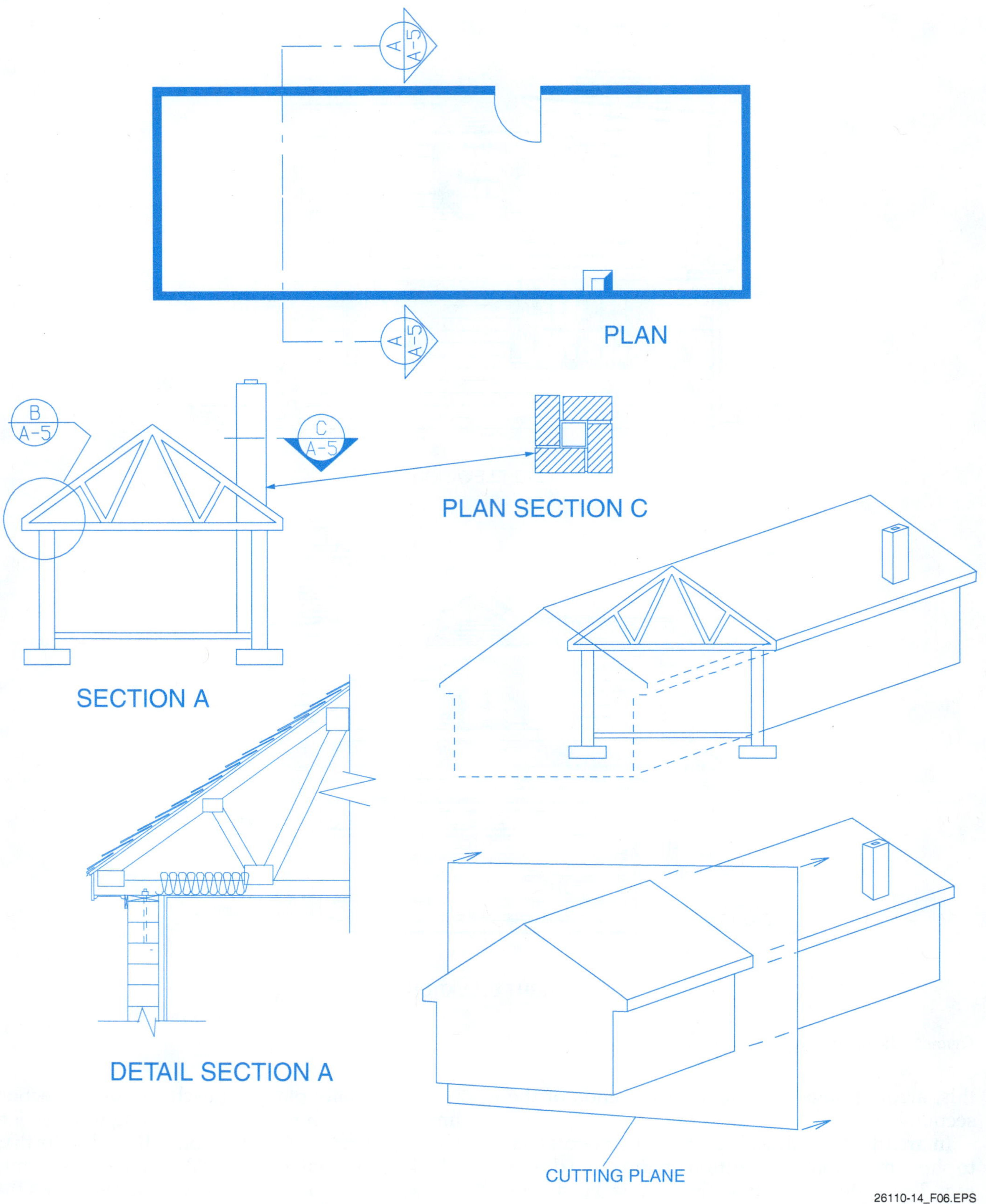

Figure 6 Sectional drawing.

A sectional view, as applied to architectural drawings, is a drawing showing the building, or portion of a building, as though it were cut through on some imaginary line. This line may be either vertical (straight up and down) or horizontal. Wall sections are nearly always made vertically so that the cut edge is exposed from top to bottom. In some ways, the wall section is one of the most important of all the drawings to construction workers, because it answers the questions as to how a structure should be built. The floor plans of a building show how each floor is arranged, but the wall sections tell how each part is constructed and usually indicate the material to be used. The electrician needs to know this information when determining wiring methods that comply with the *NEC*®.

1.5.0 Electrical Drawings

Electrical drawings show in a clear, concise manner exactly what is required of the electricians. The amount of data shown on such drawings should be sufficient, but not overdone. This means that a complete set of electrical drawings could consist of only one 8½" × 11" sheet, or it could consist of several dozen 24" × 36" (or larger) sheets, depending on the size and complexity of a given project. A **shop drawing,** for example, may contain details of only one piece of equipment, while a set of working drawings for an industrial installation may contain dozens of drawing sheets detailing the electrical system for lighting and power, along with equipment, motor controls, wiring diagrams, **schematic diagrams,** equipment **schedules,** and a host of other pertinent data.

In general, the electrical working drawings for a given project serve three distinct functions:

- They provide electrical contractors with an exact description of the project so that materials and labor may be estimated to calculate a total cost of the project for bidding purposes.
- They provide workers on the project with instructions as to how the electrical system is to be installed.
- They provide a map of the electrical system once the job is completed to aid in maintenance and troubleshooting for years to come.

Electrical drawings from consulting engineering firms will vary in quality from sketchy, incomplete drawings to neat, precise drawings that are easy to understand. Few, however, will cover every detail of the electrical system. Therefore, a good knowledge of installation practices must go hand-in-hand with interpreting electrical working drawings.

On Site

Applying Your Skills

Once you learn how to interpret construction drawings, you can apply that knowledge to any type of construction, from simple residential applications to large industrial complexes.

Sometimes electrical contractors will have electrical drafters prepare special supplemental drawings for use by the contractors' employees. On certain projects, these supplemental drawings can save supervision time in the field once the project has begun.

2.0.0 Drawing Layout

Although a strong effort has been made to standardize drawing practices in the building construction industry, the drawings or **blueprints** prepared by different architectural or engineering firms will rarely be identical. Similarities, however, will exist between most sets of drawings, and with a little experience, you should have no trouble interpreting any set of drawings that might be encountered.

Most drawings used for building construction projects will be drawn on sheets in various sizes. Each drawing sheet has border lines framing the overall drawing and one or more title blocks, as shown in *Figure 7*. The type and size of title blocks varies with each firm preparing the drawings. In addition, some drawing sheets will also contain a revision block near the title block, and perhaps an approval block. This information is normally found on each drawing sheet, regardless of the type of project or the information contained on the sheet.

2.1.0 Title Block

The architect's title block for a drawing is usually boxed in the lower right-hand corner of the drawing sheet; the size of the block varies with the size of the drawing and with the information required. See *Figure 8*.

In general, the title block of an electrical drawing should contain the following information:

- Name of the project
- Address of the project
- Name of the owner or client
- Name of the architectural firm
- Date of completion

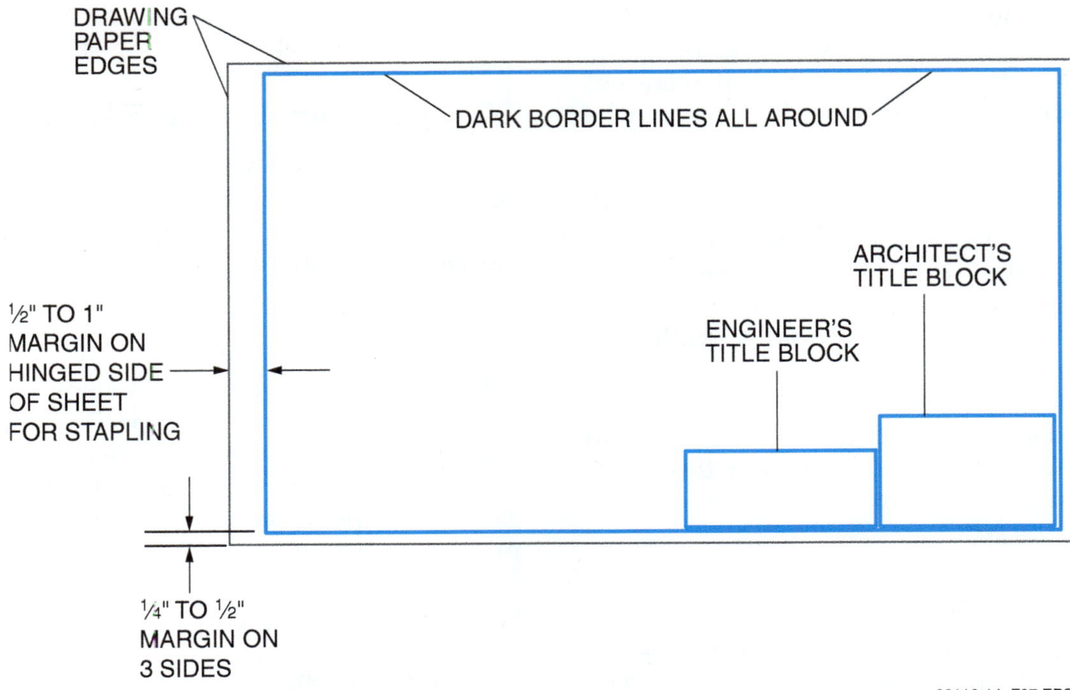

Figure 7 Typical drawing layout.

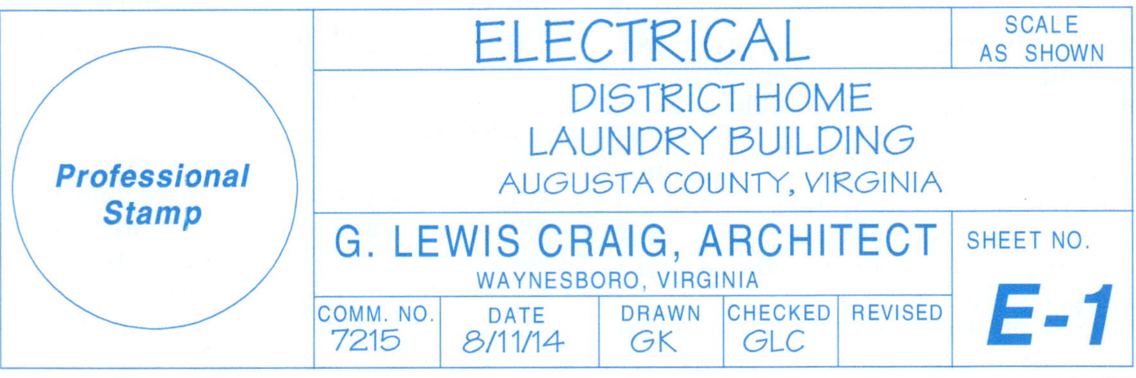

Figure 8 Typical architect's title block.

On Site

Interpreting Electrical Drawings

A good example of when an electrician must interpret the drawings is when wiring a log cabin. The drawings will show the receptacle and switch locations in branch circuits as usual, but the electrician must figure out how to route wires and install boxes where there is no hollow wall and sometimes no ceiling space.

- Scale(s)
- Initials of the drafter, checker, and designer, with dates under each
- Job number
- Sheet number
- General description of the drawing

Often, the consulting engineering firm will also be listed, which means that an additional title block will be applied to the drawing, usually next to the architect's title block. *Figure 9* shows completed architectural and engineering title blocks as they appear on an actual drawing.

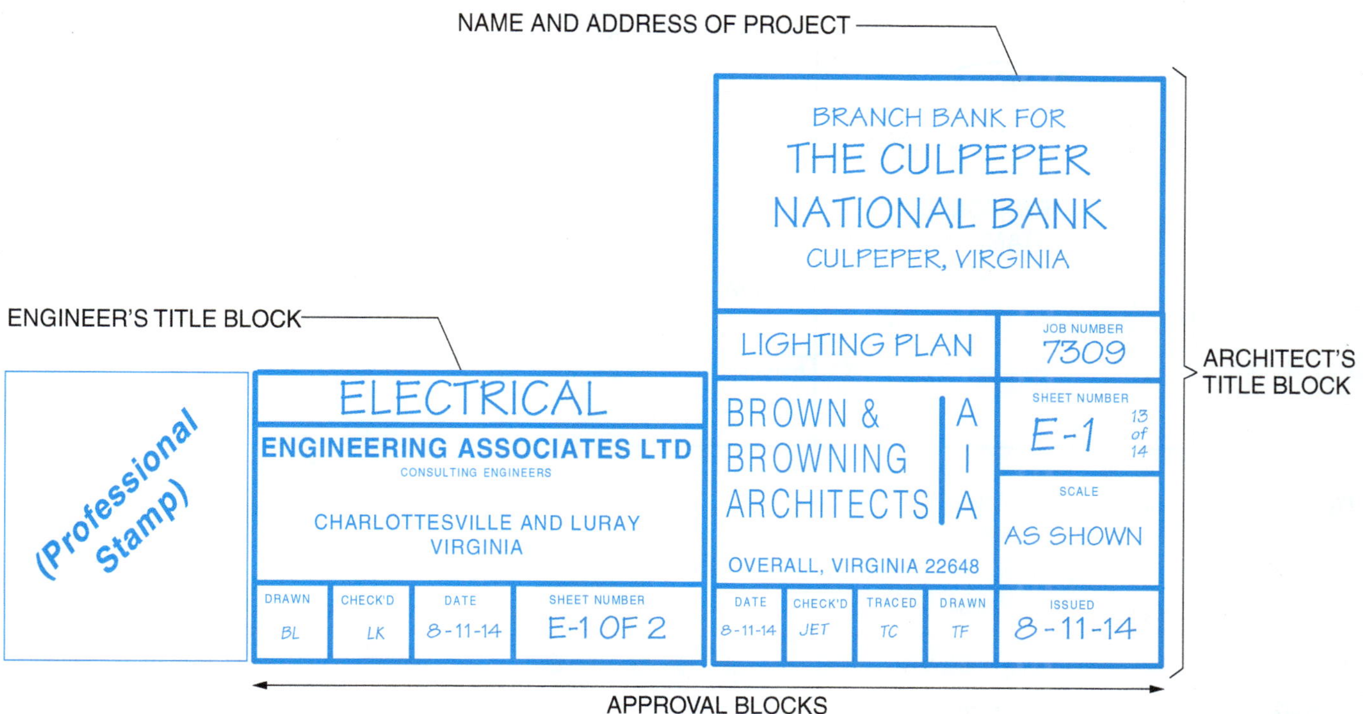

Figure 9 Title blocks.

On Site

Check the Title Block

Always refer to the title block for scale and be aware that it often changes from drawing to drawing within the same set. Also check for the current revision, and when you replace a drawing with a new revision, be sure to remove and file the older version.

2.2.0 Approval Block

The approval block, in most cases, will appear on the drawing sheet as shown in *Figure 10*. The various types of approval blocks (drawn, checked, etc.) will be initialed by the appropriate personnel. This type of approval block is usually part of the title block and appears on each drawing sheet.

On some projects, authorized signatures are required before certain systems may be installed, or even before the project begins. An approval block such as the one shown in *Figure 11* indicates that all required personnel have checked the drawings for accuracy, and that the set meets with everyone's approval. Such an approval block usually appears on the front sheet of the blueprint set and may include:

- *Professional stamp* – Registered seal of approval by the licensed architect or consulting engineer.
- *Design supervisor* – Signature of the person who is overseeing the design.
- *Drawn (by)* – Signature or initials of the person who drafted the drawing and the date it was completed.
- *Checked (by)* – Signature or initials of the person who reviewed the drawing and the date of approval.
- *Approved* – Signature or initials of the architect/engineer and the date of the approval.
- *Owner's approval* – Signature of the project owner or the owner's representative along with the date signed.

Figure 10 Typical approval block.

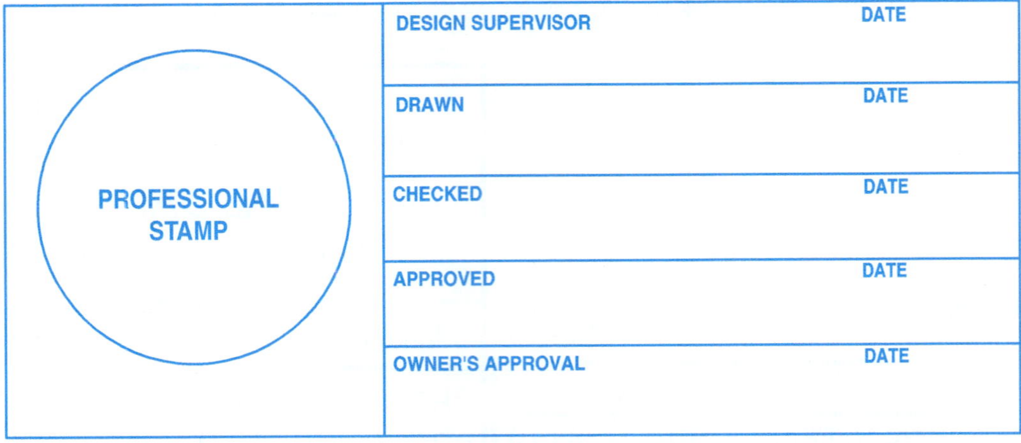

26110-14_F11.EPS

Figure 11 Alternate approval block.

On Site

Orient Yourself

When reading a drawing, find the north arrow to orient yourself to the structure. Knowing where north is enables you to accurately describe the locations of walls and other parts of the building.

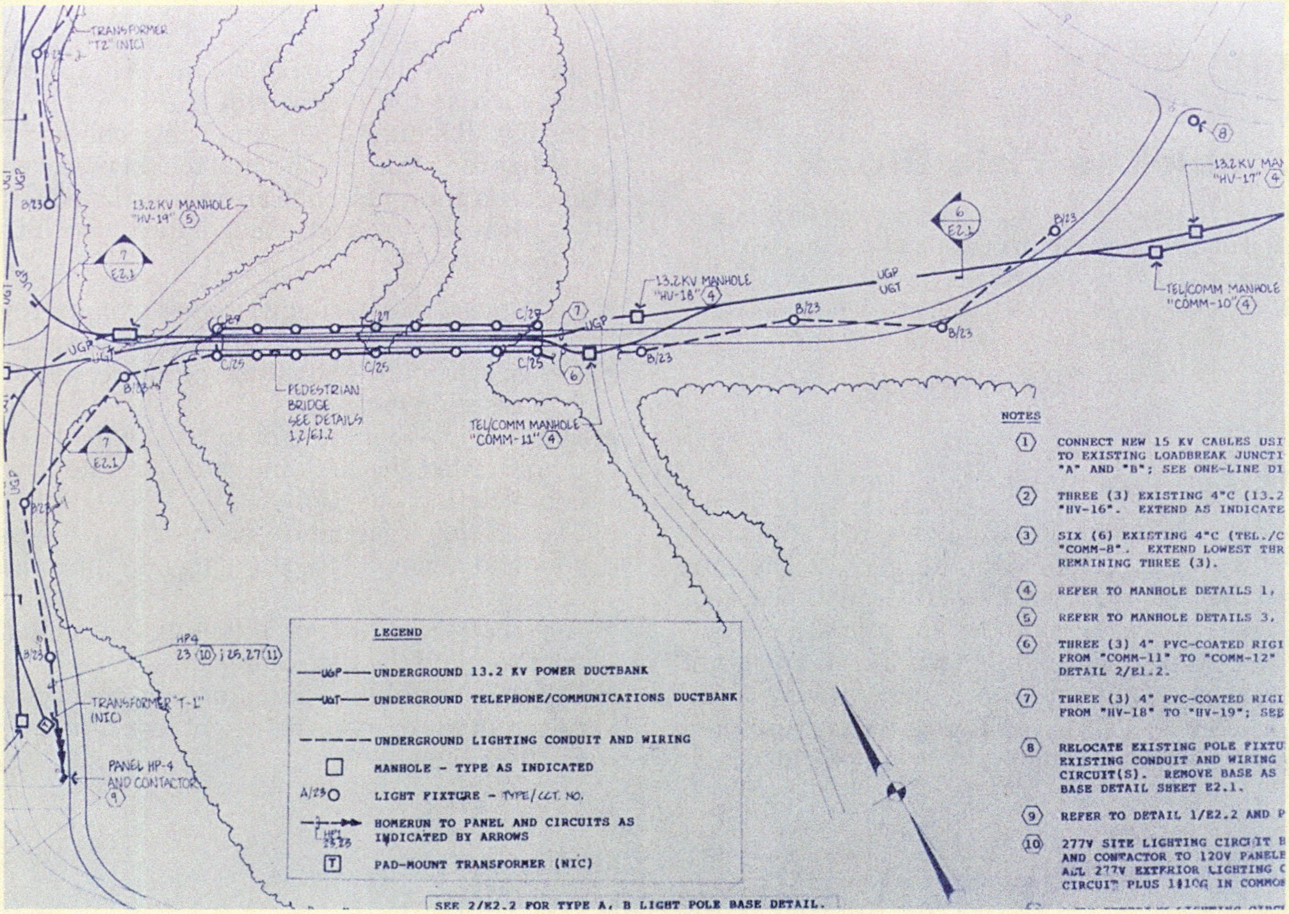

26110-14_SA01.EPS

Think About It

Using All of the Drawings

Look back over the information on floor plans, elevations, and sections. What kinds of information would an electrician get from each of these drawings? What could a sectional drawing show that a floor plan could not?

2.3.0 Revision Block

Sometimes electrical drawings will have to be partially redrawn or modified during the construction of a project. It is extremely important that such modifications are noted and dated on the drawings to ensure that the workers have an up-to-date set of drawings to work from. In some situations, sufficient space is left near the title block for dates and descriptions of revisions, as shown in *Figure 12*. In other cases, a revision block is provided (again, near the title block), as shown in *Figure 13*. The area on the drawing where the revision has been made will often be circled with a cloud shape.

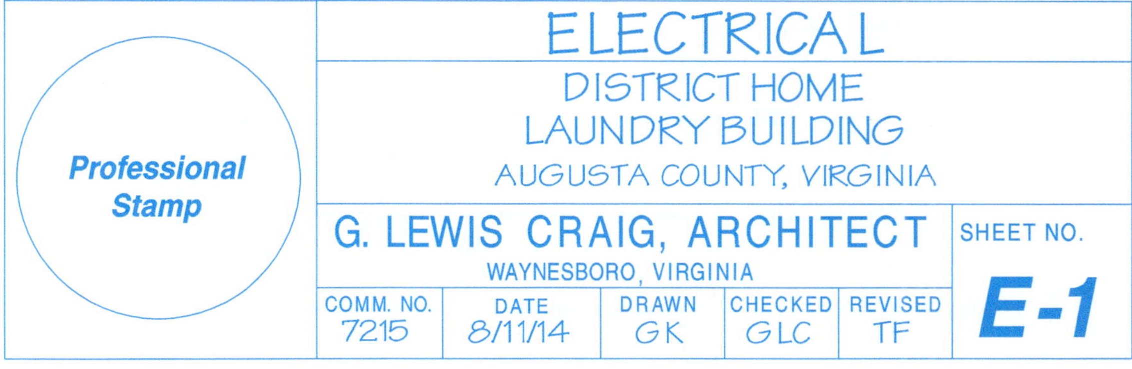

Figure 12 One method of showing revisions on working drawings.

REVISIONS

REV	DESCRIPTION	DR	APPD	DATE
1	FIXTURE NO. 3 IN. LIGHTING-FIXTURE SCHDL	GK	GLC	9/5/14

Professional Stamp

ELECTRICAL
DISTRICT HOME
LAUNDRY BUILDING
AUGUSTA COUNTY, VIRGINIA
G. LEWIS CRAIG, ARCHITECT
WAYNESBORO, VIRGINIA
SHEET NO.
E-1

COMM. NO.	DATE	DRAWN	CHECKED	REVISED
7215	8/11/14	GK	GLC	TF

26110-14_F13.EPS

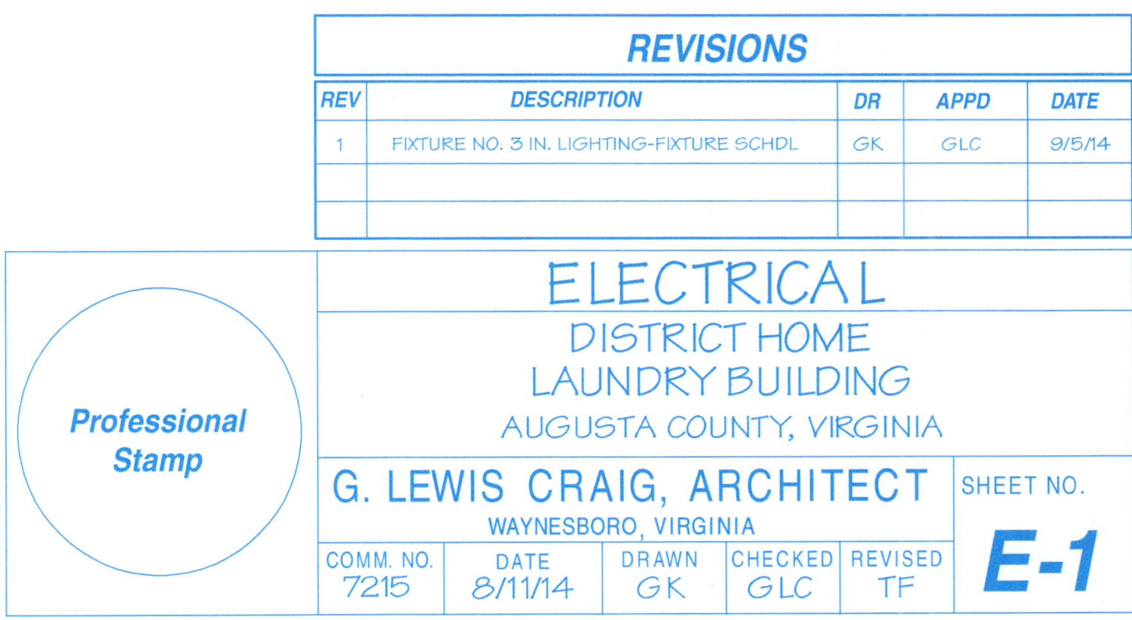

Figure 13 Alternative method of showing revisions on working drawings.

NOTE

Architects, engineers, designers, and drafters have their own methods of showing revisions, so expect to find deviations from those shown here.

CAUTION

When a set of electrical working drawings has been revised, always make certain that the most up-to-date set is used for all future layout work. Either destroy the old, obsolete set of drawings or else clearly mark on the affected sheets, *Obsolete Drawing—Do Not Use.* Also, when working with a set of working drawings and written specifications for the first time, thoroughly check each page to see if any revisions or modifications have been made to the originals. Doing so can save much time and expense to all concerned with the project.

3.0.0 Drafting Lines

You will encounter many types of drafting lines. To specify the meaning of each type of line, contrasting lines can be made by varying the width of the lines or breaking the lines in a uniform way.

Figure 14 shows common lines used on architectural drawings. However, these lines can vary. Architects and engineers have strived for a common standard for the past century, but unfortunately, their goal has yet to be reached. Therefore, you will find variations in lines and symbols from drawing to drawing, so always consult the legend or symbol list when referring to any drawing. Also, carefully inspect each drawing to ensure that line types are used consistently.

The drafting lines shown in *Figure 14* are used as follows:

- *Light full line* – This line is used for section lines, building background (outlines), and similar

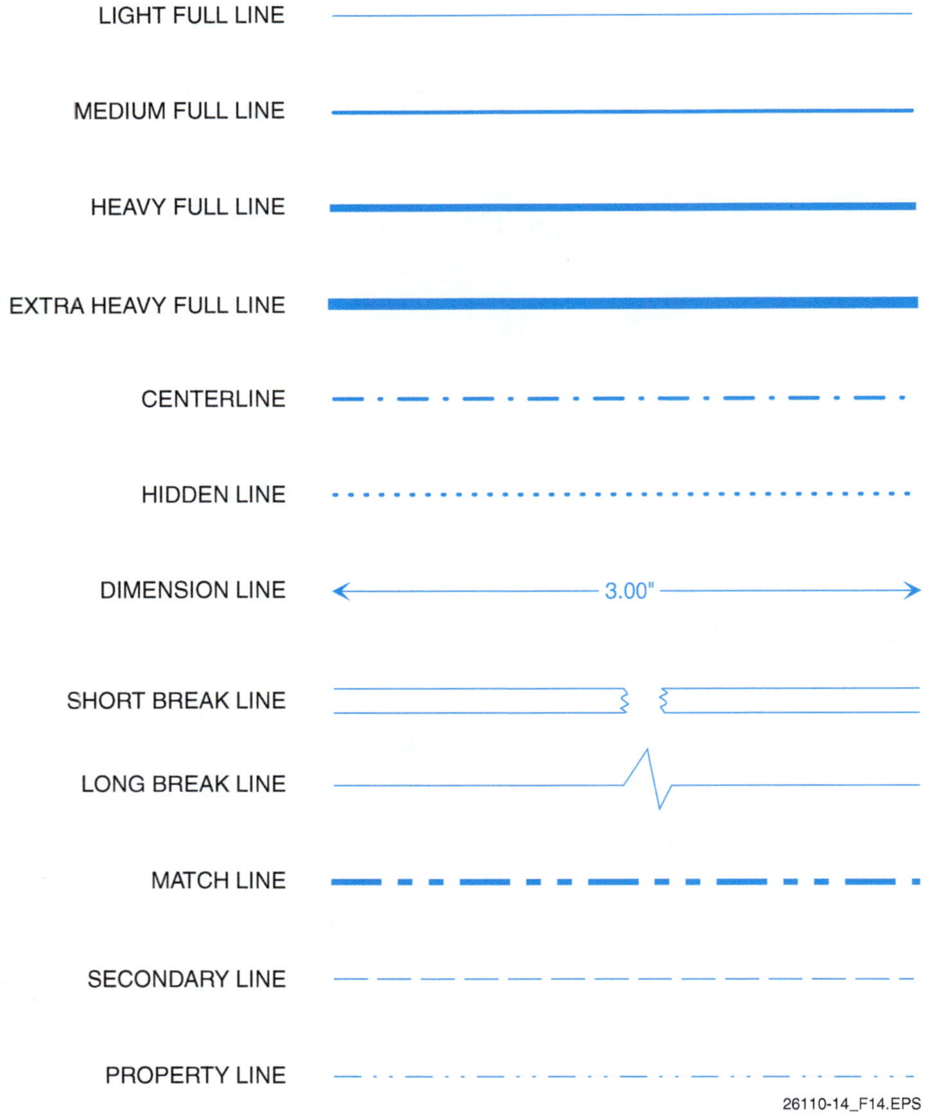

Figure 14 Typical drafting lines.

uses where the object to be drawn is secondary to the system being shown (e.g., HVAC or electrical).

- *Medium full line* – This type of line is frequently used for hand lettering on drawings. It is further used for some drawing symbols, circuit lines, etc.
- *Heavy full line* – This line is used for borders around title blocks, schedules, and for hand lettering drawing titles. Some types of symbols are frequently drawn with a heavy full line.
- *Extra heavy full line* – This line is used for border lines on architectural/engineering drawings.
- *Centerline* – A centerline is a broken line made up of alternately spaced long and short dashes. It indicates the centers of objects such as holes, pillars, or fixtures. Sometimes, the centerline indicates the dimensions of a finished floor.
- *Hidden line* – A hidden line consists of a series of short dashes that are closely and evenly spaced. It shows the edges of objects that are not visible in a particular view. The object outlined by hidden lines in one drawing is often fully pictured in another drawing.
- *Dimension line* – These are thin lines used to show the extent and direction of dimensions. The dimension is usually placed in a break inside the dimension lines. Normal practice is to place the dimension lines outside the object's outline. However, it may sometimes be necessary to draw the dimensions inside the outline.
- *Short break line* – This line is usually drawn freehand and is used for short breaks.
- *Long break line* – This line, which is drawn partly with a straightedge and partly with freehand zigzags, is used for long breaks.
- *Match line* – This line is used to show the position of the cutting plane. Therefore, it is also called the cutting plane line. A match or cutting plane line is a heavy line with long dashes alternating with two short dashes. It is used on drawings of large structures to show where one drawing stops and the next drawing starts.
- *Secondary line* – This line is frequently used to outline pieces of equipment or to indicate reference points of a drawing that are secondary to the drawing's purpose.
- *Property line* – This is a light line made up of one long and two short dashes that are alternately spaced. It indicates land boundaries on the site plan.

Other uses of the lines just mentioned include the following:

- *Extension lines* – Extension lines are lightweight lines that start about 1⁄16 inch away from the edge of an object and extend out. A common use of extension lines is to create a boundary for dimension lines. Dimension lines meet extension lines with arrowheads, slashes, or dots. Extension lines that point from a note or other reference to a particular feature on a drawing are called leaders. They usually end in either an arrowhead or a dot and may include an explanatory note at the end.
- *Section lines* – These are often referred to as cross-hatch lines. Drawn at a 45° angle, these lines show where an object has been cut away to reveal the inside.
- *Phantom lines* – Phantom lines are solid, light lines that show where an object will be installed. A future door opening or a future piece of equipment can be shown with phantom lines.

3.1.0 Electrical Drafting Lines

Besides the architectural lines shown in *Figure 14* consulting electrical engineers, designers, and drafters use additional lines to represent circuits and their related components. Again, these lines may vary from drawing to drawing, so check the symbol list or legend for the exact meaning of lines on the drawing with which you are working. *Figure 15* shows lines used on some electrical drawings.

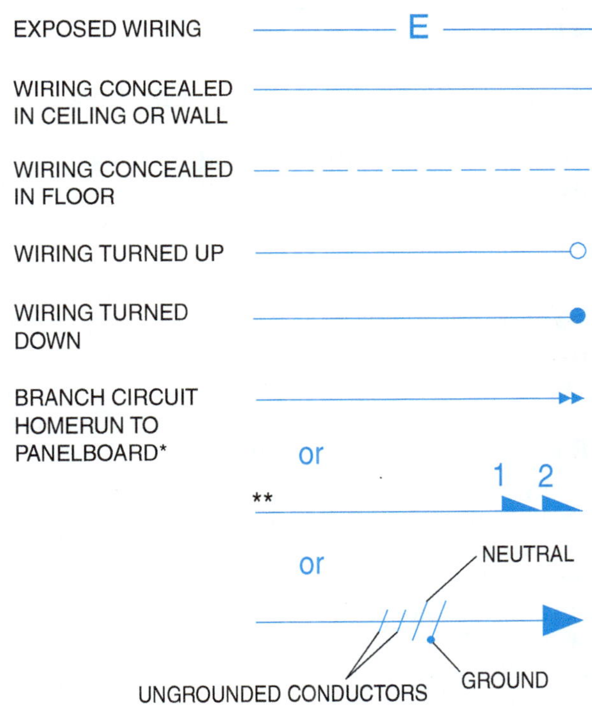

Figure 15 Electrical drafting lines.

On Site

Check the Legend

Be sure to check the legend on every drawing set. Symbols and abbreviations often vary widely from drawing set to drawing set.

4.0.0 Electrical Symbols

The electrician must be able to correctly read and understand electrical working drawings. This includes a thorough knowledge of electrical symbols and their applications.

An electrical symbol is a figure or mark that stands for a component used in the electrical system. *Figure 16* shows a list of electrical symbols that are currently recommended by the American National Standards Institute (ANSI). It is evident from this list of symbols that many have the same basic form, but, because of some slight difference, their meaning changes. For example, the receptacle symbols in *Figure 17* each have the same basic form (a circle), but the addition of a line or an abbreviation gives each an individual meaning. A good procedure to follow in learning symbols is to first learn the basic form and then apply the variations for obtaining different meanings.

It would be much simpler if all architects, engineers, electrical designers, and drafters used the same symbols; however, this is not the case. Although standardization is getting closer to a reality, existing symbols are still modified, and new symbols are created for almost every new project.

The electrical symbols described in the following paragraphs represent those found on actual electrical working drawings throughout the United States and Canada. Many are similar to those recommended by ANSI and the Consulting Engineers Council/US; others are not. Understanding how these symbols were devised will help you to interpret unknown electrical symbols in the future.

Some of the symbols used on electrical drawings are abbreviations, such as WP for weatherproof and AFF for above finished floor. Others are simplified pictographs, such as those shown in *Figure 18*.

In some cases, the symbols are combinations of abbreviations and pictographs, such as in *Figure 18* for a fusible safety switch, a nonfusible safety switch, and a double-throw safety switch. In each example, a pictograph of a switch enclosure has been combined with an abbreviation: F (fusible), DT (double-throw), and NF (nonfusible), respectively.

Lighting outlet symbols have been devised that represent incandescent, fluorescent, and high-intensity discharge lighting; a circle usually represents an incandescent fixture, and a rectangle is used to represent a fluorescent fixture. These symbols are designed to indicate the physical shape of a particular fixture, and while the circles representing incandescent lamps are frequently enlarged somewhat, symbols for fluorescent fixtures are usually drawn as close to scale as possible. The type of mounting used for all lighting fixtures is usually indicated in a lighting fixture schedule, which is shown on the drawings or in the **written specifications**.

The type of lighting fixture is identified by a numeral placed inside a triangle or other symbol, and placed near the fixture to be identified. A complete description of the fixtures identified by the symbols must be given in the lighting fixture schedule and should include the manufacturer, catalog number, number and type of lamps, voltage, finish, mounting, and any other information needed for proper installation of the fixture.

Switches used to control lighting fixtures are also indicated by symbols (usually the letter S followed by numerals or letters to define the exact type of switch). For example, S_3 indicates a three-way switch; S_4 identifies a four-way switch; and S_P indicates a single-pole switch with a pilot light. A subscript letter is often used to identify the fixtures that are controlled by that switch.

Main distribution centers, panelboards, transformers, safety switches, and other similar electrical components are indicated by electrical symbols on floor plans and by a combination of symbols and semipictorial drawings in riser diagrams.

A detailed description of the service equipment is usually given in the panelboard schedule or in the written specifications. However, on small projects, the service equipment is sometimes indicated only by notes on the drawings.

Circuit and feeder wiring symbols are getting closer to being standardized. Most circuits concealed in the ceiling or wall are indicated by a solid line; a broken line is used for circuits concealed in the floor or ceiling below; and exposed raceways are indicated by short dashes or else the letter *E* placed in the same plane with the circuit line at various intervals. The number of conductors in a conduit or raceway system may be indicated in the panelboard schedule under the appropriate column, or the information may be shown on the floor plan.

Symbols for communication and signal systems, as well as symbols for light and power,

SWITCH OUTLETS

Single-Pole Switch

Double-Pole Switch

Three-Way Switch

Four-Way Switch

Key-Operated Switch

Switch w/Pilot

Low-Voltage Switch

Switch & Single Receptacle

Switch & Duplex Receptacle

Door Switch

Momentary Contact Switch

RECEPTACLE OUTLETS

Single Receptacle

Duplex Receptacle

Triplex Receptacle

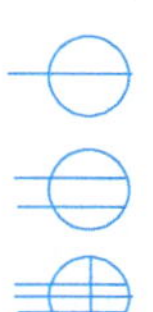

Split-Wired Duplex Recep.

Single Special Purpose Recep.

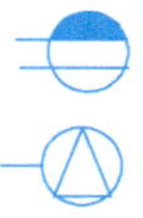

Duplex Special Purpose Recep.

Range Receptacle

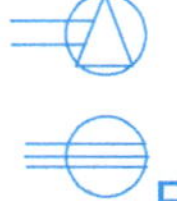

Special Purpose Connection or Provision for Connection. Subscript letters indicate Function (DW - Dishwasher; CD - Clothes Dryer, etc.)

Clock Receptacle w/Hanger

Fan Receptacle w/Hanger

Single Floor Receptacle

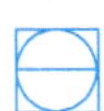

Note: A numeral or letter within the symbol or as a subscript keyed to the list of symbols indicates type of receptacle or usage.

LIGHTING OUTLETS — Ceiling / Wall

Surface Fixture

Surface Fixt. w/Pull Chain

Recessed Fixture

Surface or Pendant Fluorescent Fixture

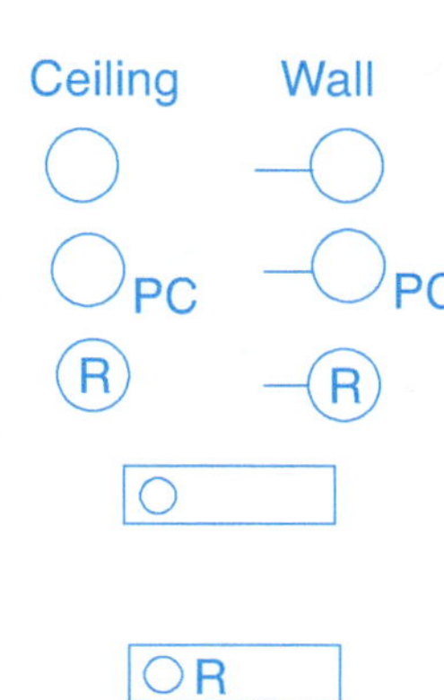

Recessed Fluor. Fixture

Surface or Pendant Continuous Row Fluor. Fixtures

Recessed Continuous Row Fluorescent Fixtures

Surface Exit Light

Recessed Exit Light

Blanked Outlet

Junction Box

CIRCUITING

Wiring Concealed in Ceiling or Wall

Wiring Concealed in Floor

Wiring Exposed

Branch Circuit Homerun to Panelboard. Number of arrows indicates number of circuits in run. Note: Any circuit without further identification is 2-wire. A greater number of wires is indicated by cross lines as shown below. Wire size is sometimes shown with numerals placed above or below cross lines.

3-Wire

4-Wire

26110-14_F16.EPS

Figure 16 ANSI electrical symbols.

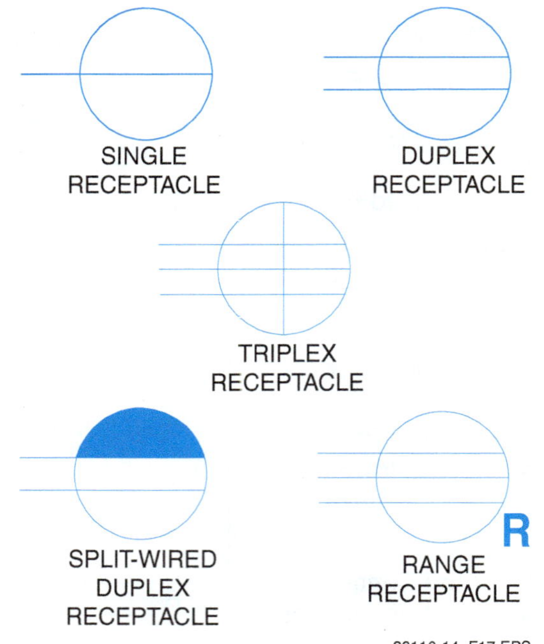

Figure 17 Various receptacle symbols used on electrical drawings.

are drawn to an appropriate scale and accurately located with respect to the building. This reduces the number of references made to the architectural drawings. Where extreme accuracy is required in locating outlets and equipment, exact dimensions are given on larger-scale drawings and shown on the plans.

Each different category in an electrical system is usually represented by a basic distinguishing symbol. To further identify items of equipment or outlets in the category, a numeral or other identifying mark is placed within the open basic symbol. In addition, all such individual symbols used on the drawings should be included in the symbol list or legend. The electrical symbols shown in *Figure 19* were modified by a consulting engineering firm for use on a small industrial electrical installation. The symbols shown in *Figure 20* are those recommended by the Consulting Engineers Council/US. You should become familiar with these symbols.

5.0.0 Scale Drawings

In most electrical drawings, the components are so large that it would be impossible to draw them actual size. Consequently, drawings are made to some reduced scale; that is, all the distances are drawn smaller than the actual dimensions of the object itself, with all dimensions being reduced in the same proportion. For example, if a floor plan of a building is to be drawn to a scale of ¼" = 1'–0", each ¼" on the drawing would equal 1 foot on the building itself; if the scale is ⅛" = 1'–0", each ⅛" on the drawing equals 1 foot on the building, and so forth.

When architectural and engineering drawings are produced, the selected scale is very important. Where dimensions must be held to extreme accuracy, the scale drawings should be made as large as practical with dimension lines added. Where dimensions require only reasonable accuracy, the object may be drawn to a smaller scale (with dimension lines possibly omitted).

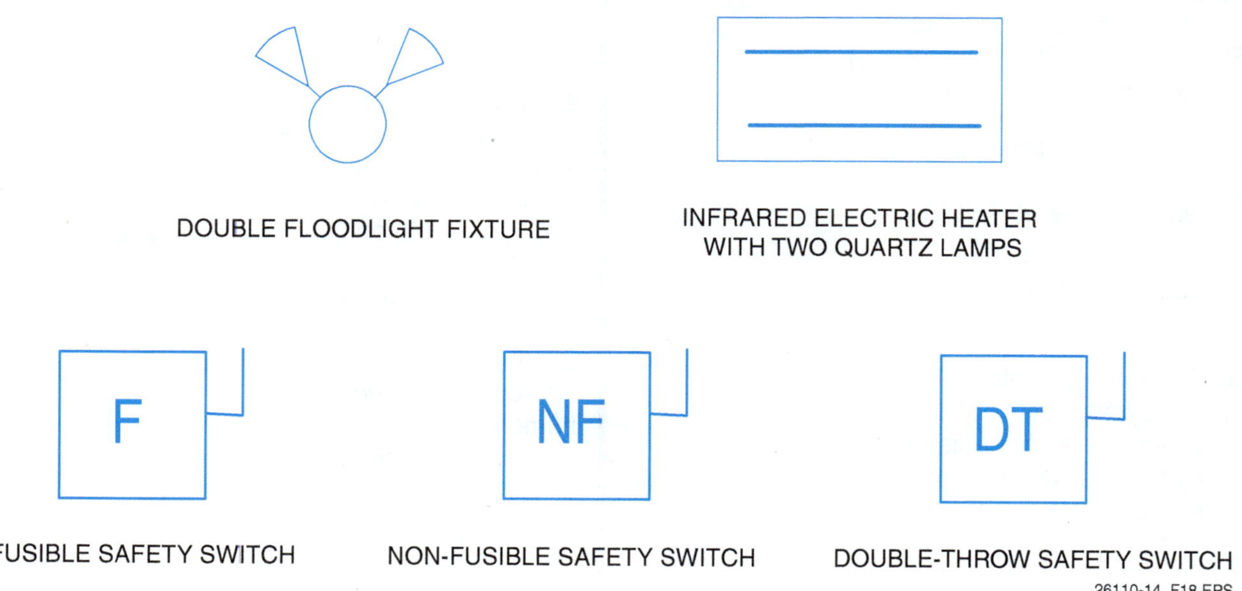

Figure 18 General types of symbols used on electrical drawings.

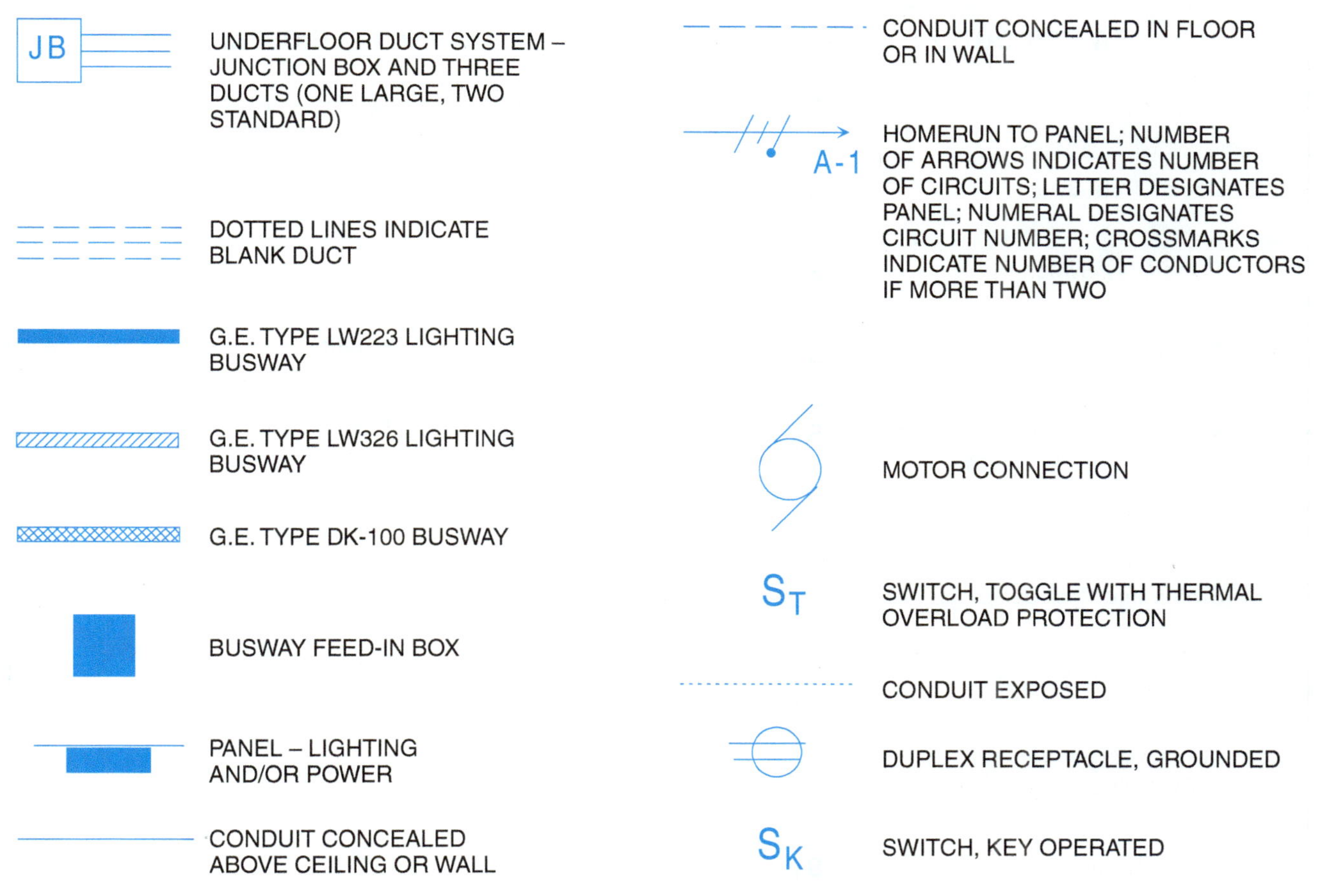

Figure 19 Electrical symbols used by one consulting engineering firm.

In dimensioning drawings, the dimensions written on the drawing are the actual dimensions of the building, not the distances that are measured on the drawing. To further illustrate this point, look at the floor plan in *Figure 21*; it is drawn to a scale of ½" = 1'–0". One of the walls is drawn to an actual length of 3½" on the drawing paper, but since the scale is ½" = 1'–0" and since 3½" contains 7 halves of an inch (7 × ½ = 3½"), the dimension shown on the drawing will therefore be 7'–0" on the actual building.

As shown in the previous example, the most common method of reducing all the dimensions (in feet and inches) in the same proportion is to choose a certain distance and let that distance represent one foot. This distance can then be divided into 12 parts, each of which represents an inch. If half inches are required, these twelfths are further subdivided into halves, etc. Now the scale represents the common foot rule with its subdivisions into inches and fractions, except that the scaled foot is smaller than the distance known as a foot and, likewise, its subdivisions are proportionately smaller.

When a measurement is made on the drawing, it is made with the reduced foot rule or scale; when a measurement is made on the building, it is made with the standard foot rule. The most common reduced foot rules or scales used in electrical drawings are the architect's scale and the engineer's scale. Drawings may sometimes be encountered that use a metric scale, but using this scale is similar to using the architect's or engineer's scales.

5.1.0 Architect's Scale

Figure 22 shows two configurations of architect's scales. The one on the top is designed so that 1" = 1'–0", and the one on the bottom has graduations spaced to represent ⅛" = 1'–0".

Note that on the one-inch scale in *Figure 23*, the longer marks to the right of the zero (with a numeral beneath) represent feet. Therefore, the distance between the zero and the numeral 1 equals one foot. The shorter mark between the zero and 1 represents ½ of a foot, or six inches.

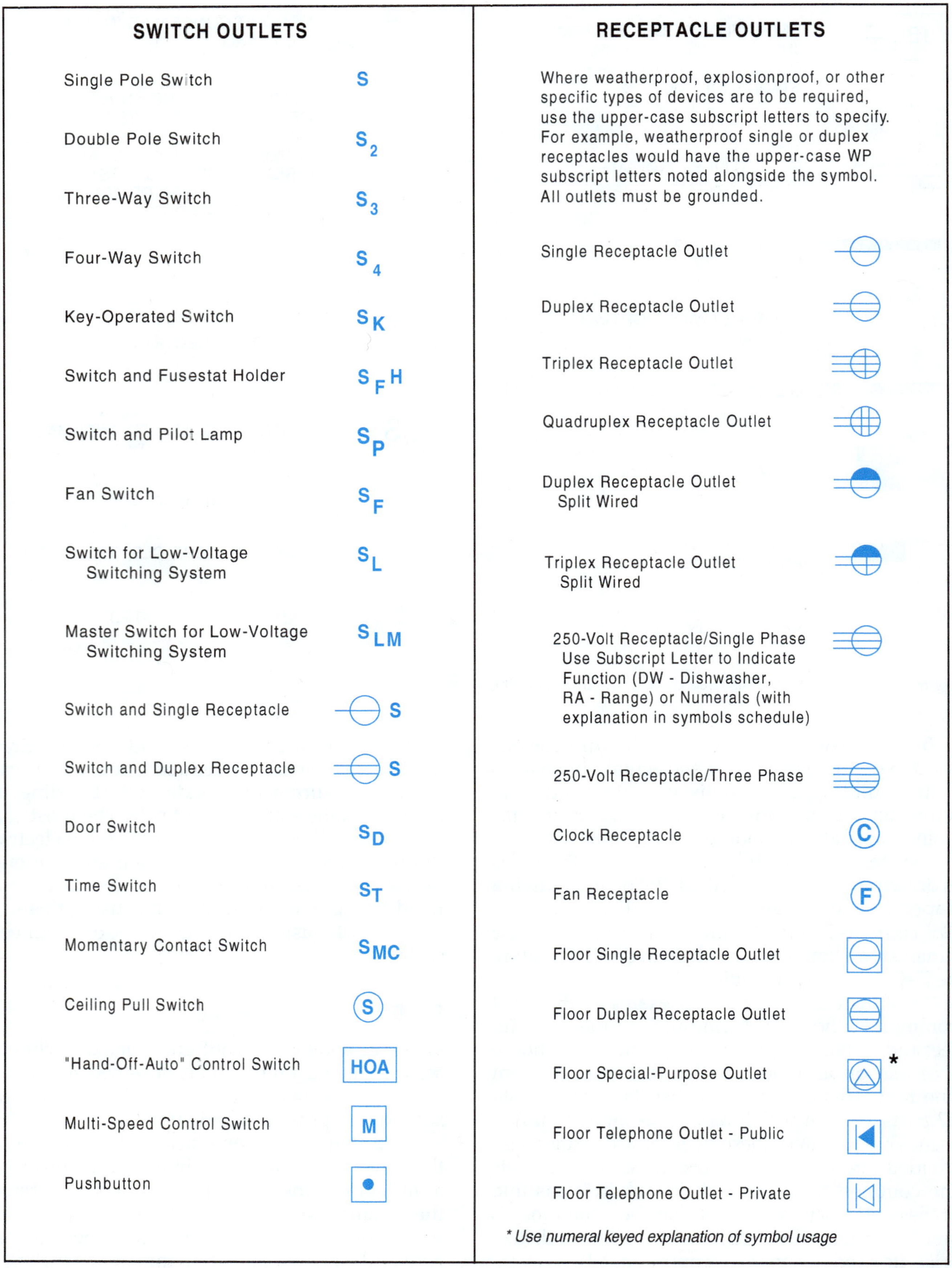

SWITCH OUTLETS	
Single Pole Switch	S
Double Pole Switch	S_2
Three-Way Switch	S_3
Four-Way Switch	S_4
Key-Operated Switch	S_K
Switch and Fusestat Holder	S_{FH}
Switch and Pilot Lamp	S_P
Fan Switch	S_F
Switch for Low-Voltage Switching System	S_L
Master Switch for Low-Voltage Switching System	S_{LM}
Switch and Single Receptacle	S
Switch and Duplex Receptacle	S
Door Switch	S_D
Time Switch	S_T
Momentary Contact Switch	S_{MC}
Ceiling Pull Switch	S
"Hand-Off-Auto" Control Switch	HOA
Multi-Speed Control Switch	M
Pushbutton	

RECEPTACLE OUTLETS

Where weatherproof, explosionproof, or other specific types of devices are to be required, use the upper-case subscript letters to specify. For example, weatherproof single or duplex receptacles would have the upper-case WP subscript letters noted alongside the symbol. All outlets must be grounded.

RECEPTACLE OUTLETS	
Single Receptacle Outlet	
Duplex Receptacle Outlet	
Triplex Receptacle Outlet	
Quadruplex Receptacle Outlet	
Duplex Receptacle Outlet Split Wired	
Triplex Receptacle Outlet Split Wired	
250-Volt Receptacle/Single Phase Use Subscript Letter to Indicate Function (DW - Dishwasher, RA - Range) or Numerals (with explanation in symbols schedule)	
250-Volt Receptacle/Three Phase	
Clock Receptacle	C
Fan Receptacle	F
Floor Single Receptacle Outlet	
Floor Duplex Receptacle Outlet	
Floor Special-Purpose Outlet	*
Floor Telephone Outlet - Public	
Floor Telephone Outlet - Private	

** Use numeral keyed explanation of symbol usage*

26110-14_F20A.EPS

Figure 20 Recommended electrical symbols (1 of 7).

Example of the use of several floor outlet symbols to identify a 2, 3, or more gang outlet:

Underfloor duct and junction box for triple, double, or single duct system as indicated by the number of parallel lines

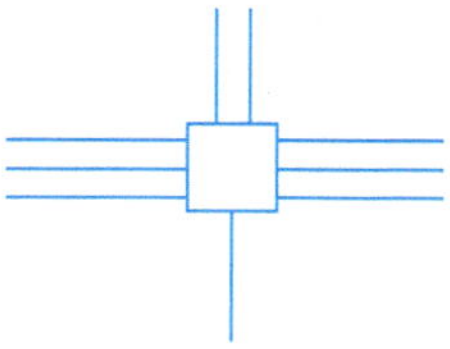

Example of the use of various symbols to identify the location of different types of outlets or connections for underfloor duct or cellular floor systems:

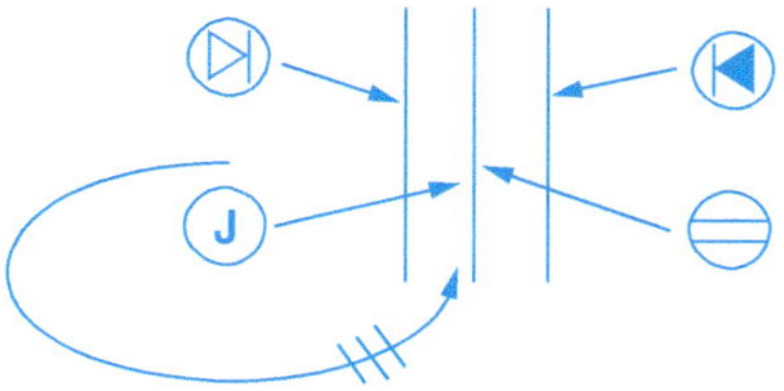

Cellular Floor Heater Duct

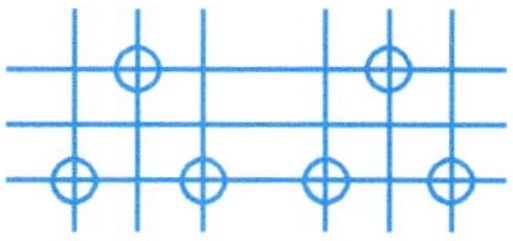

CIRCUITING

Wiring Exposed (not in conduit)

Wiring Concealed in Ceiling or Wall

Wiring Concealed in Floor

Wiring Existing*

Wiring Turned Up

Wiring Turned Down

Branch Circuit Homerun to Panelboard

Number of arrows indicates number of circuits. (A number at each arrow may be used to identify the circuit number.)**

BUS DUCTS AND WIREWAYS

Trolley Duct***

Busway (Service, Feeder or Plug-in)***

Cable Trough Ladder or Channel***

Wireway***

PANELBOARDS, SWITCHBOARDS AND RELATED EQUIPMENT

Flush Mounted Panelboard and Cabinet***

Surface Mounted Panelboard and Cabinet***

Switchboard, Power Control Center, Unit Substation (Should be drawn to scale)***

Flush Mounted Terminal Cabinet (In small scale drawings the TC may be indicated alongside the symbol)***

Surface Mounted Terminal Cabinet (In small scale drawings the TC may be indicated alongside the symbol)***

Pull Box (Identify in relation to Wiring System Section and Size)

Motor or Other Power Controller May be a starter or contactor***

Externally Operated Disconnection Switch***

Combination Controller and Disconnection Means***

*Note: Use heavy-weight line to identify service and feeders. Indicate empty conduit by notation CO.

**Note: Any circuit without further identification indicates two-wire circuit. For a greater number of wires, indicate with cross lines, e.g.:

3 wires  4 wires, etc.

Neutral and ground wires may be shown longer. Unless indicated otherwise, the wire size of the circuit is the minimum size required by the specification. Identify different functions of wiring system (e.g., signaling system) by notation or other means.

***Identify by Notation or Schedule

26110-14_F20B.EPS

Figure 20 Recommended electrical symbols (2 of 7).

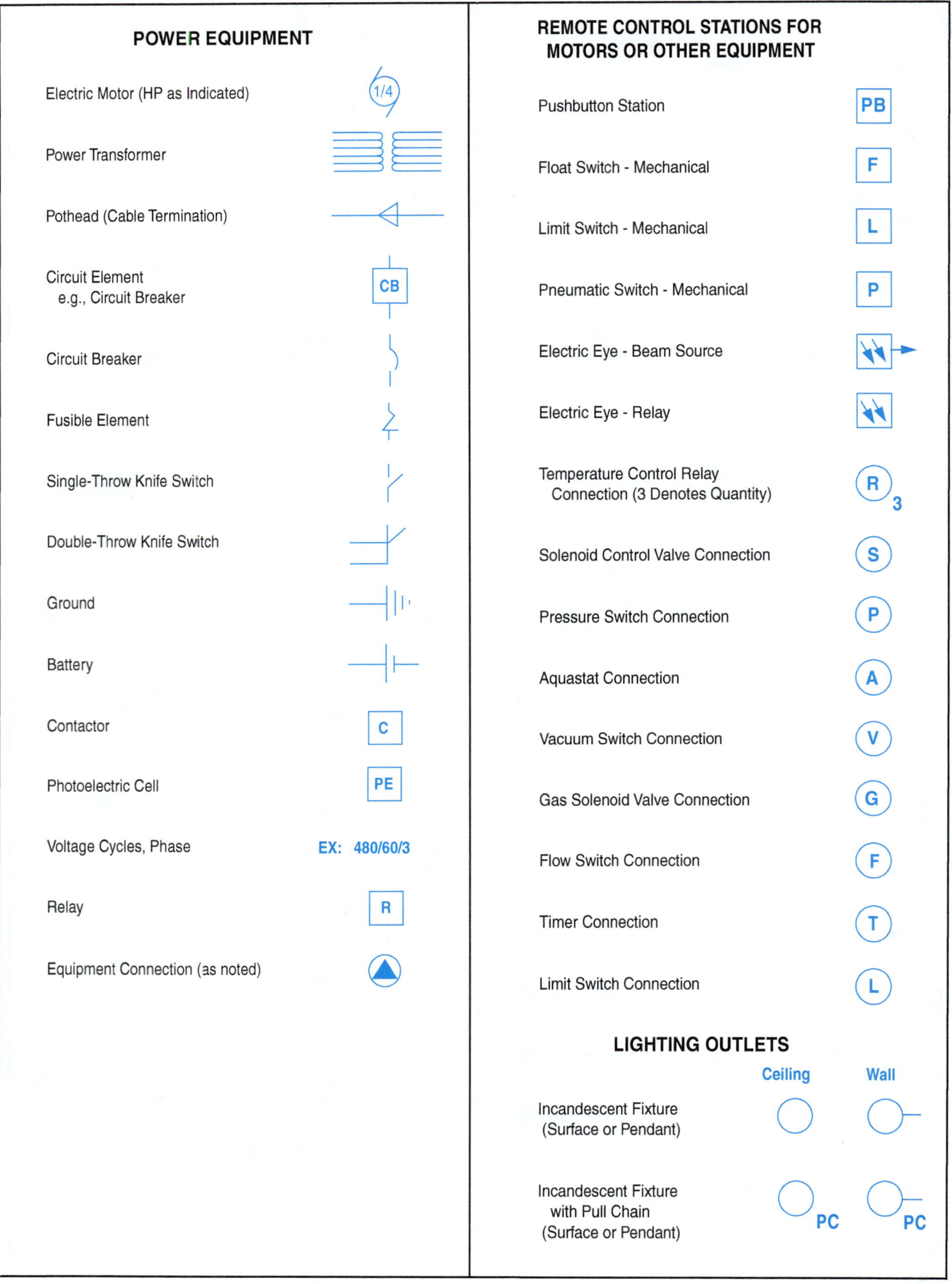

Figure 20 Recommended electrical symbols (3 of 7).

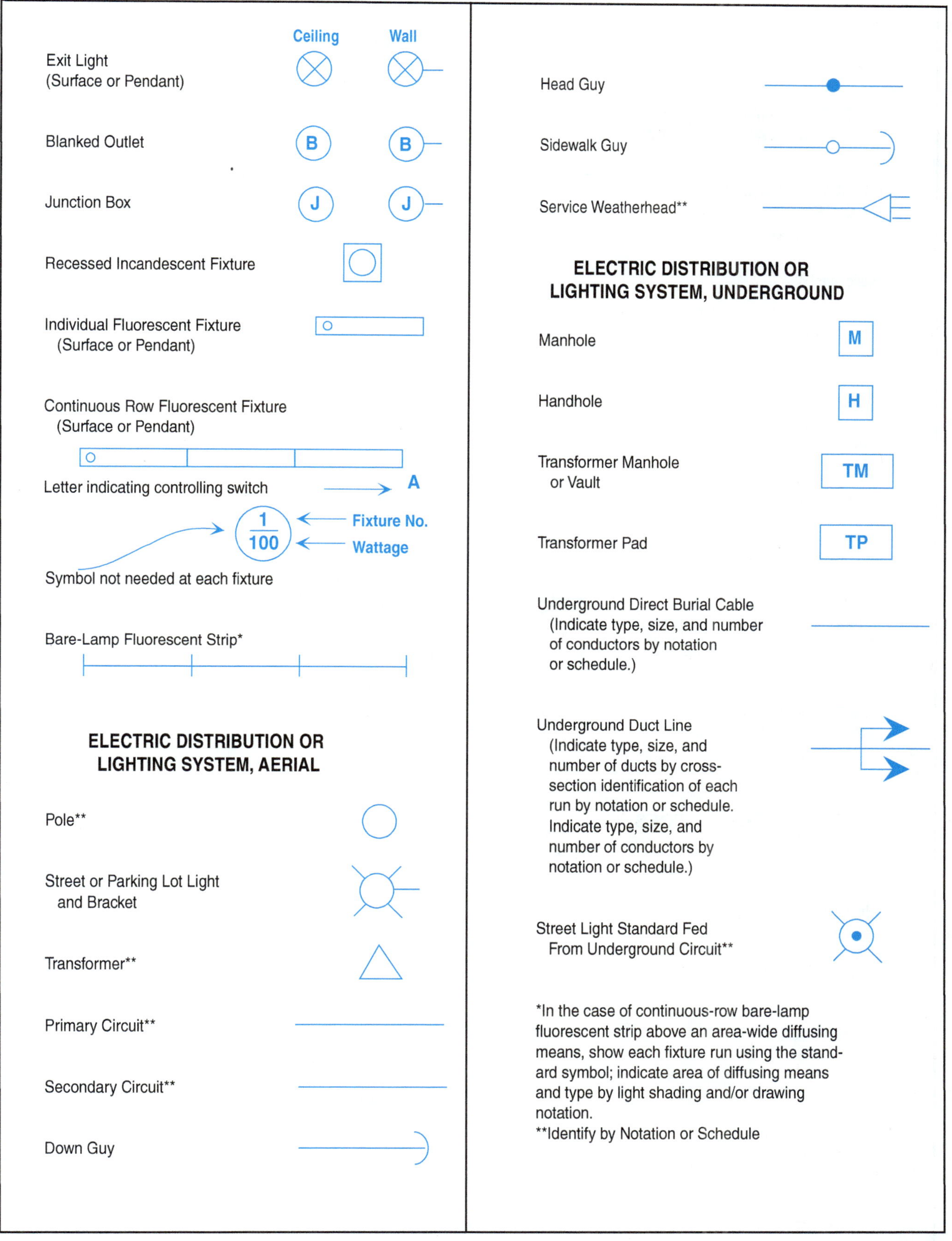

Figure 20 Recommended electrical symbols (4 of 7).

SIGNALING SYSTEM OUTLETS

INSTITUTIONAL, COMMERCIAL, AND INDUSTRIAL OCCUPANCIES

I NURSE CALL SYSTEM DEVICES (Any Type)

Basic Symbol

(Examples of Individual Item Identification Not a Part of Standard)

Nurses' Annunciator (Add a number after it as ⊢① 24 to indicate number of lamps)

Call Station, Single Cord, Pilot Light

Call Station, Double Cord, Microphone Speaker

Corridor Dome Light 1 Lamp

Transformer

Any Other Item On Same System Use Number As Required

II PAGING SYSTEM DEVICES

Basic Symbol

(Examples of Individual Item Identification Not a Part of Standard)

Keyboard

Flush Annunciator

2-Face Annunciator

Any Other Item On Same System Use Numbers As Required

III FIRE ALARM SYSTEM DEVICES (Any Type) Including Smoke and Sprinkler Alarm Devices

Basic Symbol

(Examples of Individual Item Identification. Not a Part of Standard)

Control Panel

Station

10" Gong

Pre-Signal Chime

Any Other Item On Same System Use Numbers As Required

IV STAFF REGISTER SYSTEM DEVICES (Any Type)

Basic Symbol

(Examples of Individual Item Identification. Not a Part of Standard)

Phone Operators' Register

Entrance Register - Flush

Staff Room Register

Transformer

Any Other Item On Same System Use Numbers As Required

V ELECTRIC CLOCK SYSTEM DEVICES (Any Type)

Basic Symbol

(Examples of Individual Item Identification. Not a Part of Standard)

26110-14_F20E.EPS

Figure 20 Recommended electrical symbols (5 of 7).

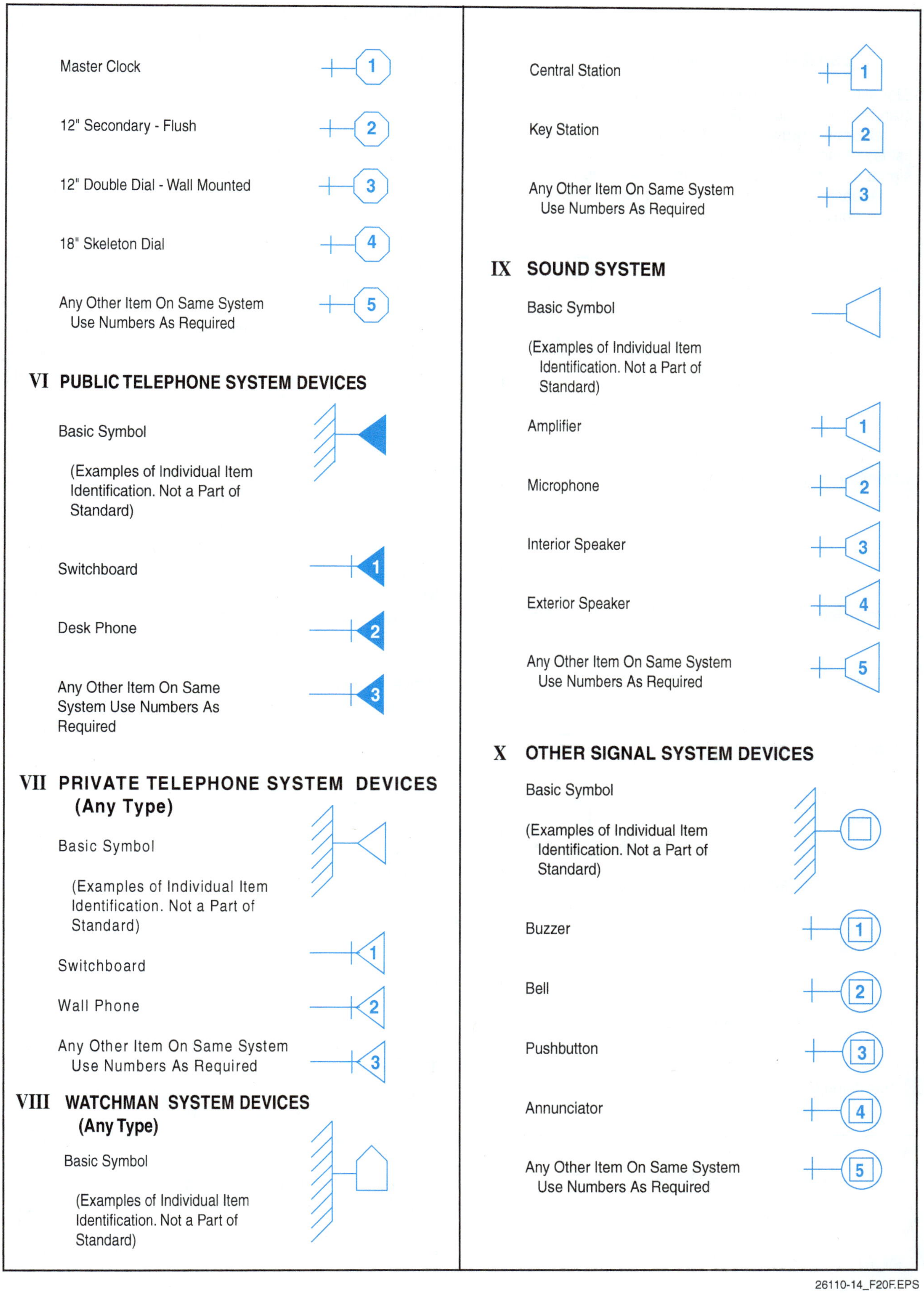

Master Clock	1
12" Secondary - Flush	2
12" Double Dial - Wall Mounted	3
18" Skeleton Dial	4
Any Other Item On Same System Use Numbers As Required	5

VI PUBLIC TELEPHONE SYSTEM DEVICES

Basic Symbol

(Examples of Individual Item Identification. Not a Part of Standard)

Switchboard	1
Desk Phone	2
Any Other Item On Same System Use Numbers As Required	3

VII PRIVATE TELEPHONE SYSTEM DEVICES (Any Type)

Basic Symbol

(Examples of Individual Item Identification. Not a Part of Standard)

Switchboard	1
Wall Phone	2
Any Other Item On Same System Use Numbers As Required	3

VIII WATCHMAN SYSTEM DEVICES (Any Type)

Basic Symbol

(Examples of Individual Item Identification. Not a Part of Standard)

Central Station	1
Key Station	2
Any Other Item On Same System Use Numbers As Required	3

IX SOUND SYSTEM

Basic Symbol

(Examples of Individual Item Identification. Not a Part of Standard)

Amplifier	1
Microphone	2
Interior Speaker	3
Exterior Speaker	4
Any Other Item On Same System Use Numbers As Required	5

X OTHER SIGNAL SYSTEM DEVICES

Basic Symbol

(Examples of Individual Item Identification. Not a Part of Standard)

Buzzer	1
Bell	2
Pushbutton	3
Annunciator	4
Any Other Item On Same System Use Numbers As Required	5

26110-14_F20F.EPS

Figure 20 Recommended electrical symbols (6 of 7).

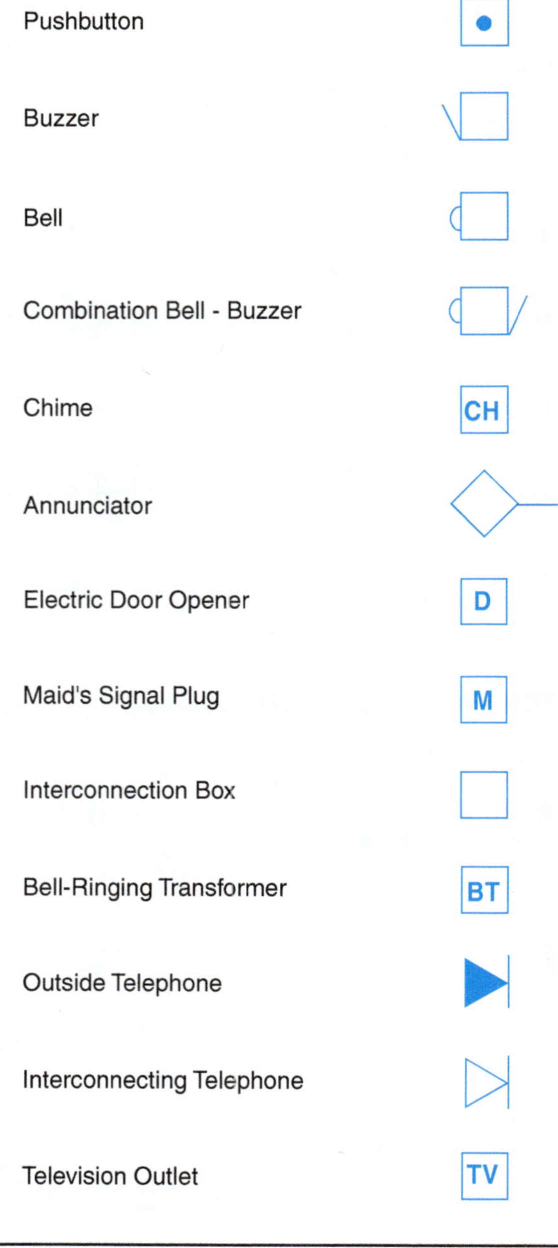

Figure 20 Recommended electrical symbols (7 of 7).

Referring again to *Figure 23*, look at the marks to the left of the zero. The numbered marks are spaced three scaled inches apart and have the numerals 0, 3, 6, and 9 for use as reference points. The other lines of the same length also represent scaled inches, but are not marked with numerals. In use, you can count the number of long marks to the left of the zero to find the number of inches, but after some practice, you will be able to tell the exact measurement at a glance. For example, the measurement A represents five inches because it is the fifth inch mark to the left of the zero; it is also one inch mark short of the six-inch line on the scale.

The lines that are shorter than the inch line are the half-inch lines. On smaller scales, the basic unit is not divided into as many divisions. For example, the smallest subdivision on some scales represents two inches.

5.1.1 Types of Architect's Scales

Architect's scales are available in several types, but the most common include the triangular scale (*Figure 24*) and the flat scale. The quality of architect's scales also varies from cheap plastic scales (costing a dollar or two) to high-quality wooden-laminated tools that are calibrated to precise standards.

The triangular scale is frequently found in drafting and estimating departments or engineering and electrical contracting firms, while the flat scales are more convenient to carry on the job site.

Triangular architect's scales have 12 different scales—two on each edge—as follows:

- Common foot rule (12 inches)
- 1⁄16" = 1'–0"
- 3⁄32" = 1'–0"
- 3⁄16" = 1'–0"
- 1⁄8" = 1'–0"
- 1⁄4" = 1'–0"
- 3⁄8" = 1'–0"
- 3⁄4" = 1'–0"
- 1" = 1'–0"
- 1⁄2" = 1'–0"
- 1 1⁄2" = 1'–0"
- 3" = 1'–0"

Two separate scales on one face may seem confusing at first, but after some experience, reading these scales becomes second nature.

In all but one of the scales on the triangular architect's scale, each face has one of the scales placed opposite to the other. For example, on the one-inch face, the one-inch scale is read from left to right, starting from the zero mark. The half-inch scale is read from right to left, again starting from the zero mark.

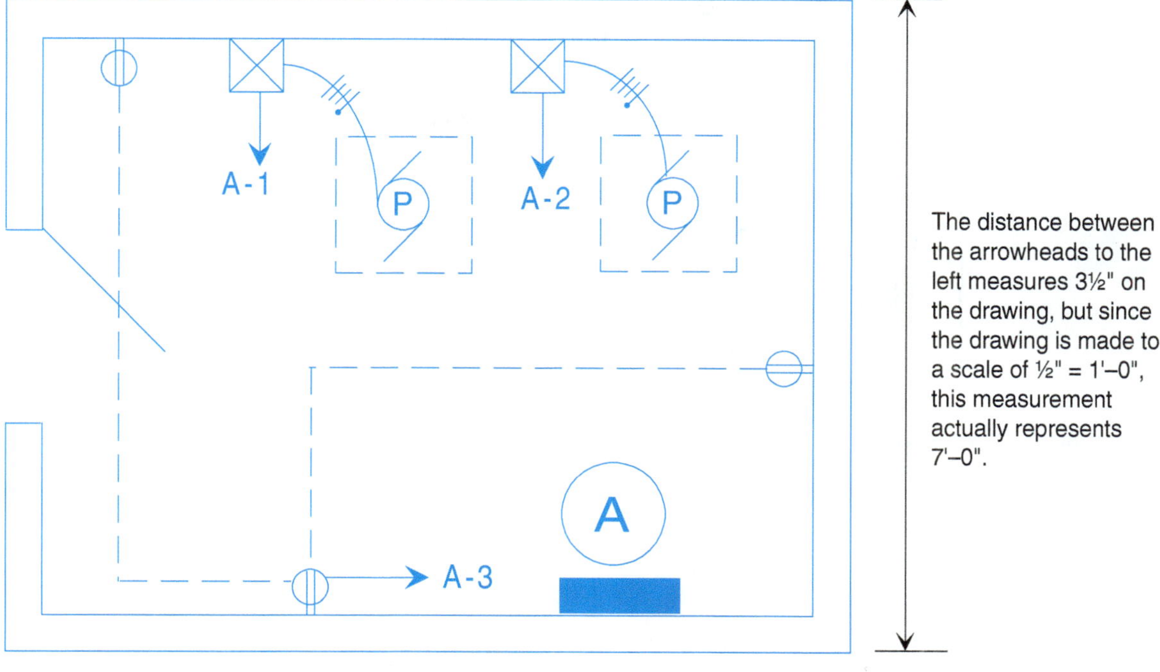

Figure 21 Typical floor plan showing drawing scale.

On the remaining foot-rule scale (1⁄16" = 1'–0") each 1⁄16" mark on the scale represents one foot.

Figure 25 shows all the scales found on the triangular architect's scale.

The flat architect's scale shown in *Figure 26* is ideal for workers on most projects. It is easily and conveniently carried in the shirt pocket, and the four scales (⅛", ¼", ½", and 1") are adequate for the majority of projects that will be encountered.

The partial floor plan shown in *Figure 26* is drawn to a scale of ⅛" = 1'–0". The dimension in question is found by placing the ⅛" architect's scale on the drawing and reading the figures. It can be seen that the dimension reads 24'–6".

Think About It

Using Electrical Symbols

Although there are many electrical symbols, you must be able to read the common ones at a glance. Looking at the simple pump house drawing in *Figure 21,* see how quickly you can explain the symbols and the circuits that they identify.

Every drawing should have the scale to which it is drawn plainly marked on it as part of the drawing title. However, it is not uncommon to have several different drawings on one blueprint sheet—all with different scales. Therefore, always check the scale of each different view found on a drawing sheet.

Figure 22 Two different configurations of architect's scales.

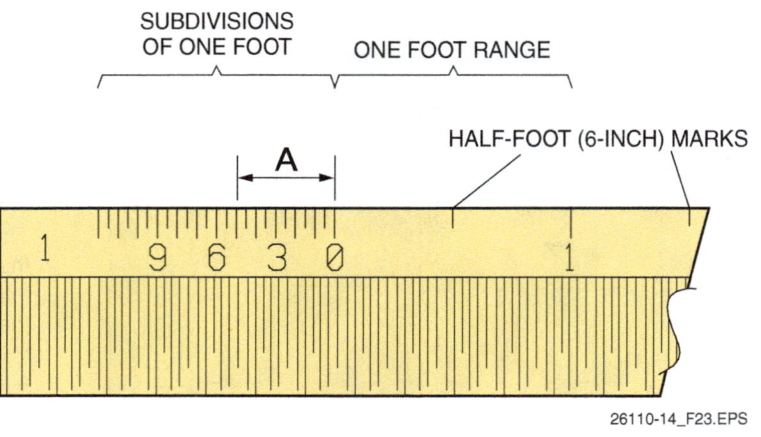

Figure 23 One-inch architect's scale.

On Site

Architect's Scale

Measurements are usually made on architectural drawings using an architect's scale rather than a standard ruler. Architect's scales, like the ones on the left, are divided into feet and inches and usually consist of several scales on one rule. Architect's scales also come in other forms such as tapes or with wheels, like the one shown on the right.

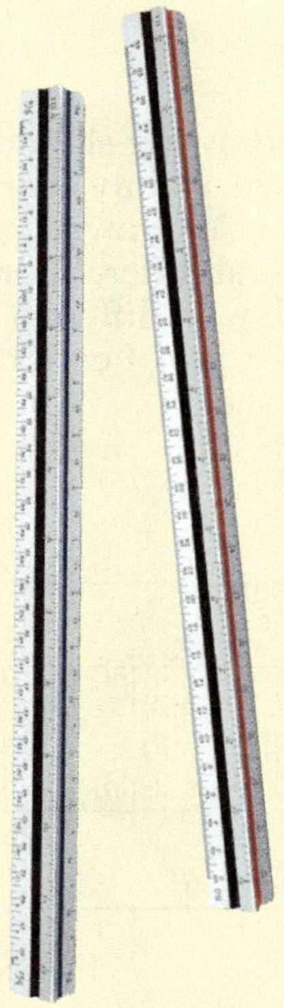

26110-14_SA02.EPS

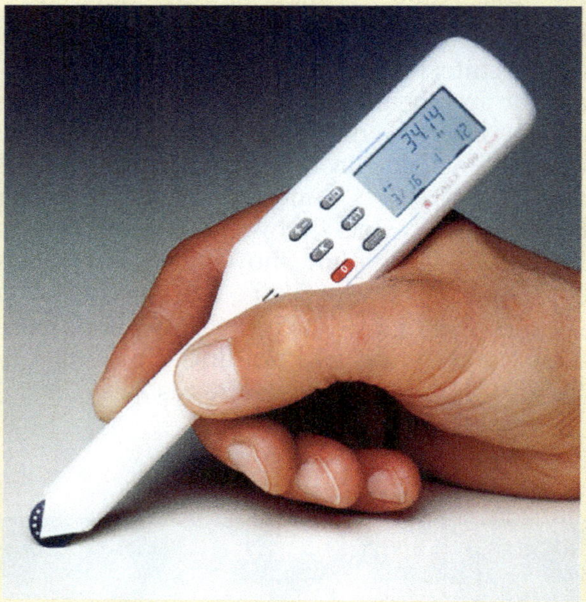

26110-14_SA03.EPS

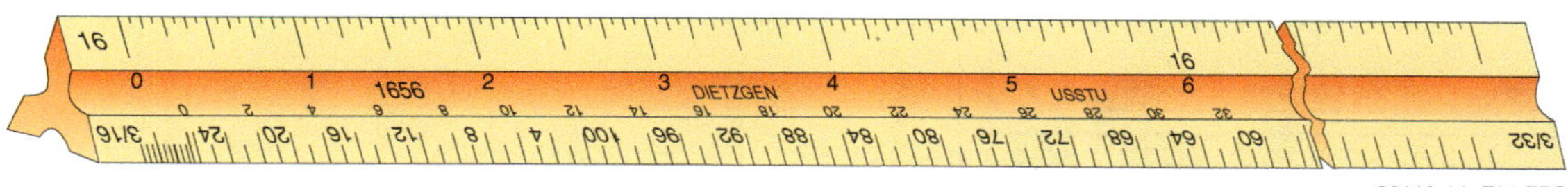

26110-14_F24.EPS

Figure 24 Typical triangular architect's scale.

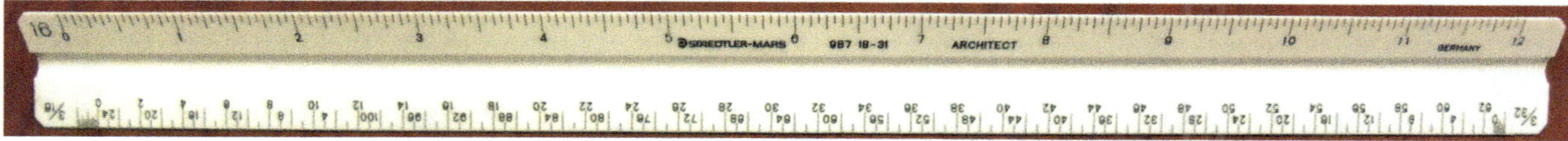

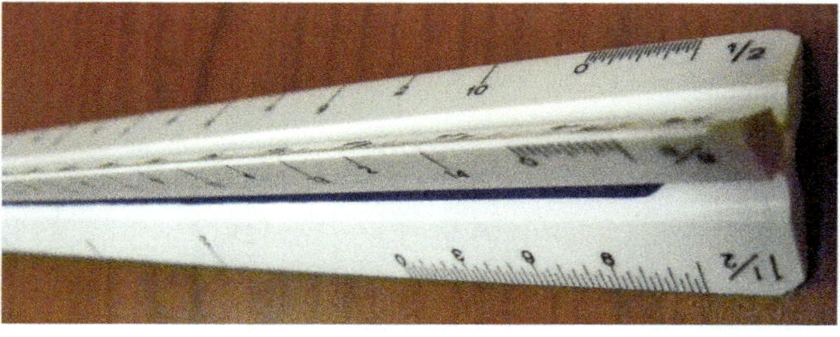

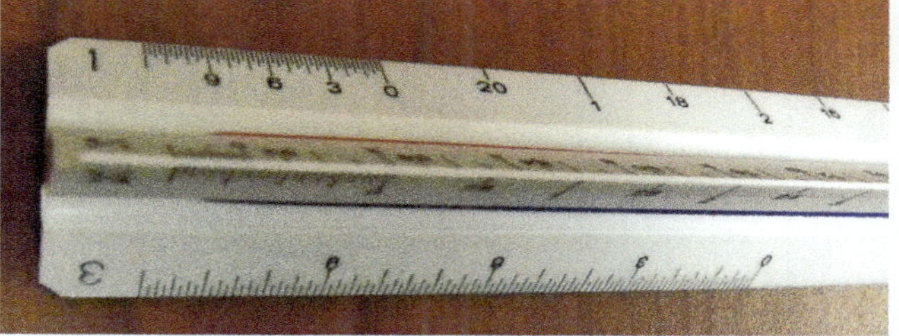

26110-14_F25.EPS

Figure 25 Various scales on a triangular architect's scale.

5.2.0 Engineer's Scale

The civil engineer's scale is used in basically the same manner as the architect's scale, with the principal difference being that the graduations on the engineer's scale are decimal units rather than feet as on the architect's scale.

The engineer's scale is used by placing it on the drawing with the working edge away from the user. The scale is then aligned in the direction of the required measurement. Then, by looking down at the scale, the dimension is read.

Civil engineer's scales commonly show the following graduations:

- 1" = 10 units
- 1" = 20 units
- 1" = 30 units
- 1" = 40 units
- 1" = 60 units
- 1" = 80 units
- 1" = 100 units

The purpose of this scale is to transfer the relative dimensions of an object to the drawing or vice versa. It is used mainly on site plans to determine distances between property lines, manholes, duct runs, direct-burial cable runs, and the like.

Site plans are drawn to scale using the engineer's scale rather than the architect's scale. On small lots, a scale of 1 inch = 10 feet or 1 inch = 20 feet is used. For a 1:10 scale, this means that one inch (the actual measurement on the drawing) is equal to 10 feet on the land itself.

On larger drawings, where a large area must be covered, the scale could be 1 inch = 100 feet or 1 inch = 1,000 feet, or any other integral power of 10. On drawings with the scale in multiples of 10, the engineer's scale marked 10 is used. If the scale is 1 inch = 200 feet, the engineer's scale marked 20 is used, and so on.

Although site plans appear reduced in scale, depending on the size of the object and the size of the drawing sheet to be used, the actual dimensions must be shown on the drawings at all times. When you are reading the drawing plans to scale, think of each dimension in its full size and not in the reduced scale it happens to be on the drawing (*Figure 27*).

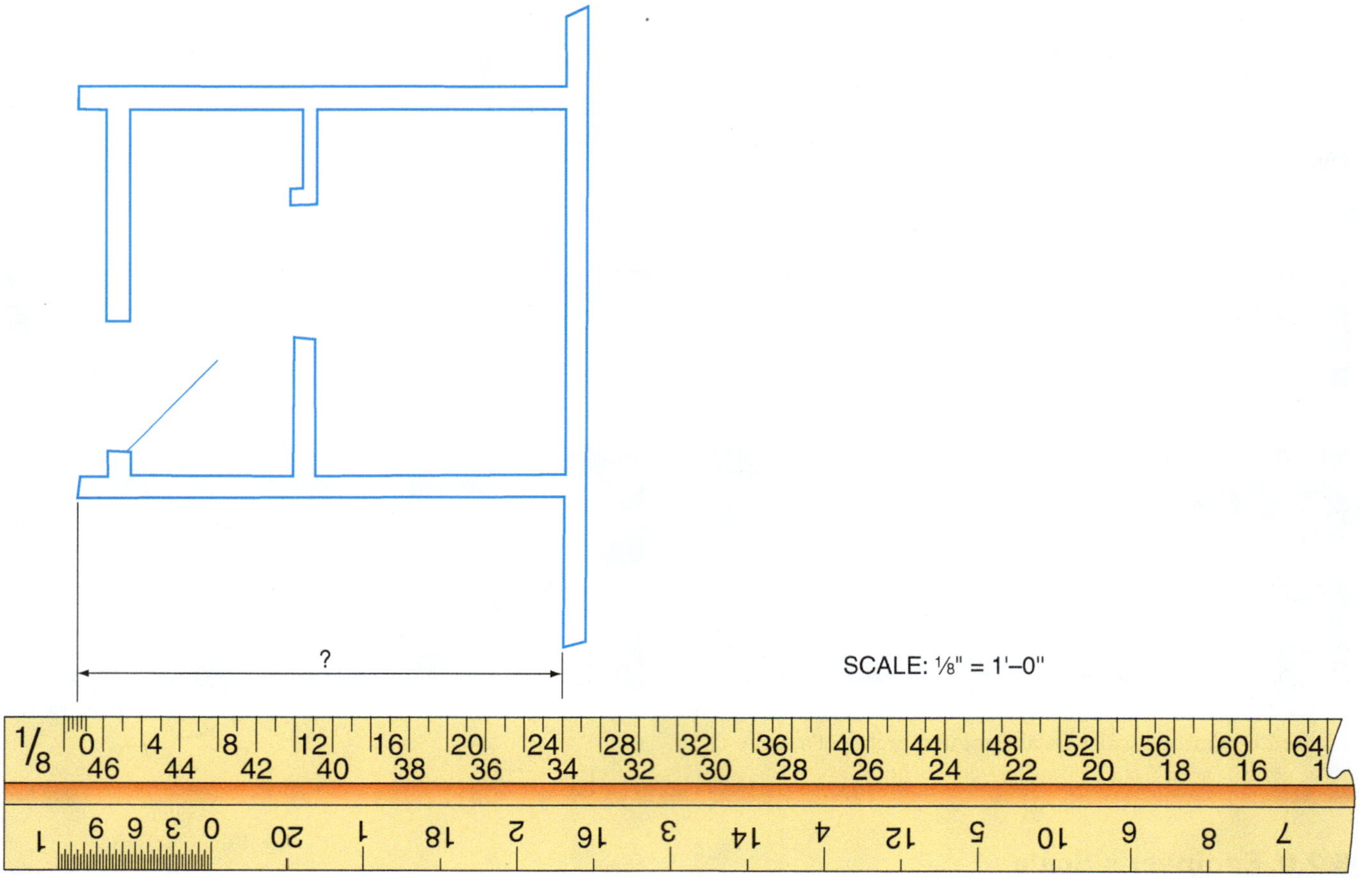

?

SCALE: ⅛" = 1'–0"

26110-14_F26.EPS

Figure 26 Using the ⅛" architect's scale to determine the dimensions on a drawing.

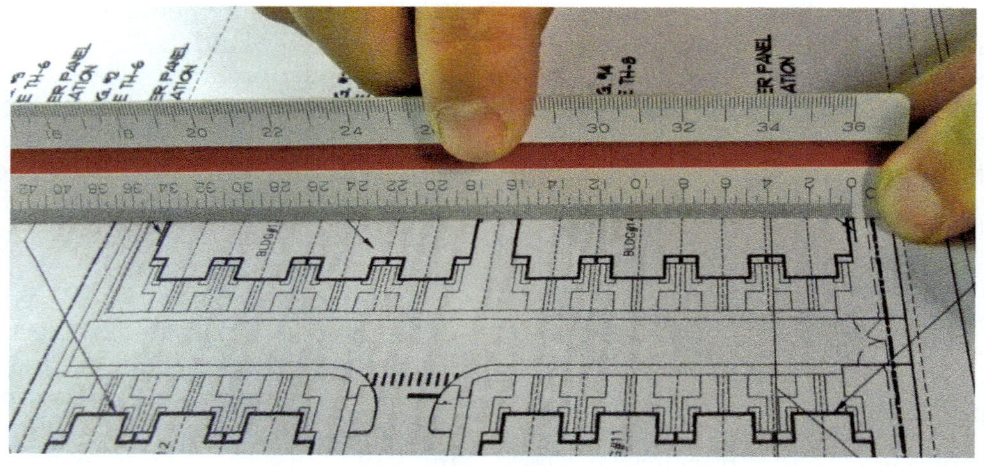

26110-14_F27.EPS

Figure 27 Practical use of the engineer's scale.

5.3.0 Metric Scale

Metric scales are calibrated in units of 10 (*Figure 28*). The two common length measurements used in the metric scale on architectural drawings are the meter and the millimeter, the millimeter being ¹⁄₁,₀₀₀ of a meter. On drawings drawn to scales between 1:1 and 1:100, the millimeter is typically used. On drawings drawn to scales between 1:200 and 1:2,000, the meter is generally used. Many contracting firms that deal in international trade have adopted a dual-dimensioning system expressed in both metric and English symbols. Drawings prepared for government projects may also require metric dimensions. A metric conversion chart is provided in the Appendix.

6.0.0 Analyzing Electrical Drawings

The most practical way to learn how to read electrical construction documents is to analyze an existing set of drawings prepared by consulting or industrial engineers.

Engineers or electrical designers are responsible for the complete layout of electrical systems for most projects. Electrical drafters then transform the engineer's designs into working drawings, using either manual drafting instruments or computer-aided design (CAD) systems. The following is a brief outline of what usually takes place in the preparation of electrical design and working drawings:

- The engineer meets with the architect and owner to discuss the electrical needs of the building or project and to discuss various recommendations made by all parties.
- After that, an outline of the architect's floor plan is laid out.
- The engineer then calculates the required power and lighting outlets for the project; these are later transferred to the working drawings.
- All communications and alarm systems are located on the floor plan, along with lighting and power panelboards.
- Circuit calculations are made to determine wire size and overcurrent protection.
- The main electric service and related components are determined and shown on the drawings.
- Schedules are then placed on the drawings to identify various pieces of equipment.
- Wiring diagrams are made to show the workers how various electrical components are to be connected.
- A legend or electrical symbol list is drafted and shown on the drawings to identify all symbols used to indicate electrical outlets or equipment.
- Various large-scale electrical details are included, if necessary, to show exactly what is required of the electricians.
- Written specifications are then made to give a description of the materials and installation methods.

6.1.0 Development of Site Plans

In general practice, it is usually the owner's responsibility to furnish the architect/engineer with property and topographic surveys, which are made by a certified land surveyor or civil engineer. These surveys show:

- All property lines
- Existing public utilities and their location on or near the property (e.g., electrical lines, sanitary sewer lines, gas lines, water-supply lines, storm sewers, manholes, telephone lines, etc.)

A land surveyor does the property survey from information obtained from a deed description of the property. A property survey shows only the property lines and their lengths, as if the property were perfectly flat.

The topographic survey shows both the property lines and the physical characteristics of the land by using contour lines, notes, and symbols. The physical characteristics may include:

- The direction of the land slope
- Whether the land is flat, hilly, wooded, swampy, high, or low, and other features of its physical nature

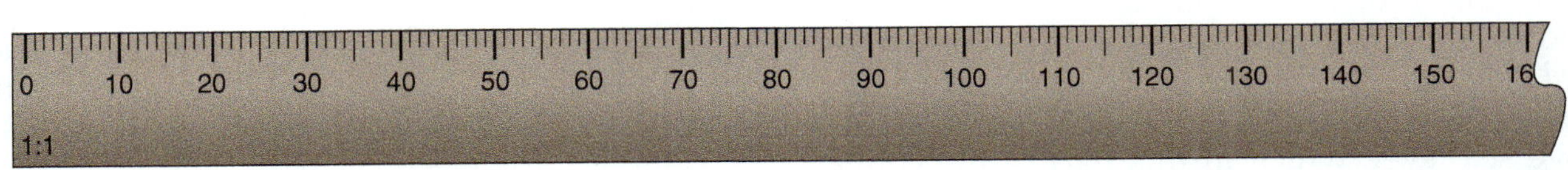

Figure 28 Typical metric scale.

All of this information is necessary so that the architect can properly design a building to fit the property. The electrical engineer also needs this information to locate existing electrical utilities and to route the new service to the building, provide outdoor lighting and circuits, etc.

Electrical site work is sometimes shown on the architect's plot plan. However, when site work involves many trades and several utilities (e.g., gas, telephone, electric, television, water, and sewage), it can become confusing if all details are shown on one drawing sheet. In cases like these, it is best to have a separate drawing devoted entirely to the electrical work, as shown in *Figure 29*. This project is an office/warehouse building for Virginia Electric, Inc. The electrical drawings consist of four 24" × 36" drawing sheets, along with a set of written specifications, which will be discussed later in this module.

The electrical site or plot plan shown in *Figure 29* has the conventional architect's and engineer's title blocks in the lower right-hand corner of the drawing. These blocks identify the project and project owners, the architect, and the engineer. They also show how this drawing sheet relates to the entire set of drawings. Note the engineer's professional stamp of approval to the left of the engineer's title block. Similar blocks appear on all four of the electrical drawing sheets.

When examining a set of electrical drawings for the first time, always look at the area around the title block. This is where most revision blocks or revision notes are placed. If revisions have been made to the drawings, make certain that you have a clear understanding of what has taken place before proceeding with the work.

Refer again to the drawing in *Figure 29* and note the north arrow in the upper left corner. A north arrow shows the direction of true north to help you orient the drawing to the site. Look directly down from the north arrow to the bottom of the page and notice the drawing title, *Plot Utilities*. Directly beneath the drawing title you can see that the drawing scale of 1" = 30' is shown. This means that each inch on the drawing represents 30 feet on the actual job site. This scale holds true for all drawings on the page unless otherwise noted.

An outline of the proposed building is indicated on the drawing along with a callout, *Proposed Bldg. Fin. Flr. Elev. 590.0*. This means that the finished floor level of the building is to be 590 feet above sea level, which in this part of the country will be about two feet above finished grade around the building. This information helps the electrician locate conduit sleeves and stub-ups to the correct height before the finished concrete floor is poured.

On Site

Reading Notes

The notes are crucial elements of the drawing set. Receptacles, for example, are hard to position precisely based on a scaled drawing alone, and yet the designer may call for exact locations. For example, the designer may want receptacles exactly 6" above the kitchen counter backsplash and centered on the sink.

The shaded area represents asphalt paving for the access road, drives, and parking lot. Note that the access road leads into a highway, which is designated Route 35. This information further helps workers to orient the drawing to the building site.

Existing manholes are indicated by a solid circle, while an open circle is used to show the position of the five new pole-mounted lighting fixtures that are to be installed around the new building. Existing power lines are shown with a light solid line with the letter E placed at intervals along the line. The new underground electric service is shown in the same way, except the lines are somewhat wider and darker on the drawing. Note that this new high-voltage cable terminates into a padmount transformer near the proposed building. New telephone lines are similar except the letter T is used to identify the telephone lines.

The direct-burial underground cable supplying the exterior lighting fixtures is indicated with dashed lines on the drawing—shown connecting the open circles. A homerun for this circuit is also shown to a time clock.

The manhole detail shown to the right of the north arrow may seem to serve very little purpose on this drawing since the manholes have already been installed. However, the dimensions and details of their construction will help the electrical contractor or supervisor to better plan the pulling of the high-voltage cable. The same is true of the cross section shown of the duct bank. The electrical contractor knows that three empty ducts are available if it is discovered that one of them is damaged when the work begins.

Although the electrical work will not involve working with gas, the main gas line is shown on the electrical drawing to let the electrical workers know its approximate location while they are installing the direct-burial conductors for the exterior lighting fixtures.

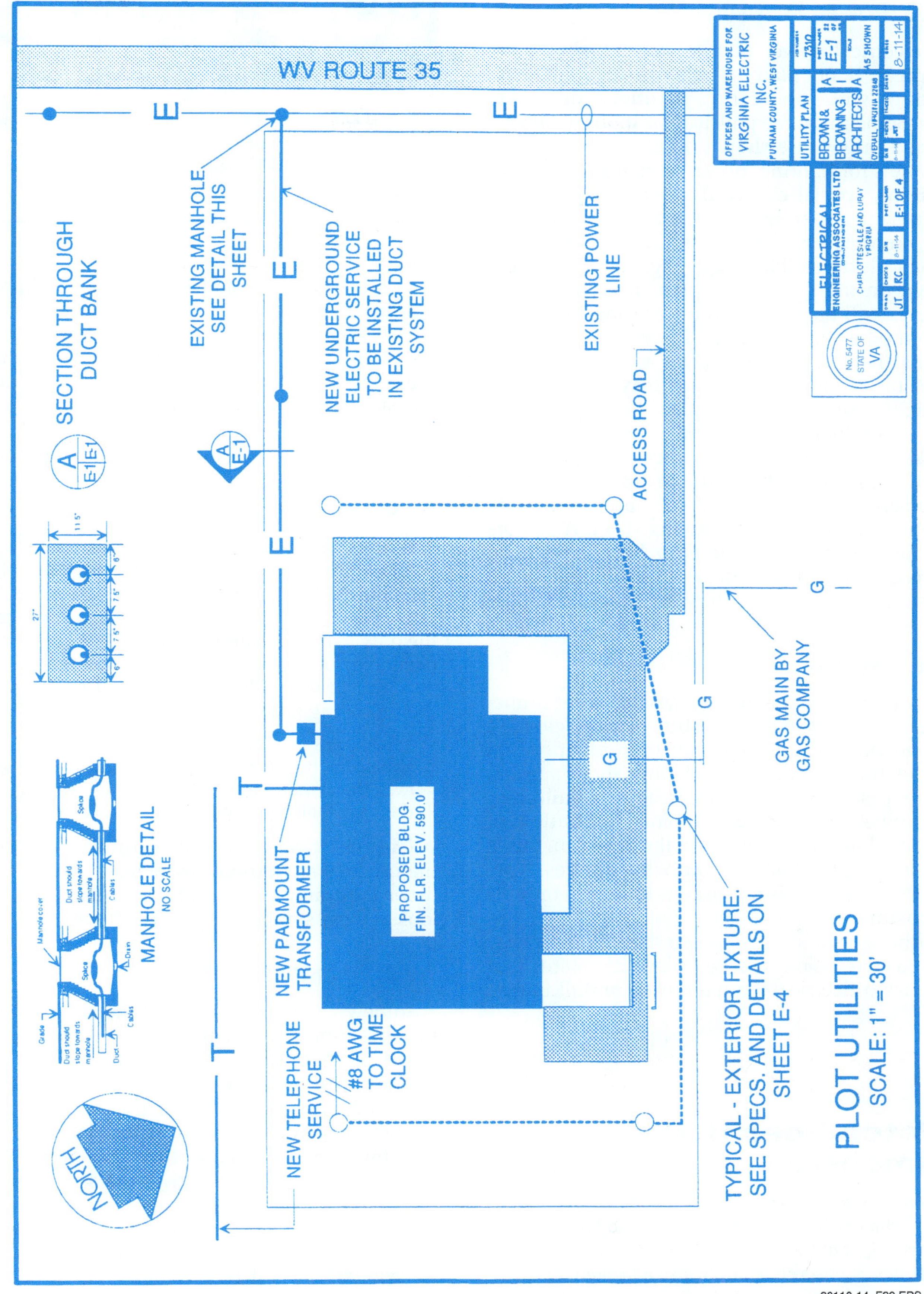

Figure 29 Typical electrical site plan.

7.0.0 Power Plans

The electrical power plan (*Figure 30*) shows the complete floor plan of the office/warehouse building with all interior partitions drawn to scale. Sometimes, the physical locations of all wiring and outlets are shown on one drawing; that is, outlets for lighting, power, signal and communications, special electrical systems, and related equipment are shown on the same plan. However, on complex installations, the drawing would become cluttered if both lighting and power were shown on the same floor plan. Therefore, most projects will have a separate drawing for power and another for lighting. Riser diagrams and details may be shown on yet another drawing sheet, or if room permits, they may be shown on the lighting or power floor plan sheets.

A closer look at this drawing reveals the title blocks in the lower right corner of the drawing sheet. These blocks list both the architectural and engineering firms, along with information to identify the project and drawing sheet. Also note that the floor plan is titled *Floor Plan "B"—Power* and is drawn to a scale of ⅛" = 1'-0". There are no revisions shown on this drawing sheet.

7.1.0 Key Plan

A key plan appears on the drawing sheet immediately above the engineer's title block (*Figure 31*). The purpose of this key plan is to identify that part of the project to which this sheet applies. In this case, the project involves two buildings: Building A and Building B. Since the outline of Building B is cross-hatched in the key plan, this is the building to which this drawing applies. Note that this key plan is not drawn to scale—only its approximate shape.

Although Building A is also shown on this key plan, a note below the key plan title states that there is no electrical work required in Building A.

On Site

Watch Specified Dimensions

When devices are to be located at heights specified above the finished floor (AFF), be sure to find out the actual height of the flooring to be installed. Some materials, such as ceramic tile, can add significantly to the height of the finished floor.

Think About It

Interpreting Site Plans

Study *Figure 29* and explain as many of its features as you can. How much can be understood using common sense? What features require special information?

On some larger installations, the overall project may involve several buildings requiring appropriate key plans on each drawing to help the workers orient the drawings to the appropriate building. In some cases, separate drawing sheets may be used for each room or area in an industrial project—again requiring key plans on each drawing sheet to identify applicable drawings for each room.

7.2.0 Symbol List

A symbol list appears on the electrical power plan (immediately above the architect's title block) to identify the various symbols used for both power and lighting on this project. In most cases, the only symbols listed are those that apply to the particular project. In other cases, however, a standard list of symbols is used for all projects with the following note:

> "These are standard symbols and may not all appear on the project drawings; however, wherever the symbol on the project drawings occurs, the item shall be provided and installed."

Only electrical symbols that are actually used for the office/warehouse drawings are shown in the list on the example electrical power plan. A close-up look at these symbols appears in *Figure 32*.

7.3.0 Floor Plan

A somewhat enlarged view of the electrical floor plan drawing is shown in *Figure 33*. However, due to the size of the drawing in comparison with the size of the pages in this module, it is still difficult to see very much detail. This illustration is meant to show the overall layout of the floor plan and how the symbols and notes are arranged.

In general, this plan shows the service equipment (in plan view), receptacles, underfloor duct system, motor connections, motor controllers, electric heat, busways, and similar details. The electric panels and other service equipment are drawn close to scale. The locations of other electrical outlets and similar components are only

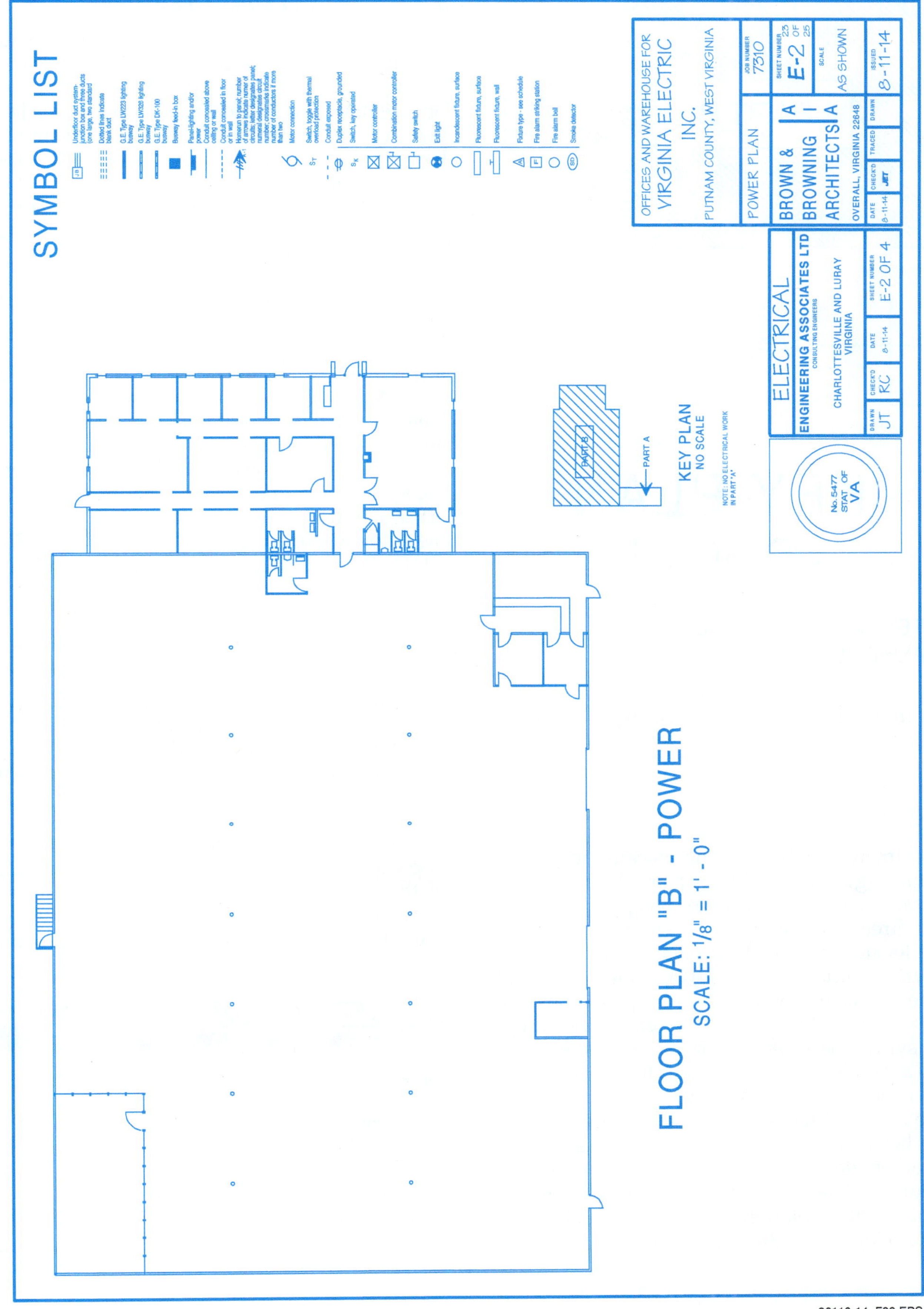

26110-14_F30.EPS

Figure 30 Electrical power plan.

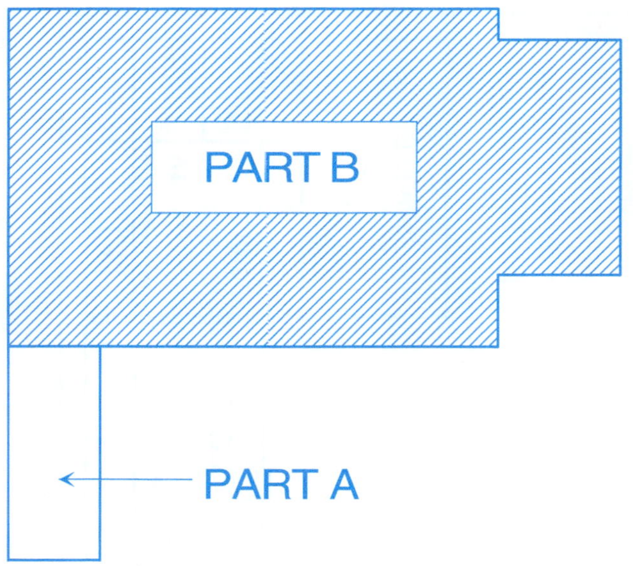

Figure 31 Key plan appearing on electrical power plan.

approximated on the drawings because they have to be exaggerated to show up on the prints. To illustrate, a common duplex receptacle is only about three inches wide. If such a receptacle were to be located on the floor plan of this building (drawn to a scale of ⅛" = 1'–0"), even a small dot on the drawing would be too large to draw the receptacle exactly to scale. Therefore, the receptacle symbol is exaggerated. When such receptacles are scaled on the drawings to determine the proper location, a measurement is usually taken to the center of the symbol to determine the distance between outlets. Junction boxes, switches, and other electrical connections shown on the floor plan will be exaggerated in a similar manner. The partial floor plan drawing in *Figure 34* allows a better view of the drawing details.

7.3.1 Notes and Building Symbols

Referring again to *Figure 33*, you will notice numbers placed inside an oval symbol in each room. These numbered ovals represent the room name or type and correspond to a room schedule in the architectural drawings. For example, room number 112 is designated as the lobby in the room schedule (not shown), room number 113 is designated as office No. 1, etc. On some drawings, these room symbols are omitted and the room names are written out on the drawings.

There are also several notes appearing at various places on the floor plan. These notes offer additional information to clarify certain aspects of the drawing. For example, only one electric heater is to be installed by the electrical contractor; this heater is located in the building's vestibule. Rather than have a symbol in the symbol list for this one heater, a note is used to identify it on the drawing. Other notes on this drawing describe how certain parts of the system are to be installed. For example, in the office area (rooms 112, 113, and 114), you will see the following note: *CONDUIT UP AND STUBBED OUT ABOVE CEILING.* This empty conduit is for telephone/communications cables that will be installed later by the telephone company.

7.3.2 Busways

The office/warehouse project utilizes three types of busways: two types of lighting busways and one power busway. Only the power busway is shown on the floor plan; the lighting busways will appear on the lighting plan.

Figure 33 shows two runs of busways: one running the length of the building on the south end (top wall on drawing), and one running the length of the north wall. The symbol list in *Figure 32* shows this busway to be designated by two parallel lines with a series of X's inside. The symbol

Think About It

Power Plans

Study *Figure 33.* Where does the power enter, and how is it distributed and controlled? What is meant by each of the symbols and lines? Is every electrical connection marked or are some left to the discretion of the electrician?

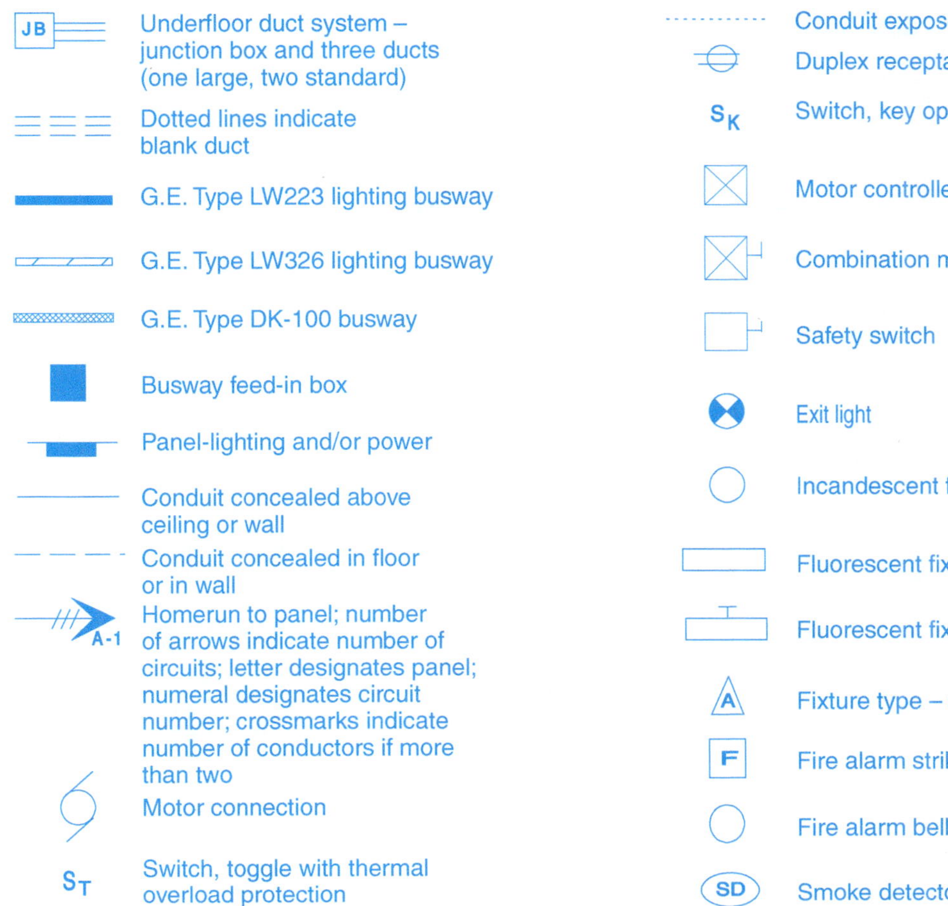

Figure 32 Sample electrical symbols list.

list further describes the busway as General Electric Type DK-100. These busways are fed from the main distribution panel (circuits MDP-1 and MDP-2) through GE No. DHIBBC41 tap boxes.

The *NEC®* defines a busway as a grounded metal enclosure containing factory-mounted, bare or insulated conductors, which are usually copper or aluminum bars, rods, or tubes.

The relationship of the busway and hangers to the building construction should be checked prior to commencing the installation so that any problems due to space conflicts, inadequate or inappropriate supporting structure, openings through walls, etc., are worked out in advance so as not to incur lost time.

For example, the drawings and specifications may call for the busway to be suspended from brackets clamped or welded to steel columns. However, the spacing of the columns may be such that additional supplementary hanger rods suspended from the ceiling or roof structure may be necessary for the adequate support of the busway. To offer more assistance to workers on the office/warehouse project, the engineer may also provide an additional drawing that shows how the busway is to be mounted.

Other details that appear on the floor plan in *Figure 34* include the general arrangement of the underfloor duct system, junction boxes and feeder conduit for the underfloor duct system, and plan views of the service and telephone equipment, along with duplex receptacle outlets. A note on the drawing requires all receptacles in the toilets to be provided with ground fault circuit interrupter (GFCI) protection. The letters EWC next to the receptacle in the vestibule designate this receptacle for use with an electric water cooler.

On Site

Understanding Contact Symbols

When a drawing shows normally open or normally closed contacts, the word *normally* refers to the condition of the contacts in their de-energized or shelf state.

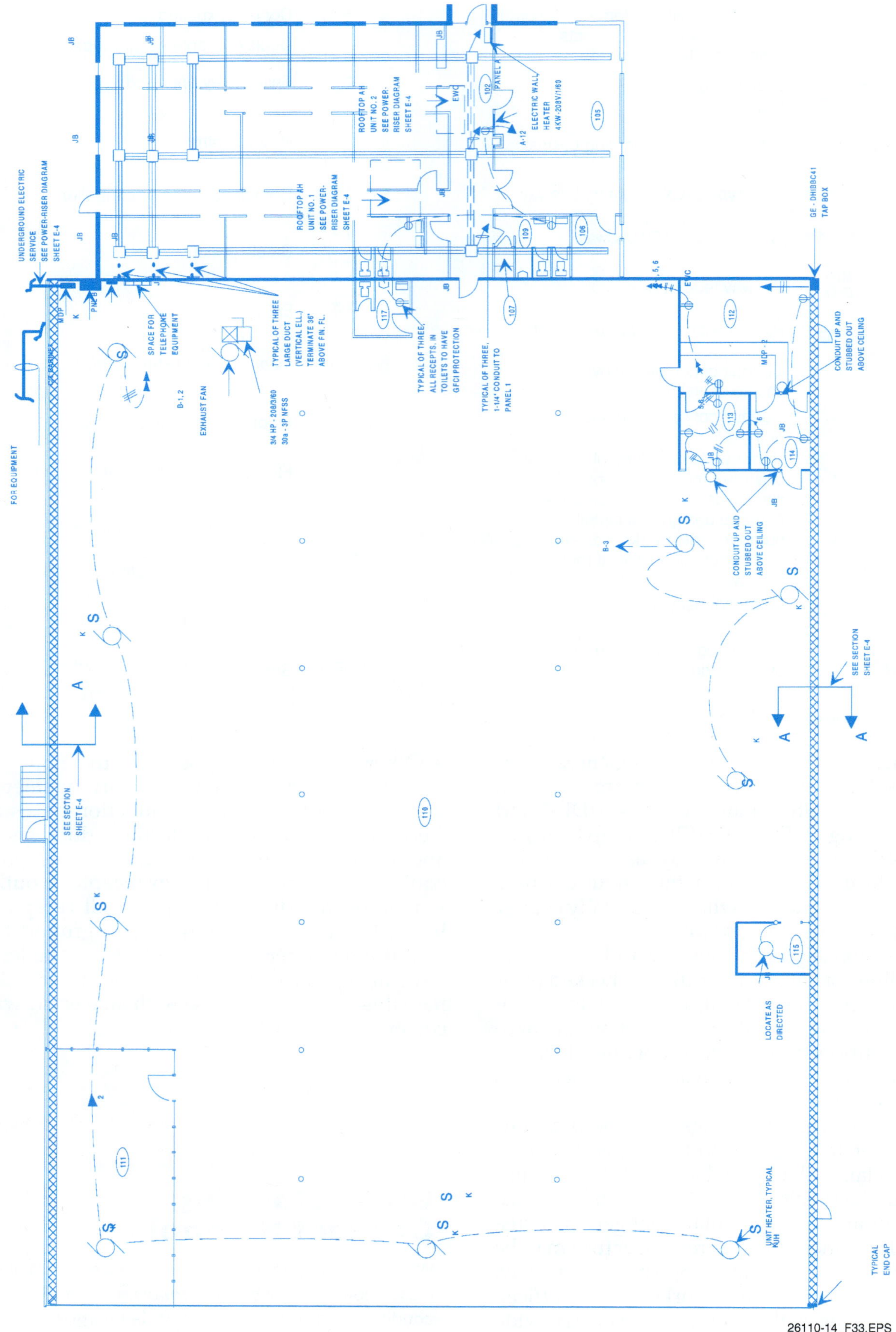

Figure 33 Floor plan for an office/warehouse building.

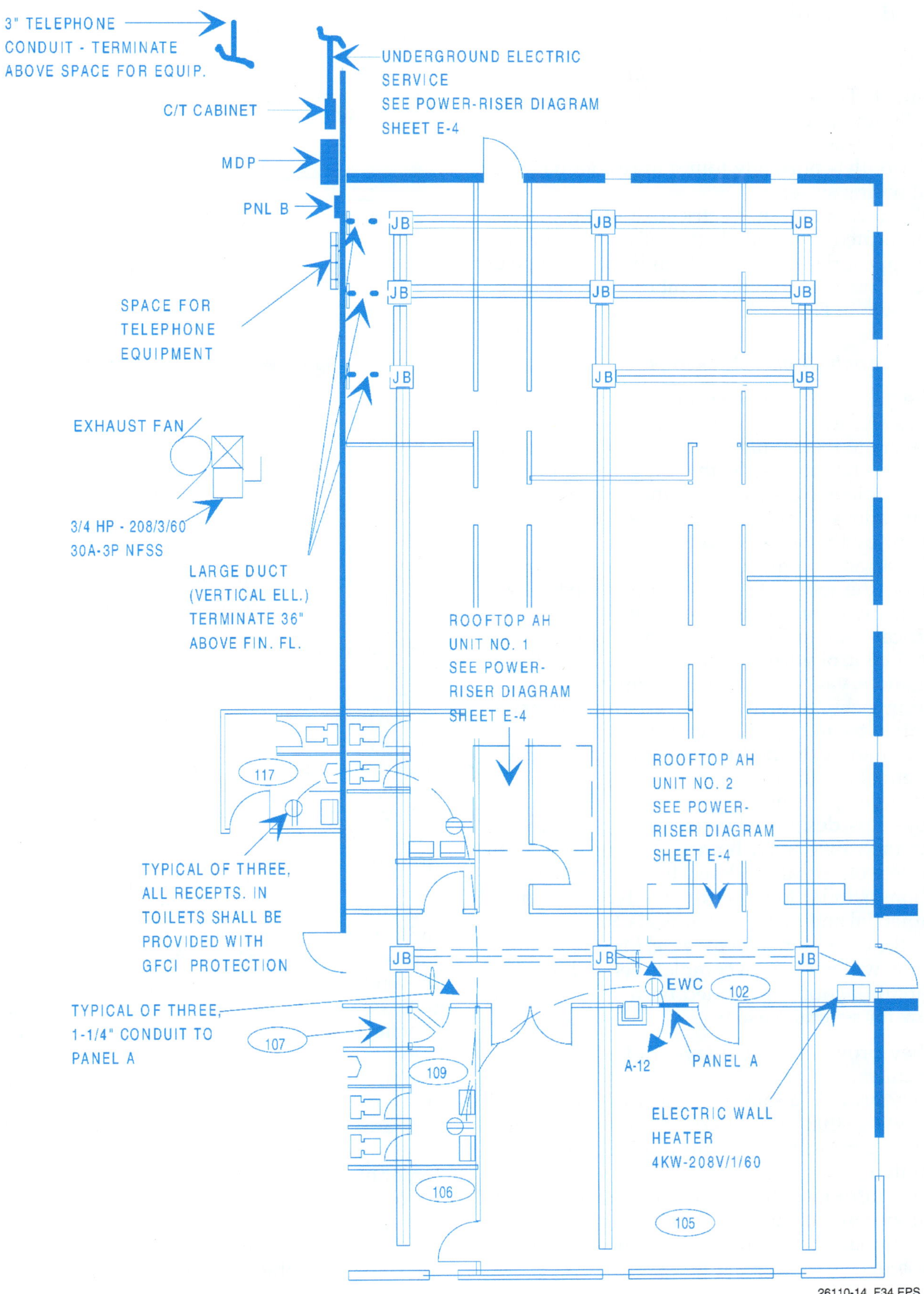

Figure 34 Partial floor plan for office/warehouse building.

7.4.0 Branch Circuit Layout for Power

The point at which electrical equipment is connected to the wiring system is commonly called an outlet. There are many classifications of outlets: lighting, receptacle, motor, appliance, and so forth. This section, however, deals with the power outlets normally found in residential electrical wiring systems.

When viewing an electrical drawing, outlets are indicated by symbols (usually a small circle with appropriate markings to indicate the type of outlet). The most common symbols for receptacles are shown in *Figure 35*.

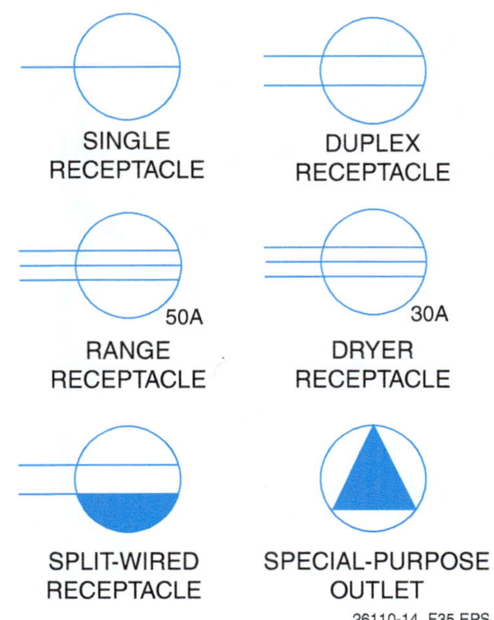

Figure 35 Typical outlet symbols appearing in electrical drawings.

7.4.1 Branch Circuit Drawings

In the past, with the exception of very large residences and tract-development houses, the size of the average residential electrical system was not large enough to justify the expense of preparing complete electrical working drawings and specifications. Such electrical systems were either laid out by the architect in the form of a sketchy outlet arrangement, or laid out by the electrician on the job as the work progressed. However, many technical developments in residential electrical use—such as electric heat with sophisticated control wiring, increased use of electrical appliances, various electronic alarm systems, new lighting techniques, and the need for energy conservation techniques—have greatly expanded the demand and extended the complexity of today's residential electrical systems.

Each year, the number of homes with electrical systems designed by consulting engineering firms increases. Such homes are provided with complete electrical working drawings and specifications, similar to those frequently provided for commercial and industrial projects. Still, these are more the exception than the rule. Most residential projects will not have a complete set of drawings.

Circuit layout is provided on the drawings to follow for several reasons:

- They provide a visual layout of house wiring circuitry.
- They provide a sample of electrical residential drawings that are prepared by consulting engineering firms, although the number may still be limited.
- They introduce the method of showing electrical systems on working drawings to provide a foundation for tackling advanced electrical systems.

Branch circuits are shown on electrical drawings by means of a single line drawn from the panelboard (or by homerun arrowheads indicating that the circuit goes to the panelboard) to the outlet, or from outlet to outlet where there is more than one outlet on the circuit.

The lines indicating branch circuits can be solid to show that the conductors are to be run concealed in the ceiling or wall; dashed to show that the conductors are to be run in the floor or ceiling below; or dotted to show that the wiring is to be run exposed. *Figure 36* shows examples of these three types of branch circuit lines.

In *Figure 36*, No. 12 indicates the wire size. The slash marks shown through the circuits in *Figure 36* indicate the number of current-carrying conductors in the circuit. Although two slash marks are shown for the current-carrying conductors (along with one slash mark for the ground), in actual practice, a branch circuit containing only two conductors usually contains no slash marks; that is, any circuit with no slash marks is assumed to have two conductors. However, three or more conductors are always indicated on electrical working drawings—either by slash marks for each conductor, or else by a note.

Never assume that you know the meaning of any electrical symbol. Although great efforts have been made in recent years to standardize drawing symbols, architects, consulting engineers, and

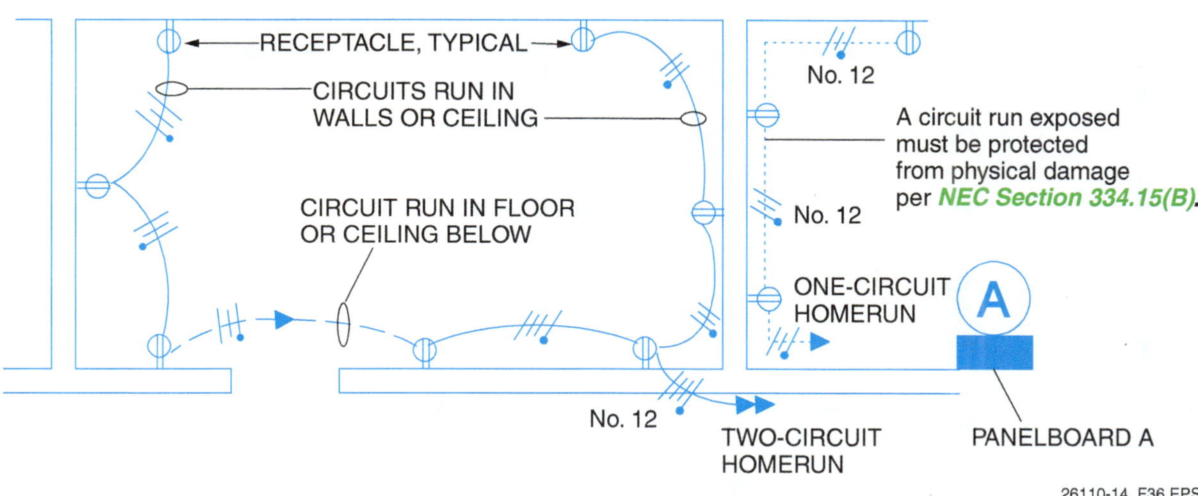

Figure 36 Types of branch circuit lines shown on electrical working drawings.

electrical drafters still modify existing symbols or devise new ones to meet their own needs. Always consult the symbol list or legend on electrical working drawings for an exact interpretation of the symbols used.

7.4.2 Locating Receptacles

NEC Section 210.52(A) states the minimum requirements for the location of receptacles in dwelling units. It specifies that in each kitchen, family room, and dining room, receptacle outlets shall be installed so that no point along the floor line in any wall space is more than 6', measured horizontally, from an outlet in that space, including any wall space 2' or more in width and the wall space occupied by fixed panels in exterior walls, but excluding sliding panels. This means that the outlets will be no more than 12' apart. When spaced in this manner, a 6' extension cord will reach a receptacle at any point along the wall line. Receptacle outlets shall, insofar as practicable, be spaced equal distances apart. Receptacle outlets in floors shall not be counted as part of the required number of receptacle outlets unless located within 18" of the wall.

The *NEC®* defines wall space as a wall that is unbroken along the floor line by doorways, fireplaces, or similar openings. Each wall space that is two feet or more in width must be treated individually and separately from other wall spaces within the room.

The purpose of *NEC Section 210.52(A)* is to minimize the use of cords across doorways, fireplaces, and similar openings.

On Site

Don't Just Check the Electrical Plan

Always review all of the drawings in a drawing set, not just the electrical plan. Several drawings in the set will have information of relevance to the electrician. For example, you should review:

- Site plans for utility lines and elevation information
- Mechanical drawings for routing, clearances, and HVAC equipment and controls
- Architectural drawings for the type of construction (block, wood, metal stud, etc.), fire ratings, and special details.
- Finish drawings (e.g., reflected ceiling plans) for locations of fixtures, fans, and other devices
- Room finish schedules for ceiling heights and floor and wall finishing details
- Plumbing drawings for pumps, water service, and sprinklers

Figure 37 shows the outlets for a sample residence. In laying out these receptacle outlets, the floor line of the wall is measured (also around corners), but not across doorways, fireplaces, passageways, or other spaces where a flexible cord extended across the space would be unsuitable.

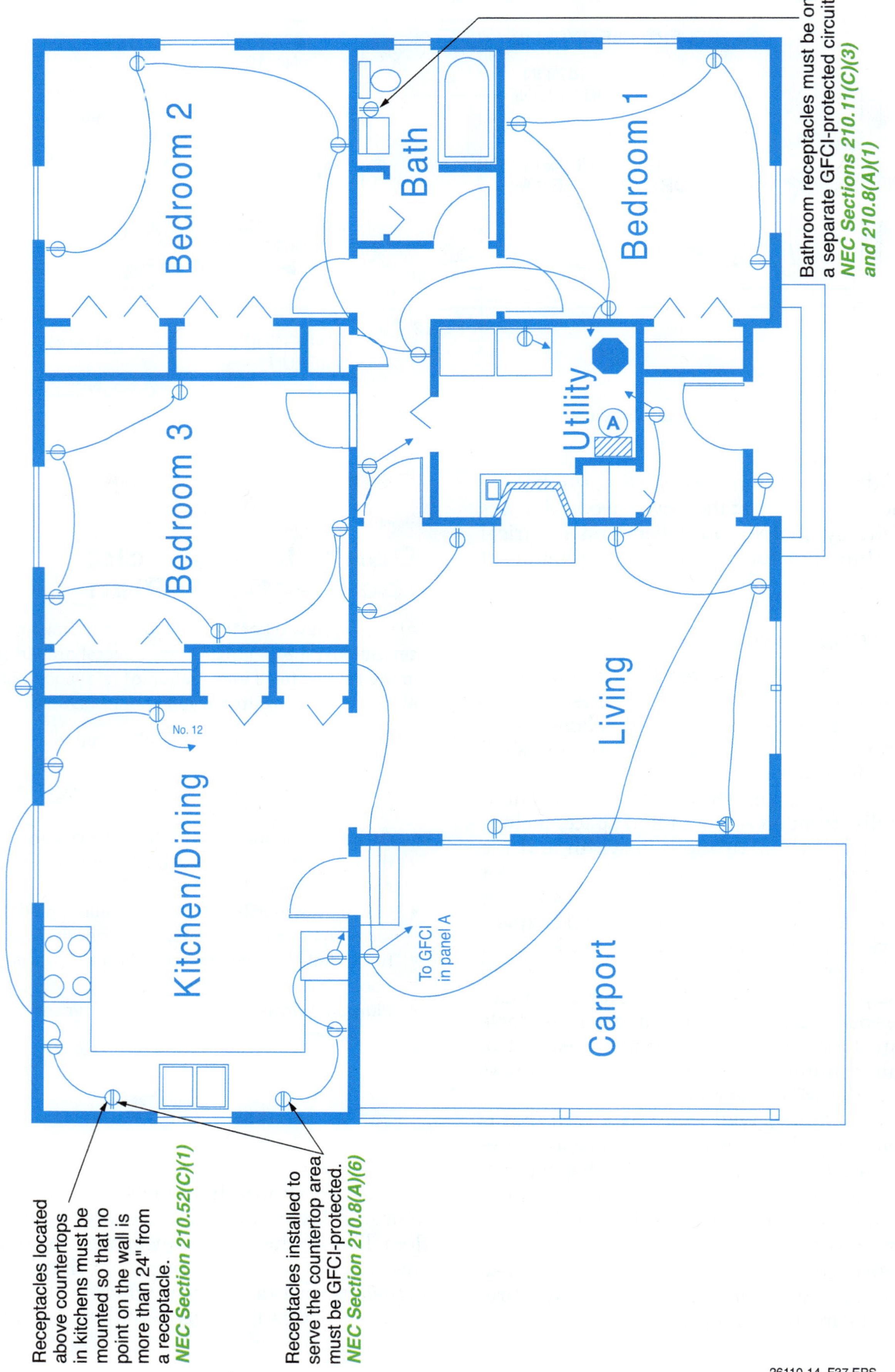

Figure 37 Floor plan of a sample residence.

8.0.0 Lighting Floor Plan

A skeleton view of a lighting floor plan is shown in *Figure 38*. Again, the architect's/engineer's title blocks appear in the lower right corner of the drawing. A key plan appears above the engineer's title block. This plan is drawn to the same scale as the power plan; that is, ⅛" = 1'–0". A lighting fixture (luminaire) schedule appears in the upper right corner of the drawing and some installation notes appear below the schedule.

The lighting outlet symbols found on the drawing for the office/warehouse building represent both incandescent and fluorescent types; a circle on most electrical drawings usually represents an incandescent fixture, and a rectangle represents a fluorescent one. All of these symbols are designed to indicate the physical shape of a particular fixture and are usually drawn to scale.

The type of mounting used for all lighting fixtures is usually indicated in a lighting fixture schedule, which in this case is shown on the drawings. On some projects, the schedule may be found only in the written specifications.

The type of lighting fixture is identified by a numeral placed inside a triangle near each lighting fixture. If one type of fixture is used exclusively in one room or area, the triangular indicator need only appear once with the word ALL lettered at the bottom of the triangle.

8.1.0 Drawing Schedules

A schedule is a systematic method of presenting notes or lists of equipment on a drawing in tabular form. When properly organized and thoroughly understood, schedules are powerful timesaving devices for both those preparing the drawings and workers on the job.

For example, the lighting fixture schedule shown in *Figure 39* lists the fixture and identifies each fixture type on the drawing by number. The manufacturer and catalog number of each type are given along with the number, size, and type of lamp for each.

At times, all of the same information found in schedules will be duplicated in the written specifications, but combing through page after page

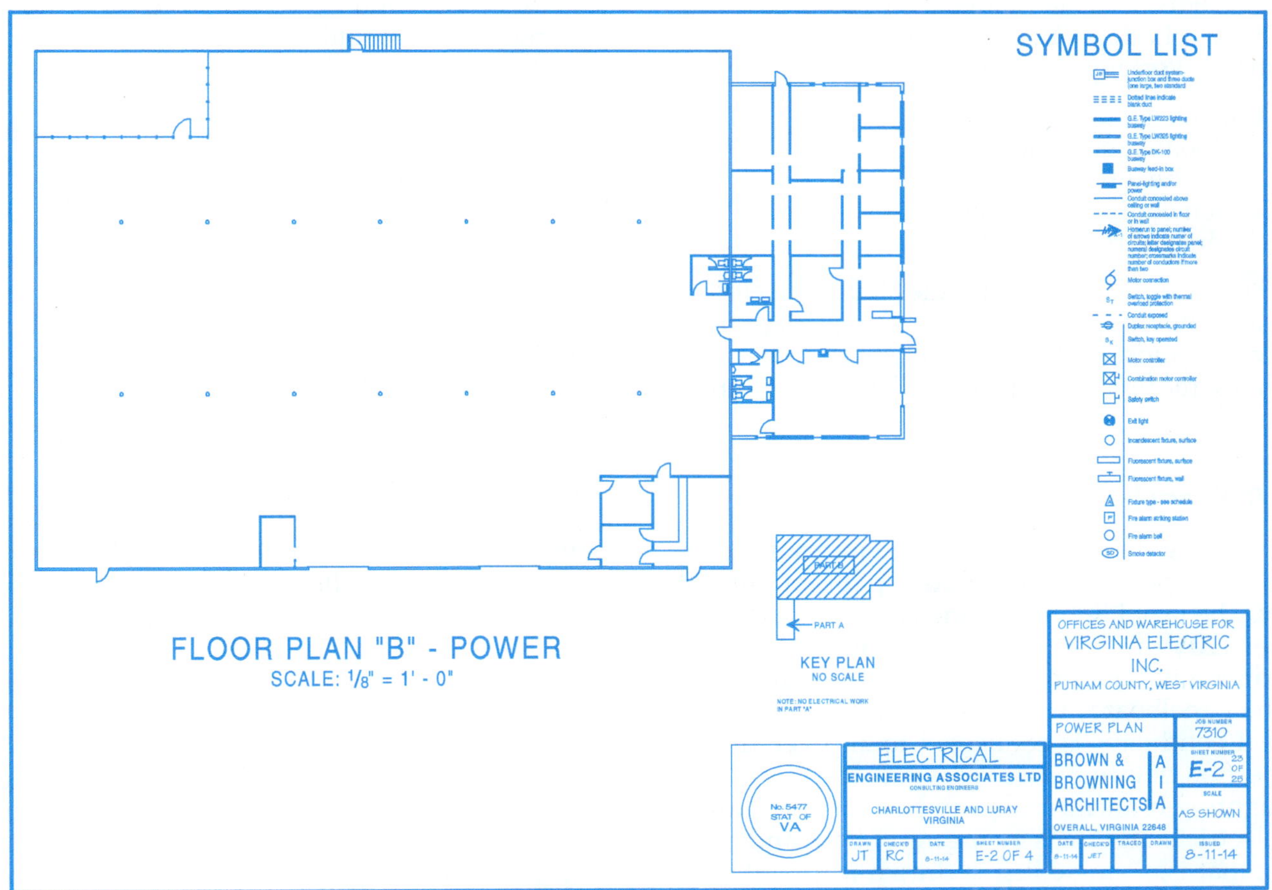

Figure 38 Sample lighting plan.

LIGHTING FIXTURE SCHEDULE				
SYMBOL	TYPE	MANUFACTURER AND CATALOG NUMBER	MOUNTING	LAMPS
	A	LIGHTOLIER 10234	WALL	2-40W T-12WWX
	B	LIGHTOLIER 10420	SURFACE	2-40W T-12 WWX
	C	ALKCO RPC-210-6E	SURFACE	2-8W T-5
	D	P 7 S AL 2936	WALL	1-100W 'A'
	E	P 7 S 110	SURFACE	1-100W 'A'

26110-14_F39.EPS

Figure 39 Lighting fixture (luminaire) schedule.

of written specifications can be time consuming. Workers do not always have access to the specifications while on the job, whereas they usually do have access to the working drawings. Therefore, the schedule is an excellent means of providing essential information in a clear and accurate manner, allowing the workers to carry out their assignments in the least amount of time.

Other schedules that are frequently found on electrical working drawings include:

- Connected load schedule
- Panelboard schedule
- Electric heat schedule
- Kitchen equipment schedule
- Schedule of receptacle types

There are also other schedules found on electrical drawings, depending upon the type of project. However, most will deal with lists of equipment such as motors, motor controllers, and similar items.

8.2.0 Branch Circuit Layout for Lighting

A simple lighting branch circuit requires two conductors to provide a continuous path for current flow. The usual lighting branch circuit operates at either 120V or 277V; the white (grounded) circuit conductor is therefore connected to the neutral bus in the panelboard, while the black (ungrounded) circuit conductor is connected to an overcurrent protection device.

Lighting branch circuits and outlets are shown on electrical drawings by means of lines and symbols; that is, a single line is drawn from outlet to outlet and then terminated with an arrowhead to indicate a homerun to the panelboard. Several methods are used to indicate the number and size of conductors, but the most common is to indicate the number of conductors in the circuit by using slash marks through the circuit lines and then indicate the wire size by a notation adjacent to these slash marks.

The circuits used to feed residential lighting must conform to standards established by the *NEC®* as well as by local and state ordinances. Most of the lighting circuits should be calculated to include the total load, although at times this is not possible because the electrician cannot be certain of the exact wattage that might be used by the homeowner. For example, an electrician may install four porcelain lampholders for the unfinished basement area, each to contain one 100-watt (100W) incandescent lamp. However, the homeowners may eventually replace the original lamps with others rated at 150W or even 200W. Thus, if the electrician initially loads the lighting circuit to full capacity, the circuit will probably become overloaded in the future.

It is recommended that no residential branch circuit be loaded to more than 80% of its rated capacity. Since most circuits used for lighting are rated at 15A, the total ampacity (in volt-amperes) for the circuit is as follows:

$$15A \times 120V = 1{,}800VA$$

Therefore, if the circuit is to be loaded to only 80% of its rated capacity, the maximum initial connected load should be no more than 1,440VA.

Figure 40 shows one possible lighting arrangement for the sample residence discussed earlier. All lighting fixtures are shown in their approximate physical location as they should be installed.

Electrical symbols are used to show the fixture types. Switches and lighting branch circuits are also shown by appropriate lines and symbols. The

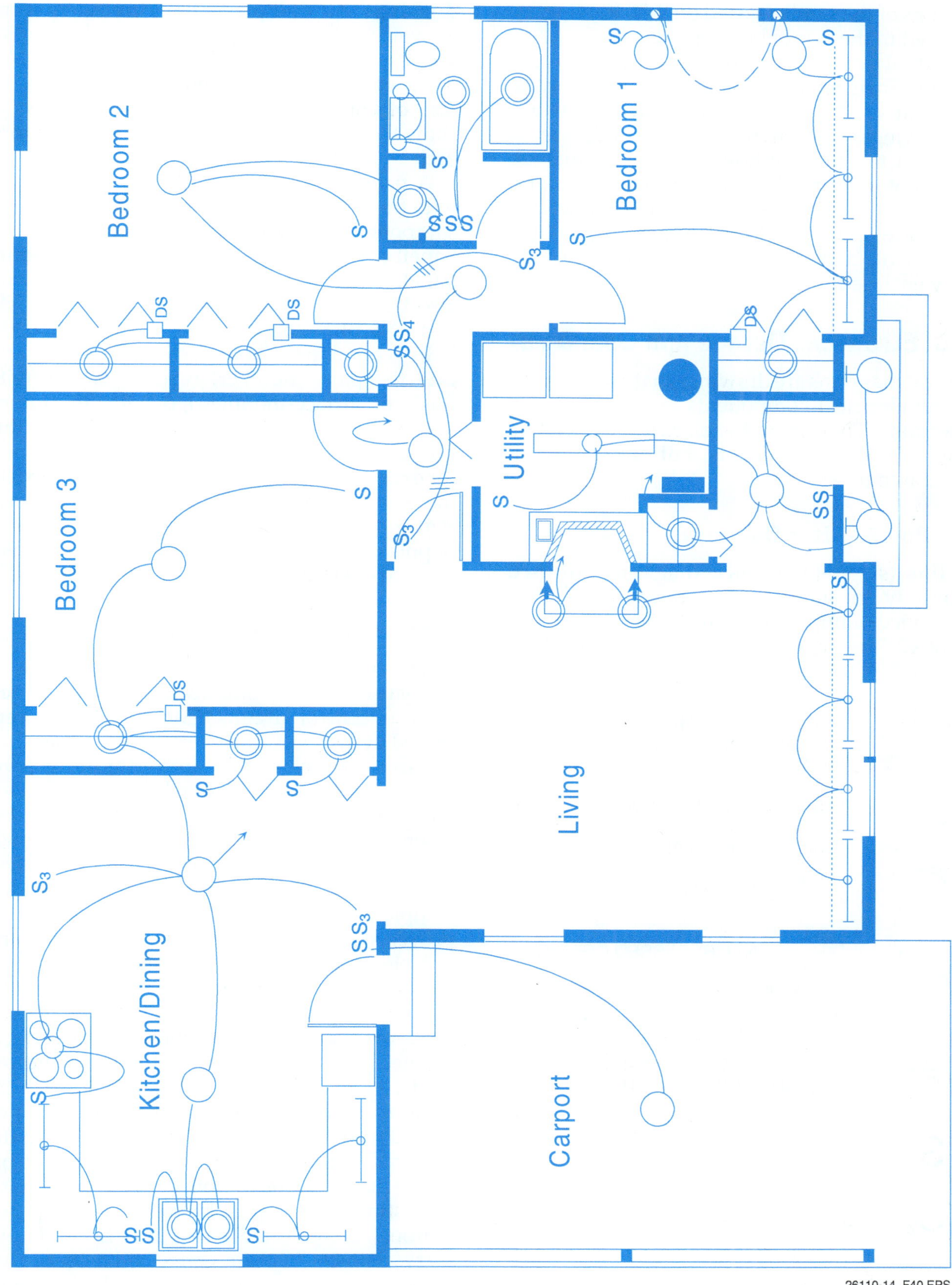

Figure 40 Lighting layout of the sample residence.

meanings of the symbols used on this drawing are explained in the symbol list in *Figure 41*.

In actual practice, the location of lighting fixtures and their related switches will probably be the extent of the information shown on working drawings. The circuits shown in *Figure 40* are meant to illustrate how lighting circuits are routed, not to imply that such drawings are typical for residential construction. If incandescent fixtures are used in a closet, they must meet the requirements of *NEC Section 410.16* and be completely enclosed.

9.0.0 Electrical Details and Diagrams

Electrical diagrams are drawings that are intended to show electrical components and their related connections. They show the electrical association of the different components, but are seldom, if ever, drawn to scale.

9.1.0 Power-Riser Diagrams

One-line (single-line) **block diagrams** are used extensively to show the arrangement of electric service equipment. **Power-riser diagrams** (*Figure 42*) are typical of such drawings. These drawings show all pieces of electrical equipment as well as the connecting lines used to indicate service-entrance conductors and feeders. Notes are used to identify the equipment, indicate the size of conduit necessary for each feeder, and show the number, size, and type of conductors in each conduit.

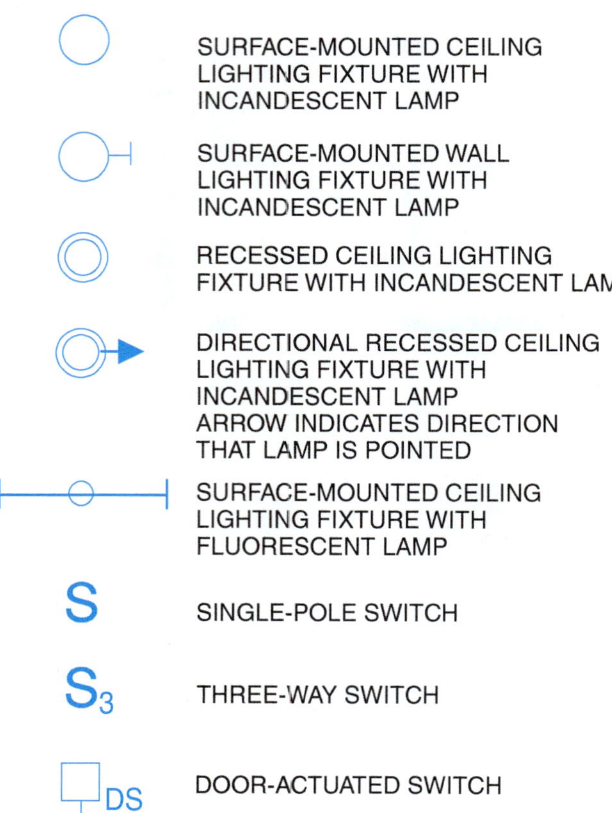

Figure 41 Electrical symbols list.

A panelboard schedule (*Figure 43*) is included with the power-riser diagram to indicate the exact components contained in each panelboard. This panelboard schedule is for the main distribution panel. On the actual drawings, schedules would also be shown for the other two panels (PNL A and PNL B).

In general, panelboard schedules usually indicate the panel number, type of cabinet (either flush- or surface-mounted), panel mains (ampere and voltage rating), phase (single- or three-phase), and number of wires. A four-wire panel, for example, indicates that a solid neutral exists in the panel. Branches indicate the type of overcurrent protection; that is, they indicate the number of poles, trip rating, and frame size. The items fed by each overcurrent device are also indicated.

9.2.0 Schematic Diagrams

Complete schematic wiring diagrams are normally used only in complicated electrical systems, such as control circuits. Components are represented by symbols, and every wire is either shown by itself or included in an assembly of several wires, which appear as one line on the drawing. Each wire should be numbered when it enters an assembly and should keep the same number when it comes out again to be connected to some electrical component in the system. *Figure 44* shows a complete schematic wiring diagram for a three-phase, AC magnetic non-reversing motor starter.

Note that this diagram shows the various devices in symbol form and indicates the actual connections of all wires between the devices. The three-wire supply lines are indicated by L_1, L_2, and L_3; the motor terminals of motor M are indicated by T_1, T_2, and T_3. Lines L_1, L_2, and L_3 each have a thermal overload protection device (OL) connected in series with normally open line contacts C_1 and C_3, respectively, which are both controlled by the magnetic starter coil, C. The control station, consisting of start pushbutton 1 and stop pushbutton 2, is connected across lines L_1 and L_2. Auxiliary contacts (C_4) are connected in series with the stop pushbutton and in parallel with the start pushbutton. The control circuit also has normally closed overload contacts (OC) connected in series with the magnetic starter coil (C).

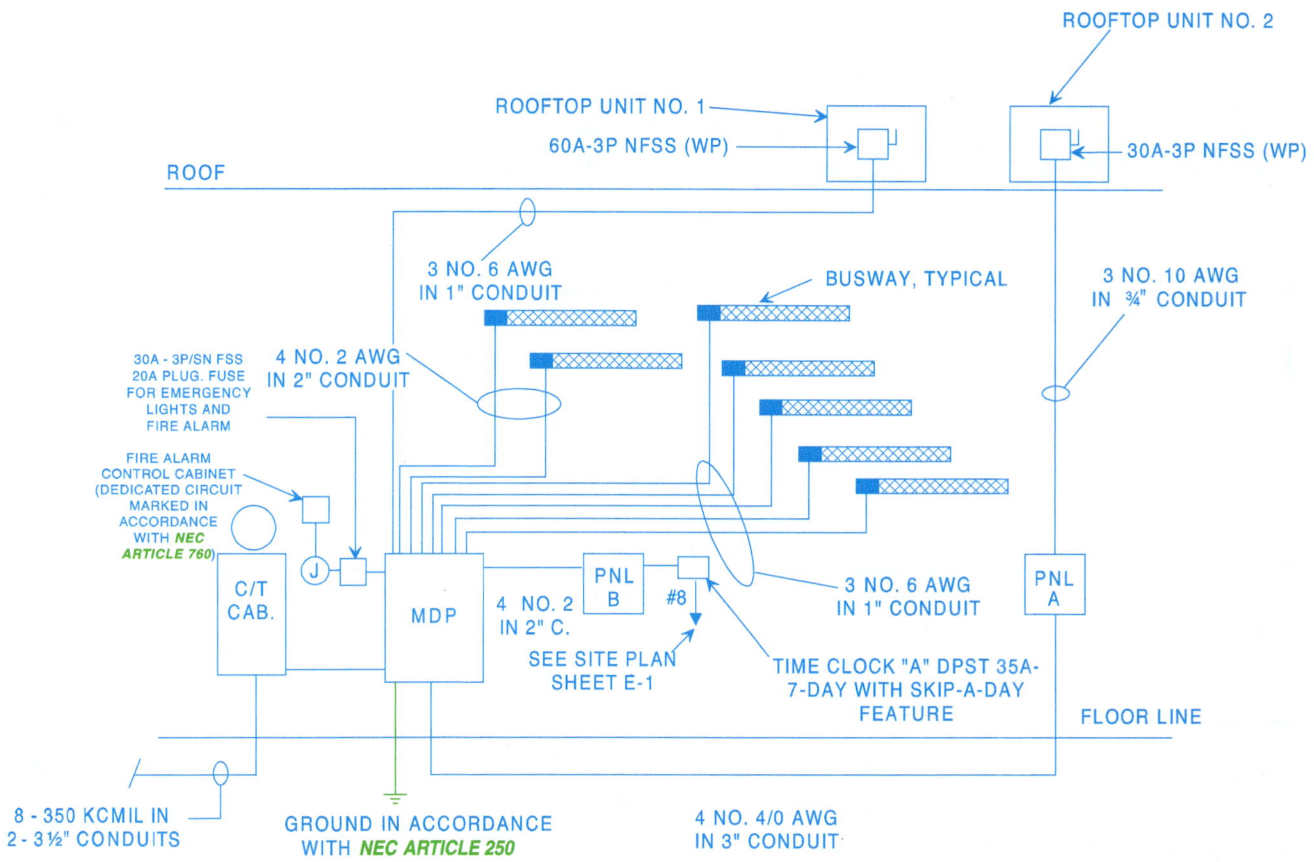

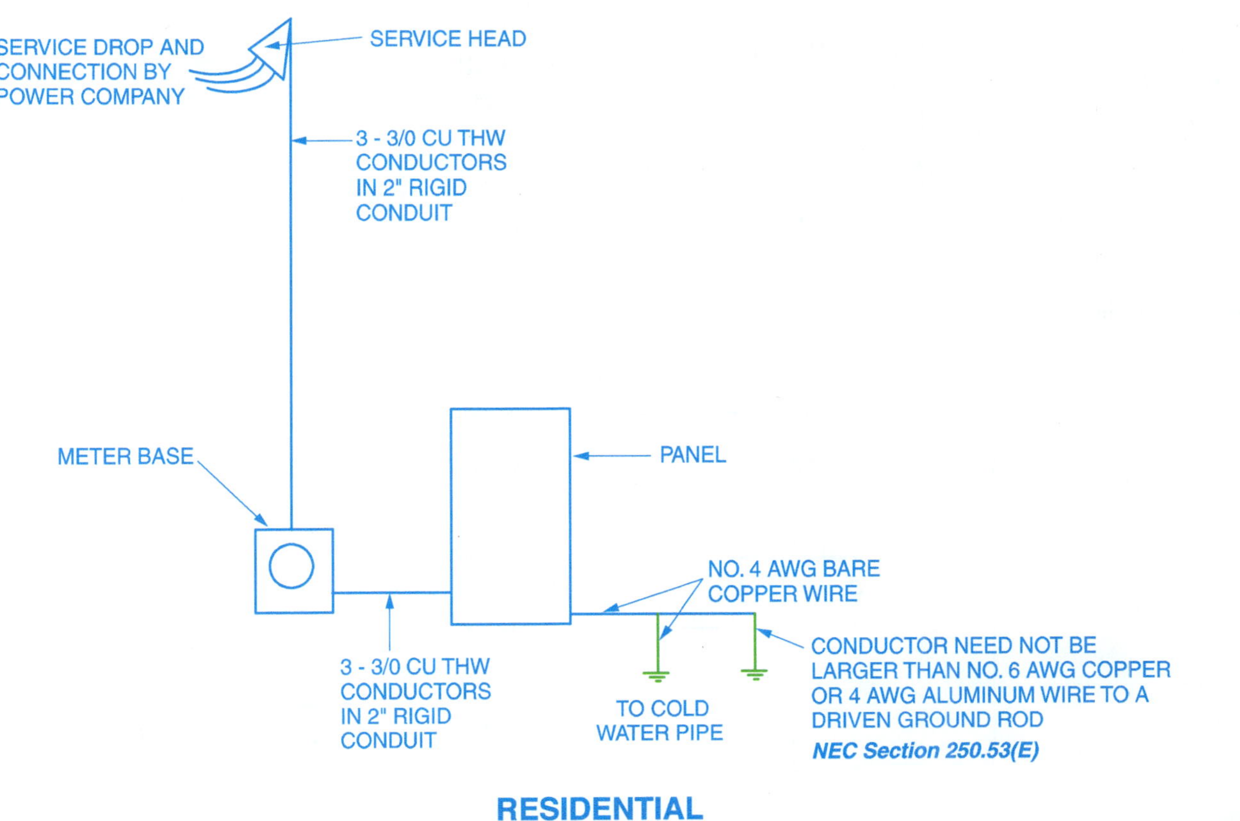

26110-14_F42.EPS

Figure 42 Typical power-riser diagrams.

PANELBOARD SCHEDULE

PANEL No.	CABINET TYPE	PANEL MAINS AMPS	PANEL MAINS VOLTS	PANEL MAINS PHASE	BRANCHES 1P	BRANCHES 2P	BRANCHES 3P	BRANCHES PROT.	BRANCHES FRAME	ITEMS FED OR REMARKS
MDP	SURFACE	600A	120/208	3ϕ,4-W	-	-	1	225A	25,000	PANEL "A"
					-	-	1	100A	18,000	PANEL "B"
					-	-	1	100A		POWER BUSWAY
					-	-	1	60A		LIGHTING BUSWAY
					-	-	1	70A		ROOFTOP UNIT #1
					-	-	1	70A		SPARE
					-	-	1	600A	42,000	MAIN CIRCUIT BRKR

26110-14_F43.EPS

Figure 43 Typical panelboard schedule.

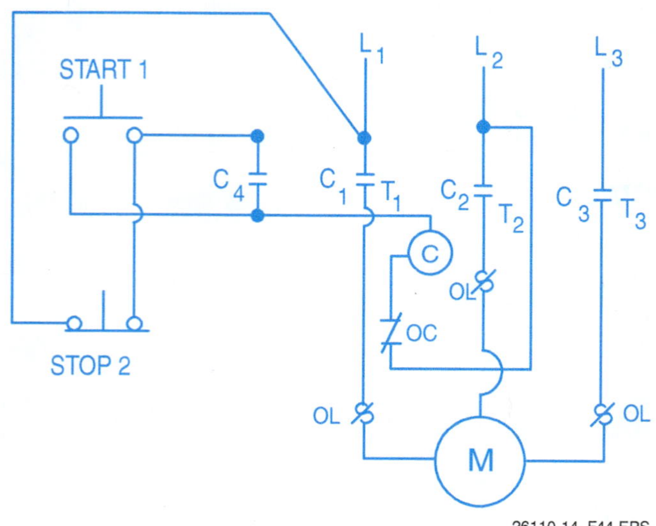

Figure 44 Wiring diagram.

Any number of additional pushbutton stations may be added to this control circuit similarly to the way in which three-way and four-way switches are added to control a lighting circuit. When adding pushbutton stations, the stop buttons are always connected in series and the start buttons are always connected in parallel. *Figure 45* shows the same motor starter circuit in *Figure 44*, but this time it is controlled by two sets of start/stop buttons.

Schematic wiring diagrams have only been touched upon in this module; there are many other details that you will need to know to perform your work in a proficient manner. Later modules cover wiring diagrams in more detail.

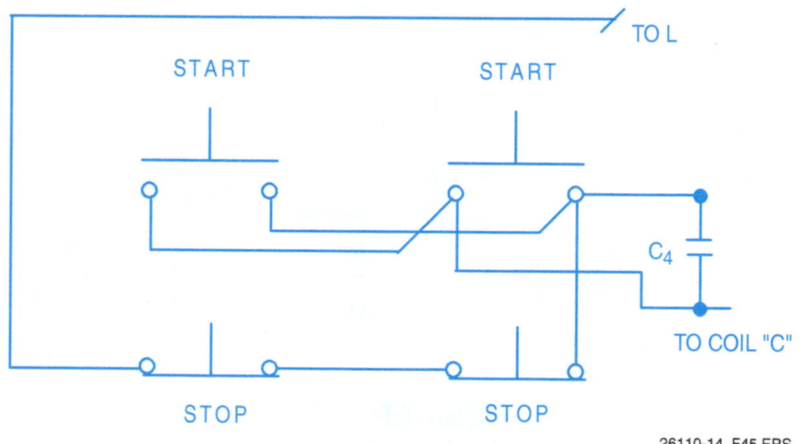

Figure 45 Circuit being controlled by two sets of start/stop buttons.

9.3.0 Drawing Details

A detail drawing is a drawing of a separate item or portion of an electrical system, giving a complete and exact description of its use and all the details needed to show the electrician exactly what is required for its installation. For example, the power plan for the office/warehouse has a sectional cut through the busduct. This is a good example of where an extra, detailed drawing is desirable.

A set of electrical drawings will sometimes require large-scale drawings of certain areas that are not indicated with sufficient clarity on the small-scale drawings. For example, the site plan may show exterior pole-mounted lighting fixtures that are to be installed by the contractor.

10.0.0 Written Specifications

The written specifications for a building or project are the written descriptions of work and duties required of the owner, architect, and consulting engineer. Together with the working drawings, these specifications form the basis of the contract requirements for the construction of the building or project. Those who use the construction drawings and specifications must always be alert to discrepancies between the working drawings and the written specifications. These are some situations where discrepancies may occur:

- Architects or engineers use standard or prototype specifications and attempt to apply them without any modification to specific working drawings.
- Previously prepared standard drawings are changed or amended by reference in the specifications only and the drawings themselves are not changed.
- Items are duplicated in both the drawings and specifications, but an item is subsequently amended in one and overlooked in the other contract document.

In such instances, the person in charge of the project has the responsibility to ascertain whether the drawings or the specifications take precedence. Such questions must be resolved, preferably before the work begins, to avoid added costs to the owner, architect/engineer, or contractor.

10.1.0 How Specifications Are Written

Writing accurate and complete specifications for building construction is a serious responsibility for those who design the buildings because the specifications, combined with the working drawings, govern practically all important decisions that are made during the construction span of every project. Compiling and writing these specifications is not a simple task, even for those who have had considerable experience in preparing such documents. A set of written specifications for a single project will usually contain thousands of products, parts, and components, and the methods of installing them, all of which must be covered in either the drawings and/or specifications. No one can memorize all of the necessary items required to describe accurately the various areas of construction. One must rely upon reference materials such as manufacturer's data, catalogs, checklists, and, most of all, a high-quality master specification.

10.2.0 Format of Specifications

For convenience in writing, speed in estimating, and ease of reference, the most suitable organization of the specifications is a series of sections dealing with the construction requirements, products, and activities that is easily understandable by the different trades. Those people who use the specifications must be able to find all the information they need without spending too much time looking for it.

The most commonly used specification-writing format used in North America is the *MasterFormat*®. This standard was developed jointly by the Construction Specifications Institute (CSI) and Construction Specifications Canada (CSC). For many years prior to 2004, the organization of construction specifications and suppliers catalogs was based on a standard with 16 sections, otherwise known as divisions. The divisions and their subsections were individually identified by a five-digit numbering system. The first two digits represented the division number and the next three individual numbers represented successively lower levels of breakdown. For example, the number 13213 represents division 13, subsection 2, sub-subsection 1, and sub-sub-subsection 3. In this older version of the standard, electrical systems, including any electronic or special electrical systems, were lumped together under Division 16 – *Electrical*. Today, specifications conforming to the 16 division format may still be in use.

In 2004, the *MasterFormat*® standard underwent a major change. What had been 16 divisions was expanded to four major groupings and 49 divisions with some divisions reserved for future expansion. *MasterFormat*® was again updated for 2010 (*Figure 46*). The first 14 divisions are essentially the same as the old format. Subjects under the old Division 15 – *Mechanical*

Applications Guide

MasterFormat GROUPS, SUBGROUPS, AND DIVISIONS

PROCUREMENT AND CONTRACTING REQUIREMENTS GROUP

Division 00 – Procurement and Contracting Requirements
- Introductory Information
- Procurement Requirements
- Contracting Requirements

SPECIFICATIONS GROUP

GENERAL REQUIREMENTS SUBGROUP

Division 01 – General Requirements

FACILITY CONSTRUCTION SUBGROUP

Division 02 – Existing Conditions
Division 03 – Concrete
Division 04 – Masonry
Division 05 – Metals
Division 06 – Wood, Plastics, and Composites
Division 07 – Thermal and Moisture Protection
Division 08 – Openings
Division 09 – Finishes
Division 10 – Specialties
Division 11 – Equipment
Division 12 – Furnishings
Division 13 – Special Construction
Division 14 – Conveying Equipment
Division 15 – Reserved for Future Expansion
Division 16 – Reserved for Future Expansion
Division 17 – Reserved for Future Expansion
Division 18 – Reserved for Future Expansion
Division 19 – Reserved for Future Expansion

FACILITY SERVICES SUBGROUP

Division 20 – Reserved for Future Expansion
Division 21 – Fire Suppression
Division 22 – Plumbing
Division 23 – Heating, Ventilating, and Air-Conditioning (HVAC)
Division 24 – Reserved for Future Expansion
Division 25 – Integrated Automation
Division 26 – Electrical
Division 27 – Communications
Division 28 – Electronic Safety and Security
Division 29 – Reserved for Future Expansion

SITE AND INFRASTRUCTURE SUBGROUP

Division 30 – Reserved for Future Expansion
Division 31 – Earthwork
Division 32 – Exterior Improvements
Division 33 – Utilities
Division 34 – Transportation
Division 35 – Waterway and Marine Construction
Division 36 – Reserved for Future Expansion
Division 37 – Reserved for Future Expansion
Division 38 – Reserved for Future Expansion
Division 39 – Reserved for Future Expansion

PROCESS EQUIPMENT SUBGROUP

Division 40 – Process Integration
Division 41 – Material Processing and Handling Equipment
Division 42 – Process Heating, Cooling, and Drying Equipment
Division 43 – Process Gas and Liquid Handling, Purification, and Storage Equipment
Division 44 – Pollution and Waste Control Equipment
Division 45 – Industry-Specific Manufacturing Equipment
Division 46 – Water and Wastewater Equipment
Division 47 – Reserved for Future Expansion
Division 48 – Electrical Power Generation
Division 49 – Reserved for Future Expansion

26110-14_F46.EPS

Figure 46 2010 *MasterFormat®*.

On Site

Specifications

Written specifications supplement the related working drawings in that they contain details not shown on the drawings. Specifications define and clarify the scope of the job. They describe the specific types and characteristics of the components that are to be used on the job and the methods for installing some of them. Many components are identified specifically by the manufacturer's model and part numbers. This type of information is used to purchase the various items of hardware needed to accomplish the installation in accordance with the contractual requirements.

have been relocated to new divisions 22 and 23. The basic subjects under old Division 16 – *Electrical* have been relocated to new divisions 26 and 27. In addition, the numbering system was changed to 6 digits to allow for more subsections in each division, which allowed for finer definition. In the new numbering system, the first two digits represent the division number. The next two digits represent subsections of the division and the two remaining digits represent the third level sub-subsection numbers. The fourth level, if required, is a decimal and number added to the end of the last two digits. For example, the number 132013.04 represents division 13, subsection 20, sub-subsection 13, and sub-sub-subsection 04. Under the new standard, the Facility Service Subgroup contains the divisions that are most important to the electrician. These include the following divisions:

- *Division 25 – Integrated Automation*
- *Division 26 – Electrical*
- *Division 27 – Communications*
- *Division 28 – Electronic Safety and Security*

Figure 47 contains a detailed breakdown of the electrical division.

NUMBER	TITLE	EXPLANATION

DIVISION 26 – ELECTRICAL

26 00 00 Electrical

may be used as division level section title.

See: 02 41 19 for selective demolition of existing electrical systems.
03 30 00 for cast-in-place concrete equipment bases.
07 84 00 for firestopping.
07 92 00 for joint sealants.
08 31 00 for access doors and panels.
09 91 00 for field painting.
31 23 33 for trenching and backfilling.

26 01 00 Operation and Maintenance of Electrical Systems

Includes: maintenance, repair, rehabilitation, replacement, restoration, preservation, etc. of electrical systems. medium voltage: 2400 V to 69 kV. low voltage: 600 V and less.

Notes: Definitions medium voltage: 2400 V to 69 kV. low voltage: 600 V and less.

Level 4 Numbering Recommendation: following numbering is recommended for the creation of Level 4 titles:
.51-.59 for maintenance.
.61-.69 for repair.
.71-.79 for rehabilitation.
.81-.89 for replacement.
.91-.99 for restoration.

26 01 10 Operation and Maintenance of Medium-Voltage Electrical Distribution
26 01 20 Operation and Maintenance of Low-Voltage Electrical Distribution
26 01 26 Maintenance Testing of Electrical Systems
26 01 30 Operation and Maintenance of Facility Electrical Power Generating and Storing Equipment
26 01 40 Operation and Maintenance of Electrical and Cathodic Protection Systems
26 01 50 Operation and Maintenance of Lighting
- 26 01 50.51 Luminaire Relamping
- 26 01 50.81 Luminaire Replacement

26 05 00 Common Work Results for Electrical

Includes: subjects common to multiple titles in Division 26. raceway and boxes includes conduit, tubing, surface raceways, and electrical boxes. medium voltage: 2400 V to 69 kV. low voltage: 600 V and less. control voltage: 50 V

311

26110-14_F47A.EPS

Figure 47 Detailed breakdown of the electrical division (1 of 13).

DIVISION **26**

NUMBER **TITLE** **EXPLANATION**

and less.

Alternate Terms/Abbreviations: EMT: electrical metallic tubing.

Notes: Definitions medium voltage: 2400 V to 69 kV. low voltage: 600 V and less. control voltage: 50 V and less.

See 01 80 00 for performance requirements of subjects common to multiple titles.
05 35 00 for raceway decking assemblies.
05 45 16 for electrical metal supports.
13 48 00 for sound, vibration, and seismic control.
25 05 13 for conductors and cables for integrated automation.
25 05 26 for grounding and bonding for integrated automation.
25 05 28 for pathways for integrated automation.
25 05 48 for vibration and seismic control for integrated automation.
25 05 53 for identification for integrated automation.
27 05 28 for pathways for communications systems.
27 05 46 for utility poles for communications systems.
27 05 48 for vibration and seismic controls for communications.
27 05 53 for identification for communications.
28 05 13 for conductors and cables for electronic safety and security.
28 05 26 for grounding and bonding for electronic safety and security.
28 05 28 for pathways for electronic safety and security.
28 05 48 for vibration and seismic controls for electronic safety and security.
28 05 53 for identification for electronic safety and security.
33 71 16 for electrical utility poles.
33 71 19 for electrical utility underground ducts and manholes.

26 05 13	Medium-Voltage Cables	
	26 05 13.13	Medium-Voltage Open Conductors
	26 05 13.16	Medium-Voltage, Single- and Multi-Conductor Cables
26 05 19	Low-Voltage Electrical Power Conductors and Cables	
	26 05 19.13	Undercarpet Electrical Power Cables
	26 05 19.23	Manufactured Wiring Assemblies
26 05 23	Control-Voltage Electrical Power Cables	
26 05 26	Grounding and Bonding for Electrical Systems	
26 05 29	Hangers and Supports for Electrical Systems	
26 05 33	Raceway and Boxes for Electrical Systems	
	26 05 33.13	Conduit for Electrical Systems
	26 05 33.16	Boxes for Electrical Systems
	26 05 33.23	Surface raceways for Electrical Systems

312

26110-14_F47B.EPS

Figure 47 Detailed breakdown of the electrical division (2 of 13).

NUMBER	TITLE	EXPLANATION
26 05 36	Cable Trays for Electrical Systems	
26 05 39	Underfloor Raceways for Electrical Systems	
26 05 43	Underground Ducts and Raceways for Electrical Systems	
26 05 46	Utility Poles for Electrical Systems	
26 05 48	Vibration and Seismic Controls for Electrical Systems	
26 05 53	Identification for Electrical Systems	
26 05 73	Overcurrent Protective Device Coordination Study	
26 05 83	Wiring Connections	
26 06 00	**Schedules for Electrical**	Notes: a schedule may be included on drawings, in the project manual, or a project book. Definitions: medium voltage: 2400 V to 69 kV. low voltage: 600 V and less. Includes: schedules of items common to multiple titles in Division 26.
26 06 10	Schedules for Medium-Voltage Electrical Distribution	
26 06 20	Schedules for Low-Voltage Electrical Distribution	
	26 06 20.13 Electrical Switchboard Schedule	
	26 06 20.16 Electrical Panelboard Schedule	
	26 06 20.19 Electrical Motor-Control Center Schedule	
	26 06 20.23 Electrical Circuit Schedule	
	26 06 20.26 Wiring Device Schedule	
26 06 30	Schedules for Facility Electrical Power Generating and Storing Equipment	
26 06 40	Schedules for Electrical and Cathodic Protection Systems	
26 06 50	Schedules for Lighting	
	26 06 50.13 Lighting Panelboard Schedule	
	26 06 50.16 Lighting Fixture Schedule	
26 08 00	**Commissioning of Electrical Systems**	Includes: commissioning of items common to multiple titles in Division 26. See: 01 91 00 for commissioning of subjects common to multiple divisions.
26 09 00	**Instrumentation and Control for Electrical Systems**	Includes: instrumentation and control associated with electrical systems. See: 13 50 00 for special instrumentation. 25 36 00 for integrated automation instrumentation and terminal devices for electrical systems. 25 56 00 for integrated automation control of electrical systems.

313

26110-14_F47C.EPS

Figure 47 Detailed breakdown of the electrical division (3 of 13).

Division **26**

Number	Title	Explanation
		25 96 00 for integrated automation control sequences for electrical systems. 33 09 70 for instrumentation and control for electrical utilities.
26 09 13	Electrical Power Monitoring	
26 09 15	Peak Load Controllers	
26 09 16	Electrical Controls and Relays	
26 09 17	Programmable Controllers	
26 09 19	Enclosed Contactors	
26 09 23	Lighting Control Devices	
		Includes: clock and calendar, photoelectric switches, occupancy sensors, and light-leveling control devices. control- and low-voltage lighting control devices connected through computers. addressable lighting control devices and lighting components (ballasts) connected through computers. See: 11 61 00 for theater and stage equipment. 26 50 00 for lighting. 26 55 61 for theatrical lighting. See Also: 11 61 00 for theatrical lighting controls.
26 09 26	Lighting Control Panelboards	
26 09 33	Central Dimming Controls	
	26 09 33.13 Multichannel Remote-Controlled Dimmers	
	26 09 33.16 Remote-Controlled Dimming Stations	
26 09 36	Modular Dimming Controls	
	26 09 36.13 Manual Modular Dimming Controls	
	26 09 36.16 Integrated Multipreset Modular Dimming Controls	
26 09 43	Network Lighting Controls	
	26 09 43.13 Digital-Network Lighting Controls	
	26 09 43.16 Addressable Fixture Lighting Control	
26 09 61	Theatrical Lighting Controls	

26 10 00 Medium-Voltage Electrical Distribution

Includes: substations, transformers, switchgear, and circuit protection devices to distribute medium-voltage electrical power from the facility service point to the point of delivery.

Notes: Definitions medium voltage: 2400 V to 69 kV.

See 26 05 13 for medium-voltage cables.
26 20 00 for low-voltage electrical distribution.
26 30 00 for facility electrical power generating and storing equipment.
33 71 00 for electrical utility distribution.

314

26110-14_F47D.EPS

Figure 47 Detailed breakdown of the electrical division (4 of 13).

NUMBER	TITLE	EXPLANATION
26 11 00	**Substations**	
		Includes: assembly of switches, circuit breakers, buses, and transformers to switch circuits and convert power from one voltage to another. See: 33 72 00 for utility substations. 34 21 16 for traction power substations.
26 11 13	Primary Unit Substations	
26 11 16	Secondary Unit Substations	
26 12 00	**Medium-Voltage Transformers**	
		Includes: transformers for medium-voltage applications. See: 26 22 00 for low-voltage transformers. 33 73 00 for utility transformers. 34 21 23 for traction power transformer-rectifier units.
26 12 13	Liquid-Filled, Medium-Voltage Transformers	
26 12 16	Dry-Type, Medium-Voltage Transformers	
26 12 19	Pad-Mounted, Liquid-Filled, Medium-Voltage Transformers	
26 13 00	**Medium-Voltage Switchgear**	
		Includes: switchgear for medium-voltage applications. See: 26 23 00 for low-voltage switchgear. 33 77 00 for medium-voltage utility switchgear. 34 21 19 for traction power switchgear.
26 13 13	Medium-Voltage Circuit Breaker Switchgear	
26 13 16	Medium-Voltage Fusible Interrupter Switchgear	
26 13 19	Medium-Voltage Vacuum Interrupter Switchgear	
26 13 23	Medium-Voltage Metal-Enclosed Switchgear	
26 13 26	Medium-Voltage Metal-Clad Switchgear	
26 13 29	Medium-Voltage Compartmentalized Switchgear	
26 16 00	**Medium-Voltage Metering**	
26 18 00	**Medium-Voltage Circuit Protection Devices**	
		Includes: circuit protection devices for medium-voltage applications. See: 26 28 00 for low-voltage circuit protective devices. 26 41 23 for lightning protection surge arresters and suppressors. 33 77 00 for medium-voltage utility circuit protection devices.
26 18 13	Medium-Voltage Cutouts	

315

26110-14_F47E.EPS

Figure 47 Detailed breakdown of the electrical division (5 of 13).

DIVISION **26**

NUMBER	TITLE	EXPLANATION
26 18 16	Medium-Voltage Fuses	
26 18 19	Medium-Voltage Lightning Arresters	
26 18 23	Medium-Voltage Surge Arresters	
26 18 26	Medium-Voltage Reclosers	
26 18 29	Medium-Voltage Enclosed Bus	
26 18 33	Medium-Voltage Enclosed Fuse Cutouts	
26 18 36	Medium-Voltage Enclosed Fuses	
26 18 39	Medium-Voltage Motor Controllers	

26 20 00 Low-Voltage Electrical Transmission

Includes: overhead power systems, transformers, switchgear, switchboards, panelboards, enclosed bus assemblies, power distribution units, controllers, wiring devices, and circuit protection devices to distribute low-voltage electrical power from the point of voltage transformation to the point of use. typical voltages: 120, 208, 230, 240, 277, 460, and 480.

Notes: Definitions low voltage: 600 V and less.

See 26 05 19 for low-voltage electrical power conductors and cables.
26 10 00 for medium-voltage electrical distribution.
26 30 00 for facility electrical power generating and storing equipment.

26 21 00 Low-Voltage Electrical Service Entrance

See: 26 05 19 for low-voltage electrical power conductors and cables.
26 05 46 for utility poles for electrical systems.
33 71 13 for site electrical transmission towers.
33 71 16 for electrical utility poles.

26 21 13 Low-Voltage Overhead Electrical Service Entrance
26 21 16 Low-Voltage Underground Electrical Service Entrance

26 22 00 Low-Voltage Transformers

Includes: transformers for low-voltage applications.

See: 26 12 00 for medium-voltage transformers.

26 22 13 Low-Voltage Distribution Transformers
26 22 16 Low-Voltage Buck-Boost Transformers
26 22 19 Control and Signal Transformers

26 23 00 Low-Voltage Switchgear

Includes: switchgear for low-voltage applications.

316

26110-14_F47F.EPS

Figure 47 Detailed breakdown of the electrical division (6 of 13).

NUMBER	TITLE	EXPLANATION
		See: 26 13 00 for medium-voltage switchgear.
26 23 13	Paralleling Low-Voltage Switchgear	
26 24 00	**Switchboards and Panelboards**	Includes: switchboards, panelboards, and control centers. See: 26 27 16 for electrical cabinets and enclosures. 26 29 13 for enclosed controllers. 26 29 23 for variable-frequency motor controllers.
26 24 13	Switchboards	
26 24 16	Panelboards	
26 24 19	Motor-Control Centers	
26 25 00	**Enclosed Bus Assemblies**	Includes: busway, step bus, and tap boxes. See: 33 72 26 for utility substation bus assemblies.
26 26 00	**Power Distribution Units**	Includes: distribution units with integral transformers, panelboards, and power conditioning components. See: 26 24 16 for panelboards.
26 27 00	**Low-Voltage Distribution Equipment**	Includes: wiring devices includes receptacles, switches, dimmers, and finish plates. See: 26 24 00 for switchboards and panelboards. 33 71 73 for utility electric meters.
26 27 13	Electricity Metering	
26 27 16	Electrical Cabinets and Enclosures	
26 27 19	Multi-Outlet Assemblies	
26 27 23	Indoor Service Poles	
26 27 26	Wiring Devices	
26 27 73	Door Chimes	
26 28 00	**Low-Voltage Circuit Protective Devices**	Includes: circuit protection devices for low-voltage applications. enclosed switches and transfer switches. See: 26 18 00 for medium-voltage circuit protection devices.

317

26110-14_F47G.EPS

Figure 47 Detailed breakdown of the electrical division (7 of 13).

DIVISION **26**

NUMBER	TITLE	EXPLANATION
26 28 13	Fuses	
26 28 16	Enclosed Switches and Circuit Breakers	
	26 28 16.13 Enclosed Circuit Breakers	
	26 28 16.16 Enclosed Switches	
26 29 00	**Low-Voltage Controllers**	Includes: contactors and motor controllers. May Include: fuses. Alternate Terms/Abbreviations: enclosed controllers: motor controllers. See: 26 24 19 for motor-control centers. 26 28 13 for fuses.
26 29 13	Enclosed Controllers	
	26 29 13.13 Across-the-Line Motor Controllers	
	26 29 13.16 Reduced-Voltage Motor Controllers	
26 29 23	Variable-Frequency Motor Controllers	
26 29 33	Controllers for Fire Pump Drivers	
	26 29 33.13 Full-Service Controllers for Fire Pump Electric-Motor Drivers	
	26 29 33.16 Limited-Service Controllers for Fire Pump Electric-Motor Drivers	
	26 29 33.19 Controllers for Fire Pump Diesel Engine Drivers	
26 30 00	**Facility Electrical Power Generating and Storing Equipment**	Includes: equipment to generate and store electrical power for a single facility. Notes: 48 10 00 for electrical power generation equipment.
26 31 00	**Photovoltaic Collectors**	Includes: solar cells to convert sunlight to electricity. See: 07 31 00 for solar collector roof shingles. 22 33 30 for residential, collector-to-tank, solar-electric domestic water heaters. 23 56 00 for solar energy heating equipment. 42 12 23 for solar process heaters. 42 13 26 for industrial solar radiation heat exchangers. 48 14 00 for solar energy electrical power generation equipment.
26 32 00	**Packaged Generator Assemblies**	Includes: generators, frequency changers, and rotary converters and uninterruptible power units.

318

26110-14_F47H.EPS

Figure 47 Detailed breakdown of the electrical division (8 of 13).

NUMBER	TITLE	EXPLANATION
		See: 23 11 00 for facility fuel piping. 23 24 00 for internal-combustion engine piping. 48 11 00 for fossil fuel plant electrical power generation equipment. 48 13 00 for hydroelectric plant electrical power generation equipment. 48 15 00 for wind energy electrical power generation equipment.
26 32 13	Engine Generators 26 32 13.13 Diesel-Engine-Driven Generator Sets 26 32 13.16 Gas-Engine-Driven Generator Sets 26 32 13.26 Gas-Turbine Engine-Driven Generators	
		Alternate Terms/Abbreviations: microturbines See: 48 11 23 Fossil Fuel Electrical Power Plant Gas Turbines
26 32 16	Steam-Turbine Generators	
26 32 19	Hydro-Turbine Generators	
26 32 23	Wind Energy Equipment	
26 32 26	Frequency Changers	
26 32 29	Rotary Converters	
26 32 33	Rotary Uninterruptible Power Units	
26 33 00	**Battery Equipment**	
		Includes: batteries, battery racks, battery chargers, static power converters, uninterruptible power supplies, and accessories. May Include: battery-operated emergency light fixtures. See: 25 36 23 for integrated automation battery monitors. 26 31 00 for photovoltaic collectors. 33 72 33 for electrical utility substation. 48 17 13 for electrical power generation batteries. See Also: 26 52 00 for emergency lighting incorporating batteries.
26 33 13	Batteries	
26 33 16	Battery Racks	
26 33 19	Battery Units	
26 33 23	Central Battery Equipment	
26 33 33	Static Power Converters	
26 33 43	Battery Chargers	
26 33 46	Battery Monitoring	

319

26110-14_F47I.EPS

Figure 47 Detailed breakdown of the electrical division (9 of 13).

DIVISION **26**

NUMBER	TITLE	EXPLANATION
26 33 53	Static Uninterruptible Power Supply	
26 35 00	**Power Filters and Conditioners**	Includes: capacitors, chokes and inductors, filters, power factor controllers, and voltage regulators. Alternate Terms/Abbreviations: EMI: electromagnetic interference. RFI: radio frequency interference. power factor correction equipment: power factor controllers. See: 08 34 46 for RFI shielding doors. 08 56 46 for RFI shielding windows. 13 49 00 for radiation protection. 26 18 23 for medium-voltage surge arresters. 28 32 00 for radiation detection and alarm. 40 91 16 for electromagnetic process measurement devices.
26 35 13	Capacitors	
26 35 16	Chokes and Inductors	
26 35 23	Electromagnetic-Interference Filters	
26 35 26	Harmonic Filters	
26 35 33	Power Factor Correction Equipment	
26 35 36	Slip Controllers	
26 35 43	Static-Frequency Converters	
26 35 46	Radio-Frequency-Interference Filters	
26 35 53	Voltage Regulators	
26 36 00	**Transfer Switches**	Includes: switches transfer from one source of electricity to another.
26 36 13	Manual Transfer Switches	
26 36 23	Automatic Transfer Switches	
26 40 00	**Electrical and Cathodic Protection**	
26 41 00	**Facility Lightning Protection**	Includes: wiring and equipment for lightning protection. See: 26 18 19 for medium-voltage lightning arresters. 33 79 00 for site grounding. 33 79 93 for site lightning protection.
26 41 13	Lightning Protection for Structures 26 41 13.13 Lightning Protection for Buildings	
26 41 16	Lightning Prevention and Dissipation	
26 41 19	Early Streamer Emission Lightning Protection	
26 41 23	Lightning Protection Surge Arresters and Suppressors	

320

26110-14_F47J.EPS

Figure 47 Detailed breakdown of the electrical division (10 of 13).

NUMBER	TITLE	EXPLANATION

26 42 00 Cathodic Protection

Includes: equipment, controls, and installation for cathodic protection of structures and underground metal construction and piping.

See: 40 46 42 for cathodic process corrosion protection.

26 42 13 Passive Cathodic Protection for Underground and Submerged Piping
26 42 16 Passive Cathodic Protection for Underground Storage Tank

26 43 00 Transient Voltage Suppression

Includes: devices to protect against voltage surges on electrical distribution systems.

26 43 13 Transient-Voltage Suppression for Low-Voltage Electrical Power Circuits

26 50 00 Lighting

Includes: luminaries, lighting equipment, ballasts, dimming controls, and lighting accessories. fluorescent, high intensity discharge, incandescent, mercury vapor, neon, and sodium vapor lighting.

Alternate Terms/Abbreviations: HID: high intensity discharge.

See: 10 84 00 for Gas Lighting.
25 36 26 for integrated automation lighting relays.
26 09 23 for lighting controls.
26 20 00 for low-voltage electrical transmission.

26 51 00 Interior Lighting

Includes: lighting for interior locations, except for emergency lighting, lighting in hazardous locations, and special purpose lighting. chandeliers, troffers.

See: 09 54 16 for luminous ceilings.
09 58 00 for integrated ceiling assemblies.
10 14 33 for illuminated panel signage.

26 51 13 Interior Lighting Fixtures, Lamps, And Ballasts

26 52 00 Emergency Lighting

Includes: equipment for exitway lighting and other emergency applications, including emergency battery units, fixtures with integral batter power supplies.

See: 26 53 00 for exit signs.

321

26110-14_F47K.EPS

Figure 47 Detailed breakdown of the electrical division (11 of 13).

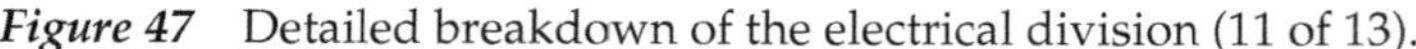

Division **26**

Number	**Title**	**Explanation**
26 53 00	**Exit Signs**	Includes: electric exit signs. See: 26 52 00 for emergency lighting.
26 54 00	**Classified Location Lighting**	Includes: lighting for application in areas classified as hazardous. See: 26 55 33 for hazard warning lighting.
26 55 00	**Special Purpose Lighting**	Includes: lighting equipment for specialized applications. Alternate Terms/Abbreviations: healthcare lighting: medical lighting. See: 11 13 26 for loading dock lights. 11 18 00 for security equipment. 11 19 00 for detention equipment. 11 59 00 for exhibit and display equipment. 11 61 00 for theater and stage equipment. 11 70 00 for healthcare equipment. 13 10 00 for swimming pools. 13 12 00 for fountains. 13 14 00 for aquatic park structures. 13 17 00 for tubs and pools. 26 54 00 for classified location lighting. 34 40 00 for transportation signals. 35 13 13 for navigation signals. See Also: 11 61 00 for theatrical lighting.
26 55 23	Outline Lighting	
26 55 29	Underwater Lighting	
26 55 33	Hazard Warning Lighting	
26 55 36	Obstruction Lighting	
26 55 39	Helipad Lighting	See: 34 43 00 Airfield Signaling and Control Equipment
26 55 53	Security Lighting	
26 55 59	Display Lighting	
26 55 61	Theatrical Lighting	
26 55 63	Detention Lighting	
26 55 70	Healthcare Lighting	

322

26110-14_F47L.EPS

Figure 47 Detailed breakdown of the electrical division (12 of 13).

NUMBER	TITLE	EXPLANATION
26 56 00	**Exterior Lighting**	

Includes: lighting equipment for exterior locations, except for special purpose and signal lighting. airfield general exterior lighting.

Alternate Terms/Abbreviations: athletic lighting: sports lighting.

See: 10 14 33 for illuminated panel signage.
11 13 26 for loading dock lights.
11 68 23 for exterior court athletic equipment.
32 94 00 for planting accessories.
34 41 13 for traffic signals.
34 42 13 for railway signals.
34 43 13 for airfield signals.
34 43 16 for airfield landing equipment.
34 71 00 for roadway construction.
34 72 00 for railway construction.
34 73 00 for airfield construction.
34 75 00 for roadway equipment.

26 56 13	Lighting Poles and Standards
26 56 16	Parking Lighting
26 56 19	Roadway Lighting
26 56 23	Area Lighting
26 56 26	Landscape Lighting
26 56 29	Site Lighting
26 56 33	Walkway Lighting
26 56 36	Flood Lighting
26 56 68	Exterior Athletic Lighting

323

26110-14_F47M.EPS

Figure 47 Detailed breakdown of the electrical division (13 of 13).

Summary

In this module, you learned the symbols and conventions used on architectural and engineering drawings. As an electrician, you need to know how to recognize the basic symbols used on electrical drawings and other drawings used in the building construction industry. You should also know where to find the meaning of symbols that you do not immediately recognize. Schedules, diagrams, and specifications often provide detailed information that is not included on the working drawings.

Building projects require detailed specifications. These written specifications are complex and detailed and need a unified format to be easily usable by the trades. The specification format most commonly used is the *MasterFormat®* developed by CSI and CSC. The *MasterFormat®* was updated in 2010 with changes to the division numbering system.

Reading architectural and engineering drawings takes practice and study. Now that you have the basic skills, take the time to master them.

Review Questions

1. A section line on a drawing shows _____.
   a. the north orientation
   b. the location of the section on the plan
   c. where to locate receptacles in that section
   d. the section scale

2. An electrical drafting line with a double arrowhead represents _____.
   a. wiring concealed in the floor
   b. wiring turned down
   c. a branch circuit homerun
   d. wiring concealed in a ceiling or wall

Questions 3 through 9 refer to the seven electrical symbols shown below. In the spaces provided, place the letter corresponding to the correct answer found in the list.

3.

4.

5. PC PC

6.

7.

8. Ceiling Wall

9.

26110-14_RQ01.EPS

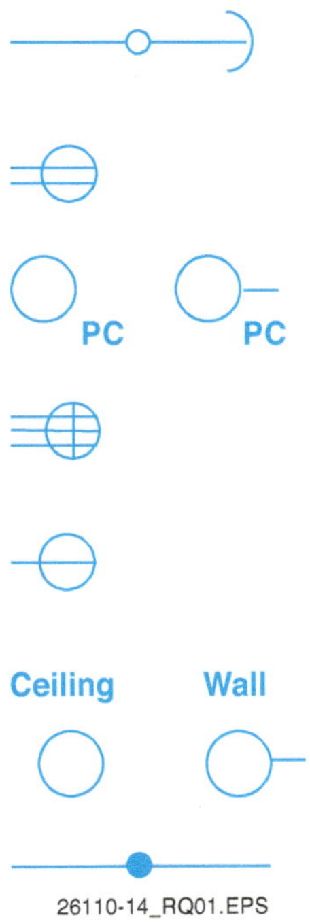

_____ a. Single Receptacle Outlet

_____ b. Duplex Receptacle Outlet

_____ c. Triplex Receptacle Outlet

_____ d. Incandescent Fixture (Surface or Pendant)

_____ e. Incandescent Fixture with Pull Chain (Surface or Pendant)

_____ f. Head Guy

_____ g. Sidewalk Guy

10. In dimension drawings, the dimensions written on the drawing are _____.
    a. for reference only
    b. on a larger scale
    c. inaccurate
    d. the actual dimensions

11. The architect's scale is designed so that one inch always equals one foot.
    a. True
    b. False

12. All views on a construction drawing are drawn to the same scale.
    a. True
    b. False

13. The *NEC*® specifies one set of electrical drawing symbols that are used in all cases.
    a. True
    b. False

14. Dotted lines used to represent a branch circuit on a drawing mean that the wiring is to be _____.
    a. concealed in the ceiling or wall
    b. run in the floor or ceiling below
    c. exposed
    d. installed in a future building expansion

15. A branch circuit line or drawing that does *not* have slashes is assumed to have two conductors.
    a. True
    b. False

16. To meet general recommendations, a residential branch circuit rated for 2,400VA should have a connected load of no more than _____.
    a. 1,680VA
    b. 1,920VA
    c. 2,040VA
    d. 2,160VA

17. Power-riser diagrams are used to show the _____.
    a. arrangement of electric service equipment
    b. branch circuit layout for power
    c. branch circuit layout for lighting
    d. panelboard schedule

18. The symbols T_1, T_2, and T_3 in a typical motor starter schematic represent _____.
    a. voltage supply lines
    b. auxiliary contacts
    c. motor terminals
    d. line contacts

19. The updated *MasterFormat*® standard _____.
    a. is specified in the *NEC*®
    b. uses a six-digit code for division content
    c. is required by OSHA
    d. allows for fewer subsections

20. The current *MasterFormat*® standard covering communications systems is under _____.
    a. Division 16
    b. Division 27
    c. Division 37
    d. Division 48

Trade Terms Quiz

Fill in the blank with the correct term that you learned from your study of this module.

1. ___________ typically include the following information: a site plan, floor plans, elevations of all exterior faces of the building, and large-scale detail drawings.
2. A(n) ___________ is an exact copy or reproduction of an original drawing.
3. A simple, single-line diagram used to show electrical equipment and related connections is a(n) ___________ diagram.
4. A(n) ___________ shows the path of an electrical circuit or system of circuits, along with the circuit components.
5. To convey a substantial amount of detailed information to installation electricians, an engineer will use a(n) ___________ drawing.
6. Shown in a separate view, a(n) ___________ view is an enlarged, detailed view taken from an area of a drawing.
7. A cutaway drawing that shows the inside of an object or building is a(n) ___________ drawing.
8. The sizes or measurements that are printed on a drawing are called ___________.
9. The relationship between an object's size in a drawing and the object's actual size is the ___________.
10. The height of the front, rear, or sides of a building is shown in a(n) ___________ drawing.
11. A building's location on the site is shown in a(n) ___________.
12. A drawing that has a top-down view of a building is a(n) ___________ plan.
13. A drawing that has a top-down view of a single object is a(n) ___________ view.
14. A(n) ___________ diagram is a single-line block diagram used to indicate the electric service equipment, service conductors and feeders, and subpanels.
15. Owners, architects, and engineers use ___________ to specify material and workmanship requirements.
16. A(n) ___________ is a systematic way of presenting equipment lists on a drawing in tabular form.
17. Complicated circuits, such as control circuits, are shown in a(n) ___________ diagram.
18. Usually developed by manufacturers, fabricators, or contractors, a(n) ___________ drawing shows specific dimensions and other information about a piece of equipment and its installation methods.

Trade Terms

Architectural drawings
Block diagram
Blueprint
Detail drawing
Dimensions
Electrical drawing
Elevation drawing
Floor plan
One-line diagram
Plan view
Power-riser diagram
Scale
Schedule
Schematic diagram
Sectional view
Shop drawing
Site plan
Written specifications

Module 26110-14
Supplemental Exercises

1. A(n) ______________________ indicates the location of the building on the property.
2. The ______________________ show the walls and partitions for each floor or level.
3. What are the three main functions of electrical drawings?

____

____

____

____

____

____

____

4. The title block of an electrical drawing should contain the following ten items:

____

____

____

____

____

____

____

____

____

____

5. Match the following names to their corresponding electrical drafting lines.

____ WIRING TURNED UP

____ BRANCH CIRCUIT HOMERUN TO PANELBOARD

____ WIRING TURNED DOWN

____ EXPOSED WIRING

____ WIRING CONCEALED IN FLOOR

____ WIRING CONCEALED IN CEILING OR WALL

26110-14_WB.EPS

6. What does the letter F stand for in reference to safety switches? ____________________

7. On a floor plan with a scale of ½" = 1'0", what would be the equivalent distance if you measured 3¾" on the drawing?____________________

8. The purpose of a(n) ____________________ is to identify that part of the project to which the sheet applies.

9. One-line block diagrams are also known as ____________________.

10. Divisions ____________________ and ____________________ of the current CSI specifications cover electrical work.

Cornerstone of Craftsmanship

Wayne Stratton

Associated Builders and Contractors

How did you choose a career in the electrical field?
Three events in my childhood created the desire to learn the electrical trade. At age six, the farmhouse we lived in was totally destroyed by fire. The cause was electrical. As a young teen, a local electrician had incorrectly wired a heating element and electrocuted several pigs. In 1973, my father hired this electrician to install a motor starter on a grain conveyor. He could not figure it out. I wanted to learn how to do this type of work and do it safely.

Tell us about your apprenticeship experience.
My education is from a technical school. I have attended several manufacturers' training sessions. I had to gain the hands-on experience after learning the trade. My observation of the apprenticeship programs is this: you get hands-on experience while you learn.

What positions have you held in the industry?
I worked as a plant industrial electrician responsible for motor control, DC motors, co-generation, and medium voltage distribution. Later, I began working for an electrical contractor who wanted to expand his business into the industrial field. I worked as a PLC technician designing and installing control systems. In 1987, I began teaching apprenticeship classes.

What would you say is the primary factor in achieving success?
The desire to learn all that I can learn, the ability to think outside the box, and the opportunities to gain a variety of experiences. All this helps me continue to learn and share with trainees.

What does your current job involve?
I teach electrical apprenticeship levels one through four at two different locations in Iowa. My other responsibilities involve task training for electrical licensing, fire alarm, and code updates.

Do you have any advice for someone just entering the trade?
Continue to learn. Completing an apprenticeship program or acquiring an electrician's license is not the end of learning. With code changes every 3 years, there is always more to learn. If you don't understand something, ask! Observe and learn from experienced individuals.

Appendix

Metric Conversion Chart

METRIC CONVERSION CHART

INCHES		METRIC	INCHES		METRIC	INCHES		METRIC
Fractional	Decimal	mm	Fractional	Decimal	mm	Fractional	Decimal	mm
.	0.0039	0.1000	.	0.5512	14.0000	.	1.8898	48.0000
.	0.0079	0.2000	9/16	0.5625	14.2875	.	1.9291	49.0000
.	0.0118	0.3000	.	0.5709	14.5000	.	1.9685	50.0000
1/64	0.0156	0.3969	37/64	0.5781	14.6844	2	2.0000	50.8000
.	0.0157	0.4000	.	0.5906	15.0000	.	2.0079	51.0000
.	0.0197	0.5000	19/32	0.5938	15.0813	.	2.0472	52.0000
.	0.0236	0.6000	39/64	0.6094	15.4781	.	2.0866	53.0000
.	0.0276	0.7000	.	0.6102	15.5000	.	2.1260	54.0000
1/32	0.0313	0.7938	5/8	0.6250	15.8750	.	2.1654	55.0000
.	0.0315	0.8000	.	0.6299	16.0000	.	2.2047	56.0000
.	0.0354	0.9000	41/64	0.6406	16.2719	.	2.2441	57.0000
.	0.0394	1.0000	.	0.6496	16.5000	2 1/4	2.2500	57.1500
.	0.0433	1.1000	21/32	0.6563	16.6688	.	2.2835	58.0000
3/64	0.0469	1.1906	.	0.6693	17.0000	.	2.3228	59.0000
.	0.0472	1.2000	43/64	0.6719	17.0656	.	2.3622	60.0000
.	0.0512	1.3000	11/16	0.6875	17.4625	.	2.4016	61.0000
.	0.0551	1.4000	.	0.6890	17.5000	.	2.4409	62.0000
.	0.0591	1.5000	45/64	0.7031	17.8594	.	2.4803	63.0000
1/16	0.0625	1.5875	.	0.7087	18.0000	2 1/2	2.5000	63.5000
.	0.0630	1.6000	23/32	0.7188	18.2563	.	2.5197	64.0000
.	0.0669	1.7000	.	0.7283	18.5000	.	2.5591	65.0000
.	0.0709	1.8000	47/64	0.7344	18.6531	.	2.5984	66.0000
.	0.0748	1.9000	.	0.7480	19.0000	.	2.6378	67.0000
5/64	0.0781	1.9844	3/4	0.7500	19.0500	.	2.6772	68.0000
.	0.0787	2.0000	49/64	0.7656	19.4469	.	2.7165	69.0000
.	0.0827	2.1000	.	0.7677	19.5000	2 3/4	2.7500	69.8500
.	0.0866	2.2000	25/32	0.7813	19.8438	.	2.7559	70.0000
.	0.0906	2.3000	.	0.7874	20.0000	.	2.7953	71.0000
3/32	0.0938	2.3813	51/64	0.7969	20.2406	.	2.8346	72.0000
.	0.0945	2.4000	.	0.8071	20.5000	.	2.8740	73.0000
.	0.0984	2.5000	13/16	0.8125	20.6375	.	2.9134	74.0000
7/64	0.1094	2.7781	.	0.8268	21.0000	.	2.9528	75.0000
.	0.1181	3.0000	53/64	0.8281	21.0344	.	2.9921	76.0000
1/8	0.1250	3.1750	27/32	0.8438	21.4313	3	3.0000	76.2000
.	0.1378	3.5000	.	0.8465	21.5000	.	3.0315	77.0000
9/64	0.1406	3.5719	55/64	0.8594	21.8281	.	3.0709	78.0000
5/32	0.1563	3.9688	.	0.8661	22.0000	.	3.1102	79.0000
.	0.1575	4.0000	7/8	0.8750	22.2250	.	3.1496	80.0000
11/64	0.1719	4.3656	.	.8858	22.5000	.	3.1890	81.0000
.	0.1772	4.5000	57/64	.89063	22.6219	.	3.2283	82.0000
3/16	0.1875	4.7625	.	.9055	23.0000	.	3.2677	83.0000
.	0.1969	5.0000	29/32	.90625	23.0188	.	3.3071	84.0000
13/64	0.2031	5.1594	59/64	.92188	23.4156	.	3.3465	85.0000
.	0.2165	5.5000	.	.9252	23.5000	.	3.3858	86.0000
7/32	0.2188	5.5563	15/16	.93750	23.8125	.	3.4252	87.0000
15/64	0.2344	5.9531	.	.9449	24.0000	.	3.4646	88.0000
.	0.2362	6.0000	61/64	.95313	24.2094	3 1/2	3.5000	88.9000
1/4	0.2500	6.3500	.	.9646	24.5000	.	3.5039	89.0000
.	0.2559	6.5000	31/32	.96875	24.6063	.	3.5433	90.0000
17/64	0.2656	6.7469	.	.9843	25.0000	.	3.5827	91.0000
.	0.2756	7.0000	63/64	.98438	25.0031	.	3.6220	92.0000
9/32	0.2813	7.1438	1	1.000	25.40	.	3.6614	93.0000
.	0.2953	7.5000	.	1.0039	25.5000	.	3.7008	94.0000
19/64	0.2969	7.5406	.	1.0236	26.0000	.	3.7402	95.0000
5/16	0.3125	7.9375	.	1.0433	26.5000	.	3.7795	96.0000
.	0.3150	8.0000	.	1.0630	27.0000	.	3.8189	97.0000
21/64	0.3281	8.3344	.	1.0827	27.5000	.	3.8583	98.0000
.	0.3346	8.5000	.	1.1024	28.0000	.	3.8976	99.0000
11/32	0.3438	8.7313	.	1.1220	28.5000	.	3.9370	100.0000
.	0.3543	9.0000	.	1.1417	29.0000	4	4.0000	101.6000
23/64	0.3594	9.1281	.	1.1614	29.5000	.	4.3307	110.0000
.	0.3740	9.5000	.	1.1811	30.0000	4 1/2	4.5000	114.3000
3/8	0.3750	9.5250	.	1.2205	31.0000	.	4.7244	120.0000
25/64	0.3906	9.9219	1 1/4	1.2500	31.7500	5	5.0000	127.0000
.	0.3937	10.0000	.	1.2598	32.0000	.	5.1181	130.0000
13/32	0.4063	10.3188	.	1.2992	33.0000	.	5.5118	140.0000
.	0.4134	10.5000	.	1.3386	34.0000	.	5.9055	150.0000
27/64	0.4219	10.7156	.	1.3780	35.0000	6	6.0000	152.4000
.	0.4331	11.0000	.	1.4173	36.0000	.	6.2992	160.0000
7/16	0.4375	11.1125	.	1.4567	37.0000	.	6.6929	170.0000
.	0.4528	11.5000	.	1.4961	38.0000	.	7.0866	180.0000
29/64	0.4531	11.5094	1 1/2	1.5000	38.1000	.	7.4803	190.0000
15/32	0.4688	11.9063	.	1.5354	39.0000	.	7.8740	200.0000
.	0.4724	12.0000	.	1.5748	40.0000	8	8.0000	203.2000
31/64	0.4844	12.3031	.	1.6142	41.0000	.	9.8425	250.0000
.	0.4921	12.5000	.	1.6535	42.0000	10	10.0000	254.0000
1/2	0.5000	12.7000	.	1.6929	43.0000	20	20.0000	508.0000
.	0.5118	13.0000	.	1.7323	44.0000	30	30.0000	762.0000
33/64	0.5156	13.0969	1 3/4	1.7500	44.4500	40	40.0000	1016.000
17/32	0.5313	13.4938	.	1.7717	45.0000	60	60.0000	1524.000
.	0.5315	13.5000	.	1.8110	46.0000	80	80.0000	2032.000
35/64	0.5469	13.8906	.	1.8504	47.0000	100	100.0000	2540.000

TO CONVERT TO MILLIMETERS, MULTIPLY INCHES X 25.4
TO CONVERT TO INCHES, MULTIPLY MILLIMETERS X 0.03937*
*FOR SLIGHTLY GREATER ACCURACY WHEN CONVERTING TO INCHES, DIVIDE MILLIMETERS BY 25.4

26110-14_A01.EPS

Trade Terms Introduced in This Module

Architectural drawings: Working drawings consisting of plans, elevations, details, and other information necessary for the construction of a building. Architectural drawings usually include:

- A site (plot) plan indicating the location of the building on the property
- Floor plans showing the walls and partitions for each floor or level
- Elevations of all exterior faces of the building
- Several vertical cross sections to indicate clearly the various floor levels and details of the footings, foundations, walls, floors, ceilings, and roof construction
- Large-scale detail drawings showing such construction details as may be required

Block diagram: A single-line diagram used to show electrical equipment and related connections. See *power-riser diagram.*

Blueprint: An exact copy or reproduction of an original drawing.

Detail drawing: An enlarged, detailed view taken from an area of a drawing and shown in a separate view.

Dimensions: Sizes or measurements printed on a drawing.

Electrical drawing: A means of conveying a large amount of exact, detailed information in an abbreviated language. Consists of lines, symbols, dimensions, and notations to accurately convey an engineer's designs to electricians who install the electrical system on a job.

Elevation drawing: An architectural drawing showing height, but not depth; usually the front, rear, and sides of a building or object.

Floor plan: A drawing of a building as if a horizontal cut were made through a building at about window level, and the top portion removed. The floor plan is what would appear if the remaining structure were viewed from above.

One-line diagram: A drawing that shows, by means of lines and symbols, the path of an electrical circuit or system of circuits along with the various circuit components. Also called a single-line diagram.

Plan view: A drawing made as though the viewer were looking straight down (from above) on an object.

Power-riser diagram: A single-line block diagram used to indicate the electric service equipment, service conductors and feeders, and subpanels. Notes are used on power-riser diagrams to identify the equipment; indicate the size of conduit; show the number, size, and type of conductors; and list related materials. A panelboard schedule is usually included with power-riser diagrams to indicate the exact components (panel type and size), along with fuses, circuit breakers, etc., contained in each panelboard.

Scale: On a drawing, the size relationship between an object's actual size and the size it is drawn. Scale also refers to the measuring tool used to determine this relationship.

Schedule: A systematic method of presenting equipment lists on a drawing in tabular form.

Schematic diagram: A detailed diagram showing complicated circuits, such as control circuits.

Sectional view: A cutaway drawing that shows the inside of an object or building.

Shop drawing: A drawing that is usually developed by manufacturers, fabricators, or contractors to show specific dimensions and other pertinent information concerning a particular piece of equipment and its installation methods.

Site plan: A drawing showing the location of a building or buildings on the building site. Such drawings frequently show topographical lines, electrical and communication lines, water and sewer lines, sidewalks, driveways, and similar information.

Written specifications: A written description of what is required by the owner, architect, and engineer in the way of materials and workmanship. Together with working drawings, the specifications form the basis of the contract requirements for construction.

Additional Resources

This module presents thorough resources for task training. The following resource material is suggested for further study.

National Electrical Code® Handbook, Latest Edition. Quincy, MA: National Fire Protection Association.

Figure Credits

AGC of America, Module opener

John Traister, Figures 6–18, 20–23, 29–31, 33–40, 43–45

Mike Powers, Figures 25, 27

MasterFormat® Numbers and Titles used in this book are from MasterFormat®, published by CSI and Construction Specifications Canada (CSC), and are used with permission from CSI. For those interested in a more in-depth explanation of MasterFormat® and its use in the construction industry visit www.masterformat.com or contact:

CSI
110 South Union Street, Suite 100
Alexandria, VA 22314
800-689-2900; 703-684-0300
www.csinet.org, Figures 46, 47

Topaz Publications, Inc., SA01

Staedtler USA, SA02

Scalex Corporation, SA03

NCCER CURRICULA — USER UPDATE

NCCER makes every effort to keep its textbooks up-to-date and free of technical errors. We appreciate your help in this process. If you find an error, a typographical mistake, or an inaccuracy in NCCER's curricula, please fill out this form (or a photocopy), or complete the online form at **www.nccer.org/olf**. Be sure to include the exact module ID number, page number, a detailed description, and your recommended correction. Your input will be brought to the attention of the Authoring Team. Thank you for your assistance.

Instructors – If you have an idea for improving this textbook, or have found that additional materials were necessary to teach this module effectively, please let us know so that we may present your suggestions to the Authoring Team.

NCCER Product Development and Revision

13614 Progress Blvd., Alachua, FL 32615

Email: curriculum@nccer.org
Online: www.nccer.org/olf

❑ Trainee Guide ❑ AIG ❑ Exam ❑ PowerPoints Other ______________________________

Craft / Level: Copyright Date:

Module ID Number / Title:

Section Number(s):

Description:

Recommended Correction:

Your Name:

Address:

Email: Phone:

Residential Electrical Services

Phoenix Fire Station No. 50

Phoenix's Fire Station No. 50 sports many environmentally protective measures in its construction, such as a roof made of recycled aluminum cans and terra-cotta colored terrazzo flooring made by grinding down the concrete structural slab. Recycled countertops are used in the kitchen and more than 80 percent of the lighting takes advantage of natural sources to save energy. The landscape is a xeriscape design that will require no irrigation after two years.

26111-14

Trainees with successful module completions may be eligible for credentialing through NCCER's National Registry. To learn more, go to **www.nccer.org** or contact us at **1.888.622.3720.** Our website has information on the latest product releases and training, as well as online versions of our *Cornerstone* magazine and Pearson's product catalog.

Your feedback is welcome. You may email your comments to **curriculum@nccer.org,** send general comments and inquiries to **info@nccer.org,** or fill in the User Update form at the back of this module.

This information is general in nature and intended for training purposes only. Actual performance of activities described in this manual requires compliance with all applicable operating, service, maintenance, and safety procedures under the direction of qualified personnel. References in this manual to patented or proprietary devices do not constitute a recommendation of their use.

26111-14
Residential Electrical Services

Objectives

When you have completed this module, you will be able to do the following:

1. Explain the role of the *National Electrical Code*® in residential wiring and describe how to determine electric service requirements for dwellings.
2. Explain the grounding requirements of a residential electric service.
3. Calculate and select service-entrance equipment.
4. Select the proper wiring methods for various types of residences.
5. Compute branch circuit loads and explain their installation requirements.
6. Explain the types and purposes of equipment grounding conductors.
7. Explain the purpose of ground fault circuit interrupters and tell where they must be installed.
8. Size outlet boxes and select the proper type for different wiring methods.
9. Describe rules for installing electric space heating and HVAC equipment.
10. Describe the installation rules for electrical systems around swimming pools, spas, and hot tubs.
11. Explain how wiring devices are selected and installed.
12. Describe the installation and control of lighting fixtures.

Performance Tasks

Under the supervision of the instructor, you should be able to do the following:

1. For a residential dwelling of a given size, and equipped with a given list of major appliances, demonstrate or explain how to:
   - Compute lighting, small appliance, and laundry loads.
   - Compute the loads for large appliances.
   - Determine the number of branch circuits required.
   - Size and select the service-entrance equipment (conductors, panelboard, and protective devices).
2. Using an unlabeled diagram of a panelboard (Performance Profile Sheet 3), label the lettered components.
3. Select the proper type and size outlet box needed for a given set of wiring conditions.

Trade Terms

Appliance
Bonding bushing
Bonding jumper
Branch circuit
Feeder
Load center
Metal-clad (MC) cable
Nonmetallic-sheathed (Type NM, NMC, NMS) cable
Romex®
Roughing in
Service drop
Service entrance
Service-entrance conductors
Service-entrance equipment
Service lateral
Switch
Switch leg

Required Trainee Materials

1. Paper and pencil
2. Copy of the latest edition of the *National Electrical Code*®
3. Appropriate personal protective equipment

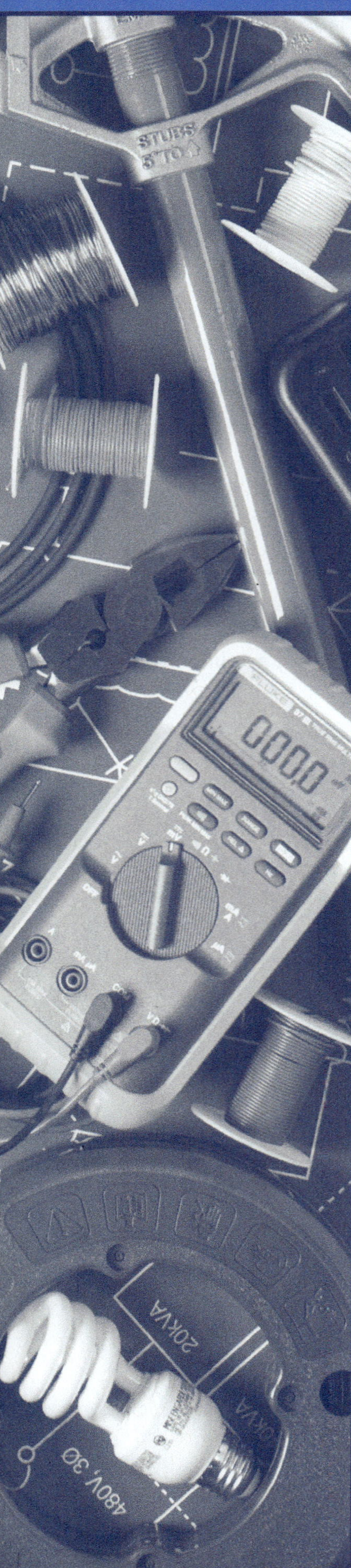

Contents

Topics to be presented in this module include:

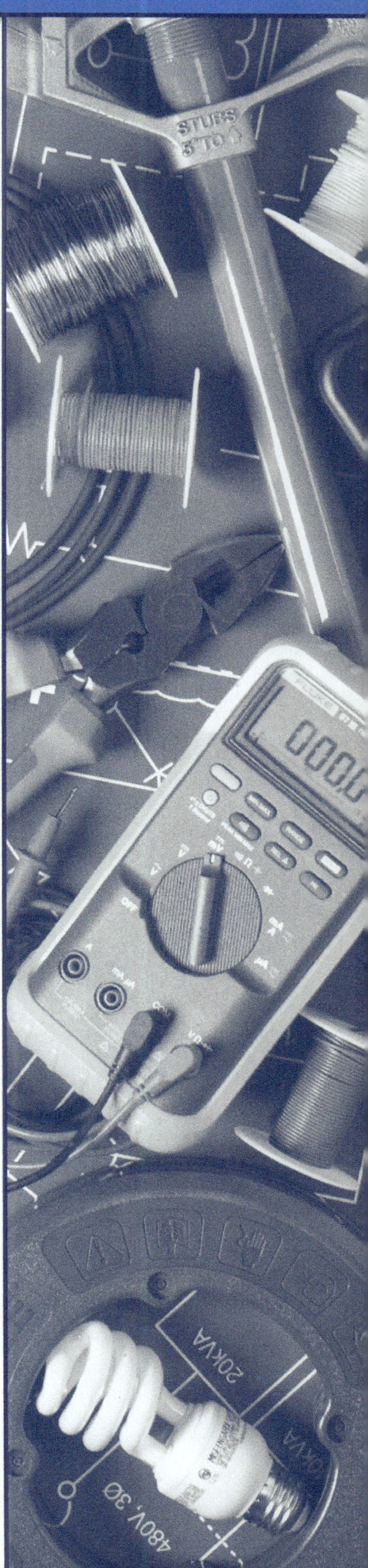

STUBS 3" TO
20KVA
480V, 3Ø

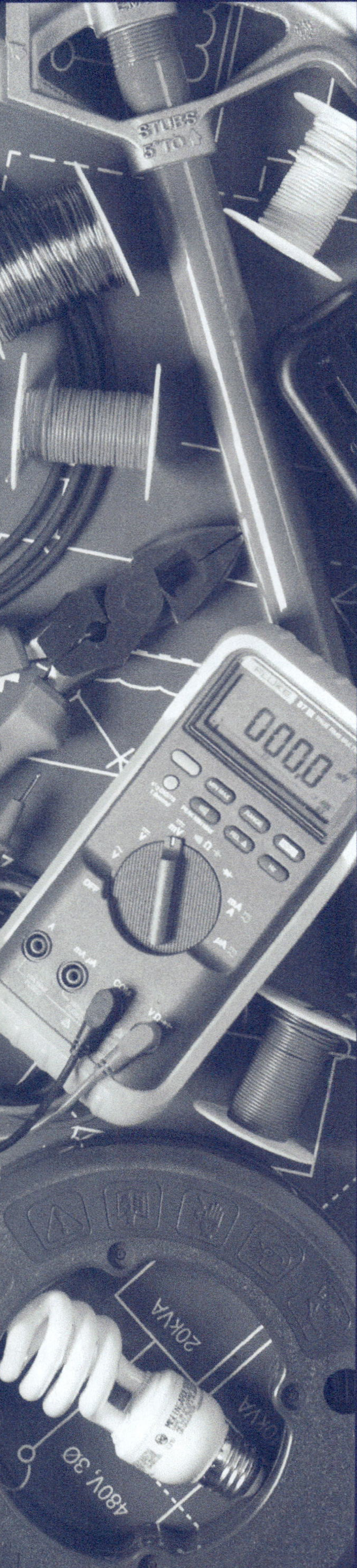

Figures

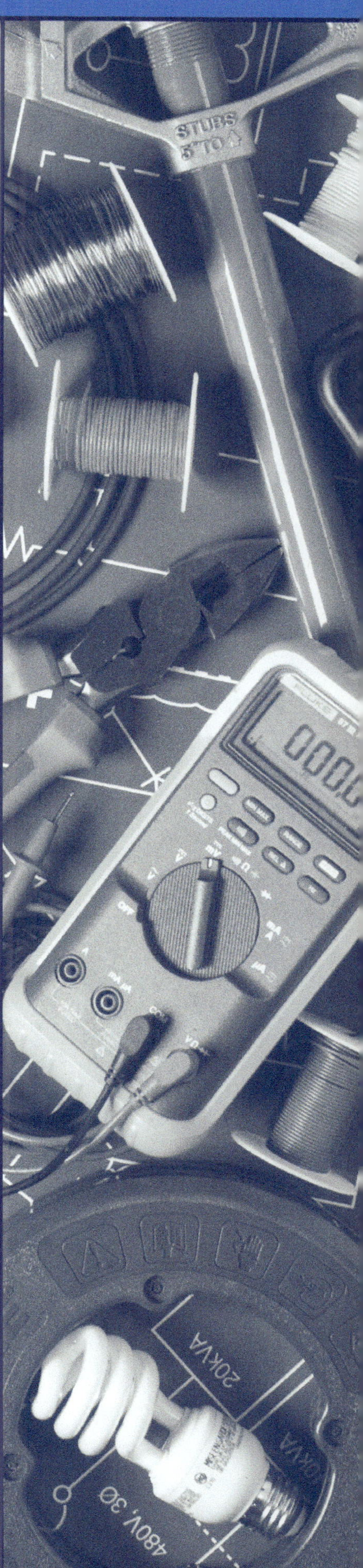
STUBS
5"TO
20KVA
480V, 3Ø

1.0.0 Introduction

The use of electricity in houses began shortly after the opening of the California Electric Light Company in 1879 and Thomas Edison's Pearl Street Station in New York City in 1882. These two companies were the first to enter the business of producing and selling electric service to the public. In 1886, the Westinghouse Electric Company secured patents that resulted in the development and introduction of alternating current; this paved the way for rapid acceleration in the use of electricity.

The primary use of early home electrical systems was to provide interior lighting, but today's uses of electricity include:

- Heating and air conditioning
- Electrical appliances
- Interior and exterior lighting
- Communications systems
- Alarm systems

When planning any electrical system, there are certain general steps to be followed, regardless of the type of construction. In planning a residential electrical system, the electrician must take certain factors into consideration. These include:

- Wiring method
- Overhead or underground electrical service
- Type of building construction
- Type of service entrance and equipment
- Grade of wiring devices and lighting fixtures
- Selection of lighting fixtures
- Type of heating and cooling system
- Control wiring for the heating and cooling system
- Signal and alarm systems
- Presence of alternative electrical systems, if any

The experienced electrician readily recognizes, within certain limits, the type of system that will be required. However, always check the local code requirements when selecting a wiring method. The *NEC*® provides minimum requirements for the practical safeguarding of persons and property from hazards arising from the use of electricity. These minimum requirements are not necessarily efficient, convenient, or adequate for good service or future expansion of electrical use. Some local building codes require electrical installations that surpass the requirements of the *NEC*®. For example, *NEC Section 230.51(A)* requires that service cable be secured by means of cable straps placed every 30 inches and within 12 inches of every service head, gooseneck, or connection to a raceway or enclosure. The electrical inspection department in one area requires these cable straps to be placed at a minimum distance of 18 inches.

If more than one wiring method may be practical, a decision as to which type to use should be made prior to beginning the installation.

See the *Appendix* for other codes and electrical standards that apply to residential electrical installations.

In a residential occupancy, the electrician should know that a 120/240-volt (V), single-phase service entrance will invariably be provided by the utility company. The electrician knows that the service and feeders will be three-wire, that the branch circuits will be either two- or three-wire, and that the safety switches, service equipment, and panelboards will be three-wire, solid neutral. On each project, however, the electrician must consult with the local utility to determine the point of attachment for overhead connections and the location of the metering equipment.

2.0.0 Sizing the Electrical Service

It may be difficult to decide at times which comes first, the layout of the outlets or the sizing of the electric service. In many cases, the service (main disconnect, panelboard, service conductors, etc.) can be sized using the *NEC*® before the outlets are actually located. In other cases, the outlets will have to be laid out first. However, in either case, the service entrance and panelboard locations will have to be determined before the circuits can be installed—so the electrician will know in which direction (and to what points) the circuit homeruns will terminate. In this module, a typical residence will be used as a model to size the electric service according to the latest edition of the *NEC*®.

2.1.0 Floor Plans

A floor plan is a drawing that shows the length and width of a building and the rooms that it contains. A separate plan is made for each floor.

Figure 1 shows how a floor plan is developed. An imaginary cut is made through the building as shown in the view on the left. The top half of this cut is removed (top right), and the resulting floor plan (bottom) is what the remaining structure looks like when viewed directly from above.

The floor plan for a small residence is shown in *Figure 2*. This building is constructed on a concrete slab with no basement or crawl space. There is an unfinished attic above the living area and an open carport just outside the kitchen entrance.

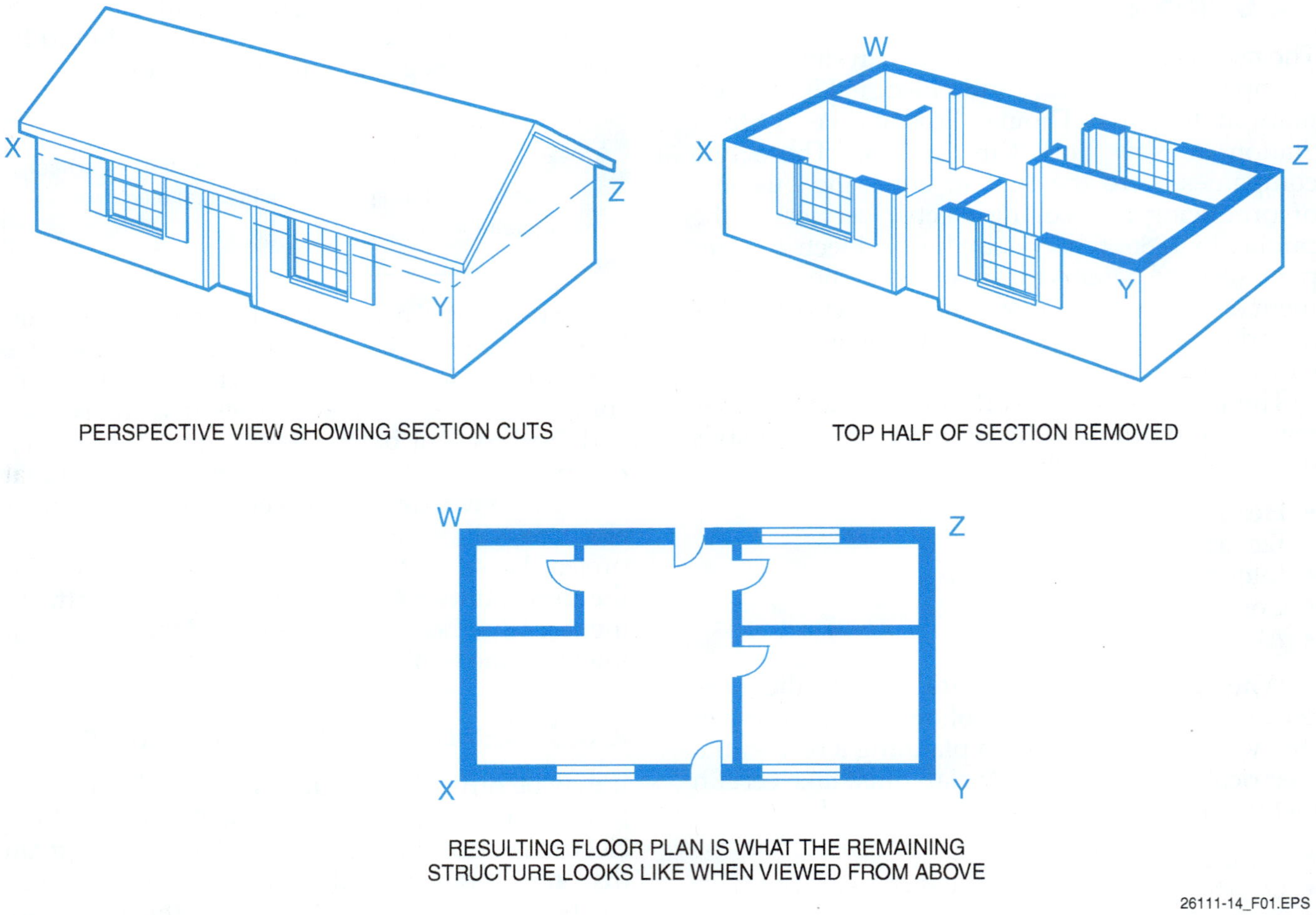

PERSPECTIVE VIEW SHOWING SECTION CUTS

TOP HALF OF SECTION REMOVED

RESULTING FLOOR PLAN IS WHAT THE REMAINING STRUCTURE LOOKS LIKE WHEN VIEWED FROM ABOVE

26111-14_F01.EPS

Figure 1 Principles of floor plan layout.

Appliances include a 12 kilovolt-ampere (kVA) electric range, a 4.5kVA water heater, a ½hp 120V disposal, and a 1.5kVA dishwasher.

There is also a washer/dryer (rated at 5.5kVA) in the utility room. A gas furnace with a ⅓hp 120V blower supplies the heating. In this module, the electrical requirements of this example building will be computed.

2.2.0 General Lighting Loads

General lighting loads are calculated on the basis of *NEC Table 220.12.* For residential occupancies, three volt-amperes (watts) per square foot of living space is the figure to use. This includes non-appliance duplex receptacles into which lamps, televisions, etc., may be connected. Therefore, the area of the building must be calculated first. If the building is under construction, the dimensions can be determined by scaling the working drawings used by the builder. If the residence is an existing building with no drawings, actual measurements will have to be made on the site.

Using the floor plan of the residence in *Figure 2* as a guide, an architect's scale is used to measure the longest width of the building (using outside dimensions). It is determined to be 33 feet. The longest length of the building is 48 feet. These two measurements multiplied together give 33 × 48 = 1,584 square feet of living area. However, there is an open carport on the lower left of the drawing. This carport area will have to be calculated and then deducted from 1,584 to give the true amount of living space. This open area (carport) is 12 feet wide by 19.5 feet long: 12 × 19.5 = 234 square feet. Subtract the carport area from 1,584 square feet: 1,584 – 234 = 1,350 square feet of living area.

When using the square-foot method to determine lighting loads for buildings, *NEC Section 220.12* requires the floor area for each floor to be computed from the outside dimensions. When calculating lighting loads for residences, the computed floor area must not include open porches, carports, garages, or unused or unfinished spaces that are not adaptable to future use.

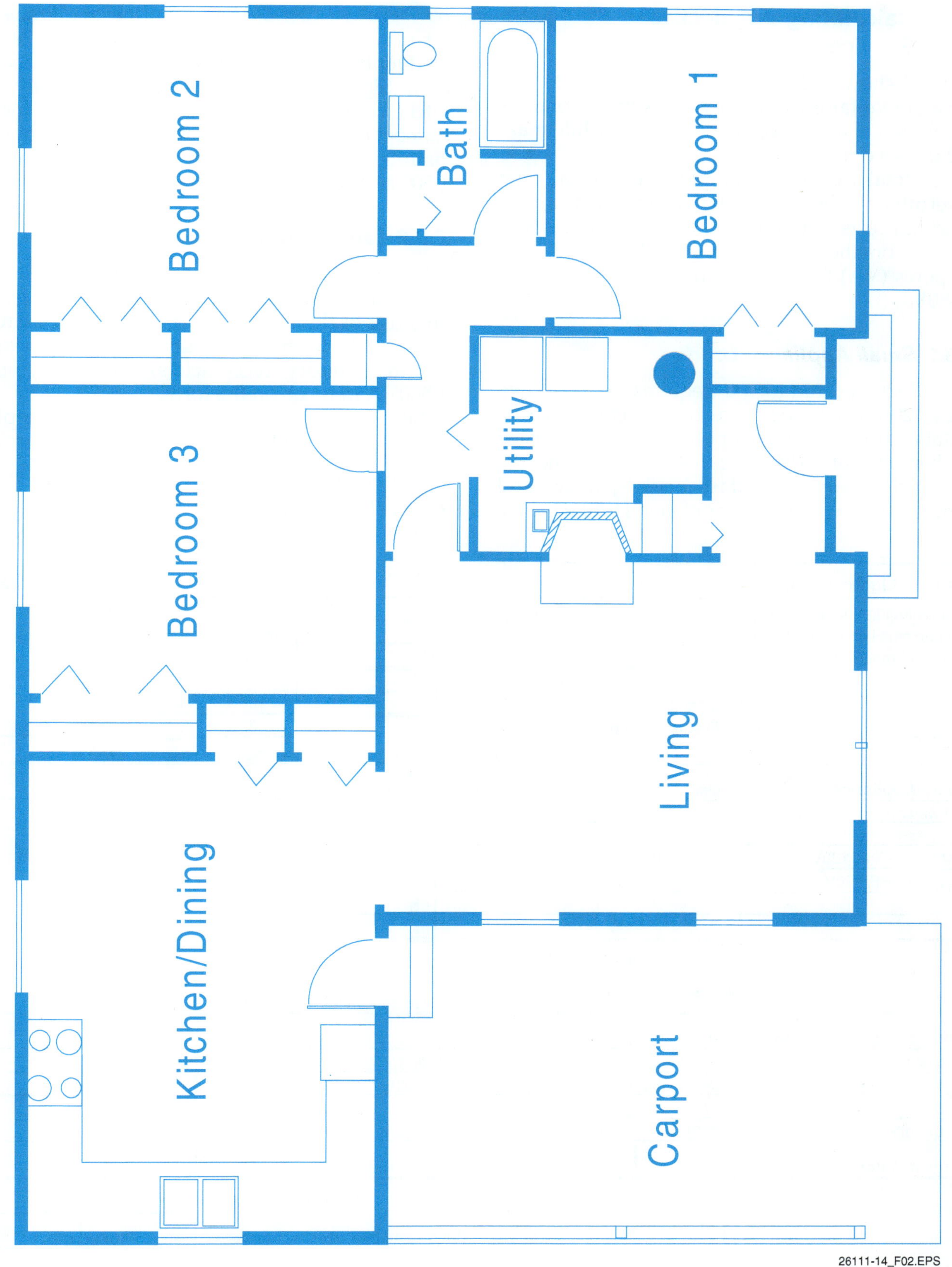

Figure 2 Floor plan of a typical residence.

2.3.0 Calculating the Electric Service Load

Figure 3 shows a standard calculation worksheet for a single-family dwelling. This form contains numbered blank spaces to be filled in while making the service calculation.

The total area of our sample dwelling has been determined to be 1,350 square feet of living space. This figure is entered in the appropriate space (Box 1) on the form and multiplied by 3 volt-amperes (VA) for a total general lighting load of 4,050VA (Box 2).

2.3.1 Small Appliance Loads

NEC Section 210.11(C)(1) requires at least two 120V, 20A small appliance branch circuits to be installed for the small appliance loads in each kitchen area of a dwelling. Kitchen areas include the dining area, breakfast nook, pantry, and similar areas where small appliances will be used.

NEC Section 220.52(A) gives further requirements for residential small appliance circuits; that is, the load for those circuits is to be computed at 1,500VA each. Since our example dwelling has only one kitchen area, the number 2 is entered in Box 3 for the number of required kitchen small appliance branch circuits. Multiply the number of these circuits by 1,500 and enter the result in Box 4.

2.3.2 Laundry Circuit

NEC Section 210.11(C)(2) requires an additional 20A branch circuit to be provided for the exclusive use of the laundry area (Box 5). This circuit must not have any other outlets connected except for the laundry receptacle(s). Therefore, enter 1,500VA in Box 6 on the form.

So far, there is enough information to complete the first portion of the service calculation form:

- General lighting 4,050VA (Box 2)
- Small appliance load 3,000VA (Box 4)

General Lighting Load					**Phase**	**Neutral**
Square footage of the dwelling	[1] 1350	× 3VA =	[2] 4050	*NEC Table 220.12*		
Kitchen small appliance circuits	[3] 2	× 1500 =	[4] 3000	*NEC Section 220.52(A)*		
Laundry branch circuit	[5] 1	× 1500 =	[6] 1500	*NEC Section 220.52(B)*		
Subtotal of gen. lighting loads			[7] 8550			
Subtract 1st 3000VA per *NEC Table 220.42*			[8] 3000	× 100% =	[9] 3000	
Remaining VA times 35% per *NEC Table 220.42*			[10] 5550	× 35% =	[11] 1943	
Total demand for general lighting loads =					[12] 4943	[13]
Fixed Appliance Loads (Nameplate or NEC FLA of motors) per *NEC Section 220.14*						
Hot water tank, 4.5kVA, 240V			[14] 4500			
Dishwasher 1.5kVA, 120V			[15] 1500			
Disposal 1/2HP, 120V per *NEC Table 430.248* = 9.8A			[16] 1176			
Blower 1/3HP, 120V per *NEC Table 430.248* = 7.2A			[17] 864			
			[18]			
			[19]			
Subtotal of fixed appliances			[20] 8040			
NEC Section 220.53 — If 3 or less fixed appliances take @ 100% =					[21]	[22]
NEC Section 220.53 — If 4 or more fixed appliances take @ 75% =					[23] 6030	[24]
Other Loads per *NEC Section 220.14*						
Electric Range per *NEC Section 220.55* [neutral @ 70% per *NEC Section 220.61(B)*]					[25] 8000	[26]
Electric Dryer per *NEC Section 220.54* [neutral @ 70% per *NEC Section 220.61(B)*]					[27] 5500	[28]
Electric Heat per *NEC Section 220.51* / Air Conditioning *NEC Section 220.82(C)*		omit smaller load per *NEC Section 220.60*			[29]	[30]
Largest Motor = 1176		× 25% (per *NEC Section 430.24*) =			[31] 294	[32]
Total VA Demand =					[33] 24767	[34]
(VA divided by 240 volts) **Amps** =					[35] **103**	[36]
Service OCD and minimum size grounding electrode conductor					[37] 125	[38]
AWG per *NEC Section 310.15(B)(7); NEC Section 220.61 and Table 310.15(B)(16)* for neutral					[39]	[40]

26111-14_F03.EPS

Figure 3 Calculation worksheet for residential requirements.

- Laundry load 1,500VA (Box 6)
- Total general lighting and appliance loads 8,550VA (Box 7)

2.3.3 Lighting Demand Factors

All residential electrical outlets are never used at one time. There may be a rare instance when all the lighting may be on for a short time every night, but even so, all the small appliances and receptacles throughout the house will never be used simultaneously. Knowing this, *NEC Section 220.42* allows a diversity or demand factor to be used when computing the general lighting load for services. Our calculation continues as follows:

- The first 3,000VA is rated at 100% 3,000VA (Box 8)
- The remaining 5,550VA (Box 10) may be rated at 35% (the allowable demand factor) Therefore, 5,550 × 0.35 = 1,943VA (Box 11)
- Net general lighting and small appliance load (rounded off) 4,943VA (Box 12)

2.3.4 Fixed Appliances

NEC Section 220.53 permits the loads for four fixed appliances in a single-family dwelling only to be computed at 75% as long as they are not electric heating, air conditioning, electric cooking, or electric clothes dryer loads. To compute the load of the fixed appliances in this dwelling, list all the fixed appliances that meet *NEC Section 220.53.* Enter the nameplate rating of the appliance or VA for motors by using *NEC Table 430.248* to find the FLA of each motor. *NEC Section 220.5(A)* tells us to use 120V (not 115V) for calculation purposes. The fixed appliances would be as follows:

- Hot water tank 4,500VA (Box 14)
- Dishwasher 1,500VA (Box 15)
- ½hp 120V disposal (9.8A × 120V) 1,176VA (Box 16)
- Gas furnace blower (7.2A × 120V) 864VA (Box 17)
- Add the loads for the fixed appliances 8,040VA (Box 20)
- Since there are four or more fixed appliances, multiply the total in Box 20 by 75% 6,030VA (Box 23)

2.3.5 Other Loads

The remaining loads of the dwelling are now computed in the Other Loads section in *Figure 3*. *NEC Section 220.14(B)* allows electric dryers to be computed as permitted in *NEC Table 220.54* and electric cooking appliances to be computed per *NEC Table 220.55.* For a single range rated over 8.75kVA, but not over 12kVA, Column C of *NEC Table 220.55* permits a demand of 8kVA for the range in this dwelling. Enter 8,000VA in Box 25.

The electric dryer must be computed at 5,000VA or the nameplate, whichever is greater, according to *NEC Section 220.54.* Up to four electric dryers must be taken at 100%. Enter 5,500VA in Box 27.

If this dwelling had electric space heating and/or air conditioning, it would be computed in this section using the larger of the two loads. Since they are typical noncoincidental loads, *NEC Section 220.60* permits the smaller of those loads to be omitted. There are no demand factors for either electric heating or air conditioning; therefore, the larger of the two loads would be computed at 100%.

The final step in this calculation is to add in 25% of the largest motor in the dwelling. This dwelling unit has two motors: the disposal at 9.8A and the blower at 7.2A. (See *NEC Section 430.17.*) In this case, the larger motor is the disposal; therefore, we must add 25% of the rating to meet the requirements of *NEC Section 430.24.* Enter 294VA (1,176 × 25%) in Box 31. Adding together the individual loads as computed, we have a minimum demand of 24,767VA (Box 33) for the phase conductors.

2.3.6 Required Service Size

The conventional electric service for residential use is 120/240V, three-wire, single-phase. Services are sized in amperes, and when the volt-amperes are known on single-phase services, amperes may be found by dividing the highest voltage into the total volt-amperes. For example:

$$24{,}767\text{VA} \div 240\text{V} = 103\text{A (Box 35)}$$

The **service-entrance conductors** have now been calculated and must be rated at a minimum of 110A, which is a standard rating for overcurrent protection. However, this is not a typical trade size; therefore, we will use the more common rating of 125A as the size of our service.

If the demand for our dwelling unit had resulted in a load of less than 100A, *NEC Section 230.79(C)* would have required that the minimum rating of the service disconnect be 100A. *NEC Section 230.42(B)* would have required the ampacity of the service conductors to be equal to the rating of the 100A disconnect as well.

2.4.0 Demand Factors

NEC Article 220, Part III provides the rules regarding the application of demand factors to certain types of loads. Recall that a demand factor is the maximum amount of volt-amp load expected at any given time compared to the total connected load of the circuit. The maximum demand of a feeder circuit is equal to the connected load times the demand factor. The loads to which demand factors apply can be found in the *NEC®* as follows:

- Receptacle loads — *NEC Table 220.14*
- Lighting loads — *NEC Table 220.42*
- Dryer loads — *NEC Table 220.54*
- Range loads — *NEC Table 220.55*

In addition to those demand factors listed in *NEC Article 220, Part III,* alternative (optional) methods for computing loads can be found in *NEC Article 220, Part IV.* They include the following:

- Dwelling unit loads — *NEC Section 220.82*
- Existing dwelling unit loads — *NEC Section 220.83*
- Multi-family dwelling unit loads — *NEC Section 220.84*

2.5.0 General Lighting and Receptacle Load Demand Factors

NEC Table 220.42 provides the demand factors allowed for dwelling units and apartment houses without provisions for tenant cooking.

2.6.0 Appliance Loads

NEC Section 210.11(C) provides the number of branch circuits required for small appliances and laundry loads. Demand factors for dryers and ranges are found in *NEC Tables 220.54 and 220.55*.

2.6.1 Small Appliance Loads

The small appliance branch circuits required by *NEC Section 210.11(C)(1)* for small appliances supplied by 15A or 20A receptacles on 20A branch circuits for each kitchen area served are calculated at 1,500VA. If a dwelling has more than one kitchen area, the *NEC®* will require two small appliance branch circuits computed at 1,500VA for each kitchen area served. Where a dwelling with only one kitchen area has more than the required two small appliance branch circuits installed to serve a single kitchen area, only the first two required circuits need be computed. Additional circuits for countertops or refrigeration provide a separation of load, not additional loads. If a dwelling has two kitchen areas, then the total small appliance branch circuits required would be four at 1,500VA each. These loads are permitted to be included with the general lighting load and subjected to the demand factors of *NEC Table 220.42.*

2.6.2 Laundry Circuit Load

A 1,500VA feeder load is added to load calculations for each two-wire laundry branch circuit installed in a home. The branch circuit is required by *NEC Section 210.11(C)(2).* This load may also be added to the general lighting load and subjected to the same demand factors provided in *NEC Section 220.42.*

2.6.3 Dryer Load

The dryer load for each electric clothes dryer is 5,000VA or the actual nameplate value of the dryer, whichever is larger. Demand factors listed in *NEC Table 220.54* may be applied for more than one dryer in the same dwelling. If two or more single-phase dryers are supplied by a three-phase, four-wire feeder, the total load is computed by using twice the maximum number connected between any two phases.

2.6.4 Range Load

Range loads and other cooking appliances are covered under *NEC Section 220.55.* The feeder demand loads for household electric ranges, wall-mounted ovens, countertop cooking units, and other similar household appliances individually rated over 1¾kW are permitted to be computed in accordance with *NEC Table 220.55.* If two or more single-phase ranges are supplied by a three-phase, four-wire feeder, the total load is computed by using twice the maximum number connected between any two phases.

Think About It

Demand Factors

Examine *NEC Table 220.55.* Why does the demand factor decrease as the number of appliances increases? Why does the demand factor decrease more for larger ranges than it does for smaller ones?

2.6.5 Demand Factors for Electric Ranges

Ranges can be computed in various ways that depend on which part of *NEC Article 220* you are using and the occupancy type for the ranges involved. Note the demand factors permitted for the following occupancy types:

- Dwelling units per Part III — *NEC Section 220.55*
- Dwelling units per Part IV — *NEC Section 220.82*
- Additions to existing dwellings per Part IV — *NEC Section 220.83*
- Multi-family dwellings per Part III — *NEC Section 220.55*
- Multi-family dwellings per Part IV — *NEC Section 220.84*

2.7.0 Demand Factors for Neutral Conductors

The neutral conductor of electrical systems generally carries only the maximum current imbalance of the phase conductors. For example, in a single-phase feeder circuit with one phase conductor carrying 50A and the other carrying 40A, the neutral conductor would carry 10A. Since the neutral in many cases will never be required to carry as much current as the phase conductors, the *NEC®* allows us to apply a demand factor. (See *NEC Section 220.61.*) Note that in certain circumstances such as electrical discharge lighting, data processing equipment, and other similar equipment, a demand factor cannot be applied to the neutral conductors because these types of equipment produce harmonic currents that increase the heating effect in the neutral conductor.

Think About It

Balanced Phase Conductors

The word *phase* is used in these modules to refer to a hot wire rather than a neutral one. Some electricians call these legs rather than phases. Why must the two phase conductors be balanced?

3.0.0 Sizing Residential Neutral Conductors

The neutral conductor in a three-wire, single-phase service carries only the unbalanced load between the two ungrounded (hot) wires or legs. Since there are several 240V loads in the above calculations, these 240V loads will be balanced and therefore reduce the load on the service neutral conductor. Consequently, in most cases, the service neutral does not have to be as large as the ungrounded (hot) conductors.

In the previous example, the water heater does not have to be included in the neutral conductor calculation, since it is strictly 240V with no 120V loads. This takes the total number of fixed appliances on the neutral conductor down to three appliances. Therefore, each of the fixed appliance loads on the neutral must be computed at 100% (dishwasher at 1,500VA, plus disposal at 1,176VA, plus the blower at 864VA). The neutral loads of the electric range and clothes dryer are permitted by *NEC Section 220.61* to be computed at 70% of the demand for the phase conductors since these appliances have both 120V and 240V loads. In this case, the largest motor is the same for the neutral conductors as it is for the phase conductors; therefore, it is computed in the same manner. Using this information, the neutral conductor may be sized accordingly:

- Net general lighting and small appliance load — 4,943VA (Box 13)
- Fixed appliance loads — 3,540VA (Box 22)
- Electric range (8,000VA × 0.70) — 5,600VA (Box 26)
- Clothes dryer (5,500VA × 0.70) — 3,850VA (Box 28)
- Largest motor — 294VA (Box 32)
- Total — 18,227VA (Box 34)

To find the total phase-to-phase amperes, divide the total volt-amperes by the voltage between phases:

$$18{,}227\text{VA} \div 240\text{V} = 75.9\text{A or } 76\text{A}$$

The service-entrance conductors have now been calculated and are rated at 125A with a neutral conductor rated for at least 76A. See *Figure 4* for a completed calculation form for the example residence.

In *NEC Section 310.15(B)(7),* special consideration is given to 120/240V, single-phase residential services and feeders. Conductor sizes are

General Lighting Load				Phase	Neutral
Square footage of the dwelling	[1] 1350	× 3VA =	[2] 4050	*NEC Table 220.12*	
Kitchen small appliance circuits	[3] 2	× 1500 =	[4] 3000	*NEC Section 220.52(A)*	
Laundry branch circuit	[5] 1	× 1500 =	[6] 1500	*NEC Section 220.52(B)*	
Subtotal of gen. lighting loads			[7] 8550		
Subtract 1st 3000VA per *NEC Table 220.42*			[8] 3000	× 100% = [9] 3000	
Remaining VA times 35% per *NEC Table 220.42*			[10] 5550	× 35% = [11] 1943	
Total demand for general lighting loads =				[12] 4943	[13] 4943
Fixed Appliance Loads (Nameplate or NEC FLA of motors) per *NEC Section 220.14*					
Hot water tank, 4.5kVA, 240V			[14] 4500		
Dishwasher 1.5kVA, 120V			[15] 1500		
Disposal 1/2HP, 120V per *NEC Table 430.248* = 9.8A			[16] 1176		
Blower 1/3HP, 120V per *NEC Table 430.248* = 7.2A			[17] 864		
			[18]		
			[19]		
Subtotal of fixed appliances			[20] 8040		
NEC Section 220.53 — If 3 or less fixed appliances take @ 100% =				[21]	[22] 3540
NEC Section 220.53 — If 4 or more fixed appliances take @ 75% =				[23] 6030	[24]
Other Loads per *NEC Section 220.14*					
Electric Range per *NEC Section 220.55* [neutral @ 70% per *NEC Section 220.61(B)*]				[25] 8000	[26] 5600
Electric Dryer per *NEC Section 220.5* [neutral @ 70% per *NEC Section 220.61(B)*]				[27] 5500	[28] 3850
Electric Heat per *NEC Section 220.51*		omit smaller load per *NEC Section 220.60*		[29]	[30]
Air Conditioning *NEC Section 220.82(C)*					
Largest Motor = 1176		× 25% (per *NEC Section 430.24*) =		[31] 294	[32] 294
Total VA Demand =				[33] 24767	[34] 18227
(VA divided by 240 volts) **Amps** =				[35] **103**	[36] **76**
Service OCD and minimum size grounding electrode conductor				[37] 125	[38] 8 AWG
AWG per *NEC Section 310.15(B)(7); NEC Section 220.61 and Table 310.15(B)(16)* for neutral				[39] 2 AWG	[40] 4 AWG

26111-14_F04.EPS

Figure 4 Completed calculation form.

shown in *NEC Table 310.15(B)(7).* Reference to this table shows that the *NEC*® allows a No. 2 AWG copper or a 1/0 AWG aluminum conductor for a 125A service. The neutral conductor is sized per *NEC Tables 310.15(B)(7) or 310.15(B)(16)* using the appropriate column for the markings on the service equipment per *NEC Section 110.14(C).* Assuming our service panel is marked as suitable for use with 75°C-rated conductors, the minimum size of the neutral would be a No. 4 AWG copper or No. 2 AWG aluminum.

When sizing the grounded conductor for services, the provisions stated in *NEC Sections 215.2, 220.61, and 230.42* must be met, along with other applicable sections.

4.0.0 Sizing the Load Center

Each ungrounded conductor in all circuits must be provided with overcurrent protection in the form of either fuses or circuit breakers. If more than six such devices are used, a means of disconnecting the entire service must be provided using either a main disconnect switch or a main circuit breaker.

To calculate the number of fuse holders or circuit breakers required in the sample residence, look at the general lighting load first. The total general lighting load of 4,050VA can be divided by 120V to find the amperage:

$$4{,}050\text{VA} \div 120\text{V} = 33.75\text{A}$$

Either 15A or 20A circuits may be used for the lighting load. Two 20A circuits (2 × 20) equal 40A, so two 20A circuits would be adequate for the lighting. However, two 15A circuits totalling only 30A and 33.75A are needed. Therefore, if 15A circuits are used, three will be required for the total lighting load. In this example, three 15A circuits will be used.

In addition to the lighting circuits, the sample residence will require a minimum of two 20A circuits

> **Think About It**
>
> ## Common Loads
>
> Which of the following devices uses the most power?
>
> - Giant-screen television
> - Typical hair dryer
> - Curling iron
> - Crockpot

for the small appliance load and one 20A circuit for the laundry. So far, the following branch circuits can be counted:

• General lighting load	Three 15A circuits
• Small appliance load	Two 20A circuits
• Laundry load	One 20A circuit
• Total	Six branch circuits

Most load centers and panelboards are provided with an even number of circuit breaker spaces or fuse holders (for example, four, six, eight, or ten). But before the panelboard can be selected, space must be provided for the remaining loads. Each 240V load will require two spaces. In some existing installations, you might find a two-pole fuse block containing two cartridge fuses being used to feed a residential electric range. Each 120V load will require one space each. Thus, the remaining number of circuits for this example is as follows:

• Hot water heater	One two-pole breaker
• Dishwasher	One single-pole breaker
• Disposal	One single-pole breaker
• Blower	One single-pole breaker
• Electric range	One two-pole breaker
• Electric dryer	One two-pole breaker

These additional appliances will therefore require an additional nine spaces in the load center or panelboard. *NEC Section 210.11(C)(3)* requires that a separate 20A branch circuit be provided for the bathroom receptacles. While this circuit requires extra space within a load center, it does not add to the demand on the service for a dwelling unit. Adding the nine spaces for the other loads in the dwelling, plus one for a bathroom circuit, to the six required for the general lighting and small appliance loads requires at least a 16-space load center to handle the circuits.

4.1.0 Ground Fault Circuit Interrupters

Under certain conditions, the amount of current it takes to open an overcurrent protective device can be critical. You should remember from the *Electrical Safety* module that when persons are subject to very low current values (less than one full ampere), it can be fatal. The overcurrent protection installed on services, feeders, and branch circuits protects only the conductors and equipment.

Because of this fact, the *NEC®* requires ground fault circuit interrupter (GFCI) protection for receptacle outlets and/or equipment in many locations and occupancies. The *NEC®* defines a GFCI as "a device intended for the protection of personnel that functions to de-energize a circuit or portion thereof within an established period of time when a current to ground exceeds the values established for a Class A device." Class A GFCIs trip when the current to ground has a value in the range of 4mA to 6mA.

For dwelling units, the majority of requirements to provide protection for 15A or 20A, 125V-rated receptacles can be found in *NEC Section 210.8(A).* Further requirements for GFCI protection at dwelling units can be found in other *NEC®* articles such as *NEC Article 590* for temporary construction sites; *NEC Article 620* for special equipment such as elevators; or in *NEC Article 680* for special equipment such as swimming pools, hot tubs, and hydromassage tubs. These articles may also expand the requirements for GFCI protection to include circuits rated at more than 20A or operating at 240V.

According to *NEC Section 210.8(A),* the 15A and 20A, 125V-rated receptacles in a dwelling that require GFCI protection must be readily accessible and include the following:

- Bathrooms
- Outdoor receptacles (except those provided on dedicated circuits for snow melting and de-icing equipment)
- Receptacles that serve the countertops in kitchens

Further requirements for GFCI protection at dwelling units are as follows:

- Receptacles within garages and accessory buildings, such as storage sheds or workshops, or similar uses that have a floor located at or below grade level
- Receptacles in unfinished basements
- Crawl spaces at or below grade level
- Receptacles that serve countertops and are within 6' of wet bar sinks, utility, or laundry sinks
- Boathouses and boat hoists

One way to provide this GFCI protection is through the use of a GFCI circuit breaker. GFCI circuit breakers require the same mounting space as standard single-pole circuit breakers and provide the same branch circuit wiring protection as standard circuit breakers. They also provide Class A ground fault protection.

Listed GFCI circuit breakers are available in single- and two-pole construction; 15A, 20A, 25A, and 30A, 50A, and 60A ratings; and have a 10,000A interrupting capacity. Single-pole units are rated at 120VAC; two-pole units are rated at 120/240VAC.

GFCI breakers can be used not only in load centers and panelboards, but they are also available factory-installed in meter pedestals and power outlet panels for recreational vehicle (RV) parks and construction sites.

The GFCI sensor continuously monitors the current balance in the ungrounded or energized (hot) load conductor and the neutral load conductor. If the current in the neutral load wire becomes less than the current in the hot load wire, then a ground fault exists, since a portion of the current is returning to the source by some means other than the neutral load wire. When a current imbalance occurs, the sensor, which is a differential current transformer, sends a signal to the solid-state circuit, which activates the ground trip solenoid mechanism and breaks the hot load connection (*Figure 5*). A current imbalance as low as four milliamps (4mA) will cause the circuit breaker to interrupt the circuit. This is indicated by the trip indicator on the front of the device.

The two-pole GFCI breaker (*Figure 6*) continuously monitors the current balance between the two hot conductors and the neutral conductor. As long as the sum of these three currents is zero, the device will not trip; that is, if the A load wire is carrying 10A of current, the neutral is carrying 5A, and the B load wire is carrying 5A, then the sensor is balanced and will not produce a signal. A current imbalance from a ground fault condition as low as 4mA will cause the sensor to produce a signal of sufficient magnitude to trip the device.

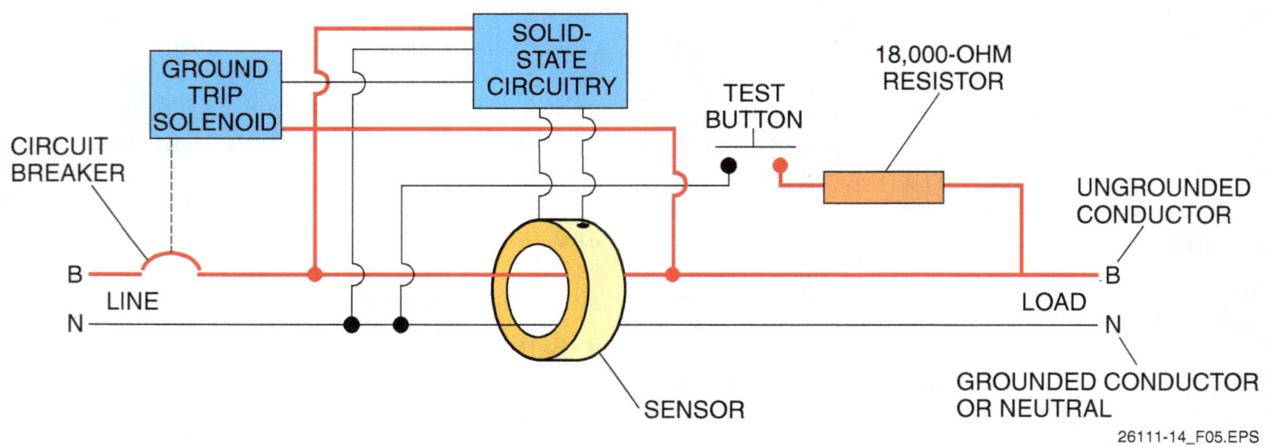

Figure 5 Operating circuitry of a typical GFCI.

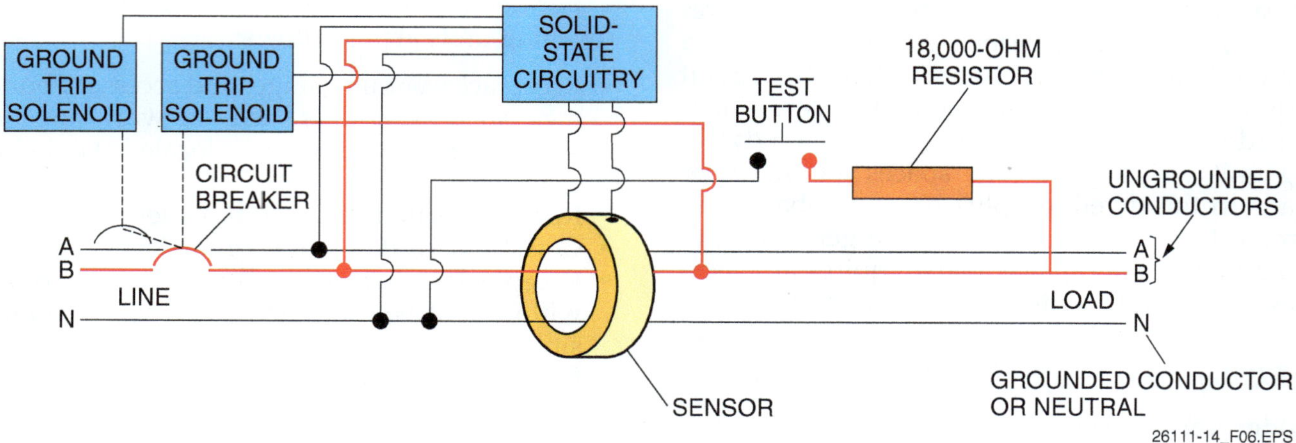

Figure 6 Operating characteristics of a two-pole GFCI.

4.1.1 Single-Pole GFCI Circuit Breakers

The single-pole GFCI breaker has two load lugs and a white wire pigtail in addition to the line side plug-on or bolt-on connector. The line side hot connection is made by installing the GFCI breaker in the panel just as any other circuit breaker is installed. The white wire pigtail is attached to the panel neutral (S/N) assembly. Both the neutral and hot wires of the branch circuit being protected are terminated in the GFCI breaker. These two load lugs are clearly marked Load Power and Load Neutral in the breaker case. Also in the case is the identifying marking for the pigtail, Panel Neutral.

> **NOTE**
> Single-pole GFCI circuit breakers cannot be used on multi-wire circuits.

Care should be exercised when installing GFCI breakers in existing panels. Be sure that the neutral wire for the branch circuit corresponds with the hot wire of the same circuit. Always remember that unless the current in the neutral wire is equal to that in the hot wire (within 4mA), the GFCI breaker senses this as being a possible ground fault (see *Figure 7*).

4.1.2 Two-Pole GFCI Circuit Breakers

A two-pole GFCI circuit breaker can be installed on a 120/240VAC single-phase, three-wire system; the 120/240VAC portion of a 120/240VAC three-phase, four-wire system; or the two phases and neutral of a 120/208VAC three-phase, four-wire system. Regardless of the application, the installation of the breaker is the same—connections are made to two hot buses and the panel neutral assembly. When installed on these systems, protection is provided for two-wire 240VAC or 208VAC circuits, three-wire 120/240VAC or 120/208VAC circuits, and 120VAC multiwire circuits.

The circuit in *Figure 8* illustrates the problems that are encountered when a common load neutral is used for two single-pole GFCI breakers. Either or both breakers will trip when a load is applied at the #2 duplex receptacle. The neutral current from the #2 duplex receptacle flows through breaker #1; this increase in neutral current through breaker #1 causes an imbalance in its sensor, thus causing it to produce a fault signal. At the same time, there is no neutral current flowing through breaker #2; therefore, it also senses a current imbalance. If a load is applied at the #1 duplex receptacle, and there is no load at the #2 duplex receptacle, then neither breaker will trip because neither breaker will sense a current imbalance.

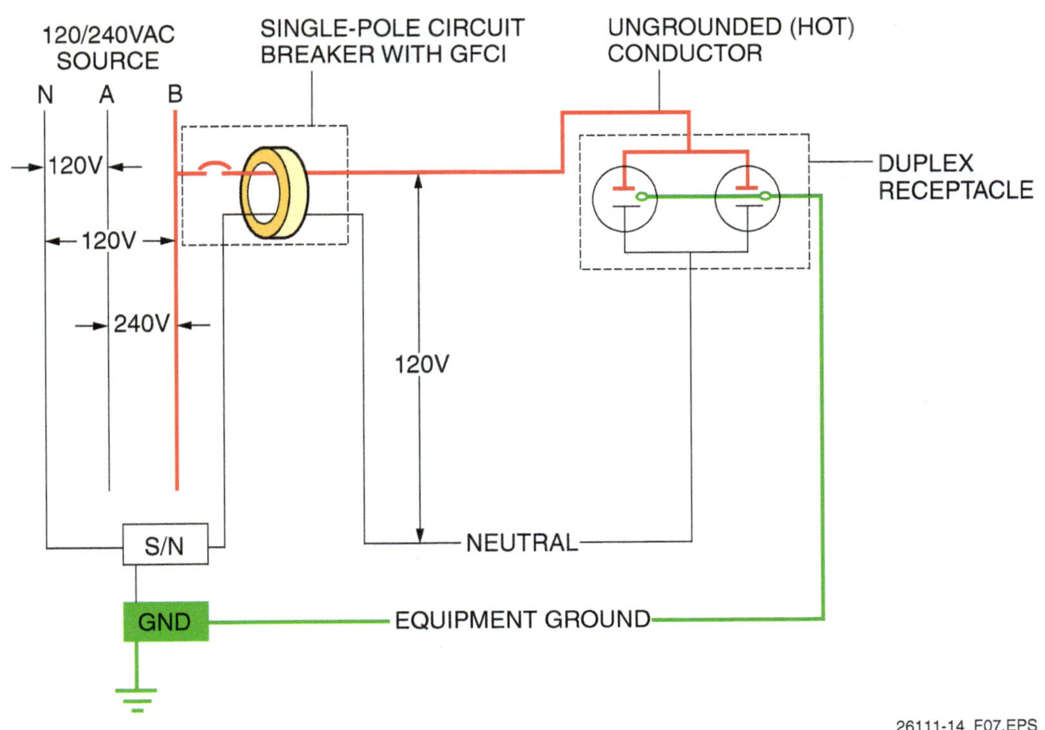

Figure 7 Operating characteristics of a single-pole circuit breaker with a GFCI.

On Site

GFCI and AFCI Circuit Breakers

GFCI breakers protect against ground faults, while AFCI breakers protect against arc faults.

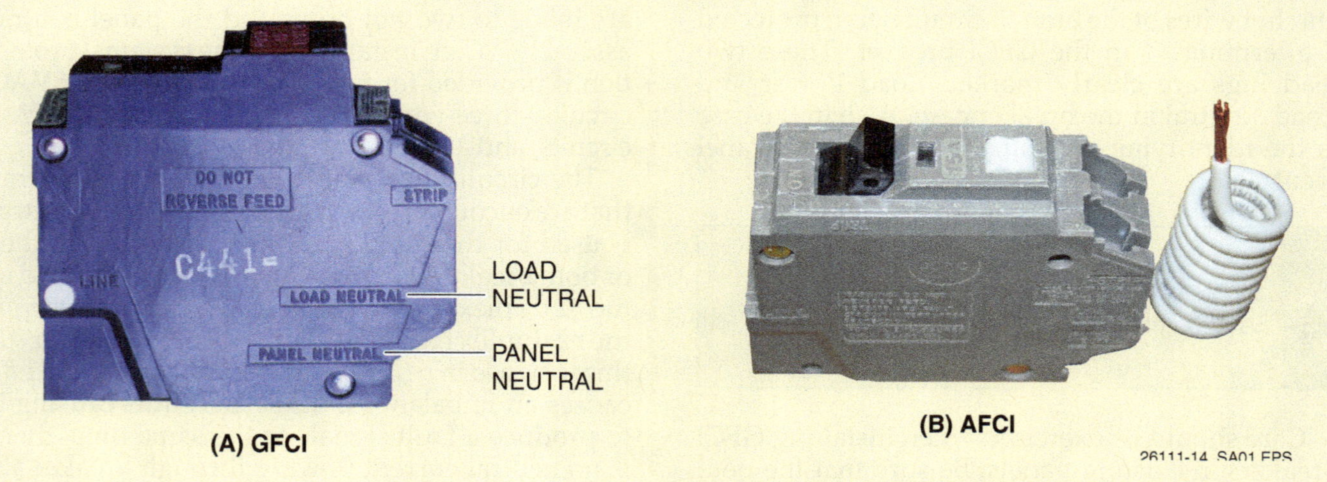

(A) GFCI

(B) AFCI

26111-14_SA01.EPS

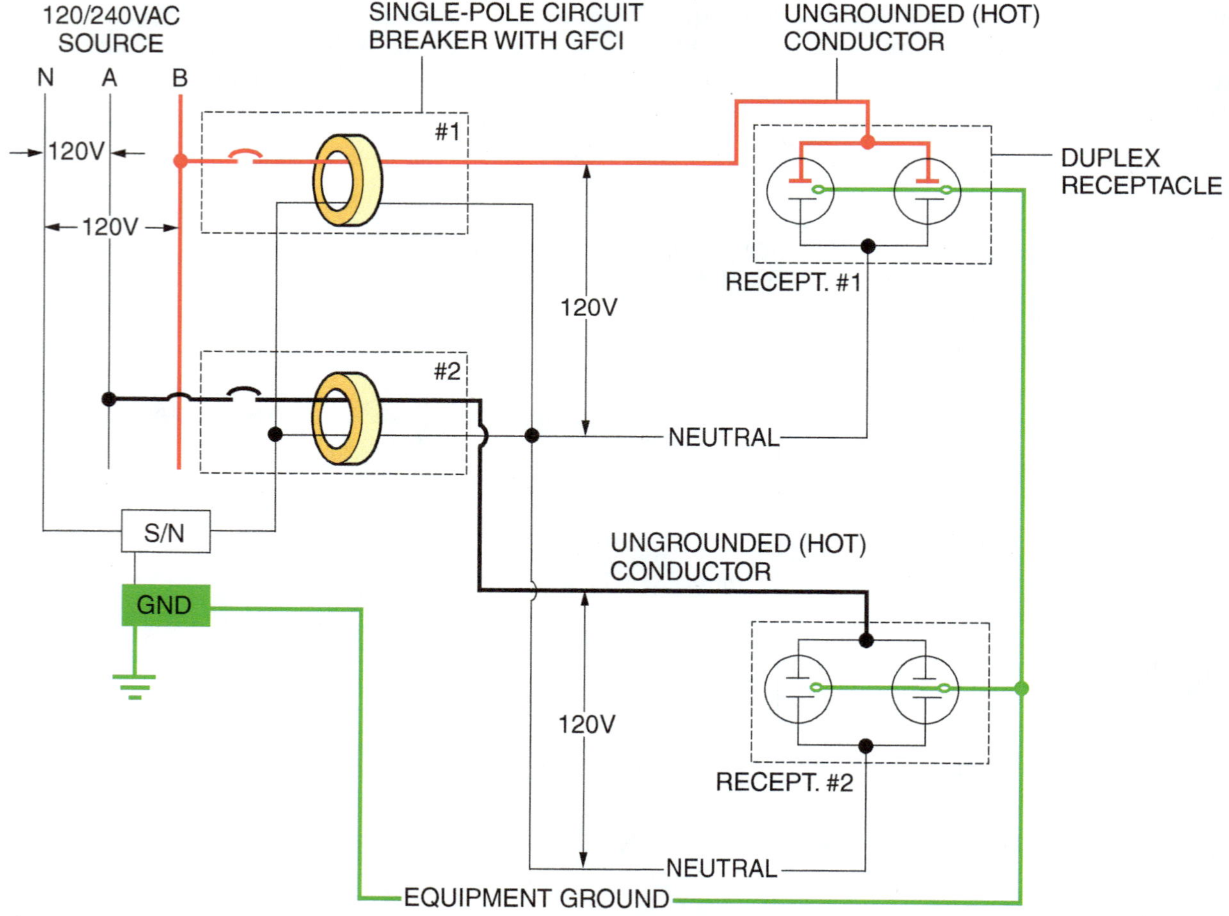

Figure 8 Circuit depicting the common load neutral.

Junction boxes can also present problems when they are used to provide taps for more than one branch circuit. Even though the circuits are not wired using a common neutral, sometimes all neutral conductors are connected together. Thus, parallel neutral paths are established, producing an imbalance in each GFCI breaker sensor, causing them to trip.

The two-pole GFCI breaker eliminates the problems encountered when trying to use two single-pole GFCI breakers with a common neutral. Because both hot currents and the neutral current pass through the same sensor, no imbalance occurs between the three currents, and the breaker will not trip.

4.1.3 Direct-Wired GFCI Receptacles

Direct-wired GFCI receptacles provide Class A ground fault protection on 120VAC circuits. They are available in both 15A and 20A arrangements. The 15A unit has a NEMA 5-15R receptacle configuration for use with 15A plugs only. The 20A device has a NEMA 5-20R receptacle configuration for use with 15A or 20A plugs. Both 15A and 20A units have a 120VAC, 20A circuit rating. This is to comply with *NEC Table 210.24,* which requires that 15A circuits use 15A receptacles but permits the use of either 15A or 20A receptacles on 20A circuits. Therefore, GFCI receptacle units that contain a 15A receptacle may be used on 20A circuits.

These receptacles have line terminals for the hot, neutral, and ground wires. In addition, they have load terminals that can be used to provide ground fault protection for other receptacles electrically downstream on the same branch circuit (*Figure 9*). All terminals will accept No. 14 to No. 10 AWG copper wire.

GFCI receptacles have a two-pole tripping mechanism that breaks both the hot and the neutral load connections.

When tripped, the RESET button pops out. The unit is reset by pushing the button back in.

GFCI receptacles have the additional benefit of noise suppression. Noise suppression minimizes false tripping due to spurious line voltages or radio frequency (RF) signals between 10 and 500 megahertz (MHz).

GFCI receptacles can be mounted without adapters in wall outlet boxes that are at least 1.5 inches deep.

Think About It

GFCIs

Explain the difference(s) in the operation of single-pole and double-pole GFCIs.

4.2.0 Arc Fault Circuit Interrupters

All branch circuits that supply the lighting and general-purpose receptacles in dwelling unit family rooms, dining rooms, living rooms, parlors, libraries, dens, bedrooms, sunrooms, recreation rooms, closets, hallways, or similar rooms or areas, as well as all circuit extensions or modifications, must have arc fault circuit interrupter protection to comply with *NEC Section 210.12.*

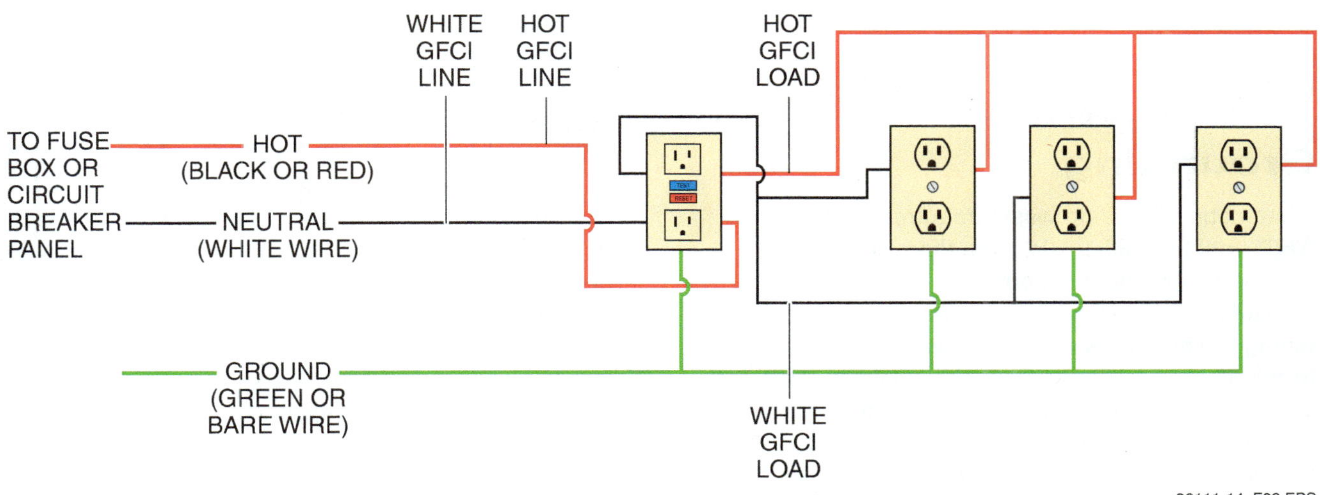

Figure 9 GFCI receptacle used to protect other outlets on the same circuit.

5.0.0 Grounding

NEC Section 250.4(A) provides the general requirements for grounding and bonding of grounded electrical systems. In order to ensure systems are properly grounded and bonded, the prescriptive requirements of *NEC Article 250* must be followed.

The grounding system is a major part of the electrical system. Its purpose is to protect people and equipment against the various electrical faults that can occur. It is sometimes possible for higher-than-normal voltages to appear at certain points in an electrical system or in the electrical equipment connected to the system. Proper grounding ensures that the electrical charges that cause these higher voltages are channeled to the earth or ground and that an effective ground fault path is provided throughout the system so that overcurrent devices will open before people are endangered or equipment is damaged.

The word *ground* refers to ground potential or earth ground. If a conductor is connected to the earth or some conducting body that serves in place of the earth, such as a driven ground rod (electrode), the conductor is said to be grounded. The neutral conductor in a three- or four-wire service, for example, is intentionally grounded, and therefore becomes a grounded conductor. This is the path back to the source of supply for all ground faults in an electrical system. This conductor is intended not only to carry the unbalanced loads of an installation, but also to provide the low-impedance path back to the source so that enough current will flow in the system to open the overcurrent devices. A wire that is used to connect this neutral conductor to a grounding electrode or electrodes is referred to as a grounding electrode conductor (GEC). Note the difference in the two meanings: one is grounded, while the other provides a means for grounding.

There are two general classifications of protective grounding:

- System grounding
- Equipment grounding

The system ground relates to the **service-entrance equipment** and its interrelated and bonded components; that is, the system and circuit conductors are grounded to limit voltages due to lightning, line surges, or unintentional contact with higher voltage and to stabilize the voltage to ground during normal operation per *NEC Sections 250.4(A)(1) and (2).*

The noncurrent-carrying conductive parts of materials enclosing electrical conductors or equipment, or forming a part of such equipment, and electrically conductive materials that are likely to become energized are all connected together to the supply source in a manner that establishes an effective ground fault path per *NEC Sections 250.4(A)(3) and (4).*

NEC Section 250.4(A)(5) defines the requirements for an effective ground path. It requires that electrical equipment and wiring and other electrically conductive materials likely to become energized shall be installed in a manner that creates a permanent, low-impedance circuit capable of safely carrying the maximum ground fault current likely to be imposed on it from any point on the wiring system where a ground fault may occur to the electrical supply source. The earth shall not be used as the sole equipment grounding conductor or effective ground fault current path.

To better understand a complete grounding system, a conventional residential system will be examined, beginning at the power company's high-voltage lines and transformer, as shown in *Figure 10*. The pole-mounted transformer is fed with a two-wire, single-phase 7,200V system, which is transformed and stepped down to a three-wired, 120/240V, single-phase electric service suitable for residential use. Note that the voltage between line A and line B is 240V. However, by connecting a third (neutral) wire on the secondary winding of the transformer—between the other two—the 240V is split in half, providing 120V between either line A or line B and the neutral conductor. Consequently, 240V is available for household appliances such as ranges, hot water heaters, and clothes dryers, while 120V is available for lights and small appliances.

Referring again to *Figure 10*, conductors A and B are ungrounded conductors, while the neutral is a grounded conductor. If only 240V loads were connected, the neutral (grounded conductor)

On Site

Grounding

Systematic grounding wasn't required by the *NEC*® until the mid-1950s; even then, electricians commonly grounded an outlet by wrapping an uninsulated wire around a cold-water pipe and taping it. Three-hole receptacles with grounding terminals became common in the 1960s, but in many older houses, you cannot assume that receptacles are grounded, even when you see a three-hole receptacle. Sometimes, new receptacles have simply been screwed onto old boxes where there is no equipment grounding conductor.

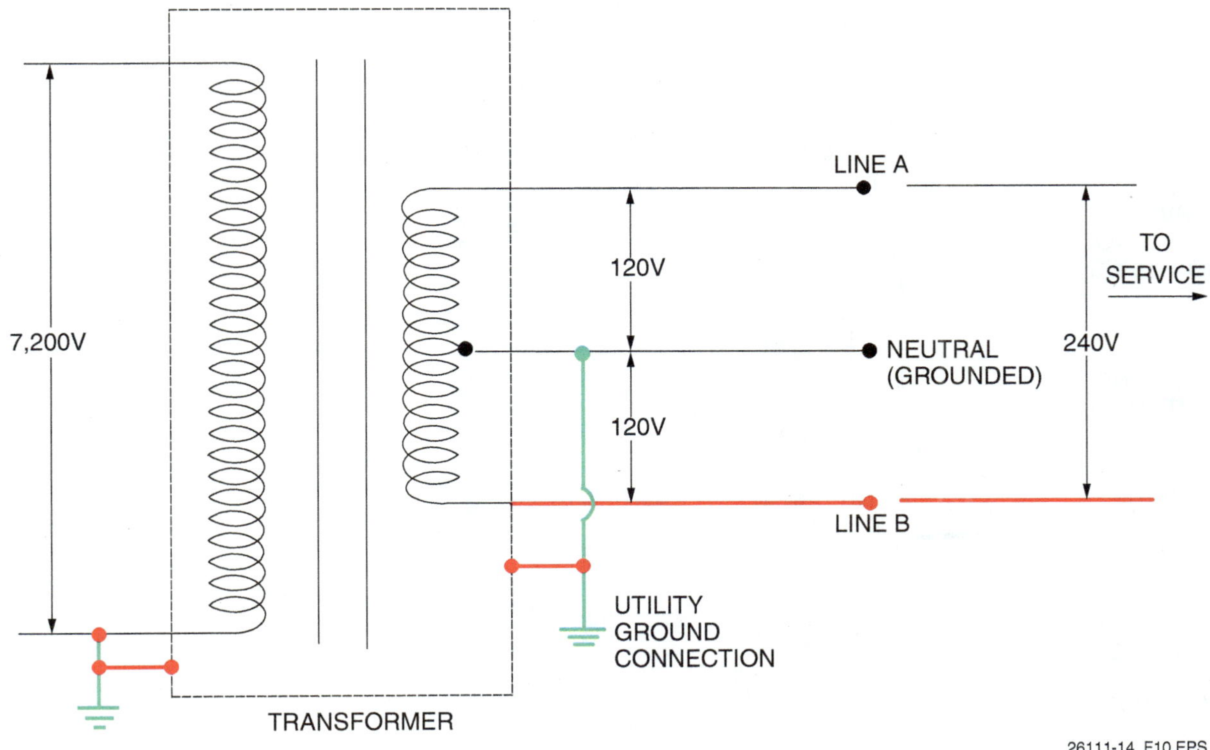

Figure 10 Wiring diagram of a 7,200V to 120/240V, single-phase transformer connection.

would carry no current. In this instance, the neutral would be used to carry any ground fault currents from the load side of the service back to the utility instead of depending on the earth as the path back to the source. However, since 120V loads are present, the neutral will carry the unbalanced load and become a current-carrying conductor. For example, if line A carries 60A and line B carries 50A, the neutral would carry only 10A (60A – 50A = 10A). This is why the *NEC®* allows the neutral conductor in an electric service to be smaller than the ungrounded conductors. However, *NEC Section 250.24(C)(1)* requires that it must be sufficient to carry fault currents back to the source and, therefore, must not be less than the required grounding electrode conductor using *NEC Table 250.66* for service conductors up to 1,100 kcmil and not less than 12.5% of the area of the service-entrance conductors (or equivalent) larger than 1,100 kcmil. The typical pole-mounted service drop conductors are normally routed by a messenger cable from a point on the pole to a point on the building being served, terminating at the point where service-entrance conductors exit a weatherhead. Service-entrance conductors are then typically routed through metering equipment into the service disconnecting means. This is the point where most services are grounded. See *Figure 11*. *NEC Section 250.24(A)(1)* requires that the grounding electrode for the structure connection to the neutral (grounded conductor) be at any accessible point from the load end of the service drop or service lateral to and including the terminal or bus to which the neutral (grounded service conductor) is connected to the service disconnecting means.

NOTE

Effectively grounded means intentionally connected to earth through one or more ground connection(s) of sufficiently low impedance and having sufficient current-carrying capacity to prevent the buildup of voltages that may result in a hazard to people or connected equipment.

5.1.0 Grounding Electrodes

NEC Article 250, Part III provides the requirements for connecting an electric service to the grounding electrode system of a building or structure. *NEC Section 250.50* requires, in general, that all of the electrodes described in *NEC Section 250.52(A)* be used (if present), and they must be bonded together to form the grounding electrode system. The electrodes listed in *NEC Section 250.52(A)* are as follows:

- Metal underground water pipe in direct contact with the earth for 10' or more and electrically continuous (or made electrically continuous

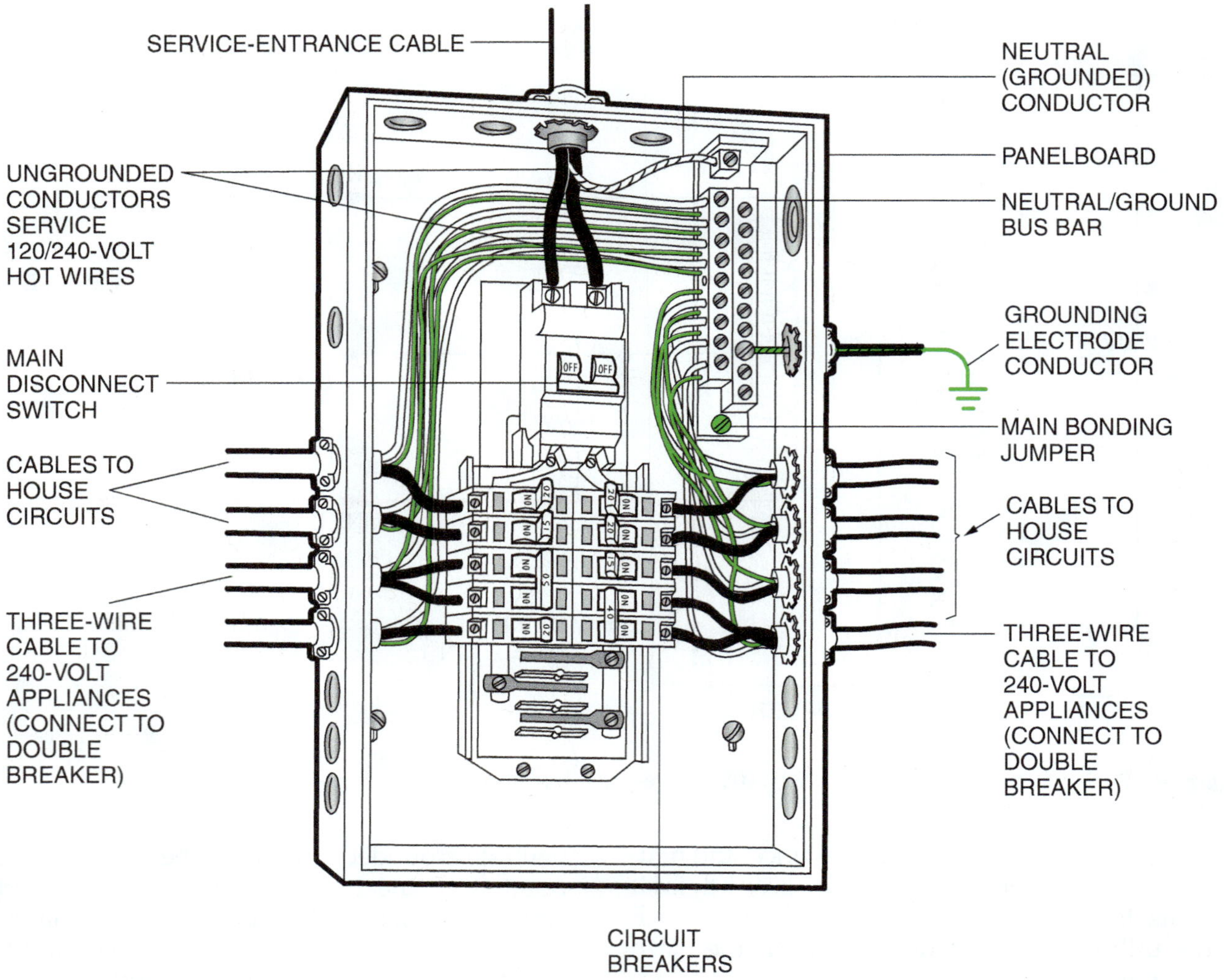

SERVICE-ENTRANCE CABLE

NEUTRAL (GROUNDED) CONDUCTOR

PANELBOARD

NEUTRAL/GROUND BUS BAR

UNGROUNDED CONDUCTORS SERVICE 120/240-VOLT HOT WIRES

GROUNDING ELECTRODE CONDUCTOR

MAIN DISCONNECT SWITCH

MAIN BONDING JUMPER

CABLES TO HOUSE CIRCUITS

CABLES TO HOUSE CIRCUITS

THREE-WIRE CABLE TO 240-VOLT APPLIANCES (CONNECT TO DOUBLE BREAKER)

THREE-WIRE CABLE TO 240-VOLT APPLIANCES (CONNECT TO DOUBLE BREAKER)

CIRCUIT BREAKERS

26111-14_F11.EPS

Figure 11 Interior view of panelboard showing connections.

by bonding around insulating joints or insulating pipe) to the points of connection of the grounding electrode conductor and the bonding conductors. Interior metal water piping located more than 5' from the point of entrance to the building shall not be used as part of the grounding electrode system or as a conductor to interconnect electrodes that are part of the grounding electrode system.

- Metal frame of the building or structure that complies with *NEC Section 250.52(A)(2).*
- An electrode encased by at least 2" of concrete may be used if it is located within and near the bottom of a concrete foundation or footing that is in direct contact with the earth. The electrode must be at least 20' long and must be made of electrically conductive coated steel reinforcing bars or rods of not less than ½" in diameter, or consisting of at least 20' of bare copper conductor not smaller than No. 4 AWG wire size.
- A ground ring encircling the building or structure, in direct contact with the earth, consisting of at least 20' of bare copper conductor not smaller than No. 2 AWG.
- Rod and pipe electrodes shall not be less than 8' in length and consist of either:
  - Pipe or conduit not smaller than trade size ¾ and, where of iron or steel, shall have the outer surface galvanized or otherwise metal-coated for corrosion protection.
  - Rods of iron or steel not smaller than ⅝" in diameter. Stainless steel rods less than ⅝" in diameter, nonferrous rods, or their equivalent shall be listed and not less than ½" in diameter.
- Plate electrodes shall expose less than two square feet of surface to exterior soil. Plates made of iron or steel shall be at least ¼" thick. Nonferrous metal plates shall be at least 0.06" thick.

- Other local metal underground systems or structures such as piping systems and underground tanks.

Often in residential construction, the only grounding electrode that is available is the metal underground water piping system. *NEC Section 250.53(D)(2)* requires that whenever water piping is used as an electrode, it must be supplemented. Any of the electrodes listed above can be used to supplement the water pipe electrode. *Figure 12* shows a typical residential electric service and the available grounding electrodes for this structure using a ground rod to supplement the water pipe electrode.

This house also has a metal underground gas piping system, but this may not be used as an electrode per *NEC Section 250.52(B).* In some cases, a water pipe electrode, building steel, and a concrete-encased electrode are not available to be used as a part of the grounding electrode system. For example, a building may be fed by plastic water piping, be constructed of wood, and an electrician may not be present at the site when the foundation for the structure is poured. When that happens, *NEC Section 250.50* requires that rod, pipe, plate, or other local metal underground structures be used.

WARNING!

A metal underground gas piping system must never be used as a grounding electrode.

Some local jurisdictions do not recognize water piping as an electrode due to the rise in the use of nonmetallic piping for both new and replacement water systems. They do not want to rely on the maintained viability of an existing metallic water service and therefore require the use of other electrodes such as the concrete-encased or rod electrodes. This means electricians must be involved with the construction prior to the foundation being poured in order to utilize concrete-encased electrodes.

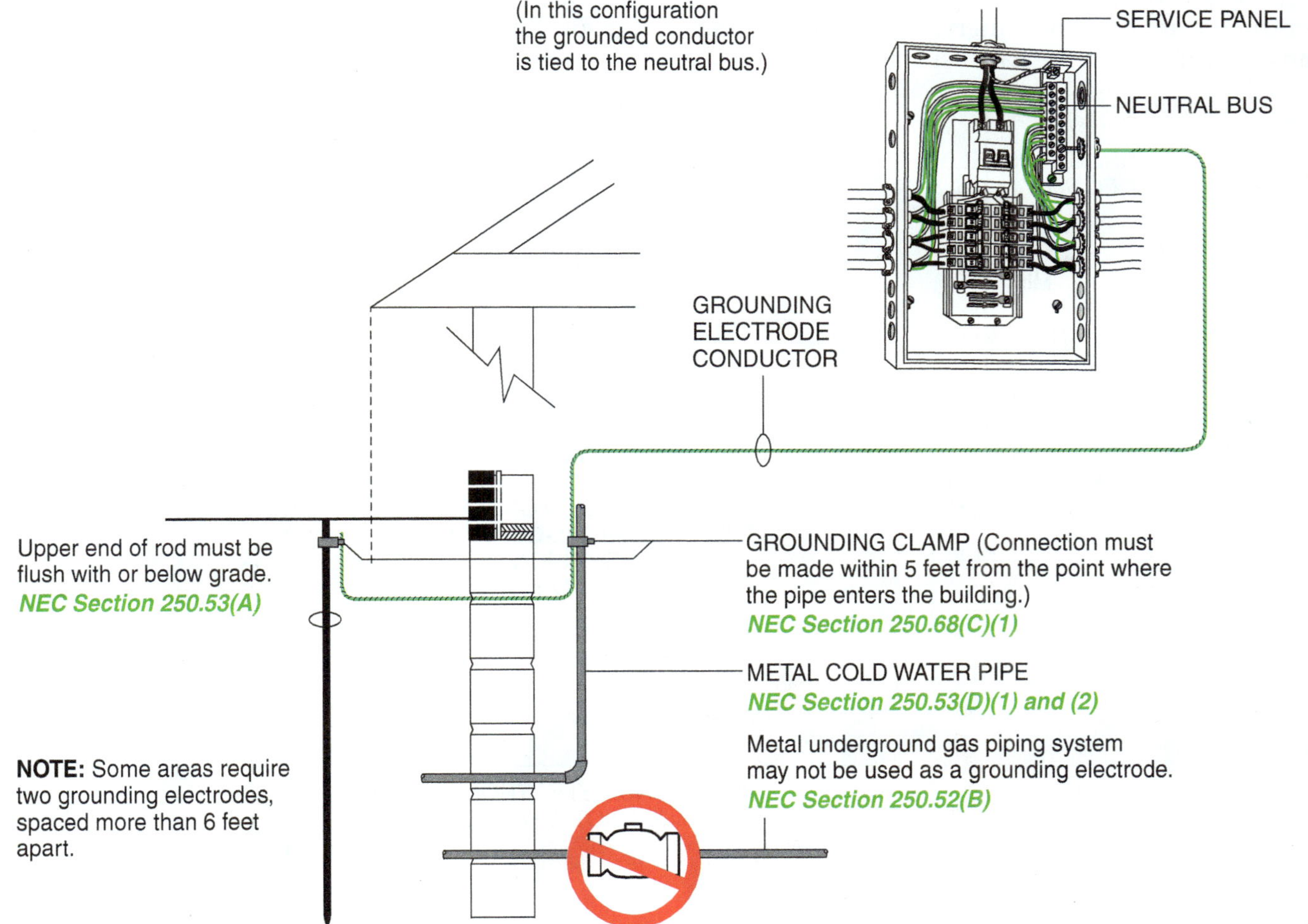

Figure 12 Components of a residential grounding system.

In most cases, the supplemental electrode used for a water pipe electrode will consist of either a driven rod or pipe electrode, the specifications for which are shown in *Figure 13*.

5.1.1 Grounding Electrode Installations

NEC Section 250.53(A)(1) requires that rod, pipe, and plate electrodes, where practical, be buried below the permanent moisture level and that they are free from any nonconductive coatings, such as paint or enamel. This section also requires that each electrode system used for a structure be at least 6' from other electrode systems, such as those for lightning protection.

NEC Section 250.53(G) permits a rod or pipe electrode to be driven at a 45-degree angle if rock bottom is encountered and prevents the rod or pipe from being driven vertically for at least 8'. Where driving a rod or pipe electrode at a 45-degree angle will not work, it is permitted to lay a rod or pipe horizontally in a trench that is at least 30" deep. For rod or pipe electrodes longer than 8', it is permitted to have the upper end above ground level if a suitable means of protection is provided for the grounding electrode conductor attachment; otherwise, the upper end must be flush with the earth surface.

NEC Section 250.53(A)(2) requires a supplemental electrode when a single rod, pipe, or plate electrode of a type specified in *NEC Sections 250.52(A)(2) through 250.52(A)(8)* is used. If a single rod or plate grounding electrode that has a resistance of 25 ohms or less is used, the supplemental electrode shall not be required. Always check with the local inspection authority, including the local utility, for rules that surpass the requirements of the *NEC*®.

- Where multiple rod, pipe, or plate electrodes are installed to meet the requirements of this section, they shall not be less than 6' apart.
- Plate electrodes must be buried at least 30" below the surface of the earth, according to *NEC Section 250.53(H).*
- Where two or more electrodes are effectively bonded together, they are treated as a single electrode system.

5.1.2 Grounding Electrode Conductors (GECs)

The grounding electrode conductor (GEC) connecting the neutral (grounded conductor of the service) at the panelboard neutral bus to the grounding electrodes must meet the requirements

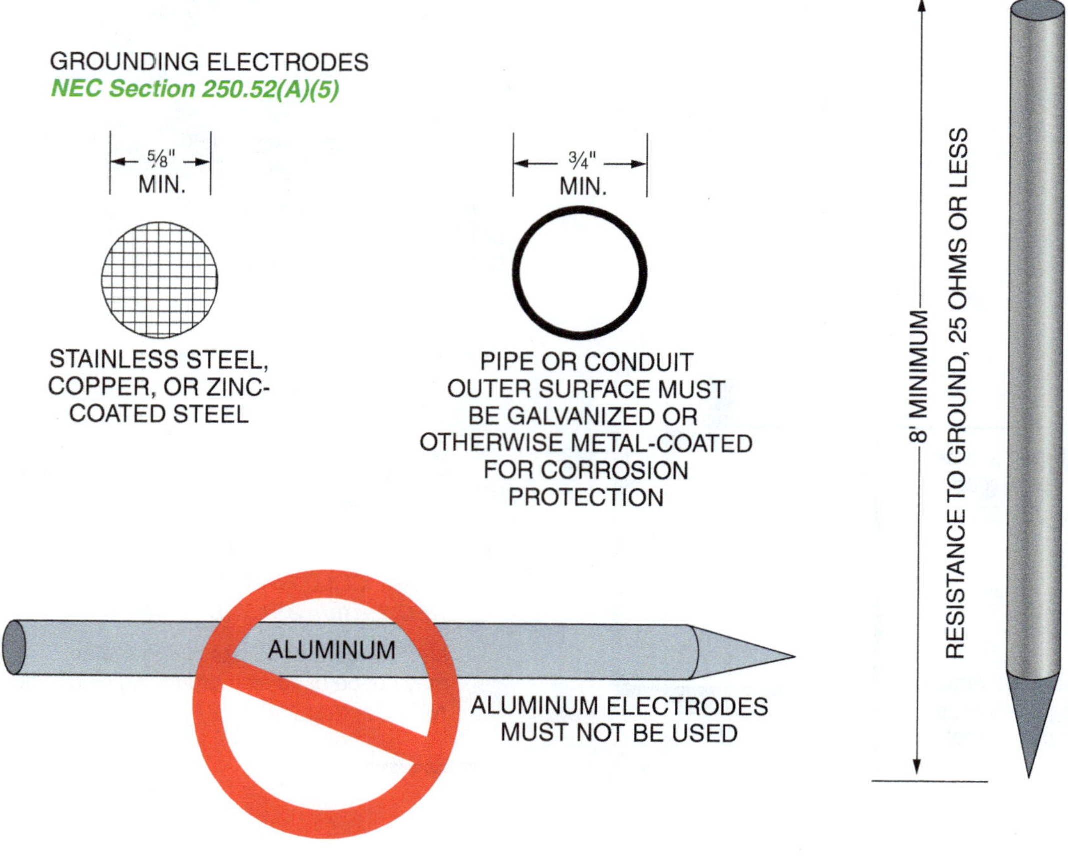

Figure 13 Specifications for rod and pipe grounding electrodes.

of *NEC Section 250.62*. This requires that it be made of copper, aluminum, or copper-clad aluminum. The material selected shall be protected against corrosion. The GEC may be either solid or stranded, covered or bare. Note that the GEC is not an equipment grounding conductor, and thus is not required to be identified by the use of the color green or green with yellow stripes, if insulated.

5.1.3 Installation of GECs

NEC Section 250.64 provides the installation requirements for GECs and does not permit bare aluminum or copper-clad aluminum grounding conductors to be used where in direct contact with masonry, the earth, where subject to corrosive conditions, or where used outside within 18" of the earth at the termination point. Other *NEC®* requirements include the following:

- A GEC or its enclosure is required to be securely fastened to the surface on which it is carried.
- A No. 4 AWG or larger copper or aluminum GEC is required to be protected if it will be exposed to severe physical damage.
- A No. 6 AWG or larger GEC that is free from exposure to physical damage is permitted to be run along the surface of the building without metal covering or protection if it is securely fastened to the building. Otherwise, it must be installed in RMC, IMC, PVC, reinforced thermosetting resin conduit (RTRC), EMT, or a cable armor.
- GECs smaller than No. 6 AWG must be protected by RMC, IMC, PVC, RTRC, EMT, or cable armor.
- The GEC shall be installed in one continuous length without a splice or joint, unless it meets the requirements of *NEC Sections 250.64(C)(1) through (C)(4)*.
- Where a service consists of more than a single enclosure, as permitted in *NEC Section 230.71(A)*, grounding electrode connections shall be made in accordance with *NEC Sections 250.64(D)(1), (D)(2), or (D)(3)*.
- Ferrous metal enclosures for the GEC are required to be electrically continuous from the point of attachment to metal cabinets or metallic equipment enclosures to the GEC. They must also be securely fastened to the ground clamp or fitting.
- Ferrous metal enclosures for the GEC that are not physically continuous from a metal cabinet or metallic equipment enclosure to the grounding electrode must be made electrically continuous by bonding each to the enclosed GEC. Bonding methods must comply with *NEC Section 250.92(B)* for installations at service equipment locations and with *NEC Sections 250.92(B)(2) through (B)(4)* for other-than-service equipment locations.
- GECs may be run to any convenient grounding electrode available in the grounding electrode system, or to one or more grounding electrode(s) individually. The GEC shall be sized for the largest grounding electrode conductor required among all the electrodes connected together.

Think About It

Grounding Conductors

This residential application has a grounding electrode connected to a rod that travels through the floor and into the ground at least 8'. In addition, it is also connected to a metal cold-water pipe (not shown). What type of grounding system is provided at your home? Does it meet *NEC®* requirements?

26111-14_SA02.EPS

5.1.4 Methods of Connecting GECs

NEC Section 250.70 requires the GEC to be connected to electrodes using exothermic welding, listed pressure connectors, listed clamps, listed lugs, or other listed means. Connections that depend on solder must never be used. To prevent corrosion, the ground clamp must be listed for the material of the grounding electrode and the GEC.

Where used on a pipe, ground rod, or other buried electrodes, the fitting must be listed for direct soil burial or concrete encasement. More than one conductor is not permitted to be connected to the grounding electrode using a single clamp or fitting unless the clamp or fitting is specifically listed for the connection of more than one conductor.

Think About It

Grounding Electrode Conductors

Which *NEC*® table would you use to size the minimum GEC required for a typical residential service?

For the connection to an electrode, you must use one of the following:

- A listed, bolted clamp of cast bronze or brass, or plain or malleable iron
- A pipe fitting, pipe plug, or other approved device that is screwed into a pipe or pipe fitting
- For indoor communication purposes only, a listed sheet metal strap-type ground clamp with a rigid metal base that seats on the electrode with a strap that will not stretch during or after installation
- An equally substantial approved means

The connection of a GEC or a **bonding jumper** to a grounding electrode must be accessible unless that connection is to the concrete-encased or buried grounding electrodes permitted in *NEC Section 250.68.* Where it is necessary to ensure the grounding path for metal piping used as a grounding electrode, effective bonding shall be provided around insulated joints and around any equipment likely to be disconnected for repairs or replacement. Bonding conductors shall be of sufficient length to permit removal of such equipment while retaining the integrity of the bond.

NOTE

The UL listing states that "strap-type ground clamps are not suitable for attachment of the grounding electrode conductor of an interior wiring system to a grounding electrode."

As required by *NEC Section 250.96*, coatings on metal piping systems must be removed to ensure that a permanent and effective grounding path is provided. Grounding electrode conductors and bonding jumpers shall be permitted to be connected at the following locations and used to extend the connections to an electrode:

- Interior metal piping located not more than 5 feet from the point of entrance of a building
- Structural frame of a building directly connected to the GEC

For the example house, the point of connection to the water piping is shown in *Figure 12* and would be required to be accessible after any wall coverings are installed. Any nonconductive coatings on the water piping would also have been scraped off or removed prior to installing the clamp on the water pipe.

5.1.5 Sizing GECs

Grounding electrode conductors must be sized per *NEC Section 250.66,* which uses the area of the largest service-entrance conductor (or equivalent area for paralleled conductors). Except as noted below, *NEC Table 250.66* will provide the minimum size GEC and any bonding jumpers used to interconnect grounding electrodes used.

- Where connected to rod, pipe, or plate electrodes, that portion of the GEC that is the sole connection to the grounding electrode shall not be required to be larger than No. 6 AWG copper or No. 4 AWG aluminum.
- Where connected to a concrete-encased electrode, that portion of the GEC that is the sole connection to the grounding electrode shall not be required to be larger than No. 4 AWG copper wire.
- Where the GEC is connected to a ground ring, that portion of the conductor that is the sole connection to the grounding electrode shall not be required to be larger than the conductor used for the ground ring.
- Where multiple sets of service-entrance conductors are used as permitted in *NEC Section 230.40, Exception 2,* the equivalent size of the largest service-entrance conductor is required to be determined by the largest sum of the areas of the corresponding conductors of each set.
- Where there are no service-entrance conductors, the GEC size is required to be determined by the equivalent size of the largest service-entrance conductor required for the load to be served.

For our sample dwelling unit, the size of the service-entrance conductors is No. 2 AWG. Using *NEC Table 250.66,* we can determine that the size of the conductor coming from the service panel to the water pipe (the GEC) must be at least No. 8 AWG copper. This No. 8 AWG may continue on without a splice to the ground rod, as shown in *Figure 12*, or a separate No. 6 AWG could be installed for the ground rod(s). This conductor would have to be connected to the service panel as an individual run or with a separate connector to any portion of the No. 8 AWG. It may not be connected directly to the water piping.

5.1.6 Air Terminals (Lightning Protection)

Air terminal conductors and driven pipes, rods, or plate electrodes used for grounding air terminals

On Site

What Does a Lightning Rod Do?

An interesting fact about grounding is that a lightning rod (air terminal) isn't meant to bring a bolt of lightning to ground. To do this, its conductors would have to be several feet in diameter. The purpose of the rod is to dissipate the negative static charge that would cause the positive lightning charge to strike the house.

are not permitted to be used in lieu of the grounding electrodes covered in *NEC Section 250.50* for grounding wiring systems and equipment. However, *NEC Section 250.106* requires that they be bonded to the wiring and equipment grounding electrode system for the structure.

5.2.0 Main Bonding Jumper

NEC Section 250.24(B) requires that an unspliced main bonding jumper (MBJ) shall be used to connect the equipment grounding conductor(s) and the service disconnect enclosure to the grounded conductor (neutral) of the system within the enclosure of each service disconnect.

The MBJ must be of copper or other corrosion-resistant material. An MBJ may be in the form of a wire, bus, screw, or similar suitable conductor.

Where an MBJ is in the form of a screw, it is required to be identified with a green finish so that the head of the screw is visible for inspection. An MBJ must be attached using exothermic welding, a listed pressure connector, listed clamp, or other listed means.

The MBJ cannot be smaller than the sizes given in *NEC Table 250.66* for grounding electrode conductors. See *NEC Section 250.28(D)* for service conductors that exceed 1,100 kcmil.

The MBJ is the means by which any ground fault in the branch circuits and feeders of the electrical system travels back to the source of supply at the utility. A ground fault will travel along the equipment conductors of the circuits back to the service disconnecting means. Where metallic raceways are used as equipment grounding conductors, there will be no connection to the grounded conductor at the service. Without the MBJ, the path back to the source would be through the grounding electrode system and the earth. This does not provide a low-impedance path, and thus will not allow enough current to flow in the circuit to let the overcurrent devices open.

For example, suppose a phase conductor makes contact with the metallic housing of a 120V, 15A appliance that was wired using EMT. Further suppose that the total combined resistance of the EMT being used as the equipment grounding conductor connected to the appliance and the resistance of the metal water piping and our ground rod in the sample house is 20Ω [less than the 25Ω permitted in *NEC Section 250.53(A)(2), Exception*]. The amount of current that could flow back to the utility source would be 120V ÷ 20 = 6A. The smallest overcurrent device in our electrical system is 15A, and would not trip. With the MBJ installed, the path back to the utility source is through the MBJ to the grounded conductor of the service. This resistance will be much less than 1Ω, and thus would allow enough current to flow to open up the overcurrent devices within the system. The MBJ provides the path back to the source for faults that occur within the service disconnect means.

5.2.1 Bonding at the Service

Electrical continuity is required at the service per *NEC Section 250.92(A),* which states that all of the following must be bonded:

- The service raceways, auxiliary gutters, or service cable armor or sheaths, except for underground metallic sheaths of continuously underground cables as noted in *NEC Section 250.84*
- All service enclosures containing service-entrance conductors, including meter fittings, boxes, or the like interposed in the service raceway or armor
- Any metallic raceway or armor enclosing a grounding electrode conductor as specified in *NEC Section 250.64(E)*

Bonding shall apply at each end and to all intervening raceways, boxes, and enclosures between the service equipment and the grounding electrode.

The items that typically require bonding include the mast and weatherhead, the meter enclosure, the armor of the SE cable (if it has armor), and the service disconnect.

5.2.2 Methods of Bonding at the Service

The electrical continuity of the service equipment, raceways, and enclosures will be ensured per *NEC Section 250.92(B)* through the use of the following methods:

- Bonding equipment to the grounded service conductor in a manner provided in *NEC Section 250.8*

- Connections utilizing threaded couplings or threaded bosses on enclosures where made up wrenchtight
- Threadless couplings and connectors where made up tight for metal raceways and metal-clad cables
- Other approved devices, such as bonding-type locknuts and **bonding bushings**

Bonding jumpers must be used around concentric or eccentric knockouts that are punched or otherwise formed so as to impair the electrical connection to ground. Standard locknuts or bushings shall not be the sole means for bonding.

5.2.3 Bonding and Grounding Requirements for Other Systems

An accessible means external to the service equipment enclosure is required for connecting intersystem bonding and grounding conductors and connections for the communications, radio and television (TV), community antenna television (CATV), and network-powered broadband communication system. The intersystem bonding termination must consist of at least three terminals and be installed in accordance with *NEC Section 250.94*. Any one of the following can be used:

- A set of listed terminals for grounding and bonding
- A bonding bar near the service-entrance enclosure, meter enclosure, or raceway for service conductors connected to an equipment grounding conductor in the enclosure or raceway with a minimum No. 6 AWG copper conductor
- A bonding bar near the grounding electrode conductor connected with a minimum No. 6 AWG copper conductor

The intersystem bonding termination must meet the following requirements per *NEC Section 250.94*:

- At the disconnecting means for a building or structure, be securely mounted and electrically connected to the metal enclosure or GEC with a minimum No. 6 AWG copper conductor
- Be accessible for connection and inspection
- Not to interfere with access to service, building, or structure disconnecting means or metering equipment
- Consist of a set of terminals with the capacity for connection of not less than three intersecting bonding conductors

5.2.4 Bonding of Water Piping Systems

Metallic water piping systems in or on a structure must be bonded as required by *NEC Section 250.104(A)*. The metallic water piping system(s) must be bonded by means of a bonding jumper sized in accordance with *NEC Table 250.66* and connected to one of the following:

- The service-entrance enclosures
- The grounded (neutral) conductor at the service
- The grounding electrode conductor where of sufficient size
- The grounding electrode(s) used

The points of attachment of the bonding jumper(s) shall be accessible. It shall be installed in accordance with *NEC Section 250.64(A), (B), and (E)*. Note that while this conductor is sized in the same manner as if the water piping system is a grounding electrode, the point of attachment to the water piping is permitted to be at any convenient point on the water piping system and not just within the first 5' of where the water enters the building.

NEC Section 250.104(A)(2) states that in multi-family dwelling units (or other multiple occupancy buildings) where the metal water piping system(s) installed in or attached to a building or structure for the individual occupancies is metallically isolated from all other occupancies by use of nonmetallic water piping, the metal water piping system(s) for each occupancy shall be permitted to be bonded to the equipment grounding terminal of the panelboard or switchboard enclosure (other than service equipment) supplying that occupancy. The bonding jumper shall be sized in accordance with *NEC Table 250.122*.

5.2.5 Bonding of Other Piping Systems

NEC Section 250.104(B) requires that other piping systems, where installed in or attached to a building or structure, including gas piping, that may become energized shall be bonded to one of the following:

- The service equipment enclosure
- The grounded conductor at the service
- The grounding electrode conductor where of sufficient size
- One or more grounding electrodes used

The bonding jumper(s) shall be sized in accordance with *NEC Table 250.122* using the rating of the circuit that may energize the piping system(s).

The equipment grounding conductor for the circuit that may energize the piping shall be permitted to serve as the bonding means. The points of attachment of the bonding jumper(s) shall be accessible.

NOTE

Bonding all piping and metal air ducts within the premises will provide additional safety.

6.0.0 Installing the Service Entrance

In practical applications, the electric service is normally one of the last components of an electrical system to be installed. However, it is one of the first considerations when laying out a residential electrical system. For instance:

- The electrician must know in which direction and to what location to route the circuit homeruns while **roughing in** the electrical wiring.
- Provisions must be made for sleeves through footings and foundations in cases where underground systems (service laterals) are used.
- The local power company must be notified as to the approximate size of service required so they may plan the best way to furnish a **service drop** to the property.

6.1.0 Service Drop Locations

The location of the service drop, electric meter, and load center should be considered first. It is always wise to consult the local power company to obtain their requirements; where you want the service drop and where they want it may not coincide. A brief meeting with the power company about the location of the service drop can prevent problems later on.

The service drop must be routed so that the service drop conductors have a clearance of not less than 3' horizontally and below windows that open, doors, porches, fire escapes, or similar locations. In addition, they must have a 10' vertical clearance that extends 3' horizontally from porches, fire escapes, balconies, and so forth, as required in *NEC Section 230.9.* Where overhead service conductors pass over rooftops, driveways, yards, and so forth, they must have clearances as specified in *NEC Section 230.24.*

A plot plan (also called a site plan) is often available for new construction. The plot plan shows the entire property, with the building or buildings drawn in their proper location on the plot of land. It also shows sidewalks, driveways, streets, and existing utilities—both overhead and underground.

A plot plan of the sample residence is shown in *Figure 14*. In reviewing this drawing, you can see that the closest power pole is located across a

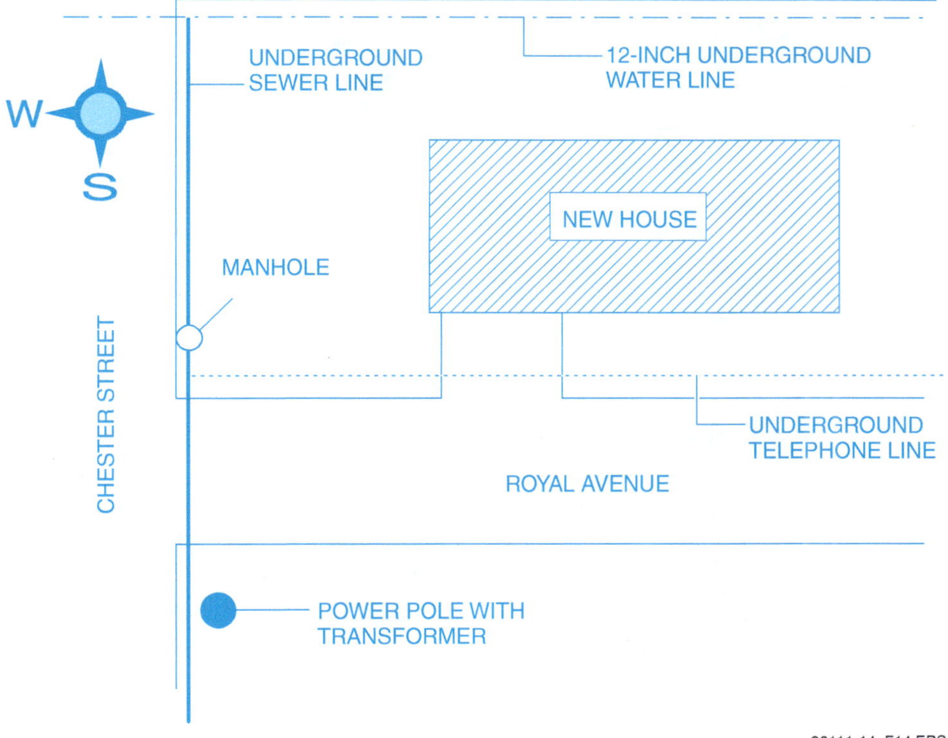

Figure 14 Plot plan of the sample residence.

public street from the house. By consulting with the local power company, it is learned that the service will be brought to the house from this pole by triplex cable, which will connect to the residence at a point on its left (west) end. The steel uninsulated conductor of triplex cable acts as both the grounded conductor (neutral) and as a support for the insulated (ungrounded) conductors. It is also suitable for overhead use.

When service-entrance cable is used, it will run directly from the point of attachment and service head to the meter base. However, since the carport is located on the west side of the building, a service mast (*Figure 15*) will have to be installed.

The *NEC®* requires a clearance of not less than 8' over rooftops, unless the roof has a slope of 4" in 12" or greater, in which case the clearance may be reduced to 3'. Where the service drop conductors pass over only the overhang (eaves) of a roof, the clearance may be reduced to 18" as long as no more than 6' of the conductors travel over no more than 4' of the overhang (eave). This minimum height requirement extends beyond the roof for a distance of not less than 3' in all directions, except the final portion of the span where the service drop conductors attach to the sides of a building.

6.2.0 Vertical Clearances of Service Drop

NEC Section 230.24(B) specifies the distances by which service drop conductors must clear the ground. These distances vary according to the surrounding conditions.

In general, the *NEC®* states that the vertical clearances of all service drop conductors that carry 600V or under are based on a conductor temperature of 60°F (15°C) with no wind and with the

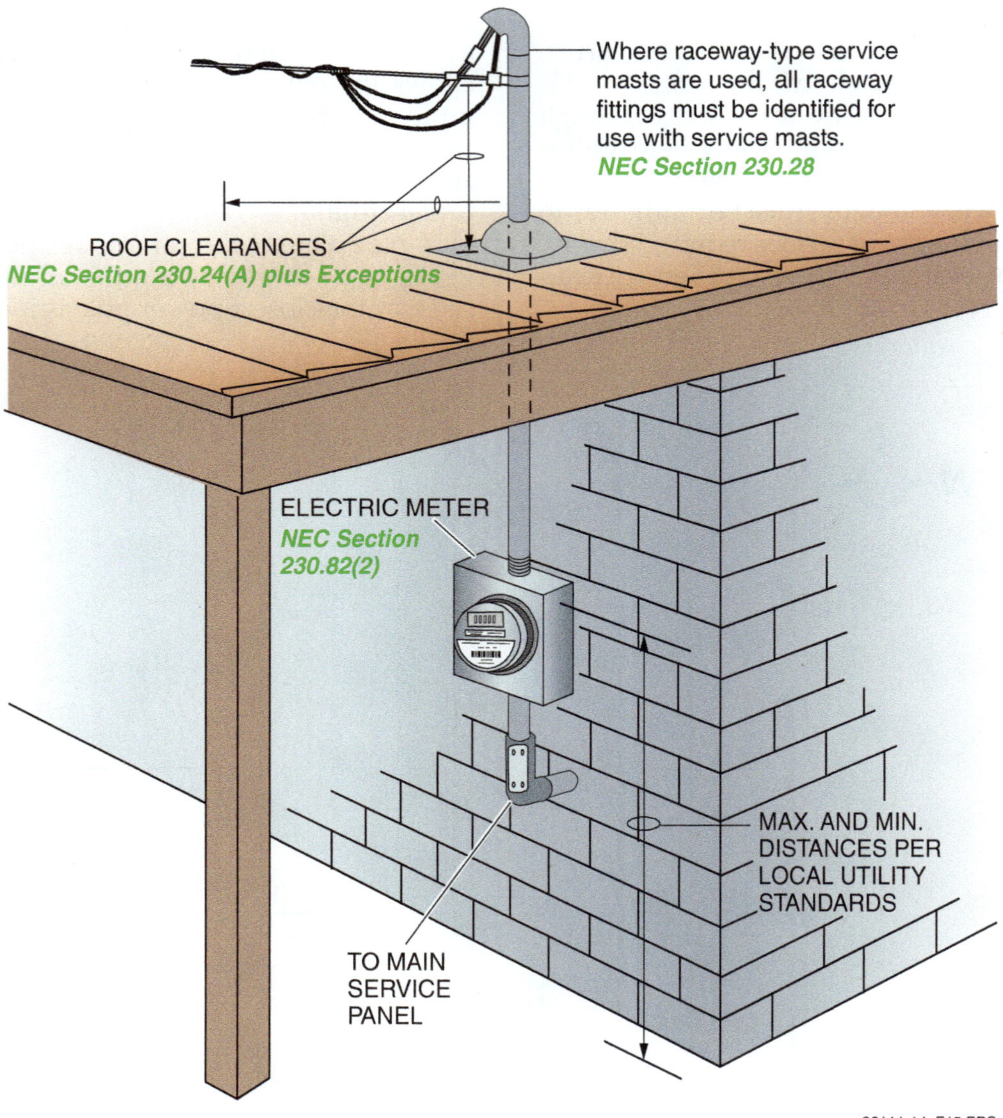

Figure 15 *NEC®* sections governing service mast installations.

final unloaded sag in the wire, conductor, or cable. Service drop conductors must be at least 10' above the ground or other accessible surfaces at all times. More distance is required under most conditions. For example, if the service conductors pass over residential property and driveways or commercial property that is not subject to truck traffic, the conductors must be at least 15' above the ground. However, this distance may be reduced to 12' when the voltage is limited to 300V to ground.

In other areas, such as public streets, alleys, roads, parking areas subject to truck traffic, driveways on other-than-residential property, the minimum vertical distance is 18'. The conditions of the sample residence are shown in *Figure 16*.

6.3.0 Service Drop (Overhead Service Conductor) Clearances for Building Openings

Service conductors that are installed as open conductors or multiconductor cable without an overall outer jacket must have a clearance of not less than 3' from windows that are designed to be opened, doors, porches, balconies, ladders, stairs, fire escapes, or similar locations (*NEC Section 230.9*). However, conductors run above the top level of a window are permitted to be less than 3' from the window opening.

The 3' of clearance is not applicable to raceways or cable assemblies that have an overall outer jacket approved for use as a service conductor. The intention of this requirement is to protect the conductors from physical damage and/or physical contact with unprotected personnel when evacuating a structure through the window opening. The exception allows service conductors, including drip loops and service drop conductors, to be located just above the window openings because they would not interfere with ladders leaning against the structure to the right, left, or below the window opening when used to evacuate people from the building.

> **Case History**
>
> **Failure to De-Energize Panelboard**
>
> A 31-year-old electrician was finishing the installation of an outdoor floodlight on a new home. He borrowed an aluminum ladder from another contractor and then proceeded with his task. He did not verify that power was removed at the panelboard, and when he used his wire strippers to remove the conductor insulation, his right thumb and index finger contacted a 110V circuit. He received a fatal shock.
>
> **The Bottom Line:** Always de-energize circuits at the panelboard before beginning any electrical task, and never use aluminum ladders while working in or near electrical devices. In this case, the ladder provided a path to ground.

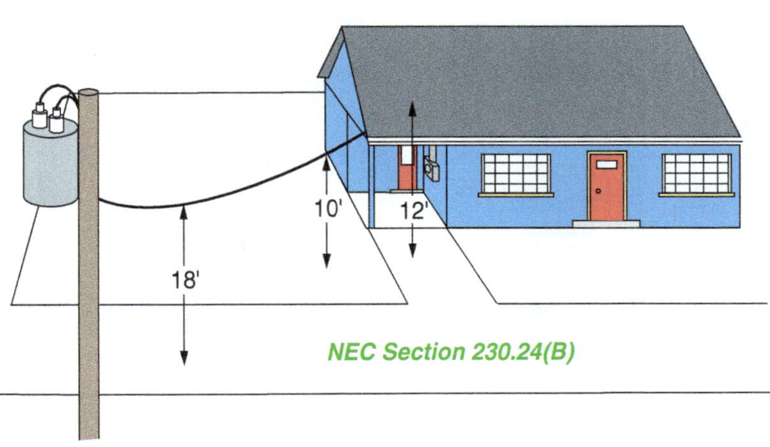

Figure 16 Vertical clearances for service drop conductors.

7.0.0 Panelboard Location

The main service disconnect or panelboard is normally located in a portion of an unfinished basement or utility room on an outside wall so that the service cable coming from the electric meter can terminate immediately into the switch or panelboard when the cable enters the building. In the example home, however, there is no basement and the utility room is located in the center of the house with no outside walls. Consequently, a somewhat different arrangement will have to be used. A load center is a type of panelboard that is normally located at the service entrance of a residential installation. The load center usually contains a main circuit breaker, which is the main disconnect. Circuit breakers are provided for equipment such as electric water heaters, ranges, dryers, air conditioning and heating units, and breakers that feed subpanels such as lighting panels.

NEC Section 230.70 requires that the service disconnecting means be installed in a readily accessible location—either outside or inside the building. If located inside the building, it must be located nearest the point of entrance of the service conductors. In the sample home, there are at least two methods of installing the panelboard in the utility room that will comply with this *NEC®* regulation, as well as the requirements in *NEC Sections 110.26 and 240.24.*

The first method utilizes a weatherproof 100A disconnect (safety switch or circuit breaker enclosure) mounted next to the meter base on the outside of the building. With this method, service conductors are provided with overcurrent protection; the neutral conductor is also grounded at this point, as this becomes the main disconnect switch. Three-wire cable with an additional grounding wire is then routed from this main disconnect to the panelboard in the utility room. All three current-carrying conductors (two ungrounded and one neutral) must be insulated with this arrangement; the equipment ground, however, may be bare. The panelboard containing overcurrent protection devices for the branch circuits, which is located in the utility room, now becomes a subpanel. See *Figure 17*.

> **NOTE**
> Local ordinances in some areas may require a disconnect at the meter base, making the panel in the utility room a subpanel.

An alternate method utilizes conduit from the meter base that is routed under the concrete slab and then up to a main panelboard located in the utility room. *NEC Section 230.6* considers conductors to be outside of a building when they are installed under not less than 2" of concrete beneath a building or installed in a conduit not less than 18" deep beneath a building. The sample residence has a 4"-thick reinforced concrete slab—well within the *NEC®* regulations. Therefore, the service conductors from the meter base that are installed under the concrete slab in conduit are considered to be outside the house, and no disconnect is required at the meter base. When this conduit emerges in the utility room, it will run straight up into the bottom of the panelboard, again meeting the *NEC®* requirement that the panel be located nearest the point of entrance of the service conductors. Always check with your local authority having jurisdiction for specific requirements about where services are to be located. Details of this service arrangement are shown in *Figure 18*.

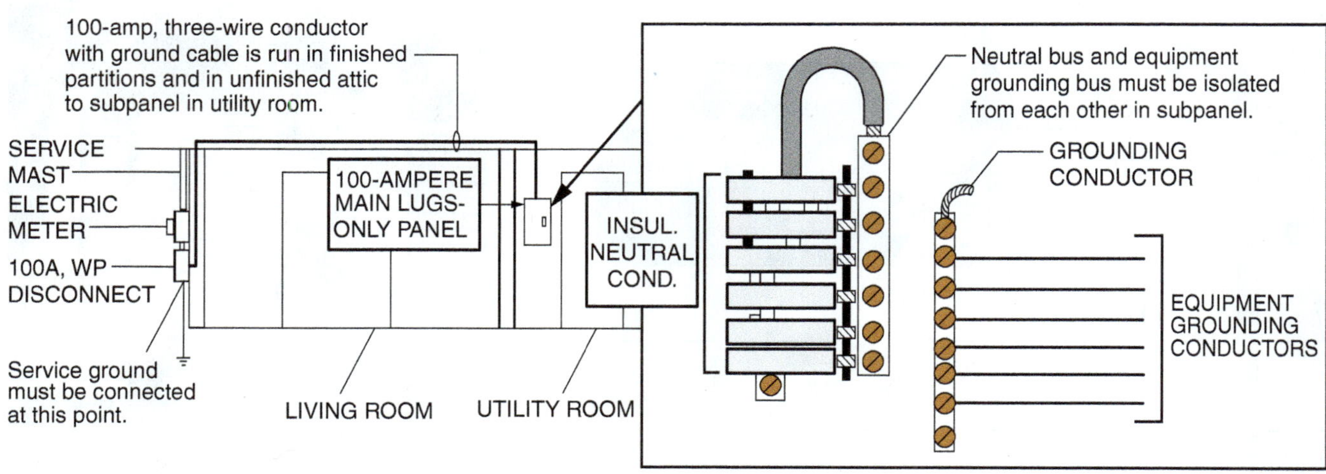

Figure 17 One method of wiring a panelboard for the sample residence.

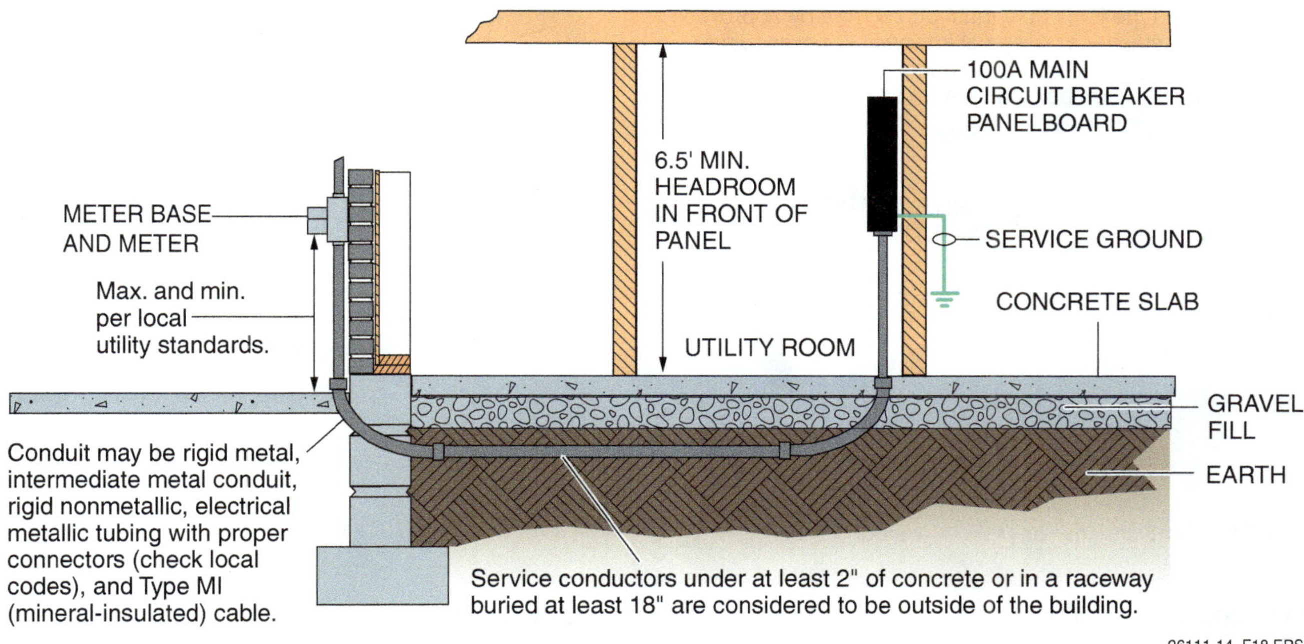

Figure 18 Alternate method of service installation for the sample residence.

8.0.0 Wiring Methods

Branch circuits and feeders are used in residential construction to provide power wiring to operate components and equipment, and control wiring to regulate the equipment. Wiring may be further subdivided into either open or concealed wiring.

In open wiring systems, the cable and/or raceways are installed on the surface of the walls, ceilings, columns, and other areas where they are in view and are readily accessible. Open wiring is often used in areas where appearance is not important, such as in unfinished basements, attics, and garages.

Concealed wiring systems are installed inside walls, partitions, ceilings, columns, and behind baseboards or moldings where they are out of view and are not readily accessible. This type of wiring is generally used in all new construction with finished interior walls, ceilings, and floors, and it is the preferred type of wiring where appearance is important.

In general, there are two basic wiring methods used in the majority of modern residential electrical systems. They are:

- Sheathed cables of two or more conductors
- Raceway (conduit) systems

The method used on a given job is determined by the requirements of the *NEC*®, any amendments made by local authorities, the type of building construction, and the location of the wiring in the building. In most applications, either of the two methods may be used, and both methods are frequently used in combination.

8.1.0 Cable Systems

Several types of cable are used in wiring systems to feed or supply power to equipment. These include nonmetallic-sheathed cable, **metal-clad (MC) cable**, underground feeder cable, and service-entrance cable.

8.1.1 Nonmetallic-Sheathed Cable

Nonmetallic-sheathed (Type NM) cable (*NEC Article 334*) is manufactured in two- or three-wire configurations with varying sizes of conductors. In both two- and three-wire cables, conductors are color-coded: one conductor is black while the other is white in two-wire cable; in three-wire cable, the additional conductor is red. Both types also have a grounding conductor, which is usually bare, but it is sometimes covered with green plastic insulation, depending upon the manufacturer. Type NMC is basically the same as NM, but the outer nonmetallic sheath is corrosion-resistant. The jacket or covering consists of rubber, plastic, or fiber. Most also have markings on this jacket giving the manufacturer's name or trademark, wire size, and number of conductors (see *Figure 19*). For example, NM 12-2 W/GRD indicates that the jacket contains two No. 12

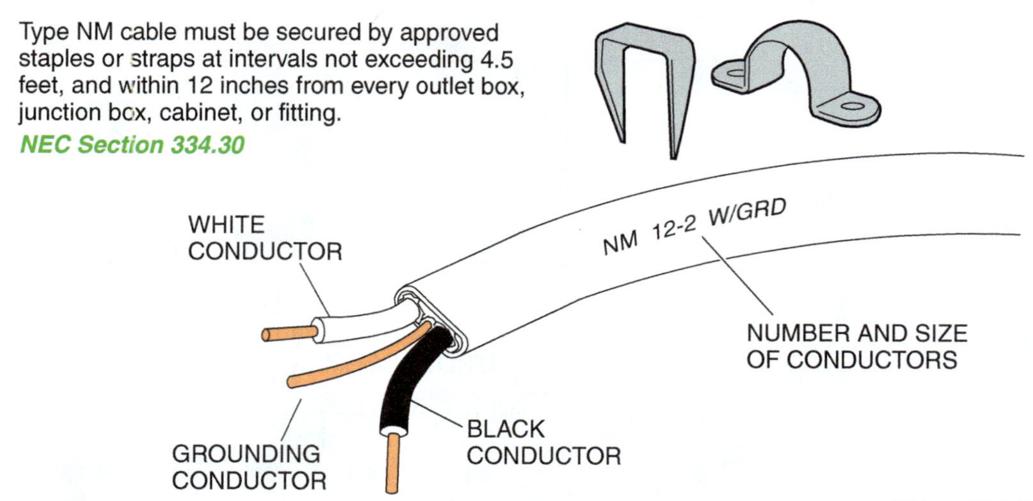

Figure 19 Characteristics of Type NM cable.

AWG conductors along with a grounding wire; NM 12-3 W/GRD indicates three conductors plus a grounding wire. Type NM cable is often referred to as **Romex®**.

NEC Section 334.10 permits Type NM, NMC, and NMS cable to be used in the following applications:

- Outdoors in dry locations
- One- and two-family dwelling units and attached or detached garages or storage buildings
- Multi-family dwellings when they are of Types III, IV, and V construction
- Other structures if concealed behind a 15-minute finish barrier and are of Types III, IV, and V construction

In addition, Type NMC can be used outdoors in dry or moist locations.

Type NMS is insulated power or control conductors with signaling, data, and communications conductors within an overall nonmetallic jacket. Type NMS cable shall be permitted as follows:

- For both exposed and concealed in normally dry locations
- To be fished in voids in masonry block or tile walls

NEC Section 334.12 prohibits the use of Type NM, NMC, and NMS cable in the following applications:

- As open runs in dropped or suspended ceilings in other-than-dwelling units
- As service-entrance cable
- In commercial garages with hazardous (classified) areas
- In theaters and similar locations, except as permitted by *NEC Section 518.4(B)*
- In motion picture studios
- In storage battery rooms
- In hoistways or on elevators or escalators
- Embedded in poured cement, concrete, or aggregate
- In hazardous (classified) areas
- Where exposed to corrosive fumes or vapors (NM and NMS)
- Embedded in masonry, adobe, fill, or plaster (NM and NMS)
- In a shallow chase in masonry, concrete, or adobe and covered with plaster, adobe, or similar finish (NM and NMS)
- Where exposed or subject to excessive moisture or dampness (NM and NMS)

Type NM cable is the most common type of cable for residential use. *Figure 20* shows additional *NEC®* regulations pertaining to the installation of Type NM cable.

8.1.2 Metal-Clad Cable

Metal-clad (MC) cable is manufactured in two-, three-, and four-wire assemblies with varying sizes of conductors, and is used in locations similar to those for Type NM, NMC, and NMS cable. Unlike Type NM, NMC, and NMS, it can also be used as service-entrance cable and in other locations permitted by *NEC Section 330.10*.

The metallic spiral covering on Type MC cable offers a greater degree of mechanical protection than Type NM cable and also provides a continuous grounding bond without the need for additional grounding conductors.

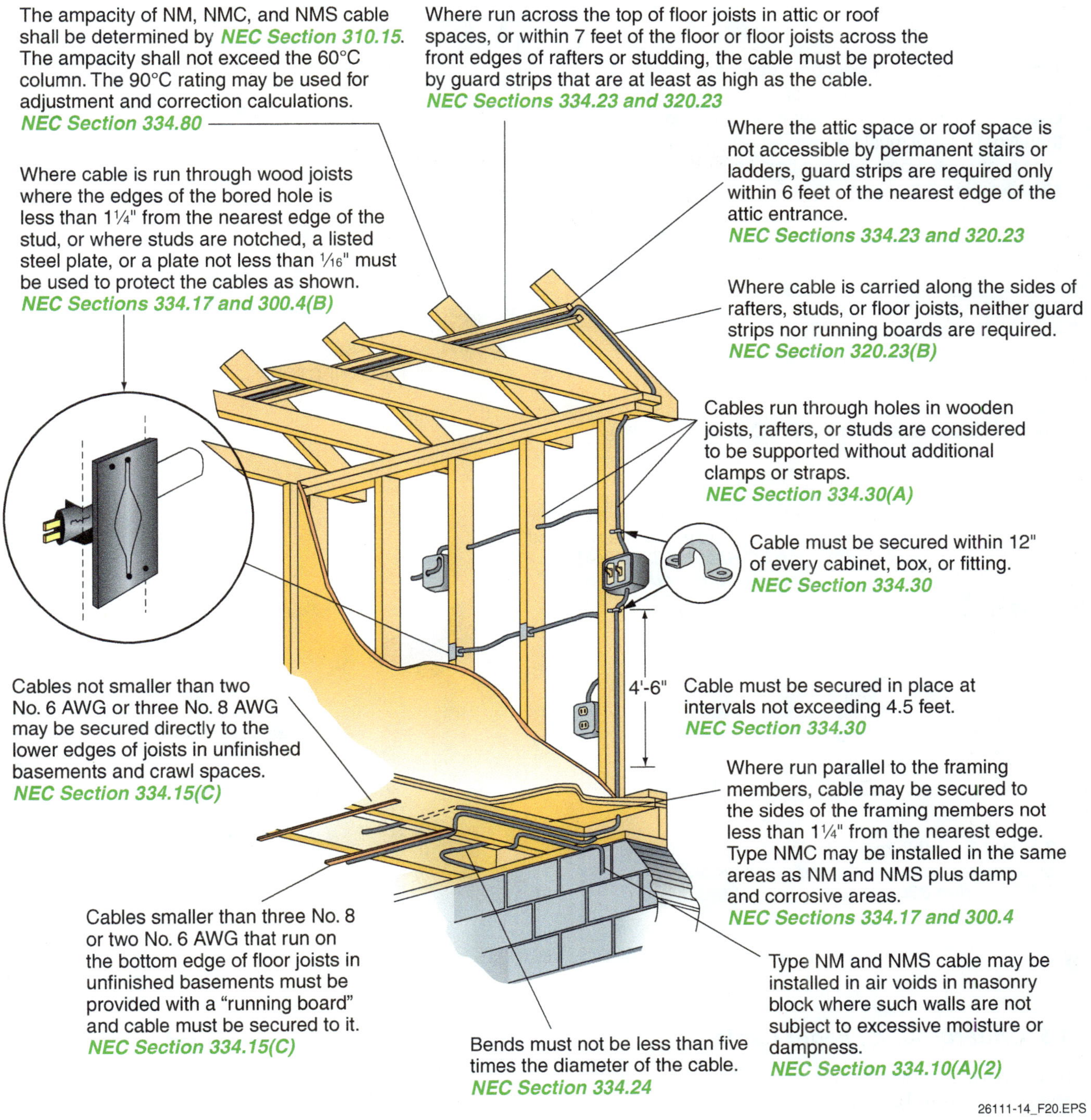

The ampacity of NM, NMC, and NMS cable shall be determined by *NEC Section 310.15*. The ampacity shall not exceed the 60°C column. The 90°C rating may be used for adjustment and correction calculations. *NEC Section 334.80*

Where run across the top of floor joists in attic or roof spaces, or within 7 feet of the floor or floor joists across the front edges of rafters or studding, the cable must be protected by guard strips that are at least as high as the cable. *NEC Sections 334.23 and 320.23*

Where the attic space or roof space is not accessible by permanent stairs or ladders, guard strips are required only within 6 feet of the nearest edge of the attic entrance. *NEC Sections 334.23 and 320.23*

Where cable is run through wood joists where the edges of the bored hole is less than 1¼" from the nearest edge of the stud, or where studs are notched, a listed steel plate, or a plate not less than $\frac{1}{16}$" must be used to protect the cables as shown. *NEC Sections 334.17 and 300.4(B)*

Where cable is carried along the sides of rafters, studs, or floor joists, neither guard strips nor running boards are required. *NEC Section 320.23(B)*

Cables run through holes in wooden joists, rafters, or studs are considered to be supported without additional clamps or straps. *NEC Section 334.30(A)*

Cable must be secured within 12" of every cabinet, box, or fitting. *NEC Section 334.30*

Cables not smaller than two No. 6 AWG or three No. 8 AWG may be secured directly to the lower edges of joists in unfinished basements and crawl spaces. *NEC Section 334.15(C)*

4'-6"

Cable must be secured in place at intervals not exceeding 4.5 feet. *NEC Section 334.30*

Where run parallel to the framing members, cable may be secured to the sides of the framing members not less than 1¼" from the nearest edge. Type NMC may be installed in the same areas as NM and NMS plus damp and corrosive areas. *NEC Sections 334.17 and 300.4*

Cables smaller than three No. 8 or two No. 6 AWG that run on the bottom edge of floor joists in unfinished basements must be provided with a "running board" and cable must be secured to it. *NEC Section 334.15(C)*

Bends must not be less than five times the diameter of the cable. *NEC Section 334.24*

Type NM and NMS cable may be installed in air voids in masonry block where such walls are not subject to excessive moisture or dampness. *NEC Section 334.10(A)(2)*

26111-14_F20.EPS

Figure 20 *NEC®* sections governing the installation of Type NM cable.

Type MC cable may be embedded in plaster finish, brick, or other masonry, except in damp or wet locations. It may also be run in the air voids of masonry block or tile walls, except where such walls are exposed or subject to excessive moisture or dampness. It may be used in wet locations if the conditions of *NEC Section 330.10(A)(11)* are met. It may not be used where subject to physical damage. See *Figures 21 and 22*.

NOTE

In the past, armored cable (Type AC), also called BX® cable, was commonly used in residential applications. Today, Type MC cable is used because it has a plastic wrapping to protect the conductors and does not require an insulating bushing at cable terminations.

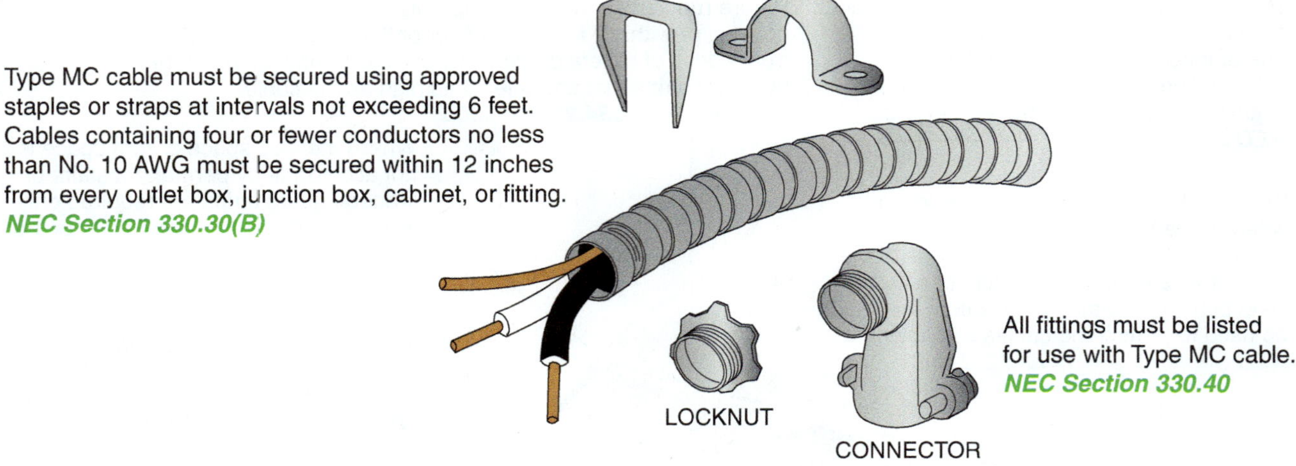

Figure 21 Characteristics of Type MC cable.

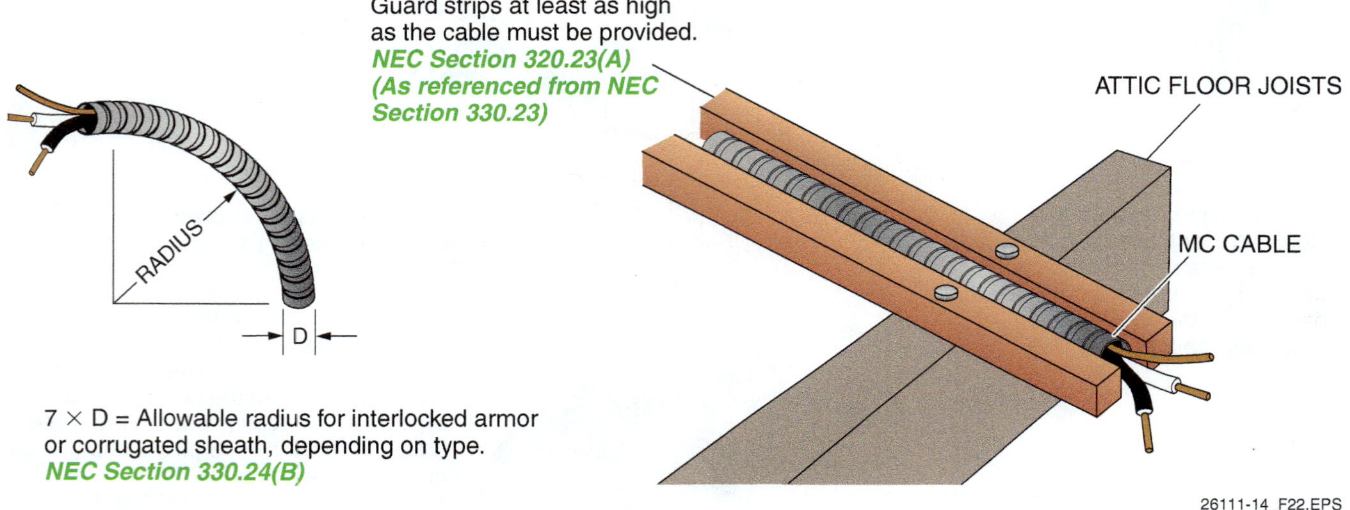

Figure 22 *NEC*® sections governing the installation of Type MC cable.

8.1.3 Underground Feeder Cable

Underground feeder (Type UF) cable *(NEC Article 340)* may be used underground, including direct burial in the earth, as a feeder or branch circuit cable when provided with overcurrent protection at the rated ampacity as required by the *NEC*®. When Type UF cable is used above grade where it will come in direct contact with the rays of the sun, its outer covering must be sun-resistant. Furthermore, where Type UF cable emerges from the ground, some means of mechanical protection must be provided. This protection may be in the form of conduit or guard strips. *NEC Section 300.5(D)(1)* requires that the protection extend from the minimum burial depth below grade to a point at least 8' above grade. *NEC Section 300.5(D)(4)* states that if conduit is used as protection, the permitted types are RMC, IMC, and Schedule 80 PVC, or equivalent. Type UF cable resembles Type NM cable; however, the jacket is constructed of weather-resistant material to provide the required protection for direct-burial wiring installations.

8.1.4 Service-Entrance Cable

Service-entrance (Type SE) and underground service-entrance (Type USE) cable, when used for electrical services, must be installed as specified in *NEC Articles 230 and 338.* Service-entrance cable is available with the grounded conductor bare for outside service conductors, and also with an insulated grounded conductor for interior wiring systems.

On Site

Cable Stripping

Special strippers are used to remove the jackets from Type NM and Type MC cable.

(A) NM CABLE RIPPER

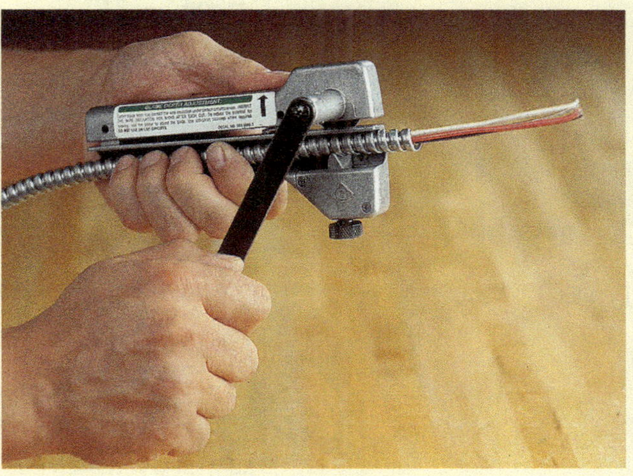

(B) MC CABLE CUTTER

26111-14 SA03.EPS

Type SE cable is permitted for use on branch circuits or feeders provided that all current-carrying conductors are insulated; this includes the grounded or neutral conductor. Where a conductor in the cable is not insulated, it is only permitted to be used as an equipment grounding conductor for branch circuits or feeders. Where used as an interior wiring method, the installation requirements of *NEC Article 334* must be followed, except for determining the ampacity of the cable. Where installed as exterior wiring, the requirements of *NEC Article 225* must be met, with the supports for the cable in accordance with *NEC Section 334.30.*

SE Style R (SER) cable is used in residential applications for subfeeds for ranges, and it is also used for service laterals in multi-family dwellings.

Figure 23 summarizes the installation rules for Type SE cable for both exterior and interior wiring.

8.2.0 Raceways

A raceway is any channel that is designed and used solely for the purpose of holding wires, cables, or busbars. Types of raceways include rigid metal conduit, intermediate metal conduit, rigid nonmetallic conduit, flexible metallic conduit, electrical metallic tubing, and auxiliary gutters. Raceways are constructed of either metal or insulating material, such as polyvinyl chloride or PVC (plastic). Metal raceways are joined using threaded, compression, or setscrew couplings; nonmetallic raceways are joined using cement-coated couplings. Where a raceway terminates in an outlet box, junction box, or other enclosure, an approved connector must be used.

Raceways provide mechanical protection for the conductors that run in them and also prevent accidental damage to insulation and the conducting material. They also protect conductors from corrosive atmospheres and prevent fire hazards to life and property by confining arcs and flames that may occur due to faults in the wiring system. Conduits or raceways are used in residential applications for service masts, underground wiring embedded in concrete, and sometimes in unfinished basements, shops, or garage areas.

Another function of metal raceways is to provide a continuous equipment grounding system throughout the electrical system. To maintain this feature, it is extremely important that all raceway systems be securely bonded together into a continuous conductive path and properly connected to the system ground. The following section explains how this is accomplished.

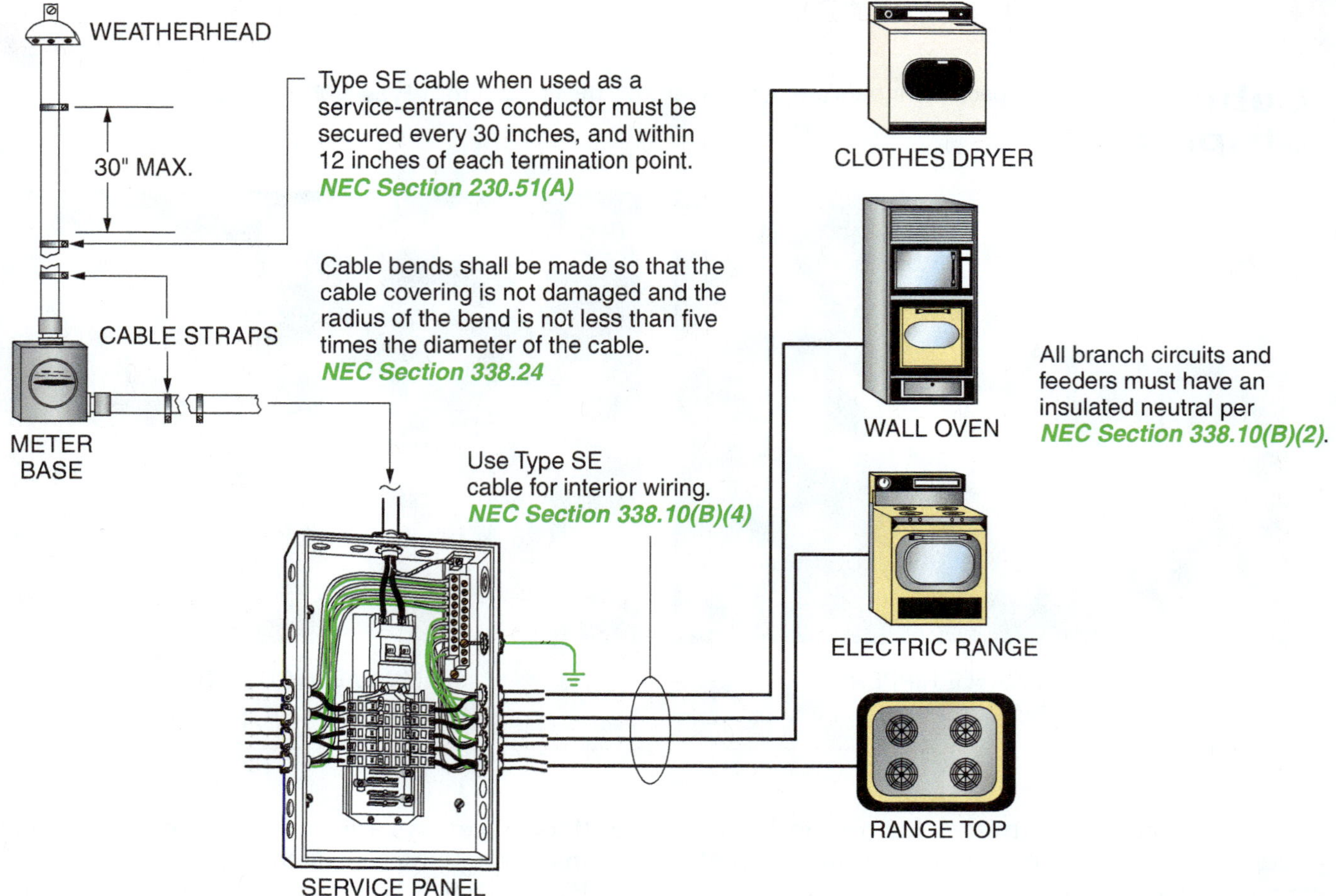

Figure 23 NEC® sections governing Type SE cable.

9.0.0 Equipment Grounding System

NEC Article 250, Part IV generally requires that all metallic enclosures, raceways, and cable armor be grounded. The exceptions in *NEC Sections 250.80 and 250.86* allow metal enclosures or short sections of raceways that are used to provide support or physical protection to be ungrounded under specific conditions.

NEC Article 250, Part VI covers equipment grounding and equipment grounding conductors. This section generally requires that the exposed noncurrent-carrying metal parts of fixed equipment likely to become energized be grounded under the following conditions:

- Where within 8' vertically or 5' horizontally of ground or grounded metal objects and subject to contact by occupants or others
- Where located in wet or damp locations
- Where in electrical contact with metal
- Where in hazardous (classified) locations as covered by *NEC Articles 500 through 517*
- Where supplied by a metal-clad, metal-sheathed, or metal raceway, or other wiring method that provides an equipment ground
- Where equipment operates with any terminal at over 150V to ground

Specific equipment that is required to be grounded regardless of the voltage is listed in *NEC Section 250.112* and includes equipment such as motors, motor controllers, and light fixtures. Types of cord- and plug-connected equipment in dwelling units that are required to be grounded are found in *NEC Section 250.114* and include equipment such as refrigerators, freezers, air conditioners, information technology equipment (computers), clothes washers, clothes dryers, and dishwashing machines.

The types of equipment grounding conductors that are acceptable to be used are found in *NEC Section 250.118.* Note that among the list of wiring methods approved for use as equipment grounding conductors, both flexible metal conduit (FMC) and liquidtight flexible metal conduit (LFMC) are permitted to be used. Listed FMC is permitted to be used as an equipment grounding conductor only when the following conditions are met:

- The conduit is terminated in fittings listed for grounding.
- The circuit conductors contained in the conduit are protected by overcurrent devices rated at 20A or less.
- The combined length of FMC, FMT, and LFMC in the same ground return path does not exceed 6'.
- The conduit is not installed for flexibility.

Type LFMC is also used in dwelling units and has slightly different requirements when used as an equipment grounding conductor:

- The conduit is terminated in fittings listed for grounding.
- For trade sizes ⅜ through ½, the circuit conductors contained in the conduit are protected by overcurrent devices rated at 20A or less.
- For trade sizes ¾ through 1¼, the circuit conductors contained in the conduit are protected by overcurrent devices rated at 60A or less and there is no FMC, FMT, or LFMC in trade sizes ⅜ through ½ in the grounding path.
- The combined length of FMC, FMT, and LFMC in the same ground return path does not exceed 6'.
- The conduit is not used for flexibility.

Where external bonding jumpers are used to provide the continuity of the fault current path, *NEC Section 250.102(E)(2)* limits the length to not more than 6', except at outside pole locations for the purposes of bonding or grounding the isolated sections of metal raceways or elbows installed in exposed risers at those pole locations. When installing an equipment grounding conductor in a raceway, *NEC Table 250.122* is used to determine the size of the equipment grounding conductor. It is permitted to install one equipment grounding conductor in a raceway that has several circuits. In that case, the size of the equipment grounding conductor is based on the rating of the largest overcurrent device protecting the circuits contained in the raceway.

NEC Section 250.148 requires that where circuit conductors are spliced within a box, or terminated on equipment within or supported by a box, separate equipment grounding conductors associated with those circuit conductors shall be spliced or joined within the box or to the box with devices suitable for the use. *Figure 24* shows several types of fittings that are suitable for this purpose.

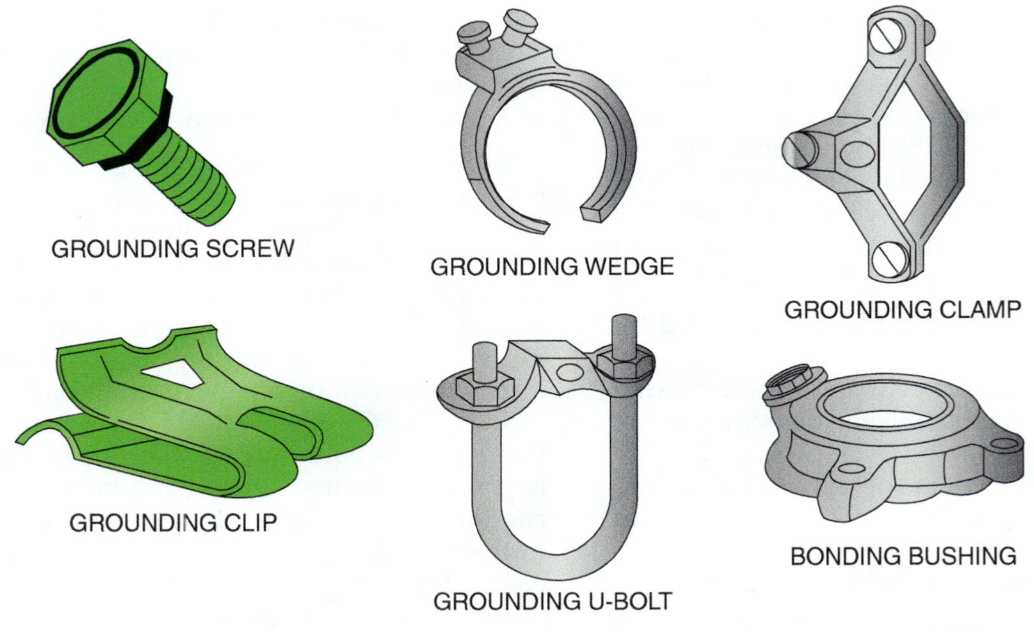

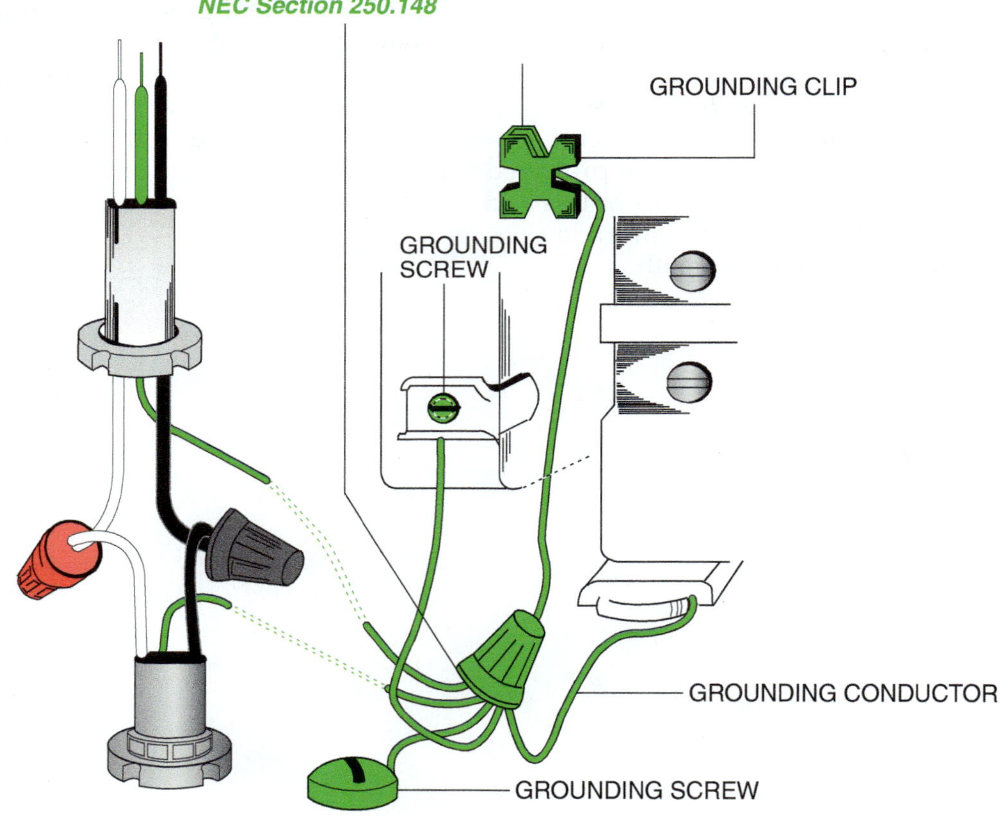

Figure 24 Equipment grounding methods.

10.0.0 Branch Circuit Layout for Power

The point at which electrical equipment is connected to the wiring system is commonly called an outlet. There are many classifications of outlets: lighting, receptacle, motor, appliance, and so forth. This section, however, deals with the power outlets normally found in residential electrical wiring systems.

When viewing an electrical drawing, outlets are indicated by symbols (usually a small circle with appropriate markings to indicate the type of outlet). The most common symbols for receptacles are shown in *Figure 25*.

10.1.0 Branch Circuits and Feeders

The conductors that extend from the panelboard to the various outlets are called branch circuits and are defined by the *NEC®* as the point of a wiring system that extends beyond the final overcurrent device protecting the circuit. See *Figure 26*.

A feeder consists of all conductors between the service equipment and the final overcurrent device. See *Figure 27*.

In general, the size of the branch circuit conductors varies depending upon the load requirements of the electrically operated equipment connected to the outlet. For residential use, most branch circuits consist of either No. 14 AWG, No. 12 AWG, No. 10 AWG, or No. 8 AWG conductors.

The basic branch circuit requires two wires or conductors to provide a continuous path for the flow of electric current, plus a third wire for equipment grounding. The usual receptacle branch circuit operates at 120V.

Fractional horsepower motors and small electric heaters usually operate at 120V and are connected to 120V branch circuits by means of a receptacle, junction box, or direct connection.

With the exception of very large residences and tract-development houses, the size of the average residential electrical system of the past has not been large enough to justify the expense of preparing complete electrical working drawings and specifications. Such electrical systems were usually laid out by the architect in the form of a sketchy outlet arrangement or else laid out by the electrician on the job, often only as the work progressed. However, many technical developments in residential electrical use—such as electric heat

Think About It

Grounding Devices

Identify each of the devices pictured here and explain their function(s). Which of these devices would not be used for equipment grounding?

26111-14_SA04.EPS

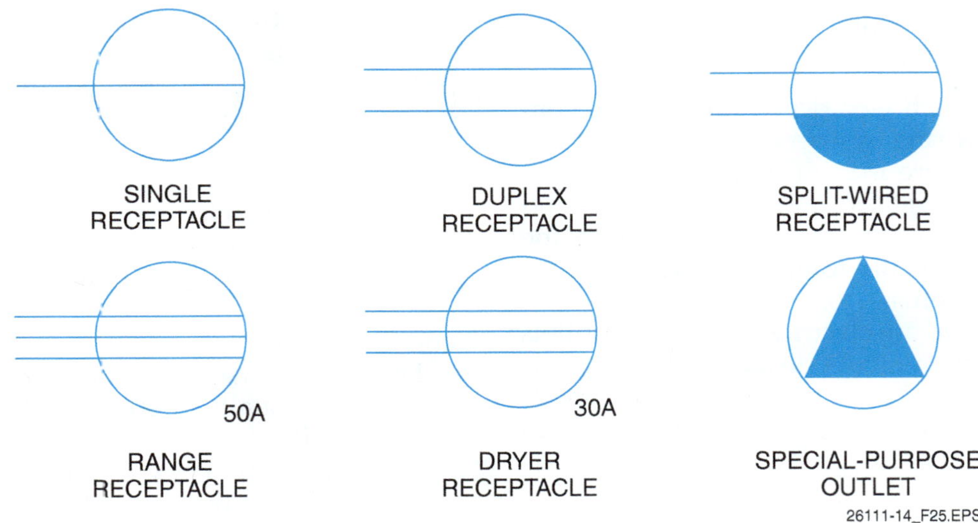

Figure 25 Typical outlet symbols appearing in electrical drawings.

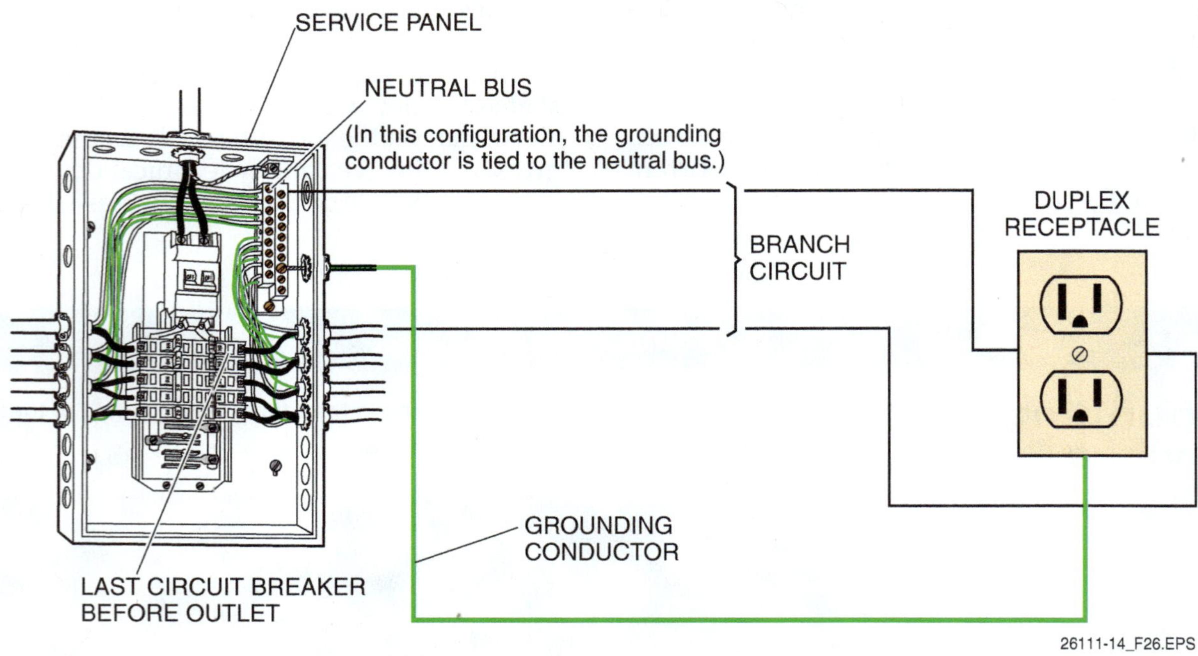

Figure 26 Components of a duplex receptacle branch circuit.

with sophisticated control wiring, increased use of electrical appliances, various electronic alarm systems, new lighting techniques, and the need for energy conservation techniques—have greatly expanded the demand and extended the complexity of today's residential electrical systems.

Each year, the number of homes with electrical systems designed by consulting engineering firms increases. Such homes are provided with complete electrical working drawings and specifications, similar to those frequently provided for commercial and industrial projects. Still, these are more the exception than the rule. Most residential projects will not have a complete set of drawings.

Circuit layout is provided on the drawings to follow for several reasons:

- They provide a visual layout of house wiring circuitry.
- They provide a sample of electrical residential drawings that are prepared by consulting engineering firms, although the number may still be limited.
- They introduce the method of showing electrical systems on working drawings to provide a foundation for tackling advanced electrical systems.

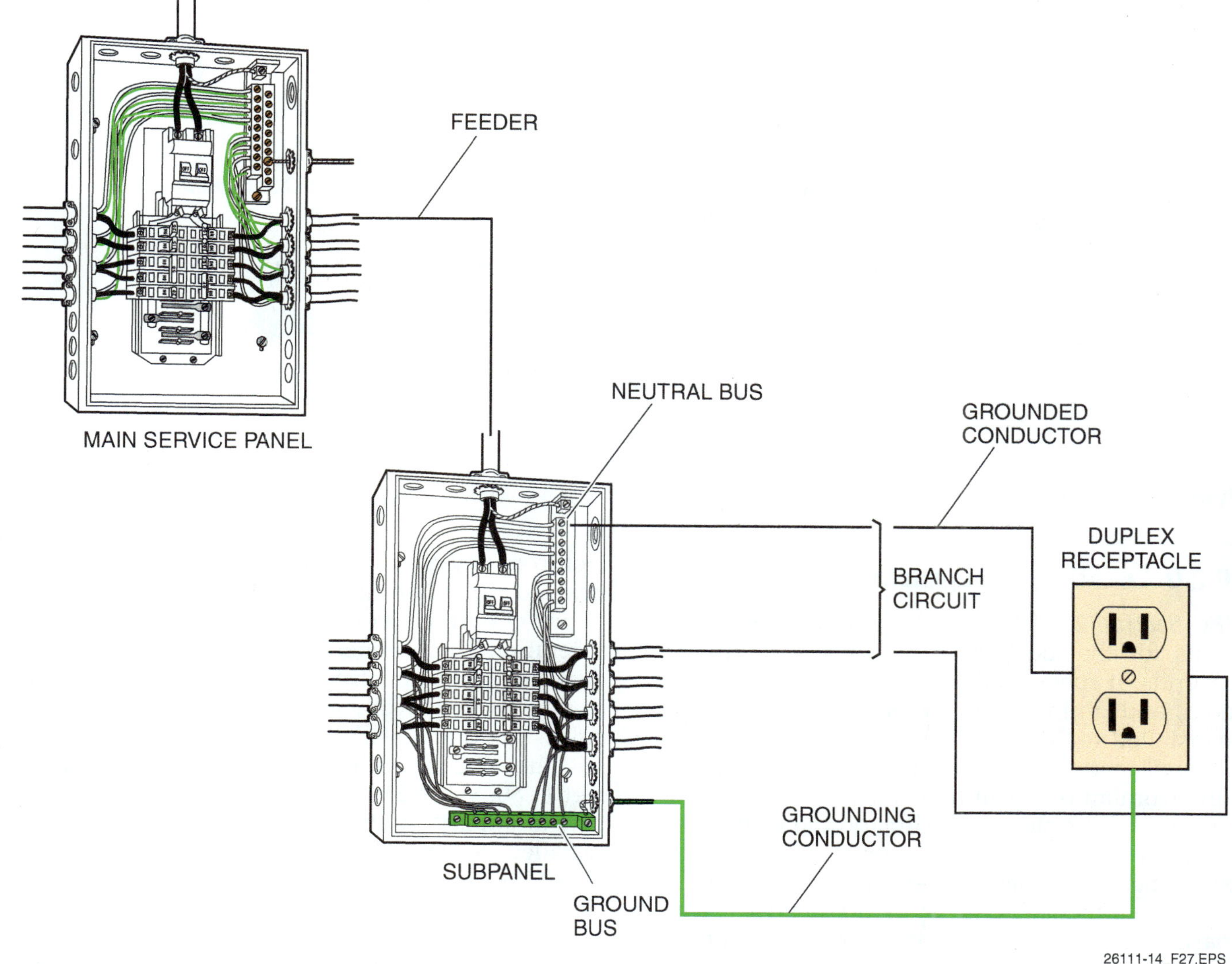

Figure 27 A feeder being used to feed a subpanel from the main service panel.

Branch circuits are shown on electrical drawings by means of a single line drawn from the panelboard (or by homerun arrowheads indicating that the circuit goes to the panelboard) to the outlet or from outlet to outlet where there is more than one outlet on the circuit.

The lines indicating branch circuits can be solid to show that the conductors are to be run concealed in the ceiling or wall; dashed to show that the conductors are to be run in the floor or ceiling below; or dotted to show that the wiring is to be run exposed. *Figure 28* shows examples of these three types of branch circuit lines.

In *Figure 28*, No. 12 indicates the wire size. The slash marks shown through the circuits in *Figure 28* indicate the number of current-carrying conductors in the circuit. Although two slash marks are shown, in actual practice, a branch circuit containing only two conductors usually contains no slash marks; that is, any circuit with no slash marks is assumed to have two conductors. However, three or more conductors are always indicated on electrical working drawings—either by slash marks for each conductor, or else by a note.

Never assume that you know the meaning of any electrical symbol. Although great efforts have been made in recent years to standardize drawing symbols, architects, consulting engineers, and electrical drafters still modify existing symbols or devise new ones to meet their own needs. Always consult the symbol list or legend on electrical working drawings for an exact interpretation of the symbols used.

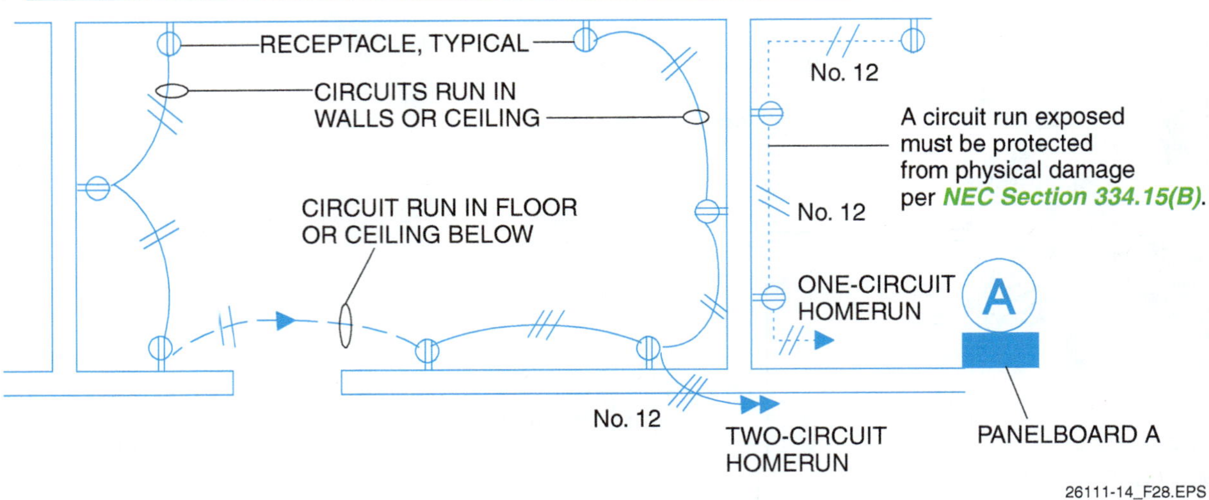

Figure 28 Types of branch circuit lines shown on electrical working drawings.

10.2.0 Locating Receptacles

NEC Section 210.52 states the minimum requirements for the location of receptacles in dwelling units. It specifies that in each kitchen, family room, and dining room, receptacle outlets shall be installed so that no point along the floor line in any wall space is more than 6', measured horizontally, from an outlet in that space, including any wall space 2' or more in width and the wall space occupied by fixed panels in exterior walls, but excluding sliding panels. Receptacle outlets shall, insofar as practicable, be spaced equal distances apart. Receptacle outlets in floors shall not be counted as part of the required number of receptacle outlets unless located within 18" of the wall.

The *NEC®* defines wall space as a wall that is unbroken along the floor line by doorways, fireplaces, or similar openings. Each wall space that is two feet or more in width must be treated individually and separately from other wall spaces within the room.

The purpose of *NEC Section 210.52* is to minimize the use of cords across doorways, fireplaces, and similar openings.

With this *NEC®* requirement in mind, outlets for our sample residence will be laid out (see *Figure 29*). In laying out these receptacle outlets, the floor line of the wall is measured (also around corners), but not across doorways, fireplaces, passageways, or other spaces where a flexible cord extended across the space would be unsuitable.

In general, duplex receptacle outlets must be no more than 12' apart. When spaced in this manner, a 6' extension cord will reach a receptacle from any point along the wall line.

Note that at no point along the wall line are any receptacles more than 12' apart or more than six feet from any door or room opening. Where practical, no more than eight receptacles are connected to one circuit. However, this is just a design consideration since general-purpose receptacles in dwelling units are sized on the basis of 3VA per square foot of dwelling space. A 15A branch circuit is rated at 1,800VA (15A × 120V = 1,800VA) and the *NEC®* requires that for every 600 square feet (1,800VA ÷ 3VA/sq. ft. = 600 sq. ft.), a circuit to supply lighting and receptacles must be installed. Always check with the local authorities about the requirements for the number of branch circuits in a dwelling.

The utility room has at least one receptacle for the laundry on a separate circuit in order to comply with *NEC Sections 210.11(C)(2) and 210.52(F).*

One duplex receptacle is located in the vestibule for cleaning purposes, such as feeding a portable vacuum cleaner or similar appliance. It is connected to the living room circuit. An additional duplex receptacle is required per *NEC Section 210.52(H)* in hallways of 10' or more.

Although this is not shown in the figure, the living room outlets could be split-wired (the lower half of each duplex receptacle is energized all the time, while the upper half can be switched on or off). The reason for this is that a great deal of the illumination for this area will be provided by portable table lamps, and the split-wired receptacles provide a means to control these lamps from several locations, such as at each entry to the living room, if desired. Split receptacles are discussed in more detail in the next section.

To comply with *NEC Sections 210.11(C)(1) and 210.52(B),* the kitchen receptacles are laid out as follows. In addition to the number of branch circuits determined previously, two or more 20A small appliance branch circuits must be provided

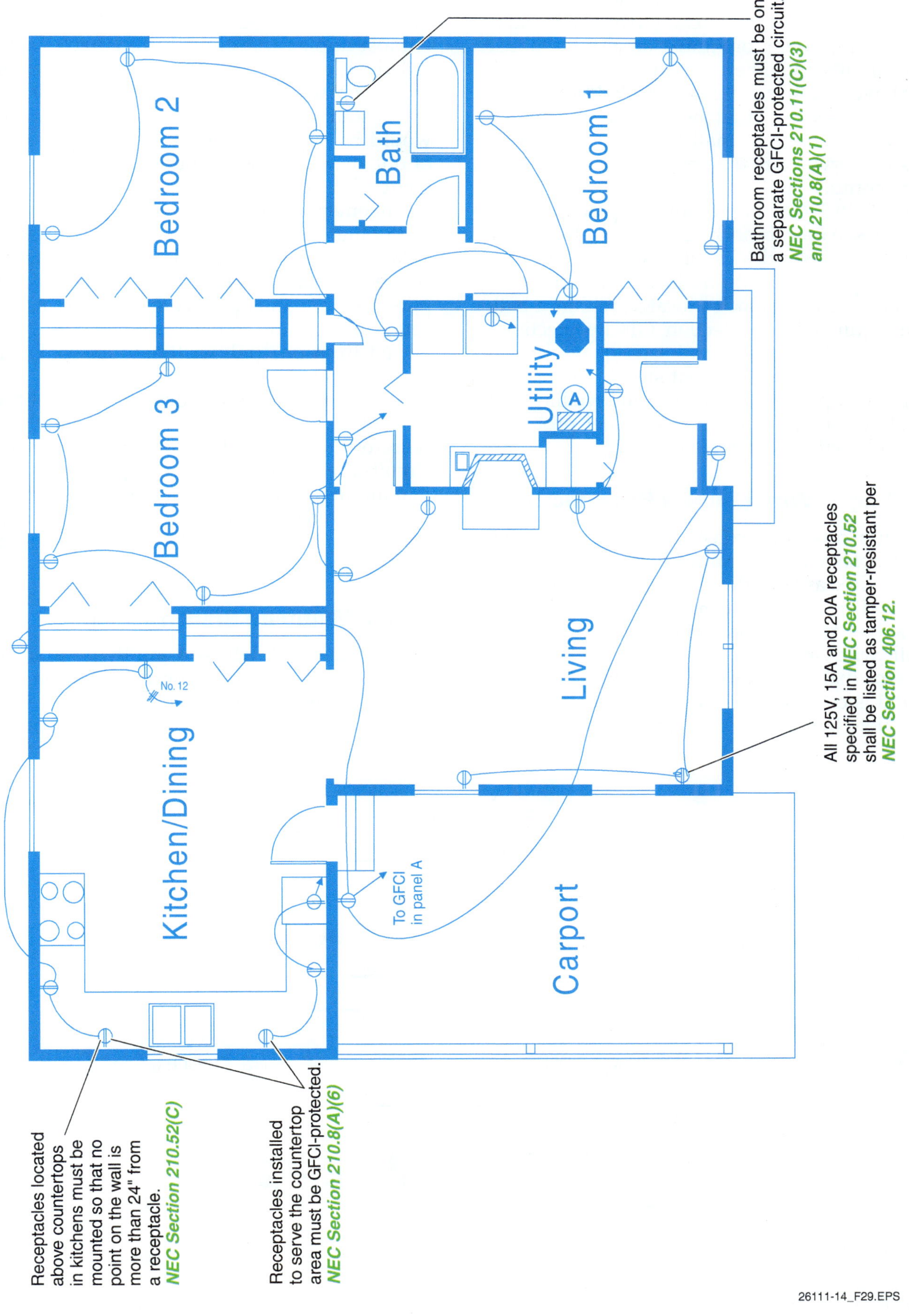

Figure 29 Floor plan of the sample residence.

to serve all receptacle outlets (including refrigeration equipment) in the kitchen, pantry, breakfast room, dining room, or similar area of the house. Such circuits, whether two or more are used, must have no other outlets connected to them. All receptacles serving a kitchen countertop require GFCI protection. No small appliance branch circuit shall serve more than one kitchen.

To comply with *NEC Sections 210.11(C)(3) and 210.52(D),* bathroom receptacle(s) must be on a separate branch circuit supplying only bathroom receptacles or on a circuit supplying a single bathroom with no loads other than that bathroom. All receptacles located within a bathroom require GFCI protection. GFCI protection is also required on garage and exterior receptacles. All other branch circuits that supply the lighting and general-purpose receptacles in dwelling units must have arc fault circuit interrupter protection to comply with *NEC Section 210.12.*

10.3.0 Split-Wired Duplex Receptacles

In modern residential construction, it is common to have duplex wall receptacles that have one of the outlets wired as a standard duplex outlet (hot all the time) and the other half controlled by a wall switch. This allows table or floor lamps to be controlled by a wall switch and leaves the other outlet available for items that are not to be switched. This wiring method is commonly referred to as a split receptacle. Note that switched receptacles are installed to provide lighting. Dimmer switches are not permitted to be used per *NEC Section 404.14(E).*

Most duplex 15A and 20A receptacles are provided with a breakoff tab that permits each of the two receptacle outlets to be supplied from a different source or polarity. For example, one outlet would be supplied from the hot leg of a series of outlets and the other outlet supplied from the switch leg of a light switch. A diagram of this arrangement is shown in *Figure 30*.

Another application of split receptacles is shown in *Figure 31*. In this example, one outlet connected from a double-pole circuit breaker supplies 240V for an appliance such as a window air conditioning unit, while the other outlet is connected from one pole of the double-pole circuit breaker and the other side is connected to the neutral or grounded conductor to supply 120V for an appliance such as a lamp. *NEC Section 210.4(B)* requires the use of a two-pole breaker when two circuits are connected to one duplex receptacle so that all ungrounded conductors of the circuit are disconnected simultaneously. This circuit and the split receptacle mentioned above are both considered multiwire branch circuits.

10.4.0 Multiwire Branch Circuits

NEC Article 100 defines a multiwire branch circuit as "two or more ungrounded conductors having a potential difference between them, and a grounded conductor having equal potential

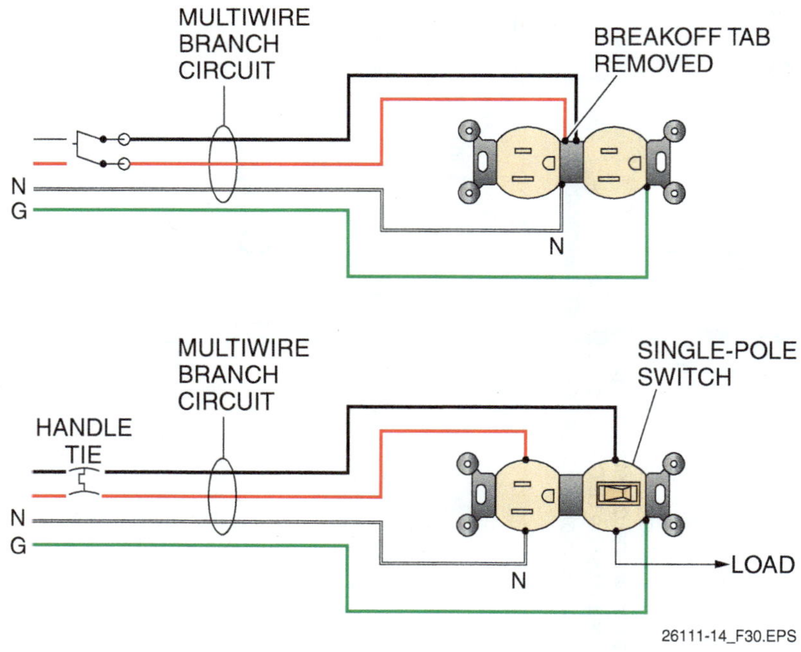

Figure 30 Two 120V receptacle outlets supplied from different sources.

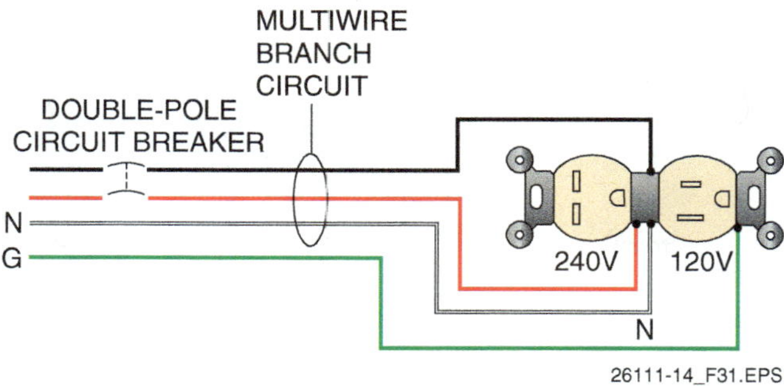

Figure 31 Combination receptacle.

difference between it and each ungrounded conductor of the circuit and that is connected to the neutral conductor of the system."

10.5.0 240-Volt Circuits

The electric range, clothes dryer, and water heater in the sample residence all operate at 240VAC. Each will be fed by a separate circuit and connected to a two-pole circuit breaker of the appropriate rating in the panelboard. To determine the conductor size and overcurrent protection for the range, proceed as follows:

Step 1 Find the nameplate rating of the electric range. This has previously been determined to be 12kVA.

Step 2 Refer to *NEC Table 220.55.* Since Column A of this table applies to ranges rated at 12kVA (12kW) and under, this will be the column to use in this example.

Step 3 Under the Number of Appliances column, locate the appropriate number of appliances (one in this case), and find the maximum demand given for it in Column A. Column A states that the circuit should be sized for 8kVA (not the nameplate rating of 12kVA).

Step 4 Calculate the required conductor ampacity as follows:

$$\frac{8{,}000\text{VA}}{240\text{V}} = 33.33\text{A}$$

The minimum branch circuit must be rated at 40A since common residential circuit breakers are rated in steps of 15A, 20A, 30A, 40A, and so forth. A 30A circuit breaker is too small, so a 40A circuit breaker is selected. The conductors must have a current-carrying capacity that is equal to or greater than the overcurrent protection. Therefore, No. 8 AWG conductors will be used.

If a cooktop and wall oven were used instead of the electric range, the circuit would be sized similarly. The *NEC®* specifies that a branch circuit for a counter-mounted cooking unit and not more than two wall-mounted ovens, all supplied from a single branch circuit and located in the same room, is computed by adding the nameplate ratings of the individual appliances and treating this total as equivalent to one range. Therefore, two appliances of 6kVA each may be treated as a single range with a 12kVA nameplate rating.

Figure 32 shows how the electric range circuit may appear on an electrical drawing. The connection may be made directly to the range junction box, but more often a 50A range receptacle is mounted at the range location and a range cord-and-plug set is used to make the connection. This facilitates moving the appliance later for maintenance or cleaning.

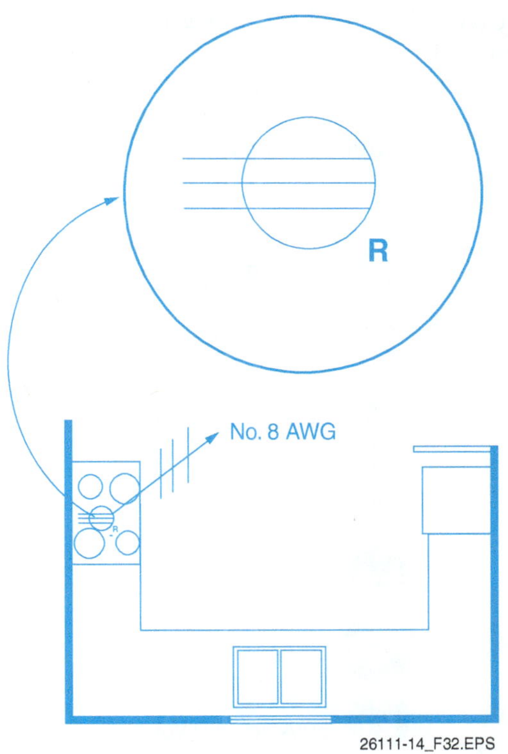

Figure 32 Range circuit shown on an electrical drawing.

> **Think About It**
>
> **240V Circuits**
>
> Calculate the ampacity required for a kitchen range with an 8kW rating. Now design the practical wiring in a labeled diagram. How will the wires be connected at the service panel and at the appliance? How will the cable be installed?

Figure 33 shows several types of receptacle configurations used in residential wiring applications. You will eventually recognize these configurations at a glance.

The branch circuit for the water heater in the sample residence must be sized for its full capacity because there is no diversity or demand factor for this appliance. Since the nameplate rating on the water heater indicates two heating elements of 4,500W each, the first inclination would be to size the circuit for a total load of 9,000W (volt-amperes). However, only one of the two elements operates at a time. See *Figure 34*. Note that each element is controlled by a separate thermostat. The lower element becomes energized when the thermostat calls for heat, and at the same time, the thermostat opens a set of contacts to prevent the upper element from operating. When the lower element's thermostat is satisfied, the lower contacts open, and at the same time, the thermostat closes the contacts for the upper element to become energized to maintain the water temperature.

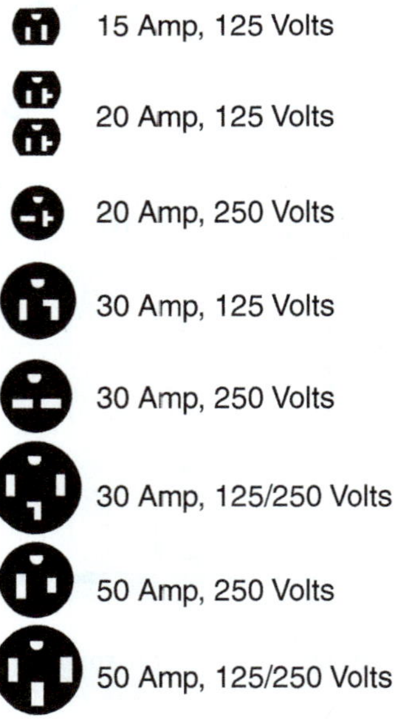

Figure 33 Residential receptacle configurations.

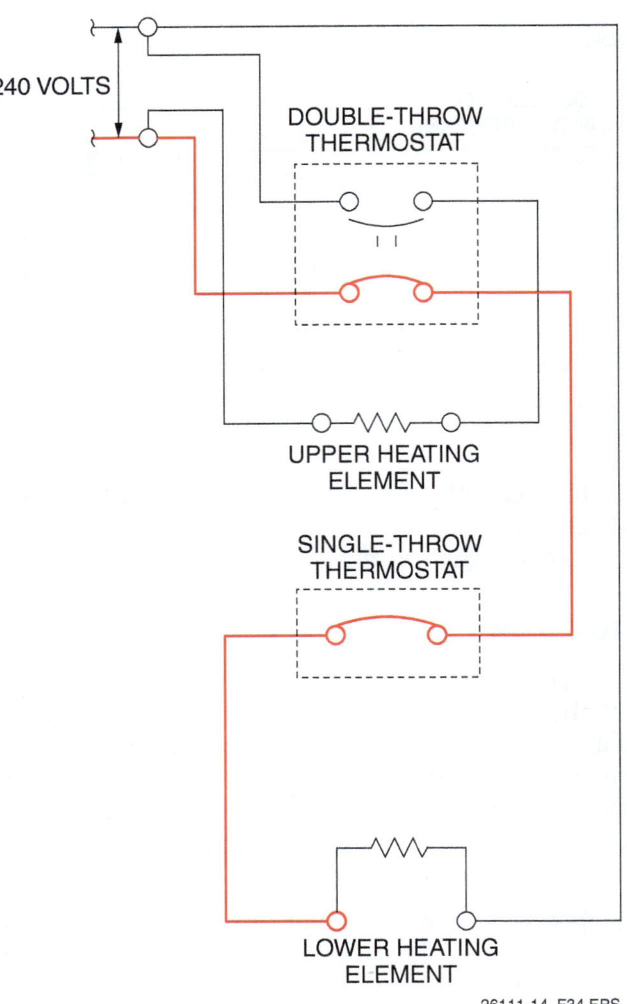

Figure 34 Wiring diagram of water heater controls.

With this information in hand, the circuit for the water heater may be sized as follows:

$$\frac{4{,}500\text{VA}}{240\text{V}} = 18.75\text{A} \times 1.25 = 23.44\text{A}$$

NEC Section 422.13 requires that the branch circuits that supply storage type water heaters having a capacity of 120 gallons or less be rated not less than 125% of the nameplate rating of the water heater. Our calculation shows this to be not less than 23A. Normally, this would require a maximum rating for the branch circuit to be not more than 25A. (See standard ratings of overcurrent devices in *NEC Section 240.6.*) However, *NEC Section 422.11(E)(3)* permits a single nonmotor-operated appliance to be protected

by overcurrent devices rated up to 150% of the nameplate rating of the appliance. In this case, 4,500VA ÷ 240V = 18.75A × 150% = 28.125A. Since the next standard rating is 30A, the water heater will be wired with No. 10 AWG conductors protected by a 30A overcurrent device.

The *NEC*® specifies that electric clothes dryers must be rated at 5kVA or the nameplate rating, whichever is greater. In this case, the dryer is rated at 5.5kVA, and the conductor current-carrying capacity is calculated as follows:

$$\frac{5{,}500\text{VA}}{240\text{V}} = 22.92\text{A}$$

A three-wire, 30A circuit will be provided (No. 10 AWG wire). It is protected by a 30A circuit breaker. The dryer may be connected directly, but a 30A dryer receptacle is normally provided for the same reasons as mentioned for the electric range.

Large appliance outlets rated at 240V are frequently shown on electrical drawings using lines and symbols to indicate the outlets and circuits. In some cases, no drawings are provided.

11.0.0 Branch Circuit Layout for Lighting

A simple lighting branch circuit requires two conductors to provide a continuous path for current flow. The usual lighting branch circuit operates at 120V; the white (grounded) circuit conductor is therefore connected to the neutral bus in the panelboard, while the black (ungrounded) circuit conductor is connected to an overcurrent protection device.

Lighting branch circuits and outlets are shown on electrical drawings by means of lines and symbols; that is, a single line is drawn from outlet to outlet and then terminated with an arrowhead to indicate a homerun to the panelboard. Several methods are used to indicate the number and size of conductors, but the most common is to indicate the number of conductors in the circuit by using slash marks through the circuit lines and then indicate the wire size by a notation adjacent to these slash marks. For example, two slash marks indicate two conductors; three slash marks indicate three conductors. Some electrical designers omit slash marks for two-conductor circuits. In this case, the conductor size is usually indicated in the symbol list or legend.

The circuits used to feed residential lighting must conform to standards established by the *NEC*® as well as by local and state ordinances. Most of the lighting circuits should be calculated to include the total load, although at times this is not possible because the electrician cannot be certain of the exact wattage that might be used by the homeowner. For example, an electrician may install four porcelain lampholders for the unfinished basement area, each to contain one 100-watt (100W) incandescent lamp. However, the homeowners may eventually replace the original lamps with others rated at 150W or even 200W. Thus, if the electrician initially loads the lighting circuit to full capacity, the circuit will probably become overloaded in the future.

It is recommended that no residential branch circuit be loaded to more than 80% of its rated capacity. Since most circuits used for lighting are rated at 15A, the total ampacity (in volt-amperes) for the circuit is as follows:

$$15\text{A} \times 120\text{V} = 1{,}800\text{VA}$$

Therefore, if the circuit is to be loaded to only 80% of its rated capacity, the maximum initial connected load should be no more than 1,440VA.

Figure 35 shows one possible lighting arrangement for the sample residence. All lighting fixtures are shown in their approximate physical location as they should be installed.

Electrical symbols are used to show the fixture types. Switches and lighting branch circuits are also shown by appropriate lines and symbols. The meanings of the symbols used on this drawing are explained in the symbol list in *Figure 36*.

In actual practice, the location of lighting fixtures (luminaires) and their related switches will probably be the extent of the information shown on working drawings. The circuits shown in *Figure 35* are meant to illustrate how lighting circuits are routed, not to imply that such drawings are typical for residential construction. If fixtures are used in a closet, they must meet the requirements of *NEC Section 410.16*.

A box used at fan outlets is not permitted to be used as the sole support for ceiling (paddle) fans, unless it is listed for the application as the sole means of support. Where a ceiling fan does not exceed 70 pounds in weight, it is permitted to be supported by outlet boxes listed and identified for such use. Boxes designed to support more than 35 pounds must be marked with the maximum weight to be supported. See *NEC Section 314.27(C)*. These boxes must be rigidly supported from a structural member of the building. *NEC Section 314.27(C)* also requires that a ceiling-fan rated box be used if a spare conductor is installed to the box and the location of the box is acceptable for a fan installation.

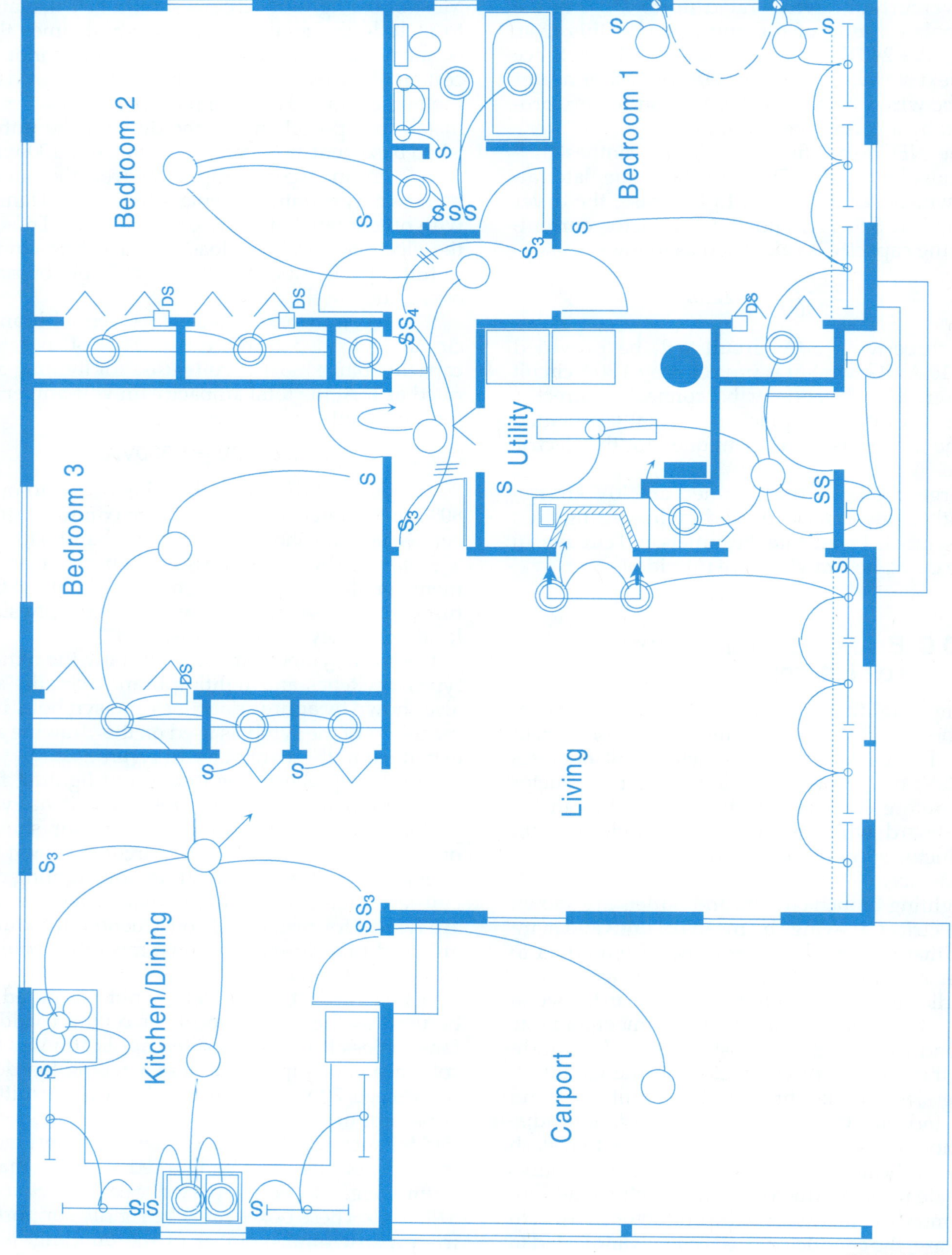

Figure 35 Lighting layout of the sample residence.

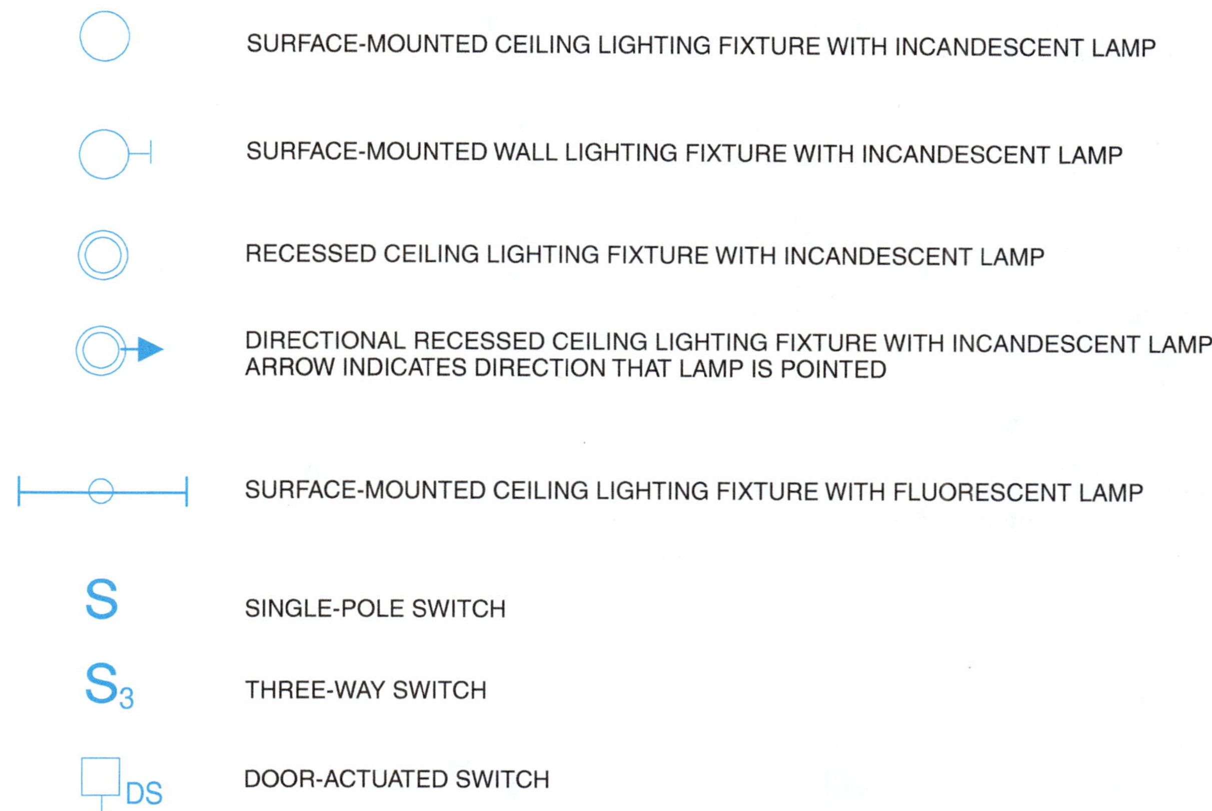

Figure 36 Symbols.

12.0.0 Outlet Boxes

Electricians installing residential electrical systems must be familiar with outlet box capacities, means of supporting outlet boxes, and other requirements of the *NEC®*. Boxes were discussed in detail in an earlier module, but a general review of the rules and necessary calculations is provided here.

The maximum numbers of conductors of the same size permitted in standard outlet boxes are listed in *NEC Table 314.16(A)*. These figures apply where no fittings or devices such as fixture studs, cable clamps, switches, or receptacles are contained in the box and where no grounding conductors are part of the wiring within the box. Obviously, in all modern residential wiring systems there will be one or more of these items contained in every outlet box installed. Therefore, where one or more of the above-mentioned items are present, the total number of conductors will be less than that shown in the table. Also, if the box contains a looped, unbroken conductor 12" or more in length, it must be counted twice.

For example, a deduction of two conductors must be made for each strap containing a wiring device entering the box (based on the largest size conductor connected to the device) such as a switch or duplex receptacle; a further deduction of one conductor must be made for one or more equipment grounding conductors entering the box (based on the largest size grounding conductor). For instance, a 3" × 2" × 2¾" box is listed in the table as containing a maximum of six No. 12 wires. If the box contains cable clamps and a duplex receptacle, three wires will have to be deducted from the total of six, providing for only three No. 12 wires. If a ground wire is used, which is always the case in residential wiring, only two No. 12 wires may be used.

For example, to size a metallic outlet box for two No. 12 AWG conductors with a ground wire, cable clamp, and receptacle, proceed as follows:

Step 1 Calculate the total number of conductors and equivalents [*NEC Section 314.16(B)*]. One ground wire plus one cable clamp plus one receptacle (two wires) plus two No. 12 conductors equals a total of six No. 12 conductors.

Step 2 Determine the amount of space required for each conductor. *NEC Table 314.16(B)* gives the box volume required for each conductor. No. 12 AWG equals 2.25 cubic inches.

Step 3 Calculate the outlet box space required by multiplying the number of cubic inches required for each conductor by the total number of conductors:

$$6 \times 2.25 = 13.5 \text{ cubic inches}$$

Step 4 Once you have determined the required box capacity, again refer to *NEC Table 314.16(A)* and note that a 3" × 2" × 2¾" box comes closest to our requirements. This box is rated for 14 cubic inches.

Now, size the box for two additional conductors. Where four No. 12 conductors enter the box with two ground wires, only the two additional No. 12 conductors must be added to our previous count for a total of 8 conductors (6 + 2 = 8). Remember, any number of ground wires in a box counts as only one conductor; any number of cable clamps also counts as only one conductor. Therefore, the box size required for use with two additional No. 12 conductors may be calculated as follows:

$$8 \times 2.25 = 18 \text{ cubic inches}$$

Again, refer to *NEC Table 314.16(A)* and note that a 3" × 2" × 3½" device box with a rated capacity of 18.0 cubic inches is the closest device box that meets *NEC*® requirements. An alternative is to use a 4" × 1¼" square box with a single-gang plaster ring, as shown in *Figure 37*. This box also has a capacity of 18.0 cubic inches.

Other box sizes are calculated in a similar fashion. When sizing boxes for different size conductors, remember that the box capacity varies as shown in *NEC Table 314.16(B)*.

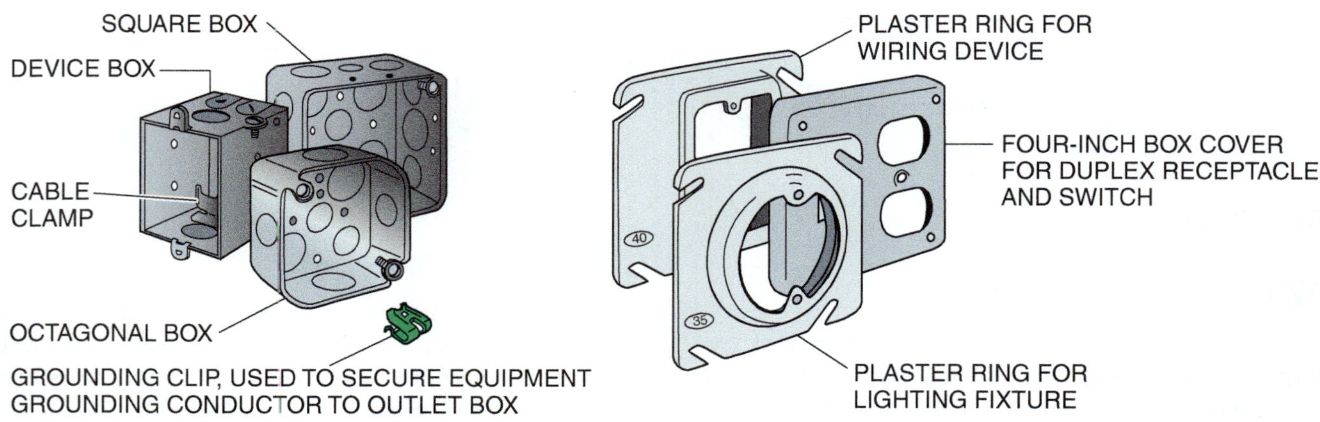

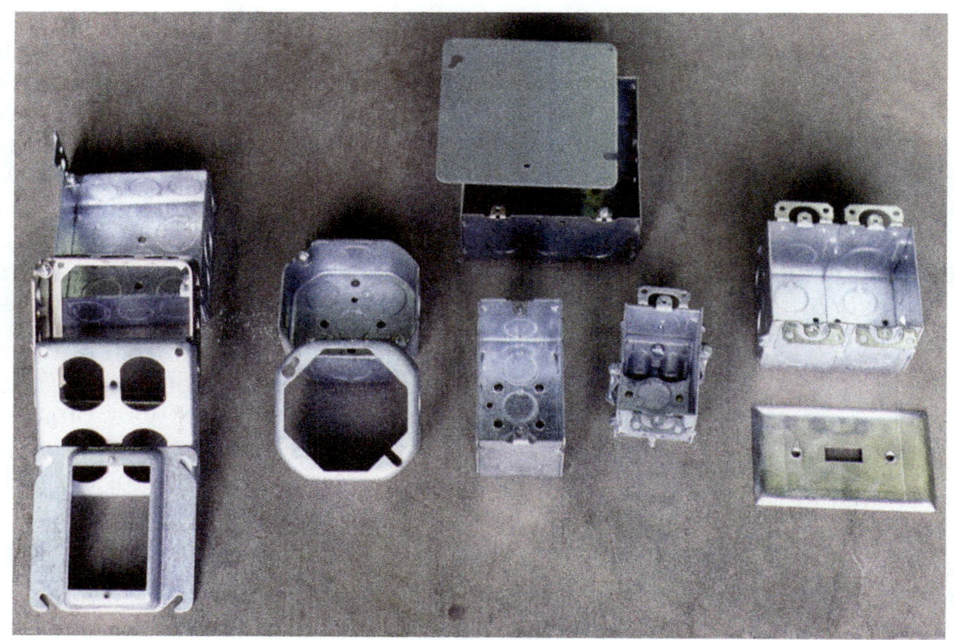

26111-14_F37B.EPS

Figure 37 Typical metallic outlet boxes with extension (plaster) rings.

On Site

Calculating Conductors

In a 4" × 4" × 1½" metal box, one 14/3 cable with ground feeds three 14/2 cables with ground wires. The red wire of the 14/3 cable feeds a receptacle, and the black wire feeds the 14/2 black wires. All of the white wires are spliced together, with one brought out to the receptacle terminal. The ground wires are all spliced, with one brought out to the grounding terminal on the receptacle and one to the ground clip on the box. All four cables are connected with box connectors rather than internal clamps. Using *NEC Section 314.16,* determine whether this wiring violates the code.

26111-14_SA05.EPS

12.1.0 Mounting Outlet Boxes

Outlet box configurations are almost endless, and if you research the various methods of mounting these boxes, you will be astonished. In this section, some common outlet boxes and their mounting considerations will be reviewed.

The conventional metallic device box, which is used for residential duplex receptacles and switches for lighting control, may be mounted to wall studs using 16d (penny) nails placed through the round mounting holes passing through the interior of the box. The nails are then driven into the wall stud. When nails are used for mounting outlet boxes in this manner, the nails must be located within ¼" of the back or ends of the enclosure.

Nonmetallic boxes normally have mounting nails fitted to the box for mounting. Other boxes have mounting brackets. When mounting outlet boxes with brackets, use either wide-head roofing nails or box nails about 1¼" in length. *Figure 38* shows various methods of mounting outlet boxes. Before mounting any boxes during the rough wiring process, first find out what type and thickness of finish will be used on the walls. This will dictate the depth to which the boxes must be mounted to comply with *NEC®* regulations. For example, the finish on plastered walls or ceilings is normally ½" thick; gypsum board or drywall is either ½" or ⅝" thick; and wood paneling is normally only ¼" thick. (Some tongue-and-groove wood paneling is ½" to ⅝" thick.)

The *NEC®* specifies the amount of space permitted from the edge of the outlet box to the finished wall. When a noncombustible wall finish (such as plaster, masonry, or tile) is used, the box may be recessed ¼". However, when combustible finishes are used (such as wood paneling), the box must be flush (even) with the finished wall or ceiling. See *Figure 39* and *NEC Section 314.20.*

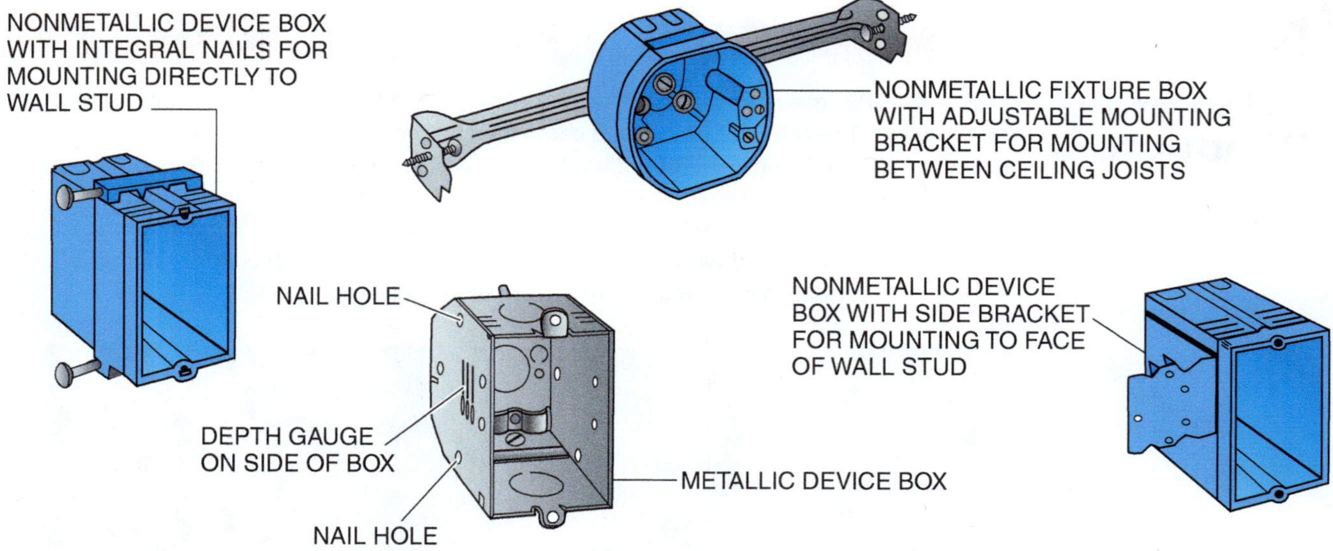

Figure 38 Several methods of mounting outlet boxes.

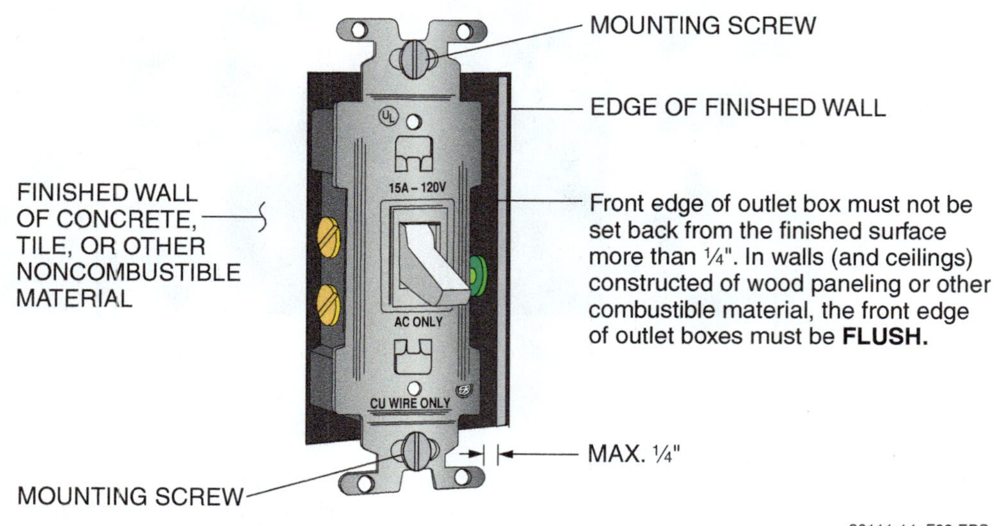

Figure 39 Outlet box installation.

On Site

Mounting Boxes

To quickly mount each box at the same height from the floor, make a simple height template (story pole) and mark it with the receptacle and switch heights. The story pole consists of an L-shaped jig made out of 2 × 2s or 2 × 4s. After installing the boxes, make sure to push the wires well back into the box so that the sheetrock installers will not damage the wires when they rout out a hole for the receptacle.

When Type NM cable is used in either metallic or nonmetallic outlet boxes, the cable assembly, including the sheath, must extend into the box by not less than ¼" [*NEC Section 314.17(C)*]. In all instances, all permitted wiring methods must be secured to the boxes by means of either cable clamps or approved connectors. The one exception to this rule is where Type NM cable is used with 2¼" × 4" (or smaller) nonmetallic boxes where the cable is fastened within eight inches of the box. In this case, the cable does not have to be secured to the box. See *NEC Section 314.17(C), Exception.*

13.0.0 WIRING DEVICES

Wiring devices include various types of receptacles and switches, the latter being used for lighting control. Switches are covered in *NEC Article 404,* while regulations for receptacles may be found in *NEC Article 406.*

13.1.0 Receptacles

Receptacles are rated by voltage and amperage capacity. *NEC Section 406.3* requires that receptacles connected to a 15A or 20A circuit have the correct voltage and current rating for the application, and be of the grounding type. *NEC Section 406.12* requires that all 15A and 20A, 125V receptacles installed in dwelling units be listed as tamper-resistant.

NEC Section 406.4(D) has several requirements for replacement receptacles:

- *NEC Section 406.4(D)(4)* – The replacement needs to be AFCI-protected if specified elsewhere.
- *NEC Section 406.4(D)(5)* – The replacement shall be tamper-resistant if specified elsewhere.
- *NEC Section 406.4(D)(4)* – The replacement must be weather-resistant if specified elsewhere.

Where there is only one outlet on a circuit, the receptacle's rating must be equal to or greater than the capacity of the conductors feeding it per *NEC Section 210.21(B)(1).* For example, if one receptacle is connected to a 20A residential laundry circuit, the receptacle must be rated at 20A or more. When more than one outlet is on a circuit, the total connected load must be equal to or less than the capacity of the branch circuit conductors feeding the receptacles.

Refer to *Figure 40* for some of the characteristics of a standard 125V, 15A duplex receptacle. Note that the terminals are color coded as follows:

- *Green* – Connection for the equipment grounding conductor
- *Silver* – Connection for the neutral or grounded conductor
- *Brass* – Connection for the ungrounded conductor

A standard 125V, 15A receptacle is also typically imprinted with the following symbols:

- *UL* – Underwriters Laboratories, Inc., listing
- *CSA* – Canadian Standards Association
- *CO/ALR* – Designed for use with both copper and aluminum wire
- *15A* – Receptacle rated for a maximum of 15A
- *125V* – Receptacle rated for a maximum of 125V

The UL label means that the receptacle has undergone testing by Underwriters Laboratories, Inc., and meets minimum safety requirements. Underwriters Laboratories, Inc., was created by the National Board of Fire Underwriters to test electrical devices and materials. The UL label is a safety rating only and does not mean that the device or equipment meets any type of quality standard. The CSA label means that the receptacle is approved by the Canadian Standards Association, the Canadian equivalent to Underwriters

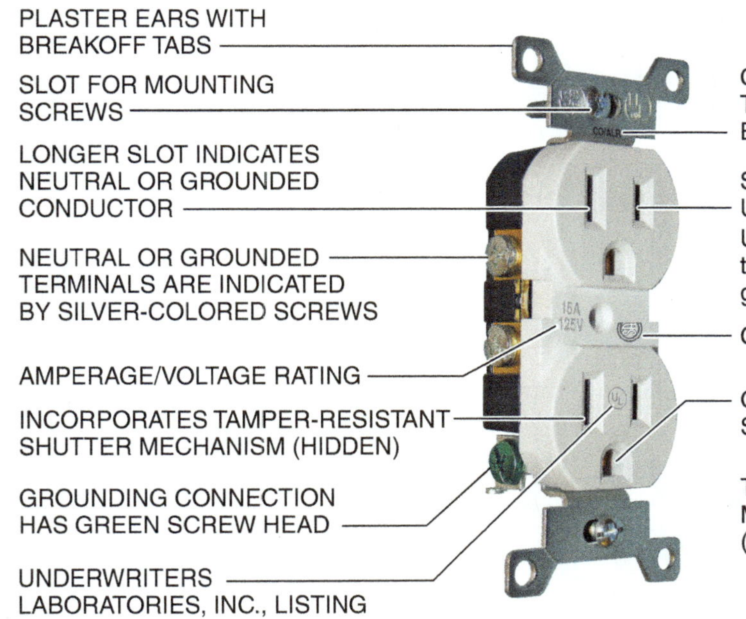

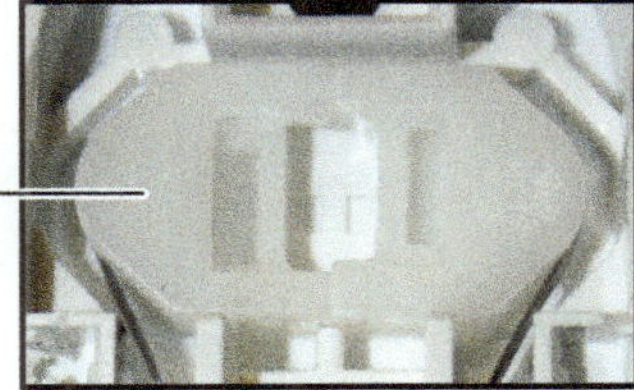

Figure 40 Standard 125V, 15A duplex receptacle.

Laboratories, Inc. The CSA label means that the receptacle is acceptable for use in Canada.

The CO/ALR symbol means that the device is suitable for use with copper, aluminum, or copper-clad aluminum wire. The CO in the symbol stands for copper while ALR stands for aluminum revised. The CO/ALR symbol replaces the earlier CU/AL mark, which appeared on wiring devices that were later found to be inadequate for use with aluminum wire in the 15A to 20A range. Therefore, any receptacle or wall switch marked with the CU/AL configuration or anything other than CO/ALR should be used only for copper wire.

These same configurations also apply to wall switches used for lighting control. These will be discussed next.

14.0.0 Lighting Control

There are many types of lighting control devices. These devices have been designed to make the best use of the lighting equipment provided by the lighting industry. They include:

- Automatic timing devices for outdoor lighting
- Dimmers for residential lighting
- Common single-pole, three-way, and four-way switches

For the purposes of this module, a switch is defined as a device that is used on branch circuits to control lighting. Switches fall into the following basic categories:

- Snap-action switches
- Quiet switches

A single-pole snap-action switch consists of a device containing two stationary current-carrying elements, a moving current-carrying element, a toggle handle, a spring, and a housing. When the contacts are open, as shown in *Figure 41*, the circuit is broken and no current flows. When the moving element is closed by manually flipping the toggle handle, the contacts complete the circuit and the lamp is energized. See *Figure 42*.

The quiet switch (*Figure 43*) is the most common switch for use in lighting applications. Its operation is much quieter than the snap-action switch.

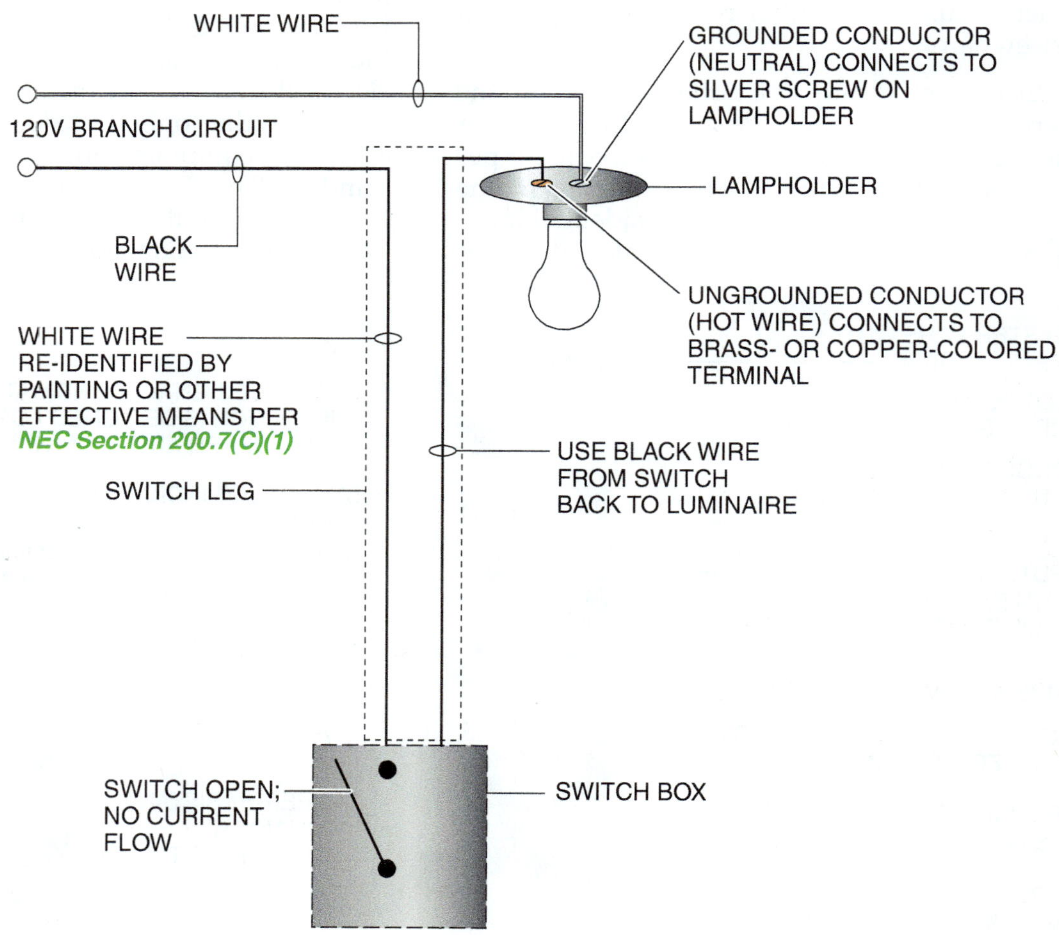

Figure 41 Switch operation, contacts open.

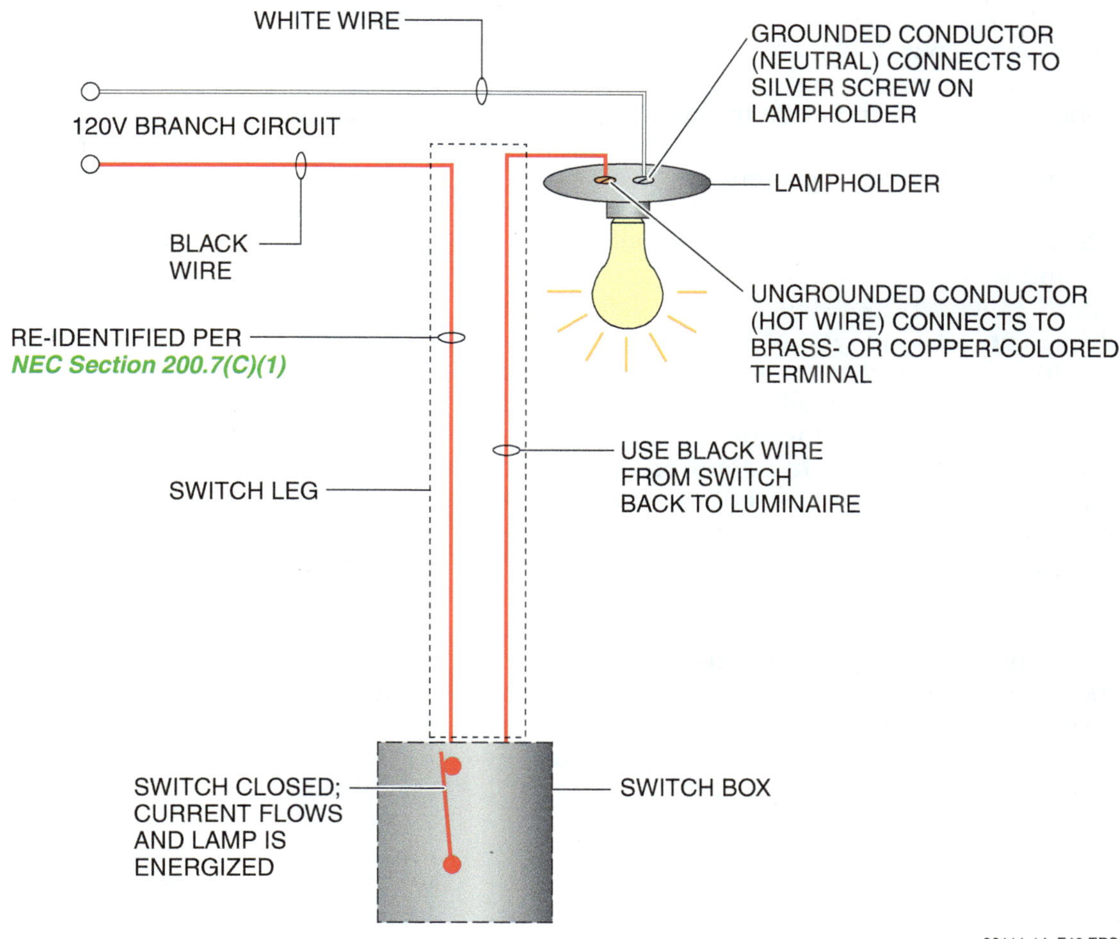

Figure 42 Switch operation, contacts closed.

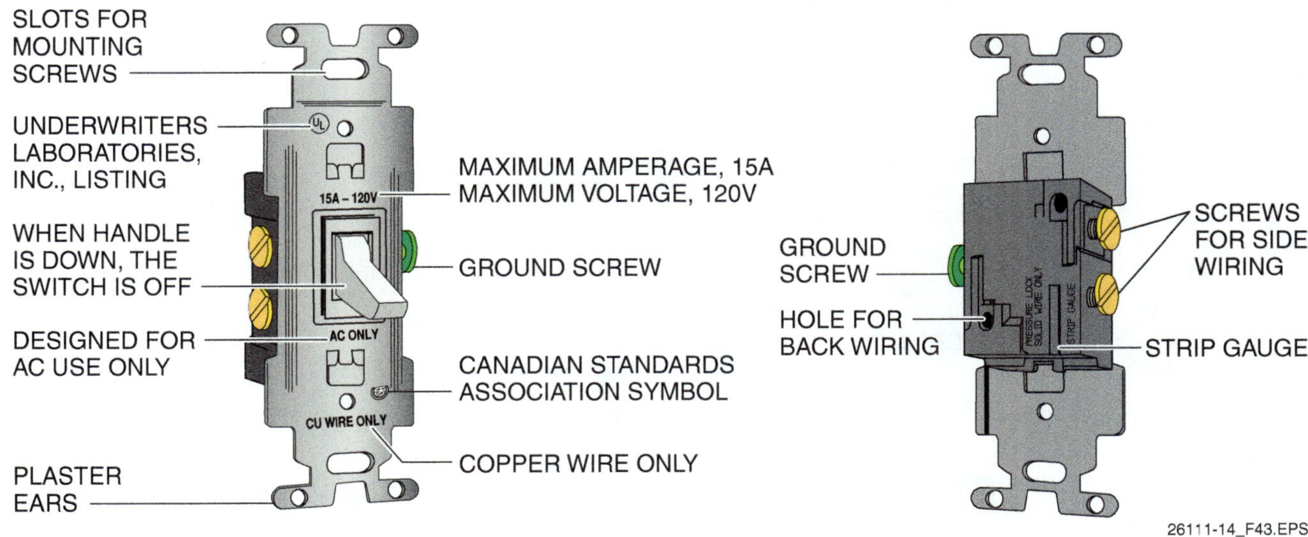

Figure 43 Characteristics of a single-pole quiet switch.

The quiet switch consists of a stationary contact and a moving contact that are close together when the switch is open. Only a short, gentle movement is required to open and close the switch, producing very little noise. This type of switch may be used only on alternating current.

Quiet switches are common for loads from 10A to 20A, and are available in single-pole, three-way, and four-way configurations.

Many other types of switches are available for lighting control. One type of switch used mainly in residential occupancies is the door-actuated switch. It is generally installed in the door jamb of a closet to control a light inside the closet. When the door is open, the light comes on; when the door is closed, the light goes out. Most refrigerator and oven lights are also controlled by door-actuated switches.

Combination switch/indicator light assemblies are available for use where the light cannot be seen from the switch location, such as an attic or garage. Switches are also made with small neon lamps in the handle that light when the switch is off. These low-current-consuming lamps make the switches easy to find in the dark.

14.1.0 Three-Way Switches

Three-way switches are used to control one or more lamps from two different locations, such as at the top and bottom of stairways, in a room that has two entrances, etc. A typical three-way switch is shown in *Figure 44*.

A three-way switch has three terminals. The single terminal at one end of the switch is called the common or hinge point. This terminal is easily identified because it is darker than the other two terminals. The feeder (hot wire) or switch leg is always connected to the common dark or black terminal. The two remaining terminals are called traveler terminals. These terminals are used to connect three-way switches together.

The connection of two three-way switches is shown in *Figure 45*. By means of the two switches, it is possible to control the lamp from two locations. By tracing the circuit, it may be seen how these three-way switches operate.

A 120V circuit emerges from the left side of the drawing. The white or neutral wire connects directly to the neutral terminal of the lamp. The hot wire carries current, in the direction of the arrows, to the common terminal of the three-way switch on the left. Since the handle is in the Up position, the current continues to the top traveler terminal and is carried by this traveler to the other three-way switch. Note that the handle is also in the Up position on this switch; this picks up the current flow and carries it to the common point, which continues to the ungrounded terminal of the lamp to make a complete circuit. The lamp is energized.

Moving the handle to a different position on either three-way switch will break the circuit, which in turn de-energizes the lamp. For example, let's say a person leaves the room at the point of the three-way switch on the left, and the switch handle is flipped down, as shown in *Figure 46*. Note that the current flow is now directed to the bottom traveler terminal, but since the handle of the three-way switch on the right is still in the Up position, no current will flow to the lamp.

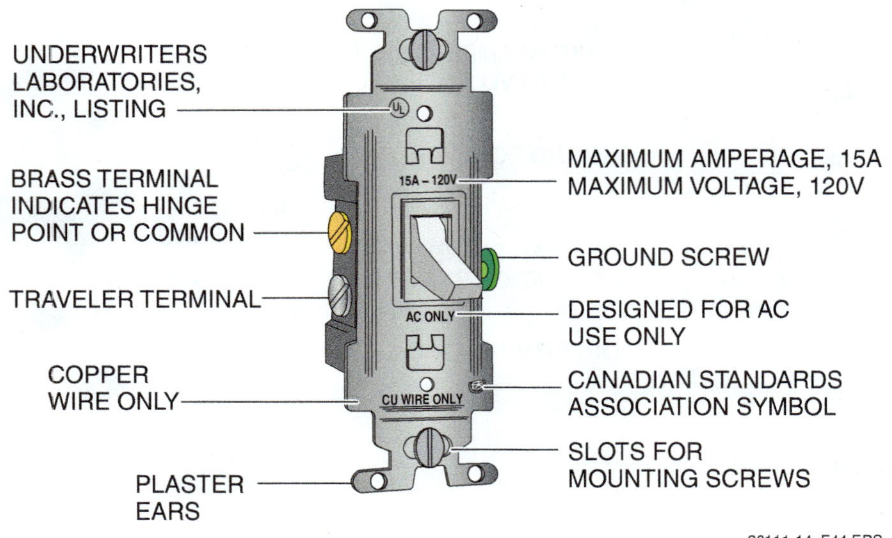

Figure 44 Typical three-way switch.

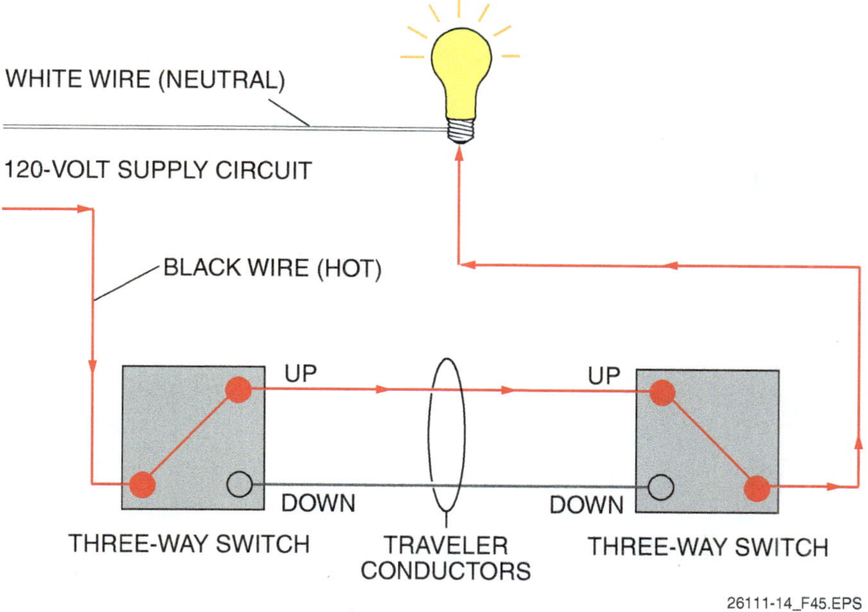

Figure 45 Three-way switches in the On position; both handles are up.

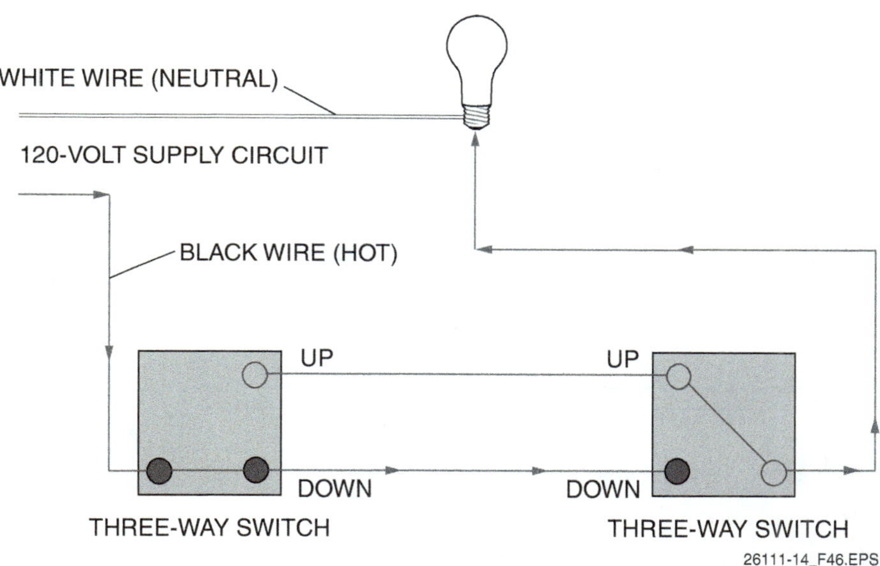

Figure 46 Three-way switches in the Off position; one handle is down, one handle is up.

If another person enters the room at the location of the three-way switch on the right, and the handle is flipped downward, as shown in *Figure 47*, this change provides a complete circuit to the lamp, which causes it to be energized. In this example, current flow is on the bottom traveler. Again, changing the position of the switch handle (pivot point) on either three-way switch will de-energize the lamp.

In actual practice, the exact wiring of the two three-way switches to control the operation of a lamp will be slightly different from the routing shown in these three diagrams. There are several ways that two three-way switches may be connected. One solution is shown in *Figure 48*. In this case, two-wire, Type NM cable is fed to the three-way switch on the left.

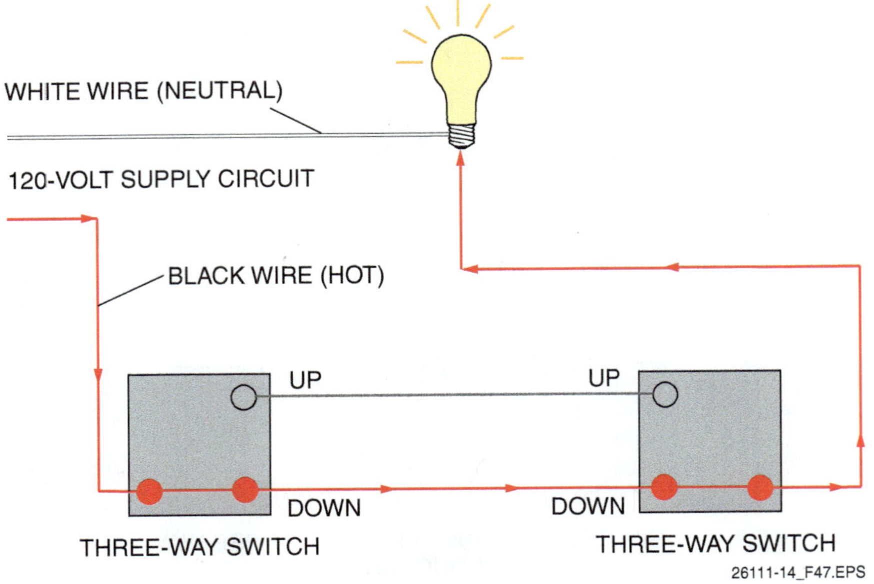

Figure 47 Three-way switches with both handles down; the light is energized.

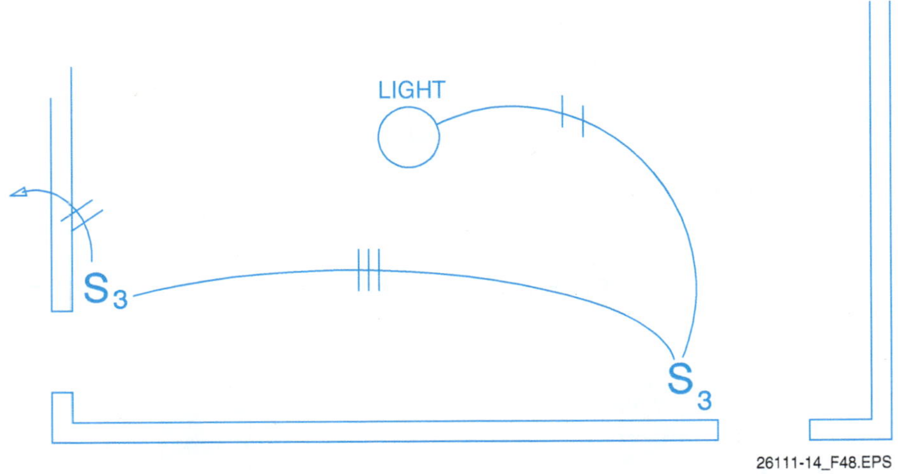

Figure 48 Method of showing the wiring arrangement on a floor plan.

The black or hot conductor is connected to the common terminal on the switch, while the white or neutral conductor is spliced to the white conductor of the three-wire, Type NM cable leaving the switch. This three-wire cable is necessary to carry the two travelers plus the neutral to the three-way switch on the right. At this point, the black and red wires connect to the two traveler terminals, respectively. The white or neutral wire is again spliced—this time to the white wire of another two-wire, Type NM cable. The neutral wire is never connected to the switch itself. The black wire of the two-wire, Type NM cable connects to the common terminal on the three-way switch. This cable, carrying the hot and neutral conductors, is routed to the lighting fixture outlet for connection to the fixture.

Another solution is to feed the lighting fixture outlet with two-wire cable. Run another two-wire cable carrying the hot and neutral conductors to one of the three-way switches. A three-wire cable is pulled between the two three-way switches, and then another two-wire cable is routed from the other three-way switch to the lighting fixture outlet.

Some electricians use a shortcut method that eliminates one of the two-wire cables in the preceding method. In this case, a two-wire cable is run from the lighting fixture outlet to one three-way switch. Three-wire cable is pulled between the two three-way switches—two of the wires for travelers and the third for the common point return. This method is shown in *Figure 49*.

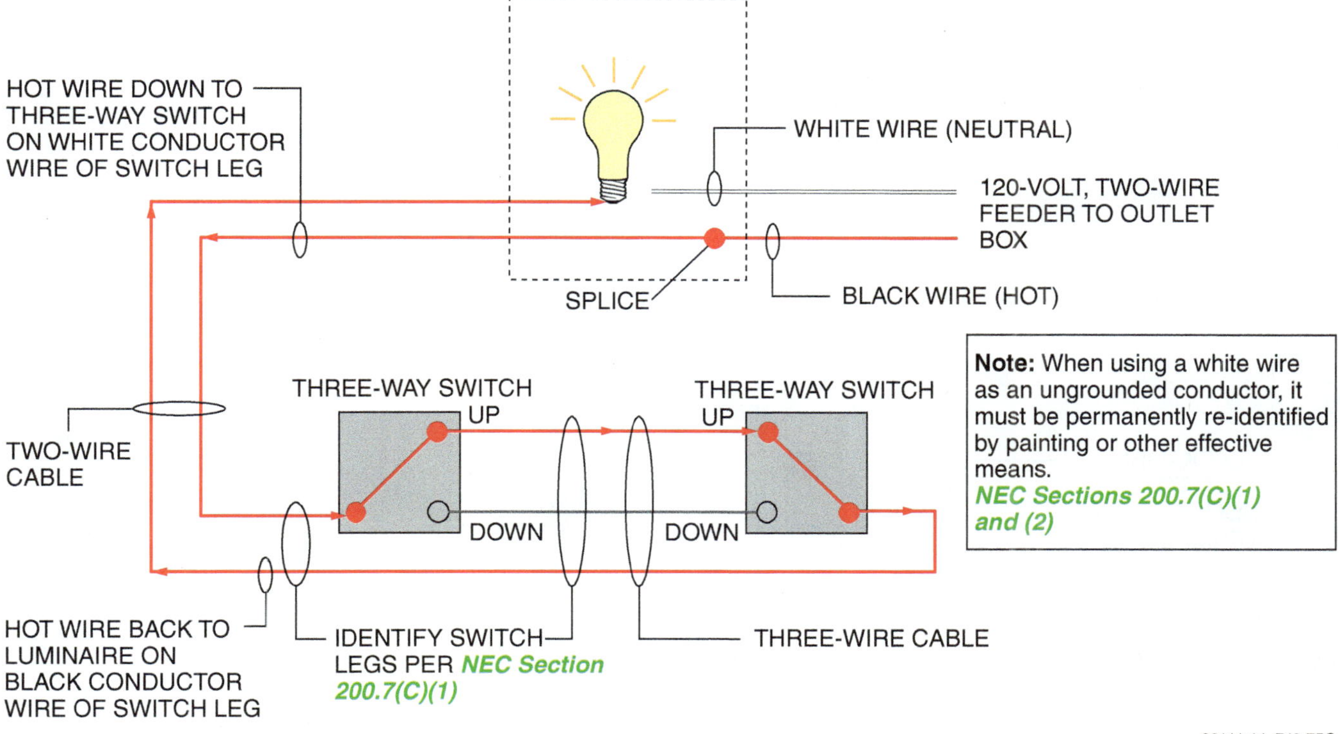

Figure 49 One way to connect a pair of three-way switches to control one lighting fixture.

Think About It

Wiring Three-Way Switches

Using a schematic drawing, explain the actual wiring of two different three-way switches, one in which the load and supply come in from different boxes, and the other in which the load and supply come in from the same box. Be specific about which wires connect to which terminals.

14.2.0 Four-Way Switches

Two three-way switches may be used in conjunction with any number of four-way switches to control a lamp, or a series of lamps, from any number of positions. When connected correctly, the actuation of any one of these switches will change the operating condition of the lamp (i.e., turn the lamp either on or off).

Figure 50 shows how a four-way switch may be used in combination with two three-way switches to control a device from three locations. In this example, note that the hot wire is connected to the common terminal on the three-way switch on the left. Current then travels to the top traveler terminal and continues on the top traveler conductor to the four-way switch. Since the handle is up on the four-way switch, current flows through the top terminals of the switch and into the traveler conductor going to the other three-way switch.

Again, the switch is in the Up position. Therefore, current is carried from the top traveler terminal to the common terminal and then to the lighting fixture to energize it.

If the position of any one of the three switch handles is changed, the circuit will be broken and no current will flow to the lamp. For example, assume that the four-way switch handle is flipped downward. The circuit will now appear as shown in *Figure 51*, and the light will be out.

Remember, any number of four-way switches may be used in combination with two three-way switches, but two three-way switches are always necessary for the correct operation of one or more four-way switches.

14.3.0 Photoelectric Switches

The chief application of the photoelectric switch is to control outdoor lighting, especially the dusk-to-dawn lights found in suburban areas. This switch has an endless number of possible uses and is a great tool for electricians dealing with outdoor lighting situations.

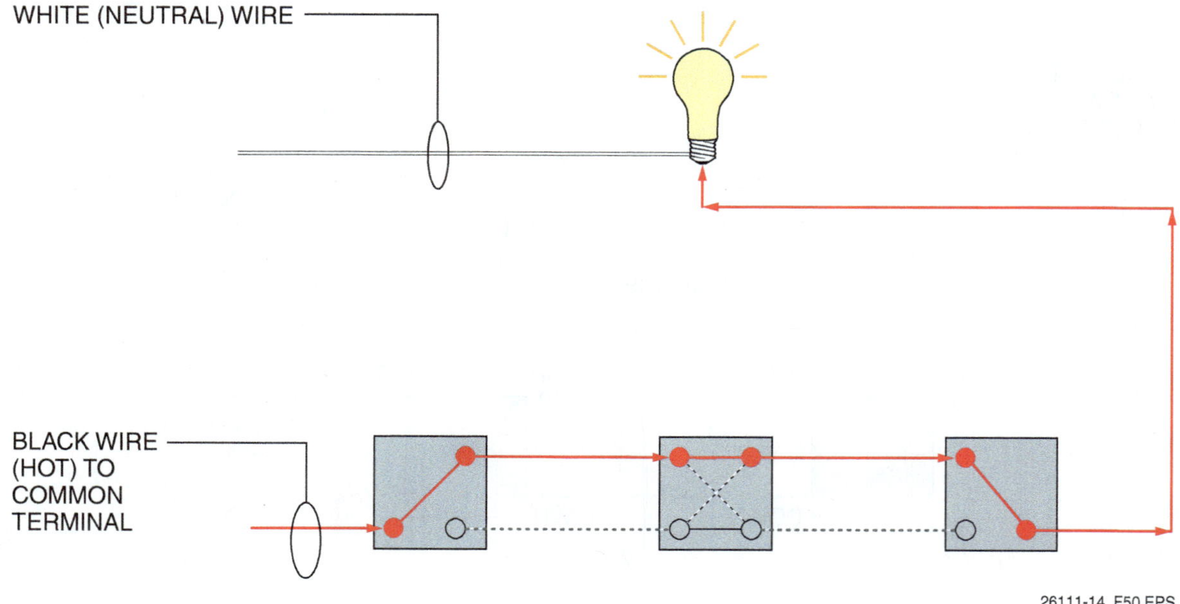

Figure 50 Three- and four-way switches used in combination; the light is on.

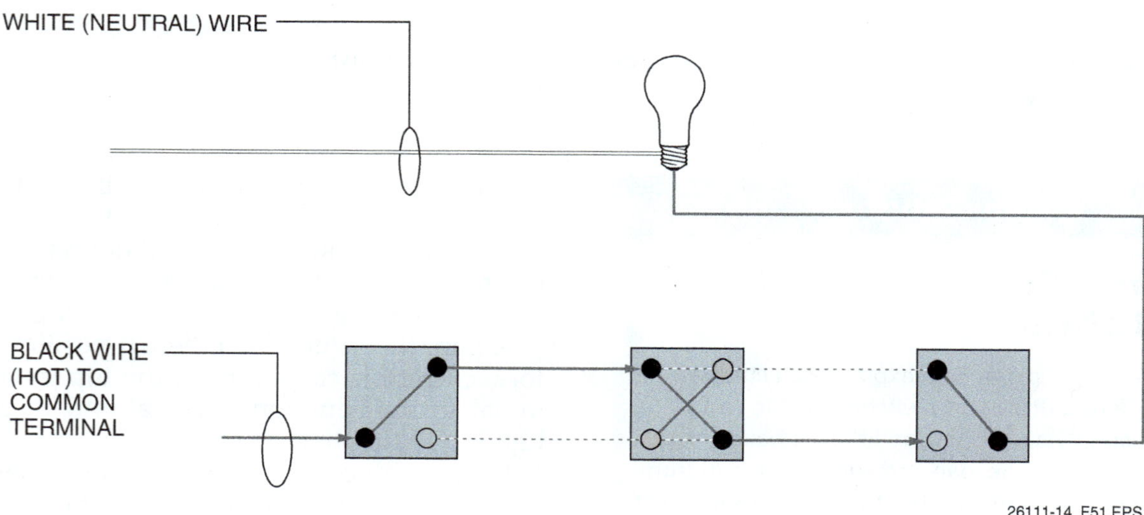

Figure 51 Three- and four-way switches used in combination; the light is off.

14.4.0 Relays

Next to switches, relays play the most important part in the control of light. However, the design and application of relays is a study in itself, and they are far beyond the scope of this module. Still, a brief mention of relays is necessary to round out your knowledge of lighting controls.

An electric relay is a device whereby an electric current causes the opening or closing of one or more pairs of contacts. These contacts are usually capable of controlling much more power than is necessary to operate the relay itself. This is one of the main advantages of relays.

One popular use of the relay in residential lighting systems is that of remote control lighting. In this type of system, all relays are designed to operate on a 24V circuit and are used to control 120V lighting circuits. They are rated at 20A, which is sufficient to control the full load of a normal lighting branch circuit, if desired.

Remote control switching makes it possible to install a switch wherever it is convenient and practical to do so or wherever there is an obvious need for a switch, no matter how remote it is from the lamp or lamps it is to control. This method enables lighting designs to achieve new advances

in lighting control convenience at a reasonable cost. Remote control switching is also ideal for rewiring existing homes with finished walls and ceilings.

One relay is required for each fixture or each group of fixtures that are controlled together. Switch locations for remote control follow the same rules as for conventional direct switching. However, since it is easy to add switches to control a given relay, no opportunities should be overlooked for adding a switch to improve the convenience of control.

Remote control lighting also has the advantage of using selector switches at central locations. For example, selector switches located in the master bedroom or in the kitchen of a home enable the owner to control every lighting fixture on the property from this location. For example, the selector switch may be used to control outside or basement lights that might otherwise be left on inadvertently.

14.5.0 Dimmers

Dimming a lighting system provides control of the quantity of illumination. It may be done to create certain moods or to blend the lighting from different sources for various lighting effects.

For example, in homes with formal dining rooms, a chandelier mounted directly above the dining table and controlled by a dimmer switch becomes the centerpiece of the room while providing general illumination. The dimmer adds versatility since it can set the mood for the activity—low brilliance (candlelight effect) for formal dining or bright for an evening of playing cards. When chandeliers with exposed lamps are used, the dimmer is essential to avoid a garish and uncomfortable atmosphere. The chandelier should be sized in proportion to the dining area.

NOTE

It is very important that dimmers be matched to the wattage of the application. Check the manufacturer's data.

14.6.0 Switch Locations

Although the location of wall switches is usually provided for convenience, the *NEC*® also stipulates certain mandatory locations for lighting fixtures and wall switches. See *NEC Section 210.70(A)* for specific switch locations in dwelling units. These locations are deemed necessary for added safety in the home for both the occupants and service personnel.

For example, the *NEC*® requires adequate light in areas where heating, ventilating, and air conditioning (HVAC) equipment is placed. Furthermore, these lights must be conveniently controlled so that homeowners and service personnel do not have to enter a dark area where they might come in contact with dangerous equipment. Three-way switches are required under certain conditions. The *NEC*® also specifies regulations governing lighting fixtures in clothes closets, along with those governing lighting fixtures that may be mounted directly to the outlet box without further support. *Figure 52* summarizes some of the *NEC*® requirements for light and switch placement in the home. For further details, refer to the appropriate sections in the *NEC*®.

14.7.0 Low-Voltage Electrical Systems

Conventional lighting systems operate and are controlled by the same system voltage, generally 120V in residential lighting circuits. The *NEC*® permits the use of low-voltage systems to control lighting circuits. There are some advantages to low-voltage systems. One advantage is that the control of lighting from several different locations is more easily accomplished, such as with the remote control system discussed earlier. For example, outside flood lighting can be controlled from several different rooms in a house. The cost of the control wiring is less in that it is rated for a lower voltage and only carries a minimum amount of current compared to a standard lighting system. When extensive or complex lighting control is required, low-voltage systems are preferred. Also, since these circuits are low-energy circuits, circuit protection is not required.

14.7.1 *NEC*® *Requirements for Low-Voltage Systems*

NEC Article 725 governs the installation of low-voltage system wiring. These provisions apply to remote control circuits, low-voltage relay switching, low-energy power circuits, and low-voltage circuits. The *NEC*® divides these circuits into three categories:

- Remote control
- Signaling
- Power-limited circuits

As mentioned earlier, circuit protection of the low-voltage circuit is not required; however, the high-voltage side of the transformer that supplies the low-voltage system must be protected. *NEC Chapter 9, Tables 11(A) and 11(B)* cover circuits that are inherently limited in power output and

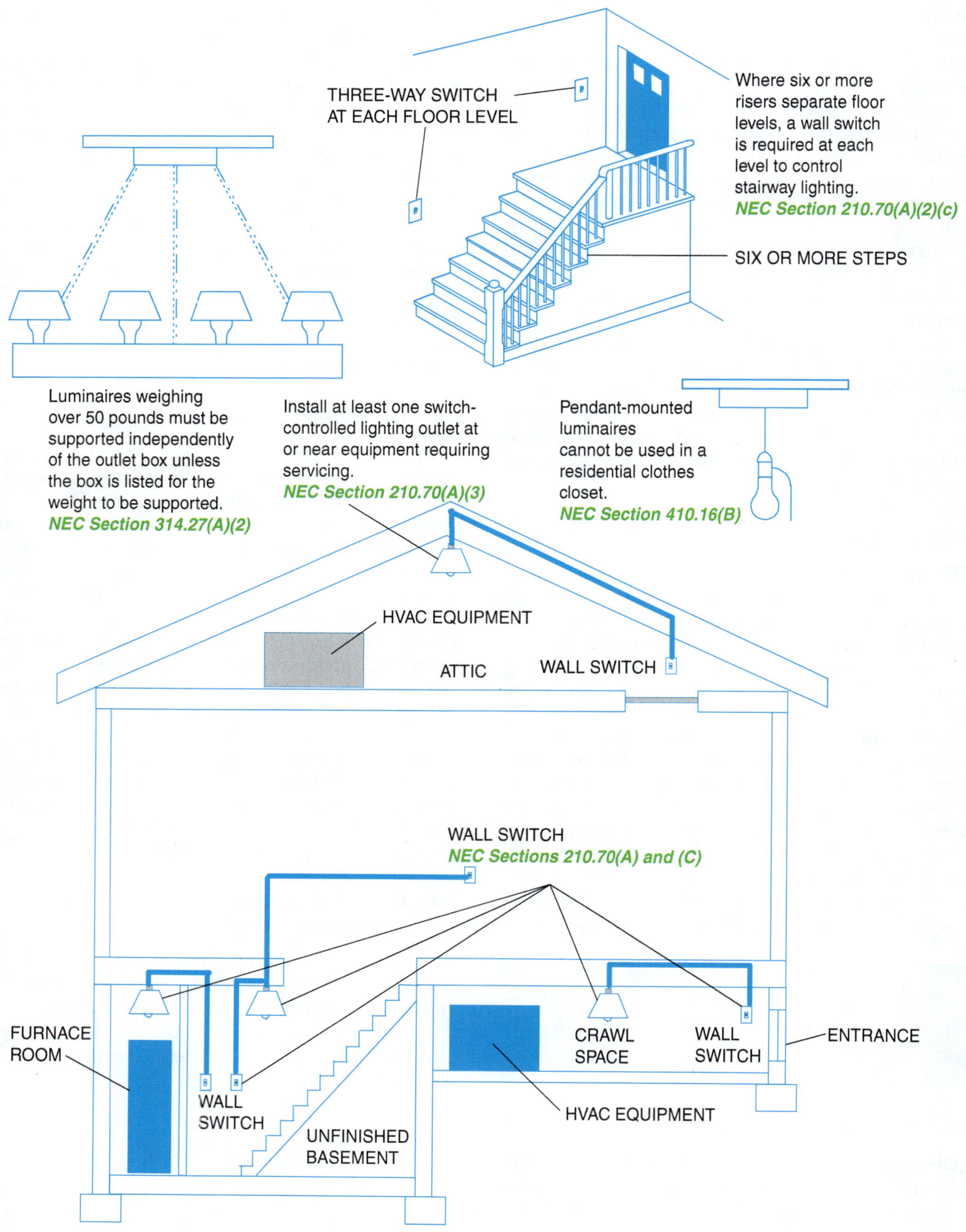

Figure 52 NEC® requirements for light and switch placement.

therefore require no overcurrent protection or are limited by a combination of power source and overcurrent protection.

There are a number of requirements of the power systems described in *NEC Chapter 9, Tables 11(A) and 11(B)* and the notes preceding the tables. You should read and study all applicable portions of the *NEC®* before installing low-voltage power systems.

Low-voltage systems are described in more detail in later modules.

15.0.0 Electric Heating

The use of electric heating in residential occupancies has risen tremendously over the past decade or so, and the practice will no doubt continue. This is due to the following advantages of electric heat over most other heating systems:

- Electric heat is noncombustible and is therefore safer than combustible fuels.
- It requires no storage space, fuel tanks, or chimneys.
- It requires little maintenance.
- The initial installation cost is relatively inexpensive when compared to other types of heating systems.
- The comfort level may be improved since each room may be controlled separately by its own thermostat.

There are also some disadvantages to using electric baseboard heat, especially in northern climates. Some of these disadvantages include:

- Electric heat is often more expensive to operate than other types of fuels.
- Receptacles must not be installed above electric baseboard heaters.
- Electric baseboard heaters tend to discolor the wall area immediately above the heater, especially if there are smokers in the home.

NOTE

It is very important to calculate the extra electric load of an electric heater installation (especially in an add-on situation). Ensure that the extra load does not exceed the maximum amperage draw of either the circuit or the panel.

The type of electric heating system used for a given residence will usually depend on the structural conditions, the kind of room, and the activities for which the room will be used. The homeowner's preference will also enter into the final decision.

Electric heating equipment is available in baseboard, wall, ceiling, kick space, and floor units; in resistance cable embedded in the ceiling or concrete floor; in forced-air duct systems similar to conventional oil- or gas-fired hot air systems; and in electric boilers for hot water baseboard heat.

Electric heat pumps have also become popular for HVAC systems in certain parts of the country. The term *heat pump*, as applied to a year-round air conditioning system, commonly denotes a system in which refrigeration equipment is used in such a manner that heat is taken from a heat source and transferred to the conditioned space when heating is desired; heat is removed from the space and discharged to a heat sink when cooling and dehumidification are desired.

A heat pump has the unique ability to furnish more energy than it consumes. This is due to the fact that under certain outdoor conditions, electrical energy is required only to move the refrigerant and run the fan; thus, a heat pump can attain a heating efficiency of two or more to one; that is, it will put out an equivalent of two or three watts of heat for every watt consumed. For this reason, its use is highly desirable for the conservation of energy.

In general, electric baseboard heating equipment should be located on the outside wall near the areas where the greatest heat loss will occur, such as under windows, etc. The controls for wall-mounted thermostats should be located on an interior wall, about 50 inches above the floor to sense the average room temperature. *Figure 53* shows an electric heating arrangement for the sample residence. *NEC®* regulations governing the installation of these units are also noted.

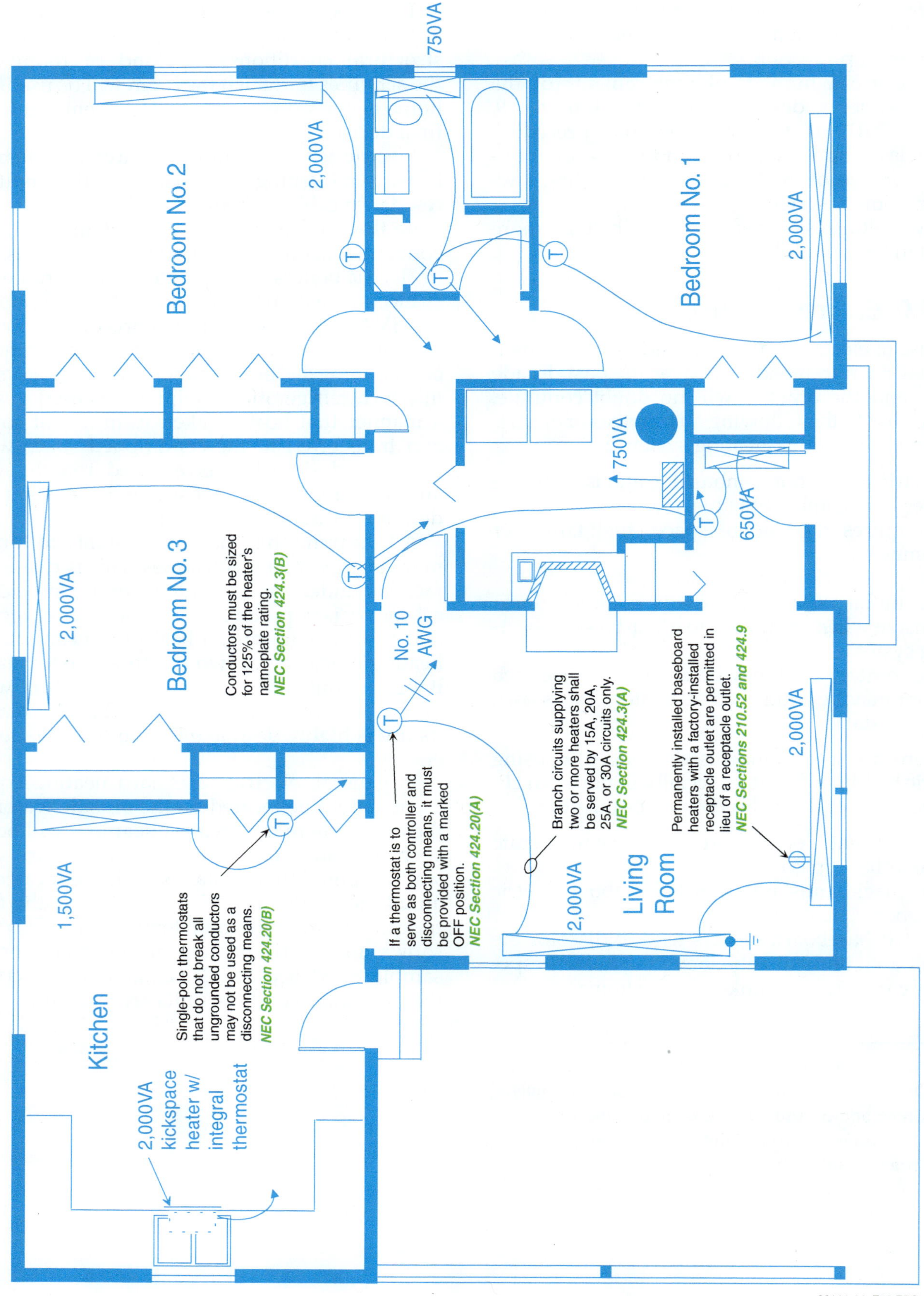

Figure 53 Electric heating arrangement for the sample residence.

16.0.0 Residential Swimming Pools, Spas, and Hot Tubs

The *NEC*® recognizes the potential danger of electric shock to persons in swimming pools, wading pools, and therapeutic pools, or near decorative pools or fountains. This shock could occur from electric potential in the water itself or as a result of a person in the water or a wet area touching an enclosure that is not at ground potential. Accordingly, the *NEC*® provides rules for the safe installation of electrical equipment and wiring in or adjacent to swimming pools and similar locations. *NEC Article 680* covers the specific rules governing the installation and maintenance of swimming pools, spas, and hot tubs.

The electrical installation procedures for hot tubs and swimming pools are too vast to be covered in detail in this module. However, the general requirements for the installation of outlets, overhead fans and lighting fixtures, and other items are summarized in *Figure 54*.

Besides *NEC Article 680* (see *Figure 55*), another good source for learning more about electrical installations in and around swimming pools is from manufacturers of swimming pool equipment, including those who manufacture and distribute underwater lighting fixtures. Many of these manufacturers offer pamphlets detailing the installation of their equipment with helpful illustrations, code explanations, and similar details. This literature is usually available at little or no cost to qualified personnel. You can write directly to manufacturers to request information about available literature, or contact your local electrical supplier or contractor who specializes in installing residential swimming pools.

Think About It

Residential Wiring

Make a mental wiring tour of your home. Picture several rooms, including the kitchen and utility/laundry room. How is each device connected to the power source, and what is the probable amperage and overcurrent protection? What other devices might or might not be included in the circuit? How many branch circuits serve each room? Later, examine the panelboard and exposed wiring to see how accurately you identified the branch circuits.

On Site

Pools

Pools are a common site for homeowners to add lights, receptacles, or heaters, and code violations are common. If you are called in for a service problem, ask the homeowner about any do-it-yourself wiring.

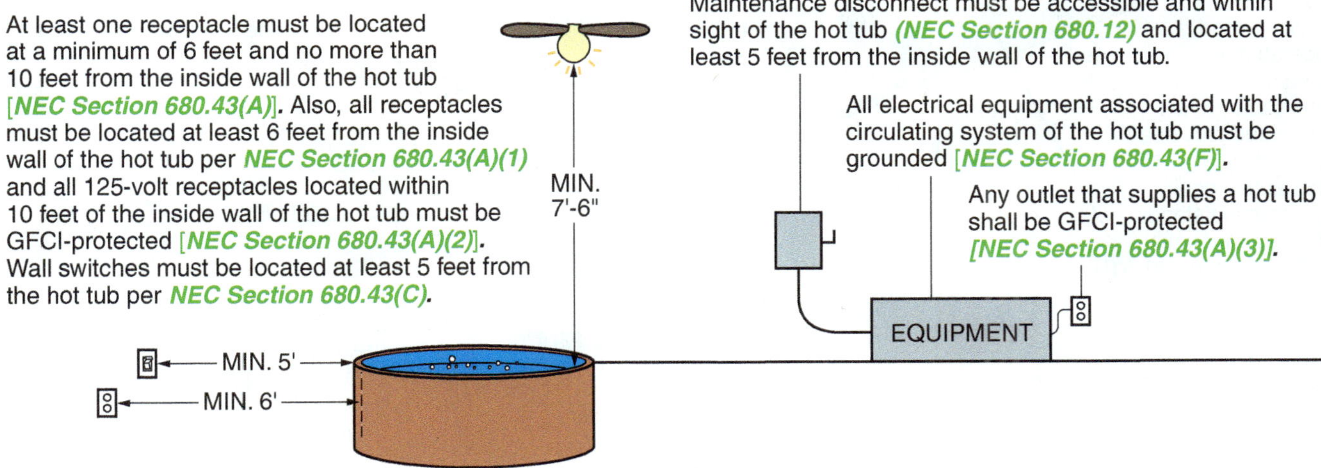

Figure 54 *NEC*® requirements for packaged indoor hot tubs.

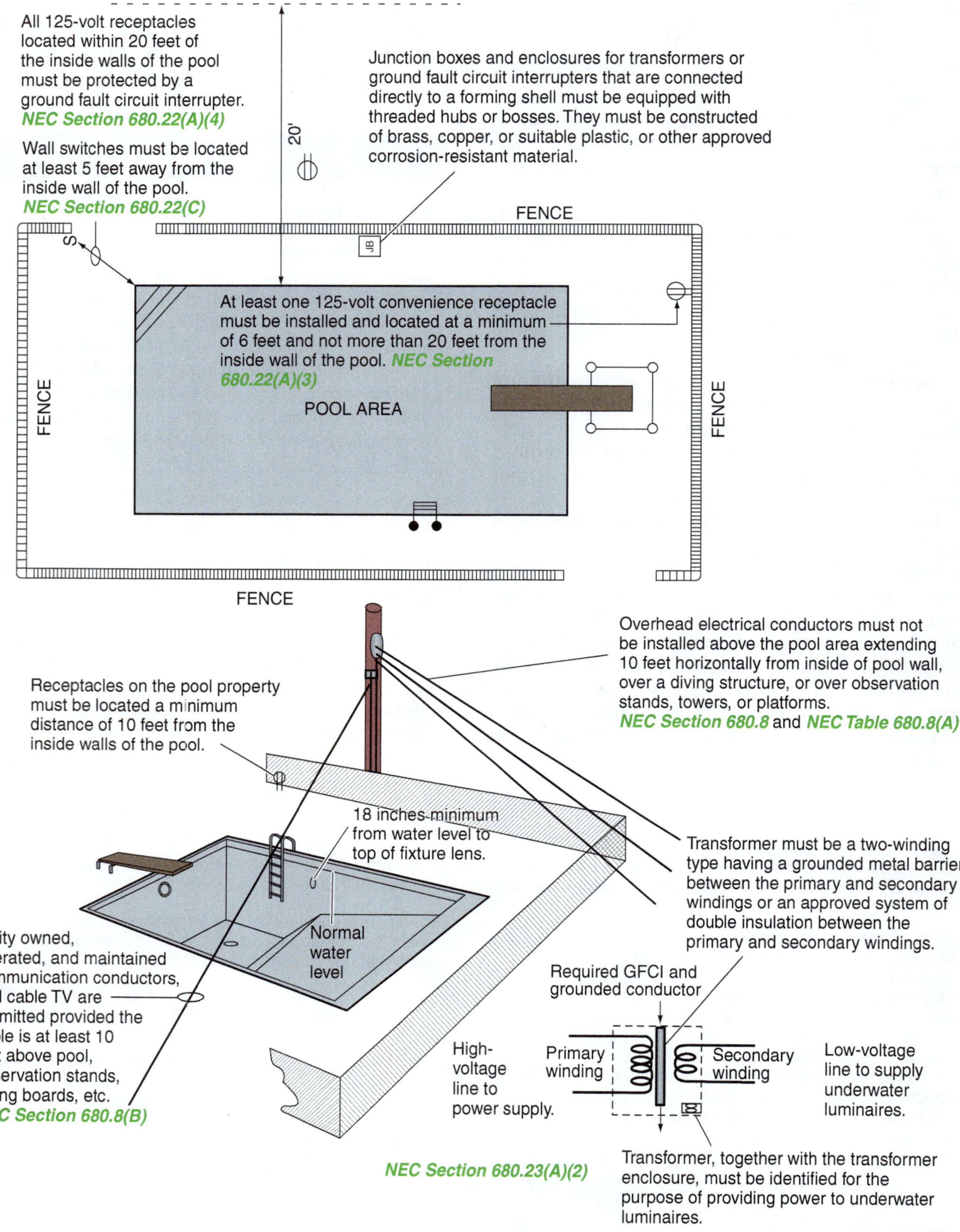

Figure 55 NEC® requirements for typical swimming pool installations.

SUMMARY

This module covered the basics of residential wiring, including load calculations and wiring devices.

Residential electrical system design begins with the floor plan. The square footage of the home is used to calculate the lighting loads. The lighting loads, along with small appliance loads, fixed appliance loads, and other loads must be calculated to determine the service size. Demand factors are applied and the neutral conductor is sized per *NEC Article 220*. GFCI and AFCI breakers must be installed in all required locations, and the system must be grounded and bonded per *NEC Article 250*. Most residences use Type NM or MC cable for branch circuits. Switches must be installed in all locations required by the *NEC*®. Special consideration must be given to electric heating units and circuits supplying swimming pools and hot tubs.

A thorough knowledge of the *NEC*® is essential to the safe and successful installation of residential wiring systems.

Review Questions

1. When sizing electrical services, at what percentage is the first 3,000VA rated?
   a. 20%
   b. 45%
   c. 80%
   d. 100%

2. What section of the *NEC*® requires that fittings be identified for use with service masts?
   a. *NEC Section 230.28*
   b. *NEC Section 230.40*
   c. *NEC Section 250.46*
   d. *NEC Section 250.83*

3. A service conductor without an overall jacket must have a clearance of not less than _____ above a window that can be opened.
   a. two feet
   b. three feet
   c. eight feet
   d. ten feet

4. *NEC Section 230.6* considers conductors installed under at least _____ inch(es) of concrete to be outside the building.
   a. one
   b. two
   c. four
   d. five

5. Type NM cable may *not* be used in _____.
   a. shallow chases of masonry, concrete, or adobe
   b. the framework of a building
   c. protective strips
   d. attic spaces

6. Type MC cable may *not* be used _____.
   a. in concrete or plaster where dry
   b. in dry masonry
   c. in attic spaces
   d. where subject to physical damage

7. Type SE cable is available with _____ for interior wiring systems.
   a. a non-insulated ground or neutral conductor
   b. an insulated grounded conductor
   c. no ground conductor
   d. guard strips

8. Type SER cable may be used _____.
   a. in overhead applications
   b. underground
   c. as a subfeed under certain conditions
   d. in hazardous locations

Identify the following receptacles by numbers shown below for each receptacle.

9. _____ Single receptacle

10. _____ Split-wired receptacle

11. _____ Dryer receptacle

12. _____ Range receptacle

13. _____ Duplex receptacle

14. _____ Special-purpose outlet

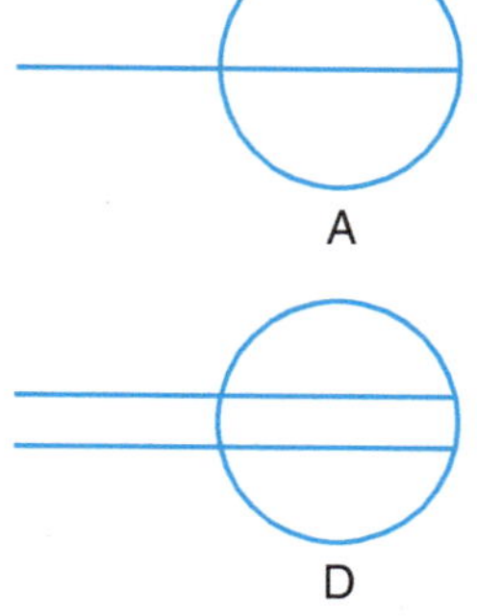

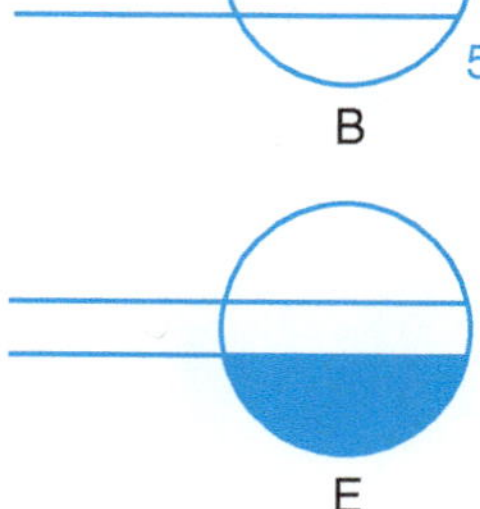

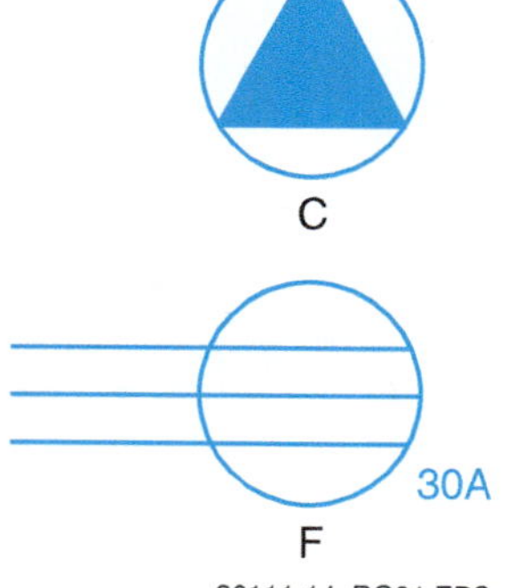

26111-14_RQ01.EPS

15. Using *NEC Table 314.16(A),* calculate the cubic inches required for the receptacle outlets shown in the table below. Then, indicate the size of the metallic box that should be used.

Number and Size of Conductors in Box	Free Space within Box for Each Conductor	Total Cubic Inches of Box Space Required	What Size Metallic Box May Be Used?
A. Six No. 12 conductors and three ground wires	2.25	____	____
B. Seven No. 12 conductors and three ground wires with one receptacle	2.25	____	____
C. Two No. 14 conductors and one ground wire	2.0	____	____
D. Four No. 14 conductors and two ground wires	2.0	____	____
E. Six No. 14 conductors and three ground wires with one receptacle	2.0	____	____

Trade Terms Quiz

Fill in the blank with the correct term that you learned from your study of this module.

1. A cable that contains insulated circuit conductors enclosed in armor made of metal is ____________ cable.
2. A factory-assembled cable with two or more insulated conductors and a nonmetallic jacket is called ____________ cable.
3. A(n) ____________ is a piece of equipment that has been designed for a particular purpose.
4. A(n) ____________ is used for turning an electrical circuit on and off.
5. The circuit that is routed to a switch box for controlling electric lights is known as a(n) ____________.
6. A(n) ____________ is equipped with a conductor terminal to take a bonding jumper.
7. A(n) ____________ is a bare or green insulated conductor used to ensure conductivity between metal parts that are required to be electrically connected.
8. The ____________ is comprised of the conductors that extend from the last power company pole to the point of connection at the service facilities.
9. The ____________ is the point where power is supplied to a building.
10. ____________ lie between the point of termination of the overhead service drop or underground service lateral and the main disconnecting device in the building.
11. ____________ mainly provides overcurrent protection to the feeder and service conductors.
12. A(n) ____________ is comprised of the underground conductors through which service is supplied between the power company's distribution facilities and their first point of connection to the building.
13. The portion of a wiring system that extends beyond the final overcurrent device is the ____________.
14. A(n) ____________ is a circuit that carries current from the service equipment to a subpanel or a branch circuit panel or to some point in the wiring system.
15. Normally located at the service entrance of a residential installation, a(n) ____________ usually contains the main disconnect.
16. Raceway, cable, wires, boxes, and other equipment are installed during ____________.

Trade Terms

Appliance
Bonding bushing
Bonding jumper
Branch circuit
Feeder
Load center
Metal-clad (Type MC) cable
Nonmetallic-sheathed (Type NM) cable
Romex®
Roughing in
Service drop
Service entrance
Service-entrance conductors
Service-entrance equipment
Service lateral
Switch
Switch leg

Module 26111-14

Supplemental Exercises

1. Because all residential electrical outlets are never used at the same time, the *NEC*® allows a diversity or ______________________ to be used when sizing the general lighting load for electric services.

2. When sizing the general lighting load for electric services, what percentage is the first 3,000VA rated at? ______________________

3. The minimum service disconnecting means for a one-family dwelling is ______________________.

4. The general lighting load in a residence is 3,600VA at 120V. What is the amperage?

   ____

5. A single-pole GFCI breaker is rated at ______________________.

6. True or False? In a residential electric service, the service can be grounded to the underground gas piping system.

7. Service drop conductors must be at least ______________________ feet above the ground or other accessible surface at all times.

8. True or False? GFCI protection is required for receptacles in residential bathrooms.

9. In a noncombustible wall, the outlet box may be recessed ______________________.

10. A duplex receptacle outlet must be installed in all hallways longer than ______________________ feet.

11. Electric ranges, clothes dryers, and water heaters that operate at 240V require a(n) ____________ circuit breaker.

12. In general, there are two basic methods used in the majority of modern residential electrical systems. What are they?

    ____

    ____

13. In a standard 125V receptacle, which wire is connected to the brass terminal?

    ____

14. Three-way switches are used to control lamps from ______________________ different locations.

15. The minimum number of small appliance circuits required by the *NEC*® in a kitchen area is ______________________.

Cornerstone of Craftsmanship

Dan Lamphear

Associated Builders and Contractors, Inc.

Like many other people, Dan Lamphear just fell into his career as an electrician. But once he discovered the electrical trade, he knew he had found a home. Since then, he has progressed from a helper to an apprentice, a journeyman, an independent contractor, an inventor, and finally, a teacher.

It was as much luck as anything else that led Dan toward a career as a professional electrician more than two decades ago. He wasn't sure what he wanted to do with his life after graduating from high school. However, after watching an electrician perform a commercial wiring job at a friend's business—and providing a helping hand—he was hooked. "It seemed like a challenging career, and I was curious to learn more about how electricity works," he recalls.

Dan was hired by that same electrician, under whom he apprenticed for several years before hearing about the NCCER program. He jumped at the chance to further his skills through the program. "Like they say, knowledge is money," he smiles.

After graduating from the program, Dan struck out on his own as an independent electrician, specializing in plant maintenance, industrial, and commercial work. His ability to diagnose and repair electrical problems in factory machinery soon made him a valuable contractor in Milwaukee's industrial sector.

He also discovered his knack for invention, and he has designed and built specialized machinery for a company that hired him as its full-time electrical maintenance supervisor. "Knowing the electrical side of machinery allowed me to understand how they operate mechanically," he says of his work as an inventor.

Dan later returned to the Associated Builders and Contractors (ABC) chapter, which trains out of a local community college, to repay the favor that helped him embark on his career. He teaches Electrical Level 2 courses for students who represent the next generation of professional electricians.

"Knowing how to use test instruments is perhaps the most important aspect of the job," he notes. "I still have some of the same meters I started out with."

Cornerstone of Craftsmanship

David Lewis

Instructor
Putnam Career & Technical Center

David Lewis started his career working in coal mines. After a few years he opened his own electrical business. Now he is an electrical instructor and works with the State Department of Education on curriculum development. He also serves on the NCCER revision team for the Electrical curriculum.

How did you first get interested in the field?
After graduating from high school in 1972 and attending college for a while, I decided that I wanted to work in the coal mines. Electricity interested me, so I became a maintenance foreman/electrician. Eventually, I started my own business.

What kind of training have you been through?
While working in the coal mines, I attended several electrical training classes and obtained my underground electrical license. While in business for myself I got my Master Electrician license and attended many update classes. Since I have started teaching, I have attended classes on PLCs and other topics. I also went back to college and obtained a bachelor of science degree in Career and Technical Education.

What work have you done in your career?
In the coal mines I worked on all types of mining equipment. After starting my own business, I worked mainly in residential and light commercial wiring.

Tell us about your present job and what you like about it.
I enjoy being an instructor in Electrical Technology. I work mostly with high school students and during the two years they are with me, it is great to see them grasp the knowledge of electricity.

What factors have contributed most to your success?
Hard work and the willingness to learn from experienced electricians.

What advice would you give to those new to the field?
Try to learn all you can. Work with an experienced electrician and learn from them. Attend any training or classes that you can. There is always something new to learn.

Trade Terms Introduced in This Module

Appliance: Equipment designed for a particular purpose (for example, using electricity to produce heat, light, or mechanical motion). Appliances are usually self-contained, are generally available for applications other than industrial use, and are normally produced in standard sizes or types.

Bonding bushing: A special conduit bushing equipped with a conductor terminal to take a bonding jumper. It also has a screw or other sharp device to bite into the enclosure wall to bond the conduit to the enclosure without a jumper when there are no concentric knockouts left in the wall of the enclosure.

Bonding jumper: A bare or green insulated conductor used to ensure the required electrical conductivity between metal parts required to be electrically connected. Bonding jumpers are frequently used from a bonding bushing to the service-equipment enclosure to provide a path around concentric knockouts in an enclosure wall, and they may also be used to bond one raceway to another.

Branch circuit: The portion of a wiring system extending beyond the final overcurrent device protecting a circuit.

Feeder: A circuit, such as conductors in conduit or a cable run, that carries current from the service equipment to a subpanel or a branch circuit panel or to some point in the wiring system.

Load center: A type of panelboard that is normally located at the service entrance of a residential installation. It usually contains the main disconnect.

Metal-clad (Type MC) cable: A factory assembly of one or more insulated circuit conductors with or without optical fiber members enclosed in an armor of interlocking metal tape, or a smooth or corrugated metallic sheath.

Nonmetallic-sheathed cable: A factory assembly of two or more insulated conductors enclosed within an overall nonmetallic jacket. Type NM contains insulated conductors enclosed within an overall nonmetallic jacket; Type NMC contains insulated conductors enclosed within an overall, corrosion-resistant, nonmetallic jacket; and Type NMS contains insulated power or control conductors with signaling, data, and communications conductors within an overall nonmetallic jacket.

Romex®: General Cable's trade name for Type NM cable; however, it is often used generically to refer to any nonmetallic-sheathed cable.

Roughing in: The first stage of an electrical installation, when the raceway, cable, wires, boxes, and other equipment are installed. This is the electrical work that must be done before any finishing work can be done.

Service drop: The overhead conductors, through which electrical service is supplied, between the last power company pole and the point of their connection to the service facilities located at the building.

Service entrance: The point where power is supplied to a building (including the equipment used for this purpose). The service entrance includes the service main switch or panelboard, metering devices, overcurrent protective devices, and conductors/raceways for connecting to the power company's conductors.

Service-entrance conductors: The conductors between the point of termination of the overhead service drop or underground service lateral and the main disconnecting device in the building.

Service-entrance equipment: Equipment that provides overcurrent protection to the feeder and service conductors, a means of disconnecting the feeders from energized service conductors, and a means of measuring the energy used.

Service lateral: The underground conductors through which service is supplied between the power company's distribution facilities and the first point of their connection to the building or area service facilities located at the building.

Switch: A mechanical device used for turning an electrical circuit on and off.

Switch leg: A circuit routed to a switch box for controlling electric lights.

Additional Resources

This module presents thorough resources for task training. The following resource material is suggested for further study.

National Electrical Code® Handbook, Latest Edition. Quincy, MA: National Fire Protection Association.

Figure Credits

Associated Builders and Contractors, Inc., Module opener

John Traister, Figure 2, Figures 5–8, Figures 10–13, Figures 19–23, Figures 25–29, Figure 35, Figure 43, Figure 44, Figure 53, Figure 55

Topaz Publications, Inc., Figure 16B, Figure 37B, 111SA01A, 111SA02, 111SA04, 111SA05

Tim Dean, Figure 40, 111SA01B

Greenlee Textron, Inc., a subsidiary of Textron Inc., 111SA03

Appendix

Other Codes and Standards That Apply to Electrical Installations

Until 2000, there were three model building codes:

- *Standard Building Code (SBC)* – Published by the Southern Building Code Congress International.
- *BOCA National Building Code (NBC)* – Published by the Building Officials and Code Administrators.
- *Uniform Building Code (UBC)* – Published by the International Conference of Building Officials.

The three code writing groups, SBCCI, BOCA, and UBC, combined into one organization called the International Code Council with the purpose of writing one nationally accepted family of building and fire codes. It is known as the *International Building Code.*

The *International Residential Code (IRC)* is adopted as part of the electrical code requirements in many areas of the country. The *IRC* covers one- and two-family dwellings of three stories or less. The *IRC* includes requirements for such things as ventilating fans for bathrooms, requirements for smoke detectors, and other items not specified by the *NEC*®. The *IRC* covers all trades, including building, plumbing, mechanical, gas, energy, and electrical.

The NFPA also publishes its own building code, *NFPA 5000.*

To be thoroughly competent in the electrical trade, you should become familiar with the contents of these codes and the terminology used in them.

NOTE

Always refer to the latest editions of codes in effect in your area.

NCCER CURRICULA — USER UPDATE

NCCER makes every effort to keep its textbooks up-to-date and free of technical errors. We appreciate your help in this process. If you find an error, a typographical mistake, or an inaccuracy in NCCER's curricula, please fill out this form (or a photocopy), or complete the online form at **www.nccer.org/olf**. Be sure to include the exact module ID number, page number, a detailed description, and your recommended correction. Your input will be brought to the attention of the Authoring Team. Thank you for your assistance.

Instructors – If you have an idea for improving this textbook, or have found that additional materials were necessary to teach this module effectively, please let us know so that we may present your suggestions to the Authoring Team.

NCCER Product Development and Revision

13614 Progress Blvd., Alachua, FL 32615

Email: curriculum@nccer.org
Online: www.nccer.org/olf

❑ Trainee Guide ❑ AIG ❑ Exam ❑ PowerPoints Other ______________________

Craft / Level: Copyright Date:

Module ID Number / Title:

Section Number(s):

Description:

Recommended Correction:

Your Name:

Address:

Email: Phone:

Electrical Test Equipment

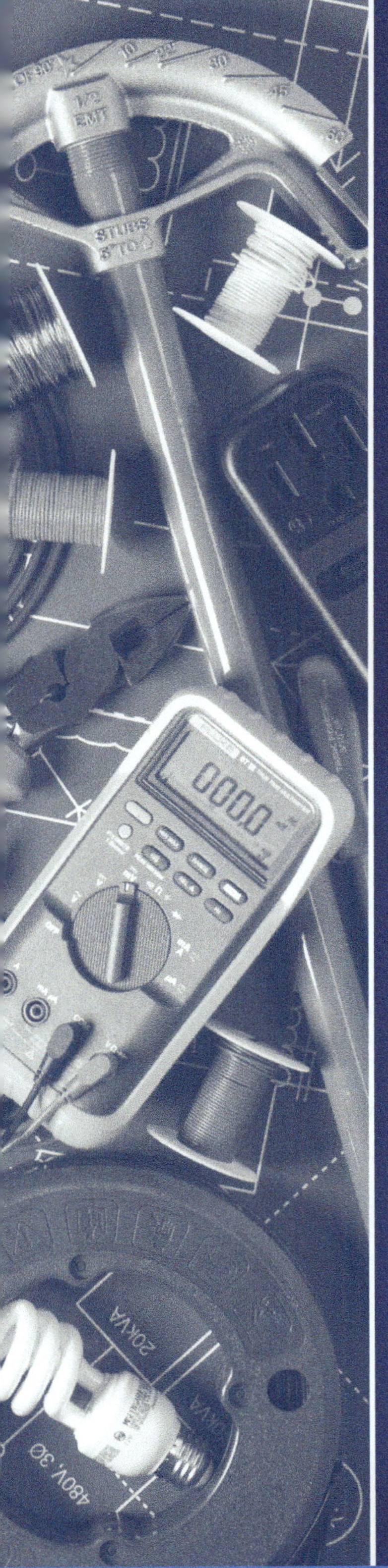

The Biodesign Institute at Arizona State University

The Biodesign Institute is the centerpiece of new growth at Arizona State University (ASU). This unique complex was built to meet the stringent demands posed by experimental programs in biotechnology and nanotechnology, and to enhance communication and collaboration between researchers with an open, shared lab design and a central atrium linking all floors.

26112-14

Trainees with successful module completions may be eligible for credentialing through NCCER's National Registry. To learn more, go to **www.nccer.org** or contact us at **1.888.622.3720.** Our website has information on the latest product releases and training, as well as online versions of our *Cornerstone* magazine and Pearson's product catalog.

Your feedback is welcome. You may email your comments to **curriculum@nccer.org,** send general comments and inquiries to **info@nccer.org,** or fill in the User Update form at the back of this module.

This information is general in nature and intended for training purposes only. Actual performance of activities described in this manual requires compliance with all applicable operating, service, maintenance, and safety procedures under the direction of qualified personnel. References in this manual to patented or proprietary devices do not constitute a recommendation of their use.

26112-14

Electrical Test Equipment

Objectives

When you have completed this module, you will be able to do the following:

1. Explain the operation of and describe the following pieces of test equipment:
   - Voltmeter
   - Ohmmeter
   - Clamp-on ammeter
   - Multimeter
   - Megohmmeter
   - Motor and phase rotation testers
2. Select the appropriate meter for a given work environment based on category ratings.
3. Identify the safety hazards associated with various types of test equipment.

Performance Tasks

Under the supervision of the instructor, you should be able to do the following:

1. Measure the voltage in your classroom from line to neutral and neutral to ground.
2. Use an ohmmeter to measure the value of various resistors.

Trade Terms

Coil
Continuity
d'Arsonval meter movement
Frequency

Required Trainee Materials

1. Pencil and paper
2. Copy of the latest edition of the *National Electrical Code®*
3. Appropriate personal protective equipment

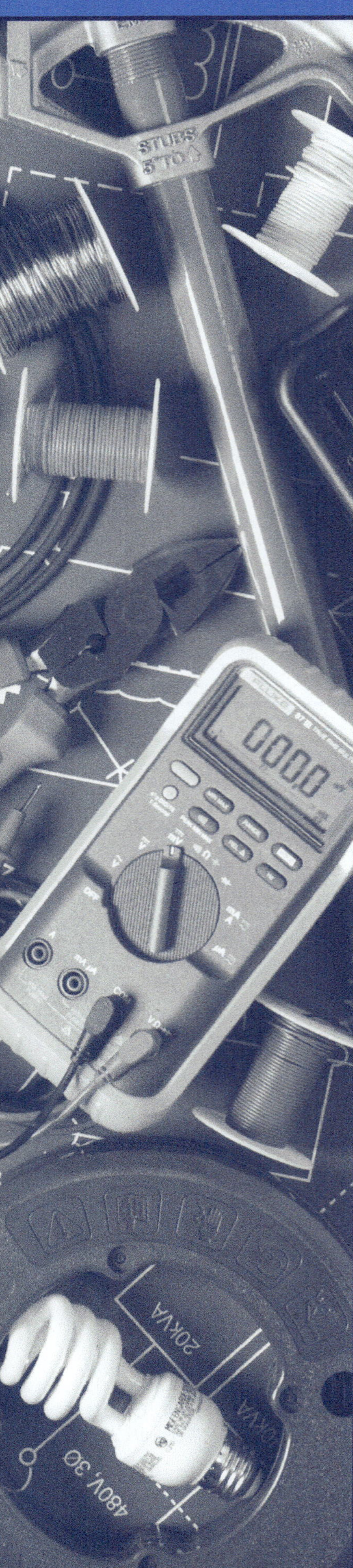

Contents

Topics to be presented in this module include:

Figures and Tables

1.0.0 Introduction

Electronic test instruments and meters are generally used for the following tasks:

- Troubleshooting electrical/electronic circuits and equipment
- Verifying proper operation of instruments and associated equipment

The test equipment selected for a specific task depends on the type of measurement and the level of accuracy required. This module will focus on some of the test equipment used by electricians. Upon completion of this module, you should be able to select the appropriate test equipment for a specific application and identify the applicable safety hazards.

2.0.0 Meters

In 1882, a Frenchman named Arsene d'Arsonval invented the galvanometer. This meter used a stationary permanent magnet and a moving **coil** to indicate current flow on a calibrated scale. The early galvanometer was very accurate but could only measure very small currents. Over the following years, many improvements were made that extended the range of the meter and increased its ruggedness. The **d'Arsonval meter movement** (*Figure 1*) is the basis for analog meters.

A moving-coil meter movement operates on the electromagnetic principle. In its simplest form, the moving-coil meter uses a coil of very fine wire wound on a light aluminum frame. A permanent magnet surrounds the coil. The aluminum frame is mounted on pivots to allow it and the coil to rotate freely between the poles of the permanent magnet. When current flows through the coil, it becomes magnetized, and the polarity of the coil is repelled by the field of the permanent magnet. This causes the coil frame to rotate on its pivots, and the distance it rotates is determined by the amount of current that flows through the coil. By attaching a pointer to the coil frame and adding a calibrated scale, the amount of current flowing through the meter can be measured. Multiplier resistors are used to extend the range of the meter movement for voltage measurements, while shunt resistors are used to extend the range of the meter movement for current measurements.

Today, most meters are solid-state digital systems; they are easier to read than mechanical (analog) meters and have no meter movement or moving parts.

Figure 1 d'Arsonval meter movement.

On Site

Phantom Readings

The sensitivity of a digital meter can sometimes produce a low reading known as a phantom or ghost reading. This is due to the induction from the electrical field around the energized conductors in close proximity to the meter.

2.1.0 Voltmeter

A voltmeter is used to measure voltage, also known as potential difference or electromotive force (emf). It is connected in parallel with the circuit or component being measured. An analog meter uses the basic d'Arsonval meter movement with internally switched resistors to measure different voltage ranges. A digital meter uses an analog-to-digital converter chip to convert the sensed values into a digital or graphic display.

Many digital voltmeters are autoranging, which means that the meter will automatically search for the correct scale. When using a voltmeter that is not autoranging, always start with the highest voltage range and work down until the indication reads somewhere between half and three-quarter scale. This will provide a more accurate reading and prevent damage to the meter. On many meters, a DC value is indicated by a straight line with three dashes beneath it, while an AC value is indicated by a sine wave.

WARNING!

Measuring voltages above 50V exposes the technician to potentially life-threatening hazards. Follow all applicable safety procedures as found in *NFPA 70E,* OSHA standards, and company/institutional policies and standards.

A voltmeter is used when the exact value of the voltage is required. However, electricians are often concerned with identifying only whether voltage is present, and if so, the general range of the voltage. In other words, is it energized, and if so, is it at 120V, 240V, or 480V? In these cases, a voltage tester is used. The range of voltage and the type of current (AC and/or DC) that a voltage tester is capable of measuring are usually indicated on the scales that display the reading. See *Figure 2*.

Advanced voltage testers offer additional features, such as a digital readout, GFCI test capability, and even the ability to switch between use as a contact and noncontact detector. See *Figure 3*.

A voltage tester must be checked before each use to make sure that it is in good condition and is operating correctly. The external check of the tester should include a careful inspection of the insulation on the leads for cracks or frayed areas. Faulty leads constitute a safety hazard, so they must be replaced. As a check to make sure that the voltage tester is operating correctly, the probes of the tester are first connected to a known energized source. The voltage indicated on the tester should match the voltage of the source. If there is no indication, the voltage tester is not operating correctly, and it must be repaired or replaced. It must also be repaired or replaced if it indicates a voltage different from the known voltage of the source.

26112-14_F02.EPS

Figure 2 Voltage tester.

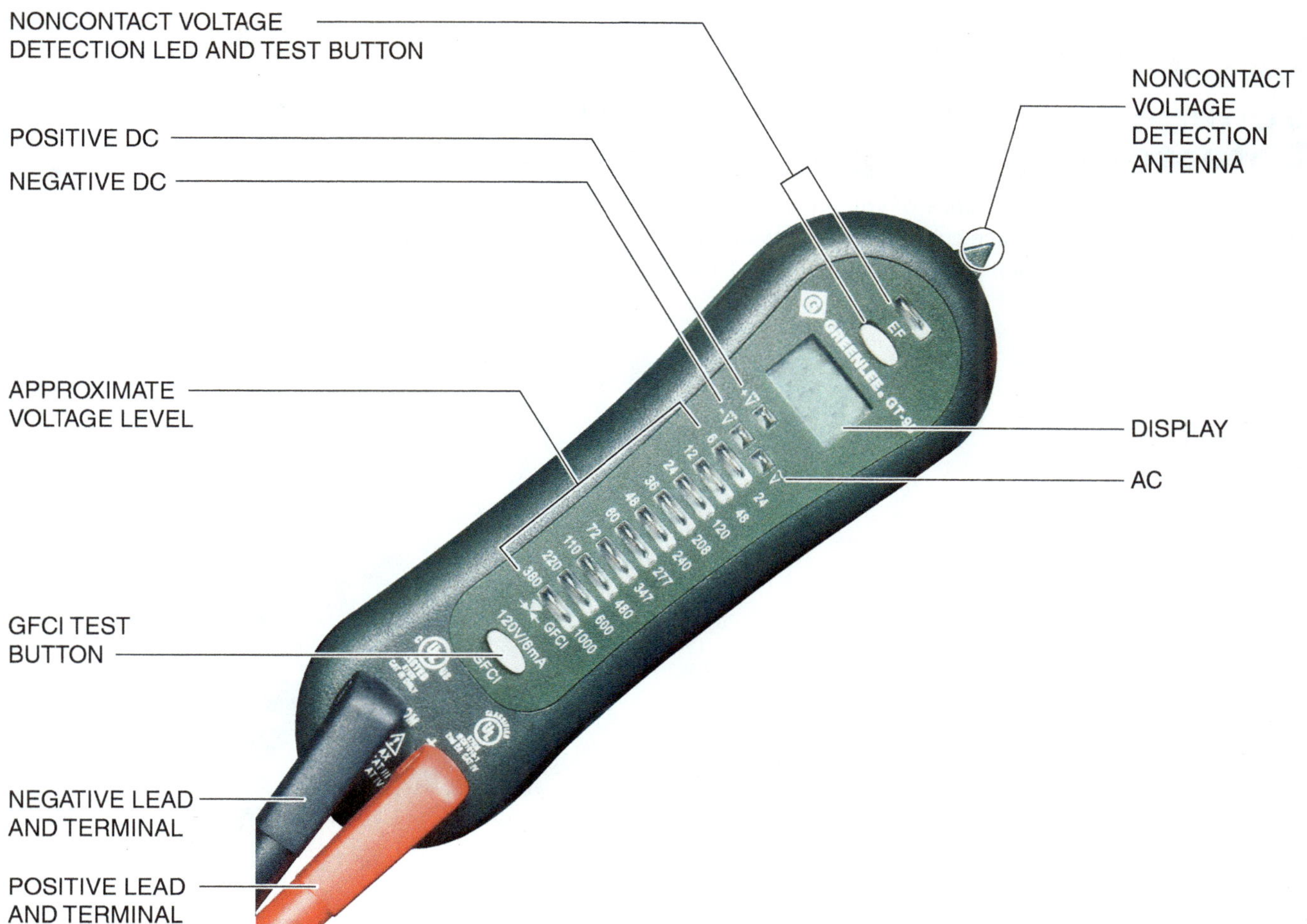

Figure 3 Multi-function voltage tester.

CAUTION

Care should be taken when placing the probes of the tester across the voltage source. Some voltage testers are designed to take quick readings and may be damaged if left in contact with the voltage source for too long.

Voltage testers are used to make sure that voltage is available when it is needed and to ensure that power has been cut off when it should have been. In a troubleshooting situation, it might be necessary to verify that power is available in order to be sure that lack of power is not the problem. For example, if there were a problem with a power tool, such as a drill, a voltage tester might be used to make sure that power is available to run the drill. A voltage tester might also be used to verify that there is power available to a three-phase motor that will not start.

WARNING!

When testing for voltage to verify that a circuit is de-energized as part of an electrical lockout, always perform a live-dead-live test. This involves first verifying operation of the test equipment on a known energized (live) source, then de-energizing and testing the target circuit to ensure it is dead, then again testing the known energized (live) source before making contact with the de-energized circuit. This is an OSHA requirement for systems above 600V but is a good practice at all voltage levels.

2.2.0 Ohmmeter

An ohmmeter measures the resistance of a circuit or component. It can also be used to locate open circuits or shorted circuits. An ohmmeter consists of a DC current meter movement, a low-voltage DC power source (usually a battery), and current-limiting resistors, all of which are connected in series with the meter (*Figure 4*).

On Site

Voltage Detectors

Simple noncontact (proximity) voltage detectors can also be used to indicate the presence of voltage within their specified range rating, but do not discriminate between ranges of values in the same way as a voltage tester. They are handy for quickly scanning for the presence of voltage in junction boxes or termination cabinets, and can even be used to trace circuits through walls. The voltage detector shown here glows in the presence of voltages between 50V and 1,000V.

26112-14_SA01.EPS

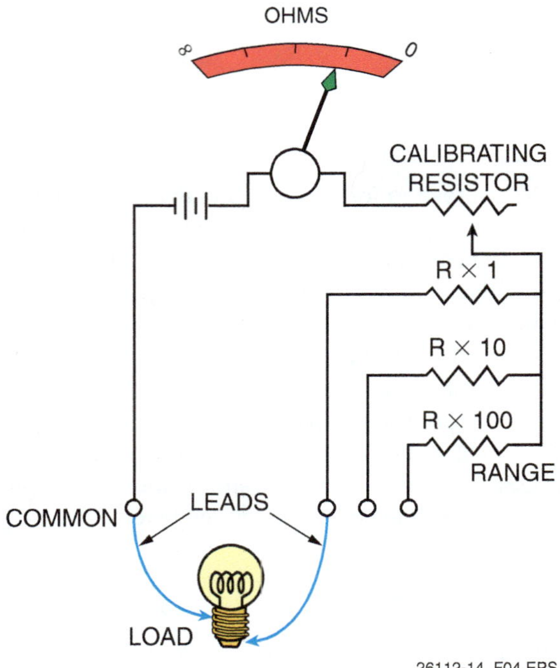

26112-14_F04.EPS

Figure 4 Ohmmeter schematic.

Before measuring the resistance of an unknown resistor or electrical circuit, connect the test leads together. This zeroes out or nulls the resistance of the leads. Some analog ohmmeters have a zero adjustment knob. With the leads connected together, turn the adjustment knob until the meter registers zero ohms. This adjustment must be made each time a different range is selected.

Many digital ohmmeters are autoranging. The correct scale is internally selected and the reading will indicate the range (ohms, K-ohms, or M-ohms). Analog ohmmeters require that you select the desired range. Most digital meters also have an audible tone when the measured value is very low or at zero ohms. This indicates a closed circuit and is useful when using the meter as a **continuity** tester. A continuity test is used to determine if a circuit is complete.

WARNING!

Prior to taking a reading with an ohmmeter, verify that both sides of the circuit are de-energized by using a voltmeter. If the circuit were energized, its voltage could cause a current to flow through the meter. This can damage the meter and/or circuit and cause personal injury.

Resistance

Why does the resistance vary when holding a resistor by pinching the meter leads against the resistor with your fingers when measuring it, versus measuring it while holding it in clips?

When making resistance measurements in circuits, each component in the circuit can be tested individually by removing the component from the circuit and connecting the ohmmeter leads across it. However, the component does not have to be totally removed from the circuit. Usually, the part can be effectively isolated by disconnecting one of its leads from the circuit. Note that this method can still be somewhat time consuming.

2.3.0 Ammeter

A clamp-on ammeter, also known as a clamp meter, can measure current without having to make contact with uninsulated wires (*Figure 5*). This type of meter operates by sensing the strength of the electromagnetic field around the wire(s).

Clamp-on ammeters measure current by using simple transformer principles. The conductor(s) being measured would be the primary and the jaws (clamp) of the meter would be the secondary. The current in the primary winding induces a current in the secondary winding. If the ratio of the primary winding to the secondary winding is 1,000, then the secondary current is $\frac{1}{1000}$ of the current flowing in the primary. The smaller secondary current is connected to the meter's input. For example, a 1A current in the conductor will produce 0.001A (1mA) in the meter.

To measure the current, open the jaws of the meter and close them around the conductor(s) that are to be measured. Make sure that the jaws are clean and close tightly. Then read the magnitude of the current on the meter display.

Many meters have a Hold function. This is useful in tight locations when it is hard to read the meter while it is clamped around the conductor(s). Just press the Hold button when the value is measured, then remove and read the meter.

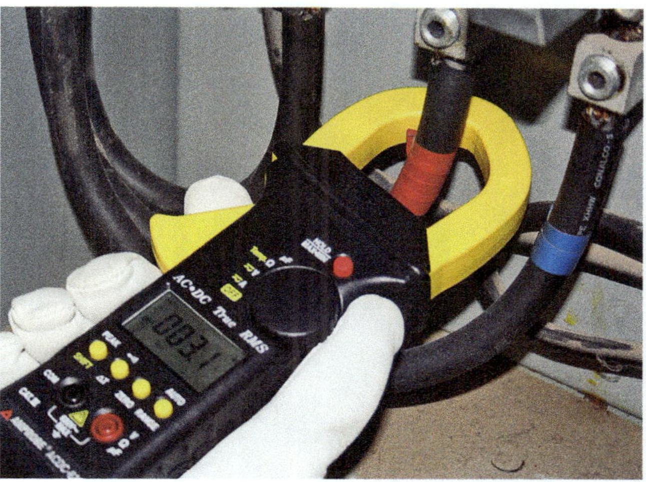

26112-14_F05.EPS

Figure 5 Clamp-on ammeter.

CAUTION

When using a clamp-on ammeter, make sure that the range of the meter is at least as high as the current to be measured. If the meter is digital and the current is too high, the display will read OL (overload). This means that the meter has been overloaded. If the meter is an analog meter, the indicating needle will peg (move) above the maximum limit on the scale, which might damage the meter.

Some meters have a Min/Max or peak function. This allows the technician to record the maximum inrush, as with a motor start.

WARNING!

Using a clamp-on ammeter may expose you to energized systems and equipment. Never use this type of meter unless you are qualified and are following NFPA, OSHA, and company/institutional safety procedures.

Think About It

Current Measurements

What happens if you loop the conductor so that the meter measures two turns instead of one?

2.4.0 Multimeter

The multimeter is also known as a volt-ohm-milliammeter (VOM). An analog VOM is shown in *Figure 6*. It is a multi-purpose instrument that combines the three previous meters discussed. When using an analog meter, you must select the proper voltage (DC or AC) and range

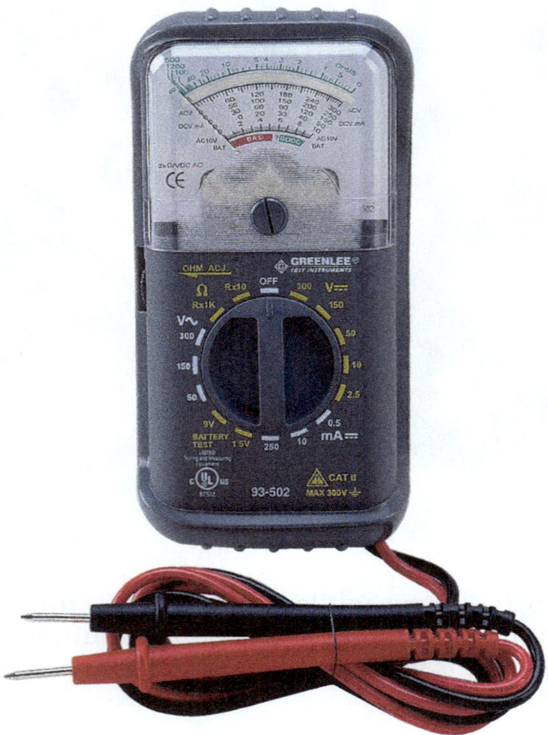

26112-14_F06.EPS

Figure 6 Analog VOM.

(in volts, amps, or ohms). When using a digital VOM (*Figure 7*), you must select the proper voltage. Most have an autoranging feature for the magnitude.

WARNING!

Before use, a VOM must be checked on a known power source.

Use of a VOM is the same as using the individual voltmeter, ohmmeter, and clamp-on ammeter. Current clamps can be used with most multimeters for measuring AC and DC currents above the milliamp level. These can be plugged into either the amp or voltage test lead connections, depending on the current clamp. Always refer to the manufacturer's instructions for the device in use. Current clamps are available in various current ranges, from 50A to several thousand amps. The jaws are available in different shapes and sizes to suit various applications, from round to rectangular, and even flexible.

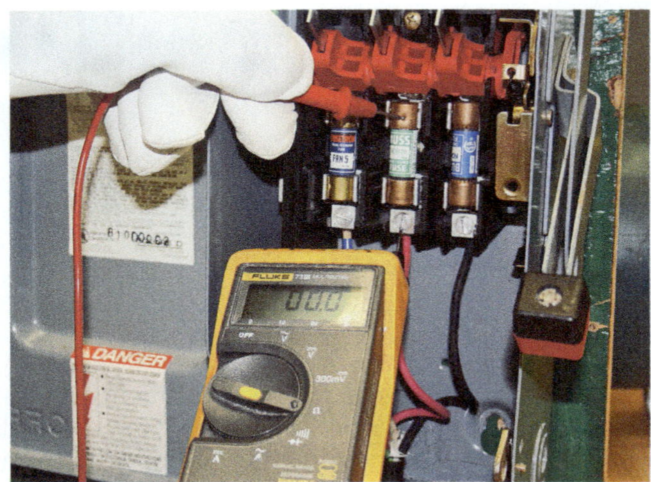

26112-14_F07.EPS

Figure 7 Digital VOM.

Some multimeters can also measure the **frequency** of an AC waveform—this function is useful for diagnosing harmonic problems in an electrical distribution system.

Another additional feature is the Min/Max memory function. It will record the minimum and maximum readings over the time period selected. Other common multimeter functions include capacitance measurement, diode and transistor testers, temperature measurement, and true RMS measurement for accurate voltage and current readings at different frequencies. Refer to the manufacturer's instructions for the meter in use. In addition to the current clamps used with standard multimeters, clamp-on multimeters are also available (*Figure 8*). They are used to measure AC current, AC and DC voltage, resistance, and other values.

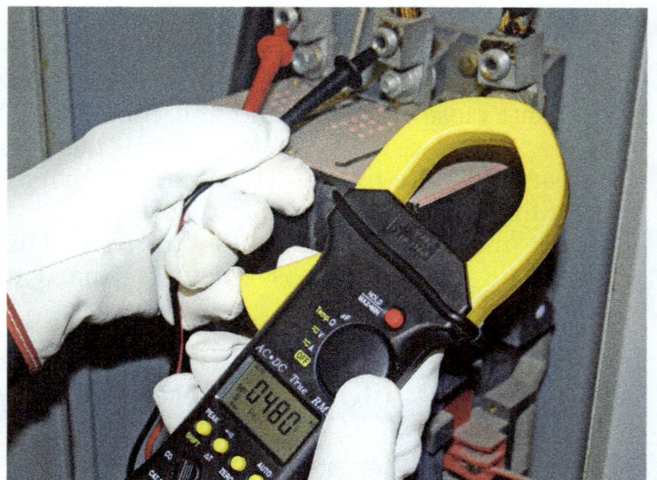

26112-14_F08.EPS

Figure 8 Clamp-on multimeter.

2.5.0 Megohmmeter

An ordinary ohmmeter cannot be used for measuring resistances of several million ohms, such as those found in conductor insulation or between motor or transformer windings. The instrument used to measure very high resistances is known as a megohmmeter. Megohmmeters are also called Meggers®, or insulation resistance testers. They can be powered by alternating current, battery (*Figure 9*), or hand cranking (*Figure 10*).

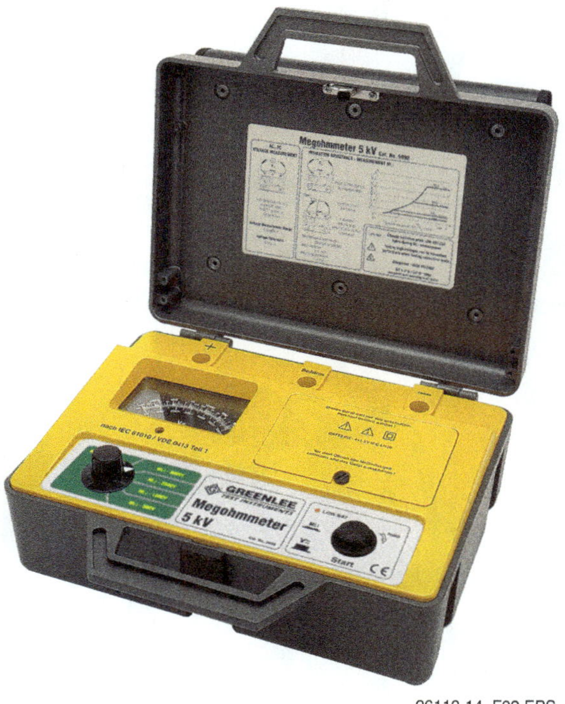

26112-14_F09.EPS

Figure 9 Battery-operated megohmmeter.

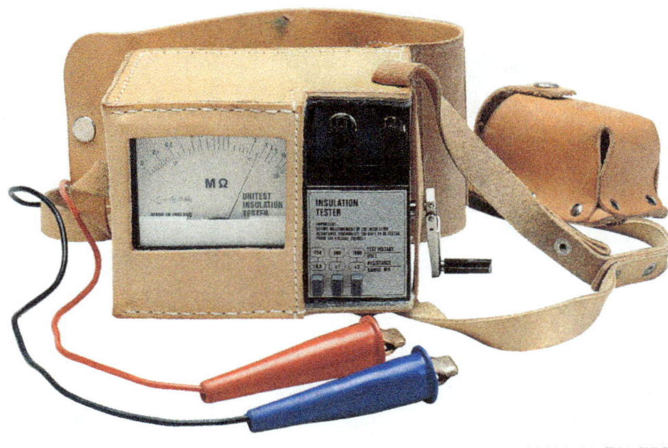

26112-14_F10.EPS

Figure 10 Hand-crank megohmmeter.

When using a megohmmeter, you could be injured or cause damage to the equipment you are working on if the following minimum safety precautions are not followed:

- High voltages are present when using a megohmmeter. For example, in 600V class systems, applied megohmmeter voltages are typically 500V and 1,000V. Only qualified individuals may use this equipment. Always wear appropriate personal protective equipment when approaching energized parts.
- De-energize and verify the de-energization of the circuit before connecting the meter. Make sure all capacitors are discharged.
- If possible, disconnect the item being checked from the other circuit components before using the meter.
- Do not exceed the manufacturer's recommended voltage test levels for the cable or equipment under test. Many manufacturers have different test levels based on the age of the cable or equipment being tested.
- Never touch the test leads when the meter is energized or powered. Meggers generate high voltage, and touching the leads could result in injury or electrical shock.
- After the test, discharge any energy that may be left in the circuit by grounding the conductor or equipment for a period of time equal to the duration of the test.
- When megging cables or busducts where you have exposed parts that are remote from your testing position, safely secure or barricade the exposed end to protect others from inadvertent contact with the test voltage.

CAUTION

If a megohmmeter is used to test switchgear, all of the electronics must be disconnected prior to testing the switchgear. The voltage produced by the megohmmeter may damage electronic equipment.

Meter manufacturers supply detailed manuals for testing various devices and equipment. Always follow these instructions.

2.6.0 Motor and Phase Rotation Testers

Before connecting a three-phase motor to a circuit, you must first match the legs or windings of the motor (T1, T2, and T3) to the phases of the circuit (L1, L2, and L3). This ensures that the motor rotates in the proper direction. A motor

On Site

Meter Care

Like all meters, a megohmmeter is a sensitive instrument. Treat it with care and keep it in its case when not in use.

26112-14_SA02.EPS

rotation tester can be used to identify the legs of the motor (*Figure 11*), while a phase rotation tester can be used to identify the phases of the circuit (*Figure 12*).

Do not connect a motor rotation tester to energized equipment. This can result in injury and equipment damage.

A motor rotation tester is a passive device; it operates on residual magnetism present in the motor after it has been run or tested by the manufacturer prior to shipping. To use a motor rotation tester, connect the three motor wires to the T1, T2, and T3 leads on the tester, then rotate the motor shaft a half-turn while pressing the Test button (the direction of rotation depends on the tester in use; always follow the manufacturer's instructions). Either the clockwise or counterclockwise LED will light up. If the required rotation is clockwise and the clockwise LED lights up, tag the motor wires to correspond to the motor rotation leads. If the required rotation is clockwise and the counterclockwise LED lights up, switch a pair of leads and retest.

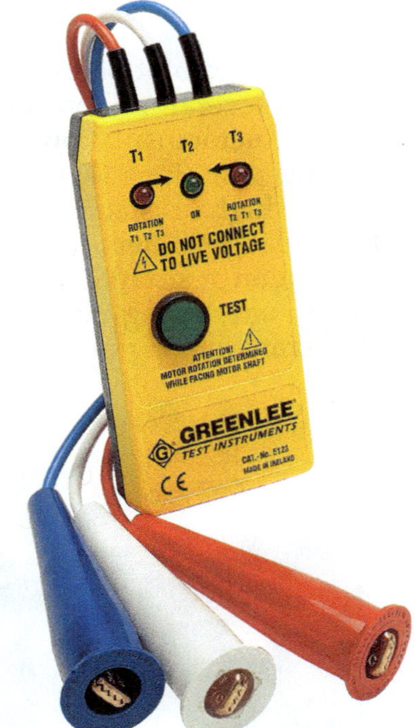

26112-14 F11.EPS

Figure 11 Motor rotation tester.

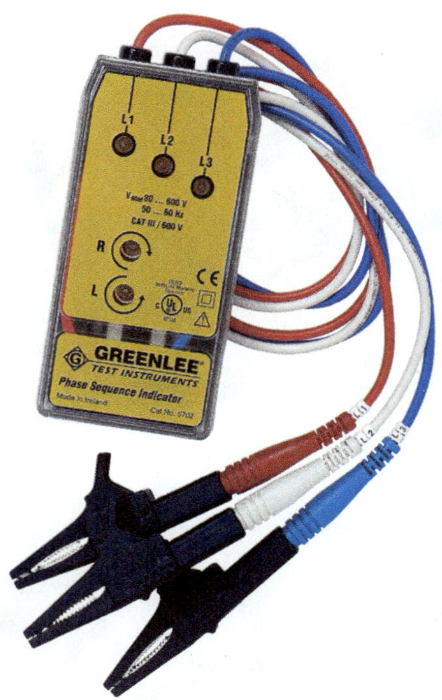

26112-14_F12.EPS

Figure 12 Phase rotation tester.

A phase rotation tester, also called a phase sequence indicator, is used on three-phase electrical systems to indicate the phase sequence rotation of the voltages. These testers typically have LEDs to indicate the phase rotation. A phase sequence is measured as clockwise or counterclockwise rotation.

On Site

Phase Rotation Tester

The leads for a phase rotation tester may not correspond to typical circuit color coding. A good idea is to put phase tape on your meter leads to correspond to the circuit color coding. This will help to ensure correct connections.

WARNING!

Phase rotation testers may only be used by qualified individuals. Always wear appropriate personal protective equipment when approaching energized parts.

A phase rotation tester is used when it is necessary to ensure the same phase rotation throughout a facility. To test phase rotation, de-energize and lock out power to the circuit, then connect the three leads of the tester to the phase conductors in the circuit. Next, safely energize the circuit and observe the meter. Make note of the color scheme of the connected leads to the system, along with the phase sequences as indicated on the meter. This is necessary to ensure that the added equipment follows the same phase rotation. De-energize and lock out the circuit before disconnecting the leads.

2.7.0 Recording Instruments

The term *recording instrument* describes many instruments that make a permanent record of measured quantities over a period of time. Recording instruments use either a paper strip or electronic accessible memory. Those using electronic memory are usually called data loggers. These instruments record electrical quantities, including potential difference, current, power, resistance, and frequency. They can also record nonelectrical quantities by electrical means, such as a temperature recorder that uses a potentiometer system to record thermocouple output.

It is often necessary to know the conditions that exist in an electrical circuit over a period of time to determine such things as peak loads, voltage fluctuations, and so on. An automatic recording instrument can be connected to take readings at specified intervals for later review and analysis. Some meters can upload data to a PC for real-time data logging and graphing. See *Figure 13*.

3.0.0 Category Ratings

Distribution systems and loads are becoming more complex, increasing the risk of transient power spikes. Lightning strikes on outdoor transmission lines and switching surges from normal switching operations can also produce dangerous high-energy transients. Motors, capacitors, variable speed drives, and power conversion equipment can also generate power spikes.

Safety systems are built into test equipment to protect electricians from transient power spikes. The International Electrotechnical Commission (IEC) developed a safety standard, *IEC 1010*, for test equipment that was adapted as *UL Standard UL3111-1*. These standards define four overvoltage installation categories, often abbreviated as CAT I, CAT II, CAT III, and CAT IV (*Table 1*). These categories identify the hazards posed by transients; the higher the category number, the greater the risk to the electrician. A higher category number refers to an installation with higher power available and higher-energy transients.

When selecting a meter, choose a meter rated for the highest category you will be working in. Then select the appropriate voltage level. In addition, make sure that your test leads are rated as high as your meter. Choose meters that are independently tested and certified by UL, CSA, or another recognized testing organization. Certified meters are marked with the category rating on the meter housing (*Figure 14*).

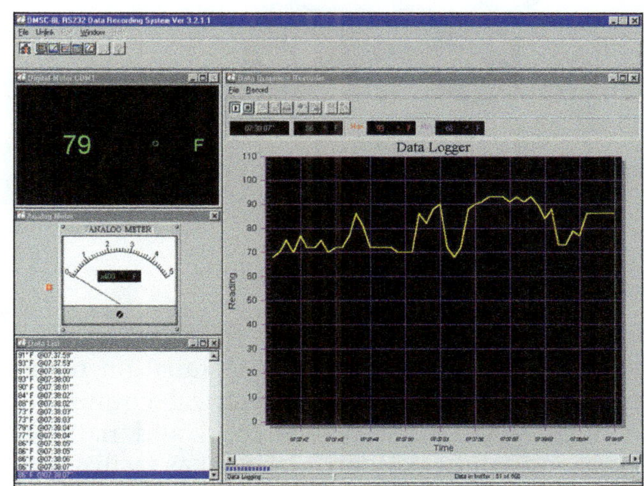

26112-14_F13.EPS

Figure 13 Data recording system.

Table 1 Overvoltage Installation Categories

Overvoltage Category	Installation Examples
CAT I	Electronic equipment and circuitry
CAT II	Single-phase loads such as small appliances and tools, outlets at more than 30 feet from a CAT III Source or 60 feet from a CAT IV source
CAT III	Three-phase motors, single-phase commercial or industrial lighting, switchgear, busduct and feeders in industrial plants
CAT IV	Three-phase power at meter, service-entrance, or utility connection, any outdoor conductors

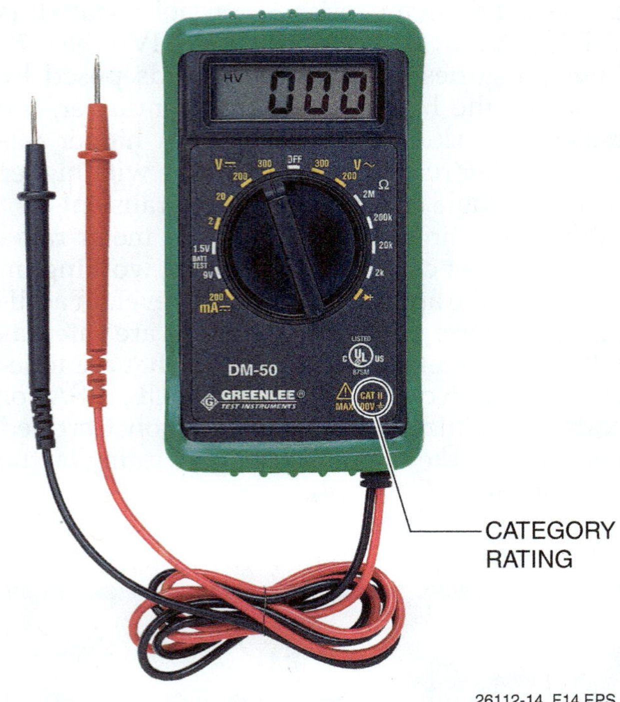

Figure 14 Category rating on a typical meter.

4.0.0 Safety

Safety must be the primary responsibility of all personnel on a job site. The safe installation, maintenance, and operation of electrical equipment requires strict adherence to local and national codes and safety standards, as well as facility and company safety policies. Carelessness can result in serious injury or death due to electrical shock, burns, falls, flying objects, etc. After an accident has occurred, investigation almost invariably shows that it could have been prevented by the exercise of simple safety precautions and procedures. It is your personal responsibility to identify and eliminate unsafe conditions and unsafe acts that cause accidents.

You must bear in mind that de-energizing main supply circuits by opening supply switches will not necessarily de-energize all circuits in a given piece of equipment. A source of danger that has often been neglected or ignored, sometimes with tragic results, is the input to electrical equipment from other sources, such as backfeeds. The rescue of a victim shocked by the power input from a backfeed is often hampered because of the time required to determine the source of power and isolate it. Always turn off all power inputs before working on equipment and lock out and tag, then check with an operating voltage tester to be sure that the equipment is safe to work on.

WARNING!

When performing lockout/tagout procedures, remember that other forms of energy may be present and must also be locked out and tagged. These include water pressure, steam, springs, gravity, and other forms of energy.

Remember that the common 120V power supply voltage is not a low, relatively harmless voltage but is a voltage that has caused more deaths than any other.

Safety can never be stressed enough. There are times when your life literally depends on it. Always observe the following precautions:

- Thoroughly inspect all test equipment before each use. Check for broken leads or knobs, damaged plugs, or frayed cords. Do not use equipment that is wet or damaged.
- Make sure the rating of any leads or accessories meets or exceeds the rating of the meter.
- Do not work with energized equipment unless you are both qualified and approved by your supervisor.
- Never shortcut safety; strictly adhere to all energized work policies and procedures.
- When testing circuits, test at higher ranges first, then work your way down to lower ranges.
- Always have a standby person present during hot work; he or she should know whom to contact in case of emergency and how to disconnect the power.

Think About It

Putting It All Together

What kind of measuring device would you select or need for the following tasks, and how would you apply the device?

- Identify a short circuit in house wiring
- Measure the secondary voltage of an AC transformer
- Identify a blown fuse in a circuit
- Identify the contact configuration of a three-way switch or multi-pole relay

SUMMARY

Meters and other devices are used to test and troubleshoot circuits and electrical equipment. One of the most important tests you will perform is verifying the absence of voltage before working on a device or circuit. This is often done using a voltage tester. Other common test equipment includes multimeters, clamp-on ammeters, megohmmeters, and motor/phase rotation testers.

You must understand the operation of and safety precautions for each piece of test equipment. In addition, you must be able to select the appropriate test equipment based on the task and category rating of the environment in which the work is to be performed. Always inspect and verify the operation of all test equipment before using it. Your life may depend on it.

Review Questions

1. Digital meters do *not* have a(n) _____.
   a. red lead wire
   b. battery
   c. meter movement
   d. autoranging setting

2. A voltmeter is used to test _____.
   a. exact voltages
   b. voltage ranges
   c. power
   d. sine waves

3. In order to ensure safety, before measuring low voltages, you should first test for _____.
   a. resistance
   b. current
   c. vibration
   d. higher voltages

4. An ammeter is used to measure _____.
   a. current
   b. voltage
   c. resistance
   d. insulation value

5. Clamp-on ammeters operate by _____.
   a. using d'Arsonval meter movement
   b. sensing the strength of the electromagnetic field around the wire
   c. measuring the high resistance end of a power transformer
   d. using a resistive shunt

6. An insulation tester is another name for a(n) _____.
   a. megohmmeter
   b. ammeter
   c. multimeter
   d. continuity tester

7. A motor rotation tester _____.
   a. tests an energized motor
   b. gets connected to the motor supply conductors
   c. works on residual magnetism
   d. works only on clockwise rotation

8. The highest level of protection is provided by instruments rated as _____.
   a. CAT I
   b. CAT II
   c. CAT III
   d. CAT IV

9. The International Electrotechnical Commission (IEC) developed a safety standard for overvoltage installation categories of CAT I, CAT II, CAT III, and CAT IV for _____.
   a. wiring
   b. electrical equipment
   c. test equipment
   d. signaling circuits

10. De-energizing the main supply circuit will always de-energize all circuits in a given piece of equipment.
    a. True
    b. False

Trade Terms Quiz

Fill in the blank with the correct term that you learned from your study of this module.

1. Used for electromagnetic effects or for providing electrical resistance, a ____________ is a number of turns of wire.
2. A ____________ uses a permanent magnet and moving coil arrangement to move a pointer across a scale.
3. Usually expressed in hertz, ____________ is the number of cycles completed each second by a given AC voltage.
4. ____________ is an uninterrupted electrical path for current flow.

Trade Terms

Coil

Continuity

d'Arsonval meter movement

Frequency

Module 26112-14
Supplemental Exercises

1. The measurement of the electromotive force of a circuit is accomplished using a(n) ______________.
   a. ammeter
   b. wattmeter
   c. voltmeter
   d. ohmmeter
2. When using a voltmeter that is not autoranging, start with the highest setting and work down until the meter reads somewhere between ______________________.
3. A(n) ______________________ is used to extend the range of a meter movement for current measurements.
4. True or False? Always connect an ohmmeter in parallel with a load.
5. The voltage range of a meter movement can be extended by adding a(n) _______________ in series.
6. Short circuits can be detected by using a(n) ______________________.
7. What type of test equipment would you use to check the resistance between motor windings?
   ____
8. The phase sequence of a circuit is identified using a(n) ______________________.
9. True or False? A voltage tester is used for precise voltage measurements.
10. A(n) ______________________ is used to take electrical readings at specified intervals.

Cornerstone of Craftsmanship

Clarence "Ed" Cockrell

HR/Safety Manager
Vector Electric & Controls, Inc.

How did you get started in the construction industry?
I worked as a summer electrical helper during high school and college and found it very rewarding. I was looking for a job that would hold my interest for more than a year. I studied electrical engineering at Louisiana State University for two years.

Who inspired you to enter the industry? Why?
My brother-in-law and father-in-law inspired me to enter the industry.

What do you enjoy most about your job?
I started off as an electrician's apprentice, which offered many potential job opportunities. I enjoyed seeing the work progress from dirt to a functional building and finding new challenges as new technology changes the way we install the electrical components. My electrical training also opened the door to the possibility of being a field superintendent, project manager, senior office manager, human resource/training manager, and then human resource/safety/training manager.

Do you think training and education are important in construction? If so, why?
Training gives an apprentice the opportunity to become a great electrician and not an electrical laborer, by that I mean not just a conduit or cable tray installer or wire puller, but a well-rounded electrician. It also allows the apprentice to advance beyond the limits of being an electrician. Almost all of our supervisors, estimators, and project managers have been trained by an apprenticeship program and followed it up with more NCCER training.

How important are NCCER credentials to your career?
I went through a state certified apprenticeship and then completed the CSST training. Without these certifications, I believe I would be no more than a second rate electrician.

How has training/construction impacted your life and you career?
It has allowed me to raise a family, buy a house and raise a child in a comfortable fashion. I have advanced many times at work, which has led to increased wages. Through my job, I have met and had dealings with many interesting people from all walks of life.

Would you suggest construction as a career to others? If so, why?
Yes. It offers a rewarding career opportunity to anyone willing to take pride and ownership in their learning and work.

How do you define craftsmanship?
I believe that craftsmanship is always delivering a great quality job in whatever you do. It takes pride and self-esteem to deliver work deemed to meet this definition of craftsmanship. This pride in their work helps build life qualities that become the building blocks of a truly honorable life. They walk through their community as a positive contributor as well as helping to build a secure and prosperous business. They teach their children through their actions how those who build contribute to the well-being of their community and country.

Trade Terms Introduced in This Module

Coil: A number of turns of wire, especially in spiral form, used for electromagnetic effects or for providing electrical resistance.

Continuity: An electrical term used to describe a complete (unbroken) circuit that is capable of conducting current. Such a circuit is also said to be closed.

d'Arsonval meter movement: A meter movement that uses a permanent magnet and moving coil arrangement to move a pointer across a scale.

Frequency: The number of cycles completed each second by a given AC voltage; usually expressed in hertz. One hertz equals one cycle per second.

Additional Resources

This module presents thorough resources for task training. The following resource material is suggested for further study.

ABCs of Multimeter Safety, Everett, WA: Fluke Corporation.

ABCs of DMMs, Multimeter Features and Functions Explained, Everett, WA: Fluke Corporation.

Clamp Meter ABCs, Everett, WA: Fluke Corporation.

Electronics Fundamentals: Circuits, Devices, and Applications, Thomas L. Floyd. New York: Prentice Hall.

Power Quality Analyzer Uses for Electricians, Everett, WA: Fluke Corporation.

Principles of Electric Circuits, Thomas L. Floyd. New York: Prentice Hall.

Figure Credits

AGC of America, Module opener

Tim Dean, Figure 1, Figure 5, Figure 7, Figure 8

Greenlee Textron, Inc., a subsidiary of Textron Inc., Figure 2, Figure 3, Figure 6, Figures 9–14, 112SA01

Topaz Publications, Inc., 112SA02

NCCER CURRICULA — USER UPDATE

NCCER makes every effort to keep its textbooks up-to-date and free of technical errors. We appreciate your help in this process. If you find an error, a typographical mistake, or an inaccuracy in NCCER's curricula, please fill out this form (or a photocopy), or complete the online form at **www.nccer.org/olf**. Be sure to include the exact module ID number, page number, a detailed description, and your recommended correction. Your input will be brought to the attention of the Authoring Team. Thank you for your assistance.

Instructors – If you have an idea for improving this textbook, or have found that additional materials were necessary to teach this module effectively, please let us know so that we may present your suggestions to the Authoring Team.

NCCER Product Development and Revision

13614 Progress Blvd., Alachua, FL 32615

Email: curriculum@nccer.org
Online: www.nccer.org/olf

❑ Trainee Guide ❑ AIG ❑ Exam ❑ PowerPoints Other ____________________

Craft / Level: Copyright Date:

Module ID Number / Title:

Section Number(s):

Description:

Recommended Correction:

Your Name:

Address:

Email: Phone:

Identifying Career Development Opportunities

Power, Structural, and Technical Agricultural Systems

Bls.gov/k12/ is an official U.S. government Web site providing links to careers that relate to your interests. It presents fun facts about the economy and jobs. Use the student-related links to explore the careers that interest you. Pay special attention to careers in the power, structural, and technical systems pathways, including:

- Agricultural engineer
- GIS specialist
- Farm equipment technician
- Land surveyor
- Welder
- Contractor/builder
- Diesel technician

Use the information you find to develop a list of potential careers, and create a table for each that identifies the average salary, education requirements, work experience requirements, development opportunities, and the job outlook.

Employees in the Power, Structural, and Technical systems pathways work with all sorts of equipment and technology, especially those in the hydraulics, engineering, pneumatics, power, electronics, and agricultural controls industries. They work to design agricultural machinery, equipment, and structures while developing methods used to help conserve as much water and soil as possible while still improving the processing of agricultural products.

Pathways

- Power
- Structural
- Technical

Sample Occupations

- Agricultural engineer
- GIS specialist
- Farm equipment technician
- Land surveyor
- Welder
- Contractor/builder
- Diesel technician

College and Career Readiness: Your Personal Strengths: Analyzing Career Requirements and Rewards

Analyzing Career Requirements and Rewards

A career is a chosen field of work in which you try to advance over time by gaining responsibility and earning more money. A job is any activity that you do in exchange for money or other payment. A job does not necessarily lead to advancement. Occupation is a word that means career or job.

Every career comes with its own requirements—things you must do or have—and rewards—things you receive in return. The specific requirements and rewards of the career you choose have a major impact on the kind of life you will lead.

- A doctor spends years in school and in training. It is expensive and leaves little time for extra activities. But, doctors help people and save lives, which can be very rewarding.
- A restaurant owner invests his own money in the business. He works long hours and must interact with temperamental chefs, kitchen staff, and customers. However, he is his own boss and can take pride in starting and running his own exciting business.

Understanding the requirements and rewards can help you choose a career that you will enjoy for many years.

Career Requirements

Career requirements are the responsibilities that you must perform in order to succeed in the career. A responsibility is something people expect you to do, or something you must accomplish. As a student, your teachers expect you to come to class and do your homework. As an employee—worker—your supervisor expects certain things from you, too.

Education is a requirement for all careers. Most employers expect employees to have diplomas from high school and college. For many careers, additional training and education is required.

The requirements of a career determine things such as the type of training and education you will need. Career requirements might also determine where you will live, how you will spend your time, and who you will be working with.

- If you choose to become an electrician, you will need education and training different from that of someone who chooses to become a chef.
- A kindergarten teacher spends most of the day with 5-year-olds.
- A lawyer must read and prepare legal documents.
- A software programmer is alone in front of a computer for hours at a time.

Career Rewards

Recall that the average person spends between 40 to 50 hours per week at work. Most people work because they require money to buy things they need, such as food, clothes, and shelter. Some people choose their careers based on how much money they can earn on the job. The money that you receive in exchange for work performed is called wages or a salary.

Money is a good reason to choose a career, but it is not the only reason. People who are happy with their careers usually do something they enjoy. Even if the job does not pay very well, people might choose a career because it gives them satisfaction, fits their values, uses their skills and abilities, and gives them a role in the community.

- A woman who starts a small jewelry business may enjoy being her own boss and setting her own hours.
- A social worker who helps homeless people may believe she is making the world a better place.
- An artist may feel he is making the world more beautiful by creating paintings or sculptures.

College and Career Readiness: Exploring the Career Clusters: Analyzing the Career Clusters: Architecture and Construction

Architecture & Construction

People in the Architecture & Construction career cluster design, plan, manage, build, and maintain the built environment. Take a look around you. The houses across the street, the roads that cars drive on, and the parks that you walk through every day were created or maintained by the people performing jobs in this cluster and its pathways.

Pathways

Construction

Design/Pre-Construction

Maintenance/Operations

Sample Occupations

Architect

Architectural and Civil Drafter

Carpenter

Civil Engineer

Civil Engineering Technician

Code Official

Computer Aided Drafter (CAD)

Concrete Finisher

Construction Worker

Cost Estimator

Drywall and Ceiling Tile Installer

Pipelayer

ENTREPRENEURIAL OPPORTUNITES

Many careers in the Architecture & Construction cluster are well-suited for someone interested in being an entrepreneur. An architect might work for a large firm for a while then start her own business when she feels that she has gained enough experience to succeed on her own. A drywall installer might prefer working independently instead of for a contractor. What are some of the benefits and drawbacks to working as an entrepreneur in the Architecture & Construction industry?

College and Career Readiness: Exploring the Career Clusters: Analyzing the Career Clusters: Science, Technology, Engineering & Mathematics

Science, Technology, Engineering & Mathematics

Can you see yourself planning, managing, and providing scientific research as a career? Are you interested in laboratory and testing services or in research and development? If so, are you also willing to earn the certification or advanced education necessary to work in this career cluster? If your answer is a "yes" to any of these questions, then take a closer look at these pathways and occupations.

Pathways

Engineering and Technology

Science and Math

Sample Occupations

Anthropologist

Biologist

Geneticist

Mathematician

Nuclear Chemist

Paleontologist

Physicist

Statistician

Technical Writer

44105-08

Construction Documents

44105-08

Construction Documents

Topics to be presented in this module include:

Overview

Documentation is vital for a successful business. Numerous studies indicate that poor documentation causes the most problems for industry, followed by poor administration.

Documentation is the who, what, where, how, and when of a construction project. It is a collection of facts that records and documents the actual history of the project. The specific documentation you'll need to maintain depends on the type of project and the needs of the owner, the architect/engineer, the contractors, and the project team members. Without exception, all projects should have some form of documentation.

Objectives

Upon completion of this module, you will be able to do the following:

1. Explain the need for documentation on a project.
2. State the various approaches for obtaining work in the construction industry.
3. Identify the parts of a typical project manual.
4. Identify the various types of drawings and format specifications.
5. Discuss the types of contracts used in the construction industry.
6. Discuss insurance requirements for a company and a project.
7. List the types of documents used on a project.
8. Describe the change order process.
9. List the documents necessary to close out a project.

Trade Terms

Addenda
Bid bond
Change order
Construction Specifications Institute (CSI)
Punch list
Scope
Specifications
Shop drawings

Prerequisites

Before you begin this module, it is recommended that you successfully complete *Project Management,* Modules 44101-08 through 44104-08.

This course map shows all of the modules in the *Project Management* curriculum. The suggested training order begins at the bottom and proceeds up. Skill levels increase as you advance on the course map. The local Training Program Sponsor may adjust the training order.

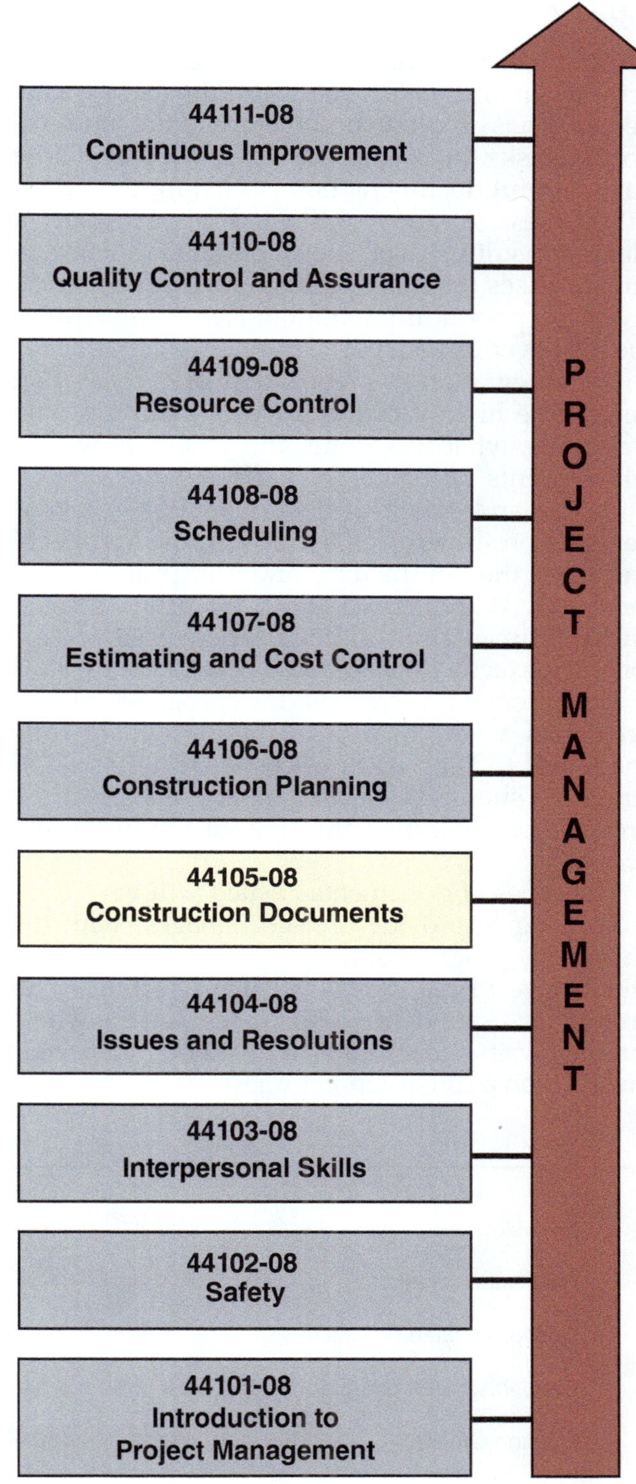

105CMAP.EPS

1.0.0 ♦ INTRODUCTION

Although documentation was once a secondary activity for the project manager, during the past decade it has grown to become one of the most important tasks the project manager will face. These days, careful documentation is vital for a successful business. A study conducted by Accenture (a global consulting firm) shows that poor documentation causes the most problems for industry, followed by poor administration. *Figure 1* illustrates the results of this survey.

Documentation is a collection of records that details the history of the project, including who was there, what was done, and where, how, and when events occurred that affected the project. The type and extent of documentation required depends on the project, the owner, the architect/engineer, the contractors, and the project team members. Without exception, all projects must have documentation, although there is some flexibility in exactly how records are kept.

Documentation records the progress of the project as a way to control the amount of time spent and to keep track of costs. Documentation strategies should be developed at the start of the project and continued throughout the life of the project.

The types of documents created will vary from a daily log to provide project managers with the day-to-day progress of the project, to photographs, which give visual details of project activities, to project correspondence, which includes the whole range of written communication among all parties involved on a construction project.

2.0.0 ♦ OBTAINING WORK IN THE CONSTRUCTION INDUSTRY

Contractors and subcontractors who want to obtain work in the construction industry must submit a bid or cost proposal to an owner or architect/engineer. There are several ways a contractor can obtain work:

- Competitive bid
- Invited bid
- Negotiation
- Design-build
- Construction management

2.1.0 Competitive Bid

The most common way of obtaining work in the construction industry is through a method known as competitive bidding. In competitive bidding, the owner will seek bids from several general contractors and will award a contract to the general contractor who has submitted the lowest responsive bid.

While preparing a bid, the general contractor may seek prices from specialized contractors (known as subcontractors), vendors, and material suppliers that have more knowledge and experience to perform the required tasks. The general contractor will review the scope of work and bids received, and may select the lowest responsive bid from subcontractors, vendors, and material suppliers to determine the price that will be submitted to the owner.

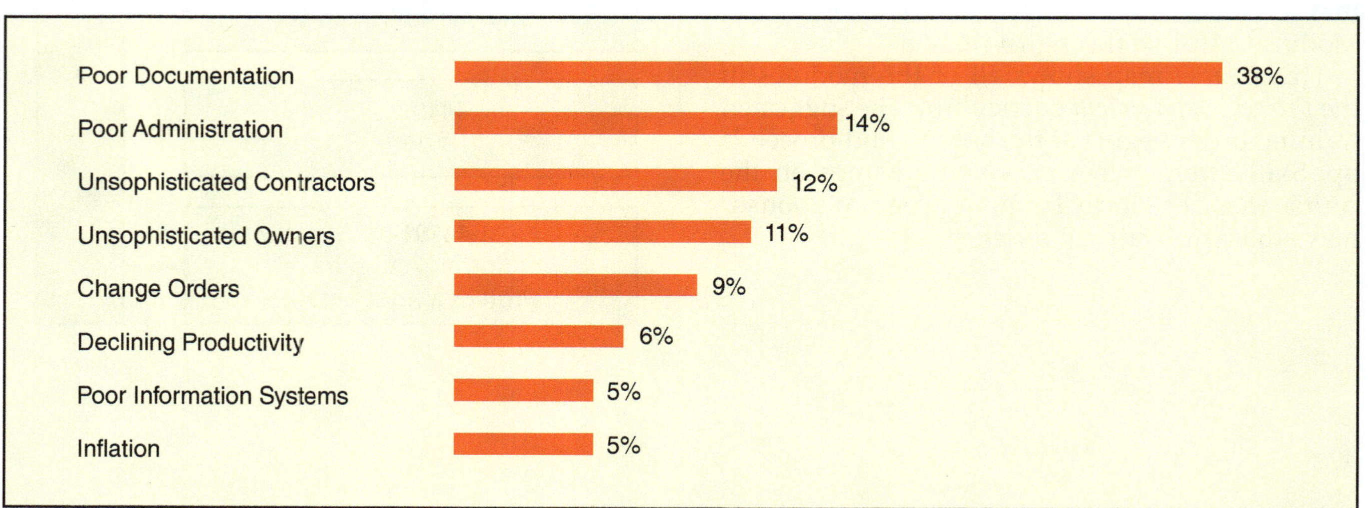

Figure 1 ♦ Construction contract hang-ups (based on a survey of 790 industry firms).

Bids are submitted to the owner and architect/engineer at a predetermined location, date, and time. On public projects and in many private projects, the bids or cost proposals are opened and read aloud. This allows all interested parties to know the results of the bids or cost proposals immediately.

2.2.0 Invited Bid

The invited bid method of obtaining work is very similar to competitive bidding, except the owner will pre-qualify the general contractors. Pre-qualified general contractors are selected based on their experience, financial strength, present and future work loads, and management team. An advantage for being selected to bid is that the contractor will be competing against other contractors with similar experience and business philosophies.

2.3.0 Negotiations

The negotiation method of obtaining work occurs when the owner selects several general contractors to prepare and submit bids or cost proposals. The owner has the opportunity to interview and discuss with each contractor its proposal, work experience, construction schedule, work sequence, and project cost before awarding a contract. The general contractor also has the same opportunity to discuss with the owner its approach and cost of the project.

2.4.0 Design-Build

The design-build method requires the owner to select a contractor that will be responsible for both the design and construction of the project. The owner and the contractor will work closely together from the very beginning of the project. Both will be involved in developing design guidelines, selecting materials, and establishing the project estimate.

2.5.0 Construction Management

Construction management is used by an owner who lacks the expertise and knowledge of construction. The owner will employ a firm that will be responsible for reviewing the contract documents and managing the construction process. The construction management firm will have experience in both the technical aspects and administration of projects.

2.6.0 Summary of Methods

Many contractors will split their methods of obtaining work between competitive bidding, invited bidding, and negotiated work. However work is obtained, each method requires the contractor to compile estimates for the project, including seeking bids from subcontractors, vendors, and material suppliers. The estimating procedure takes considerable time, company resources, and finances; therefore, contractors should examine the various approaches they use to obtain work to make the best use of their estimating and management staff.

3.0.0 ◆ PROJECT MANUAL

The term "project manual" was coined by the American Institute of Architects (AIA) in 1964 when it adopted the concept and title of *Project Manual* to replace the common title **specification** (spec) book used by the industry up to that time.

The project manual consists of written documents prepared by the architect/engineer for the owner in order to bid and construct the project. The project manual contains a variety of separate documents that a project manager must thoroughly understand. These documents include:

- Bidding documents
- Various contract forms
- Specifications
- List of drawings
- Other information required to construct the project

The project manual has been arranged into two categories: the bidding requirements for the project and the specifications. The use of bidding requirements is to assure that all bidders will have the same opportunity to review the documents, prepare a bid, and submit the bid for fair evaluation by the owner and architect/engineer.

A typical project manual will contain bidding requirement documents which include an invitation to bid, instructions to bidders, bid form, **bid bond**, performance and payment bond, form of agreement, general conditions, supplementary conditions, specifications, list of drawings, and **addenda**.

3.1.0 Invitation to Bid

The invitation to bid advises prospective bidders about a project. It contains information bidders need in order to determine if they wish to submit a bid or proposal for the project.

The following information is contained in the invitation to bid section:

- *Project identification* – Name of the owner, architect/engineer, or organization issuing the bid documents
- *Description of work* – Construction type, project type, and size of the project
- *Types of bids required* – Single contract or multiple contracts for various phases of the project
- *Time of completion* – May be either the number of calendar days or a calendar date
- *Time and location of bid* – The date, time, and location of the bid opening will be outlined, including whether it is a public or private opening
- *Review of documents* – Lists the names and addresses of sources where bidding documents may be examined and obtained
- *Bid bond or bid security* – States if a bid bond or bid security is required, in what amount, and the bond form
- *Bidder pre-qualifications* – Owner or architect/engineer may require the bidder to provide proof of experience on a similar construction project
- *Right to reject bids* – Statement that the owner reserves the right to waive irregularities or to reject all bids
- *Guaranty bonds* – States whether the owner will require a performance and payment bond from the winning bidder

3.2.0 Instructions to Bidders

This document contains information that the bidder must comply with in preparation of the bid. In addition to the information outlined in the invitation to bid, the instructions to bidders may contain the following:

- *Resolution of discrepancies and ambiguities* – Outlines the method by which issues will be resolved during the bidding process
- *Substitution of products* – Requirements, procedures, and time limit under which substitutions of products will be considered during the bidding period
- *Bid submittal* – Requirements for submitting a bid including type of form, number of copies, identification of bid, required signatures, list of subcontractors, and other related information
- *Governing laws and regulations* – Outlines applicable regulatory requirements, labor agreements, safety requirements, and hiring requirements

3.3.0 Bid Form

The bid form is the document used by each bidder to submit its price (or bid) for doing the work specified in the contract documents. The uniformity of the bid form makes it easier for the owner and architect/engineer to evaluate and compare bids.

3.4.0 Bid Bond or Bid Security

The bid bond or bid security is a legal document posted by the bidder as an assurance that the lowest bidder will enter into a contract with the owner if awarded the project. The bid bond or bid security is required at the time of the bid and is usually represented as a percentage of the total bid amount. The bid bond or bid security is usually furnished by the contractor's surety (bonding) company, which will also issue the performance and payment bond, if required by the contract documents.

Many owners and architects/engineers will use the bid bond or bid security requirements as a method for pre-qualifying contractors. Many times, only contractors who can show evidence of bond ability will be permitted to bid on a project. If a contractor's construction record or financial condition is such that a surety company will not provide a bond, the owner or architect/engineer may be reluctant to award a contract to a contractor that may not have the construction experience or the financial capacity to complete the project.

3.5.0 Performance and Payment Bond

Performance and payment bond is a document guaranteeing that the contractor will complete the work in accordance with the terms of the contract and make payments to their subcontractors and vendors. Also, the payment portion of the bond indemnifies the owner for liens filed against the project. The performance and payment bond is issued for the full amount of the contract.

If the contractor defaults (that is, becomes unable to complete the work or make payments) the surety (bonding) company becomes responsible for both the completion of the work and payment of all bills. Usually, the surety company will assume the financial responsibility and will retain another contractor to fulfill the construction portion of the project. Defaults by the contractor may result in construction delays and additional claims against the surety company from the owner and other parties involved on the project. Waiving the performance and payment bond requirement may be considered when the owner has knowledge of the contractor's reputation for performance and financial stability.

3.6.0 Form of Agreement

An example of the form of agreement (or contract) between the owner and contractor will be included in the project manual. It may vary from a standard agreement developed by a professional association, the owner, or architect/engineer. It is important that the form of agreement presented in the project manual is reviewed by both the construction management team and its legal counsel.

3.7.0 General Conditions

General conditions define the standard contractual relationships, particularly the interrelationship of the owner, architect/engineer, contractor, and subcontractors. It also details how the contract will be administered and the procedures relative to the project, including rights and responsibilities of the parties. The most commonly used general conditions for private projects is the American Institute of Architects (AIA) *A201* document. For government-related and certain large projects, the architect/engineer or owner will develop general conditions specific to the requirements of a particular project.

3.8.0 Supplementary Conditions

The supplementary conditions are modifications, deletions, and additions to the standard, or prepared, general conditions. Supplementary conditions are developed to meet the project's specific needs or to supplement the requirements of a project.

3.9.0 Specifications

The specifications include information that supports and describes the construction drawings. The role and structure of specifications will be presented in detail in another section of this module.

3.10.0 Addenda

Addenda are documents used to modify the contract prior to receiving the bid. They are issued by the architect/engineer when errors or discrepancies are discovered during the review of the bid documents. When modifications to the contract documents occur after the award of the contract, the document issued by the architect or contractor is known as a change order.

4.0.0 ◆ DRAWINGS AND SPECIFICATIONS

Those outside of the construction industry are likely bewildered by the maze of lines, dimensions, and notations associated with construction drawings. Project managers, however, must be able to read and understand construction drawings and specifications if they are to communicate the information to others.

4.1.0 Types of Drawings

The various types of construction drawings that project managers may see are:

- *Conceptual drawings* – Prepared by the architect/engineer during the initial phase of the project, these drawings provide the owner with an architect's perspective of the project, including elevations, building layout, interior finishes, and other information necessary for the owner to evaluate the project.
- *Preliminary drawings* – After the owner has evaluated the project concept drawing, the owner will authorize the architect/engineer to develop drawings that will contain construction details, project requirements, and information necessary to submit to governmental agencies for review, to financial institutes for construction loans, and for development of an estimate of the project cost. The preliminary drawings may go through several stages of development before the final drawings have been accepted by the owner.
- *Working or construction drawings* – These drawings provide the information for the contractors to prepare a bid or proposal to be submitted to the owner.

4.2.0 Reading Drawings

The first step in studying a set of drawings is to become familiar with the general features of the drawings. Project managers should review the methods the architect/engineer used for notation, legends, sections, details, and other information.

After reviewing the general features of the drawings, project managers should review the drawings systematically. One suggested method for reviewing drawings is:

Step 1 Review the site drawings for building layout and location of roadways, utilities, site and building grades, and other site-specific information.

Step 2 Review the structure drawings for footing sizes, column pads, foundation walls, reinforcement placement, floor system, superstructure, and other specific structural information.

Step 3 Review architectural drawings for floor plans, individual room layouts, elevations, wall sections, finish details, and other specific architectural information.

Step 4 Review the mechanical drawings for plumbing, heating, ventilation, air conditioning, fire sprinkler layout, equipment, and other specific mechanical information.

Step 5 Review electrical drawings for lighting layout, power distribution, equipment requirements, and other specific electrical information.

Step 6 Review all notes, details, and specific instructions.

Step 7 Review all drawings for possible interference of trades and coordination problems.

Step 8 Finally, review all drawings to ensure a complete understanding of the project and the requirements of the architect/engineer.

After becoming familiar with the drawings, project managers should review the specifications.

4.3.0 Specifications

Since there is never enough space on the drawings to include all the required technical information regarding the project, the architect/engineer will develop a set of specifications that provide important technical information about the work to be accomplished and the materials to be used.

To illustrate the need for specifications, consider the job of building custom cabinets in a particular project. The drawings would detail the locations of the units, the elevation, and the arrangement of cabinet doors, drawers, and shelving. The specifications would describe the type of wood to be used, hardware requirements, finishes, and the installation instructions, all in accordance with performance guidelines established by a trade or manufacturers' association. Therefore, in order to meet the requirements of the cabinets for the project, both drawings and specifications must be used.

Specifications serve three purposes:

- Define the responsibilities of the architect/ engineer, the owner, the contractor, and the subcontractor
- Supplement the working drawings with detailed technical information regarding the quality of work and the material to be used
- Support the drawings as part of the contract documents

The architect/engineer is responsible for developing and resolving any conflicts between the drawings and specifications. If a conflict occurs between the drawings and specifications, the technical specifications will usually take precedence.

Specifications usually follow a format created by the **Construction Specifications Institute (CSI)** that divides the technical data into divisions. Each division is divided into several sections for materials, products, and installation requirements. The project manager should be familiar with the CSI divisions, sections, subsections, and format. The current format is the MasterFormat™ 2004 Edition. *Figure 2* shows an excerpt from the MasterFormat™ 2004 Edition.

CSI has established the following format for subsections of specifications:

- *Scope* – Describes the scope of work to be performed
- *Material* – Describes the specific materials that are to be used
- *Installation* – Specifies how the material is to be applied
- *Guarantee* – States the quality parameters for which the contractor, subcontractor, and suppliers are responsible

Each division is organized in a consistent manner and contains a number of sections and subsections. For example, Division 03, Concrete, a part of the Facility Services Subgroup, has the following breakdown:

Division	03 00 00	Concrete
Section	03 06 00	Schedules for Concrete
Subsection	03 06 10	Schedules for Concrete Forming and Accessories
Subsection	03 06 20	Schedules for Concrete Reinforcing
Subsection	03 06 20.13	Concrete Beam Reinforcing Schedule
Subsection	03 06 20.16	Concrete Slab Reinforcing Schedule

MasterFormat™ 2004 Edition – Numbers & Titles

DIVISION 00 – PROCUREMENT AND CONTRACTING REQUIREMENTS

00 00 00 PROCUREMENT AND CONTRACTING REQUIREMENTS

INTRODUCTORY INFORMATION
- 00 01 01 Project Title Page
- 00 01 05 Certifications Page
- 00 01 07 Seals Page
- 00 01 10 Table of Contents
- 00 01 15 List of Drawing Sheets
- 00 01 20 List of Schedules

PROCUREMENT REQUIREMENTS

00 10 00 SOLICITATION
- **00 11 00 Advertisements and Invitations**
  - 00 11 13 Advertisement for Bids
  - 00 11 16 Invitation to Bid
  - 00 11 19 Request for Proposal
  - 00 11 53 Request for Qualifications

00 20 00 INSTRUCTIONS FOR PROCUREMENT
- **00 21 00 Instructions**
  - 00 21 13 Instructions to Bidders
  - 00 21 16 Instructions to Proposers
- **00 22 00 Supplementary Instructions**
  - 00 22 13 Supplementary Instructions to Bidders
  - 00 22 16 Supplementary Instructions to Proposers
- **00 23 00 Procurement Definitions**
- **00 24 00 Procurement Scopes**
  - 00 24 13 Scopes of Bids
    - 00 24 13.13 Scopes of Bids (Multiple Contracts)
    - 00 24 13.16 Scopes of Bids (Multiple-Prime Contract)
  - 00 24 16 Scopes of Proposals
    - 00 24 16.13 Scopes of Proposals (Multiple Contracts)
    - 00 24 16.16 Scopes of Proposals (Multiple-Prime Contract)
- **00 25 00 Procurement Meetings**
  - 00 25 13 Pre-Bid Meetings
  - 00 25 16 Pre-Proposal Meetings
- **00 26 00 Procurement Substitution Procedures**

00 30 00 AVAILABLE INFORMATION
- **00 31 00 Available Project Information**
  - 00 31 13 Preliminary Schedules

DIVISION 32 – EXTERIOR IMPROVEMENTS

32 00 00 EXTERIOR IMPROVEMENTS
- **32 01 00 Operation and Maintenance of Exterior Improvements**
  - 32 01 11 Paving Cleaning
    - 32 01 11.51 Rubber and Paint Removal from Paving
    - 32 01 11.52 Rubber Removal from Paving
    - 32 01 11.53 Paint Removal from Paving
  - 32 01 13 Flexible Paving Surface Treatment
    - 32 01 13.61 Slurry Seal (Latex Modified)
    - 32 01 13.62 Asphalt Surface Treatment
  - 32 01 16 Flexible Paving Rehabilitation
    - 32 01 16.71 Cold Milling Asphalt Paving
    - 32 01 16.72 Asphalt Paving Reuse
    - 32 01 16.73 In Place Cold Reused Asphalt Paving
    - 32 01 16.74 In Place Hot Reused Asphalt Paving
    - 32 01 16.75 Heater Scarifying of Asphalt Paving
  - 32 01 17 Flexible Paving Repair
    - 32 01 17.61 Sealing Cracks in Asphalt Paving
    - 32 01 17.62 Stress-Absorbing Membrane Interlayer
  - 32 01 19 Rigid Paving Surface Treatment
    - 32 01 19.61 Sealing of Joints in Rigid Paving
    - 32 01 19.62 Patching of Rigid Paving
  - 32 01 23 Base Course Reconditioning
  - 32 01 26 Rigid Paving Rehabilitation
    - 32 01 26.71 Grooving of Concrete Paving
    - 32 01 26.72 Grinding of Concrete Paving
    - 32 01 26.73 Milling of Concrete Paving
    - 32 01 26.74 Concrete Overlays
    - 32 01 26.75 Concrete Paving Reuse
  - 32 01 29 Rigid Paving Repair
    - 32 01 29.61 Partial Depth Patching of Rigid Paving
    - 32 01 29.62 Concrete Paving Raising
    - 32 01 29.63 Subsealing and Stabilization
  - 32 01 30 Operation and Maintenance of Site Improvements
  - 32 01 80 Operation and Maintenance of Irrigation
  - 32 01 90 Operation and Maintenance of Planting
    - 32 01 90.13 Fertilizing
    - 32 01 90.16 Amending Soils
    - 32 01 90.19 Mowing
    - 32 01 90.23 Pruning
    - 32 01 90.26 Watering
    - 32 01 90.29 Topsoil Preservation
  - 32 01 90.33 Tree and Shrub Preservation
- **32 05 00 Common Work Results for Exterior Improvements**
  - 32 05 13 Soils for Exterior Improvements
  - 32 05 16 Aggregates for Exterior Improvements
  - 32 05 19 Geosynthetics for Exterior Improvements

105F02.EPS

Figure 2 ◆ CSI MasterFormat™ 2004 Edition.

The CSI MasterFormat™ 2004 Edition has been revised to meet today's complex requirements that have evolved since the original MasterFormat™ 1995 Edition. The revised edition has increased the divisions from 16 to 49, and includes a better defined sub-classification system. The added divisions also allow for future expansion. The differences between the revised and current versions include the following:

- There are four subgroups with new topics and new sections under each subgroup.
- The first 16 divisions (00–15) cover general construction subjects, similar to the 1995 edition.
- There are separate divisions for plumbing, fire suppression, electrical, communications, integrated building systems, safety and security systems, and HVAC.
- A site infrastructure group has been added for unique civil construction (dams, roads, bridges, tunnels, utilities, and so forth) and industrial construction (power plants and factories).
- There are placeholders reserved for future information.
- The five-digit format has been replaced with a six-digit numbering system for sections to be added at a later date.

5.0.0 ♦ TYPES OF CONTRACTS

A contract, whether between an owner and general contractor or general contractor and subcontractor, is the legal agreement, either written or oral, between two parties in which one party agrees to perform some task or work for the other party in return of some type of payment or compensation.

For any contract to be considered legal and binding, it must contain the following four elements:

- An offer and acceptance
- An agreement between two or more parties
- A benefit to each party named in the contract
- A legal agreement (illegal agreements are not enforceable)

The project manager should be aware of the legal terms and specific clauses contained in each contract. Since no two contracts are ever the same, project managers should review the following clauses of a contract:

- Name of the parties
- Date of the agreement
- Scope of the work to be performed
- Start and completion dates (the schedule)
- Insurance requirements
- Bonding requirements
- Payment terms and conditions
- Indemnifications (the protection of another party against damage or loss)
- Cancellation or termination notices
- Liquidated damages or bonus/penalty clauses
- Special conditions
- Safety requirements
- Guarantees/warranties
- Change order procedures
- Employment requirements

Contracts should be fair to both parties, and each contract clause should be written so that its intent is clear to the parties. One-sided contracts, where terms are vague and benefit only one of the parties, should be reviewed carefully to determine the risk before signing.

Contracting covers a variety of situations. It is important to use the proper type of contract for specific construction requirements. Typical contracts awarded by an owner to a contractor or by a contractor to a subcontractor are:

- Fixed-price/lump sum
- Cost plus
- Target contracts
- Letter of intent or letter contracts
- Fast-tracking contract
- Subcontracts
- Purchase orders

5.1.0 Fixed-Price/Lump-Sum Contract

The most common type of construction contract is the fixed-price, also known as a lump-sum contract. Under this type of contract, the contractor agrees to perform all the work specified at a price agreed to at the time the contract is signed. As noted in the previous section, the price is usually decided through competitive bidding, invited bidding, or negotiated work.

The central characteristic of the fixed-price contract is that the contractor agrees to do the work at a pre-determined price and assumes the risk of all unforeseen conditions (for example, increased labor cost, low productivity, and adverse weather). The owner agrees to pay the price whether the contractor makes a profit or suffers a loss. The fixed-price contract provides the owner with the cost of the project before work starts.

There are circumstances under which a contractor may be entitled to an adjustment to the fixed-price contract. The contract amount may be increased or decreased by issuing change orders triggered by changes in the design or material used or unforeseen site conditions.

Many fixed-price contracts contain a liquidated damage clause requiring the contractor to pay the owner a stipulated amount, usually per day, if the work is not completed on the date stated in the contract. Other contracts will have a bonus provision, which allows the contractor to receive stated bonus payments for completion of the work before the specified contract completion date.

5.2.0 Cost-Plus Contract

A cost-plus contract is one in which the project cost is decided by the actual costs as they are incurred, plus an agreed fee to cover the contractor's overhead and profits. These contracts eliminate or at least minimize many problems associated with fixed-price contracts. For example, inclement weather, escalation of wages or material prices, or changes in work are usually handled without any controversy or claim procedures, since the cost of performing the work will be measured against the actual cost of the entire project.

Many owners favor a cost-plus contract since it creates incentives for the contractor when costs are reduced through efficiency or obtaining the best price from subcontractors or suppliers. A cost-plus contract requires documentation of all costs. In addition, the cost of reimbursable and non-reimbursable costs must be clearly defined in the contract.

Examples of typical reimbursable and non-reimbursable costs include:

- *Reimbursable cost-plus contract items* – Items such as cost of materials, cost of labor, and salaries related to the job. Other items include taxes for labor and materials, subcontractor payments, equipment costs, permits and fees, bonds, insurance, and safety costs.
- *Non-reimbursable cost-plus contract items* – These include salaries of personnel working at the contractor's principal or branch office, overhead and general expenses, capital expenses, or other costs such as disposal of materials, damage to property and other issues related to negligence.

The following are various types of cost-plus contracts that can be used to determine contractor overheads and fees (profit):

- *Cost-plus a fixed fee* – This contract reimburses contractors for their actual construction costs and their compensation for overhead and fee (profit) that is a fixed amount determined at the time the contract is signed. The fee is frequently determined by a percentage based on the original estimated cost of the project.
- *Cost-plus a percentage of cost fee* – Contractors are reimbursed for their actual construction costs and receive a fee based on an agreed percentage applied against the actual costs. The final fee will be determined when the actual cost has been specified.
- *Cost-plus an incentive fee* – Contractors are reimbursed for their actual construction costs plus compensation based on a special formula for sharing actual costs over or under the estimated costs.
- *Cost-plus with guaranteed maximum fee* – Contractors are reimbursed for their actual construction costs including overhead and profit up to but not exceeding the agreed contract amount. The contract may allow for sharing of any cost savings if the project comes in under the guaranteed maximum.

5.3.0 Target Contract

A variation of the fixed-price and cost-plus type contracts is the target contract. Although there are various types of target contracts, they all have some common elements including the following:

- An agreement between the parties on estimated target costs
- Target profit or fee
- Ceiling price
- Other combinations to establish a fair cost target.

Upon completion of the project, the final cost is decided and the contractor's fee will be adjusted if the target costs were met.

5.4.0 Unit-Price or Unit-Cost Contract

The unit-price contract, also known as the unit-cost contract, fixes the price per unit at the time the contract is signed and that price remains the same throughout the life of the contract. Several examples of unit-cost items are electrical (add ceiling outlet, add wall switch, wire and connect exhaust fans), mechanical (add water heater, add 4-inch floor drain), and finishes (add wall tile, add texture to stucco finish).

Sometimes a problem arises with a unit-price contract when there is a substantial overrun or underrun in quantities. Contractors develop their unit price based on the estimated quantities listed in the bid documents, which should be reasonably close to the final measured quantities. If the quantities vary substantially, then the contractor may be entitled to a price adjustment.

5.5.0 Letter of Intent or Letter Contract

A letter of intent or letter contract is a preliminary contract issued by the owner under which the contractor may start work prior to the execution of a formal contract. Letters of intent or letter contracts outline the liability of both parties including maximum costs, fees, scope of work, payment, and a cancellation provision if a formal contract cannot be executed.

5.6.0 Fast-Tracking Contract

A fast-tracking contract, also known as a design and construction contract, allows construction to proceed in stages or phases by issuing contracts when the design of each phase is sufficiently completed.

5.7.0 Subcontract

A subcontract is a contract between a general contractor, or prime contractor, and another contractor, usually a specialty contractor. The subcontractor agrees to do the labor and material portion of the work that the general contractor has agreed to perform under the same agreement with the owner.

A subcontract can be fixed or cost price, and is subject to the same laws that govern normal contracts.

A subcontract ordinarily does not create any contractual relationship between the subcontractor and the owner. This relationship is important in deciding rights under state and construction lien laws and with respect to payment and performance bonds.

5.8.0 Purchase Order

A purchase order is used to cover the purchase of materials from a supplier and contains provisions appropriated to such transactions that usually are governed by the law of sales, which may differ in various respects from the law of contracts.

6.0.0 ◆ INSURANCE REQUIREMENTS

An extensive amount of insurance of various types is necessary for a company to operate. Policy acquisition and maintenance is needed to protect the employees, owners, public, and the company from losses.

Insurance does not protect the company from all losses, since a deductible is often stipulated in each policy. When an accident or loss occurs, the company's premiums may increase. The project manager should understand the company's insurance requirements, be aware of the company's risk exposure, and help reduce premiums through careful planning and safety awareness.

Insurance is an extremely complicated business and the project manager must be able to recognize any potential problems related to insurance coverage. Generally the home office will handle all the administration necessary for filing a claim but relies on the project manager or field supervisor for notification and written reports. The project manager must realize one of his primary responsibilities is to minimize accidents and losses.

The basic types of insurance a company should carry include: workers' compensation, general liability, vehicle, builder's risk or installation floater, equipment floater, and property insurance.

6.1.0 Workers' Compensation Insurance

Workers' compensation insurance pays employees who are injured on company time. It is usually required by a state statute, and the company is required to maintain coverage in each state in which it works or risk financial penalty. The field supervisor must report all injuries to the project manager. Failure to report injuries may result in the injured employee losing his benefits and the company being penalized.

6.2.0 General Liability Insurance

General liability insurance provides coverage for claims that the company is obligated to pay as a result of bodily injury or property damages. General liability insurance may include the following coverage:

- Premises owned by the company
- Project construction activities
- Liability incurred as a result of injury or damage caused by a subcontractor
- Completed work
- Liability assumed under written contracts
- Libel, slander, false detention, false entry and similar personal injury claims

General liability insurance policies may not cover all possible types of losses. The project manager should be aware of the items covered by a general liability policy, as well as the standard exclusions. If an incident occurs that may involve a standard exclusion, the project manager should notify the home office immediately to either verify the coverage or advise management of a potential problem. Some standard exclusions include the following:

- Property in care, custody, or control of the company
- Losses as a result of design
- Losses from the use of watercraft, aircraft, or autos
- Losses occurring from underground work
- Losses caused by explosions
- Losses caused by collapse

The premium for general liability insurance may be very high. The premium cost is incorporated into the company overhead or project costs. As losses occur, premiums rise, increasing the company's cost of doing business and therefore making the company less competitive.

6.3.0 Vehicle Insurance

Vehicle insurance covers losses the company incurs from ownership and usage of automobiles or trucks and usually includes physical damage to company vehicles due to collision, fire, theft, and vandalism. Many companies also carry insurance on leased or employee-owned vehicles used for company business.

If an accident involving a company-insured vehicle occurs, the project manager should notify the home office. The employee involved in the accident must complete an accident report with information on the cause of the accident, names of witnesses, names of parties involved, photographs, and sheriff or police reports. All pertinent documents should be sent to the home office as soon as possible, so the insurance company can be informed.

6.4.0 Builder's Risk/Installation Floater Insurance

Builder's risk/installation floater insurance provides protection for damage to the work being performed and for losses of materials and labor due to fire, theft, vandalism, or other causes. Unless contractually provided by the owner, all builder's risk insurance is maintained by the contractor in an amount equal to the cost of the project. An installation floater policy is usually written in the amount of the subcontractor's contract or the general contract. The policy is in effect when the contractor gains possession of the materials, during temporary storage on or off site, while in transit, and during installation. Coverage ceases upon acceptance of the work performed.

Builder's risk/installation floater insurance generally provides protection against most losses of materials, labor, fixtures, and other items due to the following causes:

- Fire
- Theft
- Malicious mischief
- Vandalism
- Wind storm
- Flood and earthquake

When a loss occurs on the project, the project manager must notify the home office immediately, so the insurance company can be notified. A report that contains a list of the material lost or damaged, photographs, names of any witnesses, and a copy of the sheriff or police report should also be prepared by the project manager.

6.5.0 Equipment Floater Insurance

Equipment floater insurance provides coverage against loss or damage of company-owned or leased equipment and tools that are not covered under the builder's risk insurance policy. The policy covers only the equipment that is listed on the insurance policy schedule of equipment. It is up to the project manager and field supervisors to advise the home office of any additions of equipment so the policy can be adjusted to reflect the changes. The coverage includes damage or loss to equipment due to fire, theft, or accidents.

6.6.0 Property Insurance

Property insurance provides coverage of all company-owned and -leased property that is not already covered under another insurance policy, including the contents of the main office building and job-site offices.

7.0.0 ◆ PROJECT CORRESPONDENCE

Most projects generate a considerable amount of written correspondence, and project managers occupy the center of the correspondence flow. They determine who shall communicate to whom, proper relationship channels, and secondary parties who must be informed. The project manager is responsible for channeling information to the proper parties, so theycan quickly perform any required action or response. *Figure 3* illustrates the typical roles and authority of key personnel on a project.

Correspondence includes all drafted, written, and printed material relating to the project. To handle this vast amount of material, project managers should develop a filing system and maintain it throughout the life of the project. *Figure 4* shows a suggested filing system format; however, users will generally define their own numbering system. A successful filing system will meet two basic requirements:

- Provide easy recovery of any document at any time during the life of the project
- Provide a method of follow-up to verify that a given correspondence has achieved its intent

Correspondence that is usually prepared by the project manager includes the following:

- Formal letters
- Memorandums
- Speed messages
- Emails
- Directives

When preparing any correspondence, the project manager should ensure that it addresses the subject and reflects the company business philosophy. One poorly executed letter, a hasty email, or a poorly thought-out comment can destroy the goodwill a contractor has been building for months, or even years. When composing a letter or any other correspondence, make every effort to consider the recipient's reaction to the message. Write in a clear, concise manner, giving all essential information that will accomplish the purpose in the fewest possible words, and always remember to be courteous.

Item	Project Manager	Super-intendent	Other
Schedule			
A. Long Term	x	x	
B. Short Term		x	
Decision on Field Methods		x	
Responsible For:			
A. Subcontracts	x		
B. Submittals			x (Engineer)
C. Change Orders	x	x	x
Field Documents			
A. Quantity Reports		x	x
B. Equipment Reports		x	x
Owner/Client Relationship	x		
A/E Relationship	x		
Equipment Repairs & Maintenance		x	x
Purchase of Tools		x	x
Storage of Tools		x	
Equipment Rental	x	x	
Storage of Materials			x
Personnel Decisions	x	x	
Budget Tracking and Analysis	x	x	
Cost Control	x	x	
Review Subcontractors' Status	x	x	
Communication Procedures	x	x	

105F03.EPS

Figure 3 ◆ Project chart of authority.

7.1.0 Daily Reports

Daily reports or logs catalogue activities that occur on the project on a given day. It is a permanent written record of the project recorded by the field supervisor.

The format of a daily report will vary with the type of information each contractor requires from the field supervisor, but the basic intent of all reports is to document, for the project manager, what has taken place within the project on a given day.

The project manager should insist that the field supervisor complete the daily report by the end of each working day. The field supervisor who conscientiously writes a daily report will avoid struggling to recall (with inevitable omissions and errors) activities that occurred days or weeks earlier.

The project manager is responsible for regularly reviewing the daily reports. *Figure 5* illustrates a typical daily report and information that may be required.

7.2.0 Meetings

There are two types of formal meetings that a project manager may conduct or attend. One is the supervisor's meeting, usually held weekly, and the other is the owner or architect/engineer meeting, usually held monthly.

Each type of meeting should be well planned. The agenda will advise the attendees of the structure and objective of the meeting. *Figure 6* illustrates a typical meeting agenda.

During the meeting, minutes should be taken by at least one person, with this duty rotating to a different staff member each time. When minutes are officially recorded and distributed, they must be reviewed immediately. The accuracy of the minutes is extremely important because they become part of the project documentation. Errors in the minutes should be corrected to ensure they represent an accurate record of the meeting. *Figure 7* illustrates typical project minutes.

7.3.0 Project Photographs

Photographs are one of the best visual forms of documentation. Photographs should be taken for the following reasons:

- Show daily progress
- Verify existing conditions
- Illustrate quality of work
- Record materials and equipment stored on site
- Record construction techniques
- Communicate problems to the architect/engineer or contractor's home office

Since photographs document the project, the following information should be recorded on the reverse side of each photograph:

- Job name and job number
- Date the photograph was taken
- Name of the person taking the photograph
- Subject or purpose of the photograph
- Any other comments describing the photograph

Each job site should have a still camera, digital camera, or video camera, and it should be used on a regular basis. The subjects and number of photographs that the project managers will require field personnel to take vary from project to project.

Pre-Bid Files:

1.A. Project Solicitation
1.B. Original Specifications
1.B.1. Addenda
1.B.2. Current Amended Specifications
1.C. Original Drawings
1.C.1. Revision Drawings
1.D. Site Visit Report
1.D.1. Step-bid Site Photographs
1.D.2. Owner-furnishes Site Reports
1.D.3. Contractor-procured Site Reports
1.H. Transmittals
1.E. Owner's Finances
1.E.1. Credit Report
1.E.2. References
1.F. Subcontractor Quotations
1.G. Unit Price Calculations
1.I. Bid Sheets
1.J. Submitted Document

Pre-Construction Files:

2.A. Meeting Minutes
2.B. Modifications/Proposals
2.C. Notice of Award
2.D. Permits
2.E. Signed Contracts
2.E.1. Contractor/Subcontractor
2.E.1. Purchase Orders
2.F. Bonds (Performance and Payments)
2.F.1. Contractor/Owner
2.F.2. Subcontractor/Contractor
2.G. Insurance Certificates
2.H. Schedules
2.H.1. Contractor
2.H.2. Subcontractor
2.H.3. Suppliers
2.H.4. Schedule Update

Construction Files:

3.A. Notice to Proceed
3.B. Daily Superintendent Report
3.C. Disruption/Interference Reports
3.D. Change Order Files
3.E. Claim Files
3.F. Shop Drawing Log
3.G. Submittal Log
3.H. Transmittals
3.I. Revised Drawings
3.J. Progress Photographs
3.K. Schedule
3.K.1. Original Schedule
3.K.2. Revisions/Updates
3.K.3. Correspondence
3.L. Job Meeting Minutes
3.M. Inspection Reports
3.N. Payment Requisitions
3.P. Correspondence
3.P.1. Owner
3.P.2. Construction Manager
3.P.3. Architect
3.P.4. Subcontractors
3.P.5. Suppliers
3.Q. Intraoffice Memoranda
3.R. Accident Report
3.S. Insurance Claims

Close-Out Files

4.T.1 Lien Waivers
4.T.2. Turn over to Customer
4.T.3. Punch List
4.T.4. Architect's Approval
4.T.5. Warranties and Guarantees
4.T.6. Final Releases
4.T.7. Final Requisition

105F04.EPS

Figure 4 ◆ Correspondence filing system.

Since the advent of more user-friendly video cameras, many contractors now use these devices to record and document projects. They can provide more information than a photograph because they can record actions and verbal comments.

The project manager must maintain a master log listing each photograph or video by sequence. *Figure 8* illustrates a typical job-site photograph.

7.4.0 Schedule

The project schedule is an estimate of the time and the sequence of activities necessary to get a project completed on time. After the schedule has been approved and issued, project managers should review and update the schedule on a regular basis. All activities that affect the schedule, whether they are activities controlled by the contractor or not, should be documented on the schedule. The role

DAILY PROJECT REPORT

PROJECT		**DATE**	March 3
REPORT BY	John Doe	**REPORT NO.**	043
WEATHER AVG TEMP:	AM:50 PM:80	**Sunny, Cloudy, Rainy:**	sunny

WORKFORCE:

Plumbers	6 incl. 1 Foreman	Sheet Metal	4	Cooling Tower	
Fitters	7 incl. 1 Foreman	Insulation	1	Refrigeration	
Laborers		Temperature Controls	1	Other	
Operators		Painter			
Others					

JOB PROGRESS REPORT:
3 Plumbers working basement pump room
2 Plumbers roughing in 3rd floor
3 Pipefitters in main equipment room installing chilled and condenser water lines
2 Fitters working on HVAC risers – 4th floor
1 Fitter fabricating coil connections
4 chilled and condenser water pumps delivered today

SPECIAL DECISIONS OR INSTRUCTIONS:
General Contractor requested that we pull off 3rd floor so drywall contractor can stock floor with sheetrock. (This will affect sheet metal sub and will cause us to shift a crew of plumbers to another work area.)

ITEMS NEEDING EXPEDITING:
Air handling units were scheduled to ship February 15, but I've heard no word. Please check with supplier and expedite. Need badly or will have to cut back on pipefitter crew and will be covered up by sheetrock.

105F05.EPS

Figure 5 ◆ Sample daily project report.

of project managers in developing, maintaining, and resolving scheduling problems is discussed in the *Scheduling* module.

7.5.0 Time Cards

Time cards are important documents because they record labor and material costs as they are expended. Time cards can verify additional costs resulting from change orders or directives from the architect/engineer or owner. Time cards should record hours worked, areas or activities worked, work accomplished, and material used. Since time cards track the cost of work, project managers should verify that the field supervisors are completing the time cards properly.

7.6.0 Shop Drawings

Contract drawings and specifications provide sufficient information to estimate a project but may lack the level of detail needed to produce all the various components of a project. The information necessary to produce or provide components is contained in **shop drawings**. Shop drawings encompass drawings, schedules, lists, performance data, manufacturers' literature, samples, vendors' drawings, and submittals.

The architect/engineer normally requires that shop drawings be submitted before material or equipment is delivered to the project, and that the material or equipment meets the requirements contained in the contract documents. Consequently, shop drawings are provided for almost

XYZ Company, Inc.
Project: Mercy Hospital

Meeting Agenda
June 12

1.	Review last meeting minutes.	9:00 a.m.
2.	Review status of action items.	9:15 a.m.
3.	Discuss job safety requirements.	9:30 a.m.
4.	Update schedule, comparing actual to planned.	10:00 a.m.
5.	Develop a plan to improve schedule, recover lost time, improve productivity.	10:30 a.m.
6.	Review field observation on quality and work standard.	10:45 a.m.
7.	Review outstanding change orders.	11:00 a.m.
8.	Review submittal schedule status.	11:15 a.m.
9.	Develop goals and assign action items.	11:30 a.m.
10.	Discuss potential problems, delays, conflicts that will affect the progress of the project.	11:45 a.m.
11.	Other business.	12:15 a.m.

105F06.EPS

Figure 6 ♦ Sample meeting agenda.

every component of a project. The general contractor's shop drawings may range from reinforcing steel and millwork to finished hardware. The mechanical contractor's shop drawings may include ductwork, piping, and equipment data. The electrical contractor's shop drawings may range from electric panels and power distribution to a single light fixture.

Although the procedure for processing shop drawings may vary from contractor to contractor, there are basic steps that should be followed:

- Log in shop drawings when received.
- Verify quantities of required copies and samples.
- Review drawings and submittals for accuracy.
- Submit shop drawings to the architect/engineer.
- Review drawings when returned from the architect/engineer for notes and comments.

If shop drawings are returned approved or approved as noted, the contractor returns them to the supplier or subcontractor and distributes the information to the supervisor and other members of the project team. If the shop drawings are rejected by the architect/engineer, the contractor returns the information to the supplier or subcontractor for correction and resubmission. Rejected drawings should not be distributed to the field to avoid using inaccurate information in construction.

All shop drawings received from a supplier or subcontractor must be carefully checked against the contract drawings and specifications before being forwarded to the architect/engineer for review and approval. The architect/engineer is responsible for checking the manufacturer, style, and quality of the material or equipment to make sure that they meet project requirements, then approving or rejecting the shop drawings and returning the information to the contractor.

The contractor should note any comments or changes noted by the architect/engineer. Remember, the architect/engineer can give the only qualified approval; the contractor is ultimately responsible for verifying quantities and dimensions.

Shop drawings are important because of the information they provide. The responsibility for verifying that information should not be taken lightly. Usually the project manager or project engineer is responsible for checking shop drawings, but the superintendent should be involved when dimensions must be verified.

Over the years, the responsibility for checking shop drawings has become a major area of disagreement. Generally, contract documents stipulate that the approval of shop drawings by the architect/engineer does not relieve the contractor of responsibility for furnishing materials and equipment in accordance with the contract documents. Nor does such approval relieve the contractor from responsibility for comments not noted on the shop drawings. As a result, maintaining records of transmittal, receipt, and approval of shop drawings is critical.

Since it is the project manager's responsibility to keep track of all transactions related to shop drawings, an organized system should be developed. The shop drawing log in *Figure 9* is an example of such a system.

7.7.0 Back Charges

During the course of a project, the project manager may be requested to perform work that is outside the contractor's normal scope of work such as furnishing labor, material, or equipment to assist another trade, or to repair work damaged by another contractor. These situations may lead to back charges, companies billing each other for extra costs incurred. The project manager should recognize that back charges are a form of change order and, as such, increase or decrease the company's costs. Before a project manager can approve a back charge, the applicable documents should be checked to verify that the back charge is justifiable.

XYZ Company Minutes
May 27

Subject: Subcontractors' Meeting
Time: 9:00 A.M.

Attendees:

John Reid	Reid's Mechanical Maintenance
Jane Smith	J and R's Formwork Company
Larry Jones	Zap! Electric Company
Richard Graham	Bare Walls Drywall & Painting
Jake Robinson	Grounded Flooring, Inc.
Dan Stone	Builder's Best Company, Inc.
Walt Meyers	Walter and Sons Glass Installation

1. Safety: All subcontractors were reminded that this project is a "HARD HAT" JOB. Hard hats must be worn at all times.
2. Schedule: The total project is nine days ahead of schedule. All subcontractors must maintain their present manpower and productivity in order to continue ahead of schedule.
3. General Notes:
   a. Next window delivery is scheduled for June 2.
   b. Metal studs installation on 5th floor will be completed by June 5.
   c. Domestic water rough-in on 5th floor will be completed on June 9.
   d. Electric rough-in on 5th floor will start June 9.
   e. Flooring for top floors will start next week.
4. Change Order: The owner is revising the first floor layout. All work on the first floor is on "HOLD" until new drawings are released by the architect.
5. Clean-up: Each subcontractor must clean up its own trash. A dumpster is provided at the south end of the property for all contractors to use.

Meeting Adjourned at 9:32 A.M.
The meeting minutes are transcribed and will become part of the project record. If you disagree with the minutes, contact this office within seven days with your corrections or comments.

Minutes recorded and transcribed by:

____

105F07.EPS

Figure 7 ◆ Sample project meeting minutes.

The following list gives examples of back charges that can occur on construction projects:

- A contractor furnishes scaffolding to another contractor
- A contractor provides saw cutting and patching for another contractor
- A contractor furnishes a pump to another contractor to remove water from a trench
- A contractor provides labor and equipment to remove another contractor's trash
- A supplier furnishes materials to another contractor
- A contractor reworks material furnished by the supplier

8.0.0 ◆ CHANGE ORDERS

Few items in a project manager's daily schedule cause more work and anxiety than change orders. Change orders reflect alterations in the planning, design, or execution of a job. As authorized by the owner or architect/engineer, change orders increase or decrease the cost of the project, resulting in an increase in administrative time and cost. Dealing with change orders is unavoidable because construction projects are bound to encounter changes to the original scope of work.

The bid documents normally include contract clauses that permit the owner and architect/engineer the authority to request that the contractor carry out changes to the work. However, these

105F08.EPS

Figure 8 ◆ Sample project job-site photograph.

changes must be within the intent of the contract; a contractor cannot be required to supply craftsmen, equipment, and material not originally contemplated. When the owner requests a change, the contractor must determine if it is within the original intent of the work. If the change is outside the original scope, then the contractor should submit a request for the additional work.

8.1.0 Reasons for Creating a Change Order

During the course of the project, items may be identified by the architect/engineer or contractor that may become the subjects of change orders. Change orders may be a result of many factors to include the following:

- Document errors or omissions
- Scope of work changes requested by owner
- Adjustments to the project schedule
- Building code or regulation changes
- Changes in materials specified for the project

8.2.0 The Change Order Process

The change order process should begin with the pre-construction phase of a project. By reviewing the change order process early in the project, the manager can save time during construction and avoid disputes. The change order procedure should address how the contractor, owner, and architect/engineer will review the change order for scope of work and cost. Necessary change orders should be approved as quickly as possible since contractors are not authorized to proceed with work until the change order has been approved. *Figure 10* illustrates a typical change order.

Change order procedures should include the following information:

- The scope of work
- Justification for the change orders
- The amount requested by the contractor
- The start and completion dates of the change
- The effect the change has on the entire project

8.3.0 Change Order Review Meetings

Regular change order meetings should be scheduled to review and identify the changes to the work. Part of the meeting should involve the contractor reporting on the progress of pending, existing, and future change orders. This provides the owner with an early notification of possible additional costs and an assessment of the final construction cost.

8.4.0 Pricing of Change Orders

The cost of a change will vary depending on the phase of construction. During the beginning of a project, when only a few trades are involved, the relative cost may be lower than later in the con-

	Job Name	Warehouse Bldg.		Job No.	86006						Add'l Dwgs						
Spec Station	Contractor	No	Drawing Number or Item	Sent Dt.	Ret'd Dt.	Arct.	Sub	Files	Job	Other	Remarks	Need	File	Job	Mach'l	Elec.	Mis. Iron
2660	Utilities, Inc.	1	Water Distribution Pipe	1/4/xx	1/18/xx		xx		x		Approved						
3100	Salt	2	Construction JT Loc.	1/7/xx	1/22/xx			xx			Approved						
3200	AB Steel	3	Footing & Column Steel	1/7/xx	1/20/xx		xx		x		Approved						
3300	Concrete, Inc.	4	Concrete Mix	1/12/xx	1/20/xx		x				Not Approved						
3345	L&B Chem.	5	Curing Compound	1/13/xx	2/2/xx		xx		x		Approved						
3300	Concrete, Inc.	6	Concrete Mix-Resub.	2/1/xx	2/15/xx		xx		x		Approved						
3200	AB Steel	7	2nd Floor Steel	2/6/xx	2/16/xx		xx		x		Approved						
1210	Up/Down Elv.	8	Hydraulic	2/14/xx	3/1/xx		xx		x		Approved						
15250	Ted's Insul.	9	Pipe Insulation	2/20/xx	3/1/xx		x				Not Approved						
9340	Townsend	10	Paver Tile	2/26/xx	3/14/xx		xx		xx		Approved						

105F09.EPS

Figure 9 ◆ Sample shop drawing log.

CHANGE ORDER

PROJECT: (name, address)	Hospital USA Anywhere, USA 12345	CHANGE ORDER NUMBER: 003 INITIATION DATE: Oct. 5, 20xx
TO:	XYZ Construction PO Box 1620 Perth, NJ 04321	ARCHITECT'S PROJECT NO:_____ CONTRACT FOR: Renovation Work
		Contract Date: April 1, 20xx

You are directed to make the following changes to this contract:

Authorized extra work approved by owner per Change Order no. 24.

1. Install acoustical ceiling in utility room	$503.80
2. Install cabinet in Room 104	$3751.00
3. Install access panel Room 104	$497.00
4. Install lock & dead bolt in pair doors at Admin office	$215.00
5. Additional partition work at stair #3	$183.00
6. Revise medical records wall per C.O.R. 112	$2148.00
Total:	$7297.00

Not valid until signed by both the Owner and Architect. Signature of the contractor indicates agreement herewith, including any adjustments in the Contract Sum or Contract Time.

The original (Contract Sum) (Guaranteed Maximum Cost) was	$214,800.00
Net change by previously authorized Change Orders	$17,888.20
The (Contract Sum) (Guaranteed Maximum Cost) prior to this Change order was	$232,688.20
The (Contract Sum) (Guaranteed Maximum Cost) will be (increased) (decreased) (unchanged) by this Change Order	$7297.80
The new (Contract Sum) (Guaranteed Maximum Cost) including the Change Order will be	$239,986.00
The Contract Time will be (increased) (decreased) (unchanged) by	( ) Days

The Date of Substantial completion as of the date of this Change Order therefore is __________

Design Firm, Inc. Architect	Ajax Construction, Inc. Contractor	Hospital USA Owner
100 Design Avenue Orlando, FL 32801 Address	PO Box 1620 Perth, NJ 04331 Address	High Street Anywhere, USA 12345 Address
By __________	By __________	By __________
Date __________	Date __________	Date __________

105F10.EPS

Figure 10 ◆ Change order.

struction process, when many more trades are employed on the project.

Contract language normally states that the contractor shall not proceed with the change without written approval from the owner or architect/engineer. Some changes are discovered just prior to required implementation; therefore, timing often becomes a problem. The pricing of the change can be a lengthy process because quotations are required from suppliers and/or subcontractors.

In addition to the costs directly related to the changes, it is important to ensure that the price quote includes all the costs related to any delays, cost impacts from loss of productivity, additional temporary facilities, or additional supervision. If these costs are not included in the cost of the change order, the contractor may have to submit a claim at the completion of the project to recover the costs. A general conditions checklist can be used to identify items that may be included when preparing a change order (*Figure 11*).

The following steps should be used while reviewing a change order.

Step 1 Define the scope of work.

Step 2 Determine the quantity of materials.

Step 3 Determine acceptable unit cost for materials, labor, equipment, and any necessary subcontractors.

Step 4 Conduct an analysis of impact cost for the field/home office, idle time, sequence, or other costs.

Step 5 Determine the amount of allowable overhead.

Step 6 Determine the amount of allowable markup/fee.

Step 7 Determine bonding requirements.

8.5.0 Change Orders and Claims

Change orders often become the subject of claims and disputes. Therefore, information regarding change orders must be documented carefully. The project manager should treat each change order as though it might eventually become part of a future dispute or claim. The manager must keep written records because they may become the foundation for pursuing or defending a claim.

The project may be subject to an audit. If a dispute arises, the documents will be subject to review. In either case, well-organized, thorough, accurate records and documentation are essential. Records and documents include the following:

- Minutes of meetings
- Discussion notes
- Change order procedures
- Weather conditions
- Workforce, material, and equipment status
- Schedule updates and impacts
- Revised drawings, specifications, and sketches
- Photographs

Change order oversight is a requirement of project management. By implementing sound change order principles, the project manager can complete the change order more effectively within his or her budget, schedule, and scope of work. *Figure 12* illustrates the items that may be included in a construction claim.

9.0.0 ◆ CONTRACTOR PAYMENTS

Along with determining the right contract to meet project requirements, the project manager is usually responsible for preparing and submitting payment requests to the contractor or owner on a monthly basis. The payment documentation includes:

- Request for payment
- Waivers and affidavits
- Certificate for payment
- Final payment request

9.1.0 Request for Payment

The contractor submits a request for payment based on completed work, material, and equipment purchased up to a specific date. A date of payment is set to allow for review, approval, certification, and payment processing.

Contractors should prepare a schedule of value for their work and the work of subcontractors to facilitate progress payments before the first request for payment. The schedule of value should reflect the value of the work and should be submitted to the owner or architect/engineer for review and approval. The request for payment format can vary from a simple letter to detailed forms.

The owner and architect/engineer should review the request for payment with the contractor, including verification of the percentage of completion for each item listed in the request. In many cases, the percentage of completion is a subjective judgment and may be negotiated to a percentage acceptable to the parties involved. If the amount requested corresponds to the actual percentage of completion, and the contractor's performance has

GENERAL CONDITIONS CHECKLIST

Project ______________________________ No.___________

Change Estimate No. ________________ Owner Bulletin No._________

	Material	*Labor*	*Total*
1. Supervision	______	______	______
A. Project manager	______	______	______
B. Superintendent	______	______	______
C. Project and office engineer	______	______	______
D. Field engineer	______	______	______
E. Additional foremen	______	______	______
F. Accountant/timekeeping/material check	______	______	______
G. Home office supervision	______	______	______
H. ______________	______	______	______
2. Temporary Facilities	______	______	______
A. Field office	______	______	______
B. Material trailers/sheds	______	______	______
C. Temporary toilets	______	______	______
D. Temporary roads	______	______	______
E. Safety protection/equipment	______	______	______
F. ______________	______	______	______
3. Field Support	______	______	______
A. Office/first-aid supplies	______	______	______
B. Blueprinting/copying/photos	______	______	
C. Telephone	______	______	______
D. Fire/theft alarm	______	______	______
E. Insurance	______	______	______
F. Home office expense	______	______	______
4. Temporary Utilities	______	______	______
A. Heat	______	______	______
B. Light and power	______	______	______
C. Water	______	______	______
D. Elevators/lifting/moving	______	______	______
E. Tests/inspections	______	______	______
F. ______________	______	______	______
5. Construction Equipment	______	______	______
A. Small tools	______	______	______
B. Trash removal/light trucking	______	______	______
6. Special conditions	______	______	______
A. Winter conditions	______	______	______
B. Snow removal	______	______	______
C. Cutting and patching	______	______	______
D. Final cleanup	______	______	______
E. ______________	______	______	______

Total Change Order General Conditions $__________

105F11.EPS

Figure 11 ◆ General conditions checklist.

been satisfactory, the request for payment will be approved and forwarded to the person or lending institution (if one is involved) responsible for construction finances.

9.2.0 Waivers and Affidavits

Lien laws are designed to protect workers, materials suppliers, and, under certain conditions, general contractors and subcontractors from nonpayment by the owner or general contractor. The underlying legal principal is that of unjust enrichment. In this case, the owner or general contractor would unfairly benefit at the expense of the party that performed the work without compensation.

With each request for payment, the contractor may have to submit to the owner a waiver for the total amount received to date along with an affidavit that all suppliers and subcontractors have been paid to date. Contractors may have to submit waivers from their suppliers and subcontractors verifying the amounts have been received.

When work is completed, all final waivers must accompany the contractor's request for full and final payment. All waivers and affidavits should be signed, sealed, and notarized. They should also be examined for accuracy and compliance with the terms of the contract before the payment request is approved.

In many states contractors can file a mechanic's or construction lien, which is a legal process to secure payment for work performed and materials furnished in the improvement of land. This statutory right attaches to the land itself in much the same way as a mortgage. The purpose of a lien statute is to permit a claim on the premises where the value or condition of real property has been increased or improved and where suitable payment has not been made by the owner. Lien laws are strictly construed by the courts, and full compliance with all provisions of the local statute is mandatory.

9.3.0 Certificate for Payment

As part of a request for payment, the architect/engineer or owner completes a certificate for payment, authorizing the lending institution to make payment. The information on the certificate should match the information on the request for payment, the waivers, and affidavits.

9.4.0 Final Payment Request

Construction contracts usually allow the owner to withhold a percentage of each request for payment amount (known as retainage) to protect it against unsettled claims and uncompleted work. Retainage will usually range between 5 and 10 percent. The contractor will eventually submit a final request for payment for the total balance due, including retention, along with final waivers and affidavits from all its subcontractors and suppliers.

A. Increased cost due to:
- Labor rates escalation
- Acceleration
- Manpower
- Field overhead
- Office overhead
- Inefficiency
- Overtime

B. Cost due to the parties failure to:
- Resolve conflicts between Regulatory Agencies
- Coordination
- Meet milestone dates
- Obtain permits
- Commence on schedule
- Perform work
- Provide access
- Pay promptly

C. Cost related to changes in:
- Original scope of work
- Changed conditions
- Quality or details
- Dimension
- Nature of work
- Work schedule
- Starting or ending dates
- Contract documents

D. Cost related to third party:
- Interest rate increase
- Finance charges
- Errors & omissions
- Breach of contract
- Code of conflicts
- Bidding errors

105F12.EPS

Figure 12 ◆ Types of construction claims.

10.0.0 ◆ CLOSE-OUT

Project close-out begins at substantial completion, the point in time where the building is ready for occupancy. The purpose of project close-out is to ensure that the owner is receiving a finished product that conforms to the contract documents. The project manager should provide the documentation that provides this assurance.

10.1.0 Substantial Completion

According to the American Institute of Architects, as shown in *AIA Document A201*, substantial completion is the stage in the progress of the project when the project or a designated portion thereof is sufficiently completed in accordance with the contract documents so that the owner can occupy or use the structure for its intended purpose.

10.2.0 Close-Out Procedures

The project manager should formally organize project close-out no later than 30 days prior to substantial completion. This involves sending all subcontractors and material suppliers a copy of the project's close-out requirements. An example of a contractor's project close-out requirements is illustrated in *Figure 13*.

As all mechanical, electrical, and other systems are completed, the contractor may be required to test their operation. If a representative of the manufacturer conducts the test for proprietary systems, the project manager and the owner's representative should be present to observe. Test records should be submitted to the proper authorities, with copies going to the project manager, the architect/engineer, and the owner's representative.

Tests are time-consuming and should be scheduled in advance so that all involved parties can attend. The owner's representative should have every opportunity to ask questions about the process. Testing is not complete until it demonstrates that the system functions properly. Testing procedures are often videotaped to document the results.

10.3.0 Project Inspection

The project manager should arrange for all parties, particularly the owner, the architect/engineer, and prime contractors, to meet and conduct an inspection tour of the project well before the date of substantial completion. Written notification of the inspection tour date should be sent to all involved parties well in advance. At the conclusion of the inspection tour, the project manager should have each party sign the certificate to verify the acceptance of the work. In preparation for the inspection tour, project managers should conduct inspections about two weeks before substantial completion.

10.4.0 Punch List

As the work is completed, it should be inspected and a list of deficiencies (known as a **punch list**) is prepared. Generally, the architect/engineer conducts the inspection, but it is not unusual for the project manager and the owner to undertake their own inspections. A punch list is given to the contractors for completion prior to a stated date. A re-inspection is conducted to ensure the work has been completed. There are no established guidelines for punch lists, but they should be clear and specific.

10.5.0 Owner Acceptance

As soon as the deficiencies on the final punch list have been corrected, the project manager notifies the owner that the project is ready for acceptance. At this time, applications for required permits or certificates of occupancy should be processed. The architect/engineer prepares the certificate of substantial completion for the owner and attaches a list of any remaining deficiencies, making the substantial completion conditional on the correction of those deficiencies.

10.6.0 Certificate of Occupancy

Even if the project is accepted by the owner and the architect/engineer, the project must be certified by the local building department and must meet all health and safety requirements. Once those requirements are met, the owner may occupy the building.

10.7.0 Record Drawings

Record drawings, formerly called as-built drawings, are developed to indicate changes to the project as it was being built. When changes are made, they must be recorded on the drawings.

Normally, the architect/engineer will furnish a set of drawings (either prints or sepias) for recording any changes. These drawings must be clearly identified, kept at the project, and kept separate from other project drawings. The general contractor is usually responsible for maintaining a complete set of record drawings and verifying that all subcontractors are updating their record drawings.

RE: (Project)

SUBJECT: Project Close-out Procedures

In order to satisfy the project close-out requirements and expedite the owner's release of retainage, the following items (marked X) must be submitted to this contractor.

1. _______ Warranty and Guarantee (form enclosed)
2. _______ Specific manufacturer's guarantee
3. _______ Operating and maintenance manuals
4. _______ Maintenance agreement (per specifications requirements)
5. _______ Letter certifying instruction of owner's operating/maintenance
6. _______ Inspection and field testing reports
7. _______ Record or As-built drawings (per specifications requirements)
8. _______ Certification letter stating that your firm has complete your punch list items
9. _______ Consent of surety to final contract amount and final payment
10. _______ Final Release of Lien from your firm and all other material suppliers, vendors, or sub-subcontractors who have provided Notices to Owner
11. _______ Final Application for Payment
12. _______ Furnish spares per specs
13. _______ Other:___________________________________

A timely completion of the above marked items will require submission to this office within 30 days from the date of this letter. If for any reason any of the items marked above cannot be submitted within the requested time frame, please contact this writer immediately.

Sincerely,

John Doe
Project Manager

105F13.EPS

Figure 13 ◆ Sample contractor close-out procedures.

At project completion, the record drawings are submitted to the architect/engineer as part of the close-out documents. Often the architect/engineer or the owner will withhold final payment until all record drawings have been submitted. Therefore, the project manager must ensure that record drawings are maintained and submitted in a timely manner.

10.8.0 Operation and Maintenance Manuals

Operation and maintenance manuals must be furnished to the owner upon completion of the project. Manuals should contain all information necessary for operating, maintaining, repairing, dismantling, and assembling all equipment provided under the project contract.

10.9.0 Warranties and Guarantees

Warranties and guarantees are furnished by contractors and suppliers to guarantee workmanship and material integrity for a specified length of time. As described in the project specifications, most guarantees and warranties must be in writing and on authorized forms.

Warranties and guarantees usually go into effect at substantial completion. The project manager should advise the owner of the maintenance and security responsibilities, the cost of utilities, and services related to the equipment.

10.10.0 Consent of Surety and Final Release of Liens

Two of the most important close-out documents are the consent of surety and release of liens papers. The consent of surety releases the bonding company from further liability for the project, while the final release of liens releases the owner from all claims by the general contractors or subcontractors.

Review Questions

1. The most common way to obtain work in the construction industry is through _____.
   a. referral
   b. purchase order
   c. competitive bidding
   d. teaming arrangements

2. When the owner pre-qualifies a contractor, this is known as a(n) _____ bid.
   a. procured
   b. sole source
   c. invited
   d. negotiated

3. When an owner selects several contractors to interview for the work, this is a(n) _____ bid.
   a. procured
   b. sole source
   c. invited
   d. negotiated

4. The documents that contain the information that the bidder must comply with in bid preparation is known as a(n) _____.
   a. invitation to bid
   b. instruction to bidders
   c. general conditions document
   d. specifications sheet

5. The legal document posted by the bidder as an assurance that the apparent low bidder will enter into a contract with the owner if awarded the project is known as _____.
   a. material bond
   b. bid bond
   c. payment bond
   d. Flemish bond

6. _____ supplement the working drawings with detailed technical information not contained on the drawings.
   a. Purchase orders
   b. Change orders
   c. Specifications
   d. Daily reports

7. The revised CSI MasterFormat™ 2004 has expanded from 16 Divisions to _____.
   a. 20
   b. 29
   c. 34
   d. 49

8. The most common type of construction contract is _____.
   a. lump sum/fixed price
   b. cost-plus
   c. target
   d. unit price

Review Questions

9. Items such as labor, materials, salaries, equipment, fees, and bonds are reimbursable in a _____ contract.
   a. lump sum
   b. unit-price
   c. back charges
   d. cost-plus

10. The contractor must maintain _____ insurance to pay employees who are injured on company time.
    a. risk
    b. general liability
    c. workers' compensation
    d. vehicle

11. The type of insurance that provides coverage for claims that the company is obligated to pay as a result of bodily injury or property damages is known as _____.
    a. worker's compensation
    b. general liability
    c. builder's risk
    d. floater

12. Which item is *not* covered by general liability insurance?
    a. completed work
    b. injury or damages caused by subcontractors
    c. loss caused by explosion
    d. project construction activity

13. _____ are important documents because they identify labor and material costs.
    a. Purchase orders
    b. Job logs
    c. Change orders
    d. Time cards

14. The change order management process begins during the _____ phase.
    a. pre-bid
    b. pre-construction
    c. scope
    d. coordination

15. The purpose of record drawings is to _____.
    a. provide the owner with the original drawings
    b. provide documentation of changes during construction
    c. provide the architect with updated drawings
    d. provide stakeholders with copies

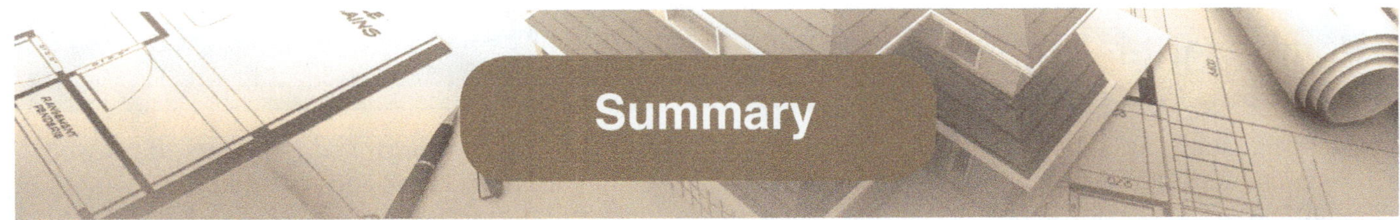

Summary

In order to be useful, good documentation must meet the following criteria:

Accuracy – The facts should be recorded the way the facts actually occurred. If the facts are misstated in one place, it may be assumed that facts have been misstated throughout the documentation.

Objectivity – Objectivity means that the facts should be stated in an unbiased way. Sometimes it is difficult for the project manager to be unbiased because they may be directly involved in the situation being documented. Facts should be stated the way they occurred.

Completeness – If the company provides the project manager or the field supervisors with specific forms for documentation, they should be filled out completely. Incomplete forms demonstrate inaccuracy or lack of knowledge of the situation.

Uniformity – Information should be recorded in a precise, defined and routine manner.

Credibility – Documents should be completed, recorded timely and without bias. The project manager should be certain that all documentation is accurate and carefully reported

Timeliness – All occurrences should be documented immediately and thoroughly.

For a successful project close-out, the project manager must plan and begin close-out activities well ahead of the date of substantial completion. He or she should develop control documents that will allow for the tracking of close-out documents and inspections. By following proper close-out procedures, the project manager reduces overhead expenses, expedites the processing of final payments, and earns the reputation of being able to finalize a project.

Notes

Trade Terms Quiz

1. The written requirements for materials, equipment, construction systems, standards and workmanship are known as ______________.
2. The work done to deliver a product with specified features and functions is known as the ______________ of the project.
3. An itemized list documenting incomplete or unsatisfactory items after the contractor has notified the owner that the tenant space is substantially complete is known as the ________.
4. ______________ are the documents used to modify the contract prior to receiving the bid.
5. A ___________ is a guarantee that the contractor will enter into a contract, if it is awarded.
6. The ______________ is an organization that maintains and advances the standardization of construction language as it pertains to building specifications.
7. Detailed drawings showing how building elements will be fabricated are called __________.
8. A ______________ is a document that modifies the original contract due to changes incurred after contract award.

Trade Terms List

Addenda
Bid bond
Change order
Construction Specifications Institute (CSI)
Punch list
Scope
Specifications
Shop drawings

Trade Terms Introduced in This Module

Addenda: Documents used to modify the contract prior to receiving the bid.

Bid bond: A guarantee that the contractor will enter into a contract, if it is awarded.

Change order: A document that modifies the original contract documents due to changes that occur after the award of the contract.

Construction Specifications Institute (CSI): An organization that maintains and advances the standardization of construction language as pertains to building specifications.

Punch list: An itemized list documenting incomplete or unsatisfactory items after the contractor has notified the owner that the tenant space is substantially complete.

Scope: The work that must be done to deliver a product with the specified features and functions.

Specifications: A part of the construction documents contained in the project manual consisting of written requirements for materials, equipment, construction systems, standards and workmanship.

Shop drawings: Detailed drawings showing how building elements will be fabricated.

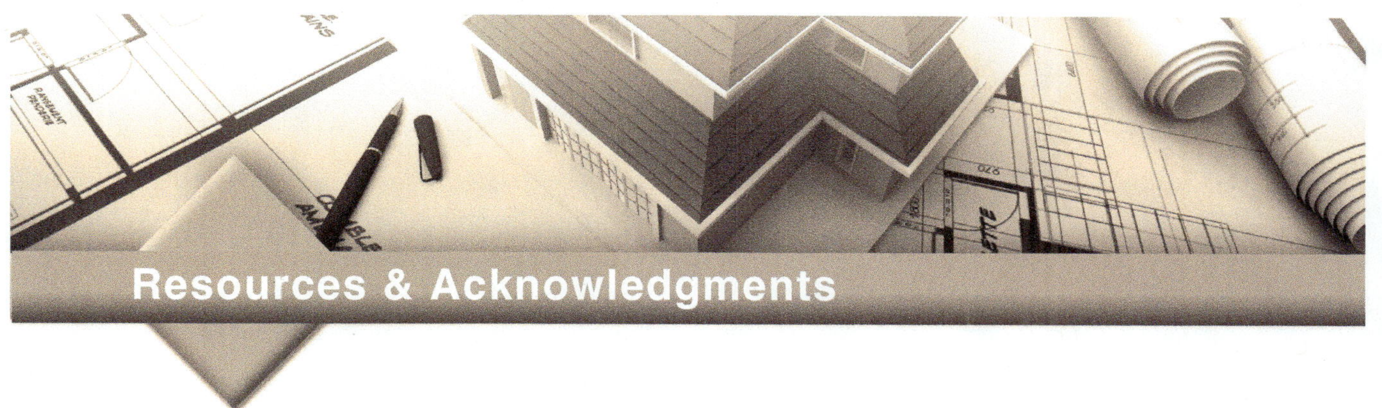

Resources & Acknowledgments

References

PMI Standards Committee, *A Guide to the Project Management Body of Knowledge*. PMI Publications, Newton Square, Pa. (2004).

The American Institute of Architects, 1735 New York Ave., NW, Washington, DC 20006-5292.

The Construction Specifications Institute, 99 Canal Center Plaza, Suite 300, Alexandria VA 22314

NCCER CURRICULA — USER UPDATE

NCCER makes every effort to keep its textbooks up-to-date and free of technical errors. We appreciate your help in this process. If you find an error, a typographical mistake, or an inaccuracy in NCCER's curricula, please fill out this form (or a photocopy), or complete the online form at **www.nccer.org/olf**. Be sure to include the exact module ID number, page number, a detailed description, and your recommended correction. Your input will be brought to the attention of the Authoring Team. Thank you for your assistance.

Instructors – If you have an idea for improving this textbook, or have found that additional materials were necessary to teach this module effectively, please let us know so that we may present your suggestions to the Authoring Team.

NCCER Product Development and Revision

13614 Progress Blvd., Alachua, FL 32615

Email: curriculum@nccer.org
Online: www.nccer.org/olf

❑ Trainee Guide ❑ Lesson Plans ❑ Exam ❑ PowerPoints Other ____________

Craft / Level: Copyright Date:

Module ID Number / Title:

Section Number(s):

Description:

Recommended Correction:

Your Name:

Address:

Email: Phone:

44101-08

Introduction to Project Management

44101-08
Introduction to Project Management

Topics to be presented in this module include:

Overview

Today's construction industry provides more than just buildings where people live, work, shop, worship, and learn. The construction industry also builds highways, bridges, airports, and tunnels that enable goods and people to move freely about the country. It builds reservoirs, dams, power stations, irrigation systems, sewer systems, and flood control networks that provide water and power and protect public health. Our lives would be significantly different without the construction industry, and considerably less comfortable.

The construction industry is expected to continue to grow, and the need for more trained construction personnel is anticipated. The skills a project manager needs are acquired and refined through experience and education. NCCER has developed this Project Management training curriculum to provide that education and to give the industry the highly qualified, highly trained managers it needs to survive, grow, and prosper.

A construction project is a short-term endeavor based on specifications and requirements that are driven by functional, budgetary, customer, and time constraints. The construction project manager needs to be sensitive to the project itself as well as the customer's desires and company's constraints that can appear between pre-construction and final completion of the project. This module provides an overview of the responsibilities and characteristics of a project manager along with phases and flow of a project.

Objectives

When you have completed this module, you will be able to do the following:

1. Define project.
2. Describe the characteristics of a project manager.
3. Describe the basic functions of project management.
4. Cite the importance of ethical approaches to project management.
5. Discuss the flow and phases of a construction project.
6. Describe the four common construction delivery systems.

Trade Terms

Accountable
Agency construction management (CM)
Bidding
Close-out
Construction
Construction management (CM) at risk
Construction manager
Delegation
Design-build
Engineering-Procurement-Construction (EPC)
Grapevine
Project manager
Span of control
Traditional contract delivery
Pre-construction
Work breakdown structure (WBS)

Prerequisites

There are no prerequisites for this module.

This course map shows all of the modules in the *Project Management* curriculum. The suggested training order begins at the bottom and proceeds up. Skill levels increase as you advance on the course map. The local Training Program Sponsor may adjust the training order.

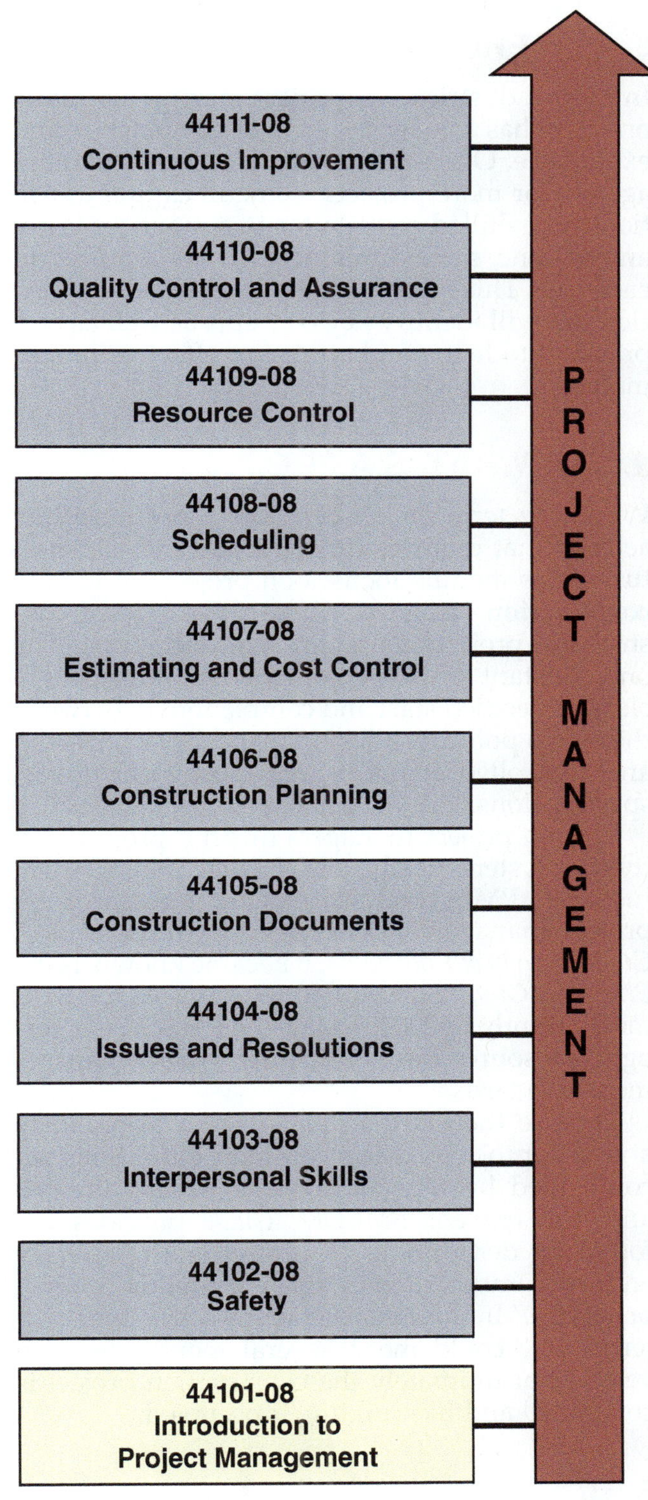

101CMAP.EPS

1.0.0 ◆ INTRODUCTION TO PROJECT MANAGEMENT

In most industries, the **project manager** is someone who has achieved a level of competency and experience. Often this person is selected to manage one or more projects. Sometimes this transition from skilled contributor to a manager is not an easy one, since different skills are required in each role. This Project Management training curriculum will identify key concepts as well as best practices to help you become an effective project manager.

2.0.0 ◆ WHAT IS A PROJECT?

While the term *project* can apply to almost any activity that requires time, people, and expenditures, this module focuses on projects in today's **construction** industry, where the range of construction projects varies in size and scope. The one constant of every construction project is a clearly specified start and completion. In between these two points is a wide range of synchronized and controlled activities that must meet project specifications.

To help project managers direct a project's activities systematically, the Project Management Institute (PMI) conducts extensive research on project management. In 1986, PMI released *A Guide to Body of Knowledge*. It became known as the *PMBOK® Guide*. Subsequent revisions were made, most recently in 2004. Today, the *PMBOK* is a recognized source for establishing project management standards.

Because there are so many ways to organize and run projects, using common definitions accompanied by repeatable steps is very important. The current *PMBOK® Guide* provides the following description: "A project is a temporary endeavor undertaken to create a unique product or service." In the construction industry, the word temporary could mean several months or even years—but ultimately the construction project is completed and the building is occupied.

3.0.0 ◆ CHARACTERISTICS AND FUNCTIONS OF A PROJECT MANAGER

In this course, project manager refers to anyone who directly manages one or more construction projects. The project manager is the person with authority and responsibility for:

- Planning, organizing, staffing, directing, and controlling one or more projects
- Reporting directly to the owner or the owner's designated representative on the status of those projects
- Completing each project on time, within budget and in full compliance with all job specifications, code requirements, drawings, and contractual agreements

There are many subsets included in the above list. The ability to plan is one of the most important characteristics a project manager needs to possess. But also consider other factors. The project manager is an important part of the project team. Through both technical experience and building relationships, the project manager can make the difference between a successful project and one with overruns, personnel attrition, and ultimately failure. What other characteristics should the project manager possess? Today, every project manager must use both technical and management skills to meet the requirements of the construction industry. These management skills include:

- Planning
- Initiating
- Directing
- Organizing
- Communicating
- Controlling

3.1.0 Planning

Planning is an important skill. In the early phases of project startup, the project manager needs to develop a **work breakdown structure (WBS)**. Using a WBS helps the project manager set up project tasks in a hierarchical arrangement that is similar to an organizational chart. Part of the planning process also involves CPM (critical path method), a technique that diagrams tasks based on their longest path through the network. At the onset of the project, either before or after the kick-off meeting, it may be necessary to brainstorm with the project team. Used effectively, brainstorming involves all project members in problem-solving and helps to prioritize tasks. Establishing a risk management plan will help to identify and analyze potential risks during construction.

Brainstorming and risk identification may help to identify resources, safety issues, budgeting constraints, staffing, and regulatory matters such as permits and codes. Most importantly, having enough information will help the project manager assemble the right staff.

Staffing involves acquiring competent employees, placing them into specific positions, and keeping those positions filled. In many companies, the project manager may be responsible for hiring supervisors, field engineers, and other field personnel. The skills of finding and keeping a high-quality staff are discussed in other modules of this training curriculum.

3.2.0 Initiating

Initiating requires building relationships and being keenly aware of customer needs. The project manager needs to know how to identify and budget resources and also how to effectively evaluate the drawings and specifications to provide accurate estimating. Another skill is the ability to effectively communicate with secondary audiences. For example, the project manager is not always involved with the team, but may have to elicit information from stakeholders and consider their input during the planning process.

3.3.0 Directing

Directing involves guiding and supervising employees in order to accomplish company objectives. This includes assigning tasks and following up to be sure they were completed. Employees cannot be forced to do their best work; instead, they must be motivated. Therefore, it is the project manager's role to develop techniques to motivate employees. The most important characteristics of a project manager are solid leadership and good communication skills.

3.4.0 Organizing

A company's staffing structure is usually presented in a diagram called an organizational chart. An organizational chart shows the horizontal segments, the vertical segments, and the lines of authority between them. By clearly showing each individual's duties and area of authority, a well-prepared organizational chart also illustrates areas where individuals' authority and responsibility overlap and areas in which no one has been assigned authority or responsibility. A typical organizational chart is presented in *Figure 1*.

In addition to the formal lines of authority and responsibility shown in the organizational chart, there are informal channels of communication and contact within every company. Informal channels among company personnel are the result of happenstance, necessity, and social relationships, and they can exist at all levels of management, supervision, and the workforce.

Like the formal company organization, the informal organization grows out of a need to accomplish company goals. For example, a proj-

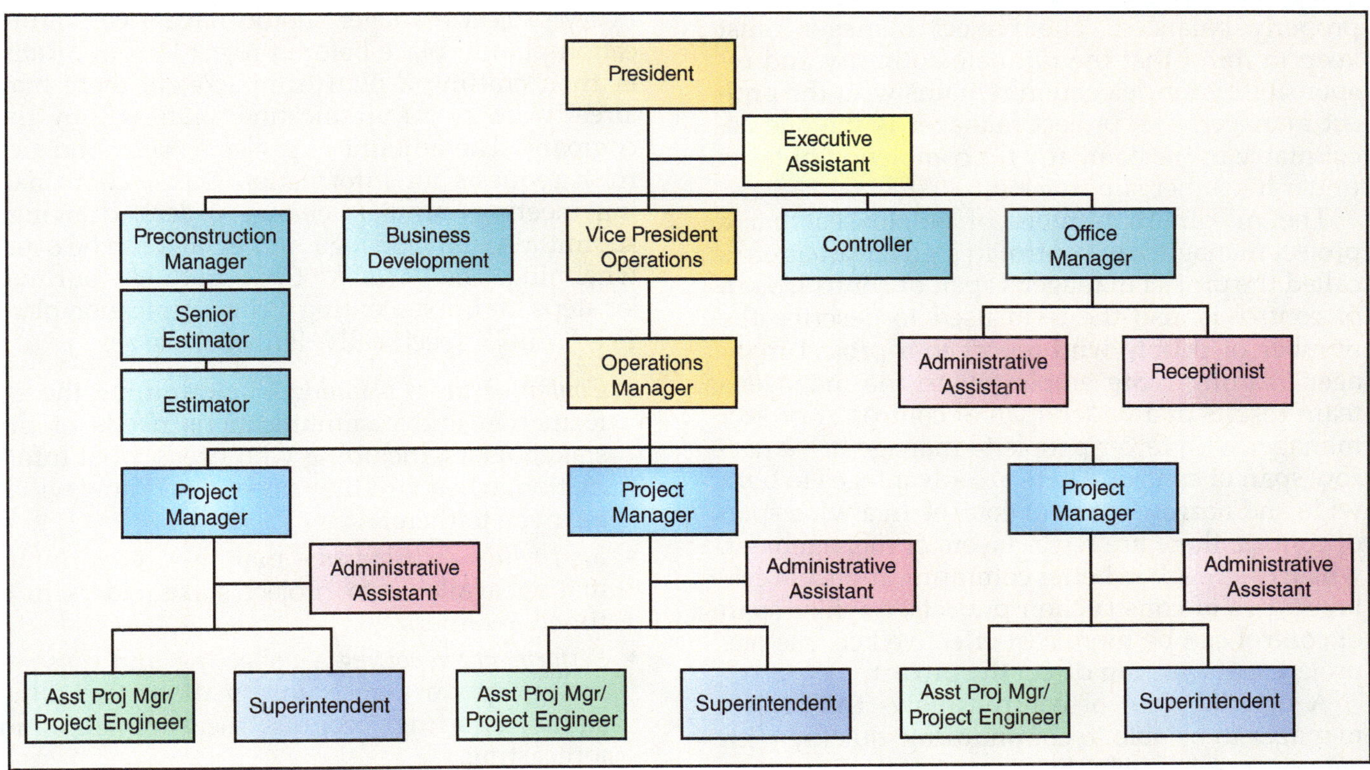

Figure 1 ◆ Typical organizational chart.

ect manager may require immediate verification that an invoice or bill has been paid. Instead of going through the formal channels outlined on the organizational chart, the project manager may simply pick up the phone and call the accounts payable department directly. The accounting staff may provide the answer immediately, and so a new channel of communication is established.

Another example of informal company organization is the **grapevine**. Based on friendship and casual contact, in most companies this network is able to originate and spread information through an office or a project faster than any company memo. Many times, damaging rumors can be stopped and incorrect information can be corrected more quickly through the grapevine than through the formal channels of communication. A company's informal organization is an important management tool, and each project manager must recognize its existence.

As an organization grows, the manager must assign duties to others in order to have time to devote to more important duties and tasks. The act of assigning duties to others is called **delegation**. In delegating activities, the project manager should assign not just the task but the authority and responsibility to complete that task.

Having authority means having the power to act or make decisions. Having responsibility means being **accountable** for the outcome of an assignment. In every act of delegation, authority and responsibility must be carefully defined and properly balanced. The project manager must keep in mind that the ultimate authority and responsibility for delegation remains with the project manager. The project manager is the one accountable to the company for completion of tasks under his or her supervision.

The maximum number of employees that a project manager can control in a given situation is called the project manager's **span of control**. Span of control is also the term used to describe the number of people who report to a project manager. Adding more employees to the manager's team results in a wider span of control. A project manager with fewer people to manage has a narrow span of control. There are advantages to both wide and narrow spans of control. In a wide span of control, there are fewer layers of management, which can result in better communication. On certain types of construction projects, a wider span of control can be more cost-effective because one project manager can direct the project.

A narrow span of control allows the project manager to be able to communicate quickly with supervisors and help direct their tasks. For example, a supervisor can provide feedback and suggestions to handle on-site tasks in a more efficient manner. However, a narrow span of control may have the potential for micro-managing.

Therefore, several factors determine a project manager's span of control. They are:

- The manager's ability and personality
- The quality of communication between the manager and the manager's supervisors
- The number and types of other duties required of the manager
- The skills and training level of the project team
- The types of projects under the manager's command

A project manager should not be responsible for managing more projects or supervising more employees than the individual can effectively monitor and control. A general rule for the number of employees a manager can effectively oversee is between four and eight.

A number of factors can contribute to a project manager losing effective control of the project. Often the reason for poor management is caused by the project manager trying to supervise more work and more employees than can be managed properly. In other words, failure comes from a project manager trying to work beyond a personal span of control.

3.5.0 Communicating

Every project manager should have a communication plan in place before project kick-off. Rather than crafting a plan from scratch, there may already be a communication plan within the company. The amount of project correspondence from requests for information (RFIs), coordinating meeting minutes, change orders, drawings submittals, daily job logs, and emails can be overwhelming. The *PMBOK® Guide* outlines four major steps for implementing a communication plan. The *PMBOK* specifically defines these as:

- *Communications planning* – determining the information and communications needs of the stakeholders, including who needs what information, when will they need it, and how will it be given to them.
- *Information distribution* – making needed information available to project stakeholders in a timely manner.
- *Performance reporting* – collecting and disseminating performance information, including status reporting, progress measurement, and forecasting.
- *Administrative closure* – generating, gathering, and disseminating information to formalize phase or project completion.

3.6.0 Controlling

Control is the means of measuring performance and correcting any apparent deviation from the project plan. The most effective control techniques involve establishing standard units for measuring progress, identifying deviations as soon as they occur, and correcting them. The elements and process of control will be discussed in more detail in modules covering cost control, resource control, and quality control in this curriculum.

4.0.0 ◆ ETHICAL APPROACHES TO PROJECT MANAGEMENT

Ethics in its simplest definition means doing the moral and right thing. A culture and practice of ethics is essential for today's corporations. Corporate boards of directors are increasingly focusing on ethics. In addition to the legal department, some firms are appointing ethics officers or hiring outside experts to assist in developing a code of ethics and practices for the organization. Practicing professional ethics serves the interests of the firm and the customer and drives business performance.

The major business focus for all construction companies is to make a profit. The business process to attain profit must be based on ethical considerations. Project managers within such environments maintain a high standard. However, there may be times when potential unethical situations arise that are somewhat in the gray area. Consider the following example:

Phil, the project manager, is reviewing estimates for copper tubing and mentions to one of his foremen that the prices are getting too high to meet the specs. The foreman lets Phil know that his cousin has connections and thinks he can get it for three cents off per foot. The foreman also mentions that he has done this before and tells Phil that he will give him all the details so Phil can prepare a product substitution. The foreman mentions that it will be a nice profit on the job.

Ask yourself – what you think your action would be? Project managers need to think through all situations that appear to stray from the plans and specifications and contractual agreements. On a larger scale, what about an owner who pressures the contractor to complete the project by a certain date because of external pressures or public perceptions? Will the construction company cut corners to save time and money? Both of these situations present problems because avoiding ethical responsibilities by the company or individuals within the company may have far-reaching implications.

The project manager must ensure that the project team understands the ethics and values of the company and puts these into practice on the job. Honesty and integrity are the foremost values that govern ethical situations. Practices that the project manager can implement include:

- Preparing ethical bids and requiring the same from suppliers
- Negotiating fairly with the customer and suppliers
- Being candid and honest with customers
- Providing the professional services consistent with best practices and the competencies of the organization
- Complying with regulations and legal procedures
- Adhering to confidentiality and proprietary information provisions and avoiding conflicts of interest

During a construction project, the project manager will encounter many situations that will require good business decisions to resolve conflict. *Table 1* lists some common conflicts that can potentially have ethics implications.

The whole idea of what is ethical and what is not ethical is difficult, and project managers are not perfect. However, the company that integrates ethics with their business model is less likely to avoid turn-over, litigation and client dissatisfaction.

5.0.0 ◆ PHASES OF A CONSTRUCTION PROJECT

A construction project can be divided into three phases or sequences that represent the project flow:

- Development
- Design
- Construction

5.1.0 Development Phase

The development phase is the first step in all construction projects. Beginning with an idea or concept, the owner prepares the required feasibility studies, locates suitable land through research, and obtains the necessary finances. The architect/engineer develops conceptual drawings and models, conceptual estimates, and begins preliminary discussions with various government agencies to examine regulating and permitting issues.

Using the conceptual estimate, the owner and architect/engineer develop the project budget, taking into account all anticipated costs and expenses.

Table 1 Conflicts with Ethical Implications

Situation	Ethical Concern(s)
Schedule conflicts resulting from lack of support, both internal and external	May impact service to the client
Restrictive administrative procedures that may constrain the project manager's authority and accountability	May impact service to the client
Priorities regarding concurrent projects within the company	The project manager is ethically bound to deliver services contracted
Problems with personnel resources	The project manager is responsible to assign qualified staff to projects
Technical issues and conflicts on how to expedite and reduce costs to obtain solutions	Possible conflicts of interest
Personality conflicts (team members, subcontractors)	Possible issues with confidentiality and conflicts of interest

101T01.EPS

In addition to analyzing the cost of the project, the owner analyzes the projected return on the investment. If the project is considered financially feasible, then the owner generally seeks financial assistance from lending institutions.

The architect/engineer and owner begin the preliminary review of the project with various government agencies. These reviews are intended to ensure that the project meets all applicable zoning laws, building restrictions, environmental concerns, road access, landscape requirements, and other considerations.

5.2.0 Design Phase

During the design phase, cost estimates are refined, project financing is secured, government regulations are met, and the marketing program is developed.

When the architect/engineer starts the preliminary drawings and specifications, other design professionals—including structural, mechanical, electrical, civil, and other specialized engineers—perform calculations, analyze technical data, and develop project details. This work is translated into drawings and specifications that illustrate all the information that the general contractors, the subcontractors, and the suppliers need to install the components that make up a project.

In addition to developing the drawings and specifications, the architect/engineer will produce all the documents necessary for the owner to obtain bids from interested contractors. With the contract documents complete, the owner selects the method that will be used to choose contractors. The most common method for selecting a contractor is through competitive **bidding**. Other methods an owner may employ to choose a contractor are discussed in a later section of this module.

5.3.0 Construction Phase

The construction phase is usually organized by a general contractor, which may perform work with its own workforce. In addition, the general contractor may seek the services of specialty contractors or subcontractors to perform certain portions of work on the project.

As construction nears completion, the architect/engineer, the owner, and government agencies start their final inspections and acceptance of the nearly completed project. If the project has been managed properly, the work has been performed satisfactorily, the architect/engineer has inspected the project regularly, and local codes have been met throughout, then final inspection and acceptance will proceed quickly. The result is a satisfied client and a profitable project for the contractors. On the other hand, if the inspection reveals faulty work, inferior materials, and code violations, the inspection and acceptance procedures can generate disputes that will undoubtedly result in a dissatisfied client and a loss of profits for the contractor.

6.0.0 ◆ CONSTRUCTION PROJECT FLOW

Understanding the project flow is an important tool for management of any project. Each of the four phases of project flow is illustrated in *Figure 2*.

- Bidding
- Pre-construction
- Construction
- Closeout

Just as the total project can be divided into phases or sequences, the construction phase can be further divided into several phases. *Figure 3* illustrates the construction flow phase of a typical project.

6.1.0 Bidding Phase

The bidding phase starts when an owner or architect/engineer requests that a general contractor or specialized contractor submit a bid or proposal on all or part of a construction project. The contractor reviews the documents to determine the extent of the work and the requirements of the project.

Usually, the contractor assigns an estimator to develop the estimate. The estimator studies the project documents; prepares a quantity survey (material take-off); and prepares a list of the various materials, components, and subcontractors necessary to complete the project. The total of all the items is summarized. The company's overhead cost and a reasonable profit are added to the cost summary. The final sum becomes the bid amount the company will submit to the owner or architect/engineer.

In a competitive bid situation, the company that submits the lowest bid usually is awarded the contract to do the work. If the project is being negotiated, the owner or architect/engineer typically may not award the project solely on the basis of the lowest bid, but will consider other factors including safety performance, contractor qualifications, financial strength, personnel, and schedule. In any case, the bidding phase ends when the project contract is awarded.

6.2.0 Pre-Construction Phase

The pre-construction phase is often the most important part of the construction flow process. During this phase, the project manager and the project team are involved in several major areas:

- Issuing subcontracts and purchase orders
- Scheduling development
- Reviewing the project estimate
- Identifying manpower and equipment requirements
- Establishing cost control
- Preparing job site mobilization plans

6.2.1 Subcontracts and Purchase Orders

The successful contractor must select subcontractors after carefully reviewing their bids for scope of work, cost, and exceptions to the contract documents. *Figure 4* lists the topics that should be discussed before any contract is awarded.

The contractor must verify that suppliers can meet the requirements of the contract documents before issuing purchase orders for materials and equipment.

6.2.2 Schedule

The project schedule is developed with cooperation and input from all members of the company's project team, the major subcontractors, the key suppliers, and others who can contribute to a realistic schedule. While developing the schedule, the highest priority is assigned to details regarding long-lead items, critical operations, sequencing, duration of activities, equipment deliveries, and labor requirements. Scheduling is discussed in greater detail in another module.

6.2.3 Estimate

The pre-construction review of the estimate is a key function of the project manager. A thorough project team review of the estimate ensures that everyone on the project has a clear understanding of the project costs. The subject of estimating will be discussed in more detail in the *Estimating and Cost Control* module.

6.2.4 Labor and Equipment

Determining the number of craftworkers for the various trades required at each stage of work gives the project team a basis for determining the availability of personnel, identifying possible labor shortage or surpluses, and controlling labor costs and cash flow. Reviewing equipment requirements ensures that construction equipment is available when required, used efficiently on the site, and removed from the job site when it is no longer needed.

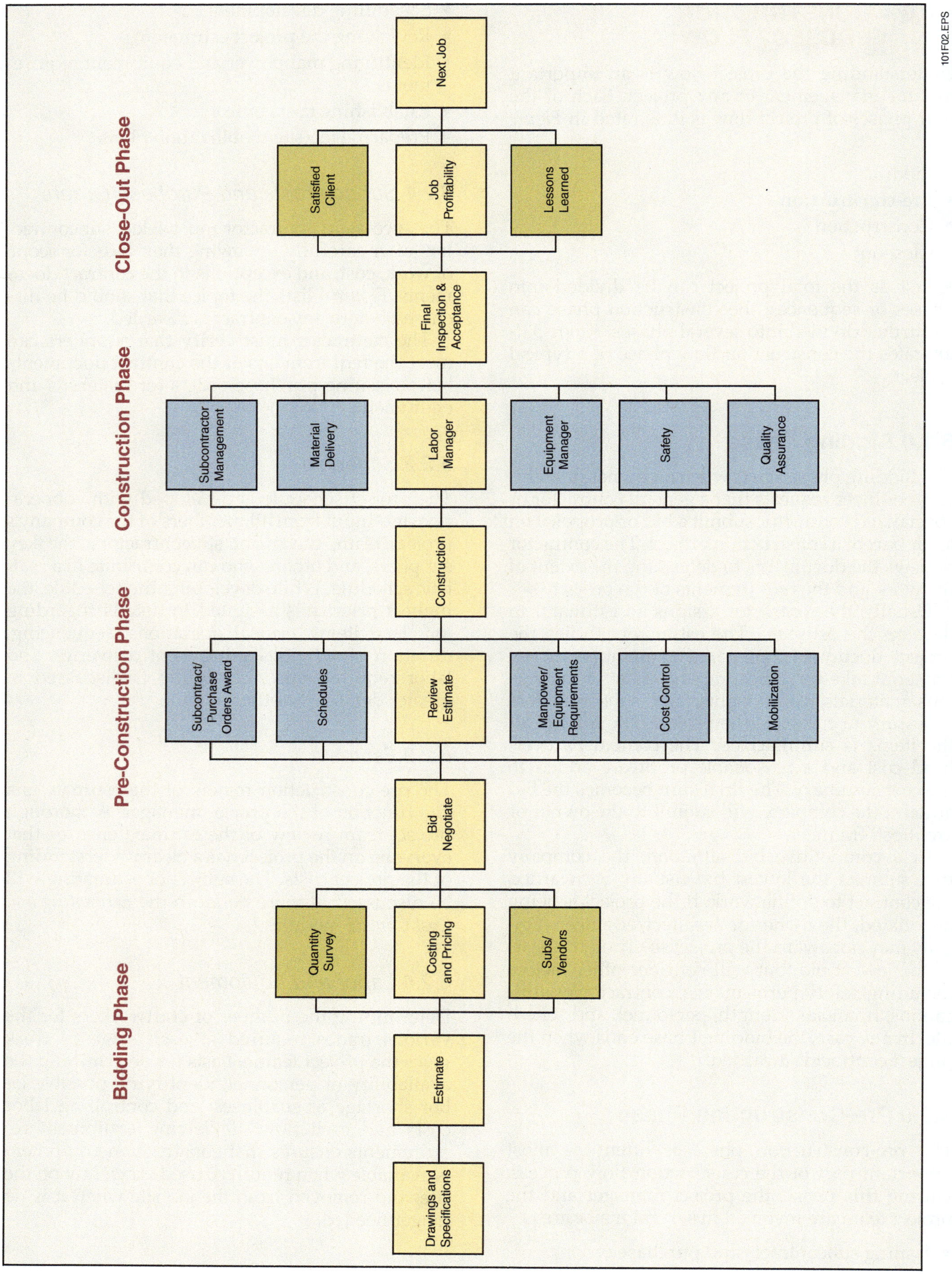

Figure 2 ◆ Flow of a traditional project.

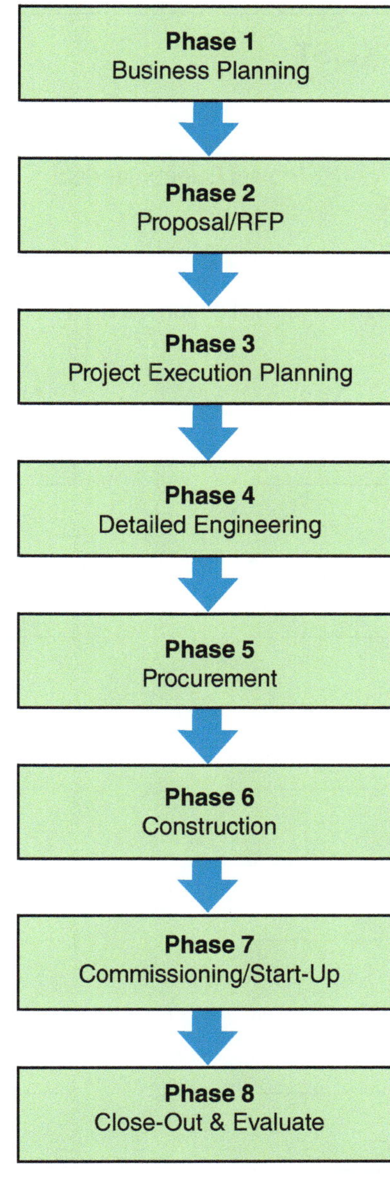

Figure 3 ◆ Construction flow of a typical project.

6.2.5 Cost and Other Controls

Every step of the project, from inception to completion, is based on cost management. The project manager should review the cost control system with every member of the project team, emphasizing the need for accurate and timely cost reporting. The project manager's role and involvement in cost control will be reviewed in detail in the *Estimating and Cost Control* module.

Other controls that need to be implemented and managed are covered in the modules on *Scheduling* and *Resource Control* (including materials, tools, and equipment).

6.2.6 Mobilization

Mobilization includes setting up field site offices and storage trailers, locating temporary facilities, establishing storage layout areas, and other tasks necessary to start the project. *Figure 5* illustrates a typical checklist that may be used to review mobilization item requirements.

6.3.0 Construction Phase

The construction phase, or the actual building of the project, involves subcontractors, equipment, labor coordination, scheduling of material delivery, safety requirements, and quality control and assurance of the project. Before construction begins, plan a pre-construction conference to discuss and distribute project-related material and information to the project team including subcontractors and major vendors and suppliers.

6.4.0 Close-Out Phase

Close-out of the project involves the final inspections and acceptance of the work by the owner, architect/engineer, and government agencies. Record drawings or as-built drawings are also submitted in accordance with the contract requirements. The closeout documents include:

- Punch list
- Guarantees and warranties
- Operation and maintenance manuals
- Release of liens
- Record or as-built drawings
- Other documents as required by the contract

7.0.0 ◆ CONSTRUCTION DELIVERY SYSTEMS

Project managers must be aware of the different construction delivery systems in today's market. The selection of a construction system depends on the project, the type of company, and the owner's preference. The most common construction project delivery systems are:

- The traditional contract delivery system
- The **design-build** system
- The construction management delivery system (CM)
- The engineering-procurement-construction (EPC) system

Each of these systems has its advantages and disadvantages which the owner must consider before making any decisions.

CONTRACT AWARD CHECKLIST

Project Name	Project Manager	Project Superintendent
Trade Contractor	Trade Contractor Superintendent	Specification Section

	Yes	N/A
Performance and payment	______	______
Bond, if required, on file	______	______
Insurance requirements on file	______	______
General conditions reviewed	______	______
Contract requirements reviewed	______	______
Drawings and specifications reviewed	______	______
Schedule reviewed	______	______
Material status reports reviewed	______	______
Permit requirements reviewed	______	______
Payment request/invoice procedure reviewed	______	______
Safety requirements reviewed	______	______
Safety program received	______	______
City, county, or state licenses on file	______	______
Change order procedures reviewed	______	______
Back charge procedures reviewed	______	______

Comments: ______________________________

101F04.EPS

Figure 4 ◆ Contract award checklist.

Mobilization Items	PM Responsibility	Superintendent Responsibility	Other (Identify)	Date Required	Date Completed
Permits					
Insurance					
Recruiting and Staffing					
Office Complex					
Temporary Utilities					
Site Security					
Fenced Storage					
Telecommunications					
Temporary Toilet					
Survey/Base Lines					
Disposal Service					
First Aid Facilities					
First Aid Training					
Survey Equipment					
Notice of Commencement					
Urgent Care Facilities					
Emergency Response Services					
Local Post Office Box					
Local Banking Services					
Job Camera/Videos					
Webcam					
Set up Local Accounts					
Verify Burning Regulations					
Copier or Copier Service					
Office Supplies					
Files					
Furniture					
Computers and Networking					
Project Signage					
Project Bulletin Board					

101F05.EPS

Figure 5 ◆ Mobilization checklist/status report.

7.1.0 Traditional Contract Delivery System

The traditional contract delivery system has been the standard approach for many years and is still used. This approach involves the owner, contractor and architect/engineer. The traditional contract delivery system reduces risk because the owner has control during the design and construction phases.

Using this approach, the owner contracts separately with the architect/engineer for the preparation of plans and specifications and assistance in the bidding stage. The architect/engineer may also be requested by the owner to oversee certain aspects of the project during construction. For example, a renovation may have requirements to maintain historical aesthetics, so the design firm would coordinate with the contractor to ensure those types of requirements are met.

Once the architect/engineer is under contract, all the design will be completed before a general contractor is selected, usually through a competitive bid process. This allows the owner to carefully analyze the bids to obtain the best price for construction. When a contractor is selected based on the firm fixed price, the owner has a high level of confidence that the estimate is accurate because several bids were reviewed during the selection process.

Once selected, the owner will negotiate a contract with the general contractor who will provide services for the construction of the project. The owner retains responsibility for overall project management. All transactions and contractual matters are handled by the owner. This approach provides checks and balances between the design and construction. Most companies are very familiar with this approach and generally see it as a good way to do business. Additionally, the owner is able to provide influential input throughout the project. Some disadvantages are that the owner must make a significant investment up front to develop the final drawings and specifications. The project design and specifications could also potentially lead to higher bids than anticipated, requiring redesign.

7.2.0 Design-Build

The design-build project delivery system is a single source procurement that allows the owner to select and enter into a contract with a design-build firm that will assume the obligation of furnishing design, supervision and construction services during the project. The design-build project delivery system is an attractive arrangement for the owner because it provides a single point of responsibility for design and construction. This arrangement keeps the owner from being in the middle of any disputes between the construction and design. The downside is that it takes away the checks and balances which occur when the design and construction phase are contracted separately. The owner also loses some of the control that it has on the project, and loses the direct advisory relationship with the architect/ engineer that the owner has in the traditional contract delivery system.

7.3.0 Construction Management (CM)

The construction management system (CM) may be a good solution for projects with unique security, requirements, regulations, codes and specifications such as medical facilities, power plants, or federal government facilities. The two common controls of CM are agency construction management (CM) and construction management (CM) at risk.

Agency CM involves providing contractor-based management at the early stages of a project to ensure that the project stays within scope during design and throughout construction. The assigned program manager reports directly to the owner and works on fee basis, but also interacts with the architect/engineer and the general contractor along with other parties assigned to mediate and solve problems. The CM program manager's primary focus is to advise the owner and to assist in providing resources from the CM firm that can support different project requirements and logistics. The following are a few examples:

- Maintain project relationships with the trade contractors
- Establish timetables and resolving conflicts
- Plan and coordinate equipment, materials, and suppliers
- Coordinate with local, state, and federal agencies regarding safety and code compliance

Since the CM program manager is involved early in the entire process, monitoring the concept, design, execution, and completion can be effectively managed.

CM at risk involves hiring a construction management firm that meets specific qualifications and often performs the work at a guaranteed maximum price. The construction manager participates with the owner in the concept and design of the project and becomes the general contractor once the price is agreed. As a part of the design team, the CM contributes expertise and recommendations early in the process. The

CM assumes the risk of subcontracting, managing schedules, adhering to contract specifications, and cost overruns.

The CM at risk approach is often more efficient than the traditional contract delivery method. The CM has the flexibility to fast track development. Mobilization can start during the design process. Subcontracts can be let for portions of the work while other portions are being designed.

Many owners find this arrangement less risky since the construction manager is responsible for all costs and overruns and any cost underrun savings remain with the owner.

7.4.0 Engineering-Procurement-Construction (EPC)

Engineering-procurement-construction (EPC) is a spin-off of the design-build approach. This solution is sometimes known as a turnkey solution because one company will provide design engineering, management and construction services. EPC is usually the model that large companies use on complex construction projects such as power plants, bridges, and refineries.

Review Questions

1. What is the definition of a project?

____

____

2. List three common functions of a project manager.
   1. ______________________________
   2. ______________________________
   3. ______________________________

3. During the planning phase, a project manager needs to develop a WBS in order to ______________________________.

4. List the four phases of a construction project.
   1. ______________________________
   2. ______________________________
   3. ______________________________
   4. ______________________________

5. When does the bidding phase begin?

____

6. Schedule development and reviewing the project estimate are part of the __________ phase.

7. Briefly describe a situation you would consider unethical (do not use the one described in this module).

____

____

____

____

8. When developing the project schedule, the project manager or estimator should _____.
   a. consult with the owner's capital projects officer
   b. modify an existing schedule that is similar to the scope
   c. consult with members of the project team

9. Select two documents that are submitted for project closeout.
   a. cost control summary report
   b. punch lists
   c. documentation of all change orders
   d. as-built drawings

10. List five factors that determine a project manager's span of control.
    1. ______________________________
    2. ______________________________
    3. ______________________________
    4. ______________________________
    5. ______________________________

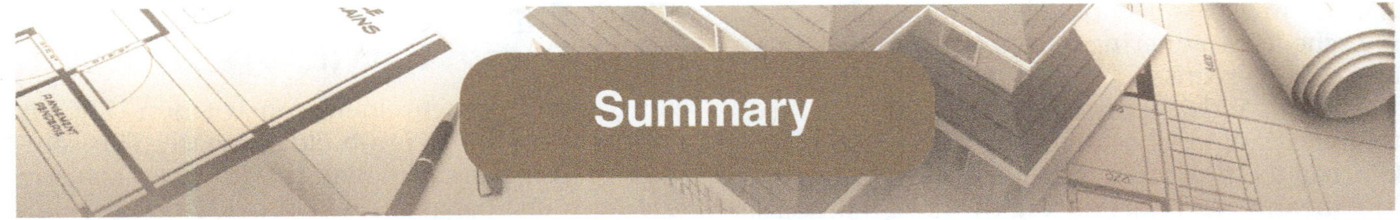

Summary

A project is a finite endeavor that requires a project manager to systematically direct its activities and the team performing the work. The project manager has a responsibility to plan, organize, and control the project and complete it within the budget and time agreed and in compliance with the specifications, codes, and contract. The project manager and the team must practice ethical standards of conduct in performing the job and in their relations with stakeholders.

There are well-defined phases to a construction project and to its flow. The major phases include development, design, and construction. Project flow includes bidding, pre-construction, construction, and closeout. There are also different approaches to a construction project that include the traditional contract delivery system, design-build, construction management, and engineering-procurement-construction.

The project manager must be involved on a daily basis in controlling the use of resources and reducing costs by such means as:

- Implementing more efficient work methods
- Using more effective tools and equipment
- Providing training and re-training of personnel
- Monitoring productivity on the job site and working to improve it
- Reviewing and following up on production records

The project manager's responsibility is to control production so that the project is constructed for the estimated cost, within the scheduled time, and to the quality standards designated in the specifications and plans.

Notes

Trade Terms Quiz

1. Making an offer to perform work described in contract documents is known as ______________.
2. ______________ is a building project that is typically divided into three phases; development, design, and construction.
3. A ______________ is the person responsible for the planning, coordination, and controlling of a project from start to finish.
4. Being ______________ means you are subject to giving an account or being answerable.
5. A hierarchical definition of the tasks and activities of a project is called the ______________.
6. ______________ is the act of empowering or the assignment of duties to others.
7. Performing final inspections and preparing documents for handing over the project is known as ______________.
8. An informal person-to-person means of circulating information or gossip is referred to as the ______________.
9. The ______________ consists of the maximum number of projects and employees that a person can control in a given situation.
10. ______________ is the planning phase that takes place before on-site work can begin.
11. ______________ is the standard approach to construction development in which the owner contracts separately with the architect/engineer and the contractor.
12. ______________ is single source procurement for an owner.
13. This delivery system is a form of design-build known as ______________.
14. A firm contracted to provide design expertise and deliver the construction project at a firm price at no risk to the owner is known as ______________.
15. ______________ involves the use of a manager as the owner's agent and advisor during a construction project.

Trade Term List

Accountable
Agency construction management (CM)
Bidding
Close-out
Construction
Construction management (CM) at risk
Construction manager
Delegation
Design-build
Engineering-procurement-construction (EPC)
Grapevine
Project manager
Span of control
Traditional contract delivery
Pre-construction
Work breakdown structure (WBS)

Trade Terms Introduced in This Module

Accountable: Subject to giving an account; answerable.

Agency construction management (CM): A firm that provides various services to owners in the course of a construction project. These services can vary from program management to design and construction.

Bidding: Making an offer to perform work described in contract documents at a specified cost.

Close-out: Performing final inspections and documents to prepare for handing the project over for the owners' acceptance and possession.

Construction: A building project that is typically divided into three phases: development, design, and construction.

Construction management (CM) at risk: The form of construction management and delivery in which the CM firm assumes the burdens and risk for the project; the CM is involved in the design phase and performs the construction delivery at a firm agreed-upon price.

Construction manager: The supervisor responsible to the project manager for day-to-day activities on the construction site.

Delegation: The act of empowering to act for another; assignment of duties to others.

Design-build: A construction delivery system whereby the contractor provides both design and construction services.

Engineering-Procurement-Construction (EPC): A variation of the design-build approach wherein there is a single point of contact with the owner throughout the project.

Grapevine: An informal person-to-person means of circulating information or gossip.

Project manager: The person responsible for the planning, coordination, and controlling of a project from inception to completion, meeting the project's requirements and ensuring completion on time, within cost, and to quality standards specified in contract documents.

Span of control: The maximum number of employees that a person can control in a given situation; also the supervisory ratio of from three-to-seven individuals, with five-to-one being established as optimum.

Traditional contract delivery: The standard approach to construction development in which the owner contracts separately with the architect/engineer and the contractor.

Pre-construction: The planning phase that takes place before on-site work begins, involving issuing contracts, schedule development, identifying labor and equipment requirements, and other preparatory work.

Work breakdown structure (WBS): A hierarchical decomposition of deliverables which breaks down the scope of work into unique and distinguishable activities and tasks.

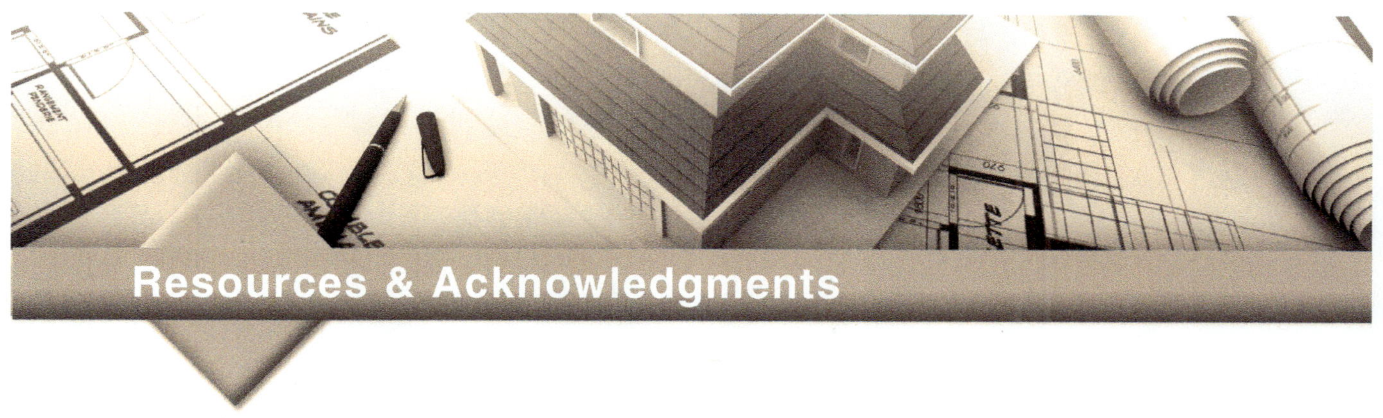

Resources & Acknowledgments

References

PMI Standards Committee, *A Guide to the Project Management Body of Knowledge*. PMI Publications, Newton Square, Pa. (2004).

NCCER CURRICULA — USER UPDATE

NCCER makes every effort to keep its textbooks up-to-date and free of technical errors. We appreciate your help in this process. If you find an error, a typographical mistake, or an inaccuracy in NCCER's curricula, please fill out this form (or a photocopy), or complete the online form at **www.nccer.org/olf**. Be sure to include the exact module ID number, page number, a detailed description, and your recommended correction. Your input will be brought to the attention of the Authoring Team. Thank you for your assistance.

Instructors – If you have an idea for improving this textbook, or have found that additional materials were necessary to teach this module effectively, please let us know so that we may present your suggestions to the Authoring Team.

NCCER Product Development and Revision

13614 Progress Blvd., Alachua, FL 32615

Email: curriculum@nccer.org
Online: www.nccer.org/olf

❏ Trainee Guide ❏ Lesson Plans ❏ Exam ❏ PowerPoints Other ______________

Craft / Level: Copyright Date:

Module ID Number / Title:

Section Number(s):

Description:

Recommended Correction:

Your Name:

Address:

Email: Phone:

00105-15

Introduction to Construction Drawings

Overview

Various types of construction drawings are used to represent actual components of a building project. The drawings provide specific information about the locations of the parts of a structure, the types of materials to be used, and the correct layout of the building. Knowing the purposes of the different types of drawings and interpreting the drawings correctly are important skills for anyone who works in the construction trades. This module introduces common types of construction drawings, their basic components, standard drawing elements, and measurement tools that are typically used when working with construction drawings.

Module Seven

Trainees with successful module completions may be eligible for credentialing through the NCCER Registry. To learn more, go to **www.nccer.org** or contact us at **1.888.622.3720**. Our website has information on the latest product releases and training, as well as online versions of our *Cornerstone* magazine and Pearson's product catalog.

Your feedback is welcome. You may email your comments to **curriculum@nccer.org,** send general comments and inquiries to **info@nccer.org**, or fill in the User Update form at the back of this module.

This information is general in nature and intended for training purposes only. Actual performance of activities described in this manual requires compliance with all applicable operating, service, maintenance, and safety procedures under the direction of qualified personnel. References in this manual to patented or proprietary devices do not constitute a recommendation of their use.

00105-15
Introduction to Construction Drawings

Objective

When you have completed this module, you will be able to do the following:

1. Identify and describe various types of construction drawings, including their fundamental components and features.
   a. Identify various types of construction drawings.
   b. Identify and describe the purpose of the five basic construction drawing components.
   c. Identify and explain the significance of various drawing elements, such as lines of construction, symbols, and grid lines.
   d. Identify and explain the use of dimensions and various drawing scales.
   e. Identify and describe how to use engineer's and architect's scales.

Performance Task

Under the supervision of your instructor, you should be able to do the following:

1. Using the floor plan supplied with this module:
   - Locate the wall common to both interview rooms.
   - Determine the overall width of the structure studio.
   - Determine the distance from the outside east wall to the center of the beam in the structure studio.
   - Determine the elevation of the slab.

Trade Terms

Architect
Architect's scale
Architectural plans
Beam
Blueprints
Civil plans
Computer-aided drafting (CAD)
Contour lines
Detail drawings
Dimension line
Electrical plans
Elevation (EL)
Elevation drawing
Engineer
Engineer's scale
Fire protection plan
Floor plan
Foundation plan
Hidden line
HVAC
Leader
Legend
Mechanical plans
Metric scale
Not to scale (NTS)
Piping and instrumentation drawings (P&IDs)
Plumbing isometric drawing
Plumbing plans
Roof plan
Scale
Schematic
Section drawing
Specifications
Structural plans
Symbol
Title block

Industry Recognized Credentials

If you are training through an NCCER-accredited sponsor, you may be eligible for credentials from NCCER's Registry. The ID number for this module is 00105-15. Note that this module may have been used in other NCCER curricula and may apply to other level completions. Contact NCCER's Registry at 888.622.3720 or go to **www.nccer.org** for more information.

Contents

Topics to be presented in this module include:

Figures

Section One

1.0.0 Construction Drawings and Their Components

Objective

Identify and describe various types of construction drawings, including their fundamental components and features.

a. Identify various types of construction drawings.
b. Identify and describe the purpose of the five basic construction drawing components.
c. Identify and explain the significance of various drawing elements, such as lines of construction, symbols, and grid lines.
d. Identify and explain the use of dimensions and various drawing scales.
e. Identify and describe how to use engineer's and architect's scales.

Performance Task

1. Using the floor plan supplied with this module:
   - Locate the wall common to both interview rooms.
   - Determine the overall width of the structure studio.
   - Determine the distance from the outside east wall to the center of the beam in the structure studio.
   - Determine the elevation of the slab.

Trade Terms

Architect: A qualified, licensed person who creates and designs drawings for a construction project.

Architect's scale: A specialized ruler used in making or measuring reduced scale drawings. The ruler is marked with a range of calibrated ratios for laying out distances, with scales indicating feet, inches, and fractions of inches. Used on drawings other than site plans.

Architectural plans: Drawings that show the design of the project. Also called architectural drawings.

Beam: A large, horizontal structural member made of concrete, steel, stone, wood, or other structural material to provide support above a large opening.

Blueprints: The traditional name used to describe construction drawings.

Civil plans: Drawings that show the location of the building on the site from an aerial view, including contours, trees, construction features, and dimensions.

Computer-aided drafting (CAD): The making of a set of construction drawings with the aid of a computer.

Contour lines: Solid or dashed lines showing the elevation of the earth on a civil drawing.

Detail drawings: Enlarged views of part of a drawing used to show an area more clearly.

Dimension line: A line on a drawing with a measurement indicating length.

Electrical plans: Engineered drawings that show all electrical supply and distribution.

Elevation (EL): Height above sea level, or other defined surface, usually expressed in feet or meters.

Elevation drawing: Side view of a building or object, showing height and width.

Engineer: A person who applies scientific principles in design and construction.

Engineer's scale: A straightedge measuring device divided uniformly into multiples of 10 divisions per inch so that drawings can be made with decimal values. Used mainly for land measurements on site plans.

Fire protection plan: A drawing that shows the details of the building's sprinkler system.

Floor plan: A drawing that provides an aerial view of the layout of each room.

Foundation plan: A drawing that shows the layout and elevation of the building foundation.

Hidden line: A dashed line showing an object obstructed from view by another object.

HVAC: Heating, ventilating, and air conditioning.

Leader: In drafting, the line on which an arrowhead is placed and used to identify a component.

Legend: A description of the symbols and abbreviations used in a set of drawings.

Mechanical plans: Engineered drawings that show the mechanical systems, such as motors and piping.

Metric scale: A straightedge measuring device divided into centimeters, with each centimeter divided into 10 millimeters. Usually used for architectural drawings and sometimes referred to as a metric architect's scale.

Not to scale (NTS): Describes drawings that show relative positions and sizes only, without scale.

Piping and instrumentation drawings (P&IDs): Schematic diagrams of a complete piping system.

Plumbing isometric drawing: A type of three-dimensional drawing that depicts a plumbing system.

Plumbing plans: Engineered drawings that show the layout for the plumbing system.

Roof plan: A drawing of the view of the roof from above the building.

Scale: The ratio between the size of a drawing of an object and the size of the actual object.

Schematic: A one-line drawing showing the flow path for electrical circuitry or the relationship of all parts of a system.

Section drawing: A cross-sectional view of a specific location, showing the inside of an object or building.

Specifications: Precise written presentation of the details of a plan.

Structural plans: A set of engineered drawings used to support the architectural design.

Symbol: A drawing that represents a material or component on a plan.

Title block: A part of a drawing sheet that includes some general information about the project.

Construction drawings are architectural or working drawings used to represent a structure or system. These were traditionally referred to as **blueprints**, because years ago the lines on a blueprint were white and the background was blue. Construction drawings are also called prints. Today, most prints are created by **computer-aided drafting (CAD)**, and they have blue or black lines on a white background.

Various kinds of construction drawings, including residential drawings, commercial drawings, landscaping plans, shop drawings, and industrial drawings, are used in construction. In this module, several of the most common types of drawings will be introduced.

Construction drawings, together with the set of **specifications** (often abbreviated as *specs*), detail what is to be built and what materials are to be used. Specifications are written statements that the architectural and engineering firm provides to the general contractors. The specs define the quality of work to be done and describe the materials to be used.

Did You Know?

How Blueprints Started

The process for making blueprints was developed in 1842 by an English astronomer named Sir John F. Herschel. The method involved coating a paper with a special chemical. After the coating dried, an original hand drawing was placed on top of the paper. Both papers were then covered with a piece of glass and set in the sunlight for about an hour. The coated paper was developed much like a photograph. After a cold-water wash, the coated paper turned blue, and the lines of the drawing remained white.

The set of construction drawings forms the basis of agreement and understanding that a building will be built as detailed in the drawings. Therefore, everyone involved in planning, supplying, and building any structure should be able to read construction drawings. For any building project, also consult the civil engineering plans for that location, including sewer, highway, and water installation plans.

1.1.0 Six Types of Construction Drawings

A set of building construction drawing plans almost always includes six major types of drawings (*Figure 1*). They include the following:

- Civil
- Architectural
- Structural
- Mechanical
- Plumbing
- Electrical
- Fire protection

This section will examine the various characteristics of each type of drawing.

1.1.1 Civil Plans

Civil plans are also called site plans, survey plans, or plot plans. They show the location of the building on the site from an aerial view (*Figure 2*). A civil plan also shows the natural contours of the earth, represented on the plan by **contour lines**. The civil plans can also include a landscape plan (*Figure 3*) that shows any trees on the property; construction features such as walks, driveways, or utilities; the dimensions of the property; and possibly a legal description of the property.

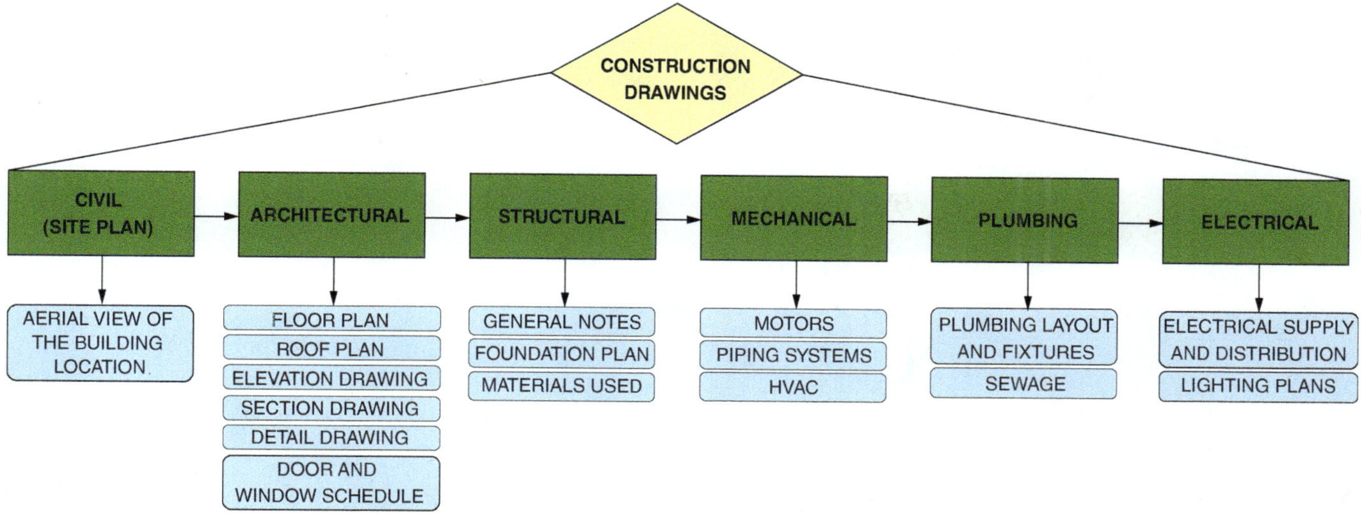

Figure 1 Types of construction drawings.

1.1.2 Architectural Plans

Architectural plans (also called architectural drawings) show the design of the project. One part of an architectural plan is a **floor plan**, also known as a plan view (refer to *Drawing 1*, First Floor Plan, in the *Appendix*). Any drawing made looking down on an object is commonly called a plan view. The floor plan is an aerial view of the layout of each room. It provides the most information about the project. It shows exterior and interior walls, doors, stairways, and mechanical equipment. The floor plan shows the floor as someone would see it from above if the upper part of the building were removed.

An architectural plan also includes a **roof plan** (*Figure 4*), which is a view of the roof from above the building. It shows the shape of the roof and the materials that will be used to finish it.

Elevation (EL) is another element of architectural drawings. **Elevation drawings** are side views that show height. On a building drawing, there are standard names for different elevations. For example, the side of a building that faces south is called the south elevation. Exterior elevations (*Figure 5*) show the size of the building; the style of the building; and the placement of doors, windows, chimneys, and decorative trim.

Building Information Modeling (BIM)

Traditionally, Building Information Modeling (BIM) was a system of paper-based documents and construction drawings designed to monitor and track specific information about a building.

In the last decade, BIM has increasingly moved toward computer technology, which allows the various trades and the building owner to track the lifecycle of the building from the start of planning, through construction, to completion. BIM allows building owners and trades to easily track and identify the various components and assemblies of individual systems (such as electrical, HVAC, and plumbing) and building materials within a specific location or locations of a building.

BIM also has the capabilities of simulating the lifecycles of various systems within the building in order to provide information on determining the wear on, and lifespan of, individual components. Additionally, BIM software allows the trades and the building owner to input and save information on the building as repairs, updates, or renovations are made to the building. Information such as part sizes, manufacturers, part numbers, and related drawings, or any other information ever researched in the past can be stored in a BIM system.

Another advantage of BIM lies in its capabilities of allowing documentation and information on a building to be easily transferred and communicated between all the members of a project.

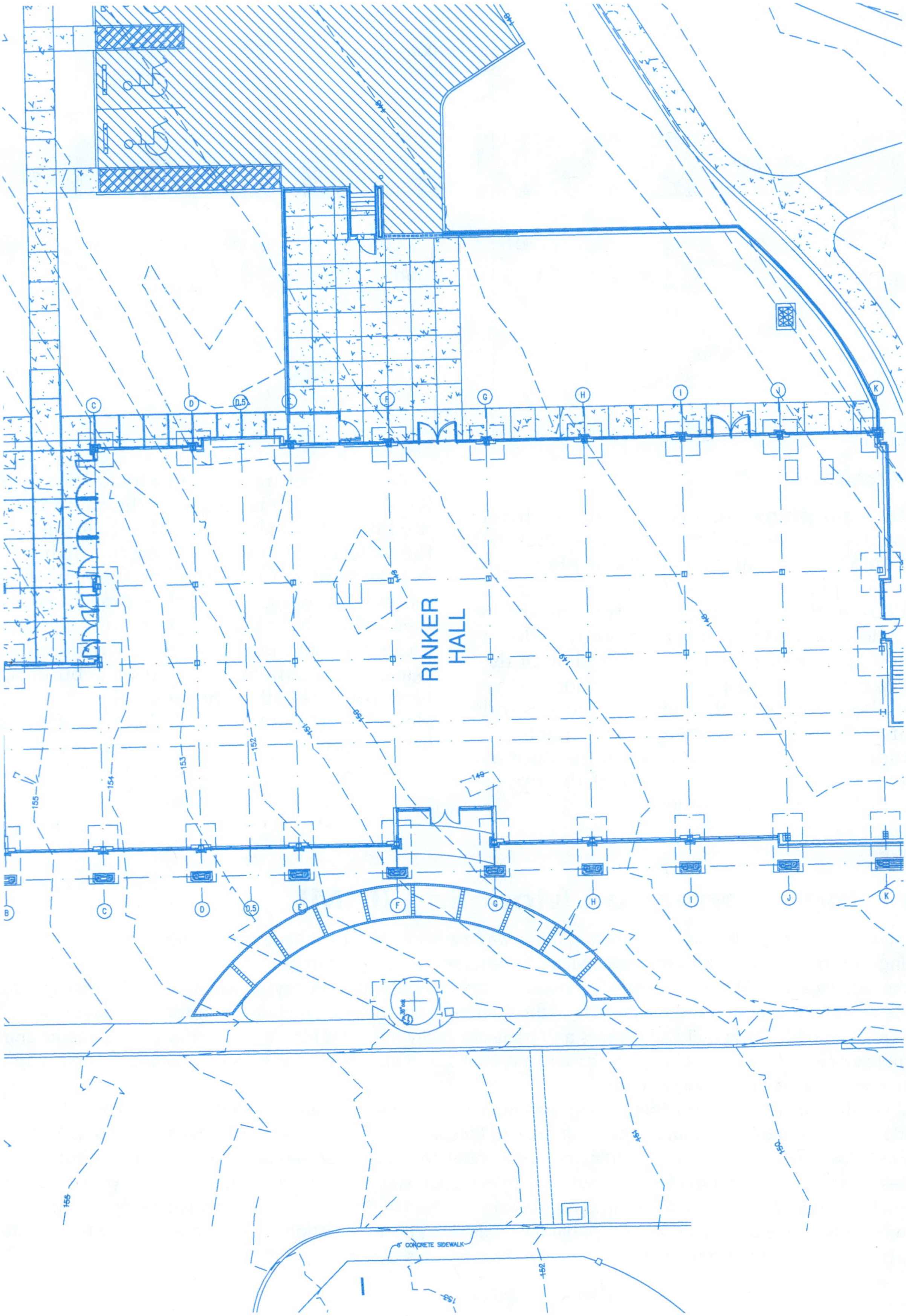

00105-15_F02.EPS

Figure 2 Civil plan, aerial view.

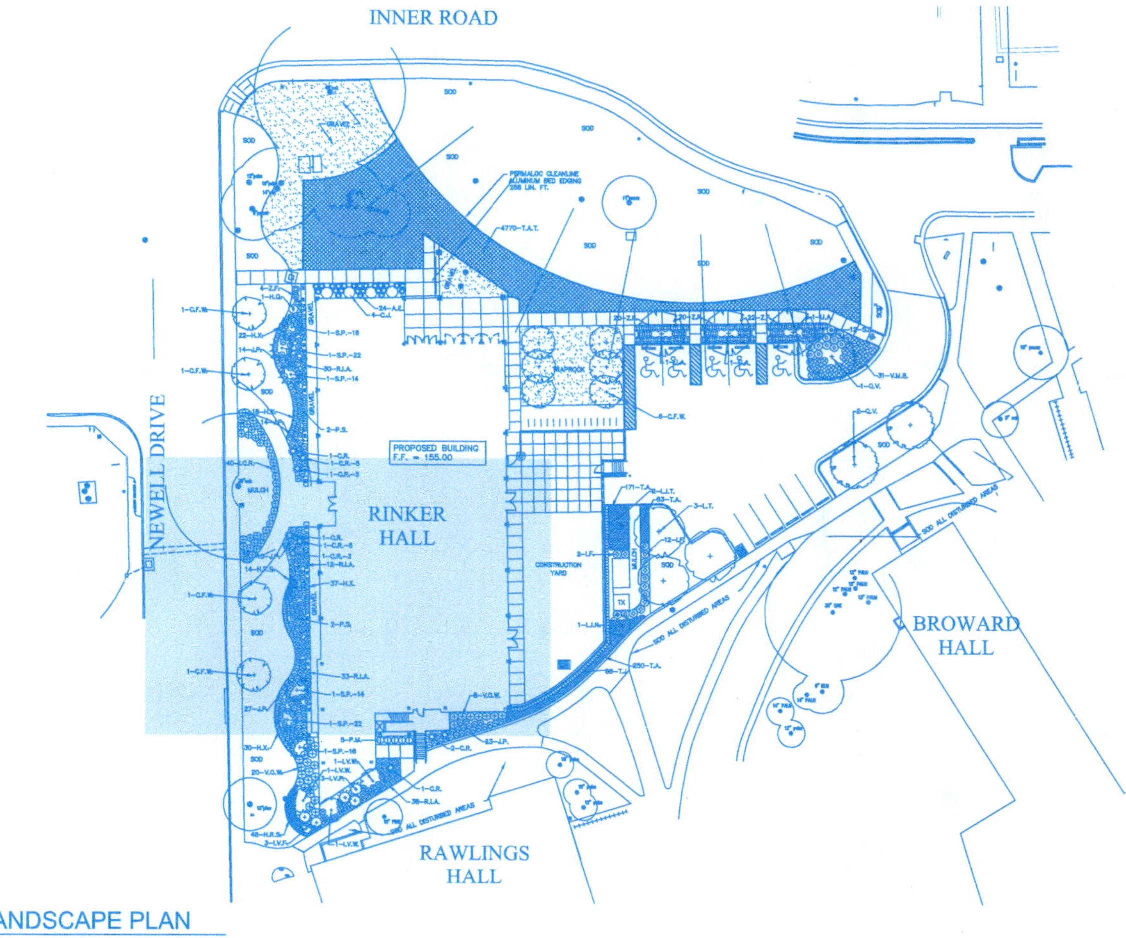

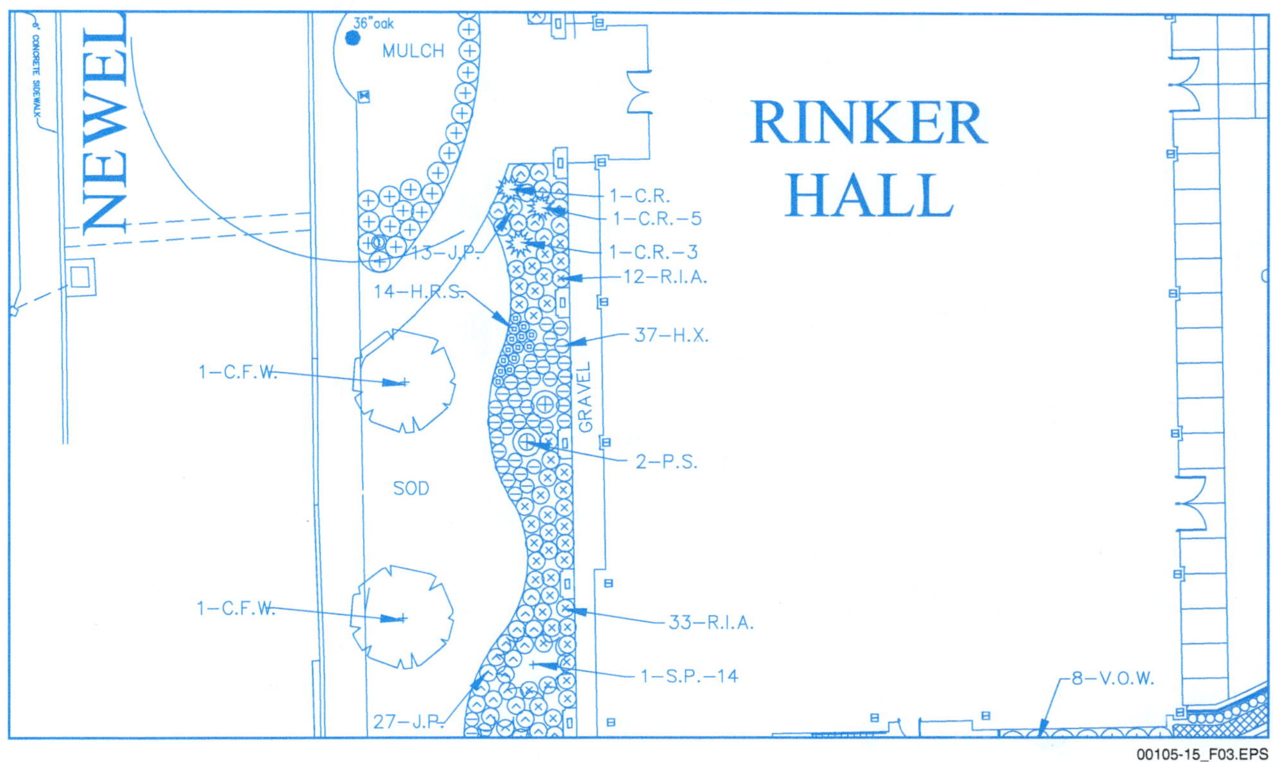

Figure 3 Landscape plan.

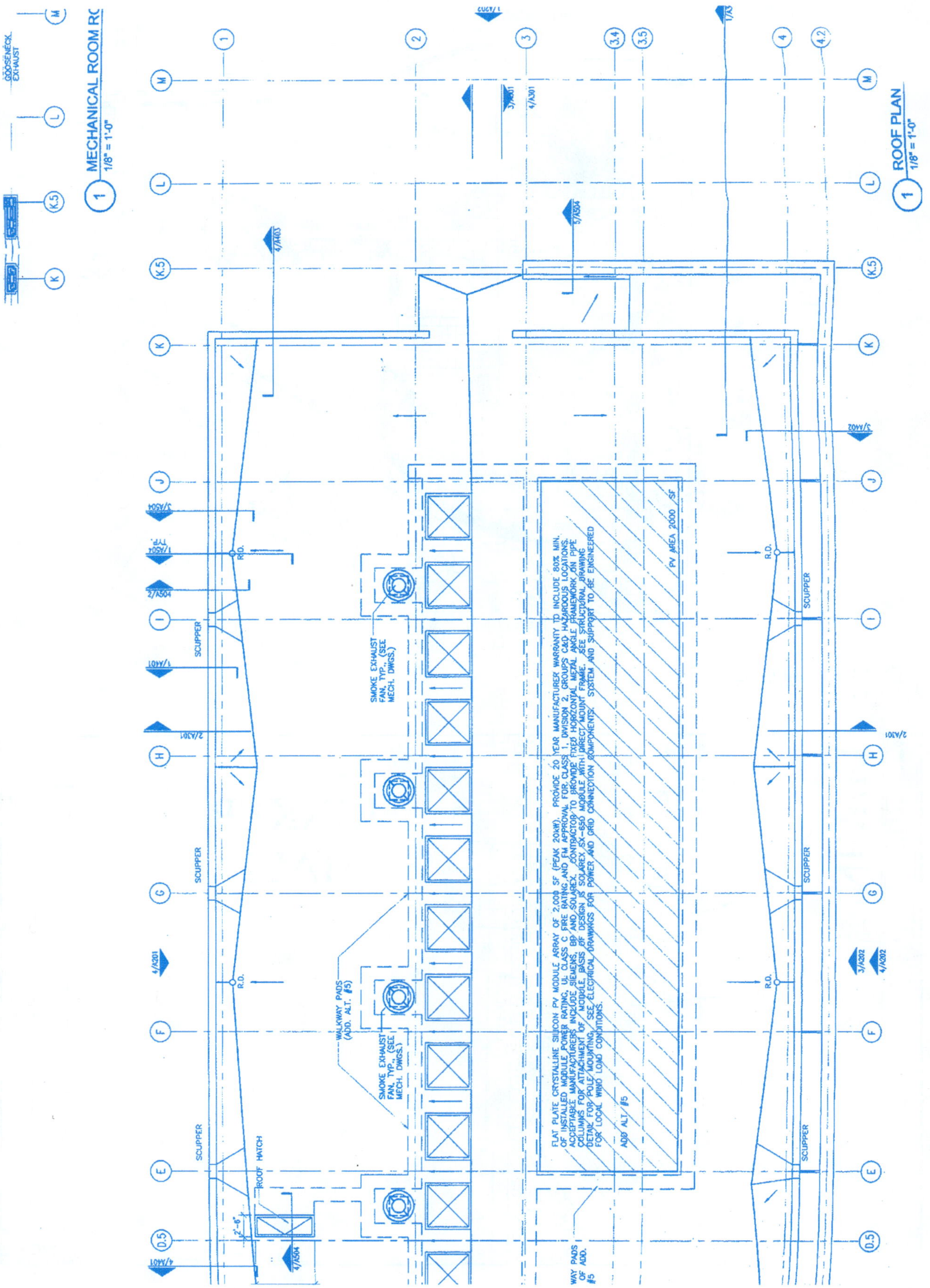

Figure 4 Roof plan.

00105-15_F04.EPS

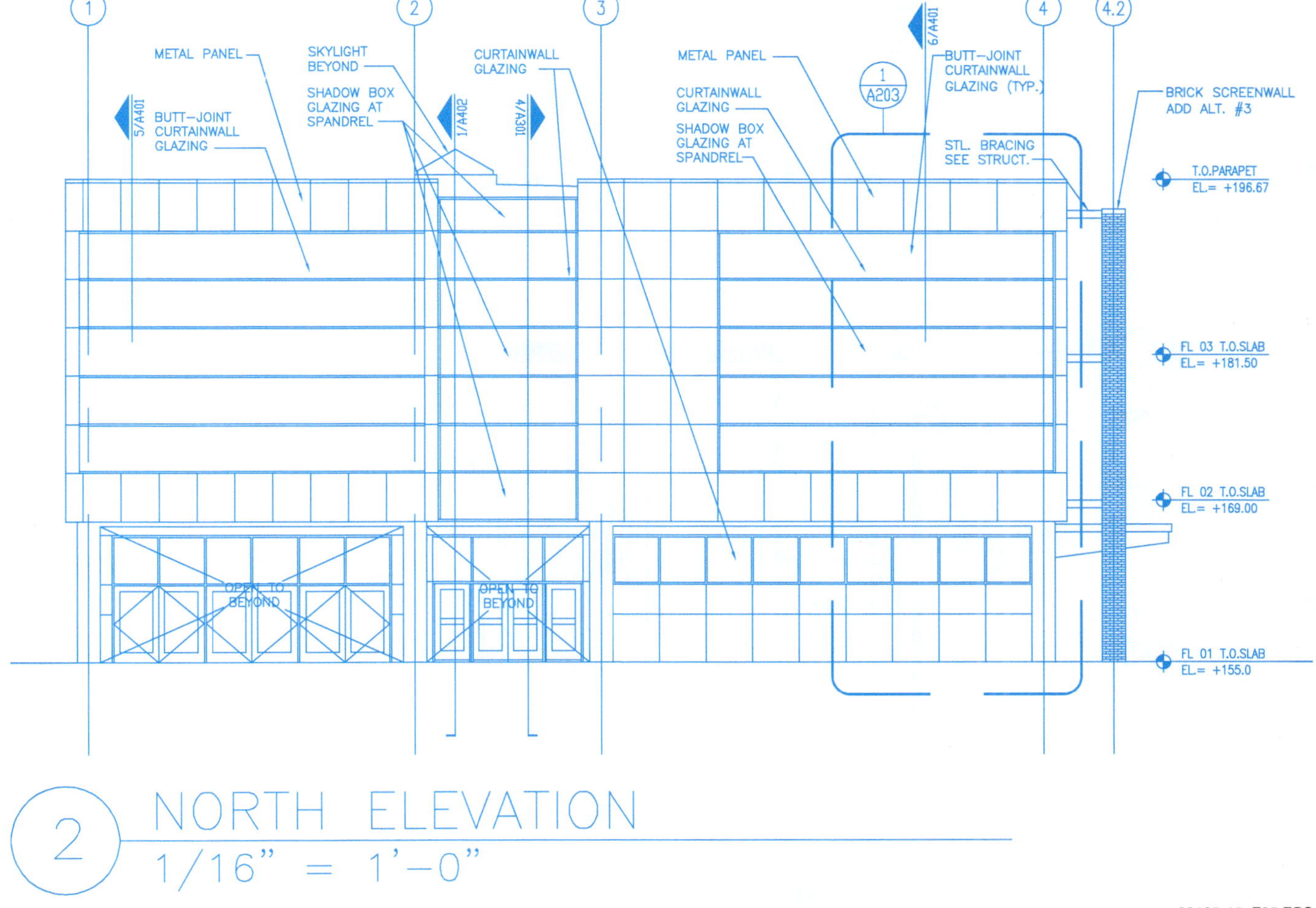

Figure 5 Exterior elevation.

Another element of the architectural plan is **section drawings**, which show how the structure is to be built. Section drawings (*Figure 6*) are cross-sectional views that show the inside of an object or building. They show what construction materials to use and how the parts of the object or building fit together. They normally show more detail than plan views. Compare the information on the drawing in *Figure 6* with the information on the drawing in *Figure 5*.

Even more detail is shown in **detail drawings** (*Figure 7*), which are enlarged views of some special features of a building, such as floors and walls. They are enlarged to make the details clearer. Often the detail drawings are placed on the same sheet where the feature appears in the plan, but sometimes they are placed on separate sheets and referred to by a number on the plan view.

The architectural plan also shows the finish schedules to be used for the doors and windows of the building. Door and window schedules, for example, are tables that list the sizes and other information about the various types of doors and windows used in the project. *Figure 8* shows an example of a window schedule and a detail drawing for the Type D1 window. Be aware that in some sets of drawings the window schedule and elevation drawings for each type of window may appear on the same sheet. Schedules may also be included for finish hardware and fixtures. These schedules are not drawings, but they are usually included in a set of working drawings.

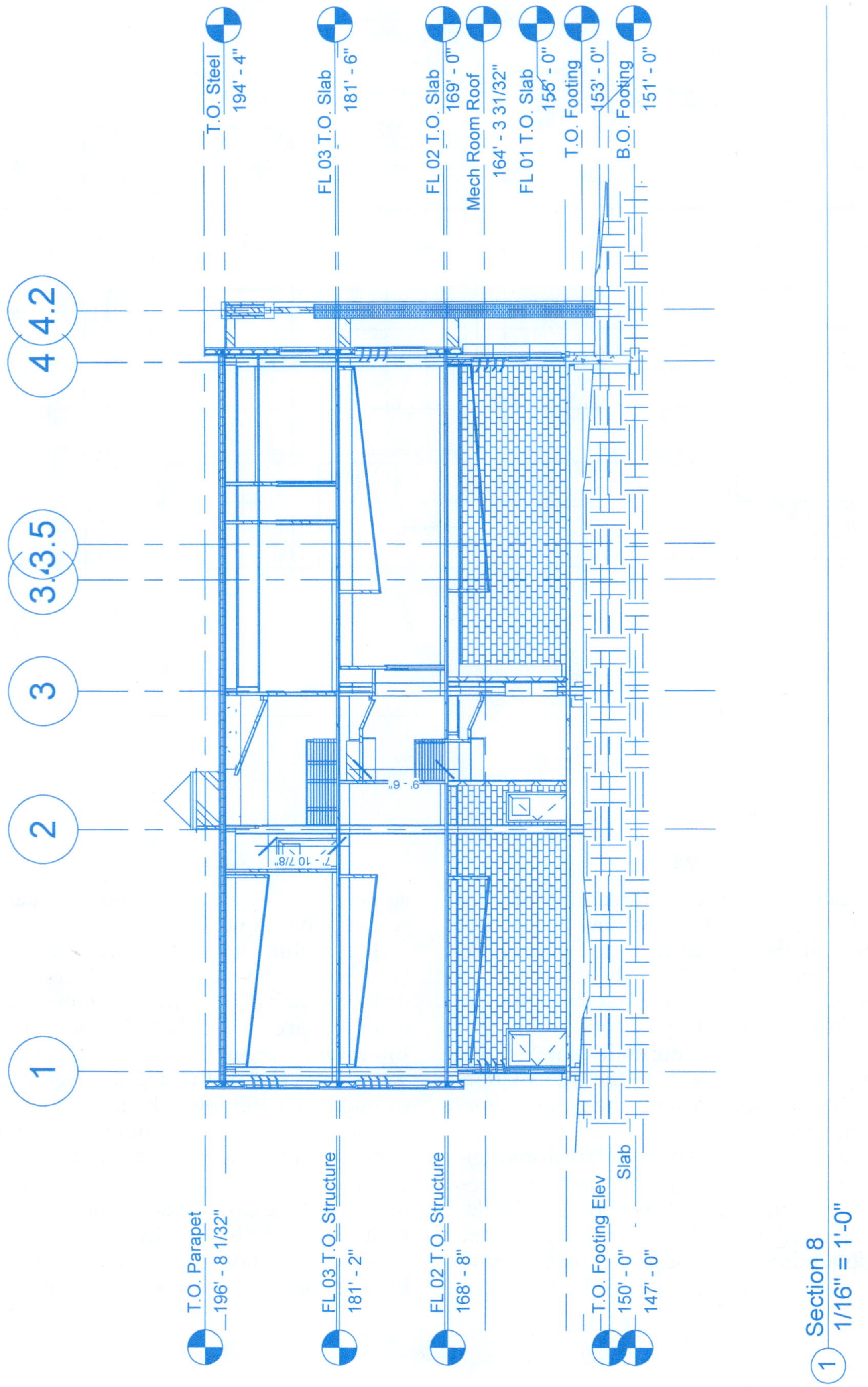

00105-15_F06.EPS

Figure 6 Section drawing (wall section).

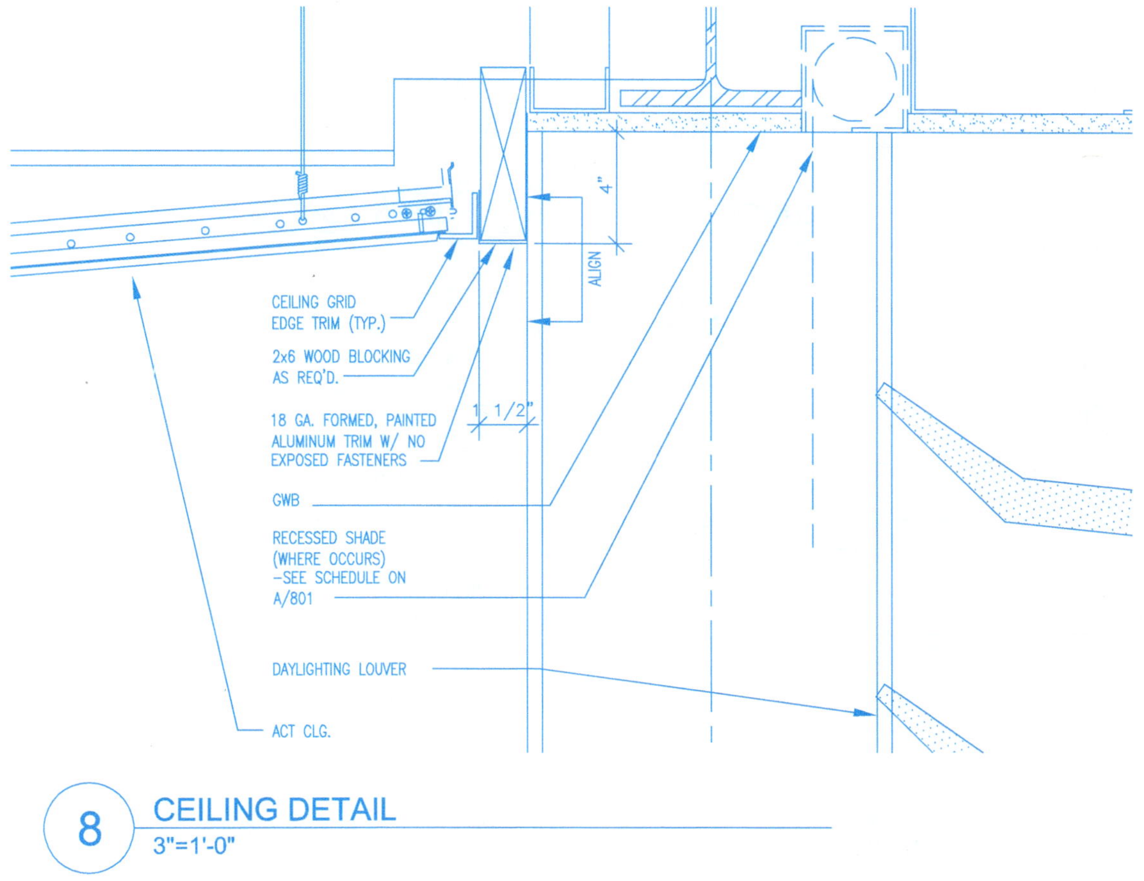

Figure 7 Detail drawing (ceiling detail).

1.1.3 Structural Plans

The **structural plans** are a set of engineered drawings used to support the architectural design. The first part of the structural plans is the general notes (*Figure 9*). These notes give details of the materials to be used and the requirements to be followed in order to build the structure that the architectural plan depicts. The notes, for instance, might specify the type and strength of concrete required for the foundation, the loads that the roof and stairs must be built to accommodate, and codes that contractors must follow. General notes may be on a separate general notes sheet or may be part of individual plan sheets.

The structural plans also include a **foundation plan** (*Figure 10*), which shows the lowest level of the building, including concrete footings, slabs, and foundation walls. They also may show steel girders, columns, or **beams**, as well as detail drawings to show where and how the foundation must be reinforced. Column and spread footing schedules, and foundation notes may be included on the foundation plan. A related element is the structural floor plan, which depicts a wood or metal joist framing and the underlayment of each floor of the structure.

Importance of Architectural Drawings

Look at architectural plans first, because all other drawings follow from them. Architectural plans are the most general; they show how all the parts of the project fit together.

WINDOW SCHEDULE				
TYPE	FRAME SIZE	FRAME FINISH	GLAZING	REMARKS
A	4' 2" x 4' 2"	CLEAR ANODIZED ALUMINUM	VIRACON VE-7-2M	
B	24' 2-1/2" x 11' 0"	CLEAR ANODIZED ALUMINUM	VIRACON VE-7-2M	
B-1	12' 10" x 11' 0"	CLEAR ANODIZED ALUMINUM	VIRACON VE-7-2M	
C	15'-11-1/2" x 8' 3-1/2"	CLEAR ANODIZED ALUMINUM	VIRACON VE-7-2M	
C-1	15'-11-1/2" x 4' 1-1/2"	CLEAR ANODIZED ALUMINUM	VIRACON VE-7-2M	
D	15 11-1/2'-0" x 11' 1-1/2"	CLEAR ANODIZED ALUMINUM	VIRACON VE-7-2M	
D-1	15 11-1/2'-0" x 11' 1-1/2"	CLEAR ANODIZED ALUMINUM	VIRACON VE-7-2M	
D-2	31' 11-1/2" x 4' 4"	CLEAR ANODIZED ALUMINUM	VIRACON VE-7-2M	
E	14' 4" x 7' 8"	CLEAR ANODIZED ALUMINUM	VIRACON VE-7-2M	
F	15'-11-1/2" x 4' 4 "	CLEAR ANODIZED ALUMINUM	VIRACON VE-7-2M	
H	11' 11-1/2" x 8' 1/4"	CLEAR ANODIZED ALUMINUM	VIRACON VE-7-2M	
H-1	11' 11-1/2" x 10' 1-3/4"	CLEAR ANODIZED ALUMINUM	VIRACON VE-7-2M	
I	15'-11-1/2" x 4' 1-1/2 "	CLEAR ANODIZED ALUMINUM	VIRACON VE-7-2M	
CW - 1	12' 0" x 27' 11-1/4"	CLEAR ANODIZED ALUMINUM	VIRACON VE-7-2M	
CW - 3	15' 11 1/2" x 37' 3 1/4"	CLEAR ANODIZED ALUMINUM	VIRACON VE-7-2M	
CW - 4	36' 10" x 20' 9 1/2"	CLEAR ANODIZED ALUMINUM	VIRACON VE-7-2M	
CW - 5	28' 11-7/8" x 20' 9-1/2"	CLEAR ANODIZED ALUMINUM	VIRACON VE-7-2M	

NOTES

1. FRAME SIZES HAVE BEEN INDICATED ASSUMING 1/2" JOINT ADJACENT TO METAL PANEL SYSTEM. ALIGNMENT OF JOINTS W/ METAL PANEL MODULE IS REQUIRED. INTERMEDIATE MULLIONS TO BE CENTERED ON ADJACENT METAL PANEL JOINTS

2. "SG" INDICATES SAFETY GLAZING (TEMPERED GLASS REQUIRED)

1

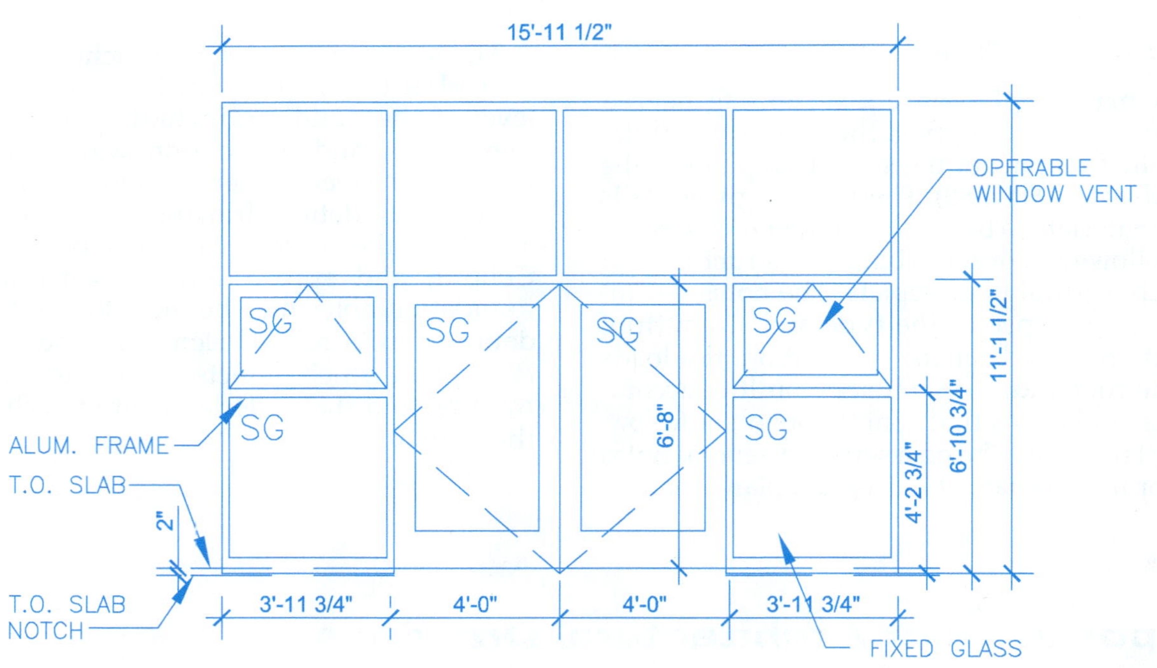

00105-15_F08.EPS

Figure 8 Window schedule and window detail.

GENERAL NOTES

I. DESIGN CRITERIA

A. GENERAL BUILDING CODE
The Contract Documents are based on the requirements of the:
1. Standard Building Code, 1997 edition.

B. DEAD LOADS
1. Partitions. An allowance of 20 PSF has been made for partitions as a uniformly distributed dead load.
2. Hanging Ceiling and Mechanical Loads. An allowance of 10 PSF has been made for hanging ceiling and mechanical equipment loads such as duct work and sprinkler pipes.

C. LIVE LOADS
1. Design live loads are based on the more restrictive of the uniform load listed below or the concentrated load listed acting over an area 2.5 feet square.

CATEGORY	UNIFORM LOAD (PSF)	CONCENTRATED LOAD (LB)
1. Roof	20	N/A
2. Elevated Floors	50	0
3. Terraces, Lobbies	100	0
4. Stairways, Exit Facilities	100	0
5. Elevator Machine Rooms	100	Assumed Eqp. Wt.
6. Mechanical Rooms, typical	150	Assumed Eqp. Wt.

NOTES:
1. Live Load Reduction. Live loads have been reduced on any member supporting more than 150 square feet, including flat slabs, except for floors in places of public assembly and for live loads greater than 100 pounds per square foot in accordance with the following formula:

R = r(A−150)

The reduction,, R, shall not exceed 40 percent for members supporting one level only, 60 percent for other members, or R as calculated in the following formula:

$$R = 23.1\left(1 + \frac{D}{L}\right)$$

R = Reduction in percent.
r = Rate of reduction equal to .08 percent for floors.
A = Area of floor supported by the member.
D = Total dead load supported by the member.
L = Total, unreduced, live load supported by the member.

2. For storage loads exceeding 100 pounds per square foot, no reduction has been made, except that design live loads on columns have been reduced 20 percent.

D. ELEVATOR LOADS
Machine Beam, Car Buffer, Counterweight Buffer, and Guide Rail Loads. Assumed elevator loads to the supporting structure are shown on the drawings, including machine beam reactions, car buffer reactions, counterweight buffer reactions, and horizontal and vertical guide rail loads. The General Contractor shall submit to the Structural Engineer final elevator shop drawings showing all loads to the structure prior to the installation of the elevators for verification of load carrying capacity.

E. MECHANICAL EQUIPMENT LOADS
The General Contractor shall submit actual weights of equipment to be used in the project to the Structural Engineer for verification of loads used in the design at least three weeks prior to fabrication and construction of the supporting structure.

F. WIND LOADS
1. Wind pressures are based on the American Society of Civil Engineers, Minimum Design Loads for Buildings and Other Structures, ASCE 7–98 with a Wind Speed = 110 MPH (3 sec. gust), Exposure C, Importance Factor 1.15.
2. Wind pressures used in the design of the cladding are shown on these Drawings.

II. FOUNDATION

A. GEOTECHNICAL REPORT
Foundation design is based on the geotechnical investigation report as follows:
1. Reports of Geotechnical exploration, M.E. Rinker Sr. Hall (Revised Location) Near the southeast Corner of Newell Drive and Inner Road, Gainesville, Alachua County, Florida. Law Engineering and Environmental Services, Inc. January 2, 2001.

The geotechnical report is available to the General Contractor upon request to the Owner. The information included therein may be used by the General Contractor for his general information only. The Architect and Engineer will not be responsible for the accuracy or applicability of such data therein.

B. FOUNDATION TYPE
1. Spread Footing.
a. Design Pressures:
1. All footings have been designed assuming an allowable bearing pressure of 4000 PSF.
Allowable pressures are increased 33% for combined gravity and wind loads.

C. SLAB–ON–GRADE
Radon resistant construction guidlines are being followed on this project. The details and specifications for slab–on–grade construction must be adhered to without deviation.
Slab–on–Grade shall be immediately underlain by a 8 mil. vapor barrier. Seams shall be lapped 12 inches and sealed with 2" wide pressure sensitive vinyl tape. All penetrations shall be sealed with tape.

D. CONSTRUCTION DEWATERING
The Contractor shall determine the extent of construction dewatering required for the excavation. The Contractor shall submit to the Geotechnical Engineer for review the proposed plan for construction dewatering, prior to beginning the excavation.

III. REINFORCED CONCRETE

A. CLASSES OF CONCRETE
All concrete shall conform to the requirements as specified in the table below unless noted otherwise on the drawings:

Usage	28 Day Comp. Strength (PSI)	Conc Type	Max Size Agg.	W/C Ratio
1. Elevated Floors	4000	NWT	3/4"	0.48
2. Spread Footings	3000	NWT	1"	0.55
3. Slab–On–Grade	4000	NWT	1"	0.48
4. Fnd. Walls & Plinths	4000	NWT	1"	0.48

All concrete shall be proportioned for a maximum allowable unit shrinkage of 0.03% measured at 28 days after curing in lime water as determined by ASTM C 157 (using air storage).

B. HORIZONTAL CONSTRUCTION JOINTS IN CONCRETE POURS
There shall be no horizontal construction joints in any concrete pours unless shown on the drawings. The Architect/Engineer shall approve all deviations or additional joints in writing.

C. REINFORCING STEEL SPECIFICATION
1. All Reinforcing Steel shall be ASTM A615 Grade 60 unless noted otherwise on the drawings or in these notes.
2. Welded Reinforcing Steel. Provide reinforcing steel conforming to ASTM A706 for all reinforcing steel required to be welded and where noted on the drawings.
3. Galvanized Reinforcing Steel. Provide reinforcing steel galvanized according to ASTM A767 Class II (2.0 oz. zinc PSF where noted on the drawings.
4. Deformed Bar Anchors. ASTM A496 minimum yield strength 70,000 PSI as noted on the drawings. Reinforcing bars shall not be substituted for deformed bar anchors.
5. Welded Wire Fabric. Welded smooth wire fabric, ASTM A 185, yield strength 65,000 PSI where noted on the drawings. Welded deformed wire fabric for, ASTM A 497, yield strength 70,000 PSI where noted on the drawings.

D. PLACEMENT OF WELDED WIRE FABRIC
Wherever welded wire fabric is specified as reinforcement, it shall be continuous across the entire concrete surface and not interrupted by beams or girders and properly lapped one cross wire spacing plus 2".

E. REINFORCEMENT IN TOPPING SLABS
Provide welded smooth wire fabric minimum 6 x 6 W2.9 x W2.9 in all topping slabs unless specified otherwise on the drawings.

F. REINFORCEMENT IN HOUSEKEEPING PADS
Provide welded smooth wire fabric 6 x 6 W2.9 x W2.9 minimum in all housekeeping pads supporting mechanical equipment whether shown on the drawings or not unless heavier reinforcement is called for on the drawings.

G. REINFORCING STEEL COVERAGE
Concrete Cover for reinforcement layer nearest to the surface unless specified otherwise on the drawings.
1. Concrete surfaces cast against and permanently exposed to earth. 3 inches
2. Concrete surfaces exposed to earth or weather or where noted on the drawings 2 inches.
3. Concrete surfaces not exposed to weather or in contact with the ground.
a. #3 to #11 bars 1 inch

H. SPLICES IN REINFORCING STEEL
1. All unscheduled splices shall be Class A tension splice.

IV. STRUCTURAL STEEL

V. STEEL DECKS

VI. CURTAIN WALL

VII. CONCRETE MASONRY

VIII. MISCELLANEOUS

IX. SUBMITTALS

X. DRAWING INTERPRETATION

00105-15_F09.EPS

Figure 9 General notes for structural plans.

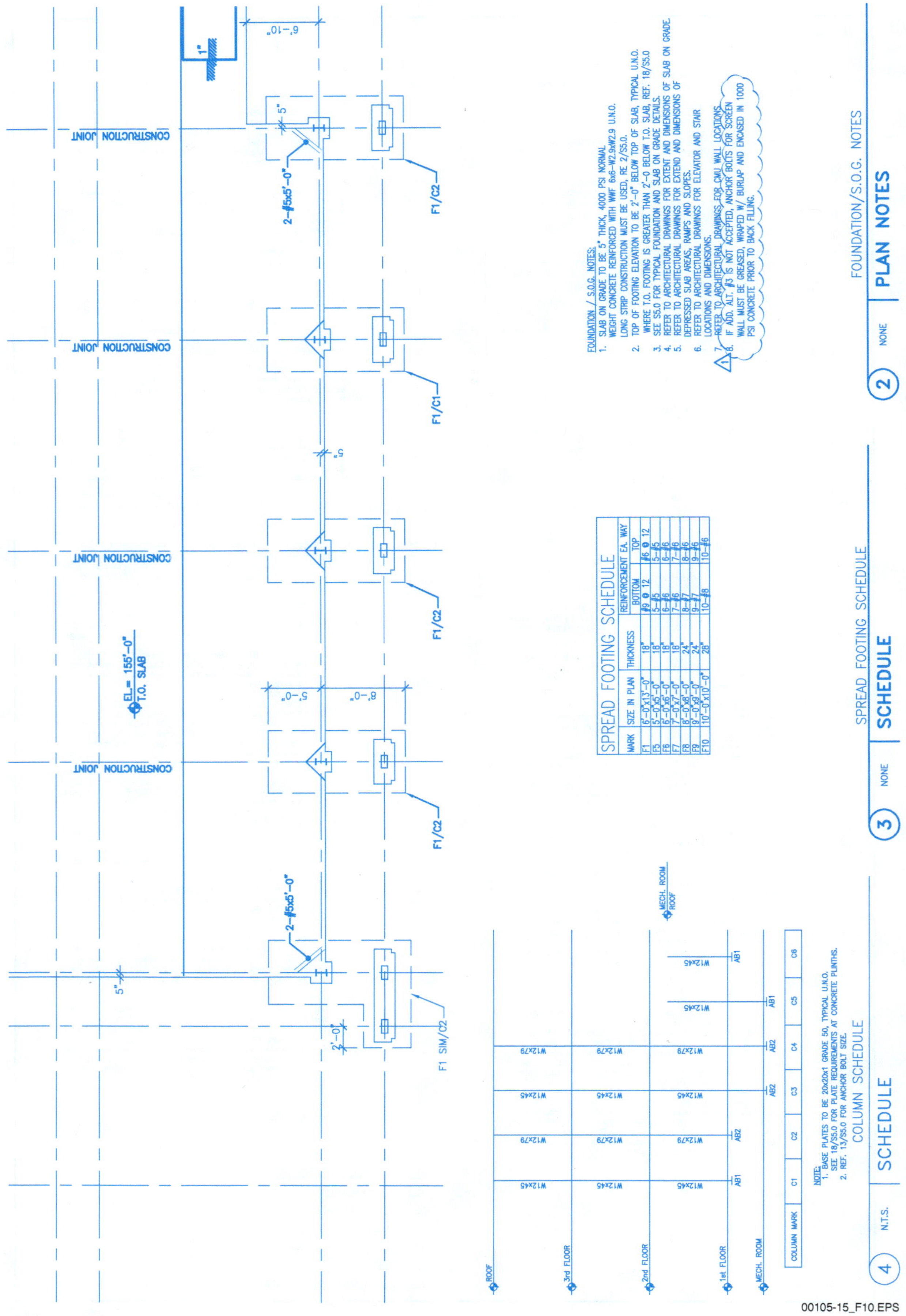

CONSTRUCTION JOINT

CONSTRUCTION JOINT

CONSTRUCTION JOINT

CONSTRUCTION JOINT

EL= 155'-0"
T.O. SLAB

1"

6'-10"

5"

2-#5x5'-0"

5'-0"

8'-0"

2'-0"

F1 SIM/C2

F1/C2

F1/C2

F1/C1

F1/C2

ROOF

3rd FLOOR

2nd FLOOR

1st FLOOR

MECH. ROOM

MECH. ROOM
ROOF

COLUMN MARK	C1	C2	C3	C4	C5	C6
Roof – 3rd Floor	W12x45	W12x79	W12x45	W12x79		
3rd Floor – 2nd Floor	W12x45	W12x79	W12x45	W12x79		
2nd Floor – 1st Floor	W12x45	W12x79	W12x45	W12x79	W12x45	W12x45
Anchor bolts	AB1	AB2	AB2	AB2	AB1	AB1

NOTE:
1. BASE PLATES TO BE 20x20x1 GRADE 50, TYPICAL U.N.O. SEE 18/S5.0 FOR PLATE REQUIREMENTS AT CONCRETE PLINTHS.
2. REF. 13/S5.0 FOR ANCHOR BOLT SIZE.

COLUMN SCHEDULE

4 N.T.S. SCHEDULE

SPREAD FOOTING SCHEDULE

MARK	SIZE IN PLAN	THICKNESS	REINFORCEMENT EA. WAY BOTTOM	REINFORCEMENT EA. WAY TOP
F1	6'-0"x13'-0"	18"	#9 @ 12	#6 @ 12
F5	5'-0"x5'-0"	18"	5-#5	5-#5
F6	6'-0"x6'-0"	18"	6-#6	6-#6
F7	7'-0"x7'-0"	18"	7-#6	7-#6
F8	8'-0"x8'-0"	24"	8-#7	8-#6
F9	9'-0"x9'-0"	24"	9-#7	9-#6
F10	10'-0"x10'-0"	28"	10-#8	10-#6

SPREAD FOOTING SCHEDULE

3 NONE SCHEDULE

FOUNDATION / S.O.G. NOTES:
1. SLAB ON GRADE TO BE 5" THICK, 4000 PSI NORMAL WEIGHT CONCRETE REINFORCED WITH WWF 6x6-W2.9xW2.9 U.N.O. LONG STRIP CONSTRUCTION MUST BE USED, RE 2/S5.0.
2. TOP OF FOOTING ELEVATION TO BE 2'-0" BELOW TOP OF SLAB, TYPICAL U.N.O. WHERE T.O. FOOTING IS GREATER THAN 2'-0 BELOW T.O. SLAB, REF. 18/S5.0
3. SEE S5.0 FOR TYPICAL FOUNDATION AND SLAB ON GRADE DETAILS.
4. REFER TO ARCHITECTURAL DRAWINGS FOR EXTENT AND DIMENSIONS OF SLAB ON GRADE.
5. REFER TO ARCHITECTURAL DRAWINGS FOR EXTEND AND DIMENSIONS OF DEPRESSED SLAB AREAS, RAMPS AND SLOPES.
6. REFER TO ARCHITECTURAL DRAWINGS FOR ELEVATOR AND STAIR LOCATIONS AND DIMENSIONS.
7. REFER TO ARCHITECTURAL DRAWINGS FOR CMU WALL LOCATIONS.
8. IF ADD. ALT. #3 IS NOT ACCEPTED, ANCHOR BOLTS FOR SCREEN WALL MUST BE GREASED, WRAPED W/ BURLAP AND ENCASED IN 1000 PSI CONCRETE PRIOR TO BACK FILLING.

FOUNDATION/S.O.G. NOTES

2 NONE PLAN NOTES

00105-15_F10.EPS

Figure 10 Foundation plan (foundation/slab-on-grade plan).

The structural plans show the materials to be used for the walls, whether concrete or masonry, and whether the framing is wood or steel. Structural plans also include a roof-framing plan, showing what kinds of ceiling joists and roof rafters are to be used and where trusses are to be placed (refer to *Drawing 2*, Roof Framing Plan, in the *Appendix*). Notes for the framing plan are usually found on the same sheet as the drawing.

The structural plans include structural section drawings (*Figure 11*), which are similar to the architectural section drawings but show only the structural requirements. Miscellaneous structural details may also be shown in these sections to provide a better understanding of such things as connections and attachments of accessories.

1.1.4 Mechanical Plans

Mechanical plans are engineered plans for motors, pumps, piping systems, and piping equipment. These plans incorporate general notes (*Figure 12*) containing specifications ranging from what the contractor is to provide to how the contractor determines the location of grilles and registers. A mechanical **legend** (*Figure 13*) defines the **symbols** used on the mechanical plans. A list of abbreviations (*Figure 14*) spells out abbreviations found on the plans.

Piping and instrumentation drawings (P&IDs) (*Figure 15*) are **schematic** diagrams of a complete piping system that show the process flow. They also show all the equipment, pipelines, valves, instruments, and controls needed to operate the system.

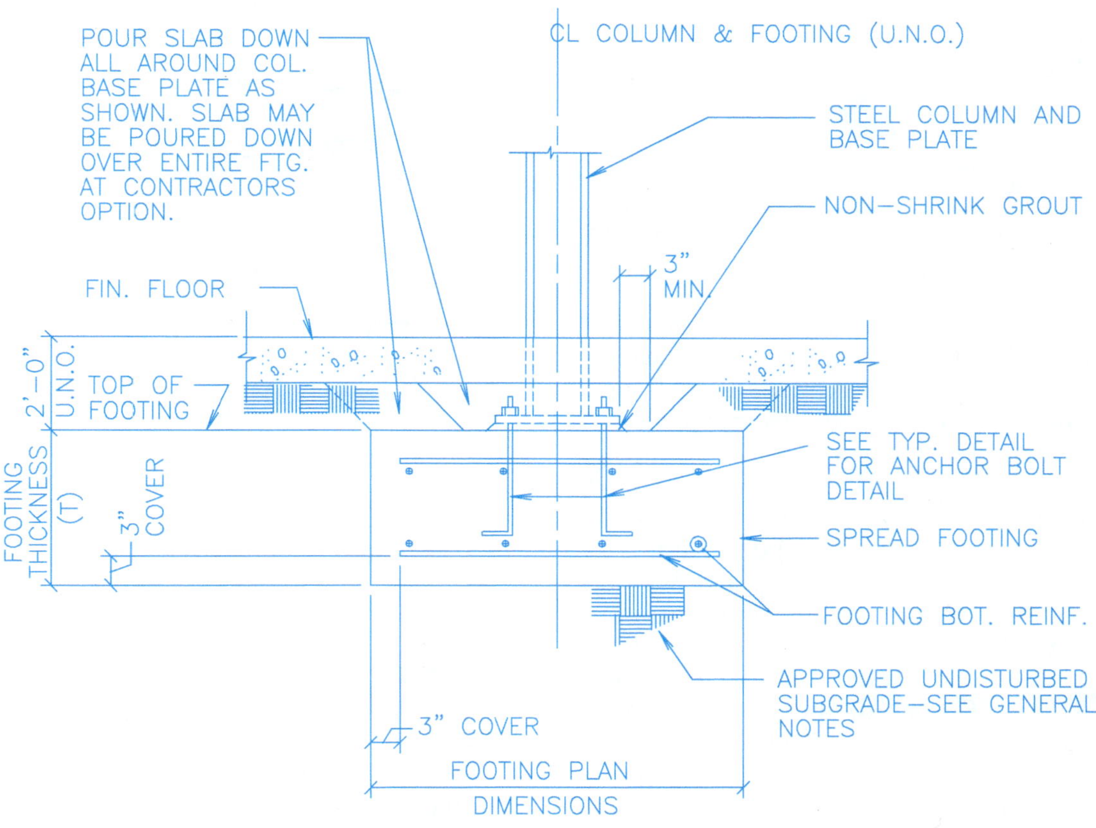

00105-15_F11.EPS

Figure 11 Structural section drawing (foundation details).

GENERAL NOTES
(FOR ALL MECHANICAL DRAWINGS)

1. CONTRACTOR IS TO PROVIDE COMPLETE CONNECTIONS TO ALL NEW AND RELOCATED OWNER FURNISHED EQUIPMENT.
2. CONTRACTOR TO COORDINATE THE LOCATION OF ALL DUCTWORK AND DIFFUSERS WITH REFLECTED CEILING PLAN AND STRUCTURE PRIOR TO BEGINNING WORK.
3. DIMENSIONS FOR INSULATED OR NON-INSULATED DUCT ARE OUTSIDE SHEET METAL DIMENSIONS.
4. DRAWINGS ARE NOT TO BE SCALED FOR DIMENSIONS. TAKE ALL DIMENSIONS FROM ARCHITECTURAL DRAWINGS, CERTIFIED EQUIPMENT DRAWINGS AND FROM THE STRUCTURE ITSELF BEFORE FABRICATING ANY WORK. VERIFY ALL SPACE REQUIREMENTS COORDINATING WITH OTHER TRADES, AND INSTALL THE SYSTEMS IN THE SPACE PROVIDED WITHOUT EXTRA CHARGES TO THE OWNER.
5. LOCATION OF ALL GRILLES, REGISTERS, DIFFUSERS AND CEILING DEVICES SHALL BE DETERMINED FROM THE ARCHITECTURAL REFLECTED CEILING PLANS.
6. THE OWNER AND DESIGN ENGINEER ARE NOT RESPONSIBLE FOR THE CONTRACTOR'S SAFETY PRECAUTIONS OR TO MEANS, METHODS, TECHNIQUES, CONSTRUCTION SEQUENCES, OR PROCEDURES REQUIRED TO PERFORM HIS WORK.
7. ALL WORK SHALL BE INSTALLED IN ACCORDANCE WITH PLRC'S SAFETY PLAN AND ALL APPLICABLE STATE AND LOCAL CODES.
8. ALL EXTERIOR WALL AND ROOF PENETRATIONS SHALL BE SEALED WEATHERPROOF. REFERENCE SPECIFICATION SECTION 15050.
9. ALL MECHANICAL WORK UNDER THIS CONTRACT IS TO FIVE (5) FEET OUTSIDE THE BUILDING.

00105-15_F12.EPS

Figure 12 Mechanical plan general notes.

P&IDs are not drawn to **scale** because they are meant only to give a representation, or a general idea, of the work to be done. Additionally, P&IDs do not indicate north, south, east, and west directions.

For more complex jobs, a separate heating, ventilating, and air conditioning **(HVAC)** plan is added to the set of plans. Piping system plans for gas, oil, or steam heat may be included in the HVAC plan. The mechanical plans include the layout of the HVAC system, showing specific requirements and elements for that system, including a floor, a reflected ceiling, or a roof. HVAC drawings (*Figure 16*) include an electrical schematic that shows the electrical circuitry for the HVAC system. HVAC plans are both mechanical and electrical drawings in one plan.

Importance of Architectural Symbols

When looking at a section drawing, pay close attention to the way different parts are drawn. Each part of the drawing represents a method of construction or a type of material. These symbols are covered in more detail later in this module.

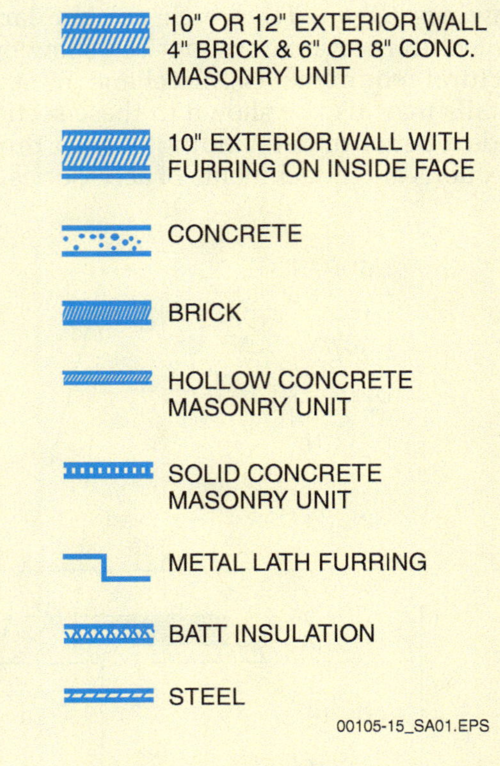

Be aware that a page with a series of mechanical detail drawings may be included in the mechanical plans. These drawings show specific details of certain components within the mechanical system. *Figure 17* is an example of a mechanical detail drawing.

Around the World

Green Buildings

Many countries around the world are pursuing sustainable building construction and design. In Brazil, for example, a Green Building Council (GBC) operates independent of the government. The efforts of the group have resulted in Brazil being ranked fourth in the world in the number of green buildings. This places Brazil behind only the United States, the United Arab Emirates, and China. Over 100 countries now have an active Green Building Council. A World Green Building Council (WGBC) has also been created; established in 2002, the WGBC helps form a network that supports all GBCs worldwide.

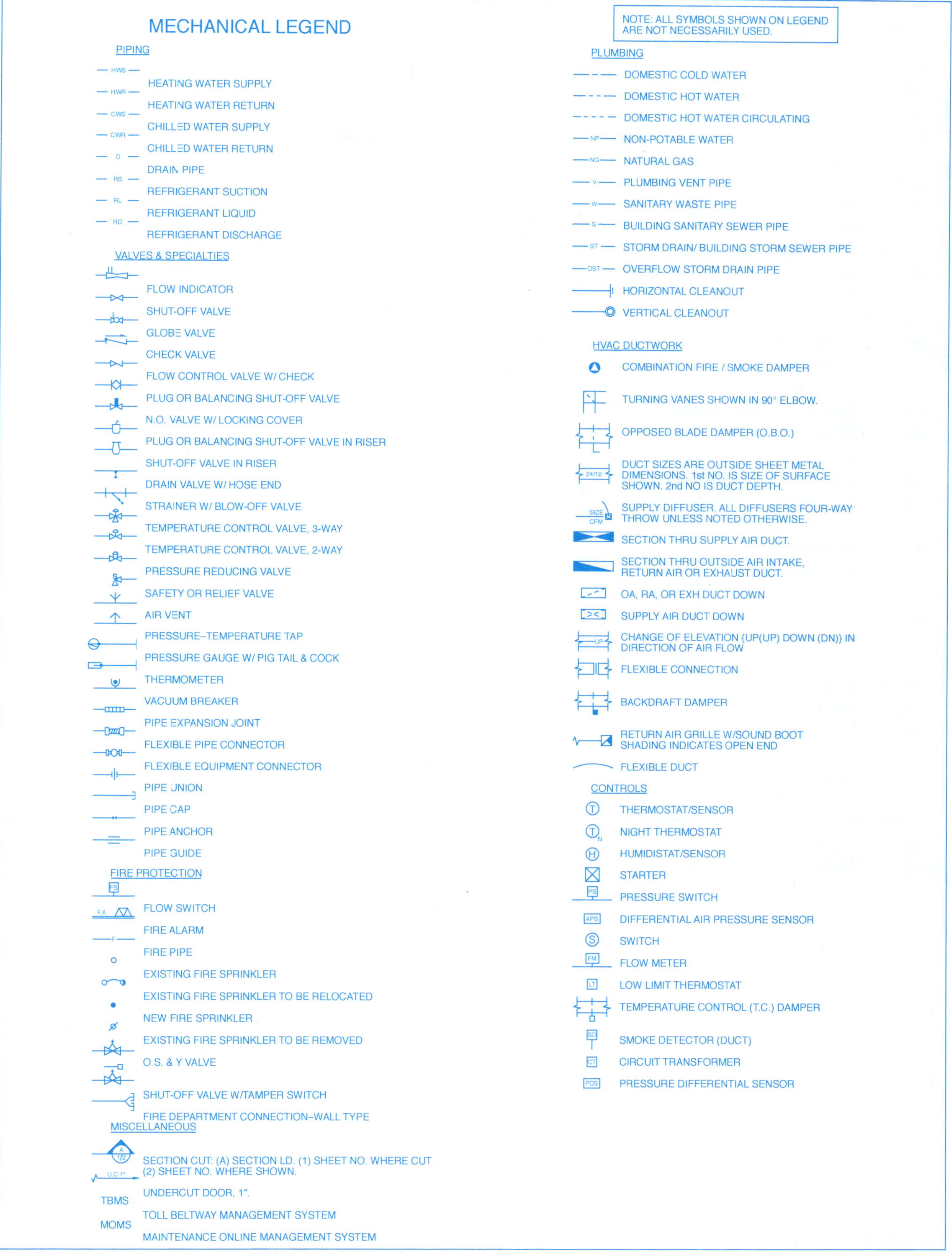

Figure 13 Mechanical plan legend.

ABBREVIATIONS

AFF	ABOVE FINISHED FLOOR
ALT	ALTITUDE
BHP	BRAKE HORSEPOWER
BTU	BRITISH THERMAL UNIT
Cv	COEFFICIENT, VALVE FLOW
CU FT	CUBIC FEET
CU IN	CUBIC INCH
CFM	CUBIC FEET PER MINUTE
SCFM	CFM, STANDARD CONDITIONS
dB	DECIBEL
DCW	DOMESTIC COLD WATER
DEG OR °	DEGREE
DHW	DOMESTIC HOT WATER
DIA	DIAMETER
DB	DRY-BULB
EAT	ENTERING AIR TEMPERATURE
EFF	EFFICIENCY
ELEV or EL	ELEVATION
ESP	EXTERNAL STATIC PRESSURE
EWT	ENTERING WATER TEMPERATURE
EXH	EXHAUST
F	FAHRENHEIT
FLA	FULL LOAD AMPS
FPM	FEET PER MINUTE
FPS	FEET PER SECOND
FT	FOOT OR FEET
FU	FIXTURE UNITS
GA	GAUGE
GAL	GALLONS
GPH	GALLONS PER HOUR
GPM	GALLONS PER MINUTE
HD	HEAD
HG	MERCURY
HGT	HEIGHT
HORZ	HORIZONTAL
HP	HORSEPOWER
HR	HOUR(S)
HWC	HOT WATER CIRCULATING (DOMESTIC)
HZ	HERTZ
ID	INSIDE DIAMETER
IE	INVERT ELEVATION
IN	INCHES
IN W.C.	INCHES WATER COLUMN
KW	KILOWATT
KWH	KILOWATT HOUR
LAT	LEAVING AIR TEMPERATURE
LBS OR #	POUNDS
LF	LINEAR FEET
LRA	LOCKED ROTOR AMPS
LWT	LEAVING WATER TEMPERATURE
MAX	MAXIMUM
MCA	MINIMUM CIRCUIT AMPS
MBH	BTU PER HOUR (THOUSAND)
MIN	MINIMUM
NC	NOISE CRITERIA
N.O.	NORMALLY OPEN
N.C.	NORMALLY CLOSED
N/A	NOT APPLICABLE
NIC	NOT IN CONTRACT
NTS	NOT TO SCALE
NO	NUMBER
OA	OUTSIDE AIR
OD	OUTSIDE DIAMETER
PPM	PARTS PER MILLION
%	PERCENT
PH OR f	PHASE (ELECTRICAL)
PSF	POUNDS PER SQUARE FOOT
PSI	POUNDS PER SQUARE INCH
PSIA	PSI ABSOLUTE
PSIG	PSI GAUGE
PRESS	PRESSURE
RA	RETURN AIR
RECIRC	RECIRCULATE
RH	RELATIVE HUMIDITY
RLA	RUNNING LOAD AMPS
RPM	REVOLUTIONS PER MINUTE
SL	SEA LEVEL
SENS	SENSIBLE
SPEC	SPECIFICATION
SQ	SQUARE
STD	STANDARD
SP	STATIC PRESSURE
SA	SUPPLY AIR
TEMP	TEMPERATURE
TD	TEMPERATURE DIFFERENCE
TSP	TOTAL STATIC PRESSURE
TSTAT	THERMOSTAT
TONS	TONS OF REFRIGERATION
VAV	VARIABLE AIR VOLUME
VEL	VELOCITY
VERT	VERTICAL
V	VOLT
VOL	VOLUME
W	WATT
WT	WEIGHT
WB	WET-BULB

00105-15_F14.EPS

Figure 14 Mechanical plan list of abbreviations.

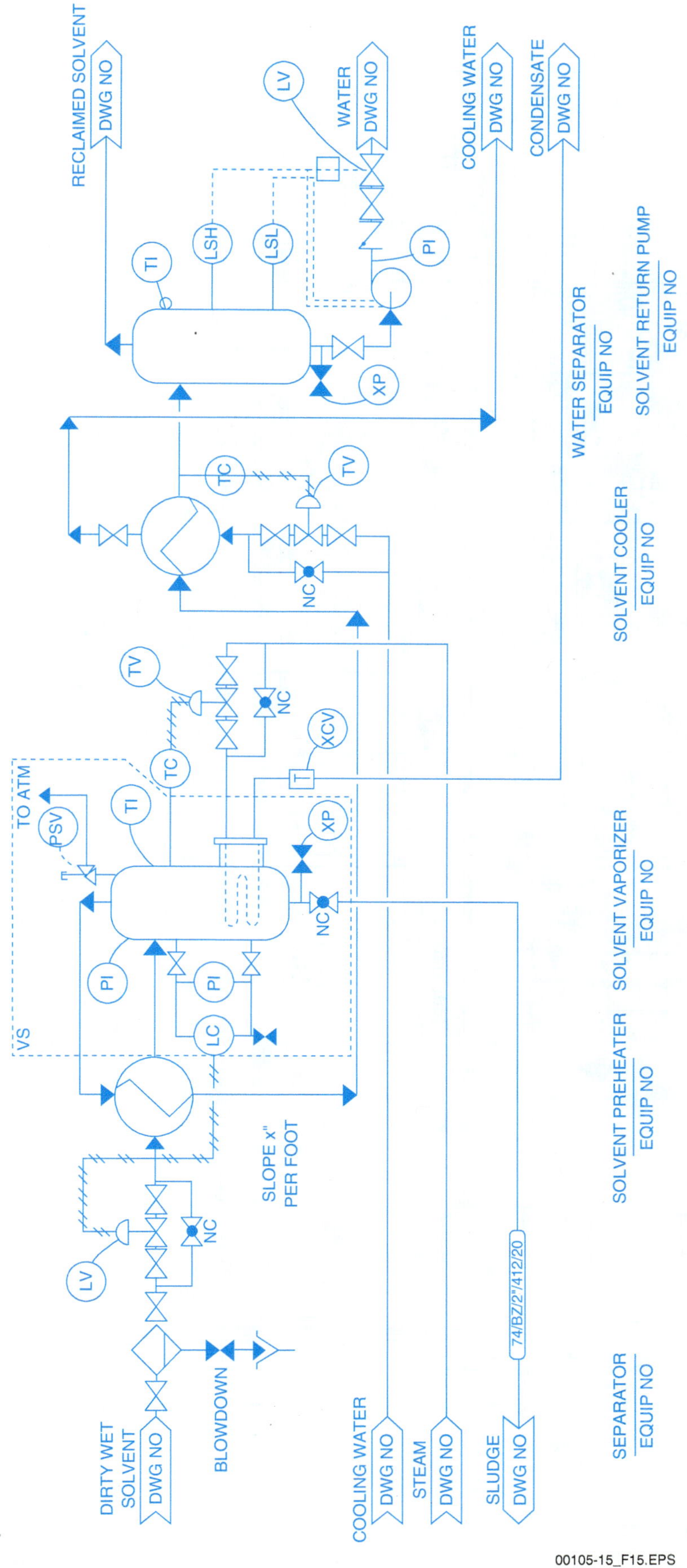

Figure 15 P&ID drawing.

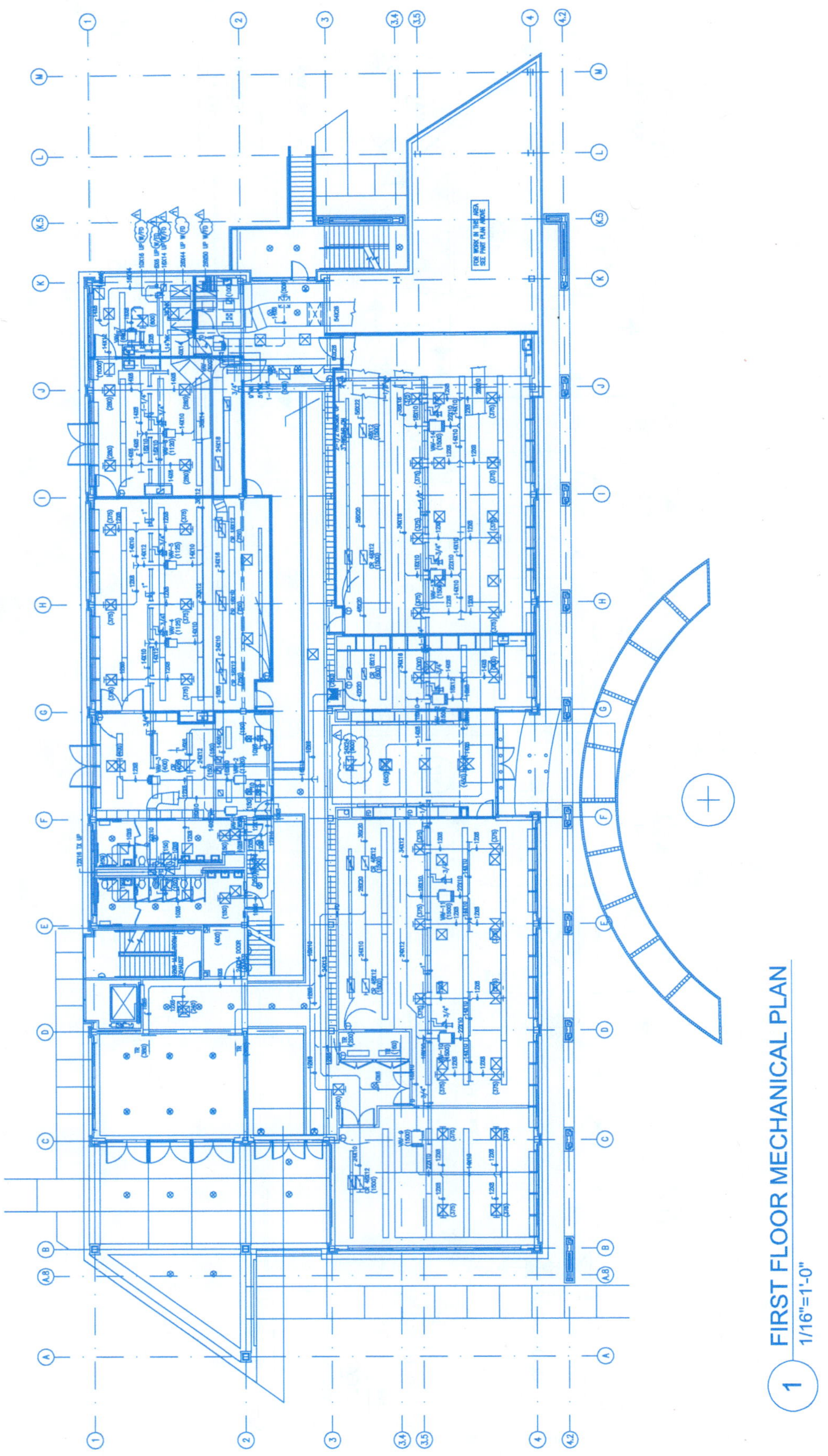

1 FIRST FLOOR MECHANICAL PLAN
1/16"=1'-0"

00105-15_F16A.EPS

Figure 16A HVAC drawing. (1 of 2)

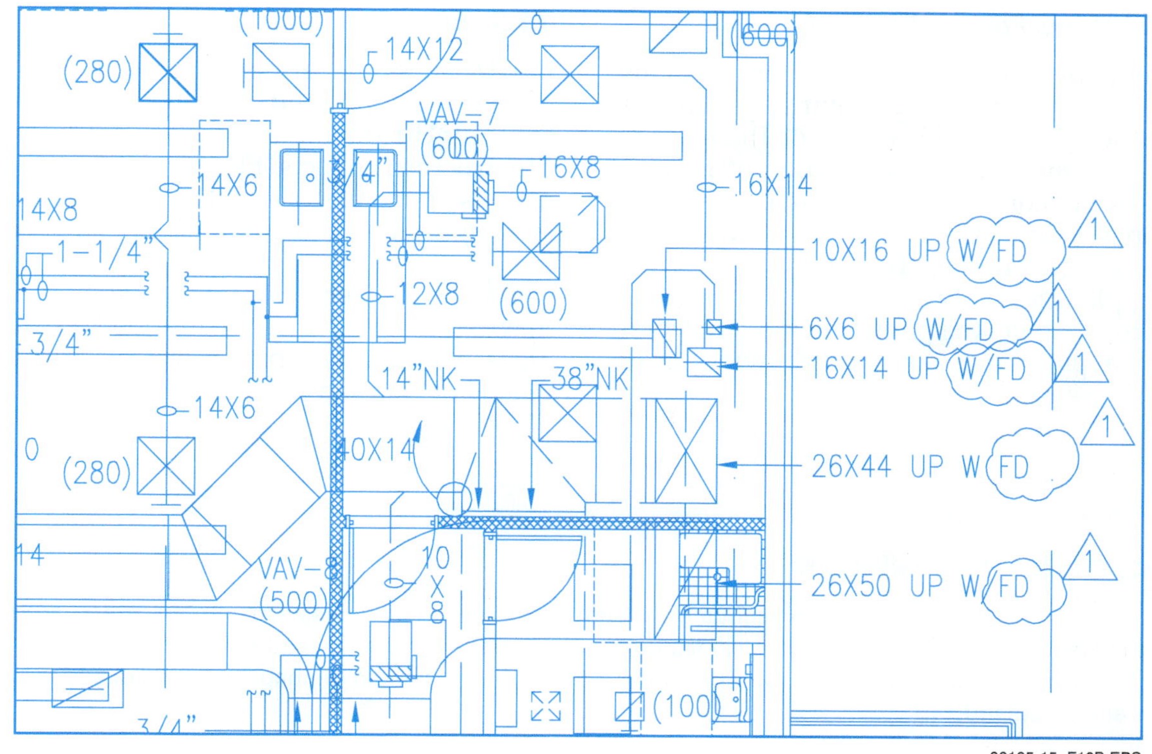

00105-15_F16B.EPS

Figure 16B HVAC drawing. (2 of2)

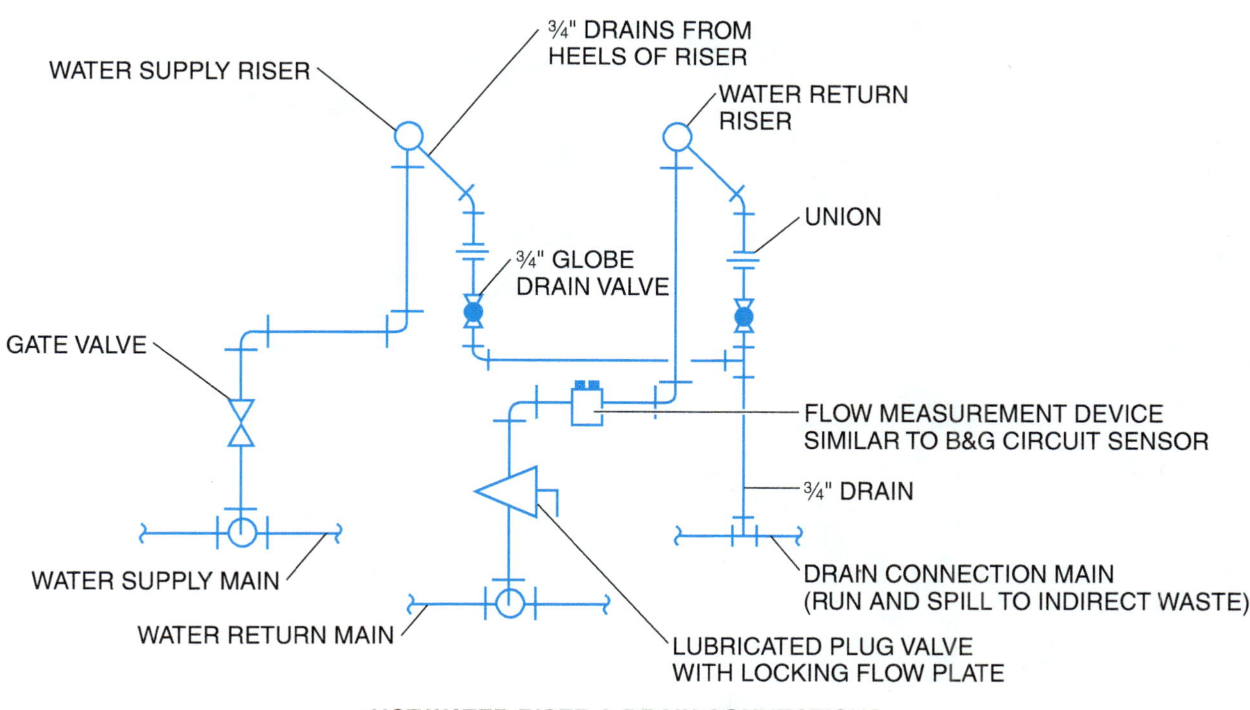

00105-15_F17.EPS

Figure 17 Mechanical detail drawing for hot water riser and drain connections.

1.1.5 Plumbing/Piping Plans

Plumbing plans (*Figure 18*) are engineered plans showing the layout for the plumbing system that supplies the hot and cold water, for the sewage disposal system, and for the location of plumbing fixtures. For commercial projects, each system may be on a separate plan.

A **plumbing isometric drawing** is part of the plumbing plan. It is a type of three-dimensional drawing that depicts the plumbing system. *Figure 19* shows a plumbing isometric drawing for a sanitary riser system.

1.1.6 Electrical Plans

Electrical plans are engineered drawings for electrical supply and distribution. These plans may appear on the floor plan itself for simple construction projects. Electrical plans include locations of the electric meter, distribution panel, switchgear, convenience outlets, and special outlets.

For more complex projects, the information may be on a separate plan added to the set of plans. This separate plan leaves out construction-related details and shows just the electrical layout.

Topographic Maps

Topographic (topo) maps provide a representation of vertical dimension that gives a feel for the shape or contour of a piece of land. Topo maps identify physical features such as mountains, lakes, and streams. These maps indicate where highways and railroads run. They often include information on drainage and land use such as orchards and woodland.

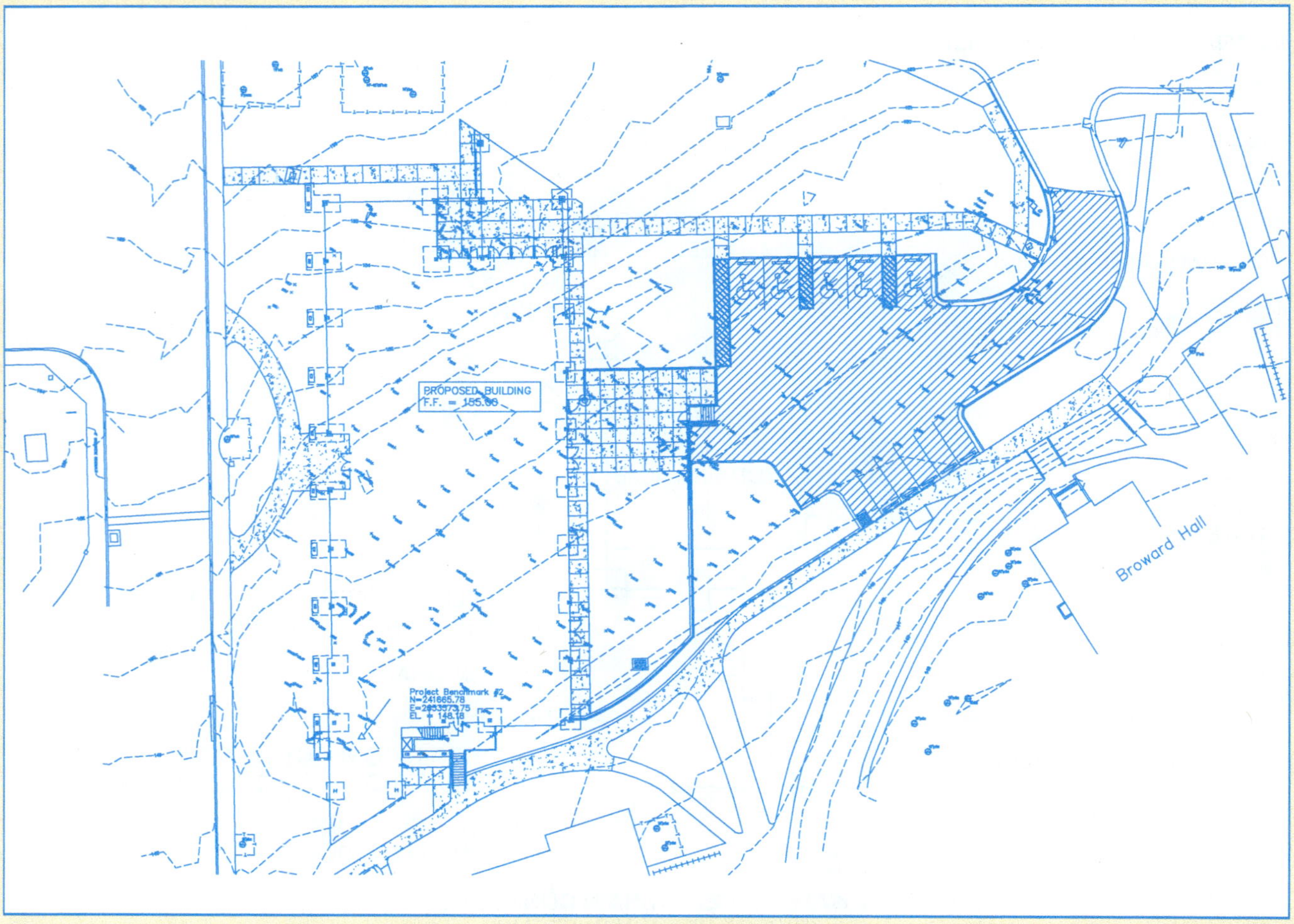

00105-15_SA02.EPS

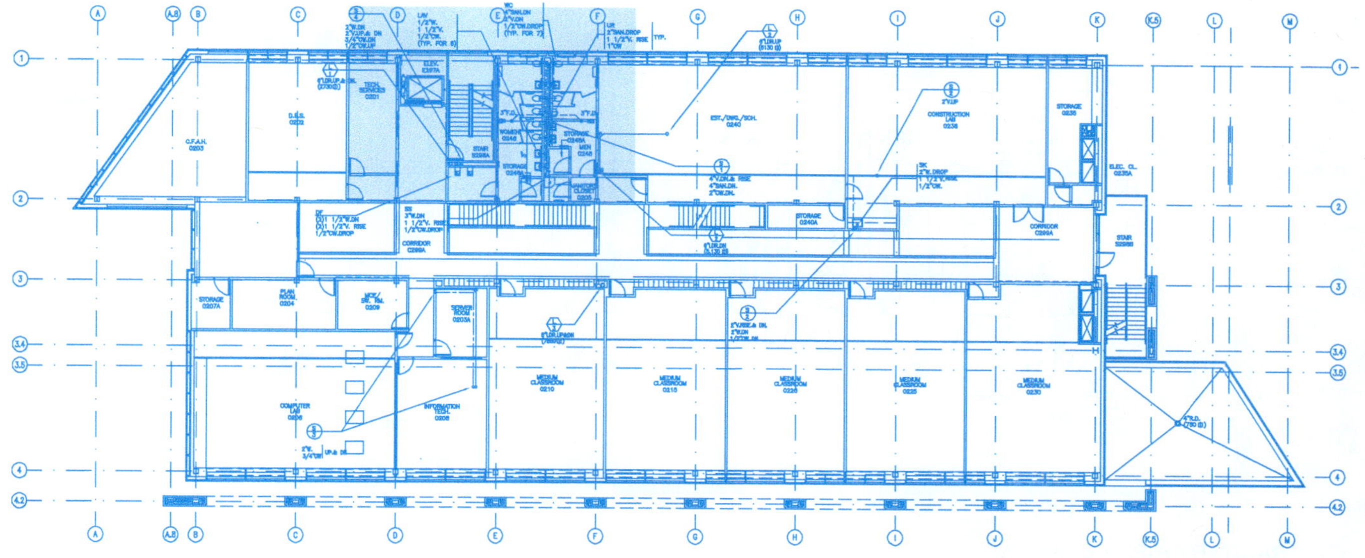

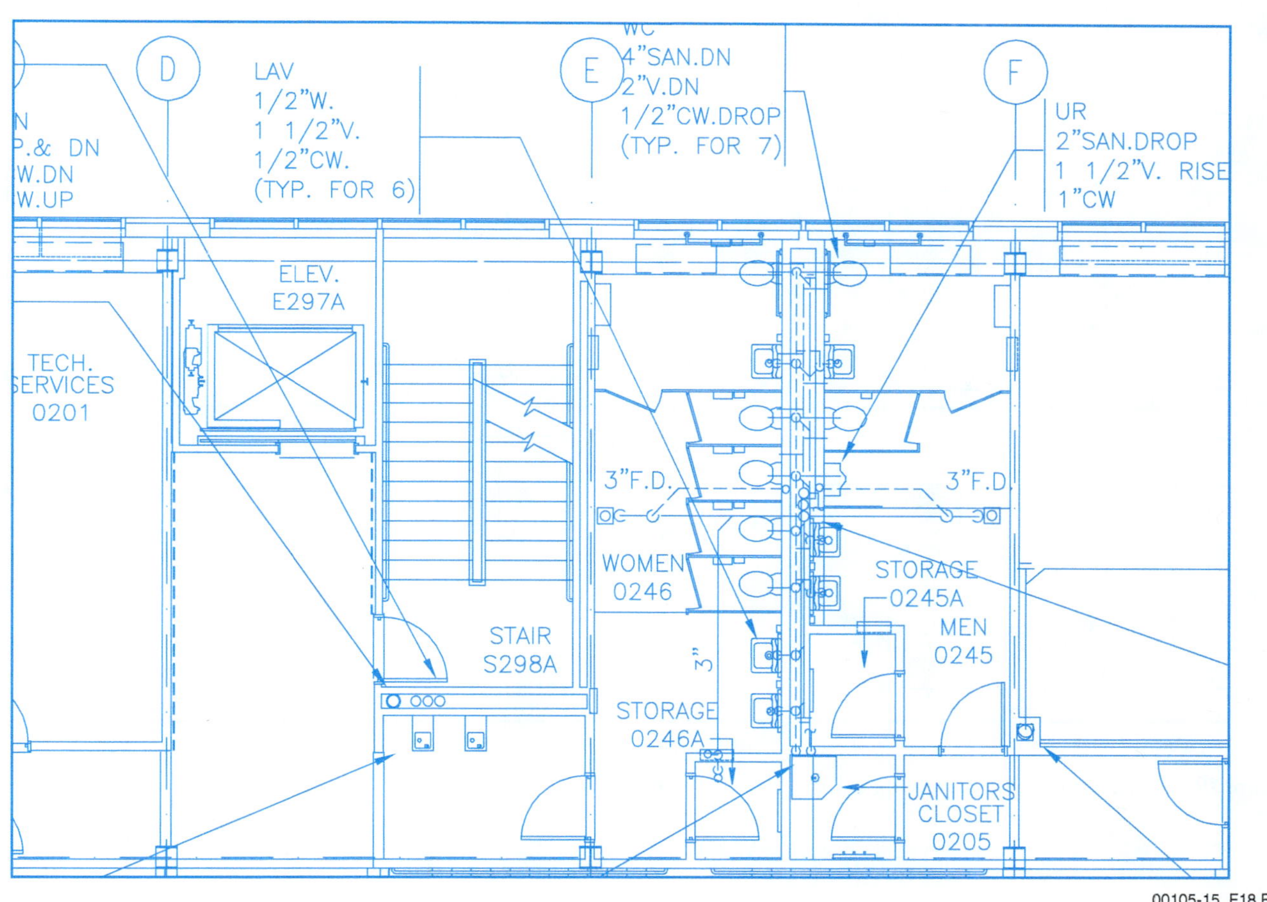

00105-15_F18.EPS

Figure 18 Plumbing plan.

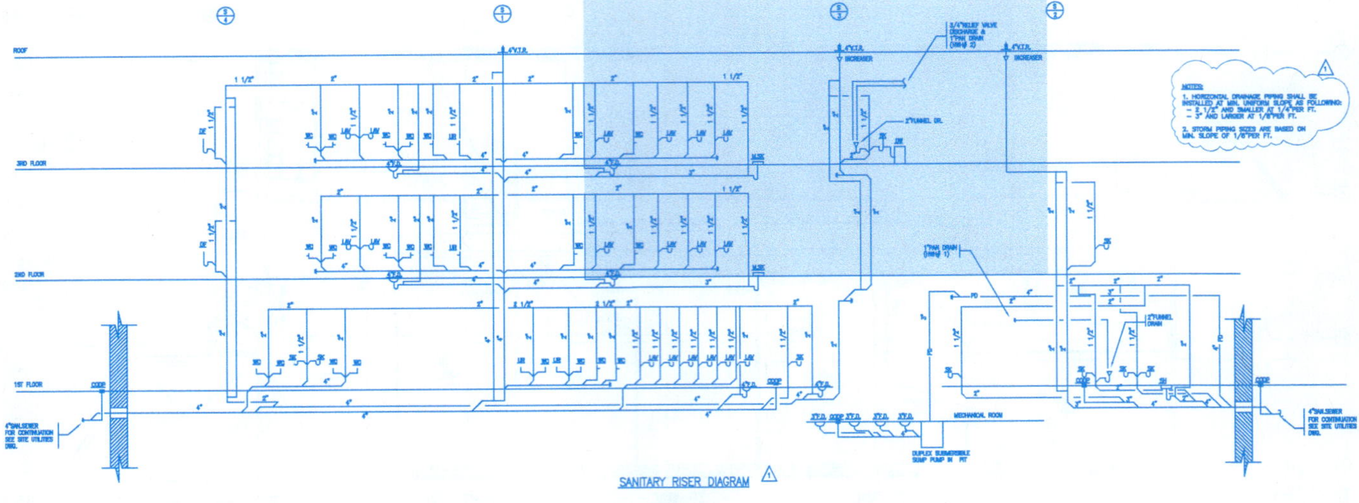

1 SANITARY RISER DIAGRAM
N.T.S.

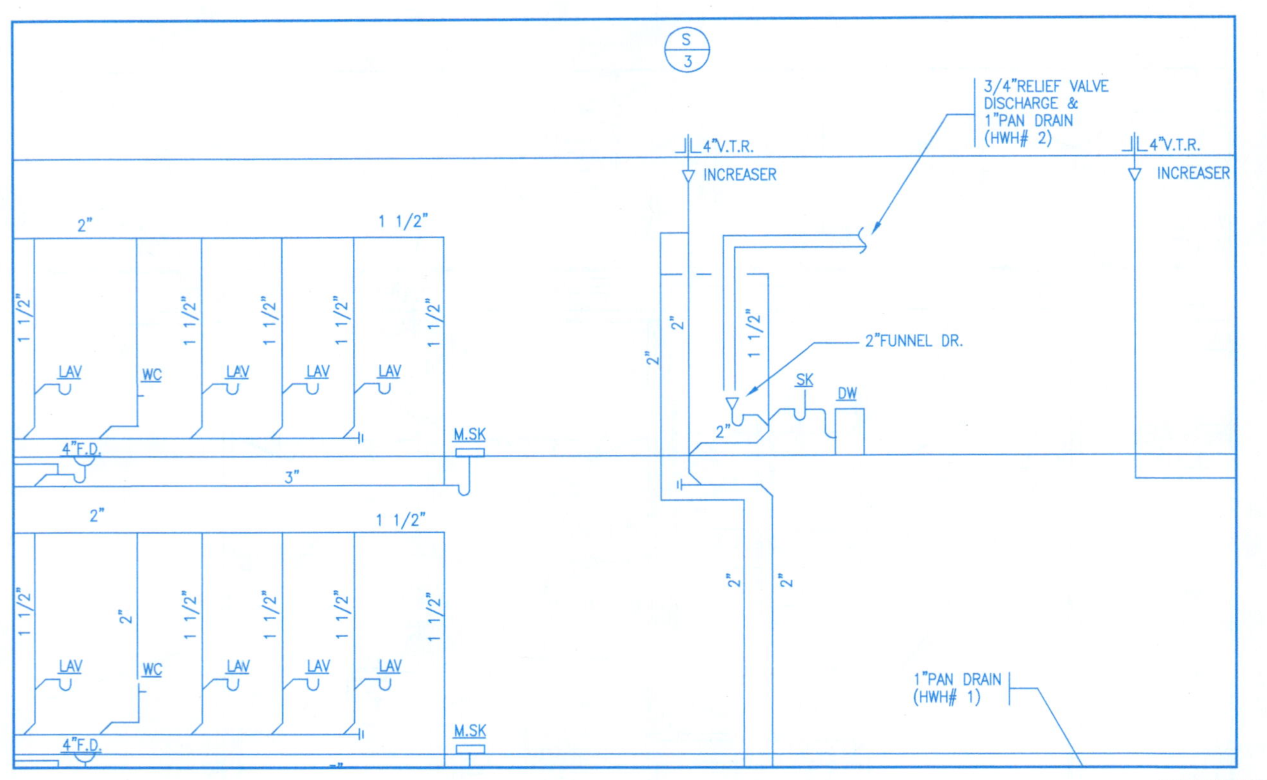

00105-15_F19.EPS

Figure 19 Plumbing isometric drawing (sanitary riser diagram).

More complex electrical plans include locations of switchgear, transformers, main breakers, and motor control centers.

The electrical plans usually start with a set of general notes (*Figure 20*). These notes cover items ranging from main transformers to the coordination of underground penetrations into the building.

The electrical plans can include lighting plans, which show the location of lights and receptacles (refer to *Drawing 3*, First Floor Lighting Plan, in the *Appendix*), power plans (*Figure 21*), and panel schedules (*Figure 22*). Electrical plans have an electrical legend, which defines the symbols (*Figure 23*) used on the plan and a key to the abbreviations (*Figure 24*) used on the plan. Depending on the size and scope of the drawings, the legend is often on a separate drawing sheet of its own, rather than on each individual drawing.

Isometric Drawings

An isometric drawing is a type of three-dimensional drawing also known as a pictorial illustration. Typically in isometric construction drawings, objects are shown at a 30-degree angle in isometric drawings to provide a three-dimensional perspective rather than a flat, two-dimensional view.

BILL OF MATERIAL

P.M.	REQ'D	SIZE	DESCRIPTION
1		1-1/2"	PIPE SCH/40 ASTM-A-120 GR.B
2		3/4"	PIPE SCH/40 ASTM-A-120 GR.B
3	5	1-1/2"	90° ELL ASTM-A-197 BW
4	1	1-1/2"	TEE ASTM-197 STD
5	2	3/4"	45° ELL ASTM-197 STD
6	1	1-1/2" × 3/4"	BELL RED. CONC.
7	2	1-1/2"	GATE VA. BW ASTM-B62
8	1	1-1/2"	CHECK VA. SWING BW 150#

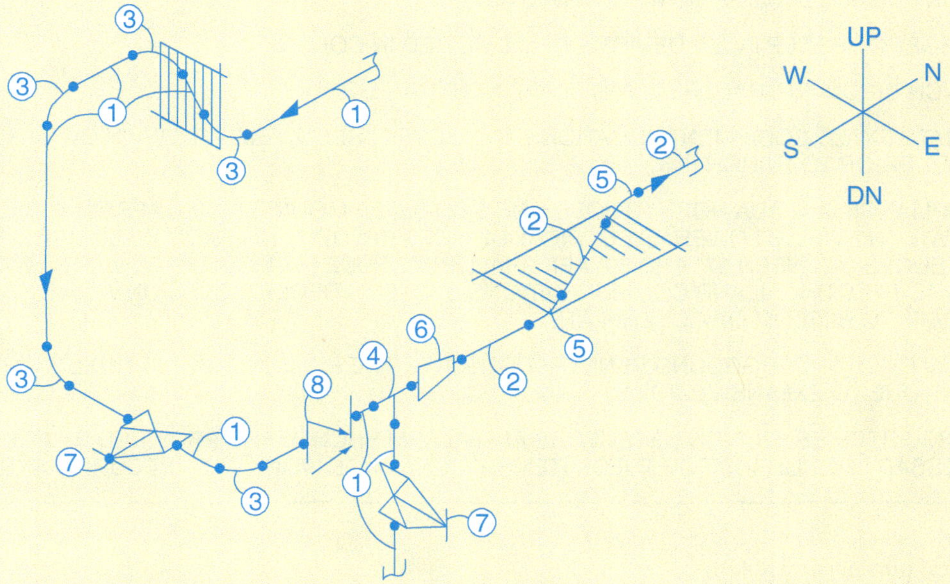

00105-15_SA03.EPS

GENERAL NOTES (FOR ALL ELECTRICAL SHEETS)

1. COORDINATE LOCATION OF LUMINARIES WITH ARCHITECTURAL REFLECTED CEILING PLANS.
2. COORDINATE LOCATION OF ALL OUTLETS WITH ARCHITECTURAL ELEVATIONS, CASEWORK SHOP DRAWINGS AND EQUIPMENT INSTALLATION DRAWINGS.
3. COORDINATE LOCATION OF MECHANICAL EQUIPMENT WITH MECHANICAL PLANS AND MECHANICAL CONTRACTOR PRIOR TO ROUGH-IN.
4. PROVIDE (1) 3/4"C WITH PULL WIRE FROM EACH TELEPHONE, DATA OR COMMUNICATION OUTLET SHOWN, TO ABOVE ACCESSIBLE CEILING, AND CAP.
5. 3-LAMP FIXTURES SHOWN HALF SHADED HAVE INBOARD SINGLE LAMP CONNECTED TO EMERGENCY BATTERY PACK FOR FULL LUMEN OUTPUT. SEE SPECIFICATIONS.
6. SITE PLAN DOES NOT INDICATE ALL OF THE UG UTILITY LINES, RE: CIVIL DRAWINGS FOR ADDITIONAL INFORMATION. CONTRACTOR TO FIELD VERIFY EXACT LOCATION OF ALL EXISTING UNDERGROUND UTILITY LINES OF ALL TRADES PRIOR TO ANY SITE WORK.
7. THE LOCATIONS OF ALL SMOKE DETECTORS SHOWN ARE CONSIDERED TO BE SCHEMATIC ONLY. THE ACTUAL LOCATIONS (SPACING TO ADJACENT DETECTORS, WALLS, ETC.) ARE REQUIRED TO MEET NFPA 72.
8. ANY ITEMS DAMAGED BY THE CONTRACTOR SHALL BE REPLACED BY THE CONTRACTOR.
9. "CLEAN POWER" AND COMMUNICATION/COMPUTER SYSTEM REQUIREMENTS SHALL BE COORDINATED WITH COMMUNICATION/COMPUTER SYSTEMS CONTRACTOR.
10. REFER TO ARCHITECTURAL PLANS, ELEVATIONS AND DIAGRAMS FOR LOCATIONS OF FLOOR DEVICES AND WALL DEVICES. LOCATION WILL INDICATE VERTICAL AND/OR HORIZONTAL MOUNTING. IF DEVICES ARE NOT NOTED OTHERWISE THEY SHALL BE MOUNTED LONG AXIS HORIZONTAL AT +16" TO CENTER.
11. ALL PLUGMOLD SHOWN SHALL BE WIREMOLD SERIES V2000 (IVORY FINISH) WITH SNAPICOIL #V20GB06 (OUTLETS 6" ON CENTER). PROVIDE ALL NECESSARY MOUNTING HARDWARE, ELBOWS, CORNERS, ENDS, ETC. REQUIRED FOR A COMPLETE SYSTEM.
12. ALL EMERGENCY RECEPTACLE DEVICES SHALL BE RED IN COLOR.
13. ALL BRANCH CIRCUITS SHALL BE 3-WIRE (HOT, NEUTRAL, GROUND).
14. COORDINATE EXACT EQUIPMENT LOCATIONS AND POWER REQUIREMENTS WITH OWNER AND ARCHITECT PRIOR TO ROUGH-INS.
15. ADA COMPLIANCE: ALL ADA HORN/STROBE UNITS SHALL BE MOUNTED +90" AFF OR 6" BELOW FINISHED CEILING, WHICH EVER IS LOWER. ELECTRICAL DEVICES PROJECTING FROM WALLS WITH THEIR LEADING EDGES BETWEEN 27" AND 80" AFF SHALL PROTRUDE NO MORE THAN 4" INTO WALKS OR CORRIDORS. ELECTRICAL AND COMMUNICATIONS SYSTEMS RECEPTACLES ON WALLS SHALL BE 15" MINIMUM AFF TO BOTTOM OF COVERPLATE.
16. COORDINATE ALL UNDERGROUND PENETRATIONS INTO THE BUILDING AND TUNNEL WITH STRUCTURAL ENGINEER, DUE TO EXPANSIVE SOILS.
17. ELECTRONIC STRIKES, MOTION DETECTORS AND ALARM SHUNTS ARE PROVIDED BY OTHERS. PROVIDE ALL NECESSARY ROUGH-INS FOR THESE ITEMS. COORDINATE WORK WITH SECURITY SYSTEM PROVIDER.

00105-15_F20.EPS

Figure 20 Electrical plan general notes.

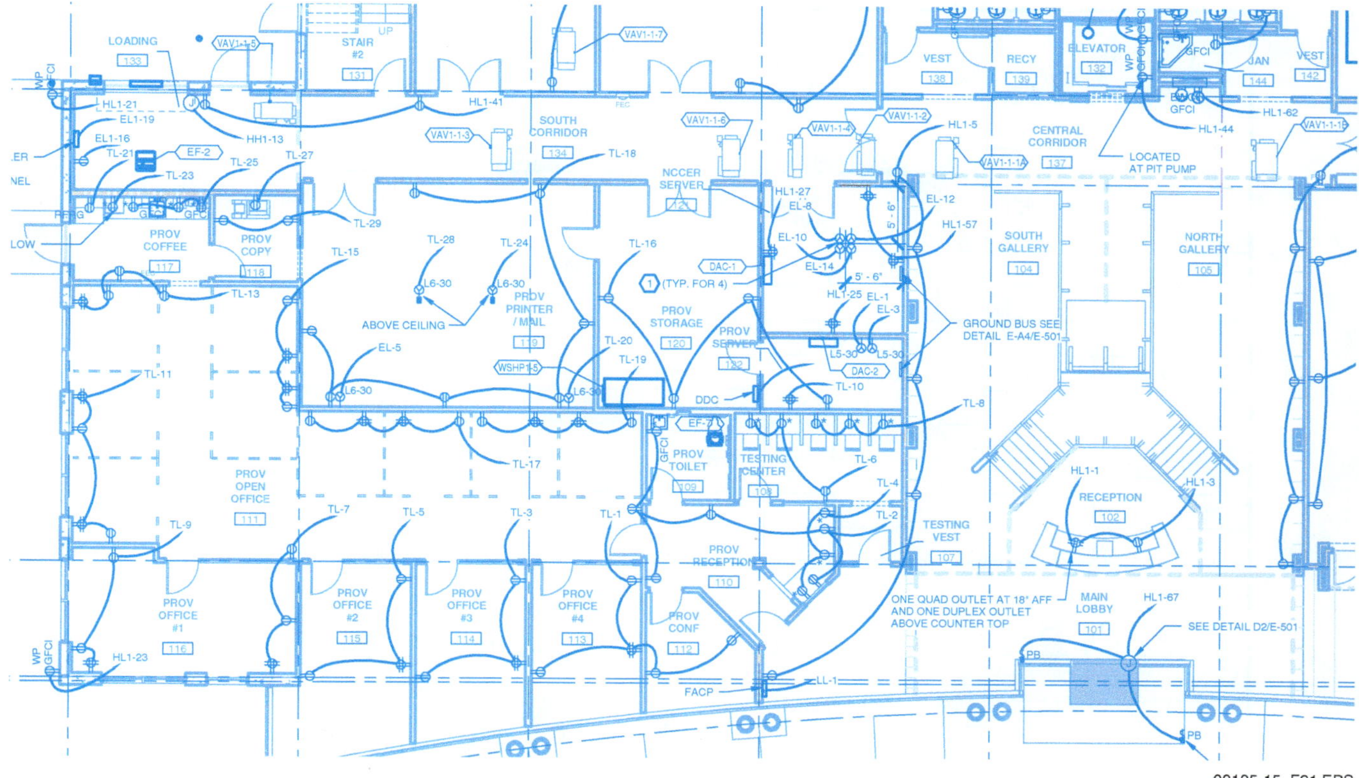

00105-15_F21.EPS

Figure 21 Portion of a power plan.

Did You Know?

The North Arrow

Most construction drawings have an arrow indicating North. It may be located near the title block or in some other conspicuous place. The North arrow is a reference point that helps to verify the locations of objects, walls, and building parts shown on the construction drawing.

Visualize Before Building

It is important for a builder to be able to visualize a finished project before starting it. Detail drawings help builders visualize different parts of the structure long before they measure a board or hammer a nail. By visualizing, builders can plan ahead and anticipate potential problems. This saves time and money on the job.

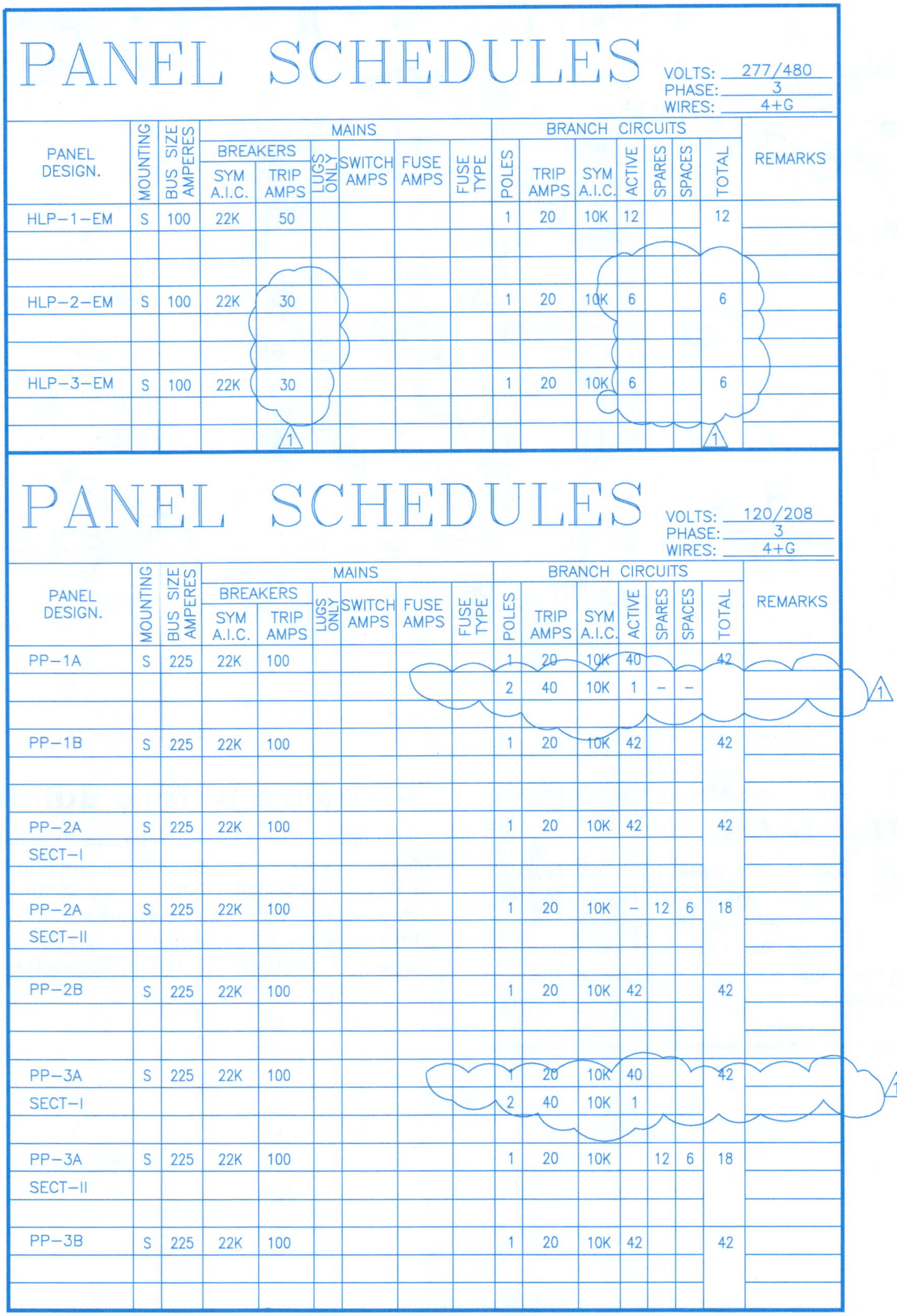

PANEL SCHEDULES

VOLTS: 277/480
PHASE: 3
WIRES: 4+G

PANEL DESIGN.	MOUNTING	BUS SIZE AMPERES	MAINS: BREAKERS SYM A.I.C.	MAINS: BREAKERS TRIP AMPS	LUGS ONLY	SWITCH AMPS	FUSE AMPS	FUSE TYPE	BRANCH CIRCUITS: POLES	TRIP AMPS	SYM A.I.C.	ACTIVE	SPARES	SPACES	TOTAL	REMARKS
HLP–1–EM	S	100	22K	50					1	20	10K	12			12	
HLP–2–EM	S	100	22K	30					1	20	10K	6			6	
HLP–3–EM	S	100	22K	30					1	20	10K	6			6	

PANEL SCHEDULES

VOLTS: 120/208
PHASE: 3
WIRES: 4+G

PANEL DESIGN.	MOUNTING	BUS SIZE AMPERES	MAINS: BREAKERS SYM A.I.C.	MAINS: BREAKERS TRIP AMPS	LUGS ONLY	SWITCH AMPS	FUSE AMPS	FUSE TYPE	BRANCH CIRCUITS: POLES	TRIP AMPS	SYM A.I.C.	ACTIVE	SPARES	SPACES	TOTAL	REMARKS
PP–1A	S	225	22K	100					1	20	10K	40			42	
									2	40	10K	1	–	–		
PP–1B	S	225	22K	100					1	20	10K	42			42	
PP–2A	S	225	22K	100					1	20	10K	42			42	
SECT–I																
PP–2A	S	225	22K	100					1	20	10K	–	12	6	18	
SECT–II																
PP–2B	S	225	22K	100					1	20	10K	42			42	
PP–3A	S	225	22K	100					1	20	10K	40			42	
SECT–I									2	40	10K	1				
PP–3A	S	225	22K	100					1	20	10K		12	6	18	
SECT–II																
PP–3B	S	225	22K	100					1	20	10K	42			42	

00105-15_F22.EPS

Figure 22 Panel schedules.

ELECTRICAL SYMBOLS LIST

J / J / J	JUNCTION BOX. CEILING/FLOOR/WALL MOUNTED
	PULL BOX SIZED AS REQUIRED
	BELL
	CHIME
	HORN (SINGLE/BI– DIRECTIONAL)
	BUZZER
	DOOR PUSHBUTTON
T S	SIGNAL TRANSFORMER
DH	MAGNETIC DOOR HOLDER
DH S	MAGNETIC TYPE DOOR CLOSER, 'S' DENOTES SMOKE DETECTOR TYPE
WF	SPRINKLER WATER FLOW SWITCH
TS	SPRINKLER SUPERVISED VALVE – TAMPER SWITCH
S R	INDICATING CLOCK – SINGLE FACE CLG/WALL MTD
S / S	SPEAKER; CLG/WALL MTD
V	VOLUME CONTROL
M / M	MICROPHONE OUTLET; CLG/WALL MTD
PA	PUBLIC ADDRESS EQUIPMENT RACK
PPP	ELECT. POWER PATCH PANEL WITH 4 RECEPTS.AND R/S SWITCH
A/V	MUTIMEDIA PATCH PANEL
W S	WINDOW ALARM SWITCH
B S	DOOR PUSHBUTTON
CCS	CLOSED CIRCUIT SURVEILLANCE CAMERA 'P' DENOTES PAN – P/T TILT
CCSM	COLSED CIRCUIT SURVEILLANCE MONITOR – NUMBERS DENOTE QUANTITY
TV TV	TELEVISION ANTENNA OUTLET
S 2 TV	TELEVISION SPLITTER NUMBER DENOTES QUANTITY OF SPLITS
TV	TELEVISION CAMERA OUTLET
TVHE	TELEVISION HEADED EQUIPMENT
	TELEPHONE OUTLET WITH BUSHED OPENING LETTERS DENOTE: 'H' – HOUSE PHONE 'J' – JACK TYPE, 'P' – PAY (PUBLIC) PHONE, 'W' – WALL TYPE PHONE
TC	TELEPHONE TERMINAL CABINET
S	TELEPHONE FLOOR BOX WITH NIPPLE (RISER) EXTENSION. 'S' DENOTES SERVICE FITTING TYPE
	DATA OUTLET WITH BUSHED OPENING WALL MOUNTED
	COMBINATION TELEPHONE/DATA OUTLET WITH BUSHED OPENING WALL MOUNTED
S	SWITCH
S	BELL
S	CHIME
S S	DOOR STRIKE
SSP	START/STOP WITH PILOT LIGHT
	POWER TRANSFORMER
	GROUND (EARTH)
	LIGHTNING ARRESTER
	CIRCUIT BREAKER
G	GROUND BUS
N	NEUTRAL BUS
CR R	CARD READER 'R' – DENOTES RECESSED TYPE
DO	POWER DOOR OPERATOR
I	INTERCOM
M	CEILING MOUNTED OCCUPANCY SENSOR
M	SURFACE MOUNTED OCCUPANCY SENSOR
P	PHOTOSENSOR

'S' – DENOTES SKELETON TYPE
'R' – DENOTES RECESSED TYPE
'D' – DENOTES DOUBLE FACED
'S' – DENOTES SURFACE MOUNTED
'D' – DENOTES DOUBLE FACED

00105-15_F23.EPS

Figure 23 Electrical symbols list.

1.1.7 Fire Protection Plans

Another important drawing that may be included in a set of drawings is the **fire protection plans** (refer to *Drawing 4*, First Floor Fire Protection Plan, in the *Appendix*). This drawing shows the piping, valves, heads, and switches that make up a building's fire sprinkler system. A fire sprinkler symbols list is usually included on a separate sheet along with the fire sprinkler specifications, details and assembly drawings, and riser diagrams.

1.2.0 Basic Components of Construction Drawings

Most construction drawings are laid out in a fairly standardized format. This section describes the following five parts of a construction drawing:

- **Title block**
- Border
- Drawing area
- Revision block
- Legend

1.2.1 Title Block

The first thing to look at on any drawing is the title block. The title block is normally in the lower right-hand corner of the drawing or across the right edge of the paper (*Figure 25*). The title block has two purposes. First, it gives information about the structure or assembly. Second, it is numbered so the print can be filed easily.

Different companies put different information in the title block. Generally, it contains the following:

- *Company logo* – Usually preprinted on the drawing.
- *Sheet title* – Identifies the project.
- *Date* – Date the drawing was checked and readied for seal, or issued for construction.
- *Drawn by* – Initials of the person who drafted the drawing.
- *Drawing number* – Code numbers assigned to a project
- *Scale* – The ratio of the size of the object as drawn to the object's actual size.

ELECTRICAL ABBREVIATIONS

A	AMPERE(S)
AC	ALTERNATING CURRENT
ACB	AIR CIRCUIT BREAKER
AFF	ABOVE FINISHED FLOOR
AFG	ABOVE FINISHED GRADE
AL	ALUMINUM
ALT	ALTERNATE
ASYM	ASYMMETRICAL
ATS	AUTOMATIC TRANSFER SWITCH
AWG	AMERICAN WIRE GAUGE
BC	BOTTOM CONDUIT
BD	BUS DUCT
BFG	BELOW FINISHED GRADE
BIL	BASIC IMPULSE LEVEL
BLDG	BUILDING
BX	ARMORED CABLE
C	CONDUIT
CATV	CABLE ANTENNA TELEVISION SYSTEM
CCAB	CONTROL CABINET
CCTV	CLOSED CIRCUIT TELEVISION
CH	CABINET HEATER
CKT	CIRCUIT
CKT BKR/CB	CIRCUIT BREAKER
CL	CLOSET
CLG	CEILING
COND	CONDUCTOR
CO	CONDUIT ONLY
CT	CURRENT TRANSFORMER
CU	COPPER
DB	DUCT BANK
DMB	DIMMER BOARD
DC	DIRECT CURRENT
DH	DUCT HEATER
DIM	DIMMER CONTROL
DISC	DISCONNECT
DM	DAMPER MOTOR
DN	DOWN
DP	DISTRIBUTION POWER PANEL(BOARD)
DT	DOUBLE THROW
DWG	DRAWING
EA	EACH
E.C.	ELECTRICAL CONTRACTOR
E.HTR	ELECTRIC HEATER
ELEV	ELEVATOR
EL	ELECTRIC
EM	EMERGENCY
EMT	ELECTRICAL METALLIC TUBING
ENT	ELECTRICAL NON-METALLIC TUBING
EWC	ELECTRIC WATER COOLER
EX	EXISTING
F	FUSE
FA	FIRE ALARM
FACP	FIRE ALARM CONTROL PANEL
FBO	FURNISHED BY OTHERS
FCC	FLAT CONDUCTOR CABLE
FCU	FAN COIL UNIT
FDR	FEEDER
FL	FLOOR
FLUOR	FLUORESCENT
F.P.C.	FIRE PROTECTION CONTRACTOR
FS	FUSIBLE SWITCH
F.S.C.	FOOD SERVICE CONTRACTOR
FT	FEET OR FOOT
G.C.	GENERAL CONTRACTOR
GEN	GENERATOR
GF	GROUND FAULT
GG	GROUND GRID
GRD	GROUND
HC	HUNG CEILING
H.I.D.	HIGH INTENSITY DISCHARGE
HP	HORSEPOWER
H.P.S.	HIGH PRESSURE SODIUM
HPU	HEAT PUMP UNIT
HT	HEIGHT
HV	HIGH VOLTAGE
HW	HEAVY WALL RIGID CONDUIT
HZ	FREQUENCY IN CYCLES PER SECOND
IC	INTERRUPTING CAPACITY
IG	ISOLATED GROUND
IMC	INTERMEDIATE METALLIC CONDUIT
INC	INCANDESCENT
JB	JUNCTION BOX
K	KEY OPERATED
kVA	KILOVOLT AMPERE(S)
kVAR	KILOVAR(S)
kW	KILOWATT(S)
kWhr	KILOWATT(S) HOUR(S)
LP	LIGHTING PANEL
L.P.S.	LOW PRESSURE SODIUM
LTG	LIGHTING
LV	LOW VOLTAGE
MATV	MASTER ANTENNA TELEVISION SYSTEM
MC	METAL CLAD CABLE
MCC	MOTOR CONTROL CENTER
MCM	THOUSAND CIRCULAR MIL(S)
MCP	MOTOR CONTROL PANEL
M.C.	MECHANICAL CONTRACTOR
MH	MANHOLE
MIC	MICROPHONE
MIN	MINIMUM
MS	MAGNETIC STARTER
MTD	MOUNTED
MTG	MOUNTING
MTR	MOTOR
MTS	MANUAL TRANSFER SWITCH
N	NEUTRAL
NA	NON-AUTOMATIC
NC	NORMALLY CLOSED
NF	NON-FUSE
N.I.C.	NOT IN CONTRACT
NL	NIGHT LIGHT
NO	NORMALLY OPEN
NP	NETWORK PROTECTOR
NTS	NOT TO SCALE
OC	ON CENTER
OL	OVERLOAD ELEMENT
P	POLE
PA	PUBLIC ADDRESS
PB	PULL BOX
P.C.	PLUMBING CONTRACTOR
PF	POWER FACTOR
Ø	PHASE
PL	PILOT (INDICATOR) LIGHT
PNL	PANEL (PANELBOARD)
PP	POWER PANEL
PRI	PRIMARY
PT	POTENTIAL TRANSFORMER
PVC	POLYVINYL CHLORIDE
PWR	POWER
R	RECESSED
RC	REMOTE CONTROL
REC	RECEPTACLE
S	SURFACE
SC	SEPARATE CIRCUIT
SDB	SUB-DISTRIBUTION BOARD
SEC	SECONDARY
SMR	SURFACE METAL RACEWAY
SP	SINGLE POLE
SPK	SPEAKER
ST	SINGLE THROW
SW	SWITCH
SWBD	SWITCHBOARD
SYM	SYMMETRICAL
T	THERMOSTAT
TEL	TELEPHONE
TB	TERMINAL BOX
TC	TOP CONDUIT
TCAB	TELEPHONE CABINET
T.C.C.	TEMPERATURE CONTROL CONTRACTOR
TP	TAMPER PROOF
TV	TELEVISION
TYP	TYPICAL
UG	UNDERGROUND
UH	UNIT HEATER
UNG	UNGROUNDED
UON	UNLESS OTHERWISE NOTED
UPS	UNINTERRUPTED POWER SYSTEM
V	VOLT(S)
VA	VOLTAMP(S)
VAR	VOLT AMPERES REACTIVE
VP	VAPORPROOF
W	WATT(S)
WP	WEATHERPROOF
WT	WATERTIGHT
XFR	TRANSFORMER
XP	EXPLOSION PROOF

00105-15_F24.EPS

Figure 24 Electrical abbreviations.

Figure 25 The title block of a construction drawing.

- *Revision blocks* – Information on revisions, including (at a minimum) the date and the initials of the person making the revision. Other information may include descriptions of the revision and a revision number.

Every company has its own system for such things as project numbers and departments. Every company also has its own placement locations for the title and revision blocks. Your supervisor should explain your company's system to you.

1.2.2 Border

The border is a clear area of approximately half an inch around the edge of the drawing area. It is there so that everything in the drawing area can be printed or reproduced on printing machines with no loss of information.

1.2.3 Drawing Area

The drawing area presents the information for constructing the project, such as the floor plan, elevations of the building, sections, and details.

1.2.4 Revision Block

A revision block is located in the drawing area, usually in the lower right corner inside the title block or near it. Different companies put the revision block in different places. This block is used to record any changes (revisions) to the drawing. It typically contains the revision number, a brief description, the date, and the initials of the person who made the revisions (*Figure 26*). All revisions must be noted in this block and dated and identified by a letter or number.

> **CAUTION**
>
> It is essential to note the revision designation on a construction drawing and to use only the latest version. Otherwise, costly mistakes may result.

1.2.5 Legend

Each line on a construction drawing has a specific design and thickness that identifies it. Note that some of the lines may be used to identify off-site utilities. The identification of these lines and other symbols is called the legend. Although a legend doesn't automatically appear on every construction drawing, when it does, it explains or defines symbols or special marks used in the drawing (*Figure 27*). Be aware that legends are specific only to the set of drawings in which they are contained.

Importance of Specifications

Specifications clarify information that cannot be shown on the drawings. Specifications are very important to the architect and owner to ensure compliance to the standards set. The figure provided shows one page of the specifications for a building's air handling units.

M.E. RINKER SR. HALL
SCHOOL OF BUILDING CONSTRUCTION
BR-191

AIR HANDLING UNITS
SECTION 15760

2.1 AIR HANDLING UNITS

A. Units shall be of the type, size and capacity as set forth in the schedule. The fan outlet velocities and coil and filter face velocities shall be within 5% of the values specified in the schedule. Units shall be double wall McQuay, or approved equal.

B. The units, as assembled, shall be complete with fans, coils, insulated casing, filters, drives and accessories. Each unit, including the fan enclosure, shall have essentially constant cross-sectional dimensions as to width and height. Internal baffles shall be provided as required to prevent bypassing of coils and filters.

C. The casing shall consist of an independent structural steel frame, properly reinforced and braced for maximum rigidity, having individually removable, flush mounted, insulated panels. The casing shall be of sectionalized construction, consisting basically of individual fan section, coil section, access sections, filter section, and drain pan. Sections shall be joined with continuous gasketing to form an air tight closure. Sections shall be so designed that the method of joining can be performed with relative ease and without damage to the insulation and vapor barrier.

The framework shall be constructed of AISC structural rolled shapes having minimum thickness of 1/8" (3 mm) or die formed sheet steel having the minimum gauges set forth in the following schedules:

Maximum Individual Casing Cross-Section	Minimum Framework Gauge
Up to 30 sq.ft. (2.8 sq.m.)	14
30.5 sq.ft. to 47 sq.ft. (2.81 to 4.4 sq.m.)	12
48 sq.ft. (4.45 sq.m.) and up	10

D. Framework shall be designed with recesses suitable to receive enclosure panels, providing neat appearance, airtight enclosure, and ease of panel removal.

E. Enclosure panels 12 sq. ft. in area and larger shall be constructed of not less than 18 gauge die formed sheet steel. Should the sides or top of a casing section exceed 20 sq. ft. in area, the panels shall be fabricated of more than one piece, with the individual panels recessed into intermediate structural members.

Protection for the insulation edges shall be provided around the perimeter of each

GEA 0300-0600

15760 - 2
Revision No. 1
Dec. 19, 2001

00105-15_SA04.EPS

PROJ	NO	REVISION	RVSD	CHKD	APPD	DATE
3483	01	RELEASED FOR CONSTRUCTION		APD	NWS	JULY 92
3483	02	DELETED PART OF LINE 12037		APD	NWS	AUG 92
3483	03	⚠ ADDED WELDING SYMBOL		APD	NWS	AUG 92

00105-15_F26.EPS

Figure 26 The revision block of a construction drawing.

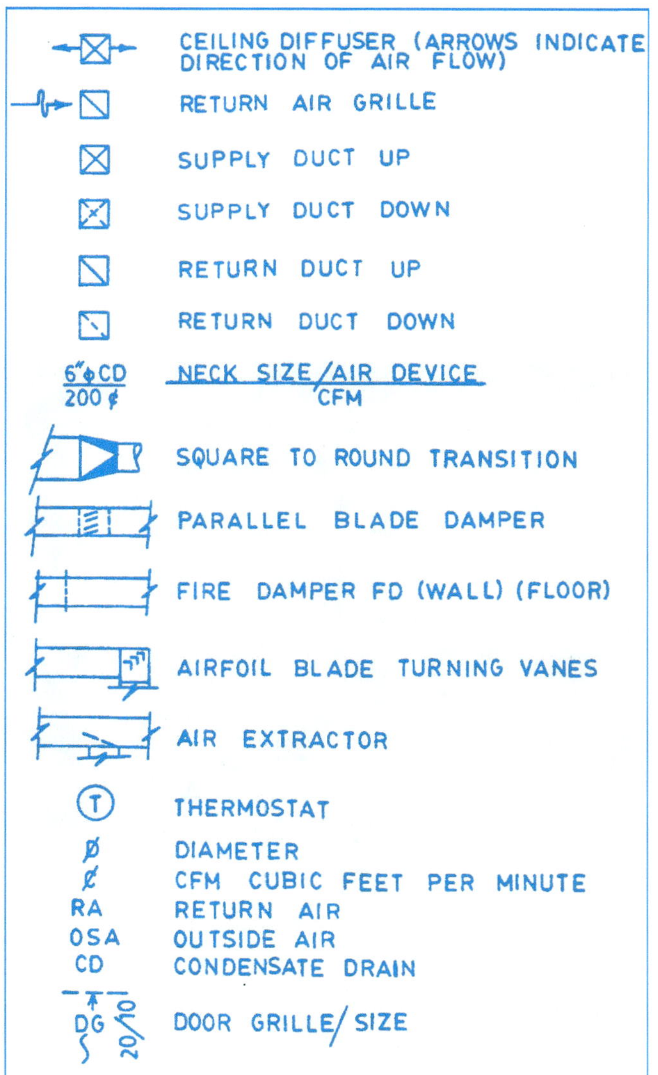

00105-15_F27.EPS

Figure 27 Sample legend.

1.3.0 Drawing Elements

Various drawing elements make it easier and quicker for people to read and understand construction drawings. This section covers several of the most common drawing elements.

1.3.1 Lines of Construction

It is very important to understand the meanings of lines on a drawing. The lines commonly used on a drawing are sometimes called the *alphabet of lines*. Here are some of the more common types of lines (*Figure 28*):

- **Dimension lines** – Establish the dimensions (sizes) of parts of a structure. These lines end with arrows (open or closed), dots, or slashes at a termination line drawn perpendicular to the dimension line.
- **Leaders** *and arrowheads* – Identify the location of a specific part of the drawing. They are used with words, abbreviations, symbols, or keynotes.
- *Property lines* – Indicate land boundaries.
- *Cut lines* – Lines around part of a drawing that is to be shown in a separate cross-sectional view.
- *Section cuts* – Show areas not included in the cutting line view.
- *Break lines* – Show where an object has been broken off to save space on the drawing.
- **Hidden lines** – Identify part of a structure that is not visible on the drawing. You may have to look at another drawing to see the part referred to by the lines.
- *Center lines* – Show the measured center of an object, such as a column or fixture.
- *Object lines* – Identify the object of primary interest or the closest object.

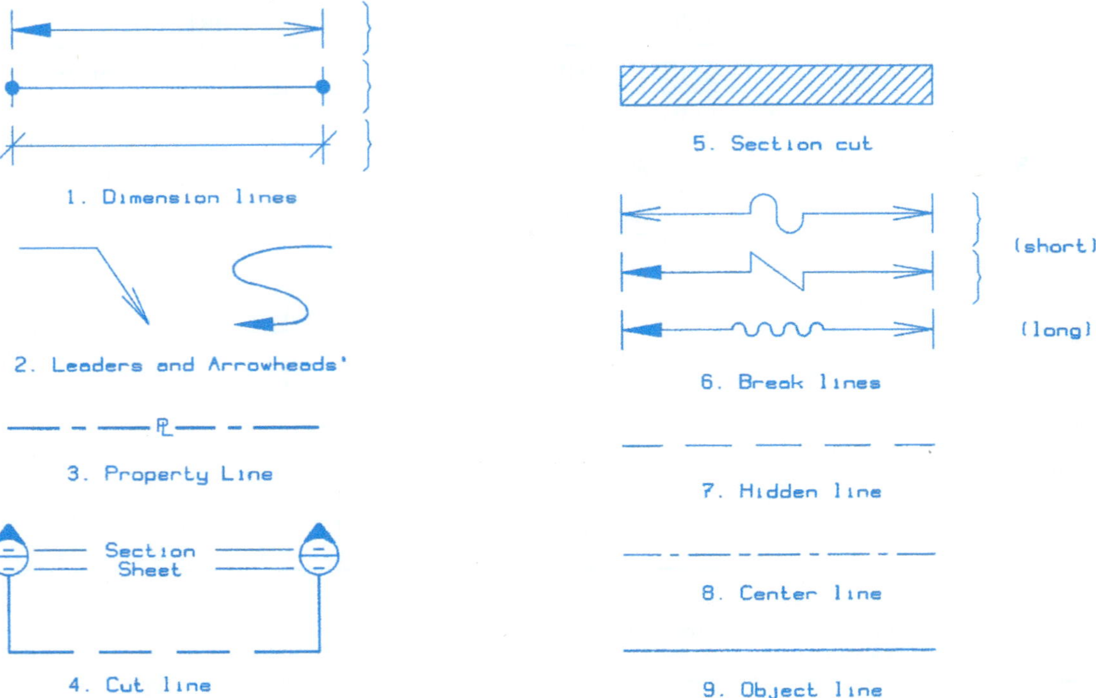

Figure 28 Lines of construction.

What Is Computer-Aided Drafting?

The use of computers is a cost-effective way to increase drafting productivity because the computer program automates much of the repetitive work. A CAD system generates drawings from computer programs. CAD has the following advantages over hand-drawn construction drawings:

- It is automated.
- The computer performs calculations quickly and easily.
- Changes can be made quickly and easily.
- Commonly used symbols can be easily retrieved.
- CAD can include three-dimensional modeling of the structure.

Regional and Company Differences

Although most symbols are standard, there can be slight variations in different regions of the country. Always check the title sheet or other introductory drawing to verify the symbols you find on the project drawings.

Your company may also use some special symbols or terms. Your instructor or supervisor will tell you about conventions that are unique to your company.

1.3.2 Abbreviations, Symbols, and Keynotes

Architects and engineers use systems of abbreviations, symbols, and keynotes to keep plans uncluttered, making them easier to read and understand. Examples of some of these items appear throughout this module. Following is additional information on these items and some variants that are commonly used.

Each trade has its own symbols, and workers should learn to recognize the symbols used by other trades. For example, an electrician should understand a carpenter's symbols, a carpenter should understand a plumber's symbols, and so on. Then, no matter what symbols are encountered on a project, the workers will understand what the symbols mean.

Did You Know?

Legality of Construction Drawings

Construction drawings are incorporated into building contracts by reference, making them part of the legal documents associated with a project. When describing the project to be completed, the legal contract between the builder and the owner refers to the accompanying construction drawings for details that would be too lengthy to write out. That makes tracking changes or revisions to construction drawings over the course of a project vitally important. If an error is made along the way, either the owner or the builder must be able to find the discrepancy by reading the drawings. Taking care of construction drawings over the course of the project makes good business sense.

Abbreviations used in construction drawings are short forms of common construction terms. For example, the term *on-center* is abbreviated O.C. Some common construction abbreviations are listed in *Figure 29*.

Abbreviations should always be written in capital letters. Abbreviations for each project should be noted on the title sheet or other introductory drawing page such as the legend page. Books that list construction abbreviations and their meanings are available. You do not need to memorize these abbreviations, as you will start to remember them as you use them.

Check All the Plans

Always double-check all the plans for a project. Be familiar with other trade work that may affect your job. Determine whether an electrical conduit is close to the plumbing pipes or whether a framing member is placed too close to an HVAC duct. By knowing all the work planned for a particular area, you can prevent errors and save time and money.

If a particular dimension is missing on the plan you will be using, check the other plans to see if it is included there. The part of the building you are working on will also be shown on other plans.

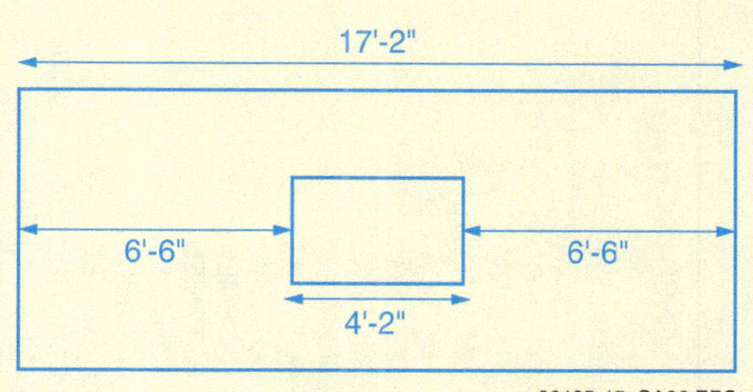

00105-15_SA06.EPS

When taking measurements from a set of plans, make sure that the dimensions of each measured section add up to the total measurement. For instance, if you are looking at a drawing of a wall with one window, add up the measurement from the left end of the wall to the left side of the window, the width of the window, and the measurement from the right side of the window to the right end of the wall. Make sure this total matches the measurement of the entire wall.

ABBREVIATIONS

A.B.	– ANCHOR BOLT
ADD'L	– ADDITIONAL
ADJ.	– ADJACENT
A.I.S.C.	– AMERICAN INSTITUTE OF STEEL CONSTRUCTION
ALT.	– ALTERNATE
ARCH.	– ARCHITECTURAL
A.S.T.M.	– AMERICAN SOCIETY FOR TESTING & MATERIALS
BLDG.	– BUILDING
BM.	– BEAM
B.O.	– BOTTOM OF
BOT.	– BOTTOM
BSMT.	– BASEMENT
BTWN.	– BETWEEN
CANT.	– CANTILEVER
CB.	– CARDBOARD
CH.	– CHAMFER
C.J.	– CONTROL/CONSTRUCTION JOINT
CLR.	– CLEAR, CLEARANCE
C.M.U.	– CONCRETE MASONRY UNIT
COL.	– COLUMN
CONC.	– CONCRETE
CONN.	– CONNECTION
CONST.	– CONSTRUCTION
CONT.	– CONTINUOUS
CONTR.	– CONTRACTOR
CTRD.	– CENTERED
DET.	– DETAIL
DIAG.	– DIAGONAL
DIAM.	– DIAMETER
DIM.	– DIMENSION
DISCONT.	– DISCONTINUOUS
DWG.	– DRAWING
EA.	– EACH
E.F.	– EACH FACE
EL.	– ELEVATION
ELECT.	– ELECTRICAL
ELEV.	– ELEVATOR
EQ.	– EQUAL
E.W.B.	– END WALL BARS
E.W.	– EACH WAY
EXIST.	– EXISTING
EXP. JNT.	– EXPANSION JOINT
EXT.	– EXTERIOR
F.D.	– FLOOR DRAIN

FDN.	– FOUNDATION
FIN.	– FINISH
FLR.	– FLOOR
F.O.B.	– FACE OF BRICK
F.O.CONC.	– FACE OF CONCRETE
F.O.W.	– FACE OF WALL
FS.	– FLAT SLAB
FT.	– FOOT
FTG.	– FOOTING
F.W.	– FILLET WELD
GA.	– GAUGE
GAL.	– GALVANIZED
G.L.	– GLU-LAM BEAM
GR.	– GRADE
GR. BM.	– GRADE BEAM
H.A.S.	– HEADED ANCHOR STUD
HORIZ.	– HORIZONTAL
H.S.B.	– HIGH STRENGTH BOLT
I.D.	– INSIDE DIAMETER
IN.	– INCH
INT.	– INTERIOR
JNT.	– JOINT
LB.	– POUND
LIN. FT.	– LINEAL FEET
L.L.V.	– LONG LEG VERTICAL
MAT'L.	– MATERIAL
MAX.	– MAXIMUM
MECH.	– MECHANICAL
MID.	– MIDDLE
MIN.	– MINIMUM
MISC.	– MISCELLANEOUS
MTL.	– METAL
N.I.C.	– NOT IN CONTRACT
NO.	– NUMBER
NOM	– NOMINAL
N.T.S.	– NOT TO SCALE
O.C.	– ON CENTER
O.D.	– OUTSIDE DIAMETER
O.H.	– OPPOSITE HAND
OPNG.	– OPENING
⅊	– PLATE
P.S.F.	– POUND PER SQUARE FOOT
P.S.I.	– POUND PER SQUARE INCH
R.	– RADIUS
REINF.	– REINFORCEMENT
REQ'D.	– REQUIRED

RM.	– ROOM
SCHED.	– SCHEDULE
SECT.	– SECTION
SHT.	– SHEET
SIM.	– SIMILAR
S.L.V.	– SHORT LEG VERTICAL
SPC.	– SPACE
SPEC.	– SPECIFICATION
SQ.	– SQUARE
STD.	– STANDARD
STIFF.	– STIFFENER
STL.	– STEEL
STOR.	– STORAGE
SYM.	– SYMMETRICAL
T.&B.	– TOP AND BOTTOM
THK.	– THICKNESS
T.O.	– TOP OF
TYP.	– TYPICAL
U.N.O.	– UNLESS NOTED OTHERWISE
VAR.	– VARIES
VERT.	– VERTICAL
V.I.F.	– VERIFY IN FIELD
WT.	– WEIGHT

SYMBOLS

℄	CENTER LINE
⌀	DIAMETER
	ELEVATION
&	AND
W/	WITH
⅊	PLATE
X	BY
#	NUMBER
◎	AT
	SQUARE
L	ANGLE

00105-15_F29.EPS

Figure 29 Abbreviations.

Request for Information (RFI)

A request for information (RFI) is used to clarify any discrepancies in the plans. If you notice a discrepancy, you should notify the foreman. The foreman will write up an RFI, explaining the problem as specifically as possible and putting the date and time on it. The RFI is submitted to the superintendent, who passes it to the general contractor, who passes it to the architect or engineer, who then resolves the discrepancy.

Always refer to specifications and the RFI when deciding how to interpret drawings.

DATE 12/07/12 RFI NO. 1

PROJECT NAME GERMANS FROM RUSSIA PROJECT NO. 15-1593

REQUEST: REF D.W.G.NO. M2 REV. DETAIL $\frac{1}{M2}$ OTHER

WILL THE 16 X 10 INTAKE AIR DUCTWORK RUNNING THROUGH RM 116 REQUIRE WALL MOUNTED FIRE DAMPERS ON ALL 4 EXIT CORRIDOR WALL PENETRATIONS AND WILL THE 14 X 10 TRANSFER DUCTWORK REQUIRE THE INSTALLATION OF A FIRE DAMPER AS WELL?

BY: LARRY MAYRE REPLY BY (DATE): 12/20/12

REPLY:

ANSWER: ALL DUCTWORK IS ABOVE CEILING, SO OK AS IS.

DATE: 12/19/12

00105-15_SA07.EPS

Symbols are used on a drawing to tell what material is required for that part of the project. A combination of these symbols, expanded and drawn to the same size, makes up the pictorial view of the plan. There are architectural symbols (*Figure 30*); civil and structural engineering symbols (*Figure 31*); mechanical symbols (*Figure 32*); plumbing symbols (*Figure 33*); and electrical symbols. Slightly different symbols may be used in different parts of the country or by different companies. The symbols used for each set of plans should be indicated on the title sheet or other introductory drawing. Many code books, manufacturers' brochures, and specifications include symbols and their meanings.

Some plans use keynotes (*Figure 34*) instead of symbols. A keynote is a number or letter (usually in a square or circle) with a leader and arrowhead that is used to identify a specific object. Part of the drawing sheet (usually on the right-hand side) lists the keynotes with their numbers or letters. The keynote descriptions normally use abbreviations.

1.3.3 Using Gridlines to Identify Plan Locations

Have you ever used a map to find a street? The map may have used a grid to make locating a detailed area easier. For example, the index might have referred you to section B-3, so you located B along the side of the map and 3 along the top. Then you located the intersection of the two to find the street.

The gridline system shown on a plan (*Figure 35*) is used like the grid on a map. On a drawing such as a floor plan, a grid divides the area into small parts called bays.

The numbering and lettering system begins in the upper left-hand corner of the floor plan. The numbers are normally across the top and the letters are along the side. To avoid confusion, certain letters and the symbol for zero are not used. Omitted from the gridline system are the letters I, O, and Q; and numbers 1 and 0.

A gridline system makes it easy to refer to specific locations on a plan. Suppose you want to refer to one outlet, but there are a dozen on a plan. Simply refer to "the outlet in bay C-8".

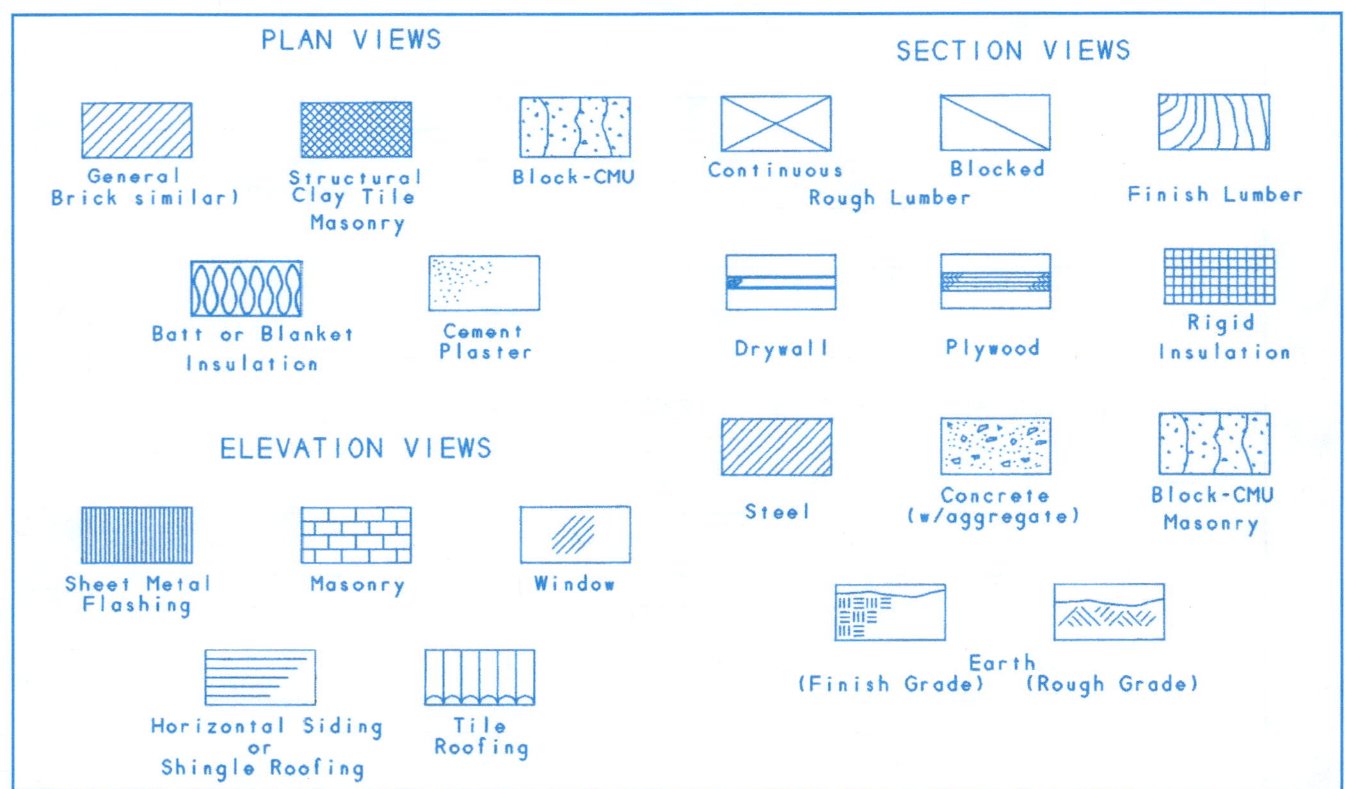

Figure 30 Architectural symbols.

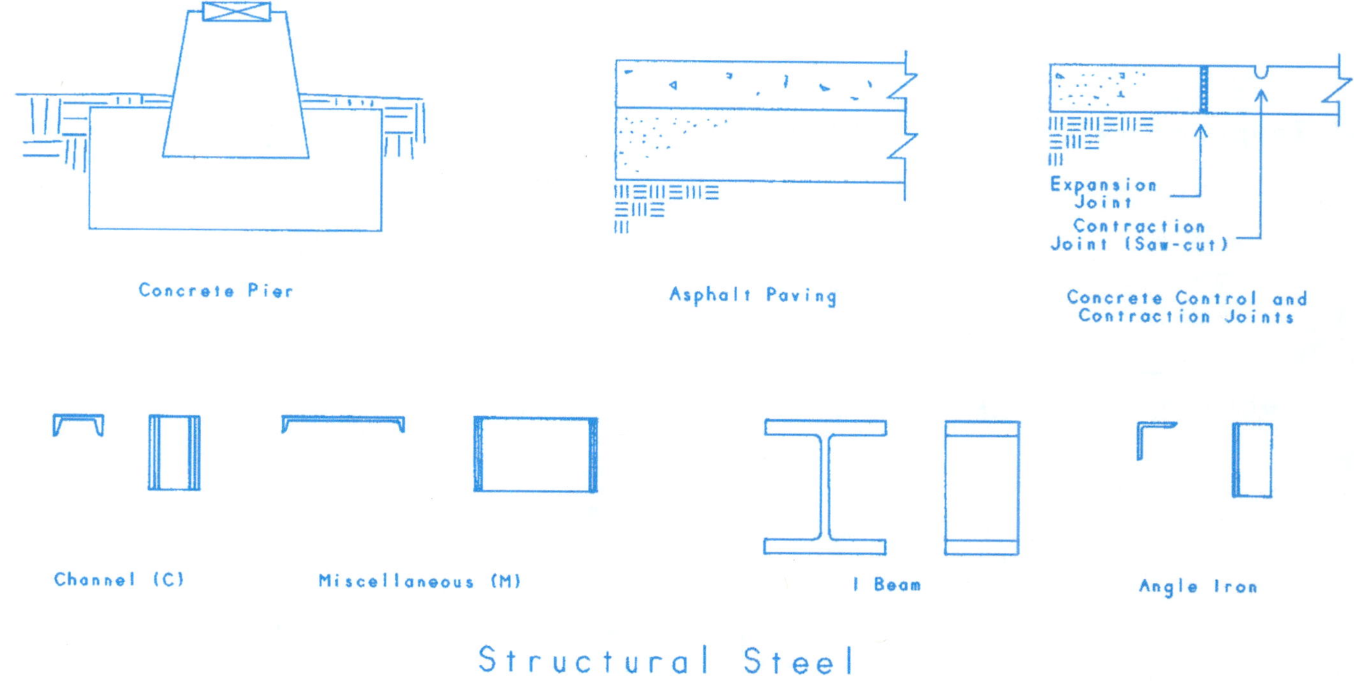

Figure 31 Civil and structural engineering symbols.

GOING GREEN

Green Construction

Green construction refers to a method of designing and building structures using materials and techniques that help minimize the stress on our natural resources and the environment.

With just under 5 percent of the world's population, the United States manages to consume about 19 percent of the world's energy (buildings account for 40 percent of this consumption) and produces 170 million tons (154,221,406 metric tons) of construction and demolition debris a year.

In response to these numbers, and a general increase in environmental awareness, organizations such as the US Green Building Council (USGBC) created environmental assessment systems, such as the LEED (Leadership in Energy and Environmental Design) Green Building Rating System. Systems like LEED provide green standards for the construction industry to follow, and they officially certify structures (nonresidential) that meet USGBC's strict criteria.

The LEED program was launched in 1988. It is a voluntary national standard that awards points for incorporating green strategies into areas such as site planning, safeguarding water quality and efficiency, efficiency in energy use and recycling, conservation of resources and materials, and the design, quality, and efficiency of the indoor environment.

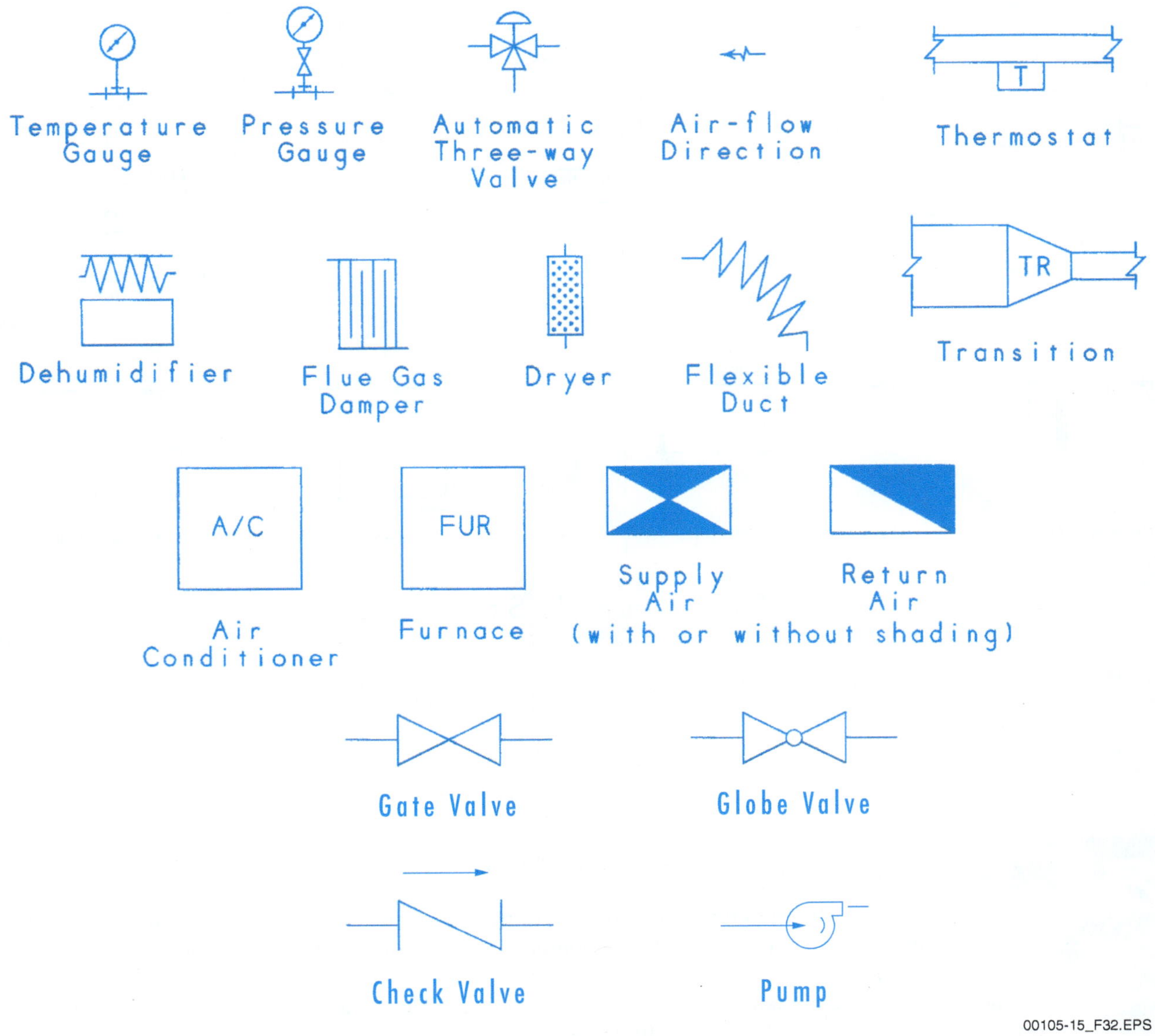

Temperature Gauge
Pressure Gauge
Automatic Three-way Valve
Air-flow Direction
Thermostat
Dehumidifier
Flue Gas Damper
Dryer
Flexible Duct
Transition
Air Conditioner
Furnace
Supply Air
Return Air
(with or without shading)
Gate Valve
Globe Valve
Check Valve
Pump

00105-15_F32.EPS

Figure 32 Mechanical symbols.

Care of Construction Drawings

Construction drawings are valuable records and must be cared for. Follow these rules when handling construction drawings:

- Never write on a construction drawing without authorization.
- Keep drawings clean. Dirty drawings are hard to read and can cause errors.
- Fold drawings so that the title block is visible.
- Fold and unfold drawings carefully to avoid tearing.
- Do not lay sharp tools or pointed objects on construction drawings.
- Keep drawings away from moisture.
- Make copies for field use; don't use originals.

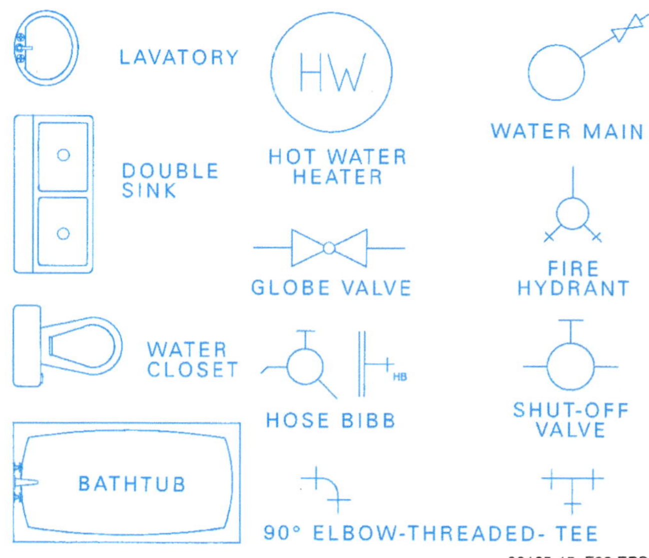

Figure 33 Plumbing symbols.

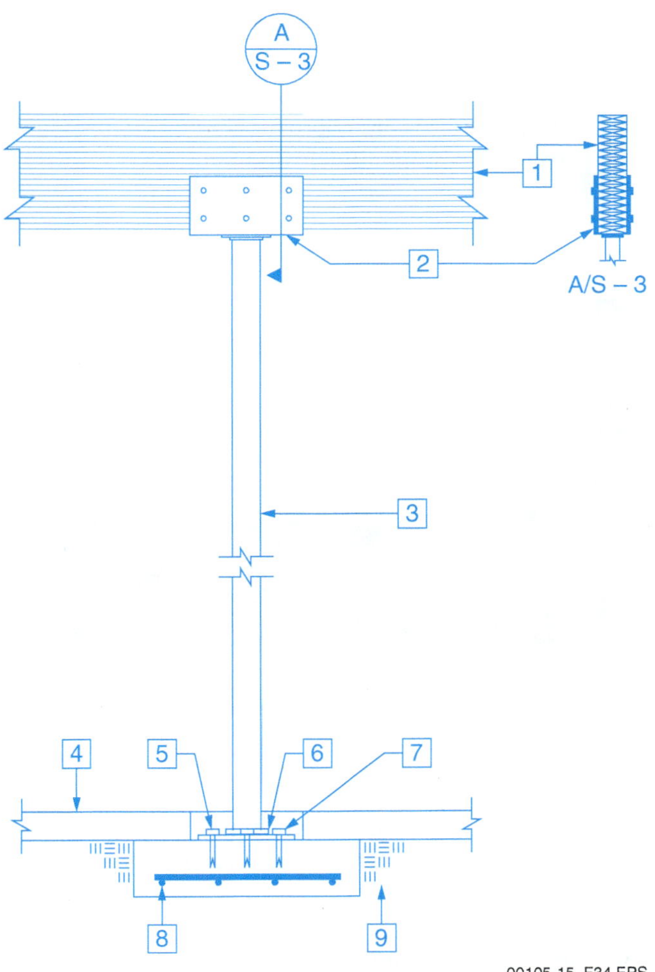

Figure 34 Keynotes.

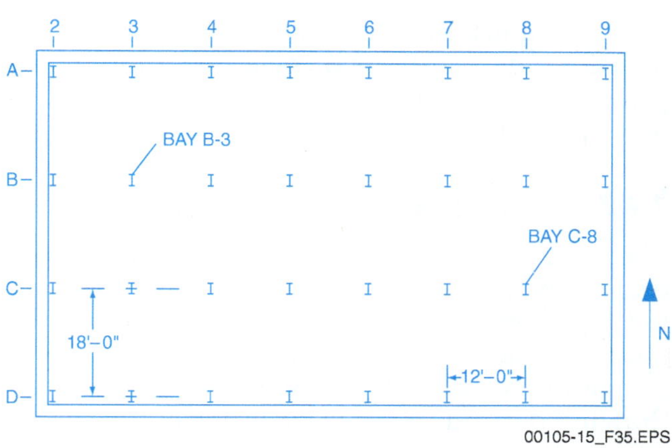

Figure 35 Grid.

1.4.0 Dimensions and Drawing Scale

Dimensions and their associated drawing scales provide information that is essential for correctly translating construction drawings into actual buildings. This section describes the basic features and purpose of dimensions and drawing scales. How to read and use different types of drawing scales will be discussed later.

1.4.1 Dimensions

Dimensions are the parts of the drawings that show the size and the placement of the objects that will be built or installed. Dimension lines can have arrowheads or slashes at both ends, with the dimension itself written near the middle of the line. The dimension is a measurement written as a number, and it may be written in inches with fractions (6½"), in feet with inches (1'-2"), or in inches with decimals (3.2"). When the metric system is used, the dimensions are usually expressed in meters, centimeters, or millimeters (9 mm).

To do accurate work, workers need to know how to read dimensions on construction drawings. This means they need to know whether the dimensions measure to the exterior or the interior of an object. To understand the difference, look at *Figure 36*, which shows a piece of pipe. There are two measurements that could be taken to get the pipe's dimensions.

The first measurement is from the pipe's exterior edge on one side directly across to its exterior edge on the other side. The second measurement is from the pipe's inside (interior) edge on one side directly across to its interior edge on the other side. Even though the difference between these two dimensions may be only a fraction of an inch (the thickness of the pipe), they are still two completely different dimensions.

This is important to remember because any dimensioning inaccuracy or miscalculation in one place will affect the accuracy of calculations in other places.

1.4.2 Drawing Scale

The scale of a drawing tells the size of the object drawn compared with the actual size of the object represented. The scale is shown in one of the spaces in the title block, beneath the drawing itself, or in both places. The type of scale used on a drawing depends on the size of the objects being shown, the space available on the paper, and the type of plan.

On a site plan, the scale may read SCALE: 1" = 20'-0". This means that every 1 inch on the drawing represents 20 feet, 0 inches. The scale used to develop site plans is an **engineer's scale**.

On a floor plan, the scale may read SCALE: ¼" = 1'-0". This means that every ¼ inch on the drawing represents 1 foot, 0 inches. Floor plans are developed using an **architect's scale**. This scale is divided into fractions of an inch. Metric floor plans typically use a ratio of 1:50, indicating that each millimeter on the drawing represents 50 meters.

Some drawings are not drawn to scale. A note on such drawings reads **not to scale (NTS)**.

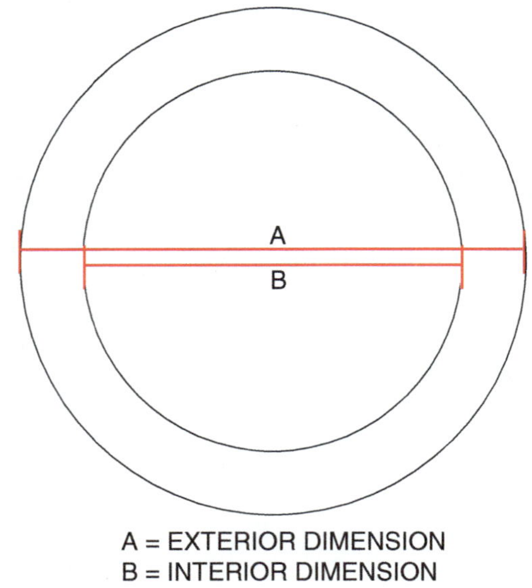

Figure 36 Exterior and interior dimensions on pipe.

CAUTION

When a plan is marked NTS, workers cannot measure dimensions on the drawing and use those measurements to build the project. Not-to-scale drawings give relative positions and sizes. The sizes are approximate and are not accurate enough for construction.

GOING GREEN

Recycling Rinker Hall, University of Florida in Gainesville

An example of a LEED Gold rated building is Rinker Hall on the campus of the University of Florida in Gainesville, Florida. During the planning phase of the building, special consideration was given to land use and the use of recycled and recyclable materials. Special attention was also given to planning the incorporation of the most recent green technologies in the areas of water consumption and conservation, heating and cooling energies, and other power needs required to run the various systems of a building.

00105-15_SA08.EPS

A few of the green strategies incorporated into the construction of Rinker Hall include:

- Orienting the structure on a pure north-south axis to allow it to use low-angle light for daytime lighting.
- Using rooftop skylights and a central skylight-covered atrium to light specific areas using natural sunlight.
- Using shade walls on the east and west sides of the building.
- Landscaping with native trees and flora that require minimal watering.
- Installing water-free urinals and fixtures that use 20 percent less water than mandated.

With the various green strategies, materials selection, and building techniques employed, Rinker Hall and all of its systems use up to 57 percent less energy than a similar structure designed in minimal compliance with the American Society of Heating, Refrigerating, and Air Conditioning Engineers (ASHRAE).

1.5.0 Measuring Scales

Standard and metric rulers and measuring tapes are usually used in a shop or in the field for a variety of measuring tasks. However, there are other types of rulers called scales used by draftsmen to produce drawings. They are also used by many workers to take scaled measurements on a drawing. These include the architect's scale, the metric scale (metric architect's scale), and the engineer's scale. Knowing how these scales work makes it easier to understand the information contained in a set of drawings.

1.5.1 Architect's Scale

The architect's scale is often used to create construction drawings. An architect's scale translates the large measurements of real structures (rooms, walls, doors, windows, duct, etc.) into smaller measurements for drawings. Architect's scales are available in several types, but the most common include the triangular and flat scales. A flat scale is shown in *Figure 37*. The triangular architect's scale is most commonly used because it can combine up to twelve different scales on one tool. Each side of the triangular form has two faces, and two scales are combined on each face. Architect's scales are available in 6- and 12-inch lengths.

Each scale on an architect's scale is designated to a different fraction of an inch that equals a foot. These fraction designations appear on the right and left corners of each scale. You read an architect's scale from left to right or right to left, depending on which scale you are reading.

Look at the point of measurement in *Figure 37*. It represents 57 feet when read from the left on the ⅛-inch scale and equals 18 feet when read from the right on the ¼-inch scale.

Now look at *Figure 38*. Using the ⅜-inch scale and reading from the right, you can determine that the section of duct is 7-feet, 5-inches long. Notice how the 0 point on an architect's scale is not at the extreme end of the measuring line. This is because numbers to the right of the 0 represent fractions of one foot (or inches).

Checking Scales on Drawings

Normally the scale of a drawing is marked directly on the drawing itself. If it is not, however, do not guess what the scale is. Rather, compare different scales by identifying a given length on the drawing (for example, an 8-foot, 6-inch pipe), and then find the scale that measures the pipe at 8-feet, 6-inches. Check at least two items on the drawing this way before proceeding.

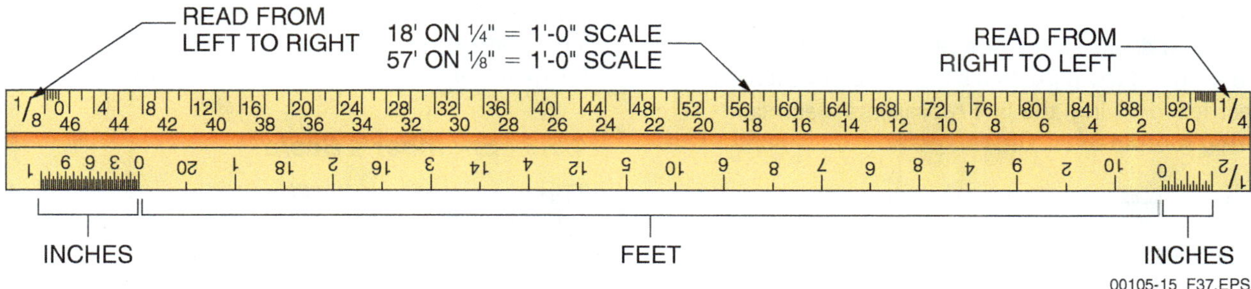

Figure 37 The architect's scale.

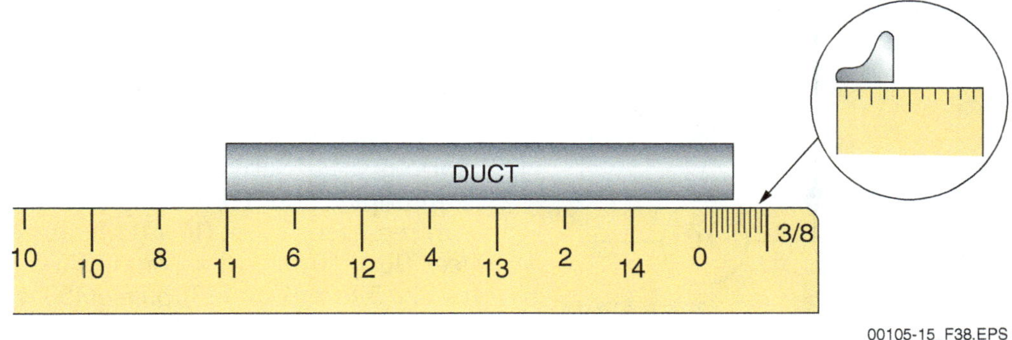

Figure 38 Measuring a section of duct with an architect's scale.

1.5.2 Metric Scale (Metric Architect's Scale)

Similar to the architect's scale is the metric scale, sometimes referred to as a metric architect's scale (*Figure 39*). Common lengths for metric scales are 30 and 60 millimeters. Like the architect's scale, a metric scale can have a number of scales on it and is used to generate drawings. *Figure 39* shows how 10 meters would be shown on a 1:200 scale. The actual length on the plans is 50 mm, equaling 10 meters of real distance on the project.

Metric scales are calibrated in units of 10. Some of the most common metric scales include 1:5, 1:10, 1:20, 1:50, 1:100, 1:200, 1:500, and 1:1,000.

The two common length measurements used with the metric scale on architectural drawings are the meter and millimeter, the millimeter being 1⁄1000 of a meter. On drawings drawn to scale between 1:1 and 1:100, the millimeter is typically used. The millimeter symbol (mm) will not be shown, but there should be a note on the drawing indicating that all dimensions are given in millimeters unless otherwise noted.

On drawings with scales between 1:200 (*Figure 39*) and 1:2,000, the meter is generally used. Again, the meter symbol (m) will not be shown, but the drawing will have a note indicating that all dimensions are in meters unless otherwise noted. Land distances shown on site and plot plans, expressed in metric units, are typically given in meters or kilometers (1,000 meters).

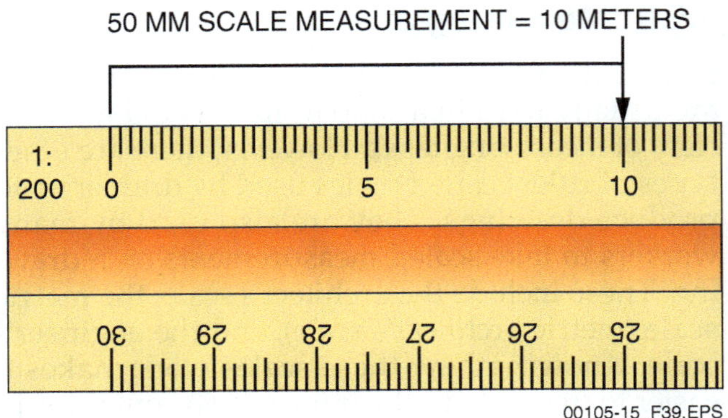

Figure 39 Metric architect's scale.

Reading a metric scale is easy once you identify the scale to use and the length that the scale increments represent on the drawing. Here's how it works.

The unit of length in *Figure 39* is the meter and the object on the drawing starts at 0 and extends out to the 10 on the scale. At a 1:200 ratio this object would be 10 meters long. This is because every millimeter on the scale represents 200 millimeters on the object. For example, if there are 50 millimeters from 0 to 10, multiply the 50 by 200, which gives 10,000 mm (50 mm × 200 = 10,000 mm). Because the unit being used is the meter and there are 1000 mm in one meter, divide the 10,000 mm by 1,000, which results in 10 meters. This same process is also used to determine lengths for other ratios on the scale.

Schematic Drawings

Most plumbing and electrical sketches are single-line drawings or schematic drawings. These drawings illustrate the scale and relationship of the project's components. In a single-line or schematic plumbing drawing, the line represents the centerline of the pipe. In a single-line or schematic electrical drawing, the line represents electrical wiring routing or circuit.

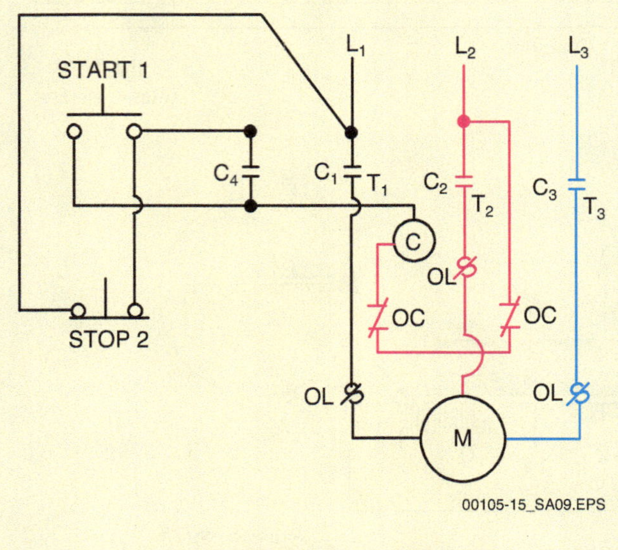

1.5.3 Engineer's Scale

The engineer's scale (*Figure 40*) is used mainly for land measurements on site plans, which means the scale must accommodate very large measurements. Each engineer's scale is set up as multiples of 10 and the measurements are taken in decimals. This is different from the architect's scale in that a unit is represented by a portion of an inch. The most common engineer's scales are 10, 20, 30, 40, 50, and 60, which can all be combined on the triangular engineer's scale.

For each scale, the measurements can represent various units derived from that scale number and a multiple of 10. For example on a 10 scale, 1 inch can represent 1 foot (10 × 0.10), 10 feet (10 × 1.0), 100 feet (10 × 10.0), and so on. On a 50 scale, an inch can be 5 feet (50 × 0.10), 50 feet (50 × 1.0), or 500 feet (50 × 10.0), and so on. This same process can also be used to determine lengths on the scales.

00105-15_F40.EPS

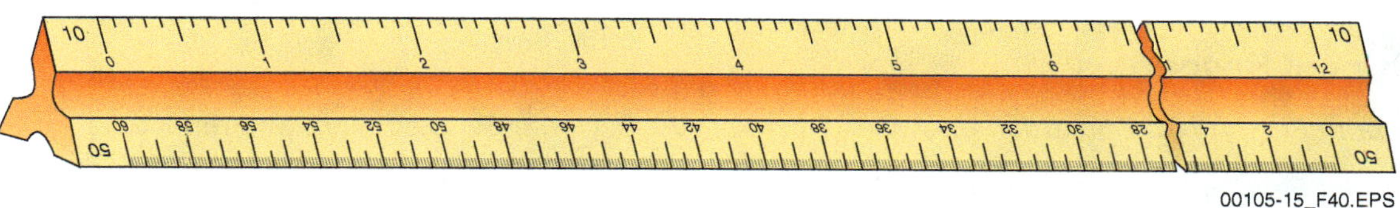

Figure 40 The engineer's scale.

Orthographic Drawings

Orthographic drawings are used for elevation drawings. They show straight-on views of the different sides of an object with dimensions that are proportional to the actual physical dimensions. In orthographic drawings, the designer draws lines that are scaled-down representations of real dimensions. Every 12 inches, for example, may be represented by ¼ inch on the drawing. Similarly, in an example using metric measurements with a ratio of 1:2, every 30 millimeters may be represented by 15 millimeters on the drawing.

00105-15_SA10.EPS

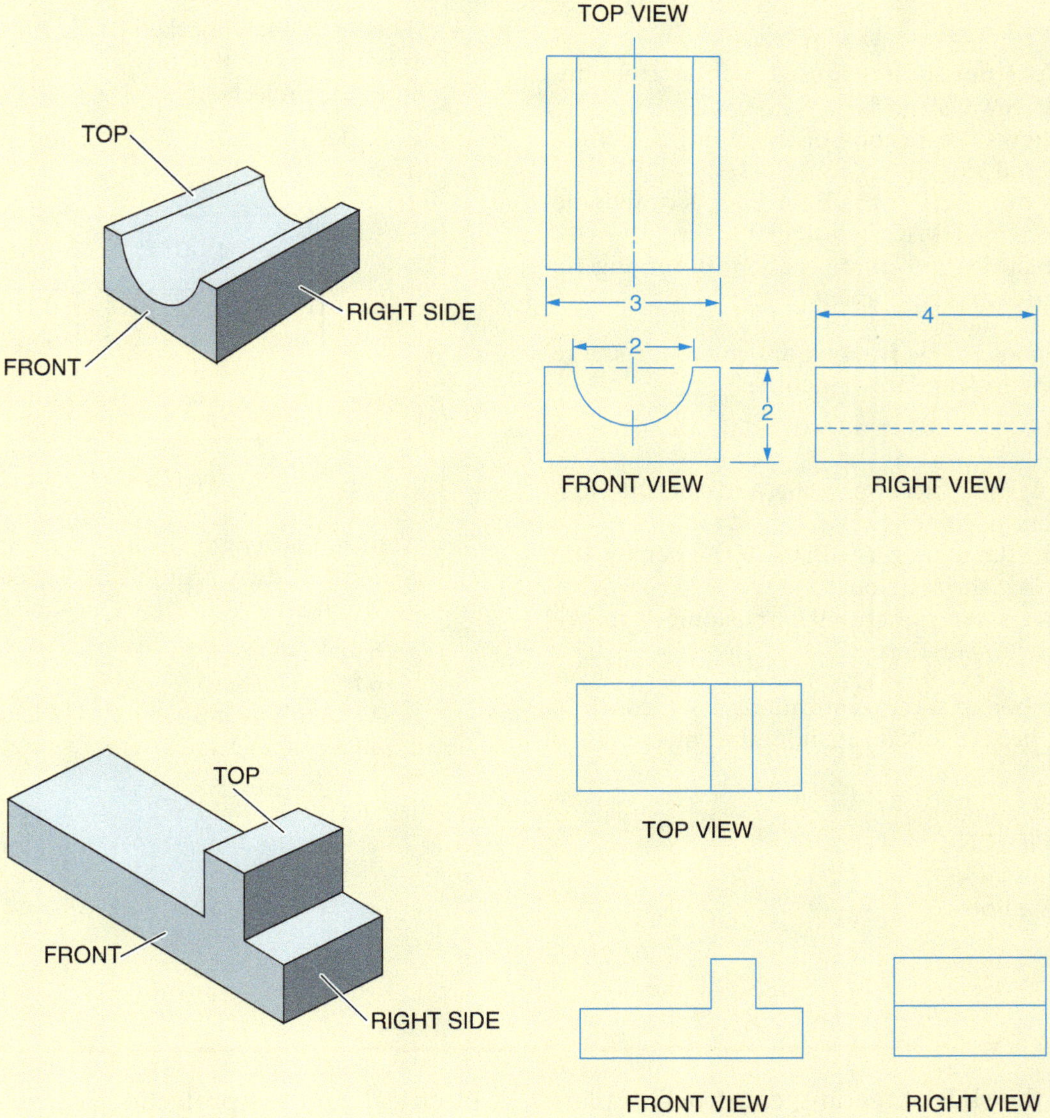

Additional Resources

Blueprint Reading for Construction, James Fatzinger. 2003. Upper Saddle River, NJ: Prentice Hall.

Blueprint Reading for the Construction Trades, Peter A. Mann. 2005. Ontario, Canada: **Micro-press.com**.

Reading Architectural Plans for Residential and Commercial Construction, Ernest R. Weidhaas. 2001. Englewood Cliffs, NJ: Prentice Hall Career & Technology.

Reading Architectural Work Drawings, Edward J. Muller and Phillip A. Grau III. 2003. Upper Saddle River, NJ: Prentice Hall.

Autodesk, 1 Market St, Suite 500, San Francisco, CA 94105, USA; 3D design, engineering and entertainment software and parent company of the AutoCAD software suite. **www.autodesk.com**.

Datacad, P.O. Box 815, Simsbury, CT 06070, USA. Windows-based CADD solutions. **www.datacad.com**.

1.0.0 Section Review

1. A site plan _____.
   a. does not show features such as trees and driveways
   b. shows the location of the building from an aerial view
   c. is drawn after the floor plan is drawn, before the HVAC is designed
   d. includes information about plumbing fixtures and equipment

2. Revisions to the drawing are entered in the revision block and must include _____.
   a. the project tag and the date the drawing was approved
   b. engineering approvals and the intended date of completion
   c. the date and the initials of the person who made the revision
   d. dates and signatures for customer approval documentation

3. Code books and specifications for a construction project often include the meanings of _____.
   a. drawings
   b. references
   c. glossaries
   d. symbols

4. When a plan is marked NTS, the dimensions as measured on the drawing cannot be used to build the project.
   a. True
   b. False

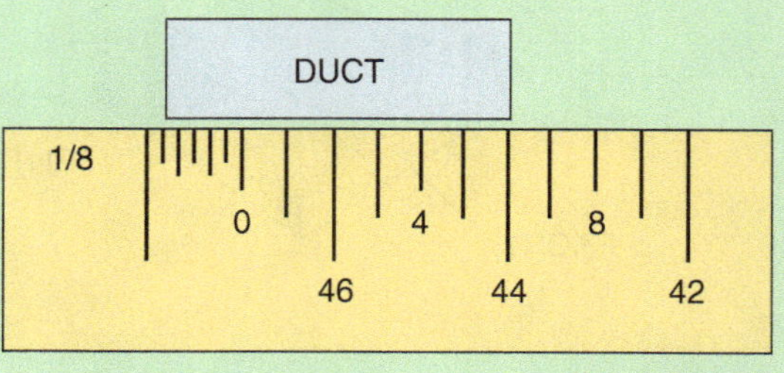

Figure 1

5. What is the length (in feet and inches) of the section of duct using the architect's scales shown in Section Review Question *Figure 1* ?
   a. 6 feet, ¾ inches
   b. 6 feet, 5 inches
   c. 44 feet, ¾ inches
   d. 48 feet, 4 inches

SUMMARY

Mastering the skill of reading construction plans requires practice, but it is a skill that anyone who works in the construction trades needs to develop. Each of the different types of construction drawings commonly found on a job site—civil, architectural, structural, mechanical, plumbing, electrical, and fire protection plans—is important for the completion of the project. By correctly interpreting the drawing elements, symbols, and scales that are used workers can visualize the entire project, detect inconsistencies or errors early, and possibly avoid costly mistakes or rework.

Review Questions

1. Which type of plan shows the layout of the HVAC system?
   a. Plumbing
   b. Structural
   c. Mechanical
   d. Foundation

2. Electrical plans can include _____.
   a. exterior elevations
   b. section drawings
   c. plumbing isometrics
   d. lighting plans

3. The title block generally contains _____.
   a. revision blocks
   b. special marks used in the drawing
   c. the mechanical plans
   d. the legend

4. The latest revision date on a set of construction drawings can be found _____.
   a. inside the detail
   b. in the schedule
   c. inside the title block
   d. in the legend

5. The Alphabet of Lines consists of _____.
   a. the line types used on a construction drawing
   b. lines that are indicated using letters of the alphabet
   c. lines that match up detail and section drawings
   d. lines that indicate land boundaries on the site plan

6. On a gridline system, a grid divides the area into small parts called _____.
   a. segments
   b. bays
   c. sections
   d. PODS

7. When the metric system is used, dimensions are written in _____.
   a. inches, feet, and yards
   b. degrees, radians, and gradians
   c. CADS, RADS, and KRADS
   d. meters, centimeters, and millimeters

8. If the scale on a site plan reads SCALE: 1" = 20'-0", then every _____.
   a. $\frac{1}{20}$th of an inch on the drawing represents 20 feet, 0 inches
   b. 20 inches on the drawing represents 1 foot, 0 inches
   c. inch on the drawing represents 20 feet, 0 inches
   d. 20 inches on the drawing represents 20 feet, 0 inches

00105-15_RQ01.EPS

Figure 1

9. What is the length in meters at the point indicated by the arrow on the metric scale in Review Question *Figure 1*? The unit of length used on the scale is the meter.
   a. 3500
   b. 350
   c. 35
   d. 3.5

10. An engineer's scale is set up in multiples of _____.
    a. 5
    b. 10
    c. 12
    d. 39.37

Trade Terms Quiz

Fill in the blank with the correct term that you learned from your study of this module.

1. A(n) ________ is a side view that shows height.
2. A(n) ________ usually has an arrowhead at both ends, with the measurement written near the middle of the line.
3. A(n) ________ is a qualified, licensed person who creates and designs drawings for a construction project.
4. ________, which show the design of the project, include many parts, such as the floor plan, roof plan, elevation drawings, and section drawings.
5. ________ support the architectural design of a building to show how the building is supported. A foundation plan would be part of this category.
6. Also called site plans or survey plans, ________ show the location of the building from an aerial view, as well as the natural contours of the earth.
7. Almost all construction drawings today are made by ________.
8. ________ are solid or dashed lines showing the elevation of the earth on a civil drawing.
9. ________ are enlarged views of some special features of a building, such as floors and walls.
10. A(n) ________ is a person who applies scientific principles in design and construction.
11. ________, or engineered drawings for electrical supply and distribution, include locations of the electric meter, switchgear, and convenience outlets.
12. A large, horizontal support made of concrete, steel, stone, or wood that may be shown in structural plans is a ________.
13. Schematic drawings called ________ show all the equipment, pipelines, valves, instruments, and controls needed to operate a piping system.
14. An element of architectural drawings, ________ refers to the height above sea level or other defined surface.
15. Architectural, civil and structural engineering, mechanical, and plumbing ________ may be used on a drawing to tell what material is required for that part of the project.
16. Also called a plan view, a(n) ________ is an aerial view of the layout of each room.
17. A(n) ________ is a one-line drawing showing the flow path for electrical circuitry or the relationship of all parts of a system.
18. Part of the structural plans, the ________ shows the lowest level of the building.
19. Plans for gas, oil, or steam heat piping may be included in the ________ plan.
20. When a plan is marked ________, it means that the drawing gives approximate positions and sizes only.
21. A(n) ________ is a dashed line on a plan showing an object obstructed from view by another object.
22. A(n) ________ is a type of three-dimensional drawing that is included in a plumbing plan.
23. The meter and millimeter are two common length measurements used with the ________ on architectural drawings.
24. In drafting, an arrowhead is placed on a(n) ________ to identify a component.
25. ________ are written statements provided by the architectural and engineering firm to define the quality of work to be done and to describe the materials to be used.
26. The ________ defines the symbols used in architectural plans.
27. ________ are engineered plans for motors, pumps, piping systems, and piping equipment.

28. A(n) ________ is a cross-sectional view that shows the inside of an object or building.
29. The triangular and flat scales are the two most common types of ________.
30. A(n) ________ shows the shape of the roof and the materials that will be used to finish it.
31. The ________ of a drawing tells the size of the object drawn compared with the actual size of the object represented.
32. ________ show the layout for the plumbing system that supplies hot and cold water, for the sewage disposal system, and for the location of plumbing fixtures.
33. Part of the construction drawing, the ________ gives information about the structure and is numbered for easy filing.
34. ________ are the traditional name for construction drawings.
35. The drawing that makes up a building's piping, valves, and switches is a(n) ________.
36. A(n) ________ is used mainly for land measurements on site plans.

Trade Terms

Architect
Architect's scale
Architectural plans
Beam
Blueprints
Civil plans
Computer-aided drafting (CAD)
Contour lines
Detail drawings
Dimension line
Electrical plans
Elevation (EL)
Elevation drawing
Engineer
Engineer's scale
Fire protection plan
Floor plan
Foundation plan
Hidden line
HVAC
Leader
Legend
Mechanical plans
Metric scale
Not to scale (NTS)
Piping and instrumentation drawings (P&IDs)
Plumbing isometric drawing
Plumbing plans
Roof plan
Scale
Schematic
Section drawing
Specifications
Structural plans
Symbol
Title block

Cornerstone of Craftsmanship

Jan Prakke

Lake Mechanical Contractors Inc.
Director of Safety and Training

How did you choose a career in the construction industry?
A friend of mine had just started his own plumbing company and needed help with a large project he landed. He called me up and asked if I would be willing to help him. I replied that I knew nothing about plumbing and asked if I would really be a help. He replied "Are you stupid?" I answered no. "Fine—I will pick you up at seven." It was my first time on a construction site and I truly enjoyed the work so much that I quit my permanent job and began a career in the plumbing industry.

Who inspired you to enter the industry?
There were a couple of people. My father-in-law was a career sheet metal worker in Chicago. And of course there was Tom Kissane, the plumber who hooked me into the trade. Both of these guys inspired craftsmanship, taught me to do the job right the first time, and to show pride in my work.

What types of training have you been through?
I have had training in the following areas:

- NCCER Apprenticeship
- NCCER Trainer
- NCCER Master Trainer
- Medical Gas Certification
- OSHA 503 Certification
- Rigging Train the Trainer Program
- Certifications earned to operate several different types of heavy equipment
- Certifications earned in several different pipe-joining methods

How important is education and training in construction?
If you want to succeed in any craft or profession it is always important to keep learning. The education I received in the past and plan to receive in the future will only make me a stronger asset to my company.

How important are NCCER credentials to your career?
The credentials I received from NCCER from my apprenticeship through the Master Trainer credential have been invaluable. It's an outstanding program with industry recognition that is unparalleled in the construction industry.

How has training/construction impacted your life?
The construction and training fields have impacted my life by allowing me to raise my family at a comfortable level. It has also helped me to instill in my children the discipline to do a job well the first time.

Tell us about your present job.
At present, I am the Director of Safety and Training for Lake Mechanical Contractors, Inc. I direct our apprenticeship program and train our field personnel on the new products and techniques that come on the market.

What do you enjoy most about your job?
I thoroughly enjoy teaching the apprenticeship classes. Trainees come in to the program knowing very little about plumbing, and about half-way through the program, the light comes on and they suddenly understand the terminology and know how to complete specific tasks. The questions then get more technical, and their hunger to learn more excites them as well as me.

Would you suggest construction as a career to others? Why?
I always tell my students about an episode of 60 Minutes on TV. The show profiled Jim Varney (the star of the movie, "Ernest Goes to Camp"). The 60 Minutes crew walked around his mansion and grounds, asking him questions about his career in acting. Once they reached the garage, they saw all of his nice cars. In the corner of the building, there was a pile of tile-setting tools, trowels, saws, etc. He explained that, prior to making it as an actor, he was a tile setter. The reporter asked why he still kept his tools. He was rich and famous—why keep them around? His reply struck home with me: "Because you never know when things will go bad. I will always have my trade to fall back on." So no matter what you do in life, you will always have a backup plan as a craftworker.

How do you define craftsmanship?
Craftsmanship is the work of a skillful person, done with great care and expertise; caring about what you are doing and showing pride in the finished product.

Trade Terms Introduced in This Module

Architect: A qualified, licensed person who creates and designs drawings for a construction project.

Architect's scale: A specialized ruler used in making or measuring reduced scale drawings. The ruler is marked with a range of calibrated ratios for laying out distances, with scales indicating feet, inches, and fractions of inches. Used on drawings other than site plans.

Architectural plans: Drawings that show the design of the project. Also called architectural drawings.

Beam: A large, horizontal structural member made of concrete, steel, stone, wood, or other structural material to provide support above a large opening.

Blueprints: The traditional name used to describe construction drawings.

Civil plans: Drawings that show the location of the building on the site from an aerial view, including contours, trees, construction features, and dimensions.

Computer-aided drafting (CAD): The making of a set of construction drawings with the aid of a computer.

Contour lines: Solid or dashed lines showing the elevation of the earth on a civil drawing.

Detail drawings: Enlarged views of part of a drawing used to show an area more clearly.

Dimension line: A line on a drawing with a measurement indicating length.

Electrical plans: Engineered drawings that show all electrical supply and distribution.

Elevation (EL): Height above sea level, or other defined surface, usually expressed in feet or meters.

Elevation drawing: Side view of a building or object, showing height and width.

Engineer: A person who applies scientific principles in design and construction.

Engineer's scale: A straightedge measuring device divided uniformly into multiples of 10 divisions per inch so that drawings can be made with decimal values. Used mainly for land measurements on site plans.

Fire protection plan: A drawing that shows the details of the building's sprinkler system.

Floor plan: A drawing that provides an aerial view of the layout of each room.

Foundation plan: A drawing that shows the layout and elevation of the building foundation.

Hidden line: A dashed line showing an object obstructed from view by another object.

HVAC: Heating, ventilating, and air conditioning.

Leader: In drafting, the line on which an arrowhead is placed and used to identify a component.

Legend: A description of the symbols and abbreviations used in a set of drawings.

Mechanical plans: Engineered drawings that show the mechanical systems, such as motors and piping.

Metric scale: A straightedge measuring device divided into centimeters, with each centimeter divided into 10 millimeters. Usually used for architectural drawings and sometimes referred to as a metric architect's scale.

Not to scale (NTS): Describes drawings that show relative positions and sizes only, without scale.

Piping and instrumentation drawings (P&IDs): Schematic diagrams of a complete piping system.

Plumbing isometric drawing: A type of three-dimensional drawing that depicts a plumbing system.

Plumbing plans: Engineered drawings that show the layout for the plumbing system.

Roof plan: A drawing of the view of the roof from above the building.

Scale: The ratio between the size of a drawing of an object and the size of the actual object.

Schematic: A one-line drawing showing the flow path for electrical circuitry or the relationship of all parts of a system.

Section drawing: A cross-sectional view of a specific location, showing the inside of an object or building.

Specifications: Precise written presentation of the details of a plan.

Structural plans: A set of engineered drawings used to support the architectural design.

Symbol: A drawing that represents a material or component on a plan.

Title block: A part of a drawing sheet that includes some general information about the project.

Additional Resources

This module presents thorough resources for task training. The following resource material is suggested for further study.

Blueprint Reading for Construction, James Fatzinger. 2003. Upper Saddle River, NJ: Prentice Hall.

Blueprint Reading for the Construction Trades, Peter A. Mann. 2005. Ontario, Canada: **Micro-press.com**.

Reading Architectural Plans for Residential and Commercial Construction, Ernest R. Weidhaas. 2001. Englewood Cliffs, NJ: Prentice Hall Career & Technology.

Reading Architectural Work Drawings, Edward J. Muller and Phillip A. Grau III. 2003. Upper Saddle River, NJ: Prentice Hall.

Autodesk, 1 Market St, Suite 500, San Francisco, CA 94105, USA; 3D design, engineering and entertainment software and parent company of the AutoCAD software suite. **www.autodesk.com**.

Datacad, P.O. Box 815, Simsbury, CT 06070, USA. Windows-based CADD solutions. **www.datacad.com**.

Figure Credits

©**iStockphoto.com/btwfoto**, Module Opener

Croxton Collaborative/Gould Evans Associates, Figures 2–11, 16–19, 22–25, SA02, SA04, Construction Drawings 1-4

Colonial Webb, SA05

Ritterbush-Ellig-Hulsing PC, SA07

Section Review Answer Key

Answer	Section Reference	Objective
Section One		
1. b	1.1.1	1a
2. c	1.2.4	1b
3. d	1.3.2	1c
4. a	1.4.2	1d
5. b	1.5.1	1e

NCCER CURRICULA — USER UPDATE

NCCER makes every effort to keep its textbooks up-to-date and free of technical errors. We appreciate your help in this process. If you find an error, a typographical mistake, or an inaccuracy in NCCER's curricula, please fill out this form (or a photocopy), or complete the online form at **www.nccer.org/olf**. Be sure to include the exact module ID number, page number, a detailed description, and your recommended correction. Your input will be brought to the attention of the Authoring Team. Thank you for your assistance.

Instructors – If you have an idea for improving this textbook, or have found that additional materials were necessary to teach this module effectively, please let us know so that we may present your suggestions to the Authoring Team.

NCCER Product Development and Revision

13614 Progress Blvd., Alachua, FL 32615

Email: curriculum@nccer.org

Online: www.nccer.org/olf

❑ Trainee Guide ❑ Lesson Plans ❑ Exam ❑ PowerPoints Other ____________________

Craft / Level: Copyright Date:

Module ID Number / Title:

Section Number(s):

Description:

Recommended Correction:

Your Name:

Address:

Email: Phone:

[illegible]

[illegible]

[illegible]

Date:

00107-15

Basic Communication Skills

Overview

The construction professional communicates constantly. The ability to communicate skillfully will help to make you a better worker and a more effective leader. This module provides guidance in listening to understand, and speaking with clarity. It explains how to use and understand written materials, and it also provides techniques and guidelines that will help you to improve your writing skills.

Module Two

Trainees with successful module completions may be eligible for credentialing through the NCCER Registry. To learn more, go to **www.nccer.org** or contact us at **1.888.622.3720**. Our website has information on the latest product releases and training, as well as online versions of our *Cornerstone* magazine and Pearson's product catalog.

Your feedback is welcome. You may email your comments to **curriculum@nccer.org,** send general comments and inquiries to **info@nccer.org**, or fill in the User Update form at the back of this module.

This information is general in nature and intended for training purposes only. Actual performance of activities described in this manual requires compliance with all applicable operating, service, maintenance, and safety procedures under the direction of qualified personnel. References in this manual to patented or proprietary devices do not constitute a recommendation of their use.

00107-15
Basic Communication Skills

Objectives

When you have completed this module, you will be able to do the following:

1. Describe the communication, listening, and speaking processes and their relationship to job performance.
   a. Describe the communication process and the importance of listening and speaking skills.
   b. Describe the listening process and identify good listening skills.
   c. Describe the speaking process and identify good speaking skills.
2. Describe good reading and writing skills and their relationship to job performance.
   a. Describe the importance of good reading and writing skills.
   b. Describe job-related reading requirements and identify good reading skills.
   c. Describe job-related writing requirements and identify good writing skills.

Performance Tasks

Under the supervision of your instructor, you should be able to do the following:

1. Perform a given task after listening to oral instructions.
2. Fill out a work-related form provided by your instructor.
3. Read and interpret a set of instructions for properly donning a safety harness and then orally instruct another person on how to don the harness.

Trade Terms

Active listening
Appendix
Body language
Bullets
Change order
Electronic signature
Font
Glossary
Graph
Index
Italics
Jargon
Memo
Nonverbal communication
Paraphrase
Permit
Punch list
Table
Table of contents

Industry Recognized Credentials

If you are training through an NCCER-accredited sponsor, you may be eligible for credentials from NCCER's Registry. The ID number for this module is 00107-15. Note that this module may have been used in other NCCER curricula and may apply to other level completions. Contact NCCER's Registry at 888.622.3720 or go to **www.nccer.org** for more information.

Contents

Topics to be presented in this module include:

Figures

SECTION ONE

1.0.0 COMMUNICATION

Objective

Describe the communication, listening, and speaking processes and their relationship to job performance.

a. Describe the communication process and the importance of listening and speaking skills.
b. Describe the listening process and identify good listening skills.
c. Describe the speaking process and identify good speaking skills.

Performance Task

1. Perform a given task after listening to oral instructions.

Trade Terms

Active listening: A process that involves respecting others, listening to what is being said, and understanding what is being said.

Body language: A person's facial expression, physical posture, gestures, and use of space, all of which communicate feelings and ideas.

Jargon: Specialized terms used in a specific industry.

Nonverbal communication: All communication that does not use words. This includes appearance, personal environment, use of time, and body language.

Paraphrase: Express something heard or read using different words.

Every construction professional learns how to use tools. Depending on your trade, the tools you use could include welding machines and cutting torches, press brakes and plasma cutters, or surveyor's levels and pipe threaders. However, some of the most important tools you will use on the job are not tools you can hold in your hand or put in a toolbox. These tools are your abilities to read, write, listen, and speak.

At first, you might think that these are not really construction tools. They are things you already learned how to do in school, so why do you have to learn them all over again? The types of communication that take place in the construction workplace are very specialized and technical, just like the communications between pilots and air traffic control. Good communications result in a job done safely—a pilot hears and understands the message to change course to avoid a storm, and a construction worker hears and understands the message to install a water heater according to the local code requirements. In a way, you are learning another language, a special language that only trained professionals know how to use. Even though you will use a professional language that other people may not understand, the same communication skills apply to all professions, whether doctors, builders, managers, or mechanics.

The following are some specific examples of why these skills are so important in the construction industry:

- *Listening* – Your supervisor tells you where to set up safety barriers, but because you did not listen carefully, you missed a spot. As a result, your co-worker falls and is injured.
- *Speaking* – You must train two co-workers to do a new task, but you mumble, use words they don't understand, and don't answer their questions clearly. Your co-workers do the task incorrectly, and all of you must work overtime to fix the mistakes.
- *Reading* – Your supervisor tells you to read the manufacturer's basic operating and safety instructions for the new drill press before you use it. You don't really understand the instructions, but you don't want to ask him. You go ahead with what you think is correct and damage the drill press.
- *Writing* – Your supervisor asks you to write up a material takeoff (supply list) for a project. You rush through the list and don't check what you've written. The supplier delivers 250 feet of PVC piping cut to your specified sizes instead of 25 feet.

As you can see, good communication on the work site has a direct effect on safety, schedules, and budgets. A good communications toolbox is a badge of honor; it lets everyone know that you have important skills and knowledge. And like a physical toolbox, the ability to communicate well verbally and in writing is something that you can take with you to any job. You will find that good communications skills can help you advance your career. This module introduces you to the techniques you will need to read, write, listen, and speak effectively on the job.

1.1.0 The Communication Process

There are two basic steps to clear communication (*Figure 1*). First, a sender sends a spoken or written message through a communication channel to a receiver (examples of communication channels include meetings, phones, two-way radios, and email). When the receiver gets the message, he or she figures out what it means by listening or reading carefully. If anything is not clear, the receiver gives the sender feedback by asking the sender for more information.

This process is called two-way communication, and it is the most effective way to make sure that everyone understands what's going on. It sounds simple, doesn't it? So why is good communication so hard to achieve? When we try to communicate, a lot of things—called noise—can get in the way. Following are some examples of communication noise:

- The sender uses work-related words, or **jargon**, that the receiver does not understand.
- The sender does not speak clearly.
- The sender's written message is disorganized or contains mistakes.
- The sender is not specific.
- The sender does not get to the point.
- The receiver is tired or distracted or just not paying attention.
- The receiver has poor listening or reading skills.
- Actual noise on the construction site makes it physically hard to hear a message.
- There is a mechanical problem with the equipment used to communicate, such as static on a phone or radio line.

1.1.1 Nonverbal Communication

It is obvious that humans communicate with their words, known technically as verbal communication. If a mechanic asks for a ratchet set, he or she expects to get it because the words have a shared meaning between the listener and speaker. However, did you know that people communicate constantly without using any words at all? This kind of communication is called **nonverbal communication**. It is hard to communicate some kinds of information with nonverbal communication. For example, try asking for that ratchet set without any words. For expressing attitudes and emotion, nonverbal communication is a very powerful form of communication.

Imagine that you are talking to a doctor in a clinic. The whole time you are talking, he is yawning, checking his cell phone messages, and cleaning his fingernails. After you finish, he writes out a prescription. Do you trust the diagnosis and the value of the prescription? His nonverbal communication told you that he was bored and inattentive. Perhaps the doctor missed the most important thing you said about your condition, including an allergy to certain medications.

Similarly, you can express feelings and attitudes in a variety of ways without intention. You may show you are nervous by fidgeting in your chair or fiddling with your hands. You may show you are angry by raising your voice, folding your arms, and furrowing your eyebrows. You may show you are happy to see someone by smiling widely and giving him or her a warm handshake. The ways for a person to communicate nonverbally are limitless.

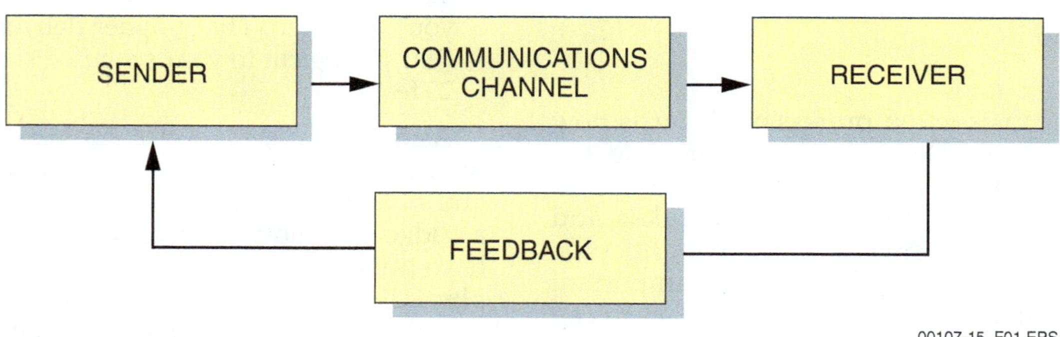

Figure 1 The communication process.

Did You Know?

Following Instructions Prevents Accidents

Many accidents are the result of not listening to or not understanding instructions. For example, according to a study by the US Occupational Safety and Health Administration (OSHA), over a 10-year period, 39 percent of crane operator deaths in the United States resulted from electrocution caused by contact with electrical power lines. This was the single largest cause of death in the study. How many of those accidents could have been prevented if the operator had heard, understood, and followed instructions? The answer is unknown, but it may have prevented all of them.

There are several basic categories of nonverbal communication:

- *Grooming* – Generally, people who maintain an attractive appearance have more successful careers. A groomed appearance communicates self-discipline and awareness. Shaving or keeping a trimmed beard could make the difference between getting promoted at work or being passed up by newcomers. Likewise, messy hair can communicate that someone is incompetent and unable to take on new responsibilities.
- *Dress* – Dress appropriately and neatly. Appropriate dress doesn't automatically mean formal dress. (Imagine working on a hot rooftop in a suit.) The best way to know how to dress in your work environment is to observe the people around you who are most successful, particularly the people in the position that you hope to obtain next. Even on occasions in which casual dress is allowed, such as a company outing to a baseball game, you must make sure that your casual clothing is clean and attractive. Casual does not mean messy. Many workplaces require a uniform. Of course, people working in these environments should always arrive at work with a neat and clean uniform.
- *Condition of one's personal environment* – People make their spaces their own by the way they arrange things even if the space does not belong to them. An office worker can show a sloppy work ethic by having binders, pens, paper clips, and candy wrappers strewn about his or her cubicle. A technician also might show a lack of responsibility by having gloves, coupons, magazines, and bags from fast food restaurants on the dashboard of his or her work vehicle. Whoever sees those environments unconsciously makes judgments about the character of the people who work there.
- *Use of time* – Clearly, people show respect and care by arriving on time or early to their scheduled events. Workers who show up late may be considered lazy or inconsiderate by those around them. Further, they may hinder the productivity of the entire organization by making others wait for them. Not only should you arrive or start on time, but when you hold meetings, you should also try to end on time. Doing so shows a humility on your part that others will appreciate.
- *Facial expressions* – One of the main ways that humans express and read emotions is through facial expressions. A smile can express interest and excitement. A frown could express displeasure or pain. An expressionless face expresses an emotion too, perhaps boredom or the desire to get away from whoever is talking. Of course, eyes are known as the window of the soul. Direct eye contact can express interest, understanding, intelligence, and confidence. On the other hand, a lack of eye contact could show inattention, a lack of confidence, or deceit. A person should not look into the eyes of another person the entire time while speaking, however. Too much eye contact could give the impression of initiating a challenge or trying to dominate. There is no solid rule for what amount of eye contact is appropriate. Different cultures have different rules. For example, in the West, authority figures expect subordinates to look at them when they are reprimanding them. In parts of Asia, however, etiquette requires the one being scolded to look down at the ground as a sign of respect. In the United States, people look at each other's eyes 50 to 60 percent of the time as they communicate.

- *Posture and gestures* – Slouching may feel comfortable, but it may also make people think of you as a sloppy person. Likewise, folding your arms in front of you may make you feel warmer on a cold day, but it will make you seem distant from others and unwilling to talk. Having a confident and powerful posture shows other people that you are confident and powerful. What if you feel timid and weak? Assume a powerful posture anyway. Research shows that people who kept a certain powerful posture for a few minutes—such as leaning slightly over a desk with their hands wide apart on the desk—experienced less stress and actually behaved more boldly when doing a task later. Those who had practiced low-power poses for a few minutes—such as folding their hands in their lap or touching their neck—showed a higher level of stress and behaved more timidly.
- *Physical distance* – Various cultures have different accepted personal distances. People of some countries stand close together while speaking, whereas others stand further apart. Also, different social situations call for a different personal range. People stand closer to their close friends and family members than they do to their business associates. Here is one rule of thumb to follow: let the more powerful person choose the distance. A manager or company owner may politely come and stand close to an employee, but an employee would be rude to do the same to his or her manager.

When it comes to safety and the accuracy of information communicated on the job, there is no doubt that what you say plays a greater role than the way you say it. However, beyond that, your nonverbal communication habits have a major effect on the quality of what is said and how it is received.

1.1.2 Listening and Speaking Skills

Every day on the job can be a learning experience. The more you learn, the more you will be able to help others learn, too (*Figure 2*). An effective method of learning and teaching is through verbal communication—that is, through speaking and listening. As a construction professional, you need to be able to state your ideas clearly. You also need to be able to listen to and understand ideas that other people express. The following are some of the ways that verbal teaching and learning take place on the job:

- *Giving and taking instructions* – One worker may read the steps in a calibration process while a second worker accomplishes the task.
- *Offering and listening to presentations* – Equipment manufacturers may visit the job site or offices to provide operating instruction for a new piece of equipment.
- *Participating in team discussions* – Safety is often discussed among teams; offer your input.

00107-15_F02.EPS

Figure 2 Teaching and learning are often accomplished by speaking and listening.

A Sense of Humor

A special tip: Maintain a sense of humor! A good sense of humor will get you through many situations. As a construction professional, you should always take yourself seriously. This means speaking well and conveying the proper professional attitude. But your listeners will always appreciate you more if you show them that you have a sense of humor and have a light side to you. Humor can diffuse tension and relieve frustration over things that aren't working properly. Remember, though, never to tell off-color or offensive jokes or make jokes at another person's expense. Also, never play practical jokes, as they can lead to accidents and injuries on the job.

Listening in the Classroom

When you are in the classroom, be aware of things that affect your ability to listen well. Take action to correct these problems. Is someone on the other side of the room speaking too softly? Ask other classmates to face the class when they speak and to speak loudly. Is there noise out in the hall? Ask permission to close the door to shut out noise from outside. Did your instructor say something you did not understand? Ask your instructor to explain things you don't understand.

- *Talking with your co-workers and your supervisor* – Listen carefully as they speak, without distracting yourself by thinking of a response too quickly. A slight pause in the conversation to prepare a response after a speaker is finished actually encourages others to listen more carefully to you.
- *Talking with clients* – Again, listening skills are critical.

Before we discuss some of the ways to become a more effective listener and speaker, evaluate your current speaking and listening skills by completing the self-assessment quizzes in *Figures 3* and *4*.

At this stage in your career, you will probably do more listening than speaking. You may be wondering why it is so important to be a good listener. The answer is simple: experience. People learn by listening, not by speaking. You are only beginning to learn how the construction industry works, and there is a lot to learn! Teachers, supervisors, and experienced workers can guide you to make sure you are learning what you need to know (*Figure 5*).

Are You a Good Listener?

Do you have good listening habits? Take the following self–assessment quiz to find out. Be sure to answer each question honestly.

	Always	Sometimes	Rarely
1. I maintain eye contact when someone is talking to me.	☐	☐	☐
2. I pay attention when someone is talking to me.	☐	☐	☐
3. I ask questions when I don't understand something I hear.	☐	☐	☐
4. I take notes when receiving instructions.	☐	☐	☐
5. I repeat instructions my supervisor has given me to make sure I understand them.	☐	☐	☐
6. I nod my head or say I understand to show others I am listening to them.	☐	☐	☐
7. I let others speak without interrupting.	☐	☐	☐
8. I move to a quieter spot or ask someone to speak up if I am in a noisy location.	☐	☐	☐
9. I put aside what I am doing when someone is speaking to me.	☐	☐	☐
10. I listen with an open mind.	☐	☐	☐

00107-15_F03.EPS

Figure 3 Listening skills self-assessment.

Are You a Good Speaker?

How good are your speaking skills? This self-assessment quiz will help you understand your speaking strengths and weaknesses.

	Always	Sometimes	Rarely
1. When giving instructions to co-workers, I explain words they might not understand.	☐	☐	☐
2. When giving instructions for a task with several steps, I organize my thoughts first, then give the instructions.	☐	☐	☐
3. I give more details when explaining a task to inexperienced co-workers.	☐	☐	☐
4. When giving instructions to others, I try to keep from sounding like a know-it-all.	☐	☐	☐
5. When giving instructions to others, I encourage them to ask questions about anything they don't understand.	☐	☐	☐
6. I am patient and will explain instructions more than once if necessary.	☐	☐	☐
7. I try to speak more carefully when giving instructions to a co-worker for whom English is a second language.	☐	☐	☐
8. When speaking on the phone or over a two-way radio, I repeat instructions and spell out words when necessary.	☐	☐	☐
9. If someone asks me a question I don't know the answer to, I admit it and then try to find the answer.	☐	☐	☐
10. When I give instructions to others, I am confident, upbeat, and encouraging.	☐	☐	☐

00107-15_F04.EPS

Figure 4 Speaking skills self-assessment.

00107-15_F05.EPS

Figure 5 Your supervisor can help you learn what you need to know.

1.2.0 Active Listening on the Job

You might think that listening just happens automatically, that someone says something and someone else hears it. However, real listening, the process not only of hearing, but of understanding what is said, is an active process. You have to be involved and pay attention to really listen. In other words, understanding begins with **active listening**. You must develop good listening skills to be able to listen actively. This section presents some tips and suggestions that you can use to develop good listening skills.

First of all, you should understand the possible consequences of not listening. Poor listening skills can cause mistakes that waste time and money (*Figure 6*), and may even result in injury or loss of life. Recognize that other people may have

00107-15_F06.EPS

Figure 6 Listen carefully and ask questions to make sure you understand.

important things to say. Even if what they are currently saying is boring, they may happen to mention something vitally important, like what to do when an alarm goes off.

Refuse to allow yourself to be distracted. If a work-related conversation or presentation is taking place, be there in the room listening, not darting around from place to place in your mind. If it helps you to stay focused, keep a pad of paper with you. When you are sidetracked by thoughts of something you need to do later, jot down a couple words about it. Then put the matter out of your mind.

One way to stay focused is to show that you are listening with your **body language**, that is your facial expressions, your posture, and your gestures (*Figure 7*). Not only does body language communicate something to other people, it affects you as well. It was mentioned earlier that behaving confidently can actually make you feel more confident. Similarly, acting like you are listening can actually help you listen better. While you listen, maintain eye contact and acknowledge what is being said by nodding or making appropriate sounds to show that you are paying attention.

While the other person is speaking, do not interrupt or criticize. Instead, wait until he or she has finished, and then ask plenty of questions, especially if he or she gave incomplete or unclear instructions. Asking questions can minimize misunderstandings and maximize your memory of the content. Of course, your mind will never remember perfectly, so take notes on what is said, especially if the content is technical in nature—dealing with numbers, times, amounts, temperatures, or other such data.

00107-15_F07.EPS

Figure 7 Body language shows whether you are paying attention.

Finally, end a conversation by **paraphrasing** what you heard back to the conversation partner. Paraphrasing is the act of repeating what you heard in your own words. By using your own words, you can check the accuracy of your understanding. Therefore, it is important not to simply repeat what the person said exactly. The original speaker must listen carefully to ensure the paraphrased information is accurate.

Imagine how many hours of wasted work you can avoid by just confirming what you think you hear. For example, your supervisor may tell you, "Recalibrate the machines according to the specifications on this paper." You can respond by saying, "So you want me to recalibrate these three machines in this building according to the chart posted here." Then he or she replies, "Oh, no! I mean, recalibrate the machines like these in the building next door, and follow the chart on the back of this paper, since they are slightly different models." Just providing a simple summary in your own words of what was said could save time, money, or even someone's life. Paraphrase instructions whenever you can. It completes the act of listening just as installing a roof completes the construction of a building.

1.2.1 Barriers to Listening

As you have learned, listening well takes some work on your part. However, even when you have mastered effective listening skills, you will still have to overcome some barriers that will keep the message from getting through. Just like the warning signs on a construction site that indicate hazards to be avoided, there are signs that indicate problems in the process of listening and understanding (*Figure 8*). Some of those barriers are listed below, along with tips to overcome them:

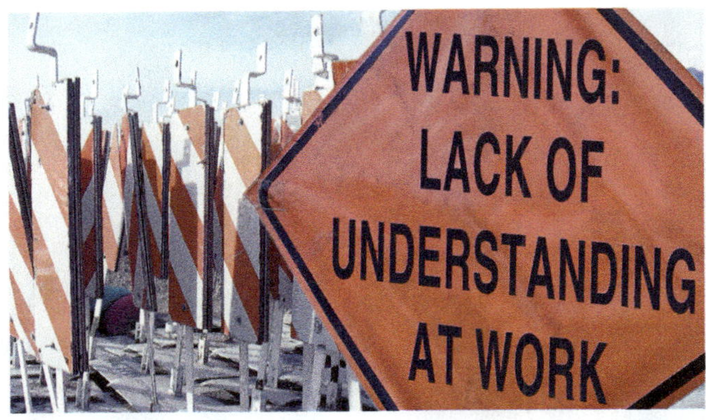

00107-15_F08.EPS

Figure 8 Learn to read the warning signs of listening problems.

- *Emotion* – When angry or upset, humans tend to stop listening actively. Try counting to 10 or asking the speaker to excuse you for a minute. Go get a drink of water and calm down. When you are ready, come back and focus your mind on the task to be done.
- *Boredom* – Maybe the speaker is dull or overbearing. Maybe you think you know it all already. There is no easy tip for overcoming this barrier. You just have to force yourself to stay focused. Keep in mind that the speaker has important information you need to hear.
- *Distractions* – Anything from too much noise and activity on the site to problems at home can steal your attention. If the problem is noise, ask the speaker to move away from it. If a personal problem is keeping you from listening, concentrate harder on staying focused. In some cases it may help if you explain to your supervisor why you are having trouble concentrating.
- *Your ego* – Do you finish people's sentences for them? Do you interrupt others a lot? Do you think about the things you are going to say next instead of listening? That is your ego putting itself squarely between you and effective listening. Be aware of your ego and try to tone it down a bit so you can get the information you need.

Ten Tips for Dealing with Conflict

Conflict can cause emotions to heat up, it can ruin relationships, and it can decrease productivity. Maybe you don't care to be friends with the person who bothers you at work. However, you still have to get along in order to accomplish your work. Allowing conflicts to get worse can only make your work less pleasurable and less effective. Here are some tips for dealing with conflicts at work:

- *Deal with it quickly.* While it is still a small problem, it will be easier to solve. Further, your frustration will not have increased to the breaking point.
- *Set aside a time to discuss the problem.* Dedicating a time later will help emotions cool and provide a neutral emotional atmosphere. You will also have more time to discuss the issue fully.
- *Discuss the problem with civility.* As the old saying goes, "A soft answer turns away wrath." Getting angry will only make the other person angry as well, escalating the situation.
- *Ask questions first.* If a co-worker had a reason for doing what made you angry, you should know what it is. Maybe just knowing why the person behaved that way will solve the problem.
- *Never say "never."* Do not say "always" either. "You never help me" is probably not true, and it will make the other person feel defensive. "You always criticize me" will also make the other person angry.
- *Always focus on specifics, not generalities.* Notice the difference between, "You are a careless person who hurts the whole team," and, "On the last two jobs we had, you forgot to bring the key for the building. We all had to wait while you went back for it. Is there anything you can do or I can do to make sure this doesn't happen again?"
- *Do not bring up things from the past that do not relate to the current situation.* Bringing up something that the co-worker did last year will only cause his or her temper to flare up.
- *Put yourself in the other person's shoes.* Understand what the other person needs in order to come out of the conflict happy. One of the main things people need is an intact sense of honor. If you publicly insult that person or force him or her with an overwhelming use of authority or even logic, you will implant a grudge that will inevitably hurt you later. Beyond giving the person the honor he or she needs, see if you can also compromise on some of the issues you disagree on.
- *Praise the person's good points.* This will show that you are not attacking his or her character, and it will help focus attention on the issue rather than the interpersonal tension.
- *Take the blame for your part of the problem.* Sincerely accepting some blame for the issue can make a huge difference. However, do not take the blame in a self-righteous or manipulative manner in order to shame the other person into submission.

1.3.0 Speaking on the Job

Although you can use the skills presented in this module to give a speech if necessary, the term *speaking skills* does not refer merely to your ability to give a speech or make a presentation to a group of people. It also refers to your ability to communicate effectively, one-on-one, to others on the job every day.

Effective listening depends on effective speaking. After all, you cannot be expected to understand what has not been made clear to you. Look at the following examples of sentences spoken by one worker to another. Which one is the clearest and most effective? Which one would you like to hear if you were the listener?

"Hand me that tool there."

"Hand me the grinder on that bench."

"Hand me the 4-inch angle grinder that's on the bench behind you."

The third example has enough information for you to identify the correct tool and its location. You do not have to ask the speaker, "Which tool? Where is it?" You will not accidentally give the other worker the wrong tool. As a result, time is not wasted trying to clear up confusion. The time it takes for someone to stop what he or she is doing and explain something again because it was not clear the first time is time that the job is not getting done. Time lost this way can add up quickly.

One of the best ways to learn to speak effectively is to listen to someone who speaks well. Think about what makes that person such an effective speaker. Is it the person's choice of words? Or perhaps their body language? Or their ability to make something complex sound simple? Keep the following things in mind when you speak, and they will make a difference for you:

- Think about what you are going to say before you say it, but not at the expense of listening actively.
- Follow the old advice about giving speeches, "Tell them what you are going to tell them. Tell them. Then tell them what you told them." This means you should introduce your topic with a brief summary, then share all of the detailed information, and finally end with a paraphrased summary of what you said.
- As with writing, take time to organize your ideas logically.
- Choose an appropriate place and time. For example, if you need to give detailed assembly instructions to your team, pick a quiet place, and do not hold the meeting just before lunch or the end of the shift.
- Encourage your listeners to take notes if necessary.
- Do not over-explain if people are already familiar with the topic.
- Always speak clearly, and maintain eye contact with the person or people you are speaking to.
- Do not talk on the phone, send text messages, or listen to music while communicating with the work crew.
- When using jargon be sure that everyone knows what the term means.
- Give your listeners enough time to ask questions, and take the time to answer questions thoroughly.
- When you are finished, make sure that everyone understands what you were saying.

Keep these things in mind when you speak, and they will make a difference for you.

1.3.1 Placing Telephone Calls

You may remember when telephones were anchored to walls and desks. To make or receive a phone call, you had to stop what you were doing and go to the telephone. Today, cell phones allow you to make and receive calls from just about anywhere. A cell phone can be a useful tool on the job site, but keep in mind the following guidelines:

- Cell phones can distract you from your job, so never make or receive personal calls while working.
- Wait until a designated break time to make or receive calls.
- Do not operate cell phones where they would pose a safety hazard, such as while operating a piece of machinery, a power tool, or driving a vehicle.

Let It Wait

Never make or receive phone calls while driving or operating heavy equipment. Distraction by cell phones is a leading cause of accidents. In 2013, a Spanish train derailed due to excess speed, killing 79 people just moments after the driver finished a work-related conversation on a cell phone. Subsequently, Spain banned cell phone use by train drivers except for emergency communication. Many areas of the United States likewise have laws banning cell phone use while operating heavy equipment, and even while behind the wheel of any motor vehicle.

- Be aware of the regulations regarding cell phone use at your workplace. While some companies allow cell phone use, many others do not even allow a phone to be in your possession. Companies make these rules to avoid accidents and wasted time.
- Recognize that cell phone cameras can be a potential threat to intellectual property. Workers may photograph secret equipment or data and sell the photographs, causing the company to lose its competitive edge. This is especially worth noting on projects for the government or military. Obviously, the military fiercely protects its secrets for good reason, and may ban employees and contractors from taking cell phones into certain areas.

When you speak to people face-to-face, you can see them and judge how they react to what you say. When you are on the telephone, you don't have these clues. Effective speaking is all the more important in such cases.

When making a call, keep the following points in mind:

- Start by identifying yourself and ask who you are speaking to.
- Speak clearly and explain the purpose of your call.
- Take notes to help you remember the conversation later (*Figure 9*).

00107-15_F09.EPS

Figure 9 Take notes to help you remember important details.

If you leave a message for someone, remember the following:

- Keep it brief.
- Prepare your message ahead of time so you will know what to say.
- Be sure to leave a number where you can be reached and the best time to reach you.

1.3.2 Receiving Telephone Calls

How you answer a phone call is just as important as how you place a phone call. Remember to be professional and courteous when answering your phone because you don't know who is going to be on the other end of the line. When you receive a phone call, remember the following guidelines:

- Don't just say "Hello." Identify yourself immediately by giving your name and the company name.
- Don't keep people on hold for long. People generally resent it. Instead, ask the caller if you can call back at a later time.
- Transfer calls courteously and introduce the caller to the recipient.
- Keep your calls brief.
- If the call is of a personal nature, continue working and do not answer. There are other opportunities throughout the day when you can return the call without interfering with the job or creating a safety hazard.
- Finally, never talk on the phone in front of co-workers, supervisors, or customers. This is usually considered rude and unprofessional.

Did You Know?

Cultural Interpretation

Communication is culturally diverse. Our experiences and surroundings, along with context, individual personality, and mood, help to form the ways we learn to speak and give nonverbal messages. For instance, in Europe the correct form for waving hello or goodbye is to keep your hand and arm stationary, palm out, with fingers moving up and down. In America, the common wave is the whole hand in motion moving back and forth. In Europe, the common American hand gesture for a wave means no, and in Greece it is considered an insult. It is important to remember that not all people communicate in the same way, so be aware of your surroundings when choosing your words and actions.

Recycling Used Cell Phones

GOING GREEN

Recycling your mobile devices (cell phones, pagers, PDAs) is a great way to save energy and protect the environment. These mobile devices contain toxic materials. Recycling keeps them out of landfills, cutting down on air and water pollution and greenhouse gas emissions. These devices contain precious metals, copper, and plastics, which require energy to manufacture and mine.

Over 100 million cell phones are retired each year. The EPA estimates that the energy saved from recycling cell phones alone could power more than 194,000 households for one year.

Your community benefits from cell phone donations. Cell phones can be reconditioned for reuse. They are often donated to charities and law enforcement, and for emergency use by individuals. Recycling is free and some organizations will even pay for your old cell phone.

Additional Resources

Tools for Success: Critical Skills for the Construction Industry, NCCER. 2009. Upper Saddle River, NJ: Pearson Education, Inc.

How to Win Friends and Influence People. Dale Carnegie. 2013. New York, NY: Simon & Schuster.

Listen Up: How to Improve Relationships, Reduce Stress, and Be More Productive by Using the Power of Listening. Larry Barker; Kittie Watson. 2000. New York: St. Martin's Press.

1.0.0 Section Review

1. Making meetings last longer than they were scheduled is a good way to show that you are a diligent person that wants to encourage others to work hard, too.
   a. True
   b. False

2. If the content of a speech or conversation is technical in nature, including lots of data such as numbers, be sure to _____.
   a. take notes
   b. ask others to summarize
   c. check to see if the information is true
   d. have the speaker repeat the key information

3. When is the best place and time to give assembly instructions to the team?
   a. In a quiet place when they are fresh and not too tired
   b. In the lunchroom while they are eating
   c. In the work area right at the end of their shift before they go home
   d. In the break room while people are getting snacks and watching TV

4. What of the following is banned in some workplaces because it can cause wasted time, accidents, and the theft of intellectual property?
   a. Pagers
   b. Watches
   c. Hearing aids
   d. Cellphones

5. When making a telephone call, start by _____.
   a. making small talk
   b. explaining the purpose of your call
   c. identifying yourself
   d. asking who you are talking to

Section Two

2.0.0 Reading and Writing

Objective

Describe good reading and writing skills and their relationship to job performance.

a. Describe the importance of good reading and writing skills.
b. Describe job-related reading requirements and identify good reading skills.
c. Describe job-related writing requirements and identify good writing skills.

Performance Tasks

2. Fill out a work-related form provided by your instructor.
3. Read and interpret a set of instructions for properly donning a safety harness and then orally instruct another person on how to don the harness.

Trade Terms

Appendix: A source of detailed or specific information placed at the end of a section, a chapter, or a book.

Bullets: Large, vertically aligned dots that highlight items in a list.

Change order: A written order by the owner of a project for the contractor to make a change in time, amount, or specifications.

Electronic signature: A signature that is used to sign electronic documents by capturing handwritten signatures through computer technology and attaching them to the document or file.

Font: The type style used for letters and numbers.

Glossary: An alphabetical list of terms and definitions.

Graph: Information shown as a picture or chart. Graphs may be represented in various forms, including line graphs and bar charts.

Index: An alphabetical list of topics, along with the page numbers where each topic appears.

Italics: Letters and numbers that lean to the right rather than stand straight up.

Memo: Informal written correspondence. Another term for memorandum (plural: *memoranda*).

Permit: A legal document that allows a task to be undertaken.

Punch list: A written list that identifies deficiencies requiring correction at completion.

Table: A way to present important text and numbers so they can be read and understood at a glance.

Table of contents: A list of book chapters or sections, usually located at the front of the book.

Imagine for a moment that you wake up one morning and head to work just like any other day, except for one difference: writing has not yet been invented. You grab a diagram (which of course has no notes on it) for installing a heating system, but you do not understand one part of the drawing. You head back to the office to ask the engineer what he meant. He starts to explain, but then he cannot clearly remember what he had discussed with the client. The discussion with the client was so long, and of course, there is no formal written contract or even an informal list of what the client wanted. Therefore, you and the engineer go to the client's office together to resolve the issue.

An existence without reading or writing may have worked for the Aztecs and Mayans, but clearly in our complex modern society, the written word is indispensable. Reading and writing is not just for academics. Almost all Americans read, if not write, dozens of times per day regardless of whether they are plumbers, professors, statisticians, or stay-at-home parents.

2.1.0 The Importance of Reading and Writing Skills

The construction industry, just like any other, depends on written materials of all kinds to carry on business, from routine office paperwork to construction drawings and building codes (*Figure 10*). Written documents allow workers to follow instructions accurately, help project managers ensure that work is on schedule, and enable the company to meet its legal obligations. As a construction professional, you need to be able to read and understand the written documents that apply to your work, whether they are printed in a book or letter, or on a computer. Further, as you gain professional experience, you will eventually take on responsibilities for writing documents.

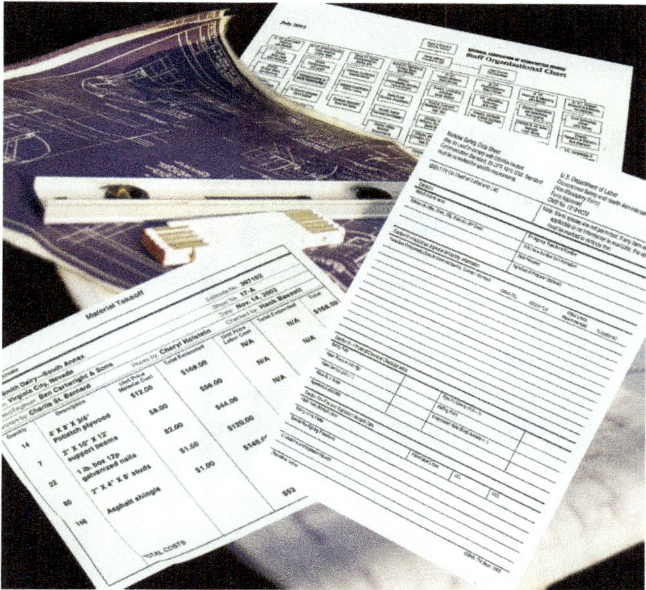

00107-15_F10.EPS

Figure 10 The written word is extremely important in the construction trade.

Other forms of written construction documents that you are very familiar with are textbooks and codebooks (*Figure 11*). A textbook is an instructional document that contains information that a reader needs to know to carry out a task or a series of tasks. A codebook is a guide that provides the codes and standards for certain areas of construction, such as electrical and building codes. Instruction manuals for tools, and installation manuals from manufacturers, are other common examples of documents used on the job. It is crucial for workers to be able to read and understand these important documents. When a person's reading skills are poor, it typically causes them to avoid reading technical data whenever possible. This can lead to errors and accidents. This section reviews some techniques that you may find helpful in reading and writing construction documents.

2.2.0 Reading on the Job

When your company issues new safety guidelines, or the latest version of the local building code contains new standards that affect how you do your job, you need to be able to understand these documents to perform your job safely and effectively. You should not always rely on someone else to tell you how to do your job. Reading is an essential skill for construction professionals.

You may think that you do not have to read on the job, but, in fact, the written word is at the center of the construction trade. The following are some typical examples of things construction workers read on the job:

- Safety instructions
- Construction drawings
- Manufacturer's installation instructions
- Materials lists
- Signs and labels
- Work orders and schedules
- **Permits**
- Specifications
- **Change orders**
- Industry magazines and company newsletters
- Emails

This section reviews some simple tips and techniques that will help you read faster and more efficiently on the job. Further, some of these tips and techniques will help you become a better writer as well. Because you have been reading for some time now, it is probably something you do without even thinking. With practice, these guidelines will become second nature, too.

You should always have a purpose in mind when you read. This will help you find the information faster. For example, say you are looking for a specific installation procedure in a manual. Do not waste time reading other parts of the manual; go straight to the section that deals with that procedure. Read slowly enough to be able to concentrate on what you are reading. Often, technical publications are packed with detailed information; you do not want to misunderstand it because you rushed.

Most books have special features that can help you locate information, including the following:

- **Table of contents**
- **Index**
- **Glossary**
- **Appendixes**
- **Tables** and **graphs**

Tables of contents (*Figure 12*) are lists of chapters or sections in a book. They are usually at the front of a book. This is the first place experienced readers look when they want to know about the book in general or when they want to find some specific information in the book.

Indexes are alphabetical listings of topics with page numbers to show where those topics appear. Indexes are also used to list information in documents other than books—for example, in construction drawings (*Figure 13*). Say you want to look up information about transistors. Whereas the table of contents will tell you where to find the chapter that focuses on transistors, the index will tell you every page that mentions transistors even if the page is about another topic and mentions transistors only briefly.

00107-15_F11.EPS

Figure 11 Codebooks are critical resources for your job.

Glossaries are alphabetical lists of terms used in a book, along with definitions for each term. A glossary is usually located at the back of a publication.

Appendixes are sources of additional information placed at the end of a section, chapter, or book. Appendixes are separate because the information in them is more detailed or specific than the information in the rest of the book. If the information were placed in the main part of the book, it could distract readers or slow them down too much. Some of the NCCER *Core Curriculum* trainee guide modules contain an Appendix.

Tables and graphs summarize important facts and figures in a way that lets the reader understand them at a glance. Tables usually contain text or numbers, and graphs use images or symbols to convey their meaning.

Given the importance of written language to your job, make sure to pay close attention when reading. This section details some techniques you can use to maximize your comprehension of written material.

The noise and activity on a construction site or in a shop can easily distract you. While you are reading, try to avoid distractions. Radios, cellphones, televisions, conversations, and nearby machinery and power tools can all affect your concentration.

Take notes or use a highlighter to help you remember and find important text. Your notes on the blank pages at the beginning of a book can be an invaluable resource. If you find the information useful now, you may want to look it up again later, so make a note of the topic and record the page number. You can find it in an instant later. This old-fashioned technique is faster than pulling out your smartphone and doing a web search. Also, highlighting a main sentence of an especially useful section will help you find the section quickly when you flip through the book.

When reading an unfamiliar book, skim or scan the chapter titles and section headings before you start to read. This will help you organize the material in your mind. Look for visual clues that indicate important material. For example, a bold **font** or ***italics*** indicate important words or information. Bold fonts are letters and numbers that are heavier and darker than the surrounding text. Italics are letters and numbers that lean to the right (like *this*), rather than stand straight up.

When reading instructions or a series of steps for performing a task, such as turning on a welding machine, imagine yourself performing the task; you may find the steps easier to remember that way. Be sure to read all the directions through before you begin to follow them. You will then be able to understand how all of the steps work together.

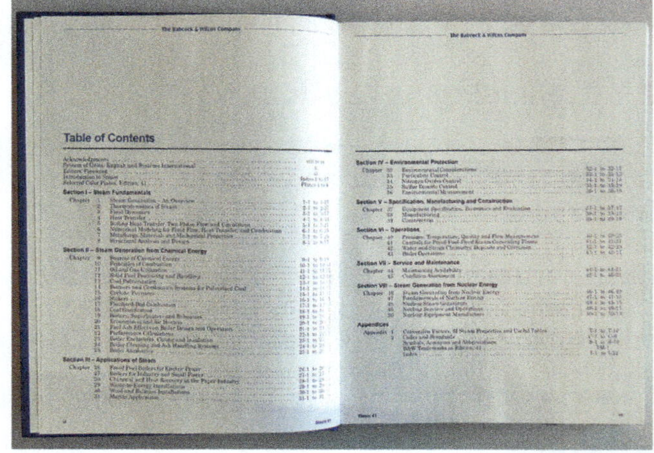

00107-15_F12.EPS

Figure 12 A table of contents lists the chapters or sections in a book.

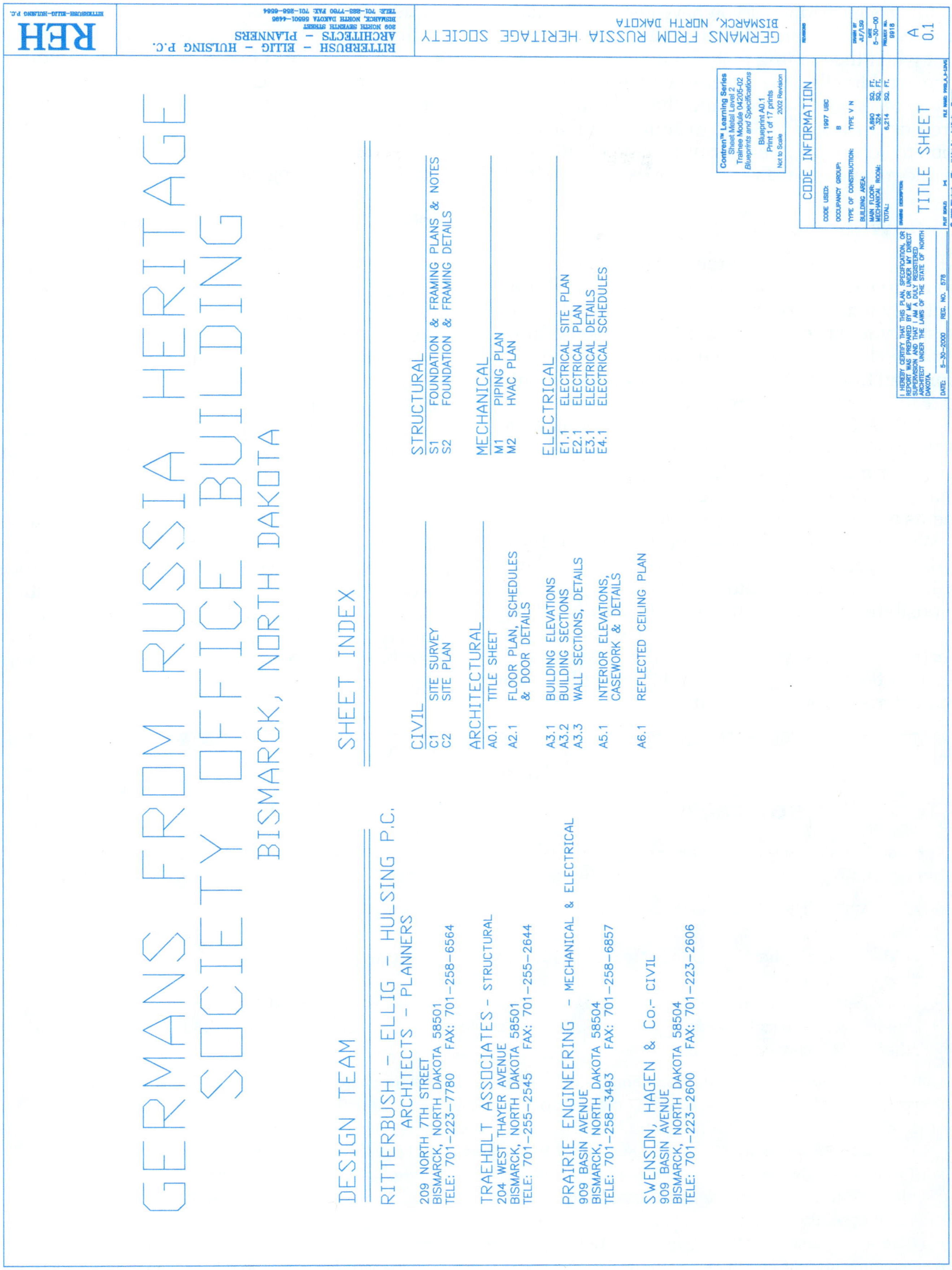

GERMANS FROM RUSSIA HERITAGE SOCIETY OFFICE BUILDING
BISMARCK, NORTH DAKOTA

DESIGN TEAM

RITTERBUSH – ELLIG – HULSING P.C.
ARCHITECTS – PLANNERS
209 NORTH 7TH STREET
BISMARCK, NORTH DAKOTA 58501
TELE: 701-223-7780 FAX: 701-258-6564

TRAEHOLT ASSOCIATES – STRUCTURAL
204 WEST THAYER AVENUE
BISMARCK, NORTH DAKOTA 58501
TELE: 701-255-2545 FAX: 701-255-2644

PRAIRIE ENGINEERING – MECHANICAL & ELECTRICAL
909 BASIN AVENUE
BISMARCK, NORTH DAKOTA 58504
TELE: 701-258-3493 FAX: 701-258-6857

SWENSON, HAGEN & Co.– CIVIL
909 BASIN AVENUE
BISMARCK, NORTH DAKOTA 58504
TELE: 701-223-2600 FAX: 701-223-2606

SHEET INDEX

CIVIL
C1 SITE SURVEY
C2 SITE PLAN

ARCHITECTURAL
A0.1 TITLE SHEET
A2.1 FLOOR PLAN, SCHEDULES & DOOR DETAILS
A3.1 BUILDING ELEVATIONS
A3.2 BUILDING SECTIONS
A3.3 WALL SECTIONS, DETAILS
A5.1 INTERIOR ELEVATIONS, CASEWORK & DETAILS
A6.1 REFLECTED CEILING PLAN

STRUCTURAL
S1 FOUNDATION & FRAMING PLANS & NOTES
S2 FOUNDATION & FRAMING DETAILS

MECHANICAL
M1 PIPING PLAN
M2 HVAC PLAN

ELECTRICAL
E1.1 ELECTRICAL SITE PLAN
E2.1 ELECTRICAL PLAN
E3.1 ELECTRICAL DETAILS
E4.1 ELECTRICAL SCHEDULES

Contren™ Learning Series
Sheet Metal Level 2
Trainee Module 04205-02
Blueprints and Specifications
Blueprint A0.1
Print 1 of 17 prints
Not to Scale 2002 Revision

CODE INFORMATION

CODE USED:	1997 UBC	
OCCUPANCY GROUP:	B	
TYPE OF CONSTRUCTION:	TYPE V N	
BUILDING AREA:		
MAIN FLOOR:	5,890	SQ. FT.
MECHANICAL ROOM:	324	SQ. FT.
TOTAL:	6,214	SQ. FT.

I HEREBY CERTIFY THAT THIS PLAN, SPECIFICATION, OR REPORT WAS PREPARED BY ME OR UNDER MY DIRECT SUPERVISION AND THAT I AM A DULY REGISTERED ARCHITECT UNDER THE LAWS OF THE STATE OF NORTH DAKOTA.
DATE: 5-30-2000 REG. NO. 578

TITLE SHEET
A 0.1

DATE 5-30-00
PROJECT NO. 9918

REH
RITTERBUSH – ELLIG – HULSING P.C.
ARCHITECTS – PLANNERS
209 NORTH SEVENTH STREET
BISMARCK, NORTH DAKOTA 58501-4496
TELE. 701-223-7780 FAX. 701-258-6564

GERMANS FROM RUSSIA HERITAGE SOCIETY
BISMARCK, NORTH DAKOTA

00107-15_F13.EPS

Figure 13 Indexes are often used in construction drawings to identify key information.

Always re-read what you have just read to make sure you understand it. This is especially important when the reading material is complex or very long. And finally, take a break every now and then. If you read slowly or you find that you are getting tired or frustrated, put the material down and do something else to give your mind a break. Reading to understand is hard work and does require effort.

2.3.0 Writing on the Job

At this stage in your career, you will probably do more reading on the job than writing. But no matter what job you have in the construction industry, you'll eventually have to write something. Writing skills are essential if you want to succeed. Construction workers write work orders, health and safety reports, **punch lists** (written lists that identify deficiencies requiring correction at completion), memoranda (informal office correspondence, or **memos** for short), emails, work orders or change orders, and a whole range of other documents as part of their job (see *Figure 14*).

Writing skills will allow you to perform these tasks, too, as you gain experience on the job. They will help you move into positions of greater responsibility. It is never too early to start brushing up on your writing skills, so that when the opportunity comes, you will be ready. This section reviews the writing process and then explains the characteristics of good writing.

First, before you begin to write, do some prewriting. Organize in your mind what you want or need to write. One prewriting technique is brainstorming, which is simply writing down a list of ideas. Do not pressure yourself to make a good list of ideas; just get ideas down on paper. Next, cross out rejected ideas and arrange the better ideas in a logical order to create an outline. After you have built a foundation of basic concepts and have a general sense of what you want to write about, do the necessary research. It may be necessary to ensure that the statements are appropriate for the situation and not something you may regret writing later.

After you have researched, write a rough draft. A rough draft is called this because it is rough, not polished. Do not attempt to write a masterpiece at this time; doing so will merely make you feel discouraged and overwhelmed. Just follow your outline, incorporate your ideas, and include your research. You can improve it at a later point.

The third step is to simply rest or do something else for a while. When you come back to the project later, your mind will be fresh and critical. Maybe you thought something was understandable when you were writing it, but when you read it a day later, you may see that it is unclear. On the other hand, you could also discover that some important information is missing.

Did You Know?

Text Presentation

Text can be modified using different styles and fonts. These elements can be changed according to the purpose of the writer. Styles can be used to make words stand out more clearly. Here are the basic styles commonly used.

- ALL CAPITAL LETTERS. Use this sparingly. Writing in capital letters is the written equivalent of shouting and can seem rude.
- **Bold.** Use to emphasize important words. Notice that in this book the trade terms are bold and in color to show their importance.
- *Italics.* Some common uses for italics are for the names of books, magazines, ships, and other vehicles.
- Underline. The uses for underlining are almost the same as for italics, so the writer should choose one or the other and be consistent in use.

In typography, font refers to the shape of the letters. The two main categories of fonts are serif and sans serif. A serif is a small line attached to the end of a stroke in a letter or symbol. The letters of sans serif fonts have no small lines. The strokes are simply straight. (The word *sans* means "without.")

Serif fonts, being easily readable, are best for text in books and printed documents. One of the most popular serif fonts is Times New Roman, but there are many other serif fonts available to use in your documents.

Sans serif fonts can also be used as body text, but they are most commonly used as headlines or titles. One of the most common sans serif fonts in America is Arial.

There are thousands of creative specialty fonts available as free internet downloads, but the construction professional should use them sparingly. Unless carefully selected by a graphic designer, they can make your document look unprofessional.

Burning - Welding - Hot Work Permit

Valid from _____ to _____, __________ Master Card No.________________

(am/pm) (am/pm) DATE

1. Work Description

Equipment Location or Area ____

Work to be done:

2. Gas Test	☐ None Required			
	☐ Instrument Check	Test Results	Other Tests	Test Results
	☐ Oxygen 20.8% Min			
	☐ Combustible % LFL			

Gas Tester Signature Date Time

3. Special Instructions ☐ None ☐ Check with issuer before beginning work

4. Hazardous Materials ☐ None What did the line / equipment last contain?

5. Personal Protection ☐ Standard Equipment: welder's hood with long sleeves; cutting goggles

☐ Goggles or Face Shield ☐ Respirator ☐ Forced Air Ventilation

☐ Standby Man ☐ Other, specify: ____________________

6. Fire Protection ☐ None Required ☐ Portable Fire Extinguisher

☐ Fire Watch ☐ Fire Blanket ☐ Other, specify: ____________________

7. Condition of Area and Equipment

Required Yes	No	THESE KEY POINTS MUST BE CHECKED
		a. Lines disconnected & blanked or if disconnecting is not possible, blinds installed?
		b. Lines steamed, purged, or otherwise properly cleared of combustibles?
		c. Area and equipment satisfactorily clean of oil or combustibles?
		d. Trenches, catch basins & sewer connections properly covered or sealed?
		e. Immediate area and/or area under the work barricaded or roped off?
		f. Adjoining equip. & operations checked to have any effect on the job?
		g. Area fire suppression (fire water and sprinkler system) in service?

Comments

00107-15_F14A.EPS

Figure 14A A hot work permit is a typical written product in a construction project. (1 of 2)

Burning - Welding - Hot Work Permit

8. Approval	Permit Authorization			Permit Acceptance		
	Area Supv.	Date	Time	Maint. Supv./Engineer Contractor Supv.	Date	Time
Issued by						
Endorsed by						
Endorsed by						

9. Individual Review

I have been instructed in the proper Hot Work Procedures

	Signed	Signed
Persons Authorized to Perform Hot Work	____________	____________
	____________	____________
	____________	____________
	____________	____________
	____________	____________
Fire Watch	____________	____________

10. Job Completion

☐ Yes ☐ No Is the work on the equipment completed?

☐ Yes ☐ No Has the worksite been cleaned and made safe?

Worker answering above questions ____________

Issuer's Acceptance ____________

Forward to Production Superintendent within 7 days of job completion

00107-15_F14B.EPS

Figure 14B A hot work permit is a typical written product in a construction project. (2 of 2)

Next, revise the paper. Come back to your writing and make major changes. You may need to rearrange some sections, add information or try to simplify them, or delete unnecessary sections. Look for wordy phrases.

Finally, proofread. Now that you have fixed the major problems in the paper, look for the smaller errors, like grammar or spelling mistakes. Watch for the errors that spell-check cannot find. If you write a different word than the one you meant to (like *four* instead of *for*) spell-check will not notice.

After you have finished writing, it may be helpful to ask a friend or a co-worker to read it. Another reader can often spot mistakes or problems that you have missed. For example, you might ask a friend to review your job application or resume. You can ask a co-worker to check over a work permit or material takeoff to make sure you haven't left out any information or made any math mistakes. Don't be embarrassed to ask for this type of help, and don't be upset if your friend or co-worker spots a mistake. Professionals are always glad to catch mistakes before those mistakes create bigger problems. Note that how you view the material may affect the accuracy of proofreading, too. Writers often find errors when they read a piece printed on paper rather than on a computer display.

Now that you understand the process of writing, we can examine the characteristics of good writing. Being a good writer is not as hard as you might think. You do not need to be a master of English composition and grammar, or have a college degree, to write well. You just need to remember these five characteristics of good writing: clear, concise, correct, complete, and considerate of the reader. Let's examine them one by one.

- *Be clear* – Make your main point easy to find, and include all the necessary details so the reader can understand what you are saying. One way to highlight important information is to list it. Lists often use **bullets**, like this one does, or numbers. Several different bullet formats can be used. When you finish writing, reread what you have written to make sure it is clear and accurate. Check for mistakes and take out words you don't need.
- *Be concise* – More words do not mean better writing. Instead of writing "In order to prevent unwanted electrical shocks, workers should make sure they disconnect the main power supply before work commences," write "To avoid being shocked, disconnect the main power supply before starting." Instead of writing "at the present time," simply write "now." It is also important to think about the reading level of the intended audience.
- *Be correct* – Accuracy is vital in the construction industry. For example, if you are filling out a form to order supplies from a manufacturer's catalog, make sure you use the correct terms and quantities. When writing material takeoffs or other documents that will be used to purchase equipment and assign workers, you need to get it right. Get in the habit of looking over your writing, whether it's a handwritten note, a supply list, a permit, or an email. Errors cost money and time.
- *Be complete* – When writing, ask yourself the following questions:
  - Have I identified myself to the reader?
  - Have I said why I'm writing this?
  - Will the reader know what to do and, if necessary, how to do it?
  - Will the reader know when to do it?
  - Will the reader know where to do it?
  - Will the reader know whom to call with questions?

You should always ask these questions when you are writing something for others to read, even if you are writing something short and simple.

The Journalist's Questions

Make sure that you cover all of the necessary information in your writing by asking yourself the following journalist's questions:

- **Who?** Who told you that? Who wants it done? Who are you?
- **What?** What is the problem? What do you need? What do you want to do?
- **When?** When did it happen? When do you want it finished?
- **Where?** Where is it taking place? Where are you going? Where can you be found?
- **Why?** Why do you want to do that? Why is this problem happening? Why did you decide to write about it now?
- **How?** How will you fix it? How can you be reached? How can they help?

- *Be considerate of the reader* – Not everyone who reads what you write will be a professional in the same trade that you are in, and not everyone will understand the same words and concepts that you do. For example, you may need to write a report to a client to describe a technical problem. In that report you must use words that intelligent, yet uninformed, people would understand. Similarly, not every professional in your field will know all of the inside information about your company or the history of the problem you are writing about.

2.3.1 Emails

Email has become the standard way to send written information, files, and pictures as long as the files are not too large. Like printed letters and memos, email is used to ask and answer questions, and to provide information quickly. However, remember that a direct telephone call may be in order if email is not providing a timely result.

There are advantages and disadvantages to email. Email delivery is much faster than paper-based delivery, and emails can be stored permanently on a computer. They can be replicated quickly and distributed to many people at the same time. However, because of their ability to be copied and sent so easily, emails are not as private as paper-based documents. Business emails should not include private, sensitive, or confidential information. They can easily be sent by accident to people who are not authorized to read them. Information such as this should be sent in a more secure fashion. Remember that emails may be stored permanently, and can represent a binding agreement or contract. Negative, aggressive emails can lead to dismissal, just like a verbal attack on someone.

Another advantage to transmitting information electronically is that it reduces the amount of paper, printing materials, and shipping or processing fees associated with written documents. In some cases, however, an email is not considered legally binding. Written documents with handwritten signatures remain the standard for contracts and other formal agreements. However, this situation is changing with the advent of **electronic signatures**.

Because email is now used daily in business to communicate with associates, clients, supervisors, and co-workers, it is important to know the proper way to write an email (see *Figures 15* and *16*). When writing an email, the nonverbal part of communication (facial expressions, body language, tone of voice) is lost, so people may misinterpret the message you are trying to convey if you are not careful with words. Many of the rules that apply to writing paper-based documents still apply to emails, but there is a certain etiquette involved. Business emailing standards exist so that emails are composed in a professional, efficient, and responsible manner.

The following are some general rules for writing a proper business email. Keep in mind that some rules will vary according to the nature of the business and company culture.

- Write business email the same way you would write a formal business letter or memo.
- Make sure you are sending the email to the correct individual(s) to maintain confidentiality.
- Always start with a clear subject line that indicates the purpose of the message.
- Begin the email by addressing the recipient.
- Try to keep the body of the email brief and to the point, typically no longer than one screen length, so that the recipient does not have to scroll when reading.

From: JQSmith@smithcontracting.com — Sent: Mon 3/17/2014 1:47 PM
To: WJones@paintersplus.com
Cc:
Subject: Quick Note
Attachments:
View As Web Page

Here are some paint colors and faucets available for your bathroom: Color #1415, Soft Jade; Color #1416, Garden Moss; and Color #1417, Forest Glen. All are available in semi-gloss or eggshell finish. There are also three faucet sets: the Meridian (single handle) $109.88; the Mermaid (dual handles) $83.50; and the Monitor (dual handles) $95.75. All are available in polished brass or polished chrome. I've included paint samples and photos of the faucets. Please tell me your choices by Friday. If you have any questions, call me at 703-555-1212.

00107-15_F15.EPS

Figure 15 Functional but unattractive email example.

From: JQSmith@smithcontracting.com
Sent: Mon 3/17/2014 1:47 PM

To: WJones@paintersplus.com

Cc:

Subject: Bathroom Paint and Faucet Options

Attachments:

View As Web Page

Dear Mr. Jones,

The paint colors and faucets available for your bathroom are listed below (photos of faucets and paint colors are attached to this email). Please let me know what you decide by 5:00 pm on Friday, March 21st. If you have any questions, please do not hesitate to contact me at 703-555-1212.

Paint Colors (Available in semi-gloss or eggshell finish)

- #1415 – Soft Jade
- #1416 – Garden Moss
- #1417 – Forest Glen

Faucet Sets (Available in polished brass or polished chrome)

Model	Price	Handle Style
• Meridian	$109.88	Single
• Mermaid	$83.50	Dual
• Monitor	$95.75	Dual

Regards,
John Q. Smith
Smith Contracting

00107-15_F16.EPS

Figure 16 Better example for an email when a sale is at stake.

- Use a concise format, such as numbers or bulleted points, to clarify what the email is about and what response or action is required of the recipient.
- Write in a positive tone and avoid using negative or blaming statements.

GOING GREEN

Reducing Your Carbon Footprint

Many companies are taking part in the paperless movement. They reduce their environmental impact by reducing the amount of paper they use. Using email helps to reduce the amount of paper used, and there are even postscripts on emails that asking you to reconsider printing the email unless necessary.

- Do not type in all capital letters, as it gives the impression of shouting.
- Use italicized, bold, or underlined text if you need to stress a point.
- Avoid sarcasm, as it can be easily misunderstood.
- Do not forward junk mail.
- Do not address private issues and concerns.
- When sending an attachment, state the name of the file, and its format.
- Double-check the email for spelling and grammatical mistakes before sending. Remember, once an email is sent it cannot be retrieved.
- Include additional contact information other than your email address, such as a phone number or a mailing address.
- Email is not a substitute for face-to-face interaction. Know when to pick up the phone or schedule a meeting if you cannot express your thoughts or concerns through an email. Never send bad news through email, and don't use email to avoid your responsibilities.

2.3.2 Texting

Like email, texting has become nearly universal. According to a 2013 Pew Research Center study, 91 percent of American adults own a cell phone. Tablet computers are also common and used for both texting and email. Texting is also an immediate form of communication. Although an email may take several hours or more to be seen, a text message is more likely to be seen right away. This depends on whether or not an individual's cell phone is set up to receive emails as well as text messages. Texting has become an important part of business and industry today. Therefore, the construction professional must be able to text effectively as well. However, how you write a text to your friend is likely quite different than a text you may write to your supervisor.

Remember that both texting and email can create a safety hazard. Safety must always be given a higher priority than a text message. Be familiar with and follow company policy on the use of phones and tablets on the job. If an immediate response is not vital, focus on the job at hand first.

You must consistently ask yourself if sending a text message is the best way to communicate. Text messages should not be used when the message is long or complex. Entering text into a cell phone is much slower than discussing the same information over the phone or even typing it into an email. The average speaking speed is 130 words per minute. Therefore, before you start a long text-message conversation, consider calling. Text messages should not be used to convey highly emotional messages, apologies, or criticism. Emotion cannot be effectively communicated this way.

Under what circumstances should text messages be used? Extremely brief messages that should be viewed immediately are best communicated with a text message. That's why text messages are also called short messages. Any snippet of information that does not require much interaction can be texted: for example, "I'm leaving now. I'll be at the factory in 10 minutes." Texts are also great for confirming information after a telephone or face-to-face conversation. For example, "Looking forward to meeting you at 3:30 Friday afternoon."

When sending a text message, be sure that it is accurate. Texting poses a special problem in the form of auto-corrections. Most smartphone systems correct typos automatically; for example, "sprll" instantly becomes "spell" after hitting the space key. This feature saves time, but it also frequently causes embarrassment. Websites commonly post autocorrect incidents that completely changed the meanings of the messages or added offensive words, but these mistakes are avoidable. Always take a moment and proofread before sending a text.

Not only should your text message be accurate, but it must be clear. Many people use shorthand in their texts in order to save time. The use of abbreviations such as LOL (laugh out loud), TTYL (talk to you later), and IDK (I don't know) is controversial. There are a few factors to consider when deciding whether to use shorthand. If the message is informal, shorthand is usually acceptable. This is especially true if your work group has its own abbreviations for certain phrases. If the person you are sending the message to is a friend or close colleague of the same level in the

He Said *What?*

Being careful about the words you use when writing is especially important when dealing with clients. Always make sure that the person you are communicating with understands the jargon that you are using. To appreciate the humor in the following story, you have to know a couple of terms used in the electrical trade. Two types of electrical wires enter a building: drops (overhead wires) and laterals (underground wires).

A power company customer called to report a power outage. The service technician looked at the problem, wrote up the work order, gave the customer a copy, and left. The customer took one look at the work order, got angry, and called the company to complain that its service technician was rude. What made the customer angry? Here is what the service technician wrote on the work order:

INVESTIGATED POWER OUTAGE – DROP DEAD

The technician described the problem accurately, but the customer took it as an insult!

Source: Adapted from *Tools for Success: Soft Skills for the Construction Industry*, Steven A. Rigolosi. 2009. Upper Saddle River, NJ: Pearson Learning.

company, you can use shorthand. If the message is to a stranger or someone in upper management, it may be better to send messages with correct spelling and grammar in order to leave a good impression. If you are sure that the person you are texting knows the shorthand, it is fine to use it. Otherwise, using abbreviations that your friend or co-worker does not understand could cause frustration.

The warnings in the previous section about making phone calls while operating machinery apply to sending texts as well. According to research by the Federal Motor Carrier Safety Administration, typing or sending a text takes a driver's eyes off the road for an average of 4.6 seconds. At 55 miles per hour, the car will travel the length of a football field in that time. Never send texts while driving or operating machinery.

Additional Resources

Successful Writing, Maxine Hairston and Michael Keene. 2003. New York, NY: W. W. Norton & Company.

The College Writer's Reference, Alan R. HayaKawa and Toby Fulwiler. 1998. Upper Saddle River, NJ: Prentice Hall.

The Elements of Style, Willian Strunk, Jr. 2015. Grammar, Inc.

2.0.0 Section Review

1. When a person's reading skills are poor, it typically causes him or her to _____.
   a. rewrite the material themselves
   b. avoid reading technical information
   c. have another worker read to them
   d. file for disability

2. Where can a reader look to discover if a book has a chapter on a particular topic?
   a. Table of contents
   b. Index
   c. Glossary
   d. Appendix

3. Bold fonts can be used to indicate that words _____.
   a. have recently been changed
   b. need to be checked for accuracy
   c. are important
   d. are illustrated below

4. After you finish writing a rough draft, you should _____.
   a. ask a colleague to proofread it for you
   b. do some research
   c. check for spelling or grammar errors
   d. have a rest and then review and revise

5. Which of the following is the best advice for sending an email?
   a. Treat a business email the same way you would treat a formal business letter.
   b. Type in all capital letters in order to emphasize the importance of your email and set it apart from others.
   c. Send bad news or emotional information via email so that the recipient can read the message in privacy.
   d. Spelling and grammar do not matter in email communication since it is informal.

Summary

Communications skills—the ability to read, write, listen, and speak effectively—are essential for success in the construction workplace. Effective communication ensures that work is done correctly, safely, and on time. The construction industry relies on a wide variety of written documents. As your responsibilities increase, you will be called on to write some of these documents yourself. This module introduced you to simple tips and techniques that you can use every day to read and write effectively.

Construction professionals state their ideas clearly, and they also listen to the ideas of others. Real listening is an active process entailing feedback, interaction, and retention of details. Not listening causes mistakes, which waste time and money. Of course, effective listening depends on effective speaking. No one can understand what is not expressed clearly. Once you master these basic communication skills and add them to your toolbox of skills, you will find that you can succeed at more things than you ever thought possible.

Review Questions

1. Good communication on the job site _____.
   a. affects safety, schedules, and budgets
   b. will make you popular
   c. takes too much time
   d. cannot be learned

2. Nonverbal communication is best for communicating _____.
   a. feelings and attitudes
   b. complex ideas
   c. what a person desires
   d. instructions for installing new equipment

3. Which of the following are examples of positive nonverbal communication?
   a. Sitting up straight, keeping a clean workspace, and looking at people who are talking to you.
   b. Giving compliments to others and encouraging them.
   c. Speaking with an even tone and looking away most of the time during a conversation.
   d. Telling people all of the information they need in order to do a task properly and answering questions.

4. Which of the following is an example of positive verbal communication on the job?
   a. Talking over or interrupting another person.
   b. Giving and taking instructions.
   c. Tuning out when someone is speaking.
   d. Ducking out of team discussions.

5. Real listening is _____.
   a. tedious
   b. an active process
   c. unnecessary
   d. an art

6. A co-worker asks you to hand her a tool, but does not specify which tool she needs. The most appropriate response would be: _____
   a. "Get it yourself."
   b. "How am I supposed to figure out which tool you need?"
   c. "I'm sorry; which tool do you need?"
   d. "I am too busy; ask someone else."

7. A supervisor wants an apprentice to insert a plug into a water supply line for a pressure test. The clearest way to instruct the apprentice would be for the supervisor to say, _____.
   a. "Put the plug in"
   b. "Insert that one plug into the pipe"
   c. "Get the water supply line ready for the pressure test"
   d. "Insert the plug into the end of the water supply line so I can do a pressure test"

8. If you are in the middle of a task and you receive a personal phone call, _____.
   a. put the caller on hold until you complete the task
   b. continue working and do not answer
   c. tell the caller you don't have time to talk
   d. hand the phone to a co-worker

9. In the construction industry, a codebook provides _____.
   a. the keypad entry codes for locked work areas
   b. codes and standards, such as building codes or electrical codes
   c. passwords for company web services, such as databases and training software
   d. keys for understanding blueprint symbols

10. A reader that wants to find any mention of a certain topic in a book can look in the _____.
    a. table of contents
    b. glossary
    c. appendixes
    d. index

11. Which is a typical example of an item a construction worker might read for work on a regular basis?
    a. Newspaper
    b. Roadmap
    c. Materials list
    d. Summons

12. When you are looking for a specific installation procedure in a manual, the best way to find the information is to _____.
    a. identify the correct section and go straight there
    b. start reading the book from the beginning
    c. study the terms in the glossary
    d. skim the book and take notes on all important information

13. Which of the following is one disadvantage to using email versus paper-based delivery?
    a. Email is an expensive way to communicate.
    b. The spell-checker is unreliable.
    c. Emails cannot easily be replicated.
    d. Emails are not as private as paper-based documents.

14. Many of the rules that apply to writing paper-based documents also apply to _____.
    a. writing emails
    b. leaving voice messages
    c. having a face-to-face conversation
    d. taking orders

15. Text messages are best used for _____.
    a. long conversations since a record of what was said will be preserved
    b. telling people how one feels
    c. detailed or complex information
    d. brief messages that should be viewed immediately

Trade Terms Quiz

Fill in the blank with the correct term that you learned from your study of this module.

1. Often, books will include additional detailed information in a(n) ________.
2. In a list, a large dot is called a(n) ________.
3. Words and numbers can be printed using many different ________, or type styles.
4. A(n) ________ includes the definition of terms used in a book.
5. Information can be presented in picture form using a(n) ________.
6. To find an alphabetical listing of topics in a book, consult the ________.
7. A type style that uses slanted print to emphasize text is called ________.
8. Construction workers write ________ to list deficiencies requiring correction at completion.
9. You can use a(n) ________ to send an informal written message to someone in your company.
10. You can find a book's chapters and headings listed in a(n) ________.
11. Numerical or written information can be presented for quick visual scanning in a(n) ________.
12. A(n) ________ is another way to sign documents.
13. Providing feedback is an essential part of the ________ process.
14. Your ________ silently communicates whether or not you are paying attention to a speaker.
15. As you gain experience, you will learn more of the special ________ that you can use to communicate with other workers in your trade.
16. A legal document that allows a task, such as constructing a building, to be undertaken is called a(n) ________.
17. The owner of a project may change the expected completion date by writing a ________.
18. After you receive instructions from people, be sure to ________ what they say back to them to make sure that you understand correctly.
19. No matter what you are doing, other people can get clues about your attitude and character from your ________.

Trade Terms

Active listening
Appendix
Body language
Bullets
Change order
Electronic signature
Font
Glossary
Graph
Index
Italics
Jargon
Memo
Nonverbal communication
Paraphrase
Permit
Punch list
Table
Table of contents

Cornerstone of Craftsmanship

Dr. Michael John Sandroussi

Craft Training Center of the Coastal Bend (CTCCB)
President

How did you choose a career in the construction industry?
As I was growing up, my career paths took me through many different construction worlds, including the petrochemical and oil field environment. Although my path ultimately led me to public education, I always believed that students that took career and technology education courses were better prepared to function in the real world.

Who inspired you to enter the industry?
I was actually inspired by economics and the need to survive in a complex world. I started working with my grandfather when I was 8 years old and developed a strong work ethic. I learned how to shake hands with adults, look them in the eye, and communicate with others in a professional manner. In working with him and other individuals, I also learned the value of ambition and learned how to set and achieve goals. I realized that a quality education was a neutralizing factor in the US.

What types of training have you been through?
In my career, I have learned plumbing, electrical, carpentry, roofing, foundation work, HVAC, flooring, heavy equipment operations, grain elevator operations, farming, pipefitting, instrumentation, oilfield flow production, wireline operations, and many others.

How important is education and training in construction?
Knowledge is power, and without a trained workforce, an organization will not be fully productive. A lack of training also results in accidents and fatalities.

How important are NCCER credentials to your career?
My NCCER credentials have empowered me to assume the responsibilities as President of a large training facility that trains adults, high school students, junior high students, and veterans in many different crafts.

How has training/construction impacted your life?
Working in construction allowed me to feed my family. Receiving training in construction kept me safe and healthy.

What kinds of work have you done in your career?
Grafting orange trees; plant production; agricultural field technician; plumbing; electrical; carpentry; roofing; masonry work; HVAC; flooring; heavy equipment operator; grain elevator/gin operations; farming; pipefitting; instrumentation; oilfield flow production; and wireline operations.

Tell us about your present job.
I am the President of the Craft Training Center of the Coastal Bend (CTCCB) and an NCCER Certified Primary Administrator. This center trains adults, high school students, junior high students and veterans. Over 1,700-plus students per year are trained in the areas of welding, pipefitting, instrumentation, scaffold-building, industrial painting, electrical, plumbing, mobile crane operations, rigging, field safety, and safety technology. The CTCCB also has an on-line assessment center that has tested over 2,500 journey-level experienced craftsmen.

What do you enjoy most about your job?
NCCER training is changing the quality of life for our students. In the real world, quality of life is what you wear, what you eat, where you live, and what you drive. To my knowledge, the CTCCB is the only training center that catapults high school students immediately into the workforce upon high school graduation. These high school graduates secured NCCER training credentials and entered the workplace as trained, educated, drug-free craftsmen and women.

What factors have contributed most to your success?
My family structure is very strong. My grandmother told me at a very young age that the only goal in her life was for her children to do better, and the only way that was going to happen was to work hard, learn as much as possible, and seek out a quality education.

Would you suggest construction as a career to others? Why?
A career in construction is very rewarding because you are actually building or constructing something that will be around for a long time to come. Your children, grandchildren, great-grandchildren and many others will be able to see what you did during your life. The "backbone" of America is the craftsman; when they are working, our nation is strong!

What advice would you give to those new to the field?
Study, learn, and strive to keep track of any changes in the industry.

Interesting career-related fact or accomplishment:
The Craft Training Center of the Coastal Bend was featured in the *Wall Street Journal* in 2014. The article focused on technical training, and the welder featured is one of our students that attended during his high school career.

How do you define craftsmanship?
Craftsmanship is skill and/or an expertise in a craft.

Trade Terms Introduced in This Module

Active listening: A process that involves respecting others, listening to what is being said, and understanding what is being said.

Appendix: A source of detailed or specific information placed at the end of a section, a chapter, or a book.

Body language: A person's facial expression, physical posture, gestures, and use of space, which communicate feelings and ideas.

Bullets: Large, vertically aligned dots that highlight items in a list.

Change order: A written order by the owner of a project for the contractor to make a change in time, amount, or specifications.

Electronic signature: A signature that is used to sign electronic documents by capturing handwritten signatures through computer technology and attaching them to the document or file.

Font: The type style used for letters and numbers.

Glossary: An alphabetical list of terms and definitions.

Graph: Information shown as a picture or chart. Graphs may be represented in various forms, including line graphs and bar charts.

Index: An alphabetical list of topics, along with the page numbers where each topic appears.

Italics: Letters and numbers that lean to the right rather than stand straight up.

Jargon: Specialized terms used in a specific industry.

Memo: Informal written correspondence. Another term for memorandum (plural: memoranda).

Nonverbal communication: All communication that does not use words. This includes tone of voice, appearance, personal environment, use of time, and body language.

Paraphrase: Express something heard or read using different words.

Permit: A legal document that allows a task to be undertaken.

Punch list: A written list that identifies deficiencies requiring correction at completion.

Table: A way to present important text and numbers so they can be read and understood at a glance.

Table of contents: A list of book chapters or sections, usually located at the front of the book.

Additional Resources

This module presents thorough resources for task training. The following resource material is suggested for further study.

How to Win Friends and Influence People. Dale Carnegie. 2013. New York, NY: Simon & Schuster.

Listen Up: How to Improve Relationships, Reduce Stress, and Be More Productive By Using the Power of Listening. Larry Barker; Kittie Watson. 2000. New York, NY: St. Martin's Press.

Successful Writing, Maxine Hairston; Michael Keene. 2003. New York, NY: W. W. Norton & Company.

The College Writer's Reference, Alan R. Hayakawa; Toby Fulwiler. 1998. Upper Saddle River, NJ: Prentice Hall.

The Elements of Style, WIlliam Strunk, Jr. 2015. Grammar, Inc.

Tools for Success: Critical Skills for the Construction Industry, NCCER. 2009. Upper Saddle River, NJ: Pearson Education, Inc.

Figure Credits

Courtesy of Oak Ridge National Laboratory, Figures 2, 8

M.C. Dean, Inc., Figure 5

Ed Gloninger, Figures 7, 10

Topaz Publications, Inc., Figures 11, 12

Ritterbush-Ellig-Hulsing PC, Figure 13

Section Review Answer Key

Answer	Section Reference	Objective
Section One		
1. b	1.1.1	1a
2. a	1.2.0	1b
3. a	1.3.0	1c
4. d	1.3.1	1c
5. c	1.3.1	1c
Section Two		
1. b	2.1.0	2a
2. a	2.2.0	2b
3. c	2.2.0	2b
4. d	2.3.0	2c
5. a	2.3.1	2c

NCCER CURRICULA — USER UPDATE

NCCER makes every effort to keep its textbooks up-to-date and free of technical errors. We appreciate your help in this process. If you find an error, a typographical mistake, or an inaccuracy in NCCER's curricula, please fill out this form (or a photocopy), or complete the online form at **www.nccer.org/olf**. Be sure to include the exact module ID number, page number, a detailed description, and your recommended correction. Your input will be brought to the attention of the Authoring Team. Thank you for your assistance.

Instructors – If you have an idea for improving this textbook, or have found that additional materials were necessary to teach this module effectively, please let us know so that we may present your suggestions to the Authoring Team.

NCCER Product Development and Revision

13614 Progress Blvd., Alachua, FL 32615

Email: curriculum@nccer.org
Online: www.nccer.org/olf

❑ Trainee Guide ❑ Lesson Plans ❑ Exam ❑ PowerPoints Other ____________________

Craft / Level: Copyright Date:

Module ID Number / Title:

Section Number(s):

Description:

Recommended Correction:

Your Name:

Address:

Email: Phone:

00101-15

Basic Safety (Construction Site Safety Orientation)

Overview

Work at construction and industrial job sites can be hazardous. Most job-site incidents are caused by at-risk behavior, poor planning, lack of training, or failure to recognize the hazards. To help prevent incidents, every company must have a proactive safety program. Safety must be incorporated into all phases of the job and involve employees at every level, including management.

Module One

Trainees with successful module completions may be eligible for credentialing through the NCCER Registry. To learn more, go to **www.nccer.org** or contact us at **1.888.622.3720**. Our website has information on the latest product releases and training, as well as online versions of our *Cornerstone* magazine and Pearson's product catalog.

Your feedback is welcome. You may email your comments to **curriculum@nccer.org,** send general comments and inquiries to **info@nccer.org** , or fill in the User Update form at the back of this module.

This information is general in nature and intended for training purposes only. Actual performance of activities described in this manual requires compliance with all applicable operating, service, maintenance, and safety procedures under the direction of qualified personnel. References in this manual to patented or proprietary devices do not constitute a recommendation of their use.

00101-15
Basic Safety (Construction Site Safety Orientation)

Objectives

When you have completed this module, you will be able to do the following:

1. Describe the importance of safety, the causes of workplace incidents, and the process of hazard recognition and control.
   a. Define incidents and the significant costs associated with them.
   b. Identify the common causes of incidents and their related consequences.
   c. Describe the processes related to hazard recognition and control, including the Hazard Communication (HAZCOM) Standard and the provisions of a safety data sheet (SDS).
2. Describe the safe work requirements for elevated work, including fall protection guidelines.
   a. Identify and describe various fall hazards.
   b. Identify and describe equipment and methods used in fall prevention and fall arrest.
   c. Identify and describe the safe use of ladders and stairs.
   d. Identify and describe the safe use of scaffolds.
3. Identify and explain how to avoid struck-by hazards.
   a. Identify and explain how to avoid struck-by and caught-in-between hazards.
   b. Identify and explain how to avoid caught-in and caught-between hazards.
4. Identify common energy-related hazards and explain how to avoid them.
   a. Describe basic job-site electrical safety guidelines.
   b. Explain the importance of lockout/tagout and describe basic procedures.
5. Identify and describe the proper use of personal protective equipment (PPE).
   a. Identify and describe the basic use of PPE used to protect workers from bodily injury.
   b. Identify potential respiratory hazards and the basic respirators used to protect workers against those hazards.
6. Identify and describe other specific job-site safety hazards.
   a. Identify various exposure hazards commonly found on job sites.
   b. Identify hazards associated with environmental extremes.
   c. Identify hazards associated with hot work.
   d. Identify fire hazards and describe basic firefighting procedures.
   e. Identify confined spaces and describe the related safety considerations.

Performance Tasks

Under the supervision of your instructor, you should be able to do the following:

1. Properly set up and climb/descend an extension ladder, demonstrating proper three-point contact.
2. Inspect the following PPE items and determine if they are safe to use:
   - Eye protection
   - Hearing protection
   - Hard hat
   - Gloves
   - Fall arrest harnesses
   - Lanyards
   - Connecting devices
   - Approved footwear
3. Properly don, fit, and remove the following PPE items:
   - Eye protection
   - Hearing protection
   - Hard hat
   - Gloves
   - Fall arrest harness
4. Inspect a typical power cord and GFCI to ensure their serviceability.

Trade Terms

Accident
Arc welding
Brazing
Combustible
Competent person
Confined spaces
Cross-bracing
Excavation
Flammable
Flash burn
Flash point
Ground
Ground fault
Ground fault circuit interrupter (GFCI)
Guarded
Hand line
Hazard communication standard (HAZCOM)
Hydraulic
Incident
Lanyards
Lockout/tagout (LOTO)
Management system
Maximum intended load
Midrail
Occupational Safety and Health Administration (OSHA)
Permit-required confined space
Personal protective equipment (PPE)
Planked
Pneumatic
Proximity work
Qualified person
Respirator
Safety culture
Safety data sheet (SDS)
Scaffold
Shielding
Shoring
Signaler
Six-foot rule
Spoil
Toeboard
Top rail
Trench
Welding curtain
Wind sock

Industry Recognized Credentials

If you are training through an NCCER-accredited sponsor, you may be eligible for credentials from NCCER's Registry. The ID number for this module is 00101-15. Note that this module may have been used in other NCCER curricula and may apply to other level completions. Contact NCCER's Registry at 888.622.3720 or go to **www.nccer.org** for more information.

Note

The successful completion of this module will award a Construction Site Safety Orientation credential.

Contents

Topics to be presented in this module include:

Contents (continued)

Figures and Tables

Figures and Tables (continued)

SECTION ONE

1.0.0 SAFETY AND HAZARD RECOGNITION

Objectives

Describe the importance of safety, the causes of workplace incidents and accidents, and the process of hazard recognition and control.

a. Define incidents and accidents and the significant costs associated with them.
b. Identify the common causes of incidents and accidents and their related consequences.
c. Describe the processes related to hazard recognition and control, including the Hazard Communication (HAZCOM) Standard and the provisions of a Safety Data Sheet (SDS).

Trade Terms

Accident: According to the US Occupational Safety and Health Administration (OSHA), an unplanned event that results in personal injury and/or property damage.

Combustible: Capable of easily igniting and rapidly burning; used to describe a fuel with a flash point at, or above, 100°F (38°C).

Competent person: A person who is capable of identifying existing and predictable hazards in the surroundings or working conditions that are unsanitary, hazardous, or dangerous to employees, and who has authorization to take prompt corrective measures to eliminate them.

Confined space: A work area large enough for a person to work, but arranged in such a way that an employee must physically enter the space to perform work. A confined space has a limited or restricted means of entry and exit. It is not designed for continuous work. Tanks, vessels, silos, pits, vaults, and hoppers are examples of confined spaces. Also see *permit-required confined space.*

Flammable: Capable of easily igniting and rapidly burning; used to describe a fuel with a flash point below 100°F (38°C).

Ground fault: Incidental grounding of a conducting electrical wire.

Hazard Communication (HAZCOM) Standard: The Occupational Safety and Health Administration standard that requires contractors to educate employees about hazardous chemicals on the job site and how to work with them safely.

Hydraulic: Powered by fluid under pressure.

Incident: Per the US Occupational Safety and Health Administration (OSHA), an unplanned event that does not result in personal injury but may result in property damage or is worthy of recording.

Management system: The organization of a company's management, including reporting procedures, supervisory responsibility, and administration.

Occupational Safety and Health Administration (OSHA): An agency of the US Department of Labor. Also refers to the Occupational Safety and Health Act of 1970, a law that applies to more than 111 million workers and 7 million job sites in the United States.

Personal protective equipment (PPE): Equipment or clothing designed to prevent or reduce injuries.

Pneumatic: Powered by air pressure, such as a pneumatic tool.

Respirator: A device that provides clean, filtered air for breathing, no matter what is in the surrounding air.

Safety culture: The culture created when the whole company sees the value of a safe work environment.

Safety data sheet (SDS): A document that must accompany any hazardous substance. The SDS identifies the substance and gives the exposure limits, the physical and chemical characteristics, the kind of hazard it presents, precautions for safe handling and use, and specific control measures.

Trench: A narrow excavation made below the surface of the ground that is generally deeper than it is wide, with a maximum width of 15 feet (4.6m). Also see *excavation.*

When you take a job, you have an obligation to your employer, co-workers, family, and yourself to work safely. You also have an obligation to make sure anyone you work with is working safely. Your employer is likewise obliged to maintain a safe workplace for all employees. The ultimate responsibility for on-the-job safety, however, rests with you; safety is part of everyone's job. In this module, you will learn to ensure your safety, and that of the people you work with, by obeying the following rules:

- Follow safe work practices and procedures, both regulatory and corporate.
- Inspect safety equipment before use.
- Use safety equipment properly.

To take full advantage of the wide variety of training, job, and career opportunities the construction industry offers, you must first understand the importance of safety. Successful completion of this module will be your first step toward achieving this goal. Later modules offer more detailed explanations of safety procedures, along with opportunities to practice them.

On a typical job site, there are often many workers from many trades in one place. These workers are all performing different tasks and operations. As a result, the job site is constantly changing and hazards are continually emerging. These hazards can jeopardize your safety. Your employer should make every effort to plan safety into each job and to provide a safe and healthful job site. Ultimately, however, your safety is in your own hands.

Safety training is provided to make you aware that hazards exist all around you every day. The time you spend learning and practicing safety procedures can save your life and the lives of others.

Safety is a learned behavior and attitude. It is a way of working that must be incorporated into the company as a culture. A **safety culture** is created when all the workers at a job site or in an organization see the value of a safe work environment and support it through their actions. Creating and maintaining a safety culture is an ongoing process that includes a sound safety structure and attitude, and relates to organizations as well as individuals. Everyone in the company, from management to laborers, must be responsible for safety every day they come to work.

There are many benefits to having a safety culture. Companies with strong safety cultures usually have the following characteristics:

- Fewer at-risk behaviors
- Lower **incident** and **accident** rates
- Less turnover
- Lower absenteeism
- Higher productivity

Did You Know?

Safety First

Safety training is required for all activities. Never operate tools, machinery, or equipment without prior training. Always refer to the manufacturer's instructions.

A strong safety culture can also improve a company's safety record, which leads to winning more bids and keeping workers employed. Contractors with poor safety records are sometimes excluded from bidding, so good safety performance is essential. Factors that contribute to a strong safety culture include the following:

- Embracing safety as a core value
- Strong leadership
- Establishing and enforcing high standards of performance
- The commitment and involvement of all employees
- Effective communication and commonly understood and agreed-upon goals
- Using the workplace as a learning environment
- Encouraging workers to have a questioning attitude and empowering them to stop work when faced with potential hazards.
- Good organizational learning and responsiveness to change
- Providing timely response to safety issues and concerns
- Continually monitoring performance
- Positive reinforcement when proper safety practices are demonstrated by employees

Around the World

GOST

While OSHA serves to protect workers by setting safety standards in the United States, other systems are used internationally. One such set of technical standards used on a regional basis is known as GOST. GOST standards are more far-reaching than OSHA standards, as they cover a much broader range of topics than worker safety alone. The first set of GOST standards were published in 1968 as state standards for the former Soviet Union. After the Soviet Union was dismantled, GOST became a regional standard used by many previous members of the Soviet Union. Although countries may also have some standards of their own, countries such as Belarus, Moldova, Armenia, and Ukraine continue to use GOST standards as well. The standards are no longer administered by Russia, however. Today, the standards are administered by the Euro-Asian Council for Standardization, Metrology and Certification (EASC).

1.1.0 Incidents and Accidents

Incidents and accidents can occur at any job site. Both at-risk behavior and poor working conditions can cause these undesirable events. You can help prevent such events by using safe work habits, understanding what causes them, and learning how to prevent them.

The terms *incident* and *accident* are often used interchangeably. However, according to the US **Occupational Safety and Health Administration (OSHA)**, an incident is an unplanned event that may or may not result in property damage. However, an incident is worthy of being documented so that steps can be taken to prevent it from recurring. When an incident occurs, no personal injury has occurred.

An accident is defined as an unplanned event that results in personal injury and/or property damage. Therefore, an event that results in property damage alone could be considered an incident or an accident. If personal injury or a fatality has occurred, the event is definitely an accident.

There are varying opinions on the use of these two terms, however. The US National Safety Council defines an incident as an unplanned, undesired event that adversely affects the completion of a task. In this definition, there is no mention of injury or property damage. Other safety organizations across the globe are likely to have their own definitions of these terms as well, or may use completely different terms.

The most important thing to understand is that both incidents and accidents are undesirable events that have a negative effect on both projects and workers. Do not be surprised when you hear the terms used interchangeably by other workers. The definitions provided by OSHA are used here to provide context for these terms as they are used throughout this module.

The lessons you will learn in this module will help you work safely. You will be able to spot and avoid hazardous conditions on the job site. By following safety procedures, you will help keep your workplace free from incidents and accidents, and protect yourself and others from injury or even death.

1.1.1 Incident and Accident Categories

Incidents and accidents are often categorized by their severity and impact, as follows:

- *Near-miss* – An unplanned event in which no one was injured and no damage to property occurred, but during which either could have happened. Near-miss incidents are warnings that should always be reported rather than overlooked or taken lightly.
- *Property damage* – An unplanned event that results in damage to tools, materials, or equipment, but no personal injuries.
- *Minor injuries* – Personnel may have received minor cuts, bruises, or strains, but the injured workers returned to full duty on their next regularly scheduled work shift.
- *Serious or disabling injuries* – Personnel received injuries that resulted in temporary or permanent disability. Included in this category would be lost-time incidents, restricted-duty or restricted-motion cases, and those that resulted in partial or total disability.
- *Fatalities* – Deaths resulting from unplanned events.

Studies have shown that for every serious or disabling injury, there were 10 injuries of a less serious nature and 30 property damage incidents. A further study showed that 600 near-miss incidents occurred for every serious or disabling injury.

There are four leading causes of death in construction work. These are often referred to as the "big four", the "fatal four", or the "focus four". They include falls; struck-by hazards; caught-in or caught-between hazards; and electrical hazards (*Figure 1*). Deaths from falls far exceed all other causes.

Here are explanations of the four leading hazard groups:

- Falls from elevation are incidents involving failure of, failure to provide, or failure to use appropriate fall protection.
- Struck-by accidents involve unsafe operation of equipment, machinery, and vehicles, as well as improper handling of materials, such as through unsafe rigging operations.
- Caught-in or caught-between accidents involve unsafe operation of equipment, machinery, and vehicles, as well as improper safety procedures at **trench** sites and in other **confined spaces**.
- Electrical shock accidents involve contact with overhead wires; use of defective tools; failure to disconnect power source before repairs; or improper **ground fault** protection.

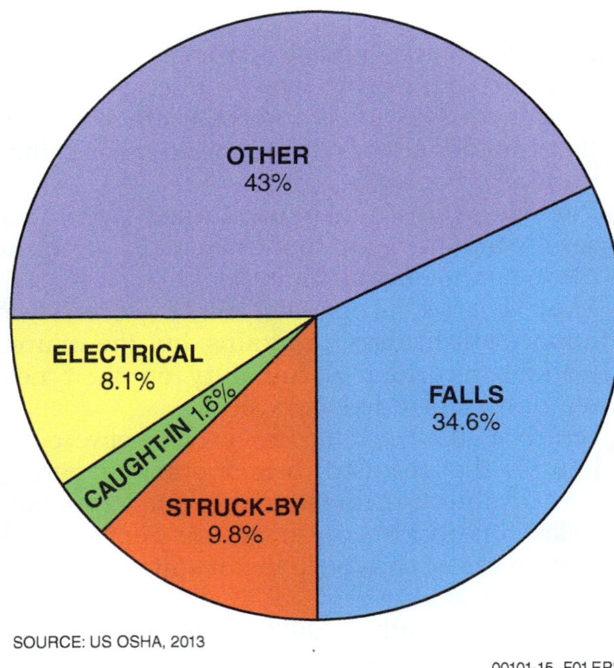

Figure 1 The four high-hazard areas.

1.1.2 Costs

Incidents and accidents cost billions of dollars each year and cause much needless suffering. The National Safety Council estimates that the organized safety movement has saved more than 4.2 million lives since it began in 1913. This section examines why incidents and accidents happen and how you can help prevent them.

Accidents that result in injury or death can have a lasting effect not only on the victims, but on their families, co-workers, and employers. An injured worker who is disabled by an accident faces potentially huge medical bills. On top of those expenses, the worker's family faces loss of the income they rely on. It's convenient to think that insurance will take care of the costs, but it may not. Workers who are injured because they violated established safety rules may have their insurance claims denied and may be dismissed from the company because of the violation. A worker who is injured in a fall because he or she was not using fall protection could be refused compensation. How these issues are handled varies dramatically among companies and countries.

Employers and co-workers can be affected because many contract awards are based, in part, on a company's safety record. Therefore, incidents and accidents can also result in the loss of future jobs, which affects the company's financial position. This can mean layoffs, hiring freezes, or inability to purchase new equipment or tools. In this way, these events affect not only injured employees and their families, but everyone on the job site.

1.2.0 Incident and Accident Causes

When an incident or accident occurs on the job site, it can often be attributed to one of the following causes:

- Failure to communicate
- At-risk work habits
- Alcohol or drug abuse
- Lack of skill
- Intentional acts
- Unsafe acts
- Rationalizing risks
- Unsafe conditions
- Housekeeping
- **Management system** failure

Each of these causes is discussed further in the sections that follow.

1.2.1 Failure to Communicate

Many incidents happen because of a lack of communication. For example, you may learn how to do things one way on one job, but what happens when you go to a new job site? You need to communicate with the people at the new job site to find out whether they do things the way you have learned to do them. If you do not communicate clearly, incidents can happen. Remember that different people, companies, and job sites do things in different ways.

Making assumptions about what other workers know and what they will do can cause incidents.

The Fatal Four

Out of 3,945 worker fatalities in US private industry during the 2012 calendar year, 775, or 19.6 percent, were in construction. The leading causes of worker deaths on construction sites were falls, followed by struck-by-object, electrocution, and caught-in/between. These fatal four were responsible for nearly three out of five (56 percent) construction worker deaths in 2012 reports, as reported by the US Bureau of Labor Statistics. Eliminating the fatal four would save the lives of 435 workers in America every year.

Half-Measures

Most workers who die from falls are wearing harnesses but failed to tie off properly. Always follow the manufacturer's instructions when wearing a harness. Know and follow your company's safety procedures when working on roofs, ladders, and other elevated locations and make sure you have an adequate anchor point at all times.

Don't assume, for example, that all workers understand what you are saying; some workers have limited language skills, especially outside of their native language. Also, don't use terms or jargon that other people may not understand.

CAUTION

Never assume anything. It never hurts to ask questions, but disaster can result if you don't ask. For example, do not assume that an electrical power source is turned off. First ask whether the power is turned off, then check it yourself to be completely safe.

All work sites have specific markings and signs to identify hazards and provide emergency information (*Figure 2*). Learn to recognize these types of signs:

- Informational
- Safety
- Caution
- Danger
- Temporary warnings

Informational markings or signs provide general information. These signs are blue. The following are considered informational signs:

- No Admittance
- No Trespassing
- For Employees Only

Toolbox Talks

Toolbox talks are one way to effectively keep all workers aware and informed of safety issues and guidelines. Toolbox talks are 5- to 10-minute meetings that review specific health and safety topics. These are very common at construction sites of all types.

INFORMATION SIGN

SAFETY SIGN

CAUTION SIGN

DANGER SIGN

00101-15_F02.EPS

Figure 2 Communication tags and signs.

Safety signs give general instructions and suggestions about safety measures. The background on these signs is white; most have a green panel with white letters. These signs tell you where to find such important areas as the following:

- First-aid stations
- Emergency eye-wash stations
- Evacuation routes
- Safety Data Sheet (SDS) stations
- Exits (usually have white letters on a red field)

Caution markings or signs tell you about potential hazards or warn against unsafe acts. When you see a caution sign, protect yourself against a possible hazard. Caution signs are yellow and have a black panel with yellow letters. They may give you the following information:

- Hearing and eye protection are required
- Respirators are required
- Smoking is not allowed

Danger markings or signs tell you that an immediate hazard exists and that you must take certain precautions to avoid an incident. Danger signs are red, black, and white. They may indicate the presence of the following:

- Defective equipment
- Flammable liquids and compressed gases
- Safety barriers and barricades
- Emergency stop button
- High voltage

Safety tags are temporary warnings of immediate and potential hazards. They are not designed to replace signs or to serve as permanent means of protection. Learn to recognize the standard incident and accident prevention signs and tags (*Table 1*).

1.2.2 At-Risk Work Habits

Some examples of at-risk work habits are procrastination, carelessness, and horseplay. Procrastination (putting things off) is a common cause of incidents. For example, delaying the repair, inspection, or cleaning of equipment and tools can cause incidents and even accidents. If you try to push machines and equipment beyond their operating capacities, you risk injuring yourself and your co-workers.

Machines, power tools, and even a pair of pliers can hurt you if you don't use them safely. It is your responsibility to be careful; tools and machines don't know the difference between wood or steel, and flesh and bone.

Work habits and work attitudes are closely related. If you resist taking orders, you may also resist listening to warnings. If you let yourself be easily distracted, you won't be able to concentrate. If you aren't concentrating, you could cause a significant problem.

Your safety is affected not only by how you do your work, but also by how you act on the job site. This is why most companies have strict policies for employee behavior. Horseplay and other inappropriate behavior are forbidden. Workers who engage in such behavior on the job site may be fired.

These strict policies are provided for the worker's protection. There are many hazards on construction sites. Each person's behavior—at work, on a break, or at lunch—must follow the principles of safety.

A person pulling a practical joke on a co-worker could consider it just having fun, but in fact, it could cause the co-worker serious, even fatal, injury. If you horse around on the job, play pranks, or don't concentrate on what you are doing, you are showing a poor work attitude that can lead to a serious incident.

Table 1 Tags and Signs

Basic Stock (background)	Safety Colors (ink)	Message(s)
White	Red panel with white or gray letters	Do Not Operate Do Not Start
White	Black square with a red oval and white letters	Danger Unsafe Do Not Use
Yellow	Black square with yellow letters	Caution
White	Black square with white letters	Out of Order Do Not Use
Yellow	Red/magenta (purple) panel with black letters and a radiation symbol	Radiation Hazard
White	Fluorescent orange square with black letters and a biohazard symbol	Biological Hazard

1.2.3 Alcohol and Drug Abuse

Alcohol and drug abuse costs the construction industry millions of dollars a year in incidents, accidents, lost time, and lost productivity. The true cost of alcohol and drug abuse is much more than just money, of course. Substance abuse can cost lives. Just as drunk driving kills thousands of people on our highways every year, alcohol and drug abuse kills on the construction site.

Using alcohol or drugs creates a risk of injury for everyone on a job site. Many states have laws that prevent workers from collecting insurance benefits if they are injured while under the influence of alcohol or illegal drugs.

Would you trust your life to a crane operator who was under the influence of drugs? Would you bet your life on the responses of a co-worker under the influence of alcohol or drugs? Alcohol and drug abuse have no place in the workplace. A person on a job site who is under the influence of alcohol or drugs is an incident or accident waiting to happen—possibly a fatal accident.

People who work while using alcohol or drugs are at risk of incident or injury; their co-workers are at risk as well. That's why most employers have a formal substance abuse policy that you should be aware of and follow. Avoid any substances that can affect your job performance; the life you save could be your own.

You do not have to be abusing drugs such as marijuana, cocaine, or heroin to create a job hazard. Many prescription and over-the-counter drugs, taken for legitimate reasons, can affect your ability to work safely. Amphetamines, barbiturates, and antihistamines are only a few of the prescription-controlled legal drugs that can affect the ability to work or operate machinery safely. The main thing is to understand and follow your company's substance abuse policy.

CAUTION

If your doctor prescribes any medication that you think might affect your job performance, ask about its effects. Your safety and the safety of your co-workers depend on everyone being alert on the job.

1.2.4 Lack of Skill

Every worker needs to learn and practice new skills under careful supervision. Never perform new tasks alone until you have been checked out by a supervisor. Lack of skill can cause incidents and accidents quickly. For example, suppose you are told to cut some boards with a circular saw, but you aren't skilled with that tool. A basic rule of circular saw operation is never to cut without a properly functioning guard. Because you haven't been trained, you don't know this. You find that the guard on the saw is slowing you down, so you jam the guard open. The result could be a serious accident. Proper training can prevent this from happening.

Never operate a power tool or machine until you have been trained to use it. Some power tools require that a worker be trained and certified in their use. You can greatly reduce the chances of incidents and accidents by learning the safety rules for each task you perform.

1.2.5 Intentional Acts

When someone purposely causes property damage or injury, it is called an intentional act. An angry or dissatisfied employee may purposely create a situation that leads to an incident or accident. If someone you are working with threatens to get even or pay back someone, let your supervisor know at once.

Did You Know?

Stress Effects

Stress creates a chemical change in your body. Although stress may heighten your hearing, vision, energy, and strength, long-term stress can harm your health. Not all stress is job-related; some stress develops from the pressures of dealing with family and friends and daily living. In the end, your ability to handle and manage your stress determines whether stress hurts or helps you. Use common sense when you are dealing with stressful situations. For example, consider the following:

- Keep daily occurrences in perspective. Not everything is worth getting upset, angry, or anxious about.
- When you have a particularly difficult workday scheduled, get plenty of rest the night before.
- Manage your time. The feeling of always being behind creates a lot of stress. Waiting until the last minute to finish an important task adds unnecessary stress.
- Talk to your supervisor. Your supervisor may understand what is causing your stress and may be able to suggest ways to manage it better.

Unfortunately, terrorism must also be a consideration on the job site. The level of concern is typically dependent on the site itself and the nature of the structure. Pipelines, for example, may represent a significant target due to their size, importance, location (sections of pipeline are in remote areas), and potential for destruction. Controversial sites also tend to attract attention, perhaps from terrorist groups operating within a country that oppose the construction or what the structure represents.

As an individual worker, the most important thing you can do is to pay attention. Become familiar with people on the job site and look for strangers that do not seem to fit. A terrorist attack can come from any direction and be delivered in many different ways. It can begin with a single individual gaining access to the job site.

1.2.6 Unsafe Acts

An unsafe act is a change from an accepted, normal, or correct procedure that can cause an incident or accident. It can be any conduct that causes unnecessary exposure to a job-site hazard or that makes an activity less safe than usual. Here are examples of unsafe acts:

- Failing to use required **personal protective equipment (PPE)**
- Failing to warn co-workers of hazards
- Lifting improperly
- Loading or placing equipment or supplies improperly
- Making safety devices (such as saw guards) inoperable
- Operating equipment at improper speeds
- Operating equipment without authority
- Servicing equipment in motion
- Taking an improper working position
- Using defective equipment
- Using equipment improperly

1.2.7 Rationalizing Risk

Everybody takes risks every day. When you get in your car to drive to work, you know there is a risk of being involved in an accident. Yet when you drive using all the safety practices you have learned, you know that there is a very good chance you will arrive at your destination safely. Driving is an acceptable risk because you have some control over your own safety and that of others.

Some risks are not acceptable. On the job, you must never take risks that endanger yourself or others just because you think you can justify doing so. This is called rationalizing risk which means ignoring safety warnings and practices. For example, because you are late for work, you might decide to run a red light. Perhaps you feel the risk is worth the time saved. This is not only a dangerous driving habit, but this kind of thinking is unacceptable on the job site.

The following are common examples of rationalized risks on the job:

- Crossing barricades or boundaries because there is no apparent activity
- Not wearing gloves because it's easier to do the job with bare hands
- Removing your hard hat because you are hot and you cannot see anyone working overhead
- Not tying off your fall protection because you only have to lean over by about a foot

Think about the job before you do it. If you think that it is unsafe, then it is unsafe. Stop working until the job can be done safely. Bring your concerns to the attention of your supervisor. Your health and safety, and that of your co-workers, make it worth taking extra care.

1.2.8 Unsafe Conditions

An unsafe condition is a physical state that is different from the acceptable, normal, or correct condition found on the job site. It usually results in an incident or accident. An unsafe condition can be anything that reduces the degree of safety normally present. The following are some examples of unsafe conditions:

- Congested workplace
- Defective tools, equipment, or supplies
- Excessive noise
- Fire and explosive hazards
- Hazardous atmospheric conditions (such as gases, dusts, fumes, and vapors)
- Inadequate supports or guards
- Inadequate warning systems
- Cluttered work area
- Poor lighting
- Poor ventilation
- Radiation exposure
- Unguarded moving parts such as pulleys, drive chains, and belts

All employees should be given the authority to stop work when an unsafe condition is observed.

1.2.9 Poor Housekeeping

Housekeeping means keeping your work area clean and free of scraps, clutter, or spills. It also means being orderly and organized. Store your materials and supplies safely and label them properly. Arranging your tools and equipment

to permit safe, efficient work practices, and easy cleaning is also important.

If the work site is indoors, make sure it is well-lit and ventilated. Don't allow aisles and exits to be blocked by materials and equipment. Make sure that flammable liquids are stored in safety cans. Oily rags must be placed only in approved, self-closing metal containers (*Figure 3*).

Remember that the major goal of housekeeping is to prevent incidents. Good housekeeping reduces the chances for slips, fires, explosions, and falling objects. Here are some good housekeeping rules:

- Remove from work areas all scrap material and lumber with nails protruding.
- Clean up spills to prevent falls.
- Remove all **combustible** scrap materials regularly.
- Make sure you have containers for the collection and separation of refuse. Containers for flammable or harmful refuse must have covers.
- Dispose of wastes often.
- Store all tools and equipment in the proper location and condition when you are finished with them.
- Keep all aisles and walkways clear of materials and tools.

Another term for good housekeeping is pride of workmanship. If you take pride in what you are doing, you won't let trash build up around you. The saying, "a place for everything and everything in its place" is the right attitude on the job site.

1.2.10 Management System Failure

Sometimes the cause of an incident or accident is failure of the management system. The management system should be designed to prevent or correct the acts and conditions that can create safety hazards. If the safety management system is not functioning properly, problems are likely to occur.

> **Tool Blades**
>
> Dull blades cause more accidents than sharp ones. If you do not keep your cutting tools sharpened, they won't cut very easily. When you have a hard time cutting, you exert more force on the tool. When that happens, something is bound to slip. And when something slips, you can be injured.

00101-15_F03.EPS

Figure 3 Container for oily rags and waste.

What traits could mean the difference between a management system that fails and one that succeeds? A company implementing a good management system will do the following:

- Put safety policies and procedures in writing
- Distribute written safety policies and procedures to each employee
- Review safety policies and procedures periodically
- Enforce all safety policies and procedures fairly and consistently
- Evaluate supplies, equipment, and services to see whether they are safe
- Provide regular, periodic safety training for employees

1.3.0 Hazard Recognition, Evaluation, and Control

The process of hazard recognition, evaluation, and control is the foundation of an effective safety program. When hazards are identified and assessed, they can be addressed quickly, reducing the hazard potential. Simply put, the more aware you are of your surroundings and the dangers in them, the less likely you are to be involved in an incident.

There is a standard rule in the United States that affects every worker in most industries. It is often referred to as HAZCOM, which is short for **Hazard Communication Standard**. The US HAZCOM also aligns with the Globally Harmonized System of Classification and Labeling of Chemicals (GHS). HAZCOM may also be called the Right-To-Know requirement. It requires all US contractors to educate their employees about the hazardous chemicals they may be exposed to on the job site. Employees must be taught how to work safely around these materials. Other countries have similar programs.

Many people think that there are very few hazardous chemicals on construction job sites. That is simply not true. In practice, the term *hazardous chemical* applies to solvents, paint, concrete, and even wood dust, along with more obvious substances such as acids.

As an employee, you have the following responsibilities under HAZCOM:

- Know where SDSs are on your job site.
- Report any hazards you spot on the job site to your supervisor.
- Know the physical and health hazards of any hazardous materials on your job site, and know and practice the precautions needed to protect yourself from these hazards.
- Know what to do in an emergency.
- Know the location and content of your employer's written hazard communication program.

The final responsibility for your safety rests with you. Your employer must provide you with information about hazards, but you must know this information and follow safety rules.

1.3.1 Hazard Recognition

There are many potential hazard indicators. The best approach in determining if a situation or equipment is potentially hazardous is to ask these questions:

- How can this situation or equipment cause harm?
- What types of energy sources are present that can cause an incident?
- What is the magnitude of the energy?
- What could go wrong to release the energy?
- How can the energy be eliminated or controlled?
- Will I be exposed to any hazardous materials?

Before you can fully answer these questions, you need to know the different types of incidents and accidents that can happen and the energy sources behind them. Some of the different types of events that can cause injuries include the following:

- Falls on the same elevations or falls from elevations
- Being caught in, on, or between equipment
- Being struck by falling objects
- Contact with acid, electricity, heat, cold, radiation, pressurized liquid, gas, or toxic substances
- Being cut by tools or equipment
- Exposure to high noise levels
- Repetitive motion or excessive vibration

You might recognize the first four high-hazard conditions as the so-called fatal four previously discussed. Remember, these four types of accidents cause more than half of all construction fatalities.

When equipment is the cause of an incident or accident, it is usually because there was an uncontrolled release of energy. The different types of energy sources that can be released include the following:

- Mechanical
- **Pneumatic**
- **Hydraulic**
- Electrical
- Chemical
- Thermal (heat or cold)
- Radioactive
- Gravitational
- Stored energy

There are a number of ways to recognize hazards and potential hazards on a job site. Some techniques are more complicated than others. In order to be effective, they all must answer this question: What could go wrong with this situation or operation? No matter what hazard recognition technique you use, answering that question in advance will save lives and prevent equipment damage.

1.3.2 Job Safety Analysis (JSA) and Task Safety Analysis (TSA)

Performing a job safety analysis (JSA), also known as job hazard analysis (JHA), is one approach to hazard recognition. Another common technique is performing a task safety analysis (TSA), also called a task hazard analysis (THA).

In a JSA, the task at hand is broken down into its individual parts or steps and then each step is analyzed for its potential hazards. Once a hazard is identified, certain actions or procedures are recommended that will correct that hazard. For example, during a JSA, it is determined that using a chain hoist to install a pump motor in a tight space would be safer than having a worker do it manually. By using the chain hoist, the chance that the worker's hand would get crushed during installation is reduced. Using the JSA process saved the worker from injury. *Figure 4* shows an example of a form used to conduct a JSA.

JOB SAFETY ANALYSIS

Job Name: ____________________
Job Number: ____________________
Supervisor: ____________________
Date: ____________________

TITLE OF JOB OR TASK

TASK	START	END	HAZARDS	CONTROLS
1.				
2.				
3.				
4.				
5.				
6.				
7.				
8.				

Required Training:

Required Personal Protective Equipment (PPE):

Weekly Vehicle Check List: ___ **Tire Pressure** ___ **Transmission Fluid**
___ **Oil** ___ **Lights**
___ **Air Filter** ___ **Wkly Mileage**

Names of Employees:

PRINT NAME	SIGN NAME	TOTAL HOURS

00101-15_F04.EPS

Figure 4 Job safety analysis form.

JSAs can also be used as pre-planning tools. This helps to ensure that safety is planned into the job. You may be asked to take part in a JSA during job planning. When JSAs are used as pre-planning tools, they contain the following information:

- Tools, materials, and equipment needs
- Staffing or manpower requirements
- Duration of the job
- Quality concerns

Task safety analysis is similar to job safety analysis in that both require workers to identify potential hazards and needed safeguards associated with a job they are about to do. The difference is the form used to report the hazards. During a TSA, a pre-printed, fill-in-the-blank checklist is often used to document any hazard found during analysis. Before work begins, the first-line supervisor or team leader should discuss the conclusions found during the TSA with the crew. Some companies require workers to sign the completed TSA forms or checklists before they start work. This helps companies document the hazards and ensure that workers have been told of the potential hazards and safety procedures.

1.3.3 Risk Assessment

Whether an action is considered safe is often a matter of evaluating risk. Risk is a measure of the probability, consequences, and exposure related to an event. Probability is the chance that a given event will occur. Consequences are the results of an action, condition, or event. Exposure is the amount of time and/or the degree to which someone or something is exposed to an unsafe condition, material, or environment.

A safe operation is one in which there is an acceptable level of risk. This means there is a low probability of an incident and that the consequences and exposure risk are all acceptable. For example, climbing a ladder has risk that is considered to be acceptable if the proper ladder is being used as intended, if it is set up correctly, and if it is in good condition. The probability of exposure to a hazard and its potential consequences are all low. If any one of these conditions were different, climbing the ladder would have an unacceptable level of risk.

1.3.4 Reporting Injuries, Incidents, and Near-Misses

All on-the-job incidents and accidents, no matter how minor, must be reported to your supervisor. Some workers think they will get in trouble if they report minor injuries, but that is not the case. Small injuries, like cuts and scrapes, can later become big problems because of infection and other complications.

US employers with more than 10 employees are required to maintain a log of significant work-related injuries and illnesses using specific OSHA forms and documents. Employee names can be kept confidential in certain circumstances. A summary of these injuries must be posted at certain intervals, although employers do not need to submit it to OSHA unless requested. Employers can calculate the total number of injuries and illnesses and compare the result with the average national rates for similar companies. By analyzing incidents and accidents, companies and OSHA can improve safety policies and procedures. By reporting an incident, you can help keep similar events from happening in the future. For details on the operation of OSHA and its important mission, refer to the *Appendix*.

1.3.5 Safety Data Sheets

SDSs are fact sheets prepared by the chemical manufacturer or importer. Each product used on a construction site must have an SDS. An SDS describes the substance, along with its hazards, safe handling, first aid, and emergency spill procedures. The HAZCOM standard requires new SDSs (formerly known as material safety data sheets or MSDSs) to be in a standardized format. The sections of the form include the following:

- *Section 1, Product identification, manufacturer contact information, recommended uses and restrictions*
- *Section 2, Hazard identification*
- *Section 3, Composition/information on ingredients*
- *Section 4, First aid measures*
- *Section 5, Firefighting information*
- *Section 6, Incidental release measures*
- *Section 7, Handling and storage*
- *Section 8, Exposure controls/personal protection*
- *Section 9, Physical and chemical properties*
- *Section 10, Stability and reactivity*
- *Section 11, Toxicological properties*
- *Section 12, Ecological properties*
- *Section 13, Disposal considerations*
- *Section 14, Transport information*
- *Section 15, Regulatory information*
- *Section 16, Other information*

Figure 5 shows a sample SDS for a PVC solvent-cement. The most important things to look for on an SDS are the specific hazards, personal protection, handling procedures, and first aid

GHS SAFETY DATA SHEET

WELD-ON® 705™ Low VOC Cements for PVC Plastic Pipe

Date Revised: **DEC 2011**
Supersedes: **FEB 2010**

SECTION I - PRODUCT AND COMPANY IDENTIFICATION

PRODUCT NAME: **WELD-ON® 705™ Low VOC Cements for PVC Plastic Pipe**
PRODUCT USE: Low VOC Solvent Cement for PVC Plastic Pipe
SUPPLIER:
MANUFACTURER: IPS Corporation
17109 South Main Street, Carson, CA 90248-3127
P.O. Box 379, Gardena, CA 90247-0379
Tel. 1-310-898-3300

EMERGENCY: Transportation: CHEMTEL Tel. 800.255-3924, 813-248-0585 (International) **Medical:** Tel. 800.451.8346, 760.602.8703 3E Company (International)

SECTION 2 - HAZARDS IDENTIFICATION

GHS CLASSIFICATION:

Health		Environmental		Physical	
Acute Toxicity:	Category 4	Acute Toxicity:	None Known	Flammable Liquid	Category 2
Skin Irritation:	Category 3	Chronic Toxicity:	None Known		
Skin Sensitization:	NO				
Eye:	Category 2B				

GHS LABEL:  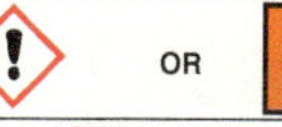 OR  **Signal Word: Danger**

WHMIS CLASSIFICATION: CLASS B, DIVISION 2

Hazard Statements	Precautionary Statements
H225: Highly flammable liquid and vapor	P210: Keep away from heat/sparks/open flames/hot surfaces – No smoking
H319: Causes serious eye irritation	P261: Avoid breathing dust/fume/gas/mist/vapors/spray
H332: Harmful if inhaled	P280: Wear protective gloves/protective clothing/eye protection/face protection
H335: May cause respiratory irritation	P304+P340: IF INHALED: Remove victim to fresh air and keep at rest in a position comfortable for breathing
H336: May cause drowsiness or dizziness	P403+P233: Store in a well ventilated place. Keep container tightly closed
EUH019: May form explosive peroxides	P501: Dispose of contents/container in accordance with local regulation

SECTION 3 - COMPOSITION/INFORMATION ON INGREDIENTS

	CAS#	EINECS #	REACH Pre-registration Number	CONCENTRATION % by Weight
Tetrahydrofuran (THF)	109-99-9	203-726-8	05-2116297729-22-0000	25 - 50
Methyl Ethyl Ketone (MEK)	78-93-3	201-159-0	05-2116297728-24-0000	5 - 36
Cyclohexanone	108-94-1	203-631-1	05-2116297718-25-0000	15 - 30

All of the constituents of this adhesive product are listed on the TSCA inventory of chemical substances maintained by the US EPA, or are exempt from that listing.
* Indicates this chemical is subject to the reporting requirements of Section 313 of the Emergency Planning and Community Right-to-Know Act of 1986 (40CFR372).
# indicates that this chemical is found on Proposition 65's List of chemicals known to the State of California to cause cancer or reproductive toxicity.

SECTION 4 - FIRST AID MEASURES

Contact with eyes: Flush eyes immediately with plenty of water for 15 minutes and seek medical advice immediately.
Skin contact: Remove contaminated clothing and shoes. Wash skin thoroughly with soap and water. If irritation develops, seek medical advice.
Inhalation: Remove to fresh air. If breathing is stopped, give artificial respiration. If breathing is difficult, give oxygen. Seek medical advice.
Ingestion: Rinse mouth with water. Give 1 or 2 glasses of water or milk to dilute. Do not induce vomiting. Seek medical advice immediately.

SECTION 5 - FIREFIGHTING MEASURES

Suitable Extinguishing Media: Dry chemical powder, carbon dioxide gas, foam, Halon, water fog.
Unsuitable Extinguishing Media: Water spray or stream.
Exposure Hazards: Inhalation and dermal contact
Combustion Products: Oxides of carbon, hydrogen chloride and smoke

	HMIS	NFPA	
			0-Minimal
Health	2	2	1-Slight
Flammability	3	3	2-Moderate
Reactivity	0	0	3-Serious
PPE	B		4-Severe

Protection for Firefighters: Self-contained breathing apparatus or full-face positive pressure airline masks.

SECTION 6 - ACCIDENTAL RELEASE MEASURES

Personal precautions: Keep away from heat, sparks and open flame.
Provide sufficient ventilation, use explosion-proof exhaust ventilation equipment or wear suitable respiratory protective equipment.
Prevent contact with skin or eyes (see section 8).
Environmental Precautions: Prevent product or liquids contaminated with product from entering sewers, drains, soil or open water course.
Methods for Cleaning up: Clean up with sand or other inert absorbent material. Transfer to a closable steel vessel.
Materials not to be used for clean up: Aluminum or plastic containers

SECTION 7 - HANDLING AND STORAGE

Handling: Avoid breathing of vapor, avoid contact with eyes, skin and clothing.
Keep away from ignition sources, use only electrically grounded handling equipment and ensure adequate ventilation/fume exhaust hoods.
Do not eat, drink or smoke while handling.
Storage: Store in ventilated room or shade below 44°C (110°F) and away from direct sunlight.
Keep away from ignition sources and incompatible materials: caustics, ammonia, inorganic acids, chlorinated compounds, strong oxidizers and isocyanates.
Follow all precautionary information on container label, product bulletins and solvent cementing literature.

SECTION 8 - PRECAUTIONS TO CONTROL EXPOSURE / PERSONAL PROTECTION

EXPOSURE LIMITS:

Component	ACGIH TLV	ACGIH STEL	OSHA PEL	OSHA STEL:
Tetrahydrofuran (THF)	50 ppm	100 ppm	200 ppm	
Methyl Ethyl Ketone (MEK)	200 ppm	300 ppm	200 ppm	
Cyclohexanone	20 ppm	50 ppm	50 ppm	

Engineering Controls: Use local exhaust as needed.
Monitoring: Maintain breathing zone airborne concentrations below exposure limits.
Personal Protective Equipment (PPE):
Eye Protection: Avoid contact with eyes, wear splash-proof chemical goggles, face shield, safety glasses (spectacles) with brow guards and side shields, etc. as may be appropriate for the exposure.
Skin Protection: Prevent contact with the skin as much as possible. Butyl rubber gloves should be used for frequent immersion.
Use of solvent-resistant gloves or solvent-resistant barrier cream should provide adequate protection when normal adhesive application practices and procedures are used for making structural bonds.
Respiratory Protection: Prevent inhalation of the solvents. Use in a well-ventilated room. Open doors and/or windows to ensure airflow and air changes. Use local exhaust ventilation to remove airborne contaminants from employee breathing zone and to keep contaminants below levels listed above.
With normal use, the Exposure Limit Value will not usually be reached. When limits approached, use respiratory protection equipment.

00101-15_F05A.EPS

Figure 5A Solvent cement SDS. (1 of 2)

GHS SAFETY DATA SHEET

WELD-ON® 705™ Low VOC Cements for PVC Plastic Pipe

Date Revised: **DEC 2011**
Supersedes: **FEB 2010**

SECTION 9 - PHYSICAL AND CHEMICAL PROPERTIES

Appearance: Clear or gray, medium syrupy liquid
Odor: Ketone
pH: Not Applicable
Melting/Freezing Point: -108.5°C (-163.3°F) Based on first melting component: THF
Boiling Point: 66°C (151°F) Based on first boiling component: THF
Flash Point: -20°C (-4°F) TCC based on THF
Specific Gravity: 0.9611 @23°C (73°F)
Solubility: Solvent portion soluble in water. Resin portion separates out.
Partition Coefficient n-octanol/water: Not Available
Auto-ignition Temperature: 321°C (610°F) based on THF
Decomposition Temperature: Not Applicable
Odor Threshold: 0.88 ppm (Cyclohexanone)
Boiling Range: 66°C (151°F) to 156°C (313°F)
Evaporation Rate: > 1.0 (BUAC = 1)
Flammability: Category 2
Flammability Limits: **LEL:** 1.1% based on Cyclohexanone; **UEL:** 11.8% based on THF
Vapor Pressure: 129 mm Hg @ 20°C (68°F)based on THF
Vapor Density: >2 (Air = 1)
Other Data: Viscosity: Medium bodied
VOC Content: When applied as directed, per SCAQMD Rule 1168, Test Method 316A,VOC content is: ≤ 510 g/l.

SECTION 10 - STABILITY AND REACTIVITY

Stability: Stable
Hazardous decomposition products: None in normal use. When forced to burn, this product gives off oxides of carbon, hydrogen chloride and smoke.
Conditions to avoid: Keep away from heat, sparks, open flame and other ignition sources.
Incompatible Materials: Oxidizers, strong acids and bases, amines, ammonia

SECTION 11 - TOXICOLOGICAL INFORMATION

Likely Routes of Exposure: Inhalation, Eye and Skin Contact
Acute symptoms and effects:
Inhalation: Severe overexposure may result in nausea, dizziness, headache. Can cause drowsiness, irritation of eyes and nasal passages.
Eye Contact: Vapors slightly uncomfortable. Overexposure may result in severe eye injury with corneal or conjunctival inflammation on contact with the liquid.
Skin Contact: Liquid contact may remove natural skin oils resulting in skin irritation. Dermatitis may occur with prolonged contact.
Ingestion: May cause nausea, vomiting, diarrhea and mental sluggishness.
Chronic (long-term) effects: None known to humans

Toxicity:

	LD_{50}	LC_{50}
Tetrahydrofuran (THF)	Oral: 2842 mg/kg (rat)	Inhalation 3 hrs. 21,000 mg/m^3 (rat)
Methyl Ethyl Ketone (MEK)	Oral: 2737 mg/kg (rat), Dermal: 6480 mg/kg (rabbit)	Inhalation 8 hrs. 23,500 mg/m^3 (rat)
Cyclohexanone	Oral: 1535 mg/kg (rat), Dermal: 948 mg/kg (rabbit)	Inhalation 4 hrs. 8,000 PPM (rat)

Reproductive Effects	Teratogenicity	Mutagenicity	Embryotoxicity	Sensitization to Product	Synergistic Products
Not Established	Not Established	Not Established	Not Established	Not Established	Not Established

SECTION 12 - ECOLOGICAL INFORMATION

Ecotoxicity: None Known
Mobility: In normal use, emission of volatile organic compounds (VOC's) to the air takes place, typically at a rate of ≤ 510 g/l.
Degradability: Biodegradable
Bioaccumulation: Minimal to none.

SECTION 13 - WASTE DISPOSAL CONSIDERATIONS

Follow local and national regulations. Consult disposal expert.

SECTION 14 - TRANSPORT INFORMATION

Proper Shipping Name: Adhesives
Hazard Class: 3
Secondary Risk: None
Identification Number: UN 1133
Packing Group: PG II
Label Required: Class 3 Flammable Liquid
Marine Pollutant: NO

EXCEPTION for Ground Shipping
DOT Limited Quantity: Up to 5L per inner packaging, 30 kg gross weight per package.
Consumer Commodity: Depending on packaging, these quantities may qualify under DOT as "ORM-D" .

TDG INFORMATION	
TDG CLASS:	FLAMMABLE LIQUID 3
SHIPPING NAME:	ADHESIVES
UN NUMBER/PACKING GROUP:	UN 1133, PG II

SECTION 15 - REGULATORY INFORMATION

Precautionary Label Information: Highly Flammable, Irritant
Symbols: F, Xi
Ingredient Listings: USA TSCA, Europe EINECS, Canada DSL, Australia AICS, Korea ECL/TCCL, Japan MITI (ENCS)

Risk Phrases:
R11: Highly flammable.
R20: Harmful by inhalation.
R36/37: Irritating to eyes and respiratory system.
R66: Repeated exposure may cause skin dryness or cracking
R67: Vapors may cause drowsiness and dizziness

Safety Phrases:
S9: Keep container in a well-ventilated place.
S16: Keep away from sources of ignition - No smoking.
S25: Avoid contact with eyes.
S26: In case of contact with eyes, rinse immediately with plenty of water and seek medical advice.
S33: Take precautionary measures against static discharges.
S46: If swallowed, seek medical advise immediately and show this container or label.

SECTION 16 - OTHER INFORMATION

Specification Information:
Department issuing data sheet: IPS, Safety Health & Environmental Affairs
E-mail address: <EHSinfo@ipscorp.com>
All ingredients are compliant with the requirements of the European Directive on RoHS (Restriction of Hazardous Substances).
Training necessary: Yes, training in practices and procedures contained in product literature.
Reissue date / reason for reissue: 12/14/2011 / Updated GHS Standard Format
Intended Use of Product: Solvent Cement for PVC Plastic Pipe

This product is intended for use by skilled individuals at their own risk. The information contained herein is based on data considered accurate based on current state of knowledge and experience. However, no warranty is expressed or implied regarding the accuracy of this data or the results to be obtained from the use thereof.

00101-15_F05B.EPS

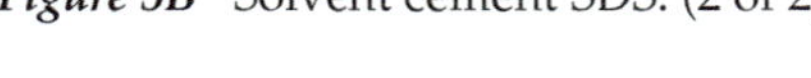

Figure 5B Solvent cement SDS. (2 of 2)

information. Most SDSs have a 24-hour emergency-response number.

Using *Figure 5*, try to find the information you would need to use the cement described on the sample SDS. First locate the hazards; section 2 of the SDS shows that the adhesive is flammable, and also an eye and skin irritant that can cause respiratory irritation, dizziness, and drowsiness.

Next, find out how to minimize these hazards. Section 7 gives general handling and storage information. It indicates that ventilation is needed to reduce hazardous vapors, which can be a fan in an open window. If ventilation is not enough, respiratory protection is needed. Section 8 tells you how to protect your eyes and skin.

Section 4 lists the first aid measures for eye contact, skin contact, or inhalation. Section 5 explains fire hazards and firefighting measures. Now you have the information you need in case of an emergency.

The SDSs must be kept in the work area and be readily accessible to all workers. The company's safety officer or **competent person** should review the SDS before a hazardous material is used. Ask your supervisor to point out where the SDSs are located if you are not sure. Have him or her point out the sections that relate to your job; the health and safety of you and your co-workers may depend on it.

Additional Resources

US Occupational Safety and Health Administration. Numerous safety videos are available on line at **www.osha.gov/video**.

Construction Safety, Jimmie W. Hinze. 2006. Upper Saddle River, NJ: Pearson Education, Inc.

DeWalt Construction Safety/OSHA Professional Reference, Paul Rosenberg; American Contractors Educational Services. 2006. DEWALT

Basic Construction Safety and Health, Fred Fanning. 2014. CreateSpace Independent Publishing Platform.

1.0.0 Section Review

1. The person primarily responsible for your safety is _____.
   a. your foreman
   b. your instructor
   c. yourself
   d. your employer

2. The color commonly used for informational signs is _____.
   a. green
   b. red
   c. yellow
   d. blue

3. The SDS for any chemical used at a job site must be available _____.
   a. at the job site
   b. on line
   c. at the contractor's office
   d. at the nearest hospital

Section Two

2.0.0 Elevated Work and Fall Protection

Objectives

Describe the safe work requirements for elevated work, including fall protection guidelines.

a. Identify and describe various fall hazards.
b. Identify and describe equipment and methods used in fall prevention and fall arrest.
c. Identify and describe the safe use of ladders and stairs.
d. Identify and describe the safe use of scaffolds.

Performance Tasks

1. Properly set up and climb/descend an extension ladder, demonstrating proper three-point contact.
2. Inspect the following PPE items and determine if they are safe to use:
   - Fall arrest harnesses
   - Lanyards
   - Connecting devices
3. Properly don, fit, and remove the following PPE:
   - Fall arrest harness

Trade Terms

Cross-bracing: Braces (metal or wood) placed diagonally from the bottom of one rail to the top of another rail that add support to a structure.

Excavation: Any man-made cut, cavity, trench, or depression in an earth surface, formed by removing earth. It can be made for anything from basements to highways. Also see *trench*.

Guarded: Enclosed, fenced, covered, or otherwise protected by barriers, rails, covers, or platforms to prevent dangerous contact.

Hand line: A line attached to a tool or object so a worker can pull it up after climbing a ladder or scaffold.

Lanyard: A short section of rope or strap, one end of which is attached to a worker's safety harness and the other to a strong anchor point above the work area.

Maximum intended load: The total weight of all people, equipment, tools, materials, and loads that a ladder can hold at one time.

Midrail: Mid-level, horizontal board required on all open sides of scaffolds and platforms that are more than 14 inches (35 cm) from the face of the structure and more than 10 feet (3 m) above the ground. It is placed halfway between the toeboard and the top rail.

Planked: Having pieces of material 2 inches (5 cm) thick or greater and 6 inches (15 cm) wide or greater used as flooring, decking, or scaffold decks.

Scaffold: An elevated platform for workers and materials.

Six-foot rule: A rule stating that platforms or work surfaces with unprotected sides or edges that are 6 feet (≈2 m) or higher than the ground or level below it require fall protection.

Toeboard: A vertical barrier at floor level attached along exposed edges of a platform, runway, or ramp to prevent materials and people from falling.

Top rail: A top-level, horizontal board required on all open sides of scaffolds and platforms that are more than 14 inches (36 cm) from the face of the structure and more than 10 feet (3 m) above the ground.

Falls from elevated areas are the leading cause of fatalities in the workplace. Falls from elevated heights account for about one-third of all deaths in the construction trade. Approximately 85 percent of the injured workers lose time from work; approximately 33 percent require hospitalization; some never return to the job.

While the risk of falls is high in construction and some other trades, there are many things that workers can do to safeguard themselves. Using the appropriate PPE; following proper safety procedures; practicing good housekeeping habits; and staying alert at all times will help you stay safe when working at an elevation. Employers are required to provide for both fall prevention and fall arrest, and to make sure workers are trained and certified in the use of fall protection equipment. Fall prevention consists of covered floor openings, climbing aids, barricades, and guardrails that are designed to protect against falls. Fall arrest consists of equipment such as body harnesses, lanyards, connection devices, lifelines, and safety nets that are intended to protect a worker in case a fall occurs. All these topics are covered in this section.

2.1.0 Fall Hazards

Falls are classified into two groups: falls from an elevation and falls from the same level. Falls from an elevation can happen during work from scaffolds, work platforms, decking, concrete forms, ladders, stairs, and work near excavations. Falls from elevation often result in death unless the fall is arrested. Falls on the same level are usually caused by tripping or slipping. Sharp edges and pointed objects, such as exposed concrete reinforcing bars (rebar), could cut and otherwise harm a worker. Other bodily injuries are also common results of tripping or slipping.

In the United States, fall protection is required for platforms or work surfaces with unprotected sides or edges that are 6 feet (≈2 meters) or above the ground or the level below. This is commonly referred to as the six-foot rule. However, some international regulations and company policies may require fall protection for heights less than 6 feet.

2.1.1 Walking and Working Surfaces

Slips, trips, and falls on walking and working surfaces cause 15 percent of all incidental deaths in the construction industry. Some incidents occur due to environmental conditions, such as snow, ice, or wet surfaces. Others happen because of poor housekeeping and careless behavior, such as leaving tools, materials, and equipment out and unattended. You can avoid slips, trips, and falls by being aware of your surroundings and following the rules on your site. Remember these general walking and working surface guidelines to avoid incidents:

- Keep all walking and working areas clean and dry. If you see a spill or ice patch, clean it up, or barricade the area until it can be properly attended to.
- Keep all walking and working surfaces clear of clutter and debris.
- Run cables, extension cords, and hoses overhead or through crossover plates so that they will not become tripping hazards.
- Do not run on scaffolds, work platforms, decking, roofs, or other elevated work areas.

2.1.2 Unprotected Sides, Wall Openings, and Floor Holes

Any opening in a wall or floor is a safety hazard. There are two types of protection for these openings: (1) they can be guarded or (2) they can be covered. Cover any hole in the floor when possible. Hole covers must be clearly marked. When it is not practical to cover the hole, use barricades. If the bottom edge of a wall opening is less than 39 inches (1 m) above the floor and would allow someone to fall 6 feet (≈2m) or more, then place guards around the opening.

The types of barriers and barricades used vary from one job site to another. There may also be different procedures for when and how barricades are put up. Learn and follow the policies at your job site. Several different types of guards are commonly used:

- Railings are used across wall openings or as a barrier around floor openings to prevent falls (*Figure 6A*).
- Warning barricades alert workers to hazards but provide no real protection (*Figure 6B*). Warning barricades can be made of plastic tape or rope strung from wire or between posts. The tape or rope is color-coded as follows:
  - *Red means danger.* No one may enter an area with a red warning barricade. A red barricade is used when there is danger from falling objects or when a load is suspended over an area.
  - *Yellow means caution.* You may enter an area with a yellow barricade, but be sure you know what the hazard is, and be careful. Yellow barricades are used around wet areas or areas containing loose dust. Yellow with black lettering warns of physical hazards such as bumping into something, stumbling, or falling.
  - *Yellow and purple together mean radiation warning.* No one may pass a yellow and purple barricade without authorization, training, and the appropriate PPE. These barricades are often used where piping welds are being X-rayed.
- Protective barricades give both a visual warning and protection from injury (*Figure 6C*). They can be wooden posts and rails, posts and chain, or steel cable. People should not be able to get past protective barricades.
- Blinking lights are placed on barricades so they can be seen at night (*Figure 6D*).
- Hole covers are used to cover open holes in a floor or in the ground (*Figure 6E*).

WARNING!

Never remove a barricade unless you have been authorized to do so. Follow your employer's procedures for putting up and removing barricades.

(A)

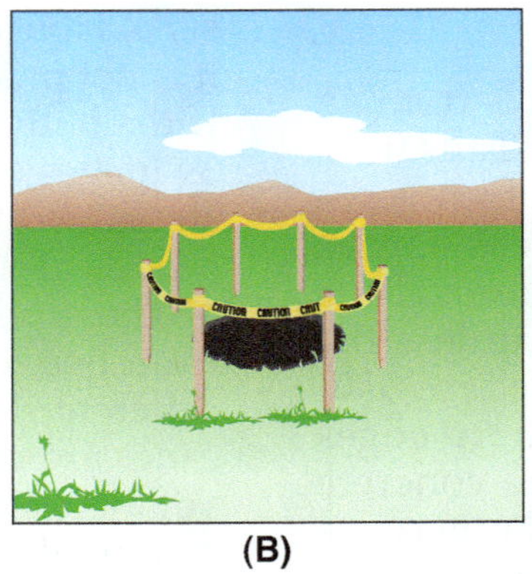
(B)

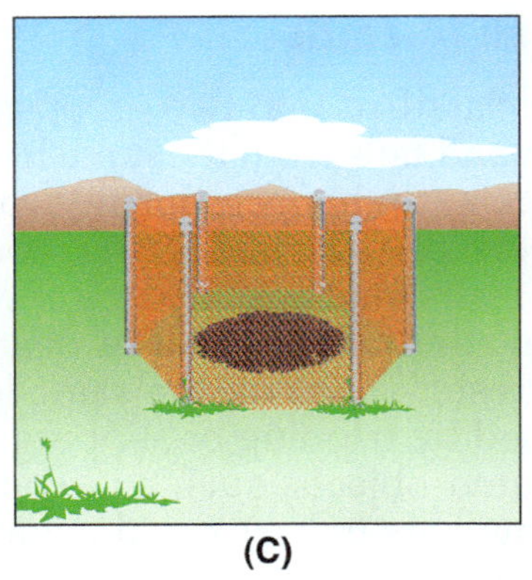
(C)

(D)

(E)

00101-15_F06.EPS

Figure 6 Common types of barriers and barricades.

Follow these guidelines when working near unprotected sides, floor holes, and wall openings:

- Hole covers must be cleated, wired, or otherwise secured to prevent them from slipping sideways or horizontally beyond the hole.
- Covers must extend adequately beyond the edge of the hole.
- Hole covers must be strong enough to support twice the weight of anything that may be placed on top of them. They must also be clearly marked. Use ¾-inch (2 cm) plywood as a hole cover, provided that one dimension of the opening is less than 18 inches (46 cm); otherwise, 2-inch (5-cm) lumber is required.
- Never store material or equipment on a hole cover.
- Guard all stairway floor openings, with the exception of the entrance, with standard railing and **toeboards**.
- Guard all wall openings from which there is a drop of more than 6 feet (≈2m) and for which the bottom of the opening is less than 39 inches (1 m) above the working surface.
- Guard all open-sided floors and platforms 6 feet (≈2m) or more above adjacent floor or ground level, using a standard railing or the equivalent.

2.2.0 Fall Arrest

The key to preventing serious injury or death should a fall occur is the personal fall arrest system (PFAS). A complete PFAS (*Figure 7*) consists of anchor points, a body harness, and connecting devices. Anchor points are related to the structure, and the type and availability of anchor points help determine what other equipment should be chosen. The body harness comprises the system of belts, rings, or hooks worn by the worker. Connecting devices and lanyards are used to maintain attachment between anchor points and the PFAS, and include lanyards and various pieces of hardware.

> **NOTE**
> This training alone does not provide any level of certification in the use of fall arrest or fall restraint equipment. Trainees should not assume that the knowledge gained in this module is sufficient to certify them to use fall arrest equipment in the field.

2.2.1 Anchor Points

There are both permanent anchor points and temporary, reusable anchor points (*Figure 8*). Anchor points must be rated at or equal to 5,000 pounds (2,267 kg) breaking or tensile strength, or twice the intended load.

Figure 7 Personal fall-arrest system.

Ideally, the fall arrest anchor point will be located directly above the back D-ring (*Figure 9*), in order to minimize any swing-zone hazards, as well as the free-fall distance. The maximum free-fall distance is 6 feet (≈2 m). To understand swing zones, picture a tied-off worker standing three feet (1 m) away from a point directly under the anchor point. If the worker falls, gravity will cause his body to swing toward the anchor point axis and momentum will cause the body to swing past that point. If there is a wall or other solid object close by, he may strike it and be injured. Swing zones are minimized when the anchor point is directly above the worker when he falls. Serious injury and damage can occur when a human body strikes an immoveable object while swinging as a pendulum. Although the PFAS may do its job by preventing the worker from falling a great distance, serious injury or death can still occur by striking an object in the swing zone.

Case Histories

Here are some examples of fatal incidents that resulted from failure to provide the proper fall protection:

- A worker taking measurements was killed when he fell backward from an unguarded balcony to the concrete below.
- A roofer handling a piece of fiberboard backed up and tripped over a 7½-inch (19 cm) parapet. He fell more than 50 feet (15 m) to the ground level and died of severe head injuries.

(A) PERMANENT ANCHOR

(B) CONCRETE ANCHOR

(C) BEAM ANCHOR

(D) TIE-BACK LANYARD

00101-15_F08.EPS

Figure 8 Anchor points.

Anchor points are often needed to secure the position of the worker, leaving the hands free to accomplish a task. Ideally, workers will select anchor points to maintain a potential fall distance of no more than 2 feet (61 cm). For example, a worker placing reinforcing bars in a concrete wall form may use two short positioning lanyards connected to the hip rings on the harness. The fall protection lanyard would be connected to the rear D-ring.

Positioning connections cannot be considered the primary anchor. Positioning anchor points are required to be rated at a 3,000 pound (1,360 kg) strength instead of the 5,000 pound (2,268 kg) rating for primary fall arrest. Remember that positioning lanyards, connected to D-rings on the harness other than the back or front chest D-ring, are fall restraints rather than fall arresting connections. A positioning lanyard does not take the place of a fall arrest lanyard or anchor point.

2.2.2 Harnesses

Full body harnesses, like the one shown in *Figure 10,* are available in sizes that are based on the height and weight of the user. The back D-ring is the only one used to connect the harness to the anchor point for primary fall arrest purposes unless you are climbing a ladder. When climbing a ladder, the front chest D-ring is the likely choice for connecting the fall arrest harness. D-rings located at the hips are used for positioning and fall restraint only. D-rings mounted to shoulders are often used for rescue situations. All of them can be used for fall restraint, but the back D-ring is the primary connection for fall arrest.

00101-15_F09.EPS

Figure 9 Back D-ring.

A body harness must fit correctly to ensure that it will provide proper protection. Do not place additional holes or openings in harness components under any circumstances. No field modifications to a body harness or lanyard should be attempted. Installation, maintenance, and inspection instructions are provided for every harness, and it is the responsibility of the worker to read and understand the details regarding his or her personal equipment. Workers must inspect their body harness each day that is in use.

Harness straps are generally designed with some stretch to help absorb some of the potential force of a fall. This means good, taut installation on the body is essential so that the worker cannot fall out of the harness in the event of a fall.

Figure 11 shows the proper procedure for donning a common full body harness. The most important adjustments to be made include the chest straps, the groin straps, and the final position of the back D-ring. The related details that follow must be considered during the fitting and wearing of a full body harness:

NOTE

These guidelines are general in nature and the instructions provided for specific equipment by the manufacturer must always take precedence. It is the worker's responsibility to be intimately familiar with the duty of each and every ring and strap on a given harness.

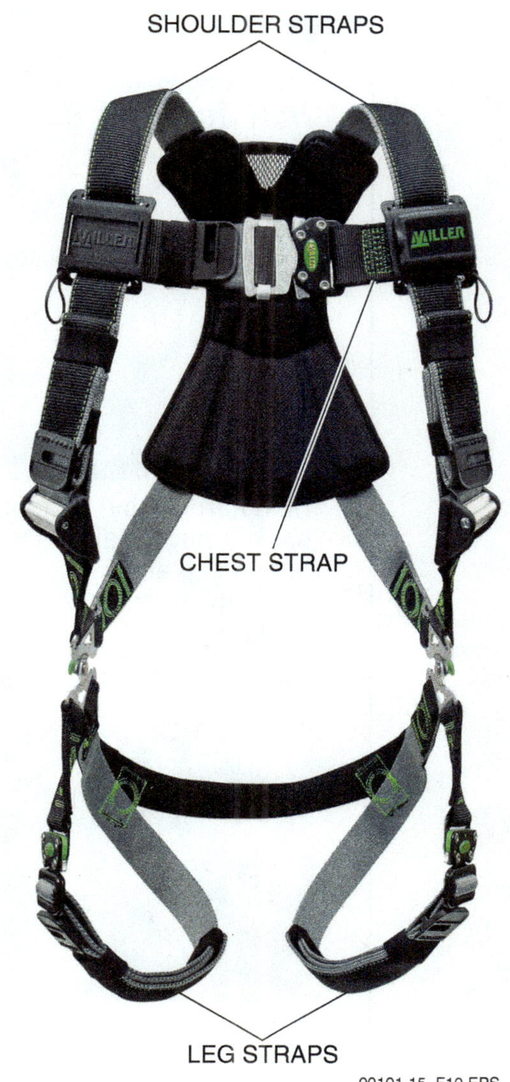

00101-15_F10.EPS

Figure 10 Full body harness.

- The back D-ring location is vital to proper fall arrest. Position this ring between the shoulder blades. If it is too low, it will tend to cause the body to hang in a more horizontal position during fall arrest, increasing pressure on the diaphragm and affecting breathing. If it is positioned too high or with too much slack, the D-ring may strike the worker's head at the base of the fall, and the shoulder straps may be pulled too tightly into the neck and restrict blood flow. The impact at the base of the fall arrest can be dramatic, so the force must be spread all around the body to prevent injury to any one portion.
- Chest straps generally form either an "H" pattern or an "X" pattern. Adjust "H" pattern straps to land between the bottom of the sternum and the belly button. This helps ensure the horizontal portion of the "H" does not contact the throat during a fall, choking the worker. Some harness designs may not allow

6 Easy Steps That Could Save Your Life

How To Don A Harness

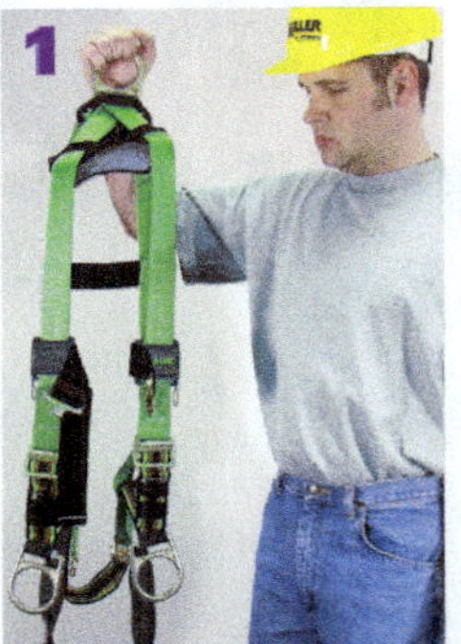

Hold harness by back D-ring. Shake harness to allow all straps to fall in place.

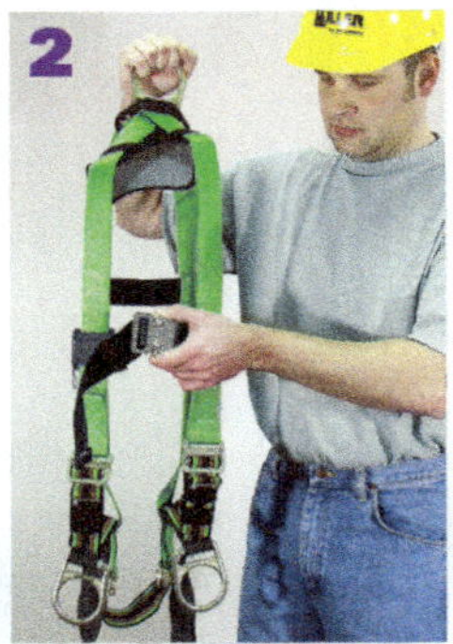

If chest, leg and/or waist straps are buckled, release straps and unbuckle at this time.

Slip straps over shoulders so **D-ring is located in middle of back between shoulder blades.**

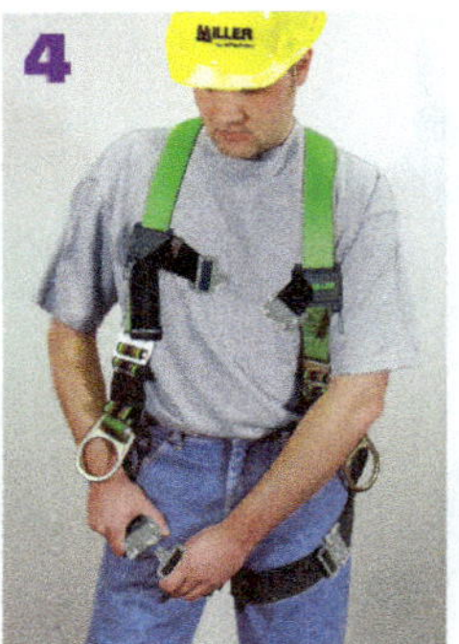

Pull leg strap between legs and connect to opposite end. Repeat with second leg strap. If belted harness, connect waist strap after leg straps.

Connect chest strap and position in midchest area. Tighten to keep shoulder straps taut.

After all straps have been buckled, **tighten all buckles so that harness fits snug but allows full range of movement.** Pass excess strap through loop keepers.

00101-15_F11.EPS

Figure 11 Installing the body harness.

for this adjustment, with the final position of the "H" being based solely on a properly sized harness.

- The position of the chest straps is also crucial for "X" pattern harnesses. Position the "X" at or just below the sternum.
- Leg straps are an integral and required part of a PFAS. Adjust the groin straps for a good, snug fit. Too much slack here will cause extreme discomfort in a fall, when the impact snatches them up tight and you are left suspended this way.
- A suspension trauma strap (*Figure 12*) is recommended as part of the PFAS gear. The suspension trauma strap is stored in a pouch connected to the harness within easy reach. This is done by either one end of the strap being permanently sewn to the harness (by the manufacturer); one end of the strap attached to the harness with a carabiner (*Figure 13*); or by choking the pouch around a harness strap or hip D-ring. The strap can then be quickly removed and used without any possibility of the user dropping it. Once connected, the strap allows

00101-15_F12.EPS

Figure 12 Suspension trauma strap use.

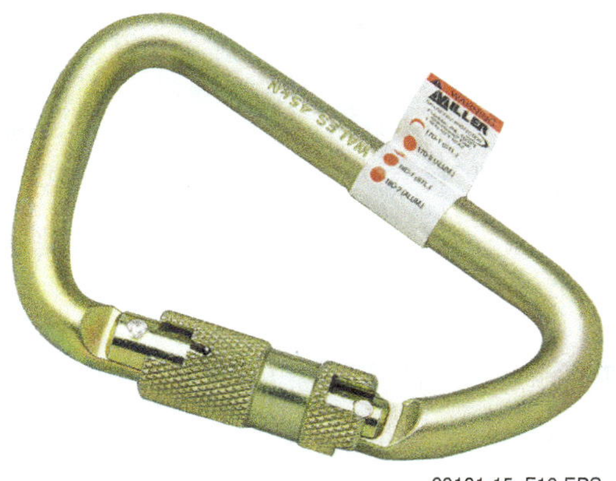

00101-15_F13.EPS

Figure 13 A carabiner.

the worker to stand up in the harness, relieving suspended weight and pressure from the hips and groin. This helps open the path for blood flow from the legs back to the heart, preventing blood from pooling in the lower extremities.

- A separate waist or tool belt, while not considered a necessary component of the fall-arrest system, must be fitted properly. It is best used for body positioning with the D-rings precisely located at the hips, rather than in the front or rear. Do not adjust it in a way that could apply pressure to the kidneys or lower back. If the waist belt is not an integral part of the harness, it must not be worn on the outside of the harness—put it on first, then add the harness over it. This is also true of any added tool belts.
- Saddles, like waist belts, are also not considered an integral or required part of the PFAS. They are optional and often detachable. They are generally used by workers to allow a seated, suspended position when the task may require long periods in the same location.

It is important to note that not all hardware will qualify as a component of a PFAS. Some hardware is to be used only for attaching tools and equipment to the worker or to structures. Hardware used as connecting devices as part of the PFAS must be drop-forged steel, and have a corrosion-resistant finish resistant to salt spray per ANSI standards. Any type of hook or carabiner must be equipped with safety gates or keepers to prevent the hooked object from being disconnected incidentally. In most cases, these safety gates will be required to be two-step, also called double action. Designs for these features vary, but those designed so that both movements required to open the gate can be done with a single hand are generally better. Using both hands to manipulate a single connector can be a hazard in itself.

Double-locking snap hooks (*Figure 14*) are usually curved and have an opening to allow connection to a line that can then be securely closed. They are usually not as consistent in appearance

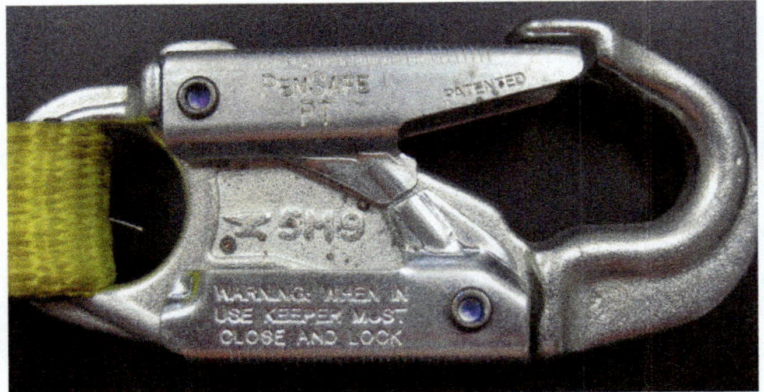

00101-15_F14.EPS

Figure 14 Double-locking snap hook.

as carabiners, and come in somewhat different shapes. When used as part of a PFAS, the security closure should be automatic. Snap hooks are designed to connect to D-rings primarily, not to each other. In most cases, snap hooks are already connected to a lanyard or rope to ensure the integrity of the connection, and are not purchased separately.

2.2.3 PFAS Inspection

Always inspect your lanyard, harness, and any other fall protection gear prior to use. Never use damaged equipment. Treat a safety harness as if your life depends on it, because it does! To maintain their service life and high performance, all belts and harnesses should be inspected frequently. Damage to fall arrest systems includes burns, hardening due to chemical contact, and excessive wear. When inspecting a harness, check that the buckles and D-ring are not bent or deeply scratched. Check the harness for any cuts or rough spots.

The PFAS should be inspected monthly by a competent person. This requirement should be established through your company's safety program. The competent person has the authority to impose prompt corrective measures to eliminate any hazards. If there is any question about a defect, no matter how small it may seem to be, err on the side of caution. Take the fall arrest component(s) out of service for testing or replacement.

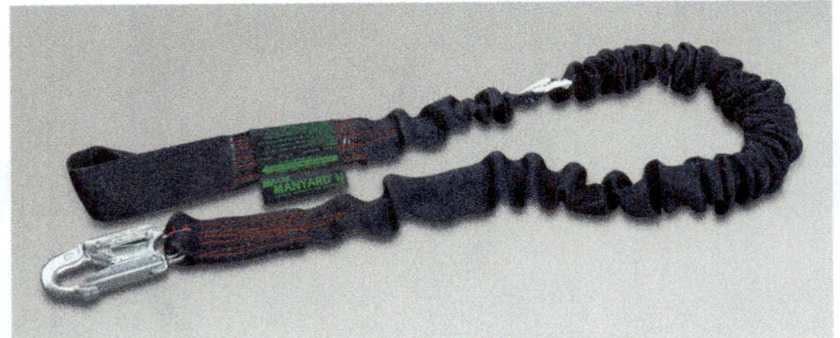

(A) SHOCK-ABSORBING

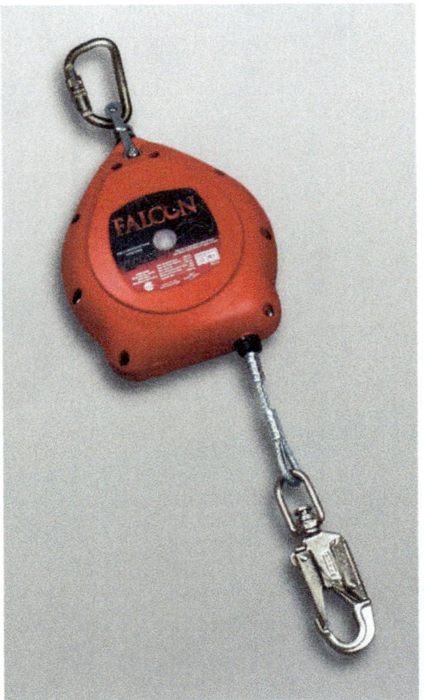

(B) RETRACTABLE

00101-15_F15.EPS

Figure 15 Fall arrest lanyards.

2.2.4 Lanyards

Lanyards consist of some primary material of construction (rope, webbing, aircraft cable, etc.) with a connecting device attached to the ends. Lanyards used for fall arrest are either shock absorbing or self-retracting (*Figure 15*). Non-shock absorbing lanyards, such as the one shown in *Figure 16*, are used for positioning and fall restraint. Lanyards used for positioning are not considered part of the fall-arrest system; they are fall restraints and will be attached to D-rings on the harness other than the back D-ring. Since fall restraint is all about preventing a fall from happening, lanyards used for positioning should not allow a fall or movement greater than 2 feet (0.61 m). Two such lanyards are usually required to permit movement from one place to another.

The retractable fall arrest lanyard shown in *Figure 15* is rapidly becoming the preferred device. Since it automatically retracts or feeds lanyard as the worker moves about, there is never a great deal of slack that can pose a risk in itself. In addition, the potential fall distance is significantly reduced.

Lanyards for fall restraint or arrest must never be field-fabricated, and must never be connected together to increase their length. Depending on the use, padding may be added during fabrication or added in the field to protect against sharp edges. Never tie a knot in a lanyard, as knots

00101-15_F16.EPS

Figure 16 Non-shock absorbing lanyard.

can severely reduce the load limit. Never wrap a lanyard around a structure and then choke it back into itself unless it is designed for this use. A special large D-ring is usually attached to the lanyard for this purpose. It is important to remember that whenever you are wearing a PFAS, 100 percent tie-off is required. A PFAS is absolutely useless unless you are tied off.

The D-ring used on the PFAS for fall arrest may only be connected to one live connection at a time. This can be challenging when trying to move from one point to another, especially horizontally. The user must be able to reach back and disconnect a lanyard, while maintaining one lanyard connected at all times. A Y-configured lanyard (*Figure 17*) can be used for this purpose. A Y-configured lanyard has a single point of attachment at the D-ring that is used to accommodate two lanyards. They are also referred to as double-leg or tieback lanyards.

Before using a shock absorbing or self-retracting lanyard, the potential fall distance is determined. Then the proper equipment is selected to meet available fall clearance (*Figure 18*). Note in the figure that the fall distance with the self-retracting lanyard is significantly shorter than it is with a shock absorbing lanyard; hence its increasing popularity. Failure to select proper equipment and calculate fall distance may result in serious personal injury or death. These calculations must be done by experienced and qualified personnel on the job site. If a personal fall arrest system is actuated due to a fall, it must be inspected by a competent person before it can be used again. In most cases, replacement will be necessary.

2.2.5 Lifelines

A lifeline is a flexible line such as a cable or rope connected vertically to an anchorage at one end (vertical lifeline), or horizontally to an anchorage at both ends (horizontal lifeline). It serves as a means for connecting other components of a personal fall-arrest system to the available anchorage.

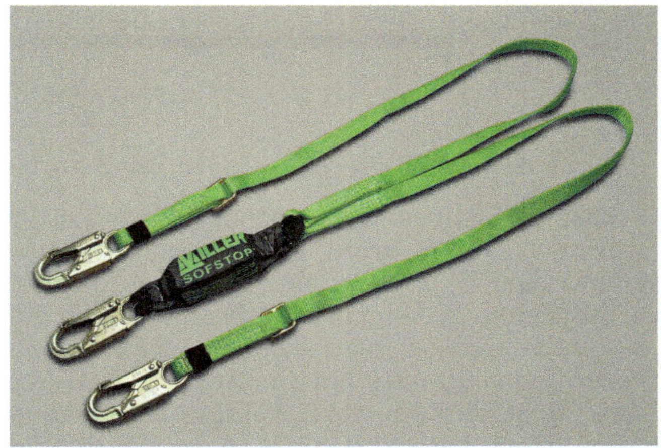

00101-15_F17.EPS

Figure 17 Y-configured shock absorbing lanyard.

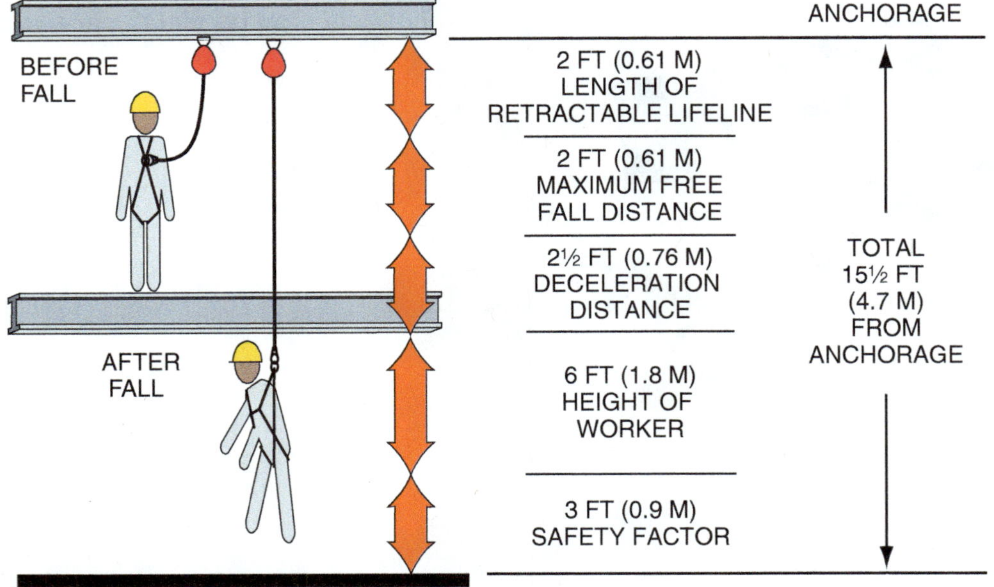

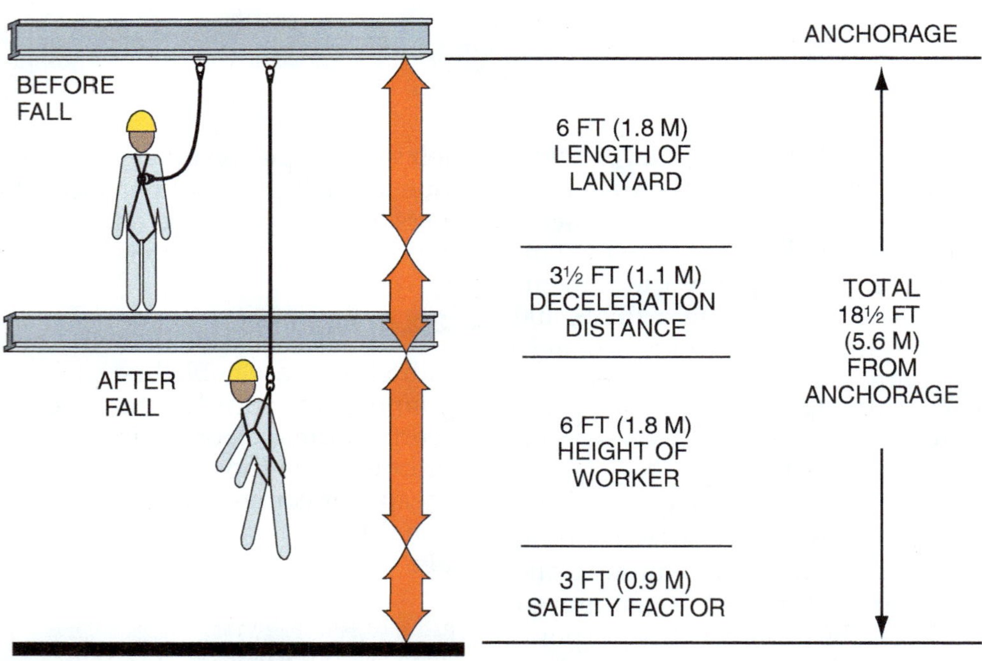

Figure 18 Potential fall distances for self-retracting and shock absorbing lanyards.

Vertical lifelines are suspended from a fixed anchorage. A fall arrest device such as a rope grab (*Figure 19*) or retractable lanyard is attached to the lifeline. A beam grab, or beamer (*Figure 20*), is sometimes used as an anchorage for a lifeline. Vertical lifelines must have a minimum breaking strength of 5,000 pounds (2,267 kg). Workers must use separate vertical lifelines. A vertical lifeline must have a termination on the end unless it extends to the ground or to the next lower level of the structure.

Horizontal lifelines (*Figure 21*) are connected between two fixed anchorages. A lanyard is attached to the lifeline. Horizontal lifelines must be designed, installed, and used under the supervision of a competent person. The required breaking strength of any horizontal lifeline is determined by the number of workers that will be attached to it.

00101-15_F19.EPS

Figure 19 Rope grab.

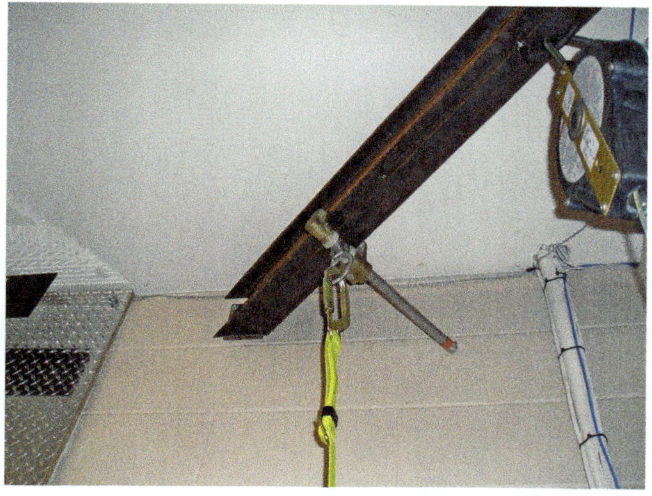

00101-15_F20.EPS

Figure 20 Beam grab.

> **WARNING!**
>
> Keep in mind that horizontal lifelines are rated for a maximum number of connected workers. Check with your supervisor or the manufacturer before connecting to a lifeline that is being used by other workers.

2.2.6 Guardrails

Guardrails (*Figure 22*) are a common type of fall prevention. They protect workers by providing a barrier between the work area and the ground or lower work areas. They may be made of wood, pipe, steel, or wire rope and must be able to support 200 pounds (90 kg) of force applied in any direction to the top rail and 150 pounds (68 kg) for the **midrail**. A guardrail must be 42" ±3" (106 ±8 cm) high to the top rail and have a toeboard that is a minimum of 4" (10 cm) high. This helps to prevent the inadvertent loss of tools or material through the bottom rail. The toeboard must be securely fastened with not more than ¼" (5 mm) clearance above the floor level.

2.2.7 Safety Nets

Safety nets are used for fall protection on bridges and similar projects. They must be installed not more than 30 feet (9 m) beneath the work area. There must be enough clearance under a safety net to prevent a worker who falls into it from hitting the surface below it. There must also be no obstruction between the work area and the net.

Depending on the actual vertical distance between the net and the work area, the net must extend 8 to 13 feet (2 to 4 m) beyond the edge of the work area. Mesh openings in the net must be limited to 36 square inches (232 sq cm) and 6 inches (15 cm) from the side. The border rope must have a 5,000-pound (2,267 kg) minimum breaking strength, and connections between net panels must be as strong as the nets themselves. Safety nets must be inspected at least once a week and after any event that might have damaged or weakened them. Worn or damaged nets must be removed from service.

2.3.0 Ladders and Stairs

Ladders are used daily to perform work in elevated locations. Any time work is performed above ground level, there is a risk of incidents. You can reduce this risk by carefully inspecting ladders before you use them and by using them properly.

Overloading means exceeding the **maximum intended load** of a ladder. Overloading can cause ladder failure, which means that the ladder could buckle, break, or topple. The maximum intended load is the total weight of all people, equipment, tools, materials, loads that are being carried, and other loads that the ladder can hold at any one time. Check the manufacturer's specifications to determine the maximum intended load. Ladders are usually given a duty rating that indicates their load capacity, as shown in *Table 2*. Note that ladders designed for the metric market are not usually direct equivalents to ladders built for the American market. Capacity ratings of 130 kg (286 lbs) and 150 kg (330 pounds) are the most common for the trades. Ladder heights will also be stated in metric units and may not be exactly the same size as those designed for the American market.

00101-15_F21.EPS

Figure 21 Horizontal lifeline.

00101-15_F22.EPS

Figure 22 Guardrails.

Drop-Testing Safety Nets

Safety nets should be drop-tested at the job site after the initial installation, whenever relocated, after a repair, and at least every six months if left in one place. The drop test consists of a 400-pound (181 kg) bag of sand of 29" to 31" (74 to 79 cm) in diameter that is dropped into the net from at least 42" (107 cm) above the highest walking/working surface at which workers are exposed to fall hazards. If the net is still intact after the bag of sand is dropped, it passed the test.

Table 2 Ladder Duty Ratings and Load Capacities

Duty Rating	Load Capacities
Type IAA	375 lbs., extra-heavy duty/ professional use
Type IA	300 lbs., extra-heavy duty/ professional use
Type I	250 lbs., heavy duty/industrial use
Type II	225 lbs., medium duty/commercial use
Type III	200 lbs., light duty/household use

There are different types of ladders to use for different jobs (*Figure 23*). Selecting the right ladder for the job at hand is important to complete a job as safely and efficiently as possible. Ladder types include portable straight ladders, extension ladders, and stepladders.

WARNING!

Use ladders only for their intended purposes. Ladders are not interchangeable; incorrect use of a ladder can result in injury or damage.

When using a ladder, be sure to maintain three-point contact with the ladder while ascending or descending. Three-point contact means that either two feet and one hand or one foot and two hands are always touching the ladder.

2.3.1 Straight Ladders

Straight ladders consist of two rails, rungs between the rails, and safety feet on the bottom of the rails (*Figure 24*). Straight ladders are generally made of aluminum, wood, or fiberglass.

Metal ladders conduct electricity and should never be used around electrical equipment. Any portable metal ladder must have "Danger! Do Not Use Around Electrical Installations" stenciled on the rails in two-inch red letters. Ladders made of dry wood or fiberglass, neither of which conducts electricity, should be used around electrical equipment. Check that any ladder, especially a wooden ladder, is completely dry before using it; even a small amount of water will conduct electricity.

Case History

A worker was climbing a 10-foot (3.05 m) ladder to access a landing, which was 9 feet (2.74 m) above the adjacent floor. The ladder slid down, and the worker fell to the floor, sustaining fatal injuries. Although the ladder had slip-resistant feet, it was not secured, and the railings did not extend 3 feet (0.91 m) above the landing.

Different types of ladders are intended for use in specific situations. Aluminum ladders are corrosion-resistant and can be used where they might be exposed to the elements. They are also lightweight and can be used where they must often be lifted and moved. Fiberglass ladders are very durable, so they are useful where some amount of rough treatment is unavoidable. Wooden ladders, which are heavier and sturdier than fiberglass or aluminum ladders, can be used where heavy loads must be moved up and down. However, wooden ladders are subject to more rapid deterioration as the wood swells and shrinks. Both fiberglass and aluminum are easier to clean than wood.

Wooden ladders should never be painted. The paint could hide cracks in the rungs or rails. Clear varnish, shellac, or a preservative oil finish will protect the wood without hiding defects.

Figure 25 shows the safety feet attached to a straight ladder. Make sure the feet are securely attached and that they are not damaged or worn down. Do not use a ladder if its safety feet are not in good working order.

It is very important to place a straight ladder at the proper angle before using it. A ladder placed at an improper angle will be unstable and could cause a fall. *Figure 26* shows a properly positioned straight ladder.

The distance between the foot of a ladder and the base of the structure it is leaning against must be one-fourth of the distance between the ground and the point where the ladder touches the structure. Stated another way, there should be a 4-to-1 ratio between the distances. For example, if the height of the wall shown in *Figure 26* is 16 feet (4.9 m), the base of the ladder should be 4 feet (1.2 m) from the base of the wall. If you are going to step off a ladder onto a platform or roof, the top of the ladder should extend at least 3 feet (0.9 m) above the point where the ladder touches the platform, roof, side rails, etc.

Ladders should be used only on stable and level surfaces unless they are secured at both the bottom and the top to prevent any incidental movement (*Figure 27*). Never try to move a ladder while you are on it. If a ladder must be placed in front of a door that opens toward the ladder, the door should be locked or blocked open. Otherwise, the door could be opened into the ladder.

Ladders are made for vertical use only. Never use a ladder as a work platform by placing it horizontally. Make sure the ladder you are about to climb or descend is properly secure before you do so. Check to make sure the ladder's feet are solidly positioned on firm, level ground. Also check

(A) ALUMINUM STEPLADDER

(B) FIBERGLASS STEPLADDER

(C) FIBERGLASS EXTENSION LADDER

(D) STRAIGHT LADDER

(E) FIBERGLASS PLATFORM LADDER

(F) ROLLING WAREHOUSE LADDER

00101-15_F23.EPS

Figure 23 Different types of ladders.

to make sure the top of the ladder is firmly positioned and in no danger of shifting once you begin your climb. Remember that your own weight will affect the ladder's steadiness once you mount it. It is important to test the ladder first by putting some of your weight on it without actually beginning to climb. This way, you can be sure that the ladder will remain steady as you climb.

When climbing a straight ladder, keep both hands on the rails or rungs (*Figure 28*). Maintain

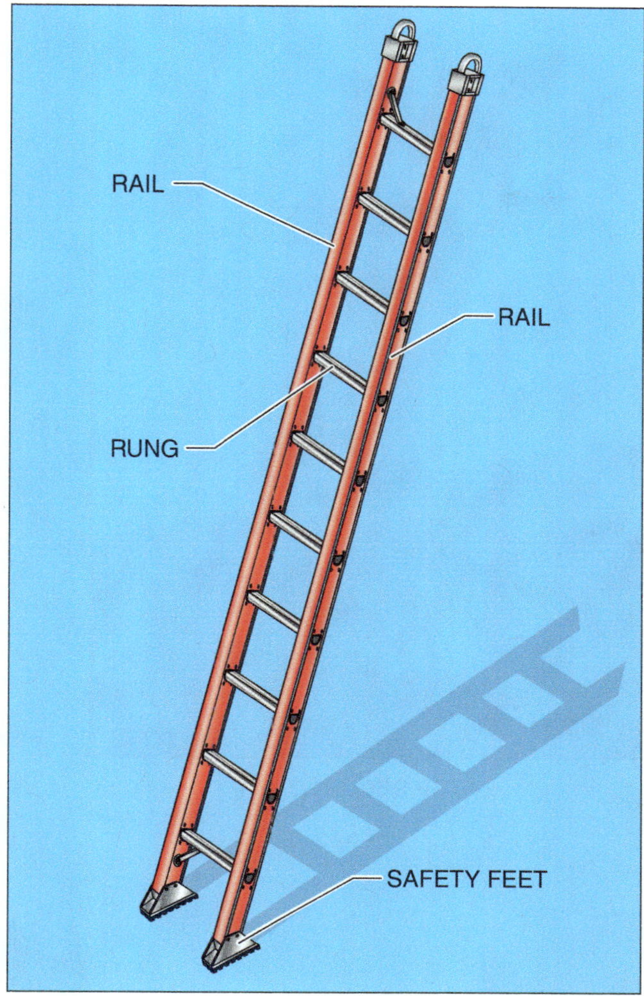

00101-15_F24.EPS

Figure 24 Portable straight ladder.

00101-15_F25.EPS

Figure 25 Ladder safety feet.

00101-15_F26.EPS

Figure 26 Proper positioning of a straight ladder.

three points of contact at all times, as shown in the figure. This can be two feet and one hand, or one foot and two hands. Always keep your body's weight in the center of the ladder between the rails. Face the ladder at all times. Never go up or down a ladder while facing away from it.

To carry a tool while you are on the ladder, use a **hand line** or tagline attached to the tool. Climb the ladder and then pull up the tool. Don't carry tools in your hands while you are climbing a ladder.

Figure 27 Securing a ladder.

Figure 28 Moving up or down a ladder.

2.3.2 Extension Ladders

An extension ladder is actually two straight ladders connected so that the overlap between them can be altered to increase or decrease the length of the ladder (*Figure 29*).

Extension ladders are positioned and secured following the same rules as straight ladders. When adjusting the length of an extension ladder, always reposition the movable section from the bottom, not the top, to ensure that the rung locks

Figure 29 Examples of extension ladders.

(*Figure 30*) are properly engaged after you make the adjustment. Check to make sure the section locking mechanisms are fully hooked over the desired rung. Also check to make sure that all ropes used for raising and lowering the extension are clear and untangled.

Extension ladders are positioned and secured following the same rules as straight ladders. There are, however, some safety rules that are unique to extension ladders:

> **WARNING!**
>
> Extension ladders have a built-in extension stop mechanism. Do not remove this mechanism. If the mechanism is removed, it could cause the ladder to collapse under a load.
>
> Haul materials up on a line rather than hand carrying them up an extension ladder.

- Make sure the extension ladder overlaps between the two sections (*Figure 31*). For ladders up to 36' (10.5 m) long, the overlap must be at least 3' (0.9 m). For ladders 36' to 48' (10.8 to 14.6 m) long, the overlap must be at least 4' (1.2 m). For ladders 48' to 60' (14.6 to 18.3 m) long, the overlap must be at least 5' (1.5 m).

00101-15_F30.EPS

Figure 30 Rung locks.

- Never stand above the highest safe standing level on a ladder. On an extension ladder, this is the fourth rung from the top. If you stand higher, you may lose your balance and fall. Some ladders have colored rungs to show where you should not stand.
- Avoid carrying anything on a ladder, because it will affect your balance and may cause you to fall. Haul materials up on a line instead.
- Keep yourself centered on the ladder. Do not over-reach, lean to one side, or try to move a ladder while standing on it.

2.3.3 Stepladders

Stepladders are self-supporting ladders made of two sections hinged at the top (*Figure 32*). The section of a stepladder used for climbing consists of rails and rungs like those on straight ladders. The other section consists of rails and braces. Spreaders are hinged arms between the sections that keep the ladder stable and keep it from folding while in use.

When positioning a stepladder, be sure that all four feet are on a hard, even surface. Otherwise, the ladder can rock from side to side or corner to corner when you climb it. With the ladder in position, be sure the spreaders are locked in the fully open position. The following safety precautions must be followed when using a stepladder:

- Never stand on the top step or the top of a stepladder. Putting your weight this high will make the ladder unstable. The top of the ladder is made to support the hinges, not to be used as a step.
- Although the rear braces may look like rungs, they are not designed to support your weight. Never use the braces for climbing or climb the back of a stepladder. However, there are specially designed two-person ladders available with steps on both sides.
- Check the load capacity of the ladder and do not exceed it.

> **WARNING!**
>
> Never stand on a step with your knees higher than the top of a stepladder. You need to be able to hold on to the ladder with your hand. Keep your body centered between the side rails.

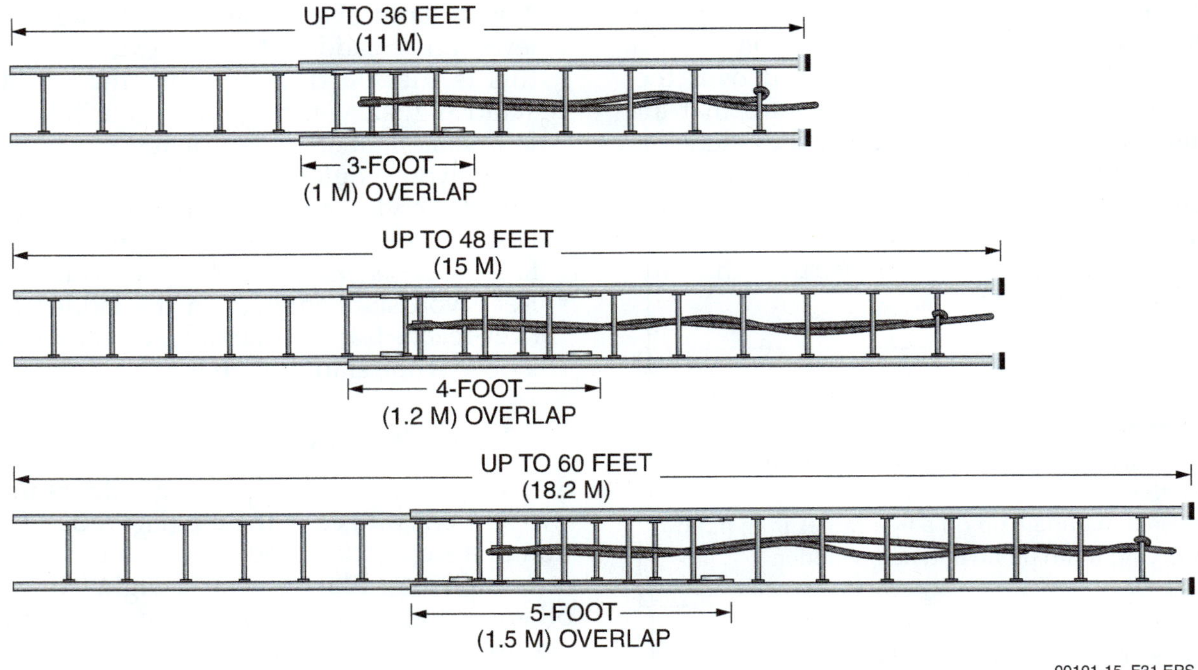

Figure 31 Overlap lengths for extension ladders.

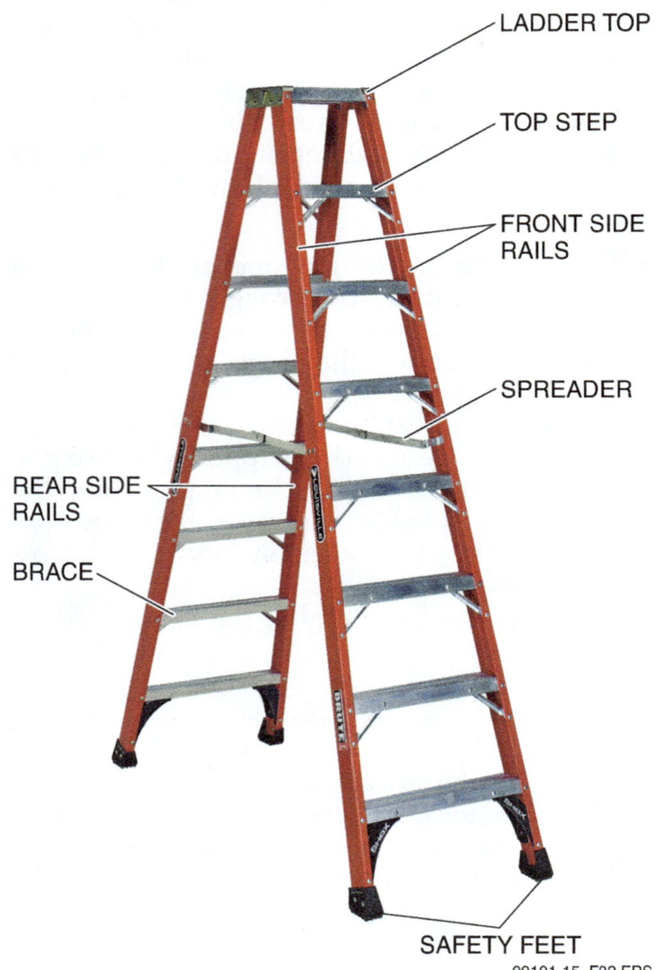

Figure 32 Typical fiberglass stepladder.

2.3.4 Inspecting Ladders

Always inspect a ladder before you use it. Check the rails and rungs for cracks or other damage. Also, check for loose rungs. If you find any damage, do not use the ladder. Check the entire ladder for loose nails, screws, brackets, or other hardware. If you find any hardware problems, tighten the loose parts or have the ladder repaired before you use it. OSHA requires regular inspections of all ladders and an inspection just before each use.

CAUTION

Wooden ladders should never be painted. The paint could hide cracks in the rungs or rails. Clear varnish, shellac, or a preservative oil finish will protect the wood without hiding defects.

The same rules for inspecting straight ladders apply to extension ladders. In addition, the rope that is used to raise and lower the movable section of the ladder should be inspected. If the rope is frayed or has worn spots, it should be replaced before the ladder is used.

The rung locks support the entire weight of the movable section and the person climbing the ladder. Inspect them for damage before each use. If they are damaged, they should be repaired or replaced before the ladder is used.

Inspect stepladders the way you inspect straight and extension ladders. Pay special attention to the hinges and spreaders to be sure they are in good repair. Also, be sure the rungs are clean. The rungs of a stepladder are usually flat, so oil, grease, or dirt can build up on them and make them slippery.

2.3.5 Stairways

Stairways are also routinely used on construction sites where there is a break in elevation of 19 inches (46 cm) or more, and no ramp, runway, sloped embankment, or personnel hoist is provided. Observe the following regulations, based on OSHA standards, when using stairways on a job site:

- Stairways having four or more risers or rising more than 30 inches (76 cm), whichever is less, must be equipped with at least one handrail and one stair railing system along each unprotected side.
- Winding and spiral stairways must be equipped with a handrail offset sufficiently to prevent walking on those portions of the stairways where the tread width is less than 6 inches (15 cm).
- Stair railings must be not less than 36 inches (91 cm) from the upper surface of the stair railing system to the surface of the tread, in line with the face of the riser at the forward edge of the tread.

To reduce the likelihood of slips, trips, or falls, keep stairways clean and clear of debris. Do not store any tools or materials on stairways, and clean up liquid spills, rain water, or mud immediately.

Stairways must have adequate lighting. This can sometimes be a problem because permanent lighting is usually installed after stairway construction is completed. If the lighting is inadequate, temporary lighting should be installed in the stairway. Each bulb should be equipped with a protective cover and the string should be inspected daily for burned out or broken bulbs.

Whenever possible, avoid using stairways to transport materials between floors. Carrying small materials and tools is fine, as long as the materials do not block your vision. Going up or down a stairway while carrying large items is physically demanding and increases the chance of injuries and falls. Use the building elevator or crane service to transport large materials from one floor to another.

2.4.0 Scaffolds

Scaffolds provide safe elevated work platforms for people and materials. They are designed and built to comply with high safety standards, but normal wear and tear or incidentally putting too much weight on them can weaken them and make them unsafe. That's why it is important to inspect every part of a scaffold before each use. Personnel who assemble scaffolds must be certified to do so.

CAUTION

Only a competent person has the authority to supervise setting up, moving, and taking down scaffolds. Only a competent person can approve the use of scaffolds on the job site after inspecting the scaffolds.

2.4.1 Types of Scaffolds

Two basic types of scaffolds—self-supporting scaffolds and suspended scaffolds—are used in the construction industry. The rules for safe use apply to both of them. Self-supporting scaffolds can be manufactured units or can be assembled at the site.

Manufactured scaffolds (*Figure 33*) are made of painted steel, stainless steel, or aluminum. They are stronger and more fire-resistant than wooden scaffolds. They are supplied in ready-made, individual units, which are assembled on site. A rolling scaffold has wheels on its legs so that it can be easily moved. The scaffold wheels have brakes so the scaffold will not move while workers are standing on it.

WARNING!

Never unlock the wheel brakes of a rolling scaffold while anyone is on it. People on a moving scaffold can lose their balance and fall.

Built-up scaffolds (*Figure 34*) are built from the ground up at a job site using steel framework sections and lumber. Swing (suspended) scaffolds (*Figure 35A*) are suspended by ropes or cables in a manner that allows it to be raised or lowered as needed. Another type of suspended scaffolding is a work cage (*Figure 35B*). A work cage is typically suspended with rigging devices that attach to I-beams with various sizes of clamps and rollers.

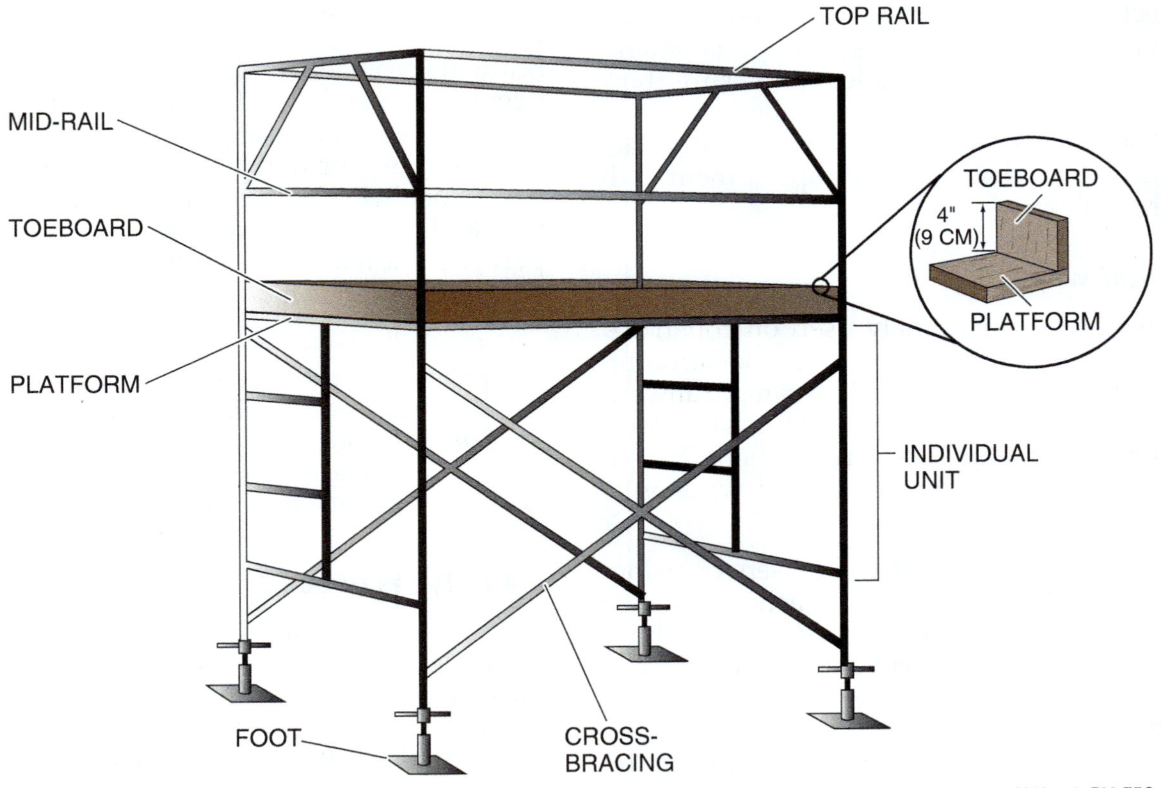

Figure 33 Typical manufactured scaffold.

2.4.2 Inspecting Scaffolds

Any scaffold that is assembled on the job site must be tagged to indicate whether the scaffold meets OSHA standards and is safe to use. Three colors of tags are used: green, yellow, and red (*Figure 36*).

- A green tag means the scaffold meets all OSHA standards and is safe to use.
- A yellow tag means the scaffold does not meet all OSHA standards. An example is a scaffold on which a railing cannot be installed because of equipment interference. To use a yellow-tagged scaffold, you must wear a safety harness attached to a lanyard. You may have to take other safety measures as well.
- A red tag means a scaffold is being put up or taken down. Never use a red-tagged scaffold.

Don't rely on the tags alone; inspect all scaffolds before you use them. Check for bent, broken, or badly rusted tubes. Also check for loose joints where the tubes are connected. Any of these problems must be corrected before the scaffold is used.

Make sure you know the weight limit of any scaffold you will be using. Compare this weight limit to the total weight of the people, tools, equipment, and material you expect to put on the scaffold. Scaffold weight limits must never be exceeded.

If a scaffold is more than 10 feet (3.1 m) high, check to see that it is equipped with **top rails**, midrails, and toeboards; otherwise, use a PFAS. All connections must be pinned. That means they must have a piece of metal inserted through a hole to prevent connections from slipping. **Cross-bracing** must be used. A handrail is not the same as cross-bracing. The walking area must be completely **planked**.

If it is possible for people to walk under a scaffold, the space between the toeboard and the top rail must be screened. This prevents objects from falling off the work platform and injuring those below.

When you examine a rolling scaffold, check the condition of the wheels and brakes. Be sure the brakes are working properly and can stop the scaffold from moving while work is in progress. Be sure all brakes are locked before you use the scaffold.

2.4.3 Using Scaffolds

Be sure that a competent person inspects the scaffold before you use it. There should be firm footing under each leg of a scaffold before putting any weight on it. If you are working on loose or soft soil, you can put planks or matting under the scaffold's legs or wheels, as shown in *Figure 34*. When moving a rolling scaffold, first unlock the brakes and then move the scaffold. Once the scaffold is repositioned, don't forget to relock the brakes.

> **WARNING!**
>
> Keep scaffolds a minimum of 3 feet (91 cm) from power lines up to 300 volts and a minimum of 10 feet (3.1 m) from power lines above 300 volts. Never move a scaffold while someone is on it.

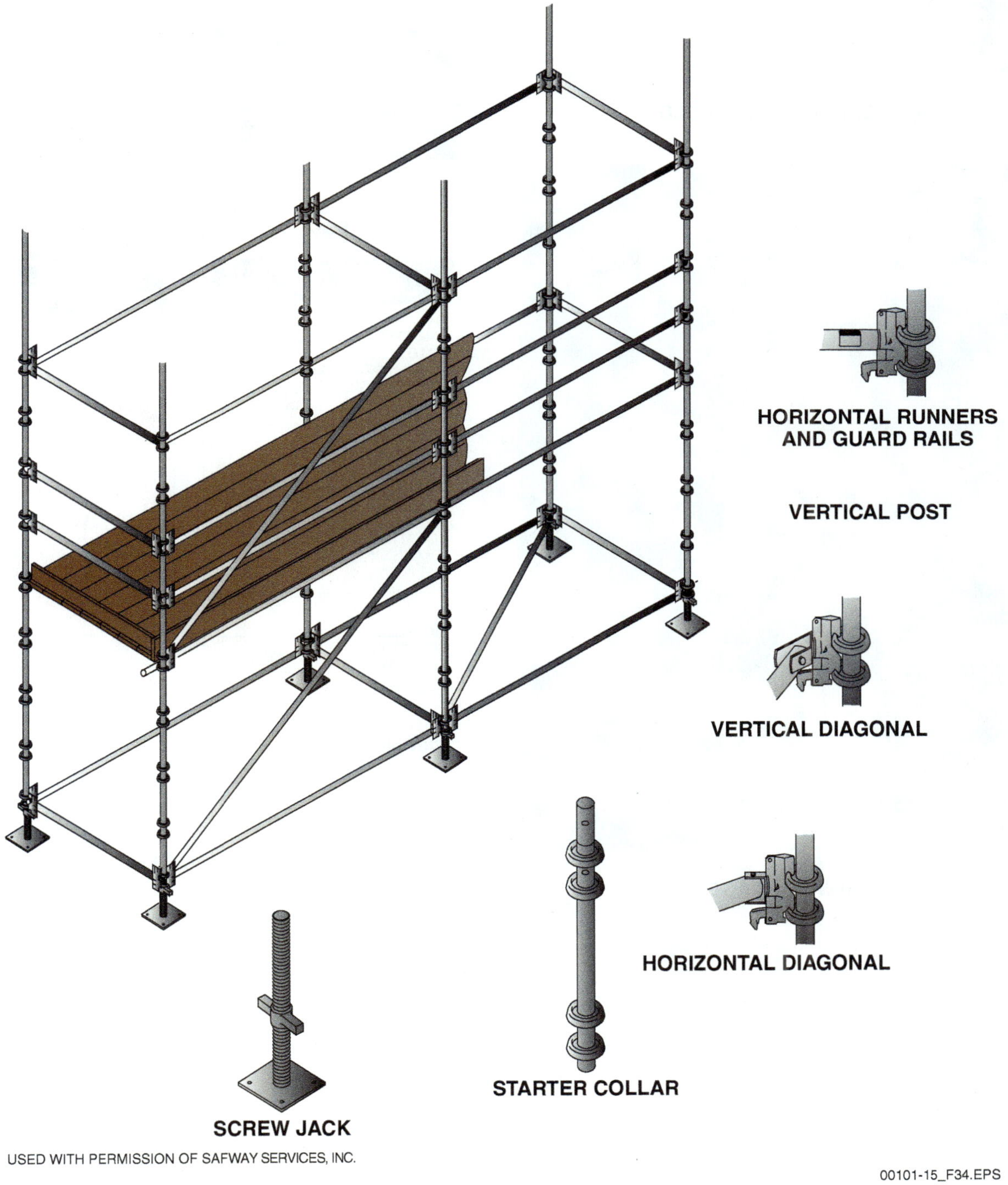

HORIZONTAL RUNNERS AND GUARD RAILS

VERTICAL POST

VERTICAL DIAGONAL

HORIZONTAL DIAGONAL

STARTER COLLAR

SCREW JACK

USED WITH PERMISSION OF SAFWAY SERVICES, INC.

00101-15_F34.EPS

Figure 34 Built-up scaffold.

(A) SUSPENDED SCAFFOLD

(B) WORK CAGE

00101-15_F35.EPS

Figure 35 Suspended scaffold and work cage.

COMPLETE

ERECTED FOR
DATE
COMPLETE
HANDRAILS
TOEBOARDS
DECK
LADDER
LIFE LINE
SCAFFOLD OVER
LOCATION

CHARGE #
ERECTION FOREMAN
SIGNATURE

DATE
LOCATION
SIGNATURE

DATE	INSPECTED BY

CAUTION

ERECTED FOR
DATE
INCOMPLETE
HANDRAILS
TOEBOARDS
DECK
LADDER
LIFE LINE
SCAFFOLD OVER
REASON INCOMPLETE

LOCATION
CHARGE #
ERECTION FOREMAN
SIGNATURE

DATE
LOCATION
SIGNATURE

DANGER

XYZ INC.

SCAFFOLD INCOMPLETE TAG

DO NOT USE OR CLIMB SCAFFOLD UNTIL COMPLETE

DATE
LOCATION
SIGNATURE

00101-15_F36.EPS

Figure 36 Typical scaffold tags.

Additional Resources

US Occupational Safety and Health Administration. Numerous safety videos are available on line at **www.osha.gov/video**.

Construction Safety, Jimmie W. Hinze. 2006. Upper Saddle River, NJ: Pearson Education, Inc.

DeWalt Construction Safety/OSHA Professional Reference. Paul Rosenberg; American Contractors Educational Services. 2006. DEWALT.

Basic Construction Safety and Health, Fred Fanning. 2014. CreateSpace Independent Publishing Platform.

2.0.0 Section Review

1. A barrier with yellow and purple markings indicates a _____.
   a. fire hazard
   b. fall hazard
   c. radiation hazard
   d. confined space hazard

2. Positioning lanyards are used only as fall restraint devices and may *not* be used as fall arrest devices.
   a. True
   b. False

3. When positioning a straight ladder against a wall, how far from the wall should the base of the ladder be?
   a. Four feet (1.2 m)
   b. One-fourth the distance from the ground to the point where the ladder touches the wall
   c. The height of the wall minus 4 feet (1.2 m)
   d. One-half the distance from the ground to the point where the ladder touches the wall

4. If a scaffold has a yellow tag, it means _____.
   a. the scaffold may only be used by one person at a time
   b. the scaffold is condemned and cannot be used
   c. a safety harness must be worn when using it
   d. the scaffold is under assembly and cannot be used

SECTION THREE

3.0.0 STRUCK-BY AND CAUGHT-IN-BETWEEN HAZARDS

Objectives

Identify and explain how to avoid struck-by and caught-in-between hazards.

a. Identify and explain how to avoid struck-by hazards.
b. Identify and explain how to avoid caught-in and caught-between hazards.

Trade Terms

Shielding: A structure used to protect workers in trenches.

Shoring: A support system designed to prevent a trench or excavation cave-in.

Signaler: A person who is responsible for directing a vehicle when the driver's vision is blocked in any way.

Spoil: Material such as earth removed while digging a trench or excavation.

There is an apparent overlap in struck-by and caught-in-between hazards that needs to be clarified, as they can involve the same kinds of equipment. *Struck-by* means being hit by a moving object, while *caught-in-between* means being trapped between a moving object and a solid surface. Assume a worker is working near a crane that is swinging a load and the load hits the worker. That is an example of a struck-by incident. However, if the worker is behind the crane and the crane backs up, pinning the worker against a wall, it becomes a caught-in-between incident.

3.1.0 Struck-By Hazards

On any job site, there is a risk of being struck by falling objects, such as tools dropped from above. Flying objects such as debris from grinding and chipping metal are another struck-by hazard. On any site where there is moving equipment, or where workers are near roadways, there is a also the danger of being struck by a vehicle.

3.1.1 Falling Objects

Workers are at risk from falling objects when they are beneath machinery and equipment such as cranes and scaffolds; where overhead work is being performed; or when working around stacked materials. To protect against struck-by injuries from falling objects, always wear an approved hard hat. Employers generally require workers to wear hard hats at all times on construction sites.

When working near machinery and equipment such as cranes, never stand or work beneath the load or in the fall zone. Barricade hazard areas where rigging equipment is in use, and post warning signs to inform other workers of falling object hazards. Inspect cranes and rigging components before use and do not exceed the rated load capacity.

When performing overhead work, be sure all tools, materials, and equipment are secured to prevent them from falling on people below. Use protective measures such as toeboards, debris nets, catch platforms, or canopies to catch or deflect falling objects. Use tool lanyards to prevent tools from falling.

Many workers are hurt or killed by falling stacks of material. Do not stack materials higher than 4:1 height-to-base ratio. Secure all loads by blocking and interlocking them. Interlocking means placing alternate layers at right angles to each other. Be aware of changing weather conditions, such as wind, that may lift and shift loads.

3.1.2 Flying Objects

There is a danger from flying objects when power tools or activities such as pushing, pulling, or prying causes objects to become airborne. Chipping, grinding, brushing, or hammering are all examples of job tasks that may cause flying objects. Tools that move at very high speed, like pneumatic and powder-actuated tools, can be very dangerous. Injuries from flying objects can range from minor abrasions to concussions, blindness, or death. The workers shown in *Figure 37* are using a pneumatic chipping hammer and a pneumatic grinder, and are properly using protective face shields.

To protect against flying object hazards, follow these guidelines:

- Use eye protection, such as safety glasses, goggles, or face shields where machines or tools may cause flying particles.
- Inspect tools and machines to ensure that protective guards are in place and in good condition.
- Make sure you are trained in the proper operation of pneumatic and powder-actuated tools.
- Use shielding devices such as welding screen or similar equipment to block flying debris.

00101-15_F37.EPS

Figure 37 Protection from flying particles.

3.1.3 Vehicle and Roadway Hazards

A common cause of incidents for workers on large job sites and highway projects are vehicle-related hazards such as being run over by vehicles or equipment (especially backing equipment) or by equipment tip-over. The most common cause of death for equipment operators is equipment rollover. If vehicle safety practices are not observed at your site, you risk being struck by swinging backhoes, moving vehicles, or swinging crane loads. If you work near public roadways (*Figure 38*), you risk being struck by passenger or commercial vehicles. When working near moving vehicles and equipment, follow these guidelines:

- Stay alert at all times and keep a safe distance from vehicles and equipment.
- Maintain eye contact with vehicle or equipment operators to ensure that they see you.
- Never get into blind spots of equipment operators.
- Keep off equipment unless authorized.
- Wear reflective or high-visibility vests or other suitable garments.
- Never stand between pieces of equipment unless they are secured.
- Never stand under loads handled by lifting or digging equipment, or near vehicles being loaded or unloaded.

Operators must also use caution when driving vehicles. The operator of any vehicle is responsible for the safety of passengers and the protection of the load. Follow these safety guidelines when you operate a vehicle on a job site:

- Always wear a seat belt.
- Be sure that each person in the vehicle has a firmly secured seat and seat belt.
- Obey all speed limits. Reduce speed in crowded areas.
- Look to the rear and sound the horn before backing up. If your rear vision is blocked, get a signaler to direct you.

Case Histories

- A worker was standing under a suspended scaffold that was hoisting a workman and three sections of ladder. Sections of that ladder became unlatched and fell 50 feet (≈15m), striking the worker in the skull. The worker was not wearing any head protection and died from his injuries.
- A carpenter was using a powder-actuated tool to anchor a plywood form in preparation for pouring a concrete wall. The nail passed through the hollow wall, traveled 72 feet (≈22m), and struck an apprentice in the head, killing him. The tool operator had never been trained in the proper use of the tool, and none of the employees in the area, including the victim, was wearing PPE.

Tool Lanyards

When working at elevations, a dropped tool can become a serious hazard. If the tool has moving parts, like battery-powered drills, the fall will likely destroy it.

Tool lanyards specifically designed for work in the elevated environment have been introduced by Snap-on®. Tethering tools to the tool belt or wrist can prevent injuries, save tools, prevent component damage, and prevent lost time in retrieval.

00101-15_SA01.EPS

- Every vehicle must have a backup alarm. Make sure the backup alarm works.
- Always turn off the engine when fueling.
- Turn off the engine and set the brakes before leaving the vehicle.
- Never stay on or in a truck that is being loaded by excavating equipment.
- Keep windshields, rearview mirrors, and lights clean and functional.
- Carry road flares, fire extinguishers, and other standard safety equipment at all times.
- Never use a cell phone while operating a motor vehicle.

00101-15_F38.EPS

Figure 38 A busy job site.

WARNING!

When a vehicle is operated indoors, the carbon monoxide in its exhaust can kill or sicken anyone in the vicinity unless there is good ventilation. Carbon monoxide is especially dangerous because you cannot see, smell, or taste it. It is never a good idea to operate a vehicle, a generator, or any other machine with an internal combustion engine indoors. Make sure there is good ventilation before you operate a motorized vehicle or equipment such as a generator indoors.

3.2.0 Caught-In and Caught-Between Hazards

Congested work sites, heavy equipment, and the presence of multiple trades can contribute to caught-in-between hazards. The primary causes of caught-in-between fatalities include trench/excavation collapse, rotating equipment, and unguarded parts.

One of the most disastrous caught-in hazards on a construction job site is being trapped by a cave-in of the walls of a trench or excavation. Other caught-in or caught-between hazards involve protective guards on power tools and machines, as well as the risk of being crushed by heavy equipment.

3.2.1 Trenches and Excavation

Trenches and excavations are common hazards, especially in construction and pipeline work. Anyone working in or around a trench or excavation must know and follow safety procedures aimed at protecting workers from cave-ins.

An excavation is any man-made cut, cavity, trench, or depression formed by removal of earth or soil. Sometimes the terms *excavation* and *trench* are used interchangeably, but there is a difference. A trench is an excavation that is deeper than it is wide, and usually not wider than 15 feet (4.6 m). Nearly all trenches are dangerous if not protected. Because trenches are narrow, workers can easily become trapped.

Did You Know?

Struck-by Fatalities

Nearly one in four struck-by vehicle deaths involve construction workers – more than any other occupation.

Hazards involved with trench and excavation work include the following:

- Cave-ins
- Water accumulation
- Falling objects
- Collapse of nearby structures
- Hazardous atmospheres produced by toxic gases in the soil

WARNING!

Just 2 to 3 feet (0.61 to 0.91 m) of soil can put enough pressure on your lungs to prevent you from breathing. In as little as 4 to 6 minutes without oxygen, you can sustain considerable brain damage.

Cave-ins are the most common and deadly hazard in excavation work. When dirt is removed from an excavation, the surrounding soil becomes unstable and gravity can force it to collapse. Cave-ins occur when soil or rock falls, or slides, into an excavation. Most cave-ins occur in trenches 5 to 15 feet (1.53 to 4.6 m) deep, and happen suddenly with little or no warning. On average, about 1,000 trench collapses occur each year in the United States.

Soil conditions can change, so they must be constantly evaluated. There are certain factors that could change the surroundings of the site, making a cave-in more likely. These factors include the following:

- Changes in weather conditions, such as freezing, thawing, or sudden heavy rain
- An excavation dug in unstable or previously disturbed soil
- Excessive vibration around the excavation
- Water accumulation in an excavation

Sometimes there are visible warning signs around the excavation that can be spotted before a cave-in occurs. Being aware of the warning signs increases your chances of getting out before a collapse occurs. Visible warning signs of a potential cave-in include the following:

- Ground settlement or narrow cracks in the sidewalls, slopes, or surface next to the excavation

Did You Know?

Excavation Fatalities

The fatality rate for excavation work is 112 percent higher than the rate for general construction.

- Flakes, pebbles, or clumps of soil separating and falling into the excavation
- Changes or bulges in the wall slope

If you notice any of these signs, get out of the excavation immediately and alert your co-workers as well.

Soil type is a key factor in determining the type of protective system needed to ensure that the trench will be safe. Solid rock is the most stable, while sandy soil is the least stable. The four types of soil are shown in *Table 3*.

To be safe, treat soil as if it is Type C soil, per *Table 3*, unless proven otherwise. It is better to over-prepare for a stronger soil than to not prepare enough for a weaker one.

A competent person must inspect excavations daily and decide whether cave-ins or failures of protective systems could occur, and whether there are any other hazardous conditions present. The competent person must conduct inspections before any work begins, as needed throughout the shift, and after every rainstorm or other hazard-increasing incident.

If the inspection reveals indications of protective system failure, hazardous atmospheres, or a possible cave-in, workers must be removed from the hazardous area and may not return until corrective action has been taken. Always ask the competent person on site or your immediate supervisor if you have questions about proper safety practices.

Once visual and manual tests are performed and the soil type is determined, a protective system must be chosen. Protective systems are required in nearly all excavations. There are various types of trench protective systems to meet each type of soil condition. Selecting a protective system for an excavation depends on soil conditions, the depth of the trench, and the environmental conditions surrounding the site.

Regardless of what type of system is used, if the excavation is more than 20 feet (6.1 m) deep, the entire excavation protective system has to be designed by a registered professional engineer. There are two basic systems of trench protection: sloping and benching systems and support systems. The method to be used is determined by the engineer and is based on the types of soil and the site conditions.

Table 3 Soil Types

Name	Type/Characteristics
Solid Rock	Excavation walls stay vertical as long as the excavation is open.
Type A Soil	Fine-grained, cohesive: clay, hardpan, and caliche. Particles too small to see with the naked eye.
Type B Soil	Angular rock, silt, and similar soil.
Type C Soil	Coarse-grained, granular: sand, gravel, and loamy sand. Particles are visible to the naked eye.

Did You Know?

Heavy Soil

One cubic yard of earth weighs approximately 2,700 pounds (1,225 kg)—that's the weight of a small car. Since the meter is longer than a yard, a cubic meter of soil weighs more — about 3,527 pounds (1,600 kg).

Sloping and benching are forms of trench protection that cut away and slant the excavation face. A sloping system is a method in which the sides of an excavation are cut back to a safe angle using relatively smooth inclines (*Figure 39*, top). A benching system is similar to a sloping system, but instead of smooth inclines, the sides of the trench wall are cut back using a series of steps (*Figure 39*, bottom). Benching systems cannot be used with Type C soil.

Trenches are often located in narrow places, so sloping and benching are not options. In these situations, support systems like **shoring** or **shielding** must be used. Shoring structures are typically made of metal or wood and are used to support the sides of a trench and prevent soil from caving in. They consist of plating held firmly in place with expandable braces (*Figure 40*). There are many types of shoring systems. Some of them are easy to install, and others require experience and engineering.

Shielding structures, also known as trench boxes, are placed inside trenches or excavations, and are strong enough to protect workers in the event of a cave-in, so long as the workers are within the confines of the box (*Figure 41*). Trench shields are used only to provide a protected space for workers. Shoring not only protects workers, but also prevents the trench walls from collapsing.

Did You Know?

Cave-In Fatalities

Each year in the United States, more than 100 people are killed and many more are seriously injured in cave-in incidents. The chances of a trapped worker being killed can be as high as 50 percent.

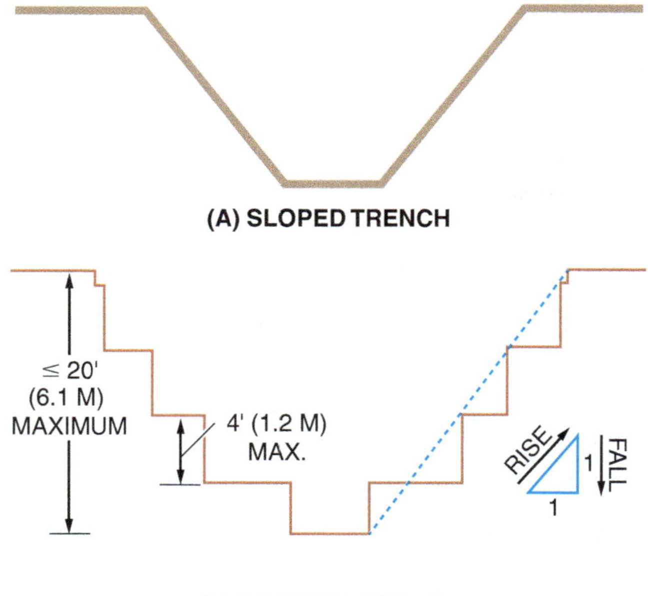

Figure 39 Sloped and benched trenches for Type B soil.

00101-15_F40.EPS

Figure 40 Shoring structure.

Case History

A worker was in a trench installing forms for concrete footers when it caved in, causing fatal injuries. The trench, which was 7½ feet (2.29 m) deep, was in loose, sandy (Type C) soil, and no inspection was conducted before the start of the shift.

Spoil piles, comprised of the material removed from an excavation, and other materials represent a hazard if not handled properly. Loose rock, soil, materials, and equipment on the face or near the excavation can fall or roll into the excavation, or overload and possibly collapse excavation walls. Keep all spoil, materials, and heavy equipment at least 2 feet (0.6 m) away from the edge of an excavation, or set up barricades to contain falling material. Scale the excavation face to remove loose material, and place spoil so that rainwater runs away from the excavation. Use a retaining device strong enough and high enough to resist expected loads. If the spoil cannot be safely stored on site, remove it to a temporary site.

When working in a trench, there must be a safe means of entry and exit for workers, such as a stairway, ladder, or ramp. There must be an exit every 25 feet (7.6 m) for every trench over 4 feet (1.2 m) deep. Lifting equipment such as loader buckets and backhoe shovels are not safe means for entering or exiting a trench. Once you are in the trench and before you begin work, take a moment to look around and find the nearest ladder so that you can plan your exit, if necessary.

00101-15_F41.EPS

Figure 41 Shielding structure.

Case Histories

- Two workers were installing pipe in a trench. No means of protection was provided in the vertical wall trench. A cave-in occurred, fatally injuring one worker and causing serious injury to the other.
- Four workers were in an excavation, boring a hole under a road. Eight-foot (2.44 m) high steel plates used as shoring were placed against the side walls of the excavation at about 30-degree angles, but were not supported by horizontal bracing. One of the plates tipped over, crushing one worker.

Your company is required to have an emergency action plan that must be communicated to every worker. If you are not sure what the emergency action plan is for your site, don't be afraid to ask questions. Your knowledge could help prevent serious injury or even death, and your supervisor wants you and everybody working with you to be as safe as possible.

Most importantly, try to prevent emergencies before they happen. When in doubt, get out! If you notice potentially dangerous conditions while working in an excavation, get yourself and your co-workers out of danger immediately, and inform your supervisor or the competent person of your concerns.

3.2.2 Tool, Machine, and Equipment Guards

Almost all tools and machines used in construction and industrial work are equipped with guards that protect workers from rotating parts. *Figure 42* show the guards on a grinding machine. All tools and machines that could harm workers must have a guard shielding the hazard. The following types of tools and machines must have guards:

- Grinding tools
- Shearing tools
- Presses
- Punches
- Cutting tools
- Rolling machines
- Tools or machines with pinch points
- Tools or machines with sharp edges

Machine guards should prevent moving parts of the machine from coming into contact with your arms, hands, or any other part of the body, while allowing you to use the machine comfortably and efficiently. Some workers find machine guards to be aggravating and try to remove them from machines. Guards should be secure and should not be easily removed. They should be maintained in good condition, made of durable material, and bolted or screwed to the machine so that tools are needed for their removal. *Figure 43* shows a coupling guard installed over the coupling that connects a motor shaft to a pump shaft. The guard is bolted in place and should only be removed when the motor is not running. The guard is there to prevent someone from being caught in the coupling, which rotates at high speed. Its highly visible markings indicate that it is a hazardous location.

Case History

A spoil pile had been placed on top of a curb, which formed the west face of a trench. A backhoe was working on top of the spoil pile. The west face of the trench collapsed on two workers who were installing sewer pipe. One worker was killed; the other received back injuries. The trench was 8 feet (2.4 m) deep with vertical walls. No cave-in protection was provided. The superimposed loads of the spoil pile and backhoe may have caused the collapse.

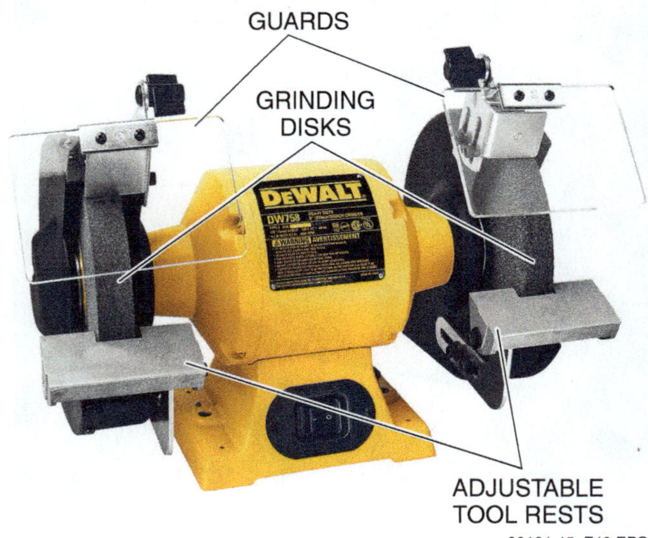

Figure 42 Bench grinder with guards.

Figure 43 Coupling guard.

Follow these guidelines for using and caring for tool and machine guards:

- Do not remove a guard from a tool or machine except for cleaning purposes or to change a blade or perform other service. Make sure the machine is turned off and tagged out.
- When you are finished with cleaning or maintenance, replace the guard immediately.
- Do not use any material to wedge a guard open.
- Guards and attachments are designed for the specific tool or machine you are using. Use only attachments that are specifically designed for that tool or machine.

3.2.3 Cranes and Heavy Equipment

Motorized equipment is used in many different jobs including construction, mining, plant maintenance and operations, road maintenance, equipment transportation, and snow removal. Working with motorized equipment can be dangerous. Workers can be crushed by falling loads, fall from equipment, be electrocuted by power lines, or be struck or trapped by vehicles. The swing radius of equipment can also be a hazard if the job is not carefully planned and properly barricaded. Most heavy equipment has pinch points. *Figure 44* shows some of the pinch point hazards on an excavator.

Dangers exist for both equipment operators and other workers on the site. In one example, a contractor was operating a backhoe when another employee attempted to walk between the swinging back end of the backhoe and a concrete wall. As the employee approached the backhoe from the operator's blind side, the back end hit the victim, crushing him against the wall. A similar problem can occur with a crane or excavator, as shown in *Figure 45*. As the cab swings, the rear end extends beyond the base of the machine. For that reason, barricades must be placed around the working perimeter of the machine.

Because working with or near motorized equipment is dangerous, you must understand that your first responsibility on a job is safety. This includes your own safety, the safety of others on the site, and the safe use of equipment on the site. You must know the hazards and safety procedures of every job you are on, regardless of the work you are doing.

00101-15_F44.EPS

Figure 44 Pinch and crush points.

00101-15_F45.EPS

Figure 45 Example of a caught-between hazard.

Case History

Two employees were attempting to adjust the brakes on a backhoe. The victim told the backhoe operator to raise the wheels off the ground with the front bucket and the outriggers so that he could get to the brakes. The victim then crawled under the machine and began to adjust the brakes. He did this without considering that there was only a 36" (0.9 m) space from the ground to the drive shaft. While adjusting the brakes, the hood of his rain jacket wrapped around the drive shaft and broke his neck. He died instantly.

The Bottom Line: Loose clothing can be caught in moving parts of machinery. You must consider all possible dangers when working on equipment.

Source: The Occupational Safety and Health Administration (OSHA)

Additional Resources

US Occupational Safety and Health Administration. Numerous safety videos are available on line at **www.osha.gov/video**.

Construction Safety Jimmie W. Hinze. 2006. Upper Saddle River, NJ: Pearson Education, Inc.

DeWalt Construction Safety/OSHA Professional Reference. Paul Rosenberg; American Contractors Educational Services. 2006. DEWALT.

Basic Construction Safety and Health, Fred Fanning. 2014. CreateSpace Independent Publishing Platform.

3.0.0 Section Review

1. Which of these activities is most likely to produce flying objects?
   a. Using a chipping hammer
   b. Using a screwdriver
   c. Painting a wall
   d. Climbing a ladder

2. The type of soil most likely to result in a trench cave-in is _____.
   a. solid rock
   b. Type A
   c. Type B
   d. Type C

SECTION FOUR

4.0.0 ENERGY RELEASE HAZARDS

Objectives

Identify common energy-related hazards and explain how to avoid them.

a. Describe basic job-site electrical safety guidelines.
b. Explain the importance of lockout/tagout and describe basic procedures.

Performance Task

4. Inspect a typical power cord and GFCI to ensure their serviceability.

Trade Terms

Ground: The conducting connection between electrical equipment or an electrical circuit and the earth.

Ground fault circuit interrupter (GFCI): A device that interrupts and de-energizes an electrical circuit to protect a person from electrocution.

Lockout/tagout (LOTO): A formal procedure for taking equipment out of service and ensuring that it cannot be operated until an authorized person has removed the lock and/or warning tag.

Proximity work: Work done near a hazard but not actually in contact with it.

Whenever possible, you will de-energize equipment before you begin working on it. Your employer will have written guidelines that you must follow to de-energize the equipment. It is important to follow all guidelines because some circuits receive power from multiple sources. To place equipment in a safe work condition, all energy sources must be removed. Once equipment is de-energized, **lockout/tagout (LOTO)** devices must be attached to all sources to prevent someone from unknowingly restoring the energy before the work has been completed.

4.1.0 Electrical Safety Guidelines

Electrical safety is a concern for all workers, not just for electricians. On many jobs, no matter what your trade, you will use or work around electrical equipment. Extension cords, power tools, portable lights, and many other pieces of equipment use electricity.

Not all electrical incidents result in death. There are different types of electrical incidents. Any of the following can happen:

- Burns
- Electric shock
- Explosions
- Falls caused by electric shock
- Fires

If the human body comes in contact with an electrically energized conductor and is in contact with a **ground** at the same time, the body becomes the path of least resistance for the electricity. The electricity flows through the body in less than the blink of an eye without warning. That's why safety precautions are so important when working with and around electrical circuits. When a body conducts electrical current, and the current is high enough, the person can be electrocuted (killed by electric shock). *Table 4* shows the effects of different amounts of electrical current on the human body and lists some common tools that operate using those currents.

Here's an example: A craft worker is operating a portable power drill while standing on damp ground. The power cord inside the drill has become frayed, and the electric wire inside the cord touches the metal drill frame. Three amps of current pass from the wire through the frame, then through the worker's body and into the ground. *Table 4* shows that this worker will probably die.

There are specific policies and procedures to keep the workplace safe from electrical hazards. You can do many things to reduce the chance of an electrical incident. If you ever have any questions about electrical safety on the job site, ask your supervisor.

> **WARNING!**
>
> Less than one amp of electrical current can kill. Always take precautions when working around electricity.

> **Did You Know?**
>
> **Electrocution**
>
> Electric shocks or burns are a major cause of incidents in the construction industry. According to the Bureau of Labor Statistics, electrocution is the fourth leading cause of death among construction workers.

Table 4 Effects of Electrical Current on the Human Body

Current	Common Item/Tool	Reaction to Current
0.001 amps	Watch battery	Faint tingle.
0.005 amps	9-volt battery	Slight shock.
0.006 – 0.025 amps (women) 0.009 – 0.030 amps (men)	Christmas tree bulb	Painful shock. Muscular control is lost.
0.050 – 0.9 amps	Small electric radio	Extreme pain. Breathing stops; severe muscular contractions occur. Death may result.
1.0 – 9.9 amps	Jigsaw (4 amps); Sawsall® or Port-a-Band® saw (6 amps); portable drill (3 – 8 amps)	Ventricular fibrillation and nerve damage occur. Death may result.
10 amps and above	ShopVac® (15-gallon); circular saw	Heart stops beating; severe burns occur. Death may result.

4.1.1 Grounding

Grounding is a method of protecting humans from electric shock; however, it is normally a secondary protective measure. The term *ground* refers to a conductive body, usually the earth. A ground is a conductive connection, whether intentional or incidental, by which an electric circuit or equipment is connected to earth or to an engineered grounding system. By grounding a tool or electrical system, a low-resistance path to the earth is intentionally created. When properly done, this path offers low resistance and has enough current-carrying capacity to prevent the buildup of voltages that could create a personnel hazard. This does not guarantee that no one will receive a shock. It will, however, greatly reduce the possibility of such incidents, especially when used in combination with your company's safety program.

Use three-wire extension cords for portable power tools and make sure they are properly connected (*Figure 46*, top). The three-wire system is one of the most common safety grounding systems used to protect you from incidental electrical shock. The third wire is connected to a ground system. If the insulation in a tool fails, the current will pass to ground through the third wire—not through your body. Double-insulated tools (*Figure 46*, bottom) are also very effective in preventing shocks. In fact, it has become more common to use double-insulated tools because they are safer than relying on a three-wire cord alone. *Figure 46* also shows the double-insulated symbol that can be found on double-insulated tools. Double-insulated tools use a two-wire power cord with no ground pin. One prong of the plug is larger than the other so it can only be connected to a polarized receptacle.

4.1.2 Ground Fault Circuit Interrupters

A **ground fault circuit interrupter (GFCI)** is a fast-acting circuit breaker that senses small imbalances in the circuit caused by current leakage to ground. A GFCI continually matches the amount of current going to an electrical device against the amount of current returning from the device. Whenever the two values differ by more than 5 milliamps, the GFCI interrupts the electric power within $\frac{1}{40}$th of a second. *Figure 47* shows an extension cord with a GFCI.

A GFCI must be used on all receptacles that are not part of the building's permanent wiring, such as temporary power and extension cords. A GFCI provides protection against a ground fault, which is the most common form of electrical shock. It also provides protection from fires, overheating, and wiring insulation deterioration.

Tripping of GFCIs—interruption of circuit flow—is sometimes caused by wet connectors and tools. Limit the amount of water that tools and connectors come into contact with by using watertight or sealable connectors. Tripping may also be caused by cumulative leakage from several tools or from extremely long circuits. GFCIs should be periodically inspected in accordance with the manufacturer's recommendations or site safety practices. GFCIs have a TEST button that can be pressed to verify that the GFCI is working. The GFCI should be tested before any use.

CAUTION

Do not plug a GFCI-protected device into a GFCI-protected circuit.

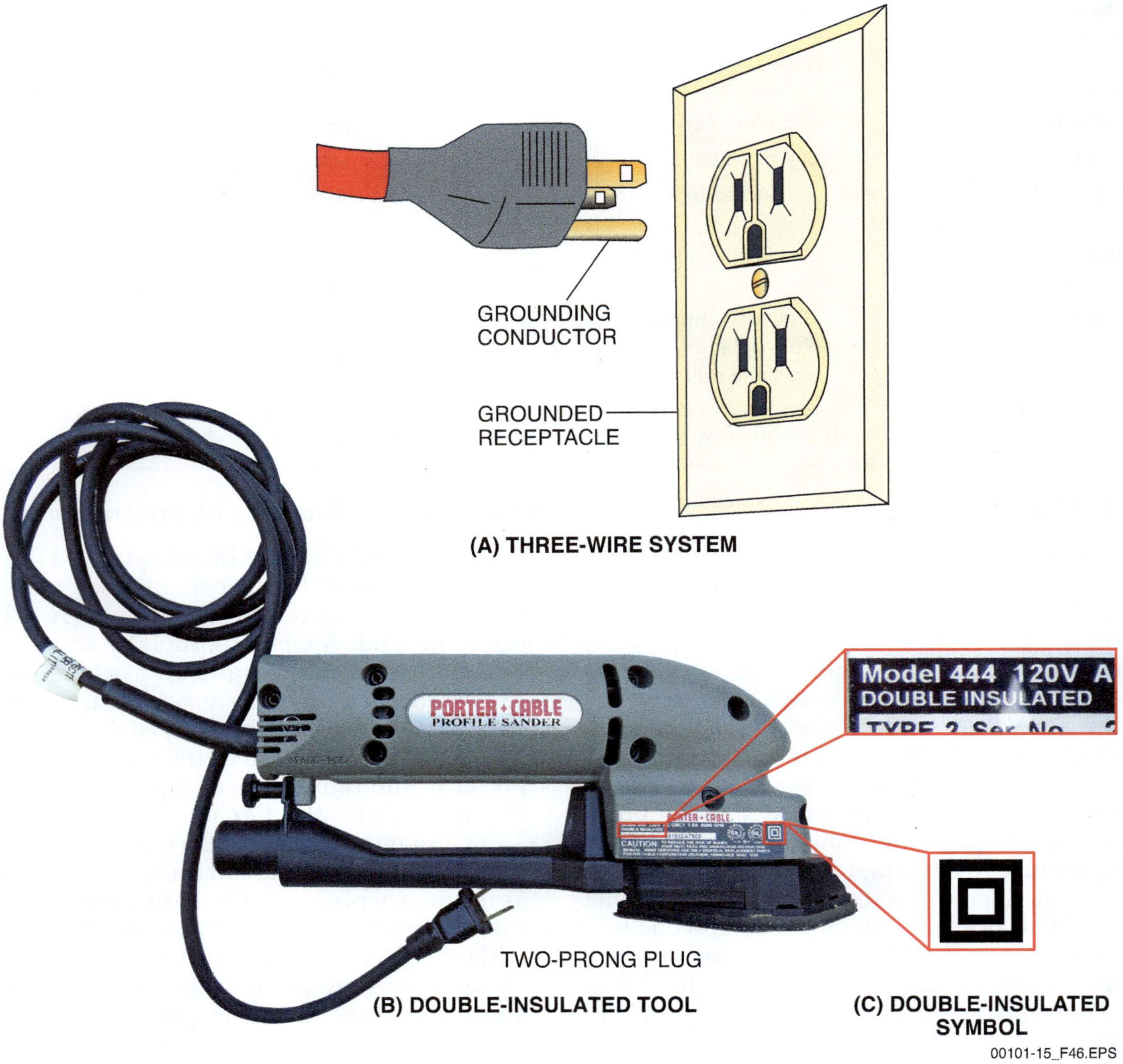

GROUNDING CONDUCTOR

GROUNDED RECEPTACLE

(A) THREE-WIRE SYSTEM

PORTER+CABLE PROFILE SANDER

Model 444 120V A
DOUBLE INSULATED

TWO-PRONG PLUG

(B) DOUBLE-INSULATED TOOL

(C) DOUBLE-INSULATED SYMBOL

00101-15_F46.EPS

Figure 46 Three-wire system, double-insulated tool, and double insulated symbol.

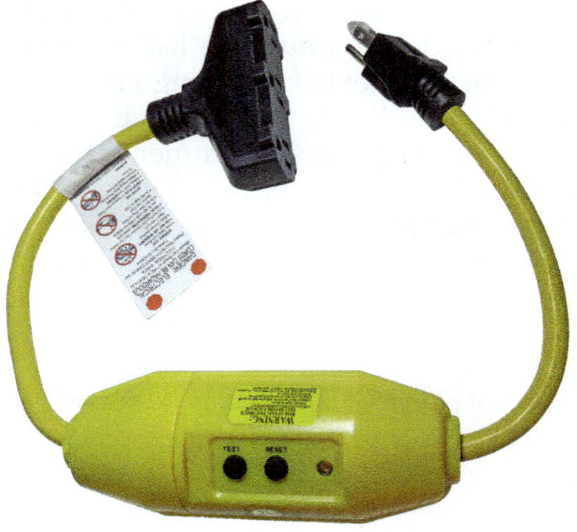

00101-15_F47.EPS

Figure 47 Extension cord with a GFCI.

Case History

A self-employed builder was using a metal cutting tool on a metal carport roof and was not using GFCI protection. The male and female plugs of his extension cord partially separated, and the active pin touched the metal roofing. When the builder grounded himself on the gutter of an adjacent roof, he received a fatal shock.

The Bottom Line: Always use GFCI protection and be aware of potential hazards.

Source: The Occupational Safety and Health Association (OSHA)

4.1.3 Summary of Electrical Safety Guidelines

Here are some basic job-site electrical safety guidelines:

- All power tools used in construction should be ground-fault protected.
- Make sure that panels, switches, outlets, and plugs are grounded.
- Never use bare electrical wire.
- Never use metal ladders near any source of electricity.
- Inspect electrical power tools before you use them.
- Never operate any piece of electrical equipment that has a danger tag or lockout device attached to it.

WARNING!

All work on electrical equipment should be done with circuits de-energized, locked out, and confirmed. All conductors, buses, and connections should be considered energized unless proven otherwise.

- Never use worn or frayed cables (*Figure 48*). If the cord is frayed or worn, disconnect the power and dispose of the cord.
- Make sure light bulbs have protective guards to prevent incidental contact (*Figure 49*).

4.1.4 Working Near Energized Electrical Equipment

No matter what your trade, your job may include working near exposed electrical equipment or conductors. This is one example of **proximity work**. Often, electrical distribution panels, switch enclosures, and other equipment must be left open during construction. This leaves the wires and components in them exposed. Some or all of the wires and components may be energized. Working

00101-15_F49.EPS

Figure 49 Work light with protective guard.

near exposed electrical equipment can be safe, but only if you keep a safe working distance.

Regulations and company policies tell you the minimum safe working distances from exposed conductors. The safe working distance can be very small or span a significant distance, depending on the voltage. The higher the voltage, the greater the required safe working distance.

You must learn the safe working distance for each situation. Make sure you never get any part of your body or any tool you are using closer to exposed conductors than that distance. You can get information on safe working distances from your instructor, your supervisor, company safety policies, and regulatory documents. This subject is also covered in greater detail in other craft curricula where it is relevant.

One of the common causes of electrical shock is coming into contact with overhead wires with metal ladders, cranes, or excavating equipment.

00101-15_F48.EPS

Figure 48 Never use damaged cords.

Case History

A fan connected to a 120-volt electrical system with an extension cord provided ventilation for a worker performing a chipping operation from an aluminum stepladder. The insulation on the extension cord was cut through, exposing bare wire, which made contact with the ladder. The ground wire was not attached on the male end of the cord's plug. When the energized conductor made contact with the ladder, the path to the ground included the worker's body, resulting in death.

A distance of at least 10 feet (3 meters) must be maintained from any conductor carrying 50,000 volts or less. Greater distances are required for higher voltages.

4.1.5 If Someone Is Shocked

If you are there when someone gets an electrical shock, you can save a life by taking immediate action. The best thing to do is to immediately shut off the power to the circuit. Do not touch the victim while he or she remains in contact with the power source. Once the circuit is disconnected, call an ambulance or the emergency number at your job site. Render first aid only if you are trained to do so.

WARNING!

Do not touch the victim or the electrical source with your hand, foot, or any other part of your body or with any object or material. You could become part of the circuit—and another victim.

4.2.0 Lockout/Tagout Requirements

Failure to disable machinery before working on it is a major cause of injury and death on job sites. Lockout/tagout procedures safeguard workers against unexpected releases from various energy sources. An energy source can be electrical, mechanical, hydraulic, pneumatic, chemical, or thermal. After the power has been turned off, energy can still be stored in a device or component. For example, even when a hydraulic system has been turned off, hydraulic pressure may remain in all or part of the system. This high-pressure energy can be released if the system is opened during service or repairs, creating an extremely hazardous situation. Additional dangers exist if the device or system contains chemicals, flammable liquids, high-temperature liquids, or gases.

When anyone is working on or around active systems, all equipment that could release energy must be shut down, drained, de-energized, or otherwise rendered harmless whenever possible. Switches, circuit breakers, valves and other components are switched off or closed, and then locks or tags (*Figure 50*) are applied so they cannot be re-energized while work is ongoing.

Generally, each lock has its own key, and the individual who applies the lock keeps the key. A variety of tags are used, depending on the circumstances and the organization. Tags (*Figure 51*) typically have the word *DANGER* on them, along with other printed information, writing space for comments, and a line for the installer's signature and date. The authorized person who applied the lock must be the one to remove it. In an emergency, a supervisor or other person authorized by the employer may remove a lockout/tagout.

Electrical Cord Safety

Electrical cords are frequently seen on construction sites, yet they are often overlooked. Use the following safety guidelines to ensure your safety and the safety of other workers.

- Every electrical cord should have an Underwriters Laboratory (UL) label attached to it. Check the UL label for specific wattage. Do not plug more than the specified number of watts into an electrical cord.
- A cord set not marked for outdoor use is to be used indoors only. Check the UL label on the cord for an outdoor marking.
- Do not remove, bend, or modify any metal prongs or pins of an electrical cord.
- Do not run a cord through doorways or through holes in ceilings, walls, and floors that might pinch the cord. Also, check to see that there are no sharp corners along the cord's path. Any of these situations will lead to cord damage.
- Extension cords are a tripping hazard. They should never be left unattended and should always be put away when not in use.

(A) ELECTRICAL LOCKOUT

(B) VALVE LOCK

(C) PNEUMATIC LOCKOUT

00101-15_F50.EPS

Figure 50 Lockout/tagout devices.

In a lockout, an energy-isolating device such as a disconnect switch or circuit breaker is placed in the Off position and a lock is applied. Padlocks are popular and versatile, but other styles (*Figure 52*) are often specifically suited for the equipment being serviced. Multiple lockout devices (*Figure 53*) are used when more than one person is accessing the equipment. When a worker needs to service a system and sees that another person's lock has been applied, it is tempting to accept this as a safe situation. However, the lock may be removed by the first worker and the circuit re-energized while the second worker is still vulnerable. In these cases, the multiple lockout devices allow several workers to apply locks, and each is ensured that the system cannot be restored until theirs is removed.

00101-15_F51.EPS

Figure 51 Typical safety tags.

In a tagout, components that control power to equipment and machinery are set to a safe position and a written warning is attached. This method may be more appropriate for valves and other energy-controlling devices that are difficult or impossible to lock, since tags alone cannot prevent a control device or switch from being repositioned.

It is important to emphasize that LOTO procedures are not solely related to electrical energy sources. Technicians cannot forget that other forms of energy may need to be rendered safe as well.

The exact procedures for LOTO will vary by organization and site. Check with your instructor or site supervisor regarding the detailed LOTO procedure at your site and be sure you are familiar with the proper steps.

4.2.1 Pressurized or High-Temperature Systems

Many jobs require that workers be close to tanks, piping systems, and pumps that contain pressurized or high-temperature fluids. Be aware that touching a container of high-temperature fluid can cause burns. Many industrial processes involve fluids that are as hot as several thousand degrees. Also, if a container holding pressurized fluids is damaged, it may leak and spray dangerous fluids. Any work around pressurized or high-temperature systems is considered proximity work. Barricades, a monitor, or both may be needed to ensure the safety of those working nearby.

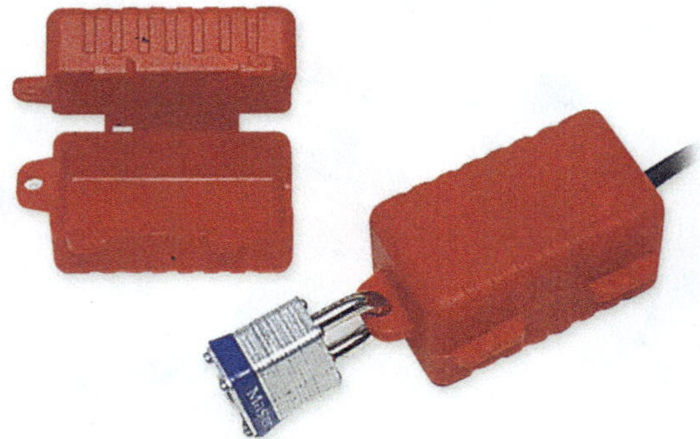

(A) ELECTRICAL PLUG LOCKOUT

(B) CIRCUIT BREAKER LOCK

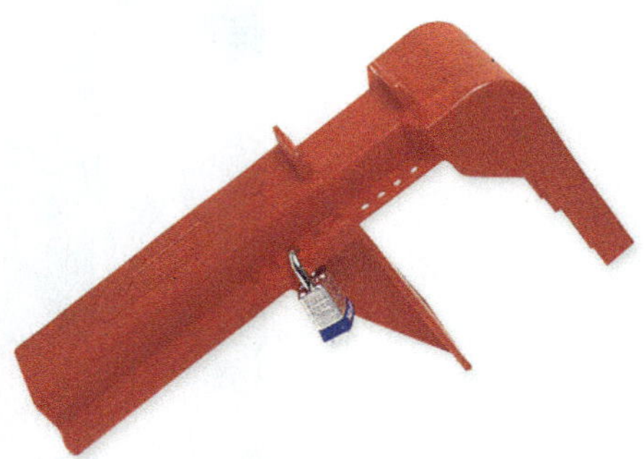

(C) BALL VALVE LOCKOUT

(C) ELECTRICAL SWITCH LOCKOUT

00101-15_F52.EPS

Figure 52 Lockout devices.

00101-15_F53.EPS

Figure 53 Multiple lockout/tagout device.

Additional Resources

US Occupational Safety and Health Administration. Numerous safety videos are available on line at **www.osha.gov/video**.

DeWalt Construction Safety/OSHA Professional Reference, Paul Rosenberg; American Contractors Educational Services. 2006. DEWALT.

Basic Construction Safety and Health, Fred Fanning. 2014. CreateSpace Independent Publishing Platform.

4.0.0 Section Review

1. The purpose of double-insulating an electrically powered tool is to _____.
   a. prevent the tool from overheating
   b. protect users from electrical shock
   c. keep the tool from getting too cold
   d. allow it to use a three-prong receptacle

2. When a lockout is performed, the key to the lock is _____.
   a. left in the lock
   b. kept by the person who applied the tag
   c. given to the foreman for safekeeping
   d. hidden in a convenient place

SECTION FIVE

5.0.0 PERSONAL PROTECTIVE EQUIPMENT

Objectives

Identify and describe the proper use of personal protective equipment.

a. Identify and describe the PPE used to protect workers from bodily injury.
b. Identify potential respiratory hazards and the basic respirators used to protect workers against those hazards.

Performance Tasks

2. Inspect the following PPE items and determine if they are safe to use:
    - Eye protection
    - Hearing protection
    - Hard hat
    - Gloves
    - Approved footwear
3. Properly don, fit, and remove the following PPE items:
    - Eye protection
    - Hearing protection
    - Hard hat
    - Gloves

Trade Terms

Arc welding: The joining of metal parts by fusion, in which the necessary heat is produced by means of an electric arc.

Every worker is responsible for wearing appropriate PPE on the job. When worn correctly, PPE is designed to protect you from injury. You must keep it in good condition and use it when you need to. When workers are injured on the job, it is often because they are not using required PPE.

You will not see all the potentially dangerous conditions just by looking around a job site. It is important to stop and consider what type of incidents could happen on any job that you are about to do. Knowing how to use PPE will greatly reduce your chance of getting hurt. Keep in mind that PPE requirements are usually established by the job site. Some job sites will have different requirements than others, so always make sure you check. For example, some sites require high-top safety boots instead of the more common standard 6-inch tops.

5.1.0 PPE Items

Remember, while PPE is the last line of defense against personal injury, using it properly and taking care of it are the first steps toward protecting yourself on the job.

The first line of defense is represented by the engineering and administrative controls used by employers to eliminate or reduce the need for some PPE. Engineering controls are implemented by making physical changes, such as reducing noise by the use of mufflers on equipment engines; installing guards on moving equipment; and making sure equipment is properly maintained. An example of an administrative control is operating noisy equipment on a shift when fewer workers are present.

The best protective equipment is of no use unless you follow these rules:

- Regularly inspect it.
- Properly care for it.
- Use it properly when it is needed.
- Never alter or modify it in any way.

The sections that follow describe protective equipment commonly used on construction sites, and tell how to use and care for each piece of equipment. Be sure to wear the equipment according to the manufacturer's specifications.

WARNING!

Your clothing must comply with good general work and safety practices. Do not wear clothing or jewelry that could get caught in machinery or otherwise cause an incident, such as loose clothing, baggy shirts, or dragging pants. You must wear a shirt at all times; some tasks will require long-sleeved shirts. Your shirt should always be tucked in unless you are performing welding.

Protective equipment commonly worn by craft workers on any job site usually includes the following:

- Hard hats
- Eye protection
- Gloves

Sharing PPE

It is not a good idea for workers to share PPE. If one person has an infection, it can be passed to another through the shared equipment.

- Safety footwear
- Hearing protection
- Respiratory protection
- High-visibility clothing

5.1.1 Hard Hats

Figure 54 shows a typical hard hat. The outer shell of the hat can protect your head from a hard blow. The webbing inside the hat keeps space between the shell and your head. Adjust the headband so that the webbing fits your head and there is at least one inch of space between your head and the shell.

Do not alter your hard hat in any way. Inspect your hard hat every time you use it. If there are any cracks or dents in the shell, or if the webbing straps are worn or torn, get a new hard hat. Wash the webbing and headband with soapy water as often as needed to keep them clean. Wear the hard hat only as the manufacturer recommends. Never wear anything but approved clothing under your hard hat.

WARNING!

No articles should be worn under the hard hat that could interfere with fit and visibility. That includes ball caps or hoodies that obscure peripheral vision. Only employer-approved gear is to be worn under the hard hat.

00101-15_F54.EPS

Figure 54 Typical hard hat.

5.1.2 Eye and Face Protection

Wear eye protection (*Figure 55*) wherever there is even the slightest chance of an eye injury. In general, eye protection is required any time you are on a job site. Face shields (*Figure 56*) are added over safety glasses for certain tasks, such as grinding. Appropriate eye or face protection is required when there is a possibility of exposure to hazards from flying particles, molten metal, liquid chemicals, acids or caustic liquids, chemical gases or vapors, or potentially injurious light radiation. The following are examples of tasks requiring eye protection:

- Grinding and chipping metal
- Using power saws and other tools/equipment that can throw out solid material

Did You Know?

Hard Hats

Hard hats were once made of metal. However, metal conducts electricity, so most hard hats are now made of reinforced plastic or fiberglass.

Case History

A training specialist traveled nearly three hours to reach a remote job site in order to observe and photograph the installation of high-voltage transmission lines. On arriving, he parked his car, put on his work boots, hard hat, reflective vest and safety glasses and walked over to the location where the work was being done. He was introduced to the project manager who shocked him by saying: "Sir, I have to ask you to leave my job site because you are not wearing electrically insulated boots." Given no choice, the training specialist headed toward his car for the long trip home. Fortunately, the project manager needed to go back to the office for a meeting and offered the loan of his boots, thus preventing six hours of wasted time, as well as the loss of a rare opportunity.

The Bottom Line: Project managers and superintendents in construction and industrial work are serious about safety. If you are not fully prepared, you may find yourself sitting on the sidelines. A worker who shows up at a job site without the required PPE could lose a day's pay and possibly face disciplinary action.

(B) TINTED SAFETY GLASSES

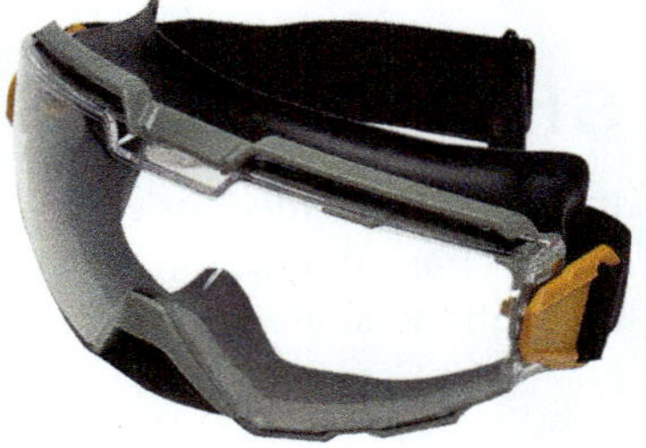

00101-15_F55.EPS

Figure 55 Typical safety glasses and goggles.

(B) TINTED FACE SHIELD

00101-15_F56.EPS

Figure 56 Full-face shields.

- Working with molten lead, tar pots, and other molten materials
- Working with chemicals, acids, and corrosive liquids
- Arc welding

On the job site, potential eye hazard areas are usually identified, but always be on the lookout for possible hazards.

> **CAUTION**
>
> Standard eyeglasses are not adequate protection in high-hazard work zones. Shatter-proof lenses and side shields are required for eye protection in those areas.

Regular safety glasses will protect you from falling objects or from objects flying toward your face. Side shields provide added protection and are required whenever there is risk of flying debris. Safety goggles provide the best protection from all directions. A face shield is required, in addition to safety glasses or goggles, when there is likely to be flying debris. Grinding and chipping activities are good examples of such activities.

Handle your safety glasses and goggles with care. If they get scratched, replace them; the scratches will interfere with your vision. Clean the lenses regularly with lens tissues or a soft cloth.

Welders must use tinted goggles or welding hoods. The tinted lenses protect the eyes from the bright welding arc or flame. Welders must use filter lenses as specified by job site requirements. Welder's helpers and all employees working in the vicinity of arc welding should not look directly at the welding process and are also required to use eye protection with the prescribed level of shading. Oxyacetylene welding and burning also require the use of a filter lens.

Follow these general precautions for eye care:

- Always report all eye injuries and suspected foreign material in your eye to your supervisor immediately. Do not try to remove foreign matter yourself.
- Keep your hands away from your eyes.

- Keep materials out of your eyes by regularly clearing debris from your hard hat brim, the top of your goggles, and your face shield. When removing a cap, hard hat, or face shield, be careful to remove it in a way that prevents accumulated dirt and debris from falling into your eyes.
- Flood your eyes with water if you feel something in them. Never rub them, as this can make the problem worse.
- Know the location of eyewash stations and how to use them.

5.1.3 Hand Protection

On many construction jobs, you must wear heavy-duty gloves to protect your hands (*Figure 57*). Construction work gloves are usually made of cloth, canvas, or leather. Make sure your work gloves are a good fit. Wearing gloves that are too big for your hands can lead to injury. Never wear gloves around rotating or moving equipment. They can easily get caught in the equipment. Replace gloves when they become worn, torn, or soaked with oil or chemicals.

Gloves help prevent cuts and scrapes when you handle sharp or rough materials. Heat-resistant gloves are needed for handling hot materials. Craftworkers should be familiar with standards governing hand protection, such as ANSI/SEA Standard 105-2011, *Hand Protection Selection Criteria* and European Standard EN388, *Protective Gloves Against Mechanical Risks*. These standards establish a rating system for gloves tested for their resistance to various hazards, including chemical burns, abrasions, cuts, punctures, resistance to heat and flame, and resistance to cold. Additional European standards govern protection from chemicals and microorganisms (EN374); thermal hazards (EN407 and EN511); and ionizing radiation and radioactive contamination (EN421).

Electricians use special rubber-insulated gloves when they work on or around live circuits. When working with solvents or other chemicals, it may be necessary to wear chemical-resistant gloves.

> **WARNING!**
>
> Only specially trained employees are allowed to use dielectrically tested rubber gloves to work on energized equipment. Never attempt this work without proper training and authorization.

5.1.4 Foot and Leg Protection

Approved safety footwear must be worn on all job sites. The best shoes to wear on a construction site are shoes with a safety toe to protect toes from falling objects (*Figure 58*). This type of shoe is generally required on every construction-related job site.

Some specialized work will require different footwear or gear. For example, safety-toed rubber boots are needed on job sites that are subject to chemically hazardous conditions or standing water. When climbing a ladder, your shoes or boots must have a well-defined heel to prevent your feet from slipping off the rungs. You will

(A) FITTED WORK GLOVES

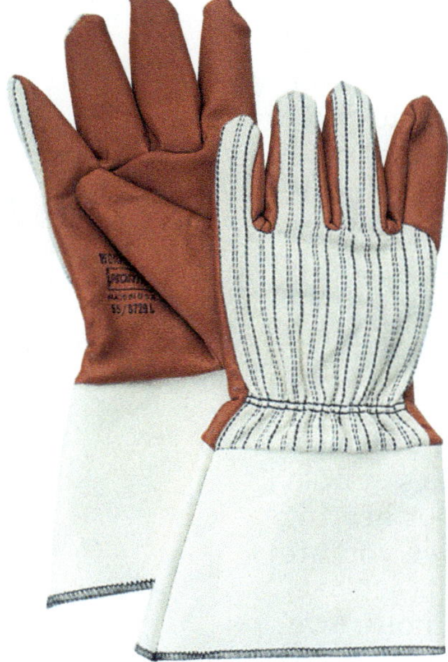

(B) GLOVES WITH CUFFS

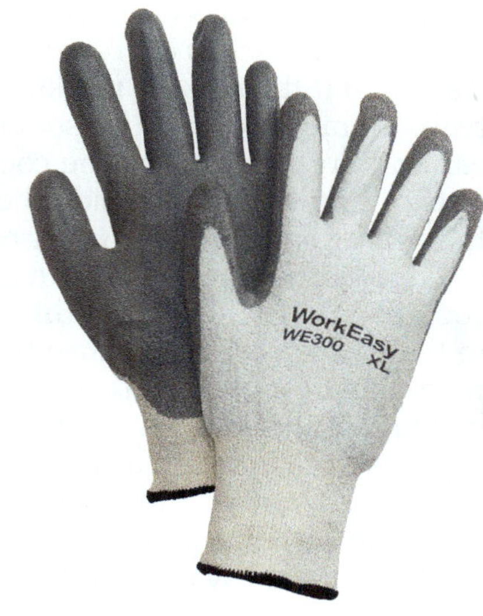

(C) RUBBERIZED COTTON

00101-15_F57.EPS

Figure 57 Work gloves.

00101-15_F58.EPS

Figure 58 Safety shoe.

need foot guards when using jackhammers and similar equipment.

Never wear canvas shoes or sandals on a construction site. They do not provide adequate protection. Always replace boots or shoes when the sole tread becomes worn or the shoes have holes, even if the holes are on top. Because of the risk of fire, do not wear oil-soaked shoes when you are welding or cutting metal.

Remember these general guidelines relating to leg protection:

- Never carry pointed tools, such as scissors or screwdrivers, in your pants pockets. Use a canvas or leather tool sheath with all sharp ends pointing down.
- When using certain special equipment, such as chainsaws and brush hooks, use shin guards.
- Consider stability when stepping into or onto locations where materials are stored. Materials may shift and pinch your legs and/or feet.

5.1.5 Hearing Protection

Unlike damage to most parts of the body, ear damage does not always cause pain. Exposure to loud noise over a long period of time can cause hearing loss, even if the noise is not loud enough to cause pain. Vibrations caused by sound waves enter the ear and are received by the cochlea, which consists of tiny hair cells inside the ear that convert the vibrations into sound signals. Exposure to intense sound waves over time can damage the cochlea and reduce the ability to hear. Damage to the cochlea can also cause tinnitus, which is a permanent condition characterized by ringing or buzzing in the ears that never stops. Hearing loss reduces your quality of life and makes simple, daily tasks more complicated. Save your hearing by using hearing protection whenever you are working in a noisy environment.

Here is a good rule to follow: If the noise level is so great that you have to raise your voice to be heard by someone who is less than 2 feet (61 cm) away, you need to wear hearing protection.

Most construction companies follow rules defined in official safety standards in deciding when hearing protection must be used. One type of hearing protection is specially designed earplugs that fit into your ears and filter out noise (*Figure 59*, top). Another type of hearing protection is earmuffs, which are large padded covers for the entire ear (*Figure 59*, bottom). The headband on earmuffs must be adjusted for a snug fit. If the noise level is very high, you may need to wear both earplugs and earmuffs.

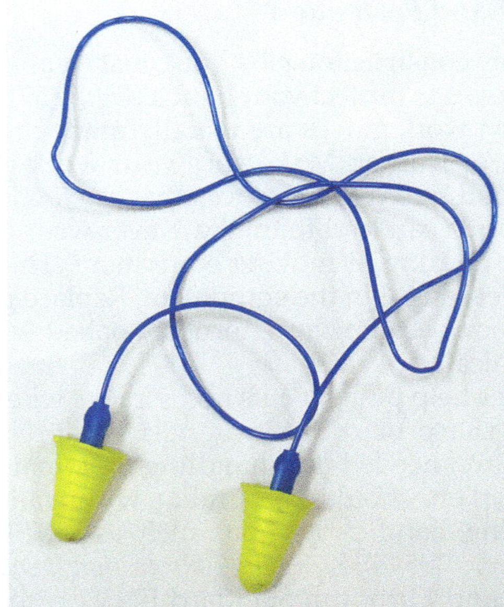

(A) EAR PLUGS

(B) EAR MUFFS

00101-15_F59.EPS

Figure 59 Hearing protection.

Noise-induced hearing loss can be prevented by using noise control measures and personal protective devices. *Table 5* shows the recommended maximum length of exposure to sound levels rated 90 decibels and higher. When noise levels exceed those outlined in *Table 5*, an effective hearing conservation program is required. A company-appointed program administrator will oversee this program. If you have questions about the hearing conservation program on your site, see your supervisor or the program administrator.

5.2.0 Respiratory Hazards and Protection

Wherever there is danger of an inhalation hazard, you must use a respirator. There are a number of job-site conditions that require workers to wear respirators, including the following:

- Dust from metal grinding
- Toxic fumes from welding or flame cutting of some metals
- Working with cleaning solvents
- Working in low-oxygen environments such as confined spaces
- Spray painting
- Sand blasting
- Drilling concrete
- Working with chemicals such as chlorine or ammonia

Silica is a mineral found in concrete, masonry, and rock. During construction, silica may be found in a dust form. Prolonged exposure to silica dust could cause silicosis and lung cancer. Silicosis is an incurable, and sometimes fatal, lung disease. The time it takes for silicosis to develop varies depending on how long the exposure lasted and how much silica the person was exposed to. When you are working in an area where silica dust is present, you must use the appropriate respiratory protection. Never work around silica dust without proper training, authorization, and PPE.

Table 5 Maximum Noise Levels

Sound Level (decibels)	Maximum Hours of Continuous Exposure per Day	Examples
90	8	Power lawn mower
92	6	Belt sander
95	4	Tractor
97	3	Hand drill
100	2	Chain saw
102	1.5	Impact wrench
105	1	Spray painter
110	0.5	Power shovel
115	0.25 or less	Hammer drill

Asbestos is another hazardous material that can be harmful to your lungs. Prolonged exposure can cause lung cancer, asbestosis (scarring of the lung tissue), and a cancer called mesothelioma. It may take more than 20 years for these diseases to develop. Smoking significantly increases the risk of lung disease. If you smoke, tar from cigarettes will stick to the asbestos fibers in the lungs, making it more difficult for your body to get rid of the asbestos.

Sources of asbestos include older insulation and other building materials, such as floor tiles, pipe insulation drywall compounds, and reinforced plaster, along with some roofing and siding materials. The United States banned production of most asbestos products in the 1970s, meaning that asbestos-containing materials (ACMs) are generally found in structures built before 1980. During renovation or maintenance operations, asbestos may be dislodged and the fibers become airborne. Asbestos fibers are particularly hazardous and must not be disturbed unless special techniques and procedures are used to remove it safely. Although you will not feel the effects immediately, this exposure can be a serious health hazard. Never handle asbestos without proper training, authorization, and PPE.

Around the World

Asbestos Use

Asbestos was identified as a health risk many years ago. However, countries have reacted to the hazard in different ways. While some countries banned its use in products and construction decades ago, other countries, such as India, continue to use asbestos without regulation of any form. Indeed, asbestos use in India has risen over 80 percent in the last decade.

The World Health Organization (WHO) estimates that 125 million people worldwide are exposed to asbestos each year in the workplace. Even in the United States, asbestos may still be encountered in old buildings, flooring, and insulation components. Be aware of the regulations concerning the use of asbestos In your region and stay alert to sources that might do you harm if not avoided.

Only workers who have been trained and licensed are authorized to perform work related to asbestos including, but not limited to, the following tasks:

- Demolition
- Removal
- Alteration
- Repair
- Maintenance
- Installation
- Cleanup
- Transportation
- Disposal
- Storage

If asbestos is encountered or suspected on the job site, a supervisor must be notified and all work must stop until the asbestos is either sealed off or removed by licensed professionals.

5.2.1 Types of Respirators

Federal law specifies which type of respirator to use for different types of hazards. There are four general types of respirators (*Figure 60*):

- Self-contained breathing apparatus (SCBA)
- Supplied air mask
- Full facepiece mask with chemical canister (gas mask)
- Half mask or mouthpiece with mechanical filter

A SCBA carries its own air supply in a compressed air tank. It is used where there is not enough oxygen or where there are dangerous gases or fumes in the air.

A supplied-air mask uses a remote compressor or air tank to provide breathable air in oxygen-deficient atmospheres. Supplied-air masks can generally be used under the same conditions as SCBAs.

A full-facepiece mask with chemical canisters is used to protect against brief exposure to dangerous gases or fumes. A half mask or mouthpiece with a mechanical filter is used in areas where you might inhale dust or other solid particles.

5.2.2 Wearing a Respirator

If you need to use a respirator, your employer must provide you with the appropriate training to select, test, wear, and maintain this equipment. Your employer is also responsible for having you medically evaluated to ensure you are fit enough to wear respiratory protection equipment without being harmed. You must also be tested to ensure a proper fit and a good seal.

Wearing a respirator generally places a burden on the employee. A self-contained breathing apparatus is heavy and may be difficult for some workers to carry; negative-pressure respirators restrict breathing; and other respirators may cause claustrophobia. For these and other reasons, workers must be medically evaluated by a physician or other licensed health care professional to determine under what conditions they may safely wear respirators.

Respirators are ineffective unless properly fit-tested to the user. To obtain the best protection from a respirator, perform positive (breathing out) and negative (breathing in) fit checks each time it is worn. These fit checks must be repeated until a good face seal is obtained. A respirator must be clean, in good condition, and all of its components must be in place in order to provide adequate protection.

When a respirator is required, a personal monitoring device is usually also required. This device samples the air to measure the concentration of hazardous chemicals.

5.2.3 Selecting a Respirator

Follow company and government procedures when choosing the type of respirator for a particular job. Also be sure that it is safe for you to wear a respirator. Under current regulations, workers must fill out a questionnaire to identify potential problems in wearing a respirator. Depending on the answers, a medical exam may be required. When a respirator is not required, workers may voluntarily use a dust or particle mask for general protection. These masks do not require fit testing or a medical examination.

Always use the appropriate respiratory protective device for the hazardous material involved and the extent and nature of the work to be performed. Before using a respirator, you must determine the type and concentration level of the contaminant, and whether the respirator can be properly fitted on your face. If you wear contact lenses, practice wearing a respirator with your contact lenses to see whether you have any problems. That way, you will identify any problems before you use the respirator under hazardous conditions.

CAUTION

All respirator instructions, warnings, and use limitations contained on each package must be read and understood by the wearer before use.

(A) SELF-CONTAINED
BREATHING APPARATUS (SCBA)

(B) SUPPLIED AIR MASK

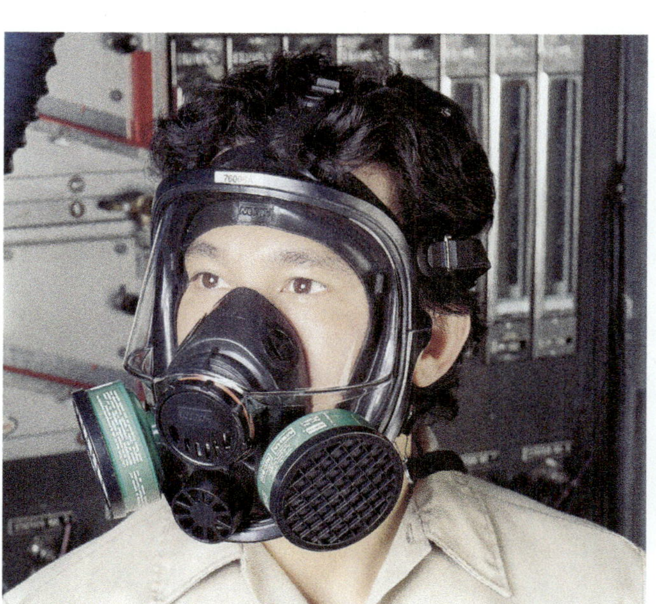

(C) FULL FACEPIECE MASK

(D) HALF MASK

00101-15_F60.EPS

Figure 60 Examples of respirators.

Additional Resources

US Occupational Safety and Health Administration. Numerous safety videos are available on line at **www.osha.gov/video**.

Construction Safety, Jimmie W. Hinze. 2006. Upper Saddle River, NJ: Pearson Education, Inc.

DeWalt Construction Safety/OSHA Professional Reference. Paul Rosenberg; American Contractors Educational Services. 2006. DEWALT.

Basic Construction Safety and Health, Fred Fanning. 2014. CreateSpace Independent Publishing Platform.

5.0.0 Section Review

1. Personnel working in the vicinity of arc welding work must wear eye protection with tinted lenses.
   a. True
   b. False

2. Drilling concrete requires use of a respirator because it produces _____.
   a. asbestos
   b. lead
   c. toxic fumes
   d. silica dust

SECTION SIX

6.0.0 JOB-SITE HAZARDS

Objectives

Identify and describe specific job-site safety hazards.

a. Identify various exposure hazards commonly found on job sites.
b. Identify hazards associated with environmental extremes.
c. Identify hazards associated with hot work.
d. Identify fire hazards and describe basic firefighting procedures.
e. Identify confined spaces and describe the related safety considerations.

Trade Terms

Brazing: A process using heat in excess of 800°F (427°C) to melt a filler metal that is drawn into a connection. Brazing is commonly used to join copper pipe.

Flash burn: The damage that can be done to eyes after even brief exposure to ultraviolet light from arc welding. A flash burn requires medical attention.

Flash point: The temperature at which fuel gives off enough gases (vapors) to burn.

Permit-required confined space: A confined space that has been evaluated and found to have actual or potential hazards, such as a toxic atmosphere or other serious safety or health hazard. Workers need written authorization to enter a permit-required confined space. Also see *confined space*.

Qualified person: A person who, by possession of a recognized degree, certificate, or professional standing, or by extensive knowledge, training, and experience, has demonstrated the ability to solve or prevent problems relating to a certain subject, work, or project.

Welding curtain: A protective screen set up around a welding operation designed to safeguard workers not directly involved in that operation.

Wind sock: A cloth cone open at both ends mounted in a high place to show which direction the wind is blowing.

It is impossible to list all the hazards that can exist on a construction or industrial job site. This section describes some of the more common hazards and explains how to deal with them. For your safety, you must know the specific hazards where you are working and how to prevent incidents and injuries. If you have questions specific to a job site, ask your supervisor.

6.1.0 Job-Site Exposure Hazards

The term *exposure* refers to contact with a chemical, biological, or physical hazard. Exposure can be chronic or acute. Chronic exposure is long-term and repeated, and may be mild or severe. Acute exposure is short-term and intense.

According to the HAZCOM standard, your employer must inform you of any hazards to which you might be exposed. In order to better protect you, your employer may require pre-employment and periodic medical examinations to ensure that your health is not being negatively affected by chemical exposure during work. Exams are usually conducted annually, and generally involve targeted organ testing. Because most medical tests cannot directly check for specific chemical exposure, blood tests are used to check the status of the particular organs (called target organs) known to be most affected by the chemicals to which a person has been exposed.

The results of these tests are compared to pre-employment test results and any other annual or periodic exams you may have had. Employers also frequently request a final screening when a worker leaves the company.

Workers can be exposed to chemicals and other hazardous materials in a variety of ways. Routes of exposure include breathing (inhalation), eating or drinking (ingestion), and skin contact (absorption). Chemical exposure is subject to permissible exposure limits (PELs). In other words, some level of exposure to certain chemicals and fumes is considered acceptable and does not represent a hazard. A PEL is the maximum concentration of a substance that a worker can be exposed to in an eight-hour shift. When a worker approaches or reaches the PEL, he or she must be removed from that environment.

There are various types of hazardous materials you may come in contact with during various types of construction work, including the following:

- Lead
- Bloodborne pathogens
- Chemicals

Taking the time to understand each of these hazards will help you stay healthy and safe. Working around these hazards requires special training and PPE. Never handle any of these materials without proper training, authorization, and PPE.

6.1.1 Lead

Lead occurs naturally in the Earth's crust and is spread through human activities, such as mining and the burning of fossil fuels. As an element, lead is indestructible—it does not break down. Once it is released into the environment, it can move from one medium to another. For example, lead in dust can be carried long distances, dissolve in slightly acidic water, and find its way into soil where it can remain for years. Lead is difficult to detect because it has no distinctive taste or smell.

Lead has many useful properties. It is soft and easily shaped, durable, resistant to some chemicals, and fairly common. Because lead is so versatile, it is used in the production of piping, batteries, and casting metals. However, lead is a toxic metal that can cause serious health problems. You can be exposed to lead by breathing air, drinking water, eating food, and swallowing or touching dust or dirt that contains lead.

You may encounter lead-based paints during demolition or renovation of structures built before 1978, when lead-based paint was banned. Dust created from sanding lead-based paints is hazardous. Protective clothing and equipment must always be used when lead levels are above the PEL. If you are unsure whether lead is present, consult your supervisor.

All waste contaminated with lead is considered hazardous. Those who handle hazardous waste must have special training. Never handle hazardous materials or waste without proper training, authorization, and PPE.

6.1.2 Bloodborne Pathogens

Bloodborne pathogens are another health hazard you may encounter on the job site. Sometimes you will hear them referred to as bloodborne infectious diseases. They are diseases that can be transmitted by contact with an infectious person's blood or other bodily fluids, the most common being HIV (the virus that causes AIDS) and hepatitis B and C.

You could be exposed to another person's blood on the job site when administering first aid to an injured person or being involved in a multi-victim incident. In order to safeguard yourself, you must know the universal precautions to prevent exposure and follow them closely:

- Always use appropriate gloves, eye protection, and a mask when administering first aid.
- If you come in contact with someone's blood, immediately wash the affected areas with soap and water. Otherwise, use antiseptic hand cleanser or antiseptic towelettes until washing with soap and running water is possible.
- Notify your supervisor of the contact right away.

You may need to seek medical attention for precautionary screening, particularly if that person has any bloodborne infectious diseases. Your employer should have a written exposure control plan and procedure for such occurrences.

6.1.3 Chemical Splashes

Chemicals such as acids and solvents can pose physical hazards, including acid reactions or burns. Acids can create toxic vapors or react violently when mixed with other chemicals. Other chemicals are flammable, combustible, or explosive. For example, solvents and compressed gases are often flammable. These materials can catch fire in their liquid or gaseous state. Some materials, such as blasting caps, are dangerous solids. These items should be kept away from open flames, sparks, intense heat, or other ignition sources to prevent fires or explosions. Know the physical hazards of the chemicals being used in order to prevent fires and chemical reactions.

Workers should always check the SDS for a chemical before working with or around it. That way, you will be prepared to take appropriate corrective action if you or another worker is exposed.

WARNING!

Chemical splashes can become medical emergencies. Before working with any chemicals, ensure that you know where the nearest shower and eyewash station are, and verify that they are functioning properly.

Remember to wear prescribed clothing and PPE on the job site at all times. This will help guard against your skin and eyes being splashed with chemicals. If you are not sure of the PPE needed on your job site or for a certain task, consult your supervisor before beginning work.

> **WARNING**
>
> Never handle any chemicals without proper training and PPE. Review the SDS for each chemical, and strictly follow SDS first aid instructions if you suspect exposure.

6.1.4 Container Labeling

On a construction site, any material in a container must have a label. Labels describe what is in a container and warn of chemical hazards. The HAZCOM Standard states that hazardous material containers must be labeled, tagged, or marked. The label must include the name of the material, the appropriate hazard warnings, and the name and address of the manufacturer. On December 1, 2015, OSHA will require that all shipments be labeled as outlined by the Global Harmonization System (GHS), which has been adopted as a worldwide standard to enhance awareness and improve the safety and health of workers exposed to these substances.

In today's global marketplace, chemicals and similar hazardous materials are in constant movement between countries. With many countries having their own hazardous material labeling system, products must often be labeled multiple times to support the systems of both the shipping and receiving countries. The GHS will solve this problem by providing some standardized labeling for various materials. There are nine pictograms that workers must become familiar with are shown here. Those that are related to flammable, combustible, and explosive materials are shown in *Figure 61*, and those that represent other hazards are shown in *Figure 62*.

Pictograms have been in use for a number of years to depict a variety of hazards or to communicate information. The advantage of pictograms is that they cross any language barrier. Regardless of language, pictograms such as those used in the GHS tell the user what type of hazard to be concerned about. Although other systems of container labeling will be encountered for years to come, the pictograms all share common features and visually represent the hazard of concern.

If a material is transferred from a labeled container to a new container, the new container must be labeled with all of the information from the original label. Make sure that any materials being worked with are labeled. Be sure that you understand your company's labels.

> **Process Safety Management**
>
> One way to avoid the hazards associated with chemical spills is to evaluate those hazards and plan ahead for dealing with unwanted releases. Forward-thinking companies adopt a method known as process safety management for this purpose. Process safety management involves training employees and contractors; making information available; establishing formal procedures; analyzing potential hazards; and taking steps to mitigate the hazards.

> **WARNING!**
>
> Never use chemicals from an unlabeled container.

FLAME

- FLAMMABLES
- PYROPHORICS
- SELF-HEATING
- EMITS FLAMMABLE GAS
- SELF-REACTIVES
- ORGANIC PEROXIDES

FLAME OVER CIRCLE

- OXIDIZERS

EXPLODING BOMB

- EXPLOSIVES
- SELF-REACTIVES
- ORGANIC PEROXIDES

00101-15_F61.EPS

Figure 61 GHS labels for flammable, combustible, and explosive materials.

HEALTH HAZARD

- CARCINOGEN
- MUTAGENICITY
- REPRODUCTIVE TOXICITY
- RESPIRATORY SENSITIZER
- TARGET ORGAN TOXICITY
- ASPIRATION TOXICITY

EXCLAMATION MARK

- IRRITANT (SKIN AND EYE)
- SKIN SENSITIZER
- ACUTE TOXICITY (HARMFUL)
- NARCOTIC EFFECTS
- RESPIRATORY TRACT IRRITANT
- HAZARDOUS TO OZONE LAYER (NON-MANDATORY)

GAS CYLINDER

- GASES UNDER PRESSURE

CORROSION

- SKIN CORROSION/ BURNS
- EYE DAMAGE
- CORROSIVE TO METALS

ENVIRONMENT (NON-MANDATORY)

- AQUATIC TOXICITY

SKULL AND CROSSBONES

- ACUTE TOXICITY (FATAL OR TOXIC)

00101-15_F62.EPS

Figure 62 GHS labels for various materials.

6.1.5 Radiation Hazards

Radioactive materials are used in construction during radiographic testing of welds in piping, vessels, and pumps. These hazards are also found at nuclear power plants. Radioactive materials must be properly labeled and warning signs must be posted in the work area. Only trained workers are allowed in these areas. If you see signs containing symbols like those shown in *Figure 63*, stay away from that area unless properly trained.

Radioactive materials are a special type of physical hazard. Excessive radiation exposure can cause skin burns, nausea, vomiting, infertility, and cancer. Employees can minimize radiation exposure by limiting the amount of time they are exposed and/or increasing their distance from the source. Always use the proper shielding or PPE. Check with your supervisor to find out if there are any radioactive hazards and how to avoid them.

6.1.6 Biological Hazards

Biological hazard signs are used to warn workers of the actual or potential presence of a biological hazard (*Figure 64*). Biological hazards, or biohazards, can be any infectious agents that create a real or potential health risk. Biological hazard signs are commonly used to identify contaminated equipment, containers, rooms, materials, and areas housing experimental animals.

00101-15_F63.EPS

Figure 63 Radioactivity warning symbols.

Background colors on signs may vary, but must contrast enough for the symbol to be easily identified. Wording is used on the sign to indicate the nature of the hazard or to identify the specific hazard.

6.1.7 Evacuation

In many work environments, specific evacuation procedures are needed. These procedures go into effect when dangerous situations arise, such as fires, chemical spills, and gas leaks. In an emergency, you must know the evacuation procedures. You must also know the signal (usually a horn or siren) that tells workers to evacuate.

When the evacuation signal sounds, follow the evacuation procedures precisely. That usually means taking a certain route to a designated assembly area and telling the person in charge that you are there. If hazardous materials are released into the air, you may have to determine which way the wind is blowing. Some sites may have a wind sock that indicates the wind direction. Different evacuation routes are planned for

00101-15_F64.EPS

Figure 64 Biological hazard symbol.

different wind directions. Taking the right route will keep you from being exposed to the hazardous material.

6.2.0 Environmental Extremes

Workers on construction and industrial sites are often required to work outdoors in extremes of heat and cold. It is important to know how to protect yourself in such environments and to recognize the symptoms of conditions that could be caused by environmental extremes.

6.2.1 Heat Stress

Heat stress occurs when abnormally hot air and/or high humidity, or extremely heavy exertion, prevents your body from cooling itself fast enough. When this happens, you may suffer heat cramps, heat exhaustion, or a heat stroke. To prevent heat stress, take the following precautions:

- Drink plenty of water.
- Avoid alcoholic or caffeinated drinks.
- When possible, perform the most strenuous work during cooler parts of the day.
- Wear lightweight, light-colored clothing.
- Wear loose-fitting cotton clothing if it does not create a hazard.
- Keep your head covered and your face shaded.
- Take frequent, short breaks.
- Rest in the shade whenever possible.

Workers are encouraged to drink 4 cups (≈1 liter) of water per hour when the heat index is 103°F (39.5°C) or higher when working in hot conditions. Start by having one or two glasses of water before beginning work, and then drink one cup (237 milliliters) every 15 to 20 minutes when working in hot conditions. It is possible to drink too much water as well. Water intake should not exceed 6 cups (1.4 l) per hour or 3 gallons (11.4 l) per day. Drinking water should be cool but not cold.

6.2.2 Heat Cramps

Heat cramps are muscular pains and spasms caused by heavy exertion. Any muscles can be affected, but most often it is the muscles that have been used the most. Loss of water and electrolytes from heavy sweating causes these cramps. Symptoms of heat cramps include the following:

- Painful muscle spasms and cramping
- Pale, sweaty skin
- Normal body temperature
- Abdominal pain
- Nausea

As noted above, body temperature is not normally elevated during the onset of heat cramps. An increase in body temperature usually indicates that the problem is escalating to a more serious stage. If you experience heat cramps, take the following steps: move to a cool area, drink some water, and gently stretch and massage cramped muscles.

6.2.3 Heat Exhaustion

Heat exhaustion typically occurs when people exercise heavily or work in a warm, humid place where body fluids are lost through heavy sweating. When it is humid, sweat does not evaporate fast enough to cool the body properly.

Symptoms of heat exhaustion include the following:

- Cool, pale, and moist skin
- Heavy sweating
- Headache, nausea, vomiting
- Dilated pupils
- Dizziness
- Possible fainting
- Fast, weak pulse
- Slight elevation in body temperature

If someone is showing the symptoms of heat exhaustion, get the victim to a shaded area and have him or her lie down. Try to cool the victim by applying cold wet cloths, fanning, removing heavy clothing, and giving small amounts of cool water if the victim is conscious. If the victim's condition does not improve within a few minutes, call emergency medical services.

6.2.4 Heat Stroke

Heat stroke is life threatening. The body's temperature-control system, which produces sweat to cool the body, stops working. Body temperature can rise so high that brain damage and death may result if the body is not cooled quickly.

> **WARNING!**
> If you suspect someone has heat stroke, call emergency medical services immediately.

The symptoms of heat stroke include the following:

- Hot, dry, or spotted skin
- Extremely high body temperature
- Very small pupils
- Mental confusion
- Headache
- Vision impairment
- Convulsions
- Loss of consciousness

While waiting for help to arrive, try to cool the victim using the methods described for heat exhaustion. Ice packs placed in the armpits and the groin should be used, if possible.

6.2.5 Cold Stress

When your body temperature drops even a few degrees below normal, which is about 98.6°F (37°C), you can begin to shiver uncontrollably and become weak, drowsy, disoriented, unconscious, or even fatally ill. This loss of body heat is known as cold stress, or hypothermia.

People who work outdoors during the winter need to learn how to protect against loss of body heat. The following guidelines can help you keep your body warm and avoid the dangerous consequences of hypothermia, frostbite, and overexposure to the cold.

> **WARNING!**
> Always seek immediate medical attention if you suspect hypothermia or frostbite.

Outdoors, indoors, in mild weather, or in cold, it is a good idea to dress in layers. Layering your clothes allows you to adjust what you are wearing to suit changing temperature conditions. In cold weather, wear cotton, polypropylene, or lightweight wool next to your skin, and wool layers over your undergarments. For outdoor activities, choose clothing made of waterproof, wind-resistant fabrics such as nylon. Since a great deal of body heat is lost through the head, always wear a hat for added protection.

Water chills the body far more rapidly than air or wind. Always take along an extra set of clothing whenever working outdoors. If clothes become damp, change to dry clothes to prevent your body temperature from dropping in cold weather. Wear waterproof boots in damp or snowy weather, and always pack raingear even if the forecast calls for sunny skies.

6.2.6 Frostbite

Frostbite is a dangerous condition that can have lifelong effects on your body. It usually affects the hands, fingers, feet, toes, ears, and nose. Symptoms of frostbite include a pale, waxy-white skin color and hard, numb skin.

Remember the following when providing first aid for frostbite:

- Never rub the affected area. This can damage the skin and tissue.
- After the affected area has been warmed, it may become puffy and blister.
- The affected area(s) may have a burning feeling or numbness. When normal feeling, movement, and skin color have returned, the affected area(s) should be dried and wrapped to keep it warm.
- If there is a chance the affected area may get cold again, do not warm the skin. If the skin is warmed and then becomes cold again, it will cause severe tissue damage. Seek medical attention as soon as possible. First aid for frostbite includes moving the victim to a warm area, removing wet clothing, and applying warm water of about 105°F (41°C) to the affected area. Do not pour the warm water directly onto the area, however.

6.2.7 Hypothermia

Hypothermia is a serious, potentially fatal condition caused by loss of body heat. You do not have to be in below-freezing temperatures to be at risk for hypothermia. Hypothermia can happen on land or in water.

The effects of hypothermia can be gradual and often go unnoticed until it is too late. If you know you'll be working outdoors for an extended period of time, work with a buddy. At the very least, let someone know where you'll be and what time you expect to return. Ask your buddy to check you for overexposure to the cold, and do the same for your buddy. Check for shivering, slurred speech, mental confusion, drowsiness, and weakness. If anyone shows any of these signs, call for emergency medical assistance and see that the victim is moved indoors as soon as possible to warm up.

WARNING!

If a co-worker exhibits uncontrolled shivering, slurred speech, clumsy movements, fatigue, and confused behavior, immediately call for medical assistance.

Symptoms of hypothermia include the following:

- A drop in body temperature
- Fatigue or drowsiness
- Uncontrollable shivering
- Slurred speech
- Clumsy movements
- Irritable, irrational, confused behavior

If a person is showing symptoms of hypothermia from exposure to cold air, immediately call for medical assistance. The victim should be moved to a warm, dry area. Any wet clothing should be removed and replaced with dry

Around the World

Temperature Extremes

Some places in the world reach incredibly hot and cold temperatures. Although work may come to a halt in the worst conditions, the job goes on between the extremes. In Chita, Russia for example, the average low temperature in January is roughly –30°C (≈–22°F). The average high during this period is around –17°C (≈1°F). Since these are average temperatures, people in Chita must adapt and continue working. The admiration of the world's workers should go to those who must work in temperature extremes like this without hesitation.

clothing or blankets. If possible, give the victim warm sweet drinks; avoid drinks with caffeine, as well as alcohol. Have the victim move his arms and legs to create muscle heat. If he is unable to do this, place warm bottles or hot packs around the armpits, groin, neck, and head. Do not rub the victim's body or place him in a warm bath. This could cause additional harm.

Body heat is lost up to 25 times faster in water than on land and the methods used to help victims while waiting for medical help to arrive is different. If you are a victim, first of all, do not remove any clothing. Close up and tighten all clothing to create a layer of trapped water close to the body. This provides insulation that slows heat loss. Keep your head out of the water and put on a hat or hood if possible. Get out of the water as quickly as possible, or climb onto a floating object. Do not attempt to swim unless you can reach a floating object or another person. Swimming or other physical activity uses the body's heat and reduces survival time by about 50 percent.

If it is not possible to get out of the water, wait quietly, and conserve your body heat by folding your arms across your chest, keeping your thighs together, bending your knees and crossing your ankles. If another person is in the water, huddle together with your chests pressed tightly together.

6.3.0 Hot Work Hazards

Welding and torch cutting of metals is done using heat that is high enough to melt steel. In addition, most flame cutting and some welding processes are done with flammable gases. For those reasons, such processes are classified as hot work and are subject to special safety precautions. Grinding with a pneumatic or electric grinding machine (*Figure 65*) is also classified as hot work, as it gives off sparks that could cause a fire or explosion in an environment containing flammable gases or concentrated dust. Hot work always carries a risk of severe burns as well as fire hazards. Being aware of these risks is of great importance and will help keep you safe.

6.3.1 Arc Welding Hazards

Arc welding is a process in which metals are joined using a high-intensity electric arc (*Figure 66*) at temperatures in the range of 1,500 to 3,000°F (815 to 1,648°C). The arc melts and fuses the base metals. In most cases, a filler metal is melted along with the base metal to strengthen the joint. Welders are required to wear specialized PPE

00101-15_F65.EPS

Figure 65 Grinding creates sparks.

00101-15_F66.EPS

Figure 66 Arc welding.

that protects their eyes from damage and protects their bodies from burns caused by sparks and molten metal. Such eye protection includes tinted goggles as well as a full face shield with a tinted viewing window. Anyone working in the vicinity of arc welding must also be protected, especially from damage to their eyes caused by looking at the welding arc. Never look at an arc welding operation without wearing the proper tinted eye protection, because the ultraviolet light from the

arc will burn your eyes. Even a reflected arc can harm your eyes. It is extremely important to follow proper safety procedures at and around all welding operations. Serious eye injury or even blindness, as well as burns, can result from unsafe conditions.

When people are working near a welding operation, **welding curtains** (*Figure 67*) must be set up and everyone in the vicinity must wear tinted protective eyewear.

Even a brief exposure to the ultraviolet light from arc welding can cause a **flash burn** and damage your eyes badly. You may not notice the symptoms until sometime after the exposure. Here are some symptoms of flash burns to the eye:

- Headache
- Feeling of sand in your eyes
- Red or weeping eyes
- Trouble opening your eyes
- Impaired vision
- Swollen eyes

Welded material is dangerously hot. Mark it with a sign and stay clear for a while after the welding has been completed.

6.3.2 Oxyfuel Cutting, Welding, and Brazing

Oxygen and acetylene (oxyfuel) gases are combined to produce a flame with a temperature high enough to melt steel. So-called oxyfuel is commonly used to cut and weld steel and to join copper pipe by **brazing**.

Figure 68 shows a worker using an oxyfuel torch to cut steel. Persons working with or around welding, brazing, or metal cutting equipment can be severely burned from the intense heat and flames generated in these processes. In addition, the gases used in these processes pose an additional danger because they are stored under pressure in metal cylinders (*Figure 69*).

Many hazards are involved in the handling, storage, and use of compressed gases. It takes energy to compress and confine the gas. That energy is stored until purposely released to perform useful work or until incidentally released by container failure or other causes.

00101-15_F67.EPS

Figure 67 Welding curtain.

00101-15_F68.EPS

Figure 68 Oxyfuel cutting.

00101-15_F69.EPS

Figure 69 Compressed-gas cylinders used for oxyfuel cutting.

Some gases, such as acetylene, are highly flammable. Oxygen is an explosion hazard if it comes into contact with grease or oil. Flammable compressed gases have additional stored energy besides simple compression-released energy. In case of a fire, for example, escaping oxygen will make the fire more intense. Other compressed gases, such as nitrogen, can cause asphyxiation simply because they displace oxygen.

The cylinders containing oxygen and acetylene must be transported, stored, and handled very carefully. Always follow these safety guidelines:

- Keep the work area clean and free from potentially hazardous items such as combustible materials and petroleum products.

WARNING!

Keep oxygen away from sources of flame and combustible materials, especially substances containing oil, grease, or other petroleum products. Compressed oxygen mixed with oil or grease will explode. Never use petroleum-based products around fittings that serve compressed oxygen lines.

- Use great caution when you handle compressed gas cylinders.
- Store cylinders in an upright position where they will not be struck, where they will be away from corrosives, and where they cannot tip over or fall. Cylinders must be stored in the upright position in a secure container or secured to a wall with chains. When in storage, oxygen and fuel cylinders must be separated by 20 feet (6.1 m) or by a 5-foot (1.5 m) high ½-hour fire-rated barrier (*Figure 70*).

WARNING!

Do not remove the protective cap unless a cylinder is secured. If the cylinder falls over and the nozzle breaks off, the cylinder could shoot off like a rocket, injuring or killing anyone in its path.

6.3.3 Transporting and Securing Cylinders

Always handle cylinders with care. They are under high pressure and should never be dropped, knocked over, rolled, or exposed to heat in excess of 140°F (60°C). When moving cylinders, always be certain that the valve caps are in place. Cylinders must be transported to the workstation in the upright position on a hand truck or bottle cart such as the one shown in *Figure 69*. Oxygen and cutting gases should not be transported on the same cart unless the cart has a divider, as shown. Be aware that some job sites will not permit oxygen and cutting gases to be transported on the same cart under any circumstances.

Never attempt to lift a cylinder using the holes in a safety cap; use an approved lifting cage (*Figure 71*). Make sure that the cylinder is secured in the cage. Cages of various sizes are available for high-pressure cylinders and cylinders containing liquids.

6.3.4 Hot Work Permits

A hot work permit (*Figure 72*) is an official authorization from the site manager to perform work that may pose a fire hazard. The permit includes information such as the time, location, and type of work being done. The hot work permit system promotes the development of standard fire safety guidelines. Permits also help managers keep records of who is working where and at what time. This information is essential in the event of an emergency at times when personnel need to be evacuated.

Most sites require the use of hot work permits and fire watches. When these requirements are violated, severe penalties may be imposed because of the risk it represents. Before a hot work permit is issued, a competent person must inspect the work area to ensure that no flammable or combustible materials are present. This inspection includes the areas above and below where the work will be performed, as well as the surrounding area.

A fire watch is posted when welding or cutting work is being done in many areas. One person other than the welding or cutting operator must constantly scan the work area for fires. Fire watch personnel should have ready access to fire extinguishers and alarms and know how to use them. Welding and cutting operations should never be performed without a fire watch. The area where welding is done must be monitored afterwards until there is no longer a risk of fire. The person on fire watch must be dedicated to that activity and may not perform other work during the time when hot work is being performed.

6.4.0 Fire Hazards and Firefighting

Fire is always a hazard on construction job sites. Many of the materials used in construction are flammable. In addition, welding, grinding, and many other construction activities create heat or sparks that can cause a fire. Fire safety involves two elements: fire prevention and firefighting.

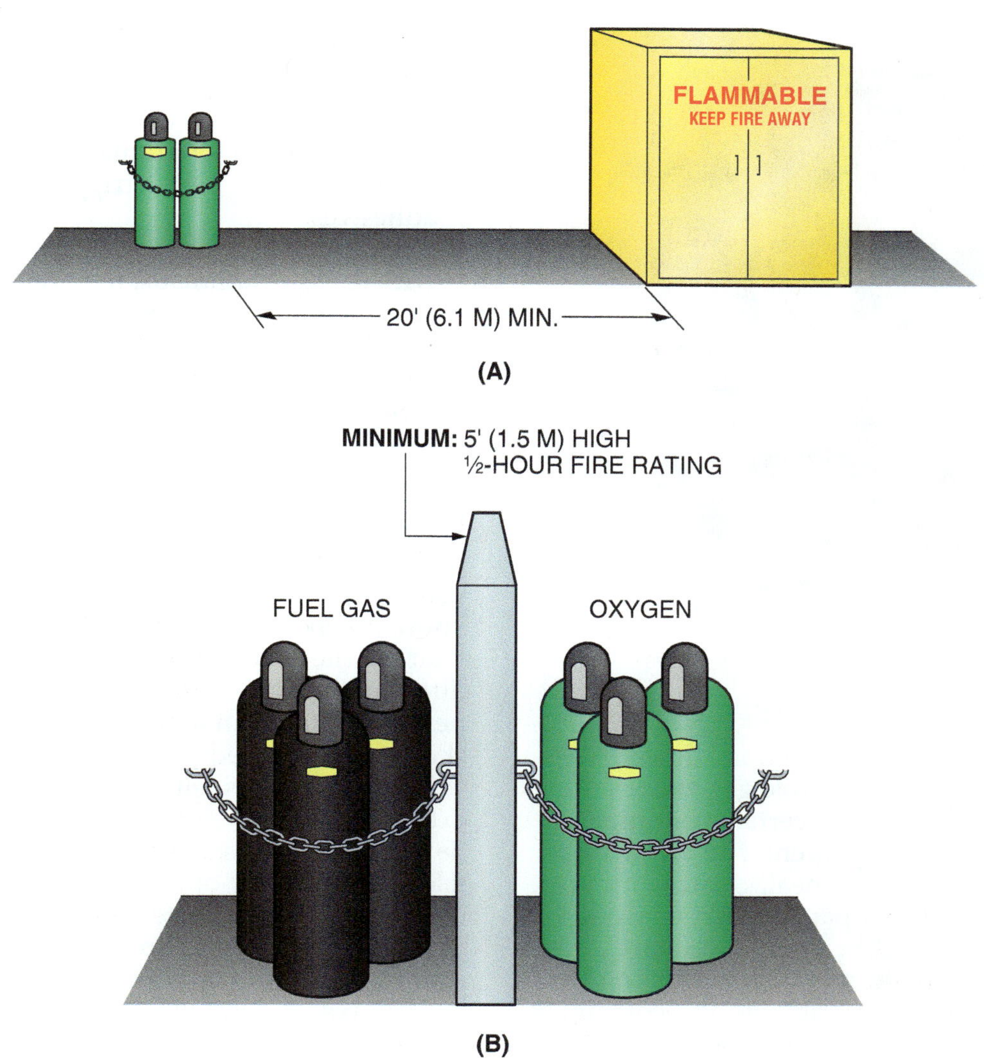

Figure 70 Cylinder storage.

6.4.1 How Fires Start

For a fire to start, three things are needed in the same place at the same time: fuel, heat, and oxygen. If one of these three is missing, a fire will not start.

Fuel is anything that will combine with oxygen and heat to burn. Oxygen is always present in the air. When pure oxygen is present, such as near a leaking oxygen hose or fitting, material that would not normally be considered fuel (including some metals) will burn.

Heat refers to a source of ignition. It may be as simple as a single spark. Heat is anything that will raise a fuel's temperature to the flash point. The flash point is the temperature at which a fuel is encouraged to burn. The flash points of some fuels are quite low—room temperature or less. When the burning gases raise the temperature of a fuel to the point at which it ignites, the fuel itself will burn—and keep burning—even if the original source of heat is removed.

What is needed for a fire to start can be shown as a fire triangle (*Figure 73*). If one element of the triangle is missing, a fire cannot start. If a fire has started, removing any one element from the triangle will put it out.

6.4.2 Combustibles

Combustibles are categorized as liquid, gas, or ordinary combustibles. The term *ordinary combustibles* means paper, wood, cloth, and similar fuels. Liquids can be flammable or combustible. Flammable liquids have a flash point below 100°F (38°C). Combustible liquids have a flash point at or above 100°F (38°C). Flammable gases used on construction sites include acetylene, hydrogen, ethane, and propane. To save space, these gases are compressed so that a large amount is stored in a small cylinder or bottle. As long as the gas is kept in the cylinder, oxygen cannot get to it and start a fire. The cylinders should be stored away

00101-15_F71.EPS

Figure 71 Cylinder lifting cage.

from sources of heat. If oxygen is allowed to escape and mix with a flammable gas, the resulting mixture will explode under certain conditions.

The easiest way to prevent fire in ordinary combustibles is to keep a neat, clean work area. If there are no scraps of paper, cloth, or wood lying around, there will be no fuel to start a fire. Establish and maintain good housekeeping habits. Use approved storage cabinets and containers for all waste and other ordinary combustibles.

6.4.3 Fire Prevention

The best way to ensure fire safety is to prevent a fire from starting. Fire can be prevented by the following actions:

- *Removing the fuel* – Liquid does not usually burn. What generally burns are the gases (vapors) given off as the liquid evaporates. Keeping liquids in an approved, sealed container prevents evaporation. If there is no evaporation, there is no fuel to burn.
- *Removing the heat* – If the liquid is stored or used away from a heat source, it will not be able to ignite.
- *Removing the oxygen* – The vapor from a liquid will not burn if oxygen is not present. Keeping safety containers tightly sealed prevents oxygen from coming into contact with the fuel.

The best way to prevent a fire is to make sure that fuel, oxygen, and heat are never present in the same place at the same time.

The following are some basic safety guidelines for fire prevention:

- Always work in a well-ventilated area, especially when using flammable materials such as shellac, lacquer, paint stripper, or construction adhesives.
- Never smoke or light matches when working with or near flammable materials.
- Keep oily rags in approved, self-closing metal containers.
- Store combustible materials only in approved containers.

6.4.4 Basic Firefighting

You are not expected to be an expert firefighter, but you may have to deal with a fire to protect your safety and the safety of others. You need to know the locations of firefighting equipment on your job site as well as which equipment to use on different types of fires. However, only a **qualified person** is placed in a position where firefighting skills are a required activity.

Most companies tell new employees where fire extinguishers are kept. If you have not been told, be sure to ask. Also, ask how to report fires. The telephone number of the nearest fire department should be clearly posted in your work area. If your company has a fire brigade, learn how to contact them. Learn your company's fire safety procedures. Know what type of extinguisher to use for different kinds of fires and how to use them. Make sure all extinguishers are fully charged. Never remove the tag from an extinguisher—it shows the date the extinguisher was last serviced and inspected.

A fire watch is required to have a fire extinguisher while on duty. A portable extinguisher such as the one shown in *Figure 74* is commonly provided for that purpose.

The function of a fire extinguisher is to remove one of the three elements (oxygen, heat, fuel) needed to sustain a fire. A fire extinguisher cannot remove fuel, so it is designed to remove either heat or oxygen. Heat is removed by using a coolant such as water; oxygen is removed by smothering the fire with dry chemicals or CO_2.

Fire extinguishers are rated for specific types of fires. Class A extinguishers, for example, are intended for fires involving ordinary combustibles such as paper, wood, and fabric. An extinguisher that is rated only for Class A fires might contain water, which could not be used on electrical or grease fires. Using water on a Class B fire (flammable liquids, grease, or gases) could spread the fire or splash burning fuel. Using water on a

HOT WORK PERMIT

For Cutting, Welding, or Soldering with Portable Gas or ARC Equipment

(References: 1997 Uniform Fire Code Article 49 & National Fire Protection Association Standard NFPA 51B.)

Job Date____________ Start Time____________ Expiration____________ WO #____________

Name of Applicant______________________ Company____________ Phone____________

Supervisor__________________________ Phone____________

Location / Description of work ____

IS FIRE WATCH REQUIRED?

1. ______ (yes or no) Are combustible materials in building construction closer than 35 feet to the point of operation?
2. ______ (yes or no) Are combustibles more than 35 feet away but would be easily ignited by sparks?
3. ______ (yes or no) Are wall or floor openings within a 35 foot radius exposing combustible material in adjacent areas, including concealed spaces in floors or walls?
4. ______ (yes or no) Are combustible materials adjacent to the other side of metal partitions, walls, ceilings, or roofs which could be ignited by conduction or radiation?
5. ______ (yes or no) Does the work necessitate disabling a fire detection, suppression, or alarm system component?

YES to any of the above indicates that a qualified fire watch is required.

Fire Watcher Name(s) ________________________________ Phone____________

NOTIFICATIONS

Notify the following groups at least 72 hours prior to work and 30 minutes after work is completed.
Write in names of persons contacted.

Notify in person OR by phone ONLY if question #5 above is answered "yes":

- Facilities Management Fire Alarm Supervisor

Notify by phone or in person: (If by phone, write down name of person and send them a completed copy of this permit.)

- Facilities Management Fire Protection Group
- Environmental Health & Safety Industrial Hygiene Group

SIGNATURES REQUIRED

University Project Manager____________________________ Date __________ Phone____________

I understand and will abide by the conditions described in this permit. I will implement the necessary precautions which are outlined on both sides of this permit form. Thirty minutes after each hot work session, I will reinspect work areas and adjacent areas to which spark and heat might have spread to verify that they are fire safe, and contact Facilities Management Alarm Technicians to have any disabled fire protection systems reactivated.

____________________ ______________________ Date __________ Phone____________
Permit Applicant Company or Department

1/17/03

00101-15_F72.EPS

Figure 72 Hot work permit.

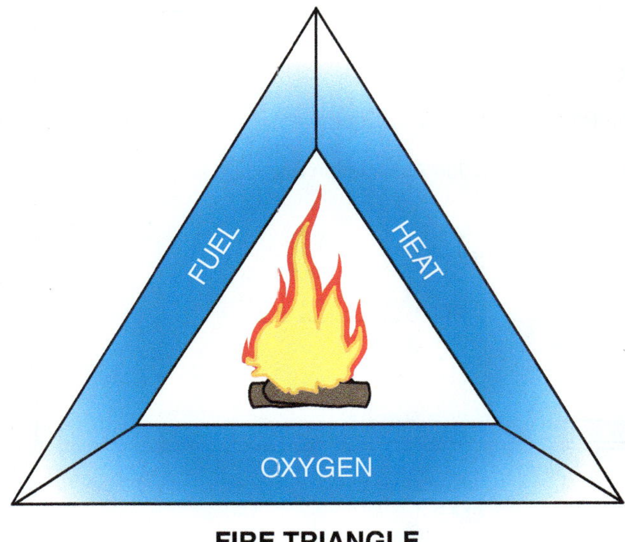

Figure 73 The fire triangle.

Class C (electrical) fire could cause you to be electrocuted because water conducts electricity. Class D fire extinguishers are required for fires involving reactive metals such as sodium, potassium, magnesium, titanium, zirconium, and the metal hydrides. A Class D fire extinguisher contains a powder designed to coat the metal to either smother the fire or keep oxygen from reaching the fire. If the powder coat fails to extinguish the fire, the best solution is to reapply the powder to keep the fire from spreading.

A pictogram label (*Figure 75*) on the extinguisher denotes the type(s) of fire for which the extinguisher is intended. The extinguisher shown in the figure is a dry-chemical extinguisher, which can be used on a Class A, B, or C fire. This type is the most common because it can be used to fight most fires. Some CO_2 extinguishers can also be used on Class A, B, or C fires, but others are designed for use on only class B and C fires. This can be seen in *Figure 76*.

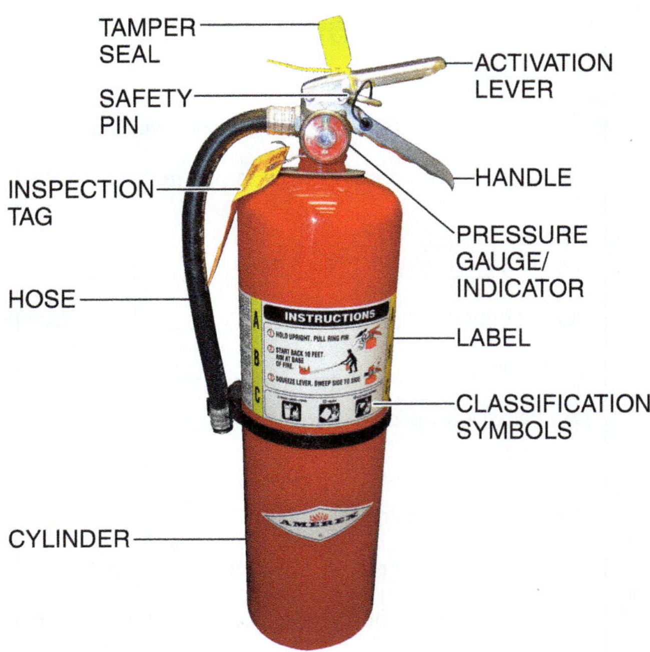

Figure 74 A portable fire extinguisher.

CO_2 is heavier than oxygen, so it smothers the fire by displacing the available oxygen. It also cools as it expands, so it can serve as a coolant. Some companies permit only CO_2 extinguishers to be used because the dry chemical in a dry-chemical extinguisher is a corrosive that can damage electrical systems. In addition, the fine powder emitted by the material in a dry-chemical extinguisher can cause eye and respiratory irritation when it becomes airborne. One advantage of the dry-chemical extinguisher is that its working range is 10 to 20 feet (3.1 to 6.1 m), while the range of the CO_2 extinguisher is 3 to 8 feet (0.9 to 2.4 m). A pressurized-water extinguisher, which is used only for Class A fires, has a range of about 50 feet (15 m).

Prevention and Preparation Are the Keys to Fire Safety

Any fire in the workplace can cause serious injury or property damage. When chemicals are involved, the risks are even greater. Prevention is the key to eliminating the hazards of fire in the workplace. Preparation is the key to controlling any fires that do start. Take the following precautions to ensure safety from fire in the workplace:

- Keep work areas clean and clutter-free.
- Know how to handle and store chemicals.
- Know what you are expected to do in case of a fire emergency.
- Should a fire start, call for professional help immediately. Don't let a fire get out of control.
- Know what chemicals you work with. You might have to tell firefighters at a chemical fire what kinds of hazardous substances are involved.
- Make sure you are familiar with your company's emergency action plan for fires.
- Use caution when using power tools near flammable substances.

00101-15_F75.EPS

Figure 75 Fire extinguisher pictograms.

WARNING!

Do not use a CO_2 extinguisher on a Class D fire, as it can cause the fire to spread. Also note that CO_2 is heavier than air, so it will concentrate in low areas, displacing oxygen. For that reason, a CO_2 extinguisher must never be taken into, or used in, a confined space. Before it can be used in a confined space, all personnel in the space must first be evacuated. After the extinguisher has been used in the confined space, the space must be cleared by a competent person before it can be re-entered.

Finally, CO_2 is discharged from the extinguishers at a temperature of −109°F (−78°C), so it can cause frostbite if it contacts the skin. Never discharge it while it is pointed at someone and do not hold it by the activation lever unless the activation lever is insulated.

WARNING!

When using a CO_2 extinguisher do not hold the activation lever with your bare hand, as it will become extremely cold.

Fire extinguishers must be periodically inspected to make sure they are properly charged. The needle on the extinguisher's pressure gauge must register in the green area (*Figure 77*). The inspection must be documented on the tag attached to the extinguisher. Whenever you check out an extinguisher, make sure its inspection tag (*Figure 78*) is up to date and that the pressure gauge registers the correct charge. Not all extinguishers have gauges, however. Extinguishers without gauges must be weighed to determine if they are fully charged.

6.4.5 Using a Fire Extinguisher

Use the PASS method (*Figure 79*) when attacking a fire:

- **P**ull the pin from the handle, breaking the tamper seal in the process (*Figure 80*).
- **A**im the nozzle at the base of the fire while 8 to 10 feet away.
- **S**queeze the discharge handle.
- **S**weep the nozzle back and forth at the base of the fire.

Note that the method of attacking the fire can vary from one extinguisher to another. The instructions for a specific fire extinguisher are printed on the label, as shown in *Figure 81*. Be sure you are familiar with the type of extinguishers that are commonly provided on your current job site.

As the fire is extinguished, move in closer until it is completely out. Stop shooting every three or four sweeps to check progress. Try not to use all of the contents on the initial attack, in case the fire flares up. Once the extinguisher has been used, it must be re-charged.

EXTINGUISHER TYPE	TYPE OF FIRE

Ordinary Combustibles

Fires in paper, cloth, wood, rubber, and many plastics require a water-type extinguisher labeled A.

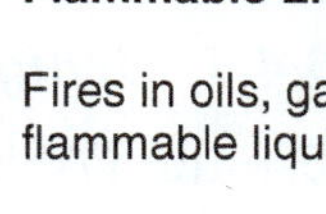

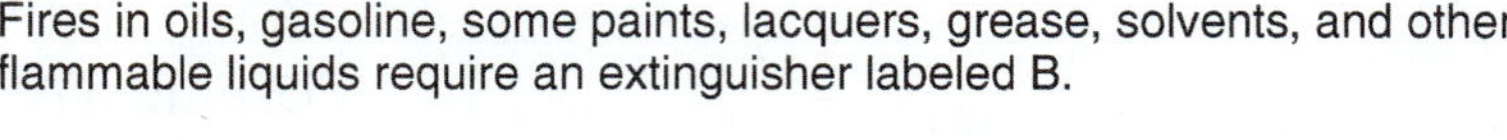

CO_2

OR

Flammable Liquids

Fires in oils, gasoline, some paints, lacquers, grease, solvents, and other flammable liquids require an extinguisher labeled B.

Dry Chemical

Electrical Equipment

Fires in liquids, fuse boxes, energized electrical equipment, computers, and other electrical sources require an extinguisher labeled C.

Ordinary Combustibles, Flammable Liquids, or Electrical Equipment

Multi-purpose dry chemical extinguishers are suitable for use on Class A, B, and C fires.

Metals

Combustible metals such as magnesium and sodium require special extinguishers labeled D.

00101-15_F76.EPS

Figure 76 Fire extinguisher applications.

00101-15_F77.EPS

Figure 77 Fire extinguisher pressure gauge.

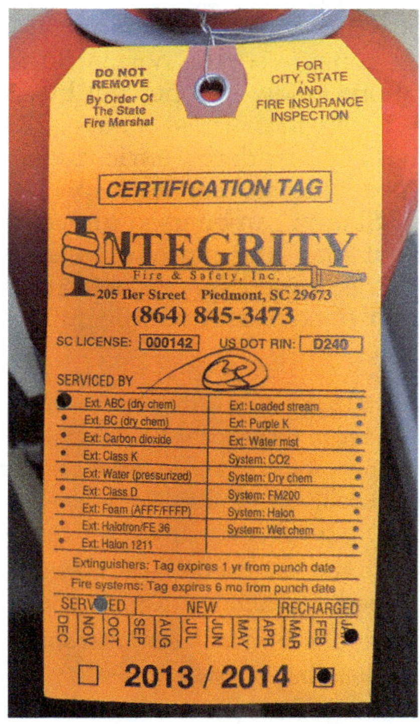

00101-15_F78.EPS

Figure 78 Fire extinguisher inspection tag.

00101-15_F79.EPS

Figure 79 PASS Method.

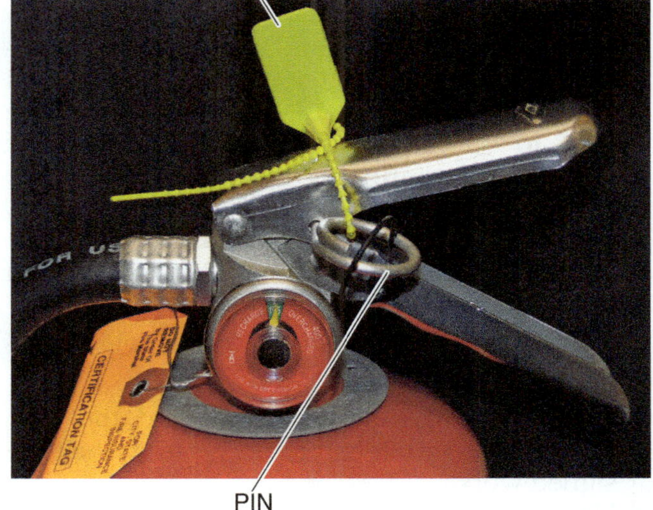

00104-15_F80.EPS

Figure 80 Pin and tamper seal.

Figure 81 Example of fire extinguisher instructions.

6.5.0 Confined Spaces

Construction and maintenance work is not always done outdoors; a lot of it is done in confined spaces. A confined space is a space that is large enough to work in but that has limited means of entry or exit. A confined space is not designed for human occupancy and it has limited ventilation. Examples of confined spaces are tanks, vessels, silos, storage bins, hoppers, vaults, pits, and certain compartments on ships and barges (*Figure 82*).

OSHA defines a confined space as a space that has the following characteristics:

- Is large enough and so configured that employees can bodily enter and perform their assigned work.
- Has a limited or restricted means of entry or exit.
- Is not designed for continuous employee occupancy.

OSHA further defines a **permit-required confined space** as a space that has one or more of the following characteristics:

- Contains or has the potential to contain a hazardous atmosphere.
- Contains a material that has the potential for engulfing an entrant.
- Has an internal configuration such that an entrant could be trapped or asphyxiated by inwardly converging walls or by a floor that slopes downward and tapers to a smaller cross-section.
- Contains any other recognized serious safety or health hazard.

Atmospheric hazards are a concern in a confined space. The air in the space can contain flammable or explosive vapors or toxic gases. The space can also contain too much or too little oxygen. For safe working conditions, the oxygen level in a confined space atmosphere must range between 19.5 and 23.5 percent by volume as measured with an oxygen analyzer, with 21 percent being considered the normal level. Oxygen concentrations below 19.5 percent by volume are considered deficient; those above 23.5 percent by volume are considered enriched. Special meters are available to test atmospheric hazards. Forced ventilation can be used to overcome these hazards, but that does not eliminate the need to satisfy other confined space safety requirements, including inspection and permitting.

WARNING!

If too much oxygen is introduced into a confined space, it can be absorbed by a worker's clothing and ignite. If too little oxygen is present, it can lead to death in minutes. For this reason, the following precautions apply:

- Make sure confined spaces are ventilated properly for cutting or welding purposes.
- Never use oxygen in confined spaces for ventilation purposes.
- Always remain aware of the work going on around you, as conditions can change.

A permit-required confined space is a type of confined space that has been evaluated by a qualified person and found to have actual or potential hazards. Written authorization, commonly known as an entry permit, is required in order to enter a permit-required confined space. Once the entry permit is issued, it must be posted at the entry to the confined space.

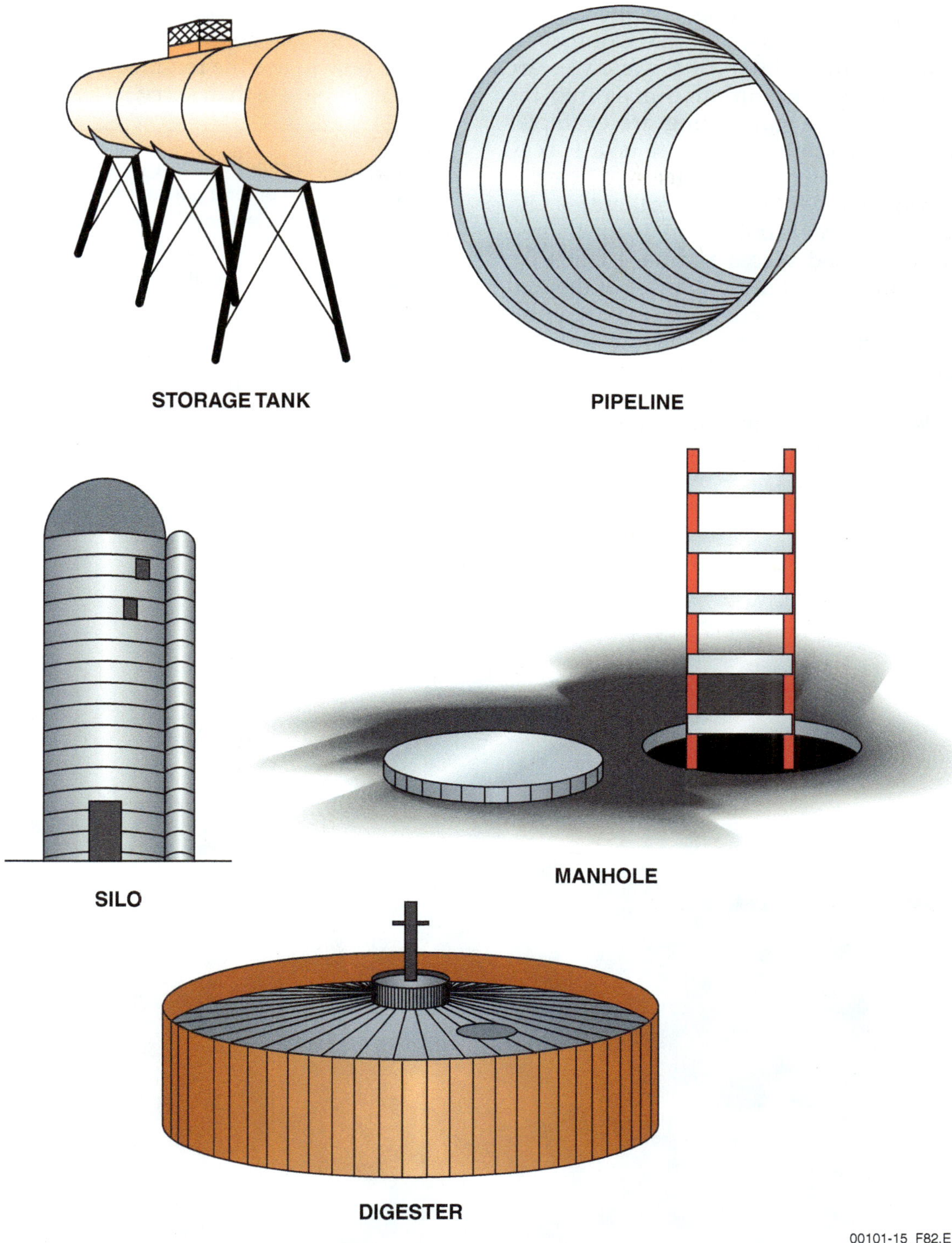

Figure 82 Examples of confined spaces.

When equipment is operating, confined spaces may contain hazardous gases or fluids, or may be oxygen-deficient. In addition, the work you are doing may introduce hazardous fumes into the space. Welding and metal cutting are examples of such work. For safety, you must take special precautions both before you enter and leave a confined space, and while you work there.

Until you have been trained to work in permit-required confined spaces and have taken the needed precautions, you must stay out of them. If you are not sure whether a confined space requires a permit, ask your supervisor. You must always follow your employer's procedures and your supervisor's instructions. Confined space procedures may include getting clearance from a safety representative before starting the work. You will be told what kinds of hazards are involved and what precautions you need to take. You will also be shown how to use the required PPE. Remember, it is better to be safe than sorry, so ask!

WARNING!

Without proper training, no employee is allowed to enter a permit-required or non-permit-required confined space. Employers are required to have programs to control entry to, and hazards in, both types of confined spaces.

Never work in a confined space without an attendant. An attendant must remain outside a permit-required confined space. The attendant monitors entry, work, and exit (*Figure 83*).

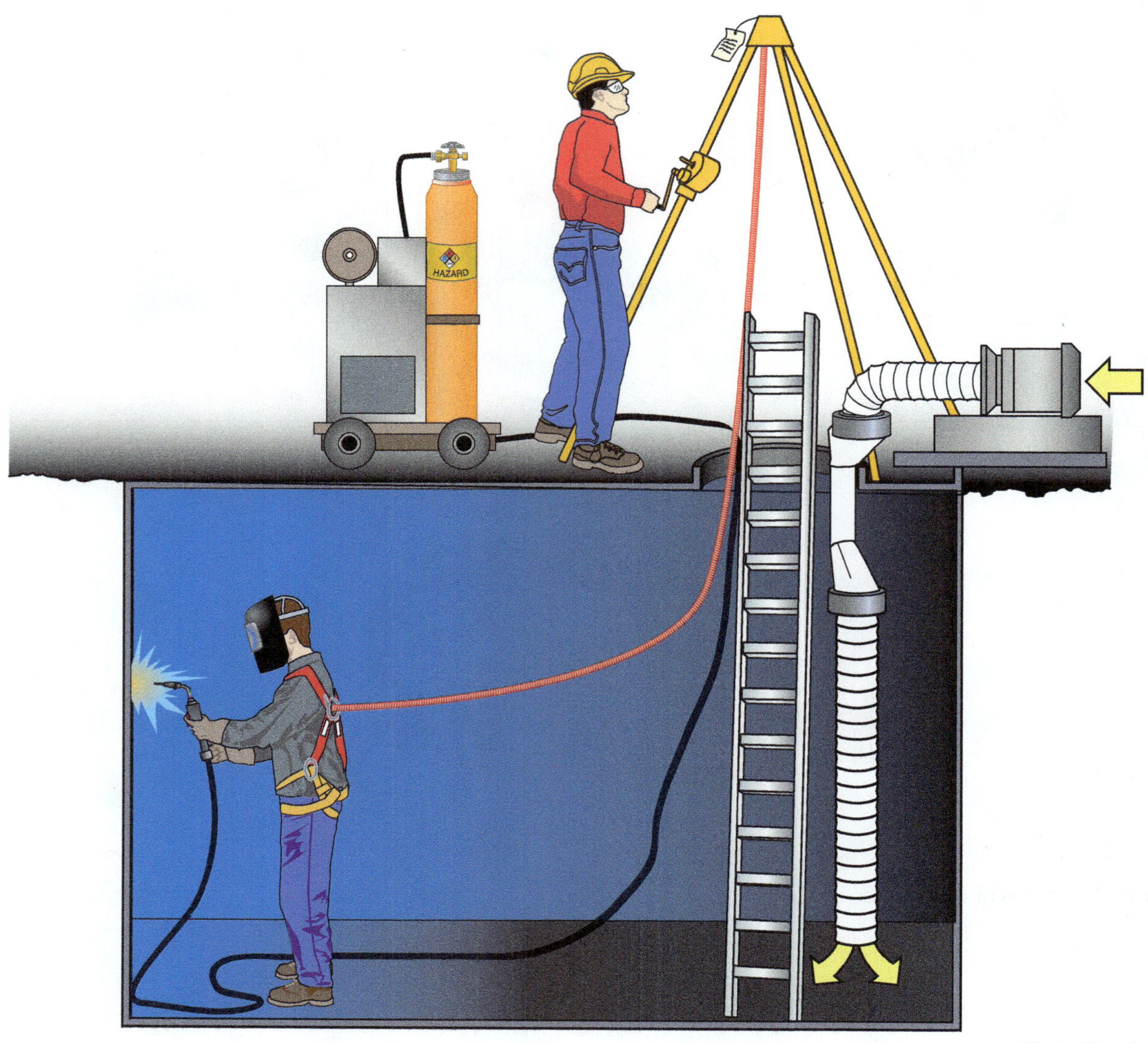

Figure 83 Permit-required confined space.

Additional Resources

US Occupational Safety and Health Administration. Numerous safety videos are available on line at **www.osha.gov/video**.

Construction Safety, 2006. Jimmie W. Hinze. Upper Saddle River, NJ: Pearson Education, Inc.

DeWalt Construction Safety/OSHA Professional Reference, Paul Rosenberg; American Contractors Educational Services. 2006. DEWALT.

Basic Construction Safety and Health, Fred Fanning. 2014. CreateSpace Independent Publishing Platform.

6.0.0 Section Review

1. The term PEL refers to a worker's _____.
   a. personal energy level
   b. maximum exposure limit
   c. the amount of lead in the bloodstream
   d. product efficiency level

2. A condition in which body temperature drops a few degrees below normal is known as _____.
   a. heat stress
   b. cold stroke
   c. cold stress
   d. heat exhaustion

3. Swollen eyes and a sandy feeling in the eyes are symptoms of _____.
   a. a flash burn
   b. heat stroke
   c. asbestos exposure
   d. frostbite

4. The three elements needed for a fire to start are fuel, combustible liquid, and oxygen.
   a. True
   b. False

5. A confined space is an enclosed space that has _____.
   a. only one door
   b. limited access
   c. hazardous gases
   d. locked doors

Summary

Although the typical job site has many hazards, it does not have to be a dangerous place to work. Your employer has programs to deal with potential hazards. Basic rules and regulations help protect you and your co-workers from unnecessary risks.

This module has presented many of the basic guidelines you must follow to ensure your safety and the safety of your co-workers. These guidelines fall into the following categories:

- Following safe work practices and procedures
- Inspecting safety equipment before use
- Using safety equipment properly

The basic approach to safety is to eliminate hazards in the equipment and the workplace; to learn the rules and procedures for working safely with and around the remaining hazards; and to apply those rules and procedures. The information covered here provides the foundation for a safe, productive, and rewarding career.

Review Questions

1. The four leading causes of death in the construction industry include electrical incidents, struck-by incidents, caught-in or caught-between incidents, and _____.
   a. vehicular incidents
   b. falls
   c. radiation exposure
   d. chemical burns

2. A sign that has a white background with a green panel with white lettering is a _____.
   a. general information sign
   b. safety instruction sign
   c. caution sign
   d. danger sign

3. To properly dispose of oily rags, they must be _____.
   a. stored in a container designed for the purpose
   b. washed thoroughly and returned to use
   c. taken outdoors and thrown into a dumpster
   d. burned at the end of the shift

4. Keeping your work area clean and free of scraps or spills is referred to as _____.
   a. managing
   b. organizing
   c. housekeeping
   d. stacking and storing

5. HAZCOM classifies all paint, concrete, and wood dust as _____.
   a. hazardous materials
   b. common materials
   c. inexpensive materials
   d. nonhazardous materials

6. Under HAZCOM, if you spot a hazard on your job site you must _____.
   a. report it to your supervisor
   b. leave immediately
   c. notify your co-workers
   d. correct the problem

7. Which of the following must be reported to your supervisor?
   a. Only major injuries
   b. Only incidents and major injuries
   c. All injuries, incidents, and incidents
   d. Only incidents in which a death occurred

8. Metal ladders should *not* be used near _____.
   a. stairways
   b. scaffolds
   c. electrical equipment
   d. windows

9. If you lean a straight ladder against the top of a 16-foot (4.8 m) wall, the base of the ladder should be _____.
   a. 3 feet (0.9 m) from the base of the wall
   b. 4 feet (1.2 m) from the base of the wall
   c. 5 feet (1.5 m) from the base of the wall
   d. 6 feet (1.8 m) from the base of the wall

10. The two basic types of scaffolds are _____.
    a. self-supporting and suspended scaffolds
    b. fixed and portable scaffolds
    c. metal and wooden scaffolds
    d. assembled and deliverable scaffolds

11. Interlocking stacked material is done by _____.
    a. applying a chain and padlock to it
    b. driving stakes around the stack
    c. securing the objects with rope
    d. placing the objects at right angles

12. To reduce the risk of workers being hurt or killed by falling materials, the maximum height-to-base ratio of a stack of materials should be _____.
    a. 2:1
    b. 4:1
    c. 8:1
    d. 10:1

13. The most common cause of death for equipment operators is _____.
    a. hit and run
    b. equipment rollover
    c. brake malfunction
    d. head-on collisions

14. Most cave-ins happen suddenly with little or no warning and occur in trenches 5 feet (1.5 m) to _____ .
    a. 10 feet (3.4 m) deep
    b. 15 feet (4.6 m) deep
    c. 20 feet (6.1 m) deep
    d. 25 feet (7.6 m) deep

15. In every trench over 4 feet (1.2 m) deep, there must be an exit every _____.
    a. 10 feet (3.4 m)
    b. 12 feet (3.7 m)
    c. 18 feet (5.5 m)
    d. 25 feet (7.6 m)

16. The minimum distance that a spoil pile must be located from the edge of an excavation is _____.
    a. 6 inches (15 cm)
    b. 2 feet (61 cm)
    c. 5 feet (1.5 m)
    d. 100 yards (91.4 m)

17. The type of trench protection designed to prevent trench wall cave-ins is _____.
    a. trench shield
    b. trench box
    c. spoil pile
    d. shoring

18. Protective guards are provided on power tools and machines in order to keep _____.
    a. dirt out of the tool or machine
    b. workers from being caught in rotating or moving parts
    c. the tool or machine from being struck by moving equipment
    d. unauthorized personnel from using the tool or machine

19. The minimum safe working distance from exposed electrical conductors _____.
    a. depends on the voltage
    b. is 6 inches (15 cm)
    c. is one foot (30 cm)
    d. is unlimited

20. Work that is performed near a hazard but not in direct contact with it is called _____.
    a. close call work
    b. near miss work
    c. proximity work
    d. barricade work

21. Circuit breakers and disconnect switches are examples of _____.
    a. energy-isolating devices
    b. energy-removal devices
    c. lockout/tagout devices
    d. multiple lockout devices

22. Which of the following provide the best eye protection?
    a. Welding hoods
    b. Face shields
    c. Safety goggles
    d. Strap-on glasses

23. The type of respirator that has its own clean air supply is the _____.
    a. half mask
    b. mouthpiece with mechanical filter
    c. self-contained breathing apparatus
    d. full facepiece mask

24. A worker can be exposed to hazardous materials by inhalation, ingestion, and _____.
    a. osmosis
    b. absorption
    c. radiation
    d. proximity

25. Hypothermia is a condition brought on by _____.
    a. excessive alcohol consumption
    b. prolonged exposure to cold
    c. excessive sweating
    d. prolonged exposure to heat

26. When in storage, oxygen and fuel cylinders used in oxyfuel cutting must be separated by _____.
    a. 20 feet (6.1 m)
    b. a metal barrier
    c. a 20-foot (6.1 m) wall
    d. 10 feet (3.4 m)

27. Grease and oil must be kept away from oxygen tanks because _____.
    a. grease or oil will contaminate the oxygen
    b. oxygen causes oil to freeze
    c. grease or oil can cause oxygen to explode
    d. oxygen will contaminate the oil or grease

28. Which of these must be present in the same place at the same time for a fire to occur?
    a. Oxygen, carbon dioxide, and heat
    b. Oxygen, heat, and fuel
    c. Hydrogen, oxygen, and wood
    d. Grease, liquid, and heat

29. A fire extinguisher labeled C would be used to fight a(n) _____.
    a. electrical fire
    b. magnesium fire
    c. paper fire
    d. gasoline fire

30. Which of the following characteristics is typical of a confined space?
    a. It has a limited amount of ventilation.
    b. There is no means of escape.
    c. It is too small to work in.
    d. It may be entered by untrained employees.

Trade Terms Quiz

Fill in the blank with the correct term that you learned from your study of this module.

1. The formal procedure for taking equipment out of service and ensuring it cannot be operated until an authorized person has returned it to service is ________.
2. A(n) ________ is any man-made cut, cavity, trench, or depression in an earth surface, formed by removing earth.
3. A(n) ________ identifies unsanitary, hazardous, or dangerous working conditions and has the authority to correct or eliminate them.
4. A(n) ________ is large enough to work in but has limited means of entry or exit.
5. Liquids that are ________ must be stored in safety cans to avoid the risk of fire.
6. To save lives, prevent injuries, and protect the health of America's workers, ________ publishes rules and regulations that employees and employers must follow.
7. Even a brief exposure to the ultraviolet light from arc welding can damage the eyes, causing a(n) ________.
8. ________ is an OSHA rule requiring all contractors to educate their employees about the hazardous chemicals they may be exposed to on the job site.
9. The temperature at which a fuel gives off enough gases to burn is called the ________.
10. A(n) ________ is the conducting connection between electrical equipment or an electrical circuit and the earth.
11. When climbing a ladder or scaffold, a tagline or ________ should be used to pull up tools.
12. If a work area is ________, that means it has pieces of material at least 2 inches (5 cm) thick and 6 inches (15 cm) wide used as flooring, decking, or scaffolds.
13. When doing work more than 6 feet (≈2 m) above the ground, a worker must wear a safety harness with a(n) ________ that is attached to a strong anchor point.
14. A structure that prevents trench walls from collapsing is known as ________.
15. Information on how to handle hazardous substances is located in the ________.
16. Overloading, which means exceeding the ________ of a ladder, can cause ladder failure.
17. If you are operating a vehicle on a job site and cannot see to your rear, get a(n) ________ to direct you.
18. Before working in a(n) ________ you must be trained, obtain written authorization, and take the necessary precautions.
19. A(n) ________ is a narrow excavation made below the surface of the ground that is generally deeper than it is wide.
20. The ________ on a scaffold is placed halfway between the toeboard and the top rail.
21. ________ for welding includes a face shield, ear plugs, and gloves.
22. A(n) ________ has proven his or her extensive knowledge, training, and experience and has successfully demonstrated the ability to solve problems relating to the work.
23. A(n) ________ provides clean air for breathing.
24. A(n) ________ is an elevated working platform for workers and material.
25. Working near a hazard, but not actually in contact with it is known as ________.
26. A vertical barrier called a(n) ________ is used at floor level on scaffolds to prevent materials from falling.
27. A(n) ________ shows which way the wind is blowing.
28. A horizontal board called a(n) ________ is used at top-level on all open sides of scaffolds and platforms.

29. The ________ is a general construction rule that states that fall protection is required any time you are working 6 feet (≈2 m) above a lower level.
30. ________ is the company organization that includes reporting procedures and supervisory responsibility.
31. The process of joining metal parts using an electric arc is known as ________.
32. An equipment or process powered by compressed air is ________.
33. Wood or metal braces placed diagonally from the bottom of one rail to the top of another rail in order to provide support are referred to as ________.
34. When a whole company sees the value of safety it is said to have a(n) ________.
35. Material such as soil removed from a trench or excavation is known as ________.
36. An area that is enclosed, fenced, covered, or otherwise protected by barriers, rails, covers, or platforms to prevent dangerous contact is said to be ________.
37. A(n) ________ is an unintentional, electrically conducting connection between an ungrounded conductor of an electrical circuit and the normally noncurrent-carrying conductors, metal objects, or the earth.
38. A(n) ________ is a device that interrupts and de-energizes an electrical circuit to protect a person from electrocution.
39. The type of equipment that is powered by fluid pressure is ________.
40. An unplanned event that results in personal injury is referred to as a(n) ________.
41. A protective screen set up around a welding operation designed to safeguard workers not directly involved in that operation is a(n) ________.
42. A process using heat in excess of 800°F (427°C) to melt base metal, along with a filler metal in order to join copper pipe is called ________.
43. A material that is capable of easily igniting and rapidly burning with a flash point at or above 100°F (38°C) is considered ________.
44. A structure used to protect workers in trenches, but lacking the ability to prevent cave-ins is ________.
45. A near miss that occurs but does not result in personal injury is considered to be a(n) ________.

Trade Terms

Accident
Arc welding
Brazing
Combustible
Competent person
Confined space
Cross-bracing
Excavation
Flammable
Flash burn
Flash point
Ground
Ground fault
Ground fault circuit interrupter (GFCI)
Guarded
Hand line
Hazard communication Standard (HAZCOM)
Hydraulic
Incident
Lanyard
Lockout/tagout (LOTO)
Management system
Maximum intended load
Midrail
Occupational Safety and Health Administration (OSHA)
Permit-required confined space
Personal protective equipment (PPE)
Planked
Pneumatic
Proximity work
Qualified person
Respirator
Safety culture
Safety data sheet (SDS)
Scaffold
Shielding
Shoring
Signaler
Six-foot rule
Spoil
Toeboard
Top rail
Trench
Welding curtain
Wind sock

Cornerstone of Craftsmanship

Bob Fitzgerald

Southern Company
Manager – Project Safety and Health

How did you choose a career in the Safety and Health field?
After many years working as an emergency medical technician (EMT) in the public and private sectors, I was hired as a medic on a construction site. What sparked my interest in safety and health was the notion of not just treating the injured, but actually being able to do something to prevent the incident from happening in the first place. When working in the medical field, you have to really care and have compassion for people. I think that is one of the major attributes of a successful safety professional.

What types of training have you been through?
I have taken all sorts of safety and health classes through the years, including trenching, shoring, electrical, fall protection, respiratory protection, and more. I was privileged to attend the NCCER Safety Academy at Clemson University early in my safety career and I think that was a turning point in my philosophy and thinking about safety management. Since then I have earned my Bachelor's of Science and a Master's degree in Safety and Health. I also hold the Certified Safety Professional and Construction Health and Safety Technician designation.

What kinds of work have you done in your career?
I have had two jobs in my professional career: EMT and Safety Professional.

Tell us about your present job.
I serve as the Manager for Project Safety & Health for Southern Company Operations Engineering and Construction Services, presently based in Birmingham, Alabama. Southern Company is the premier energy company serving the Southeast through its subsidiaries. Our firm provides engineering and project management services for most major capital work at Southern Company Electric Generating facilities as well as new generation projects.

What do you enjoy most about your job?
I enjoy helping people and making a difference in their lives. I feel very gratified to know that something I may say or some influence I may have could impact a person getting home safely to their families. That truly humbles me and is my motivation.

What factors have contributed most to your success?
You really have to care about people. Safety is a people business! My early experiences in the medical field helped me to have empathy for people and to see the pain and suffering that could happen if things go wrong.

What advice would you give to those new to Safety and Health field?
Realize that the value for safety and health is fundamental in all crafts or whatever role you perform in construction or industrial work. Watch out for each other and don't be afraid to speak up if you see an at-risk situation or action. Take the risk of intervening to help someone. Learn all you can about safety and put it to use.

Interesting career-related fact or accomplishment:
In 1995, I had the privilege to work with a wonderful team of construction professionals and craftworkers to earn the OSHA Voluntary Protection Program (VPP) STAR designation at a site in Stevenson, Alabama. This particular construction site was the first in the state of Alabama to earn the OSHA STAR. The effort that went into achieving OSHA STAR still echoes today in folks that I meet that worked on that site. They have instilled a culture of safety and a lifelong safety commitment. It was, and is, great!

Appendix

OSHA Mission and Standards

The mission of the Occupational Safety and Health Administration (OSHA) is to save lives, prevent injuries, and protect the health of America's workers. To accomplish this, federal and state governments work in partnership with millions of working men and women who are covered by the Occupational Safety and Health Act (OSH Act) of 1970.

Nearly every worker in the US comes under OSHA's jurisdiction. There are some exceptions, such as miners, transportation workers, many public employees, and the self-employed. These specific groups are not covered by OSHA standards, but most are covered by standards developed by the specific industry.

The Code of Federal Regulations

The *Code of Federal Regulations (CFR)* Part 1910 covers the OSHA standards for general industry. *CFR* Part 1926 covers the OSHA standards for the construction industry. Either or both may apply to you, depending on where you are working and what you are doing. If a job-site condition is covered in the CFR book, then that standard must be used. However, if a more stringent requirement is listed in *CFR* 1910, it should also be met. Check with your supervisor to find out which standards apply to your job.

29 *CFR* 1926 is divided into subparts A through Z. As you progress in task-specific training, you will learn about all the subparts applicable to your work. *Subpart C of* 29 *CFR* 1926 applies to all construction and maintenance work. It outlines the general safety and health provisions for the construction industry. It covers the following topics:

- Safety training and education
- Injury reporting and recording
- First aid and medical attention
- Housekeeping
- Illumination
- Sanitation
- PPE
- Standards incorporated by reference
- Definitions
- Access to employee exposure and medical records
- Means of egress
- Employee emergency action plans

For assistance in identifying parts, sections, paragraphs, and subparagraphs of an OSHA standard, refer to Appendix *Figure 1.*

All of OSHA's safety requirements in the Code of Federal Regulations apply to residential as well as commercial construction. In the past, OSHA enforced safety only at commercial sites. The increasing rate of incidents at residential sites led OSHA to enforce safety guidelines for the building of houses and townhomes. Today, however, OSHA still focuses its enforcement efforts on commercial construction.

The General Duty Clause

If a standard does not specifically address a hazard, the general duty clause must be invoked. Failing to adhere to the general duty clause can result in heavy fines for your employer. The general duty clause reads as follows:

> In practice, OSHA, court precedent, and the review commission have established that if the following elements are present, a general duty clause citation may be issued:
>
> - The employers failed to keep the workplace free of a hazard to which employees of that employer were exposed.
> - The hazard was recognized. (Examples might include: through your safety personnel, employees, organization, trade organization, or industry customs.)
> - The hazard was causing or was likely to cause death or serious physical harm.
> - There was a feasible and useful method to correct the hazard.

Employee Rights and Responsibilities

While it is the employer's responsibility to keep workers safe by complying with the General Duty Clause and all other OSHA regulations, workers have certain rights and responsibilities on the job site as well. First and foremost, workers must follow their employers' safety rules. While workers cannot be cited or fined by OSHA, they can be disciplined for violating their employer's safety rules. Workers must also wear the provided personal protective equipment. Workers should also inform their foreman about health and safety concerns on the job.

An OSHA Standard reference may look like this:

29 CFR 1926.501 (a)(1)(i)(A)

and breaks down like this:

29	=	Title (Labor)
CFR	=	Code of Federal Regulations
1926	=	Part (Construction)
.501	=	Section
(a)	=	Paragraph
(1)	=	Subparagraph
(i)	=	Subparagraph
(A)	=	Subparagraph

00101-15_A01.EPS

Figure 1 Reading OSHA standards.

Section 11(c) of the OSH Act prohibits employers from disciplining or discriminating against any worker for practicing their rights under OSHA, including filing a complaint. You have the right to file a complaint if you do not think that your employer is protecting your health and safety at work. You may submit a written request to OSHA asking for an inspection of your worksite. Workers who file a complaint have the right to have their names withheld from their employers, and OSHA will not reveal this information.

Workers who would like an on-site inspection must submit a written request. You have the following rights when job site inspection is conducted:

- You must be informed of imminent dangers. An OSHA inspector must tell you if you are exposed to an imminent danger. An imminent danger is one that could cause death or serious injury now or in the near future. The inspector will also ask your employer to stop any dangerous activity.
- You have the right to accompany the OSHA inspector in the walk-around inspection. Walk-around activities include all opening and closing conferences related to the conduct of the inspection.
- You have the right to be told about citations issued at your workplace. Notices of OSHA citations must be posted in the workplace near the site where the violation occurred and must remain posted for three days or until the hazard is corrected, whichever is longer.

After an inspection has been performed, OSHA will give the employer a date by which any hazards cited must be fixed. Employers can appeal these dates, and appeals must be filed within 15 days of the citation. Workers have the right to meet privately with the OSHA inspector to discuss the results of the inspection.

If you have been discriminated against for asserting your OSHA rights, you have the right to file a complaint with the OSHA area office within 30 days of the incident. Make sure you file your complaint as soon as possible, as the time limit is strictly enforced.

You also have the right to see and copy any medical records about you that the employer has obtained. Your employer is required by OSHA 29 *CFR* 1926.33 and OSHA 29 *CFR* 1910.1020 to maintain your medical records for 30 years after you leave employment. If you are employed for less than one year, the employer can maintain your records or give them to you when you leave the job.

Inspections

OSHA conducts six types of inspections to determine if employers are in compliance with standards:

- *Imminent danger inspections* – OSHA's top priority for inspection, conducted when workers face an immediate risk of death or serious physical harm.
- *Catastrophe inspections* – Performed after an incident that requires hospitalization of three or more workers. Employers are required to report fatalities and catastrophes to OSHA within eight hours.
- *Worker complaint and referral inspections* – Conducted due to complaints by workers or a worker representative, or a referral from a recognized professional.
- *Programmed inspection* – Aimed at high-risk areas based on OSHA's targeting and priority methods.
- *Follow-up inspection* – Completed after citations to assure employer has corrected violations.
- *Monitoring inspection* – Used for long-term abatement follow-up or to assure compliance with variances.

Before beginning an inspection, OSHA staff must be able to determine from the complaint that there are reasonable grounds to believe that a violation of an OSHA standard or a safety or health hazard exists. If OSHA has information indicating the employer is aware of the hazard and is correcting it, the agency may not conduct an inspection after obtaining the necessary documentation from the employer.

Complaint inspections are typically limited to the hazards listed in the complaint, although other violations in plain sight may be cited as well. The inspector may decide to expand the inspection based on professional judgment or conversations with workers.

Complaints are not necessarily inspected in first-come, first-served order. OSHA ranks complaints based on the severity of the alleged hazard and the number of workers exposed. That is why lower-priority complaints can often be handled more quickly using the phone/fax method than through on-site inspections.

Inspections are typically performed by conducting a walk-around. During a walk-around inspection, the inspector typically does the following:

- Observes conditions of the job site.
- Talks to workers.
- Inspects records.
- Examines posted hazard warnings and signs.
- Points out hazards and suggests ways to reduce or eliminate them.

After the walk-around inspection, there is typically a closing conference held between the inspector and the site contractor or company managers. During this conference, inspectors discuss their findings, citing specific violations and suggested abatement methods. Inspectors may also conduct interviews with the employers, workers, and representatives at this point.

Violations

Employers who violate OSHA regulations can be fined. The fines are not always high, but they can harm a company's reputation for safety. Fines for serious safety violations can cost up to $7,000. Fines for each violation that was done willfully can cost up to $70,000. In 2012, roughly $260,000,000 in fines were issued against employers. In the decade prior to 2012, annual fines averaged approximately $150,000,000. Significant increases and decreases in annual fines tend to be dependent on the current federal administration and OSHAs budget provided by federal lawmakers.

Compliance

Just as employers are responsible to OSHA for compliance, employees must comply with their company's safety policies and rules. Employers are required to identify hazards and potential hazards within the workplace and eliminate them, control them, or provide protection from them. This can only be done through the combined efforts of the employer and employees. Employers must provide written programs and training on hazards, and employees must follow the procedures. You, as the employee, must read and understand the OSHA poster at your job site explaining your rights and responsibilities. If you are unsure where the OSHA poster is, ask your supervisor.

To help employers provide a safe workplace, OSHA requires companies to provide a competent person to ensure the safety of the employees. In *OSHA 29 CFR* 1926, OSHA defines a competent person as follows:

> Competent person means one who is capable of identifying existing and predictable hazards in the surroundings or working conditions which employees, and who has authorization to take prompt corrective measures to eliminate them.

In comparison, *OSHA* 29 *CFR* 1926 defines a qualified person as follows:

> Someone who, by possession of a recognized degree, certificate, or professional standing, or who by extensive knowledge, training, and experience, has successfully demonstrated his ability to solve or resolve problems relating to the subject matter, work, or the project.

In other words, a competent person is experienced and knowledgeable about the specific operation and has the authority from the employer to correct the problem or shut down the operation until it is safe. A qualified person has the knowledge and experience to handle problems. A competent person is not necessarily a qualified person.

These terms will be an important part of your career. It is important for you to know who the competent person is on your job site. OSHA requires a competent person for many of the tasks you may be assigned to perform, such as confined space entry, ladder use, and trenching. Different individuals may be assigned as a competent person for different tasks, according to their expertise. To ensure safety for you and your co-workers, work closely with your competent person and supervisor.

Trade Terms Introduced in This Module

Accident: Per the US Occupational Safety and Health Administration (OSHA), an unplanned event that results in personal injury or property damage.

Arc welding: The joining of metal parts by fusion, in which the necessary heat is produced by means of an electric arc.

Brazing: A process using heat in excess of 800°F (427°C) to melt a filler metal that is drawn into a connection. Brazing is commonly used to join copper pipe.

Combustible: Capable of easily igniting and rapidly burning; used to describe a fuel with a flash point at or above 100°F (38°C).

Competent person: A person who is capable of identifying existing and predictable hazards in the surroundings or working conditions that are unsanitary, hazardous, or dangerous to employees, and who has authorization to take prompt corrective measures to eliminate them.

Confined space: A work area large enough for a person to work in, but with limited means of entry and exit and not designed for continuous occupancy. Tanks, vessels, silos, pits, vaults, and hoppers are examples of confined spaces. Also see *permit-required confined space*.

Cross-bracing: Braces (metal or wood) placed diagonally from the bottom of one rail to the top of another rail to add support to a structure.

Excavation: Any man-made cut, cavity, trench, or depression in an earth surface, formed by removing earth. It can be made for anything from basements to highways. Also see *trench*.

Flammable: Capable of easily igniting and rapidly burning; used to describe a fuel with a flash point below 100°F (38°C).

Flash burn: The damage that can be done to eyes after even brief exposure to ultraviolet light from arc welding. A flash burn requires medical attention.

Flash point: The temperature at which fuel gives off enough gases (vapors) to burn.

Ground: The conducting connection between electrical equipment or an electrical circuit and the earth.

Ground fault: An unintentional, electrically conducting connection between an ungrounded conductor of an electrical circuit and the normally noncurrent-carrying conductors, metal objects, or the earth.

Ground fault circuit interrupter (GFCI): A device that interrupts and de-energizes an electrical circuit to protect a person from electrocution.

Guarded: Enclosed, fenced, covered, or otherwise protected by barriers, rails, covers, or platforms to prevent dangerous contact.

Hand line: A line attached to a tool or object so a worker can pull it up after climbing a ladder or scaffold.

Hazard Communication Standard (HAZCOM): The standard that requires contractors to educate employees about hazardous chemicals on the job site and how to work with them safely.

Hydraulic: Powered by fluid under pressure.

Incident: Per the US Occupational Safety and Health Administration (OSHA), an unplanned event that does not result in personal injury but may result in property damage or is worthy of recording.

Lanyard: A short section of rope or strap, one end of which is attached to a worker's safety harness and the other to a strong anchor point above the work area.

Lockout/tagout: A formal procedure for taking equipment out of service and ensuring that it cannot be operated until a qualified person has removed the lock and/or warning tag.

Management system: The organization of a company's management, including reporting procedures, supervisory responsibility, and administration.

Maximum intended load: The total weight of all people, equipment, tools, materials, and loads that a ladder can hold at one time.

Midrail: Mid-level, horizontal board required on all open sides of scaffolds and platforms that are more than 14 inches (35 cm) from the face of the structure and more than 10 feet (3.05 m) above the ground. It is placed halfway between the toeboard and the top rail.

Occupational Safety and Health Administration (OSHA): An agency of the US Department of Labor. Also refers to the Occupational Safety and Health Act of 1970, a law that applies to more than more than 111 million workers and 7 million job sites in the country.

Permit-required confined space: A confined space that has been evaluated and found to have actual or potential hazards, such as a toxic atmosphere or other serious safety or health hazard. Workers need written authorization to enter a permit-required confined space. Also see *confined space*.

Personal protective equipment (PPE): Equipment or clothing designed to prevent or reduce injuries.

Planked: Having pieces of material 2 inches (5 cm) thick or greater and 6 inches (15 cm) wide or greater used as flooring, decking, or scaffold decks.

Pneumatic: Powered by air pressure, such as a pneumatic tool.

Proximity work: Work done near a hazard but not actually in contact with it.

Qualified person: A person who, by possession of a recognized degree, certificate, or professional standing, or by extensive knowledge, training, and experience, has demonstrated the ability to solve or prevent problems relating to a certain subject, work, or project.

Respirator: A device that provides clean, filtered air for breathing, no matter what is in the surrounding air.

Safety culture: The culture created when the whole company sees the value of a safe work environment.

Safety data sheet (SDS): A document that must accompany any hazardous substance. The SDS identifies the substance and gives the exposure limits, the physical and chemical characteristics, the kind of hazard it presents, precautions for safe handling and use, and specific control measures.

Scaffold: An elevated platform for workers and materials.

Shielding: A structure used to protect workers in trenches but lacking the ability to prevent cave-ins.

Shoring: Using pieces of timber, usually in a diagonal position, to hold a wall in place temporarily.

Signaler: A person who is responsible for directing a vehicle when the driver's vision is blocked in any way.

Six-foot rule: A rule stating that platforms or work surfaces with unprotected sides or edges that are 6 feet (≈2 m) or higher than the ground or level below it require fall protection.

Spoil: Material such as earth removed while digging a trench or excavation.

Toeboard: A vertical barrier at floor level attached along exposed edges of a platform, runway, or ramp to prevent materials and people from falling.

Top rail: A top-level, horizontal board required on all open sides of scaffolds and platforms that are more than 14 inches (36 cm) from the face of the structure and more than 10 feet (3 m) above the ground.

Trench: A narrow excavation made below the surface of the ground that is generally deeper than it is wide, with a maximum width of 15 feet (4.6 m). Also see *excavation*.

Welding curtain: A protective screen set up around a welding operation designed to safeguard workers not directly involved in that operation.

Wind sock: A cloth cone open at both ends mounted in a high place to show which direction the wind is blowing.

Additional Resources

This module presents thorough resources for task training. The following reference material is suggested for further study.

US Occupational Safety and Health Administration. Numerous safety videos are available on line at **www.osha.gov/video.**

Construction Safety, Jimmie W. Hinze. 2006. Upper Saddle River, NJ: Pearson Education, Inc.

DeWalt Construction Safety/OSHA Professional Reference, Paul Rosenberg; American Contractors Educational Services. 2006. DEWALT.

Basic Construction Safety and Health, Fred Fanning. 2014. CreateSpace Independent Publishing Platform.

Figure Credits

Honeywell Safety Products, Module opener, Figures 8B–D, 21, 50B, 52, 57C

Accuform Signs, Figures 2, 51

Courtesy of Justrite Mfg. Co. LLC, Figure 3

L.J. LeBlanc, Figure 4

Weld-On Adhesives, Inc., a division of IPS Corporation, Figure 5

Fall protection materials provided courtesy of Miller Fall Protection, Franklin, PA, Figures 7, 8A, 9-11, 13, 15–17, 19

Photo Courtesy of Capital Safety (DBI-SALA and PROTECTA), Figure 12

Topaz Publications Inc., Figures 14, 20, 37, 43, 46B, 47–49, 50A, 57A, 58, 59A, 65, 75, 77, 78, 80

Guardian Fall Protection, Figure 22

Courtesy of Louisville Ladders, Figures 23, 32

Werner Co., Figures 29, 30

Courtesy of Safway Services LLC, Figure 34

Spider Staging, Figure 35

Courtesy of Snap-on Industrial - Tools at Height Program, SA01

Carolina Bridge Co., Figure 38

Kundel Industries, Figures 40, 41

DeWALT Industrial Tool Co., Figure 42

Courtesy of Atlas Copco, Figure 44

Panduit Corp., Figure 50B

Photo courtesy of Bullard, Figure 54

MSA The Safety Company, Figures 55, 56, 59B, 60A-B, 60D

North Safety Products USA, Figures 57B, 60C

The Lincoln Electric Company, Cleveland, OH, USA, Figures 66, 68

Sellstrom Manufacturing, Figure 67

Vestil Manufacturing, Figure 69

Courtesy of Saf-T-Cart, Figure 71

Courtesy of Amerex Corporation, Figures 74, 76 (photos), 79

Badger Fire Protection, Figure 81

Section Review Answer Key

Answer	Section Reference	Objective
Section One		
1. c	1.0.0	1a
2. d	1.2.1	1b
3. a	1.3.5	1c
Section Two		
1. c	2.1.2	2a
2. a	2.2.1	2b
3. b	2.3.1	2c
4. c	2.4.2	2d
Section Three		
1. a	3.1.2	3a
2. d	3.2.1	3b
Section Four		
1. b	4.1.1	4a
2. b	4.2.0	4b
Section Five		
1. a	5.1.2	5a
2. d	5.2.0	5b
Section Six		
1. b	6.1.0	6a
2. c	6.2.5	6b
3. a	6.3.1	6c
4. b	6.4.1	6d
5. b	6.5.0	6e

NCCER CURRICULA — USER UPDATE

NCCER makes every effort to keep its textbooks up-to-date and free of technical errors. We appreciate your help in this process. If you find an error, a typographical mistake, or an inaccuracy in NCCER's curricula, please fill out this form (or a photocopy), or complete the online form at **www.nccer.org/olf**. Be sure to include the exact module ID number, page number, a detailed description, and your recommended correction. Your input will be brought to the attention of the Authoring Team. Thank you for your assistance.

Instructors – If you have an idea for improving this textbook, or have found that additional materials were necessary to teach this module effectively, please let us know so that we may present your suggestions to the Authoring Team.

NCCER Product Development and Revision
13614 Progress Blvd., Alachua, FL 32615
Email: curriculum@nccer.org
Online: www.nccer.org/olf

❑ Trainee Guide ❑ Lesson Plans ❑ Exam ❑ PowerPoints Other ______________________

Craft / Level: Copyright Date:

Module ID Number / Title:

Section Number(s):

Description:

Recommended Correction:

Your Name:

Address:

Email: Phone:

00108-15

Basic Employability Skills

Overview

Becoming gainfully employed in the construction industry takes more preparation than simply filling out a job application. It is essential to understand how the construction industry and potential employers operate. Your trade skills are extremely important, but all employers are also looking for those who are eager to advance and demonstrate positive personal characteristics. This module discusses the skills needed to pursue employment successfully.

Module One

Trainees with successful module completions may be eligible for credentialing through the NCCER Registry. To learn more, go to **www.nccer.org** or contact us at **1.888.622.3720**. Our website has information on the latest product releases and training, as well as online versions of our *Cornerstone* magazine and Pearson's product catalog.

Your feedback is welcome. You may email your comments to **curriculum@nccer.org,** send general comments and inquiries to **info@nccer.org**, or fill in the User Update form at the back of this module.

This information is general in nature and intended for training purposes only. Actual performance of activities described in this manual requires compliance with all applicable operating, service, maintenance, and safety procedures under the direction of qualified personnel. References in this manual to patented or proprietary devices do not constitute a recommendation of their use.

00108-15 BASIC EMPLOYABILITY SKILLS

Objectives

When you have completed this module, you will be able to do the following:

1. Describe the opportunities in the construction businesses and how to enter the construction workforce.
   a. Describe the construction business and the opportunities offered by the trades.
   b. Explain how workers can enter the construction workforce.
2. Explain the importance of critical thinking and how to solve problems.
   a. Describe critical thinking and barriers to solving problems.
   b. Describe how to solve problems using critical thinking.
   c. Describe problems related to planning and scheduling.
3. Explain the importance of social skills and identify ways good social skills are applied in the construction trade.
   a. Identify good personal and social skills.
   b. Explain how to resolve conflicts with co-workers and supervisors.
   c. Explain how to give and receive constructive criticism.
   d. Identify and describe various social issues of concern in the workplace.
   e. Describe how to work in a team environment and how to be an effective leader.

Performance Tasks

This is a knowledge-based module; there are no Performance Tasks.

Trade Terms

Absenteeism
Amphetamine
Barbiturate
Bullying
Cannabinoids
Compromise
Confidentiality
Constructive criticism
Hallucinogen
Harassment
Initiative
Leadership
Methamphetamine
Mission statement
Opiates
Professionalism
Reference
Self-presentation
Sexual harassment
Synthetic drugs
Tactful
Tardiness
Work ethic
Zero tolerance

Industry Recognized Credentials

If you are training through an NCCER-accredited sponsor, you may be eligible for credentials from NCCER's Registry. The ID number for this module is 00108-15. Note that this module may have been used in other NCCER curricula and may apply to other level completions. Contact NCCER's Registry at 888.622.3720 or go to **www.nccer.org** for more information.

Contents

Topics to be presented in this module include:

Figures

SECTION ONE

1.0.0 OPPORTUNITIES IN THE CONSTRUCTION INDUSTRY

Objective

Describe the opportunities in the construction business and how to enter the construction workforce.

a. Describe the construction business and the opportunities offered by the trades.
b. Explain how workers can enter the construction workforce.

Trade Terms

Mission statement: A statement of how a company does business.

Reference: A person who can confirm to a potential employer that you have the skills, experience, and work habits that are listed in your resume.

Whether a laborer (blue collar) or professional (white collar), the construction business offers an abundance of opportunities. It is up to each individual to take the necessary steps to advance their own career.

To establish a career in your trade of choice, it is important to know where to look for the opportunities and be prepared to act when opportunities arise.

1.1.0 The Construction Business

The construction industry is made up of a wide variety of specialized skills (*Figure 1*). This diverse industry is made up of both laborers and professionals working together. Industry opportunities have a global presence; skilled tradespeople are needed in areas such as war zones and major construction sites around the world. The trades are not limited solely to building construction. There are also global opportunities for many craftworkers such as wind turbine technicians, welders, powerline workers, crane operators, and instrumentation technicians that are not directly related to new construction.

> **Did You Know?**
>
> **Collar Color**
>
> The difference between blue collar and white collar jobs began in the 1920s when office workers started wearing white shirts, almost exclusively with a tie. The term *blue collar* was coined because manual laborers typically wore dark colors that did not show soiled areas.

The construction industry consists of independent companies of all sizes that specialize in one or more types of work. For example, a smaller company might install HVAC systems in residences, whereas a larger mechanical contractor might be responsible for the HVAC and piping systems associated with a hospital.

The following are some examples of the workforce needed to build a large-scale project:

- Architects to create and design the shape, color, and spaces of the structure.
- Civil engineers to analyze the architects' drawings to determine how to make the construction design possible and suggest modifications if it is not. They are also responsible for determining suitable building materials.
- Project managers to plan, coordinate, and control a project's progress or a portion thereof.
- Heavy equipment operators to excavate the site.
- Concrete pourers to pour the foundation.
- Crane operators to lift the steel beams.
- Ironworkers to cut and fit the steel beams.
- Welders to secure the framework of buildings and other metal structures.
- Electricians to install an array of electrical devices and wiring.
- Plumbers to properly install fixtures and other plumbing equipment.
- Heating, ventilation, and air conditioning (HVAC) technicians to install building comfort systems.
- Pipefitters to install chilled and hot water systems, as well as process piping.
- Carpenters to frame the walls.
- Instrumentation technicians to install measuring and control instruments.
- Landscape designers to decide on the aesthetics of the project.

(A) WIND TURBINE TECHNICIAN

(B) IRONWORKER

(C) WELDER

(D) HVAC TECHNICIAN

(E) SURVEYOR

(F) CREW LEADER/SUPERVISOR

00108-15_F01.EPS

Figure 1 The construction industry offers many career options.

Many companies have a mission statement, which explains how a company does business. The company may describe its philosophy in an employee handbook or other materials during a new-hire orientation. However a company describes its role in the construction industry, it is very important to become familiar with your company's policies and procedures to know what is expected of you. In fact, the time to learn about a potential employer is before an application is ever submitted. The more you know about a company and what is important to them, the better you can plan your responses for an interview.

1.2.0 Entering the Construction Workforce

Upon completion of your training, it will be time to find a job where your newly aquired skills can be put to use. The first step of a job search is writing a resume that will get an employer's attention.

A good resume can make the task of finding a job much easier. The following are some guidelines for resume writing:

- Use a readable font, such as Times New Roman or Arial, for the text of the resume.
- Write your objective using key words found in the job description to create a relationship between the two.
- Format your resume chronologically, putting your current or most recent job first.
- Use short sentences and bulleted lists to describe your skills, again using key words found in the job description.
- Include certifications and training credentials in specific trade areas. Credentials could include such items as NCCER training certificates or wallet cards, or an automated external defibrillator (AED) certification.
- Make sure that all contact information is current, accurate, and easy to find on the resume.
- Personal information, such as hobbies or volunteer work and commitments can be listed. Employers often appreciate employees who have such interests, as they indicate a well-rounded individual. However, such information should be kept brief.
- Check your resume for inaccurate or missing dates. The timeline of your career should be easy to follow. Large gaps of undocumented time on a resume may require a reasonable explanation.
- Have someone proofread your resume to make sure it is free of spelling errors, typos, and poor grammar.
- Provide clear personal references or indicate that such a list can be provided upon request.

Did You Know?

The High Cost of Benefits

While the pay rates for craftworkers vary widely, the actual cost of an employee to a company averages 35 percent more than the employee's wage. For example, if a worker's hourly wage is $25.00, the company will spend another $8.75 per hour to pay for benefits such as healthcare and vacation time. Even higher percentages are not uncommon.

Look for a job that will match your skill level. Typically, a job posting will include a summary of the position, the required qualifications, and an overview of benefits. Applying for positions well above your skill level usually leads to disappointment.

Most employers advertise openings on the Internet and require online applications. To perform an Internet search, use a website dedicated to job searching or go directly to a company's website. If a computer is not available, libraries and unemployment offices generally have computer stations for public use. Job openings can also be found in local newspapers and trade magazines.

Today, one of the most important and productive methods to identify job opportunities is through networking. Studies indicate that as many as 70 percent of available positions are not publicly advertised. At the same time, job seekers are spending roughly 70 percent of their time looking through public job postings. Obviously, there are many more opportunities available than job seekers are able to identify through postings. There are many ways, including social media, to share your interest in finding a specific type of job. Identify companies that you would like to work with, regardless of job postings, and then look for contacts within the company that could help you identify opportunities. Express your interest in finding a job to family and friends. Networking is quite often the most productive method of finding and landing a job.

Ask your current supervisor, your co-workers, and friends that know you well if they would act as references for you. Current and accurate contact information for references must be provided. Both personal and professional references are recommended.

Ensure that your resume is up-to-date, well organized, and easy to read (*Figure 2*). Apply the effective writing techniques that you learned in *Basic Communication Skills* when writing a resume.

John Q. Smith
123 Main Street
Anytown, MD 67890
(555) 111-2233 – Home phone
(555) 333-4455 – Cell phone
johnqsmith@email.com

EMPLOYMENT OBJECTIVE

- To obtain a position within a construction company where I can use my carpentry and custom woodworking skills.

SKILLS AND QUALIFICATIONS

- Able to work independently and with a crew.
- Practical knowledge of carpentry hand and power tools, measuring tools, and woodworking tools.
- Trained and certified to use powder-actuated tools.
- Qualified to read residential and commercial construction drawings.
- Experience with energy efficient and conservation building techniques.
- Knowledge of LEED standards.

WORK EXPERIENCE

Journeyman Carpenter — *May 2011 – present*
LJL Construction, 123 Hammer Heights Road, Fairfax, VA
800-222-2222

- Built custom cabinets and closet systems
- Framed single family homes
- Trained apprentice carpenters

Carpenter — *June 2005 – May 2011*
Hammer Company, 456 Lathe Lane, Philadelphia, PA
866-555-1234

- Built closets and storage systems for commercial storage units

Carpenter — *February 2002 – June 2005*
Blue Ridge Carpenters, 789 Mountain Road, Harrisburg, PA
888-123-4567

- Framed condo units
- Built fences
- Raise wood walkways

CONSTRUCTION EDUCATION

- Carpentry apprenticeship. MK Builders Inc., Gettysburg, PA. Certificate, 2002.
- Professional carpentry certification program. ABC Training Corp., Philadelphia, PA. Certificate, 2000.
- Powder-actuated tools certification program. All States Community College, Marietta, PA. Certificate, 2000.

00108-15_F02.EPS

Figure 2 Resume.

By carefully selecting jobs that you are qualified for and by submitting an accurate, well-written resume, you improve your chances of being called for interviews (*Figure 3*). The effective communication skills that you learned in *Basic Communication Skills* will help you present yourself well.

Before accepting an offer, ask yourself the following questions to ensure the position will be the right fit:

- Is the salary enough to meet my needs?
- Does the company offer a benefits package that covers what I need?
- Will the work be interesting and challenging, but not more than I can handle?
- Does the company have a good reputation in the industry?
- Do the people appear to be nice to work with?
- Is travel required and am I able to agree to that?
- Does the company offer training?
- Are advancement opportunities within the organization evident?
- What is the company's safety record?

Before an offer is made, some trades require credentials or qualification testing to assess the applicant's skill level. To ensure workplace safety, many companies also require drug testing as part of the pre-employment process. Appropriate identification, such as a driver license, US Social Security card, national identification card, and/or a valid passport must be given to the employer. If a position outside of your country of citizenship is applied for, check the hiring and documentation requirements for that specific country and what may be required of non-citizens.

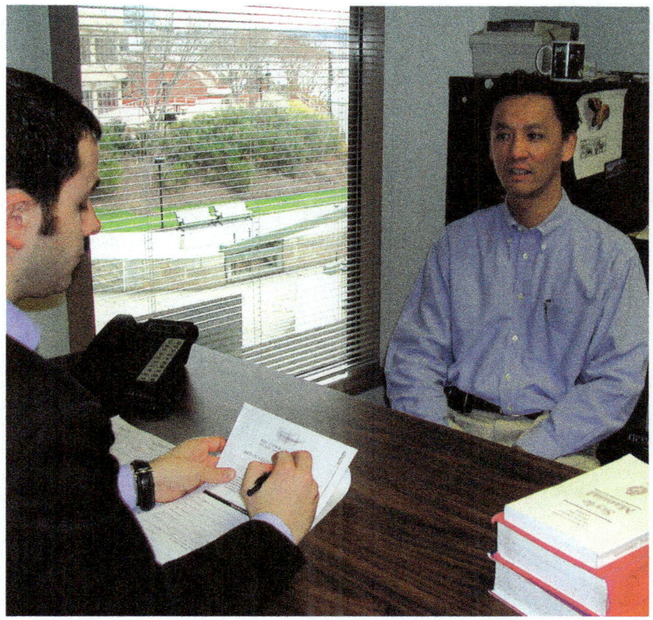

00108-15_F03.EPS

Figure 3 Job interviews are an important part of the hiring process.

After accepting a position, and even before the interview process begins, it is wise to obtain an organizational chart (*Figure 4*). Organizational charts are used to help all employees understand the structure of the company and what position and/or individual has responsibility for various areas. Understanding the organizational structure and being able to recall some important details about a prospective employer during an interview can make a strong impression.

GOING GREEN

Green-Collar Jobs

In recent years, there has been a big shift to environmentally friendly construction. This has created many new sectors in the US economy, including everything from recycling to energy retrofits; green building; solar installation/maintenance; water retrofits; and whole house performance (HVAC). These jobs are in green businesses and all involve work that directly affects environmental quality. Any job that involves the design, manufacture, installation, operation, and/or maintenance of renewable energy and energy-efficient technologies are considered green-collar jobs. Many of these jobs have opened new doors or removed barriers to employment. In the construction industry, opportunities exist in the following areas:

- Installation, construction, maintenance, and repair of energy retrofits, HVAC, solar panels, and whole home performance
- Installation, construction, maintenance, and repair of water conservation and adaptive grey water reuse
- Construction, carpentry, and demolition in green building and recycling

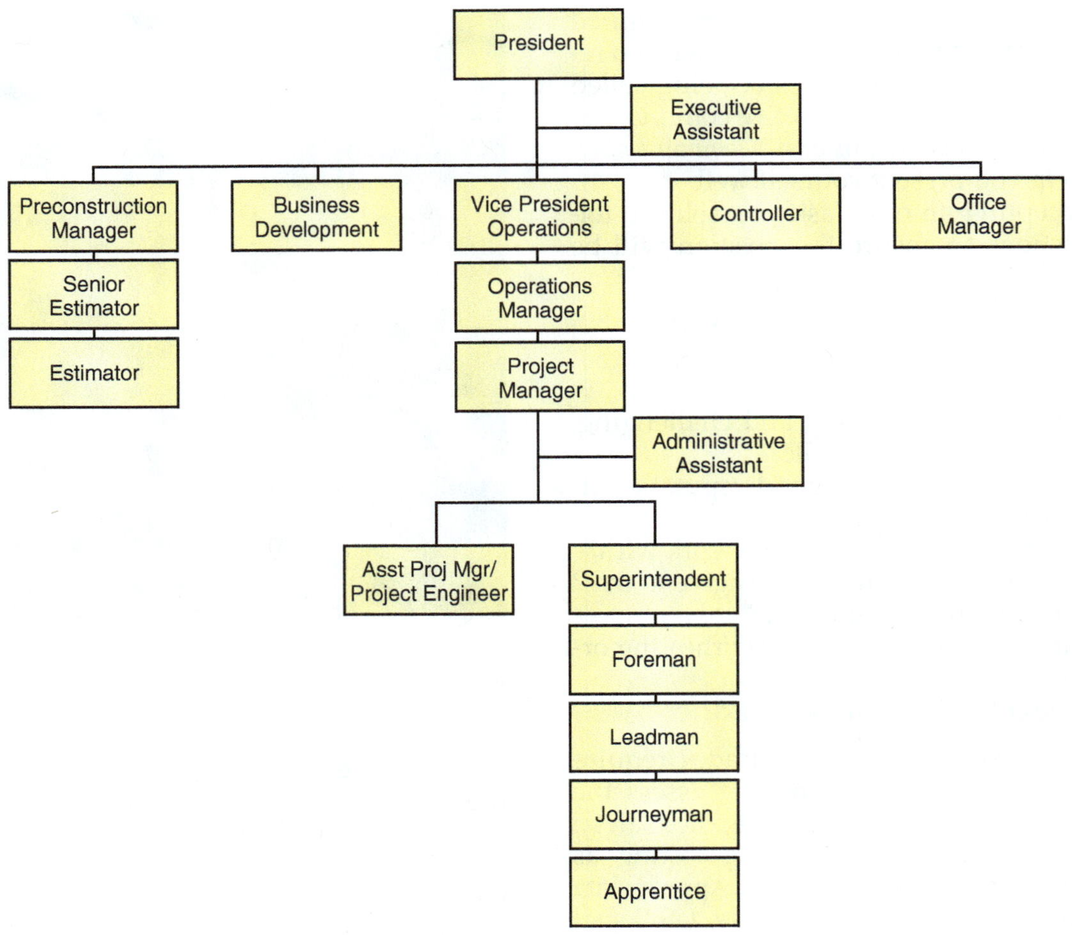

Figure 4 An organizational chart.

TWIC

In the United States, the Transportation Security Administration (TSA) and the US Coast Guard have worked together to create the guidelines for the Transportation Worker Identification Credential (TWIC). The tamper-resistant, biometric-based card is required for all workers who require unescorted access into secure areas of port facilities, outer continental shelf facilities, and all vessels regulated under the Maritime Transportation Security Act of 2002, as well as for all US Coast Guard-credentialed merchant mariners. An integrated circuit chip on the card stores the holder's information and biometric data for positive identification. This credential will likely expand to other forms of transportation in the future.

Completing an Application: Then and Now

Before computers and the Internet became part of daily life, job applicants approached the job market in a very different way. Instead of searching websites and sending their resumes off with the click of a mouse, more footwork was involved. Reading through the classifieds of the local newspaper or going door-to-door to fill out applications were arduous, time-consuming tasks.

Additional Resources

Knock 'em Dead Resumes: A Killer Resume Gets More Job Interviews! Martin Yate. 2014. Avon, MA: Adams Media.

Knock 'em Dead: The Ultimate Job Search, Martin Yate. 2014. Avon, MA: Adams Media.

1.0.0 Section Review

1. Who plans, coordinates, and controls a project's progress, or a portion thereof?
   a. Civil engineer
   b. Architect
   c. Project manager
   d. The funding bank

2. Which of the following is a valid statement about a resume?
   a. Always use a highly creative and unusual font to set your resume apart from all others.
   b. Format your resume chronologically, putting your current or last job first.
   c. Do not allow others to proofread your resume, as they will likely be critical of it.
   d. Resumes should not include certifications; present them only if called for an interview.

SECTION TWO

2.0.0 CRITICAL THINKING AND PROBLEM SOLVING

Objective

Explain the importance of critical thinking and how to solve problems.

a. Describe critical thinking and barriers to solving problems.
b. Describe how to solve problems using critical thinking.
c. Describe problems related to planning and scheduling.

Trade Terms

Absenteeism: A consistent failure to show up for work.

Tardiness: Arriving late for work.

Having the ability to solve problems using critical thinking skills is a valuable skill in the workplace. There will be situations in which timely job completion is at risk because a problem suddenly arises. Using the tools of critical thinking skills can get the project moving forward and will also demonstrate your problem-solving capabilities.

2.1.0 Critical Thinking and Barriers

Throughout your construction career, you will encounter a variety of problems that must be solved. Critical thinking skills allow you to solve such problems effectively. Critical thinking means evaluating information and then using it to reach a conclusion or to make a decision. Critical thinking allows you to draw sound conclusions and make good decisions when you use the following approach:

- Do not let personal feelings get in the way of fairly evaluating information; put them aside and remain objective.
- Determine the cause (why it happened) and effect (what happened as a result).
- Think about the expertise or experience of people who are sources of information. Ask experts and people you trust for their advice. There is no reason to think through a problem in isolation.
- Compare new information with what you already know. If it does not fit, question it and try to separate fact from fiction.
- Weigh the merits of each option and alternative, and consider if one or more can be justified.

2.1.1 Barriers to Problem Solving

When searching for the solution to a problem, it is easy to fall into a trap that prevents you from making the best possible decision. The following are the most common barriers to effective problem solving:

- Closed-mindedness
- Personality conflicts
- General fear of change

To be closed-minded is to distrust any new ideas. Closed-minded people may also resist any suggested changes. Effective problem solving, however, requires you to be open to new ideas. Sometimes the best solution is one that you would have never considered on your own. Remember that other people have good ideas, too. You must be willing to listen and evaluate their input fairly.

Sometimes you may fail to appreciate the value of information or advice simply because you do not get along with the person offering it. One of the most important skills one can master is the ability to separate the message from the messenger. Weigh the value of the information separately from your feelings about the individual. This ability will show people that you are a true professional.

People often fear change when they believe it threatens them somehow, while it is often the lack of change that is the problem. In addition, it is often difficult to clearly see the path to implementing change or to envision the final result. When the path toward, or results of, a proposed change is difficult to see, the fear of change is combined with a fear of the unknown. If you embrace the concept of change, you will never stop finding new ways to solve problems. An ancient Greek philosopher named Heraclitus is credited for first speaking the words "change is the only constant in life."

2.2.0 Solving Problems Using Critical Thinking

Problems arise when there is a difference between the way something is and the way it should (or needs to) be. You might feel frustrated or intimidated by a problem, or you might feel that you do not have enough time to solve it. Your reaction might be to simply ignore the problem. Instead, try to look at it as an opportunity to demonstrate your skills. By actively seeking solutions to problems, you will demonstrate to your colleagues and supervisors that you are responsible and capable. Encountering and conquering a challenging problem can have significant effect on your reputation, in addition to building your confidence and level of experience.

To solve problems that arise on the job, use the following five-step process (*Figure 5*). Reviewing each of these steps in detail will help you understand how to apply them.

Step 1 – Define the problem. Before you can solve a problem, you need to know exactly what it is. This step might seem obvious, but people often find themselves trying to solve the wrong problem or one that doesn't even exist.

Step 2 – Analyze and explore alternatives. Once you have defined the problem, consider different ways to solve it. Collect information from a wide range of sources, and ask people for their opinions. Perhaps you have a solution in mind, or maybe just one step towards the solution. It can be very productive to seek opinions on solving small portions of a problem rather than seeking opinions on the big picture.

Teamwork is an important part of this step. The more suggestions you get from experts and co-workers, the better your chances are of finding the right answer. Identify and compare the alternatives. Look for solutions that will be the most cost-effective, take the least time, and ensure the highest-quality result. Try to identify the short- and long-term consequences, both good and bad, of each possible solution. Eventually, you will have created several possible options and given each one some thought.

Step 3 – Choose a solution and plan its implementation. Move forward by choosing the solution that you think will work best. Next, develop a plan to put the solution into action. The plan should include all necessary tools and materials. It should also specify all the tasks involved and who is responsible for them, and estimate how long each task will take.

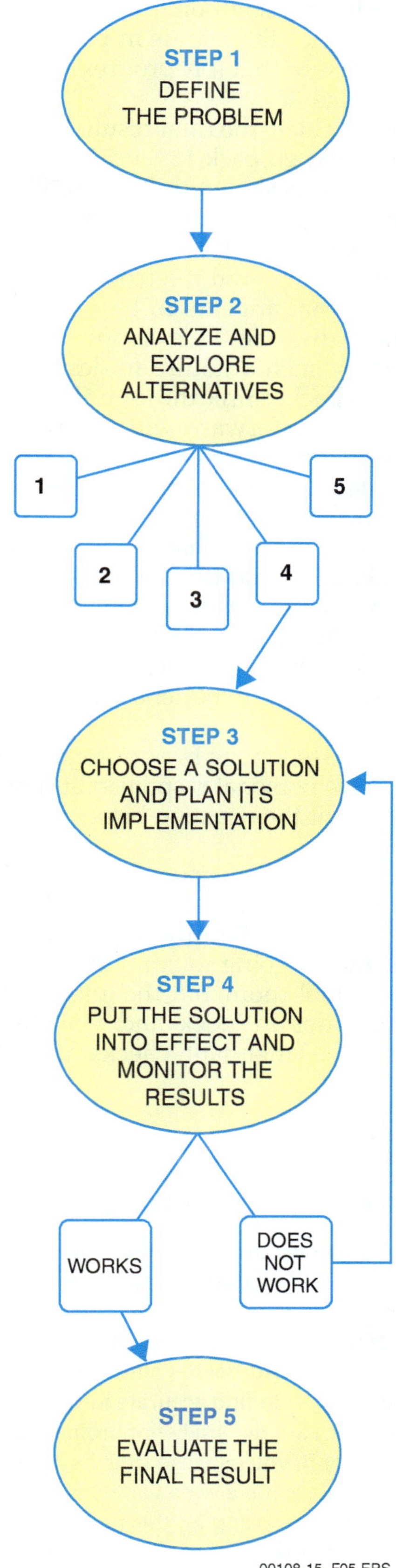

Figure 5 The five-step problem-solving process.

Step 4 – Put the solution into effect and monitor the results. Once the plan is in progress, follow it closely to ensure that it is progressing toward the desired results.

Step 5 – Evaluate the final result. If your solution does not work, go back to Step 3 and choose another solution, as shown in *Figure 5*. Then develop and implement a plan suited to the new solution.

Remember that not every solution will be successful. In fact, life and the workplace have a way of turning what appears to be a logical solution into a new disaster. This happens to everyone; do not allow such a failure to destroy your confidence. Go back, think through the options once again, and move forward with a new solution.

When the problem is finally solved, look back over the steps you took and see what lessons can be learned from the process. Perhaps there was something you could have done better. Maybe you could have saved time or money by changing a small part of the plan. When you are satisfied with the solution, remember what worked and what did not. When you face a similar problem in the future, your experience will help you find a solution.

Review the following hypothetical case. Imagine that you have personal responsibility for finding a way to prevent the situation below from repeating itself.

While on a job site, your co-workers ask you to retrieve more welding electrodes. This particular welding task must be finished by the end of the day, and that is going to cut it close. You walk to the nearest tool room, but the attendant only has sixty electrodes in stock and you need several hundred. The attendant checks with the other tool room and they are fully stocked with the electrode you need. It will take some time to get to the other tool room, though. What can be done to solve the immediate problem? What can be done to ensure that the shortage doesn't happen again?

The following sections describe an example approach you may take in this situation.

2.2.1 Defining the Problem

The proper type of welding electrodes must be readily available in order to keep the job running smoothly. It is critical that the task is complete by the end of the shift because the next shift needs to start the next phase. If the welding is not complete, it will impact the schedule and budget. There is a short-term shortage of the proper welding electrode.

2.2.2 Analyzing the Alternatives

You could take the sixty electrodes in stock, walk back to the task site, and then take a bicycle to the other tool room for the remaining electrodes. Or you could take a bicycle now to the other tool room, but it would take at least 30 minutes to complete the trip.

You could also contact the other tool room attendant that has the electrodes and explain the predicament. Then you could ask the attendant to bring the electrodes to you. This will save you (and the welders) the time involved in making the trip. While you are waiting, you can take the sixty electrodes to your co-workers so the job does not come to a halt.

What are some of the benefits and drawbacks of each alternative?

Information on the Internet

Everybody knows that the Internet provides access to information about almost everything. The trick is to know whether the answers you find there are the right ones for you. How do you evaluate resources that you find on the Internet?

One of the easiest and most effective ways is to look at the two- or three-letter extension at the end of the website address. Addresses ending in .gov or .us are official local, state, or federal government web pages. Use those pages to find accurate information about codes and regulations that apply to your work. Addresses ending in .org are generally nonprofit organizations. If the organization is affiliated with your industry or sets the standards for it, you can also trust its information.

Manufacturers put a wide variety of information, such as product specifications and catalogs, on their websites. When looking for this type of information, go straight to the manufacturer's website. The websites of companies that sell products from many different manufacturers may not be as complete or as up-to-date as the manufacturer's own site.

2.2.3 Choose a Solution and Plan

After weighing each option, you decide to contact the other tool room attendant, ask for assistance, and then deliver the sixty electrodes to your co-workers. This will keep the job going and will save some of your time away from the task by having the electrodes delivered. However, the delivery requires the attendant to close the stock room while making the trip, which could affect others needing important materials.

To keep the tool rooms properly stocked with all consumable materials from now on, you suggest to your supervisor that the local stock attendants should take an inventory at the end of each shift and restock as appropriate.

2.2.4 Implement and Monitor

At the end of each shift, the attendants now inventory consumable materials to ensure a full stock of materials for the next day. They previously took inventory every three days. After a few days, someone in authority returns to the local stock rooms to see how the change is working and how the attendants feel about the new process. Perhaps they might have fresh input and suggest additional improvements to the process.

2.2.5 Evaluate the Final Result

Since implementation, no tool room has run low on consumables. No one has had to scramble and search for materials, and an attendant does not have to close the tool room to deliver materials.

Remember, the word *critical* means important or indispensable. Critical thinking skills are important and indispensable tools in your personal tool box (*Figure 6*). As with any other tool, you must learn how to use these skills correctly and safely. When you do, you will find that one of the most rewarding experiences you can have as a trade professional is to face a problem and solve it yourself.

2.3.0 Planning and Scheduling Problems

Every task has a logical position in the overall project schedule. Each task must occur before or after other tasks, or at the same time as other tasks. Before beginning each task, ensure that it is understood what you are expected to do and what is required to accomplish the task. You must know where your work begins and ends and which materials to install. Divide the task into individual steps and determine the sequence for performing the work.

00108-15_F06.EPS

Figure 6 Critical thinking is just as important as any tool you carry.

No matter how carefully you plan ahead, unexpected problems can suddenly appear. Usually, extra time is built into project schedules so that simple problems will not cause a delay. However, more complex problems can throw the project off track.

Project managers and/or site planners (when the position is assigned) are generally responsible for ensuring that a project stays on schedule; therefore, it is common for them to use project management software (*Figure 7*). Nevertheless, as a team member who has been working on a particular job, you may be called upon to evaluate and help solve problems. Schedules must change to accommodate all types of problems that arise. It is important to be familiar with the types of problems that can disrupt a job, including regulatory issues. Typically, problems will fall under one or more of the following categories:

- Materials
- Equipment
- Tools
- Labor

2.3.1 Materials

The materials required for a job are identified during the estimating stages of a project. Once the contract has been completed, they are then ordered from suppliers and delivered according to a prearranged schedule. Many job sites do not have sufficient room to store all the needed materials for a structure at one time. Problems with materials can include errors of quantity or type, delays in delivery, and unavailability due to backorders or even to abrupt business closure. A shortage of materials due to waste, theft, or vandalism is another source of project delays.

In many cases, material problems lead to changes in the material to be used. If the new material was not listed as an option in the contract specifications, a significant amount of time and effort may be required for the change request to make its way through the approval process.

2.3.2 Equipment

Major construction equipment is usually selected and scheduled well before the need arises. If a site planner is a member of the construction team, he or she may be responsible for ensuring that equipment is on site at the right time. Otherwise, project managers and other supervisors may be assigned this responsibility. This generally depends on the scope of the project.

Often, multiple pieces of equipment must be scheduled at the same time: backhoes to excavate a foundation and dump trucks to haul away the excavated dirt, for example. Problems with equipment can include unavailability on the scheduled day(s), lack of qualified operators, mechanical breakdown, and extended maintenance.

The lack of major equipment can create significant delays. In many cases, multiple crafts and subcontractors are relying on the equipment to move forward with their responsibilities. For example, if a crane is required to lift steel but it is not available for any reason, ironworkers, welders, and possibly other crafts find themselves at a standstill.

2.3.3 Tools

Workers and supervisors alike are responsible for ensuring that the appropriate tools are available. Workers are often expected to have common but specific tools when they are employed. Special tools that are expensive or unique are usually the responsibility of the employer. Both parties must ensure that they do their part to provide and maintain proper and functional tools.

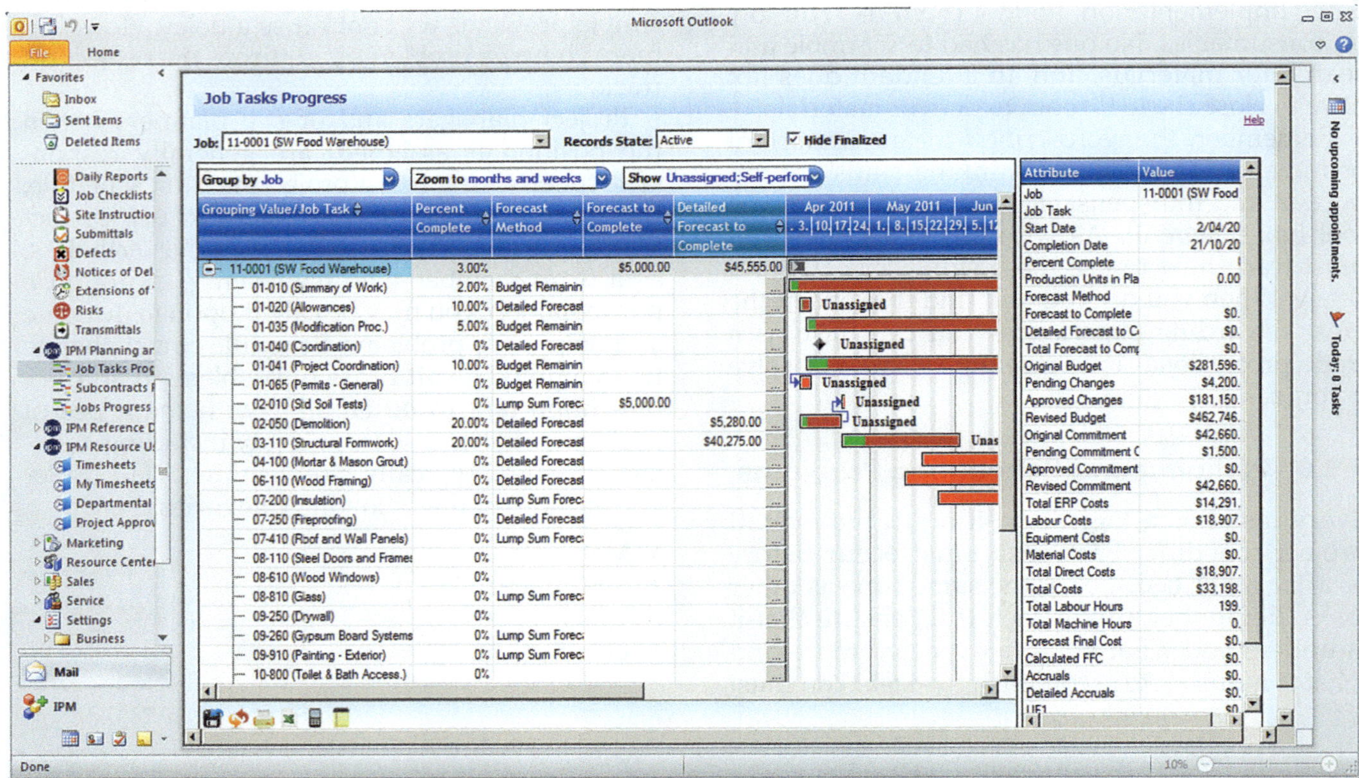

00108-15_F07.EPS

Figure 7 Project management software.

Imagine how a house project would stall if all the carpenters arrived for a day's work without hammers. On the other hand, if pneumatic nail guns are being used extensively, but the supervisor for the contractor does not ensure an air compressor is on site, the project comes just as quickly to a standstill. Other common problems related to tools include breakage or damage, loss or theft, or a lack of skilled users.

2.3.4 Labor

Labor—the men and women who perform the work on the job site—is the most important component of a project. Labor is often one of the most expensive project components as well, depending upon the country, project, and type of workforce needed. Project planners estimate the number of people needed for each day and identify the range of skills required of those workers. Foremen ensure that the right people are available and working on the right tasks. Common problems related to labor include **tardiness** and **absenteeism**, lack of experience or qualifications to perform a given task, and not enough available workers. Idle time due to a lack of guidance or laziness can also contribute to delays.

2.3.5 Handling Delays

Always be aware of possible delays as you carry out your assigned tasks. Look as far forward into the portion of a project that you are responsible for, and consider the types of delays that could happen. If you see a potential cause of delay, notify your supervisor immediately. Having a plan in place for delays, especially delays that occur regularly or happen often, enables a solution to be applied promptly. Always know clearly what your responsibilities are in such situations. Be prepared to solve the problem yourself if that is part of your assignment. Even then, always keep your supervisors updated on delicate situations involving delays.

GOING GREEN

Banner Bank Building in Boise

What makes the Banner Bank Building one of the greenest buildings in the US? It was designed to do more with less. It uses many modular components that can be reused over and over again as changes are made. Modular wall systems, along with modular power and data components, are easy to relocate without waste. The pressurized raised floor system delivers air at a higher temperature and a lower velocity than common overhead systems. It also eliminates a great deal of ductwork, and allows the location of supply air grilles to be changed at will. The structural components were designed to use less steel and concrete than most buildings in its class. Materials with a high level of recycled components were used where available and when quality would not be compromised.

00108-15_SA01.EPS

The building was intentionally constructed close to public transportation nodes to encourage their use by employees. It also has inside storage for bicycles and showers to accommodate those who want to take green living a step further. Water is reclaimed and a geothermal heating system was installed. The building is even equipped with backup generators that run on biodiesel derived from used vegetable oil.

Additional Resources

The Re-Discovery of Common Sense – A Guide to the Lost Art of Critical Thinking. Chuck Clayton. 2007. Lincoln, NE: iUniverse, Inc.

2.0.0 Section Review

1. In critical thinking, if new information you receive about an issue does not fit with what you already know, _____.
   a. question it and try to separate fact from fiction
   b. change what you previous held as factual in your mind
   c. berate the person and let them know you don't appreciate that kind of input
   d. thank them and forget about the information they offered

2. Why is it important to follow a plan through once it is implemented?
   a. To be sure there are no other options
   b. To be sure no one asks questions
   c. To look for other opinions
   d. To be sure the results are what you are looking for

3. If material that was not specified in contract documents must be substituted for another type, the change will likely need to _____.
   a. pass through an approval process
   b. pass through a testing procedure
   c. be agreeable to all other trades
   d. be considered after the project is done

SECTION THREE

3.0.0 RELATIONSHIP AND SOCIAL SKILLS

Objective

Explain the importance of social skills and identify ways good social skills are applied in the construction trade.

a. Identify good personal and social skills.
b. Explain how to resolve conflicts with co-workers and supervisors.
c. Explain how to give and receive constructive criticism.
d. Identify and describe various social issues of concern in the workplace.
e. Describe how to work in a team environment and how to be an effective leader.

Trade Terms

Amphetamine: A class of drugs that causes mental stimulation and feelings of euphoria.

Barbiturate: A class of drugs that induces relaxation, slowing the body's ability to react.

Bullying: Unwanted, aggressive behavior that involves a real or perceived power imbalance. This form of harassment may include offensive, persistent, insulting, or physically threatening behavior directed at an individual.

Cannabinoids: A diverse category of chemical substances that repress neurotransmitter releases in the brain. Cannabinoids have a variety of sources; some are created naturally by the human body, while others come from cannabis (marijuana). Still others are synthetic.

Compromise: When people involved in a disagreement make concessions to reach a solution that everyone agrees on.

Confidentiality: Privacy of information.

Constructive criticism: A positive offer of advice intended to help someone correct mistakes or improve actions.

Hallucinogen: A class of drugs that distort the perception of reality and cause hallucinations.

Harassment: A type of discrimination that can be based on race, age, disabilities, sex, religion, cultural issues, health, or language barriers.

Initiative: The ability to work without constant supervision and solve problems independently.

Leadership: The ability to set an example for others to follow by exercising authority and responsibility.

Methamphetamine: A highly addictive crystalline drug, derived from amphetamines, that affects the central nervous system.

Opiates: A narcotic painkiller derived from the opium poppy plant or synthetically manufactured. Heroin is the most commonly used opiate.

Professionalism: Integrity and work-appropriate manners.

Self-presentation: The way a person dresses, speaks, acts, and interacts with others.

Sexual harassment: A type of discrimination that results from unwelcome sexual advances, requests for sexual favors, or other verbal or physical behavior with sexual overtones.

Synthetic drugs: A drug with properties and effects similar to known substances but having a slightly altered chemical structure. Such drugs are often not illegal since they are somewhat different than well-defined restricted or illegal substances. The two typical categories are cannabinoids (lab-produced THC or marijuana substitutes) and cathinones, which are designed to mimic the effects of cocaine or methamphetamines.

Tactful: Being aware of the effects of your statements and actions on others.

Work ethic: Work habits that are the foundation of a person's ability to do his or her job.

Zero tolerance: The policy of applying laws or penalties to even minor infringements of a code in order to reinforce its overall importance, typically related to drug and alcohol abuse when applied to the workplace.

A relationship results from the process of interacting with another person, or a group of people. Relationships are affected both by real actions and the perceptions of individuals. Every day, you interact with co-workers, supervisors, and members of the public who see you working. No one wants to work with, hire, or spend time with an unprofessional person.

Craftworkers need to be aware of the appropriate professional conduct for work situations, and follow that conduct at all times. Your actions reflect on your own professional status, that of your colleagues, the company you work for, and the image of your profession as seen by the public.

3.1.0 Personal and Social Skills

Proper **self-presentation**—the way you dress, speak, act, and interact—is a vital part of any successful work relationship. Self-presentation involves developing good personal and work habits. Personal habits apply to your appearance and general behavior. Work habits, also known as your **work ethic**, apply to how you do your job. Your personal habits and work ethic make powerful impressions on your colleagues, supervisors, and potential employers. Something as simple as filling out a job application with no errors and good penmanship will give an employer a good first impression.

When you introduce yourself, look the other person in the eyes and stand with confidence. In the United States and a number of other countries, a firm and sincere handshake has long been considered an indicator of a person's character and personality, especially among men. Recent studies indicate that there actually may be a connection. A firm grip (but not so firm as to imply a competition is at hand) and making eye contact at the appropriate moment make a good impression, regardless of gender. Once formed, first impressions are hard to change, so make sure that the first impression you make is a good one.

3.1.1 *Personal Habits*

Co-workers and supervisors like people who are dependable. Co-workers know that they can trust dependable people to pull their own weight, take their responsibilities seriously, and look after one another's safety. Supervisors know that dependable people will do their best to finish a job correctly and on schedule. When dependable workers say they will do a task, they follow through on that commitment. It also means showing up for work on time every day and not stretching out lunch hours and breaks.

Organizational skills are important as well. Keep your tools and your work areas clean and organized. Know which tools you are responsible for and keep them in good working order. This applies to your personal tools as well as those belonging to your employer. Always plan what you need to do each day before you begin. Although your best plan is likely to change as the day unfolds, a failure to plan is a plan for failure. Approach your work in an organized fashion and follow the schedule that you have been assigned.

The War Against Absenteeism

During World War II, factories across America were humming along, many producing products needed for the war effort. To combat absenteeism in the workplace, employers like Remington Arms related the issue to patriotism and the crucial role that America's workers played in the war effort.

00108-15_SA02.EPS

Ensure that you are technically qualified to perform your duties and specific tasks. This means that you know how to use tools, equipment, and machines safely. Take advantage of opportunities to expand your technical knowledge through classes, books and trade periodicals, and mentoring by experienced colleagues. Being technically qualified also means that you will not attempt a task that you are unfamiliar with when any level of safety hazard exists.

Offer to pitch in and help whenever you can. The best workers are willing to take on new tasks and learn new ideas. Supervisors notice employees who are willing, but be careful not to take on more tasks than you can handle; you will not impress anyone if you cannot fulfill what you promised.

Honesty is one of the most important personal habits you can have. Do not abuse your company's system by calling in sick just to take a day off or by leaving early and asking someone else to punch you off the clock. Never steal from the company or from your co-workers; this includes everything from tools and equipment to simple office supplies. If you are struggling with a problem, don't hide it; speak with your supervisor about it truthfully. Be willing to look actively for a solution to problems you are facing.

Professionalism means that you approach your work with integrity and a professional manner. As you learned in *Basic Safety*, there is no place for horseplay or irresponsible behavior on the construction site. Employers want workers who respect the rules, and who understand that rules exist to keep people safe and projects on schedule. A professional employee always respects company **confidentiality**. In some cases, workers are required to sign documents indicating that they understand the importance of confidentiality and that they agree not to discuss certain information that is shared with them.

Most companies have a dress code, and often they have additional special requirements for specific jobs. Always follow these requirements. Pushing the limits of such policies never impresses anyone of importance. Also, do not forget to attend to the basics of good grooming. People who pay attention to their appearance and develop positive personal habits are more likely to be considered for a job over people who do not have good grooming habits. On the job site, you are a reflection of your trade and employer, as well as yourself.

3.1.2 Work Ethic

Along with good personal habits, a strong work ethic is essential for getting hired and being promoted. Employers look for people who they believe will give them a fair day's work for a fair day's pay. Having a strong work ethic means that you enjoy working and that you always try to do your best on each task. When work is important to you, you believe that you can make a positive contribution to any project you are working on.

Construction work requires people who can work without constant supervision. When there is a problem that you can solve, solve it without waiting for someone to tell you to do it. If you finish your task ahead of time, look for another task that can be finished in the remaining time or ask co-workers if they need help. This type of positive action is called taking **initiative**. Colleagues and supervisors will respect you more when you demonstrate initiative on the job.

An important part of taking initiative, however, is to know when and how to implement it. Suppose a supervisor tells you to perform a task. He demonstrates five steps required to complete the task. Later, as you perform the work, you realize that one of the steps is unnecessary. Leaving out the step could save time—but should you? No; that would not be the right initiative. Instead, tell your supervisor about your discovery, and ask what you should do. Bringing the options to your supervisor is showing initiative. Your supervisor may realize that you are right and allow you to leave out the extra step. After that, every time anyone performs that task, it will take less time and save the company money. Or, your supervisor may explain why the step is important. Then you will have learned something more about the work you are doing. The result of taking the initiative in either case is a positive one.

3.1.3 Tardiness and Absenteeism

The two most common problems supervisors face on the job are tardiness and absenteeism. Tardiness is when a worker habitually shows up late to work. Absenteeism is when a worker consistently fails to show up for work at all, with or without excuses. People with a strong work ethic are rarely late or absent.

To make a profit and stay in business, construction companies operate under tight schedules. These schedules are primarily built around workers. If you are late or do not show up regularly, expensive adjustments become necessary. Your employer may decide that the best way to save money is to stop wasting it on you.

Consider the following suggestions for improving or maintaining your record of punctuality and attendance:

- Think about what would happen if everyone on the job were late or absent frequently.
- Think about how being late or absent affects your co-workers.
- Know and follow your company's policy for reporting legitimate absences or lateness.
- Keep your supervisor informed if you need to be out for more than a day.
- Allow yourself enough time to get to work.
- Explain a late arrival to your supervisor as soon as you get to work.
- Do not abuse lunch and break-time privileges.

3.2.0 Conflict Resolution

Conflict resolution is an important relationship skill, because conflict can happen anywhere, anytime. Conflicts between you and your co-workers can arise because of disagreements over work habits; different attitudes about the job or the company; differences in personality, appearance, culture, or age; or distractions caused by problems at home. Conflicts between you and your supervisor can happen because of a disagreement over workload; lateness or absenteeism; or criticism of mistakes and inefficiencies. Significant disagreements with a supervisor should be reported to the Human Resources (HR) office of your employer.

Most of the time, people are not trying to turn disagreements into conflicts. People are often unaware of the effects of their behavior. Before reacting negatively to a co-worker's behavior, remember to be **tactful**. This means considering how the other person will feel about what you say or do. Do not accuse, embarrass, or threaten the person. This behavior rarely leads to an acceptable solution to anything.

Never let a disagreement or conflict affect job performance, team morale, or—most importantly—site safety. The goal is to keep events from turning into conflicts in the first place. If that is not possible, then the next best thing is to address the conflict quickly and resolve it professionally. If a disagreement with a co-worker is getting out of hand, try one of the following techniques to cool the situation down:

- Think before you react.
- Walk away, explaining that you need some time to think through the situation.
- Try not to take it personally.
- Avoid being drawn into others' disagreements.

Do not let a conflict simmer and then boil over before you take action. Not only is it unprofessional, tempers may flare unexpectedly and make the situation worse. Being proactive when an issue becomes uncomfortable is usually better than being reactive. This is true unless tempers are already out of bounds. In that case, it may best to allow some time to pass. If, despite your best efforts, the conflict escalates, walk away and notify your supervisor immediately.

Keep in mind, however, that there are important differences between the way you resolve conflicts with your co-workers and the way you resolve conflicts with your supervisor. The following sections discuss how to handle these types of conflicts.

3.2.1 Resolving Conflicts with Co-Workers

Remember to have respect for the people you disagree with. After all, they believe they are right, too. Be clear, rational, respectful, and open-minded at all times. Begin by admitting to each other that there is a conflict. Then analyze and discuss the problem. Allow everyone to describe his or her own perception of the conflict. You may realize that the whole problem was simply miscommunication. Begin by asking the following questions:

- How did the conflict start?
- What is keeping the conflict going?
- Is the conflict based on personality issues or a specific event?
- Has this problem been building up for a while, or did it start suddenly?
- Did the conflict start because of a difference in expectations?
- Could the problem have been prevented?
- Do both sides have the same perception of what is happening?

Once you have analyzed the situation, discuss the possible solutions. You will probably have to **compromise** to find a solution that everyone agrees on. When you agree on the solution, act on it, and see if it works. If it does not, then consult your supervisor for help in resolving the conflict. Notice that this process is similar to the problem-solving techniques discussed earlier. If you find that the conflict happened because you were wrong, apologize. An apology is not a sign of weakness; it is a sign of respect.

Once an agreement or compromise has been reached, stand by it in every possible way, and do so with a positive attitude.

3.2.2 Resolving Conflicts with Supervisors

You can usually approach co-workers as equals, because you are working on the same job. However, on the job, your supervisor is in charge of you and your work. This means that you must use a different approach to resolve conflicts with your supervisor.

Before going to your supervisor, take some time to think about the cause of the conflict. Consider writing down your thoughts; organizing the information this way often puts things into perspective. When you approach your supervisor, do so with respect. Wait until your supervisor has a free moment, and then ask if you could arrange a time to talk about something

important; or leave a message or note for your supervisor asking to meet. Be willing to meet at a time that is convenient for your supervisor. Remember, supervisors have many responsibilities. They do not have much free time during the regular workday, and you may have to meet before or after work.

When meeting with your supervisor, speak calmly and clearly. Do not be emotional, sarcastic, or accusatory. Do not confront your supervisor in a threatening or angry way. State only the facts as you see them; never say anything that you cannot prove. Do not mention the names of your co-workers unless they are directly involved. If you want to suggest changes or solutions, explain them clearly and discuss how they will benefit the people involved.

Once you have made your case, allow your supervisor to make a decision. You should accept and respect your supervisor's final decision. It may not be the one you wanted, but it will be the one that must be followed.

3.3.0 Giving and Receiving Criticism

As a construction professional, you will always be learning something new about your job. New technologies, materials, and methods appear all the time. For example, talking to an experienced construction worker can help you learn a new way to perform a task. As your skills improve with practice, you will be able to use tools and methods that you were not able to use before. This is a common way of learning on the job (*Figure 8*).

Another way to learn on the job is through **constructive criticism**. Constructive criticism is advice designed to help you correct a mistake or improve an action. Constructive criticism can improve your job performance and relations with co-workers. You have probably heard the word *criticism* used in a negative way to indicate fault or blame; that is not the type of criticism discussed here. Constructive criticism does not mean that colleagues and supervisors think little of you; in fact, it means exactly the opposite. Colleagues and supervisors who offer constructive criticism do so because they believe it is worth their time to help you improve your skills.

As you gain experience on the job, you will be able to give constructive criticism as well as receive it. To make sure that someone does not mistake your constructive criticism for blame, you need to know how to offer it in a positive manner. The following sections offer some general advice on offering and receiving constructive criticism.

00108-15_F08.EPS

Figure 8 A good employee never stops learning on the job.

3.3.1 Offering Constructive Criticism

When you are training a less-experienced person or working with co-workers, you might find yourself offering some constructive criticism. How should you offer it? Before you say anything, think about the rules of effective speaking that you learned in *Basic Communication Skills*. Use positive, supportive words and offer facts, not opinions.

Constructive criticism works best when offered occasionally. Do not constantly comment on another person's work or methods; they may block out your criticism or even become angry. Never criticize people in front of their co-workers or supervisors; they may feel embarrassed and will likely resent you. Constructive criticism should include suggested alternatives. Do not criticize how a co-worker does something unless you can suggest another way. Above all, limit your comments to the person's work or methods, without targeting them as a person. There is a great deal of difference in how people react to a statement such as "your pipefitting work is unacceptable" compared to "this threaded joint is unacceptable."

Point out improper behavior or incorrect work techniques the first time they occur. If not addressed, the behavior or technique could become a bad habit. Bad habits, in turn, can lead to accidents. Gently and firmly offer constructive criticism for improper behavior as soon as it is noticed.

Remember to compliment the person you are criticizing. Compliments have a genuinely positive effect on the person receiving them, and they are easy to offer. You do not have to wait to compliment someone until you are prepared to offer constructive criticism. However, if a compliment is to be offered along with constructive criticism, offer the compliment first.

Try to offer compliments on a regular basis. Your appreciative and respectful approach will help keep team spirits high and your supervisor will appreciate your professional attitude.

It's All In How You Say It

Constructive criticism should be given in a positive, non-confrontational way. People can be sensitive, and when approached about how they do their jobs, it becomes very personal to them.

When delivering constructive criticism, it is not a good idea to start the sentence off with "You" because it will immediately put the person on the defense. Instead, try starting out with "In the future." This approach will sound less like an attack on their abilities and they will react and listen better. Talk about the work, not the person.

3.3.2 Receiving Constructive Criticism

Not only can constructive criticism mean the difference between success and failure, but in cases where workplace safety is involved, it can also be the difference between life and death. Think of constructive criticism as a chance to learn and to improve your skills. Take constructive criticism from your supervisor seriously. Treat constructive criticism from experienced co-workers with the same respect; even though they may not be your supervisor, their experience gives them a level of expertise. When someone presents you with constructive criticism in the proper manner, learn from their approach as well.

Always take responsibility for your actions. Never be overly defensive or dispute criticism. If you respond negatively, you might offend a person who is simply trying to help you. As a result, that person will either lose respect for you or will no longer be willing to help. If that happens, the quality of your work, and that of everybody who works with you, will suffer.

When someone offers you constructive criticism, demonstrate your positive personal habits and work ethic. If the criticism is vague, ask the person to suggest specific changes that you should try to make. If you do not understand the criticism, ask for more details. The deepest respect you can show is to improve your performance based on their advice.

You have a right to disagree with criticism that is unacceptable or incorrect. Someone might honestly misunderstand your situation and offer incorrect advice. Or a co-worker might criticize you simply because he thinks he performs a task better than you do. In such cases, clearly and respectfully give your reasons for disagreeing, and explain why you believe that you are correct.

3.3.3 Destructive Criticism

Constructive criticism has positive effects. However, you may also encounter destructive criticism. Destructive criticism, as its name suggests, is designed to hurt, not help, the person receiving it. The destructive criticism could actually be about something that needs improvement, but because it is offered negatively, there is no desire to listen.

If you receive destructive criticism, the professional thing to do is to stay calm. Do not get into a fight by replying in a negative tone. Find a way to let the person know how the criticism made you feel. If there is a legitimate criticism beneath a destructive tone, ask the person to offer positive suggestions for ways you can improve. By taking a positive approach to negative criticism, you can set a good example for others.

3.4.0 Social Issues in the Workplace

The modern construction workplace is a cross-section of our society (*Figure 9*). A typical construction project involves men and women from all walks of life, of many ethnic and racial backgrounds, and often from many different countries. Many workers speak more than one language. Workers have grown up in, and currently live in, many different income brackets.

00108-15_F09.EPS

Figure 9 Today's construction workplace is a cross-section of our society.

Construction workers face a wide range of mental and physical demands every day on the job. These demands can sometimes feel overwhelming, and people may seek to escape pressure through illegal drugs or alcohol abuse. As a construction professional, you need to be aware of these issues. You also need to know what to do if someone is behaving inappropriately.

You may encounter the following issues in the workplace:

- **Harassment** or **bullying**
- Stress
- Drug and alcohol abuse

3.4.1 Harassment

Harassment is a broad term describing negative social behavior that can be based on race, age, disabilities, gender, religion, cultural issues, health, or language barriers. Harassment often takes the form of ethnic slurs, racial jokes, offensive or derogatory comments, and other verbal or physical conduct. It can create an intimidating, hostile, or offensive working environment. It can also interfere with an individual's work performance.

One type of harassment commonly reported and talked about is **sexual harassment**. While other forms of harassment may not have been talked about as much or as openly in the past, this issue has made headlines for many years.

When someone makes unwelcome sexual advances, or requests or exhibits verbal or physical behavior with sexual overtones, he or she is guilty of sexual harassment. Contrary to popular belief, sexual harassment can happen between members of the opposite sex or even those of the same gender. The harasser may threaten to fire the victim to keep him or her quiet. Sexual harassment is illegal, and if you experience it, you should report it to your supervisor immediately. Usually, sexual harassment is a pattern of behavior repeated over a period of time, so if you do not report it, you run the risk of experiencing it over and over again.

Bullying may be more common than either sexual harassment or racial discrimination on the job, but it is not reported as often. The term *bullying* is perceived by most people to be related to school-age children. The United States government defines bullying as the unwanted, aggressive behavior among school-aged children that involves a real or perceived power imbalance. That definition covers a lot of ground. When bullying is discussed among adults, it is typically referred to as hazing, stalking, or simply harassment. Most adults are uncomfortable using the term *bullying* when reporting such behavior, as they also perceive it to be related to school-age children. Using the term makes adult victims feel childish to report it as a result. Bullies may target those that appear to be a threat to them and feel the need to control the threat. In some cases, it occurs simply because the individual feels stronger and wishes to exert control to build on their feeling of superiority.

Workers should understand that dealing with an adult bully can be very much like trying to work things out with a school-age bully. Adult bullies are often lifetime bullies, having started the practice at an earlier age. They are not typically interested in compromise or improvement. Bullying makes them feel important and in control. Adult bullies are hard to change through simple interaction and positive discussions.

Workplace bullying can come from an individual or even a group. Bullying, just as any other form of harassment, can cause serious emotional and physical illnesses such as anxiety, depression, and hypertension among victims.

With the use of today's electronic technology, bullies can even cause harm through social media and other public formats; this is referred to as cyberbullying. Once emailed rumors, malicious text messages, embarrassing pictures, videos, or fake profiles are posted on social media, the damage is difficult to undo. While the most common target for cyberbullying is school children, adults are not immune to it.

Once a bullying situation has been reported to your employer, the situation may not change immediately. Your supervisor and/or employer must work within the confines of the law. Careful documentation of events is very important, as it is in all forms of harassment. Clear documentation allows the situation to be handled properly by those in authority. Since there are laws prohibiting such behavior and your co-workers are not school children, legal action will often occur based on careful documentation.

3.4.2 Drug and Alcohol Abuse

Drinking is a common way to avoid or forget about stress. Alcohol is an accepted, legal part of modern social life. The moderate use of alcohol is socially acceptable and is generally believed to cause little or no harm to most people. Alcohol abuse, or habitually drinking to excess, not only is socially unacceptable, but also poses a serious health and safety risk to everyone involved. Although some countries and US states have legalized or decriminalized the recreational and/or medical use of marijuana—and others are contemplating it—it too can be abused and have the same effects as alcohol abuse. Regardless of whether alcohol and marijuana are considered socially acceptable, there is no room in the workplace for either substance. Among employers, this is referred to as **zero tolerance** (*Figure 10*).

Illegal drug use is never acceptable, nor is the misuse of legal prescription drugs. Many companies require a clean drug test before offering a job or allowing an employee to begin work. In fact, many employers in the construction industry are often required to share records related to drug testing as part of their obligations to their clients. Drug testing can also be ordered randomly, to test the current workforce and ensure compliance. Further, an accident in the workplace often leads to immediate drug testing to determine if illegal substances were a factor. There are a number of situations that can lead to a drug test beyond a pre-employment situation. This matter is taken very seriously by the industry as a whole.

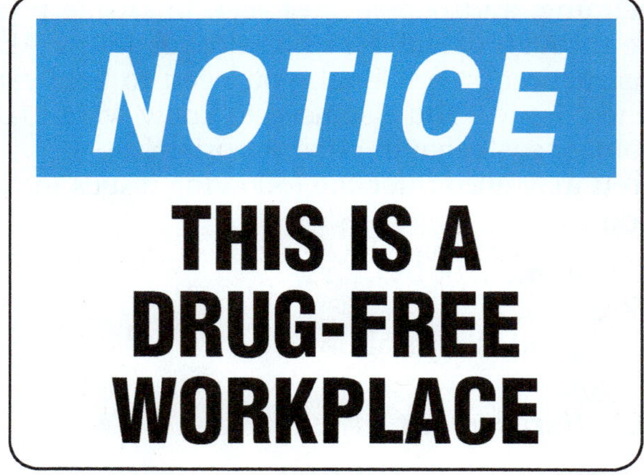

00108-15_F10.EPS

Figure 10 Signs like this are common in the construction workplace.

WARNING!

Working under the influence of drugs or alcohol can have serious consequences. It can lead to fatal accidents and serious injuries on the job site affecting you and your co-workers, and can cause serious errors affecting the integrity of the work. In addition, causing harm to people or property in the workplace while under the influence of substances can also lead to personal liability for the results of a poor choice.

Types of drugs that are sometimes used inappropriately by workers include the following:

- **Amphetamines**
- **Methamphetamines**
- **Barbiturates**
- **Hallucinogens**
- **Opiates**
- **Synthetic drugs**

Amphetamines and methamphetamines are addictive stimulants that affect the central nervous system. Also called uppers, they are used to prolong wakefulness and endurance, and they produce feelings of euphoria, or excitement. They can disturb vision; cause dizziness and an irregular heartbeat; cause the loss of coordination; and can even cause physical collapse or unconsciousness. Cocaine, crack, and crystal meth are examples of illegal amphetamine drugs.

Barbiturates are a sedative, which basically means they cause you to relax. They are also called downers. They can create slurred speech; slow reactions; cause mood swings; and cause a loss of inhibition, which means that they make a person feel less shy or self-conscious. Many abused barbiturates are controlled-substance prescription drugs.

Hallucinogens distort the perception of reality to the point where people see things that are not there (hallucinations). The effects of hallucinogens can range from ecstasy to terror. When someone is hallucinating, they put themselves and others at great risk because they cannot react to situations the right way. Hallucinogens can also cause chills, nausea, trembling, and weakness. Mescaline and LSD are two examples of illegal hallucinogens.

Derived from opium, opiates are used to kill pain. Morphine is the primary psychoactive chemical found in opium, and therefore the basis of most opiates. While opiate use creates a feeling of euphoria and relieves pain, it can cause muscle spasms, cramps, anxiety, nausea, fever, and diarrhea over time. Some prescription-drug forms of opiates are Oxycontin and Vicodin.

Synthetic drugs are relatively new. They are defined as drugs with properties and effects similar to well-defined and familiar restricted or illegal substances, but having a slightly altered chemical structure. Such drugs are often not illegal since they are somewhat different from the illegal substances. However, due to the health risks and other problems associated with them, legislation has begun in some areas to control them. The two typical categories are **cannabinoids** (lab-produced THC or marijuana substitutes) and cathinones, which are designed to mimic the effects of cocaine or methamphetamines. The cathinones are more commonly known as bath salts (but they are not used for this purpose). Synthetic drugs have quickly become readily available for retail sale. They are often deceptively packaged and do not generally list all the ingredients. Synthetic drugs can be extremely hazardous and even deadly, as users have no idea what their true composition is. Lab testing has shown the potency of some formulas can be as much as 500 times as potent as marijuana.

If you abuse drugs or alcohol, you will probably not be able to get or keep the job you want. If you are discovered using illegal drugs, you can be fired on the spot because of the safety risks related to drug use. Addiction affects your well-being and state of mind, and you cannot do your best on the job or at home. It can also jeopardize your relationships with friends, family, and co-workers. If people you know are addicted to alcohol or drugs, seek help for them or encourage them to get help for themselves. If you are having trouble with alcohol or drugs, talk to your supervisor. Many companies provide substance abuse counseling and will support you with periodic drug testing to help you remain sober and drug-free. The *Appendix* contains a list of organizations that can help people cope with alcohol and drug issues.

3.5.0 Teamwork and Leadership

Every day, you will interact with members of your work crew and your supervisor. Although the ability to get along with people is an important skill, it is not quite enough. The success of any team depends on all of its members filling their roles competently. Everyone must contribute to the effort to ensure that the team achieves its goals. This cooperation is referred to as teamwork.

From the moment your career begins, you will be assigned to different teams for different reasons. Teams can be as small as two people and as large as your entire company. Teams are often made up of people from different trades who work together to complete a task. For example, a team that is assigned the task of burying water and electrical lines might include a backhoe operator to dig the trench, pipefitters to lay and join the pipe, and electricians to run the electrical cable.

Always show respect for the other members of your team. Although you have a specific job to do, always be willing to help a co-worker. Support your co-workers when they need it and they will support you in return.

Good team members are goal-oriented, and those goals are shared among team members. Being goal-oriented means making sure that all activities benefit the team's final objective. Whatever the end result is—fabricating fittings, laying pipe, or framing a house—everyone on the team knows the desired result in advance and concentrates on achieving it.

As a team member, use your skills and strengths to the team's advantage. At the same time, accept your personal limitations and the limitations of others. To be a good team member, strive to do the following:

- Follow your team leader's and/or supervisor's directions.
- Accept that others might be better at some tasks than you.
- Keep a positive attitude when you work with other people.
- Recognize that the work you do is for the benefit of the entire company, not for you personally.
- Learn to work with people who work at different speeds.
- Accept goals that are set by someone else, not by you.
- Trust other members of the team to perform their tasks, just as you perform yours.
- Appreciate the work of others as much as you appreciate your own.

Keep in mind that you are not the only person working hard. Everyone on your team should be focusing on the goal, too. Offer praise and encouragement to your co-workers, and they will do the same for you. This mutual respect will help you feel confident that your team will reach its goal. Share the credit for good work, and be willing to take responsibility for your mistakes and errors. These actions will help you earn the respect of your co-workers.

Did You Know?

Discrimination is Prohibited

Several federal laws prohibit job discrimination:

- **Title VII of the Civil Rights Act of 1964** – Prohibits employment discrimination based on race, color, religion, sex, or national origin.
- **Equal Pay Act of 1963** – Protects men and women who perform substantially equal work in the same establishment from sex-based wage discrimination.
- **Age Discrimination in Employment Act of 1967** – Protects individuals who are 40 years of age or older from workplace discrimination.
- **Title I and Title V of the Americans with Disabilities Act of 1990** – Prohibits employment discrimination against qualified individuals with disabilities in the private sector and in state and local governments.
- **Sections 501 and 505 of the Rehabilitation Act of 1973** – Prohibits discrimination against qualified individuals with disabilities who work in the federal government.
- **Civil Rights Act of 1991** – Provides monetary damages in cases of intentional employment discrimination.

Source: US Equal Employment Opportunity Commission

If goal-oriented teamwork is practiced, the team should be able to meet deadlines and keep projects on schedule. This translates into time and money saved by your company that can be used for pay increases. There will be times when your co-workers or other trades cannot start their tasks until you have completed yours. Out of respect for them and your company's reputation, always finish your work in a timely manner. The success of a team reflects on you, just as your personal behavior and characteristics reflect on the team.

An important part of teamwork is training. As an apprentice, you will receive guidance and advice from more experienced colleagues. As you gain experience, you will be able to help less-experienced colleagues. Training also offers an excellent opportunity to build good relationships.

Being asked to teach someone is an honor. It means that someone believes that you do your job well enough to teach it to others. Such confidence should inspire you to be the best teacher you can be. Be patient with the person you are teaching, and teach by example. Offer encouragement and give constructive criticism, as you have learned in this module. Teach people in the same manner that you would like to be taught. Keep in mind that your teachers in the workplace are not likely to be trained educators. As a result, their approach may be less than perfect. Discount their limitations and focus on the skill or knowledge they are sharing with you.

3.5.1 Leadership Skills

As you gain experience and earn credentials, you will assume positions of greater responsibility. These positions are earned through hard work and dedication. Starting as an apprentice, you can work your way up to team leader, foreman, supervisor, and project manager. Someday, with enough hard work, dedication, and ambition, you may even choose to run your own company.

To progress steadily through your career, you will need to develop leadership skills and learn how to use them. Leaders set an example for others to follow. Because of their skills, leaders are trusted not only with the authority to make decisions, but also with the responsibility to carry them out.

You can become a leader at any stage in your career. As an apprentice, you have the authority to perform the task given to you by your supervisor, and your supervisor expects you to be a responsible worker. By carrying out your task quickly, correctly, and independently, you are setting an example for others to follow. In doing so, you are demonstrating leadership skills.

People with the ability to become leaders often exhibit the following characteristics:

- They lead by example.
- They have a high level of drive, determination, and persistence.
- They are effective communicators.
- They can motivate their team to do its best work.
- They are organized planners.
- They have self-confidence.

The functions of a leader vary with the environment, the group of workers being led, and the tasks to be performed. However, certain functions are common to all situations. Some of these functions include the following:

- Organizing, planning, staffing, directing, and controlling.
- Empowering team members with authority and responsibility.
- Resolving disagreements before they become problems.
- Enforcing company policies and procedures.
- Accepting responsibility for failures as well as for successes.
- Representing the team to different trades, clients, and others.

Leadership styles vary. If leadership styles are classified according to the way a leader makes decisions, the result is three broad categories of leadership: autocratic, democratic, and hands-off. An autocratic leader makes all decisions independently, without seeking recommendations or suggestions from the team. A democratic leader involves the team in the decision-making process; such a leader takes team members' recommendations and suggestions into account before making a decision. A hands-off leader leaves all decision making to the team members themselves.

How Do Your Co-Workers See You?

Do you exhibit any of the following unpopular behaviors on the job? If so, it is likely they are keeping you from being an effective team member. Develop an action plan to deal with them. Your action plan should be a two- or three-step process that will allow you to correct the behavior. Be honest with yourself!

- Being a loner
- Taking yourself too seriously
- Being uptight
- Always needing to be the best at everything
- Holding a grudge
- Arriving late to work
- Being inconsiderate
- Taking breaks that are too long
- Gossiping about others
- Sticking your nose into other people's business
- Acting as a spy and reporting on the behavior of others to your supervisor
- Saying or doing things to create tension or unhappiness
- Complaining constantly
- Bad-mouthing the company or your boss
- Being unable to take a joke
- Bragging
- Taking credit for others' work
- Being sarcastic
- Refusing to listen to other people's ideas
- Looking down on other people
- Being unwilling to pitch in and help
- Being stingy with a compliment
- Having a chip on your shoulder
- Horsing around when others are trying to work
- Thinking you work harder than everyone else
- Manipulating people

Select a leadership style that is appropriate to the situation. Leading with a single style at all times generally results in some real problems. You will need to consider your authority, experience, expertise, and personality each time a style is chosen. Leaders need to have the respect of people on the team; otherwise, they will be unable to set an example others are willing to follow.

Leaders have to make sure that the decisions they make are ethical. Success in the construction industry demands the highest standards of ethical conduct. The three types of ethics that you will encounter on the job are business or legal ethics, professional ethics, and situational ethics. Business or legal ethics involve adhering to all relevant laws and regulations. Professional ethics involve being fair to everybody, and situational ethics involve appropriate responses to a particular event or situation.

Effective leaders motivate, or inspire, people to do their best. People are motivated by different things at different times. The following are some common ways to motivate people on the job:

- Recognize and praise a job well done.
- Allow people to feel a sense of accomplishment.
- Provide opportunities for advancement.
- Encourage people to feel that their job is important.
- Provide opportunities for change to prevent boredom.
- Reward people for their efforts.

Leaders who can motivate people are more likely to have a team with high morale and a positive work attitude. Morale and attitude are key components of a successful company with satisfied workers, and such a company is more likely to be successful.

Action Plan for Improvement

Example:

Problem: Gossiping about others

Action Plan: 1. Walk away when people start gossiping or bad-mouthing others.

2. Don't repeat what I hear.

3. Focus on my own work, not on that of others.

Problem: ____

Action Plan: ______________________________________

____

Problem: ____

Action Plan: ______________________________________

____

Problem: ____

Action Plan: ______________________________________

____

00108-15_SA03.EPS

Building a Team

In this exercise, you will practice building a team. Consider the following hypothetical situation, and then select an appropriate team. Discuss your selections with your instructor and with the other trainees.

Situation

You are the team leader on a job site where a new single-family home is going to be built. Shortly before construction begins, the architect changes the plans to add a sink to one of the rooms. When you review the plans, you see that an electrical conduit is located where the drain for the new sink should be. In addition, a cabinet needs to be built for the new sink. You may choose up to five of the following construction professionals to be on your team. Whom would you choose and why?

Mason	Painter
Carpenter	Roofer
Plumber	Landscape designer
Welder	Secretary
Electrician	Cabinetmaker

GOING GREEN

Green Employers

Some employers are doing their part to reduce their environmental footprint. Since 2004, the United Parcel Service (UPS) has revised delivery routes of its drivers to minimize left-hand turns. Doing so has made making the deliveries safer (they no longer have to cross traffic) and quicker (due to the ability to make a right on red and green turn arrows). As a result, the company has shaved 30 million miles off its deliveries in 2007 and thus saved the cost of 3 million gallons of gas. It also reduced UPS truck emissions by 32,000 metric tons (equivalent to the emissions of 5,300 passenger cars).

Additional Resources

Bullying and Harassment in the Workplace: Developments in Theory, Research, and Practice, Stale Einarsen and Helge Hoel. 2010. Boca Raton, FL: CRC Press.

The 7 Habits of Highly Effective People: Powerful Lessons in Personal Change, Stephen R. Covey. 2013. New York, NY: Simon & Schuster.

3.0.0 Section Review

1. If you finish a task ahead of time, looking for another task that can be finished in the remaining time is referred to as _____.
   a. stealing someone else's job
   b. being tardy
   c. showing off
   d. taking initiative

2. To find a solution everyone can agree on, you may have to _____.
   a. compromise
   b. discuss it with your supervisor
   c. let someone else solve the problem
   d. come up with more than one solution

3. When does constructive criticism work best?
   a. When offered often
   b. When offered occasionally
   c. When offered by a member of management
   d. When offered by someone of another trade

4. To put an end to personal bullying, one of the most critical things an employee can do is _____.
   a. confront the person one-on-one
   b. stage an intervention
   c. carefully document the events when they happen
   d. avoid the person in spite of work assignments

5. Before displaying their skills as a leader, all craftworkers must wait until they have a great deal of experience in their chosen craft.
   a. True
   b. False

SUMMARY

This module reviewed some of the important non-technical skills that you must learn to be successful as a construction professional. Whatever positions you may achieve throughout your career, these skills are vital.

Whether you are just starting out or moving on to new opportunities and challenges, you should be able to write a top-notch resume, prepare for interviews, and select a job for which you are qualified. Eventually, many construction professionals become entrepreneurs and start their own businesses.

The ability to solve problems using critical thinking skills is important for any employer and employee. Critical thinking involves evaluating and using information to reach conclusions or make decisions.

Good self-presentation skills, including personal habits, work ethic, and honesty, are characteristics of professionalism. Developing your skills in conflict resolution, teamwork, and leadership will help you to advance in your career.

The construction industry demands the best from its people and has great opportunity for personal and career growth. Only you can decide where your career will take you.

Review Questions

1. Someone who can vouch for your skills, experience, and work habits is called a(n) _____.
   a. mission statement
   b. entrepreneur
   c. interviewer
   d. reference

2. One of the most important and productive methods to identify job opportunities is by _____.
   a. networking with family and friends
   b. posting a "Seeking Employment" ad at the grocery store
   c. trying to call Human Resource offices directly
   d. checking the Help Wanted section of the newspaper daily

3. A common barrier to effective problem solving includes _____.
   a. overwork
   b. fear of change
   c. inadequate supervision
   d. procrastination

4. After putting a solution to a problem into effect, it is important to _____.
   a. stop considering other solutions
   b. insist it become a company procedure
   c. monitor the results
   d. develop a business plan for it

5. Depending on the type of project and region of work, one of the most expensive components of a project is often _____.
   a. equipment rental
   b. power tools
   c. hand tools
   d. labor

6. Positive or negative interactions affect _____.
   a. managerial skills
   b. tool use
   c. relationships
   d. work schedules

7. A person who works without constant supervision is showing _____.
   a. initiative
   b. fortitude
   c. respect
   d. self-presentation

8. The process of solving disagreements between co-workers is called _____.
   a. tactfulness
   b. conflict resolution
   c. relationship crafting
   d. addressing conflict

9. Constructive criticism should not be given unless _____.
   a. someone asks your opinion
   b. the supervisor tells you to do so
   c. you feel it is your business
   d. you can also be complimentary

10. A broad term identifying negative social behavior that can be based on race, age, disabilities, sex, religion, cultural issues, health, or language barriers is _____.
    a. absenteeism
    b. racism
    c. cyberbullying
    d. harassment

11. Zero tolerance refers to an employer's policy regarding _____.
    a. being sick
    b. training
    c. alcohol and drug abuse
    d. overtime

12. Barbiturates typically cause a person to _____.
    a. hallucinate
    b. accelerate their reaction times
    c. slow their reaction times
    d. be deprived of oxygen

13. When everyone in a group is focused on the final objective, it is called _____.
    a. being goal-oriented
    b. being team players
    c. taking initiative
    d. being leaders

14. To progress steadily through your career, you will need to develop and learn how to use _____.
    a. leadership skills
    b. classroom skills
    c. studying skills
    d. diversity skills

15. The three categories of leadership are hands-off, democratic, and _____.
    a. automatic
    b. liberal
    c. autocratic
    d. conservative

Trade Terms Quiz

Fill in the blank with the correct term that you learned from your study of this module.

1. A company uses a(n) ________to state how it does business.
2. Someone who can vouch for your skills, experience, and work habits is a(n) ________.
3. Substances known as ________ can be manufactured by the human body, found in specific plants, or produced in a lab.
4. The way you act, speak, and dress is called ________.
5. Another term for work habit is ________.
6. Although ________ is closely related to self-presentation, it relies more on your integrity and manner.
7. When you break ________, you share with other people information that belongs to the company.
8. You will gain the respect of your supervisor if you regularly take the ________ to complete additional tasks when your work is done.
9. A(n) ________ approach should be used when approaching a co-worker about negative behavior.
10. When a worker has a problem with ________, they are always running late.
11. In order to resolve some situations, there must be a ________.
12. Advice that is given to point out or correct a mistake is called ________.
13. Reinforcing the importance of minor infringements is known as ________.
14. When someone is harassing a person in an offensive or threatening manner, often based on a real or perceived imbalance of power, it can likely be considered ________.
15. A person who sets an example for other people to follow is generally demonstrating good ________ skills.
16. Racial jokes are considered a form of ________.
17. Unwelcome sexual advances or requests are considered ________.
18. Uppers are another word for ________.
19. Drugs that cause you to feel sedated are known as ________.
20. The drug LSD is considered a(n) ________.
21. The crystalline derivative of amphetamines is called ________.
22. Heroin is known to be the most commonly used ________.
23. When a person has a problem with ________, co-workers cannot depend on them.
24. Since ________ are different from specific illegal or restricted substances, they may not be illegal.

Trade Terms

Absenteeism
Amphetamine
Barbiturate
Bullying
Cannabinoids
Compromise
Confidentiality
Constructive criticism
Hallucinogen
Harassment
Initiative
Leadership
Methamphetamine
Mission statement
Opiates
Professionalism
Reference
Self-presentation
Sexual harassment
Synthetic drugs
Tactful
Tardiness
Work ethic
Zero tolerance

Cornerstone of Craftsmanship

Mark Bonda

Indian Land High School
Building Construction Instructor

How did you get started in the construction industry?
My construction career began in high school. I attended Southeastern Regional Vocational Technical High School in Easton, Massachusetts. At Southeastern, I majored in the electrical trade. We learned residential, commercial, and industrial electrical applications. After high school, I moved to South Carolina and earned a Bachelor of Science degree in Industrial Technology Education from Clemson University. I have taken many classes, workshops, and seminars in the construction field over the years.

Who or what inspired you to enter the industry?
My grandfather was a major influence in my life and helped mold my view and understanding of construction. He was the Chief Electrical Inspector for the city of Boston. I remember, while growing up as a kid, learning how to work with wood in his basement shop. That's where my love of construction was awakened.

What do you enjoy most about your career?
The most rewarding and enjoyable thing about my job is exposing young people to different careers in construction. It is exciting working with teens and helping them learn the basic skills required in different trades. A lot of young people are not aware of the opportunities they have before them. Together, we explore these opportunities and help them find a craft they like.

Why do you think training and education are important in construction?
Training and education are extremely important to construction. With an ever-changing environment, there are always revisions to existing practices and new innovations. To keep up with the demand for skilled workers, we need training that meets industry standards. In addition, we need skilled workers who have credentials acceptable to the industry.

Why do you think credentials are important in construction?
Credentials give employers a baseline of what their employees and job applicants can do and what they know. This is also important to employees because it offers them better wages and advancement in a company. Credentials are also valuable to ensure workers are receiving consistent training across the country. With a mobile construction workforce, uniform credentials provide a means of unifying the industry regardless of location.

How has training/construction impacted your life and career?
Training in construction has impacted my life tremendously. I have been able to travel, meet new people, and advance my career within the construction field. It has also helped me learn skills like time management, leadership, and budgeting.

Would you recommend construction as a career to others? Why?
I would recommend a career in construction for several reasons. It is a very rewarding profession. You get to see a project transform from start to finish. You get to go and see places most people would not have the chance to visit. There is lots of room for growth professionally and many opportunities to explore.

What does craftsmanship mean to you?
Craftsmanship is a vital part of a skilled tradesperson's body of work. It means taking pride in your craft and paying attention to detail. Even though times have changed, it is important to produce the best possible result you can. Craftsmanship is an extension of oneself and your capabilities as a leader.

Trade Terms Introduced in This Module

Absenteeism: Consistent failure to show up for work.

Amphetamine: A class of drugs that causes mental stimulation and feelings of euphoria.

Barbiturate: A class of drugs that induces relaxation, slowing the body's ability to react.

Bullying: Unwanted, aggressive behavior that involves a real or perceived power imbalance. This form of harassment may include offensive, persistent, insulting, or physically threatening behavior directed at an individual.

Cannabinoids: A diverse category of chemical substances that repress neurotransmitter releases in the brain. Cannabinoids have a variety of sources; some are created naturally by the human body, while others come from cannabis (marijuana). Still others are synthetic.

Compromise: When people involved in a disagreement make concessions to reach a solution that everyone agrees on.

Confidentiality: Privacy of information.

Constructive criticism: A positive offer of advice intended to help someone correct mistakes or improve actions.

Hallucinogen: A class of drugs that distort the perception of reality and cause hallucinations.

Harassment: A type of discrimination that can be based on race, age, disabilities, sex, religion, cultural issues, health, or language barriers.

Initiative: The ability to work without constant supervision and solve problems independently.

Leadership: The ability to set an example for others to follow by exercising authority and responsibility.

Methamphetamine: A highly addictive crystalline drug, derived from amphetamines, that affects the central nervous system.

Mission statement: A statement of how a company does business.

Opiates: A narcotic painkiller derived from the opium poppy plant or synthetically manufactured. Heroin is the most commonly used opiate.

Professionalism: Integrity and work-appropriate manners.

Reference: A person who can confirm to a potential employer that you have the skills, experience, and work habits that are listed in your resume.

Self-presentation: The way a person dresses, speaks, acts, and interacts with others.

Sexual harassment: A type of discrimination that results from unwelcome sexual advances, requests, or other verbal or physical behavior with sexual overtones.

Synthetic drugs: A drug with properties and effects similar to known substances but having a slightly altered chemical structure. Such drugs are often not illegal since they are somewhat different than well-defined restricted or illegal substances. The two typical categories are cannabinoids (lab-produced THC or marijuana substitutes) and cathinones, which are designed to mimic the effects of cocaine or methamphetamines.

Tactful: Being aware of the effects of your statements and actions on others.

Tardiness: Habitually showing up late for work.

Work ethic: Work habits that are the foundation of a person's ability to do his or her job.

Zero tolerance: The policy of applying laws or penalties to even minor infringements of a code in order to reinforce its overall importance, typically related to drug and alcohol abuse when applied to the workplace.

Additional Resources

This module presents thorough resources for task training. The following resource material is suggested for further study.

Bullying and Harassment in the Workplace: Developments in Theory, Research, and Practice, Stale Einarsen and Helge Hoel. 2010. Boca Raton, FL: CRC Press

Knock 'em Dead Resumes: A Killer Resumé Gets More Job Inteviews!, Martin Yate. 2014. Avon, MA: Adams Media.

Knock 'em Dead: The Ultimate Job Search. Martin Yate. 2014. Avon, MA: Adams Media.

The 7 Habits of Highly Effective People: Powerful Lessons in Personal Change, Stephen R. Covey. 2013. New York, NY: Simon & Schuster.

The Re-Discovery of Common Sense–A Guide to the Lost Art of Critical Thinking, Chuck Clayton. 2007. Lincoln, NE: iUniverse, Inc.

Figure Credits

©iStockphoto.com/NAN104, Module Opener

Mike Casey, National Science Foundation, Figure 1A

Topaz Publications, Inc., Figure 1B

Lincoln Electric Company, Cleveland, OH, USA, Figure 1C

Courtesy of Mark S. Knoke, Figure 1D

Trimble Navigation Limited, Figure 1E

Roanoke Land Surveying, Figure 1F

Ed Gloninger, Figures 3, 6, 9

IPM Global, Figure 7

The Banner Bank Building, SA01

National Archives and Records Administration, SA02

Section Review Answer Key

Answer	Section Reference	Objective
Section One		
1. c	1.1.0	1a
2. b	1.2.0	1b
Section Two		
1. a	2.1.0	2a
2. d	2.2.0	2b
3. a	2.3.1	2c
Section Three		
1. d	3.1.2	3a
2. a	3.2.1	3b
3. b	3.3.1	3c
4. c	3.4.1	3d
5. b	3.5.1	3e

NCCER CURRICULA — USER UPDATE

NCCER makes every effort to keep its textbooks up-to-date and free of technical errors. We appreciate your help in this process. If you find an error, a typographical mistake, or an inaccuracy in NCCER's curricula, please fill out this form (or a photocopy), or complete the online form at **www.nccer.org/olf**. Be sure to include the exact module ID number, page number, a detailed description, and your recommended correction. Your input will be brought to the attention of the Authoring Team. Thank you for your assistance.

Instructors – If you have an idea for improving this textbook, or have found that additional materials were necessary to teach this module effectively, please let us know so that we may present your suggestions to the Authoring Team.

NCCER Product Development and Revision

13614 Progress Blvd., Alachua, FL 32615

Email: curriculum@nccer.org
Online: www.nccer.org/olf

❏ Trainee Guide ❏ Lesson Plans ❏ Exam ❏ PowerPoints Other ____________

Craft / Level: Copyright Date:

Module ID Number / Title:

Section Number(s):

Description:

Recommended Correction:

Your Name:

Address:

Email: Phone: